a **LANGE** medical book

1988

Harper's Biochemistry

Twenty-first Edition

a **LANGE** medical book

1988
Harper's
Biochemistry
Twenty-first Edition

Robert K. Murray, MD, PhD
Professor of Biochemistry
University of Toronto

Daryl K. Granner, MD
Professor and Chairman
Department of Molecular Physiology and Biophysics
Professor of Medicine
Vanderbilt University
Nashville, Tennessee

Peter A. Mayes, PhD, DSc
Reader in Biochemistry
Royal Veterinary College
University of London

Victor W. Rodwell, PhD
Professor of Biochemistry
Purdue University
West Lafayette, Indiana

APPLETON & LANGE
Norwalk, Connecticut/San Mateo, California

0-8385-3648-4

Prentice-Hall of Australia, Pty. Ltd., Sydney
Prentice-Hall Canada, Inc.
Prentice-Hall Hispanoamericana, S.A., Mexico
Prentice-Hall of India Private Limited, New Delhi
Prentice-Hall International (UK) Limited, London
Prentice-Hall of Japan, Inc., Tokyo
Prentice-Hall of Southeast Asia (Pte.) Ltd., Singapore
Whitehall Books Ltd., Wellington, New Zealand
Editora Prentice-Hall do Brasil Ltda., Rio de Janeiro

ISSN: 0734-9866
ISBN: 0-8385-3648-4

PRINTED IN THE UNITED STATES OF AMERICA

Table of Contents

Preface ix

1. Biochemistry & Medicine 1
   Robert K. Murray, MD, PhD

2. Biomolecules & Biochemical Methods 4
   Robert K. Murray, MD, PhD

I. STRUCTURES & FUNCTIONS OF PROTEINS & ENZYMES

3. Amino Acids 13
   Victor W. Rodwell, PhD

4. Peptides 24
   Victor W. Rodwell, PhD

5. Proteins: Structure & Properties 32
   Victor W. Rodwell, PhD

6. Proteins: Myoglobin & Hemoglobin 40
   Victor W. Rodwell, PhD

7. Enzymes: General Properties 50
   Victor W. Rodwell, PhD

8. Enzymes: Kinetics 61
   Victor W. Rodwell, PhD

9. Enzymes: Mechanisms of Action 75
   Victor W. Rodwell, PhD

10. Enzymes: Regulation of Activities 82
    Victor W. Rodwell, PhD

II. BIOENERGETICS & THE METABOLISM OF CARBOHYDRATES & LIPIDS

11. Bioenergetics 93
    Peter A. Mayes, PhD, DSc

12. Biologic Oxidation 100
    Peter A. Mayes, PhD, DSc

13. Oxidative Phosphorylation & Mitochondrial Transport Systems 108
    Peter A. Mayes, PhD, DSc

14. Carbohydrates of Physiologic Significance 120
    Peter A. Mayes, PhD, DSc

15. Lipids of Physiologic Significance 130

Peter A. Mayes, PhD, DSc

16. Overview of Intermediary Metabolism 142

Peter A. Mayes, PhD, DSc

17. The Citric Acid Cycle: The Catabolism of Acetyl-CoA 149

Peter A. Mayes, PhD, DSc

18. Glycolysis & the Oxidation of Pyruvate 158

Peter A. Mayes, PhD, DSc

19. Metabolism of Glycogen 165

Peter A. Mayes, PhD, DSc

20. Gluconeogenesis & the Pentose Phosphate Pathway 172

Peter A. Mayes, PhD, DSc

21. Metabolism of Important Hexoses 180

Peter A. Mayes, PhD, DSc

22. Regulation of Carbohydrate Metabolism 186

Peter A. Mayes, PhD, DSc

23. Oxidation & Biosynthesis of Fatty Acids 198

Peter A. Mayes, PhD, DSc

24. Metabolism of Unsaturated Fatty Acids & Eicosanoids 210

Peter A. Mayes, PhD, DSc

25. Metabolism of Acylglycerols & Sphingolipids 218

Peter A. Mayes, PhD, DSc

26. Lipid Transport & Storage 226

Peter A. Mayes, PhD, DSc

27. Cholesterol Synthesis, Transport, & Excretion 241

Peter A. Mayes, PhD, DSc

28. Regulation of Lipid Metabolism & Tissue Fuels 253

Peter A. Mayes, PhD, DSc

III. METABOLISM OF PROTEINS & AMINO ACIDS

29. Biosynthesis of Amino Acids 265

Victor W. Rodwell, PhD

30. Catabolism of Amino Acid Nitrogen 271

Victor W. Rodwell, PhD

31. Catabolism of the Carbon Skeletons of Amino Acids 281

Victor W. Rodwell, PhD

32. Conversion of Amino Acids to Specialized Products 306

Victor W. Rodwell, PhD

33. Porphyrins & Bile Pigments 319

Robert K. Murray, MD, PhD

IV. STRUCTURE, FUNCTION, & REPLICATION OF INFORMATIONAL MACROMOLECULES

34. Nucleotides 335
Victor W. Rodwell, PhD

35. Metabolism of Purine & Pyrimidine Nucleotides 344
Victor W. Rodwell, PhD

36. Recombinant DNA Technology 362
Daryl K. Granner, MD

37. Nucleic Acid Structure & Function 377
Daryl K. Granner, MD

38. DNA Organization & Replication 387
Daryl K. Granner, MD

39. RNA Synthesis & Processing 404
Daryl K. Granner, MD

40. Protein Synthesis & the Genetic Code 414
Daryl K. Granner, MD

41. Regulation of Gene Expression 428
Daryl K. Granner, MD

V. BIOCHEMISTRY OF EXTRACELLULAR & INTRACELLULAR COMMUNICATION

42. Membranes: Structure, Assembly, & Function 445
Daryl K. Granner, MD

43. Characteristics of Hormone Systems 463
Daryl K. Granner, MD

44. Hormone Action 472
Daryl K. Granner, MD

45. Pituitary & Hypothalamic Hormones 482
Daryl K. Granner, MD

46. Thyroid Hormones 496
Daryl K. Granner, MD

47. Hormones That Regulate Calcium Metabolism 502
Daryl K. Granner, MD

48. Hormones of the Adrenal Cortex 511
Daryl K. Granner, MD

49. Hormones of the Adrenal Medulla 524
Daryl K. Granner, MD

50. Hormones of the Gonads 530
Daryl K. Granner, MD

51. Hormones of the Pancreas 547
Daryl K. Granner, MD

52. Gastrointestinal Hormones 564
Daryl K. Granner, MD

VI. SPECIAL TOPICS

53. Nutrition, Digestion, & Absorption.. 571
   Peter A. Mayes, PhD, DSc

54. Glycoproteins & Proteoglycans 589
   Robert K. Murray, MD, PhD

55. Blood Plasma & Clotting 607
   Robert K. Murray, MD, PhD

56. Contractile & Structural Proteins... 618
   Victor W. Rodwell, PhD

57. Cancer, Oncogenes, & Growth Factors 635
   Robert K. Murray, MD, PhD

Appendix 649

Abbreviations Encountered in Biochemistry 665

Index... 667

Preface

Harper's Biochemistry provides concise yet comprehensive treatment of the principles of biochemistry and molecular biology. By also offering insights into the pathologic applications of biochemistry, it gives medical students an introduction to the clinical disorders resulting from or associated with biochemical derangements.

Changes in the Twenty-first Edition

In preparing this new edition, the authors had two main goals: to reflect the latest advances in biochemistry that are important to medicine and to provide medical students with a book that would be more interesting and easier to study from. The following features reflect these objectives:

- The book has been divided into six major sections in order to make the logic behind its organization more obvious. Very long chapters have been subdivided to make them more digestible, and some chapters have been moved to more logical positions in the text.
- Two new introductory chapters have been added. Chapter 1, "Biochemistry and Medicine," is designed to illustrate the relevance of biochemistry to medicine by acquainting students with some interesting and important ways biochemistry has been used in medical science. Chapter 2, "Biomolecules and Biochemical Methods," gives a brief overview of the major molecules found in cells and the methods used to study them and thus sets the stage for the rest of the book. Significant findings in biochemistry are described briefly, and key areas of current research are mentioned.
- Each chapter opens with an introduction that summarizes its contents and their overall significance.
- Each chapter introduction is followed by a short section on "biomedical importance" that emphasizes the significance of certain of the chapter topics to medicine.
- A new chapter entitled "Overview of Intermediary Metabolism" (Chapter 16) gives thematic unity to all of the chapters on metabolism.
- A new chapter entitled "Recombinant DNA Technology" (Chapter 36) sets out the basic concepts and techniques of this topic and illustrates its powerful impact on biology and medicine.
- A special chapter entitled "Cancer, Oncogenes, and Growth Factors" (Chapter 57) introduces students to the problem of cancer and shows how biochemical information acquired from the study of oncogenes and growth factors is shedding light on its fundamental nature.
- An Appendix summarizing the major biochemical laboratory tests used in clinical medicine has been added. It contains a summary of the important concepts of sensitivity and specificity as applied to laboratory tests. This Appendix is intended to help students bridge the gap between basic and clinical biochemistry by showing the diagnostic uses of many biochemical estimations.

In addition, all chapters have been revised to reflect the latest scientific knowledge. Particular discussions of interest include the following:

- Membranes (in Chapter 43).
- Nutrition, Digestion, and Absorption (in Chapter 53).
- Glycoproteins (in Chapter 54).

Organization of This Book

This text is divided into two introductory chapters followed by six main sections.

Section I is devoted to proteins and enzymes, the workhorses of the body. Because almost every reaction in the human cell is catalyzed by an enzyme, it is vital to understand the properties of enzymes before moving on to other topics.

Section II sets out how various cellular reactions either utilize or produce energy and traces the pathways by which carbohydrates and lipids are synthesized and degraded. The functions of the various members of these two classes of molecules are also described.

Section III deals with the amino acids and their fates and shows how the metabolism of amino acids also uses and yields energy.

Section IV describes the structure and functions of the nucleic acids, covering the representation DNA → RNA → protein. This section also contains a new chapter describing the principles of recombinant DNA technology, a topic with tremendous implications for biomedical science.

Section V discusses hormones and the key roles they play in intercellular communication and metabolic regulation. In order to affect cells, hormones must interact with the plasma membranes of cells; thus, membrane structure and function are addressed initially in this section.

Section VI consists of five special topics, including a new chapter on "Cancer, Oncogenes, and Growth Factors."

The **Appendix** briefly discusses the interpretation of biochemical laboratory tests and also presents their major diagnostic applications.

Acknowledgments

The authors extend their appreciation to Dr. David W. Martin, Jr., for the many contributions that he made to *Harper's Biochemistry* during his tenure as senior author. Considerable sections of the present edition reflect Dr. Martin's contributions. His place in the present edition has been taken by Dr. Robert K. Murray, who will attempt to match the energy and biochemical knowledge that Dr. Martin displayed.

The authors wish to extend their gratitude to professional colleagues and friends throughout the world who have conveyed suggestions for improvement and corrections to us. We wish to encourage their continued effort and interest. Comments from students are especially welcome.

The authors are most gratified by the broad base of acceptance and support this book has received all over the world. Several editions of the English language version have been reprinted in Japan, Lebanon, Taiwan, the Philippines, and Korea. In addition, there are now translations in Italian, Spanish, French, Portuguese, Japanese, Polish, German, Indonesian, Serbo-Croatian, and Greek.

—RKM
—DKG
—PAM
—VWR

October, 1987

Biochemistry & Medicine

Robert K. Murray, MD, PhD

INTRODUCTION

Biochemistry is the science concerned with studying the various molecules, chemical reactions, and processes that occur in living cells and organisms. Those who acquire a sound knowledge of biochemistry will be in a position to tackle the 2 central concerns of the biomedical sciences: (1) the understanding and maintenance of health and (2) the understanding and effective treatment of disease.

BIOCHEMISTRY & HEALTH

The World Health Organization (WHO) defines health as a state of "complete physical, mental, and social well-being and not merely the absence of disease and infirmity." From a strictly biochemical viewpoint, health may be considered as that situation in which all of the many thousands of intra- and extracellular reactions that occur in the body are proceeding under conditions and at rates commensurate with its maximal survival in the physiologic (as opposed to pathologic) state.

BIOCHEMISTRY, NUTRITION, PREVENTIVE MEDICINE, & TREATMENT

One major prerequisite for the maintenance of health is that there be optimal dietary intake of a number of chemicals; the chief of these are vitamins, certain amino acids, certain fatty acids, various minerals, and water. Much of the subject matter of biochemistry and nutrition is concerned with the study of various aspects of these chemicals. Consequently, there is an intimate relationship between these 2 sciences. Moreover, it appears probable that as attempts are made to curb the rising costs of medical care, more emphasis will be placed on systematic attempts to maintain health and forestall disease—ie, on preventive medicine. Thus, nutritional approaches to, for example, the prevention of atherosclerosis and cancer are likely to receive increasing emphasis. Understanding nutrition depends to a great extent on a knowledge of biochemistry.

BIOCHEMISTRY & DISEASE

All diseases are manifestations of abnormalities of molecules, chemical reactions, or processes. The major factors responsible for causing diseases in animals and humans are listed in Table 1–1. All of them affect one or more critical chemical reactions or molecules in the body.

The uses of biochemical investigations in relation to diseases may be summarized as follows. They can (1) reveal their causes; (2) suggest rational and effective treatments; (3) make available screening tests for early diagnosis; (4) assist in monitoring progress; and (5) help in assessing response to therapy. An appendix to this text describes the most important routine biochemical diagnostic tests used to investigate various diseases. It will serve as a useful reference when the biochemical diagnosis of diseases (eg, myocardial infarction, acute pancreatitis) is discussed in this text or in lectures.

The following 3 examples briefly illustrate the relationship of biochemistry to the prevention and treatment of disease; additional examples are given later in this chapter.

(1) It is well known that humans must ingest a number of complex organic molecules called **vitamins** in order to maintain normal health. Vitamins are generally converted in the body to more complex molecules (coenzymes) that play key roles in many cellular reactions. If a particular vitamin is deficient in the diet, a deficiency disease may result, such as scurvy or rickets (due to lack of intake of vitamins C and D, respectively). The elucidation of the key roles played by the

Table 1–1. The major causes of diseases. All of the causes listed act by influencing various biochemical mechanisms in the cell or in the body.*

1. **Physical agents:** Mechanical trauma, extremes of temperature, sudden changes in atmospheric pressure, radiation, electric shock.
2. **Chemical agents and drugs:** Certain toxic compounds, therapeutic drugs, etc.
3. **Biologic agents:** Viruses, rickettsiae, bacteria, fungi, higher forms of parasites.
4. **Oxygen lack:** Loss of blood supply, depletion of the oxygen-carrying capacity of the blood, poisoning of the oxidative enzymes.
5. **Genetic:** Congenital, molecular.
6. **Immunologic reactions:** Anaphylaxis, autoimmune disease.
7. **Nutritional imbalances:** Nutritional deficiencies, nutritional excesses.

*Adapted, with permission, from Robbins SL, Cotram RS, Kumar V: *The Pathologic Basis of Disease*, 3rd ed. Saunders, 1984.

vitamins or their biologically active derivatives has been a major concern of biochemists and nutritionists since the turn of the century.

(2) The condition known as **phenylketonuria (PKU),** if untreated, may lead to severe mental retardation in infancy. The biochemical basis of PKU has been known for approximately 30 years; the disorder is due to low or absent activity of the enzyme that converts the amino acid phenylalanine to the amino acid tyrosine. As a consequence of the enzyme deficiency, phenylalanine and some of its metabolites, such as ketones, accumulate in the tissues and apparently damage the developing central nervous system. When the nature of the biochemical lesion in PKU was revealed, it became rational to treat the disease by placing affected infants on a diet low in phenylalanine. Once biochemical screening tests for diagnosing PKU at birth became available, effective treatment could be started immediately.

(3) **Cystic fibrosis** is an inherited disease of the exocrine glands and of the eccrine sweat glands. The cause is unknown. Cystic fibrosis is one of the most common genetic disorders in North America. It is characterized by abnormally viscous secretions that plug up the secretory ducts of the pancreas and the bronchioles. Its victims often die at an early age because of lung infections. Because the molecular basis of this disease is unknown, only symptomatic therapy is possible. However, the molecular basis may be at least partly revealed in the near future using the techniques of recombinant DNA, and a more effective treatment may become evident.

FORMAL DEFINITIONS OF BIOCHEMISTRY

Biochemistry, as the name implies (Gk *bios* "life"), is the chemistry of life. It can be defined more formally as the science concerned with the chemical basis of life.

The cell is known to be the structural unit of living systems. Consideration of this concept leads to an alternative definition of biochemistry as the science concerned with the chemical constituents of living cells and with the reactions and processes they undergo. By this definition, biochemistry encompasses wide areas of **cell biology** and all of **molecular biology.**

OBJECTIVES OF BIOCHEMISTRY

The major objective of biochemistry is the complete understanding at the molecular level of all of the chemical processes associated with living cells.

To achieve this objective, biochemists have attempted to **isolate** the numerous molecules found in cells, determine their **structures,** and analyze how they **function.** To give one example, the efforts of many biochemists to understand the molecular basis of contractility—a process associated primarily, but not exclusively, with muscle cells—have entailed purification of many molecules, of both simple and complex structure, followed by detailed structure-function studies. Through these efforts, some of the features of the molecular basis of muscle contraction have, in fact, been revealed.

An additional objective of biochemistry is to attempt to understand how life began. Knowledge of this fascinating subject is still embryonic.

AREAS OF STUDY

The scope of biochemistry is as wide as life itself. Wherever there is life, chemical processes are occurring. Biochemists study the chemical processes that occur in microorganisms, plants, insects, fish, birds, lower and higher mammals, and human beings. The student in the biomedical sciences will be particularly interested in the biochemistry of the 2 latter groups. However, it would be shortsighted to exclude an appreciation of the biochemistry of some other forms of life, which, in many cases, is essential for understanding situations of direct relevance to humans.

BIOCHEMISTRY & MEDICINE

The interrelationship of biochemistry and medicine is a wide, 2-way street. Biochemical studies have illuminated many aspects of disease, and the study of certain diseases has opened up new areas of biochemistry.

Biochemical Studies Illuminating Disease Mechanisms
In addition to the instances given earlier, 4 examples are provided here in order to illustrate the wide-ranging nature of this topic. (1) Analysis of the mechanism of action of the bacterial toxin that causes cholera has provided important insights concerning how the clinical manifestations (diarrhea, dehydration) of this disease are brought about. (2) The fact that many plants in Africa are deficient in one or more essential amino acids helps explain the debilitating malnutrition suffered by those who depend on such plants as major dietary sources of protein. (3) The finding that mosquitoes which transmit malaria parasites in humans can develop biochemically based resistance to the action of insecticides has important consequences for attempts to eradicate this disease. (4) Greenland Eskimos consume large quantities of fish oils rich in certain polyunsaturated fatty acids and are known to have low plasma levels of cholesterol and a low incidence of atherosclerosis. These observations have stimulated interest in the use of polyunsaturated fatty acids to reduce plasma levels of cholesterol.

Diseases Illuminating Biochemistry
The initial observations made by the English physi-

cian Sir Archibald Garrod on a small group of **inborn errors of metabolism** in the early 1900s stimulated the investigation of the biochemical pathways affected in these conditions. Efforts to understand the basis of the genetic disease known as **familial hypercholesterolemia,** which results in severe atherosclerosis at an early age, have led to dramatic progress in knowledge of cell receptors and of mechanisms of uptake of cholesterol into cells. The ongoing studies of oncogenes in **cancer cells** have directed considerable attention to the molecular mechanisms involved in the control of cell growth.

Studies of Lower Organisms & Viruses

At an even more basic level, the results of studies on certain lower organisms and on viruses have yielded important information useful in the design of biochemical studies on humans. For instance, contemporary theories on the **regulation of the activities of genes and of enzymes** emanate from pioneering studies on bread molds and bacteria. The field of **recombinant DNA** emerged from studies on bacteria and their viruses. The major properties of bacteria and viruses that have made them so suitable for biochemical studies are their **rapid multiplication times** and their suitability for **genetic** analyses and manipulations. Knowledge gained from the study of viral genes (**viral oncogenes**) responsible for certain types of cancer in animals provided insights into how human cells become cancerous.

BIOCHEMISTRY & OTHER BIOSCIENCES

The biochemistry of the nucleic acids lies at the heart of **genetics;** in turn, the use of genetic approaches has been critical for the elucidation of many areas of biochemistry. **Physiology,** the study of body function, overlaps with biochemistry almost completely. **Immunology** employs numerous biochemical techniques, and many immunologic approaches have found wide use by biochemists. **Pharmacology** and **pharmacy** rest on a sound knowledge of biochemistry and physiology; in particular, most drugs are metabolized by enzyme-catalyzed reactions. Poisons act on biochemical reactions or processes; this is the subject matter of **toxicology.** It was stated earlier that all diseases are chemical in nature; thus, increasingly, biochemical approaches are being used to study basic aspects of **pathology,** such as inflammation, cell injury, and cancer. Many workers in **zoology** and **botany** employ biochemical approaches almost exclusively. These relationships are not surprising, because life as we know it depends on biochemical reactions and processes. In fact, the old barriers among the biosciences are breaking down, and biochemistry is increasingly becoming their common language.

Biomolecules & Biochemical Methods

Robert K. Murray, MD, PhD

INTRODUCTION

This chapter has 5 objectives. The first is to indicate **the composition of the body and the major classes of molecules** found in it. These molecules make up the principal components of this text.

The **cell** is the major structural and functional unit of biology. Most chemical reactions occurring in the body take place within cells. Thus, the second objective is to give a concise account of the **components of cells** and how they may be isolated; the details of how these components function form much of the fabric of this text.

The third objective concerns the fact that biochemistry is an **experimental science.** It is important to have an understanding and appreciation of the experimental approach and methods used in biochemistry, lest its study become an exercise in rote learning. Moreover, biochemistry is not an immutable corpus of knowledge but a constantly evolving field. Advances, like those in other biomedical areas, depend on the experimental approach and technologic innovation.

The fourth objective is to briefly summarize the **principal achievements** that have been made in biochemistry. The bird's-eye view of the science that will be presented will assist in imparting to the reader a sense of the overall direction of the remainder of the text.

The fifth objective is to indicate **how little is known about certain areas,** eg, about development, differentiation, brain function, cancer, and many other human diseases. Perhaps this will serve as a stimulus to some readers to contribute to research in these areas.

ELEMENTARY COMPOSITION OF THE HUMAN BODY

The elementary composition of the human body has been determined, and the major findings are listed in Table 2–1. **Carbon, oxygen, hydrogen,** and **nitrogen** are the major constituents of most biomolecules. **Phosphate** is a component of the nucleic acids and other molecules and is also widely distributed in its ionized form in the human body. **Calcium** plays a key role in innumerable biologic pro-

Table 2–1. Approximate elementary composition of the human body (dry weight basis).*

Element	Percent	Element	Percent
Carbon	50	Potassium	1
Oxygen	20	Sulfur	0.8
Hydrogen	10	Sodium	0.4
Nitrogen	8.5	Chlorine	0.4
Calcium	4	Magnesium	0.1
Phosphorus	2.5	Iron	0.01
		Manganese	0.001
		Iodine	0.00005

*Reproduced, with permission, from West ES, Todd WR: *Textbook of Biochemistry,* 3rd ed. Macmillan, 1961.

cesses and is the focus of much current research. The elements listed in column 3 of the table fulfill diverse roles. Most are encountered on an almost daily basis in medical practice in dealing with patients with electrolyte imbalances (K^+, Na^+, Cl^-, and Mg^{2+}), iron-deficiency anemia (Fe^{2+}), and thyroid diseases (I^-).

THE MAJOR CLASSES OF BIOMOLECULES FOUND IN NATURE

As shown in Table 2–2, the major **complex biomolecules** found in the cells and tissues of higher animals (including humans) are **DNA, RNA, proteins, polysaccharides,** and **lipids.** The complex molecules are constructed from **simple biomolecules,** which are also listed. The building blocks of DNA and RNA (collectively known as the nucleic acids) are **deoxynucleotides** and **ribonucleotides,** respectively. The building blocks of proteins are **amino acids.** Polysaccharides are built up from simple carbohydrates; in the case of **glycogen** (the principal polysaccharide found in human tissues), the carbohydrate is **glucose. Fatty acids** may be considered to be the building blocks of lipids, although lipids are not polymers of fatty acids. DNA, RNA, proteins, and polysaccharides are referred to as **biopolymers** because they are composed of repeating units of their building blocks (the monomers). These complex molecules essentially make up the "stuff of life"; most of this text will be largely concerned with descriptions

Table 2–2. The major complex organic biomolecules of cells and tissues. The nucleic acids, proteins, and polysaccharides are biopolymers, constructed from the building blocks shown. The lipids are not generally biopolymers, and not all lipids have fatty acids as building blocks.

Biomolecule	Building Block	Major Functions
DNA	Deoxynucleotide	Genetic material
RNA	Ribonucleotide	Template for protein synthesis
Protein	Amino acids	Numerous; usually they are the molecules of the cell that carry out work (eg, enzymes, contractile elements)
Polysaccharide (glycogen)	Glucose	Short-term storage of energy as glucose
Lipids	Fatty acids	Numerous, eg, membrane components and long-term storage of energy as triacylglycerols

of various biochemical features of both biopolymers and monomers. The same complex molecules are also generally found in lower organisms, although the building blocks in certain cases may differ from those shown in Table 2–2. For instance, bacteria do not contain glycogen or triacylglycerols, but they do contain other polysaccharides and lipids.

CHEMICAL COMPOSITION OF THE HUMAN BODY

The elementary composition of the human body is given above. Its chemical composition is shown in Table 2–3; **protein, fat, carbohydrate, water,** and **minerals** are the chief components. Water constitutes the major component, although its amount varies widely among different tissues. Its polar nature and ability to form hydrogen bonds render water ideally suited for its function as the solvent of the body.

THE CELL

The cell was established as the fundamental unit of biologic activity by Schleiden and Schwann and other pioneers such as Virchow in the 19th century. However, in the years immediately after World War II, 3 developments helped usher in a period of unparalleled activity in biochemistry and cell biology. These were (1) the increasing availability of the **electron microscope;** (2) the introduction of methods permitting **disruption of cells** under relatively mild conditions that

preserved function; and (3) the increasing availability of the high-speed, refrigerated **ultracentrifuge,** capable of generating centrifugal forces sufficient to separate the constituents of disrupted cells from one another without overheating them. Use of the electron microscope revealed many previously unknown or poorly observable cellular components, while disruption and ultracentrifugation permitted their isolation and analysis in vitro.

The Hepatocyte of the Rat

A diagram of the structure of a liver cell (hepatocyte) of a rat is shown in Fig 2–1; this cell is probably

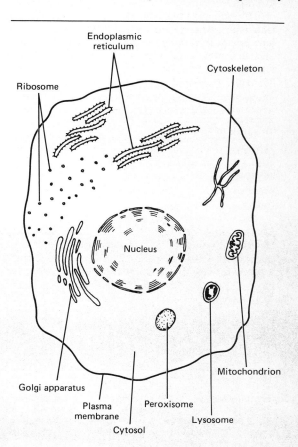

Figure 2–1. Schematic representation of a rat liver cell with its major organelles.

Table 2–3. Normal chemical composition for a man weighing 65 kg.*

	Kg	Percent
Protein	11	17.0
Fat	9	13.8
Carbohydrate	1	1.5
Water†	40	61.6
Minerals	4	6.1

*Reproduced, with permission, from Davidson SD, Passmore R, Brock JF: *Human Nutrition and Dietetics,* 5th ed. Churchill Livingstone, 1973.
†The value for water can vary widely among different tissues, being as low as 22.5% for marrow-free bone. The percentage comprised by water also tends to diminish as body fat increases.

the most studied of all cells from a biochemical viewpoint, partly because of its availability in relatively large amounts, suitability for fractionation studies, and diversity of functions. The hepatocyte contains the major **organelles** found in eukaryotic cells (Table 2–4); these include the nucleus, mitochondria, endoplasmic reticulum, free ribosomes, Golgi apparatus, lysosomes, peroxisomes, plasma membrane, and certain cytoskeletal elements.

Subcellular Fractionation

In order to study the function of any organelle in depth, it is first necessary to isolate it in relatively pure form, free of significant contamination by other organelles. The usual process by which this is achieved is called subcellular fractionation and generally entails 3 procedures: extraction, homogenization, and centrifugation. Much of the pioneering work in this area was done using rat liver.

A. Extraction: As a first step toward isolating a specific organelle (or molecule), it is necessary to extract it from the cells in which it is located. Most organelles and many biomolecules (eg, proteins) are quite **labile** and subject to loss of biologic activities. Accordingly, they must be extracted using **mild conditions** (ie, employment of aqueous solutions and avoidance of extremes of pH and osmotic pressure and of high temperatures). In fact, most procedures for isolating organelles are performed at about 0–4 °C (eg, in a cold room or using material kept on ice). Significant losses of activity can occur at room temperature, partly owing to the action of various digestive enzymes (proteases, nucleases, etc) liberated when cells are disrupted. A common solution for extraction of organelles consists of sucrose, 0.25 mol/L (isosmotic), adjusted to pH 7.4 by TRIS (tris[hydroxymethyl] aminomethane)-hydrochloric acid buffer, 0.05 mol/L, containing K^+ and Mg^{2+} ions at near physiologic concentrations; this solution is conveniently called STKM. Not all solvents used for extraction are as mild as STKM; eg, **organic solvents** are used for the extraction of lipids and of nucleic acids.

B. Homogenization: To extract an organelle (or biomolecule) from cells, it is first necessary to disrupt the cells under mild conditions. Organs (eg, liver, kidney, brain) and their contained cells may be conveniently disrupted by the process of homogenization, in which a manually operated or motor-driven pestle is rotated within a glass tube of suitable dimensions containing minced fragments of the organ under study and a suitable homogenizing medium, such as STKM. The controlled rotation of the pestle exerts mechanical shearing forces on cells and disrupts them, liberating their constituents into the sucrose. The resulting suspension, containing many intact organelles, is known as a **homogenate.**

C. Centrifugation: Subfractionation of the contents of a homogenate by differential centrifugation has been a technique of central importance in biochemistry. The classic method uses a series of 3 different centrifugation steps at successively greater speeds

Table 2–4. Major intracellular organelles and their functions. Only the major functions associated with each organelle are listed. In a number of instances, many other pathways, processes, or reactions may occur in the organelle.

Organelle or Fraction*	Marker	Major Functions
Nucleus	DNA	Site of chromosomes Site of DNA-directed RNA synthesis (**transcription**)
Mitochondrion	Glutamic dehydrogenase	Citric acid cycle, production of ATP
Ribosome*	High content of RNA	Site of protein synthesis (**translation** of mRNA into protein)
Endoplasmic reticulum	Glucose-6-phosphatase	Membrane-bound ribosomes are a major site of protein synthesis Synthesis of various lipids Oxidation of many xenobiotics (cytochrome P-450)
Lysosome	Acid phosphatase	Site of many hydrolases (enzymes catalyzing degradative reactions)
Plasma membrane	Na^+/K^+-ATPase 5′-Nucleotidase	Transport of molecules in and out of cells Intercellular adhesion and communication
Golgi apparatus	Galactosyl transferase	Intracellular sorting of proteins Glycosylation reactions Sulfation reactions
Peroxisome	Catalase Uric acid oxidase	Degradation of certain fatty acids and amino acids Production and degradation of hydrogen peroxide
Cytoskeleton*	No specific enzyme markers†	Microfilaments, microtubules, intermediate filaments
Cytosol*	Lactate dehydrogenase	Enzymes of glycolysis, fatty acid synthesis

*An organelle can be defined as a subcellular entity that is membrane-limited and is isolated by centrifugation at high speeds. According to this definition, ribosomes, the cytoskeleton, and the cytosol are not organelles. However, they are considered in this table along with the organelles because they also are usually isolated by centrifugation. They can be considered as subcellular entities or fractions. An organelle as isolated by one cycle of differential centrifugation is rarely pure; to obtain a pure fraction usually requires at least a number of cycles.
†The cytoskeletal fractions can be recognized by electron microscopy or by analysis by electrophoresis of characteristic proteins that they contain.

(Fig 2–2), each yielding a pellet and a supernatant. The supernatant from each step is subjected to centrifugation in the next step. This procedure provides 3 pellets, named the nuclear, mitochondrial, and microsomal fractions. None of these fractions are composed of absolutely pure organelles. However, it has been well established by the use of the electron microscope and by measurements of suitable "marker" enzymes and chemical components (eg, DNA and RNA) that the major constituents of each of these 3 fractions are nuclei, mitochondria, and microsomes, respectively. A "marker" enzyme or chemical is one that is almost exclusively confined to one particular organelle, eg, acid phosphatase to lysosomes and DNA to the nucleus (Table 2–4). The marker can thus serve to indicate the presence or absence in any particular fraction of the organelle in which it is contained. The **microsomal fraction (microsomes)** contains mostly a mixture of smooth endoplasmic reticulum, rough endoplasmic reticulum (ie, endoplasmic reticulum with attached ribosomes), and free ribosomes. The contents of the final supernatant correspond approximately to those of the **cell sap (cytosol)**. Modifications of this basic approach, using different homogenization media or different protocols or methods of centrifugation (eg, the use of gradients—either continuous or discontinuous—of sucrose), have permitted the isolation in more or less pure form of all of the organelles illustrated in Fig 2–1 and listed in Table 2–4. The scheme described above is applicable in general terms to most organs and cells; however, cell fractionations of this type must be assessed by the use of measurements of marker enzymes and chemicals and by the electron microscope until the overall procedure can be considered to be standardized.

The importance of subcellular fractionation studies in the development of biochemistry and cell biology cannot be overemphasized. It has been one of the major components of the experimental approach (see below), and—largely because of its application—the functions of the organelles indicated in Table 2–4 have been elucidated. The information on function summarized in this table represents one of the major achievements of biochemical research (see below).

THE EXPERIMENTAL APPROACH USED IN BIOCHEMISTRY

There are 3 major components to the experimental approach: (1) **isolation of biomolecules** and organelles (see Subcellular Fractionation, above) found in cells; (2) **determination of the structures of biomolecules;** and (3) **analyses, using various preparations, of the function and metabolism** (ie, synthesis and degradation) **of biomolecules.**

A. Isolation of Biomolecules: As in the case of organelles, elucidation of the function of any biomolecule requires that it first be isolated in pure form. Table 2–5 lists the major methods that are used to separate and purify biomolecules. No details of these methods will be given here; certain of them are described briefly at various places in this text. A combination of the successive use of several of them is almost always needed to purify a biomolecule to **homogeneity** (freedom from contamination by any other biomolecule).

It is important to appreciate that advances in biochemistry are dependent upon the development of new methods of analysis, purification, and structural deter-

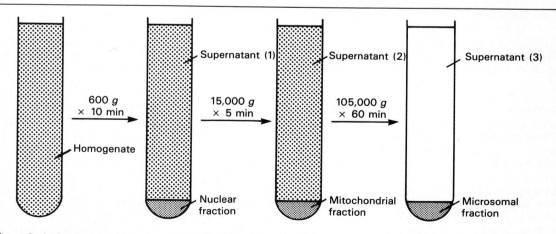

Figure 2–2. Scheme of separation of subcellular fractions by differential centrifugation. The homogenized tissue (eg, liver) is first subjected to low-speed centrifugation, which yields the nuclear fraction (containing both nuclei and unruptured cells) and supernatant (*1*). The latter is decanted and subjected to centrifugation at an intermediate speed, yielding the mitochondrial fraction (containing mitochondria, lysosomes, and peroxisomes) and supernatant (*2*). The latter is next decanted and subjected to high-speed centrifugation, yielding the microsomal fraction (containing a mixture of free ribosomes and smooth and rough endoplasmic reticulum) and the final clear solution, supernatant (*3*). The latter corresponds approximately to the cytosol or cell sap. By various modifications of this basic approach, it is generally possible to isolate each cell organelle in relatively pure form.

Table 2–5. Major methods used to separate and purify biomolecules. Most of these methods are suitable for analysis of the components present in cell extracts and other biochemical materials. The sequential use of a number of these techniques will generally permit purification of most biomolecules. The reader is referred to texts on methods of biochemical research for details regarding each.

Salt fractionation (eg, precipitation with ammonium sulfate)
Chromatography
    Paper
    Ion exchange (anion and cation exchange)
    Affinity
    Thin-layer
    Gas-liquid
    High-pressure liquid
Gel filtration
Electrophoresis
    Paper
    High-voltage
    Agarose
    Cellulose acetate
    Starch gel
    Polyacrylamide
    SDS-polyacrylamide
Ultracentrifugation

Table 2–6. Principal methods used for the determination of the structures of biomolecules.

Elemental analysis
Ultraviolet, visible, infrared, and NMR spectroscopy
Use of acid or alkaline hydrolysis to degrade the biomolecule under study into its basic constituents
Use of a battery of enzymes of known specificity to degrade the biomolecule under study (eg, proteases, nucleases, glycosidases)
Mass spectrometry
Specific sequencing methods (eg, for proteins and nucleic acids)
X-ray crystallography

mination. For instance, the field of lipid biochemistry was turned on its heels by the introduction of **gas-liquid** and **thin-layer chromatography.** The analysis of membrane and many other proteins was extremely difficult until the introduction of **sodium dodecyl sulfate-polyacrylamide gel electrophoresis** (SDS-PAGE); the introduction of the detergent SDS permitted the "solubilization" for electrophoresis of many proteins that were previously rather insoluble. The development of methods for **sequencing and cloning DNA** has had a revolutionary impact on the study of the nucleic acids and of biology in general.

B. Determination of the Structures of Biomolecules: Once a biomolecule has been purified, it is then necessary to determine its structure. This should allow detailed correlations to be made between structure and function. The major methods used to analyze the structures of biomolecules are listed in Table 2–6. They will be familiar to the reader who has some knowledge of organic chemistry. The known specificity of certain enzymes makes them very powerful tools in the elucidation of the structural features of certain biomolecules. Improvements in resolution afforded by theoretic and technologic advances are increasingly making **mass spectrometry** and **NMR spectrometry** the routine methods of choice for structural determination. For instance, the structures of the extremely complex carbohydrate chains found in certain biomolecules, such as glycoproteins, can now often be elucidated by high-resolution NMR spectrometry. The most detailed information about the structure of biomolecules is provided by **x-ray diffraction and crystallography.** Their use was critical in revealing the detailed structures of various proteins and enzymes and the double helical nature of DNA.

C. Analyses, Using Various Preparations, of the Function and Metabolism of Biomolecules: Initial biochemical research on humans and

animals was performed at the level of the whole animal. Examples were studies of respiration and of the fate of ingested compounds. It soon became apparent that the whole animal was too complex to permit definitive answers to be given to many questions. Accordingly, simpler in vitro preparations were developed that removed many of the complications experienced at the level of the whole animal. Table 2–7 summarizes the various types of preparations that are now available to study biochemical processes; most of the facts presented in this text have been obtained by their use. The items in the list are arranged in decreasing order of complexity. Just as the use of whole animals has its drawbacks, the other preparations also have limitations. Erroneous results (artifacts) can be derived from the use of in vitro approaches; eg, homogenization of cells can liberate enzymes that may partially digest cellular molecules.

A General Guide to the Strategy Used to Elucidate Biochemical Processes

Much of this text will be concerned with complex biochemical processes (eg, protein synthesis and muscle contraction), including metabolic pathways. A **metabolic pathway** is a series of reactions responsible for the **synthesis** of a more complex compound from one or more simple compounds, or for the **degradation** of a compound to its end product. The existence of a complex biochemical process or of certain metabolic pathways can be inferred from observations made at the level of the whole animal. For instance, direct observations on our fellow humans indicate that muscles contract. We know that glucose serves as a source of energy for humans and other animals and can thus infer that it must be degraded (metabolized) in the body to yield energy. However, to fully understand how glucose is metabolized by human cells—and knowledge of this is still far from complete—requires analyses at a variety of levels. Fig 2–3 shows the various types of observations and analyses that are required in order to attempt to comprehend biochemical processes, such as the initial breakdown of glucose to yield energy (a process known as **glycolysis**). The scheme shown applies at a general level to all of the major biochemical processes discussed in this text and thus represents an overall strategy for elucidating biochemical processes; it should be borne in mind when

Table 2–7. Hierarchy of preparations used to study biochemical processes.

Method	Comment
Studies at the whole-animal level	This can include (1) removal of an organ (eg, hepatectomy). (2) alterations of diet (eg, fasting-feeding). (3) administration of a drug (eg, phenobarbital). (4) administration of a toxin (eg, carbon tetrachloride). (5) use of an animal with a specific disease (eg, diabetes mellitus). (6) use of sophisticated techniques such as NMR spectroscopy and positron emission tomography.
Isolated perfused organ	Liver, heart, and kidney are particularly suitable.
Tissue slice	Liver slices have been especially used.
Use of whole cells	(1) Particularly applicable to blood cells, which can be purified relatively easily. (2) Use of cells in tissue culture is indispensable in many areas of biology.
Homogenate	(1) Ensures a cell-free preparation. (2) Specific compounds can be added or removed (eg, by dialysis) and their effects studied. (3) Can be subfractionated by centrifugation to yield individual cell organelles.
Isolated cell organelles	Extensively used to study the function of mitochondria, the endoplasmic reticulum, ribosomes, etc.
Subfractionation of organelles	Extensively used, eg, in studies of mitochondrial function.
Isolation and characterization of metabolites and enzymes	A vital part of the analysis of any chemical reaction or pathway.
Cloning of genes for enzymes and proteins	Isolation of the cloned gene is vital for studying the details of its structure and regulation; it can also reveal the amino acid sequence of the enzyme or protein for which it codes.

Inference of existence of biochemical process or metabolic pathway from observations made at the whole-animal level
↓
Analyses of its control mechanisms in vivo
↓
Analyses of effects on it of specific diseases (eg, inborn errors of metabolism, cancer, etc)
↓
Its localization to one or more organs
↓
Its localization to one or more cellular organelles or subcellular fractions
↓
Delineation of the number of reactions involved in it
↓
Purification of its individual substrates, products, enzymes, and cofactors or other components
↓
Analyses of its control mechanisms in vitro
↓
Establishment of reaction mechanisms involved in it
↓
Its reconstitution
↓
Studies of it at the gene level by the methods of recombinant DNA

Figure 2–3. Scheme of the general strategy used to analyze a biochemical process or metabolic pathway. The approaches used need not necessarily be performed in the precise sequence listed here. However, it is by their employment that the details of a biochemical process or pathway are usually elucidated. The scheme thus applies at a general level to all of the major metabolic pathways discussed in subsequent chapters of this text.

each of the major biochemical processes described in this text (eg, glycolysis, fatty acid oxidation) is under consideration, although not every point listed will always be relevant.

Comments on the Preparations & Strategy Used to Elucidate Biochemical Processes

A number of important points concerning Table 2–7 and Fig 2–3 merit discussion. (1) Despite the possibility of artifacts, it is absolutely necessary to **isolate** and **identify** each of the components of a biochemical process in pure form in order to understand the process at a molecular level. Numerous examples of this will be encountered subsequently. (2) It is also important to be able to **reconstitute** the process under study in vitro by the systematic reassembly of its individual components. If the process does not function when its parts are reassembled, one explanation is that some critical component has escaped identification and has not been added back. (3) Recent technologic advances (eg, in NMR spectroscopy and positron emission tomography [PET scanning]) have permitted detection of certain biomolecules at the whole-organ level and monitoring of changes of their amounts with time. Such developments indicate that it is becoming possible to make sophisticated analyses of many biochemical processes at the in vivo level. (4) When the results obtained using many different levels of approach are consistent, then one is justified in concluding that real progress has been made in understanding the biochemical process under study. If major inconsistencies are obtained using the various approaches, then their causes must be investigated until rational explanations are obtained. (5) The preparations and levels of analyses outlined can be used to study biochemical alterations in animals with **altered metabolic states** (eg, fasting, feeding) or **specific diseases** (eg, diabetes mellitus, cancer). (6) Most of the methods and approaches indicated can be applied to studies of normal or diseased human cells or tissues. However, care must be taken to obtain such material in a fresh condition, and particular attention must be paid to those ethical considerations that apply to human experimentation.

The Use of Isotopes in Biochemistry

The introduction of the use of isotopes into biochemistry in the 1930s had a dramatic impact; consequently, their use deserves special mention. Prior to their employment, it was very difficult to "tag" biomolecules so that their metabolic fates could be monitored conveniently. Pioneering studies, particularly by Schoenheimer and colleagues, applied the use of certain stable isotopes (eg, D^2, N^{15}) combined with their detection by mass spectrometry to many biochemical problems. For instance, certain amino acids, sugars, and fatty acids could be synthesized containing a suitable stable isotope and then administered to an animal or added to an in vitro preparation to follow their metabolic fates (eg, half-lives, conversion to other biomolecules). Compounds labeled with suitable stable isotopes were used to investigate many aspects of the metabolism of proteins, carbohydrates, and lipids. From such studies, it became apparent that metabolism is a very active process, with most of the compounds in the cell being continuously synthesized and degraded although at widely differing rates. These findings were epitomized by Schoenheimer as "the dynamic nature of metabolism."

The subsequent introduction of **radioactive isotopes** and of instruments enabling their measurement was also extremely important. The principal stable and radioactive isotopes used in biologic systems are listed in Table 2–8. The use of isotopes, both stable and radioactive, is critical to the development of every area of biochemistry. Investigations of complex and simple biomolecules rely heavily on their use, either in vivo or in vitro. The tremendous progress made recently in sequencing nucleic acids and in measuring extremely small amounts of compounds found in biologic systems by radioimmunoassay has also depended on their employment.

PRINCIPAL ACHIEVEMENTS OF BIOCHEMISTRY

Following is a summary of the principal achievements made in the field of biochemistry, particularly in relation to human biochemistry.

- The overall chemical composition of cells, tissues, and the body has been determined, and the principal compounds present have been isolated and their structures established.
- The functions of many of the simple biomolecules are understood, at least at a general level, and are described later in this text. The functions of the major complex biomolecules have also been established. Of central interest is that DNA is known to be the genetic material and to transmit its information to one type of RNA (messenger RNA), which in turn dictates the linear sequence of amino acids in proteins. The flow of information from DNA may be conveniently represented as DNA → RNA → protein.
- The principal organelles of animal cells have been isolated and their major functions established.
- Almost all of the reactions occurring in cells have

Table 2–8. Principal isotopes used in biochemical research.

Stable Isotopes	Radioactive Isotopes
D^2	H^3
N^{15}	C^{14}
O^{18}	P^{32}
	S^{35}
	Ca^{35}
	I^{125}
	I^{131}

been found to be catalyzed by enzymes; many enzymes have been purified and studied, and the broad features of their mechanisms of action have been revealed.

- The metabolic pathways involved in the synthesis and degradation of the major simple and complex biomolecules have been delineated. In general, the pathway of synthesis of a compound has been found to be distinct from its pathway of degradation.
- A number of aspects of the regulation of metabolism have been clarified.
- The broad features of how cells conserve and utilize energy have been recognized.
- Many aspects of the structure and function of the various membranes found in cells are understood; proteins and lipids are their major components.
- Considerable information is available at a general level on how the major hormones act.
- Biochemical bases for a considerable number of diseases have been discovered.

HOW LITTLE IS KNOWN

While it is important to know that much biochemical information has been accumulated, it is just as important to appreciate how little is known in many areas. Probably the 2 major problems to be solved concern establishing the biochemical bases of **development and differentiation** and of **brain function.** Although the chemical nature of the genetic material is now well established, almost nothing is known about the mechanisms that turn genes on and off during development. Understanding **gene regulation** is also a key area in learning how cells differentiate and how they become cancerous. Knowledge of **cell division and growth**—both normal and malignant—and of their regulation is very primitive. Virtually nothing is known concerning the biochemical bases of **complex neural phenomena** such as consciousness and memory. Only very limited information is available concerning mechanisms of **cell secretion.** Despite some progress, the molecular bases of **most major genetic diseases** are unknown, but approaches provided by recombinant DNA technology suggest that remarkable progress will be made in this area during the next few years. The human genome may be sequenced by the turn of the next century or earlier; however, it is not clear whether such an approach is the best way to use available human and financial resources.

REFERENCES

Freifelder D: *Physical Biochemistry: Applications to Biochemistry and Molecular Biology.* Freeman, 1982.
Fruton JS: *Molecules and Life: Historical Essays on the Interplay of Chemistry and Biology.* Wiley-Interscience, 1972.
Weinberg RA: The molecules of life. *Sci Am* (Oct) 1985; **253:** 48. [Entire issue is of interest.]

Section I.
Structures & Functions of Proteins & Enzymes

Amino Acids

Victor W. Rodwell, PhD

3

INTRODUCTION

Living cells produce an impressive variety of **macromolecules (proteins, nucleic acids, polysaccharides)** that serve as structural components, biocatalysts, hormones, receptors, or repositories of genetic information. These macromolecules are **biopolymers** constructed of **monomer units,** or **building blocks.** For nucleic acids, the monomer units are **nucleotides;** for complex polysaccharides, the monomer units are **sugar derivatives;** and for proteins, the monomer units are L-α-**amino acids.**

While proteins may also contain substances other than amino acids, the 3-dimensional structure and thus the biologic properties of proteins are determined largely by the **kinds of amino acids present,** the **order in which they are linked together** in a polypeptide chain, and thereby the **spatial relationship of one amino acid to another.**

Amino acids play many important roles in cells; some of the many compounds of biologic importance that arise from amino acids are indicated in Tables 3–4 and 3–5.

BIOMEDICAL IMPORTANCE

In addition to being the building blocks of peptides and proteins, amino acids are important in other ways. Certain amino acids appear to be involved in the transmission of impulses in the nervous system; glycine and glutamic acid are examples. Essential amino acids must be supplied in the diet, since our bodies cannot synthesize them in amounts adequate to support growth (infants) or to maintain health (adults). The metabolism of amino acids gives rise to many compounds of biomedical importance. For instance, decarboxylation of certain amino acids produces the corresponding amines, some of which (eg, histamine and γ-aminobutyrate [GABA]) have important biologic functions. A number of diseases are due to abnormali-

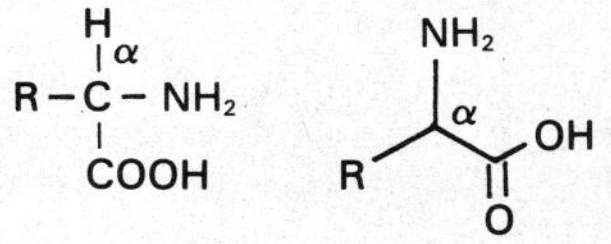

Figure 3–1. Two representations of an α-amino acid.

ties of the transport of amino acids into cells. Many of these conditions are characterized by the presence of greatly increased amounts of one or more amino acids in the urine, and for this reason they are often referred to as the aminoacidurias.

AMINO ACIDS

Amino acids contain both amino and carboxylic acid functional groups. In an α-amino acid, both are attached to the same (α) carbon atom (Fig 3–1).

Although about 300 amino acids occur in nature, only 20 of these occur in proteins. Complete hydrolysis* of proteins produces 20 L-α-amino acids (Table 3–2). The same 20 amino acids are present in proteins from all forms of life—plant, animal, or microbial. The reason for this becomes apparent when the universality of the genetic code is discussed (see Chapter 30). However, some proteins contain amino acid derivatives that are generated after incorporation of the amino acid into the protein molecule (see Table 5–4).

With the exception of glycine, for which R = a hydrogen atom (Fig 3–1), all 4 groups linked to the α-

---

*Hydrolysis = rupture of a covalent bond with addition of the atoms of water.

carbon atom of amino acids are different. This tetrahedral orientation of 4 different groups about the α-carbon atom confers optical activity (the ability to rotate the plane of plane-polarized light) on amino acids. Although some amino acids found in proteins are dextrorotatory and some levorotatory at pH 7.0, all have the **absolute configurations** of L-glyceraldehyde and hence are **L-α-amino acids.**

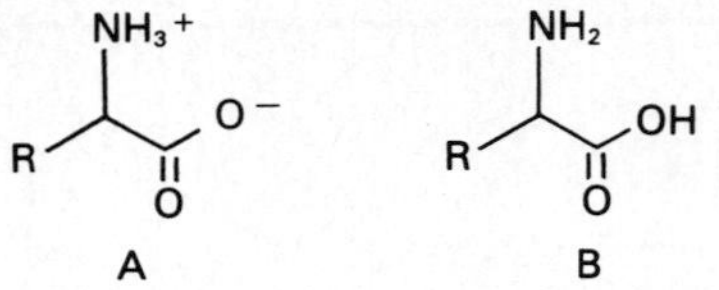

IONIC FORMS OF AMINO ACIDS

Amino acids bear at least **2 ionizable weak acid groups, a $-COOH$ and an $-NH_3{}^+$.** In solution, 2 forms of these groups, one charged and one uncharged, exist in protonic equilibrium:

$$R-COOH \rightleftharpoons R-COO^- + H^+$$

$$R-NH_3{}^+ \rightleftharpoons R-NH_2 + H^+$$

$R-COOH$ and $R-NH_3{}^+$ represent the **protonated** or **acidic** partners in these equilibria. $R-COO^-$ and $R-NH_2$ are the **conjugate bases** (ie, proton acceptors) of the corresponding acids. Although both $R-COOH$ and $R-NH_3{}^+$ are weak acids, $R-COOH$ is a several thousand times stronger acid than is $R-NH_3{}^+$. At the pH of blood plasma or the intracellular space (7.4 and 7.1, respectively), carboxyl groups exist almost entirely as **carboxylate ions, $R-COO^-$.** At these pH values, most amino groups are predominantly in the associated (protonated) form, $R-NH_3{}^+$. The prevalent ionic species of amino acids present in blood and most tissues should be represented as shown in Fig 3–2A. Structure B (Fig 3–2) cannot exist at *any* pH. At a pH sufficiently low to protonate the carboxyl group, the

Figure 3–2. Ionically correct structure for an amino acid at or near physiologic pH (*A*). The uncharged structure shown as (*B*) cannot exist at any pH but may be used as a convenience when discussing the chemistry of amino acids.

more weakly acidic amino group would also be protonated. The approximate pK_a values for α-carboxyl and α-amino groups of an α-amino acid are 2 and 10, respectively (Table 3–1). At a pH below its pK_a, an acid will be predominantly protonated, and at a pH 2 units below its pK_a, it will be approximately 99% protonated. If the pH is gradually raised, the proton from the carboxylic acid will be lost long before that from the $R-NH_3{}^+$. At any pH sufficiently high for the uncharged conjugate base of the amino group to predominate, a carboxyl group is present as the carboxylate ion ($R-COO^-$). However, for convenience the B representation is used for many equations not involving protonic equilibria.

The relative acid strengths of weak acids may be expressed in terms of their acid dissociation constant, K_a, or of their pK_a, the negative log of the dissociation constant, ie,

$$pK_a = -\log K_a{}^*$$

*For convenience, the a subscript in K_a and pK_a will be implied but dropped hereafter from the notation.

Table 3–1. Weak acid groups of the amino acids present in proteins.

	Conjugate Acid	Conjugate Base	Approximate pK_a
α-Carboxyl	$R-COOH$	$R-COO^-$	2.1 ± 0.5
Non-α-carboxyl (aspartate, glutamate)	$R-COOH$	$R-COO^-$	4.0 ± 0.3
Imidazolium (histidine)	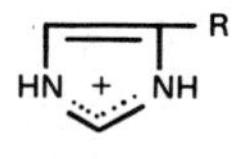	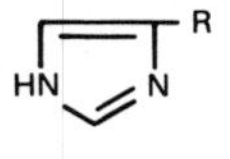	6.0
α-Amino	$R-NH_3{}^+$	$R-NH_2$	9.8 ± 1.0
ϵ-Amino (lysine)	$R-NH_3{}^+$	$R-NH_2$	10.5
Phenolic OH (tyrosine)	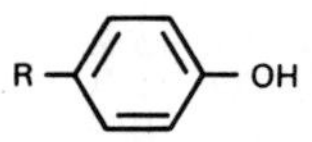	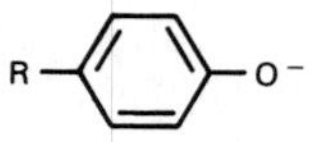	10.1
Guanidinium (arginine)			12.5
Sulfhydryl (cysteine)	$R-SH$	$R-S^-$	8.3

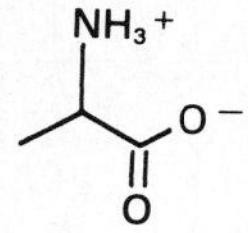

$$NH_3^+$$

Figure 3–3. Isoelectric or "zwitterionic" structure of alanine. Although charged, the zwitterion bears no *net* charge and hence does not migrate in a direct current electrical field.

Table 3–2. Classification of the L-α-amino acids present in proteins on the basis of the relative polarities of their R groups. A nonpolar group is one which has little or no charge difference from one region to another, whereas a polar group has a relatively large charge difference in different regions.

Nonpolar	Polar
Alanine	Arginine
Isoleucine	Asparagine
Leucine	Aspartic acid
Methionine	Cysteine
Phenylalanine	Glutamic acid
Proline	Glutamine
Tryptophan	Glycine
Valine	Histidine
	Lysine
	Serine
	Threonine
	Tyrosine

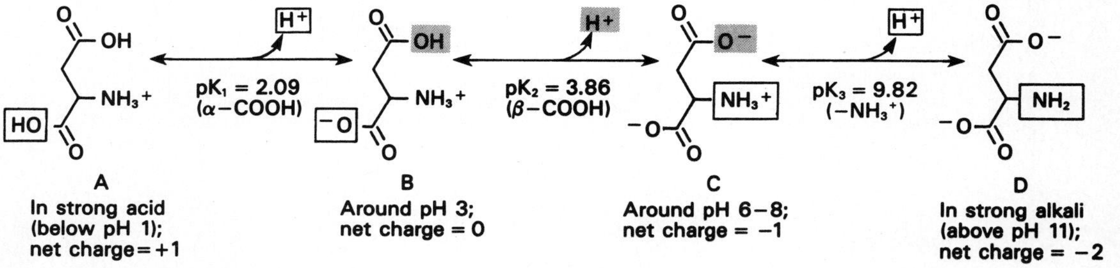

Table 3–1 lists the acidic groups and their pK values for the functional groups present on the 20 amino acids of proteins.

The **net charge** (the algebraic sum of all the positively and negatively charged groups present) of an amino acid **depends upon the pH, or proton concentration, of the surrounding solution.** The ability to alter the charge on amino acids or their derivatives by manipulating the pH facilitates the physical separation of amino acids, peptides, and proteins.

The pH at which an amino acid **bears no net charge** and hence does not move in a direct current electrical field is termed its **isoelectric pH (pI).** For an aliphatic amino acid such as alanine, the isoelectric species is the form shown in Fig 3–3.

The isoelectric pH is the pH midway between pK values on either side of the isoelectric species. For an amino acid with only 2 dissociating groups, there can be no possible ambiguity, as is shown below in the calculation of the pI for alanine. Since pK_1 (R–COOH) = 2.35 and pK_2 (R–NH_3^+) = 9.69, the isoelectric pH (pI) of alanine is:

$$pI = \frac{pK_1 + pK_2}{2} = \frac{2.35 + 9.69}{2} = 6.02$$

Calculation of pI for a compound with more than 2 dissociable groups carries more possibility for error. For example, from consideration of Fig 3–4, what would be the isoelectric pH (pI) for aspartic acid? To answer such a query unambiguously, write out all possible ionic structures for a compound in the order in which they occur as one proceeds from strongly acidic to basic solution (eg, as for aspartic acid in Fig 3–4). Next, identify the isoionic, zwitterionic, or neutral representation (as in Fig 3–4B). pI is the pH at the

midpoint between the pK values on either side of the isoionic species. In this example,

$$pI = \frac{2.09 + 3.86}{2} = 2.98$$

This approach works equally well for amino acids with additional dissociating groups, eg, lysine or histidine. After writing the formulas for all possible charged species of the basic amino acids lysine and arginine, observe that

$$pI = \frac{pK_2 + pK_3}{2}$$

For lysine, pI is 9.7; for arginine, pI is 10.8. The student should determine the pI for histidine.

The above approach, that of determining by inspection of charged structures the two pK values on either side of the zwitterion, is by no means limited in applicability to amino acids. It may be applied to calculation of the charge on a molecule with any number of dissociating groups. The ability to perform calculations of this type is of value in the clinical laboratory to predict the mobility of compounds in electrical fields and to select appropriate buffers for separations. For example, a buffer at pH 7.0 would suffice to separate 2 molecules with a pI of 6 and 8, respectively, because

Figure 3–4. Protonic equilibria of aspartic acid.

Table 3–3. L-α-Amino acids present in proteins.*

Name	Symbol	Structural Formula
With Aliphatic Side Chains		
Glycine	Gly [G]	H—CH—COO⁻ , with ⁺NH₃
Alanine	Ala [A]	CH₃—CH—COO⁻ , with ⁺NH₃
Valine	Val [V]	(H₃C)₂CH—CH—COO⁻ , with ⁺NH₃
Leucine	Leu [L]	(H₃C)₂CH—CH₂—CH—COO⁻ , with ⁺NH₃
Isoleucine	Ile [I]	CH₃—CH₂—CH(CH₃)—CH—COO⁻ , with ⁺NH₃
With Side Chains Containing Hydroxylic (OH) Groups		
Serine	Ser [S]	CH₂(OH)—CH—COO⁻ , with ⁺NH₃
Threonine	Thr [T]	CH₃—CH(OH)—CH—COO⁻ , with ⁺NH₃
Tyrosine	Tyr [Y]	See below.
With Side Chains Containing Sulfur Atoms		
Cysteine†	Cys [C]	CH₂(SH)—CH—COO⁻ , with ⁺NH₃
Methionine	Met [M]	CH₂—CH₂(S—CH₃)—CH—COO⁻ , with ⁺NH₃
With Side Chains Containing Acidic Groups or Their Amides		
Aspartic acid	Asp [D]	⁻OOC—CH₂—CH—COO⁻ , with ⁺NH₃
Asparagine	Asn [N]	H₂N—C(=O)—CH₂—CH—COO⁻ , with ⁺NH₃

*Except for hydroxylysine (Hyl) and hydroxyproline (Hyp), which are incorporated into polypeptide linkages as lysine and proline and subsequently hydroxylated (see Chapters 29 and 54), specific transfer RNA molecules exist for all the amino acids listed in Table 3–3. Their incorporation into proteins is thus under direct genetic control.

†Cystine consists of 2 cysteine residues linked by a disulfide bond:

$$^-OOC-\underset{\underset{NH_3^+}{|}}{CH}-CH_2-S-S-CH_2-\underset{\underset{NH_3^+}{|}}{CH}-COO^-$$

Table 3–3 (cont'd). L-α-Amino acids present in proteins.

Name	Symbol	Structural Formula
Glutamic acid	Glu [E]	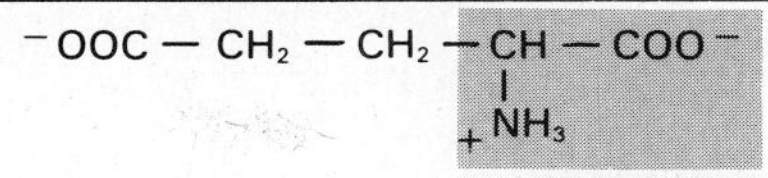 $^-OOC - CH_2 - CH_2 - CH - COO^-$ (with $^+NH_3$ on the α-carbon)
Glutamine	Gln [Q]	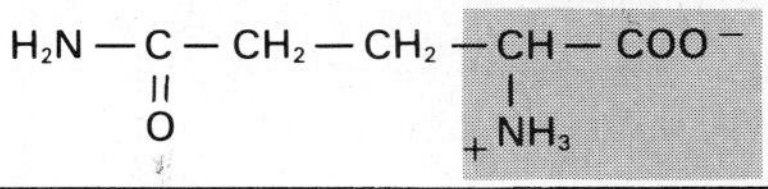 $H_2N - \underset{\underset{O}{\parallel}}{C} - CH_2 - CH_2 - CH - COO^-$ (with $^+NH_3$ on the α-carbon)

With Side Chains Containing Basic Groups

Name	Symbol	Structural Formula
Arginine	Arg [R]	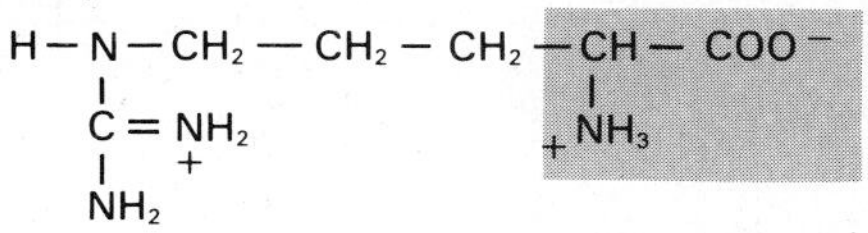 $H - N - CH_2 - CH_2 - CH_2 - CH - COO^-$ $\overset{\mid}{C} = NH_2$ $\underset{NH_2}{\overset{+}{\mid}}$ (with $^+NH_3$ on the α-carbon)
Lysine	Lys [K]	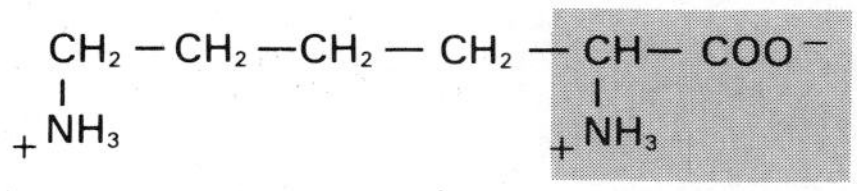 $CH_2 - CH_2 - CH_2 - CH_2 - CH - COO^-$ $^+NH_3$ (with $^+NH_3$ on the α-carbon)
Histidine	His [H]	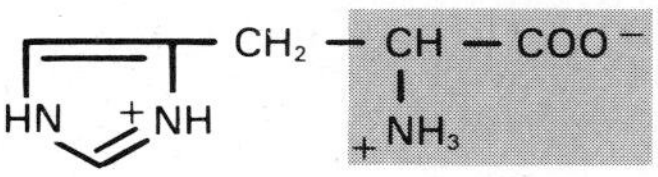 imidazole ring ($HN \quad ^+NH$) $- CH_2 - CH - COO^-$ (with $^+NH_3$ on the α-carbon)

Containing Aromatic Rings

Name	Symbol	Structural Formula
Histidine	His [H]	See above.
Phenylalanine	Phe [F]	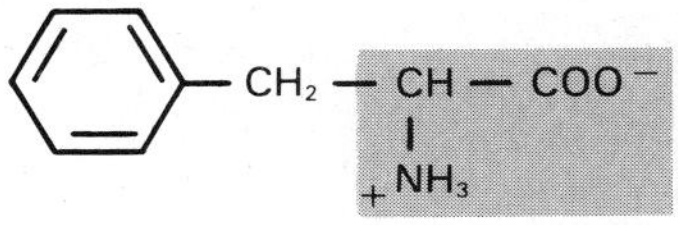 benzene ring $- CH_2 - CH - COO^-$ (with $^+NH_3$ on the α-carbon)
Tyrosine	Tyr [Y]	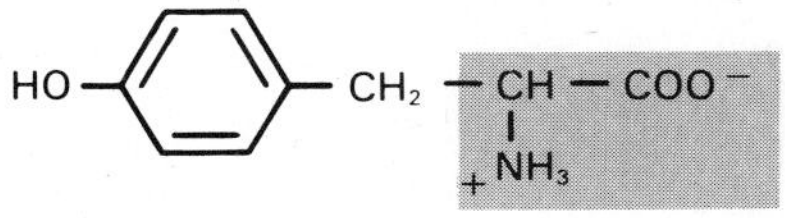 $HO -$ benzene ring $- CH_2 - CH - COO^-$ (with $^+NH_3$ on the α-carbon)
Tryptophan	Trp [W]	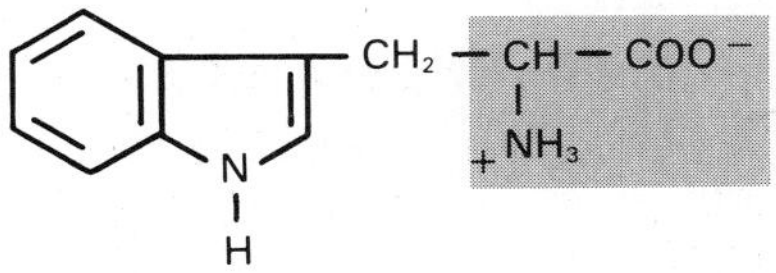 indole ring $- CH_2 - CH - COO^-$ (with $^+NH_3$ on the α-carbon)

Imino Acids

Name	Symbol	Structural Formula
Proline	Pro [P]	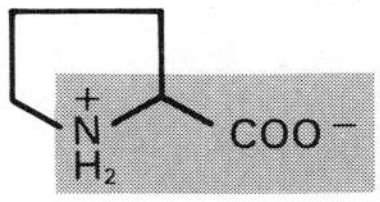 pyrrolidine ring with $\overset{+}{\underset{H_2}{N}}$ and COO^-

the molecule with pI = 6 will have a larger net negative charge at pH 7.0 than the molecule with pI = 8.

STRUCTURES OF AMINO ACIDS

The amino acids present in proteins may be divided into 2 broad groups on the basis of whether the R groups attached to the α-carbon atoms are polar or nonpolar (Table 3–2).

Table 3–3 gives, in addition, the 3-letter and single-letter abbreviations in common use among protein chemists. The single-letter abbreviations are used to represent extremely long sequences of amino acids (eg, for listing the complete sequence of amino acids in a protein). Tables 3–4 and 3–5 list important amino acids that occur in various natural products but not in proteins.

Amino acids that occur in free or combined states (but not in proteins) fulfill important roles in metabolic processes (Tables 3–4 and 3–5). For example, the amino acids ornithine, citrulline, and argininosuccinate (see Table 3–4) participate in the metabolism of urea. Over 20 D-amino acids occur naturally. These in-

Table 3–4. Selected examples of α-amino acids that do not occur in proteins but perform essential functions in mammalian metabolism.

Common and Systematic Names	Formula at Neutral pH	Significance
Homocysteine (2-amino-4-mercaptobutanoic acid)	$CH_2 - CH_2 - CH - COO^-$ | | SH $_+NH_3$	Intermediate in methionine biosynthesis (see Chapter 29).
Cysteine sulfinic acid (2-amino-3-sulfinopropanoic acid)	$CH_2 - CH - COO^-$ | | $SO_2^- {}_+NH_3$	Intermediate in cysteine catabolism (see Chapter 31).
Homoserine (2-amino-4-hydroxybutanoic acid)	$CH_2 - CH_2 - CH - COO^-$ | | OH $_+NH_3$	An intermediate in threonine, aspartate, and methionine metabolism (see Chapter 31).
Ornithine (2,5-bisaminopentanoic acid)	$CH_2 - CH_2 - CH_2 - CH - COO^-$ | | $_+NH_3$ $_+NH_3$	Intermediate in threonine, aspartate, and methionine metabolism (see Chapter 31).
Citrulline (2-amino-5-ureidopentanoic acid)	$CH_2 - CH_2 - CH_2 - CH - COO^-$ | | NH $_+NH_3$ | $C = O$ | NH_2	Intermediate in the biosynthesis of urea (see Chapter 31).
Argininosuccinic acid	$\overset{+}{NH}$ $CH_2 - CH_2 - CH_2 - CH - COO^-$ $\parallel$ | $HN - C - NH$ $_+NH_3$ | $^-OOC - CH_2 - C - COO^-$	Intermediate in the biosynthesis of urea (see Chapter 31).
Dopa (3,4-dihydroxyphenylalanine)	HO⬡$CH_2 - CH - COO^-$ with HO and $_+NH_3$	Precursor of melanin (see Chapter 32).
3-Monoiodotyrosine	HO⬡$CH_2 - CH - COO^-$ with I and $_+NH_3$	Precursor of thyroid hormones.
3,5-Diiodotyrosine	HO⬡$CH_2 - CH - COO^-$ with I, I and $_+NH_3$	Precursor of thyroid hormones.
3,5,3'-Triiodothyronine (T$_3$)	HO⬡$- O -$⬡$CH_2 - CH - COO^-$ with I's and $_+NH_3$	Precursor of thyroid hormones.
Thyroxine (3,5,3',5'-tetraiodothyronine; T$_4$)	HO⬡$- O -$⬡$CH_2 - CH - COO^-$ with I's and $_+NH_3$	Precursor of thyroid hormones.

Table 3–5. Selected examples of amino acids with non-α-amino groups that perform important functions in mammalian metabolism.

Common and Systematic Names	Formula at Neutral pH	Significance
β-Alanine (3-aminopropanoate)	$CH_2 - CH_2 - COO^-$ $\quad\underset{+}{\vert}NH_3$	Part of coenzyme A and of the vitamin pantetheine (see Chapter 17).
Taurine (2-aminoethylsulfonate)	$CH_2 - CH_2 - SO_3^-$ $\quad\underset{+}{\vert}NH_3$	Occurs in bile combined with bile acids (see Chapter 27).
γ-Aminobutyrate (GABA) (4-aminobutanoate)	$CH_2 - CH_2 - CH_2 - COO^-$ $\vert$ $\underset{+}{}NH_3$	Neurotransmitter formed from glutamate in brain tissue (see Chapter 32).
β-Aminoisobutyrate (2-methyl-3-aminopropanoate)	$H_3N^+ - CH_2 - CH - COO^-$ $\qquad\qquad\quad\vert$ $\qquad\qquad\quad CH_3$	End product of pyrimidine catabolism in urine of some persons (see Chapter 35).

clude the D-alanine and D-glutamate of certain bacterial cell walls and a variety of D-amino acids in antibiotics.

SOLUBILITY OF AMINO ACIDS

Since multiple charged groups are present on amino acids, they are readily solvated by and hence soluble in polar solvents such as water and ethanol but insoluble in nonpolar solvents such as benzene, hexane, and ether. Their high melting points (>200 °C) reflect the presence of charged groups—ie, the high energy needed to disrupt the ionic forces maintaining the crystal lattice.

GENERAL CHEMICAL REACTIONS

The carboxyl and amino groups of amino acids exhibit all the expected reactions of these functions, eg, salt formation, esterification, and acylation.

Color Reactions
Ninhydrin (Fig 3–5) oxidatively decarboxylates α-amino acids to CO_2, NH_3, and an aldehyde with one less carbon atom than the parent amino acid. The reduced ninhydrin then reacts with the liberated ammonia, forming a blue complex that maximally absorbs light of wavelength 570 nm. This blue color forms the basis of a **quantitative test for α-amino acids** that can detect as little as 1 μg of amino acid. Amines other than α-amino acids also react with ninhydrin, forming a blue color but without evolving CO_2. The evolution of CO_2 is thus indicative of an α-amino acid. NH_3 and peptides also react but more slowly than α-amino acids. Proline and 4-hydroxyproline produce a yellow color with ninhydrin.

Fluorescamine (Fig 3–6), an even more sensitive reagent, can detect nanogram quantities of an amino acid. Like ninhydrin, fluorescamine forms a complex with amines other than amino acids.

Formation of Peptide Bonds
The most important reaction of amino acids is the formation of the **peptide bond.** In principle, peptide bond formation involves removal of 1 mol of water between the α-amino group of one amino acid and the α-carboxyl group of a second amino acid (Fig 3–7). This reaction does not, however, proceed as written, since the equilibrium constant strongly favors peptide bond hydrolysis. To synthesize peptide bonds between 2 amino acids, the carboxyl group must first be **activated.** Chemically, this may involve prior conversion to an acid chloride. **Biologically, activation involves initial condensation with ATP** (see Chapter 30.)

PROPERTIES OF INDIVIDUAL AMINO ACIDS

Glycine, the smallest of the amino acids, can fit into regions of the 3-dimensional structure of proteins inaccessible to other amino acids.

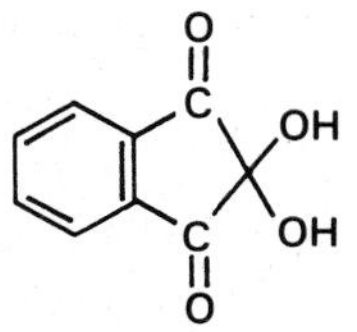

Figure 3–5. Ninhydrin.

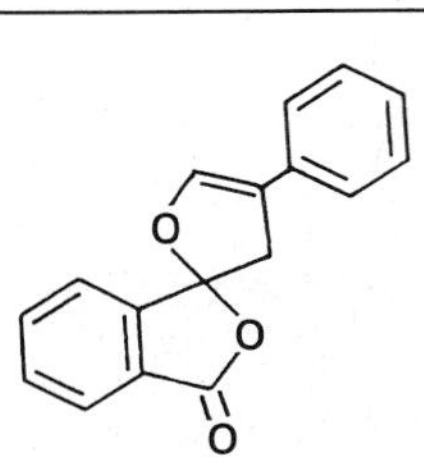

Figure 3–6. Fluorescamine.

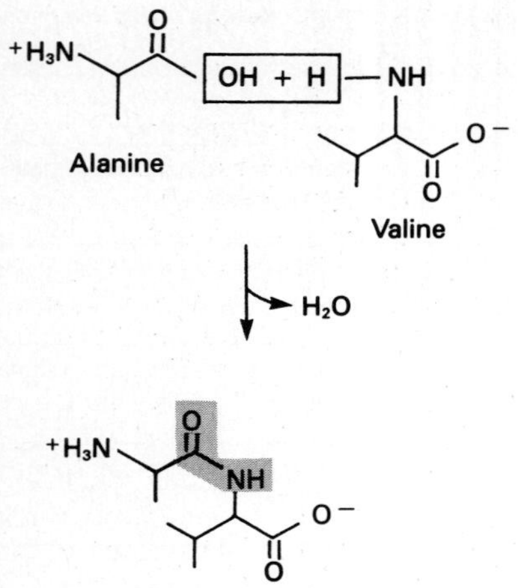

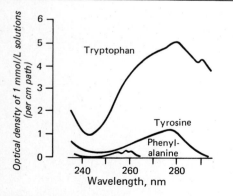

Figure 3–7. Amino acids united by a peptide bond (shaded).

Table 3–6. Phase relationships for chromatographic systems important in biochemistry.

Form of Chromatography	Stationary Phase	Mobile Phase
Partition chromatography on paper sheets, thin layers of cellulose powder, or columns of inert supports coated with thin layers of liquid; gel filtration.	Liquid	Liquid
Ion exchange; adsorption on thin layers or particles in columns.	Solid	Liquid
Partition chromatography between thin layer of liquid on support and mobile gas.	Liquid	Gas

The aliphatic R groups of alanine, valine, leucine, and isoleucine and the aromatic R groups of phenylalanine, tyrosine, and tryptophan are hydrophobic, a property that has important consequences for the ordering of water molecules in proteins in their immediate neighborhood.

The charged R groups of both the basic and acidic amino acids perform key roles in stabilization of specific protein conformations via formation of salt bonds. In addition, and together with histidine, amino acids with positively or negatively charged R groups function in the "charge relay" systems that transmit charges across considerable distances during enzymatic catalysis. Finally, histidine occupies a unique and important place in enzymatic catalysis, since the pK of its imidazole proton permits it, at pH 7.0, to function alternatively as either a base or an acid catalyst.

The primary alcohol group of serine and the primary thioalcohol (−SH) group of cysteine are excellent nucleophiles and may function as such during enzymatic catalysis. While the secondary alcohol group of threonine also is a good nucleophile, it is not known to perform this role in catalysis. In addition to its catalytic role, the −OH of serine functions in regulation of the activity of certain key metabolic enzymes whose catalytic activity depends upon the phosphorylation state of specific seryl residues.

Amino acids do not absorb visible light (ie, they are colorless) and, with the exceptions of the aromatic amino acids tryptophan, tyrosine, phenylalanine, and histidine, do not absorb ultraviolet light of a wavelength above 240 nm. As shown in Fig 3–8, above 240 nm, **most of the ultraviolet absorption of proteins is due to their tryptophan content.**

SEPARATORY TECHNIQUES FOR AMINO ACIDS

Chromatography

In all chromatographic separations, molecules are **partitioned between a stationary and a mobile phase** (Table 3–6). **Separation depends on the relative tendencies of molecules in a mixture to associate more strongly with one or the other phase.**

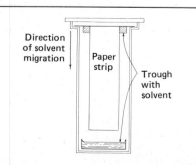

Figure 3–9. Apparatus for descending paper chromatography.

Figure 3–8. The ultraviolet absorption spectra of tryptophan, tyrosine, and phenylalanine.

While these separatory techniques are discussed principally with respect to amino acids, their use is by no means restricted to these molecules.

Paper Chromatography

While to a large extent supplanted by more sophisticated techniques, paper chromatography still finds application in amino acid separations. Samples are applied at a marked point about 5 cm from the end of a filter paper strip. The strip is then suspended in a sealed vessel that contains the chromatographic solvent (Fig 3–9).

For amino acid separations, solvents are polar binary, ternary, or more complex mixtures of water, alcohols, and acids or bases. The more polar components of the solvent associate with the cellulose and form the stationary phase. The less polar components constitute the mobile phase. This is **normal partition chromatography.** For **reversed phase partition chromatography,** the polarities of the mobile and stationary phases are reversed (eg, by first dipping the paper in a solution of a silicone). Reversed phase partition chromatography is used to separate nonpolar peptides or lipids, not polar compounds such as amino acids. The solvent may migrate up or down the paper (ascending or descending chromatography). When it has migrated almost to the end, the strip is dried and treated to allow visualization of the molecules of interest (eg, for amino acids, with 0.5% ninhydrin in acetone followed by heating at 90–110 °C for a few minutes). Amino acids with large nonpolar side chains (Leu, Ile, Phe, Trp, Val, Met, Tyr) migrate farther than those with shorter nonpolar side chains (Pro, Ala, Gly) or with polar side chains (Thr, Glu, Ser, Arg, Asp, His, Lys, Cys). (See Fig 3–10.) This reflects the greater relative solubility of polar molecules in the hydrophilic stationary phase and of nonpolar molecules in organic solvents. Note that, for a nonpolar series (Gly, Ala, Val, Leu), increasing length of the nonpolar side chain, which increases nonpolar character, results in increased mobility.

The ratio of the distance traveled by an amino acid to that traveled by the solvent front, both measured from the marked point of application of the amino acid mixture, is called the R_f **value** (mobility relative to the solvent front) for that amino acid. R_f values for a given amino acid vary with experimental conditions, eg, the solvent used. Although it is possible to identify tentatively an amino acid by its R_f value alone, it is preferable to chromatograph known amino acid standards simultaneously with the unknown mixture. In this case, mobility may be expressed relative to that of a standard (eg, as R_{ala} rather than as R_f). Mobilities expressed relative to a standard vary less than R_f values from experiment to experiment.

Quantitation of amino acids may be accomplished by cutting out each spot, eluting with a suitable solvent, and performing a quantitative colorimetric (ninhydrin) analysis. Alternatively, the paper may be sprayed with ninhydrin and the color densities of the spots measured with a recording transmittance or reflectance photometer.

For **2-dimensional paper chromatography,** sample is applied to one corner of a square sheet of paper or other suitable medium and chromatographed in one solvent mixture. The sheet is then removed, dried, turned through 90 degrees, and chromatographed in a second solvent (Fig 3–11).

Thin-Layer Chromatography

There are 2 distinct classes of thin-layer chromatography (TLC). Partition TLC (PTLC) closely re-

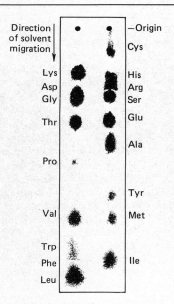

Figure 3–10. Identification of amino acids present in proteins. After descending paper chromatography in butanolacetic acid, spots were visualized with ninydrin.

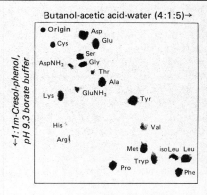

Figure 3–11. Two-dimensional chromatogram of protein amino acids. (Redrawn, slightly modified, from Levy AI, Chung D: Two-dimensional chromatography of amino acids on buffered papers. *Anal Chem* 1953;25:396. Copyright © 1953 by American Chemical Society; reproduced with permission.)

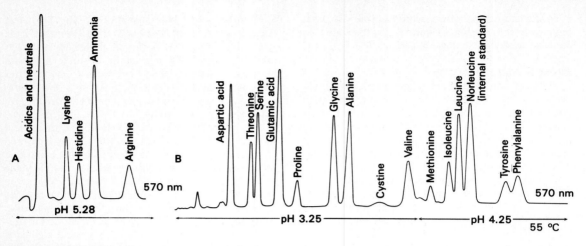

Figure 3–12. Automated analysis of an acid hydrolysate of corn endosperm on Moore-Stein Dowex 50 columns (at 55 °C). *A:* A short (5.0 × 0.9 cm) column used to resolve basic amino acids by elution at pH 5.28. Time required = 60 minutes. *B:* A longer (55 × 0.9 cm) column used to resolve neutral and acidic amino acids by elution first with pH 3.25 and then with pH 4.25 buffer. An internal standard of norleucine is included for reference. Basic amino acids remain bound to the column. Time required = 180 minutes. Emerging samples are automatically reacted with ninhydrin and the optical density of samples recorded at 570 nm and 440 nm. The latter wavelength is used solely to detect proline and hydroxyproline (absent from corn endosperm). Ordinate = optical density plotted on a log scale. Abscissa = time in minutes. (Courtesy of Professor ET Mertz, Purdue University.)

sembles partition chromatography on paper. Adsorption TLC (ATLC) bears no similarity to paper chromatography and is based on entirely different principles.

For PTLC on powdered cellulose or other relatively inert supports, the solvent systems and detection reagents used for paper chromatography are fully applicable. Reversed phase PTLC also is possible.

For ATLC, chromatography depends on the ability of the solvent (which need not be binary or more complex) to elute sample components from adsorption sites on an activated sorbent such as heated silica gel. ATLC is applicable to nonpolar materials such as lipids and hence not to amino acids or most peptides.

Automated Ion Exchange Chromatography

While amino acids may be separated by various techniques, analysis of amino acid residues after hydrolysis of a polypeptide generally involves **automated ion exchange chromatography.** Complete separation, identification, and quantitation require less than 3 hours. The procedure of Moore and Stein uses a short and long column containing the Na^+ form of a sulfonated polystyrene resin. When acid hydrolysate at pH 2 is applied to the columns, the amino acids bind via cation exchange with Na^+. The columns are then eluted with sodium citrate under preprogrammed conditions of pH and temperature. The short column requires a single elution buffer; the long column, two. Eluted material is reacted with ninhydrin reagent, and color densities are monitored in a flow-through colorimeter. Data are displayed on a strip chart recorder

that may incorporate computer-linked integration of peak areas (Fig 3–12).

High-Voltage Electrophoresis (HVE) on Inert Supports

Separations of amino acids, polypeptides, and other ampholytes (molecules whose net charge depends on the pH of the surrounding medium) in an imposed direct current field have extensive applications in biochemistry. For amino acids, paper sheets or thin layers of powdered cellulose are most frequently used as inert supports. Separations in a 2000- to 5000-volt direct current field for 0.5–2 hours depend upon the net charge on the ampholyte and its molecular weight. For molecules with identical charge, those of lower molecular weight migrate farther. Net charge is, however, the more important factor in determining separation. Applications include amino acids, low-molecular-weight polypeptides, certain proteins, nucleotides, and phosphosugars. Samples are applied to the support, which is then moistened with buffer of an appropriate pH and connected to buffer reservoirs by paper wicks. The paper may be covered by a glass plate or immersed in a hydrocarbon coolant. When current is applied, molecules with a net negative charge at the selected pH migrate toward the anode and those with a net positive charge toward the cathode. For visualization, the dried **electropherogram** is treated with ninhydrin (amino acids, peptides) or exposed to ultraviolet light (nucleotides), etc.

The choice of pH is dictated by the pK values of the dissociating groups on the molecules in the mixture. At pH 6.4, glutamate and aspartate bear a net charge of about -1, move toward the anode, and are readily separated on the basis of their difference in molecular weight. Lysine, arginine, and histidine move in the opposite direction, whereas all of the other protein amino acids remain at or near the point of application.

For separation of the peptides resulting from enzymatic digestion of a protein, a pH of 3.5 will induce a greater cationic charge on the molecules and thus provide better resolution.

REFERENCES

Barrett GC: *Chemistry and Biochemistry of the Amino Acids.* Chapman & Hall, 1985.

Cooper TG: *The Tools of Biochemistry.* Wiley, 1977.

Friedman M: *The Chemistry and Biochemistry of the Sulfhydryl Group in Amino Acids, Peptides, and Proteins.* Pergamon Press, 1973.

Greenstein JP, Winitz M: *Chemistry of the Amino Acids.* 3 vols. Wiley, 1961.

Heftman E: *Chromatography: A Laboratory Handbook of Chromatographic and Electrophoretic Methods,* 3rd ed. Van Nostrand, 1975.

Touchstone JC: *Practice of Thin Layer Chromatography.* Wiley-Interscience, 1978.

Zweig G, Sherma J: *Handbook of Chromatography.* 2 vols. CRC Press, 1972.

4

Peptides

Victor W. Rodwell, PhD

INTRODUCTION

When the amino and carboxyl groups of amino acids combine to form peptide bonds, the constituent amino acids are termed amino acid residues. **A peptide consists of 2 or more amino acid residues linked by peptide bonds.** Peptides of more than 10 amino acid residues are termed **polypeptides.**

BIOMEDICAL IMPORTANCE

The peptides are of immense biomedical interest, particularly in relation to endocrinology. Many of the major hormones of the body are peptides. The peptide hormones are frequently given to patients to correct corresponding deficiency states. The administration of insulin to patients with diabetes mellitus is the most prevalent example. Certain antibiotics are peptides (eg, valinomycin and gramicidin A), as are a few antitumor agents (eg, bleomycin). Recently developed methods permitting their rapid chemical synthesis have facilitated the manufacture of substantial amounts of peptide hormones, many of which are present in the body in relatively minute concentrations and thus difficult to isolate in quantities sufficient for therapy. The same technology allows the synthesis of other peptides, also obtainable from natural sources in only small amounts (eg, certain viral peptides and proteins), for use in vaccines.

PEPTIDE STRUCTURES

Representation of Peptide Structures

Fig 4–1 shows a tripeptide made up of the amino acid residues alanine, cysteine, and valine. Note that **a tripeptide is one with 3 residues, not 3 peptide bonds.** By convention, peptide structures are written with the **N-terminal residue** (the residue with a free α-amino group) **at the left** and with the **C-terminal residue** (the residue with a free α-carboxyl group) **at the right.** This peptide has a **single** free α-amino group and a **single** free α-carboxyl group (circled). This is true for all polypeptides comprised solely of amino acid residues linked by peptide bonds formed between α-amino and α-carboxyl groups. In some peptides, the terminal amino or carboxyl groups may

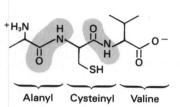

Figure 4–1. Structural formula for a tripeptide. Peptide bonds shaded for emphasis.

be derivatized (eg, an N-formyl amine or an amide of the carboxyl group) and thus not free.

Writing Structural Formulas of Peptides

What follows is a simple way to write peptide structures. First, draw its "backbone" of linked α-NH$_2$, α-COOH, and α-carbon atoms. These alternate along the backbone. Next, insert the appropriate side chains on the α-carbon atoms. This is illustrated below.

(1) Write a zig-zag of arbitrary length, and insert the N-terminal amino group:

(2) Insert the α-carbon, α-carboxyl, and α-amino groups:

(3) Add the appropriate R groups (shaded) and α-hydrogens to the α-carbon atoms:

Glu-Ala-Lys-Gly-Tyr-Ala

E A K G Y A

Figure 4–2. Use of 3-letter and one-letter abbreviations for amino acid residues to represent the primary structure of a hexapeptide with glutamate (Glu, E) at the N terminus and alanine (Ala, A) at the C terminus.

Table 4–1. pK values for glycine and glycine peptides.

	pK (COOH)	pK (NH$_3^+$)
Gly	2.34	9.60
Gly-Gly	3.12	8.17
Gly-Gly-Gly	3.26	7.91

Primary Structure of Peptides

The linear sequence of amino acid residues in a polypeptide constitutes its **primary structure.** When the **number, chemical structure,** and **order** of all of the amino acid residues in a polypeptide are known, its primary structure has been determined.

Since polypeptides (proteins) may contain 100 or more residues, it is often inconvenient to use conventional structural formulas to represent primary structure. The "chemical shorthand" used is either the 3-letter or one-letter abbreviation for the amino acids shown in column 2 of Table 3–3 (Fig 4–2). Note that **peptides are named as derivatives of the C-terminal amino acid residue.**

Three-letter abbreviations for amino acid residues linked by straight lines represent the primary structure that is known and unambiguous. These lines are omitted for single-letter abbreviations. Where there is uncertainty about the precise **order** of the amino acid residues of a portion of a polypeptide, the questionable residues are enclosed in brackets and separated by commas (Fig 4–3).

Physiologic Consequences of Changes in Primary Structure

Substitution of a single amino acid for another in a linear sequence of possibly 100 or more amino acids may reduce or abolish biologic activity, with potentially serious consequences (eg, sickle cell disease; see Chapter 6). Indeed, many inherited metabolic errors may involve no more than a single change of this type. The introduction of powerful new methods to determine protein and DNA structure has greatly increased our understanding of the biochemical basis for many inherited metabolic diseases.

IONIC FORMS OF PEPTIDES

The peptide (amide) bond is uncharged at any pH of physiologic interest. Formation of peptides from amino acids at pH 7.4 is therefore accompanied by a net loss of one positive and one negative charge per peptide bond formed. Peptides are, however, charged molecules at physiologic pH owing to the charges on the C- and N-terminal groups and on functional groups present in polar amino acid residues attached to the α-carbon atoms (see Table 3–1).

Polypeptides, like amino acids and other charged molecules, may be isolated by techniques (eg, electrophoresis, ion exchange chromatography) that separate on the basis of charge. The pK value for the C-terminal carboxyl group of a polypeptide is higher than that of the α-carboxyl group in the corresponding amino acid (ie, the peptide COOH is a weaker acid). Conversely, the N-terminal amino group is a stronger acid (has a lower pK) than the amino acid from which it was derived (Table 4–1).

CONFORMATION OF PEPTIDES IN SOLUTION

A large number of conformations (spatial arrangements) are possible for a polypeptide. The available evidence suggests, however, that in solution a narrow range of conformations tends to predominate. These favored conformations result from factors such as steric hindrance, coulombic interactions, H bonding, and hydrophobic interactions (see Chapter 5). As is the case for proteins, specific conformations are required for physiologic activity of polypeptides such as angiotensin and vasopressin (see Chapters 45 and 48).

Physiologically Active Peptides

Animal, plant, and bacterial cells contain a wide variety of low-molecular-weight polypeptides (3–100 amino acid residues) having profound physiologic activity. Some, including most mammalian polypeptide hormones, contain only peptide bonds formed between α-amino and α-carboxyl groups of the 20 L-α-amino acids present in proteins. However, additional amino acids or derivatives of the protein amino acids may also be present in polypeptides (though not in proteins). Shown below are a few selected examples.

The short polypeptides bradykinin and kallidin are smooth muscle hypotensive agents liberated from specific plasma proteins by proteolysis. Since they are derived from proteins, these peptides contain only the amino acids of proteins.

Glu-Lys-(Ala,Gly,Tyr)-His-Ala

Figure 4–3. A heptapeptide containing a region of uncertain primary structure.

Arg-Pro-Pro-Gly-Phe-Ser-Pro-Phe-Arg

Bradykinin

Lys-Arg-Pro-Pro-Gly-Phe-Ser-Pro-Phe-Arg

Kallidin

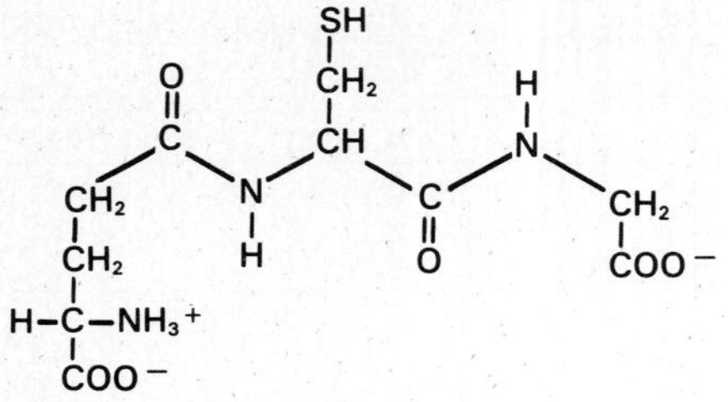

Figure 4–4. Glutathione (γ-glutamyl-cysteinyl-glycine).

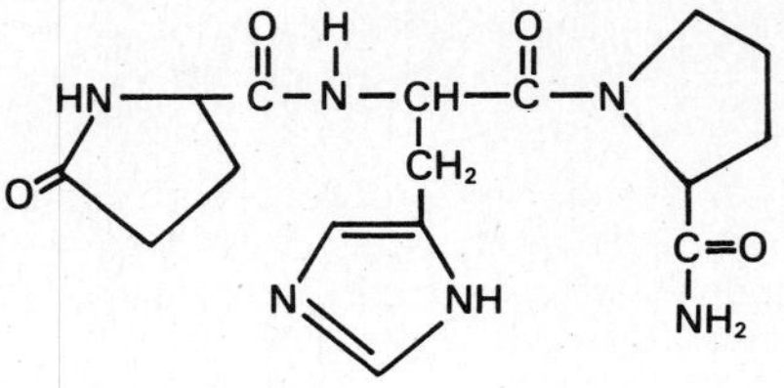

Figure 4–5. TRH (pyroglutamylhistidylprolinamide).

Glutathione (Fig 4–4), an atypical tripeptide in which the N-terminal glutamate is linked to cysteine via a non-α-peptidyl bond, is present in all forms of life. In humans and other animals, glutathione is required for the action of several enzymes. It is believed that glutathione and the enzyme glutathione reductase participate in the formation of the correct disulfide bonds of many proteins and polypeptide hormones (see Chapter 5).

Polypeptide antibiotics elaborated by fungi frequently contain both D- and L-amino acids and amino acids not present in proteins. Examples include tyrocidine and gramicidin S, cyclic polypeptides that contain D-phenylalanine, and the nonprotein amino acid ornithine. These polypeptides are not synthesized on ribosomes.

Thyrotropin-releasing hormone (TRH) (Fig 4–5) illustrates yet another variant. The N-terminal glutamate is cyclized to pyroglutamic acid, and the C-terminal prolyl carboxyl is amidated.

A mammalian polypeptide may contain more than one physiologically potent polypeptide. Within the primary structure of β-lipotropin—a hypophyseal hormone that stimulates the release of fatty acids from adipose tissue—are sequences of amino acids that are common to several other polypeptide hormones with diverse physiologic activities (Fig 4–6). The large polypeptide is a precursor of the smaller polypeptides.

SEPARATORY TECHNIQUES FOR PEPTIDES

Chromatography & High-Voltage Electrophoresis (HVE)

These techniques (see Chapter 3) are applicable to low-molecular-weight polypeptides as well as to amino acids.

Gel Filtration

Automated sequencing utilizes small numbers of large (30- to 100-residue) peptides. However, many denatured, high-molecular-weight polypeptides may be insoluble owing to exposure during denaturation of previously buried hydrophobic residues. While insolubility can be overcome by urea, alcohols, organic acids, or bases, these restrict the subsequent use of ion exchange techniques for peptide purification. Gel

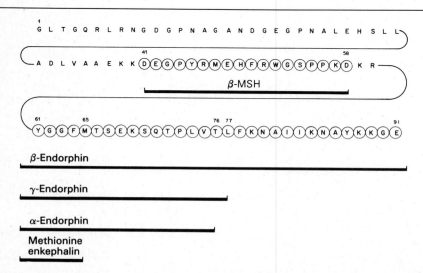

Figure 4–6. Primary structure of β-lipotropin. Residues 41–58 are melanocyte-stimulating hormone (β-MSH). Residues 61–91 contain the primary structures of the indicated endorphins.

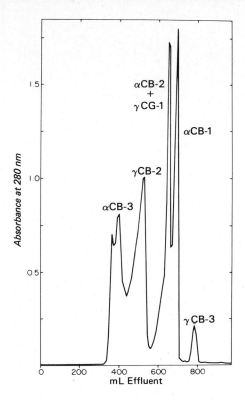

Figure 4–7. Gel filtration chromatography of cyanogen bromide (CB) fragments of human fetal globin. Chromatography was on G-50 Sephadex equilibrated and eluted with 1% HCOOH. Designations refer to arbitrarily numbered fragments from the α or γ chains. (Courtesy of JD Pearson et al, Department of Biochemistry, Purdue University.)

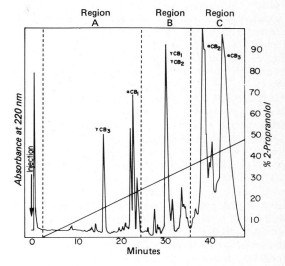

Figure 4–8. RPHPLC elution profile for cyanogen bromide (CB) fragments of human fetal globin. Designations are for arbitrarily numbered fragments from the α and γ chains. (Courtesy of JD Pearson et al, Department of Biochemistry, Purdue University.)

filtration of large hydrophobic peptides may, however, be performed in 1–4 molar formic or acetic acid (Fig 4–7).

Reversed Phase High-Pressure Liquid Chromatography (RPHPLC)

A powerful technique for purification of high-molecular-weight nonpolar peptides is high-pressure liquid chromatography on nonpolar materials with elution by polar solvents (RPHPLC). Fig 4–8 illustrates resolution of the same cyanogen bromide fragments of human fetal globin as shown in Fig 4–7 by this technique. Gel filtration and RPHPLC are used in conjunction to purify complex mixtures of peptides that result from partial digestion of proteins. New stationary phases have been developed for resolution of large peptides.

High-Voltage Electrophoresis (HVE) on Molecular Sieves

Molecular sieving may be superimposed on charge separation to facilitate separation. While starch and agarose are used, most commonly the support is a cross-linked polymer of acrylamide ($CH_2 =$

$CH \cdot CONH_2$). For **polyacrylamide gel electrophoresis (PAGE),** protein solutions are applied to buffered tubes or slabs of polyacrylamide cross-linked 2–10% by inclusion of methylene bisacrylamide ("bis") or similar cross-linking reagents. Direct current is then applied. Visualization is by staining with Coomassie blue or Ag^+ (polypeptides), ethidium bromide (polynucleotides), etc. A popular variant is PAGE under denaturing conditions. Proteins are boiled and subsequently electrophoresed in the presence of the denaturing agents, urea or sodium dodecyl sulfate (SDS), to produce conditions that favor separation based strictly on molecular size. SDS-PAGE is widely used to establish subunit molecular weights of proteins by comparison of mobilities with those of standards of known molecular weight.

DETERMINATION OF THE AMINO ACID COMPOSITION OF PEPTIDES

The peptide bonds linking the amino acids are first broken by hydrolysis. Since peptide bonds are stable at neutral pH, catalysis by acid or base is employed. Enzymatic catalysis is relatively unsuitable for complete hydrolysis. No procedure completely hydrolyzes proteins to constituent amino acids without partial loss of certain amino acid residues. The method of choice generally is hydrolysis in 6 N HCl at 110 °C in a sealed evacuated tube. Under these conditions, all of the tryptophan and cysteine and most of the cystine are destroyed. If metals are present, methionine and tyrosine are partially lost. Glutamine and asparagine are quantitatively deamidated to glutamate and aspartate. Recovery of serine and threonine is incomplete and decreases with increasing time of hydrolysis. Finally, certain bonds between neutral residues (Val-Val, Ile-Ile, Val-Ile, Ile-Val) are only 50% hydrolyzed after 20 hours. Typically, replicate samples are hydrolyzed for 24, 48, 72, and 96 hours. Serine and threonine data are then plotted on semilog paper and extrapolated back to zero time of hydrolysis. Valine and isoleucine are taken from 96-hour data. Dicarboxylic acids and their amides are determined together and are reported collectively as "Glx" or "Asx." Cysteine and cystine are converted to an acid-stable derivative (eg, cysteic acid) prior to hydrolysis. Base-catalyzed hydrolysis, which destroys serine, threonine, arginine, and cysteine and racemizes all amino acids, is employed to analyze for tryptophan. Following hydrolysis, amino acid composition may be determined by **automated ion exchange chromatography** (see Fig 3–12) or by high-pressure liquid chromatography.

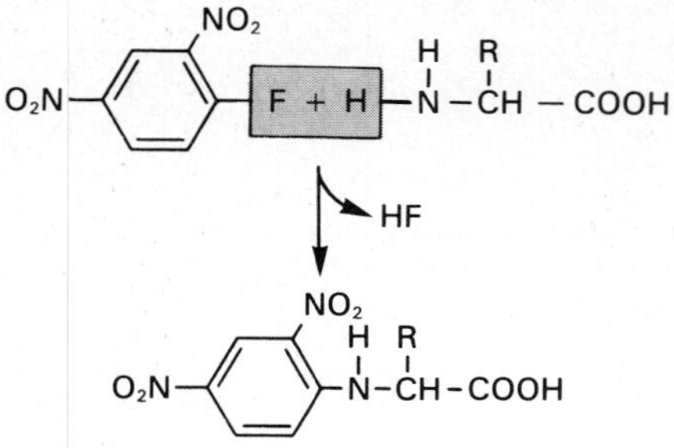

DETERMINATION OF THE PRIMARY STRUCTURE OF POLYPEPTIDES

Almost 3 decades have elapsed since Sanger amazed the biochemical world by applying chemical and enzymatic techniques to determine the complete primary structure of the polypeptide hormone insulin. Sanger's approach was first to separate the 2 polypeptide chains, A and B, of insulin and then to convert them by specific enzymatic cleavage into smaller peptides that contained regions of overlapping sequence. Using the reagent 1-fluoro-2,4-dinitrobenzene (Fig 4–9), he then removed and identified, one at a time, the N-terminal amino acid residues of these peptides. By comparison of the sequences of overlapping peptides, he was able to deduce an unambiguous primary structure for both the A and the B chains.

While the overall strategy of Sanger's approach remains valid to the present day, the ensuing decades have witnessed the introduction of 2 techniques that have revolutionized determination of the primary structures of polypeptides (proteins). The first of these was the introduction, in 1967, of an automated procedure for the sequential removal and identification of N-terminal amino acid residues as their phenylthiohydantoin derivatives. The second was the independent introduction by Sanger and by Maxam and Gilbert of techniques for the rapid and unambiguous sequencing of the DNA of the gene that codes for the protein in question. At present, optimal strategy is to utilize both

Figure 4–9. Reaction of an amino acid with 1-fluoro-2,4-dinitrobenzene (Sanger's reagent). The reagent is named for the Nobel laureate (1958) biochemist Frederick Sanger, who used it to determine the primary structure of insulin. This quantitatively arylates all free amino groups, producing intensely yellow 2,4-dinitrophenyl amino acids. These derivatives are readily quantitated by spectrophotometry. In addition to the N-terminal residue, the ϵ-amino groups of lysine, the imidazole of histidine, the OH of tyrosine, and the SH of cysteine also react with fluorodinitrobenzene. Since the dinitrophenyl group is resistant to removal by acid hydrolysis, it was used to determine the N-terminal amino acid of polypeptides.

approaches simultaneously. The automated Edman technique, while rapid compared to the manual methods used by Sanger, is slow relative to DNA sequencing methods, and difficulties may be encountered. On the other hand, DNA sequencing techniques do not infallibly yield unambiguous primary structures for the protein of interest. A major complication in sequencing eukaryotic genes is the presence of **introns** (see Chapter 38) within the gene that are not expressed in the mature protein. A major advantage of DNA sequencing is, however, the relative ease of detection and sequencing of regions of precursor molecules that may have eluded detection by the automated Edman technique owing to maturation of the protein prior to its isolation. DNA sequencing and the automated Edman technique thus are complementary techniques that have revolutionized and vastly expanded our knowledge of the primary structures of proteins. The former is discussed in Chapter 38; the latter is described below.

AUTOMATED EDMAN DETERMINATION OF POLYPEPTIDE STRUCTURES

Resolution Into Linear Polypeptides

Since many proteins consist of more than one polypeptide chain associated by noncovalent forces or disulfide bridges, the first step may be to dissociate and resolve individual polypeptide chains. Denaturing agents (urea, guanidine hydrochloride) that disrupt hydrogen bonds dissociate noncovalently associated polypeptides. Oxidizing and reducing agents disrupt

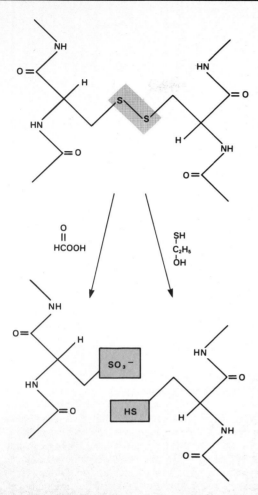

Figure 4–10. Representation of adjacent polypeptide chains linked by a disulfide bond (shaded). Oxidative cleavage by performic acid (left) or reductive cleavage by β-mercaptoethanol (right) forms 2 peptides that incorporate cysteic acid residues or cysteinyl residues, respectively.

disulfide bridges (Fig 4–10). Polypeptides are then separated by chromatographic techniques.

Cleavage of Polypeptides Into Fragments Suitable for Automated Sequencing

Automated sequencing instruments (sequenators) operate most efficiently on polypeptides 20–60 residues long. Automated techniques have therefore greatly influenced selection of the techniques for initial cleavage of polypeptides and for purification of the resulting fragments. Emphasis has shifted from production of large numbers of small fragments suitable for manual sequencing to small numbers of large (30- to 100-residue) fragments. Highly specific and complete cleavage at a restricted number of sites is therefore desired. Cleavage with cyanogen bromide

(CNBr), trypsin, or *o*-iodosobenzene meets these requirements.

A. CNBr: Cysteine residues are first modified with iodoacetic acid. CNBr then cleaves Met-peptide bonds specifically and, in most instances, quantitatively, on the COOH side of Met. Since Met is comparatively rare in polypeptides, this usually generates peptide fragments of the desired size range.

B. Trypsin: Trypsin cleaves on the COOH side of Lys and Arg residues. To restrict the number of cleavage sites, Lys residues are first derivatized with citraconic anhydride (a reversible reaction) to change the charge on Lys residues from positive to negative. Derivatization of Arg residues is less useful because of the relative abundance of Lys residues. It is, however, useful for subsequent cleavage of CNBr fragments.

C. *o*-Iodosobenzene: The reagent *o*-iodosobenzene cleaves specifically and quantitatively at the comparatively rare Trp-X residues. It requires no prior protection of other residues.

D. Hydroxylamine: Hydroxylamine cleaves comparatively rare Asn-Gly bonds, although generally not in quantitative yield.

E. Protease: *Staphylococcus aureus* protease V8 cleaves Glu-X-peptide residues with a preference for situations where X is hydrophobic. Glu-Lys resists cleavage. This reaction is useful for subsequent degradation of CNBr fragments.

F. Mild Acid Hydrolysis: This cleaves the rare Asp-Pro bond.

Two or 3 digests of the original polypeptide, normally at Met, Trp, Arg, and Asn-Gly, combined with appropriate subdigests of the resulting fragments, usually will permit determination of the entire primary structure of the polypeptide. Barring unusual difficulties in purification of fragments, this can—with care—be accomplished with a few micromoles of polypeptide.

Fragment purification is achieved chiefly by gel filtration in acetic or formic acid (Fig 4–7), by RPHPLC (Fig 4–8), or by ion exchange chromatography on phosphocellulose or sulfophenyl Sephadex in solutions of phosphoric acid.

The Edman Reagent & the Edman Reaction

The role performed initially by Sanger's reagent is replaced in automated sequencing by phenylisothiocyanate (Edman reagent) and a sequence of reactions that result in the removal of the N-terminal residue as its phenylthiohydantoin derivative (the Edman reaction, Fig 4–11).

The heart of the instrument is a spinning cup reaction chamber in which reactions occur in a thin film of solution on the cup wall. This facilitates extractions and subsequent removal of solvents. Several companies now market fully automated apparatus for the determination of polypeptide sequences of up to 30–40 residues (or, in exceptional cases, up to 60 or even 80 residues) in one continuous operation. The apparatus is programmed to perform sequential Edman degrada-

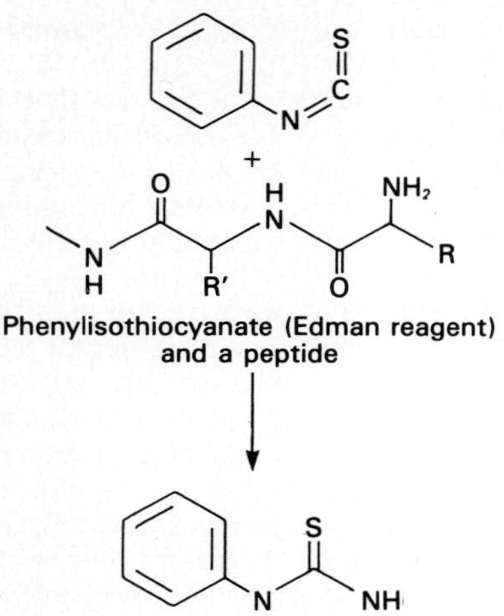

Phenylisothiocyanate (Edman reagent)
and a peptide

A phenylthiohydantoic acid

H^+, nitro-
methane $\longrightarrow H_2O$

A phenylthiohydantoin and a peptide
shorter by one residue

Figure 4–11. Conversion of an amino acid (or of the N-terminal residue of a polypeptide) to a phenylthiohydantoin. Phenylisothiocyanate reacts with the amino groups of amino acids and peptides, yielding phenylthiohydantoic acids. On treatment with acid in nonhydroxylic solvents, these cyclize to phenylthiohydantoins. The principal use of this reaction, which identifies the N-terminal residues of a peptide, is in automated sequencing of polypeptides.

tions on the N-terminal residue of a polypeptide. After the initial N-terminal amino acid has been removed, separated, and identified, an Edman derivative (Fig 4–11) of the next one in the sequence is formed, etc. Separation of the phenylthiohydantoin derivatives is accomplished by high-pressure liquid chromatography. The apparatus materially extends the sequence that can be determined by manual techniques and is incomparably faster.

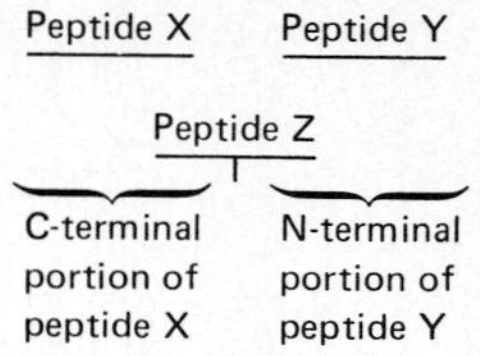

Figure 4–12. Use of the overlapping peptide Z to deduce that peptides X and Y are present in the original protein in the order X→Y, not Y→X.

Deduction of the Complete Primary Structure by Comparison of the Residue Sequences of Overlapping Peptides

The final step is to deduce the order of the constituent peptides derived from the unmodified protein. For this purpose, it is essential to have sequenced peptides produced by techniques (eg, digestion with trypsin and chymotrypsin) that cleave the protein at different locations. These are then compared to establish an unequivocal primary structure (Fig 4–12).

SYNTHESIS OF PEPTIDES BY AUTOMATED TECHNIQUES

While classic chemical techniques were adequate for synthesis of the octapeptides vasopressin and oxytocin, and later of bradykinin, the yields of final product are too low to permit synthesis of long polypeptides or proteins. This has been achieved by the automated, solid-phase synthesis technique developed by Merrifield. In this process, an automated synthesis is carried out in a single vessel by a machine programmed to add reagents, remove products, etc, at timed intervals. The steps involved are as follows:

(1) The amino acid that ultimately will form the C-terminal end of the polypeptide is attached to an insoluble resin particle.

(2) The second amino acid bearing an appropriately blocked amino group is introduced and the peptide bond is formed in the presence of the dehydrating agent dicyclohexylcarbodiimide.

(3) The blocking group is removed with acid, forming gaseous products, which are removed.

(4) Steps 2 and 3 are repeated with the next amino acid in sequence, then the next, until the entire polypeptide attached to the resin particle has been synthesized.

(5) The polypeptide is cleaved from the resin particle.

The process proceeds rapidly and with excellent yields. About 3 hours are required per peptide bond synthesized. By this technique, the A chain of insulin (21 residues) was synthesized in 8 days and the B chain (30 residues) in 11 days. The crowning achievement to date has been the total synthesis of pancreatic ribonuclease (124 residues; see Fig 5–10) in 18%

overall yield, the first total synthesis of an enzyme. It foreshadows a new era not only in confirmation of protein structures but in related areas such as immunology and vaccine production and perhaps in the treatment of inborn errors of metabolism. A variety of physiologically important peptides, prepared synthetically from L-amino acids by routes that involve no racemization, have full physiologic activity. These include the octapeptides oxytocin and vasopressin, ACTH, and melanocyte-stimulating hormone (see Chapter 45).

REFERENCES

Cantor CR, Schimmel PR: *Biophysical Chemistry, Part I: The Conformation of Macromolecules*. Freeman, 1980.

Chin CCQ, Wold F: Separation of peptides on phosphocellulose and other cellulose ion exchangers. *Methods Enzymol* 1977;**47**:204.

Cooper TG: *The Tools of Biochemistry*. Wiley, 1977.

Craig LC, Cowburn D, Bleich H: Methods for the study of small polypeptide hormones and antibiotics in solution. *Annu Rev Biochem* 1975;**44**:509.

Dayhoff M (editor): *Atlas of Protein Sequence and Structure*. Vol 5. National Biomedical Research Foundation, Washington, DC, 1972. Suppl 1, 1973; Suppl 2, 1976; Suppl 3, 1979.

Hash JH (editor): Antibiotics. In: *Methods in Enzymology*. Vol 43. Academic Press, 1975.

Heftman E: *Chromatography: A Laboratory Handbook of Chromatographic and Electrophoretic Methods,* 3rd ed. Van Nostrand, 1975.

Jauregui-Adell J, Marti J: Acidic cleavage of the aspartyl-proline bond and the limitations of the reaction. *Anal Biochem* 1975;**69**:468.

Mahoney WC, Hermodson MA: High-yield cleavage of tryptophanyl peptide bonds by *o*-iodosobenzoic acid. *Biochemistry* 1979;**18**:3810.

Mahoney WC, Smith PK, Hermodson MA: Fragmentation of proteins with *o*-iodosobenzoic acid: Chemical mechanism and identification of *o*-iodosobenzoic acid as a reactive contaminant that modifies tyrosyl residues. *Biochemistry* 1981;**20**:443.

Marglin A, Merrifield RB: Chemical synthesis of peptides and proteins. *Annu Rev Biochem* 1970;**39**:841.

Needelman SB (editor): *Protein Sequence Determination*. Springer-Verlag, 1970.

Patthy L, Smith EL: Reversible modification of arginine residues: Application to sequence studies by restriction of tryptic hydrolysis to lysine residues. *J Biol Chem* 1975;**250**:557.

Pearson JD et al: Reversed-phase supports for the resolution of large denatured protein fragments. *J Chromatogr* 1981;**207**:325.

Regnier FE, Gooding KM: High performance liquid chromatography of proteins. *Anal Biochem* 1980;**103**:1.

Snyder SH, Innes RB: Peptide neurotransmitters. *Annu Rev Biochem* 1979;**48**:755.

Stewart JM, Young JD: *Solid Phase Peptide Synthesis*. Freeman, 1969.

Storm DR, Rosenthal KS, Swanson PE: Polymyxin antibiotics. *Annu Rev Biochem* 1977;**46**:723.

Zweig G, Sherma J: *Handbook of Chromatography*. 2 vols. CRC Press, 1972.

Proteins: Structure & Properties

Victor W. Rodwell, PhD

INTRODUCTION

All proteins are **high-molecular-weight polypeptides.** While arbitrary, the dividing line between large polypeptides and small proteins is customarily drawn between MW 8000 and 10,000. **Simple proteins** contain only amino acids. **Complex proteins** contain additional, non-amino acid materials such as heme, vitamin derivatives, lipid, or carbohydrate. This chapter deals with the properties of simple proteins. The unique properties of specific complex proteins such as heme proteins (see Chapter 6), glycoproteins (see Chapter 54), and lipoproteins (see Chapter 26) are considered elsewhere, as are the properties of simple proteins of highly individual structure such as collagen and contractile proteins (see Chapter 56).

BIOMEDICAL IMPORTANCE

Proteins play a central role in cell function (for example, as enzymes) and cell structure. Analyses of certain proteins and enzymes of the blood are widely used for diagnostic purposes. In particular, electrophoretic analysis of the plasma albumin:globulin ratio is an integral part of the diagnostic workup of liver disease. Analyses of plasma lipoproteins and of plasma immunoglobulins by electrophoresis and other methods are commonly employed for the diagnosis of specific types of hyperlipoproteinemias and of specific types of immune disorders, respectively. Normal human urine is usually free of protein; thus, detection by appropriate laboratory tests of significant proteinuria is generally an important indicator of renal diseases, such as the various forms of nephritis.

CLASSIFICATION OF PROTEINS

Since no universally satisfactory system of protein classification exists, several mutually contradictory protein classification systems persist in current use. All are of comparatively limited value in assisting our understanding of many key properties of proteins. The persistence of these systems and terms—particularly in the clinical laboratory—dictates, however, brief consideration. Discussed below are salient features of protein classification systems based on solubility, shape, function, physical properties, and 3-dimensional structure.

Table 5–1. Classification of p
solubilities

Albumins	Soluble in water and tive amino acids.
Globulins	Sparingly soluble in solutions. No distinc
Prolamines	Soluble in 70–80% water and absolute
Histones	Soluble in salt solut
Scleroproteins	Insoluble in water o Ala, Pro.

Solubility

A classification system ba
oped in 1907–1908 is still in u
clinical biochemistry (Table
marcation between the classe
example, a clear distinction between albumins and globulins cannot be made solely on the basis of their solubilities in water or salt solutions. Globulins were therefore subdivided into pseudoglobulins, which are freely water-soluble, and euglobulins, which are insoluble in salt-free water.

Overall Shape

Two broad classes of proteins may be distinguished on the basis of their **axial ratios** (ratios of length to breadth). **Globular proteins,** which have axial ratios less than 10 and generally not over 3–4, are characterized by compactly folded and coiled polypeptide chains. Examples include insulin, plasma albumins and globulins, and many enzymes. **Fibrous proteins,** which have axial ratios greater than 10, are characterized by groups of polypeptide chains coiled in a spiral or helix and cross-linked covalently or by hydrogen bonds. Examples include keratin, myosin, collagen, and fibrin.

Function

Proteins may be classified according to their biologic functions—for example, as structural, catalytic, or transport proteins (Table 5–2). Catalytic proteins (enzymes), which comprise the majority of protein types, are themselves classified by the type of reaction they catalyze (see Chapter 7).

Physical Properties

For certain proteins of medical interest, there are

Table 5–2. Principal functions of the proteins.

Function	Protein (Examples)
Catalytic role	Enzymes
Contraction	Actin, myosin
Gene regulation	Histones, nonhistone nuclear proteins
Hormonal role	Insulin
Protection	Fibrin, immunoglobulins, interferon
Regulatory role	Calmodulin
Structural role	Collagen, elastin, keratins
Transport	Albumin (of bilirubin, fatty acids, etc), hemoglobin (oxygen), lipoproteins (various lipids), transferrin (iron)

specialized systems of classification that distinguish between closely related proteins. For example, two systems of nomenclature for plasma lipoproteins are in wide use, and a third is under consideration. One system distinguishes lipoproteins by their behavior in electrical or gravitational fields as "origin," α_1-, α_2-, β-, and γ-lipoproteins, on the basis of their electrophoretic mobility at pH 8.6. Lipoproteins are also classified on the basis of their hydrated densities as chylomicrons, VLDL, LDL, HDL, and VHDL. Yet another way in which they might be classified is based on the primary structure of the apoproteins. Six broad classes of plasma lipoproteins might be differentiated, based on whether apoprotein A, B, C, D, E, or F is present. The apoproteins may be differentiated by immunologic criteria.

Three-Dimensional Structure

Proteins may be distinguished on the basis of whether or not they possess quaternary structure (see below). In addition, similarities in structure, revealed primarily by x-ray crystallography, provide a potentially valuable basis for protein classification. For instance, proteins that bind nucleotides share a "nucleotide-binding domain" of tertiary structure, and these proteins may be evolutionarily related.

BONDS RESPONSIBLE FOR PROTEIN STRUCTURE

The primary structure of proteins derives from covalent linkage of L-α-amino acids by α-peptide bonds. While this was deduced long ago by multiple lines of evidence, the most convincing proof was the total chemical synthesis of insulin and ribonuclease solely by linking amino acids via peptide bonds.

Most protein structures are stabilized by 2 classes of strong bonds (peptide and disulfide) and 3 classes of weak bonds (hydrogen, hydrophobic, and electrostatic or salt).

Rigidity of the Peptide Bond

While peptides are written with a single bond con-

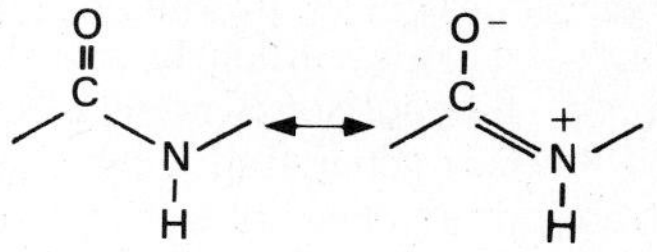

Figure 5–1. Resonance stabilization of the peptide bond confers partial double bond character, and hence rigidity, on the C–N bond.

necting α-carboxyl and α-nitrogen atoms, the carbon-nitrogen bond in fact has partial double bond character (Fig 5–1). There is no freedom of rotation about the bond that connects the C and N atoms, and all 4 of the atoms shown in Fig 5–1 lie in the same plane (ie, are coplanar). There is, by contrast, ample freedom of rotation about the remaining bonds of the polypeptide backbone. These concepts are summarized in Fig 5–2 where the bonds having freedom of rotation are shown circled by arrows and the coplanar atoms are shown in shaded boxes. This semirigidity has important consequences for orders of protein structure above the primary level.

Interchain & Intrachain Cross-Linking of Polypeptides by Disulfide Bonds

The disulfide bond formed between 2 cysteine residues links 2 portions of polypeptide chains through

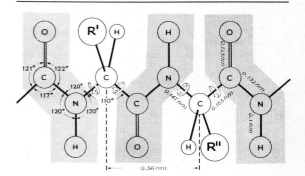

Figure 5–2. Dimensions of a fully extended polypeptide chain. The 4 atoms in shaded boxes, which are **coplanar**, comprise the polypeptide bond. The unshaded atoms are the α-carbon atom, the α-hydrogen atom, and the α-R group of the particular amino acid. Free rotation can occur about the bonds connecting the α-carbon with the α-nitrogen and α-carbonyl functions (white arrows). The extended polypeptide chain is thus a semirigid structure with two-thirds of the atoms of the backbone held in a fixed planar relationship one to another. The distance between adjacent α-carbon atoms is 0.36 nm. The interatomic distances and bond angles, which are not equivalent, are also shown. (Redrawn and reproduced, with permission, from Pauling L, Corey LP, Branson HR: The structure of proteins: Two hydrogen-bonded helical configurations of the polypeptide chain. *Proc Natl Acad Sci USA* 1951; 37:205.)

cysteine residues. The cystine bond is resistant to usual conditions for protein denaturation. Performic acid (oxidizes S–S bonds) or β-mercaptoethanol (reduces the S–S bonds generating 2 cysteine residues) separates polypeptide chains linked by disulfide bonds without affecting primary structure (see Fig 4–10).

Interchain & Intrachain Stabilization of Polypeptides by Hydrogen Bonds

Hydrogen bonds formed between bonding residues present in the side chains of peptide-linked amino acids, those formed between the hydrogen and oxygen atoms of the peptide bonds themselves, and those formed between polar residues on the surface of proteins and water all play important roles in the maintenance of protein structure above the primary order (see α-Helix and β-Pleated Sheet, below).

Hydrophobic Interactions

The nonpolar side chains of neutral amino acids tend to associate in proteins. Since the relationship is not stoichiometric, no true bond exists. Nonetheless, these interactions play a significant role in maintaining protein structure.

Electrostatic Bonds

These are salt bonds formed between oppositely charged groups in the side chains of amino acids. For example, the epsilon-amino group of lysine bears a net charge of +1 at physiologic pH and the non-α-carboxyl of aspartate and glutamate a net charge of −1. These may therefore interact electrostatically to stabilize a protein structure.

Bond Stabilities

During denaturation of proteins (eg, by reagents such as urea, sodium dodecyl sulfate [SDS], mild H^+, or OH^-), hydrogen, hydrophobic, and electrostatic bonds—but not peptide or disulfide bonds—are broken.

ORDERED CONFORMATIONS OF POLYPEPTIDES

The α Helix

The discovery that polypeptide chains may exist in highly ordered conformations maintained by hydrogen bonds formed between peptide residues constituted a major conceptual advance in our understanding of protein structure. Although the existence of these highly ordered structures was subsequently amply confirmed by high-resolution x-ray crystallography of crystalline proteins, their existence was first proposed from purely theoretical considerations.

X-ray data obtained in the early 1930s indicated that hair and wool α-keratins possessed repeating units spaced 0.5–0.55 nm along their longitudinal axis. However, no dimension of the extended polypeptide chain appears to measure 0.5–0.55 nm

(Fig 5–2). This apparent anomaly was resolved by Pauling and Corey, who proposed that the polypeptide chain of α-keratin is arranged as an α helix (Fig 5–3). In this structure, the R groups on the α-carbon atoms protrude outward from the center of the helix (Fig 5–4). There are 3.6 amino acid residues per turn of the helix, and the distance traveled per turn is 0.54 nm—a reasonable approximation of the 0.5- to 0.55-nm spacing observed by x-ray diffraction. The spacing per amino acid residue is 0.15 nm, which also corresponds with x-ray data. The main features of the α helix (Fig 5–5) are as follows:

(1) The α helix is stabilized by inter-residue hydrogen bonds formed between the H atom attached to a peptide N and the carbonyl O of the residue fourth in line behind in the primary structure.

(2) Each peptide bond participates in the H-bonding. This confers maximum stability.

(3) All of the main chain peptide N and carbonyl O residues are hydrogen-bonded, thus greatly reducing

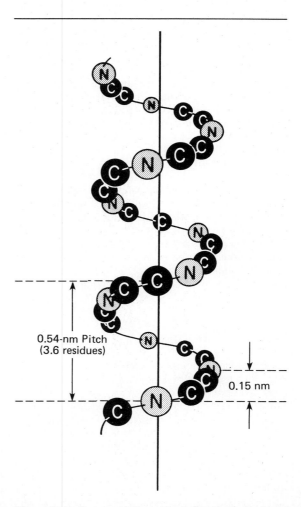

Figure 5–3. Representation of the helical (spiral) orientation of the main chain of a peptide about the axis of an α helix.

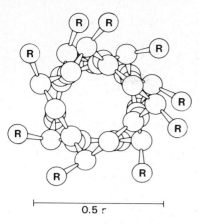

Figure 5-4. Cross-sectional view of an α helix. The side chains (R) are on the outside of the helix. The van der Waals radii of the atoms are larger than shown here; hence, there is almost no free space inside the helix. (Slightly modified and reproduced, with permission, from Stryer L: *Biochemistry,* 2nd ed. Freeman, 1981. Copyright © 1981 by W.H.Freeman and Co.)

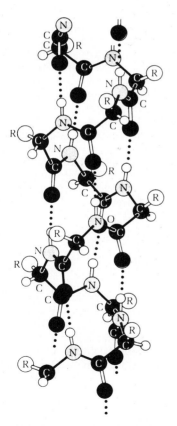

Figure 5-5. Representation of the R substituents on the α carbons and the H and O atoms involved in the hydrogen bonds (dots) that hold a polypeptide in an α-helical conformation. (Reprinted with permission, from Haggis GH et al: *Introduction to Molecular Biology.* Wiley, 1964.)

the hydrophilic (increasing the hydrophobic) nature of the α-helical region.

(4) An α helix forms spontaneously, since it is the lowest energy, most stable conformation for a polypeptide chain.

(5) When the residues are L-amino acids, the right-handed helix that occurs in proteins is significantly more stable than the left-handed helix.

Certain amino acids tend to disrupt the α helix. Among these are proline (the N-atom is part of a rigid ring and no rotation of the N–C bond can occur) and amino acids with charged or bulky R groups that either electrostatically or physically interfere with helix formation (Table 5–3).

The β-Pleated Sheet

Pauling and Corey also proposed a second ordered structure, the β-pleated sheet (β because it was their second structure, the α helix being the first). Whereas in the α helix the polypeptide chain is condensed, in the β-pleated sheet it is almost fully extended (Fig 5–6). When the adjacent polypeptide chains in a β-pleated sheet run in opposite directions (N to C terminus), the structure is termed an **antiparallel** β-pleated sheet (shown in Fig 5–6). When the chains run in the same direction, it is termed **parallel** (not shown).

Regions of β-pleated structure are present in many proteins, and both parallel and antiparallel forms occur. From 2 to 5 adjacent strands of polypeptide may combine to form these structures. Fig 5–7 illustrates a region of ribonuclease in which 3 sections of polypeptide chain form a β-pleated sheet structure. Both α helix and β-pleated sheet structures commonly occur in many proteins (see Fig 5–7).

While the α helix is stabilized by hydrogen bonding between peptide bonds 4 residues apart in a primary structural sense, stabilization of the β-pleated sheet results from formation of hydrogen bonds between peptides far removed from one another in a primary structural sense. This is illustrated in Fig 5–7.

The Random Coil

Regions of proteins that are not identifiably organized as helices or pleated sheets are said to be present in random coil conformation. As shown in Fig 5–7, a

Table 5-3. Effect of various amino acid residues on α helix formation.

Promote α Helix	Destabilize α Helix	Terminate α Helix
Ala	Arg	Pro
Asn	Asp	Hyp
Cys	Glu	
Gln	Gly	
His	Lys	
Leu	Ile	
Met	Ser	
Phe	Thr	
Trp		
Tyr		
Val		

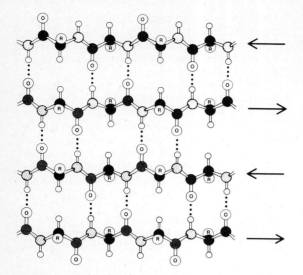

Figure 5–6. Antiparallel β-pleated sheet. Adjacent strands run in opposite directions. Hydrogen bonds between NH and CO groups of adjacent strands stabilize the structure. The side chains (R) are above and below the plane of the sheet. ● Carbon atoms; ◍ nitrogen atoms; ○ hydrogen atoms. (Modified and reproduced, with permission, from Stryer L: *Biochemistry,* 2nd ed. Freeman, 1981. Copyright © 1981 by W.H. Freeman and Co.)

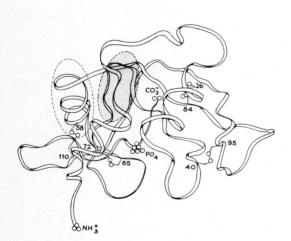

Figure 5–7. Representation of the main chain folding of bovine pancreatic ribonuclease, a single chain of 124 residues cross-linked at 4 places by disulfide bridges. A region of α helix is indicated by the dotted oval, and a region of pleated sheet is shaded. Other portions of the molecule are predominantly random coil. The active site (see Chapter 8) is indicated by the phosphate ion (PO_4^{3-}). (Adapted from Kartha G, Bello J, Harker D: Tertiary structure of ribonuclease. *Nature* 1967;213:862.) The protein has been chemically synthesized in its entirety.

considerable portion of a protein may be present in this conformation. The term "random" is unfortunate, since it may imply less biologic significance than more highly repeating regions. In terms of biologic function, regions of random coil are of equal importance to those of α helix or β-pleated sheet.

ORDERS OF PROTEIN STRUCTURE

Primary

Primary structure, already familiar from peptides (see Chapter 4), refers to the order of the amino acids in the polypeptide chain or chains and the location of disulfide bonds, if these are present.

Secondary

Secondary structures, the steric relationship of amino acids close together in a pimary structural sense, may be regular (eg, α helix, β-pleated sheet), or it may exhibit few regularities (eg, random coil).

Tertiary

The overall arrangement and interrelationship of the various regions, or domains, and individual amino acid residues of a single polypeptide chain is called the tertiary structure of the protein. While the division between secondary and tertiary structure is not clear-cut, tertiary structure considers the steric relationship of amino acid residues that are, in general, far apart in a primary structural sense.

Quaternary

Proteins are said to possess quaternary structure if they consist of 2 or more polypeptide chains **united by forces other than covalent bonds** (ie, not peptide or disulfide bonds). The forces that stabilize these aggregates are hydrogen bonds and electrostatic (or salt) bonds formed between residues on the surfaces of the polypeptide chains. Such proteins are termed **oligomers,** and the individual polypeptide chains of which they are composed are variously termed **protomers, monomers,** or **subunits.**

Many oligomeric proteins contain 2 or 4 protomers and are termed dimers or tetramers, respectively. Oligomers containing more than 4 protomers are also common, particularly among regulated enzymes (eg, aspartate transcarbamoylase). Oligomeric proteins play special roles in intracellular regulation, because the protomers can assume different spatial orientations relative to one another with resulting changes in the properties of the oligomer. The best-studied example is hemoglobin (see Chapter 16).

Role of Primary Structures in Determining Higher Levels of Protein Structure

The secondary and tertiary structures of a protein are themselves determined by the primary structure of the polypeptide chain. Once the chain has been formed, the chemical groups that extend from the α carbons direct the specific regional folding (secondary structure) and specific aggregation of the regions (tertiary structure). For example, treatment of the monomeric enzyme ribonuclease with a mild reducing agent (β-mercaptoethanol) and a denaturing agent (urea or guanidine; see below) inactivates it as it assumes a random coil conformation. Slow removal of the denaturing agent and gentle reoxidation to re-form the S–S bonds lead to almost complete restoration of enzymatic activity. **It is not necessary to postulate independent genetic control of orders of protein structure above the primary level, since the primary structure specifies the secondary, tertiary, and (when present) quarternary structure (ie, conformation) of a protein.** The native conformation of a protein such as ribonuclease appears to be that which is thermodynamically most stable for a given environment, eg, a hydrophilic versus hydrophobic one.

The structure of a protein may be modified during posttranslational processing, such as the conversion of a preproenzyme to the catalytically active form or removal of the "leader peptide" that directs exported proteins through membranes (see Chapter 42).

Macromolecular Protein Complexes

Aggregation of different functional proteins—each of which alone has all 4 orders of structure—into multifunctional macromolecular complexes is encountered in electron transport (see Chapter 12), in fatty acid biosynthesis (see Chapter 23), and in pyruvate metabolism (see Chapter 18).

DENATURATION

The comparatively weak forces responsible for maintaining secondary, tertiary, and quaternary structure of proteins are readily disrupted with resulting loss of biologic activity. This disruption of native structure is termed denaturation. Physically, denaturation may be viewed as randomizing the conformation of a polypeptide chain without affecting its primary structure. For a protomer, the process may be represented as shown in Fig 5–8.

For an oligomeric protein, denaturation may involve dissociation of the protomers with or without accompanying changes in protomer conformation.

The biologic activity of most proteins is destroyed by exposure to strong mineral acids or bases, heat, ionic detergents (amphipaths), chaotropic agents (urea, guanidine), heavy metals (Ag, Pb, Hg), or organic solvents. Denatured proteins generally are less soluble in water, and they often precipitate from aqueous solution. This property is used to advantage in the clinical laboratory. Blood or serum samples to be analyzed for small molecules (eg, glucose, uric acid, drugs) generally are first treated with trichloroacetic, phosphotungstic, or phosphomolybdic acid to precipitate proteins. These are removed by centrifugation, and the protein-free supernatant liquid is then analyzed.

The heat, acid, and protease lability of most enzymes provides a preliminary test to determine whether a reaction is enzyme-catalyzed. If a cell extract having catalytic activity loses this activity when boiled, acidified and reneutralized, or treated with a protease, the catalyst probably was an enzyme.

Frequently, the denaturation of an enzyme is influenced by the presence of its substrate. The effect is attributed to a conformational change in the enzyme structure occurring when substrate is bound. The new conformation may be either more stable or less stable than before.

METHODS FOR DETERMINATION OF PRIMARY STRUCTURE

Complex proteins are first treated to remove prosthetic groups (eg, heme), and disulfide bonds are oxidized to yield linear polypeptides (see Fig 4–10). The methods used to sequence these polypeptides are discussed in Chapter 4. While most proteins contain only the amino acids listed in Table 3–3, derivatives of these amino acids also occur in certain proteins (Tables 5–4 and 5–5). While discussion of the methods used to identify these amino acid derivatives lies be-

Table 5–4. Modified α-COOH and α-NH$_2$ groups of proteins.*

Group Modified			
α-COOH	α-NH$_2$		
Amide	N-Formyl	N-Acetyl	N-Methyl
Asp		Ala	Ala
Glu		Asp	Asp
Gly	Gly	Gly	Gly
His			
Met	Met	Met	Met
Phe			
Pro			
		Ser	Ser
		Thr	Thr
Tyr			
Val		Val	

*Modified and reproduced, with permission, from Uy R, Wold F: Posttranslational covalent modification of proteins. *Science* 1977;**198**:890. Copyright © 1977 by the American Association for the Advancement of Science.

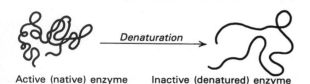

Active (native) enzyme ⟶ Inactive (denatured) enzyme

Figure 5–8. Representation of denaturation of a protomer.

Table 5–5. Modified non-α functional groups of proteins.*

–OH	Non-α-N		
PO₃H₂	**N-Methyl**	**N-Dimethyl**	**N-Trimethyl**
	Arg	Arg	
	His		
	Lys	Lys	Lys
Ser			
Thr			
Tyr			

*Modified and reproduced, with permission, from Uy R, Wold F: Posttranslational covalent modification of proteins. *Science* 1977;**198**:890. Copyright © 1977 by the American Association for the Advancement of Science.

yond the scope of this chapter, their presence can complicate the determination of primary structure.

Primary Structures of Insulin & Ribonuclease

Insulin consists of 2 polypeptide chains linked covalently by disulfide bonds (Fig 5–9). The A chain has an N-terminal Gly and a C-terminal Asn; the B chain has Phe and Ala as the N- and C-terminal residues, respectively. When insulin is oxidized with performic acid, the disulfide bonds linking the A and B chains are ruptured. Both chains are biosynthesized as a single polypeptide chain, **proinsulin,** which after formation undergoes proteolytic processing, forming insulin. (see Chapter 51).

Ribonuclease consists of a single chain of 124 residues with Lys as the N terminus and Val as the C terminus. Eight cysteine residues are joined by disulfide bonds, forming 4 cross-linkages in the protein (Fig 5–10).

DETERMINATION OF SECONDARY & TERTIARY STRUCTURE BY X-RAY CRYSTALLOGRAPHY

While various techniques (eg, optical rotatory dispersion, tritium exchange of labile protons) formerly were much used to infer the presence of helical structures in proteins, these have been largely supplanted by the powerful technique of x-ray crystallography.

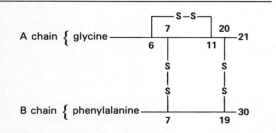

Figure 5–9. Relationship of the A and B chains of human insulin.

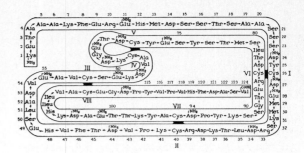

Figure 5–10. Structure of bovine ribonuclease. Two-dimensional schematic diagram showing the arrangement of the disulfide bonds and the sequence of the amino acid residues. Arrows indicate the direction of the peptide chain starting from the amino end. (Reproduced, with permission, from Smyth DG, Stein WH, Moore S: The sequence of amino acid residues in bovine pancreatic ribonuclease: Revisions and confirmations. *J Biol Chem* 1963;**238**:227.)

X-rays are directed at a crystal of protein and generally also at a derivative of that protein which contains an added heavy metal ion. The rays are scattered in a pattern that depends upon the electron densities in different parts of the protein. Images, collected on a photographic plate, are translated into electron density maps which, when superimposed one on another, permit the crystallographer to construct a faithful model of the protein in question. Although time-consuming, expensive, and requiring highly specialized training, x-ray crystallography reveals detailed, precise views of the orientations of all the amino acids in many proteins. Its contributions to our present-day concepts of protein structure can hardly be overstated.

DETERMINATION OF QUATERNARY STRUCTURE

Determining the quaternary structure of oligomeric proteins encompasses determining the number and kind of protomers present, their mutual orientation, and the interactions that unite them.

As long as oligomers do not undergo denaturation during the procedure used to determine their molecular weight, many methods can yield molecular weight data for oligomers. These same techniques may be used to determine protomer molecular weight if the oligomer is first denatured.

Ultracentrifugation

Developed by Svedberg, this method measures sedimentation rate in an ultracentrifugal field of around $10^5 \times g$. It has tended in recent years to be replaced by less complex techniques.

Sucrose Density Gradient Centrifugation

Protein standards and unknowns are layered over a 5–20% sucrose gradient in a plastic tube and cen-

Table 5–6. Quaternary structures of selected enzymes.*

Enzyme (Oligomer)	Number of Protomers	Molecular Weight of Protomer
Chicken heart aspartate transaminase (L-aspartate:2-oxoglutarate aminotransferase, E.C. 2.6.1.1)	2	50,000
Pigeon liver fatty acid synthase	2	230,000
Rabbit liver fructose diphosphatase (D-fructose-1,6-diphosphate 1-phosphohydrolase, E.C. 3.1.3.11)	2†	29,000
	2†	37,000
Rat liver ornithine transaminase (L-ornithine:2-oxoacid aminotransferase, E.C. 2.6.1.13)	4	33,000
Pig heart propionyl-CoA carboxylase (propionyl-CoA:CO_2 ligase [ADP], E.C. 6.4.1.2)	4	175,000
Beef heart, liver, or muscle LDH (L-lactate:NAD oxidoreductase, E.C. 1.1.1.27)	4†	35,000
Beef heart mitochondrial ATPase (ATP phosphohydrolase, E.C. 3.6.1.3)	10	26,000
E coli glutamine synthase (L-glutamate:NH_3 ligase [ADP], E.C. 6.3.1.2)	12	48,500
Chicken liver acetyl-CoA carboxylase (acetyl-CoA:CO_2 ligase [ADP], E.C. 6.4.1.2)	2†	4,100,000
	10†	409,000

*Adapted from Klotz IM, Langerman NR, Darnall DW: Quaternary structure of enzymes. *Annu Rev Biochem* 1970;**39**:25.
†Nonidentical subunits.

trifuged overnight at around $10^5 \times g$. A small hole is then punched in the bottom of the tube, the contents collected in a set of small tubes, the relative position of the proteins in the gradient determined, and their mobilities computed.

Filtration Through Molecular Sieves

Columns of Sephadex or similar materials are calibrated using proteins of known molecular weight. The molecular weight of an unknown protein is then calculated from its mobility relative to these standards. Large errors may result if the protein is highly asymmetric or interacts strongly with the materials from which the molecular sieve was manufactured.

Polyacrylamide Gel Electrophoresis (PAGE)

Protein standards are separated by electrophoresis in 5–15% cross-linked gels of varying porosity. Gels are stained for protein, generally with Coomassie blue stain or silver, and the molecular weight is estimated relative to the mobility of the standards. The most common application of this technique is to determine protomer molecular weight by first denaturing the oligomer (eg, by boiling in a detergent in the presence of β-mercaptoethanol) and separating on gels that contain the ionic detergent sodium dodecyl sulfate (SDS).

Electron Photomicrography

Pictures of small objects at magnifications as high as 100,000 diameters can be obtained with the electron microscope. This permits the visualization of proteins of high molecular weight, such as virus particles, enzyme complexes, and oligomeric proteins.

Table 5–6 lists examples of the numbers and molecular weights of protomers contributing to the quaternary structures of selected enzymes.

REFERENCES

Advances in Protein Chemistry. Academic Press, 1944–1987. [Annual publication.]

Aisen P, Listowsky I: Iron transport and storage proteins. *Annu Rev Biochem* 1980;**49**:357.

Baldwin RL: Intermediates in protein folding. *Annu Rev Biochem* 1975;**44**:453.

Chou PY, Fassman GD: Empirical predictions of protein structure. *Annu Rev Biochem* 1978;**47**:251.

Croft LR: *Handbook of Amino Acid Sequences of Proteins.* Joynson-Bruvvers Ltd (Oxford, England), 1973.

Gurd FN, Rothgeb TM: Motions in proteins. *Adv Protein Chem* 1979;**33**:74.

Haschemeyer RH, deHarven E: Electron microscopy of enzymes. *Annu Rev Biochem* 1974;**43**:279.

Lennarz WJ (editor): *The Biochemistry of Glycoproteins and Proteoglycans.* Plenum Press, 1980.

Lijas A, Rossmann MG: X-ray studies of protein interactions. *Annu Rev Biochem* 1974;**43**:475.

Neurath H, Hill RL (editors): *The Proteins,* 3rd ed. Academic Press, 1975.

Osborne JC Jr, Brewer HB Jr: The plasma lipoproteins. *Adv Protein Chem* 1977;**31**:253.

Smith LC, Pownall HJ, Gotto AM Jr: The plasma lipoproteins: Structure and metabolism. *Annu Rev Biochem* 1978;**47**:751.

6 Proteins: Myoglobin & Hemoglobin

Victor W. Rodwell, PhD

INTRODUCTION

The proteins discussed below are of cardinal biologic significance. They are also of great significance for the insights they provide into the diverse ways in which the structures of proteins dictate their biologic functions.

BIOMEDICAL IMPORTANCE

Heme proteins function in oxygen binding, oxygen transport, electron transport, and photosynthesis. Detailed study of hemoglobin and myoglobin illustrates structural themes common to many proteins. In a sense, their greatest medical significance is that this knowledge eloquently illustrates protein structure-function relationships. In addition, it provides insight into the molecular basis of genetic diseases such as sickle cell disease (a result of altered surface properties of the hemoglobin β-subunit), and the thalassemias (chronic, familial hemolytic diseases characterized by defective synthesis of hemoglobin). Cyanide and carbon monoxide kill because they disrupt the physiologic function of the heme proteins cytochrome oxidase and hemoglobin, respectively. Finally, stabilization of the quaternary structure of deoxyhemoglobin by 2,3-bisphosphoglycerate (DPG) is central to understanding the mechanisms of high-altitude sickness and of adaptation to high altitudes.

THE HEME PROTEINS MYOGLOBIN & HEMOGLOBIN

Myoglobin and hemoglobin provide clear insights into the relationships between structure and function for proteins in general and for globular proteins in particular. These 2 complex proteins possess as their prosthetic group **heme, a cyclic tetrapyrrole** that accounts both for their red color and for their abilities to store oxygen (myoglobin) and to transport oxygen (hemoglobin). Tetrapyrroles consist of 4 molecules of pyrrole (Fig 6–1) linked in a planar ring by four α-methylene bridges. The β substituents determine whether a tetrapyrrole is heme or a related compound. In heme, these are methyl (M), vinyl (V), and propionate (Pr) groups arranged in the order M,V,M, V,M,Pr,Pr,M (Fig 6–2). One atom of ferrous iron (Fe^{2+}) is present at the center of this planar ring. Other proteins with tetrapyrrole prosthetic groups (and their associated metal ions) include cytochromes (Fe^{2+} and

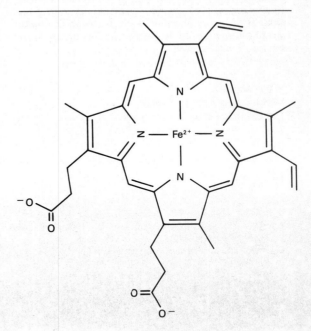

Figure 6–2. Heme. The pyrrole rings and methylene bridge carbons are coplanar, and the iron atom (Fe^{2+}) resides in almost the same plane. The fifth and sixth coordination positions of Fe^{2+} are directed perpendicular to, and directly above and below, the plane of the heme ring. Observe the nature of the substituent groups on the β carbons of the pyrrole rings, the central iron atom, and the location of the polar side of the heme ring (at about 7 o'clock) that faces the surface of the myoglobin molecule.

Figure 6–1. Pyrrole. The α carbons are linked by methylene bridges in a tetrapyrrole. The β carbons bear the substituents characteristic of a specific tetrapyrrole such as heme.

Fe^{3+}), certain enzymes (eg, catalase, tryptophan pyrrolase), and chlorophyll (Mg^{2+}). In cytochromes, oxidation and reduction of the iron atom are essential to their biologic function (electron transport; see Chapter 12). By contrast, **oxidation of the Fe^{2+} of myoglobin or hemoglobin destroys their biologic activity.**

MYOGLOBIN

Biologic Function of Myoglobin

Myoglobin of red muscle tissue stores oxygen. Under conditions of oxygen deprivation (eg, severe exercise), this oxygen is released for use by muscle mitochondria for oxygen-dependent synthesis of ATP (oxidative phosphorylation; see Chapter 13).

Primary Structure & Amino Acid Distribution of Myoglobin

Myoglobin, a single polypeptide chain of MW 17,000, is unexceptional with respect to the structures of its 153 aminoacyl residues. Clear differences are, however, apparent in their spatial distribution. **The surface is polar and the interior nonpolar,** a pattern characteristic of globular proteins. Residues with both polar and nonpolar regions (eg, Thr, Trp, and Tyr) orient their nonpolar regions inward. Apart from 2 histidyl residues that function in oxygen binding, **the interior of myoglobin contains only nonpolar residues** (eg, Leu, Val, Phe, and Met).

Secondary-Tertiary Structure of Myoglobin

X-ray crystallographic analysis revealed myoglobin to be a compact, roughly spherical molecule measuring $4.5 \times 3.5 \times 2.5$ nm (Fig 6–3). Its conformation is, however, atypical. Approximately 75% of the residues are present in 8 right-handed α helices from 7 to 20 residues in length. Starting at the N terminus, these are termed helices A through H. Interhelical regions are identified by the letters of the 2 helical regions they connect. Individual residues are designated by a letter for the helix in which they reside and a number that indicates their distance from the N terminus of that helix. For example, "His F8" refers to the eighth residue in the F helix and identifies it as a histidyl residue. Residues far distant in a primary structural sense (eg, in different helices) may nevertheless be spatially close together, eg, the F8 (proximal) and E7 (distal) histidyl residues (Fig 6–3).

Several lines of evidence suggest that the secondary-tertiary structure of myoglobin in solution closely resembles that of crystalline myoglobin. They exhibit virtually identical absorption spectra; crystalline myoglobin binds oxygen; and the amount of α helix in solution (estimated by optical rotatory dispersion and circular dichroism) closely approximates that revealed by x-ray analysis.

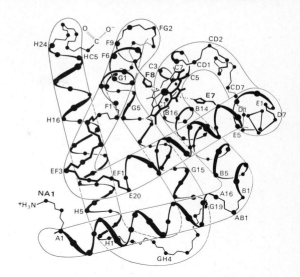

Figure 6–3. A model of myoglobin at low resolution. Only the α-carbon atoms are shown. (Based on Dickerson RE in: *The Proteins,* 2nd ed. Vol 2. Neurath H [editor]. Academic Press, 1964. Reproduced with permission.)

Effect of Heme on Myoglobin Conformation

When apomyoglobin (myoglobin with the heme removed) is prepared by lowering the pH to 3.5, its α-helical content decreases dramatically. Subsequent addition of urea to apomyoglobin at neutral pH reduces the α-helical content to about 0%. If the urea is then removed by dialysis and heme is added, full α-helical content is restored, and addition of Fe^{2+} restores full biologic (oxygen-binding) activity. The primary structural information implicit in apomyoglobin thus can, in the presence of heme, specify folding of the protein to its native, biologically active conformation. This important concept extends to other proteins: **the primary structure of a protein dictates its secondary-tertiary conformation.**

Spatial Orientation of the Heme Iron & of the Proximal & Distal Histidyl Residues of Myoglobin

The heme of myoglobin, which resides in a crevice between helices E and F, is oriented with its polar propionate substituents on the surface and the remainder in the interior surrounded by nonpolar residues except for His F8 and His E7. The fifth coordination position of the iron atom is linked to a ring nitrogen of the **proximal histidine,** His F8 (Fig 6–4). While not linked to the sixth coordination position of the iron, the **distal histidine** (His E7) lies on the side of the heme ring across from His F8 (Fig 6–4).

Location of the Heme Iron

In unoxygenated myoglobin, the heme iron resides about 0.03 nm outside the plane of the ring in the direction of His F8. In oxygenated myoglobin, an oxy-

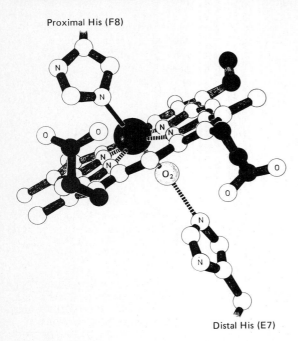

Proximal His (F8)

Distal His (E7)

Figure 6–4. Addition of oxygen to heme iron in oxygenation. Shown also are the imidazole side chains of the 2 important histidine residues of globin that attach to the heme iron. (Reproduced, with permission, from Harper HA et al: *Physiologische Chemie.* Springer-Verlag, 1975.)

gen atom occupies the sixth coordination position of the iron atom, which then lies only about 0.1 nm outside the plane of the heme. Oxygenation of myoglobin is therefore accompanied by movement of the iron atom, and consequently movement of His F8 and residues covalently linked to His F8, toward the plane of the ring. This motion brings about a new conformation of portions of the protein.

Ligands

When oxygen bonds to myoglobin, the bond between one oxygen atom and the Fe^{2+} forms perpendicular to the plane of the heme ring. The second oxygen atom bonds at an angle of 121 degrees to the plane of the heme and oriented away from the distal histidine (Fig 6–5).

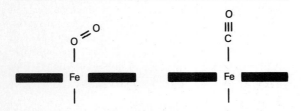

Figure 6–5. Preferred angles for bonding of oxygen and of carbon monoxide to the iron atom of free heme (solid line).

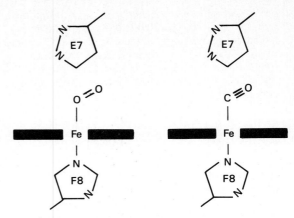

Figure 6–6. Angles for bonding of oxygen and carbon monoxide to the heme iron of myoglobin. The distal E7 histidine hinders bonding of CO at the preferred (180-degree) angle to the plane of the heme ring.

Carbon monoxide (CO) bonds to isolated heme about 25,000 times more strongly than does oxygen. Since the atmosphere contains traces of CO and normal catabolism of heme itself forms small quantities of CO, why then does not CO (rather than O_2) occupy the sixth coordination position of the heme iron of myoglobin? The answer lies in the **hindered environment** of heme in myoglobin. The preferred orientation for CO bonded to heme iron is with all 3 atoms (Fe, C, O) perpendicular to the heme ring (Fig 6–5). While this orientation is possible for isolated heme, in myoglobin **the distal histidine sterically hinders bonding of CO** at this angle (Fig 6–6). This forces CO to bond in a less favored configuration and reduces the strength of the heme-CO bond by over 2 orders of magnitude to about 200 times that of the heme-O_2 bond. A small portion (about 1%) of myoglobin nevertheless normally is present in the form of myoglobin-CO.

Kinetics of Myoglobin Oxygenation

Why is myoglobin unsuitable as an oxygen transport protein but effective as an oxygen storage protein? The quantity of oxygen bound to myoglobin (expressed as "percent saturation") depends upon the oxygen concentration (expressed as P_{O_2}, or partial pressure of oxygen) in the immediate environment of the heme iron. The relationship between P_{O_2} and the quantity of oxygen bound may be expressed graphically as an oxygen saturation (oxygen dissociation) curve. **For myoglobin, the shape of the oxygen adsorption isotherm is hyperbolic** (Fig 6–7). Since P_{O_2} in the lung capillary bed is 100 mm Hg, myoglobin could effectively load oxygen in the lungs. However, the P_{O_2} of venous blood is 40 mm Hg and that of active muscle about 20 mm Hg. Since myoglobin cannot deliver a large fraction of its bound oxygen even at 20 mm Hg, it cannot serve as an effective vehicle for delivery of oxygen from lungs to peripheral tissues.

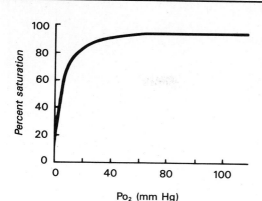

Figure 6–7. Oxygen saturation curve for myoglobin. Observe the relationship between percent saturation and the partial pressures representative of lungs (100 mm Hg), tissues (20 mm Hg), and actively working muscle (5 mm Hg).

However, in the oxygen deprivation that accompanies severe physical exercise, the P_{O_2} of muscle tissue may decrease to as little as 5 mm Hg. At 5 mm Hg, myoglobin readily releases its bound oxygen for the support of oxidative synthesis of ATP by muscle mitochondria.

HEMOGLOBINS

Biologic Functions of Hemoglobins

Hemoglobins are structurally related proteins of vertebrate erythrocytes that perform 2 major biologic functions: (1) transport of O_2 from the respiratory organ to peripheral tissues, and (2) transport of CO_2 and protons from peripheral tissues to the respiratory organ for subsequent excretion. While the comparative biochemistry of vertebrate hemoglobins provides fascinating insights, we shall here be concerned solely with human hemoglobins.

Primary Structure of Hemoglobin A

Unlike myoglobin, which lacks quaternary structure, hemoglobins are tetrameric proteins comprised of 2 each of 2 different polypeptides or monomer units (termed α, β, γ, δ, S, etc). While similar in overall length, the α (141 residue) and β (146 residue) polypeptides of hemoglobin A (HbA) are encoded by different genes and have different primary structures. By contrast, the primary structures of β-, γ-, and δ-chains of human hemoglobins have highly conserved primary structures.

Secondary-Tertiary Structure of Hemoglobin A

Despite differences in the type and number of amino acids present in myoglobin and the β polypeptide of HbA, they exhibit almost identical secondary-

tertiary structures. This striking similarity, which extends to the location of the heme and the 8 helical regions, results in part from the substitution of amino acids of similar properties at equivalent points in the primary structures of myoglobin and of the β subunit of HbA. The β polypeptide also closely resembles myoglobin despite the presence of 7 rather than 8 helical regions. As for myoglobin, hydrophobic residues are internal and (again with the exception of 2 His residues per subunit) hydrophilic residues are surface features of both the α and β subunits of HbA.

Quaternary Structure of Hemoglobin A

The properties of individual hemoglobins are inextricable consequences of their quaternary (as well as of their secondary-tertiary) structure. The tetrameric structures of common hemoglobins are: HbA (normal adult hemoglobin) = $\alpha_2\beta_2$, HbF (fetal hemoglobin) = $\alpha_2\gamma_2$, HbS (sickle cell hemoglobin) = α_2S_2, and HbA_2 (a minor adult hemoglobin) = $\alpha_2\delta_2$. The quaternary structure of hemoglobin confers striking additional properties (absent from myoglobin) that adapt it to its unique biologic roles and permit precise regulation of its properties. The **allosteric** (Gk *allos* "other," *steros* "space") properties of hemoglobin provide, in addition, a model for understanding other allosteric proteins.

Kinetics of Hemoglobin Oxygenation

Hemoglobins bind 4 oxygen atoms per tetramer (one per subunit heme), and in further contrast to myoglobin, oxygen saturation curves for hemoglobins are sigmoidal (Fig 6–8). The facility with which O_2 binds to hemoglobin thus depends on whether other O_2 atoms are present on the same tetramer. If O_2 is already present, binding of subsequent O_2 atoms occurs more readily. Hemoglobin thus exhibits **cooperative binding kinetics,** a property that permits it to bind a maximal quantity of O_2 at the respiratory organ and to deliver a maximal quantity of O_2 at the prevailing P_{O_2} of the peripheral tissues. Compare, for example, these quantities at the P_{O_2} of human lungs (100 mm Hg) and tissues (20 mm Hg) with myoglobin (Fig 6–8).

We compare the affinities of hemoglobins for O_2 by the quantity P_{50}, **the** P_{O_2} **which half-saturates a hemoglobin with** O_2. Depending on the organism, P_{50} can vary widely but in all instances is above the peripheral tissue P_{O2} in the organism in question. Human fetal hemoglobin (HbF) provides an illustrative example. For HbA, P_{50} = 26 mm Hg; for HbF, P_{50} = 20 mm Hg. This difference permits HbF to extract oxygen from the HbA of placental blood. Postpartum, HbF is, however, unsuitable, since its high affinity for O_2 dictates that it can deliver less O_2 to the tissues.

Gross Conformational Changes That Accompany Oxygenation of Hemoglobin

Binding of O_2 is accompanied by the rupture of salt

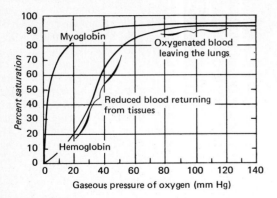

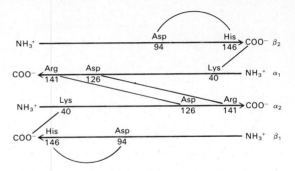

Figure 6–8. Oxygen-binding curves of hemoglobin and myoglobin. Arterial oxygen tension is about 100 mm Hg, mixed venous oxygen tension is about 40 mm Hg, capillary (active muscle) oxygen tension is about 20 mm Hg; and the minimum oxygen tension required for the cytochrome enzymes is about 5 mm Hg. The figure illustrates that association of chains into a tetrameric structure (hemoglobin) results in much greater oxygen delivery than would be possible with single chains. (Myoglobin and individual hemoglobin chains have about the same oxygen affinity and similar hyperbolic saturation curves.) (Modified, with permission, from Stanbury JB, Wyngaarden JB, Fredrickson DS [editors]: *The Metabolic Basis of Inherited Disease,* 4th ed. McGraw-Hill, 1978.)

Figure 6–9. Salt-links between different subunits in deoxyhemoglobin. These noncovalent, electrostatic interactions are disrupted on oxygenation. (Slightly modified and reproduced, with permission, from Stryer L: *Biochemistry,* 2nd ed. Freeman, 1981.)

The quaternary structure of partially oxygenated hemoglobin is described as the "T" (taut) state and that of oxygenated hemoglobin (HbO_2) as the "R" (relaxed) state (Fig 6–12). R and T are also used to describe the quaternary structures of allosteric enzymes, where the T state has the lower substrate affinity.

Conformational Changes in the Neighborhood of the Heme Groups

As for myoglobin, oxygenation of hemoglobin is accompanied by additional structural changes near the heme groups. On oxygenation, the iron atoms of deoxyhemoglobin (which lie about 0.06 nm beyond the plane of the heme ring) move into the plane of the heme ring (Fig 6–13). This motion is transmitted to the proximal (F8) histidine, which moves toward the plane of the ring, and to residues attached to His F8.

bonds between the carboxyl termini of all 4 subunits (Fig 6–9). This facilitates binding of additional O_2, since it involves rupture of fewer salt bonds. These changes also profoundly alter hemoglobin's secondary-tertiary and quaternary structures. One pair of α/β subunits rotates with respect to the other α/β pair, compacting the tetramer and increasing the affinity of the hemes for O_2 (Figs 6–10 and 6–11).

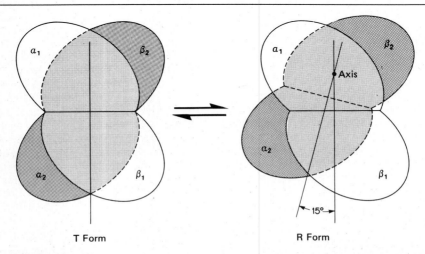

T Form

R Form

Figure 6–10. During the transition of the T form to the R form of hemoglobin, there occurs a rotation of one pair of rigid subunits (α_2/β_2) through 15 degrees relative to the other rigid pair of subunits (α_1/β_1). The axis of rotation is eccentric, and the α_2/β_2 pair also shifts toward the axis somewhat. In the diagram, the α_1/β_1 pair is unshaded and held fixed, while the shaded α_2/β_2 pair rotates and shifts.

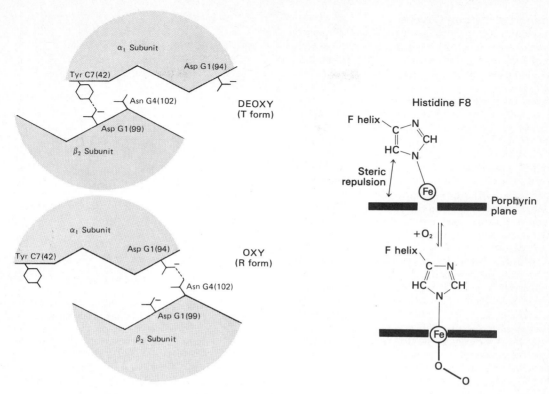

Figure 6–11. Changes at the α_1/β_2 contact on oxygenation. The contact "clicks" from one dovetailing area to another, involving a switch from one hydrogen bond to a second. The other bonds are nonpolar. (Reproduced, with permission, from Perutz MF: Molecular pathology of human hemoglobin: Stereochemical interpretation of abnormal oxygen affinities. *Nature* 1971;**232**:408.)

Figure 6–13. The iron atom moves into the plane of the heme on oxygenation because its diameter becomes smaller. Histidine F8 is pulled along with the iron atom. (Slightly modified and reproduced, with permission, from Stryer L: *Biochemistry*, 2nd ed. Freeman, 1981.)

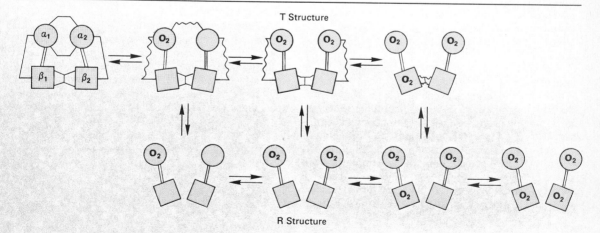

Figure 6–12. Transition from the T structure to the R structure increases in likelihood as each of the 4 heme groups is oxygenated. In this model, salt bridges (thin lines) linking the subunits in the T structure break progressively as oxygen is added, and even those salt bridges that have not yet ruptured are progressively weakened (wavy lines). The transition from T to R does not take place after a fixed number of oxygen molecules have been bound, but it becomes more probable with each successive oxygen bound. The transition between the 2 structures is influenced by several factors, including protons, carbon dioxide, chloride, and DPG. The higher their concentration, the more oxygen must be bound to trigger the transition. Fully oxygenated molecules in the T structure and fully deoxygenated molecules in the R structure are not shown, because they are too unstable to exist in significant numbers. (Modified and redrawn, with permission, from Perutz MF: Hemoglobin structure and respiratory transport. *Sci Am* [Dec] 1978;**239**:92.)

Transport of Carbon Dioxide

In addition to transporting oxygen from the lungs to peripheral tissues, hemoglobin facilitates the **transport of CO_2** from tissues to the lungs for exhalation. Hemoglobin can bind CO_2 directly when oxygen is released, and about 15% of the CO_2 carried in blood is carried directly on the hemoglobin molecule. However, as CO_2 is absorbed in blood, the carbonic anhydrase in erythrocytes catalyzes the formation of carbonic acid (Fig 6–14). Carbonic acid rapidly dissociates into bicarbonate and a proton; the equilibrium is toward the dissociation. To avoid the extreme danger of increasing the acidity of blood, there must exist a buffering system to absorb this excess proton. **Hemoglobin binds 2 protons for every 4 oxygen molecules lost** and thus provides a major buffering capacity of blood (Fig 6–15). In the lungs, the process is reversed—ie, **as oxygen binds to the deoxygenated hemoglobin, protons are released** and bind with the bicarbonate to drive the bicarbonate toward carbonic acid. With the aid of the very efficient carbonic anhydrase, the carbonic acid forms CO_2, which is exhaled. Thus, the **binding of oxygen forces the exhalation of CO_2.** This reversible phenomenon is called the **Bohr effect.** The Bohr effect is a property of the tetrameric hemoglobin and is dependent upon its heme-heme interaction or cooperative effects. Myoglobin does not exhibit any Bohr effect.

The Molecular Basis of the Bohr Effect

The protons responsible for the Bohr effect are generated by the breaking of salt bridges during the binding of oxygen to the T structure. When oxygen binds to the T structure, salt bridges are broken, and the protons are released from the N atoms of beta chain His residues HC3 (146). These released protons drive the equilibrium with bicarbonate toward carbonic acid, which is then released as CO_2 in alveolar blood (Fig 6–15.)

Conversely, upon the release of oxygen, the T structure and its salt bridges are re-formed, requiring protons to bind to the beta chain HC3 residues. Thus, the presence of protons from peripheral tissues favors the formation of salt bridges by protonating the terminal His residue of the beta subunits. Re-formation of the salt bridges forces the release of oxygen from oxygenated (R form) hemoglobin. Overall, **an increase in protons causes oxygen release,** while **an increase in oxygen causes proton release.** The former can be represented in an oxygen dissociation curve by a right-ward shift in the dissociation curve upon increasing hydrogen ions (protons).

Regulation by 2,3-Bisphosphoglycerate

In peripheral tissues, an oxygen shortage causes an increased accumulation of 2,3-bisphosphoglycerate (diphosphoglycerate; DPG) (Fig 6–16). This compound is formed from the glycolytic intermediate 1,3-diphosphoglycerate. One molecule of DPG is bound per hemoglobin tetramer in a central cavity formed by

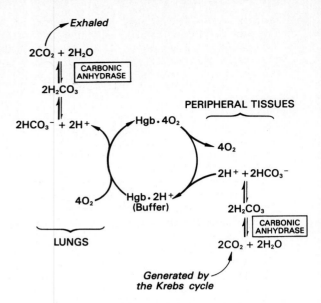

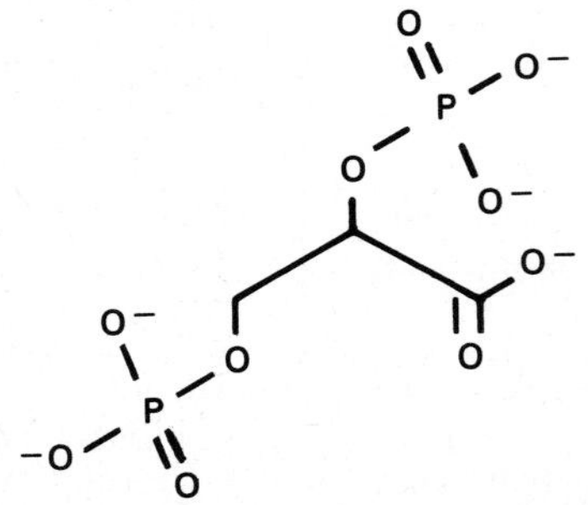

Figure 6–15. The Bohr effect. The carbon dioxide generated in peripheral tissues combines with water to form carbonic acid which dissociates into protons and bicarbonate ions. The deoxygenated hemoglobin acts as a buffer by binding protons and delivering them to the lungs. In the lungs, the binding of oxygen by hemoglobin forces the protons off the hemoglobin. The protons combine with bicarbonate ion, generating carbonic acid, which with the aid of carbonic anhydrase, becomes carbon dioxide. The carbon dioxide is exhaled from the lungs.

$$CO_2 + H_2O \xrightleftharpoons[\text{CARBONIC ANHYDRASE}]{} H_2CO_3 \xrightleftharpoons[]{} HCO_3^- + H^+$$

Carbonic *(Spontaneous)* acid

Figure 6–14. The formation of carbonic acid by erythrocyte carbonic anhydrase and the dissociation of carbonic acid to bicarbonate ion and a proton.

Figure 6–16. Structure of 2,3-bisphosphoglycerate (DPG).

residues of all 4 subunits. The central cavity is of sufficient size for the entry of DPG only when the hemoglobin molecule is in the **T form,** ie, when the space between the H helices of the beta chains is wide enough. The DPG is bound by salt bridges between its oxygen atoms and both beta chains via their N-terminal amino groups (Val NA1), Lys EF6, and **His H21** residues (Fig 6–17). Thus, **DPG stabilizes the T or deoxygenated form of hemoglobin by cross-linking the beta chains** and contributing additional salt bridges that must be broken for the T form to click into the R form of hemoglobin.

DPG binds more weakly to **fetal hemoglobin** than adult hemoglobin because the **H21** residue of the gamma chain of fetal hemoglobin is **Ser rather than His** and cannot contribute to the salt bridges that hold the DPG in the central cavity. Therefore, **DPG has a less profound effect on the stabilization of the T form of fetal hemoglobin** and is responsible for fetal hemoglobin appearing to have **a higher affinity** for oxygen than does adult hemoglobin.

The **trigger** for the transition between the R and T forms of hemoglobin is the **movement of the iron in and out of the plane of the porphyrin ring.** Both steric and electrostatic factors mediate this trigger with a free energy of about 3000 calories per mole. Thus, a minimal change in the position of Fe^{2+} relative to the porphyrin ring induces significant switching of the conformations of hemoglobin and crucially affects its biologic function in response to an environmental signal.

Mutant Human Hemoglobins

Mutations in the genes that code for the α or β chains potentially can affect the biologic function of hemoglobin. Of the several hundred known mutant human hemoglobins (most of them benign) several in which biologic function is altered are described below. When biologic function is altered owing to a mu-

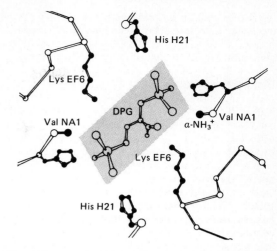

Figure 6–17. Mode of binding of 2,3-bisphosphoglycerate to human deoxyhemoglobin. DPG interacts with 3 positively charged groups on each β chain. (Based on Arnone A: X-ray diffraction study of binding of 2,3-diphosphoglycerate to human deoxyhemoglobin. *Nature* 1972;**237**:146. Reproduced with permission.)

tation in hemoglobin, the condition is known as a **hemoglobinopathy.**

In hemoglobin M variants, tyrosyl residues replace the proximal or distal His residues of α or β subunits. Heme iron is stabilized in the Fe^{3+} state, since it forms a tight ionic complex with the phenolate anion of Tyr. **Methemoglobinemia** results, since ferri-heme cannot bind O_2. In α-chain hemoglobin M variants, the R–T equilibrium favors the T form. Oxygen affinity is reduced, and a Bohr effect is absent. β-Chain hemoglobin M variants exhibit R–T switching, and a Bohr effect is therefore present.

Mutations (eg, hemoglobin Chesapeake) that favor

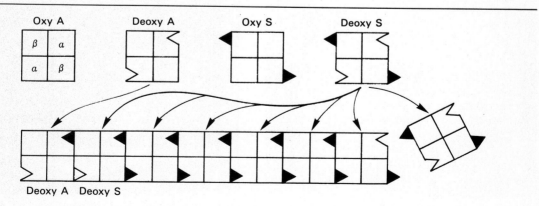

Figure 6–18. Diagrammatic representation of the sticky patch (▲) on hemoglobin S and the sticky patch "receptor" (△) present on deoxyhemoglobin A and deoxyhemoglobin S. The complementary surfaces allow deoxyhemoglobin S to polymerize into a fibrous structure, but the presence of deoxyhemoglobin A will terminate the polymerization by failing to provide sticky patches. (Modified and reproduced, with permission, from Stryer L: *Biochemistry,* 2nd ed. Freeman, 1981.)

the R form exhibit **increased oxygen affinity.** They therefore fail to deliver adequate oxygen to peripheral tissues. The resulting tissue hypoxia leads to **polycythemia** (increased concentration of erythrocytes).

Sickle Hemoglobin

Hemoglobin S is a substitution of Glu A2(6)β by a Val residue. The A2 residue, whether it be Glu or Val, can be seen to be on the surface of the hemoglobin molecule, exposed to water. The substitution in hemoglobin S replaces the polar glutamate residue with a nonpolar one and thereby generates a **"sticky patch"** on the surface of the beta chain. The sticky patch is present on oxygenated and deoxygenated hemoglobin S but not on hemoglobin A. On the surface of **deoxygenated hemoglobin,** there exists a **complement to the sticky patch,** but in oxygenated hemoglobin this complementary site is masked (Fig 6–18). When hemoglobin S is deoxygenated, the sticky patch of hemoglobin S can bind to the complementary patch on another deoxygenated hemoglobin molecule. This binding causes a **polymerization of deoxyhemoglobin S, forming long fibrous precipitates** that mechanically distort (sickle) the red cell, causing lysis and multiple secondary clinical effects. Thus, if hemoglobin S can be maintained in an oxygenated state or at least if the concentration of deoxygenated hemoglobin S can be kept at a minimum, formation of these polymers of deoxygenated hemoglobin S will not occur and "sickling" can be prevented. Clearly, it is the **T form of hemoglobin S that is subject to polymerization.** It is interesting but of no practical use to note that the ferric ion in methemoglobin A remains in the plane of the porphyrin ring and thus stabilizes the R form of hemoglobin. The same occurs in sickle hemoglobin; ie, hemoglobin S in the ferric state (methemoglobin S) will not polymerize into fibers, since it is stabilized in the R form.

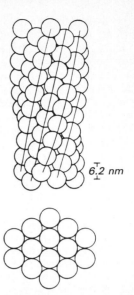

Figure 6–19. Proposed helical structure of a fiber of aggregated deoxyhemoglobin S. (Reproduced, with permission, from Maugh T II: A new understanding of sickle cell emerges. *Science* 1981;**211**:265. Copyright © 1981 by the American Association for the Advancement of Science.)

Although deoxyhemoglobin A contains the receptor sites for the sticky patch present on oxygenated or deoxygenated hemoglobin S (Fig 6–18), the binding of sticky hemoglobin S to deoxyhemoglobin A cannot extend the polymer, since the latter does not itself have a sticky patch to promote binding to still another hemoglobin molecule. Therefore, the **binding of deoxyhemoglobin A to either the R or the T form of hemoglobin S will terminate the polymerization.**

The polymerization of deoxyhemoglobin S forms a

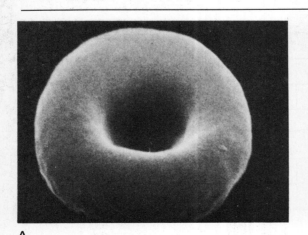

A

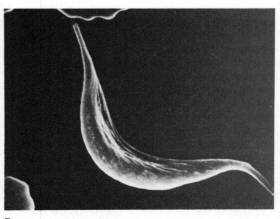

B

Figure 6–20. Scanning electron micrograph of normal *(A)* and sickle *(B)* red blood cells. The change of the β-globin molecule that causes this structural alteration results from a single base mutation in DNA, T to A, which results in the substitution of valine for glutamate in the β-globin molecule (see Chapter 36).

helical fibrous structure, each hemoglobin molecule making contact with 4 neighbors in a tubular helix (Fig 6–19). The formation of these tubular fibers is responsible for the mechanical distortion of the erythrocyte containing them, so that they take on the shape of a sickle (Fig 6–20) and are vulnerable to lysis as they penetrate the interstices of the splenic sinusoids.

Thalassemias

Another important group of conditions involving hemoglobin are the thalassemias. In these conditions, the synthesis of either the alpha (α-thalassemias) or beta (β-thalassemias) chains of hemoglobin is reduced. This results in anemia, which may be very severe. Considerable progress has been made in recent years in elucidating the molecular mechanisms responsible for the production of the thalassemias (see Chapter 36).

REFERENCES

Dean J, Schechter AN: Sickle-cell anemia: Molecular and cellular basis of therapeutic approaches. (3 parts.) *N Engl J Med* 1978;**299:**752, 804, 863.

Klotz IM, Haney DN, King LC: Rational approaches to chemotherapy: Antisickling agents. *Science* 1981;**213:**724.

Perutz MF: Hemoglobin structure and respiratory transport. *Sci Am* (Dec) 1978;**239:**92.

Perutz MF: The regulation of oxygen-affinity of hemoglobin: Influence of structure of globin on heme iron. *Annu Rev Biochem* 1979;**48:**327.

Stamatoyannopoulos G: The molecular basis of hemoglobin disease. *Annu Rev Genet* 1972;**6:**47.

Winslow RM, Anderson WF: The hemoglobinopathies. Page 1666 in: *The Metabolic Basis of Inherited Disease,* 5th ed. Stanbury JB et al (editors). McGraw-Hill, 1983.

Enzymes: General Properties

Victor W. Rodwell, PhD

INTRODUCTION

Catalysts, substances that accelerate chemical reactions, undergo physical change during a reaction but revert to their original state when the reaction is complete. **Enzymes** are **protein catalysts** for biochemical reactions, most of which would occur extremely slowly were it not for catalysis by enzymes. In contrast to nonprotein catalysts (H^+, OH^-, metal ions), each enzyme catalyzes a small number of reactions, frequently only one. Enzymes are thus **reaction-specific** catalysts. **Essentially all biochemical reactions are enzyme-catalyzed.**

BIOMEDICAL IMPORTANCE

Enzymes make life on Earth possible. It is no surprise that enzymes impinge on many fields of the biomedical sciences. Many diseases (the inborn errors of metabolism) are due to genetically determined abnormalities in the synthesis of enzymes. When cells are injured (eg, by impairment of blood supply or by inflammation), certain enzymes leak into the plasma. Measurement of the activity of such enzymes has become an integral part of the diagnosis of a number of important medical disorders (eg, myocardial infarction). Diagnostic enzymology is the area of medicine involving the use of enzymes to assist in diagnosis and management. Enzymes can also be used in therapy.

ENZYME CLASSIFICATION & NOMENCLATURE

Enzymes initially were named by adding the suffix **-ase** to the name of the substrate on which they acted. Thus, enzymes that hydrolyzed starch (amylon) were termed amylases; those that hydrolyzed fat (lipos), lipases; and those that hydrolyzed proteins, proteinases. Somewhat later, enzymes that catalyzed similar reactions were given names that indicated the type of chemical reaction catalyzed. These were termed dehydrogenases, oxidases, decarboxylases, acylases, etc. Many of these names remain in current use.

The **International Union of Biochemistry (IUB) Nomenclature System,** at first glance cumbersome and complex, is unambiguous. Its underlying principle, that of naming and classifying enzymes on the basis of chemical reaction type and reaction mechanism, can materially facilitate integration of information from widely divergent areas of metabolism. The major features of the IUB system are as follows:

(1) Reactions and the enzymes that catalyze them are divided into 6 classes, each with from 4 to 13 subclasses.

(2) The enzyme name has 2 parts. The first names the substrate or substrates. The second, ending in **-ase,** indicates the **type of reaction catalyzed.**

(3) Additional information, if needed to clarify the reaction, may follow in parentheses. For example, the enzyme catalyzing L-malate + NAD^+ = pyruvate + CO_2 + NADH + H^+ is designated 1.1.1.37 L-malate:NAD^+ oxidoreductase (decarboxylating).

(4) Each enzyme has a code number (E.C.) that characterizes the reaction type as to class (first digit), subclass (second digit), and subsubclass (third digit). The fourth digit is for the specific enzyme. Thus, E.C. 2.7.1.1 denotes class 2 (a transferase), subclass 7 (transfer of phosphate), subsubclass 1 (an alcohol functions as the phosphate acceptor). The final digit denotes the enzyme, hexokinase, or ATP:D-hexose 6-phosphotransferase, an enzyme catalyzing phosphate transfer from ATP to the hydroxyl group on carbon 6 of glucose.

The 6 classes of enzymes with some illustrative examples are given on p 51. The name in brackets is the recommended name.

COENZYMES

Many enzymes catalyze reactions of their substrates only in the presence of a specific heat-stable, low-molecular-weight organic molecule, a coenzyme. Where coenzymes are required, the **holoenzyme** (complete catalytic entity) consists of the **apoenzyme** (protein part) plus the bound **coenzyme.** A particular coenzyme may bind covalently or noncovalently to the apoenzyme. The term "prosthetic group" denotes a covalently bonded coenzyme. Reactions that require coenzymes include oxidoreductions, group transfer and isomerization reactions, and reactions that form covalent bonds (IUB classes 1, 2, 5, and 6). Lytic re-

1. Oxidoreductases. Enzymes catalyzing oxidoreductions between 2 substrates, S and S′.

$$S_{reduced} + S'_{oxidized} = S_{oxidized} + S'_{reduced}$$

Enzymes catalyzing oxidoreductions of CH–OH, CH–CH, C=O, CH–NH₂, and CH=NH groups. Representative subclasses:

1.1 Enzymes acting on the CH–OH group as electron donor. *For example:*

1.1.1.1 Alcohol:NAD⁺ oxidoreductase [alcohol dehydrogenase].

$$Alcohol + NAD^+ = Aldehyde \; or \; ketone + NADH + H^+$$

1.4 Enzymes acting on the CH–NH₂ group as electron donor. *For example:*

1.4.1.3 L-Glutamate:NAD(P)⁺ oxidoreductase (deaminating) [glutamic dehydrogenase of animal liver]. NAD(P)⁺ means that either NAD⁺ or NADP⁺ acts as the electron acceptor.

$$L\text{-}Glutamate + H_2O + NAD(P)^+ =$$
$$\alpha\text{-}Ketoglutarate + NH_4^+ + NAD(P)H + H^+$$

2. Transferases. Enzymes catalyzing a transfer of a group, G (other than hydrogen), between a pair of substrates S and S′.

$$S\text{-}G + S' = S'\text{-}G + S$$

Enzymes catalyzing the transfer of one-carbon groups, aldehyde or ketone residues, and acyl, alkyl, glycosyl and phosphorus- or sulfur-containing groups. Representative subclasses:

2.3 Acyltransferases. *For example:*

2.3.1.6 Acetyl-CoA:choline O-acetyltransferase [choline acyltransferase].

$$Acetyl\text{-}CoA + Choline = CoA + O\text{-}Acetylcholine$$

2.7 Enzymes catalyzing transfer of phosphorus-containing groups. *For example:*

2.7.1.1 ATP:D-hexose 6-phosphotransferase [hexokinase].

$$ATP + D\text{-}Hexose = ADP + D\text{-}Hexose \; 6\text{-}phosphate$$

3. Hydrolases. Enzymes catalyzing hydrolysis of ester, ether, peptide, glycosyl, acid-anhydride, C–C, C-halide, or P–N bonds. *For example:*

3.1 Enzymes acting on ester bonds. *For example:*

3.1.1.8 Acylcholine acylhydrolase [pseudocholinesterase].

$$An \; acylcholine + H_2O = Choline + An \; acid$$

3.2 Enzymes acting on glycosyl compounds. *For example:*

3.2.1.23 β-D-Galactoside galactohydrolase [β-galactosidase].

$$A \; \beta\text{-}D\text{-}Galactoside + H_2O = An \; alcohol + D\text{-}Galactose$$

3.4 Enzymes acting on peptide bonds. Classification (11 subclasses) distinguishes peptidases from proteases, whether dipeptides or longer peptides are substrates, whether one or more amino acids are removed, and whether attack is from the C- or the N-terminal end. Proteinases are further distinguished by their catalytic mechanism as serine, –SH, or metalloenzyme proteinases. *For example:*

3.4.21 Serine proteinases. *For example:* Chymotrypsin, trypsin, plasmin, coagulation factors IXa and XIa.

3.4.23 Carboxyl (acid) proteinases. *For example:* Pepsin A, B, and C.

4. Lyases. Enzymes that catalyze removal of groups from substrates by mechanisms other than hydrolysis, leaving double bonds.

$$\overset{X \quad Y}{\underset{}{C\text{-}C}} = X\text{-}Y + C=C$$

Enzymes acting on C–C, C–O, C–N, C–S, and C-halide bonds. Representative subgroups:

4.1.2 Aldehyde-lyases. *For example:*

4.1.2.7 Ketose-1 phosphate aldehyde-lyase [aldolase].

$$A \; ketose \; 1\text{-}phosphate = Dihydroxyacetone$$
$$phosphate + An \; aldehyde$$

4.2 Carbon-oxygen lyases. *For example:*

4.2.1.2 L-Malate hydro-lyase [fumarase].

$$L\text{-}Malate = Fumarate + H_2O$$

5. Isomerases. Includes all enzymes catalyzing interconversion of optical, geometric, or positional isomers. Two subclasses:

5.2 *Cis-trans* isomerases. *For example:*

5.2.1.3 All-*trans*-retinene 11-*cis-trans* isomerase [retinene isomerase].

$$All \; trans\text{-}retinene = 11\text{-}cis\text{-}retinene$$

5.3 Enzymes catalyzing interconversion of aldoses and ketoses. *For example:*

5.3.1.1 D-Glyceraldehyde-3-phosphate ketolisomerase [triosephosphate isomerase].

$$D\text{-}Glyceraldehyde \; 3\text{-}phosphate =$$
$$Dihydroxyacetone \; phosphate$$

6. Ligases. (Lat *ligare* "to bind.") Enzymes catalyzing the linking together of 2 compounds coupled to the breaking of a pyrophosphate bond in ATP or a similar compound. Included are enzymes catalyzing reactions forming C–O, C–S, C–N, and C–C bonds. Representative subclasses are:

6.3 Enzymes catalyzing formation of C–N bonds. *For example:*

6.3.1.2 L-Glutamate-ammonia ligase (ADP) [glutamine synthetase].

$$ATP + L\text{-}Glutamate + NH_4^+ =$$
$$ADP + Orthophosphate + L\text{-}Glutamine$$

6.4 Enzymes catalyzing formation of C–C bonds. *For example:*

6.4.1.2 Acetyl-CoA:CO₂ ligase (ADP) [acetyl-CoA carboxylase].

$$ATP + Acetyl\text{-}CoA + CO_2 = ADP + P_i + Malonyl\text{-}CoA$$

actions, including hydrolytic reactions such as those catalyzed by digestive enzymes, do not require coenzymes (IUB classes 3 and 4).

Coenzymes as Second Substrates

The coenzyme may be regarded as a second substrate or **cosubstrate** for 2 reasons. First, **the chemical changes in the coenzyme exactly counterbalance those taking place in the substrate.** For example, in oxidoreduction (dehydrogenase) reactions, one molecule of substrate is oxidized (dehydrogenated) and one molecule of coenzyme is reduced (hydrogenated) (Fig 7–1).

Similarly, in transamination reactions pyridoxal phosphate acts as a second substrate in 2 concerted reactions and as carrier for transfer of an amino group between different α-keto acids.

A second reason to give equal emphasis to the reactions of the coenzyme is that this aspect of the reaction may be of greater fundamental physiologic significance. For example, the importance of the ability of muscle working anaerobically to convert pyruvate to lactate does not reside in pyruvate or lactate. The reaction serves merely to convert NADH to NAD^+. Without NAD^+, glycolysis cannot continue and anaerobic ATP synthesis (and hence work) ceases. Under anaerobic conditions, reduction of pyruvate to lactate reoxidizes NADH and permits synthesis of ATP. Other reactions can serve this function equally well. This point can best be appreciated by consideration of life forms other than animals. In bacteria or yeast growing anaerobically, substances derived from pyruvate serve as oxidants for NADH and are themselves reduced (Table 7–1).

Role of Coenzymes as Group Transfer Reagents in Intermediary Metabolism

Biochemical group transfer reactions of the type

$$D–G + A \rightleftharpoons A–G + D$$

in which a functional group, G, is transferred from a donor molecule, D–G, to an acceptor molecule, A,

usually involve a coenzyme either as the ultimate acceptor (eg, dehydrogenation reactions) or as an intermediate group carrier (eg, transamination reactions). The reaction

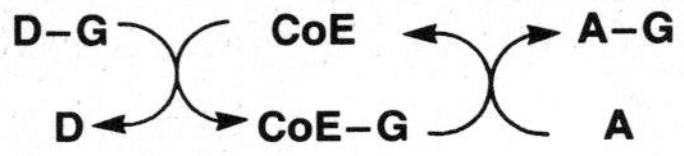

illustrates the latter concept. While this suggests formation of a single CoE–G complex during the course of the overall reaction, several intermediate CoE–G complexes may be involved in a particular reaction (eg, transamination).

When the group transferred is hydrogen, it is customary to represent only the left "half reaction":

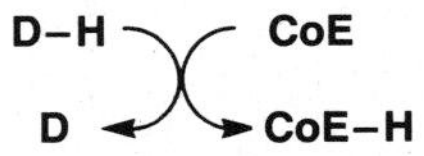

That this actually represents only a special case of general group transfer can best be appreciated in terms of the reactions that occur in intact cells (Table 7–1). These can be represented as follows:

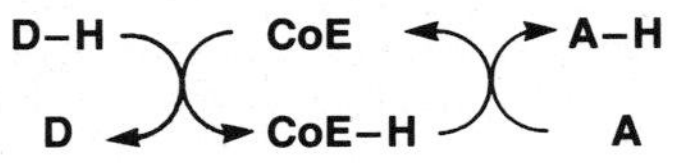

Based on this concept, we might classify coenzymes as follows:

For transfer of groups other than H
Sugar phosphates
CoA · SH
Thiamin pyrophosphate
Pyridoxal phosphate
Folate coenzymes
Biotin
Cobamide (B_{12}) coenzymes
Lipoic acid
For transfer of H
NAD^+, $NADP^+$
FMN, FAD
Lipoic acid
Coenzyme Q

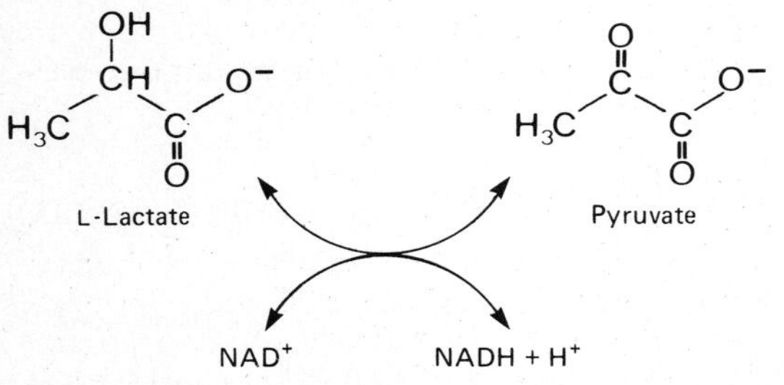

Figure 7–1. NAD^+ acting as cosubstrate in an oxidoreduction reaction.

Table 7–1. Mechanisms for anaerobic regeneration of NAD^+.

Oxidant	Reduced Product	Life Form
Pyruvate	Lactate	Muscle, homo-lactic bacteria
Acetaldehyde	Ethanol	Yeast
Dihydroxyacetone phosphate	α-Glycerophosphate	*E coli*
Fructose	Mannitol	Heterolactic bacteria

Coenzymes as B-Vitamin & AMP Derivatives

B vitamins form part of the structure of many coenzymes. For example, many enzymes of amino acid metabolism require vitamin B_6. The B vitamins **nicotinamide, thiamin, riboflavin,** and **pantothenic acid** are essential constituents of coenzymes for biologic oxidations and reductions, and **folic acid** and **cobamide** coenzymes function in one-carbon metabolism.

A structural feature common to many coenzymes is an adenine ring joined to D-ribose and inorganic phosphate. Many coenzymes may therefore be regarded as derivatives of adenosine monophosphate (AMP) (see Table 34–1). The structures of NAD^+ and $NADP^+$ are shown in Fig 7–2.

THREE-POINT ATTACHMENT OF SUBSTRATES TO ENZYMES

Most substrates form at least 3 bonds with enzymes. This **"3-point attachment"** can confer asymmetry on an otherwise symmetric molecule. To explain this, we shall represent the region of an enzyme that binds a substrate as a planar surface, although as we shall shortly see, the substrate-binding sites of enzymes are rarely, if ever, planar. Fig 7–3 shows a substrate molecule, represented as a carbon atom having 3 different groups, about to attach at 3 points to a planar enzyme site. If the site can be approached only from

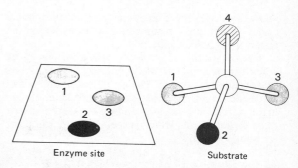

Figure 7–3. Representation of 3-point attachment of a substrate to a planar active site of an enzyme.

one side and only complementary atoms and sites can interact (both valid assumptions for actual enzymes), the molecule can bind in only one way. The reaction itself may be confined to the atoms bound at sites 1 and 2 even though atoms 1 and 3 are identical. By mentally turning the substrate molecule in space, note that it can attach at 3 points to one side of the planar site with only one orientation. Consequently, atoms 1 and 3, although identical, become distinct when the substrate is attached to the enzyme. A chemical change thus can involve atom 1 but not atom 3, or vice versa. Extension of this line of reasoning can explain why the enzyme-catalyzed reduction of the optically inactive pyruvate molecule results in formation of L- and not D,L-lactate.

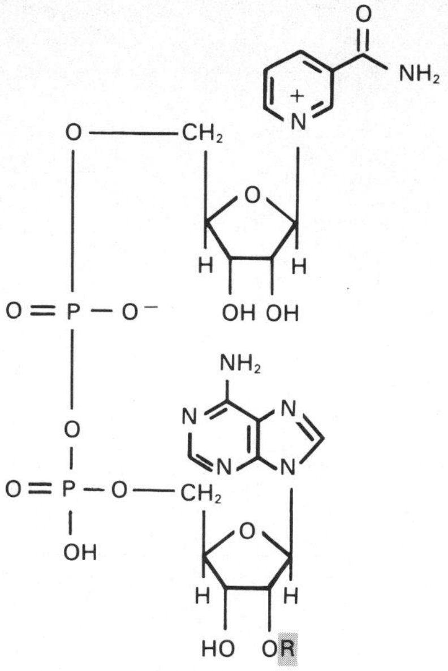

ENZYME SPECIFICITY

The ability of an enzyme to catalyze one specific reaction and essentially no others is perhaps its most significant property. The rates of specific metabolic processes may thus be regulated by changes in the catalytic efficiency of specific enzymes. However, most enzymes catalyze the same type of reaction (phosphate transfer, oxidation-reduction, etc) with a small number of structurally related substrates. Reactions with alternative substrates tend to take place if these are present in high concentration. Whether all of the possible reactions will occur in living organisms depends on the relative concentration of alternative substrates in the cell and the relative affinities of the enzyme for those substrates. Some general aspects of enzyme specificity are given below.

Optical Specificity of Enzymes

With the exception of epimerases (racemases), which catalyze interconversion of optical isomers, **enzymes generally exhibit absolute optical specificity for at least a portion of a substrate molecule.** Thus, enzymes of the glycolytic and direct oxidative pathways catalyze the interconversion of D- but not L-phos-

Figure 7–2. NAD(P)$^+$. In NAD$^+$, R = H; in NADP$^+$, R = $^-OPO_3{}^{2-}$.

phosugars. With few exceptions (eg, kidney D-amino acid oxidase), most mammalian enzymes act on the L-isomers of amino acids.

Optical specificity may extend to a portion of the substrate molecule or to its entirety. Glycosidases illustrate both extremes. These catalyze hydrolysis of glycosidic bonds between sugars and alcohols, are highly specific for the sugar portion and for the linkage (α or β), but are relatively nonspecific for the aglycone (alcohol portion).

Group Specificity of Enzymes

Lytic enzymes act on specific chemical groupings, eg, glycosidases on glycosides, pepsin and trypsin on peptide bonds, and esterases on esters. A large number of substrates may be attacked, thus lessening the number of digestive enzymes that might otherwise be required. Many proteases catalyze hydrolysis of esters also. While the ability of proteases to hydrolyze esters is of limited physiologic importance, the use of esters as synthetic substrates has been central to the study of the mechanism of action of proteases.

Certain lytic enzymes exhibit a higher order of group specificity. Chymotrypsin preferentially hydrolyzes peptide bonds in which the carboxyl group is contributed by the aromatic amino acids phenylalanine, tyrosine, or tryptophan. Carboxypeptidases and aminopeptidases split off amino acids one at a time from the carboxy- or amino-terminal end of polypeptide chains, respectively.

Although some oxidoreductases utilize either NAD$^+$ or NADP$^+$ as electron acceptor, most use exclusively one or the other. As a broad generalization, **oxidoreductases functional in biosynthetic processes in mammalian systems (eg, fatty acid or sterol synthesis) tend to use NADPH as reductant, while those functional in degradative processes (eg, glycolysis, fatty acid oxidation) tend to use NAD$^+$ as oxidant.**

QUANTITATIVE MEASUREMENT OF ENZYME ACTIVITY

The extremely small quantities of enzymes present in cells pose problems in determining the amount of an enzyme in tissue extracts or fluids quite different from those of determining the concentration of more abundant organic or inorganic substances. Fortunately, **the catalytic activity of an enzyme provides a sensitive and specific probe for its own measurement.**

To measure the amount of an enzyme in a sample of tissue extract or other biologic fluid, the **rate of the reaction** catalyzed by the enzyme in the sample is measured. Under appropriate conditions, **the measured rate is proportionate to the quantity of enzyme present.** Since it is difficult to determine the number of molecules or mass of enzyme present, results are expressed in **enzyme units.** Relative amounts of enzyme in different extracts may then be compared. Enzyme units are best expressed in micromoles (μmol; 10^{-6} mol), nanomoles (nmol; 10^{-9} mol), or picomoles (pmol; 10^{-12} mol) of substrate reacting or product produced per minute. The corresponding International Enzyme Units are μU, nU, and pU.

Example of Quantitative Analysis of Enzyme Activity: Assay of a Dehydrogenase

In reactions involving NAD$^+$ or NADP$^+$ (dehydrogenases), advantage is taken of the property of NADH or NADPH (but not NAD$^+$ or NADP$^+$) to absorb light of wavelength 340 nm (Fig 7–4).

When NADH is oxidized to NAD$^+$ (or vice versa), the optical density (OD) at 340 nm changes. Under specified conditions, the rate of change in OD depends directly on the enzyme activity (Fig 7–5).

A calibration curve (Fig 7–6) is prepared by plotting the slopes of the lines (velocities) in Fig 7–5 versus the volume of enzyme preparation added. The quantity of enzyme present in an unknown solution may then be calculated from the observed rate of change in OD at 340 nm.

Coupled Enzymatic Analyses

In the above example, the rate of formation of a product (NADH) was measured to determine enzyme activity. The activity of enzymes other than dehydrogenases may also be determined by measuring the rate of appearance of a product (or, less commonly, the rate of disappearance of a substrate). The physicochemical properties of the product or substrate determine the specific method selected for quantitation. It

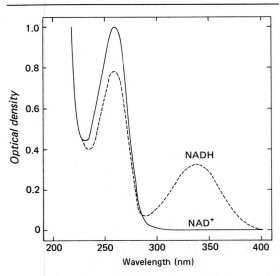

Figure 7–4. Absorption spectra of NAD$^+$ and NADH. Densities are for a 44 mg/L solution in a cell of 1-cm light path. NADP$^+$ and NADPH have spectra analogous to those of NAD$^+$ and NADH, respectively.

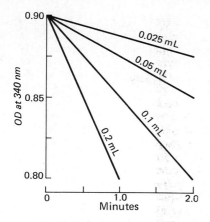

Figure 7–5. Assay of an NADH- or NADPH-dependent dehydrogenase. The rate of change in OD at 340 nm due to conversion of reduced to oxidized coenzyme is observed. Oxidized substrate (S), reduced coenzyme (NADH), and buffer are added to a cuvette. Light of 340-nm wavelength is then passed through it. Initially, the OD is high, since NADH (or NADPH) absorbs at 340 nm. On addition of 0.025–0.2 mL of a standard enzyme solution, the OD decreases.

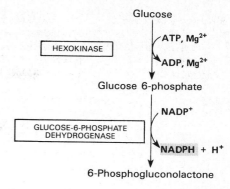

Figure 7–7. Coupled assay for hexokinase activity. The reaction is coupled to that catalyzed by glucose-6-phosphate dehydrogenase. Glucose-6-phosphate dehydrogenase, glucose, ATP, Mg^{2+}, and $NADP^+$ all are added in excess. The quantity of hexokinase present then determines the rate of the overall coupled reaction and therefore the rate of formation of NADPH, which can be measured at 340 nm.

often is convenient to "couple" the product of a reaction to a dehydrogenase for which this product is a substrate (Fig 7–7).

ISOLATION OF ENZYMES

Knowledge concerning the individual reactions of and chemical intermediates in metabolic pathways and of the regulatory mechanisms that operate at the level of catalysis are derived to a great extent from studies of purified enzymes. Reliable information concerning the kinetics, cofactors, active sites, structure, and mechanism of action also requires highly purified enzymes.

The objective of enzyme purification is to isolate a specific enzyme from a crude cell extract containing many other components. Small molecules may be removed by dialysis or gel filtration, nucleic acids by precipitation with the antibiotic streptomycin, etc. The problem is to separate the desired enzyme from hundreds of chemically and physically similar proteins.

Classic Purification Procedures

Useful classic purification procedures include precipitation with varying salt concentrations (generally ammonium or sodium sulfate) or solvents (acetone or ethanol), differential heat or pH denaturation, differential centrifugation, gel filtration, and electrophoresis. Selective adsorption and elution of proteins from the cellulose anion exchanger diethylaminoethylcellulose and the cation exchanger carboxymethylcellulose have also been extremely successful for extensive and rapid purification. Separation of proteins on molecular sieves such as Sephadex that segregate proteins on the basis of their size is also widely used. These methods are, however, relatively unselective, since they do not, except in combination, resolve a single protein from all others. This is more readily achieved by affinity chromatography.

The progress of a typical classic enzyme purification for a liver enzyme with good recovery and 490-fold overall purification is shown in Table 7–2. Note how specific activity and recovery of initial activity are calculated. The aim is to achieve the maximum specific activity (enzyme units per milligram of protein) with the best possible recovery of initial activity.

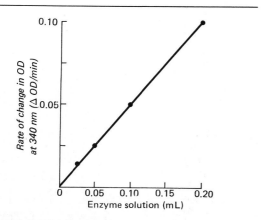

Figure 7–6. Calibration curve for enzymatic analysis. The slopes of the lines in Fig 7–5 are plotted versus the quantity of enzyme.

Table 7–2. Summary of a typical enzyme purification scheme.

Enzyme Fraction	Total Activity (pU)	Total Protein (mg)	Specific Activity (pU/mg)	Overall Recovery (%)
Crude liver homogenate	100,000	10,000	10	(100)
100,000 × g supernatant liquid	98,000	8,000	12.2	98
40–50% $(NH_4)_2SO_4$ precipitate	90,000	1,500	60	90
20–35% acetone precipitate	60,000	250	240	60
DEAE column fractions 80–110	58,000	29	2,000	58
43–48% $(NH_4)_2SO_4$ precipitate	52,000	20	2,600	52
First crystals	50,000	12	4,160	50
Recrystallization	49,000	10	4,900	49

Affinity Chromatographic Techniques

The salient feature of this purification technique is its ability to remove selectively from a complex protein mixture one particular protein or, at most, a small number of particular proteins. The technique employs an immobilized ligand that interacts specifically with the enzyme whose purification is desired. When the protein mixture is exposed to this immobilized ligand, the only proteins that bind are those which interact strongly with the ligand. After the unwanted protein is removed, the desired protein is eluted from the immobilized ligand, generally with a high concentration of salt or of the soluble form of the ligand. When successful, the purifications achieved by affinity chromatographic techniques are extremely impressive, often surpassing that possible by successive application of numerous classic techniques.

Since enzymes exhibit generally high specificity with respect to their substrates and their coenzymes, the most favored ligands are substrate and coenzyme derivatives covalently attached to a support such as Sephadex. Attachment may be direct or via a linker molecule 3–8 carbon atoms in length. Difficulties may arise if the mode of ligand attachment precludes its ability to interact with the enzyme—a difficulty that linker molecules may help to circumvent. The introduction of a hydrophobic "linker" may, however, complicate the separation by introducing an element of hydrophobic ligand chromatography (see below). Examples of successful affinity chromatography include purification of many different dehydrogenases on NAD^+ affinity supports. While many dehydrogenases may be bound and may be eluted together when the column is treated with soluble NAD^+, subsequent application of substrate (rather than coenzyme) affinity supports or elution with an "abortive ternary mixture" of one product and one substrate have proved successful in many instances.

Dye ligand chromatography on supports such as blue-, green-, or red-Sepharose and hydrophobic ligand chromatography on supports such as octyl- or phenyl-Sepharose are techniques closely related to affinity chromatography. The former employs as the immobilized ligand an organic dye that serves as an analog of a substrate, coenzyme, or allosteric effector. Elution generally is achieved using salt gradients.

In hydrophobic ligand chromatography, an alkyl or aryl hydrocarbon is attached to a support such as Sephadex. Retention of proteins on these supports involves hydrophobic interactions between the alkyl chain and hydrophobic regions on the protein. Proteins are applied in solutions that contain a high concentration of a salt (eg, $[NH_4]_2SO_4$) and are eluted with *decreasing* gradients of the same salt.

Determination of Protein Homogeneity by Polyacrylamide Gel Electrophoretic (PAGE) Techniques

Protein homogeneity is best assessed by polyacrylamide gel electrophoresis (PAGE) under several sets of conditions. One-dimensional PAGE of the native protein will, if sufficient sample is applied, reveal the presence of major and minor protein contaminants. In 2-dimensional (O'Farrell) PAGE, the first dimension separates denatured proteins on the basis of their pI values by equilibrating them in an electrical field that contains urea and a pH gradient maintained by polymerized ampholytes. The second dimension then separates proteins, after denaturation with SDS, on the basis of the molecular sizes of their protomer units (if present).

INTRACELLULAR DISTRIBUTION OF ENZYMES

The spatial arrangement and compartmentalization of enzymes, substrates, and cofactors within the cell (see Chapter 2) are of cardinal significance. In liver cells, for example, the enzymes of glycolysis are located in the cytoplasm, whereas enzymes of the citric acid cycle are in the mitochondria.

The distribution of enzymes among subcellular organelles may be studied following fractionation of cell homogenates by high-speed centrifugation. The enzyme content of each fraction is then examined (see Chapter 2).

Localization of a particular enzyme in a tissue or cell in a relatively unaltered state is frequently accomplished by histochemical procedures ("histoenzymology"). Thin (2- to 10-μm) frozen sections of tissue are treated with a substrate for a particular enzyme. Where the enzyme is present, the product of the enzyme-catalyzed reaction is formed. If the product is colored and insoluble, it remains at the site of formation and localizes the enzyme. Histoenzymology provides a graphic and relatively physiologic picture of patterns of enzyme distribution.

ISOZYMES

While terms such as "malate dehydrogenase" or "glucose-6-phosphatase" appear to describe a single catalytic entity, they encompass all proteins which catalyze the oxidation of malate to oxaloacetate or the hydrolysis of glucose 6-phosphate to glucose and P_i, respectively. When techniques for purification of enzymes were applied to, for example, malate dehydrogenase from different sources (eg, rat liver and *Escherichia coli*), it became apparent that while rat liver and *E coli* malate dehydrogenase both catalyze the same reaction, their physical and chemical properties exhibited many significant differences. Physically distinct forms of the same catalytic activity may also be present in different tissues of the same organism, in different cell types within a tissue, or even within a prokaryotic organism such as *E coli*. This discovery followed from the application of electrophoretic separation procedures to separation of electrophoretically distinct forms of a particular enzymatic activity.

While the term "isozyme" (isoenzyme) encompasses all the above examples of physically distinct forms of a given catalytic activity, in practice, and particularly in clinical medicine, "isozyme" has a more restricted meaning, namely, the physically distinct and separable forms of a given enzyme present in different cell types of a specific eukaryote such as the human being. Isozymes are common in sera and tissues of all vertebrates, insects, plants, and unicellular organisms. Both the kind and the number of enzymes involved are equally diverse. Isozymes of numerous dehydrogenases, oxidases, transaminases, phosphatases, transphosphorylases, and proteolytic enzymes are known. Different tissues may contain different isozymes, and these isozymes may differ in their affinity for substrates.

DIAGNOSTIC SIGNIFICANCE OF ISOZYMES

Medical interest in isozymes was stimulated by the discovery that **human sera contained several lactate dehydrogenase isozymes and that their relative proportions changed significantly in certain pathologic conditions.** Subsequently, many additional examples of changes in isozyme proportions as a result of disease were described.

Serum lactate dehydrogenase isozymes may be visualized by subjecting a serum sample to electrophore-

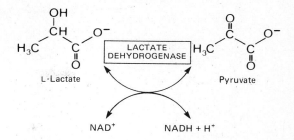

Figure 7–8. The L-lactate dehydrogenase reaction.

sis, usually at pH 8.6, on a starch, agar, or polyacrylamide gel support. The isozymes have different charges at this pH and migrate to 5 distinct regions of the electropherogram. Isozymes are then localized by means of their ability to catalyze reduction of a colorless dye to an insoluble, colored form.

A typical dehydrogenase isoenzyme assay reagent contains the following:

(1) Reduced substrate (eg, lactate).

(2) Coenzyme (NAD⁺).

(3) Oxidized dye (eg, nitroblue tetrazolium salt [NBT]).

(4) An intermediate electron carrier to transport electrons between NADH and the dye (eg, phenazine methosulfate [PMS]).

(5) Buffer; activating ions if required.

Lactate dehydrogenase catalyzes transfer of 2 electrons and one H^+ from lactate to NAD^+ (Fig 7–8). The reaction proceeds at a measurable rate only in the presence of lactate dehydrogenase. When the assay mixture is spread on the electropherogram and incubated at 37 °C, concerted electron transfer reactions take place only in those regions where lactate dehydrogenase is present (Fig 7–9). The relative intensities of the colored bands may then be quantitated by a scanning photometer (Fig 7–10). The most negative isoenzyme is termed I_1.

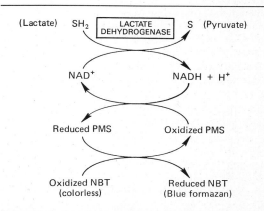

Figure 7–9. Coupled reactions in detection of lactate dehydrogenase activity on an electropherogram.

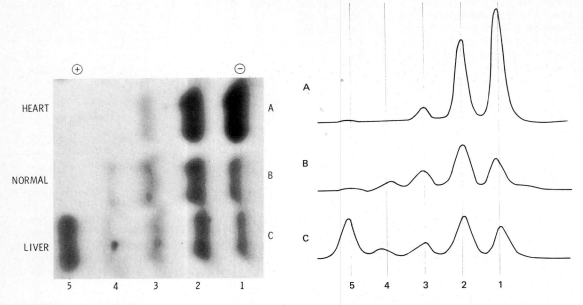

Figure 7–10. Normal and pathologic patterns of lactate dehydrogenase (LDH) isozymes in human serum. LDH isozymes of serum were separated on cellulose acetate at pH 8.6 and stained for enzyme. The photometer scan shows the relative proportion of the isozymes. Pattern A is serum from a patient with a myocardial infarct, B is normal serum, and C is serum from a patient with liver disease. (Courtesy of Dr Melvin Black and Mr Hugh Miller, St Luke's Hospital, San Francisco.)

PHYSICAL BASIS FOR ISOZYMES

Oligomeric enzymes with dissimilar protomers can exist in several forms. Frequently, one tissue produces one protomer predominantly and another tissue a different protomer. If these can combine in various ways to construct an active enzyme (eg, a tetramer), **isozymes** of that enzymatic activity are formed.

Lactate dehydrogenase isozymes differ at the level of quaternary structure. The oligomeric lactate dehydrogenase molecule (MW 130,000) consists of 4 protomers of 2 types, H and M (MW about 34,000). Only the tetrameric molecule possesses catalytic activity. If order is unimportant, these protomers might be combined in the following 5 ways:

$$
\begin{array}{c}
HHHH \\
HHHM \\
HHMM \\
HMMM \\
MMMM
\end{array}
$$

Markert used conditions known to disrupt and reform quaternary structure to clarify the relationships between the lactate dehydrogenase isozymes. Splitting and reconstitution of lactate dehydrogenase-I_1 or lactate dehydrogenase-I_5 produced no new isozymes. These therefore consist of a single type of protomer. When a mixture of lactate dehydrogenase-I_1 and lactate dehydrogenase-I_5 was subjected to the same treatment, lactate dehydrogenase-I_2, -I_3, and -I_4 were generated. The proportions of the isozymes found are

those which would result if the relationship were:

Lactate Dehydrogenase Isozyme	Subunits
I_1	HHHH
I_2	HHHM
I_3	HHMM
I_4	HMMM
I_5	MMMM

Syntheses of H and M subunits are controlled by distinct genetic loci that are differentially expressed in different tissues, eg, the heart and skeletal muscle.

ENZYMES IN CLINICAL DIAGNOSIS

Distinction Between Functional & Nonfunctional Plasma Enzymes

Certain enzymes, proenzymes, and their substrates are present at all times in the circulation of normal individuals and perform a physiologic function in blood. Examples of **functional plasma enzymes** include lipoprotein lipase, pseudocholinesterase, and the proenzymes of blood coagulation and of blood clot dissolution. They generally are synthesized in the liver but are present in blood in equivalent or higher concentrations than in tissues.

As the name implies, **nonfunctional plasma enzymes** perform no known physiologic function in blood. Their substrates frequently are absent from plasma, and the enzymes themselves are present in the blood of normal individuals at levels up to a million-fold lower than in tissues. Their presence in plasma at levels elevated above normal values suggests an increased rate of tissue destruction. Measurement of these nonfunctional plasma enzyme levels can thus provide the physician with valuable diagnostic and prognostic clinical evidence.

Nonfunctional plasma enzymes include those present in exocrine secretions and true intracellular enzymes. Exocrine enzymes—pancreatic amylase, lipase, bile alkaline phosphatase, and prostatic acid phosphatase—diffuse passively into the plasma. True intracellular enzymes normally are absent from the circulation.

Origin of Nonfunctional Plasma Enzymes

Low levels of nonfunctional enzymes found ordinarily in plasma arise apparently from the routine, normal destruction of erythrocytes, leukocytes, and other cells. With accelerated cell death, soluble enzymes enter the circulation. Although elevated plasma enzyme levels are generally interpreted as evidence of cellular necrosis, vigorous exercise also releases significant quantities of muscle enzymes.

DIAGNOSTIC & PROGNOSTIC VALUE OF SPECIFIC ENZYMES

Practicing physicians have long made use of quantitation of the levels of certain nonfunctional plasma enzymes. This valuable diagnostic and prognostic information is in most instances obtained on fully automated equipment. Table 7–3 contains a list of the principal enzymes employed in the field of diagnostic enzymology. Further details on the uses of these enzymes are given in the Appendix, including discussions of the important concepts of the sensitivity and specificity of diagnostic tests.

DIAGNOSTIC APPLICATIONS OF RESTRICTION ENDONUCLEASES

The diagnosis of genetic diseases has received tremendous impetus from recent developments in recombinant DNA technology. While all molecular diseases have long been known to be a consequence of altered DNA, techniques for direct examination of DNA sequences have only recently become available. For example, development of hybridization probes for DNA fragments (Southern, 1975) has led to techniques of sensitivity sufficient for prenatal screening for hereditary disorders by restriction enzyme mapping of DNA derived from fetal cells in the amniotic fluid.

DNA probes can, in principle, be constructed for the diagnosis of most genetic diseases. For example, for prenatal detection of thalassemias (characterized by defects in the synthesis of hemoglobin subunits; see Chapter 6), a probe constructed against a portion of the

Table 7–3. Principal serum enzymes used in clinical diagnosis. Many of the enzymes are not specific for the disease listed; further details of the range of conditions in which the activities of these enzymes are affected are given in the Appendix.

Serum Enzyme	Major Diagnostic Use
Aminotransferases Aspartate aminotransferase (AST, SGOT)	Myocardial infarction
Alanine aminotransferase (ALT, SGPT)	Viral hepatitis
Amylase	Acute pancreatitis
Ceruloplasmin	Hepatolenticular degeneration (Wilson's disease)
Creatine phosphokinase	Muscle disorders and myocardial infarction
γ-Glutamyl transpeptidase	Various liver diseases
Lactate dehydrogenase (isozymes)	Myocardial infarction
Lipase	Acute pancreatitis
Phosphatase, acid	Metastatic carcinoma of the prostate
Phosphatase, alkaline (isozymes)	Various bone disorders, obstructive liver diseases

gene for a normal hemoglobin subunit can detect a shortened or absent restriction fragment arising from a deletion in that gene, as occurs in some α-thalassemias and certain rare types of β- and β,δ-thalassemias (Dozy, Forman, and Abuelo, 1979; Kan, Chang, and Dozy, 1982). Alternatively, a synthetic cDNA probe has been constructed that hybridizes to a β-globin sequence that contains a nonsense mutation present in certain β-thalassemias, but not to the normal β-globin gene (Pirastu et al, 1984). Absence of the plasma protease inhibitor α_1-antitrypsin is associated with emphysema and with infantile liver cirrhosis. The presence of inactive α_1-antitrypsin has been detected by a probe constructed against the inactive allele that contains a point mutation in the α_1-antitrypsin gene (Kidd et al, 1983).

Hybridization probes may also be used to detect genetic alterations that lead to the loss of a restriction endonuclease site (see Chapter 38). For example, the point mutation of the Glu codon (GAG) to the Val codon (GTG) characteristic of sickle cell disease can be detected in the β-globin gene from cells in as little as 10 mL of amniotic fluid using the restriction endonuclease Mst II or Sau I (Orkin et al, 1982).

DNA probes may also be used to detect DNA sequences tightly linked to, but not actually within, the gene of interest. The analyses may be extended to detection of chromosome-specific variations (differences in sequence between homologous chromosomes). Digestion of the DNA with a restriction endonuclease generates different restriction maps (patterns of DNA fragments) from homologous genes that contain different base sequences. This phenomenon is termed "restriction fragment length polymorphism" (RFLP). For a genetic disease linked to a restriction length polymorphism, a human carrier of the disease will bear one chromosome with a normal gene and one with the defective gene. When the restriction fragments are resolved and probed, 2 hybridizing bands are detected (as distinct from a single band where both genes are identical). Offspring who inherit the disease-bearing chromosome exhibit only a single hybridizing band that differs from the band produced by normal chromosomes. The phenomenon of associated RFLP has been applied to the analysis of sickle cell trait (based on an associated Hpa I RFLP) and of β-thalassemia (linked to a Hind III and a Bam HI RFLP) (Little et al, 1980; Woo et al, 1983).

A screen based on RFLPs has been developed for the detection of infant phenylketonuria (Woo et al, 1983) (see Chapter 31). Note, however, that since the RFLP does not cause the disease but is merely located near the defective gene, this general approach is not infallible. The future importance of RFLPs appears to be as points of departure for identification of the gene responsible for linked diseases. This approach has already been applied to a screen for the mutational events involved in formation of retinoblastoma tumors (Cavenee et al, 1983) and Huntington's disease (Gusella et al, 1984).

Further examples of the uses of restriction enzymes for diagnosis are given in Chapter 36.

REFERENCES

General Enzymology

Boyer PD, Lardy H, Myrbäck K (editors): *The Enzymes,* 3rd ed. 7 vols. Academic Press, 1970–1973.

Nord FF (editor): *Advances in Enzymology.* Interscience. [Issued annually.]

Enzyme Structure

Fersht A: *Enzyme Structure and Mechanism,* 2nd ed. Freeman, 1985.

Hirs CHW, Timascheff SN (editors): Enzyme structure. Parts A–H in: *Methods in Enzymology.* Vol 11, 1967; Vols 25 and 26, 1972; Vol 27, 1973; Vol 47, 1977; Vols 48 and 49, 1978; Vol 49, 1979. Academic Press.

Coenzymes

McCormick DB, Wright LD (editors): Vitamins and coenzymes. Parts A–F in: *Methods in Enzymology.* Vol 18A, 1970; Vols 18B and 18C, 1971; Vol 62, 1979; Vols 66 and 67, 1980. Academic Press.

Nomenclature

Enzyme Nomenclature 1978. Recommendations of the Nomenclature Committee of the International Union of Biochemistry on the Nomenclature and Classification of Enzymes. Academic Press, 1979.

Assay & Purification of Enzymes

Bergmeyer H-U (editor): *Methods of Enzymatic Analysis,* 2nd English ed. 4 vols. Academic Press, 1974.

Boyer PD, Lardy H, Myrbäck K (editors): *The Enzymes,* 3rd ed. 7 vols. Academic Press, 1970–1973.

Colowick SP, Kaplan NO (editors): *Methods in Enzymology.* 69 vols. Academic Press, 1955–1987.

Hoffmann-Ostenhoff O et al: *Affinity Chromatography.* Pergamon Press, 1978.

Jacoby WB (editor): Enzyme purification and related techniques. In: *Methods in Enzymology.* Vol 22. Academic Press, 1971.

Jacoby WB, Wilchek M (editors): Affinity techniques. In: *Methods in Enzymology.* Vol 34, 1974; Vol 46, 1977. Academic Press.

Mosbach K (editor): Immobilized enzymes. In: *Methods in Enzymology.* Vol 44. Academic Press, 1976.

Intracellular Distribution of Enzymes

DePierre JW, Ernster L: Enzyme topology of intracellular membranes. *Annu Rev Biochem* 1977;**46:**201.

Clinical Enzymology

Bergmeyer HU: Aspartate aminotransferase. *Test of the Month 6* 1980; No. 2.

Bergström K: Determination of serum alkaline phosphatase activity. *Test of the Month 1* 1974; No. 22.

Cavanee WK et al: Expression of recessive alleles by chromosomal mechanisms in retinoblastomas. *Nature* 1983; **305:**779.

Dozy AM, Forman EN, Abuelo DN: Prenatal diagnosis of homozygous α-thalassemia. *JAMA* 1979;**241:**1610.

Fishinger AF: Creatine phosphokinase and its isoenzymes. *Test of the Month 2* 1976; No. 6.

Gusella JF et al: DNA markers for nervous system diseases. *Science* 1984;**225:**1320.

Kan YW, Chang J, Dozy AM: Pages 275–283 in: *Thalassemia: Recent Advances in Detection and Treatment.* Cao A, Carcassi U, Rowley P (editors). AR Liss, 1982.

Kidd VJ et al: α_1-Antitrypsin deficiency detection by direct analysis of the mutation in the gene. *Nature* 1983; **304:**230.

Little PFR et al: Model for antenatal diagnosis of β-thalassemia and other monogenic disorders by molecular analysis of linked DNA polymorphisms. *Nature* 1980; **285:**144.

McNair RD: Lactate dehydrogenase. *Test of the Month 2* 1976; No. 3.

Orkin SH et al: Improved detection of the sickle mutation by DNA analysis: Application to prenatal diagnosis. *N Engl J Med* 1982;**307:**32.

Pirastu M et al: Multiple mutations produce $\delta\beta^0$-thalassemia in Sardinia. *Science* 1984;**223:**929.

Southern EM: Detection of specific sequences among DNA fragments separated by gel electrophoresis. *J Mol Biol* 1975;**98:**503.

Wilkinson JH: Clinical applications of isozymes. *Clin Chem* 1970;**16:**733.

Wilkinson JH: Clinical significance of enzyme activity measurements. *Clin Chem* 1970;**16:**882.

Woo SLC et al: Cloned human phenylalanine hydroxylase gene allows prenatal diagnosis and carrier detection of classical phenylketonuria. *Nature* 1983;**306:**151.

Enzymes: Kinetics

8

Victor W. Rodwell, PhD

INTRODUCTION

Preceding chapters have reviewed the physical and chemical properties of proteins and the relationship between protein structure and function. Chapters 5 and 11 discuss the general properties of enzymes and the energy changes that generally accompany biochemical reactions, most of which are catalyzed by specific enzymes. In this chapter we consider the chemical nature of the catalysis carried out by enzymes and the nature of the enzyme-substrate interaction responsible for the reaction specificity of these biologic catalysts.

BIOMEDICAL IMPORTANCE

Of the major factors (enzyme and substrate concentration, temperature, pH, and inhibitors) that affect assays of enzyme activity, the last 3 are of particular clinical interest. Because enzyme activities increase as temperature increases, the rates of metabolic processes increase significantly during fevers. If fevers continued indefinitely, the results would be fatal, in part because of metabolic overdemand. On the other hand, lowering body temperature (hypothermia)—and consequently the activities of most enzymes—proves useful when it is important to reduce overall metabolic demand (eg, during open-heart surgery or transportation of organs for transplantation surgery). A cardinal biologic principle is that of homeostasis—maintenance of the internal milieu of the body very close to its normal conditions. Relatively small changes in pH can affect the activities of many enzymes or proteins. Most drugs act by affecting enzyme-catalyzed reactions. Many of these compounds resemble natural substrates and thus act as competitive inhibitors of enzyme activity. Understanding much of pharmacology and toxicology depends on a thorough knowledge of the basics of enzyme inhibition.

FORMATION & DECAY OF TRANSITION STATES

Shown below is a displacement reaction in which an entering group, Y, displaces a leaving group, X.

$$Y + R{-}X \rightleftharpoons Y{-}R + X$$

This overall reaction proceeds via 2 half reactions: (1) formation of a **transition state** in which Y and X both are attached to R, and (2) decay of the transition state to form products.

$$Y + R{-}X \rightleftharpoons \underbrace{Y\cdots R\cdots X}_{\substack{\textbf{Transition}\\\textbf{state}}} \rightleftharpoons Y{-}R + X$$

As with all chemical reactions, characteristic changes in free energy are associated with each half reaction. We define ΔG_F as the change in free energy associated with **formation** of the transition state and ΔG_D as the change in free energy associated with **decay** of the transition state to form products

$$\Delta G_F = \Delta G_F^0 + RT \ln \frac{[Y\cdots R\cdots X]}{[Y][R{-}X]}$$

$$\Delta G_D = \Delta G_D^0 + RT \ln \frac{[Y{-}R][X]}{[Y\cdots R\cdots X]}$$

where the change in free energy for the **overall** reaction, ΔG, is the sum of the changes in free energy for both half reactions.

$$\Delta G = \Delta G_F + \Delta G_D$$

As for any equation with 2 terms, it is not possible by inspection of the algebraic sign and magnitude of ΔG to infer the sign or magnitude of either ΔG_F or ΔG_D. Stated another way, we are unable simply from consideration of the change in free energy for the overall reaction, ΔG, to infer anything whatever concerning the free energy changes associated with formation and decay of transition states. Since catalysis is intimately associated with ΔG_F and ΔG_D, it follows that the thermodynamics of the overall reaction (ΔG) can tell us nothing of the **path** a reaction follows (ie, its mechanism). This, as we shall see, is the task of kinetics.

FREE ENERGY CHANGES ASSOCIATED WITH FORMATION & DECAY OF TRANSITION STATES

The above concepts are represented graphically in Figs 8–1 and 8–2 as "reaction profiles" that illustrate

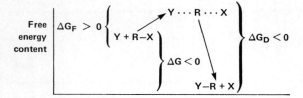

Figure 8–1. Reaction profile for a displacement reaction with a negative overall change in free energy, ie, $\Delta G < 0$.

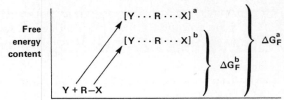

Figure 8–3. Reaction profile for formation of 2 different transition states, $[Y \cdots R \cdots X]^a$ and $[Y \cdots R \cdots X]^b$, and their associated free energies of formation, ΔG_F^a and ΔG_F^b.

the relationship between ΔG, ΔG_F, and ΔG_D. Note that whereas in Fig 8–1 ΔG is negative ($\Delta G < 0$) and in Fig 8–2 ΔG is positive ($\Delta G > 0$), in both instances, ΔG_F is positive ($\Delta G_F > 0$) and ΔG_D is negative ($\Delta G_D < 0$). Therefore, as stated above, **from the sign and magnitude of ΔG we cannot infer the sign or magnitude of either ΔG_F or ΔG_D.**

ROLE OF CATALYSTS IN FORMATION OF PRODUCTIVE TRANSITION STATES

Consider the reaction profiles for 2 different transition states in the same overall reaction (Fig 8–3).

In both instances, the magnitude of the free energy of formation of the transition state represents the energy barrier for the overall reaction. The energy barrier for the reaction that proceeds via the transition state $[Y \cdots R \cdots X]^b$ is thus lower than the energy barrier for the reaction that proceeds via the transition state $[Y \cdots R \cdots X]^a$. **Catalysts alter the free energy content of the transition state.** In the above example, $[Y \cdots R \cdots X]^a$ represents the transition state for the **noncatalyzed** reaction, while $[Y \cdots R \cdots X]^b$ represents the transition state for the **catalyzed** reaction. All catalysts, including enzymes, lower the free energy of formation, ΔG_F, of the transition state. Note further that since catalysis has no effect on ΔG, **the change in free energy for the overall reaction is independent of catalysis.** Since the equilibrium constant for a chemical reaction is a function of the standard free energy change for a reaction

$$\Delta G^0 = -RT \ln K_{eq}$$

it follows that **enzymes and other catalysts have no effect on the equilibrium constant for a reaction.**

THE COLLISION THEORY OF REACTION

The **kinetic**, or **collision, theory** for chemical reactions incorporates 2 key concepts:

(1) In order to react, molecules must collide (ie, be within bond-forming distance of one another).

(2) For a collision to be productive (ie, to result in a reaction), the reacting molecules must possess sufficient energy to overcome the energy barrier for the reaction.

It follows that if the reacting molecules have sufficient energy to react, anything that increases the frequency of collision between molecules will increase their rate of reaction. Conversely, factors that decrease either collision frequency or kinetic energy will decrease the rate of reaction.

If some molecules in the population have insufficient energy to react, increased temperature, which increases kinetic energy, will increase the rate of the reaction. These concepts are illustrated diagrammatically in Fig 8–4. In A none, in B a portion, and in C all of the molecules have sufficient kinetic energy to overcome the energy barrier for reaction.

In the absence of enzymatic catalysis, many chemical reactions proceed exceedingly slowly at the temperature of living cells. However, even at this temperature molecules are in motion and undergo collisions. **They fail to react rapidly, because most possess in-**

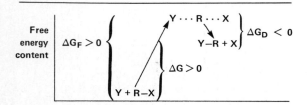

Figure 8–2. Reaction profile for a displacement reaction with a positive overall change in free energy, ie, $\Delta G > 0$.

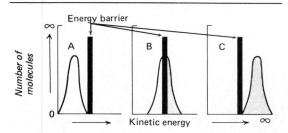

Figure 8–4. The energy barrier for chemical reactions.

sufficient kinetic energy to overcome the energy barrier for reaction. At a considerably higher temperature (and higher kinetic energy), the reaction will occur more rapidly. That the reaction takes place at all shows that it is spontaneous ($\Delta G < 0$). At the lower temperature, it is spontaneous but slow; at the higher temperature, spontaneous and fast. **Enzymes make spontaneous reactions proceed rapidly under the conditions prevailing in living cells.**

ROLE OF ENZYMES IN THE RUPTURE & FORMATION OF COVALENT BONDS

Most chemical reactions of biochemical interest involve breakage or formation of covalent bonds. For example, consider the group transfer reactions introduced in Chapter 7

$$D-G + A \rightleftharpoons A-G + D$$

in which a group, G, is transferred from a donor, D–G, to an acceptor, A. The overall reaction involves both rupture of the D–G bond and formation of a new A–G bond. Enzyme-catalyzed group transfer reactions are, however, better represented as follows:

$$\begin{array}{ccc} D-G & Enz & A-G \\ & Enz-G & \\ D & & A \end{array}$$

This representation emphasizes 3 important features of enzyme-catalyzed group transfer reactions:

(1) Each half reaction involves both the rupture and formation of a covalent bond.

(2) The enzyme is a reactant coequal with D–G and A.

(3) Whereas in the overall reaction the enzyme acts catalytically (ie, is required only in trace quantities and may be recovered unchanged when the reaction is complete), for each of the half reactions, the enzyme is a stoichiometric reactant (ie, it is required in a 1:1 molar ratio with the other reactants).

Many additional biochemical reactions may be considered as special cases of group transfer in which D, A, or both may be absent. Isomerization reactions (eg, the interconversion of glucose 6-phosphate and glucose 1-phosphate) might be represented as reactions in which both D and A are absent:

$$\begin{array}{ccc} S & Enz & P \\ & Enz-S & \end{array}$$

These representations fail, however, to emphasize yet another key feature of enzyme-catalyzed reactions—**participation in the overall reaction of 2 or more intermediate forms of EnzS complex and the consequent participation of a set of several sequential half reactions.** A representation of a group trans-

fer reaction that emphasizes these features might be

$$\begin{array}{ccccc} D-G & & Enz & & A-G \\ & & & & \\ D & & & & A \\ & Enz-G & & Enz-G** & \\ & & Enz-G* & & \end{array}$$

in which EnzG, EnzG*, and EnzG** represent successive EnzS complexes in the overall reaction.

It is clear from all of the above representations that for the reaction to occur, all of the reactants must come within bond-forming (or bond-breaking) distance of one another. That is, the reactants must collide. For homogeneous solution chemistry in the absence of catalysts, the concentrations of the reacting molecules are constant throughout the solution. This condition no longer obtains following introduction of a catalyst. To be effective, a catalyst must have surface domains that bind the reacting molecules. While this binding is a reversible process, the overall equilibrium constant for binding strongly favors the bound rather than the free forms of the reacting molecules. Qualitatively, we might represent this as follows:

Reactant + Catalyst $\rightleftharpoons$ Reactant-catalyst complex

Quantitatively, we might express the tightness of association between a reactant, R, and a catalyst, C, in terms of the dissociation constant, K_d, for the R–C complex, or the equilibrium constant for the reaction:

$$R-C \rightleftharpoons R + C$$

$$K_d = \frac{[R]\ [C]}{[R-C]}$$

A low value for K_d thus represents a tight R–C complex.

One important consequence is that when a **reactant binds to a catalyst this raises the concentration of the reactant in a localized area of the solution** well above that of its concentration in free solution. Thus, we are no longer dealing with homogeneous but with heterogeneous solution chemistry.

If the catalyst for a bimolecular (2-reactant) reaction binds both reactants, the local concentration of each reactant is increased by a factor that depends on its individual affinity (K_d value) for the catalyst. Since the rate of the overall bimolecular reaction

$$A + B \rightarrow A-B$$

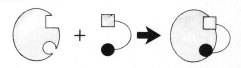

Figure 8–5. Representation of formation of an EnzS complex according to the Fischer template hypothesis.

Figure 8–6. Representation of sequential adsorption of a coenzyme (CoE) and of 2 substrates (S_1 and S_2) to an enzyme in terms of the template hypothesis. The coenzyme is assumed to bear a group essential for binding the first substrate (S_1), which in turn facilitates binding of S_2.

is, as we shall see below, proportionate to the concentrations of **both** A and B, binding of both A and B by the catalyst can result in an enormous (several thousand-fold) increase in overall reaction rate.

One key factor in the ability of enzymes to act as catalysts is their ability to bind effectively one or (more frequently) both reactants in a bimolecular reaction with an accompanying increase in local reactant concentration and hence in local reaction rate. That enzymes are, relative to most nonprotein catalysts, both extremely efficient and highly selective catalysts requires further explanation. To understand these distinctive properties of enzymes, we must introduce the concept of the "active" or "catalytic" site.*

THE CATALYTIC SITE

The large size of proteins relative to substrates led to the concept that a restricted region of the enzyme was concerned with catalysis. This region we refer to as the **catalytic site.** Initially, it was puzzling why enzymes were so large when only a portion of their structure appeared to be required for substrate binding and catalysis. Today, we recognize from 3-dimensional models of enzymes that a far greater portion of the protein interacts with the substrate than was formerly supposed. When the need for allosteric sites of equal size also arises (see Chapter 10), the size of enzymes should no longer be surprising.

The "Lock & Key" or "Template" Model

The original model of a catalytic site, proposed by Emil Fischer, visualized interaction between substrate and enzyme in terms of a "lock and key" analogy. This **lock and key,** or **rigid template, model** (Fig 8–5), is still useful for understanding certain properties of enzymes—for example, the ordered binding of 2 or more substrates (Fig 8–6) or the kinetics of a simple substrate saturation curve.

---

*While many texts equate the active and catalytic sites of enzymes, there are on enzymes other "active" sites concerned with regulation of enzyme activity rather than with the intermediary enzymology of the catalytic process per se. We therefore use the term "catalytic site" to avoid ambiguity.

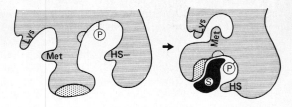

Figure 8–7. Representation of an induced fit by a conformational change in the protein structure. Note the relative positions of key residues before and after the substrate is bound. (After Koshland.)

The "Induced Fit" Model

An unfortunate feature of the Fischer model is the implied rigidity of the catalytic site. A more general model is the **"induced fit" model** of Koshland. This model has considerable experimental support. An essential feature is the flexibility of the catalytic site. In the Fischer model, the catalytic site is presumed to be preshaped to fit the substrate. In the induced fit model, the substrate induces a conformational change in the enzyme. This aligns amino acid residues or other groups on the enzyme in the correct spatial orientation for substrate binding, catalysis, or both. At the same time, other amino acid residues may become buried in the interior of the enzyme.

In the example (Fig 8–7), hydrophobic groups (hatched) and charged groups (stippled) both are involved in substrate binding. A phosphoserine ($-P$) and the $-SH$ of a cysteine residue are involved in catalysis. Other residues involved in neither process are represented by Lys and Met residues. In the absence of substrate, the catalytic and the substrate-binding groups are several bond distances apart. Approach of the substrate induces a conformational change in the enzyme protein, aligning the groups correctly for substrate binding and for catalysis. At the same time, the spatial orientations of other regions are also altered—the Lys and Met are now closer together (Fig 8–7).

Substrate analogs may cause some, but not all, of the correct conformational changes (Fig 8–8). On attachment of the true substrate (A), all groups (shown as closed circles) are brought into correct alignment. Attachment of a substrate analog that is too "bulky" (Fig 8–8B) or too "slim" (Fig 8–8C) induces incor-

---

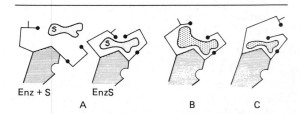

Figure 8–8. Representation of conformational changes in an enzyme protein when binding substrate (A) or inactive substrate analogs (B, C). (After Koshland.)

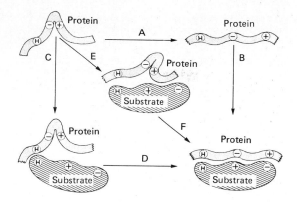

Figure 8–9. Representation of alternative reaction paths for a substrate-induced conformational change. The enzyme may first undergo a conformational change (*A*), then bind substrate (*B*). Alternatively, substrate may first be bound (*C*), whereupon a conformational change occurs (*D*). Finally, both processes may occur in a concerted manner (*E*) with further isomerization to the final conformation (*F*). (Adapted from Koshland.)

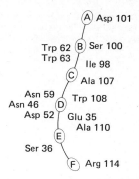

Figure 8–10. Schematic representation of the catalytic site in the cleft region of lysozyme. A to F represent the glycosyl moieties of a hexasaccharide. Some residues in the cleft region are shown with their numbers in the lysozyme sequence. (Adapted from Koshland.)

rect alignment. One final feature is the site shown as a small notch on the right. One may visualize a regulatory molecule attaching at this point and "holding down" one of the polypeptide arms bearing a catalytic group. Substrate binding, but not catalysis, might then occur.

The exact sequence of events in a substrate-induced conformational change remains to be established. Several possibilities exist (Fig 8–9).

Even when the complete primary structure of an enzyme is known, it usually is still difficult to decide exactly which residues constitute the catalytic site. As illustrated by the induced fit model, these may be distant one from another in the primary structure but spatially close in the 3-dimensional (tertiary) structure.

In the representation of a catalytic site, several regions of a polypeptide chain each contribute residues to the site. Furthermore, these residues generally are not all sequential within a polypeptide chain, as described for hemoglobin in Chapter 6 and for chymotrypsin in Chapter 9.

Catalytic Site of Lysozyme

Lysozyme, present in tears, nasal mucus, sputum, tissues, gastric secretions, milk, and egg white, catalyzes the hydrolysis of β-1,4-linkages of N-acetyl-neuraminic acid (see Chapters 13 and 33) in proteoglycans and glycosaminoglycans. It performs the function, in tears and nasal mucus, of destroying the cell walls of many airborne gram-positive bacteria. Lysozyme (MW about 15,000) consists of a single polypeptide chain of 129 residues. Since there is no coenzyme or metal ion, catalysis, specificity, and 3-dimensional structure are determined solely by these amino acid residues. There are small regions of pleated sheet, little α-helix, and large regions of random coil. For a model of lysozyme and its substrate,

photographed in 3-dimensional color, see *J Biol Chem* 1968;**243:**1663. The molecule bears a deep central cleft which harbors a catalytic site with 6 subsites (Fig 8–10) that bind various substrates or inhibitors. The residues responsible for bond cleavage are thought to lie between sites D and E close to the carboxyl groups of Asp 52 and Glu 35. Glu 35 apparently protonates the acetal bond of the substrate, while the negatively charged Asp 52 stabilizes the resulting carbonium ion from the back side.

Catalytic Site of Ribonuclease

Unlike the case for lysozyme, considerable information about the catalytic site of ribonuclease was available prior to solution of the 3-dimensional structure. The conclusions based on chemical investigations were largely confirmed by crystallography. The structure contains a cleft similar to that of lysozyme across which lie 2 residues, His 12 and His 119. These previously were implicated by chemical evidence as being at the catalytic site. Both residues are near the binding site for uridylic acid (Fig 8–11).

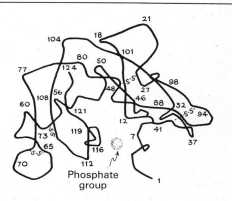

Figure 8–11. Structure of ribonuclease as determined by x-ray diffraction. Numbers refer to specific residues. See also Fig 5–7.

Table 8–1. Amino acid sequences in the neighborhood of the catalytic sites of several bovine proteases. Regions shown are those on either side of the catalytic site seryl (S) and histidyl (H) residues. For explanation of single letter abbreviations for amino acids, see Chapter 3. (Reproduced, with permission, from Dayhoff MO [editor]: *Atlas of Protein Sequence and Structure.* Vol 5. National Biomedical Research Foundation, 1972.)

Enzyme	Sequence Around Serine Ⓢ		Sequence Around Histidine Ⓗ
Trypsin	D S C Q D G Ⓢ G G P V V C S G K		V V S A A Ⓗ C Y K S G
Chymotrypsin A	S S C M G D Ⓢ G G P L V C K K N		V V T A A Ⓗ G G V T T
Chymotrypsin B	S S C M G D Ⓢ G G P L V C Q K N		V V T A A Ⓗ C G V T T
Thrombin	D A C E G D Ⓢ G G P F V M K S P		V L T A A Ⓗ C L L Y P

Amino Acid Sequences at Catalytic Sites

The primary structures of active sites of hydrolytic enzymes exhibit many similarities (Table 8–1). This implies that the number of bond-breaking mechanisms operating in biologic systems is relatively small. In view of these similarities, it is perhaps not surprising that the amino acid sequences near the catalytic sites of the same enzyme from different species bear even greater similarity.

FACTORS AFFECTING ENZYME ACTIVITY

Major factors that affect enzyme activity include enzyme and substrate concentration, temperature, pH, and inhibitors.

TEMPERATURE

Over a limited range of temperatures, the velocity of enzyme-catalyzed reactions increases as temperature rises. The factor by which the velocity increases for a 10 °C temperature rise is the **Q_{10}, or temperature coefficient.** The velocity of many biologic reactions roughly doubles with a 10 °C rise in temperature ($Q_{10} = 2$), and is halved if the temperature is decreased by 10 °C. Many physiologic processes—eg, the rate of contraction of an excised heart—consequently exhibit a Q_{10} of about 2.

For rates of enzyme-catalyzed reactions measured at several temperatures, Fig 8–12 is typical. There is an optimal temperature at which the reaction is most rapid. Above this, the reaction rate decreases sharply, mainly owing to denaturation of the enzyme by heat.

The increase in rate below optimal temperature results from increased kinetic energy of the reacting molecules. As the temperature is raised still further, the kinetic energy of the enzyme molecule exceeds the energy barrier for breaking the secondary bonds that maintain its native, catalytically active state. There is consequently a loss of secondary and tertiary structure and a parallel loss of catalytic activity.

For most enzymes, optimal temperatures are at or above those of the cells in which they occur. Enzymes from microorganisms adapted to growth in natural hot springs may exhibit optimal temperatures close to the boiling point of water.

pH

Moderate pH changes affect the **ionic state of the enzyme** and frequently that of the substrate also. When enzyme activity is measured at several pH values, optimal activity typically is observed between pH values of 5.0 and 9.0. However, a few enzymes, eg, pepsin, are active at pH values well outside this range.

The shape of pH-activity curves is determined by the following factors:

(1) Enzyme denaturation at extremely high or low pH.

(2) Effects on the charged state of the substrate or enzyme. For the enzyme, charge changes may affect activity either by changing structure or by changing the charge on a residue functional in substrate binding or catalysis. To illustrate, consider a negatively charged enzyme (Enz^-) reacting with a positively charged substrate (SH^+):

$$Enz^- + SH^+ \rightarrow EnzSH$$

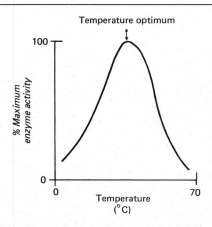

Figure 8–12. Effect of temperature on the velocity of a hypothetical enzyme-catalyzed reaction.

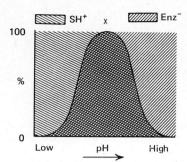

Figure 8–13. Effect of pH on enzyme activity.

At low pH, Enz^- protonates and loses its negative charge:

$$Enz^- + H^+ \rightarrow EnzH$$

Similarly, at high pH, SH^+ ionizes and loses its positive charge:

$$SH^+ \rightarrow S + H^+$$

Since the only forms that will interact are SH^+ and Enz^-, extreme pH values will lower the effective concentration of Enz^- and SH^+, thus lowering the reaction velocity (Fig 8–13). Only in the cross-hatched area are both Enz and S in the appropriate ionic state, and the maximal concentrations of Enz and S are correctly charged at X.

Enzymes may also undergo changes in conformation when the pH is varied. A charged group distal to the region where the substrate is bound may be necessary to maintain an active tertiary or quaternary structure, as described for hemoglobin (see Chapter 6). As the charge on this group is changed, the protein may unravel, become more compact, or dissociate into protomers—all with resulting loss of activity.

REACTANT CONCENTRATION

At high reactant concentrations, both the number of molecules with sufficient energy to react and their frequency of collision are high. This is true whether all or only a fraction of the molecules have sufficient energy to react. Consider reactions involving 2 different molecules, A and B,

$$A + B \rightarrow AB$$

Doubling the concentration either of A or of B will double the reaction rate. Doubling the concentration of both A and B will increase the probability of collision 4-fold. The reaction rate therefore increases 4-fold. **The reaction rate is proportionate to the concentrations of the reacting molecules.** Square

brackets ([]) are used to denote molar concentrations;* $\propto$ means "proportionate to." The rate expression is

$$Rate \propto [reacting\ molecules]$$

or

$$Rate \propto [A]\ [B]$$

For the situation represented by

$$A + 2B \rightarrow AB_2$$

the rate expression is

$$Rate \propto [A]\ [B]\ [B]$$

or

$$Rate \propto [A]\ [B]^2$$

For the general case where n molecules of A react with m molecules of B

$$nA + mB \rightarrow A_nB_m$$

the rate expression is

$$Rate \propto [A]^n[B]^m$$

THE EQUILIBRIUM CONSTANT

Since all chemical reactions are reversible, for the reverse reaction where n molecules of A react with m molecules of B

$$A_nB_m \rightarrow nA + mB$$

the appropriate rate expression is

$$Rate \propto [A_nB_m]$$

We represent reversibility by double arrows,

$$nA + mB \rightleftharpoons A_nB_m$$

This expression reads: "n molecules of A and m molecules of B are in equilibrium with A_nB_m." We may replace the "proportionate to" symbol ($\propto$) with an equality sign by inserting a proportionality constant, k, characteristic of the reaction under study. For the general case

$$nA + mB \rightleftharpoons A_nB_m$$

expressions for the rates of the forward reaction ($Rate_1$) and back reaction ($Rate_{-1}$) are

$$Rate_1 = k_1[A]^n[B]^m$$

---

*Strictly speaking, molar activities rather than concentrations should be used.

and

$$Rate_{-1} = k_{-1}[A_nB_m]$$

When the rates of the forward and back reactions are equal, the system is said to be **at equilibrium**, ie,

$$Rate_1 = Rate_{-1}$$

Then

$$k_1[A]^n[B]^m = k_{-1}[A_nB_m]$$

and

$$\frac{k_1}{k_{-1}} = \frac{[A_nB_m]}{[A]^n[B]^m} = K_{eq}$$

The ratio of k_1 to k_{-1} is termed the **equilibrium constant, K_{eq}.** The following important properties of a system at equilibrium should be kept in mind.

(1) The equilibrium constant is the ratio of the reaction rate **constants** k_1/k_{-1}.

(2) At equilibrium, the reaction **rates** (not the reaction rate constants) of the forward and back reactions are equal.

(3) **Equilibrium is a dynamic state.** Although no **net** change in concentration of reactant or product molecules occurs at equilibrium, A and B are continually being converted to A_nB_m and vice versa.

(4) **The equilibrium constant may be given a numerical value if we know the concentrations of A, B, and A_nB_m at equilibrium.**

Relationship Between Equilibrium Constant & Standard Free Energy Change (ΔG^0) for the Overall Reaction

The equilibrium constant is related to ΔG^0 as follows:

$$\Delta G^0 = -RT \ln K_{eq}$$

R is the gas constant and T the absolute temperature. Since these are known, **knowledge of the numerical value of K_{eq} permits one to calculate a value for ΔG^0.** If the equilibrium constant is greater than 1, the reaction is spontaneous; ie, the reaction as written (from left to right) is favored. If it is less than 1, the opposite is true; ie, the reaction is more likely to proceed from right to left. Note, however, that although the equilibrium constant for a reaction indicates the **direction** in which a reaction is spontaneous, it does not indicate whether it will take place **rapidly.** That is, it does not tell us anything about the **magnitude of the energy barrier** for the reaction (ie, ΔG_F; see above). This follows because K_{eq} determines ΔG^0, previously shown to concern only initial and final states. **Reaction rates depend on the magnitude of the energy barrier, not on the magnitude of ΔG^0.**

Most factors affecting the velocity of enzyme-catalyzed reactions do so by **changing local reactant concentration.**

ENZYME CONCENTRATION

In many situations, it is useful to know not only whether a given enzyme is present but also *how much* is present. Under appropriate conditions, the velocity of an enzyme-catalyzed reaction will be directly proportionate to the amount of the enzyme present. (See Chapter 7.)

That the rate is not always proportionate to enzyme concentration may be seen by considering the forward reaction at equilibrium. Although the forward reaction is proceeding, the rate of the reverse reaction equals it. Accordingly, it would *appear* that the forward reaction velocity is zero. However, when the enzyme-catalyzed reaction *initiates* the conversion of one or more substrates to product (P), there is no P for the reverse reaction to occur. Furthermore, at the *initiation* of the (forward) reaction, the concentration of S will not have been depleted at all. Therefore, at the initiation of the reaction, the velocity, ie, the **initial velocity (v_i), will be directly proportionate to the enzyme concentration [Enz]** (see Chapter 7).

The enzyme is a reactant that combines with substrate to form an **enzyme-substrate complex, EnzS,** which decomposes to form a product, P, and free enzyme. In its simplest form, this may be represented as

$$Enz + S \underset{k_{-1}}{\overset{k_1}{\rightleftharpoons}} Enz + P$$

Note that although the rate expressions for the forward, back, and *overall* reactions include the term [Enz],

$$Enz + S \underset{k_{-1}}{\overset{k_1}{\rightleftharpoons}} Enz + P$$

$$Rate_{-1} = k_1 [Enz] [S]$$

$$Rate_{-2} = k_{-2} [Enz] [P]$$

in the expression for the *overall* equilibrium constant, [Enz] cancels out.

$$K_{eq} = \frac{k_1}{k_{-1}} = \frac{[Enz] [P]}{[Enz] [S]} = \frac{[P]}{[S]}$$

The enzyme concentration thus has no effect on the equilibrium constant. Stated another way, since enzymes affect rates, not rate constants, they cannot affect K_{eq}, which is a ratio of rate constants. **The K_{eq} of a reaction is the same regardless of whether equilibrium is approached with or without enzymatic catalysis** (recall ΔG^0). Enzymes change the reaction path but do not affect the initial and final equilibrium concentrations of the reactants and products, the factors that determine K_{eq} and ΔG^0.

SUBSTRATE CONCENTRATION

In the following discussion, enzyme reactions are treated as if they had a single substrate and a single

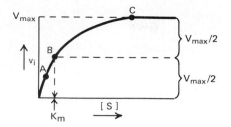

Figure 8–14. Effect of substrate concentration on the velocity of an enzyme-catalyzed reaction.

product. While this is the case for some enzyme-catalyzed reactions, most enzyme-catalyzed reactions have 2 or more substrates and products. This consideration does not, however, invalidate the discussion. What holds for a single substrate is true also for 2 substrates.

If the concentration of a substrate [S] is increased while all other conditions are kept constant, the **measured initial velocity,** v_i (the velocity measured when very little substrate has reacted), increases to a maximum value, V_{max}, and no further (Fig 8–14).

The velocity increases as the substrate concentration is increased up to a point where the enzyme is said to be "saturated" with substrate. The measured initial velocity reaches a maximal value and is unaffected by further increases in substrate concentration, because substrate is present in large molar excess over the enzyme. For example, if an enzyme with a molecular weight of 100,000 acts on a substrate with a molecular weight of 100 and both are present at a concentration of 1 mg/mL, there are 1000 mol of substrate for every mole of enzyme. More realistic figures might be

$$[Enz] = 0.1 \ \mu g/mL = 10^{-9} \ molar$$

$$[S] = 0.1 \ mg/mL = 10^{-3} \ molar$$

giving a 10^6 molar excess of substrate over enzyme. Even if [S] is decreased 100-fold, substrate is still present in 10,000-fold molar excess over enzyme.

The situations at points A, B, and C in Fig 8–14 are illustrated in Fig 8–15. At points A and B only a portion of the enzyme present is combined with substrate, even though there are many more molecules of substrate than of enzyme. This is because the equilibrium constant for the reaction Enz + S ⇌ EnzS (formation of the EnzS complex) is not infinitely large. **At point A or B, increasing or decreasing [S] will therefore increase or decrease the amount of Enz associated with S as EnzS, and v_i will thus depend on [S].** At C, essentially all the enzyme is combined with substrate, so that a further increase in [S], although it increases the frequency of collision between Enz and S, cannot result in increased rates of reaction, since no free enzyme is available to react.

Case B depicts a situation of major theoretical interest where exactly half the enzyme molecules are "saturated with" substrate. The velocity is accordingly **half the maximal velocity** ($V_{max}/2$) attainable at that particular enzyme concentration.

THE MICHAELIS-MENTEN EQUATION

Graphic Evaluation of the Michaelis Constant, K_m

The substrate concentration that produces half-maximal velocity, termed the K_m value or Michaelis constant, may be determined experimentally by graphing v_i as a function of [S] (Fig 8–14). Note that K_m has the dimensions of molar concentration.

When [S] is approximately equal to the K_m, v_i is very responsive to changes in [S], and the enzyme is working at precisely half-maximal velocity. In fact, many enzymes possess K_m values that are approximate to the physiologic concentration of their substrates.

The Michaelis-Menten expression

$$v_i = \frac{V_{max}[S]}{K_m + [S]}$$

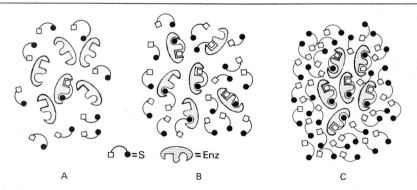

Figure 8–15. Representation of an enzyme at low (A), at high (C), and at the K_m concentration of substrate (B). Points A, B, and C correspond to those of Fig 8–14.

describes the behavior of many enzymes as substrate concentration is varied. The dependence of the initial velocity of an enzyme-catalyzed reaction on [S] and on K_m may be illustrated by evaluating the Michaelis-Menten equation as follows:

(1) When [S] is very much less than K_m (point A in Figs 8–14 and 8–15). Adding [S] to K_m in the denominator now changes its value very little, so that the [S] term can be dropped from the denominator. Since V_{max} and K_m are both constants, we can replace their ratio by a new constant, K.

$$v_i = \frac{V_{max}[S]}{K_m + [S]}; v_i \sim \frac{V_{max}[S]}{K_m} \sim \frac{V_{max}}{K_m}[S] \sim K[S]$$

[$\sim$ means "approximately equal to."]

In other words, **when the substrate concentration is considerably below that required to produce half-maximal velocity (the K_m value), the initial velocity, v_i, depends upon the substrate concentration, [S].**

(2) When [S] is very much greater than K_m (point C in Figs 8–14 and 8–15). Now adding K_m to [S] in the denominator changes the value of the denominator very little, so that the term K_m can be dropped from the denominator.

$$v_i = \frac{V_{max}[S]}{K_m + [S]}; v_i \sim \frac{V_{max}[S]}{[S]} \sim V_{max}$$

This states that **when the substrate concentration [S] far exceeds the K_m value, the initial velocity, v_i, is maximal, V_{max}.**

(3) When [S] = K_m (point B in Figs 8–14 and 8–15),

$$v_i = \frac{V_{max}[S]}{K_m + [S]}; v_i = \frac{V_{max}[S]}{[S] + [S]} = \frac{V_{max}[S]}{2[S]} = \frac{V_{max}}{2}$$

This states that **when the substrate concentration is equal to the K_m value, the initial velocity v_i, is half-maximal.** It also tells how **to evaluate K_m,** namely, to **determine experimentally the substrate concentration at which the initial velocity is half-maximal.**

Since many enzymes give saturation curves that do not readily permit evaluation of V_{max} (and hence of K_m) when v_i is plotted versus [S], it is convenient to rearrange the Michaelis-Menten expression to simplify evaluation of K_m and V_{max}. The Michaelis-Menten equation may be inverted and factored as follows:

$$v_i = \frac{V_{max}[S]}{K_m + [S]}$$

Invert:

$$\frac{1}{v_i} = \frac{K_m + [S]}{V_{max}[S]}$$

Factor:

$$\frac{1}{v_i} = \frac{K_m}{V_{max}} \cdot \frac{1}{[S]} + \frac{[S]}{V_{max}[S]}$$

Simplify:

$$\frac{1}{v_i} = \frac{K_m}{V_{max}} \cdot \frac{1}{[S]} + \frac{1}{V_{max}}$$

This is the equation for a **straight line**

$$y = a \cdot x + b$$

where

$$y = \frac{1}{v_i} \text{ and } x = \frac{1}{[S]}$$

If y, or $1/v_i$, is plotted as a function of x, or $1/[S]$, the y-intercept, b, is $1/V_{max}$, and the slope, a, is K_m/V_{max}. The negative x-intercept may be evaluated by setting y = 0. Then

$$x = -\frac{b}{a} = -\frac{1}{K_m}$$

Such a plot is called a double reciprocal plot; ie, the reciprocal of v_i ($1/v_i$) is plotted versus the reciprocal of [S] ($1/[S]$).

K_m may be estimated from the **double-reciprocal or Lineweaver-Burk plot** (Fig 8–16) using either the slope and y-intercept or the negative x-intercept. Since [S] is expressed in molarity, **the dimensions of K_m are molarity or moles per liter.** Velocity, v_i, may be expressed in any units, since **K_m is independent of [Enz].** The double-reciprocal treatment requires relatively few points to define K_m and is the method most often used to determine K_m.

Experimentally, use of the Lineweaver-Burk approach to evaluate K_m can give rise to unwarranted emphasis on data gathered at low substrate concentrations. This is the case if the substrate concentrations selected for study differ by a constant increment. This shortcoming may be circumvented by selecting substrate concentrations the reciprocals of which differ by constant increments.

An alternative approach to the experimental evaluation of K_m and of V_{max} is that of Eadie and Hofstee. The Michaelis-Menten equation may be rearranged to

$$\frac{v_i}{[S]} = -v_i \cdot \frac{1}{K_m} + \frac{V_{max}}{K_m}$$

To evaluate K_m and V_{max}, plot $v_i/[S]$ (y-axis) versus v_i

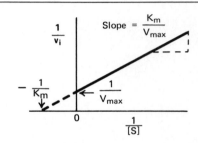

Figure 8–16. Double-reciprocal or Lineweaver-Burk plot of $1/v_i$ versus $1/[S]$ used for graphic evaluation of K_m and V_{max}.

(x-axis). The y-intercept is then V_{max}/K_m, and the x-intercept is V_{max}. The slope is $-1/K_m$.

While both the Lineweaver-Burk and Eadie-Hofstee approaches are useful in selected instances, rigorous determination of K_m and of V_{max} requires statistical treatment.

Apart from their usefulness in interpretation of the mechanisms of enzyme-catalyzed reactions, K_m values are of considerable practical value. At a substrate concentration of 100 times K_m, the enzyme will act at essentially maximum rate, and therefore the **maximal velocity (V_{max}) will reflect the amount of active enzyme present.** This situation is generally desirable in the assay of enzymes. The **K_m value tells how much substrate to use in order to measure V_{max}.** Double-reciprocal treatments also find extensive application in the evaluation of enzyme inhibitors.

Relationship of K_m to K_d, the Dissociation Constant for the Enzyme-Substrate Complex

The **affinity** of an enzyme for its substrate is equal to the **inverse of the dissociation constant, K_d, for EnzS.**

$$Enz + S \underset{k_{-1}}{\overset{k_1}{\rightleftharpoons}} EnzS$$

$$K_d = \frac{k_{-1}}{k_1}$$

That is, the less the tendency of the substrate and enzyme to dissociate, the greater is the affinity of the enzyme for the substrate.

The K_m value of an enzyme for its substrate may also serve as a measure of its K_d. However, in order for this to be true, an assumption included in the derivation of the Michaelis-Menten expression must be valid. In the derivation, it was assumed that the first step of the enzyme-catalyzed reaction

$$Enz + S \underset{k_{-1}}{\overset{k_1}{\rightleftharpoons}} EnzS$$

is fast and always at equilibrium. In other words, the rate of dissociation of EnzS to Enz + S must be much faster than its dissociation to enzyme + product

$$EnzS \underset{k_{-2}}{\overset{k_2}{\rightleftharpoons}} Enz + P$$

In the Michaelis-Menten expression, the [S] that gives $v_i = V_{max}/2$ is

$$[S] = \frac{k_2 + k_{-1}}{k_1} = K_m$$

But when $$k_{-1} >> k_2$$

then $$k_2 + k_{-1} \sim k_{-1}$$

and $$[S] = \frac{k_{-1}}{k_1} \sim K_d$$

Under these conditions, $1/K_m = 1/K_d = $ **affinity.** If

$k_2 + k_{-1} \not\sim k_{-1}$, then $1/K_m$ underestimates the affinity, $1/K_d$.

LIMITATIONS OF THE MICHAELIS-MENTEN MODEL

Certain enzymes and other ligand-binding proteins, such as hemoglobin (see Chapters 6 and 10), do not exhibit classic Michaelis-Menten saturation kinetics. When [S] is plotted versus v_i, the saturation curve is sigmoid (Fig 8–17). This generally indicates cooperative binding of substrate to multiple sites. Binding at one site affects binding at the others, as described in Chapter 6 for hemoglobin.

For sigmoid substrate saturation kinetics, the methods of graphic evaluation of the substrate concentration that produces half-maximal velocity discussed above are invalid (straight lines are not produced). To evaluate sigmoid saturation kinetics, we employ a graphic representation of the Hill equation, an equation originally derived to describe the cooperative binding of O_2 to hemoglobin (see Chapter 6). Written in the form of a straight line, the Hill equation is

$$\log \frac{v_i}{V_{max} - v_i} = n\log [S] - \log k'$$

where k' is a complex constant. The equation states that, when [S] is low compared to k', the reaction velocity increases as the nth power of [S]. Fig 8–18 illustrates a Hill plot of kinetic data for an enzyme with cooperative binding kinetics. A plot of $v_i/V_{max} - v_i$ versus $\log [S]$ yields a straight line with slope = n, where n is an empirical parameter whose value depends on the number of substrate-binding sites and the number and type of interactions between these binding sites. When $n = 1$, the binding sites act independently of one another. If $n > 1$, the sites are cooperative; and the greater the value of n, the stronger is the cooperativity and thus the more "sigmoid" are the saturation kinetics. If $n < 1$, the sites are said to exhibit negative cooperativity.

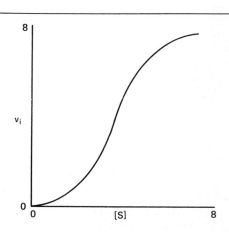

Figure 8–17. Sigmoid saturation kinetics.

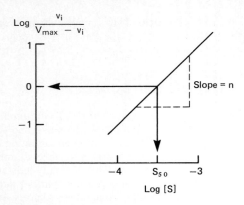

Figure 8–18. Graphic evaluation of the Hill equation to determine the substrate concentration that produces half-maximal velocity when substrate saturation kinetics are sigmoid.

At half-maximal velocity $(v_i = V_{max}/2)$, $v_i/(V_{max} - v_i) = 1$, and hence $\log v_i/(V_{max} - v_i) = 0$. Thus, to determine S_{50} (the concentration of substrate that produces half-maximal velocity), drop a perpendicular line to the x-axis from the point where $\log v_i/(V_{max} - v_i) = 0$.

INHIBITION OF ENZYME ACTIVITY

We distinguish 2 broad classes of inhibitors of enzyme activity—competitive and noncompetitive—depending on whether the inhibition is (competitive) or is not (noncompetitive) relieved by increasing the substrate concentration. In practice, many inhibitors do not exhibit the idealized properties of pure competitive or noncompetitive inhibition discussed below. An alternative way to classify inhibitors is by their site of action. Some bind to the enzyme at the same site as does the substrate (the catalytic site); others bind at some site (an allosteric site) away from the catalytic site.

Competitive Inhibition by Substrate Analogs

Classic competitive inhibition occurs at the substrate-binding (catalytic) site. The chemical structure of a substrate analog inhibitor (I) generally resembles that of the substrate (S). It may therefore combine reversibly with the enzyme, forming an enzyme inhibitor (EnzI) complex rather than an EnzS complex. When both the substrate and this type of inhibitor are present, they compete for the same binding sites on the enzyme surface. A much studied case of competitive inhibition is that of malonate (I) with succinate (S) for succinate dehydrogenase.

Succinate dehydrogenase catalyzes formation of fumarate by removal of one hydrogen atom from each α-carbon atom of succinate (Fig 8–19).

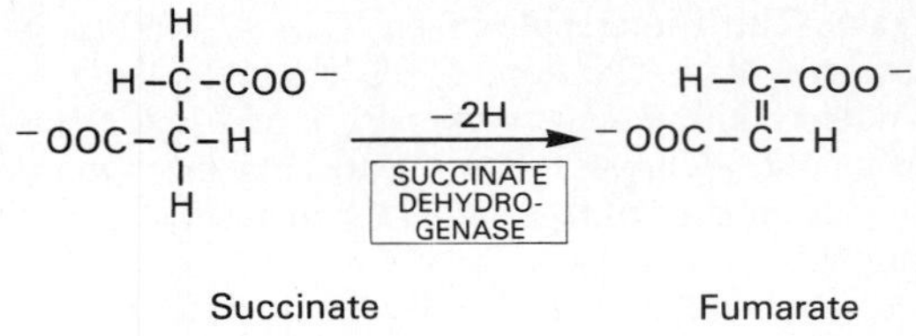

Figure 8–19. The succinate dehydrogenase reaction.

Malonate ($^-OOC–CH_2–COO^-$) can combine with the dehydrogenase, forming an EnzI complex. This cannot be dehydrogenated, since there is no way to remove even one H atom from the single α-carbon atom of malonate without forming a pentavalent carbon atom. The only reaction the EnzI complex can undergo is decomposition back to free enzyme plus inhibitor. For the reversible reaction,

$$\text{EnzI} \underset{k_{-1}}{\overset{k_1}{\rightleftharpoons}} \text{Enz} + \text{I}$$

the equilibrium constant, K_i, is

$$K_i = \frac{[\text{Enz}]\,[\text{I}]}{[\text{EnzI}]} = \frac{k_1}{k_{-1}}$$

The action of competitive inhibitors may be understood in terms of the following reactions:

$$\text{Enz} \underset{\pm\,S}{\overset{\pm\,I}{\rightleftarrows}} \begin{array}{l} \text{EnzI (inactive)} \rightarrow\!\!\times\!\!\rightarrow \text{Enz} + \text{P} \\ \text{EnzS (active)} \rightarrow \text{Enz} + \text{P} \end{array}$$

The rate of product formation, which is what generally is measured, depends solely on the concentration of EnzS. Suppose I binds very tightly to the enzyme $(K_i = $ a small number). There now is little free enzyme (Enz) available to combine with S to form EnzS and eventually Enz + P. The reaction rate (formation of P) will thus be slow. For analogous reasons, an equal concentration of a less tightly bound inhibitor $(K_i = $ a larger number) will not decrease the rate of the catalyzed reaction so markedly. Suppose that, at a fixed concentration of I, more S is added. This increases the probability that Enz will combine with S rather than with I. The ratio of EnzS/EnzI and the reaction rate also rise. At a sufficiently high concentration of S, the concentration of EnzI should be vanishingly small. If so, the rate of the catalyzed reaction will be the same as in the absence of I (Fig 8–20).

Graphic Evaluation of Competitive Inhibition Constants

Fig 8–20 represents a typical case of competitive inhibition shown graphically in the form of a Lineweaver-Burk plot. The reaction velocity (v_i) at a fixed concentration of inhibitor was measured at various concentrations of S. The lines drawn through the

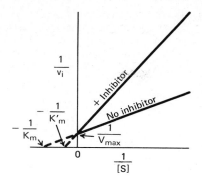

Figure 8-20. Lineweaver-Burk plot of classic competitive inhibition. Note the complete relief of inhibition at high [S] (low 1/[S]).

experimental points coincide at the y-axis. Since the y-intercept is $1/V_{max}$, this states that **at an infinitely high concentration of S** $(1/S = 0)$, v_i **is the same as in the absence of inhibitor.** However, the intercept on the x-axis (which is related to K_m) varies with inhibitor concentration and becomes a larger number ($-1/K'_m$ is smaller than $-1/K_m$) in the presence of the inhibitor. Thus, **a competitive inhibitor raises the apparent K_m (K'_m) for the substrate.** Since K_m is the substrate concentration at which the concentration of free enzyme is equal to the concentration of enzyme present as EnzS, substantial free enzyme is available to combine with inhibitor. For simple competitive inhibition, the intercept on the x-axis is

$$x = \frac{1}{K_m\left(1 + \frac{[I]}{K_I}\right)}$$

K_m may be evaluated in the absence of I, and K_i evaluated using the above equation. If the number of moles of I added is much greater than the number of moles of enzyme present, [I] may generally be taken as the added (known) concentration of inhibitor. The K_i values for a series of substrate analog (competitive) inhibitors indicate which are most effective. **At a low concentration, those with the lowest K_i values will cause the greatest degree of inhibition.**

Many clinically efficacious drugs act as competitive inhibitors of important enzyme activities in microbial and animal cells.

Reversible Noncompetitive Inhibition

As the name implies, in this case no competition occurs between S and I. The inhibitor usually bears little or no structural resemblance to S and may be assumed to bind to a different domain on the enzyme. **Reversible noncompetitive inhibitors lower the maximum velocity attainable with a given amount of enzyme (lower V_{max}) but usually do not affect K_m.** Since I and S may combine at different sites, formation of both EnzI and EnzIS complexes is possible.

Since EnzIS may break down to form product at a slower rate than does EnzS, the reaction may be slowed but not halted. The following competing reactions may occur:

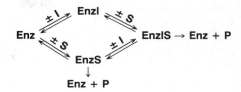

If S has equal affinity both for Enz and for EnzI (I does not affect the affinity of Enz for S), the results shown in Fig 8-21 are obtained when $1/v_i$ is plotted against $1/[S]$ in the presence and absence of inhibitor. (It is assumed that there has been no significant alteration of the conformation of the active site when I is bound.)

Irreversible Noncompetitive Inhibition

A variety of enzyme "poisons," eg, iodoacetamide, heavy metal ions (Ag^+, Hg^{2+}), oxidizing agents, etc, reduce enzyme activity. Since these inhibitors bear no structural resemblance to the substrate, an increase in substrate concentration generally does not relieve this inhibition. The presence of one or more substrates or products may, however, protect the enzyme against inactivation. Kinetic analysis of the type discussed above may not distinguish between enzyme poisons and true reversible noncompetitive inhibitors. Reversible noncompetitive inhibition is, in any case, rare. Unfortunately this is not always appreciated, since both reversible and irreversible noncompetitive inhibition exhibit similar kinetics.

MODULATORS OF ENZYME ACTIVITY

The flow of carbon and energy in metabolism is profoundly influenced both by enzyme synthesis and

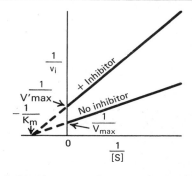

Figure 8-21. Lineweaver-Burk plot for reversible noncompetitive inhibition.

by activation of proenzymes. However, these processes are irreversible. Like all mammalian proteins, enzymes are degraded to amino acids (protein turnover). In bacteria, the activity can be rapidly diluted out among daughter cells on successive divisions. Although both mechanisms effectively reduce enzyme concentration and hence catalytic activity, they are slow, wasteful of carbon and energy, and rather like turning out a light by smashing the bulb and then inserting a new one when light is needed. An "on-off" switch for enzymes clearly would be advantageous. It thus is not surprising that the **catalytic activity** of certain key enzymes can be reversibly decreased or increased by low-molecular-weight intermediary metabolites (see Chapter 6). Small molecule **modulators** that decrease catalytic activity are termed **negative modulators;** those which increase activity are called **positive modulators.** These are discussed further in Chapter 10 and in subsequent chapters.

REFERENCES

Christensen HN: *Dissociation, Enzyme Kinetics, Bioenergetics.* Saunders, 1975.

Engle PC: *Enzyme Kinetics.* Wiley, 1977.

Piszkiwicz D: *Kinetics of Chemical and Enzyme-Catalyzed Reactions.* Oxford Univ Press, 1977.

Segel IH: *Enzyme Kinetics.* Wiley, 1975.

Sigman DS, Mooser G: Chemical studies of enzyme active sites. *Annu Rev Biochem* 1975;**44:**889.

Van Tamlen EE (editor): *Bioorganic Chemistry.* Vol 1, *Enzyme Action,* 1977. Vol 2, *Macro- and Multimolecular Systems,* 1977. Vol 3, *Substrate Behavior,* 1978. Academic Press.

Enzymes: Mechanisms of Action

9

Victor W. Rodwell, PhD

INTRODUCTION

This chapter uses a detailed description of catalysis by the proteolytic enzyme chymotrypsin to illustrate principles of catalysis that apply to enzymes in general.

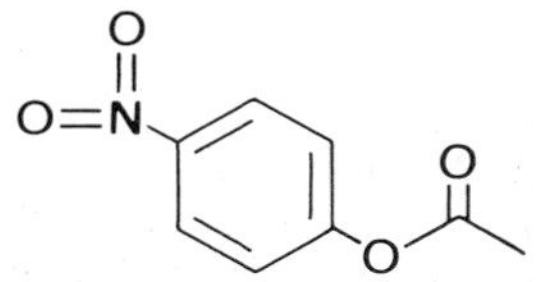

BIOMEDICAL IMPORTANCE

How enzymes achieve their remarkable catalytic efficiencies and specificities is a fundamental scientific problem. However, certain aspects of this problem have applied relevance. The conversion of inactive proenzymes to their active catalytic states within cells under certain pathologic conditions (eg, acute pancreatitis) can result in intracellular digestion and damage. As fundamental knowledge of enzyme action grows, it should be possible, using the techniques of recombinant DNA technology and site-directed mutagenesis, to systematically synthesize enzymes that display increased catalytic efficiency or new specificities. Some of these "synthetic" enzymes may prove to be powerful therapeutic agents. A knowledge of how metals affect biologic systems also depends on an understanding of how these compounds participate in enzyme action.

MECHANISM OF CATALYSIS BY CHYMOTRYPSIN

Nature & Specificity of the Overall Reaction

Chymotrypsin catalyzes hydrolysis of peptide bonds in which the carboxyl group is contributed by an aromatic amino acid (Phe, Tyr, or Trp) or by one with a bulky nonpolar R group (Met).

Like many other proteases, chymotrypsin also catalyzes the hydrolysis of certain esters. The ability of chymotrypsin to catalyze ester hydrolysis is of no physiologic significance. Rather, it facilitates mechanistic experiments that reveal details of the catalytic mechanism.

p-Nitrophenylacetate, a Useful Synthetic Substrate

The synthetic substrate *p*-nitrophenylacetate (Fig 9–1) as substrate facilitates colorimetric analysis of

Figure 9–1. *p*-Nitrophenylacetate.

chymotrypsin activity because hydrolysis of *p*-nitrophenylacetate releases *p*-nitrophenol. In alkali, this converts to the yellow *p*-nitrophenylate anion.

"Stop-Flow" Kinetics

The kinetics of chymotrypsin hydrolysis of *p*-nitrophenylacetate can be studied in a "stop-flow" apparatus. These stop-flow experiments use substrate quantities of enzyme (roughly equimolar quantities of enzyme and of substrate) and measure events that occur in the first few milliseconds after enzyme and substrate are mixed. The stop-flow apparatus has 2 syringes: one for chymotrypsin and the other for *p*-nitrophenylacetate. Instantaneous mixing of enzyme and substrate is achieved by a mechanical device that rapidly and simultaneously expels the contents of both syringes into a single narrow tube that passes through a spectrophotometer. The optical density as a function of time after mixing is displayed on an oscilloscopic screen.

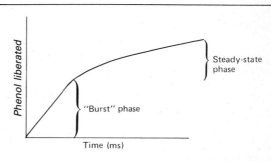

Figure 9–2. Representation of the kinetics of release of *p*-nitrophenylate anion when chymotrypsin hydrolyzes *p*-nitrophenylacetate in a stop-flow apparatus. In this representation, "phenol liberated" has been calculated from optical density.

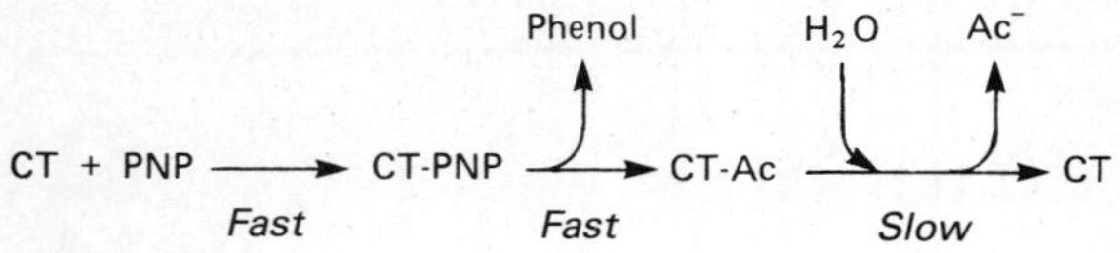

$$CT + PNP \xrightarrow{\text{Fast}} CT\text{-}PNP \xrightarrow[\text{Fast}]{\text{Phenol}} CT\text{-}Ac \xrightarrow[\text{Slow}]{H_2O \quad Ac^-} CT$$

Figure 9–3. Intermediate steps in catalysis of the hydrolysis of p-nitrophenylacetate by chymotrypsin. CT, chymotrypsin; PNP, p-nitrophenylacetate; CT-PNP, chymotrypsin-p-nitrophenylacetate complex; CT-Ac, chymotrypsin-acetate complex; phenol, p-nitrophenylate anion; Ac⁻, acetate anion. Formation of the CT-PNP and CT-Ac complexes is fast relative to the hydrolysis of the CT-Ac complex.

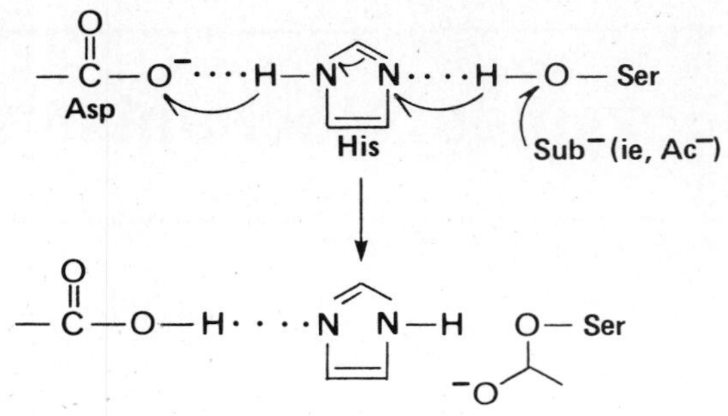

Figure 9–5. Operation of the proton shuttle of chymotrypsin during acylation of Ser 195 by the substrate (Sub⁻).

Biphasic Character of p-Nitrophenylacetate Hydrolysis

Release of p-nitrophenylate anion takes place in 2 distinct phases (Fig 9–2): (1) a "burst" phase, characterized by rapid liberation of p-nitrophenylate anion; and (2) a subsequent, slower release of additional p-nitrophenylate anion.

Formation & Decay of the Enzyme-Substrate Complex

The biphasic character of the release of p-nitrophenylate anion is comprehensible in terms of the successive steps in catalysis shown in Fig 9–3.

The slow step in overall catalysis is hydrolysis of the chymotrypsin-acetate (CT-Ac) complex. Once all of the available chymotrypsin has been converted to CT-Ac, no further release of p-nitrophenylate anion can occur until more free chymotrypsin is liberated by the slow, hydrolytic removal of acetate anion from the CT-Ac complex (Fig 9–3). The "burst" phase of p-nitrophenylate anion release (Fig 9–2) corresponds to the conversion of all of the available free chymotrypsin to the CT-Ac complex with simultaneous release of p-nitrophenylate anion. The subsequent release of p-nitrophenylate anion that follows the burst phase results from the slow liberation of free chymotrypsin by hydrolysis of the CT-Ac complex. This free chymotrypsin then is available for further formation of the CT-PNP and CT-Ac complexes with attendant release of p-nitrophenylate anion. Indeed, the magnitude of the "burst" phase (ie, moles of p-nitro-

phenylate anion released) is directly proportionate to the number of moles of chymotrypsin present initially.

Role of Ser 195 in the Catalytic Mechanism

The acyl group of the acyl-CT intermediate is linked to a highly reactive seryl residue—serine 195—of chymotrypsin. The essential nature and high reactivity of Ser 195 are evidenced by its ability (but not by the ability of the remaining 27 seryl residues of chymotrypsin) to react with diisopropylphosphofluoridate (DIPF) (Fig 9–4). Analogous reactions occur with other serine proteases.

Derivatization of Ser 195 inactivates chymotrypsin. Many other proteases are inactivated by DIFP by an analogous mechanism. These are termed "serine proteases."

The "Charge Relay Network"

A "charge relay network" functions as a proton shuttle during catalysis by chymotrypsin. The charge relay network of chymotrypsin involves 3 aminoacyl residues that are far apart in a primary structural sense but within bond-forming distance of one another in a tertiary structural sense. These residues are Asp 102, His 57, and Ser 195. While most of the charged residues of chymotrypsin are at the surface of the molecule, those of the charge relay network are

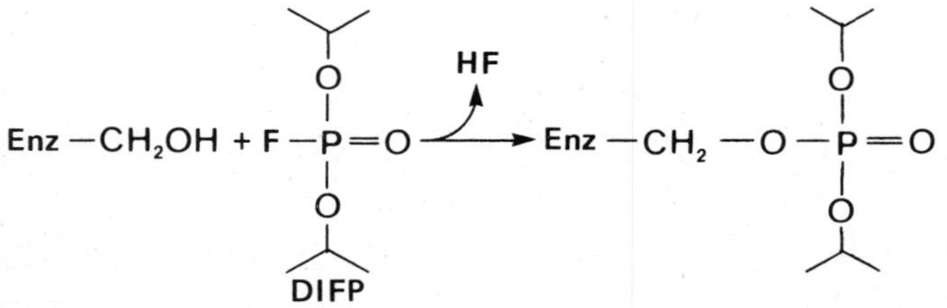

Figure 9–4. Reaction of the primary hydroxyl of Ser 195 of chymotrypsin with diisopropylphosphofluoridate (DIFP).

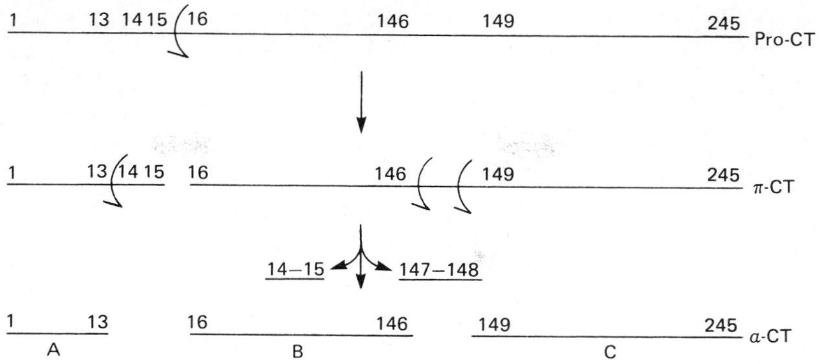

```
1        13 14 15 ⟨16              146        149                  245
─────────────────────────────────────────────────────────────────── Pro-CT
                ↘

                                    ↓

1        13⟨14 15   16              146   ⟨   ⟨ 149                 245
─────────────    ──────────────────────       ─────────────────────── π-CT
           ↘

              14—15  ◄━━▲━━►  147—148
                        ↓

1        13         16              146        149                  245
───────         ──────────────────────    ──────────────────────────── α-CT
    A                     B                         C
```

Figure 9–6. Representation of the conversion of prochymotrypsin (pro-CT) to π-chymotrypsin (π-CT) and subsequently to the mature, catalytically active enzyme α-chymotrypsin (α-CT).

"buried" in the otherwise nonpolar interior of the molecule. The 3 residues are aligned in the order Asp 102—His 57—Ser 195.

Recall that Ser 195 is the residue that is acylated during catalysis by chymotrypsin. The approach of the acetate anion (derived from p-nitrophenylacetate) to the oxygen atom on the R group of Ser 195 triggers sequential proton shifts that "shuttle" protons from Ser 195 through His 57 to Asp 102 (Fig 9–5).

During deacylation of the acyl–Ser 195 intermediate, protons shuttle in the reverse direction. An analogous series of proton shifts is believed to accompany hydrolysis of a physiologic chymotrypsin substrate such as a peptide.

ROLE OF SELECTIVE PROTEOLYSIS IN CREATION OF THE CATALYTIC SITES OF ENZYMES

Conversion of Prochymotrypsin to Chymotrypsin

The conversion of prochymotrypsin (pro-CT), a 245-aminoacyl residue polypeptide, to the active enzyme α-chymotrypsin involves 3 proteolytic clips and the formation of an active intermediate known as π-chymotrypsin (π-CT)(Fig 9–6).

In α-chymotrypsin, the A, B, and C chains (Fig 9–6) remain associated owing to the presence in α-CT of 2 interchain disulfide bonds (Fig 9–7).

Proenzymes

Many proteins are manufactured and secreted in the form of inactive precursor proteins known as **"proproteins."** When the proteins are enzymes, the proproteins are termed **"proenzymes"** or **"zymogens."** Conversion of a proprotein to the mature protein involves selective proteolysis. This converts the proprotein by one or more successive proteolytic "clips" to a form in which the characteristic activity of the mature protein (its enzymatic activity) is expressed. Examples of proteins manufactured as proproteins include the hormone insulin (proprotein = proinsulin), the digestive enzymes pepsin, trypsin, and chymotrypsin (proproteins = pepsinogen, trypsinogen, and chymotrypsinogen, respectively), several factors of the blood clotting and of the blood clot dissolution cascades (see Chapter 55), and the connective tissue protein collagen (proprotein = procollagen).

Why are certain proteins secreted in an inactive form? Certain proteins are needed at essentially all times. Others (for example, the enzymes of blood clot formation and dissolution) are needed only intermittently. Furthermore, when these intermittently needed enzymes are required, they are frequently needed rapidly. Certain physiologic processes such as digestion are intermittent but fairly regular and predictable (although this may not have been the case for primitive humans). Others (for example, blood clot formation, clot dissolution, and tissue repair) need only to be brought "on line" in reponse to pressing physiologic or pathophysiologic need. It may be readily appreciated that the processes of blood clot formation and dissolution must be temporally coordinated to achieve homeostasis. In addition, the synthesis of proteases as catalytically inactive precursor proteins serves to pro-

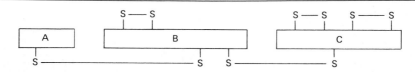

Figure 9–7. Representation of the intra- and interchain polypeptide bonds of α-chymotrypsin (α-CT).

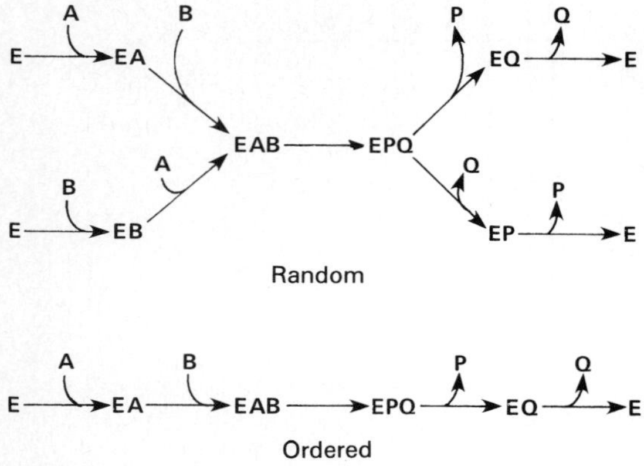

Figure 9–8. Random and ordered addition of substrates A and B and dissociation of products P and Q from an enzyme, E.

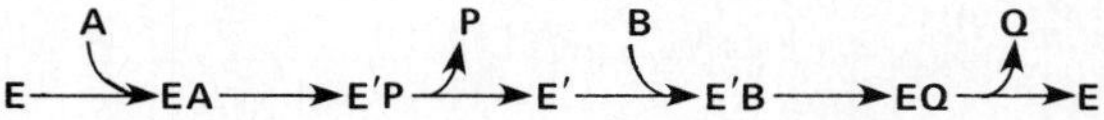

Figure 9–9. Generalized "ping-pong" mechanism for enzymatic catalysis.

tect the tissue of origin (eg, the pancreas) from autodigestion. (Autodigestion can occur in pancreatitis.)

It might be thought that de novo synthesis of the required proteins might be sufficiently rapid to respond to a pressing pathophysiologic demand such as the loss of blood. However, an adequate and complete pool of the precursor amino acids must be available. Furthermore, the secretion process may be slow relative to the physiologic demand.

The example of conversion of a proprotein to its mature, physiologically active form discussed below illustrates the following general principles of proprotein to protein conversions:

(1) The process involves selective proteolysis, which in some instances requires only a single proteolytic clip.

(2) The polypeptide products may separate or may remain associated in the mature protein.

(3) The process may (or may not) be attended by a significant change in molecular weight.

(4) A major consequence of selective proteolysis is the attainment of a new conformation.

(5) If the proprotein is an enzyme, the above conformational change generates the catalytic site of the enzyme. Indeed, selective proteolysis of a proenzyme may be viewed as a process that triggers essential conformational changes that "create" the catalytic site.

Note that while His 57 and Asp 102 reside on the B peptide of α-chymotrypsin, Ser 195 resides on the C peptide (Fig 9–6). The selective proteolysis of

prochymotrypsin (chymotrypsinogen) thus facilitates approximation of the 3 residues concerned with the charge relay network. This illustrates how selective proteolysis can give rise to the catalytic site. Note also that contact and catalytic residues can be located on different peptide chains but still be within bond-forming distance of bound substrate.

ORDERED & RANDOM BINDING OF SUBSTRATES

Most enzymes catalyze a reaction between 2 or more substrates, yielding one or more products. For some enzymes, all substrates must be present simultaneously for the reaction to occur. For others, the enzyme first alters one substrate and then catalyzes its reaction with a second substrate. The order in which an enzyme binds its substrates may be **random** or **ordered** (Fig 9–8).

Many reactions that require coenzymes proceed by "ping-pong" mechanisms (so termed because the enzyme alternates between forms E and E') (Figs 9–9 and 9–10).

While a coenzyme frequently may be regarded as a second substrate, certain coenzymes (eg, pyridoxal phosphate) are covalently bonded to the enzyme or bound noncovalently so tightly that dissociation rarely occurs (eg, thiamin pyrophosphate). In these cases, we regard the enzyme-coenzyme complex as the enzyme.

ENZYMES AS GENERAL ACID OR GENERAL BASE CATALYSTS

Once the substrate has bound at the catalytic site, the charged (or chargeable) functional groups of the side chains of nearby aminoacyl residues may participate in catalysis by functioning as acidic or basic catalysts.

We recognize 2 broad categories of acid-base catalysis by enzymes: **general** acid (or base) catalysis and **specific** acid (or base) catalysis.

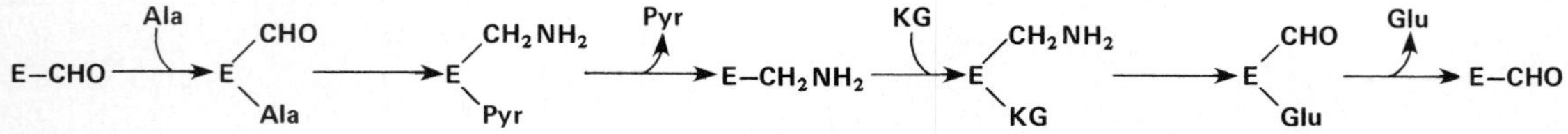

Figure 9–10. "Ping-pong" mechanism for transamination. E–CHO and E–CH₂NH₂ represent the enzyme-pyridoxal phosphate and enzyme-pyridoxamine phosphate complexes, respectively. (Ala, alanine; Pyr, pyruvate; KG, α-ketoglutarate; Glu, glutamate.)

Reactions whose rates vary in response to changes in H^+ or H_3O^+ concentration but are independent of the concentrations of other acids or bases present in the solution are said to be subject to **specific acid** or **specific base catalysis.** Reactions whose rates are responsive to all the acids (proton donors) or bases (proton acceptors) present in solution are said to be subject to **general acid** or **general base catalysis.**

To determine whether a given enzyme-catalyzed reaction is subject to general or specific acid or base catalysis, one measures the rate of the reaction under 2 sets of conditions: (1) at various pH values but at a constant buffer concentration; (2) at constant pH but at various buffer concentrations.

If the rate of the reaction changes as a function of pH at constant buffer concentration, the reaction is said to be **specific base-catalyzed** (if the pH is above 7) or **specific acid-catalyzed** (if the pH is below 7). If the reaction rate at constant pH increases as the buffer concentration increases, the reaction is said to be subject to **general base catalysis** (if the pH is above 7) or **general acid catalysis** (if the pH is below 7).

As an example of specific acid catalysis, consider the conversion of a substrate (S) to a product (P). This occurs in 2 steps—a rapid, reversible proton transfer step,

$$S + H_3O^+ \rightleftharpoons SH^+ + H_2O$$

followed by a slower, and therefore rate-determining, step of rearrangement of the protonated substrate to product

$$SH^+ + H_2O \rightarrow P + H_3O^+$$

Increasing the concentration of hydronium ion $[H_3O^+]$ increases the reaction rate by elevating the concentration of SH^+, the conjugate acid of the substrate, which is the substrate for the rate-determining step in the overall reaction. Stated mathematically,

$$\text{Rate} = \frac{d[P]}{dt} = k[SH^+]$$

where P = the product, t = time, k = the specific rate constant, and $[SH^+]$ = the concentration of the conjugate acid of the substrate.

Since the concentration of SH^+ depends upon both the concentration of S and the concentration of H_3O^+, the general rate expression for specific acid-catalyzed reactions is

$$\frac{d[P]}{dt} = k'[S][H_3O^+]$$

Note that it is a requirement of specific acid catalysis that the rate expression contain *only* terms for S and for H_3O^+.

Next consider that, in addition to the specific acid catalysis described above, there is also catalysis by imidazolium ion of an imidazole buffer. Since imidazole is a weak acid (pK_a about 7), it is a poor proton donor; hence the reaction

$$S + \text{Imidazole} \cdot H^+ \rightarrow SH^+ + \text{Imidazole}$$

is slow and is rate-determining for the overall reaction. Note that the fast and slow steps are reversed when the mechanism changes from specific to general acid catalysis. The rate expressions for general acid catalysis frequently are complex and for this reason are not discussed here.

METAL IONS

Over 25% of all enzymes contain tightly bound metal ions or require them for activity. The functions of these metal ions are studied by x-ray crystallography, nuclear magnetic resonance (NMR), and electron spin resonance (ESR). Coupled with knowledge of the formation and decay of metal complexes and of reactions within the coordination spheres of metal ions, this provides insight into the roles of metal ions in enzymatic catalysis. These roles are considered below.

Metalloenzymes & Metal-Activated Enzymes

Metalloenzymes contain a definite quantity of functional metal ion that is retained throughout purification. Metal-activated enzymes bind metals less tightly but require added metals. The distinction between metalloenzymes and metal-activated enzymes thus rests on the affinity of a particular enzyme for its metal ion. The mechanisms whereby metal ions perform their functions appear to be similar both in metalloenzymes and metal-activated enzymes.

Ternary Enzyme-Metal-Substrate Complexes

For ternary (3-component) complexes of the catalytic site (Enz), a metal ion (M), and substrate (S) that exhibit 1:1:1 stoichiometry, 4 schemes are possible:

Enz–S–M	**M–Enz–S**
Substrate-bridge complex	**Enzyme-bridge complex**

$$\text{Enz} \Big\langle \begin{matrix} M \\ \mid \\ S \end{matrix}$$

Enz–M–S	
Simple metal-bridge complex	**Cyclic metal-bridge complex**

All 4 are possible for metal-activated enzymes. Metalloenzymes cannot form the EnzSM complex, because they retain the metal throughout purification (ie, are already as EnzM). Three generalizations can be stated:

(1) Most but not all kinases (ATP:phosphotransferases) form substrate-bridge complexes of the type Enz-nucleotide-M.

(2) Phosphotransferases using pyruvate or phosphoenolpyruvate as substrate, enzymes catalyzing other reactions of phosphoenolpyruvate, and carboxylases form metal-bridge complexes.

(3) A given enzyme may form one type of bridge

complex with one substrate and a different type with another.

Enzyme-Bridge Complexes (MEnzS)

The metals in enzyme-bridge complexes are presumed to perform structural roles maintaining an active conformation (eg, glutamine synthase) or to form a metal bridge to a substrate (eg, pyruvate kinase). In addition to its structural role, the metal ion in pyruvate kinase appears to hold one substrate (ATP) in place and to activate it.

$$\text{Pyruvate kinase} \diagdown \begin{array}{c} \text{M} \\ | \\ \text{ATP} \end{array} \diagdown \text{Creatine}$$

Substrate-Bridge Complexes (EnzSM)

The formation of ternary substrate-bridge complexes of nucleoside triphosphates with enzyme, metal, and substrate appears attributable to displacement of H_2O from the coordination sphere of the metal by ATP:

$$ATP^{4-} + M(H_2O)_6{}^{2+} \rightleftharpoons ATP-M(H_2O)_3{}^{2-} + 3H_2O$$

Substrate then binds, forming the ternary complex:

$$ATP-M(H_2O)_3{}^{2-} + Enz \rightleftharpoons Enz-ATP-M(H_2O)_3{}^{2-}$$

In phosphotransferase reactions, metal ions are thought to activate the phosphorus atoms and form a rigid, polyphosphate-adenine complex of appropriate conformation in the active, quaternary complex.

Metal-Bridge Complexes

$$Enz-M-S \text{ or } Enz \diagdown \begin{array}{c} \text{M} \\ | \\ \text{S} \end{array}$$

Crystallographic and sequencing data have established that a His residue is concerned with metal binding at the active site of many proteins (eg, carboxypeptidase A, cytochrome c, rubredoxin, metmyoglobin, and methemoglobin; see Chapter 6). For binary (2-component) EnzM complexes, the rate-limiting step is in many cases the departure of water from the coordination sphere of the metal ion. For many peptidases, activation by metal ions is a slow process requiring many hours. The slow reaction probably is conformational rearrangement of the binary EnzM complex to an active conformation, eg,

Metal binding:

$$Enz + M(H_2O)_6 \xrightarrow{\text{Rapid}} Enz-M(H_2O)_{6-n} + nH_2O$$

Rearrangement to active conformation (Enz*):

$$Enz-M(H_2O)_{6-n} \xrightarrow{\text{Slow}} Enz^*-M(H_2O)_{6-n}$$

For metalloenzymes, however, the ternary metal-bridge complex must be formed by combination of the substrate (S) with the binary EnzM complex:

$$Enz-M + S \rightleftharpoons Enz-M-S \text{ or } Enz \diagdown \begin{array}{c} \text{M} \\ | \\ \text{S} \end{array}$$

Role in Catalysis

Metal ions may participate in each of the 4 mechanisms by which enzymes are known to accelerate the rates of chemical reactions: (1) general acid-base catalysis, (2) covalent catalysis, (3) approximation of reactants, and (4) induction of strain in the enzyme or substrate. Other than iron and manganese, which function in heme proteins, the metal ions most commonly concerned in enzymatic catalysis are Mg^{2+}, Mn^{2+}, and Ca^{2+}, although other metal ions (eg, K^+) are important for the activity of certain enzymes.

Metal ions, like protons, are Lewis acids (electrophiles) and can share an electron pair forming a sigma bond. Metal ions may also be considered "super acids," since they exist in neutral solution, frequently have a positive charge of >1, and may form pi bonds. In addition (and unlike protons), metals can serve as 3-dimensional templates for orientation of basic groups on the enzyme or substrate.

Metal ions can also accept electrons via sigma or pi bonds to activate electrophiles or nucleophiles (general acid-base catalysis). By donating electrons, metals can activate nucleophiles or act as nucleophiles themselves. The coordination sphere of a metal may bring together enzyme and substrate (approximation) or form chelate-producing distortion in either the enzyme or substrate (strain). A metal ion may also "mask" a nucleophile and thus prevent an otherwise likely side-reaction. Finally, stereochemical control of the course of an enzyme-catalyzed reaction may be achieved by the ability of the metal coordination sphere to act as a 3-dimensional template to hold reactive groups in a specific steric orientation (Table 9–1).

Table 9–1. Selected examples of the roles of metal ions in the mechanism of action of enzymes.*

Enzyme	Role of Metal Ion
Histidine deaminase	Masking a nucleophile
Kinases, lyases, pyruvate decarboxylase	Activation of an electrophile
Carbonic anhydrase	Activation of a nucleophile
Cobamide enzymes	Metal acts as a nucleophile
Pyruvate carboxylase, carboxypeptidase, alcohol dehydrogenase	π-Electron withdrawal
Nonheme iron proteins	π-Electron donation
Pyruvate kinase, pyruvate carboxylase, adenylate kinase	Metal ion gathers and orients ligands
Phosphotransferase, D-xylose isomerase, hemoproteins	Strain effects

*Adapted from Mildvan AS: Metals in enzyme catalysis. Vol 2. Page 456 in: *The Enzymes.* Boyer PD, Lardy H, Myrbäck K (editors). Academic Press, 1970.

REFERENCES

Crane F: Hydroquinone dehydrogenases. *Annu Rev Biochem* 1977;**46**:439.

Fersht A: *Enzyme Structure and Mechanism,* 2nd ed. Freeman, 1985.

Kraut J: Serine proteases: Structure and mechanism of catalysis. *Annu Rev Biochem* 1977;**46**:331.

Mildvan AS: Mechanism of enzyme action. *Annu Rev Biochem* 1974;**43**:357.

Purich DL (editor): Enzyme kinetics and mechanisms. Parts A and B in: *Methods in Enzymology*. Vol 63, 1979; Vol 64, 1980. Academic Press.

Wimmer MJ, Rose IA: Mechanisms of enzyme-catalyzed group transfer reactions. *Annu Rev Biochem* 1978;**47**:1031.

Wood HG, Barden RE: Biotin enzymes. *Annu Rev Biochem* 1977;**46**:385.

Enzymes: Regulation of Activities

Victor W. Rodwell, PhD

INTRODUCTION

In this chapter, mechanisms by which metabolic processes are regulated via enzymes are illustrated by selected examples. The intent is to characterize overall patterns of regulation. Throughout this book, reference is made to many other specific examples to illustrate these diverse features of metabolic regulation.

BIOMEDICAL IMPORTANCE

The mechanisms by which cells and intact organisms regulate and coordinate overall metabolism are of concern to workers in areas of the biomedical sciences as diverse as cancer, heart disease, aging, microbial physiology, differentiation, metamorphosis, hormone action, and drug action. In all of these areas, important examples of normal or abnormal regulation of enzymes are to be found. To take the first as an example, analyses on experimental tumors show that many cancer cells exhibit abnormalities in the regulation of their enzyme complement (lack of induction or repression). This speaks to the well-established conclusion that alterations of gene control are fundamental events in cancer cells. Again, certain oncogenic viruses contain a gene that codes for a tyrosine-protein kinase; when this kinase is expressed in host cells, it can phosphorylate many proteins and enzymes that are normally not phosphorylated and thus lead to dramatic changes in cell phenotype. A change of this nature appears to lie at the heart of certain types of viral oncogenic transformation. The last area cited above, drug action, provides another important example involving enzyme regulation. Specifically, the administration of some drugs is known to increase the synthesis of certain enzymes (these drugs act as enzyme inducers). Phenobarbital, for instance, leads to a marked (3- to 5-fold) increase (induction) of the microsomal enzyme cytochrome P-450, which plays a central role in the metabolism of phenobarbital itself and of many other drugs. One such drug is the anticoagulant warfarin. If a patient has been taking warfarin and is then started on phenobarbital therapy, the dosage of the former will have to be increased substantially in order to take into account its increased rate of degradation brought about by induction of cytochrome P-450. Enzyme induction is one important biochemical cause of a drug interaction, the situation in which the administration of one drug results in a significant change in the metabolism of another.

METABOLIC REGULATION

Homeostasis

The concept of homeostatic regulation of the internal milieu advanced by Claude Bernard in the late 19th century stressed the ability of animals to maintain the constancy of their intracellular environments. This implies that all the necessary enzyme-catalyzed reactions proceed at rates responsive to changes in the internal and external environment. A cell or organism might be defined as **diseased** when it responds inadequately or incorrectly to an internal or external stress. Knowledge of factors affecting the rates of enzyme-catalyzed reactions is essential both to understand the mechanism of homeostasis in normal cells and to comprehend the molecular basis of disease.

All chemical reactions, including enzyme-catalyzed reactions, are to some extent reversible.* Within living cells, however, reversibility may not obtain, because reaction products are promptly removed by additional enzyme-catalyzed reactions. Metabolite flow in living cells is analogous to the flow of water in a pipe. Although the pipe can transfer water in either direction, in practice the flow is unidirectional. Metabolite flow in living cells also is largely unidirectional. True equilibrium, far from being characteristic of life, is approached only when cells die. The living cell is a dynamic steady-state system maintained by a unidirectional flow of metabolites (Fig 10–1). In mature cells the mean concentrations of metabolites remain relatively constant over considerable periods of time.† The flexibility of the steady-state system is well illustrated in the delicate shifts and balances by which organisms maintain the constancy of the internal environment despite wide variations in food, water, and mineral intake, work output, or external temperature.

---

*A readily reversible reaction has a small numerical value of ΔG. One with a large negative value for ΔG might be termed "effectively irreversible" in most biochemical situations.
†Short-term oscillations of metabolite concentrations and of enzyme levels do occur, however, and are of profound physiologic importance.

Scope of Metabolic Regulation

For life to proceed in orderly fashion, metabolite flow through anabolic and catabolic pathways must be regulated. All requisite chemical events must proceed at rates consistent with the requirements of the intact organism in relation to its environment. ATP production, synthesis of macromolecular precursors, transport, secretion, and tubular reabsorption all must respond to subtle changes in the environment of the cell, organ, or intact animal. These processes must be coordinated and must respond to short-term changes in the external environment (eg, addition or removal of a nutrient) as well as to periodic intracellular events (eg, DNA replication). Until recently, the molecular details of regulation were best understood in bacteria, which lack the complexities of hormonal or neural control and in which genetic studies can readily be conducted to analyze molecular events. Our understanding of molecular regulation in animal cells is, however, presently in a state of rapid expansion.

While knowledge of cellular regulatory processes in humans is central to an understanding and therapy of metabolic diseases, the molecular events in regulation of many metabolic processes in mammals are still poorly understood. It is clear that metabolic regulation in mammals differs significantly from superficially similar phenomena in bacteria. Regulation of metabolic processes in bacteria will be discussed because it provides a conceptual framework for considering regulation in humans.

Available Options for Regulation of Enzymes

Net flow of carbon through any enzyme-catalyzed reaction might be influenced (1) by changing the absolute quantity of enzyme present, (2) by altering the pool size of reactants other than enzyme, and (3) by altering the catalytic efficiency of the enzyme. All 3 options are exploited in most forms of life.

REGULATION OF ENZYME QUANTITY BY CONTROL OF THE RATES OF ENZYME SYNTHESIS & DEGRADATION

General Principles

The absolute quantity of an enzyme present is determined by its rate of synthesis (k_s) and rate of degradation (k_{deg}) (Fig 10–2). The quantity of an enzyme in a cell may be raised either by an increase in its rate of synthesis (increase in k_s), by a decrease in its rate of degradation (decrease in k_{deg}), or by both. Similarly, a lower quantity of enzyme can result from a decrease in k_s, an increase in k_{deg}, or both. Examples of changes in both k_s and k_{deg} occur in human subjects. In all forms of life, enzyme (protein) synthesis from amino acids and enzyme (protein) degradation to amino acids are distinct processes catalyzed by entirely different sets of enzymes. Independent regulation of enzyme syn-

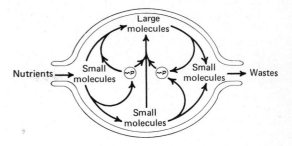

Figure 10–1. An idealized cell in steady state.

thesis and enzyme degradation is thus readily achieved.

Enzyme Synthesis a Result of Information Stored in DNA

The primary structure of an enzyme, like that of all proteins, is dictated by the trinucleotide (triplet) code of its messenger RNA (mRNA). The sequence of nucleotide bases of the mRNA is in turn dictated by a complementary base sequence in a DNA template or gene (see Chapters 38 and 40). Information for protein synthesis, stored in DNA, thus determines a cell's ability to synthesize a particular enzyme.

Mutations alter the nucleotide sequence of DNA and result in synthesis of proteins with modified primary structures. This may alter structure at higher levels of organization if the new amino acid has properties significantly different from the original. Mutations may cause partial or complete loss of catalytic activity or, rarely, enhanced catalytic activity. Since mutations at various genetic loci can produce enzymes with impaired activity, a large number of human molecular diseases are possible.

Enzyme Induction

Cells can synthesize specific enzymes in response to the presence of specific low-molecular-weight inducers. Enzyme induction is illustrated by the following experiment: *Escherichia coli* grown on glucose will not ferment lactose owing to the absence of the enzyme β-galactosidase, which hydrolyzes lactose to galactose and glucose. If lactose or certain other β-galactosides are added to the growth medium, synthesis of the β-galactosidase is induced and the culture can now ferment lactose.

Figure 10–2. Enzyme quantity is determined by the net balance between enzyme synthesis and enzyme degradation.

The inducer (lactose) is a substrate for the induced protein (β-galactosidase). Although many inducers are substrates for the enzymes they induce, compounds structurally similar to the substrate may be inducers but not substrates. These are termed **gratuitous inducers.** Conversely, a compound may be a substrate but not an inducer. Frequently, a compound induces several enzymes of a catabolic pathway (eg, β-galactoside permease and β-galactosidase are both induced by lactose). Where the structural genes that specify a group of catabolic enzymes comprise an **operon,** all enzymes of that operon are induced by a single inducer **(coordinate induction).** The ability to regulate the synthesis of enzymes dependent upon the availability of a nutrient permits the bacterium to use its available nutrients to maximum advantage; ie, it does not synthesize "unnecessary enzymes."

Enzymes whose concentration in a cell is independent of added inducer are termed **constitutive.** A particular enzyme may be constitutive in one strain, inducible in another, and absent in a third. Cells capable of being induced for a particular enzyme or other protein usually contain a small measurable **basal level** of that protein even when grown in the absence of added inducer. The extent to which a particular organism responds to an inducer is genetically determined (see Chapter 41). Increases in enzyme content from 2- to 1000-fold may be observed on induction in different strains. The genetic heritage of the cell thus determines both the nature and magnitude of the response to an inducer. The terms "constitutive" and "inducible" are therefore relative terms, like "hot" and "cold," that represent the extremes of a spectrum of responses.

Enzyme induction also occurs in eukaryotes. Examples of inducible enzymes in animals are tryptophan pyrrolase, threonine dehydrase, tyrosine-α-ketoglutaric transaminase, invertase, enzymes of the urea cycle, HMG-CoA reductase, and cytochrome P-450.

Enzyme Repression & Derepression

In bacteria capable of synthesizing a biosynthetic metabolite, its presence in the medium may curtail new synthesis of that metabolite via **repression.** A small molecule such as a purine or amino acid, acting as a **corepressor,** can ultimately block synthesis of the enzymes involved in its own biosynthesis. For example, in *Salmonella typhimurium,* addition of histidine (His) represses synthesis of all the enzymes of His biosynthesis, and addition of leucine (Leu) represses synthesis of the first 3 enzymes unique to Leu biosynthesis. In both cases, these biosynthetic enzymes comprise **operons; coordinate repression** occurs following addition of the end products His and Leu. Coordinate repression is not general for all biosynthetic pathways. Following removal or exhaustion of an essential biosynthetic intermediate from the medium, enzyme biosynthesis again occurs. This constitutes **derepression.** Derepression may be coordinate or noncoordinate.

The above examples illustrate **product feedback repression** characteristic of biosynthetic pathways in bacteria. **Catabolite repression,** a related phenomenon, refers to the ability of an intermediate in a sequence of **catabolic** enzyme-catalyzed reactions to repress synthesis of catabolic enzymes. This effect was first noted in cultures of *E coli* growing on a carbon source (X) other than glucose. Addition of glucose repressed synthesis of the enzymes concerned with catabolism of X. This phenomenon was initially termed the "glucose effect." Since oxidizable nutrients other than glucose produce similar effects, the term **catabolite repression** was adopted. Catabolite repression is mediated by cAMP. The molecular mechanisms of induction, repression, and derepression are discussed in Chapter 41.

In multiple-branched biosynthetic pathways such as those generating the branched-chain amino acids or the aspartate family of amino acids, early enzymes function in the biosynthesis of several amino acids (Fig 10–3). Following addition of lysine (Lys) to the medium of growing bacteria, synthesis of the enzymes unique to Lys biosynthesis (Enz_L) are repressed. Repression of the enzymes unique to threonine (Thr) biosynthesis (Enz_T) follows addition of Thr to the medium. These effects illustrate simple product feedback repression. Enzymes Enz_1 and Enz_2, however, function both in Lys and Thr biosynthesis. Product feedback repression of their synthesis by Lys or Thr alone would starve the bacterium of the other amino acid. If, however, both Lys and Thr are added to the medium, Enz_1 and Enz_2 become redundant, and re-

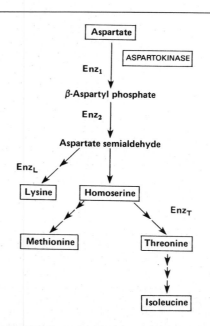

Figure 10–3. The aspartate family of amino acids. Enz_L and Enz_T denote groups of enzymes involved in lysine and threonine biosynthesis, respectively. Multiple arrows indicate multiple reactions.

pression of synthesis of Enz_1 and Enz_2 could be advantageous to survival, since it would permit more efficient use of available nutrients.

In the presence of all necessary end products of a branched or multiple-branched biosynthetic pathway, **multivalent repression** may occur. This occurs only when all end products of a particular set of biosynthetic enzymes are present in ample supply. Complete repression of aspartokinase (Enz_1) should therefore require methionine (Met) and isoleucine (Ile) in addition to Lys and Thr.

Enzyme Turnover

In rapidly growing bacteria, the overall rate of protein degradation is about 2% per hour, and control of enzyme levels is achieved primarily by increases or decreases in k_s. This is not true for starving bacteria or for bacteria transferred to fresh medium providing a poorer source of carbon for growth ("stepdown culture"). Under these conditions, bacteria degrade protein at 7–10% per hour.

The combined processes of enzyme synthesis and degradation constitute **enzyme turnover.** While turnover occurs both in bacteria and mammals, the importance of enzyme degradation as a device by which enzyme levels are regulated in bacteria has received little emphasis. Turnover of protein was recognized as a characteristic property of all mammalian cells long before it was shown also to occur in bacteria. The existence of protein (enzyme) turnover in humans was deduced from dietary experiments well over a century ago. It was, however, Schoenheimer's classic work, just prior to and during World War II, that conclusively established that turnover of cellular protein occurred throughout life. By measuring the rates of incorporation of ^{15}N-labeled amino acids into protein and the rates of loss of ^{15}N from protein, Schoenheimer deduced that body proteins are in a state of "dynamic equilibrium," a concept since extended to other body constituents, including lipids and nucleic acids.

Regulation of Enzyme Synthesis & Degradation

While the major events in protein synthesis are well understood, those in enzyme degradation are not. Enzyme degradation involves hydrolysis by proteolytic enzymes, but little is known of the processes by which proteolytic activity is regulated other than that it may require ATP. The susceptibility of an enzyme to proteolytic degradation depends upon its conformation. The presence or absence of substrates, coenzymes, or metal ions, which can alter protein conformation, alters proteolytic susceptibility. The concentrations of substrates, coenzymes, and possibly ions in cells may thus determine the rates at which specific enzymes are degraded. Arginase and tryptophan oxygenase (tryptophan pyrrolase) illustrate these concepts. Regulation of liver arginase levels can involve a change either in k_s or in k_{deg}. After a protein-rich diet is ingested, liver arginase levels rise owing to an increased rate of arginase synthesis. Liver arginase levels also rise in starved animals. Here, however, it is arginase degradation that is decreased, while k_s remains unchanged. In a second example, injection of glucocorticoids and ingestion of Trp both elevate levels of tryptophan oxygenase in mammals. The hormone raises the rate of oxygenase synthesis (raises k_s). Trp, however, has no effect on k_s but lowers k_{deg} by stabilizing the oxygenase toward proteolytic digestion. Contrast these 2 examples with enzyme induction in bacteria. For arginase, the increased intake of nitrogen on a high-protein diet may elevate liver arginase levels (see Chapter 30). The increased rate of arginase synthesis thus superficially resembles that of substrate induction in bacteria. For tryptophan pyrrolase, however, even though Trp may act as an inducer in bacteria (affects k_s), its effect in mammals is solely on the enzyme degradative process (lowers k_{deg}).

Enzyme levels in mammalian tissues may be altered by a wide range of physiologic, hormonal, or dietary manipulations. Examples are known for a variety of tissues and metabolic pathways (Table 10–1), but our knowledge of the molecular details that account for these changes is fragmentary.

Glucocorticoids increase the concentration of tyrosine transaminase by stimulating k_s. This was the first clear case of a hormone regulating the synthesis of a mammalian enzyme. Insulin and glucagon—despite their mutually antagonistic physiologic effects—both independently increase k_s 4- to 5-fold. The effect of glucagon probably is mediated via cAMP, which mimics the effect of the hormone in organ cultures of rat liver.

Conversion of Proenzymes to Active Enzymes

Enzyme activity can be regulated by converting an inactive proenzyme to a catalytically active form. To become catalytically active, the proenzyme must undergo limited proteolysis, a process accompanied by conformational changes that either reveal or "create" the catalytic site (see Chapter 8). Synthesis as a catalytically inactive proenzyme is characteristic of digestive enzymes and enzymes of blood coagulation and of blood clot dissolution (see Chapter 55).

REGULATION OF THE CATALYTIC EFFICIENCY OF ENZYMES

Definition of Terms

If a physiologic manipulation alters the level of enzyme activity, we have no way of knowing whether the quantity of enzyme has changed or whether the enzyme is a more efficient or less efficient catalyst. **We shall refer to all changes in enzyme activity that occur without change in the quantity of enzyme present as "effects on catalytic efficiency."**

Availability of Reactants

The kinetic and regulatory properties of enzymes

Table 10–1. Selected examples of rat liver enzymes that adapt to an environmental stimulus by changes in activity.*

Enzyme	$t_{1/2}$ (hours)	Stimulus	Fold Change
Amino acid metabolism			
Arginase	100–120	Starvation or glucocorticoids.	+2
		Change from high- to low-protein diet.	−2
Serine dehydratase	20	Glucagon or dietary amino acids.	+100
Histidase	60	Change from low- to high-protein diet.	+20
Carbohydrate metabolism			
Glucose-6-phosphate dehydrogenase	15	Thyroid hormone or change from fasting state to ingestion of a high-carbohydrate diet.	+10
α-Glycerophosphate dehydrogenase	100	Thyroid hormone.	+10
Fructose 1,6-phosphatase		Glucose.	+10
Lipid metabolism			
Citrate cleavage enzyme		Change from starvation to ingestion of a high-carbohydrate, low-fat diet.	+30
Fatty acid synthase		Starvation.	−10
		Change from starvation to ingestion of a fat-free diet.	+30
HMG-CoA reductase	2–3	Fasting or 5% cholesterol diet.	−10
		Twenty-four-hour diurnal variation.	±5
		Insulin or thyroid hormone.	+2 to 10
Purine or pyrimidine metabolism			
Xanthine oxidase		Change to high-protein diet.	−10
Aspartate transcarbamoylase	60	One percent orotic acid diet.	+2
Dihydroorotase	12	One percent orotic acid diet.	+3

*Data, with the exception of those for HMG-CoA reductase, from Schimke RT, Doyle D: Control of enzyme levels in animal tissues. *Annu Rev Biochem* 1970;**39**:929.

provide insights into physiologic processes in intact cells, tissues, and organisms. However, most information was obtained by studying enzymes in vitro under conditions that differ substantially from those in living cells. Application of this knowledge to the in vivo situation therefore requires considerable caution. For instance, the concentrations of substrates studied in vitro differ significantly from those in vivo.

Enzyme Compartmentation

The importance of compartmentation of metabolic processes in eukaryotic cells, including those of mammals, cannot be overemphasized. Localization of specific metabolic processes in the cytosol or in cellular organelles facilitates regulation of these processes independent of processes proceeding elsewhere. The extensive compartmentalization of metabolic processes characteristic of higher forms of life thus confers the potential for finely tuned regulation of metabolism. At the same time, it poses problems with respect to translocation of metabolites across compartmental barriers. This is achieved via "shuttle mechanisms" that convert the metabolite to a form permeable to the compartmental barrier. This is followed by transport and conversion back to the original form on the other side of the barrier. Consequently, these interconversions require, for example, cytosolic and mitochondrial forms of the same catalytic activity. Since

these 2 forms of the enzyme are physically separated, their independent regulation is facilitated. The role of shuttle mechanisms in achieving equilibration of metabolic pools of reducing equivalents, of citric acid cycle intermediates, and of other amphibolic intermediates is discussed in Chapter 17.

Macromolecular Complexes

Organization of a set of enzymes that catalyze a protracted sequence of metabolic reactions as a macromolecular complex coordinates the enzymes and channels intermediates along a metabolic path. Appropriate alignment of the enzymes can facilitate transfer of product between enzymes without prior equilibration with metabolic pools. This permits a finer level of metabolic control than is possible with the isolated components of the complex. In addition, conformational changes in one component of the complex may be transmitted by protein-protein interactions to other enzymes of the complex. Amplification of regulatory effects thus is possible.

Effective Concentrations of Substrates, Coenzymes, & Cations

The **mean** intracellular concentration of a substrate, coenzyme, or metal ion may have little meaning for the in vivo behavior of an enzyme. Information on the concentrations of essential metabolites **in the** immediate neighborhood of the enzyme in question is needed. However, even measuring metabolite concentrations in different cellular compartments does not account for local discontinuities in metabolite concentrations within compartments brought about by factors such as proximity to the site of entry or production of a metabolite. Finally, little consideration generally is given to the discrepancy between total and free metabolite concentrations. For example, while the total concentration of 2,3-bisphosphoglycerate in erythrocytes is extremely high, the concentration of free bisphosphoglycerate is comparable to that of other tissues. Erythrocytes contain approximately 5 mmol of hemoglobin, which binds 1 mol of bisphosphoglycerate per mole of deoxygenated tetramer. A **total** concentration of 4 mmol of bisphosphoglycerate would therefore result in a minuscule concentration of **free** bisphosphoglycerate in venous erythrocytes. Similar considerations apply to other metabolites in the presence of proteins that bind them effectively and reduce their concentrations in the free state.

An assumption of the Michaelis-Menten kinetic approach was that the concentration of total substrate was essentially equal to the concentration of free substrate. As noted above, this assumption may well be invalid in vivo, where concentrations of free substrates often are of the same order of magnitude as those of enzyme concentrations.

Metal ions, which perform catalytic and structural roles in over one-fourth of all known enzymes (see Chapter 9), may also fulfill regulatory roles, particularly for reactions where ATP is a substrate. Where the ATP-metal ion complex is the substrate for the reaction, maximal activity typically is observed at molar ratio of ATP to metal of about unity. Excess metal or excess ATP is inhibitory. Since nucleoside di- and triphosphates form stable complexes with divalent cations, intracellular concentrations of the nucleotides can influence intracellular concentrations of free metal ions and hence the activity of certain enzymes. For example, in the absence of metal ions, *E coli* glutamine synthase assumes a "relaxed" configuration that is catalytically inactive. Mg^{2+} or Mn^{2+} converts the synthase to the active, "tightened" form. In addition, adenylylation of the synthase changes the divalent cation specificity from Mg^{2+} to Mn^{2+}. The activity of the adenylylated enzyme is, furthermore, sensitive to the $ATP:Mg^{2+}$ ratio, whereas that of the unadenylylated form is not.

ALLOSTERIC REGULATION

Principles

The catalytic activity of certain **regulatory enzymes** is modulated by low-molecular-weight **allosteric effectors** that generally have little or no structural similarity to the substrates or coenzymes for the regulated enzyme. **Feedback inhibition** refers to the inhibition of the activity of an enzyme in a biosynthetic pathway by an end product of that pathway. For

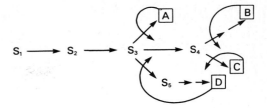

Figure 10–4. Sites of feedback inhibition in a branched biosynthetic pathway. S_1–S_5 are intermediates in the biosynthesis of end products A–D. Straight arrows represent enzymes catalyzing the indicated conversions. Curved arrows represent feedback loops and indicate probable sites of feedback inhibition by specific end products.

biosynthesis of D from A, catalyzed by enzymes Enz_1 through Enz_3,

$$A \xrightarrow{Enz_1} B \xrightarrow{Enz_2} C \xrightarrow{Enz_3} D$$

a high concentration of D typically inhibits conversion of A to B. This involves not simple "backing up" of intermediates but the ability of D to bind to and inhibit Enz_1. D thus acts as a **negative allosteric effector** or **feedback inhibitor** of Enz_1. Feedback inhibition of Enz_1 by D therefore regulates the synthesis of D. Typically, D binds to the sensitive enzyme at an **allosteric site** remote from the catalytic site.

The kinetics of feedback inhibition may be competitive, noncompetitive, partially competitive, uncoupled, or mixed. Feedback inhibition is commonest in biosynthetic pathways. **Frequently the feedback inhibitor is the last small molecule before a macromolecule** (eg, amino acids before proteins, nucleotides before nucleic acids). **Feedback regulation generally occurs at the earliest functionally irreversible* step unique to a particular biosynthetic sequence.**

Examples of feedback inhibition in microorganisms include inhibition by His of phosphoribosyl:ATP pyrophosphorylase, by Trp of anthranilate synthase, and by CTP of aspartate transcarbamoylase. In each case the regulated enzyme is involved in biosynthesis of a single end product—His, Trp, or CTP.

Frequently, a biosynthetic pathway is branched, with the initial portion serving for synthesis of 2 or more essential metabolites. Fig 10–4 shows probable sites of simple feedback inhibition in a branched biosynthetic pathway (eg, for amino acids, purines or pyrimidines). S_1, S_2, and S_3 are precursors of all 4 end products (A, B, C, and D), S_4 is a precursor of B and C, and S_5 a precursor solely of D. The sequences:

---

*One strongly favored (in thermodynamic terms) in a single direction, ie, one with a large negative ΔG.

$$S_3 \longrightarrow A$$

$$S_4 \longrightarrow B$$

$$S_4 \longrightarrow C$$

$$S_3 \longrightarrow S_5 \longrightarrow D$$

thus constitute linear reaction sequences that might be expected to be feedback-inhibited by their end products.

Multiple feedback loops (Fig 10–5) provide additional fine control. For example, if B is present in excess, the requirement for S_2 decreases. The ability of B to decrease production of S_2 thus confers a biologic advantage. However, if excess B inhibits not only the portion of the pathway unique to its own synthesis but also portions common to that for synthesis of A, C, or D, excess B should curtail synthesis of all 4 end products. Clearly, this is undesirable. Mechanisms have, however, evolved to circumvent this difficulty.

In **cumulative feedback inhibition,** the inhibitory effect of 2 or more end products on a single regulatory enzyme is strictly additive.

In **concerted** or **multivalent feedback inhibition,** complete inhibition occurs only when 2 or more end products both are present in excess.

In **cooperative feedback inhibition,** a single end product present in excess inhibits the regulatory enzyme, but **the inhibition when 2 or more end products are present far exceeds the additive effects of cumulative feedback inhibition.**

The aspartate family provides yet another variant—**multiple enzymes** or isozymes each with distinct regulatory characteristics. *E coli* produces 3 aspartokinases. One (AK_L) is specifically and completely inhibited by Lys, a second (AK_T) by Thr, and the third (AK_H) by homoserine, a precursor of Met, Thr, and Ile (Fig 10–6). In the presence of excess Lys, AK_L is inhibited and β-aspartyl phosphate production decreases. This alone would not suffice to channel

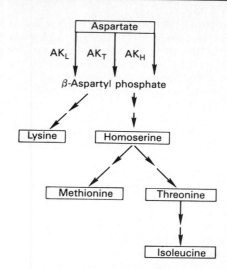

Figure 10–6. Regulation of aspartokinase (AK) activity in *E coli.* Multiple enzymes are subject to end product inhibition by lysine (AK_L), threonine (AK_T), or homoserine (AK_H).

metabolites toward synthesis of homoserine and its products. Channeling is achieved by feedback inhibition at secondary sites farther along the pathway. Lys also inhibits the first enzyme in the sequence leading from β-aspartyl phosphate to Lys. This facilitates unrestricted synthesis of homoserine, and hence of Thr and Ile. Additional control points exist at the branch point where homoserine leads both to Met and to Thr and Ile.

That all these variations can regulate metabolism is suggested by the persistence in different bacteria of distinctive patterns of feedback inhibition of a single biosynthetic pathway (Table 10–2).

The most extensively studied allosteric enzyme, **aspartate transcarbamoylase,** catalyzes the first reaction unique to pyrimidine biosynthesis (Fig 10–7). Aspartate transcarbamoylase (ATCase) is **feedback-inhibited by cytidine triphosphate (CTP).** Following treatment with mercurials, ATCase loses its sensitivity to inhibition by CTP but retains its full activity for carbamoyl aspartate synthesis. This suggests that CTP is bound at a different (allosteric) site from either substrate. ATCase consists of 2 catalytic and 3 or 4 regulatory protomers. Each catalytic protomer contains 4 aspartate (substrate) sites and each regulatory

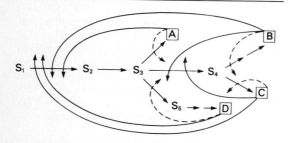

Figure 10–5. Multiple feedback inhibition in a branched biosynthetic pathway. Superimposed on simple feedback loops (dashed, curved arrows) are multiple feedback loops (solid, curved arrows) that regulate enzymes common to biosynthesis of several end products.

Table 10–2. Patterns of allosteric regulation of aspartokinase.

Organism	Feedback Inhibitor	Repressor
E coli (kinase I)	Homoser	. . .
E coli (kinase II)	Lys	Lys
E coli (kinase III)	Thr	. . .
R rubrium	Thr	. . .
B subtilis	Thr + Lys	. . .

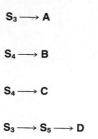

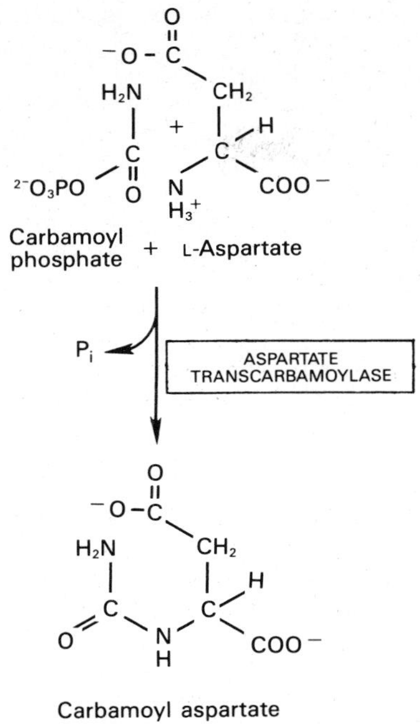

O
‖
$^-$O - C
H$_2$N   CH$_2$
| + | H
C   C
‖ |
$^{2-}$O$_3$PO   O   N   COO$^-$
H$_3^+$

Carbamoyl   +   L-Aspartate
phosphate

P$_i$ ↙ [ASPARTATE TRANSCARBAMOYLASE]

O
‖
$^-$O - C
H$_2$N   CH$_2$
| | H
C   C
O =   N   COO$^-$
H

Carbamoyl aspartate

Figure 10-7. The aspartate transcarbamoylase (ATCase) reaction.

protomer at least 2 CTP (regulatory) sites. Each type of protomer is subject to independent genetic control, as shown by the production of mutants lacking normal feedback control of CTP and, from these, of revertants with essentially normal regulatory properties.

Evidence for Allosteric Sites on Regulated Enzymes

About 1963, Monod noted the lack of structural similarity between a feedback inhibitor and the substrate for the enzyme whose activity it regulated. Since the effectors are not isosteric with a substrate but **allosteric** ("occupy another space"), he proposed that enzymes whose activity is regulated by **allosteric effectors** (eg, feedback inhibitors) bind the effector at an **allosteric site** that is physically distinct from the catalytic site. **Allosteric enzymes** thus are enzymes whose activity at the catalytic site may be modulated by the presence of allosteric effectors at an allosteric site. Lines of evidence that support the existence of physically distinct allosteric sites on regulated enzymes include the following:

(1) Regulated enzymes modified by chemical or physical techniques frequently become insensitive to their allosteric effectors without alteration of their catalytic activity. Selective denaturation of allosteric sites has been achieved by treatment with mercurials, urea, x-rays, proteolytic enzymes, extremes of ionic strength or pH, aging at 0–5 °C, by freezing, or by heating.

(2) Allosteric effectors frequently protect the **catalytic** site from denaturation under conditions where the substrates themselves do not protect. Since it seems unlikely that an effector bound at the catalytic site would protect when substrates do not, this suggests a second, allosteric site elsewhere on the enzyme molecule.

(3) In certain bacterial and mammalian cell mutants, the regulated enzymes have altered regulatory properties but identical catalytic properties to those of the wild-type from which the mutant derived. The structures of the allosteric and catalytic sites thus are genetically distinct.

(4) Binding studies of substrates and of allosteric effectors to regulated enzymes show that each may bind independently of the other.

(5) In certain cases (eg, ATCase), the allosteric site is present on a different protomer from the catalytic site.

Kinetics of Allostery

Fig 10–8 illustrates the rate of a reaction catalyzed by a typical allosteric enzyme measured at several concentrations of substrate in the presence and absence of an allosteric inhibitor. In the absence of the allosteric inhibitor, hyperbolic saturation kinetics are observed. In its presence, the substrate saturation curve is distorted from a hyperbola into a sigmoid, which at high substrate concentrations may merge with the hyperbola. Note the analogy to the relationship between the O$_2$ saturation curves for myoglobin and hemoglobin (see Chapter 6).

On kinetic analysis, feedback inhibition may appear to be competitive, noncompetitive, partially competitive, or of other types. If, at high concentrations of S, comparable activity is observed in the presence or absence of the allosteric inhibitor, the kinetics superficially resemble those of competitive inhibition.

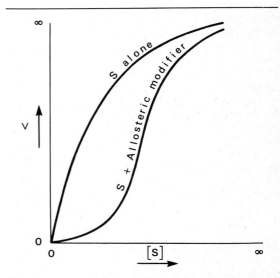

Figure 10–8. Sigmoid saturation curve for substrate in the presence of an allosteric inhibitor.

However, since the substrate saturation curve is sigmoid rather than hyperbolic, it is not possible to obtain meaningful results by graphing data for the allosteric inhibition by the double-reciprocal technique. This method of analysis was developed for substrate competitive inhibition **at the catalytic site.** Since allosteric inhibitors act at a different (allosteric) site, that kinetic model is invalid.

The sigmoid character of the V versus S curve in the presence of an allosteric inhibitor reflects the phenomenon of **cooperativity.** At low concentrations of S, the activity in the presence of the inhibitor is low relative to that in its absence. However, as S is increased, the extent of inhibition becomes relatively less severe. The kinetics are consistent with the presence of 2 or more interacting substrate-binding sites, where the presence of a substrate molecule at one catalytic site facilitates binding of a second substrate molecule at a second site. Cooperativity of substrate binding has been described in Chapter 6 for hemoglobin. The sigmoid O_2 saturation curve results from cooperative interactions between four O_2 binding sites located on different protomers.

Models of Allostery

Reference to the kinetics of allosteric inhibition as "competitive" or "noncompetitive" with substrate carries mechanistic implications which are misleading. We refer instead to 2 classes of regulated enzymes, K-series and V-series enzymes. For K-series allosteric enzymes, the substrate saturation kinetics are competitive in the sense that K_m is raised (decreased affinity for substrate) without effect on V_{max}. For V-series allosteric enzymes, the allosteric inhibitor lowers V_{max} (lowered catalytic efficiency) without affecting the apparent K_m. Alterations in K_m or V_{max} probably result from conformational changes at the catalytic site induced by binding of the allosteric effector at the allosteric site. For a K-series allosteric enzyme, this conformational change may weaken the bonds between substrate and substrate-binding residues. For a V-series allosteric enzyme, the primary effect may be to alter the orientation of catalytic residues so as to lower V_{max}. Intermediate effects on K_m and V_{max} may, however, be observed consequent to these conformational changes.

While various models have been proposed for regulation of allosteric enzymes, it is unlikely that a single model can explain the behavior of all regulatory enzymes. Since sigmoidicity of the substrate saturation curve confers a regulatory advantage, any mutation that gives rise to sigmoidicity should tend to be retained. To expect that these mutations would involve identical mechanisms is unrealistic. The presence of sigmoid kinetics does not, therefore, imply a particular mechanism of inhibition.

Physiologic Consequences of Cooperativity

The consequences of cooperative substrate-binding kinetics are analogous to those resulting from the co-operative binding of O_2 to hemoglobin. At low substrate concentrations, the allosteric effector is an effective inhibitor. It thus regulates most effectively at the time of greatest need, ie, when intracellular concentrations of substrates are low. As more substrate becomes available, stringent regulation is less necessary. As substrate concentration rises, the degree of inhibition therefore lessens, and more product is formed. As with hemoglobin, the sigmoid substrate saturation curve in the presence of inhibitor also means that relatively small changes in substrate concentration result in large changes in activity. Sensitive control of catalytic activity thus is achieved by small changes in substrate concentration. Finally, by analogy with the differing O_2 saturation curves of hemoglobins from different species, regulatory enzymes from different sources may have sigmoid saturation curves shifted to the left or right to accommodate to the range of prevailing in vivo concentrations of substrate.

Feedback Regulation in Mammalian Cells

In both mammalian and bacterial cells, end products "feed back" and control their own synthesis. In some instances (eg, ATCase), this involves feedback inhibition of an early biosynthetic enzyme. We must, however, distinguish between **feedback regulation,** a phenomenologic term devoid of mechanistic implications, and **feedback inhibition,** a mechanism for regulation of many bacterial and mammalian enzymes. For example, dietary cholesterol restricts the synthesis of cholesterol from acetate in mammalian tissues. This feedback regulation does not, however, appear to involve feedback inhibition of an early enzyme of cholesterol biosynthesis. An early enzyme (HMG-CoA reductase) is affected, but the mechanism involves curtailment by cholesterol or a cholesterol metabolite of the expression of the genes that code for the formation of HMG-CoA reductase. Cholesterol added directly to HMG-CoA reductase has no effect on its catalytic activity.

COVALENT MODIFICATION OF ENZYMES

General Principles

Reversible modulation of the catalytic activity of enzymes can occur by covalent attachment of a phosphate group (predominates in mammals) or a nucleotide (predominates in bacteria). Enzymes that undergo covalent modification with attendant modulation of their activity are termed "interconvertible enzymes" (Fig 10–9).

Interconvertible enzymes exist in 2 activity states, one of high and the other of low catalytic efficiency. Depending on the enzyme concerned, the phospho- or the dephosphoenzyme may be the more active catalyst (Table 10–3).

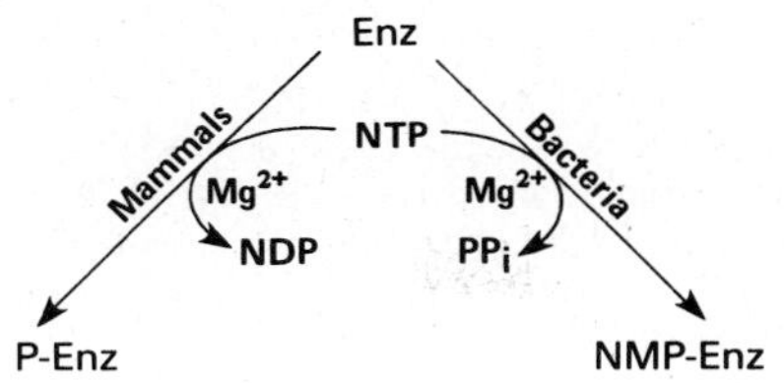

Figure 10-9. Regulation of enzyme activity by covalent modification. *Left:* phosphorylation. *Right:* nucleotidylation. For both processes the nucleoside triphosphate (NTP) generally is ATP.

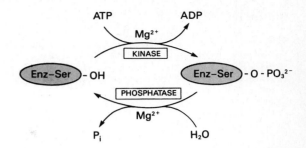

Figure 10-10. Covalent modification of a regulated enzyme by phosphorylation-dephosphorylation of a Ser residue.

Phosphorylation Sites

A specific Ser residue is phosphorylated, forming O-phosphoseryl residue, or, in rarer cases, a tyrosyl residue is phosphorylated to form O-phosphotyrosyl. While an interconvertible enzyme may contain many Ser or Tyr residues, phosphorylation is highly selective and occurs at only a small number (1–3) of possible sites. These sites probably do not form part of the catalytic site, at least in a primary structural sense, and thus constitute another example of an allosteric site.

Converter Proteins

Phosphorylation and dephosphorylation are catalyzed by protein kinases and protein phosphatases (converter proteins), respectively (Fig 10–10). In specific instances, the converter proteins themselves may be interconvertible enzymes (Table 10–3). Thus, there are protein kinase kinases and protein kinase phosphatases that catalyze the interconversion of these converter proteins. Evidence that protein phosphatases are also interconvertible proteins is less convincing, although their activity is regulated. The activity of both protein kinases and protein phosphatases is under hormonal and neural control, although the precise details by which these agents act are in most instances far from clear.

Energetics

The reactions shown in Fig 10–10 resemble those for interconversion of glucose and glucose 6-phos-

phate or of fructose 6-phosphate and fructose 1,6-bisphosphate (see Chapter 18). The net result of phosphorylating and then dephosphorylating 1 mol or substrate (enzyme or sugar) is the hydrolysis of 1 mol of ATP.

The activities of the kinases (catalyzing reactions 1 and 3) and of the phosphatases (catalyzing reactions 2 and 4) may themselves be regulated, for if not, they would act together to catalyze uncontrolled hydrolysis of ATP.

1. $Glucose + ATP \longrightarrow ADP + Glucose\ 6\text{-}P$

2. $H_2O + Glucose\ 6\text{-}P \longrightarrow P_i + Glucose$

Net: $\qquad H_2O + ATP \longrightarrow ADP + P_i$

3. $Enz\text{—}Ser\text{–}OH + ATP \longrightarrow ADP + Enz\text{–}Ser\text{–}O\text{–}P$

4. $H_2O + Enz\text{–}Ser\text{–}O\text{–}P \longrightarrow P_i + Enz\text{–}Ser\text{–}OH$

Net: $\qquad H_2O + ATP \longrightarrow ADP + P_i$

Analogies to Feedback Inhibition

Regulation of enzyme activity by phosphorylation-dephosphorylation has analogies to regulation by feedback inhibition. Both provide for short-term regulation of metabolite flow in response to specific physiologic signals; both act without altering gene expression. Both act on early enzymes of a protracted (often biosynthetic) metabolic sequence; and both act at allosteric rather than catalytic sites. Feedback inhibition, however, involves a single protein and lacks hormonal and neural features. By contrast, regulation of mammalian enzymes by phosphorylation-dephosphorylation involves several proteins and ATP or another nucleoside triphosphate and is under direct neural and hormonal control.

Table 10-3. Examples of mammalian enzymes whose catalytic activity is altered by covalent phosphorylation-dephosphorylation. E, dephosphoenzyme; EP, phosphoenzyme.

Enzyme	Activity State	
	Low	High
Acetyl-CoA carboxylase	EP	E
Glycogen synthase	EP	E
Pyruvate dehydrogenase	EP	E
HMG-CoA reductase	EP	E
Glycogen phosphorylase	E	EP
Citrate lyase	E	EP
Pyruvate dehydrogenase	E	EP
Phosphorylase b kinase	E	EP
HMG-CoA reductase kinase	E	EP

REFERENCES

Gumaa KA, McLean P, Greenbaum AL: Compartmentation in relation to metabolic control in liver. *Essays Biochem* 1971;**7:**39.

Kun E, Grisolia S: *Biochemical Regulatory Mechanisms in Eukaryotic Cells.* Wiley, 1972.

Nestler EJ, Greengard P: Protein phosphorylation in the brain. *Nature* 1983;**305:**583.

Newsholme EA, Stuart C: *Regulation in Metabolism.* Wiley, 1973.

Schimke RT, Doyle D: Control of enzyme levels in animal tissues. *Annu Rev Biochem* 1970;**39:**929.

Soderling TR: Role of hormones and protein phosphorylation in metabolic regulation. *Fed Proc* 1982;**41:**2615.

Sols A, Marco R: Concentrations of metabolites and binding sites: Implications in metabolic regulation. *Curr Top Cell Regul* 1970;**2:**227.

Stanbury JB et al (editors): *The Metabolic Basis of Inherited Disease,* 5th ed. McGraw-Hill, 1983.

Umbarger HE: Amino acid biosynthesis and its regulation. *Annu Rev Biochem* 1978;**47:**533.

Weber G (editor): *Advances in Enzyme Regulation.* Vols 1–9. Pergamon Press, 1963–1987.

Section II.
Bioenergetics & the Metabolism of Carbohydrates & Lipids

Bioenergetics

11

Peter A. Mayes, PhD, DSc

INTRODUCTION

Bioenergetics, or biochemical thermodynamics, is the study of the energy changes accompanying biochemical reactions. It provides the underlying principles to explain why some reactions may occur while others do not. Nonbiologic systems may utilize heat energy to perform work, but biologic systems are essentially isothermic and use chemical energy to power the living processes.

BIOMEDICAL IMPORTANCE

Suitable fuel is required to provide the energy that enables the animal to carry out its normal processes. How the organism obtains this energy from its food is basic to the understanding of normal nutrition and metabolism. Death from starvation occurs when available energy reserves are depleted, and certain forms of malnutrition are associated with energy imbalance (kwashiorkor, marasmus). The rate of energy release, measured by the metabolic rate, is controlled by the thyroid hormones. Storage of surplus energy results in obesity, one of the most common diseases of occidental society.

FREE ENERGY & THE LAWS OF THERMODYNAMICS

Change in free energy (ΔG) is that portion of the total energy change in a system which is available for doing work; ie, it is the useful energy, also known in chemical systems as the **chemical potential.**

The first law of thermodynamics states that **"the total energy of a system, plus its surroundings, remains constant."** This is also the law of conservation of energy. It implies that within the total system, energy is neither lost nor gained during any change. However, within that total system, energy may be transferred from one part to another or may be transformed into another form of energy. For example, chemical energy may be transformed into heat, electrical energy, radiant energy, or mechanical energy.

The second law of thermodynamics states that **"the total entropy of a system must increase if a process is to occur spontaneously." Entropy** represents the extent of disorder or randomness of the system and becomes maximum in a system as it approaches true equilibrium. Under conditions of constant temperature and pressure, the relationship between the free energy change (ΔG) of a reacting system and the change in entropy (ΔS) is given by the following equation which combines the 2 laws of thermodynamics:

$$\Delta G = \Delta H - T\Delta S$$

where ΔH is the change in **enthalpy** (heat) and T is the absolute temperature.

Under the conditions of biochemical reactions, because ΔH is approximately equal to ΔE, the total change in internal energy of the reaction, the above relationship may be expressed in the following way:

$$\Delta G = \Delta E - T\Delta S$$

If ΔG is negative in sign, the reaction proceeds spontaneously with loss of free energy; ie, it is **exergonic.** If, in addition, ΔG is of great magnitude, the reaction goes virtually to completion and is essentially irreversible. On the other hand, if ΔG is positive, the reaction proceeds only if free energy can be gained; ie, it is **endergonic.** If, in addition, the magnitude of ΔG is great, the system is stable with little or no tendency for a reaction to occur. If **ΔG is zero,** the system is at **equilibrium** and no net change takes place.

When the reactants are present in concentrations of 1.0 mol/L, ΔG^0 is known as the **standard free energy change.** For biochemical reactions, a standard state is defined as having a pH of 7.0. The standard free energy change at this standard state is denoted by $\Delta G^{0\prime}$.

The standard free energy change can be calculated from the equilibrium constant K'_{eq}

$$\Delta G^{0\prime} = -2.303 \text{ RT log } K'_{eq}$$

where R is the gas constant and T is the absolute temperature (see p 68). It is important to note that **ΔG may be larger or smaller than $\Delta G^{0\prime}$ depending on the concentrations of the various reactants.**

In a biochemical reaction system, it must be appreciated that an enzyme only speeds up the attainment of equilibrium; **it never alters the final concentrations of the reactants at equilibrium.**

THE COUPLING OF ENDERGONIC TO EXERGONIC PROCESSES

The vital processes—eg, synthetic reactions, muscular contraction, nerve impulse conduction, and active transport—obtain energy by chemical linkage, or **coupling,** to oxidative reactions. In its simplest form, this type of coupling may be represented as shown in Fig 11–1.

The conversion of metabolite A to metabolite B occurs with release of free energy. It is coupled to another reaction, in which free energy is required to convert metabolite C to metabolite D. As some of the energy liberated in the degradative reaction is transferred to the synthetic reaction in a form other than heat, the normal chemical terms exothermic and endothermic cannot be applied to these reactions. Rather, the terms **exergonic** and **endergonic** are used to indicate that a process is accompanied by loss or gain, respectively, of free energy, regardless of the form of energy involved. In practice, an endergonic process cannot exist independently but must be a com-

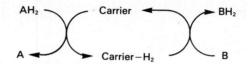

Figure 11–2. Coupling of dehydrogenation and hydrogenation reactions by an intermediate carrier.

ponent of a coupled exergonic/endergonic system where the **overall net change is exergonic.** The exergonic reactions are termed **catabolism** (the breakdown or oxidation of fuel molecules), whereas the synthetic reactions that build up substances are termed **anabolism.** The total of all of the catabolic and anabolic processes is **metabolism.**

If the reaction shown in Fig 11–1 is to go from left to right, then the overall process must be accompanied by loss of free energy as heat. One possible mechanism of coupling could be envisaged if a common obligatory intermediate (I) took part in both reactions, ie,

$$A + C \longrightarrow I \longrightarrow B + D$$

Some exergonic and endergonic reactions in biologic systems are coupled in this way. It should be appreciated that this type of system has a built-in mechanism for biologic control of the rate at which oxidative processes are allowed to occur, since the existence of a common obligatory intermediate for both the exergonic and endergonic reactions allows the rate of utilization of the product of the synthetic path (D) to determine by mass action the rate at which A is oxidized.

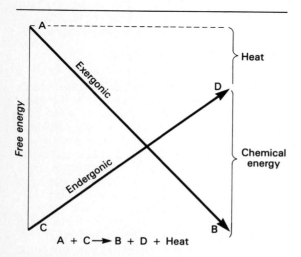

Figure 11–1. Coupling of an exergonic to an endergonic reaction.

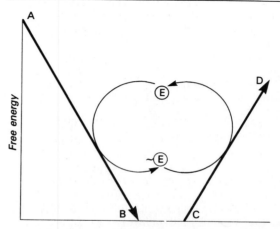

Figure 11–3. Transfer of free energy from an exergonic to an endergonic reaction via a high-energy intermediate compound.

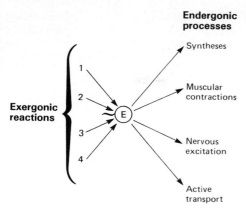

Figure 11–4. Transduction of energy through a common high-energy compound to energy-requiring (endergonic) biologic processes.

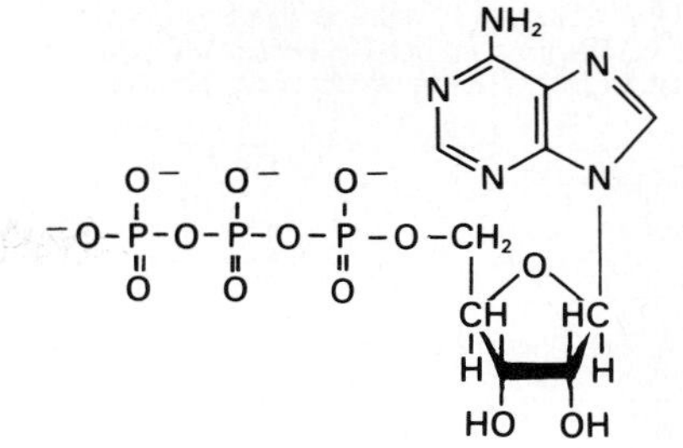

Figure 11–5. Adenosine triphosphate (ATP).

Indeed, these relationships supply a basis for the concept of **respiratory control,** the process that prevents an organism from burning out of control. An extension of the coupling concept is provided by dehydrogenation reactions, which are coupled to hydrogenations by an intermediate carrier (Fig 11–2).

An alternative method of coupling an exergonic to an endergonic process is to synthesize a compound of high-energy potential in the exergonic reaction and to incorporate this new compound into the endergonic reaction, thus effecting a transference of free energy from the exergonic to the endergonic pathway (Fig 11–3).

In Fig 11–3, ~Ⓔ is a compound of high potential energy and Ⓔ is the corresponding compound of low potential energy. The biologic advantage of this mechanism is that Ⓔ, unlike I in the previous system, need not be structurally related to A, B, C, or D. This would allow Ⓔ to serve as a transducer of energy from a wide range of exergonic reactions to an equally wide range of endergonic reactions or processes, as shown in Fig 11–4.

In the living cell, the principal high-energy intermediate or carrier compound (designated ~Ⓔ) is **adenosine triphosphate (ATP).**

ROLE OF HIGH-ENERGY PHOSPHATES IN BIOENERGETICS & ENERGY CAPTURE

In order to maintain living processes, all organisms must obtain supplies of free energy from their environment. **Autotrophic** organisms couple their metabolism to some simple exergonic process in their surroundings; eg, green plants utilize the energy of sunlight, and some autotrophic bacteria utilize the reaction $Fe^{2+} \rightarrow Fe^{3+}$. On the other hand, **heterotrophic**

organisms obtain free energy by coupling their metabolism to the **breakdown of complex organic molecules** in their environment. In all of these processes, **ATP** plays a central role in the transference of free energy from the exergonic to the endergonic processes (Figs 11–3 and 11–4). As can be seen from Fig 11–5, ATP is a specialized nucleotide containing adenine, ribose, and 3 phosphate groups. In its reactions in the cell, it functions as the Mg^{2+} complex (Fig 11–6).

The importance of phosphates in intermediary metabolism became evident with the discovery of the chemical details of glycolysis and of the role of ATP, adenosine diphosphate (ADP), and inorganic phosphate (P_i) in this process. ATP was considered to be a means of transferring phosphate radicals in the process of phosphorylation. The role of ATP in biochemical energetics was indicated in experiments demonstrating that ATP and creatine phosphate were broken down during muscular contraction and that their resynthesis depended on supplying energy from oxidative processes in the muscle. It was not until Lipmann introduced the concept of "high-energy phosphates" and the "high-energy phosphate bond" that the role of these compounds in bioenergetics was clearly appreciated.

The Free Energy of Hydrolysis of ATP & Other Organophosphates

The standard free energy of hydrolysis of a number of biochemically important phosphates is shown in Table 11–1. An estimate of the comparative tendency

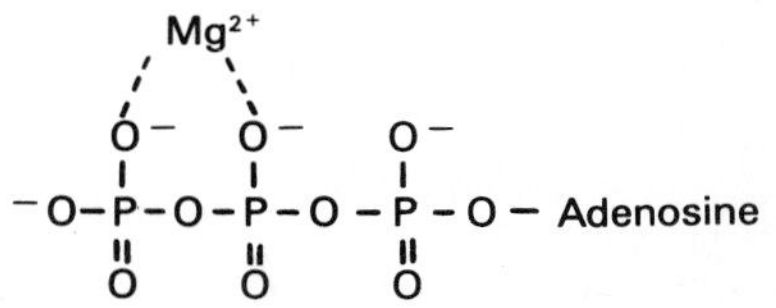

Figure 11–6. The magnesium complex of ATP. (Mg-ADP is similar.)

Table 11–1. Standard free energy of hydrolysis of some organophosphates of biochemical importance.*†

Compound	ΔG⁰′ kJ/mol	ΔG⁰′ kcal/mol
Phosphoenolpyruvate	−61.9	−14.8
Carbamoyl phosphate	−51.4	−12.3
1,3-Bisphosphoglycerate (to 3-phosphoglycerate)	−49.3	−11.8
Creatine phosphate	−43.1	−10.3
ATP → ADP + Pᵢ	−30.5	−7.3
ADP → AMP + Pᵢ	−27.6	−6.6
Pyrophosphate	−27.6	−6.6
Glucose 1-phosphate	−20.9	−5.0
Fructose 6-phosphate	−15.9	−3.8
AMP	−14.2	−3.4
Glucose 6-phosphate	−13.8	−3.3
Glycerol 3-phosphate	−9.2	−2.2

*Pᵢ, inorganic orthophosphate.
†Values for ATP and most others taken from Krebs and Kornberg (1957).

of each of the phosphate groups to transfer to a suitable acceptor may be obtained from the $\Delta G^{0'}$ of hydrolysis (measured at 37 °C). It may be seen from the table that the value for the hydrolysis of the terminal phosphate of ATP of −30.5 kJ per mole divides the list into 2 groups. One group of "low-energy phosphates," exemplified by the ester phosphates found in the intermediates of glycolysis, has $\Delta G^{0'}$ values which are smaller than that of ATP, while in the other group, designated "high-energy phosphates," the value is higher than that of ATP. The components of this latter group, in-

cluding ATP and ADP, are usually anhydrides (eg, the 1-phosphate of 1,3-bisphosphoglycerate), enolphosphates (eg, phosphoenolpyruvate), and phosphoguanidines (eg, creatine phosphate, arginine phosphate). Other biologically important compounds that are classed as "high-energy compounds" are thiol esters involving coenzyme A (eg, acetyl-CoA), acyl carrier protein, amino acid esters involved in protein synthesis, S-adenosylmethionine (active methionine), and UDPGlc (uridine diphosphate glucose).

High-Energy Phosphates

To indicate the presence of the high-energy phosphate group, Lipmann introduced the symbol ~Ⓟ, indicating **high-energy phosphate bond.** The symbol indicates that the group attached to the bond, on transfer to an appropriate acceptor, results in transfer of the larger quantity of free energy. For this reason, the term **group transfer potential** is preferred by some to "high-energy bond." Thus, ATP contains 2 high-energy phosphate groups and ADP contains one, whereas the phosphate bond in AMP (adenosine monophosphate) is of the low-energy type, since it is a normal ester link (Fig 11–7).

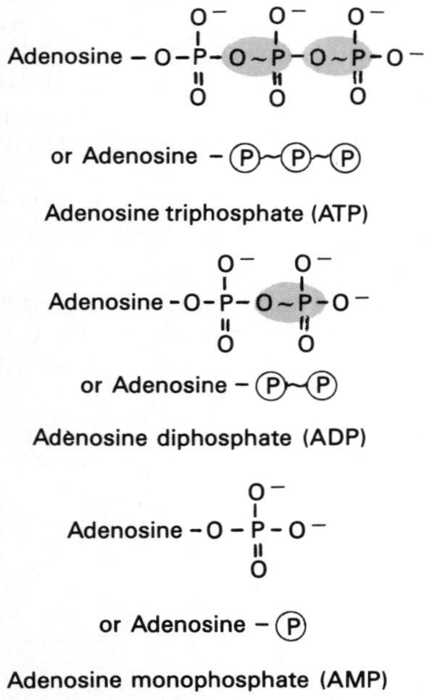

Adenosine – O–P–O~P–O~P–O⁻

or Adenosine – Ⓟ~Ⓟ~Ⓟ

Adenosine triphosphate (ATP)

Adenosine –O–P–O~P–O⁻

or Adenosine – Ⓟ~Ⓟ

Adenosine diphosphate (ADP)

Adenosine –O – P – O⁻

or Adenosine – Ⓟ

Adenosine monophosphate (AMP)

Figure 11–7. Structure of ATP, ADP, and AMP showing the position and the number of high-energy bonds (~).

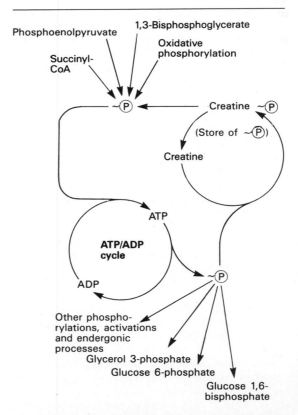

Figure 11–8. Role of ATP/ADP cycle in transfer of high-energy phosphate. Note that ~Ⓟ does not exist in a free state but is transferred in the reactions shown.

Role of High-Energy Phosphates as the "Energy Currency" of the Cell

As a result of its position midway down the list of standard free energies of hydrolysis (Table 11–1), ATP is able to act as a donor of high-energy phosphate to those compounds below it in the table. Likewise, provided the necessary enzymatic machinery is available, ADP can accept high-energy phosphate to form ATP from those compounds above ATP in the table. In effect, an **ATP/ADP cycle** connects these processes which **generate** ~Ⓟ to those processes that **utilize** ~Ⓟ (Fig 11–8).

There are 3 major sources of ~Ⓟ taking part in **energy conservation** or **energy capture:** (1) **Oxidative phosphorylation.** This is the greatest quantitative source of ~Ⓟ in aerobic organisms. The free energy to drive this process comes from respiratory chain oxidation within mitochondria (see p 110). (2) **Glycolysis.** A net formation of 2 ~Ⓟ results from the formation of lactate from one molecule of glucose (see Fig 18–2), generated in 2 reactions catalyzed by phosphoglycerate kinase and pyruvate kinase, respectively (see Fig 11–9). (3) **The citric acid cycle.** One ~Ⓟ is generated directly in the cycle at the succinyl thiokinase step (see Fig 17–3).

Another group of compounds **(phosphagens)** represented in Table 11–1 act as storage forms of high-energy phosphate. These include creatine phosphate, occurring in vertebrate muscle and brain, and arginine phosphate, occurring in invertebrate muscle.

Under physiologic conditions, phosphagens permit ATP concentrations to be maintained in muscle when ATP is rapidly being utilized as a source of energy for muscular contraction. On the other hand, when ATP is plentiful, its concentration can build up sufficiently to cause the reverse reaction to occur and allow the concentration of creatine phosphate to increase substantially so as to act as a store of high-energy phosphate (Fig 11–10). When ATP acts as a phosphate donor to form those compounds of lower free energy of hydrolysis (Table 11–1), the phosphate group is invariably converted to one of low energy, eg,

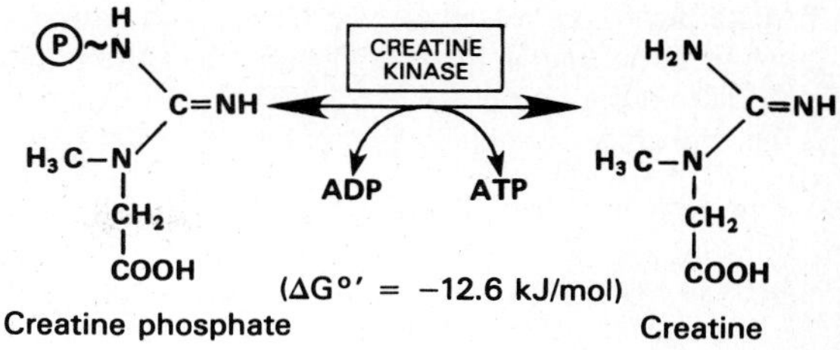

Figure 11–10. Transfer of high-energy phosphate between ATP and creatine.

Glycerol + Adenosine – Ⓟ~Ⓟ~Ⓟ $\xrightarrow{\text{GLYCEROL KINASE}}$ **Glycerol – Ⓟ + Adenosine – Ⓟ~Ⓟ**

Bioenergetics of Coupled Reactions

We can now consider in more detail the energetics of coupled reactions, as depicted in Fig 11–1 or 11–3. Such a reaction is the first in the glycolysis pathway (see Fig 18–2), the phosphorylation of glucose to glucose 6-phosphate, which is highly endergonic and would not proceed as such under physiologic conditions.

(1) Glucose + P_i ⟶ Glucose 6-phosphate + H_2O

($\Delta G^{0\prime}$ = +13.8 kJ/mol)

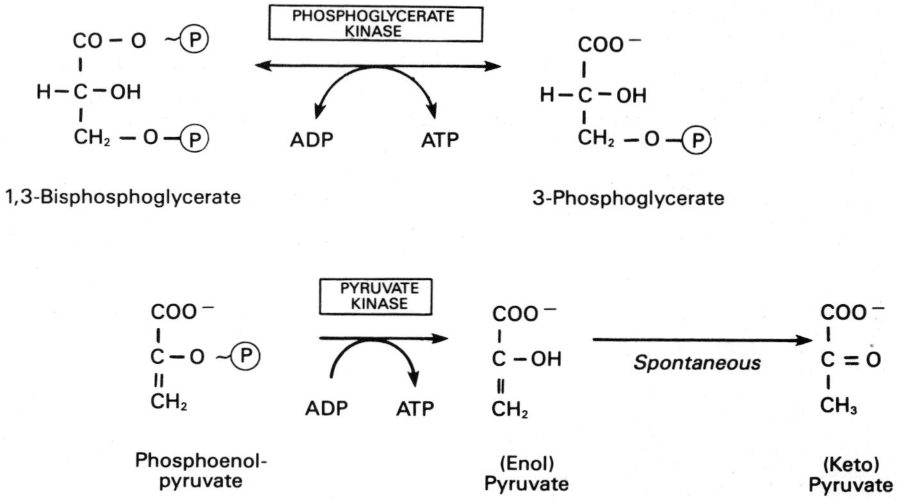

Figure 11–9. Transfer of high-energy phosphate from intermediates of glycolysis to ADP.

To take place, the reaction must be coupled with another reaction that is more exergonic than the phosphorylation of glucose is endergonic. Such a reaction is the hydrolysis of the terminal phosphate of ATP.

$$(2)\ ATP \longrightarrow ADP + P_i\ (\Delta G^{0'} = -30.5\ kJ/mol)$$

When (1) and (2) are coupled in a reaction catalyzed by hexokinase, phosphorylation of glucose readily proceeds in a highly exergonic reaction that under physiologic conditions is far from equilibrium and thus irreversible for practical purposes.

Glucose + ATP $\xrightarrow{\text{HEXOKINASE}}$

Glucose 6-phosphate + ADP

$$(\Delta G^{0'} = +13.8\ kJ/mol)$$

Many "activation" reactions follow this pattern.

Interconversion of Adenine Nucleotides

The enzyme adenylate kinase (myokinase) is present in most cells. It catalyzes the interconversion of ATP and AMP on the one hand and ADP on the other:

Adenosine − ℗~℗~℗ + Adenosine − ℗

    (ATP)   (AMP)

$\xrightarrow[\text{KINASE}]{\text{ADENYLATE}}$ **2 Adenosine − ℗~℗**

                                (2 ADP)

This reaction has 3 functions: (1) it allows high-energy phosphate in ADP to be used in the synthesis of ATP. (2) It allows AMP, formed as a consequence of several activating reactions involving ATP, to be recovered by rephosphorylation to ADP. (3) It allows AMP to increase in concentration when ATP becomes depleted and act as a metabolic (allosteric) signal to increase the rate of catabolic reactions, which in turn leads to the generation of more ATP (see p 191).

METABOLISM OF PYROPHOSPHATE

When ATP reacts to form AMP, inorganic pyrophosphate (PP$_i$) is formed, as occurs, for example, in the activation of long-chain fatty acids:

ATP + CoA · SH + R · COOH $\xrightarrow{\text{ACYL-CoA}\atop\text{SYNTHETASE}}$

AMP + PP$_i$ + R · CO~SCoA

This reaction is accompanied by loss of free energy as heat, which ensures that the activation reaction will go to the right; this is further aided by the hydrolytic splitting of PP$_i$, catalyzed by **inorganic pyrophosphatase,** a reaction that itself has a large $\Delta G^{0'}$ of -4.6 kcal/mol. Note that activations via the pyrophosphate

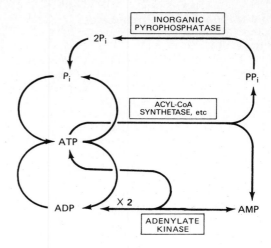

Figure 11–11. Phosphate cycles and interchange of adenine nucleotides.

pathway result in the loss of 2 ~℗ rather than one ~℗, as occurs when ADP and P$_i$ are formed.

PP$_i$ + H$_2$O $\xrightarrow{\text{INORGANIC}\atop\text{PYROPHOSPHATASE}}$ **2 P$_i$**

A combination of the above reactions makes it possible for phosphate to be recycled and the adenine nucleotides to interchange (Fig 11–11).

Nucleoside Phosphates Related to ATP & ADP

By means of the enzyme **nucleoside diphosphate kinase,** nucleoside triphosphates similar to ATP but containing an alternative base to adenine can be synthesized from their diphosphates, eg,

ATP + UDP $\xleftrightarrow{\text{NUCLEOSIDE}\atop\text{DIPHOSPHATE}\atop\text{KINASE}}$ **ADP + UTP**
                                (uridine triphosphate)

ATP + GDP $\longleftrightarrow$ **ADP + GTP**
                                (guanosine triphosphate)

ATP + CDP $\longleftrightarrow$ **ADP + CTP**
                                (cytidine triphosphate)

All of these triphosphates take part in phosphorylations in the cell. Similarly, **nucleoside monophosphate kinases,** specific for each purine or pyrimidine nucleoside, catalyze the formation of nucleoside diphosphates from the corresponding monophosphates

ATP + Nucleoside − ℗ $\xleftrightarrow{\text{SPECIFIC NUCLEOSIDE}\atop\text{MONOPHOSPHATE KINASE}}$

ADP + Nucleoside − ℗~℗

Thus, adenylate kinase is a specialized monophosphate kinase.

REFERENCES

Ernster L (editor): *Bioenergetics*. Elsevier, 1984.

Harold FM: *The Vital Force: A Study of Bioenergetics*. Freeman, 1986.

Klotz IM: *Introduction to Biomolecular Energetics*. Academic Press, 1986.

Krebs HA, Kornberg HL: *Energy Transformations in Living Matter*. Springer, 1957.

Lehninger AL: *Bioenergetics: The Molecular Basis of Biological Energy Transformations*, 2nd ed. Benjamin, 1971.

12 Biologic Oxidation

Peter A. Mayes, PhD, DSc

INTRODUCTION

Chemically, **oxidation is defined as the removal of electrons** and **reduction as the gain of electrons,** as illustrated by the oxidation of ferrous to ferric ion.

$$Fe^{2+} \xrightarrow{\quad e^- \text{ (electron)} \quad} Fe^{3+}$$

It follows that oxidation is always accompanied by **reduction of an electron acceptor.** This principle of oxidation-reduction applies equally to biochemical systems and is an important concept underlying understanding of the nature of biologic oxidation. It will be appreciated that many biologic oxidations can take place without the participation of molecular oxygen, eg, dehydrogenations.

BIOMEDICAL IMPORTANCE

Although certain bacteria (anaerobes) survive in the absence of oxygen, the life of higher animals is absolutely dependent upon a supply of oxygen. The principal use of oxygen is in respiration, which may be defined as the process by which cells derive energy in the form of ATP from the controlled reaction of hydrogen with oxygen to form water. In addition, molecular oxygen is incorporated into a variety of substrates by enzymes designated as oxygenases; many drugs, pollutants, and chemical carcinogens (xenobiotics, foreign compounds) are metabolized by enzymes of this class, known as the cytochrome P-450 system. Administration of oxygen can be lifesaving in the treatment of patients with respiratory or circulatory failure. Moreover, in a relatively small number of clinical situations, administration of oxygen at high pressure (hyperbaric oxygen therapy) has proved of value. However, both acute and chronic administration of the gas at high pressure can result in oxygen toxicity.

OXIDATION-REDUCTION EQUILIBRIA; REDOX POTENTIAL

In reactions involving oxidation and reduction, the free energy exchange is proportionate to the tendency of reactants to donate or accept electrons. Thus, in addition to expressing free energy change in terms of

Table 12–1. Some redox potentials of special interest in mammalian oxidation systems.

System	E_o' volts
Oxygen/water	+0.82
Cytochrome a; Fe^{3+}/Fe^{2+}	+0.29
Cytochrome c; Fe^{3+}/Fe^{2+}	+0.22
Ubiquinone; ox/red	+0.10
Cytochrome b; Fe^{3+}/Fe^{2+}	+0.08
Fumarate/succinate	+0.03
Flavoprotein-old yellow enzyme; ox/red	−0.12
Oxaloacetate/malate	−0.17
Pyruvate/lactate	−0.19
Acetoacetate/β-hydroxybutyrate	−0.27
Lipoate; ox/red	−0.29
$NAD^+/NADH$	−0.32
H^+/H_2	**−0.42**
Succinate/α-ketoglutarate	−0.67

$\Delta G^{0'}$ (see Chapter 11), it is possible, in an analogous manner, to express it numerically as an **oxidation-reduction** or **redox potential** (E_o'). It is usual to compare the redox potential of a system (E_o) against the potential of the hydrogen electrode, which at pH 0 is designated as 0.0 volts. However, for biologic systems it is normal to express the redox potential (E_o') at pH 7.0, at which pH the electrode potential of the hydrogen electrode is −0.42 volts. The redox potentials of some redox systems of special interest in mammalian physiology are shown in Table 12–1. The list of redox potentials shown in the table allows prediction of the direction of flow of electrons from one redox couple to another.

ENZYMES & COENZYMES INVOLVED IN OXIDATION & REDUCTION

All enzymes concerned in oxidative processes are designated **oxidoreductases.** In the following account, they are classified into 5 groups.

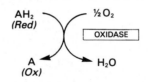

Figure 12–1. Oxidation of a metabolite catalyzed by an oxidase.

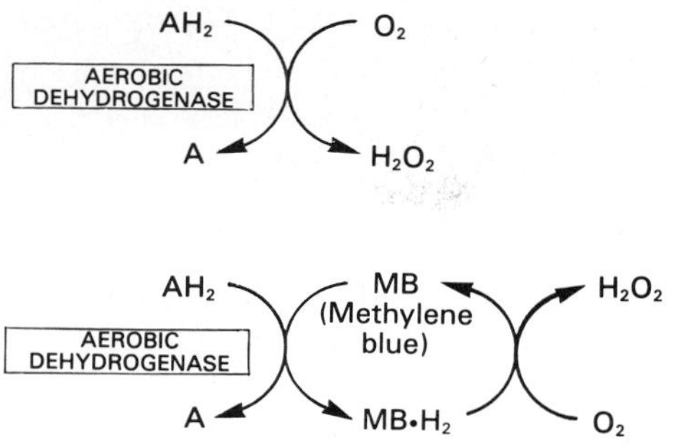

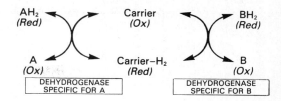

Figure 12–3. Oxidation of a metabolite catalyzed by anaerobic dehydrogenases, not involving a respiratory chain.

oxidized at the expense of another, is particularly useful in enabling oxidative processes to occur in the **absence of oxygen.**

(b) As components in a **respiratory chain** of electron transport from substrate to oxygen (Fig 12–4).

(4) Hydroperoxidases: Enzymes utilizing hydrogen peroxide or an organic peroxide as a substrate. Two types of enzymes fall into this category: **peroxidases,** found in milk, plants, leukocytes, platelets, and erythrocytes, etc; and **catalase,** found in animals and plants.

(5) Oxygenases: Enzymes that catalyze the direct transfer and **incorporation of oxygen** into a substrate molecule.

Oxidases

Cytochrome oxidase is a hemoprotein widely distributed in many plant and animal tissues. It is the terminal component of the chain of respiratory carriers found in mitochondria and is therefore responsible for the reaction whereby electrons resulting from the oxidation of substrate molecules by dehydrogenases are transferred to their final acceptor, oxygen. The enzyme is poisoned by carbon monoxide, cyanide, and hydrogen sulfide. It has also been termed cytochrome a_3. It was formerly assumed that cytochrome a and cytochrome a_3 were separate compounds, since each has a distinct spectrum and different properties with respect to the effects of carbon monoxide and cyanide. More recent studies show that the 2 cytochromes are combined with the same protein, and the complex is known as **cytochrome aa_3.** It contains 2 molecules of heme, each having one Fe atom that oscillates between Fe^{3+} and Fe^{2+} during oxidation and reduction. Also, 2 atoms of Cu are present, each associated with a heme unit.

Figure 12–2. Oxidation of a metabolite catalyzed by an aerobic dehydrogenase.

(1) Oxidases: True oxidases that catalyze the removal of hydrogen from a substrate but use **only oxygen** as a hydrogen acceptor.* They invariably contain **copper** and form water as a reaction product (with the exception of uricase and monoamine oxidase, which form H_2O_2) (Fig 12–1).

(2) Aerobic dehydrogenases: Enzymes catalyzing the removal of hydrogen from a substrate but which, as distinct from oxidases, can use **either oxygen or artificial substances** such as methylene blue as hydrogen acceptor. Characteristically, these dehydrogenases are **flavoproteins.** Hydrogen peroxide rather than water is formed as a product (Fig 12–2).

(3) Anaerobic dehydrogenases: Enzymes catalyzing the removal of hydrogen from a substrate but **not able to use oxygen** as hydrogen acceptor. There are a large number of enzymes in this class. They perform 2 main functions:

(a) Transfer of hydrogen from one substrate to another in a coupled oxidation-reduction reaction (Fig 12–3). These dehydrogenases are specific for their substrates but often utilize the same coenzyme or hydrogen carrier as other dehydrogenases. As the reactions are reversible, these properties enable reducing equivalents to be freely transferred within the cell. This type of reaction, which enables a substrate to be

---

*Sometimes the term "oxidase" is used collectively to denote all enzymes that catalyze reactions involving molecular oxygen.

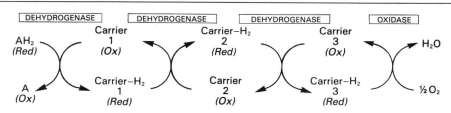

Figure 12–4. Oxidation of a metabolite by anaerobic dehydrogenases and finally by a true oxidase in a respiratory chain.

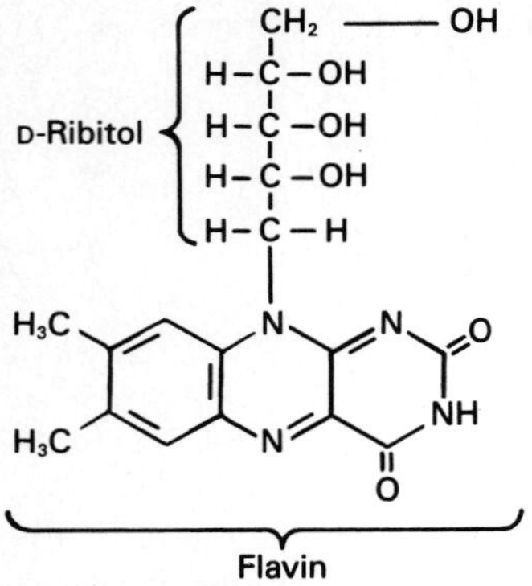

Figure 12–5. Riboflavin.

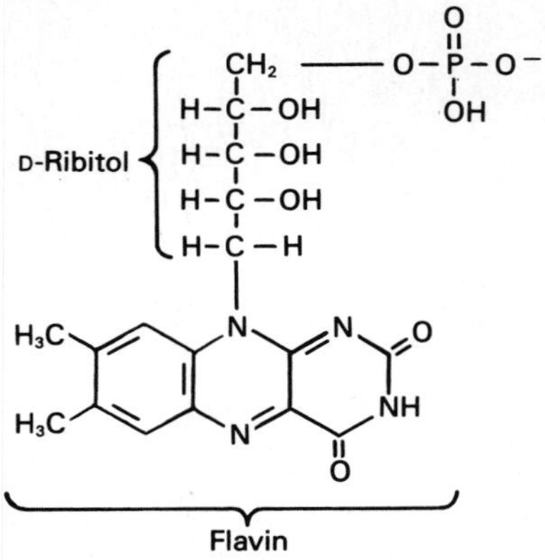

Figure 12–6. Riboflavin phosphate (flavin mononucleotide, FMN).

Phenolase (tyrosinase, polyphenol oxidase, catechol oxidase) is a copper-containing enzyme of broad specificity. It is able to convert monophenols or o-diphenols to o-quinones. Copper has been claimed to be present in a number of other enzymes such as **uricase,** which catalyzes the oxidation of uric acid to allantoin, and **monoamine oxidase,** an enzyme that oxidizes epinephrine and tyramine in mitochondria.

Aerobic Dehydrogenases

Aerobic dehydrogenases are flavoprotein enzymes containing **flavin mononucleotide (FMN)** or **flavin adenine dinucleotide (FAD)** as prosthetic groups. **FMN and FAD are formed in the body from the vitamin riboflavin** (Fig 12–5).

Riboflavin (vitamin B₂) cannot be synthesized by mammalian tissues but is synthesized by plants and microorganisms. It is therefore an **essential** component of the diet. Riboflavin is formed from the sugar alcohol ribitol and the heterocyclic flavin. It is converted in the tissues to FMN (Fig 12–6) by ATP-dependent phosphorylation and to FAD (Fig 12–7) by transfer of AMP from another molecule of ATP.

FMN and FAD are usually tightly—but not covalently—bound to their respective apoenzyme protein. Many flavoprotein enzymes contain one or more metals as essential cofactors and are known as **metalloflavoproteins.**

Enzymes belonging to this group of aerobic dehydrogenases include L-**amino acid dehydrogenase** (L-amino acid oxidase), an FMN-linked enzyme found in kidney with general specificity for the oxidative deamination of the naturally occurring L-amino acids. **Xanthine dehydrogenase** (xanthine oxidase) has a wide distribution, occurring in milk, small intestine, kidney, and liver. It contains molybdenum and plays an important role in the conversion of purine bases to uric acid. It is of particular significance in the liver and kidney of birds, which excrete uric acid as the main nitrogenous end product, not only of purine metabolism but also of protein and amino acid catabolism.

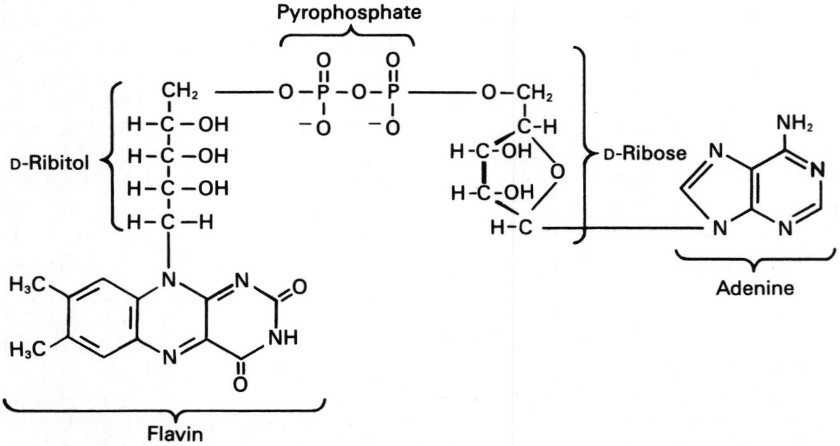

Figure 12–7. Flavin adenine dinucleotide (FAD).

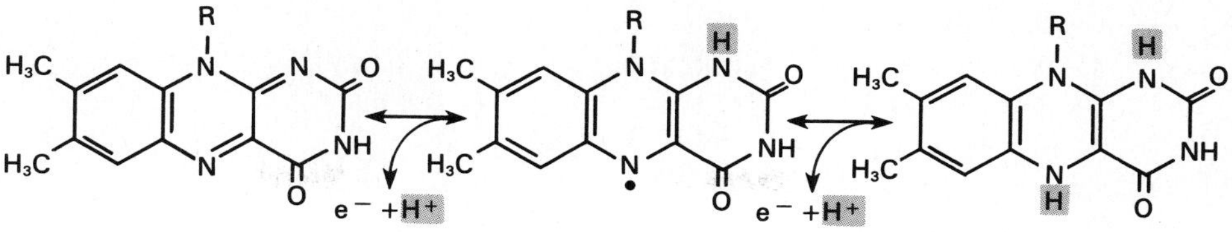

Figure 12–8. Reduction of isoalloxazine ring in flavin nucleotides.

Aldehyde dehydrogenase is an FAD-linked enzyme present in mammalian livers. It is a metalloflavoprotein containing molybdenum and nonheme iron and acts upon aldehydes and N-heterocyclic substrates.

Of interest because of its use in estimating glucose is **glucose oxidase,** an FAD-specific enzyme prepared from fungi.

The mechanisms of oxidation and reduction of these enzymes are complex. However, evidence points to reduction of the isoalloxazine ring taking place in 2 steps via a semiquinone (free radical) intermediate (Fig 12–8).

Anaerobic Dehydrogenases

A. Dehydrogenases Dependent on Nicotinamide Coenzymes: A large number of dehydrogenases fall into this category. They are specific for either **nicotinamide adenine dinucleotide** (NAD$^+$) or **nicotinamide adenine dinucleotide phosphate** (NADP$^+$) as coenzymes (Fig 12–9). However, some dehydrogenases can use either NAD$^+$ or NADP$^+$.

NAD$^+$ and NADP$^+$ are formed in the body from the vitamin niacin. Niacin comprises both **nicotinic acid** and **nicotinamide,** and either may act as a source of the vitamin in the diet. Nicotinate, not nicotinamide, is the form of niacin required for the synthesis of NAD$^+$ and NADP$^+$ by enzymes present in the cytosol of most cells. The nicotinamide moiety of NAD$^+$ is formed from nicotinate while nicotinate is in combination as a nucleotide, the amido group coming from glutamine (Fig 12–9). There is evidence of biosynthesis of NAD$^+$ from nicotinamide in mitochondria. The coenzymes are reduced by the specific substrate of the dehydrogenase and reoxidized by a suitable electron acceptor (Fig 12–10). They may freely and reversibly dissociate from their respective apoenzymes.

Generally, **NAD-linked dehydrogenases** catalyze oxidoreduction reactions in the **oxidative pathways** of metabolism, particularly in glycolysis, in the citric acid cycle, and in the respiratory chain of mitochondria. **NADP-linked dehydrogenases** are found characteristically in **reductive syntheses,** as in the extramitochondrial pathway of fatty acid synthesis and steroid synthesis. They are also to be found as coenzymes to the dehydrogenases of the pentose phosphate pathway. Some nicotinamide coenzyme-dependent dehydrogenases have been found to contain zinc, notably alcohol dehydrogenase from liver and glyceraldehyde-3-phosphate dehydrogenase from skeletal muscle. The zinc ions are not considered to take part in the oxidation and reduction.

B. Dehydrogenases Dependent on Riboflavin: The flavin groups associated with these dehydrogenases are similar to those of the aerobic dehydrogenase group, namely FMN and FAD. They are generally more tightly bound to their apoenzymes than are the nicotinamide coenzymes. Most of the riboflavin-linked anaerobic dehydrogenases are concerned with electron transport in (or to) the respiratory chain. **NADH dehydrogenase** is a member of the respiratory chain acting as a carrier of electrons between NADH and the more electropositive components. Other dehydrogenases such as **succinate dehydrogenase, acyl-CoA dehydrogenase,** and **mitochondrial glycerol-3-phosphate dehydrogenase** transfer reducing equivalents directly from the substrate to the respiratory chain. Another role of the flavin-dependent dehydrogenases is in the dehydrogenation (by dihydrolipoyl dehydrogenase) of reduced lipoate, an intermediate in the oxidative decarboxylation of pyruvate and α-ketoglutarate (see Fig 18–5). In this particular instance, owing to the low redox potential, the flavoprotein (FAD) acts as a hydrogen carrier from reduced lipoate to NAD$^+$. The **electron-transferring flavoprotein** is an intermediary carrier of electrons between acyl-CoA dehydrogenase and the respiratory chain (see Fig 13–3).

C. The Cytochromes: Except for cytochrome oxidase (previously described), the cytochromes are classified as anaerobic dehydrogenases. Their identification and study are facilitated by the presence in the reduced state of characteristic absorption bands that disappear on oxidation. In the respiratory chain, they are involved as **carriers of electrons from flavoproteins on the one hand to cytochrome oxidase on the other.** The cytochromes are iron-containing hemoproteins in which the iron atom oscillates between Fe^{3+} and Fe^{2+} during oxidation and reduction. Several identifiable cytochromes occur in the respiratory chain, ie, cytochromes b, c$_1$, c, a, and a$_3$ (cytochrome oxidase). Of these, only cytochrome c is soluble. Besides the respiratory chain, cytochromes are found in other locations, eg, the endoplasmic reticulum (cytochromes P-450 and b$_5$), plant cells, bacteria, and yeasts.

Hydroperoxidases

A. Peroxidase: Although originally considered to be plant enzymes, peroxidases are found in milk and

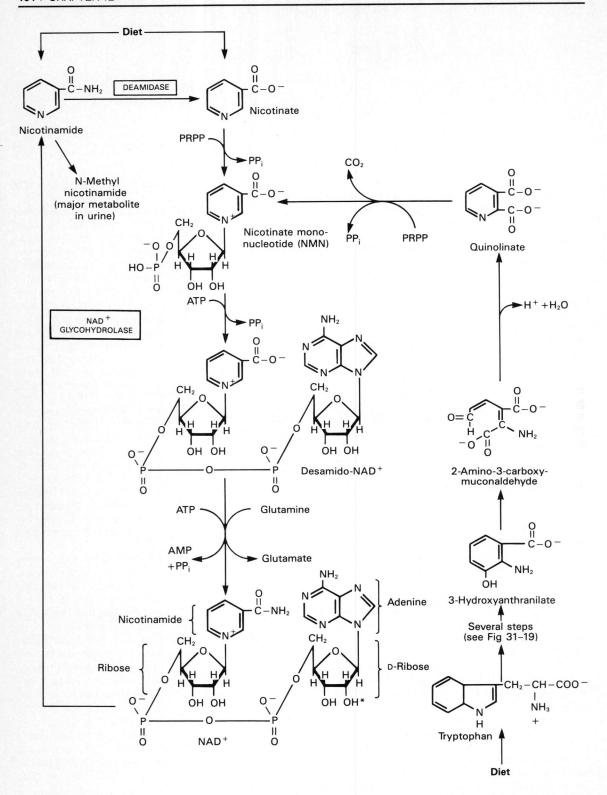

Figure 12–9. The synthesis and breakdown of nicotinamide adenine dinucleotide (NAD^+). The 2'-hydroxyl group (*) of the adenosine moiety is phosphorylated in nicotinamide dinucleotide phosphate ($NADP^+$). Humans, but not cats, can provide all of their niacin requirement from tryptophan if there is a sufficient amount in the diet. Normally, about two-thirds comes from this source. PRPP, 5-phosphoribosyl-1-pyrophosphate.

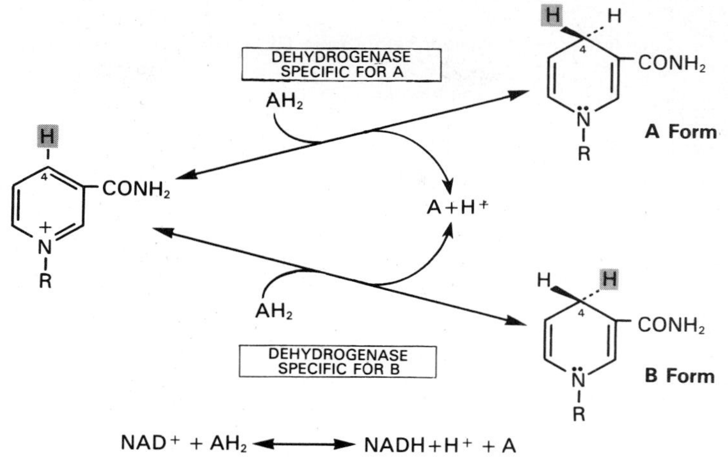

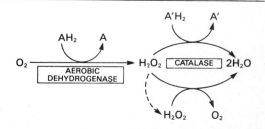

Figure 12–10. Mechanism of oxidation and reduction of nicotinamide coenzymes. There is stereospecificity about position 4 of nicotinamide when it is reduced by a substrate AH_2. One of the hydrogen atoms is removed from the substrate as a hydrogen nucleus with 2 electrons (hydride ion, H^-) and is transferred to the 4 position, where it may be attached in either the A- or B-position according to the specificity determined by the particular dehydrogenase catalyzing the reaction. The remaining hydrogen of the hydrogen pair removed from the substrate remains free as a hydrogen ion.

in leukocytes, platelets, and other tissues involved in eicosanoid metabolism (see p 213). The prosthetic group is protoheme, which, unlike the situation in most hemoproteins, is only loosely bound to the apoprotein. In the reaction catalyzed by peroxidase, hydrogen peroxide is reduced at the expense of several substances that will act as electron acceptors, such as ascorbate, quinones, and cytochrome c. The reaction catalyzed by peroxidase is complex, but the overall reaction is as follows:

$$H_2O_2 + AH_2 \xrightarrow{\boxed{\text{PEROXIDASE}}} 2H_2O + A$$

In erythrocytes, the enzyme **glutathione peroxidase,** containing **selenium** as a prosthetic group, catalyzes the destruction of H_2O_2 and lipid hydroperoxides by reduced glutathione, protecting membrane lipids and hemoglobin against oxidation by peroxides (see p 179).

B. Catalase: Catalase is a hemoprotein containing 4 heme groups. In addition to possessing peroxidase activity, it is able to use one molecule of H_2O_2 as a substrate electron donor and another molecule of H_2O_2 as oxidant or electron acceptor. Under most conditions in vivo, the peroxidase activity of catalase seems to be favored.

$$2H_2O_2 \xrightarrow{\boxed{\text{CATALASE}}} 2H_2O + O_2$$

Catalase is found in blood, bone marrow, mucous membranes, kidney, and liver. Its function is assumed to be the **destruction of hydrogen peroxide** formed by the action of aerobic dehydrogenases. Microbodies or **peroxisomes** are found in many tissues, including liver. They are rich in aerobic dehydrogenases and in catalase, which suggests that there may be a biologic

advantage in grouping the enzymes which produce H_2O_2 with the enzyme that destroys it (Fig 12–11). In addition to the peroxisomal enzymes, mitochondrial and microsomal electron transport systems must be considered as sources of H_2O_2.

Oxygenases

Oxygenases are concerned with the synthesis or degradation of many different types of metabolites rather than taking part in reactions that have as their purpose the provision of energy to the cell. Enzymes in this group catalyze the incorporation of oxygen into a substrate molecule. This takes place in 2 steps: (1) oxygen binding to the enzyme at the active site, and (2) the reaction in which the bound oxygen is reduced or transferred to the substrate. Oxygenases may be divided into 2 subgroups:

A. Dioxygenases (Oxygen Transferases, True Oxygenases): These enzymes catalyze the incorporation of both atoms of oxygen into the substrate:

$$A + O_2 \longrightarrow AO_2$$

Figure 12–11. Role of catalase in oxidative reactions.

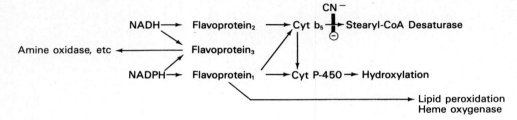

Figure 12–12. Electron transport chain in microsomes. Cyanide (CN^-) inhibits the indicated step.

Examples of this type include enzymes that contain iron such as **homogentisate dioxygenase** and **3-hydroxyanthranilate dioxygenase** from the supernatant fraction of the liver, and enzymes utilizing heme such as L-**tryptophan dioxygenase** (tryptophan pyrrolase) from the liver.

B. Monooxygenases (Mixed Function Oxidases, Hydroxylases): These enzymes catalyze the incorporation of only one atom of the oxygen molecule into a substrate. The other oxygen atom is reduced to water, an additional electron donor or cosubstrate being necessary for this purpose.

$$A–H + O_2 + ZH_2 \longrightarrow A–OH + H_2O + Z$$

Microsomal Cytochrome P-450 Monooxygenase Systems

The enzymes involved in the metabolism of many drugs by hydroxylations belong to this group. They are found in the microsomes of the liver together with cytochrome P-450 and cytochrome b_5. Both NADH and NADPH donate reducing equivalents for the re-

duction of these cytochromes (Fig 12–12), which in turn are oxidized by substrates in a series of enzymatic reactions collectively known as the hydroxylase cycle (Fig 12–13).

$$\boxed{\text{HYDROXYLASE}}$$
$$\text{DRUG–H} + O_2 + 2Fe^{2+} + 2H^+ \longrightarrow$$
$$\text{(P-450)}$$

$$\text{DRUG–OH} + H_2O + 2Fe^{3+}$$
$$\text{(P-450)}$$

Among the drugs metabolized by this system are benzpyrene, aminopyrine, aniline, morphine, and benzphetamine. Many drugs such as phenobarbital have the ability to induce the formation of microsomal enzymes and of cytochrome P-450.

Mitochondrial Cytochrome P-450 Monooxygenase Systems

These systems are found in steroidogenic tissues such as adrenal cortex, testis, ovary, and placenta and are concerned with the biosynthesis of steroid hormones from cholesterol (hydroxylation at C_{22} and C_{20}

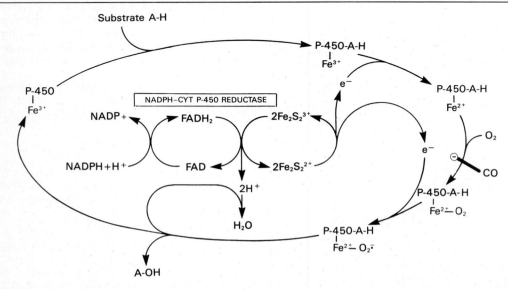

Figure 12–13. Cytochrome P-450 hydroxylase cycle in microsomes. The system shown is typical of steroid hydroxylases of the adrenal cortex. Liver microsomal cytochrome P-450 hydroxylase does not require the iron-sulfur protein Fe_2S_2. Carbon monoxide (CO) inhibits the indicated step.

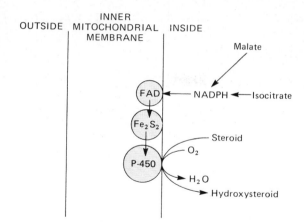

Figure 12–14. Mitochondrial cytochrome P-450 mono-oxygenase system. Fe_2S_2, iron-sulfur protein (adrenodoxin). Note that because NADP(H) cannot penetrate the mitochrondrial membrane, sources of reducing equivalents are confined to substrates such as malate and isocitrate for which there are intramitochondrial NADP-specific dehydrogenases.

in side-chain cleavage and at 11β and 18 positions). Renal systems catalyze 1α- and 24-hydroxylations of 25-hydroxycholecalciferol, and the liver catalyzes 26-hydroxylation in bile acid biosynthesis. In the adrenal cortex, mitochondrial cytochrome P-450 is 6 times more abundant than cytochromes of the respiratory chain. The monooxygenase system consists of 3 components situated on the inside of the inner mitochondrial membrane: an NADP-specific FAD containing flavoprotein, an Fe_2S_2 protein (adrenodoxin), and cytochrome P-450 (Fig 12–14).

Superoxide Metabolism

Oxygen is a potentially toxic substance, the toxicity of which has hitherto been attributed to the formation of H_2O_2. Recently, however, the ease with which oxygen can be reduced in tissues to the superoxide anion free radical (O_2^{-}) and the occurrence of **superoxide dismutase** in aerobic organisms (although not in obligate anaerobes) have suggested that the toxicity of oxygen is due to its conversion to superoxide. However, no direct evidence of superoxide toxicity has yet been obtained.

Superoxide is formed when reduced flavins, present, for example, in xanthine dehydrogenase, are reoxidized univalently by molecular oxygen. It is also formed during univalent oxidations with molecular oxygen in the respiratory chain.

$$EnzH_2 + O_2 \longrightarrow EnzH + O_2^{-} + H^+$$

Superoxide can reduce oxidized cytochrome c

$$O_2^{-} + Cyt\ c\ (Fe^{3+}) \longrightarrow O_2 + Cyt\ c\ (Fe^{2+})$$

or be removed by the presence of the specific enzyme superoxide dismutase.

$$O_2^{-} + O_2^{-} + 2H^+ \xrightarrow{\text{SUPEROXIDE DISMUTASE}} H_2O_2 + O_2$$

In this reaction, superoxide acts as both oxidant and reductant. The chemical effects of superoxide in the tissues are amplified by free-radical chain reactions. It has been proposed that O_2^{-} bound to cytochrome P-450 is an intermediate in the activation of oxygen in hydroxylation reactions (Fig 12–13).

The function of superoxide dismutase seems to be that of protecting aerobic organisms against the potential deleterious effects of superoxide. The enzyme occurs in several different compartments of the cell. The cytosolic enzyme is composed of 2 similar subunits, each one containing one equivalent of Cu^{2+} and Zn^{2+}, whereas the mitochondrial enzyme contains Mn^{2+}, similar to the enzyme found in bacteria. This finding supports the hypothesis that mitochondria have evolved from a prokaryote that entered into symbiosis with a protoeukaryote. The dismutase is present in all major aerobic tissues. Although exposure of animals to an atmosphere of 100% oxygen causes an adaptive increase of the enzyme, particularly in the lungs, prolonged exposure leads to lung damage and death. Antioxidants, eg, α-tocopherol (vitamin E), also act as scavengers of free radicals such as O_2^{-} and reduce the toxicity of oxygen.

REFERENCES

Bonnett R: Oxygen activation and tetrapyrroles. *Essays Biochem* 1981;**17**:1.

Ernster L (editor): *Bioenergetics*. Elsevier, 1984.

Fleisher S, Packer L (editors): Biological oxidations, microsomal, cytochrome P-450, and other hemoprotein systems. In: *Methods in Enzymology*. Vol 52. Biomembranes, part C. Academic Press, 1978.

Friedovich I: Superoxide dismutases. *Annu Rev Biochem* 1975;**44**:147.

Salemme FR: Structure and function of cytochromes c. *Annu Rev Biochem* 1977;**46**:299.

Schenkman JB, Jansson I, Robie-Suh KM: The many roles of cytochrome b_5 in hepatic microsomes. *Life Sci* 1976;**19**:611.

Tolbert NE: Metabolic pathways in peroxisomes and glyoxysomes. *Annu Rev Biochem* 1981;**50**:133.

Tyler DD, Sutton CM: Respiratory enzyme systems in mitochondrial membranes. Page 33 in: *Membrane Structure and Function*. Vol 5. Bittar EE (editor). Wiley, 1984.

White RE, Coon MJ: Oxygen activation by cytochrome P-450. *Annu Rev Biochem* 1980;**49**:315.

Oxidative Phosphorylation & Mitochondrial Transport Systems

Peter A. Mayes, PhD, DSc

INTRODUCTION

The mitochondrion has appropriately been termed the "powerhouse" of the cell, since it is within this organelle that most of the capture of energy derived from respiratory oxidation takes place. The system in mitochondria whereby respiration is coupled to the generation of the high-energy intermediate, ATP, is termed oxidative phosphorylation.

BIOMEDICAL IMPORTANCE

Oxidative phosphorylation enables aerobic organisms to capture a far greater proportion of the available free energy of respiratory substrates than anaerobic organisms. The chemiosmotic theory offers an insight into how this is accomplished. A number of drugs (eg, amobarbital) and poisons (eg, cyanide, carbon monoxide) inhibit oxidative phosphorylation, usually with fatal consequences. Oxidative phosphorylation is so basic and vital that disturbances of its function are incompatible with life. This explains why so few genetic abnormalities involving it have been reported.

THE RESPIRATORY CHAIN

All of the useful energy liberated during the oxidation of fatty acids and amino acids and nearly all of that released from the oxidation of carbohydrate is made available within the mitochondria as reducing equivalents (−H or electrons). The mitochondria contain the series of catalysts known as the respiratory chain that collect and transport reducing equivalents and direct them to their final reaction with oxygen to form water. Also present is the machinery for trapping the liberated free energy as high-energy phosphate. Mitochondria also contain the enzyme systems responsible for producing most of the reducing equivalents in the first place, ie, the enzymes of β-oxidation and of the citric acid cycle. The latter is the final common metabolic pathway for the oxidation of all the major foodstuffs. These relationships are shown in Fig 13–1.

ORGANIZATION OF THE RESPIRATORY CHAIN IN MITOCHONDRIA

The major components of the respiratory chain (Fig 13–2) are arranged sequentially in order of increasing redox potential (see Table 12–1). Hydrogen or electrons flow through the chain in a stepwise manner **from the more electronegative components to the more electropositive oxygen** through a redox span of 1.1 volts from $NAD^+/NADH$ to $O_2/2H_2O$.

The main respiratory chain in mitochondria pro-

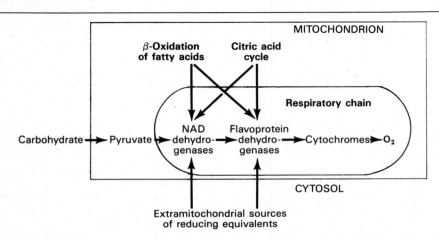

Figure 13–1. The major sources of reducing equivalents and their relationship to the mitochondrial respiratory chain. The main extramitochondrial source is NADH formed in glycolysis.

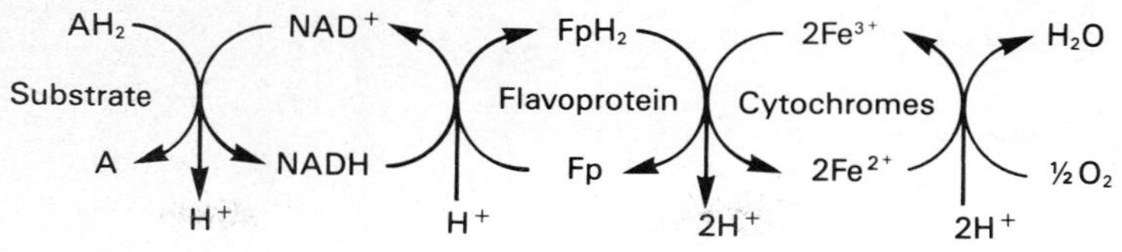

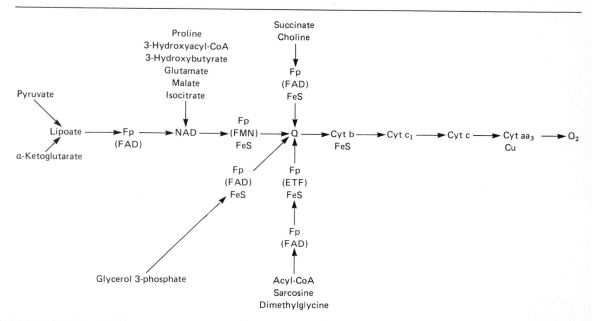

Figure 13-2. Transport of reducing equivalents through the respiratory chain.

ceeds from the NAD-linked dehydrogenase systems on the one hand, through flavoproteins and cytochromes, to molecular oxygen on the other. Not all substrates are linked to the respiratory chain through NAD-specific dehydrogenases; some, because their redox potentials are more positive (eg, fumarate/succinate; see Table 12-1), are linked directly to flavoprotein dehydrogenases, which in turn are linked to the cytochromes of the respiratory chain (Fig 13-3).

In recent years, it has become clear that an additional carrier is present in the respiratory chain linking the flavoproteins to cytochrome b, the member of the cytochrome chain of lowest redox potential. This substance, which has been named **ubiquinone** or **Q (coenzyme Q;** see Fig 13-4), exists in mitochondria in the oxidized quinone form under aerobic conditions and in the reduced quinol form under anaerobic conditions. Q is a constituent of the mitochondrial lipids; the other lipids are predominantly phospholipids that constitute part of the mitochondrial membrane. The structure of Q is very similar to vitamin K and vitamin E. It is also similar to plastoquinone, found in chloroplasts. All of these substances are characterized by the pos-

session of a polyisoprenoid side chain. In mitochondria, there is a large stoichiometric excess of Q compared with other members of the respiratory chain; this suggests that Q is a mobile component of the respiratory chain and that it collects reducing equivalents from the more fixed flavoprotein complexes and passes them on to the cytochromes.

An additional component found in respiratory chain preparations is the **iron-sulfur protein (FeS;** nonheme iron). It is associated with the flavoproteins (metalloflavoproteins) and with cytochrome b. The sulfur and iron are thought to take part in the oxidoreduction mechanism, which involves only a single e^- change (Fig 13-5).

A current view of the sequence of the principal components of the respiratory chain is shown in Fig 13-3. At the electronegative end of the chain, dehydrogenase enzymes catalyze the transfer of electrons from substrates to NAD of the chain. Several differences exist in the manner in which this is carried out. The α-keto acids pyruvate and ketoglutarate have complex dehydrogenase systems involving lipoate and FAD prior to the passage of electrons to NAD of the respiratory chain. Electron transfers from other dehy-

Figure 13-3. Components of the respiratory chain in mitochondria. FeS occurs in the sequences on the O_2 side of Fp or Cyt b. Cyt, cytochrome; ETF, electron-transferring flavoprotein; FeS, iron-sulfur protein; Fp, flavoprotein; Q, ubiquinone.

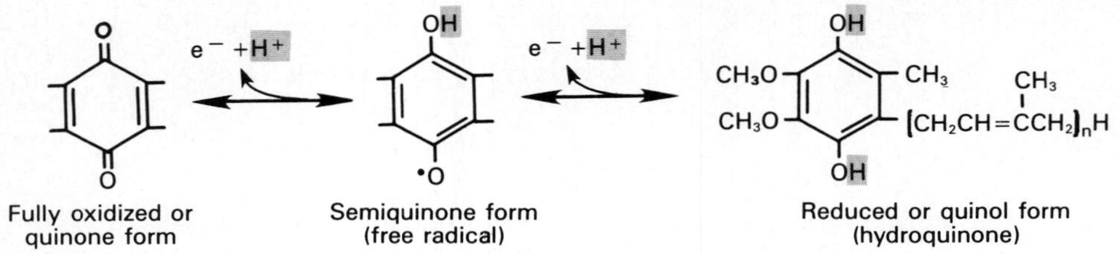

Figure 13-4. Structure of ubiquinone (Q). n = Number of isoprenoid units, which varies from 6 to 10, ie, Q_{6-10}.

drogenases such as L(+)-3-hydroxyacyl-CoA, D(−)-3-hydroxybutyrate, proline, glutamate, malate, and isocitrate dehydrogenases appear to couple directly with NAD of the respiratory chain.

The reduced NADH of the respiratory chain is in turn oxidized by a metalloflavoprotein enzyme—**NADH dehydrogenase.** This enzyme contains FeS and FMN and is tightly bound to the respiratory chain. Q is the collecting point in the respiratory chain for reducing equivalents derived from other substrates that are linked directly to the respiratory chain through flavoprotein dehydrogenases. These substrates include succinate, choline, glycerol 3-phosphate, sarcosine, dimethylglycine, and acyl-CoA (Fig 13-3). The flavin moiety of all these dehydrogenases appears to be FAD.

Electrons flow from Q, through the series of cytochromes shown in Fig 13-3, to molecular oxygen. The cytochromes are arranged in order of increasing redox potential. The terminal cytochrome aa₃ (cytochrome oxidase) is responsible for the final combination of reducing equivalents with molecular oxygen. It has been noted that this enzyme system contains copper, an essential component of true oxidase enzymes. Cytochrome oxidase has a very high affinity for oxygen, which allows the respiratory chain to

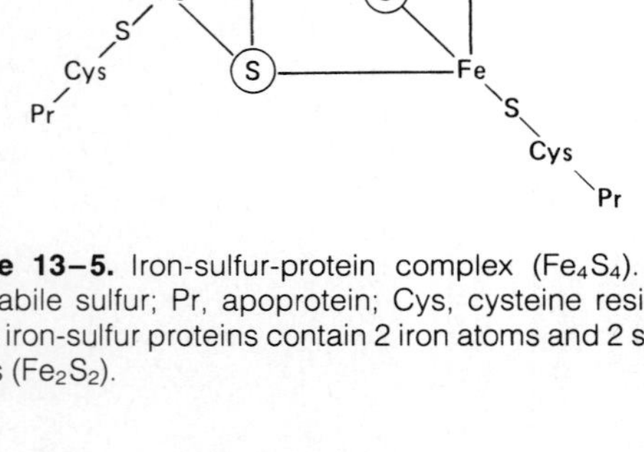

Figure 13-5. Iron-sulfur-protein complex (Fe₄S₄). Ⓢ, acid-labile sulfur; Pr, apoprotein; Cys, cysteine residue. Some iron-sulfur proteins contain 2 iron atoms and 2 sulfur atoms (Fe₂S₂).

function at the maximum rate until the tissue has become virtually depleted of O_2. Since this is an irreversible reaction (the only one in the chain), it gives direction to the movement of reducing equivalents in the respiratory chain and to the production of ATP, to which it is coupled.

The structural organization of the respiratory chain has been the subject of considerable speculation. Of significance is the finding of nearly constant molar proportions between the components. Functionally and structurally, the components of the respiratory chain are present in the inner mitochondrial membrane as 4 protein-lipid **respiratory chain complexes.** These findings have suggested that these components have a definite spatial orientation in the membranes. Cytochrome c is the only soluble cytochrome and, together with Q, seems to be a more mobile component of the respiratory chain connecting the fixed complexes (Fig 13-6).

ROLE OF THE RESPIRATORY CHAIN IN ENERGY CAPTURE

ADP is a molecule that captures, in the form of **high-energy phosphate,** some of the free energy released by catabolic processes. The resulting **ATP** passes on this free energy to drive those processes requiring energy. Thus, ATP has been called the **energy "currency"** of the cell (see Fig 11-8).

There is a net capture of 2 high-energy phosphate groups directly in the glycolytic reactions, equivalent to approximately 61 kJ/mol of glucose (see Table 18-1). Since 1 mol of glucose yields approximately 2780 kJ on complete combustion, the energy captured by phosphorylation in glycolysis is small. The reactions of the citric acid cycle, the final pathway for the complete oxidation of glucose, include one phosphorylation step, the conversion of succinyl-CoA to succinate, which allows the capture of only 2 more high-energy phosphates per mole of glucose. All of the phosphorylations described occur **at the substrate level.** Examination of intact respiring mitochondria reveals that when substrates are oxidized via an NAD-linked dehydrogenase and the respiratory chain, 3 mol of inorganic phosphate are incorporated into 3 mol of ADP to form 3 mol of ATP per ½ mol of O_2 consumed; ie, the P:O ratio = 3 (Fig 13-6). On the other hand,

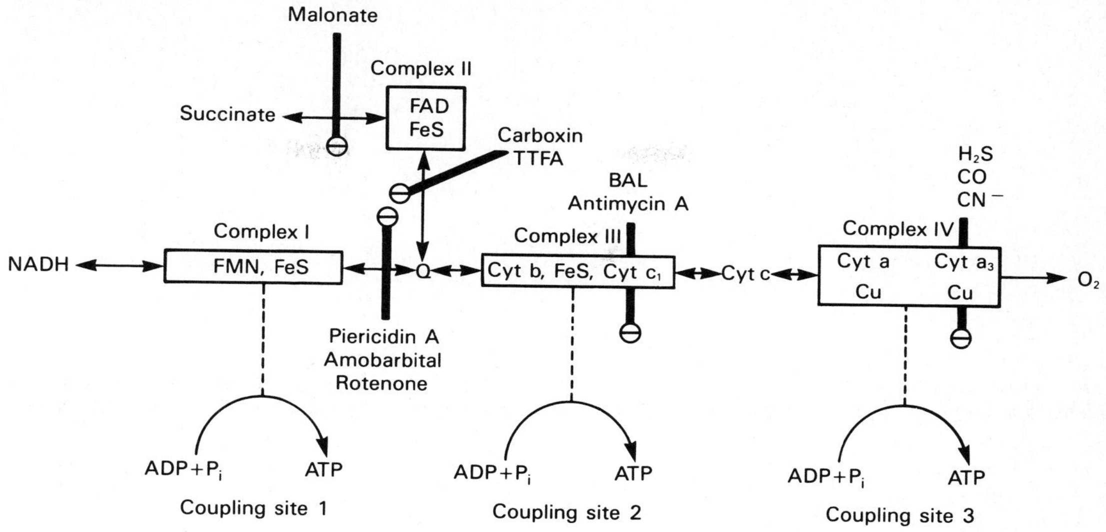

Figure 13−6. Proposed sites of inhibition ($\ominus$) of the respiratory chain by specific drugs, chemicals, and antibiotics. The sites that appear to support phosphorylation are indicated. BAL, dimercaprol. TTFA is an Fe-chelating agent. Complex I, NADH:ubiquinone oxidoreductase; complex II, succinate:ubiquinone oxidoreductase; complex III, ubiquinol: ferricytochrome c oxidoreductase; complex IV, ferrocytochrome c:oxygen oxidoreductase. Other abbreviations as in Fig 13−3.

when a substrate is oxidized via a flavoprotein-linked dehydrogenase, only 2 mol of ATP are formed; ie, P:O = 2. These reactions are known as **oxidative phosphorylation at the respiratory chain level.** Dehydrogenations in the pathway of catabolism of glucose in both glycolysis and the citric acid cycle, plus phosphorylations at the substrate level, can now account for nearly 42% of the free energy resulting from the combustion of glucose, captured in the form of high-energy phosphate. It is evident that the **respiratory chain is responsible for a large proportion of total ATP formation.**

RESPIRATORY CONTROL

The rate of respiration of mitochondria can be controlled by the concentration of ADP. This is because **oxidation and phosphorylation are tightly coupled;** ie, oxidation cannot proceed via the respiratory chain without concomitant phosphorylation of ADP. Chance and Williams have defined 5 conditions that can control the rate of respiration in mitochondria (Table 13−1).

Table 13−1. States of respiratory control.

	Conditions Limiting the Rate of Respiration
State 1	Availability of ADP and substrate
State 2	Availability of substrate only
State 3	The capacity of the respiratory chain itself, when all substrates and components are present in saturating amounts
State 4	Availability of ADP only
State 5	Availability of oxygen only

Generally, most cells in the resting state are in state 4, and respiration is controlled by the availability of ADP. When work is performed, ATP is converted to ADP, allowing more respiration to occur, which in turn replenishes the store of ATP (Fig 13−7). It would appear that under certain conditions the concentration of inorganic phosphate could also affect the rate of functioning of the respiratory chain. As respiration increases (as in exercise), the cell approaches state 3 or state 5 when either the capacity of the respiratory chain becomes saturated or the P_{O_2} decreases below the K_m for cytochrome a_3. There is also the possibility that the ADP/ATP transporter (see p 118), which facilitates entry of cytosolic ADP into the mitochondrion, becomes rate-limiting.

Thus, the manner in which biologic oxidative processes allow the free energy resulting from the oxidation of foodstuffs to become available and to be captured is stepwise, efficient (40−45%), and controlled—rather than explosive, inefficient, and uncontrolled. The remaining free energy that is not captured as high-energy phosphate is liberated as **heat.** This need not be considered as "wasted," since in the warm-blooded animal it contributes to maintenance of body temperature.

INHIBITORS OF THE RESPIRATORY CHAIN & OF OXIDATIVE PHOSPHORYLATION

Much information about the respiratory chain has been obtained by the use of inhibitors, and their proposed loci of action are shown in Fig 13−6. For de-

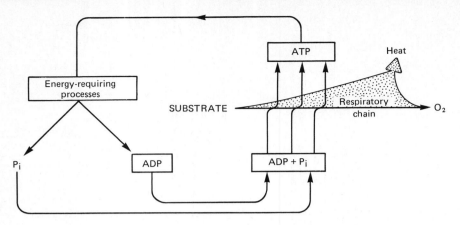

Figure 13–7. The role of ADP in respiratory control.

scriptive purposes, they may be divided into inhibitors of the respiratory chain proper, inhibitors of oxidative phosphorylation, and uncouplers of oxidative phosphorylation.

Inhibitors that arrest respiration by blocking the respiratory chain appear to act at 3 loci. The first is inhibited by **barbiturates** such as **amobarbital,** by the antibiotic **piericidin A,** and by the fish poison **rotenone.** These inhibitors prevent the oxidation of substrates that communicate directly with the respiratory chain via an NAD-linked dehydrogenase, eg, 3-hydroxybutyrate.

Dimercaprol and **antimycin A** inhibit the respiratory chain between cytochrome b and cytochrome c. The classic poisons **H_2S, carbon monoxide,** and **cyanide** inhibit cytochrome oxidase. **Carboxin** and **TTFA** specifically inhibit transfer of reducing equivalents from succinate dehydrogenase to Q, whereas **malonate** is a competitive inhibitor of succinate dehydrogenase.

The antibiotic **oligomycin** completely blocks oxidation and phosphorylation in intact mitochondria. However, in the additional presence of the uncoupler **dinitrophenol,** oxidation proceeds without phosphorylation, indicating that oligomycin does not act directly on the respiratory chain but subsequently on a step in phosphorylation (Fig 13–8).

Atractyloside inhibits oxidative phosphorylation that is dependent on the transport of adenine nucleotides across the inner mitochondrial membrane. It is considered to inhibit the transporter of ADP into the mitochondrion and of ATP out of the mitochondrion (Fig 13–16).

The action of **uncouplers** is to dissociate oxidation in the respiratory chain from phosphorylation. This results in respiration becoming uncontrolled, since the concentration of ADP or P_i no longer limits the rate of respiration. The uncoupler that has been used most frequently is 2,4-dinitrophenol, but other compounds act in a similar manner, including dinitrocresol, pentachlorophenol, and CCCP (*m*-chlorocarbonyl

cyanide phenylhydrazone). The latter, compared with dinitrophenol, is about 100 times as active.

Energy-Linked Transhydrogenase

There is evidence for an energy-linked transhydrogenase that can catalyze the transfer of hydrogen from NADH to NADP. The energy required for the reaction is provided either directly by the respiratory chain in an oligomycin-insensitive process or by ATP, in which case it is blocked by oligomycin.

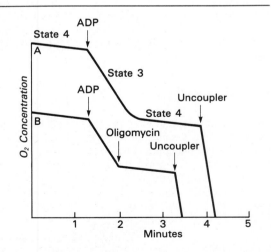

Figure 13–8. Respiratory control in mitochondria. Experiment A shows the basic state of respiration in state 4 that is accelerated upon addition of ADP. When the exogenous ADP has been phosphorylated to ATP, respiration reverts to state 4. The addition of uncoupler, eg, dinitrophenol, releases respiration from phosphorylation. In experiment B, addition of oligomycin blocks phosphorylation of added ADP and therefore of respiration as well. Addition of uncoupler again releases respiration from phosphorylation.

MECHANISM OF OXIDATIVE PHOSPHORYLATION

Two principal hypotheses have been advanced to account for the coupling of oxidation and phosphorylation. The **chemical hypothesis** postulated direct chemical coupling at all stages of the process, as in the reactions that generate ATP in glycolysis. An energy-rich intermediate (I~X) linking oxidation with phosphorylation was envisaged. Because this has never been isolated, the hypothesis has become discredited and will no longer be described in detail in this text. For an account of the chemical hypothesis, see Harper, Rodwell, and Mayes (1979). The **chemiosmotic theory** postulates that oxidation of components in the respiratory chain generates hydrogen ions, which are ejected to the outside of a coupling membrane in the mitochondrion. The electrochemical potential difference resulting from the asymmetric distribution of the hydrogen ions (protons, H$^+$) is used to drive the mechanism responsible for the formation of ATP.

Other hypotheses have been advanced in which it is envisaged that energy from oxidation is conserved in conformational changes of molecules, which in turn lead to the generation of high-energy phosphate bonds.

The Chemiosmotic Theory

According to Mitchell, the primary event in oxidative phosphorylation is the translocation of protons (H$^+$) to the exterior of a coupling membrane (ie, the mitochondrial inner membrane), driven by oxidation in the respiratory chain. It is also postulated that the membrane is impermeable to ions in general but particularly to protons which accumulate outside the membrane, creating an **electrochemical potential difference across the membrane** ($\Delta\mu_{H^+}$). This consists of a chemical potential (difference in pH) and an electrical potential. The electrochemical potential difference is used to drive a **membrane-located ATP synthase** (or the reversal of a membrane-located ATP hydrolase) which in the presence of P$_i$ + ADP forms ATP (Fig 13–9). Thus, there is no high-energy intermediate that is common to both oxidation and phosphorylation as in the chemical hypothesis.

It is proposed that the respiratory chain is folded into 3 oxidation/reduction (**o/r**) loops in the membrane, each loop corresponding to respiratory chain complex I, III, and IV, respectively. An idealized single loop consisting of a hydrogen carrier and an electron carrier is shown in Fig 13–10. A possible configuration of the respiratory chain folded into 3 functional o/r loops is shown in Fig 13–11.

In this scheme, each electron pair transferred from

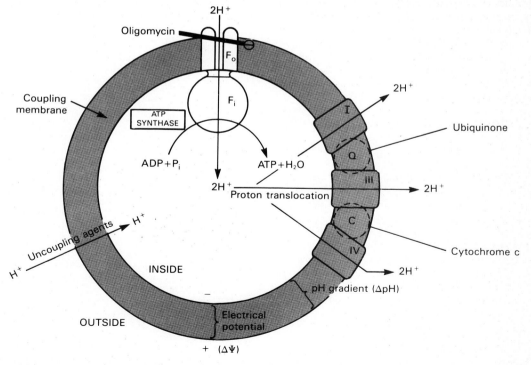

Figure 13–9. Principles of the chemiosmotic theory of oxidative phosphorylation. F$_i$, F$_o$, protein subunits responsible for phosphorylation. The main proton circuit is created by the coupling of oxidation to proton translocation from the inside to the outside of the membrane, driven by the respiratory chain complexes I, III, and IV, each of which acts as a proton pump. Uncoupling agents such as dinitrophenol allow leakage of H$^+$ across the membrane, thus collapsing the electrochemical proton gradient. Oligomycin specifically blocks conduction of H$^+$ through F$_o$.

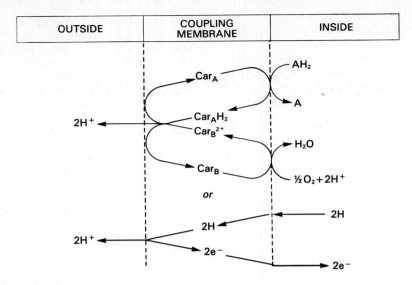

Figure 13–10. Proton-translocating oxidation/reduction (o/r) loop (chemiosmotic theory).

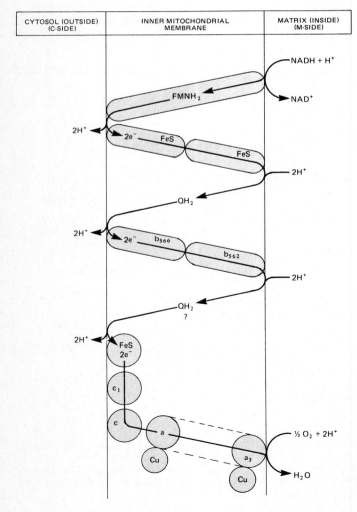

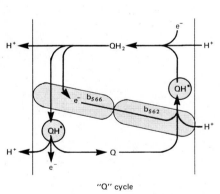

"Q" cycle

Figure 13–11. Possible configuration of o/r loops in the respiratory chain (chemiosmotic theory). Much of this scheme is still tentative, particularly around the Q/cytochrome b region, where the exact nature and relative positions of the intermediates are not known with certainty. It is possible that the semiquinone ($QH^•$) is involved in a proton-motive "Q" cycle as indicated on the right. $QH^•$ is anchored on each side of the membrane by attachment to a Q-binding protein, whereas QH_2 and Q are mobile. Cytochromes are shown respectively as b, c_1, c, a, and a_3. (The last is part of cytochrome aa_3, which traverses the membrane.) FeS, iron-sulfur protein. More recent work suggests that cytochrome c oxidoreductase acts as a proton pump as in Fig13–9.

NADH to oxygen causes 6 protons to be translocated from the inside to the outside of the mitochondrial membrane. NADH first donates one proton and 2 electrons, which, together with another proton from the internal medium, reduce FMN to $FMNH_2$. FMN is part of a large protein complex that is considered to extend the full width of the membrane, enabling it to release 2 protons to the outside of the membrane and then to return 2 electrons to the inside surface via FeS proteins that become reduced. Each reduced FeS complex donates one electron to a ubiquinone (Q) molecule which, upon taking up a proton from inside the membrane, forms QH_2. Being lipid-soluble and a small molecule, QH_2 is free to move to the outside of the membrane, where it discharges a proton pair into the cytosol and donates 2 electrons to 2 molecules of the next carrier in the respiratory chain, cytochrome b. This electron carrier is thought to span the mitochondrial membrane (as cytochromes b_{566} and b_{562}), enabling the electrons to join another molecule of ubiquinone together with 2 more protons from the internal medium. The resulting QH_2 shuttles to the outer surface, where 2 protons are liberated and 2 electrons passed to 2 molecules of cytochrome c. The more recently proposed "Q" cycle, for which there is convincing evidence, obviates the necessity of QH_2 operating the H-transporting limbs of 2 separate o/r loops. The

electrons then pass through the remainder of the cytochrome chain, traversing the membrane to cytochrome a_3, which lies on the inside of the membrane. At this site, 2 electrons combine with two H^+ from the internal medium and an oxygen atom to form water.

The inner membrane contains the enzyme proteins of the respiratory chain arranged in a sided manner as indicated in Fig 13–11. Scattered over the surface of the inner membrane are the phosphorylating subunits responsible for the production of ATP (Fig 13–12). These consist of several proteins, collectively known as an F_i subunit, which project into the matrix and which contain the ATP synthase (Fig 13–9). These subunits are attached, possibly by a stalk, to a membrane protein subunit known as F_o, which probably extends through the membrane (Fig 13–9). For every proton pair passing through the F_o–F_i complex, one ATP molecule is formed from ADP and P_i. It is of interest that similar phosphorylating units are found inside the plasma membrane of bacteria but outside the thylakoid membrane of chloroplasts. It is significant that the proton gradient is from outside to inside in mitochondria and bacteria but in the reverse direction in chloroplasts.

The mechanism of coupling of proton translocation to the anisotropic (vectorial) ATP synthase system is

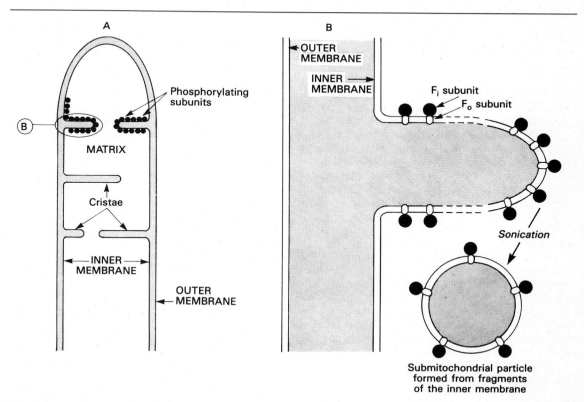

Figure 13–12. Structure of the mitochondrial membranes. Submitochondrial particles are "inside out" and allow study of an enclosed membrane system where the phosphorylating subunits are on the outside and where the proton gradient is reversed.

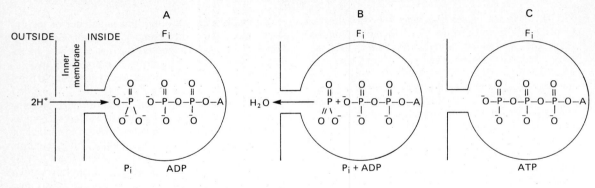

Figure 13–13. Proton-translocating ATP synthase (Mitchell).

conjectural. One model suggested by Mitchell is shown in Fig 13–13. A proton pair attacks one oxygen of P_i to form H_2O and an active form of P_i, which immediately combines with ADP to form ATP. Other studies have suggested that ATP synthesis is not the main energy-requiring step—rather it is the release of ATP from the active site. This may involve conformational changes in the F_1 subunit.

The following experimental findings support the chemiosmotic theory:

(1) Addition of protons (acid) to the external medium of mitochondria leads to the generation of ATP.

(2) Oxidative phosphorylation does not occur in soluble systems where there is no possibility of a vectorial ATP synthase. A closed membrane must be present in order to obtain oxidative phosphorylation (Fig 13–9).

(3) The respiratory chain contains components organized in a sided manner (transverse asymmetry) as required by the chemiosmotic theory (Fig 13–11).

(4) The $P:H^+$ (transported out) quotient of the ATP synthase is 1:2, and the H^+ (transported out):O quotients for succinate and 3-hydroxybutyrate oxidation are 4 and 6, respectively, conforming with the expected P:O ratios of 2 and 3, respectively. These ratios are compatible with the existence of 3 o/r loops in the respiratory chain.

The chemiosmotic theory can account for the following:

(1) The phenomenon of respiratory control. The electrochemical potential difference across the membrane, once established as a result of proton translocation, inhibits further transport of reducing equivalents through the respiratory chain unless discharged by back-translocation of protons across the membrane through the vectorial ATP synthase. This in turn depends on availability of ADP and P_i.

(2) The action of uncouplers. These components (eg, dinitrophenol) are amphipathic (see p 140) and increase the permeability of mitochondria to protons (Fig 13–9), thus reducing the electrochemical potential and short-circuiting the ATP synthase. Thus, oxidation can proceed without phosphorylation.

(3) The existence of mitochondrial exchange transporter systems (see below). These are a consequence of the coupling membrane, which must be impermeable to protons and other ions in order to maintain the electrochemical gradient. Exchange diffusion systems are present in the membrane for exchange of anions against OH^- ions and cations against H^+ ions. Such systems are necessary for uptake and output of ionized metabolites while preserving electrical and osmotic neutrality.

MITOCHONDRIAL TRANSPORT SYSTEMS

Mitochondrial Membranes & the Location of Important Enzymes

Mitochondria have an outer membrane that is permeable to most metabolites, an inner membrane which is selectively permeable and which is thrown into folds or cristae, and a matrix within the inner membrane (Fig 13–12). The outer membrane may be removed by treatment with digitonin and is characterized by the presence of monoamine oxidase and a few other enzymes (eg, acyl-CoA synthetase, glycerophosphate acyltransferase, monoacyl glycerophosphate acyltransferase, phospholipase A_2). Adenylate kinase and creatine kinase are found in the intermembrane space. The phospholipid cardiolipin is concentrated in the inner membrane.

The soluble enzymes of the citric acid cycle and the enzymes of β-oxidation of fatty acids are found in the matrix, necessitating mechanisms for transporting metabolites and nucleotides across the inner membrane. Succinate dehydrogenase is found on the inner surface of the inner mitochondrial membrane, where it transports reducing equivalents into the respiratory chain at ubiquinone, bypassing the first o/r loop. 3-Hydroxybutyrate dehydrogenase is also bound to the matrix side of the inner mitochondrial membrane. Glycerol-3-phosphate dehydrogenase is found on the outer surface of the inner membrane, where it is suitably located to participate in the glycerophosphate shuttle (Fig 13–14).

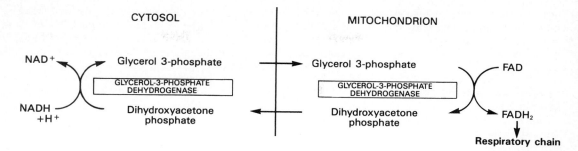

Figure 13–14. Glycerophosphate shuttle for transfer of reducing equivalents from the cytosol into the mitochondrion.

Oxidation of Extramitochondrial NADH by Substrate Shuttles

NADH cannot penetrate the mitochondrial membrane, but it is produced continuously in the cytosol by 3-phosphoglyceraldehyde dehydrogenase, an enzyme in the glycolysis sequence (see Fig 18–2). However, under aerobic conditions, extramitochondrial NADH does not accumulate and is presumed to be oxidized by the respiratory chain in mitochondria. Several possible mechanisms have been considered to permit this process. These involve transfer of reducing equivalents through the mitochondrial membrane via substrate pairs, linked by suitable dehydrogenases. It is necessary that the specific dehydrogenase be present on both sides of the mitochondrial membrane. The mechanism of transfer using the **glycerophosphate shuttle** is shown in Fig 13–14. It is to be noted that since the mitochondrial enzyme is linked to the respiratory chain via a flavoprotein rather than NAD, only 2 rather than 3 mol of ATP are formed per atom of oxygen consumed. In some species, the activity of the FAD-linked enzyme decreases after thyroidectomy and in-

creases after administration of thyroxine. Although this shuttle is present in insect flight muscle and in white muscle and might be important in liver, in other tissues (eg, heart muscle) the mitochondrial glycerol-3-phosphate dehydrogenase is deficient. It is therefore believed that a transport system involving malate and cytosolic and mitochondrial malate dehydrogenase is of more universal utility. The **malate "shuttle"** system is shown in Fig 13–15. The complexity of this system is due to the impermeability of the mitochondrial membrane to oxaloacetate, which must react with glutamate and transaminate to aspartate and α-ketoglutarate before transport through the mitochondrial membrane and reconstitution to oxaloacetate in the cytosol.

Energy-Linked Ion Transport in Mitochondria

Actively respiring mitochondria in which oxidative phosphorylation is taking place maintain or accumulate cations such as K^+, Na^+, Ca^{2+}, and Mg^{2+}, and P_i. Uncoupling with dinitrophenol leads to loss of ions

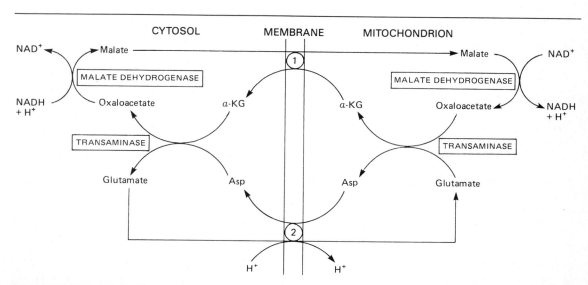

Figure 13–15. Malate shuttle for transfer of reducing equivalents from the cytosol into the mitochondrion. 1, Ketoglutarate transporter; 2, glutamate-aspartate transporter (note the proton symport).

from the mitochondria, but the ion uptake is not inhibited by oligomycin, suggesting that the energy need not be supplied by phosphorylation of ADP. It is envisaged that a primary proton pump drives cation exchange.

Transporter Systems
(See Fig 13–16.)

The inner bilipoid mitochondrial membrane is freely permeable to uncharged small molecules, such as oxygen, water, CO_2, and NH_3, and to monocarboxylic acids, such as 3-hydroxybutyric, acetoacetic, and acetic. Long-chain fatty acids are transported into mitochondria via the carnitine system (see Fig 23–1), and there is also a special carrier for pyruvate involving a symport that utilizes the H^+ gradient from outside to inside the mitochondrion. However, dicarboxylate and tricarboxylate anions and amino acids require specific transporter or carrier systems to facilitate their transport across the membrane. It appears that monocarboxylate anions penetrate more readily, because of the lesser degree of dissociation of these acids. It is the undissociated and more lipid-soluble acid that is thought to be the molecular species that penetrates the lipid membrane.

The transport of di- and tricarboxylate anions is closely linked to that of inorganic phosphate, which penetrates readily as the $H_2PO_4^-$ ion in exchange for OH^-. The net uptake of malate by the dicarboxylate transporter requires inorganic phosphate for exchange in the opposite direction. The net uptake of citrate, isocitrate, or *cis*-aconitate by the tricarboxylate transporter requires malate in exchange. α-Ketoglutarate transport also requires an exchange with malate. Thus, by the use of exchange mechanisms, osmotic balance is maintained. It will be appreciated that citrate transport across the mitochondrial membrane depends not only on malate transport but on the transport of inorganic phosphate as well. The adenine nucleotide transporter allows the exchange of ATP and ADP but not AMP. It is vital in allowing ATP exit from mitochondria to the sites of extramitochondrial utilization and in allowing the return of ADP for ATP production within the mitochondrion (Fig 13–17). Na^+ can be exchanged for H^+, driven by the proton gradient. It is believed that active uptake of Ca^{2+} by mitochondria occurs with a net charge transfer of 1 (Ca^+ uniport), possibly through a Ca^{2+}/H^+ antiport. Calcium release

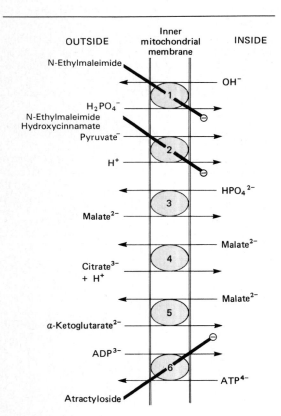

Figure 13–16. Transporter systems in the mitochondrial membrane. 1, Phosphate transporter; 2, pyruvate symport; 3, dicarboxylate transporter; 4, tricarboxylate transporter; 5, α-ketoglutarate transporter; 6, adenine nucleotide transporter. N-Ethylmaleimide, hydroxycinnamate, and atractyloside inhibit ($\ominus$) the indicated systems. Also present (but not shown) are transporter systems for glutamate/aspartate (Fig 13–15), glutamine, ornithine, and carnitine (Fig 23–1).

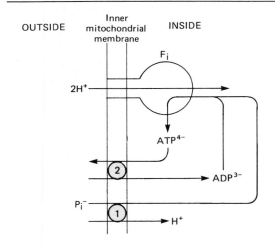

Figure 13–17. Combination of phosphate transporter (1) with the adenine nucleotide transporter (2) in ATP synthesis. The H^+/P_i symport shown is equivalent to the P_i/OH^- antiport shown in Fig 13–16. Three protons are taken into the mitochondrion for each ATP exported. However, only 2 protons are taken in when ATP is used inside the mitochondrion. This model conforms with a stoichiometry of 3 protons transported per electron pair at each coupling site instead of 2 protons (Cross), or it agrees with the original Mitchell hypothesis (Fig 13–11) of 2 protons per electron pair at each coupling site and revised values for the P/O ratio of 2 for NADH-linked oxidations and 1.3 for succinate oxidation (Hinkle).

from mitochondria is facilitated by exchange with Na^+.

Action of Ionophores

These substances are so termed because of their ability to complex specific cations and facilitate their transport through biologic membranes. This property of ionophoresis is due to their lipophilic character, which allows penetration of lipoid membranes such as the mitochondrial membrane. An example is the antibiotic **valinomycin,** which allows penetration of K^+ through the mitochondrial membrane and then discharges the membrane potential between the inside and the outside of the mitochondrion. **Nigericin** also

acts as an ionophore for K^+ but in exchange for H^+. It therefore abolishes the pH gradient across the membrane. In the presence of both valinomycin and nigericin, both the membrane potential and the pH gradient are eliminated, and phosphorylation is therefore completely inhibited. The classic uncouplers such as dinitrophenol are, in fact, proton ionophores.

Dysfunction of the Respiratory Chain

The condition of **fatal infantile mitochondrial myopathy and renal dysfunction** involves severe diminution or absence of most oxidoreductases of the respiratory chain.

REFERENCES

Cross RL: The mechanism and regulation of ATP synthesis by F_i-ATPases. *Annu Rev Biochem* 1981;**50:**681.

Harper HA, Rodwell VW, Mayes PA: Page 276 in: *Review of Physiological Chemistry,* 17th ed. Lange, 1979.

Hatefi Y: The mitochondrial electron transport and oxidative phosphorylation system. *Annu Rev Biochem* 1985;**54:**1015.

Hinkle PC, McCarty RE: How cells make ATP. *Sci Am* (March) 1978;**238:**104.

Hinkle PC, Yu ML: The phosphorus/oxygen ratio of mitochondrial oxidative phosphorylation. *J Biol Chem* 1979;**254:**2450.

Mitchell P: Keilin's respiratory chain concept and its chemiosmotic consequences. *Science* 1979;**206:**1148.

Nicholls DG: *Bioenergetics: An Introduction to the Chemiosmotic Theory.* Academic Press, 1982.

Tyler DD: The mitochondrial ATP synthase. Page 117 in: *Membrane Structure and Function.* Vol 5. Bittar EE (editor). Wiley, 1984.

Tyler DD, Sutton CM: Mitochondrial transporting systems. Page 181 in: *Membrane Structure and Function.* Vol 5. Bittar EE (editor). Wiley, 1984.

Carbohydrates of Physiologic Significance

Peter A. Mayes, PhD, DSc

INTRODUCTION

Carbohydrates are widely distributed in plants and animals, where they fulfill both structural and metabolic roles. In plants, glucose is synthesized from carbon dioxide and water by photosynthesis and stored as starch or is converted to the cellulose of the plant framework. Animals can synthesize some carbohydrate from fat and protein, but the bulk of animal carbohydrate is derived ultimately from plants.

BIOMEDICAL IMPORTANCE

A knowledge of the structure and properties of the carbohydrates of physiologic significance is essential to understanding their fundamental role in the economy of the mammalian organism. The sugar glucose is the most important carbohydrate. It is as glucose that the bulk of dietary carbohydrate is absorbed into the bloodstream or into which it is converted in the liver, and it is from glucose that all other carbohydrates in the body can be formed. Glucose is a major fuel of the tissues of mammals (except ruminants) and a universal fuel of the fetus. It is converted to other carbohydrates having highly specific functions, eg, glycogen for storage; ribose in nucleic acids; galactose in lactose of milk, in certain complex lipids, and in combination with protein in glycoprotein and proteoglycans. Diseases associated with carbohydrates include diabetes mellitus, galactosemia, glycogen storage diseases, and milk intolerance.

CLASSIFICATION OF CARBOHYDRATES

Carbohydrates may be defined chemically as aldehyde or ketone derivatives of the polyhydric (more than one OH group) alcohols or as compounds that yield these derivatives on hydrolysis.

(1) Monosaccharides are those carbohydrates that cannot be hydrolyzed into a simpler form. They may be subdivided into **trioses, tetroses, pentoses, hexoses, heptoses,** or **octoses,** depending upon the number of carbon atoms they possess; and as **aldoses** or **ketoses,** depending upon whether the aldehyde or ketone group is present. Examples are

		Aldoses	Ketoses
Trioses	$(C_3H_6O_3)$	Glycerose	Dihydroxyacetone
Tetroses	$(C_4H_8O_4)$	Erythrose	Erythrulose
Pentoses	$(C_5H_{10}O_5)$	Ribose	Ribulose
Hexoses	$(C_6H_{12}O_6)$	Glucose	Fructose

(2) Disaccharides yield 2 molecules of the same or of different monosaccharide(s) when hydrolyzed. Examples are sucrose, lactose, and maltose.

(3) Oligosaccharides yield 3–6 monosaccharide units on hydrolysis. Maltotriose* is an example.

(4) Polysaccharides yield more than 6 molecules of monosaccharides on hydrolysis. Examples of polysaccharides, which may be linear or branched, are the starches and dextrins. These are sometimes designated as hexosans or pentosans, or homopolysaccharides or heteropolysaccharides, depending upon the identity of the monosaccharides they yield on hydrolysis.

MONOSACCHARIDES

Structure of Glucose

The straight-chain structural formula (aldohexose, Fig 14–1A) can account for some of the properties of glucose, but a cyclic structure is favored on thermodynamic grounds and accounts completely for its chemical properties. For most purposes, the structural formula may be represented as a simple ring in perspective as proposed by Haworth (Fig 14–1B). X-ray diffraction analysis shows that the 6-membered ring containing one oxygen atom is actually in the form of a chair (Fig 14–1C).

Isomerism in Sugars

Compounds that have the same structural formula but differ in spatial configuration are known as **stereoisomers.** The presence of asymmetric carbon atoms (carbon atoms attached to 4 different atoms or groups) allows the formation of isomers. The number of possible isomers of a compound depends on the number of asymmetric carbon atoms (n) and is equal to 2^n. Glucose, with 4 asymmetric carbon atoms, therefore has 16 isomers. The more important types of isomerism found with glucose are as follows:

---

*Note that this is not a true triose but a trisaccharide containing 3 α-glucose residues.

A

$$
\begin{array}{c}
\overset{1}{C} - H \\
\parallel \\
O
\end{array}
$$

```
        O
        ‖
       ¹C —H
  H— ²C —OH
 HO— ³C —H
  H— ⁴C —OH
  H— ⁵C —OH
       ⁶CH₂OH
```

B

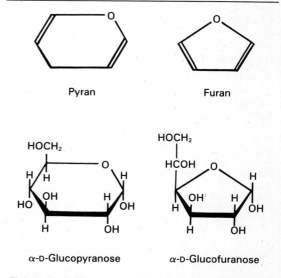

α-D-glucose.

C

Figure 14–1. α-D-glucose.

```
        O                          O
        ‖                          ‖
       ¹C —H                      ¹C —H
 HO— ²C —H               H— C — OH
       ³CH₂OH                    CH₂OH

   L-Glycerose                 D-Glycerose
 (L-glyceraldehyde)        (D-glyceraldehyde)


        O                          O
        ‖                          ‖
       ¹C —H                       C —H
 HO— ²C —H               H—C —OH
  H— ³C —OH             HO—C —H
 HO— ⁴C —H               H—C —OH
 HO— ⁵C —H               H—C — OH
       ⁶CH₂OH                    CH₂OH

   L-Glucose                   D-Glucose
```

Figure 14–2. D- and L- isomerism of glycerose and glucose.

(1) D and L: The designation of an isomer as the D form or of its mirror image as the L form is determined by its spatial relationship to the parent compound of the carbohydrate family, the 3-carbon sugar glycerose. The L and D forms of this sugar are shown in Fig 14–2 together with the corresponding isomers of glucose. The orientation of the –H and –OH groups around the carbon atom **adjacent** to the terminal primary alcohol carbon (eg, carbon atom 5 in glucose) determines whether the sugar belongs to the D or L series. When the –OH group on this carbon is on the right (as seen in Fig 14–2), the sugar is a member of the D series; when it is on the left, it is a member of the L series. Most of the monosaccharides occurring in mammals are of the D configuration, and enzymes responsible for their metabolism are specific for this configuration.

The presence of asymmetric carbon atoms also confers **optical activity** on the compound. When a beam of plane-polarized light is passed through a solution of an **optical isomer,** it will be rotated either to the right, dextrorotatory $(+)$, or to the left, levorotatory $(-)$. A compound may be designated $D(-)$, $D(+)$, $L(-)$, or $L(+)$, indicating structural relationship to D or L glycerose but not necessarily exhibiting the same optical rotation. For example, the naturally occurring form of fructose is the $D(-)$ isomer.

When equal amounts of D and L isomers are present, the resulting mixture has no optical activity, since the activities of each isomer cancel one another.

Such a mixture is said to be a **racemic**—or DL—mixture. Synthetically produced compounds are necessarily racemic because the opportunities for the formation of each optical isomer are identical.

(2) Pyranose and Furanose Ring Structures: This terminology is based on the fact that the stable ring structures of monosaccharides are similar to the ring structures of either pyran or furan (Fig 14–3). Ketoses may also show ring formation (eg, D-fructofura-

Pyran

Furan

```
  HOCH₂                      HOCH₂
                               |
                             HCOH
            O                        O
                                          H
   OH         H          OH       H
HO         OH          H       OH
   H          OH          H        OH
```

α-D-Glucopyranose α-D-Glucofuranose

Figure 14–3. Pyranose and furanose forms of glucose.

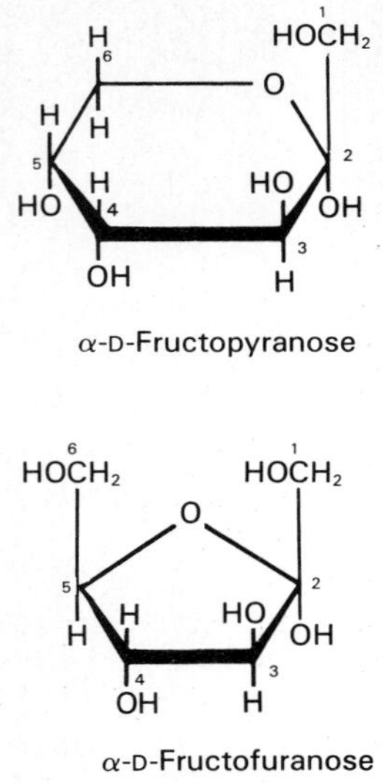

α-D-Fructopyranose

α-D-Fructofuranose

Figure 14–4. Pyranose and furanose forms of fructose.

nose or D-fructopyranose) (Fig 14–4). In the case of glucose in solution, more than 99% is in the pyranose form; thus, less than 1% is in the furanose form.

(3) α and β Anomers: The ring structure of an aldose is a hemiacetal, since it is formed by combination of an aldehyde and an alcohol group (Fig 14–5). Similarly, the ring structure of a ketose is a hemiketal. Crystalline glucose is α-D-glucopyranose. The cyclic structure is retained in solution, but isomerism takes place about position 1, the carbonyl or **anomeric carbon atom,** to give a mixture of α-glucopyranose (36%) and β-glucopyranose (63%), the remaining 1% represented mainly by α- and β-anomers of glucofuranose. This equilibration is accompanied by optical rotation **(mutarotation)** as the hemiacetal ring opens

and re-forms with change of position of the −H and −OH groups on carbon 1. The change probably takes place via a hydrated straight-chain acyclic molecule, although polarography has indicated that glucose exists only to the extent of 0.0025% in the acyclic form. The optical rotation of glucose in solution is dextrorotatory; hence, the alternative name of **dextrose,** often used in clinical practice.

(4) Epimers: Isomers differing as a result of variations in configuration of the −OH and −H on carbon atoms 2, 3, and 4 of glucose are known as epimers. Biologically, the most important epimers of glucose are mannose and galactose, formed by epimerization at carbons 2 and 4, respectively (Fig 14–6).

(5) Aldose-Ketose Isomerism: Fructose has the same molecular formula as glucose but differs in its structural formula, since there is a potential keto group in position 2 of fructose, whereas there is a potential aldehyde group in position 1 of glucose (Figs 14–3 and 14–4).

Monosaccharides of Physiologic Importance

Derivatives of trioses are formed in the course of the metabolic breakdown of glucose by the glycolysis pathway. Derivatives of trioses, tetroses, and pentoses and of a 7-carbon sugar (sedoheptulose) are formed in the breakdown of glucose via the pentose phosphate pathway. Pentose sugars are important constituents of nucleotides, nucleic acids, and many coenzymes (Table 14–1). Of the hexoses, glucose, galactose, fructose, and mannose are physiologically the most important (Table 14–2).

The structures of the aldo sugars of biochemical significance are shown in Fig 14–7. Five keto sugars which are important in metabolism are shown in Fig 14–8.

Of additional significance are carboxylic acid derivatives of glucose such as D-glucuronate (impor-

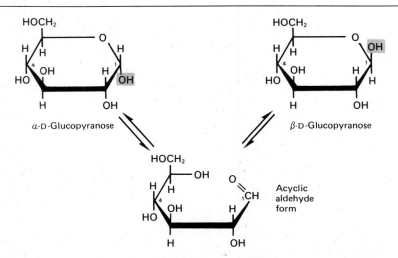

α-D-Glucopyranose   β-D-Glucopyranose

Acyclic aldehyde form

Figure 14–5. Mutarotation of glucose.

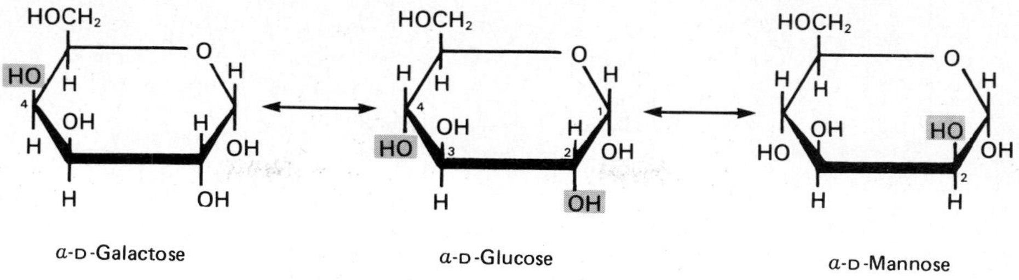

Figure 14-6. Epimerization of glucose.

tant in glucuronide formation and present in glycosaminoglycans) and its metabolic derivatives, L-iduronate (present in glycosaminoglycans) (Fig 14–9) and L-gulonate (a member of the uronic acid pathway; see Fig 21–1).

Glycosides

Glycosides are compounds formed from a condensation between a monosaccharide, or monosaccharide residue, and the hydroxyl group of a second compound that may, or may not (in the case of an **agly-cone**), be another monosaccharide. The **glycosidic bond** is an **acetal** link because it results from a reaction between a hemiacetal group (formed from an aldehyde and an −OH group) and another −OH group. If the hemiacetal portion is glucose, the resulting compound is a **glucoside;** if galactose, a **galactoside,** etc.

Glycosides are found in many drugs and spices and in the constituents of animal tissues. The aglycone may be methanol, glycerol, a sterol, or a phenol. The glycosides which are important in medicine because of

Table 14-1. Pentoses of physiologic importance.

Sugar	Where Found	Biochemical Importance	Clinical Significance
D-Ribose	Nucleic acids.	Structural elements of nucleic acids and coenzymes, eg, ATP, NAD, NADP, flavoproteins. Intermediate in pentose phosphate pathway.	
D-Ribulose	Formed in metabolic processes.	Intermediate in pentose phosphate pathway.	
D-Arabinose	Gum arabic. Plum and cherry gums.	Constituent of glycoproteins.	
D-Xylose	Wood gums, proteoglycans, glycosaminoglycans.	Constituent of glycoproteins.	
D-Lyxose	Heart muscle.	A constituent of a lyxoflavin isolated from human heart muscle.	
L-Xylulose	Intermediate in uronic acid pathway.		Found in urine in essential pentosuria.

Table 14-2. Hexoses of physiologic importance.

Sugar	Source	Importance	Clinical Significance
D-Glucose	Fruit juices. Hydrolysis of starch, cane sugar, maltose, and lactose.	The "sugar" of the body. The sugar carried by the blood, and the principal one used by the tissues.	Present in the urine (glycosuria) in diabetes mellitus owing to raised blood glucose (hyperglycemia).
D-Fructose	Fruit juices. Honey. Hydrolysis of cane sugar and of inulin (from the Jerusalem artichoke).	Can be changed to glucose in the liver and intestine and so used in the body.	Hereditary fructose intolerance leads to fructose accumulation and hypoglycemia.
D-Galactose	Hydrolysis of lactose.	Can be changed to glucose in the liver and metabolized. Synthesized in the mammary gland to make the lactose of milk. A constituent of glycolipids and glycoproteins.	Failure to metabolize leads to galactosemia and cataract.
D-Mannose	Hydrolysis of plant mannans and gums.	A constituent of many glycoproteins.	

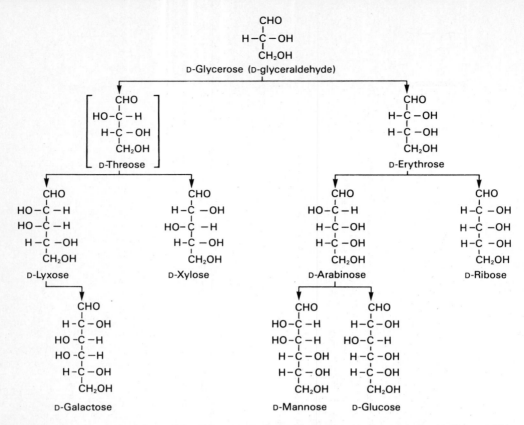

Figure 14–7. The structural relations of the aldoses, D series. D-Threose is not of physiologic significance. The series is built up by the theoretical addition of a CH_2O unit to the —CHO group of the sugar.

their action on the heart **(cardiac glycosides)** all contain steroids as the aglycone component. These include derivatives of digitalis and strophanthus such as **ouabain,** an inhibitor of the Na^+/K^+-ATPase of cell membranes. Other glycosides include antibiotics such as **streptomycin** (Fig 14–10).

Deoxy Sugars

Deoxy sugars are those in which a hydroxyl group attached to the ring structure has been replaced by a hydrogen atom. They are obtained on hydrolysis of certain substances that are important in biologic processes. An example is the **deoxyribose** (Fig 14–11) occurring in nucleic acids (DNA).

Also found as a carbohydrate of glycoproteins is L-fucose (Fig 14–17), and of importance as an inhibitor of glucose metabolism is 2-deoxyglucose.

Amino Sugars (Hexosamines)

Examples of amino sugars are D-glucosamine (Fig

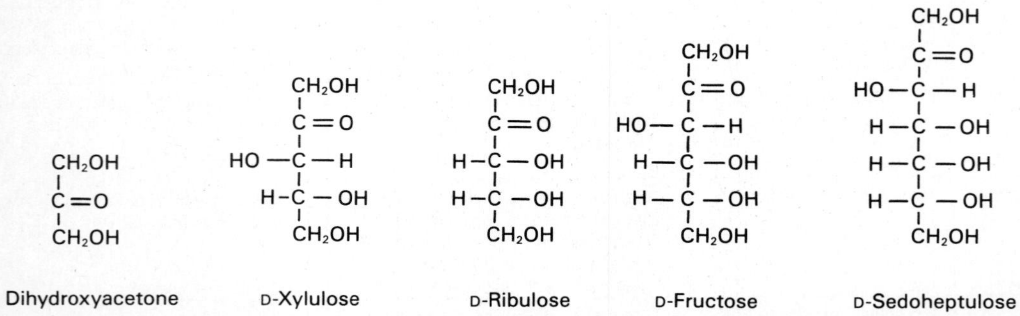

Figure 14–8. Examples of ketoses.

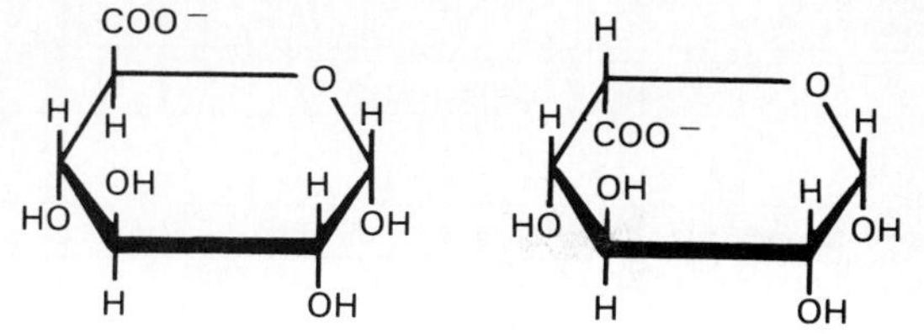

Figure 14-9. α-D-Glucuronate *(left)* and β-L-iduronate *(right).*

14-12), D-galactosamine, and D-mannosamine, all of which have been identified in nature. Glucosamine is a constituent of hyaluronic acid. Galactosamine (chondrosamine) is a constituent of chondroitin (see Chapter 54).

Several antibiotics (erythromycin, carbomycin) contain amino sugars. The amino sugars are believed to be related to the antibiotic activity of these drugs.

DISACCHARIDES

The disaccharides are sugars composed of 2 monosaccharide residues united by a glycosidic linkage (Fig 14-13). Their chemical name reflects their component monosaccharides. The physiologically important disaccharides are maltose, sucrose, lactose, and trehalose (Table 14-3).

Hydrolysis of sucrose yields a crude mixture called "invert sugar" because the strongly levorotatory fructose thus produced changes (inverts) the previous dextrorotatory action of the sucrose.

POLYSACCHARIDES

Polysaccharides include the following physiologically important carbohydrates.

Starch is formed of an α-glucosidic chain. Such a compound, yielding only glucose on hydrolysis, is a homopolymer called a **glucosan** or **glucan.** It is the most important food source of carbohydrate and is found in cereals, potatoes, legumes, and other vegetables. The 2 chief constituents are **amylose** (15-20%), which is a nonbranching helical structure (Fig 14-14), and **amylopectin** (80-85%), which consists of branched chains composed of 24-30 glucose residues united by 1→4 linkages in the chains and by 1→6 linkages at the branch points.

Glycogen (Fig 14-15) is the storage polysaccharide of the animal body. It is often called animal starch. It is a more highly branched structure than amylopectin and has chains of 11-18-α-D-glucopyranose residues (in α[1→4]-glucosidic linkage) with branching by means of α(1→6)-glucosidic bonds.

Inulin is a starch found in tubers and roots of dahlias, artichokes, and dandelions. It is hydrolyzable to fructose, and hence it is a fructosan. This starch, unlike potato starch, is easily soluble in warm water and has been used in physiologic investigation for determination of the rate of glomerular filtration.

Dextrins are substances formed in the course of the hydrolytic breakdown of starch. Limit dextrins are the first formed products as hydrolysis reaches a certain degree of branching.

Cellulose is the chief constituent of the framework of plants. It is not soluble in ordinary solvents and consists of β-D-glucopyranose units linked by β (1→4) bonds to form long, straight chains strengthened by

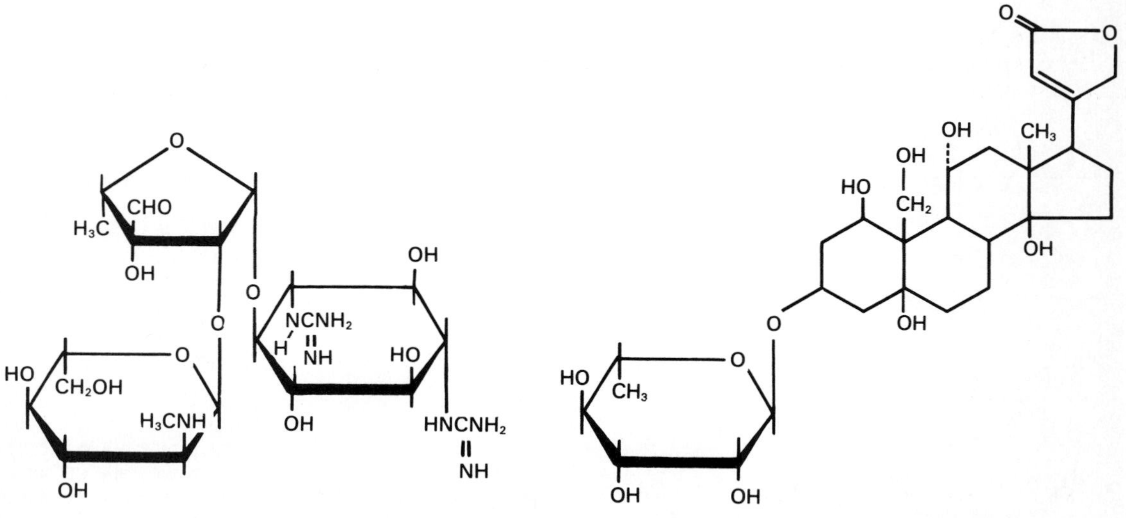

Figure 14-10. Streptomycin *(left)* and ouabain *(right).*

Table 14–3. Disaccharides.

Sugar	Source	Clinical Significance
Maltose	Digestion by amylase or hydrolysis of starch. Germinating cereals and malt.	
Lactose	Milk. May occur in urine during pregnancy.	In lactase deficiency, malabsorption leads to diarrhea and flatulence.
Sucrose	Cane and beet sugar. Sorghum. Pineapple. Carrot roots.	In sucrase deficiency, malabsorption leads to diarrhea and flatulence.
Trehalose	Fungi and yeasts. The major sugar of insect hemolymph.	

Figure 14–11. 2-Deoxy-D-ribofuranose (β form).

Figure 14–12. Glucosamine (2-amino-D-glucopyranose) (α form). Galactosamine is 2-amino-D-galactopyranose. Both glucosamine and galactosamine occur as N-acetyl derivatives in more complex carbohydrates, eg, glycoproteins.

Maltose

O-α-D-Glucopyranosyl-(1→4)-α-D-glucopyranose

Trehalose

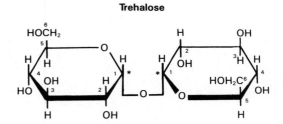

O-α-D-Glucopyranosyl-(1→1)-α-D-glucopyranoside

Sucrose

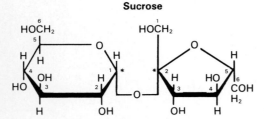

O-α-D-Glucopyranosyl-(1→2)-β-D-fructofuranoside

Cellobiose

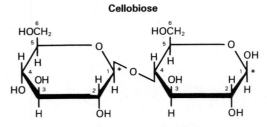

O-β-D-Glucopyranosyl-(1→4)-β-D-glucopyranose

Lactose

Figure 14–13. Structures of representative disaccharides. The -α and -β refer to the configuration at the anomeric carbon atom (*). When the anomeric carbon of the second residue takes part in the formation of the glycosidic bond, the residue becomes a glycoside known as a furanoside or pyranoside.

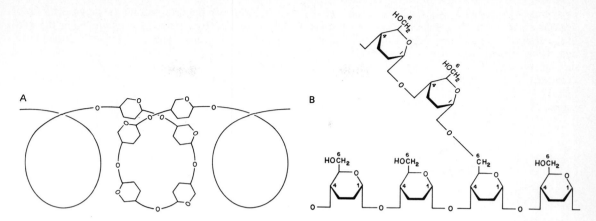

Figure 14–14. Structure of starch. *A:* Amylose, showing helical coil structure. *B:* Amylopectin, showing 1→6 branch point.

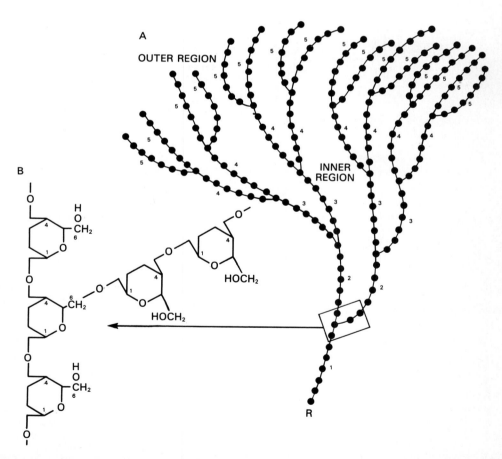

Figure 14–15. The glycogen molecule. *A:* Enlargement of structure at a branch point. *B:* Structure. The numbers refer to equivalent stages in the growth of the macromolecule. R, primary glucose residue. The branching is more variable than shown, the ratio of 1→4 to 1→6 bonds being from 12 to 18.

cross-linked hydrogen bonds. Cellulose cannot be digested by many mammals, including humans, because of the absence of a hydrolase that attacks the β linkage. Thus, it is an important source of "bulk" in the diet. In the gut of ruminants and other herbivores, there are microorganisms that can attack the β linkage, making cellulose available as a major calorigenic source.

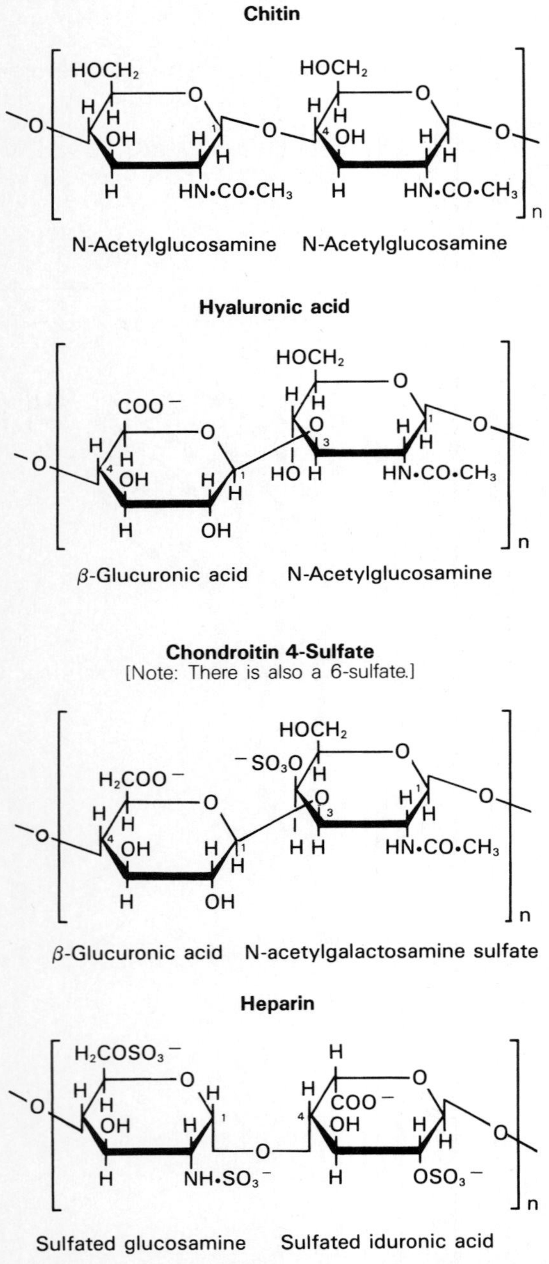

Chitin

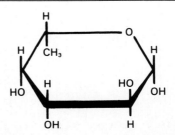

Chitin is an important structural polysaccharide of invertebrates. It is found, for example, in the exoskeletons of crustaceans and insects. Structurally, chitin consists of N-acetyl-D-glucosamine units joined by $\beta(1\rightarrow4)$-glycosidic linkages (Fig 14–16).

Glycosaminoglycans (mucopolysaccharides) consist of chains of complex carbohydrates characterized by their content of amino sugars and uronic acids. When these chains are attached to a protein molecule, the compound is known as a **proteoglycan.** As the ground or packing substance, they are associated with the structural elements of the tissues such as bone, elastin, and collagen. Their property of holding large quantities of water and occupying space, thus cushioning or lubricating other structures, is assisted by the large number of –OH groups and negative charges on the molecules, which, by repulsion, keep the carbohydrate chains apart. Examples are **hyaluronic acid, chondroitin sulfate,** and **heparin** (Fig 14–16), discussed in detail in Chapter 54.

Glycoproteins (mucoproteins) occur in many different situations in fluids and tissues, including the cell membranes (see Chapters 42 and 54). They are proteins containing carbohydrates in varying amounts attached as short or long (up to 15 units) branched or unbranched chains. Such chains are usually called oligosaccharide chains. Constituent carbohydrates include

Hexoses
 Mannose (Man)   Galactose (Gal)

Acetyl hexosamines
 N-Acetylglucosamine   N-Acetylgalactosamine
 (GlcNAc)   (GalNAc)

Pentoses
 Arabinose (Ara)   Xylose (Xyl)

Methyl pentose
 L-Fucose (Fuc; see Fig 14–17)

Sialic acids
 N-Acyl derivatives of neuraminic acid, eg, N-acetylneuraminic acid (NeuAc; see Fig 14–18),
 the predominant sialic acid

Glucose is not found in mature glycoproteins apart from collagen, and, in contrast to the glycosaminoglycans and proteoglycans, uronic acids are absent.

Figure 14–16. Structure of some complex polysaccharides.

Figure 14–17. β-L-Fucose (6-deoxy-β-L-galactose).

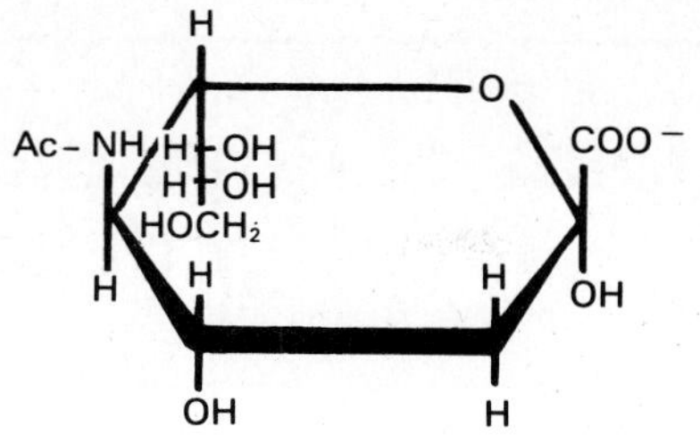

Figure 14–18. Structure of N-acetylneuraminic acid, a sialic acid (Ac = CH_3–CO–).

The **sialic acids** are N- or O-acyl derivatives of neuraminic acid (Fig 14–18). **Neuraminic acid** is a 9-carbon sugar derived from mannosamine (an epimer of glucosamine) and pyruvate. Sialic acids are constituents of both **glycoproteins** and **gangliosides.**

CARBOHYDRATES OF CELL MEMBRANES

The lipid structure of the cell membrane is described in Chapters 15 and 42. However, analysis of mammalian cell membrane components indicates that approximately 5% are carbohydrates, present in glycoproteins and glycolipids. Their presence on the outer surface of the plasma membrane (the **glycocalyx**) has been shown with the use of plant **lectins,** protein agglutinins that bind specifically with certain glycosyl residues. For example, **concanavalin A** has a specificity toward α-glucosyl and α-mannosyl residues.

Glycophorin is a major integral membrane glycoprotein of human erythrocytes. It has 130 amino acid residues and spans the lipid membrane, having free polypeptide portions outside both the external and internal (cytoplasmic) surfaces. Carbohydrate chains are only attached to the N-terminal portion outside the external surface (see Chapter 42).

REFERENCES

Advances in Carbohydrate Chemistry. Academic Press, 1945–current.

Collins PM (editor): *Carbohydrates.* Chapman & Hall, 1987.

Cook GMW, Stoddart RW: *Surface Carbohydrates of the Eukaryotic Cell.* Academic Press, 1973.

Ferrier RJ, Collins PM: *Monosaccharide Chemistry.* Penguin Books, 1972.

Hughes RC: The complex carbohydrates of mammalian cell surfaces and their biological roles. *Essays Biochem* 1975;**11**:1.

Lindahl U, Höök M: Glycosaminoglycans and their binding to biological macromolecules. *Annu Rev Biochem* 1978; **47**:385.

Pigman WW, Horton D (editors): *The Carbohydrates.* Vols 1A and 1B. Academic Press, 1972.

Sharon N: Lectins. *Sci Am* (June) 1977;**236**:108.

15

Lipids of Physiologic Significance

Peter A. Mayes, PhD, DSc

INTRODUCTION

The lipids are a heterogeneous group of compounds related, either actually or potentially, to the fatty acids. They have the common property of being (1) relatively insoluble in water and (2) soluble in nonpolar solvents such as ether, chloroform, and benzene. Thus, the lipids include fats, oils, waxes, and related compounds.

BIOMEDICAL IMPORTANCE

Lipids are important dietary constituents not only because of their high energy value but also because of the fat-soluble vitamins and the essential fatty acids contained in the fat of natural foods. In the body, fat serves as an efficient source of energy—both directly, and potentially when stored in adipose tissue. It serves as a thermal insulator in the subcutaneous tissues and around certain organs, and nonpolar lipids act as electrical insulators allowing rapid propagation of depolarization waves along myelinated nerves. The fat content of nerve tissue is particularly high. Combinations of fat and protein (lipoproteins) are important cellular constituents, occurring both in the cell membrane and in the mitochondria within the cytoplasm, and serving also as the means of transporting lipids in the blood. A knowledge of lipid biochemistry is important in understanding many current biomedical areas of interest, eg, obesity, atherosclerosis, and the role of various polyunsaturated fatty acids in nutrition and health.

CLASSIFICATION OF LIPIDS

The following classification of lipids is modified from Bloor:

A. Simple Lipids: Esters of fatty acids with various alcohols.

1. Fats—Esters of fatty acids with glycerol. A fat in the liquid state is known as an oil.

2. Waxes—Esters of fatty acids with higher molecular weight monohydric alcohols.

B. Complex Lipids: Esters of fatty acids containing groups in addition to an alcohol and a fatty acid.

1. Phospholipids—Lipids containing, in addition to fatty acids and an alcohol, a phosphoric acid residue. They frequently have nitrogen-containing bases and other substituents.

a. Glycerophospholipids—The alcohol is glycerol.

b. Sphingophospholipids—The alcohol is sphingosine.

2. Glycolipids (glycosphingolipids)—Lipids containing a fatty acid, sphingosine, and carbohydrate.

3. Other complex lipids—Lipids such as sulfolipids and aminolipids. Lipoproteins may also be placed in this category.

C. Precursor and Derived Lipids: These include fatty acids, glycerol, steroids, alcohols in addition to glycerol and sterols, fatty aldehydes, and ketone bodies (see Chapter 28), hydrocarbons, lipid-soluble vitamins, and hormones.

Because they are uncharged, acylglycerols (glycerides), cholesterol, and cholesteryl esters are termed **neutral lipids.**

FATTY ACIDS

Fatty acids are aliphatic carboxylic acids mostly obtained from the hydrolysis of natural fats and oils. Fatty acids that occur in natural fats usually contain an **even number** of carbon atoms, because they are synthesized from 2-carbon units, and are straight-chain derivatives. The chain may be saturated (containing no double bonds) or unsaturated (containing one or more double bonds).

Nomenclature

The most frequently used systematic nomenclature is based on naming the fatty acid after the hydrocarbon with the same number of carbon atoms, **-oic** being substituted for the final **e** in the name of the hydrocarbon (Genevan system). Thus, saturated acids end in **–anoic,** eg, octanoic acid, and unsaturated acids with double bonds end in **-enoic,** eg, octadecenoic acid (oleic acid). Carbon atoms are numbered from the carboxyl carbon (carbon No. 1). The carbon atom adjacent to the carboxyl carbon (No. 2) is also known as the α-carbon. Carbon atom No. 3 is the β-carbon, and

18:1;9 or Δ⁹ 18:1

$$\overset{18}{CH_3}(CH_2)_7\overset{10}{CH}=\overset{9}{CH}(CH_2)_7\overset{1}{COOH}$$

or

ω9,C18:1 *or* n−9, 18:1

$$\overset{\omega}{\underset{n}{CH_3}}\,\overset{2}{CH_2}\,\overset{3}{CH_2}\,\overset{4}{CH_2}\,\overset{5}{CH_2}\,\overset{6}{CH_2}\,\overset{7}{CH_2}\,\overset{8}{CH_2}\,\overset{9}{CH}=CH(CH_2)_7\,COOH$$

Figure 15−1. Oleic acid. n − 9, n minus 9.

the end methyl carbon is known as the ω-carbon or n-carbon. Various conventions are in use for indicating the number and position of the double bonds; eg, Δ⁹ indicates a double bond between carbon atoms 9 and 10 of the fatty acid. The ω9 indicates a double bond on the ninth carbon counting from the ω-carbon atom. Widely used conventions to indicate the number of carbon atoms, the number of double bonds, and the positions of the double bonds are shown in Fig 15−1.

In animals, additional double bonds are introduced only **between the existing double bond (eg, ω9, ω6, or ω3) and the carboxyl carbon,** leading to 3 series of fatty acids known as the ω9, ω6, and ω3 families, respectively.

Saturated Fatty Acids

Saturated fatty acids may be envisaged as based on acetic acid as the first member of the series. Examples are shown in Table 15−1.

Table 15−1. Saturated fatty acids.

Common Name	Number of C atoms	
Formic*	1	Takes part in the metabolism of "C₁" units (formate)
Acetic	2	Major end product of carbohydrate fermentation by rumen organisms
Propionic	3	An end product of carbohydrate fermentation by rumen organisms
Butyric	4	In certain fats in small amounts (especially butter). An end product of carbohydrate fermentation by rumen organisms.
Valeric	5	
Caproic	6	
Caprylic (octanoic)	8	In small amounts in many fats (including butter), especially those of plant origin
Capric (decanoic)	10	
Lauric	12	Spermaceti, cinnamon, palm kernel, coconut oils, laurels
Myristic	14	Nutmeg, palm kernel, coconut oils, myrtles
Palmitic	16	Common in all animal and plant fats
Stearic	18	
Arachidic	20	Peanut (arachis) oil
Behenic	22	Seeds
Lignoceric	24	Cerebrosides, peanut oil

* Strictly, not an alkyl derivative.

Other higher members of the series are known to occur, particularly in waxes. A few branched-chain fatty acids have also been isolated from both plant and animal sources.

Unsaturated Fatty Acids
(See Table 15−2.)

These may be further subdivided according to degree of unsaturation.

A. Monounsaturated (Monoethenoid, Monoenoic) Acids.

B. Polyunsaturated (Polyethenoid, Polyenoic Acids.

C. Eicosanoids: These compounds, derived from eicosa- (20-C) polyenoic fatty acids, comprise the **prostanoids** and **leukotrienes (LT)**. Prostanoids include **prostaglandins (PG), prostacyclins (PGI),** and **thromboxanes (TX)**. The term "prostaglandins" is often used loosely to include all prostanoids.

Prostaglandins were originally discovered in seminal plasma but are now known to exist in virtually every mammalian tissue and have important physiologic and pharmacologic activities. They are synthesized in vivo by cyclization of the center of the carbon chain of 20-C (eicosanoic) polyunsaturated fatty acids (eg, arachidonic acid) to form a cyclopentane ring (Fig 15−2). A related series of compounds, the **thromboxanes**, discovered in platelets, have the cyclopentane ring interrupted with an oxygen atom (oxane ring) (Fig 15−3). Three different eicosanoic fatty acids give rise to 3 groups of eicosanoids characterized by the number of double bonds in the side chains, eg, PG₁, PG₂, PG₃. Variations in the substituent groups attached to the rings give rise to different types in each series of prostaglandins and thromboxanes, labeled A, B, etc. For example, the "E" type of prostaglandin (as in PGE₂) has a keto group in position 9, whereas the "F" type has a hydroxyl group in this position. The **leukotrienes** are a third group of eicosanoid derivatives formed via the lipoxygenase pathway rather than cyclization of the fatty acid chain (Fig 15−4). First described in leukocytes, they are characterized by the presence of 3 conjugated double bonds.

Figure 15−2. Prostaglandin E₂ (PGE₂).

Figure 15−3. Thromboxane A₂ (TXA₂).

Table 15–2. Unsaturated fatty acids of physiologic and nutritional significance.

Number of C Atoms and Number and Position of Double Bonds	Series	Common Name	Systematic Name	Occurrence
Monoenoic acids (one double bond)				
16:1;9	$\omega7$	Palmitoleic	*cis*-9-Hexadecenoic	In nearly all fats.
18:1;9	$\omega9$	Oleic	*cis*-9-Octadecenoic	Possibly the most common fatty acid in natural fats.
18:1;9	$\omega9$	Elaidic	*trans*-9-Octadecenoic	Hydrogenated and ruminant fats.
22:1;13	$\omega9$	Erucic	*cis*-13-Docosenoic	Rape and mustard seed oils.
24:1;15	$\omega9$	Nervonic	*cis*-15-Tetracosenoic	In cerebrosides.
Dienoic acids (2 double bonds)				
18:2;9,12	$\omega6$	Linoleic	all-*cis*-9,12-Octadecadienoic	Corn, peanut, cottonseed, soybean, and many plant oils.
Trienoic acids (3 double bonds)				
18:3;6,9,12	$\omega6$	γ-Linolenic	all-*cis*-6,9,12-Octadecatrienoic	Some plants, eg, oil of evening primrose; minor fatty acid in animals.
18:3;9,12,15	$\omega3$	α-Linolenic	all-*cis*-9,12,15-Octadecatrienoic	Frequently found with linoleic acid but particularly in linseed oil.
Tetraenoic acids (4 double bonds)				
20:4;5,8,11,14	$\omega6$	Arachidonic	all-*cis*-5,8,11,14-Eicosatetraenoic	Found with linoleic acid particularly in peanut oil; important component of phospholipids in animals.
Pentaenoic acids (5 double bonds)				
20:5;5,8,11,14,17	$\omega3$	Timnodonic	all-*cis*-5,8,11,14,17-Eicosapentaenoic	Important component of fish oils, eg, cod liver oil.
22:5;7,10,13,16,19	$\omega3$	Clupanodonic	all-*cis*-7,10,13,16,19-Docosapentaenoic	Fish oils, phospholipids in brain.
Hexaenoic acids (6 double bonds)				
22:6;4,7,10,13,16,19	$\omega3$	Cervonic	all-*cis*-4,7,10,13,16,19-Docosahexaenoic	Fish oils, phospholipids in brain.

D. Others: Many other fatty acids have been detected in biologic material. Various structures, such as hydroxy groups (ricinoleic acid) or cyclic groups, have been found.

Cis-Trans Isomerism in Unsaturated Fatty Acids

The carbon chains of saturated fatty acids form a zigzag pattern when extended, as at low temperatures. At higher temperatures, some bonds rotate, causing chain shortening, which explains why biomembranes become thinner with increase in temperature. A type of **geometric isomerism** occurs in unsaturated fatty acids, depending on the orientation of atoms or groups around the axis of double bonds. If the acyl chains are on the same side of the bond, it is *cis-*, as in oleic acid; if on opposite sides, *trans-*, as in elaidic acid, the isomer of oleic acid (Fig 15–5). Naturally occurring un-

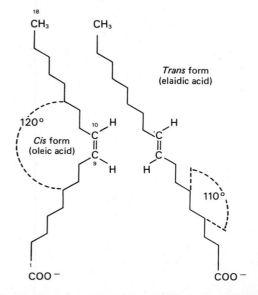

Figure 15–5. Geometric isomerism of Δ⁹, 18:1 fatty acids (oleic and elaidic acids).

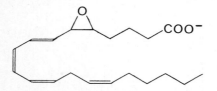

Figure 15–4. Leukotriene A₄ (LTA₄).

saturated long-chain fatty acids are nearly all of the *cis* configuration, the molecules being "bent" 120 degrees at the double bond. Thus, oleic acid has an L-shape, whereas elaidic acid remains "straight" at its *trans* double bond. Increase in the number of *cis* double bonds in a fatty acid leads to a variety of possible spatial configurations of the molecule. This may have profound significance on molecular packing in membranes and on the positions occupied by fatty acids in more complex molecules such as phospholipids. The presence of *trans* double bonds will alter these spatial relationships. *Trans* fatty acids are present in certain foods. Most arise as a by-product during the saturation of fatty acids in the process of hydrogenation, or "hardening," of natural oils in the manufacture of margarine. An additional small contribution comes from the ingestion of ruminant fat that contains *trans* fatty acids arising from the action of microorganisms in the rumen.

Alcohols

Alcohols associated with lipids include glycerol, cholesterol, and higher alcohols (eg, cetyl alcohol, $C_{16}H_{33}OH$), usually found in the waxes, and the polyisoprenoid alcohol dolichol (Fig 15–27).

Fatty Aldehydes

The fatty acids may be reduced to fatty aldehydes. These compounds are found either combined or free in natural fats.

Physiologically Relevant Properties of Fatty Acids

The physical properties of body lipids depend to a large extent on the lengths of the carbon chains and degree of unsaturation of their constituent fatty acids. Thus, the melting points of even-numbered-carbon fatty acids increase with chain length and decrease according to unsaturation. A triacylglycerol containing all saturated fatty acids of 12 C or more is solid at body temperature, whereas if all 3 fatty acid residues are 18:2, it is liquid to below 0 °C. In practice, natural acylglycerols contain a mixture of fatty acids tailored to suit their functional roles. The membrane lipids, which must be fluid, are more unsaturated than storage lipids. Lipids in tissues that are subject to cooling, eg, in hibernators or in the extremities of animals, are more unsaturated.

TRIACYLGLYCEROLS*
(Triglycerides)

The triacylglycerols, or so-called neutral fats, are esters of the alcohol glycerol and fatty acids. In natu-

*According to the current standardized terminology of the International Union of Pure and Applied Chemistry (IUPAC) and the International Union of Biochemistry (IUB), the monoglycerides, diglycerides, and triglycerides are to be designated monoacylglycerols, diacylglycerols, and triacylglycerols, respectively.

rally occurring fats, the proportion of triacylglycerol molecules containing the same fatty acid residue in all 3 ester positions is very small. They are nearly all **mixed acylglycerols.**

In Fig 15–6, if all 3 fatty acids represented by R were stearic acid, the fat would be known as tristearin, since it consists of 3 stearic acid residues esterified with glycerol. Examples of mixed acylglycerols are shown in Figs 15–7 and 15–8.

Nomenclature

When it is required to number the carbon atoms of glycerol unambiguously, the -*sn*- (stereochemical numbering) system is used, eg, 1,2-distearyl-3-palmityl-*sn*-glycerol (as below, or more generally as a projection formula shown in Fig 15–9). It is important to realize that carbons 1 and 3 of glycerol are not identical when viewed in 3 dimensions. Enzymes readily distinguish between them and are nearly always specific for one or the other carbon; eg, glycerol is always phosphorylated on *sn*-3 by glycerokinase to give glycerol 3-phosphate and not glycerol 1-phosphate.

Partial acylglycerols consisting of mono- and diacylglycerols wherein a single fatty acid or 2 fatty acids are esterified with glycerol are also found in the tis-

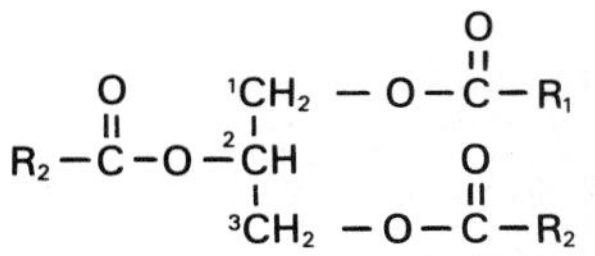

Figure 15–6. Triacylglycerol.

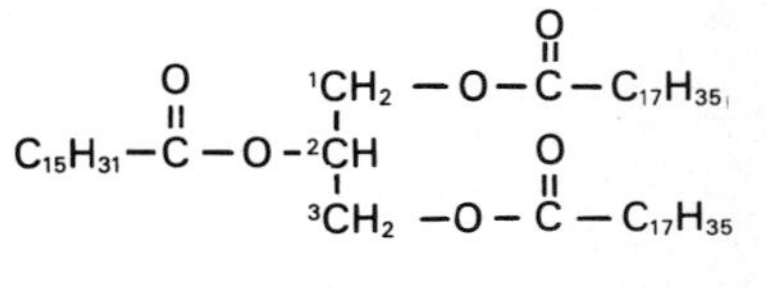

Figure 15–7. 1,3-Distearopalmitin.

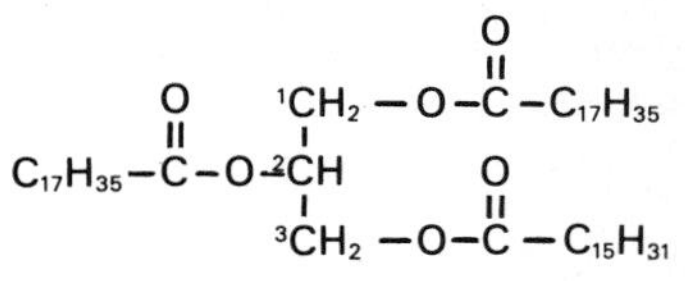

Figure 15–8. 1,2-Distearopalmitin.

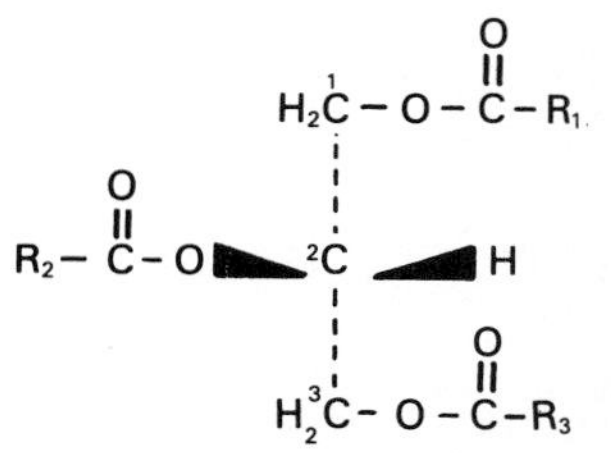

Figure 15–9. Triacyl-*sn*-glycerol.

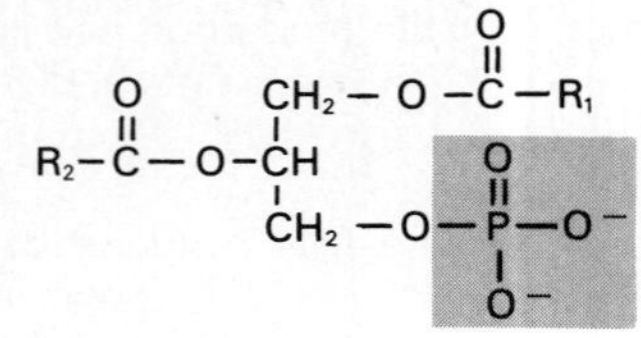

Figure 15–10. Phosphatidic acid.

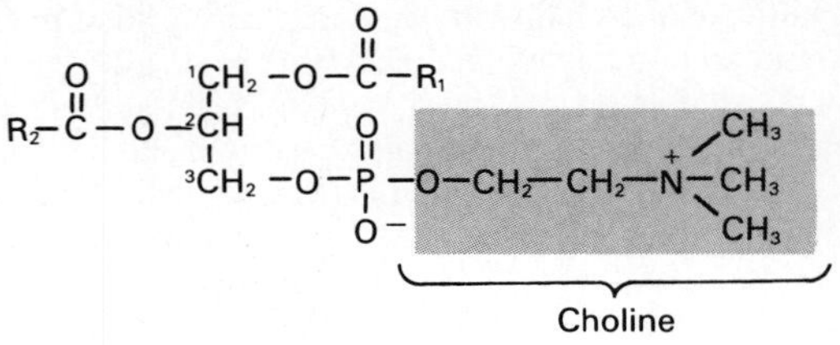

Figure 15–12. 3-Phosphatidylcholine.

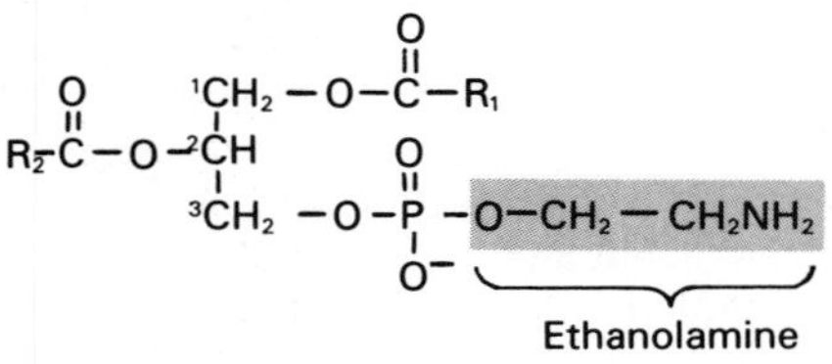

Figure 15–13. 3-Phosphatidylethanolamine.

sues. These are of particular significance in the synthesis and hydrolysis of triacylglycerols.

PHOSPHOLIPIDS

The phospholipids include the following: (1) phosphatidic acid and phosphatidylglycerols, (2) phosphatidylcholine, (3) phosphatidylethanolamine, (4) phosphatidylinositol, (5) phosphatidylserine, (6) lysophospholipids, (7) plasmalogens, and (8) sphingomyelins.

Phosphatidic Acid & Phosphatidylglycerols

Phosphatidic acid is important as an intermediate in the synthesis of triacylglycerols and phospholipids but is not found in any great quantity in tissues (Fig 15–10).

Cardiolipin is a phospholipid that is found in membranes of mitochondria. It is formed from phosphatidlyglycerol (Fig 15–11).

Phosphatidylcholine (Lecithin)

The lecithins contain glycerol and fatty acids, as do the simple fats, but they also contain phosphoric acid and choline. The lecithins are widely distributed in the cells of the body, having both metabolic and structural functions in membranes. Dipalmityl lecithin is a very effective surface active agent, preventing adherence, due to surface tension, of the inner surfaces of the lungs. Its absence from the lungs of premature infants causes respiratory distress syndrome. However,

most phospholipids have a saturated acyl radical in the C_1 position but an unsaturated radical in the C_2 position (Fig 15–12).

Phosphatidylethanolamine (Cephalin)

The cephalins differ from lecithins only in that ethanolamine replaces choline (Fig 15–13).

Phosphatidylinositol

The inositol is present as the stereoisomer, myo-inositol (Fig 15–14). Phosphatidylinositol 4,5-bisphosphate is an important constituent of cell membrane phospholipids; upon stimulation by a suitable hormone agonist, it is cleaved into diacylglycerol and isositol triphosphate, both of which act as internal signals or second messengers.

Phosphatidylserine

A cephalinlike phospholipid, phosphatidylserine, which contains the amino acid serine rather than ethanolamine, has been found in tissues (Fig 15–15). In addition, phospholipids containing threonine have been isolated.

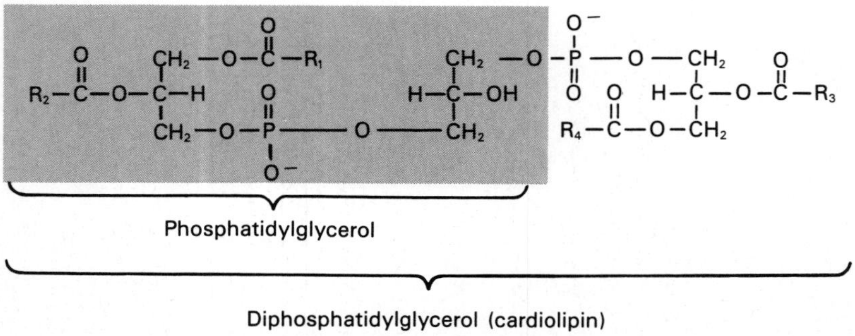

Phosphatidylglycerol

Diphosphatidylglycerol (cardiolipin)

Figure 15–11. Diphosphatidylglycerol (cardiolipin).

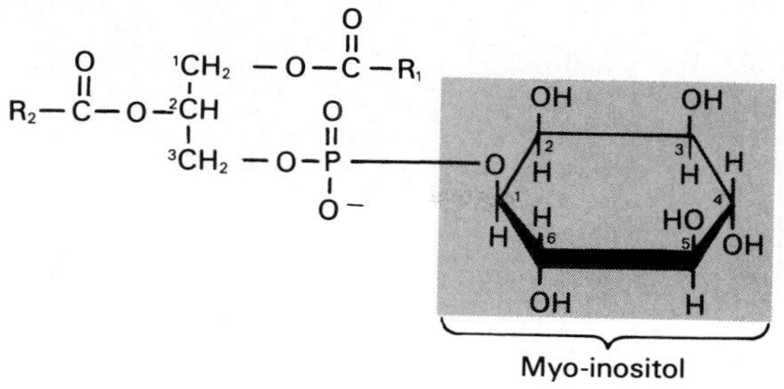

Figure 15–14. 3-Phosphatidylinositol.

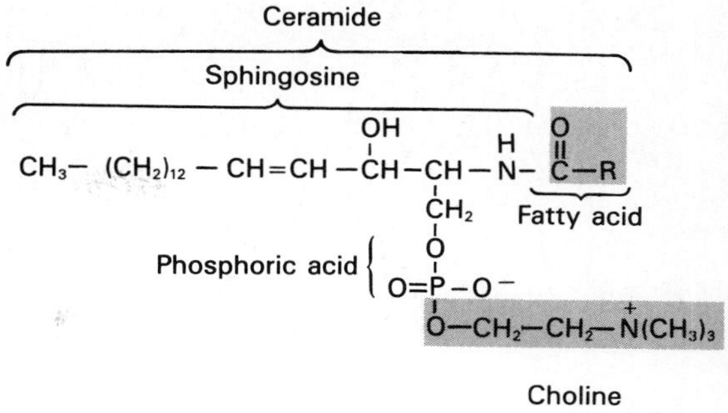

Figure 15–18. A sphingomyelin.

Lysophospholipids

These are phosphoacylglycerols containing only one acyl radical, eg, lysolecithin, important in the metabolism of phospholipids (Fig 15–16).

Plasmalogens

These compounds constitute as much as 10% of the phospholipids of brain and muscle. Structurally, the plasmalogens resemble phosphatidylethanolamine but possess an ether link on the C_1 carbon instead of the normal ester link found in most acylglycerols. Typically, the alkyl radical is an unsaturated alcohol (Fig 15–17).

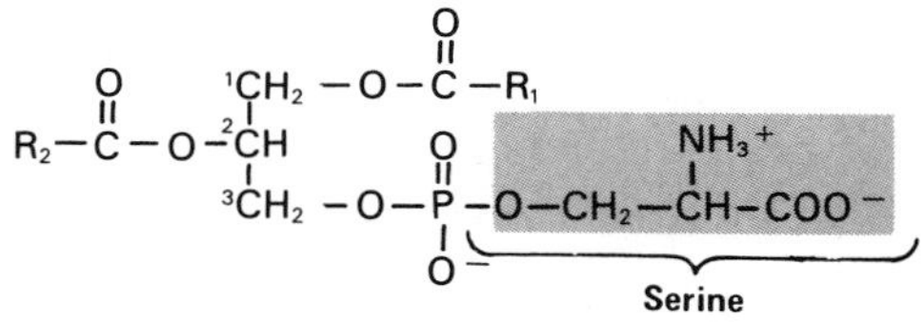

Figure 15–15. 3-Phosphatidylserine.

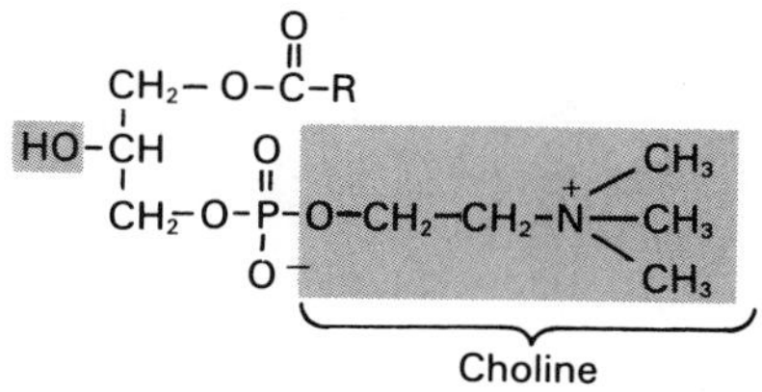

Figure 15–16. Lysolecithin.

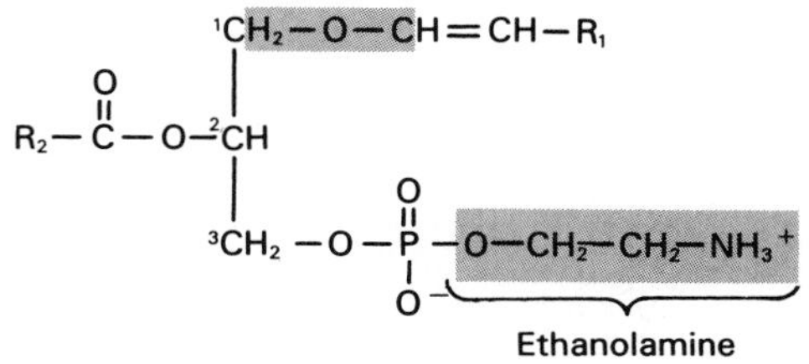

Figure 15–17. Plasmalogen (phosphatidal ethanolamine).

In some instances, choline, serine, or inositol may be substituted for ethanolamine.

Sphingomyelins

Sphingomyelins are found in large quantities in brain and nerve tissue. On hydrolysis, the sphingomyelins yield a fatty acid, phosphoric acid, choline, and a complex amino alcohol, **sphingosine** (Fig 15–18). No glycerol is present. The combination sphingosine plus fatty acid is known as **ceramide,** a structure also found in the glycolipids (see below).

GLYCOLIPIDS
(Glycosphingolipids)

Glycolipids are widely distributed in every tissue of the body, particularly in nervous tissue such as brain. They occur particularly in the outer leaflet of the plasma membrane where they contribute to **cell surface carbohydrates.**

The major glycolipids found in animal tissues are glycosphingolipids. They contain ceramide and one or more sugars. The two simplest are **galactosylceramide** and **glucosylceramide.** Galactosylceramide is a major glycosphingolipid of brain and other nervous tissue, but it is found in relatively low amounts elsewhere. It contains a number of characteristic C_{24} fatty acids. Galactosylceramide (Fig 15–19) can be converted to sulfogalactosylceramide (classic **sulfatide**), which is present in high amounts in myelin. Glucosylceramide is the predominant simple glycosphingolipid of extraneural tissues, but it also occurs in brain in small amounts. The more complex glycosphingolipids are **gangliosides** derived from glucosylceramide. A ganglioside is a glycosphingolipid that contains in addition one or more molecules of a sialic acid. Neuraminic acid (NeuAc; see Chapter 14) is the principal sialic acid found in human tissues. Gangliosides are also present in nervous tissues in high concentration. They appear to have receptor and other functions. The simplest ganglioside found in tissues is G_{M3}, which contains ceramide, one molecule of glucose, one molecule of galactose, and one molecule of NeuAc. In

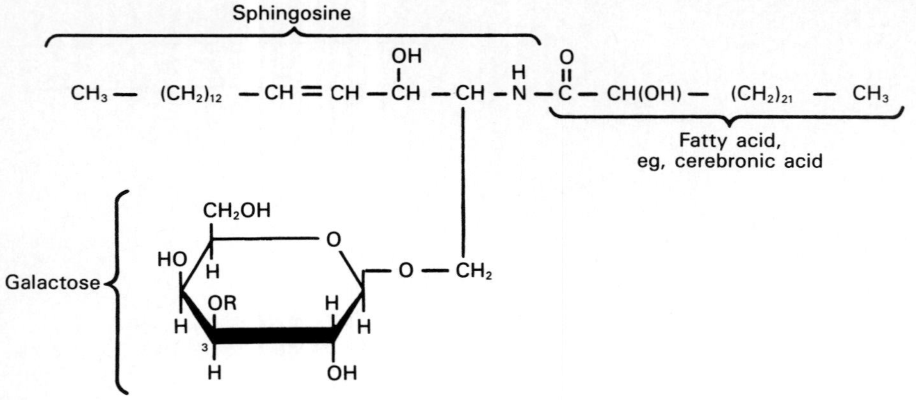

Sphingosine

$$CH_3 — (CH_2)_{12} — CH = CH — \underset{|}{CH} — \underset{|}{CH} — \underset{|}{N} — \overset{O}{\underset{\|}{C}} — CH(OH) — (CH_2)_{21} — CH_3$$

OH   H

Fatty acid,
eg, cerebronic acid

Galactose { [galactose ring structure with CH₂OH, HO, H, OR, H, O—CH₂, H, OH, labeled position 3]

Figure 15–19. Structure of galactosylceramide (galactocerebroside, R=H), and sulfogalactosylceramide (a sulfatide, R=SO₄²⁻).

Ceramide–Glucose–Galactose–N-Acetylgalactosamine–Galactose
   (Acyl-
  sphingo- |
    sine)   NeuAc

or

Cer–Glc–Gal–GalNAc–Gal
               |
         NeuAc

Figure 15–20. G_{M1} ganglioside, a monosialoganglioside.

the shorthand nomenclature used, G represents ganglioside; M is a monosialo-containing species; and the subscript 3 is an arbitrary number assigned on the basis of chromatographic migration. The structure of a more complex ganglioside derived from G_{M3}, named G_{M1}, is shown in Fig 15–20. G_{M1} is a compound of considerable biologic interest as it is known to be the receptor in human intestine for cholera toxin. Other gangliosides can contain anywhere from one to 5 molecules of sialic acid, giving rise to di-, trisialogangliosides, etc.

STEROIDS

The steroids are often found in association with fat. They may be separated from the fat after saponification* in the "unsaponifiable residue." All of the steroids have a similar cyclic nucleus resembling phenanthrene (rings A, B, and C) to which a cyclopentane ring (D) is attached. The carbon positions on the steroid nucleus are numbered as shown in Fig 15–21.

It is important to realize that in structural formulas of steroids, a simple hexagonal ring denotes a completely saturated 6-carbon ring with all valences

---

*Hydrolysis of a fat by alkali is called **saponification**. The products are glycerol and the alkali salts of the fatty acids, which are called **soaps**.

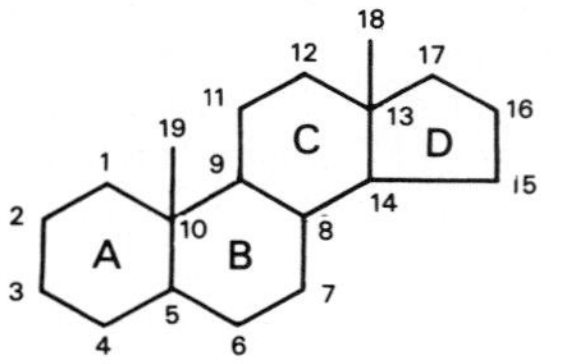

[steroid nucleus diagram with rings A, B, C, D and carbon positions numbered 1–19]

Figure 15–21. The steroid nucleus.

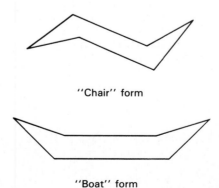

"Chair" form

"Boat" form

Figure 15–22. Conformations of stereoisomers.

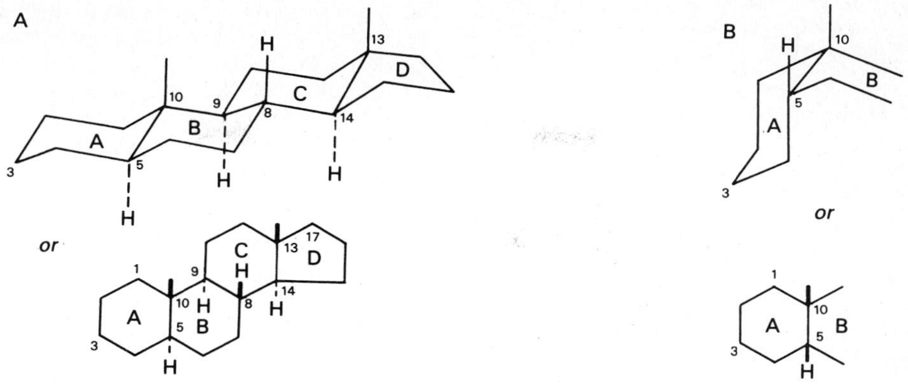

Figure 15–23. Generalized steroid nucleus, showing *(A)* an all-*trans* configuration between adjacent rings and *(B)* a *cis* configuration between rings A and B.

satisfied by hydrogen bonds unless shown otherwise; ie, it is not a benzene ring. All double bonds are shown as such. Methyl side chains are shown as single bonds unattached at the farther (methyl) end. These occur typically at positions 10 and 13 (constituting C atoms 19 and 18). A side chain at position 17 is usual (as in cholesterol). If the compound has one or more hydroxyl groups and no carbonyl or carboxyl groups, it is a **sterol**, and the name terminates in -ol.

Stereochemical Aspects

Because of their complexity and the possibilites of assymmetry in the molecule, steroids have many potential stereoisomers. Each of the 6-carbon rings of the steroid nucleus is capable of existing in the 3-dimensional conformation either of a "chair" or a "boat" (Fig 15–22).

In naturally occurring steroids, virtually all the rings are in the "chair" form, which is the more stable conformation. With respect to each other, the rings can be either -*cis* or -*trans* (Fig 15–23).

The junction between the A and B rings can be -*cis* or -*trans* in naturally occurring steroids. That between B and C is -*trans* and the C/D junction is -*trans* except in cardiac glycosides and toad poisons. Bonds attaching substituent groups above the plane of the rings are shown with bold solid lines (β), whereas those bonds attaching groups below are indicated with broken lines (α). The A ring of a 5α steroid is always -*trans* to the B ring, whereas it is -*cis* in a 5β steroid. The methyl groups attached to C_{10} and C_{13} are invariably in the β configuration.

Cholesterol

Cholesterol is widely distributed in all cells of the body, but particularly in nervous tissue. It is a major constituent of the plasma membrane and of plasma lipoproteins. It is often found in combination with fatty acids as cholesteryl ester and is the parent compound of all steroids synthesized in the body. It occurs in animal fats but not in plant fats. Cholesterol is designated as 3-hydroxy-5,6-cholestene (Fig 15–24).

Ergosterol

Ergosterol occurs in plants and yeast and is important as a precursor of vitamin D (Fig 15–25). When irradiated with ultraviolet light, it acquires antirachitic properties consequent to the opening of ring B.

Coprosterol

Coprosterol (coprostanol) occurs in feces as a result of the reduction of the double bond of cholesterol between C_5 and C_6 by bacteria in the intestine.

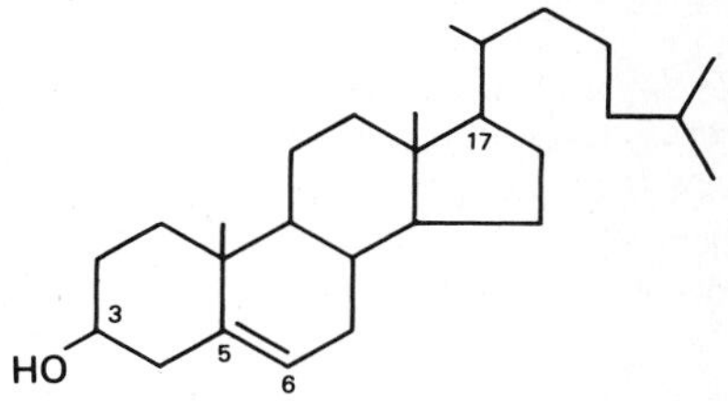

Figure 15–24. Cholesterol.

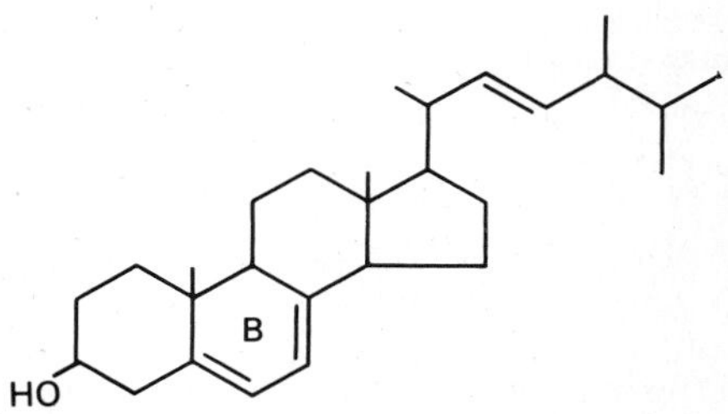

Figure 15–25. Ergosterol.

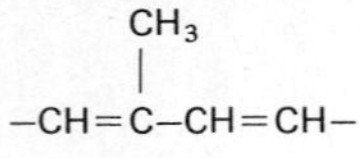

$$CH_3$$
$$-CH=C-CH=CH-$$

Figure 15–26. Isoprene unit.

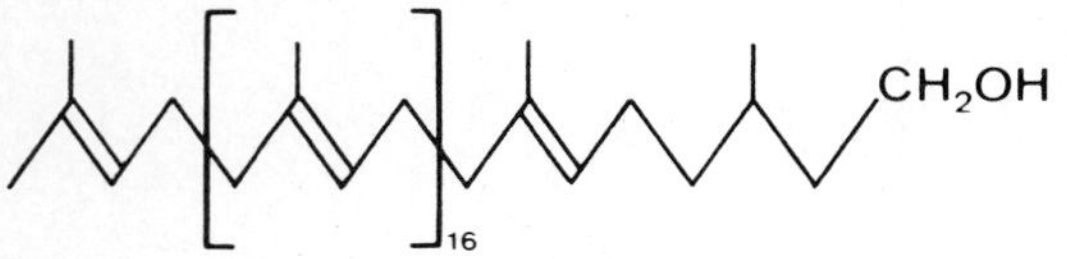

Figure 15–27. Dolichol—a C_{95} alcohol.

Other Important Sterols & Steroids

These include the bile acids, adrenocortical hormones, sex hormones, D vitamins, cardiac glycosides, sitosterols of the plant kingdom, and some alkaloids.

Polyprenoid Compounds

Although not steroids, these compounds are related because they are synthesized, like cholesterol (see Fig 27–3), from 5-carbon isoprene units (Fig 15–26). They include **ubiquinone** (see p 109), a member of the respiratory chain in mitochondria, and the long-chain alcohol **dolichol** (Fig 15–27), which takes part in glycoprotein synthesis by transferring carbohydrate residues to asparagine residues of the polypeptide (see Chapter 54). Plant-derived isoprenoid compounds include rubber, camphor, the fat-soluble vitamins A, D, E, and K, and β-carotene (provitamin A).

LIPID PEROXIDATION

Peroxidation (**auto-oxidation**) of lipids exposed to oxygen is responsible not only for deterioration of foods (**rancidity**) but also for damage to tissues in vivo, where it may be a cause of cancer. The deleterious effects are initiated by free radicals ($ROO^{\cdot}$, $RO^{\cdot}$,

$OH^{\cdot}$) produced during peroxide formation from fatty acids containing methylene-interrupted double bonds, ie, those found in the naturally occurring polyunsaturated fatty acids (Fig 15–28). Lipid peroxidation is a chain reaction providing a continuous supply of free radicals that initiate further peroxidation. The whole process can be depicted as follows:

(1) Initiation: Production of $R^{\cdot}$ from a precursor.

$$ROOH + metal^{(n)+} \rightarrow ROO^{\cdot} + metal^{(n-1)+} + H^+$$

(2) Propagation:

$$R^{\cdot} + O_2 \rightarrow ROO^{\cdot}$$
$$ROO^{\cdot} + RH \rightarrow ROOH + R^{\cdot}, etc.$$

(3) Termination:

$$ROO^{\cdot} + ROO^{\cdot} \rightarrow ROOR + O_2$$
$$ROO^{\cdot} + R^{\cdot} \rightarrow ROOR$$
$$R^{\cdot} + R^{\cdot} \rightarrow RR$$

Since the molecular precursor for the initiation process is generally the hydroperoxide product ROOH, lipid peroxidation is a branching chain reaction with potentially devastating effects. To control and reduce lipid peroxidation both humans and nature invoke the use of **antioxidants.** Propyl gallate, butylated hydroxyanisole (BHA), and butylated hydroxytoluene (BHT) are antioxidants used as food additives. Naturally occurring antioxidants include vitamin E (tocopherol), which is lipid-soluble, and urate and vitamin C, which are water-soluble. β-Carotene is an antioxidant at low P_{O_2}. Antioxidants fall into 2 classes: (1) preventive antioxidants, which reduce the rate of chain initiation, and (2) chain-breaking antioxidants, which interfere with chain propagation. Preventive antioxidants include catalase and other peroxidases and react with ROOH and chelators of metal ions such as DTPA (diethylenetriaminepentaacetate) and EDTA (ethylenediaminetetraacetate). Chain-breaking antioxidants are often phenols or aromatic amines. In vivo, the principal chain-breaking antioxidants are su-

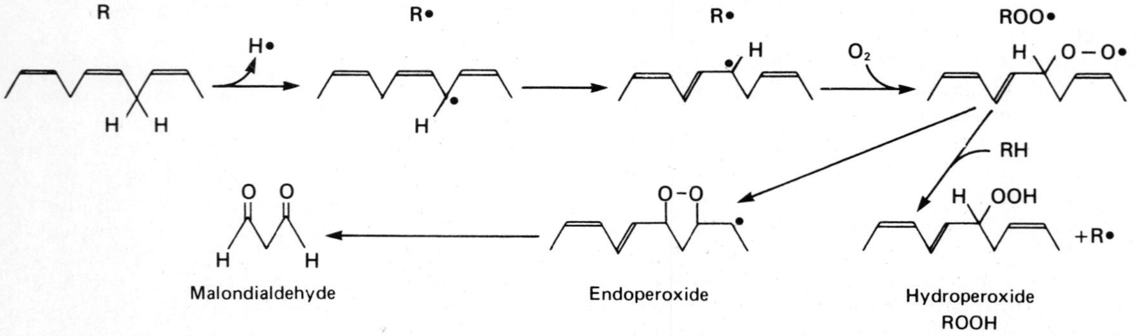

Figure 15–28. Lipid peroxidation. The reaction is initiated by light or by metal ions. Malondialdehyde is only formed by fatty acids with 3 or more double bonds and is used as a measure of lipid peroxidation together with ethane from the terminal 2-carbon of ω3 fatty acids and pentane from the terminal 5-carbon of ω6 fatty acids.

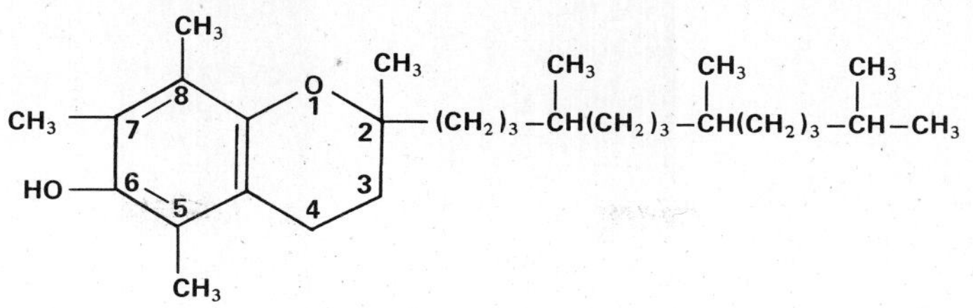

Figure 15–29. α-Tocopherol.

peroxide dismutase (see p 107), which acts in the aqueous phase to trap superoxide free radicals (O_2^-); perhaps urate; and vitamin E, which acts in the lipid phase to trap ROO˙ radicals.

Peroxidation is also catalyzed in vivo by heme compounds and by **lipoxygenases** found in platelets and leukocytes, etc.

Vitamin E
(α-Tocopherol)

There are several naturally occurring tocopherols. All are isoprenoid substituted 6-hydroxychromanes or tocols (Fig 15–29). α-Tocopherol has the widest natural distribution and the greatest biologic activity as a vitamin.

Vitamin E has at least 2 metabolic roles: It acts as nature's most potent fat-soluble **antioxidant,** and it plays a specific but incompletely understood role in **selenium metabolism.**

Vitamin E appears to be the first line of defense against peroxidation of cellular and subcellular membrane phospholipids. The phospholipids of mitochondria, endoplasmic reticulum, and plasma membranes possess specific affinities for α-tocopherol, and the vitamin appears to concentrate at these sites. The tocopherols act as chain-breaking antioxidants as a result of their ability to transfer a phenolic hydrogen to a peroxyl radical (Fig 15–30). The phenoxy radical is resonant-stabilized and relatively unreactive, except toward other peroxyl radicals. Thus, α-tocopherol does not readily engage in reversible oxidation; the chromane ring and the side chain of α-tocopherol are oxidized to produce the nonradical product shown in Fig 15–31. This oxidation product is conjugated with glucuronic acid via the 2-hydroxyl group and excreted in bile. The antioxidant effect of tocopherol is effective at high oxygen concentrations, and thus it is not surprising that vitamin E tends to be concentrated in those lipid regions which are exposed to the highest

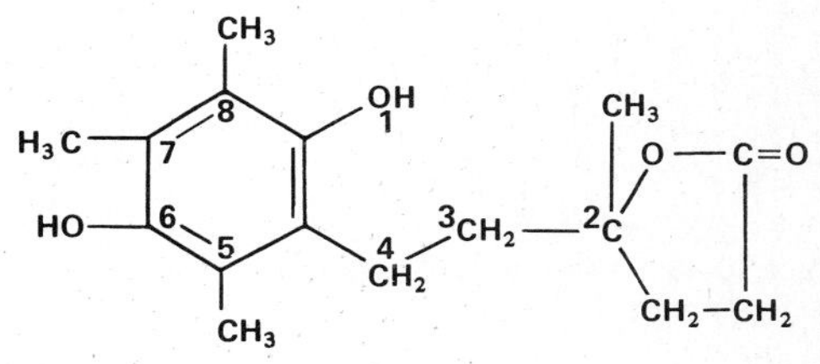

Figure 15–31. The oxidation product of α-tocopherol. The numbers allow one to relate the atoms to those in the parent compound.

partial pressures of oxygen, such as the erythrocyte membrane and membranes of the respiratory tree.

However, even in the presence of adequate vitamin E, some peroxides are formed. **Glutathione peroxidase,** of which selenium is an integral component, provides a second line of defense to destroy the peroxides before they cause damage to the membranes (see p 179). Thus, the biochemical action of vitamin E and selenium seems to be prevention of peroxidative damage to cellular and subcellular elements, which thereby preserves the organelles necessary to cope with disease, physical and chemical environmental insults, and other stresses.

METHODS FOR SEPARATING & IDENTIFYING LIPIDS IN BIOLOGIC MATERIAL

The older methods of separation and identification of lipids, based on classic chemical procedures of crystallization, distillation, and solvent extraction, have now been largely supplanted by chromatographic procedures. Particularly useful for the separation of the various lipid classes is **thin-layer chromatography** (TLC) and for the separation of the individual fatty acids, **gas-liquid chromatography** (GLC; Fig 15–32). Before these techniques are applied to wet tissues, the lipids are extracted by a solvent system based usually on a mixture of chloroform and methanol (2:1).

Gas-liquid chromatography involves the physical separation of a moving gas phase by adsorption

Figure 15–30. The chain-breaking antioxidant activity of tocopherols (TocOH) toward peroxyl radicals (ROO˙).

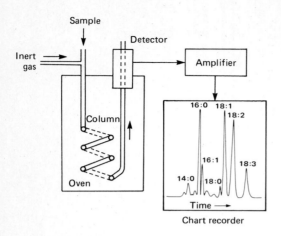

Figure 15–32. Diagrammatic representation of a gas-liquid chromatography apparatus and the separation of long-chain fatty acids (as methyl esters). (A section of the record of a chromatogram is shown at right.)

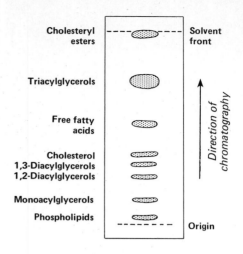

Figure 15–33. Separation of major lipid classes by thin-layer chromatography. A suitable solvent system for the above would be hexane-diethyl ether-formic acid (80:20:2 v/v/v).

onto a stationary phase consisting of an inert solid such as silica gel or inert granules of ground firebrick coated with a nonvolatile liquid (eg, lubricating grease or silicone oils). In practice, a glass or metal column is packed with the inert solid, and a mixture of the methyl esters of fatty acids is evaporated at one end of the column, the entire length of which is kept at temperatures of 170–225 °C (Fig 15–32). A constantly flowing stream of an inert gas such as argon or helium keeps the volatilized esters moving through the column. As with other types of chromatography, separation of the vaporized fatty acid esters is dependent upon the different affinities of the components of the gas mixture for the stationary phase. Gases that are strongly attracted to the stationary phase move through the column at a slower rate and therefore emerge at the end of the column later than those that are relatively less attracted. As the individual fatty acid esters emerge from the column, they are detected by physical or chemical means and recorded automatically as a series of peaks that appear at different times according to the tendency of each fatty each ester to be retained by the stationary phase (Fig 15–32). The area under each peak is proportionate to the concentration of a particular component of the mixture. The identity of each component is established by comparison with the gas chromatographic pattern of a related standard mixture of known composition. A detector of radioactivity may also be incorporated into the gas stream, together with the mass detector. Thus, a measure of the specific radioactivity of each component separated is obtained.

The advantages of gas-liquid chromatography are its extreme sensitivity, which allows very small quantities of mixtures to be separated, and the fact that the columns may be used repeatedly. Application of the

technique has shown that natural fats contain a wide variety of hitherto undetected fatty acids.

Thin-layer chromatography (TLC) is carried out on glass plates coated with a thin slurry of adsorbent, usually silica gel. This is allowed to dry and is then heated in an oven at a standard temperature and for a standard time. After cooling, the "activated" plate is "spotted" with the lipid mixture contained in a suitable solvent. The solvent is evaporated, the edge of the plate nearest the spots is dipped in an appropriate solvent mixture, and the plate is run inside a closed tank until the solvent front arrives near the top edge of the plate. The plate is dried of solvent, and the position of the spots is determined by "charring" (spraying with sulfuric acid followed by heating) or by fluorescence (with dichlorofluorescein) or by reacting with iodine vapor (Fig 15–33).

AMPHIPATHIC LIPIDS

Membranes, Micelles, Liposomes, & Emulsions

In general, lipids are insoluble in water, since they contain a predominance of nonpolar (hydrocarbon) groups. However, fatty acids, phospholipids, sphingolipids, bile salts, and, to a lesser extent, cholesterol contain polar groups. Therefore, part of the molecule is **hydrophobic,** or water-insoluble, and part is **hydrophilic,** or water-soluble. Such molecules are described as **amphipathic** (Fig 15–34). They become oriented at oil-water interfaces with the **polar group in the water phase** and the **nonpolar group in the oil phase.** A bilayer of such polar lipids has been regarded as a basic structure in biologic membranes (see

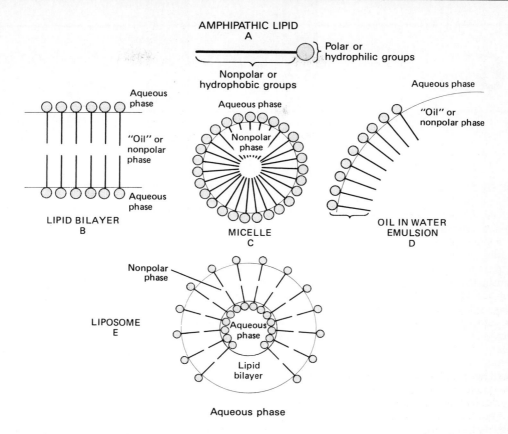

Figure 15–34. Formation of lipid membranes, micelles, emulsions, and liposomes from amphipathic lipids, eg, phospholipids.

Chapter 42). When a critical concentration of polar lipids is present in an aqueous medium, they form **micelles.** Aggregations of bile salts into micelles and the formation of mixed micelles with the products of fat digestion are important in facilitating absorption of lipids from the intestine. **Liposomes** are formed by sonicating an amphipathic lipid in an aqueous medium. They consist of spheres of lipid bilayers that enclose part of the aqueous medium. They are of potential clinical use, particularly when combined with tissue-specific antibodies, as carriers of drugs in the circulation, targeted to specific organs. **Emulsions** are much larger particles, formed usually by nonpolar lipids in an aqueous medium. These are stabilized by emulsifying agents such as polar lipids (eg, lecithin), which form a surface layer separating the main bulk of the nonpolar material from the aqueous phase (Fig 15–34).

REFERENCES

Christie WW: *Lipid Analysis*. Pergamon Press, 1973.

Frankel EN: Chemistry of free radical and singlet oxidation of lipids. *Prog Lipid Res* 1985;**23**:197.

Gunstone FD, Harwood JL, Padley FB: *The Lipid Handbook*. Chapman & Hall, 1986.

Gurr AI, James AT: *Lipid Biochemistry: An Introduction,* 3rd ed. Wiley, 1980.

Hawthorne JN, Ansell GB (editors): *Phospholipids*. Elsevier, 1982.

Johnson AR, Davenport JB: *Biochemistry and Methodology of Lipids*. Wiley, 1971.

Sevanian A, Hochstein P: Mechanisms and consequences of lipid peroxidation in biological systems. *Annu Rev Nutr* 1985;**5**:365.

Vance DE, Vance JE (editors): *Biochemistry of Lipids and Membranes*. Benjamin/Cummings, 1985.

16

Overview of Intermediary Metabolism

Peter A. Mayes, PhD, DSc

INTRODUCTION

The fate of dietary components after digestion and absorption constitutes intermediary metabolism. Thus, it encompasses a wide field that not only seeks to describe the metabolic pathways taken by individual molecules but also attempts to understand their interrelationships and the mechanisms that regulate the flow of metabolites through the pathways. Metabolic pathways fall into 3 categories (Fig 16–1): (1) **Anabolic pathways** are those involved in the synthesis of the compounds constituting the body's structure and machinery. Protein synthesis is such a pathway. The free energy required for these processes comes from the next category. (2) **Catabolic pathways** involve oxidative processes that release free energy, usually in the form of high-energy phosphate or reducing equivalents, eg, the respiratory chain and oxidative phosphorylation. (3) **Amphibolic pathways** have more than one function and occur at the "crossroads" of metabolism, acting as links between the anabolic and catabolic pathways, eg, the citric acid cycle.

BIOMEDICAL IMPORTANCE

A knowledge of metabolism in the normal animal is a prerequisite to a sound understanding of many diseases. Normal metabolism includes the variations and adaptation in metabolism due to periods of starvation, exercise, pregnancy, and lactation. Abnormal metabolism results from, for example, nutritional deficiency, enzyme deficiency, or abnormal secretion of hormones. An important example of a disease resulting from abnormal metabolism (a "metabolic disease") is diabetes mellitus.

THE BASIC METABOLIC PATHWAYS

The nature of the diet sets the basic pattern of metabolism in the tissues. Mammals such as humans need to process the absorbed products of digestion of dietary carbohydrate, lipid, and protein. These are mainly glucose, triacylglycerol, and amino acids, re-

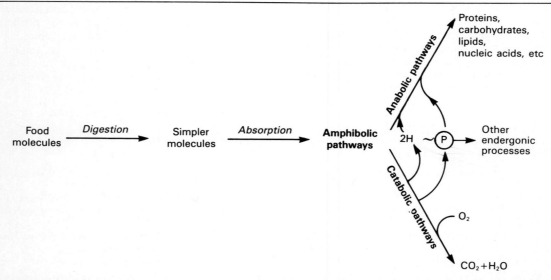

Figure 16–1. The 3 major categories of metabolic pathways. Catabolic pathways release free energy in the form of reducing equivalents (2H) or high-energy phosphate (~P) to power the anabolic pathways. Amphibolic pathways act as links between the other 2 categories of pathways.

142

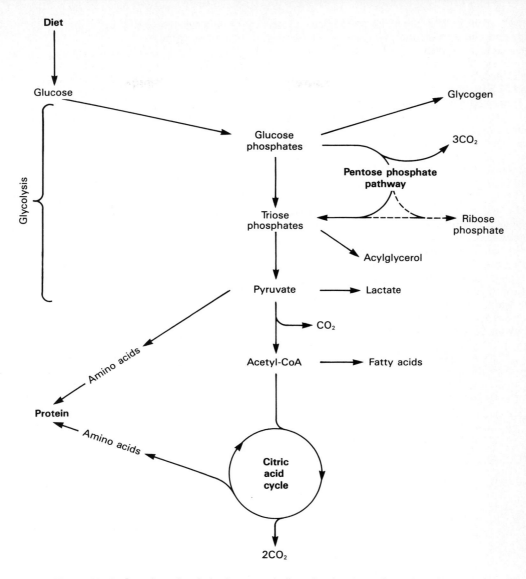

Figure 16–2. Overview of carbohydrate metabolism showing the major end products.

spectively. In ruminants (and to a lesser extent other herbivores), cellulose in the diet is digested by symbiotic microorganisms to lower fatty acids (acetic, propionic, butyric), and tissue metabolism in these animals is adapted to utilize lower fatty acids as major substrates.

Carbohydrate Metabolism
(See Fig 16–2.)

Glucose is metabolized to pyruvate and lactate by all mammalian cells by the pathway of **glycolysis.** Phosphorylation is necessary for glucose to enter this pathway. Glycolysis can occur in the absence of oxygen (anaerobic), when the end product is lactate only. Tissues that can utilize oxygen (aerobic) are able to metabolize pyruvate to acetyl-CoA, which can enter

the **citric acid cycle** for complete oxidation to CO_2 and H_2O, with liberation of much free energy as ATP in the process of **oxidative phosphorylation** (see Fig 17–2). Thus, glucose is a major fuel of many tissues. But it (and certain of its metabolites) also takes part in other processes, as follows: (1) Conversion to its storage polymer, **glycogen,** particularly in skeletal muscle and liver. (2) The **pentose phosphate pathway,** which arises from intermediates of glycolysis. It is a source of reducing equivalents (2H) for biosynthesis—eg, of fatty acids—and it is also the source of ribose, which is important for nucleotide and nucleic acid formation. (3) Triose phosphate gives rise to the **glycerol moiety** of acylglycerols (fat). (4) Pyruvate and intermediates of the citric acid cycle provide the carbon skeletons for the synthesis of **amino acids,** and acetyl-CoA is the building block for long-chain **fatty acids** and **cholesterol,** the precursor of all steroids synthesized in the body.

Lipid Metabolism
(See Fig 16–3.)

The source of long-chain fatty acids is either de novo synthesis from acetyl-CoA derived from carbohydrate, or from dietary lipid. In the tissues, fatty acids may be oxidized to acetyl-CoA (**β-oxidation**) or esterified to acylglycerols, where as triacylglycerol (fat) they constitute the body's main caloric reserve. Acetyl-CoA formed by β-oxidation has several important fates.

(1) As in the case of acetyl-CoA derived from carbohydrate, it is **oxidized completely** to $CO_2 + H_2O$ via the citric acid cycle. Fatty acids yield considerable energy both in β-oxidation and in the citric acid cycle and are therefore very effective tissue fuels.

(2) It is a source of the carbon atoms in **cholesterol.**

(3) In the liver it forms acetoacetate, the parent **ketone body.** Ketone bodies are alternative water-soluble tissue fuels, which become important sources of energy under certain conditions (eg, starvation).

Amino Acid Metabolism
(See Fig 16–4.)

The amino acids are necessary for protein synthesis. Some must be supplied specifically in the diet (the **essential amino acids**), since the tissues are unable to synthesize them. The remainder, or **nonessential amino acids,** are also supplied in the diet, but they can be formed from intermediates by **transamination** using the amino nitrogen from other surplus amino acids. After **deamination,** excess amino nitrogen is removed as **urea,** and the carbon skeletons that remain after transamination (1) are oxidized to CO_2 via the citric acid cycle, (2) form glucose (gluconeogenesis), or (3) form ketone bodies.

In addition to their requirement for protein synthesis, the amino acids are also the precursors of many other important compounds, eg, purines, pyrimidines, and hormones such as epinephrine and thyroxine.

THE LOCATION OF METABOLIC PATHWAYS

As indicated in Fig 2–3, metabolic pathways may be studied at many levels of organization. It is conve-

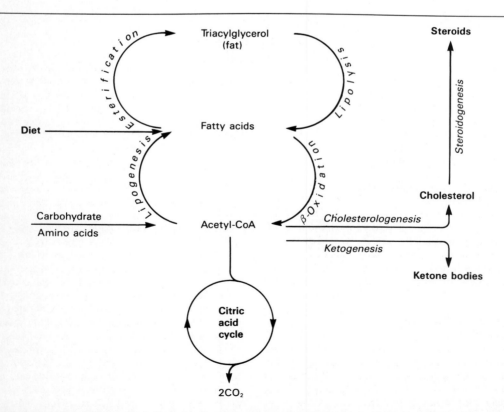

Figure 16–3. Overview of lipid metabolism showing the major end products. Ketone bodies comprise the substances acetoacetate, 3-hydroxybutyrate, and acetone.

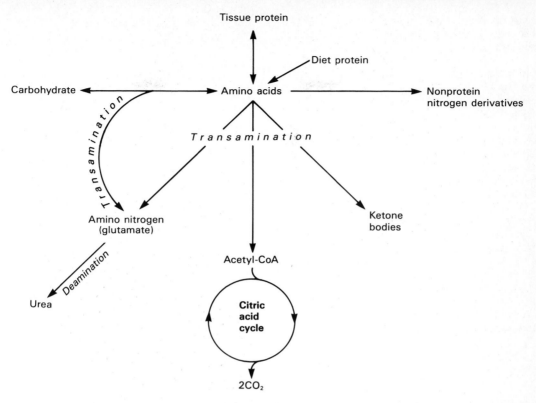

Figure 16–4. Overview of amino acid metabolism showing the major end products.

nient to divide these levels into 2 major groups: (1) **At the tissue and organ level**—the nature of the substrates entering and metabolites leaving tissues and organs is defined, and their overall fate is described. (2) **At the subcellular level**—each cell organelle (eg, the mitochondrion) or compartment (eg, the cytosol) carries out specific biochemical roles that form part of a subcellular pattern of metabolic pathways.

Intermediary Metabolism at the Tissue & Organ Level

Amino acids resulting from the digestion of dietary protein and glucose resulting from the digestion of carbohydrate share a common route of absorption via the **hepatic portal vein.** This ensures that both of these metabolites and other water-soluble products of digestion are initially directed to the liver (Fig 16–5). The **liver** has the primary metabolic function of regulating the blood concentration of most metabolites, particularly glucose and amino acids. In the case of glucose this is achieved by taking up excess glucose and converting it to glycogen (**glycogenesis**) or to fat (**lipogenesis**). Between meals, it can draw upon its glycogen stores to replenish glucose in the blood (**glycogenolysis**) or, in company with the kidney, to convert non-carbohydrate metabolites such as lactate, glycerol, and amino acids to glucose (**gluconeogenesis**). The maintenance of an adequate concentration of

blood glucose is vital for certain tissues in which it is an obligatory fuel, eg, brain and erythrocytes. The liver also has the task of **synthesizing the major plasma proteins** (eg, albumin) and of **deaminating amino acids** that are in excess of requirements with formation of urea, which is transported via the blood to the kidney and excreted.

Skeletal muscle utilizes glucose as a fuel, forming both lactate and CO_2. It stores glycogen as a fuel for its use in muscular contraction and synthesizes muscle protein from plasma amino acids. Muscle accounts for approximately 50% of body mass and consequently represents a considerable store of protein that can be drawn upon to supply plasma amino acids, particularly during dietary shortage.

Lipids (Fig 16–6) upon digestion form monoacylglycerols and fatty acids. These are recombined in the intestinal cells with protein and secreted initially into the lymphatic system and then into the circulation as a **lipoprotein** known as a **chylomicron.** All hydrophobic, lipid-soluble products of digestion (eg, cholesterol) form lipoproteins, which facilitates their transport between tissues in an aqueous environment—the plasma. Unlike glucose and amino acids, chylomicron triacylglycerol is not taken up by the liver. It is metabolized by extrahepatic tissues possessing the enzyme **lipoprotein lipase,** which hydrolyzes the triacylglycerol, releasing fatty acids that are incorporated into tis-

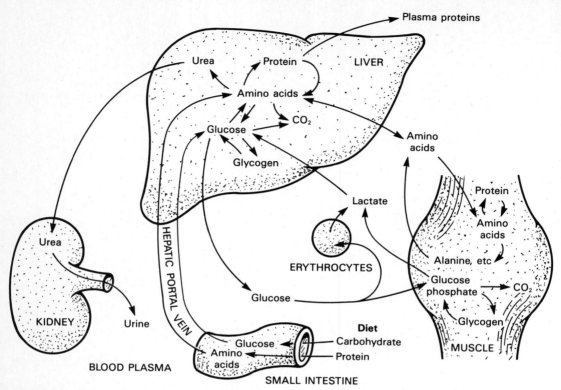

Figure 16–5. Transport and fate of major carbohydrate and amino acid substrates and metabolites. Note that there is little free glucose in muscle, since it is rapidly phosphorylated upon entry.

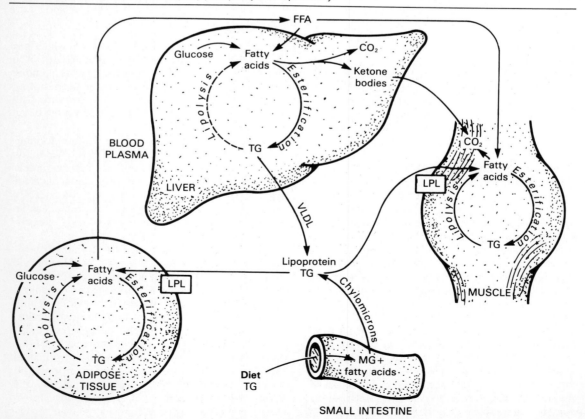

Figure 16–6. Transport and fate of major lipid substrates and metabolites. FFA, free fatty acids; LPL, lipoprotein lipase; MG, monoacylglycerol; TG, triacylglycerol; VLDL, very low density lipoprotein.

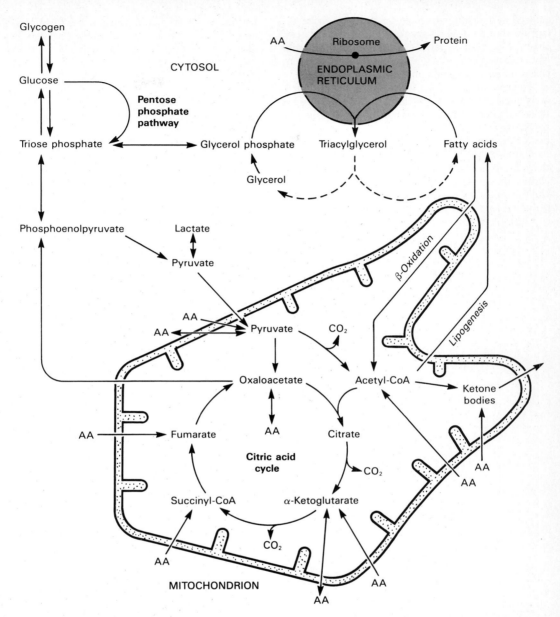

Figure 16–7. Intracellular location and integration of major metabolic pathways in a liver parenchymal cell. AA →, metabolism of one or more essential amino acids; AA ⟷, metabolism of one or more nonessential amino acids.

sue lipids or oxidized as fuel. The other major source of long-chain fatty acid is synthesis **(lipogenesis)** from carbohydrate, mainly in adipose tissue and the liver.

Adipose tissue triacylglycerol is the main fuel reserve of the body. Subsequent to its hydrolysis **(lipolysis),** fatty acids are released into the circulation. Free fatty acids are taken up by most tissues (but not brain or erythrocytes) and esterified to acylglycerols or oxidized as a major fuel to CO_2. Two pathways of additional importance occur in liver: (1) Surplus triacylglycerol arising from both lipogenesis and free fatty acids is secreted into the circulation as **very low den-**

sity lipoprotein (VLDL). This triacylglycerol undergoes a fate similar to that of chylomicrons. (2) Partial oxidation of free fatty acid leads to ketone body production **(ketogenesis).** Ketone bodies are transported to extrahepatic tissues where they act as another major fuel source.

Intermediary Metabolism at the Subcellular Level

A summary of the main biochemical functions of the subcellular components and organelles of the cell is given in Table 2–4. However, most cells are specialized in their functions and tend to emphasize certain metabolic pathways and relegate others. Fig 16–7 depicts the major metabolic pathways and their integration in a hepatic parenchymal cell, with special emphasis on their intracellular location.

The central role of the **mitochondrion** is immediately apparent, since it acts as the focus and crossroad of carbohydrate, lipid, and amino acid metabolism. In particular, it houses the enzymes of the citric acid cycle, of the respiratory chain and ATP synthase, of β-oxidation of fatty acids, and of ketone body production. In addition, it is the collecting point for the carbon skeletons of amino acids after transamination and for providing these skeletons for the synthesis of the nonessential amino acids.

Glycolysis, the pentose phosphate pathway, and fatty acid synthesis all occur in the **cytosol.** It will be noticed that in gluconeogenesis, even substances such as lactate and pyruvate that are formed in the cytosol must enter the mitochondrion and form oxaloacetate before conversion to glucose.

The membranes of the **endoplasmic reticulum** contain the enzyme system for acylglycerol synthesis, and the ribosomes are responsible for protein synthesis.

It will be appreciated that the transport of metabolites of varying size, charge, and solubility through the membranes separating organelles involves complex mechanisms. Some have been discussed in relation to the mitochondrial membrane (see Chapter 13), and others will be discussed in succeeding chapters.

The Citric Acid Cycle: The Catabolism of Acetyl-CoA

17

Peter A. Mayes, PhD, DSc

INTRODUCTION

The citric acid cycle (Krebs cycle, tricarboxylic acid cycle) is a series of reactions in mitochondria that bring about the catabolism of acetyl residues, liberating hydrogen equivalents, which, upon oxidation, lead to the release of most of the free energy of tissue fuels. The acetyl residues are in the form of **acetyl-CoA** ($CH_3-CO\sim S-CoA$, active acetate), an ester of coenzyme A. CoA contains the vitamin pantothenic acid.

BIOMEDICAL IMPORTANCE

The major function of the citric acid cycle is to act as the final common pathway for the oxidation of carbohydrate, lipids, and protein, since glucose, fatty acids, and many amino acids are all metabolized to acetyl-CoA or intermediates of the cycle. It also plays a major role in gluconeogenesis, transamination, deamination, and lipogenesis. While several of these processes are carried out in most tissues, the liver is the only tissue in which all occur. The repercussions are therefore profound when, for example, large numbers of hepatic cells are damaged or replaced by connective tissue, as in acute hepatitis and cirrhosis, respectively. A mute testimony to the vital importance of the citric acid cycle is the fact that very few if any genetic abnormalities of its enzymes have been reported in humans; such abnormalities are presumably incompatible with normal development.

CATABOLIC ROLE OF THE CITRIC ACID CYCLE

Essentially, the cycle comprises the combination of a molecule of acetyl-CoA with the 4-carbon dicarboxylic acid oxaloacetate, resulting in the formation of a **6-carbon tricarboxylic acid, citrate.** There follows a series of reactions in the course of which 2 molecules of CO_2 are released and oxaloacetate is re-

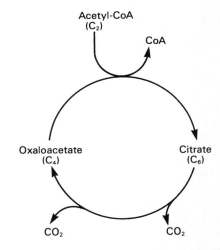

Figure 17–1. Citric acid cycle, illustrating the catalytic role of oxaloacetate.

generated (Fig 17–1). Since only a small quantity of oxaloacetate is needed to facilitate the conversion of a large quantity of acetyl units to CO_2, oxaloacetate may be considered to play a **catalytic role.**

The citric acid cycle is the mechanism by which much of the free energy liberated during the oxidation of carbohydrate, lipids, and amino acids is made available. During the course of oxidation of acetyl-CoA in the cycle, reducing equivalents in the form of hydrogen or of electrons are formed as a result of the activity of specific dehydrogenases. These reducing equivalents then enter the respiratory chain, where large amounts of ATP are generated in the process of oxidative phosphorylation (Fig 17–2; see also Chapter 13).

The enzymes of the citric acid cycle are located in the **mitochondrial matrix,** either free or attached to the inner surface of the inner mitochondrial membrane, which facilitates the **transfer of reducing equivalents to the adjacent enzymes of the respiratory chain,** situated in the inner mitochondrial membrane.

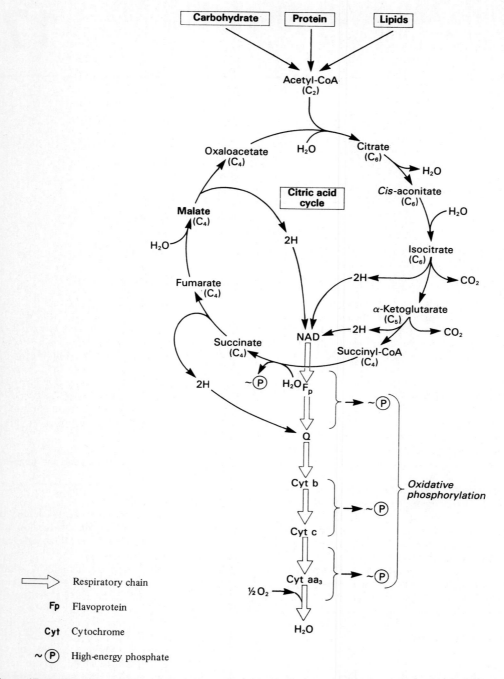

Figure 17–2. The citric acid cycle: the major catabolic pathway for acetyl-CoA in aerobic organisms. Acetyl-CoA, the product of carbohydrate, protein, and lipid catabolism, is taken up into the cycle, together with H_2O, and oxidized to CO_2 with the release of reducing equivalents (2H). Subsequent oxidation of 2H in the respiratory chain leads to coupled phosphorylation of ADP to ATP. For one turn of the cycle, 11 ~$\textcircled{P}$ are generated via oxidative phosphorylation and one ~$\textcircled{P}$ arises at substrate level from the conversion of succinyl-CoA to succinate. ⇨, Respiratory chain; F_p, flavoprotein; Cyt, cytochrome; ~$\textcircled{P}$, high-energy phosphate.

REACTIONS OF THE CITRIC ACID CYCLE
(See Fig 17–3.)*

The initial condensation of acetyl-CoA with oxaloacetate to form citrate is catalyzed by a condensing enzyme, **citrate synthase,** which effects synthesis of a carbon-to-carbon bond between the methyl carbon of acetyl-CoA and the carbonyl carbon of oxaloacetate. The condensation reaction, which forms citryl-CoA, is followed by hydrolysis of the thioester bond of CoA, accompanied by considerable loss of free energy as heat, ensuring that the reaction goes to completion.

Acetyl–CoA + Oxaloacetate + H$_2$O ⟶

Citrate + CoA·SH

Citrate is converted to isocitrate by the enzyme **aconitase** (aconitate hydratase), which contains iron in the Fe^{2+} state. This conversion takes place in 2 steps: dehydration to *cis*-aconitate, some of which remains bound to the enzyme, and rehydration to isocitrate.

Citrate ⟷ *Cis*-aconitate ⟷ Isocitrate
(enzyme bound)
H$_2$O   H$_2$O

The reaction is inhibited by **fluoroacetate,** which, in the form of fluoroacetyl-CoA, condenses with oxaloacetate to form fluorocitrate. The latter inhibits aconitase, causing citrate to accumulate.

Experiments using ^{14}C-labeled intermediates indicate that aconitase reacts with citrate in an asymmetric manner, with the result that aconitase always acts on that part of the citrate molecule that is derived from oxaloacetate. This was puzzling, since citric acid appeared to be a symmetric compound. However, it is now realized (when the molecule is viewed in 3 dimensions) that the two –CH$_2$COOH groups are not identical in space with respect to the –OH and –COOH groups. The consequences of the asymmetric action of aconitase may be appreciated by reference to the fate of labeled acetyl-CoA in the citric acid cycle as shown in Fig 17–3. It is possible that *cis*-aconitate may not be an obligatory intermediate between citrate and isocitrate but may in fact be a side branch from the main pathway.

Isocitrate undergoes dehydrogenation in the presence of **isocitrate dehydrogenase** to form oxalosuccinate. Three different isocitrate dehydrogenases have been described. One, which is NAD$^+$-specific, is found only in mitochondria. The other 2 enzymes are NADP$^+$-specific and are found in the mitochondria and the cytosol, respectively. Respiratory chain-linked oxidation of isocitrate proceeds almost completely through the NAD$^+$-dependent enzyme.

Isocitrate + NAD$^+$ ⟷ Oxalosuccinate ⟷
(enzyme bound)

α-Ketoglutarate + CO$_2$ + NADH + H$^+$

There follows a decarboxylation to α-ketoglutarate, also catalyzed by isocitrate dehydrogenase. Mn^{2+} (or Mg^{2+}) is an important component of the decarboxylation reaction. It would appear that oxalosuccinate remains bound to the enzyme as an intermediate in the overall action.

Next, α-ketoglutarate undergoes **oxidative decarboxylation** in a manner analogous to the oxidative decarboxylation of pyruvate (see Fig 18–5), both substrates being α-keto acids.

α-Ketoglutarate + NAD$^+$ + CoA·SH ⟶

Succinyl-CoA + CO$_2$ + NADH + H$^+$

The reaction, catalyzed by an **α-ketoglutarate dehydrogenase** complex, also requires identical cofactors—eg, thiamin diphosphate, lipoate, NAD$^+$, FAD, and CoA—and results in the formation of succinyl-CoA, a thioester containing a high-energy bond. The equilibrium of this reaction is so much in favor of succinyl-CoA formation that the reaction must be considered as physiologically unidirectional. As in the case of pyruvate oxidation (see p 163), arsenite inhibits the reaction, causing the substrate, **α-ketoglutarate,** to accumulate.

To continue the cycle, succinyl-CoA is converted to succinate by the enzyme **succinate thiokinase (succinyl-CoA synthetase).**

Succinyl-CoA + P$_i$ + GDP ⟷

Succinate + GTP + CoA·SH

This reaction requires GDP or IDP, which is converted in the presence of inorganic phosphate to either GTP or ITP. This is the only example in the citric acid cycle of the **generation of a high-energy phosphate at the substrate level** and arises because the release of free energy from the oxidative decarboxylation of α-ketoglutarate is sufficient to generate a high-energy bond in addition to the formation of NADH (equivalent to 3 ~℗). By means of a phosphokinase, ATP may be formed from either GTP or ITP,

eg, GTP + ADP ⟷ GDP + ATP

An alternative reaction in extrahepatic tissues, which is catalyzed by **succinyl-CoA-acetoacetate-CoA transferase (thiophorase),** is the conversion of succinyl-CoA to succinate coupled with the conversion of acetoacetate to acetoacetyl-CoA (see p 255). In liver

---

*From Circular No. 200 of the Committee of Editors of Biochemical Journals Recommendations (1975): "According to standard biochemical convention, the ending *ate* in, eg, palmitate, denotes any mixture of free acid and the ionized form(s) (according to pH) in which the cations are not specified." The same convention is adopted in this text for all carboxylic acids.

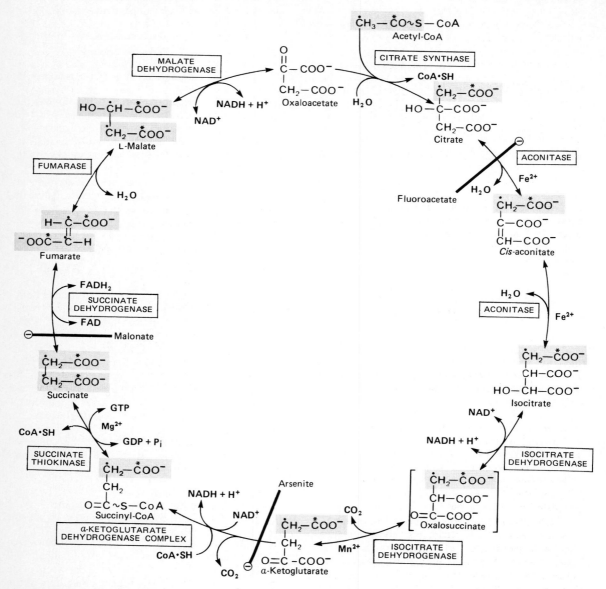

Figure 17–3. The citric acid (Krebs) cycle. Oxidation of NADH and FADH₂ in the respiratory chain leads to the generation of ATP via oxidative phosphorylation. In order to follow the passage of acetyl-CoA through the cycle, the 2 carbon atoms of the acetyl radical are shown labeled on the carboxyl carbon (using the designation [*]) and on the methyl carbon (using the designation [•]). Although 2 carbon atoms are lost as CO_2 in one revolution of the cycle, these atoms are not derived from the acetyl-CoA that has immediately entered the cycle but from that portion of the citrate molecule which derived from oxaloacetate. However, on completion of a single turn of the cycle, the oxaloacetate that is regenerated is now labeled, which leads to labeled CO_2 being evolved during the second turn of the cycle. Because succinate is a symmetric compound and because succinate dehydrogenase does not differentiate between its 2 carboxyl groups, "randomization" of label occurs at this step such that all 4 carbon atoms of oxaloacetate appear to be labeled after one turn of the cycle. During gluconeogenesis, some of the label in oxaloacetate is incorporated into glucose and glycogen (see Fig 20–1). For a discussion of the stereochemical aspects of the citric acid cycle, see Greville (1968). The sites of inhibition (⊖) by fluoroacetate, malonate, and arsenite are indicated.

there is also deacylase activity, causing some hydrolysis of succinyl-CoA to succinate plus CoA.

Succinate is metabolized further by undergoing a dehydrogenation followed by the addition of water, and subsequently by a further dehydrogenation which regenerates oxaloacetate.

$$\text{Succinate} + \text{FAD} \longleftrightarrow \text{Fumarate} + \text{FADH}_2$$

The first dehydrogenation reaction is catalyzed by **succinate dehydrogenase,** which is bound to the inner surface of the inner mitochondrial membrane. It is the only dehydrogenation in the citric acid cycle that involves the **direct transfer of hydrogen from the substrate to a flavoprotein without the participation of NAD⁺.** The enzyme contains FAD and iron-sulfur (Fe:S) protein. Fumarate is formed as a result of the dehydrogenation. Isotopic experiments have shown that the enzyme is stereospecific for the *trans* hydrogen atoms of the methylene carbons of succinate. Addition of malonate or oxaloacetate inhibits succinate dehydrogenase competitively, resulting in succinate accumulation.

Fumarase (fumarate hydratase) catalyzes the addition of water to fumarate to give malate.

$$\text{Fumarate} + \text{H}_2\text{O} \longleftrightarrow \text{L-Malate}$$

In addition to being specific for the L-isomer of malate, fumarase catalyzes the addition of the elements of water to the double bond of fumarate in the *trans* configuration. Malate is converted to oxaloacetate by **malate dehydrogenase,** a reaction requiring NAD⁺.

$$\text{L-Malate} + \text{NAD}^+ \longleftrightarrow \text{Oxaloacetate} + \text{NADH} + \text{H}^+$$

Although the equilibrium of this reaction strongly favors malate, the net flux is toward the direction of oxaloacetate because this compound, together with the other product of the reaction (NADH), is removed continuously in further reactions.

The enzymes of the citric acid cycle, except for the α-ketoglutarate and succinate dehydrogenases, are also found outside the mitochondria. While they may catalyze similar reactions, some of the enzymes, eg, malate dehydrogenase, may not in fact be the same proteins as the mitochondrial enzymes of the same name.

ENERGETICS OF THE CITRIC ACID CYCLE

As a result of oxidation catalyzed by dehydrogenase enzymes of the citric acid cycle, **3 molecules of NADH** and **one of FADH₂** are produced for each molecule of acetyl-CoA catabolized in one revolution of the cycle. These reducing equivalents are transferred to the respiratory chain in the inner mitochondrial membrane (Fig 17–2). During passage along the chain, reducing equivalents from NADH generate 3

high-energy phosphate bonds by the esterification of ADP to ATP in the process of oxidative phosphorylation (see Chapter 13). However, FADH₂ produces only 2 high-energy phosphate bonds because it transfers its reducing power to Q, thus bypassing the first site for oxidative phosphorylation in the respiratory chain (see Fig 13–6). A further high-energy phosphate is generated at the level of the cycle itself (ie, at substrate level) during the conversion of succinyl-CoA to succinate. Thus, **12 new high-energy phosphate bonds are generated for each turn of the cycle** (Table 17–1).

ROLE OF VITAMINS IN THE CITRIC ACID CYCLE

Four of the soluble vitamins of the B complex have precise roles in the functioning of the citric acid cycle. They are (1) **Riboflavin** in the form of **flavin adenine dinucleotide (FAD),** a cofactor in the α-ketoglutarate dehydrogenase complex and in succinate dehydrogenase. (2) **Niacin** in the form of **nicotinamide adenine dinucleotide (NAD),** the coenzyme for 3 dehydrogenases in the cycle, **isocitrate dehydrogenase, α-ketoglutarate dehydrogenase,** and **malate dehydrogenase.** (3) **Thiamin (B₁),** as thiamin diphosphate, the coenzyme for decarboxylation in the α-ketoglutarate dehydrogenase reaction. (4) **Pantothenic acid** as part of **coenzyme-A,** the cofactor attached to "active" acyl residues such as acetyl-CoA and succinyl-CoA. (See p 103 for further information on riboflavin and niacin.)

Thiamin (Vitamin B₁)

The structure of thiamin is shown in Fig 17–4. An ATP-dependent thiamin pyrophosphotransferase present in brain and liver is responsible for the conversion of thiamin to its active form, thiamin diphosphate (pyrophosphate).

Thiamin diphosphate serves as a coenzyme in reactions transferring an activated aldehyde unit. There are 2 types of such reactions: (1) **oxidative decarboxyla-**

Table 17–1. Generation of high-energy phosphate bonds by the citric acid cycle.

Reaction Catalyzed By	Method of ~Ⓟ Production	Number of ~Ⓟ Formed
Isocitrate dehydrogenase	Respiratory chain oxidation of NADH	3
α-Ketoglutarate dehydrogenase	Respiratory chain oxidation of NADH	3
Succinate thiokinase	Oxidation at substrate level	1
Succinate dehydrogenase	Respiratory chain oxidation of FADH₂	2
Malate dehydrogenase	Respiratory chain oxidation of NADH	3
		Net 12

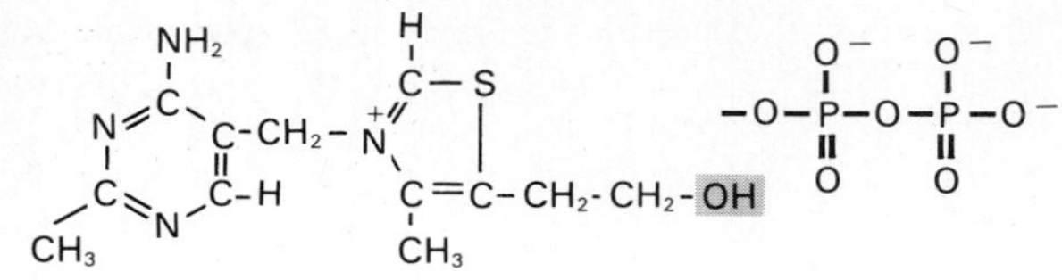

2,5,Dimethyl-
6-aminopyrimidine

4-Methyl-5-hydroxy-
ethylthiazole

Figure 17–4. Thiamin. In thiamin diphosphate, the —OH group is replaced by pyrophosphate.

tions of α-keto acids (eg, α-ketoglutarate, pyruvate, and the α-keto analogs of leucine, isoleucine, and valine); and (2) **transketolase** reactions (eg, in the pentose phosphate pathway). All of these reactions are inhibited in **thiamin deficiency** (see Chapter 53).

The role of thiamin diphosphate in the mechanism of action of pyruvate dehydrogenase complex is shown in Fig 18–5; its action in α-ketoglutarate dehydrogenase activity is analogous.

Pantothenic Acid

Pantothenic acid is an amide of pantoic acid and β-alanine (Fig 17–5). It is absorbed readily in the intestines and subsequently phosphorylated by ATP to form 4'-phosphopantothenate (Fig 17–6). On the path of conversion to the active coenzyme, **coenzyme A,** cysteine is added to the phosphopantothenate, and then the carboxyl group of cysteine is removed, resulting in the net addition of thioethanolamine, generating 4'-phosphopantetheine. Like the active coenzymes of so many other water-soluble vitamins, the active coenzyme of pantothenate contains an adenine nucleotide. Thus, 4'-phosphopantetheine is adenylated by ATP to form dephospho-coenzyme A. The final phosphorylation occurs with ATP adding phosphate to the 3'-hydroxyl group of the ribose moiety to generate coenzyme A. The **thiol group acts as a carrier of acyl groups** in reactions involving fatty acid oxidation and synthesis, acetylation reactions, and (as discussed above) oxidative decarboxylations in which thiamin diphosphate also participates. The acyl-sulfur bond is a **high-energy bond,** equivalent to the high-energy bond of ATP. Formation of these high-energy bonds therefore requires a source of energy, either from a coupled exergonic reaction or from the transfer of energy from a high-energy phosphate or a high-energy sulfur bond. It is customary to abbreviate the structure of the free (ie, reduced) coenzyme A as CoA · SH, in which the reactive SH group of the coenzyme is designated.

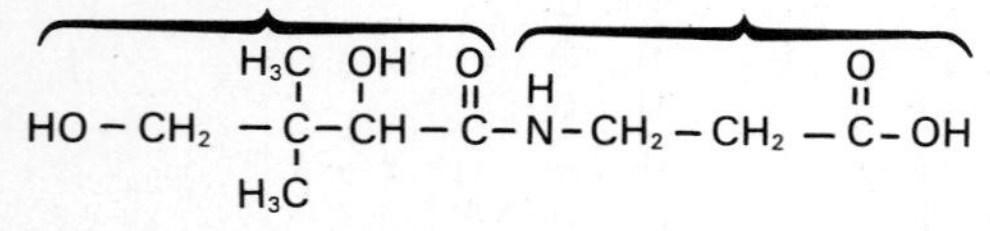

AMPHIBOLIC ROLE OF THE CITRIC ACID CYCLE

Some metabolic pathways end in a constituent of the cycle while other pathways originate from the cycle. These pathways concern the processes of gluconeogenesis, transamination, deamination, and fatty acid synthesis. Although these will be discussed in greater detail in subsequent chapters, their relationships with the cycle are summarized below.

Gluconeogenesis, Transamination, & Deamination

All major members of the cycle, from citrate to oxaloacetate, are potentially glucogenic, since they can give rise to a net production of glucose in the liver or kidney, the organs that contain a complete set of enzymes necessary for gluconeogenesis (see p 172). The key enzyme that facilitates the net transfer out of the cycle into the main pathway of gluconeogenesis is **phosphoenolpyruvate carboxykinase,** which catalyzes the decarboxylation of oxaloacetate to phosphoenolpyruvate, GTP acting as the source of high-energy phosphate (Fig 17–7).

Oxaloacetate + GTP $\longrightarrow$

Phosphoenolpyruvate + CO$_2$ + GDP

Net transfer into the cycle occurs as a result of several different reactions. Among the most significant is the formation of oxaloacetate by the carboxylation of pyruvate, catalyzed by **pyruvate carboxylase.**

ATP + CO$_2$ + H$_2$O + Pyruvate $\longrightarrow$

Oxaloacetate + ADP + P$_i$

This reaction is considered important in maintaining adequate concentrations of oxaloacetate for the condensation reaction with acetyl-CoA. If acetyl-CoA accumulates, it acts as an allosteric activator of pyruvate carboxylase, thereby ensuring a supply of oxaloacetate. Lactate, an important substrate for gluconeogenesis, enters the cycle via conversion to pyruvate and oxaloacetate.

Pantoic acid β-Alanine

$$HO-CH_2 - \underset{\underset{H_3C}{|}}{\overset{\overset{H_3C\ OH}{|\ \ |}}{C}}-CH - \overset{O}{\overset{||}{C}}-\overset{H}{\overset{|}{N}}-CH_2-CH_2-\overset{O}{\overset{||}{C}}-OH$$

Figure 17–5. Pantothenic acid.

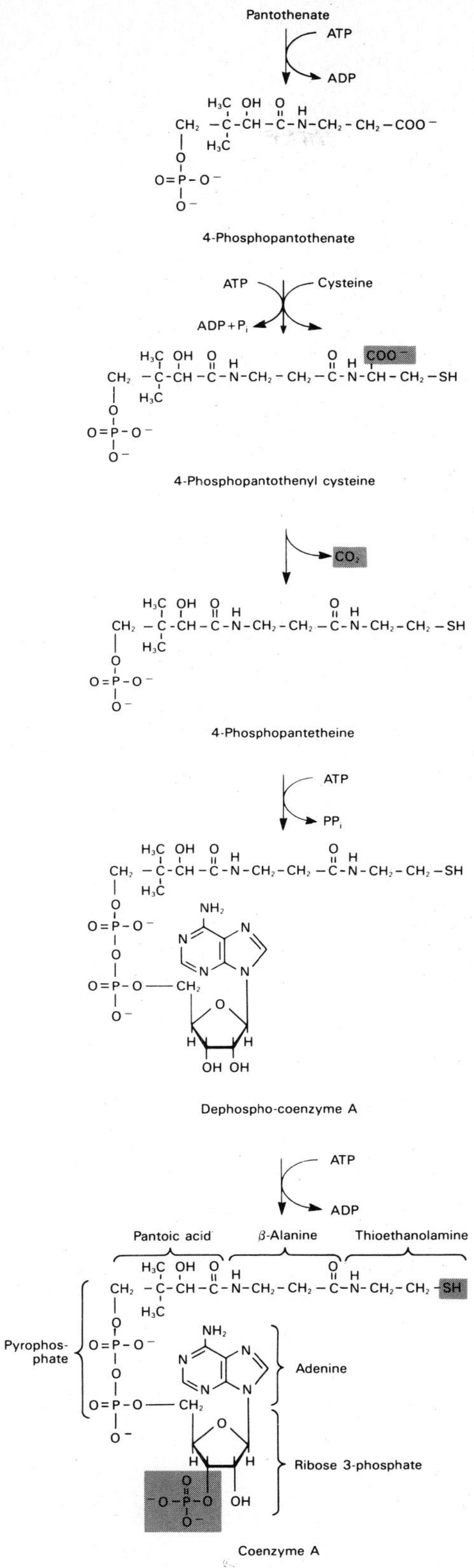

Pantothenate

ATP

ADP

4-Phosphopantothenate

ATP — Cysteine

ADP + P_i

4-Phosphopantothenyl cysteine

CO_2

4-Phosphopantetheine

ATP

PP_i

Dephospho-coenzyme A

ATP

ADP

Pantoic acid β-Alanine Thioethanolamine

Pyrophosphate

Adenine

Ribose 3-phosphate

Coenzyme A

Figure 17–6. The synthesis of coenzyme A from pantothenic acid.

Transaminase reactions produce pyruvate from alanine, oxaloacetate from aspartate, and α-ketoglutarate from glutamate. Because these reactions are reversible, the cycle also serves as a source of carbon skeletons for the synthesis of nonessential amino acids, eg,

Aspartate + Pyruvate ⟷ Oxaloacetate + Alanine

Glutamate + Pyruvate ⟷

α-Ketoglutarate + Alanine

Other amino acids contribute to gluconeogenesis because all or part of their carbon skeletons enter the citric acid cycle after deamination or transamination. Examples are alanine, cysteine, glycine, hydroxyproline, serine, threonine, and tryptophan, which form pyruvate; arginine, histidine, glutamine, and proline, which form α-ketoglutarate via glutamate; isoleucine, methionine, and valine, which form succinyl-CoA; and tyrosine and phenylalanine, which form fumarate (see Fig 17–7). Substances forming pyruvate have the option of complete oxidation to CO_2 if they follow the pyruvate dehydrogenase pathway to acetyl-CoA, or they may follow the gluconeogenic pathway via carboxylation to oxaloacetate.

Of particular significance to ruminants is the conversion of propionate, the major glucogenic product of rumen fermentation, to succinyl-CoA via the methylmalonyl-CoA pathway (see Fig 20–2).

Fatty Acid Synthesis
(See Fig 17–8.)

Acetyl-CoA, formed from pyruvate by the action of pyruvate dehydrogenase, is the major building block for long-chain fatty acid synthesis in nonruminants. (In ruminants, acetyl-CoA is derived directly from acetate.) As pyruvate dehydrogenase is a mitochondrial enzyme and the enzymes responsible for fatty acid synthesis are extramitochondrial, the cell needs to transport acetyl-CoA through the mitochondrial membrane, which is impermeable to acetyl-CoA. This is accomplished by allowing **acetyl-CoA to form citrate** in the citric acid cycle, **transporting citrate** out of the mitochondria, and finally making acetyl-CoA available in the cytosol by **cleaving citrate** in a reaction catalyzed by the enzyme **ATP-citrate lyase.**

Citrate + ATP + CoA ⟶

Acetyl-CoA + Oxaloacetate + ADP + P_i

Regulation of the Citric Acid Cycle

This is discussed in Chapter 22.

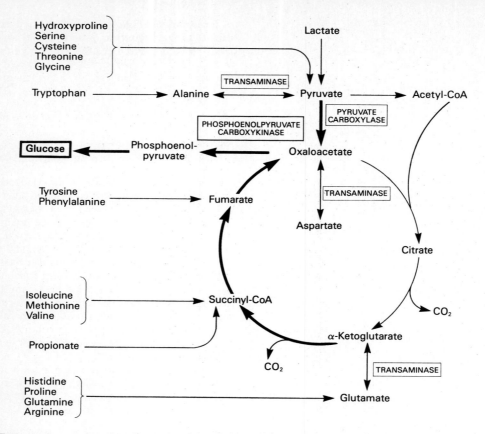

Figure 17–7. Involvement of the citric acid cycle in transamination and gluconeogenesis. The bold arrows indicate the main pathway of gluconeogenesis.

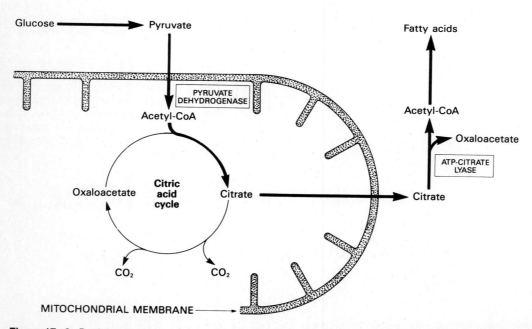

Figure 17–8. Participation of the citric acid cycle in fatty acid synthesis from glucose. See also Fig 23–9.

REFERENCES

Boyer PD (editor): *The Enzymes*, 3rd ed. Academic Press, 1971.

Goodwin TW (editor): *The Metabolic Roles of Citrate*. Academic Press, 1968.

Greville GD: Vol 1, p 297, in: *Carbohydrate Metabolism and Its Disorders*. Dickens F, Randle PJ, Whelan WJ (editors). Academic Press, 1968.

Lowenstein JM: Vol 1, p 146, in: *Metabolic Pathways,* 3rd ed.

Greenberg DM (editor). Academic Press, 1967.

Lowenstein JM (editor): *Citric Acid Cycle: Control and Compartmentation*. Dekker, 1969.

Lowenstein JM (editor): *Citric Acid Cycle*. Vol 13 in: *Methods in Enzymology*. Academic Press, 1969.

Srere PA: The enzymology of the formation and breakdown of citrate. *Adv Enzymol* 1975;**43:**57.

18 Glycolysis & the Oxidation of Pyruvate

Peter A. Mayes, PhD, DSc

INTRODUCTION

There is a minimal requirement for glucose in all tissues, and in some (eg, brain and erythrocytes) the requirement is substantial. Glycolysis is the major pathway for the utilization of glucose and is found in all cells. It is a unique pathway, since it can utilize oxygen if available (aerobic), or it can function in the absence of oxygen (anaerobic).

BIOMEDICAL IMPORTANCE

Glycolysis is not only the principal route for glucose metabolism leading to the production of acetyl-CoA and oxidation in the citric acid cycle, but it also provides the main pathway for the metabolism of fructose and galactose derived from the diet. Of crucial biomedical significance is the ability of glycolysis to provide ATP in the absence of oxygen, because this allows skeletal muscle to perform at very high levels when aerobic oxidation becomes insufficient and it allows tissues with significant glycolytic ability to survive anoxic episodes. Conversely, heart muscle, which is adapted for aerobic performance, has relatively poor glycolytic ability and poor survival under conditions of ischemia. A small number of diseases occur in which enzymes of glycolysis (eg, pyruvate kinase) are deficient in activity; these conditions are mainly manifested as hemolytic anemias. In fast-growing cancer cells, glycolysis proceeds at a much higher rate than is required by the citric acid cycle. Thus, more pyruvate is produced than can be metabolized. This in turn results in excessive production of lactate, which favors a relatively acid local environment in the tumor, a situation that may have implications for certain types of cancer therapy. Lactic acidosis also results from pyruvate dehydrogenase deficiency.

THE PATHWAY OF GLYCOLYSIS

The Nature of Anaerobic Glycolysis

At an early period in the course of investigations on carbohydrate metabolism it was realized that the process of fermentation in yeast was similar to the breakdown of glycogen in muscle. The early investigations of the glycolytic pathway were carried out on these 2 systems.

In studies on the biochemical changes that occur during muscular contraction, it was noted that when a muscle contracts in an anaerobic medium, ie, one from which oxygen is excluded, **glycogen disappears** and **pyruvate and lactate appear** as the principal end products. When oxygen is admitted, aerobic recovery takes place and glycogen reappears, while pyruvate and lactate disappear. However, if contraction takes place under aerobic conditions, lactate does not accumulate and pyruvate is oxidized further to CO_2 and water. As a result of these observations, it has been customary to separate carbohydrate metabolism into anaerobic and aerobic phases. However, this distinction is arbitrary, since the reactions in glycolysis are the same in the presence of oxygen as in its absence,

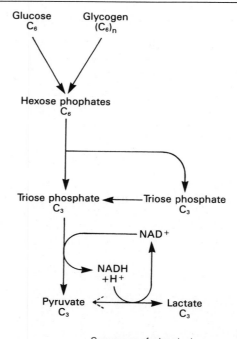

Summary of glycolysis.

Figure 18–1. Summary of glycolysis.

158

except in extent and end products. When oxygen is in short supply, reoxidation of NADH formed from NAD during glycolysis is impaired. Under these circumstances, NADH is reoxidized by coupling to the reduction of pyruvate to lactate, and the NAD so formed allows further glycolysis to proceed (Fig 18–1). Thus, glycolysis can take place under anaerobic conditions, but this has a price, for it limits the amount of energy liberated per mole of glucose oxidized. Consequently, **to provide a given amount of energy, more glucose must undergo glycolysis under anaerobic as compared with aerobic conditions.**

SEQUENCE OF REACTIONS IN GLYCOLYSIS

The overall equation for glycolysis to lactate is

Glucose + 2ADP + 2P$_i$ $\longrightarrow$

2L-Lactate + 2ATP + 2H$_2$O

All of the enzymes of the glycolysis pathway (Fig 18–2) are found in the extramitochondrial soluble fraction of the cell, the cytosol. They catalyze the reactions involved in the glycolysis of glucose to pyruvate and lactate, as follows:

Glucose enters into the glycolytic pathway by phosphorylation to glucose 6-phosphate. This is accomplished by the enzyme **hexokinase** and in liver parenchymal cells by **glucokinase,** whose activity is inducible and affected by changes in the nutritional state. ATP is required as phosphate donor, and as in many reactions involving phosphorylation, it reacts as the Mg-ATP complex. One high-energy phosphate bond of ATP is utilized, and ADP is produced. The reaction is accompanied by considerable loss of free energy as heat and therefore, under physiologic conditions, may be regarded as irreversible. Hexokinase is inhibited in an allosteric manner by the product, glucose 6-phosphate.

α-D-Glucose + ATP $\xrightarrow{\text{Mg}^{2+}}$

α-D-Glucose 6-phosphate + ADP

Hexokinase, present in all cells except those of the liver parenchyma, has a high affinity (low K$_m$) for its substrate, glucose. Its function is to ensure a supply of glucose for the tissues, even in the presence of low blood glucose concentrations, by phosphorylating all the glucose that enters the cell, thereby maintaining a large glucose concentration gradient between the blood and the intracellular environment. It acts on both the α- and β-anomer of glucose and will also catalyze the phosphorylation of other hexoses but at a much slower rate than glucose.

The function of glucokinase is to remove glucose from the blood following a meal. In contrast to hexokinase, it has a high K$_m$ for glucose and operates opti-

mally at blood glucose concentrations above 100 mg/dL (see Fig 22–7). It is specific for glucose.

Glucose 6-phosphate is an important compound at the junction of several metabolic pathways (glycolysis, gluconeogenesis, the pentose phosphate pathway, glycogenesis, and glycogenolysis) (Fig 18–2). In glycolysis it is converted to fructose 6-phosphate by **phosphohexose isomerase,** which involves an aldose-ketose isomerization. Only the α-anomer of glucose 6-phosphate is acted upon.

α-D-Glucose 6-phosphate $\longleftrightarrow$

α-D-Fructose 6-phosphate

This reaction is followed by another phosphorylation with ATP catalyzed by the enzyme **phosphofructokinase (phosphofructokinase-1)** to produce fructose 1,6-bisphosphate. Phosphofructokinase is another inducible enzyme whose activity is considered to play a major role in the regulation of the rate of glycolysis. The phosphofructokinase reaction is another that may be considered to be functionally irreversible under physiologic conditions.

D-Fructose 6-phosphate + ATP $\longrightarrow$

D-Fructose 1,6-bisphosphate

Fructose 1,6-bisphosphate is split by **aldolase** (fructose 1,6-bisphosphate aldolase) into 2 triose phosphates, glyceraldehyde 3-phosphate and dihydroxyacetone phosphate.

D-Fructose 1,6-bisphosphate $\longleftrightarrow$

D-Glyceraldehyde 3-phosphate
+ Dihydroxyacetone phosphate

Several different aldolases have been described, all of which contain 4 subunits. Aldolase A occurs in most tissues, and, in addition, aldolase B occurs in liver and kidney. The fructose phosphates exist in the cell mainly in the furanose form, but they react with phosphohexose isomerase, phosphofructokinase, and aldolase in the open chain configuration.

Glyceraldehyde 3-phosphate and dihydroxyacetone phosphate are interconverted by the enzyme **phosphotriose isomerase.**

D-Glyceraldehyde 3-phosphate $\longleftrightarrow$

Dihydroxyacetone phosphate

Glycolysis proceeds by the oxidation of glyceraldehyde 3-phosphate to 1,3-bisphosphoglycerate, and, because of the activity of phosphotriose isomerase, the dihydroxyacetone phosphate is also oxidized to 1,3-diphosphoglycerate via glyceraldehyde 3-phosphate.

D-Glyceraldehyde 3-phosphate + NAD$^+$ + P$_i$ $\longleftrightarrow$

1,3-Bisphosphoglycerate + NADH + H$^+$

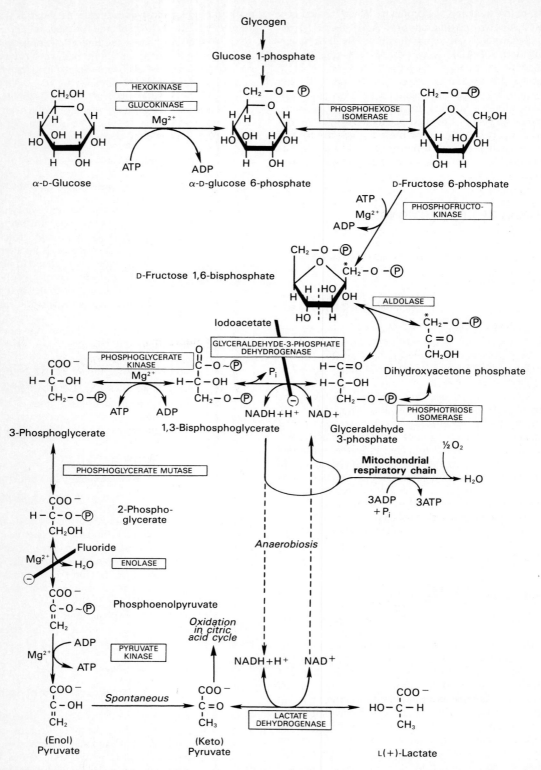

Figure 18–2. The pathway of glycolysis. Ⓟ, $-PO_3^{2-}$; P_i, $HOPO_3^{2-}$; ⊖, inhibition. *Carbon atoms 1–3 of fructose bisphosphate form dihydroxyacetone phosphate, whereas carbons 4–6 form glyceraldehyde 3-phosphate.

The enzyme responsible for the oxidation, **glyceraldehyde-3-phosphate dehydrogenase,** is NAD-dependent. Structurally, it consists of 4 identical polypeptides (monomers) forming a tetramer. Four −SH groups are present on each polypeptide, probably derived from cysteine residues within the polypeptide chain. One of the −SH groups is found at the active site of the enzyme. It is believed that the −SH group participates in the reaction in which glyceraldehyde 3-phosphate is oxidized. The substrate initially combines with a cysteinyl moiety on the dehydrogenase forming a thiohemiacetal that is converted to a thiol ester by oxidation; the hydrogens removed in this oxidation are transferred to NAD bound to the enzyme. The NADH produced on the enzyme is not so firmly bound to the enzyme as is NAD. Consequently, NADH is easily displaced by another molecule of NAD. Finally, by phosphorolysis, inorganic phosphate (P_i) is added, forming 1,3-bisphosphoglycerate, and the free enzyme with a reconstituted −SH group is liberated (Fig 18−3). Energy released during the oxidation is retained by the formation of a high-energy sulfur bond that becomes, after phosphorolysis, a high-energy phosphate bond in position 1 of 1,3-bisphosphoglycerate. This high-energy phosphate is captured as ATP in a further reaction with ADP catalyzed by **phosphoglycerate kinase,** leaving 3-phosphoglycerate.

1,3-Bisphosphoglycerate + ADP $\longleftrightarrow$

3-Phosphoglycerate + ATP

Since 2 molecules of triose phosphate are formed per molecule of glucose undergoing glycolysis, 2 molecules of ATP are generated at this stage per molecule of glucose, an example of phosphorylation "at the substrate level."

If arsenate is present, it will compete with inorganic phosphate (P_i) in the above reactions to give 1-arseno-3-phosphoglycerate, which hydrolyzes spontaneously to give 3-phosphoglycerate plus heat, without generating ATP. This is an important example of the ability of arsenate to accomplish uncoupling of oxidation and phosphorylation.

3-Phosphoglycerate arising from the above reactions is converted to 2-phosphoglycerate by the enzyme **phosphoglycerate mutase.** It is likely that 2,3-bisphosphoglycerate (diphosphoglycerate, DPG) is an intermediate in this reaction.

3-Phosphoglycerate $\longleftrightarrow$ 2-Phosphoglycerate

The subsequent step is catalyzed by **enolase** and involves a dehydration and redistribution of energy within the molecule, raising the phosphate on position 2 to the high-energy state, thus forming phosphoenolpyruvate. Enolase is inhibited by **fluoride,** a prop-

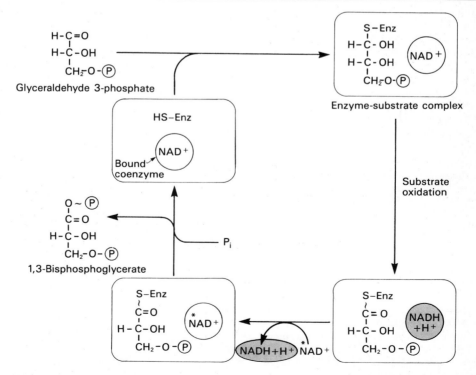

Figure 18−3. Oxidation of glyceraldehyde 3-phosphate. Enz, glyceraldehyde-3-phosphate dehydrogenase. The enzyme is inhibited by the −SH poison **iodoacetate,** which is thus able to inhibit glycolysis.

erty that can be made use of when it is required to prevent glycolysis prior to the estimation of blood glucose. The enzyme is also dependent on the presence of either Mg^{2+} or Mn^{2+}.

2-Phosphoglycerate $\longleftrightarrow$

Phosphoenolpyruvate + H_2O

The high-energy phosphate of phosphoenolpyruvate is transferred to ADP by the enzyme **pyruvate kinase** to generate, at this stage, 2 mol of ATP per mole of glucose oxidized. Enolpyruvate formed in this reaction is converted spontaneously to the keto form of pyruvate. This is another nonequilibrium reaction that is accompanied by considerable loss of free energy as heat and must be regarded as physiologically irreversible.

Phosphoenolpyruvate + **ADP** $\longrightarrow$ **Pyruvate** + **ATP**

The redox state of the tissue now determines which of 2 pathways is followed. If **anaerobic** conditions prevail, the reoxidation of NADH by transfer of reducing equivalents through the respiratory chain to oxygen is prevented. Pyruvate is reduced by the NADH to lactate, the reaction being catalyzed by **lactate dehydrogenase.** Several isozymes of this enzyme have been described and have clinical significance (see p 58).

Pyruvate + **NADH** + **H$^+$** $\longleftrightarrow$ L-**Lactate** + **NAD$^+$**

The reoxidation of NADH via lactate formation allows glycolysis to proceed in the absence of oxygen by **regenerating sufficient NAD$^+$** for another cycle of the reaction catalyzed by glyceraldehyde-3-phosphate dehydrogenase. Thus, tissues that function under **hypoxic circumstances tend to produce lactate** (Fig 18–2). This is particularly true of skeletal muscle, where the rate at which the organ performs work is not limited by its capacity for oxygenation. The additional quantities of lactate produced may be detected in the tissues and in the blood and urine. **Glycolysis in erythrocytes,** even under aerobic conditions, **always terminates in lactate,** because mitochondria that contain the enzymatic machinery for the aerobic oxidation of pyruvate are absent. The mammalian erythrocyte is unique in that about 90% of its total energy requirement is provided by glycolysis. Besides skeletal muscle and erythrocytes, other tissues that normally derive most of their energy from glycolysis and produce lactate include brain, gastrointestinal tract, renal medulla, retina, and skin. The liver, kidneys, and heart usually take up lactate but will produce it under hypoxic conditions.

Although most of the glycolytic reactions are reversible, 3 of them are markedly exergonic and must therefore be considered physiologically irreversible. These reactions are catalyzed by **hexokinase** (and glucokinase), **phosphofructokinase,** and **pyruvate kinase** and are the major sites of regulation of glycoly-

sis. Cells that are capable of effecting a net movement of metabolites in the synthetic direction of the glycolytic pathway (gluconeogenesis) do so because of the presence of different enzyme systems which provide alternative routes around the irreversible reactions catalyzed by the above-mentioned enzymes. These will be discussed under gluconeogenesis.

2,3-Bisphosphoglycerate Cycle

In the erythrocytes of many mammalian species, the step catalyzed by phosphoglycerate kinase is bypassed by a process that effectively dissipates as heat the free energy associated with the high-energy phosphate of 1,3-bisphosphoglycerate (Fig 18–4). An additional enzyme, **bisphosphoglycerate mutase,** catalyzes the conversion of 1,3-bisphosphoglycerate to 2,3-bisphosphoglycerate. The latter is converted to 3-phosphoglycerate by **2,3-bisphosphoglycerate phosphatase,** an activity also attributed to phosphoglycerate mutase. The loss of a high-energy phosphate, which means that there is no net production of ATP when glycolysis takes this route, may be of advantage to the economy of the red cell, since it would allow glycolysis to proceed when the need for ATP was minimal. However, 2,3-bisphosphoglycerate combines

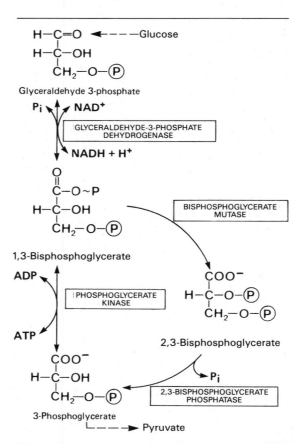

Figure 18–4. 2,3-Bisphosphoglycerate cycle in erythrocytes.

with hemoglobin, causing a decrease in affinity for oxygen and a displacement of the oxyhemoglobin dissociation curve to the right. Thus, its **presence in the red cells helps oxyhemoglobin to unload oxygen** (see Chapter 6).

OXIDATION OF PYRUVATE TO ACETYL-CoA

Before pyruvate can enter the citric acid cycle, it must be transported into the mitochondrion via a special pyruvate transporter that aids its passage across the inner mitochondrial membrane. This involves a symport mechanism whereby one proton is cotransported (see Fig 13–15). Within the mitochondrion, pyruvate is oxidatively decarboxylated to acetyl-CoA. This reaction is catalyzed by several different enzymes working sequentially in a multienzyme complex. They

are collectively designated as the **pyruvate dehydrogenase** complex and are analogous to the α-ketoglutarate dehydrogenase complex of the citric acid cycle (see p 151). Pyruvate is decarboxylated in the presence of thiamin diphosphate to a hydroxyethyl derivative of the thiazole ring of enzyme-bound thiamin diphosphate, which in turn reacts with oxidized lipoamide to form acetyl lipoamide (Fig 18–5). In the presence of **dihydrolipoyl transacetylase,** acetyl lipoamide reacts with coenzyme A to form acetyl-CoA and reduced lipoamide. The cycle of reaction is completed when the latter is reoxidized by a flavoprotein in the presence of **dihydrolipoyl dehydrogenase.** Finally, the reduced flavoprotein is oxidized by NAD, which in turn transfers reducing equivalents to the respiratory chain.

Pyruvate + NAD⁺ + CoA ⟶

Acetyl-CoA + NADH + H⁺ + CO₂

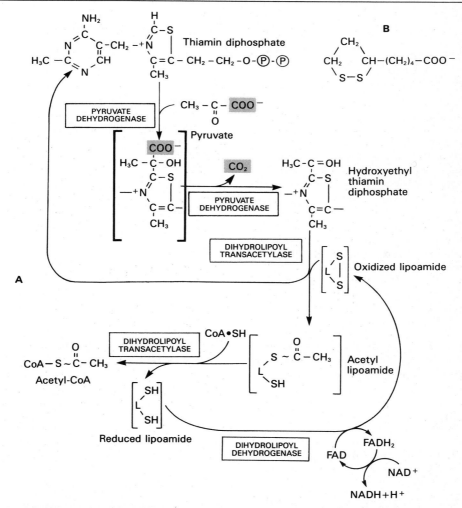

Figure 18–5. *A:* Oxidative decarboxylation of pyruvate by the pyruvate dehydrogenase complex. *B:* Lipoic acid. Lipoic acid is joined by an amide link to a lysine residue of the transacetylase component of the enzyme complex.

Table 18–1. Generation of high-energy phosphate bonds in the catabolism of glucose.

Pathway	Reaction Catalyzed By	Method of ~P Production	Number of ~P Formed per Mole of Glucose
Glycolysis	Glyceraldehyde-3-phosphate dehydrogenase	Respiratory chain oxidation of 2 NADH	6*
	Phosphoglycerate kinase	Oxidation at substrate level	2
	Pyruvate kinase	Oxidation at substrate level	2
			10
	Allow for consumption of ATP by reactions catalyzed by hexokinase and phosphofructokinase		−2
			Net 8
Citric acid cycle	Pyruvate dehydrogenase	Respiratory chain oxidation of 2 NADH	6
	Isocitrate dehydrogenase	Respiratory chain oxidation of 2 NADH	6
	α-Ketoglutarate dehydrogenase	Respiratory chain oxidation of 2 NADH	6
	Succinate thiokinase	Oxidation at substrate level	2
	Succinate dehydrogenase	Respiratory chain oxidation of 2 FADH$_2$	4
	Malate dehydrogenase	Respiratory chain oxidation of 2 NADH	6
			Net 30
	Total per mole of glucose under aerobic conditions		38
	Total per mole of glucose under anaerobic conditions		2

*It is assumed that NADH formed in glycolysis is transported into mitochondria via the malate shuttle (see Fig 13–15). If the glycerophosphate shuttle is used, only 2 ~P would be formed per mole of NADH, the total net production being 36 instead of 38. The calculation ignores the small loss of ATP due to a transport of H^+ into the mitochondrion with pyruvate and a similar transport of H^+ in the operation of the malate shuttle, totaling about 1 mol of ATP.

The pyruvate dehydrogenase complex consists of about 29 mol of pyruvate dehydrogenase and about 8 mol of flavoprotein (dihydrolipoyl dehydrogenase) distributed around 1 mol of transacetylase. Movement of the individual enzymes appears to be restricted, and the metabolic intermediates do not dissociate freely but remain bound to the enzymes.

It is to be noted that the pyruvate dehydrogenase system is sufficiently electronegative with respect to the respiratory chain that, in addition to generating a reduced coenzyme (NADH), it also generates a high-energy thio ester bond in acetyl-CoA.

Clinical Aspects of Pyruvate Metabolism

Arsenite or mercuric ions complex the –SH groups of lipoic acid and inhibit pyruvate dehydrogenase, as does a dietary deficiency of thiamin, allowing pyruvate to accumulate. Nutritionally deprived alcoholics are thiamin-deficient and if administered glucose exhibit rapid accumulation of pyruvate and **lactic acidosis,** which is frequently lethal. Patients with inherited pyruvate dehydrogenase deficiency present with a similar lactic acidosis, particularly after glucose load. Mutations have been reported for virtually all of the enzymes of carbohydrate metabolism, each associated with human disease.

Energetics of Carbohydrate Oxidation

When 1 mol of glucose is combusted in a calorimeter to CO_2 and water, approximately 2780 kJ are liberated as heat. When oxidation occurs in the tissues, some of this energy is not lost immediately as heat but is "captured" in high-energy phosphate bonds. On the order of 38 high-energy phosphate bonds are generated per molecule of glucose oxidized to CO_2 and water. Assuming each high-energy bond to be equivalent to 30.5 kJ, the total energy captured in ATP per mole of glucose oxidized is 1159 kJ, or approximately 41.7% of the energy of combustion. Most of the ATP is formed as a consequence of oxidative phosphorylation resulting from the reoxidation of reduced coenzymes by the respiratory chain. The remainder is generated by phosphorylation at the "substrate level." (See Chapter 13.) Table 18–1 indicates the reactions responsible for the generation of high-energy phosphate during oxidation of glucose and the net production under aerobic and anaerobic conditions.

REFERENCES

Blass JP: Disorders of pyruvate metabolism. *Neurology* 1979; **29:** 280.

Boyer PD (editor): *The Enzymes,* 3rd ed. Vols 5–9. Academic Press, 1972.

Dickens F, Randle PJ, Whelan WJ (editors): *Carbohydrate Metabolism and Its Disorders.* 2 vols. Academic Press, 1968.

Greenberg DM (editor): *Metabolic Pathways,* 3rd ed. Vol 1. Academic Press, 1967.

Randle PJ, Steiner DF, Whelan WJ (editors): *Carbohydrate Metabolism and Its Disorders.* Vol 3. Academic Press, 1981.

Veneziale CM (editor): *The Regulation of Carbohydrate Formation and Utilization in Mammals.* University Park Press, 1981.

Metabolism of Glycogen

Peter A. Mayes, PhD, DSc

INTRODUCTION

Glycogen is the major storage form of carbohydrate in animals and corresponds to starch in plants. It occurs mainly in liver (up to 6%) and muscle, where it rarely exceeds 1% (Table 19–1). Like starch, it is a branched polymer of α-glucose (see Fig 14–15).

BIOMEDICAL IMPORTANCE

The function of muscle glycogen is to act as a readily available source of hexose units for glycolysis **within the muscle itself.** Liver glycogen is largely concerned with export of hexose units for maintenance of the **blood glucose,** particularly between meals. After 12–18 hours of fasting, the liver becomes almost totally depleted of glycogen. Muscle glycogen is only depleted significantly after prolonged vigorous exercise. Higher concentrations of muscle glycogen can be induced by feeding high-carbohydrate diets after depletion by exercise. "Glycogen storage diseases" are a group of inherited disorders characterized by deficient mobilization of glycogen and deposition of abnormal forms of glycogen.

GLYCOGENESIS

The Pathway of Glycogen Biosynthesis (See Fig 19–1.)

Glucose is phosphorylated to glucose 6-phosphate, a reaction that is common to the first reaction in the pathway of glycolysis from glucose. This reaction is catalyzed by **hexokinase** in muscle and **glucokinase** in liver. Glucose 6-phosphate is converted to glucose

1-phosphate in a reaction catalyzed by the enzyme **phosphoglucomutase.** The enzyme itself is phosphorylated, and the phospho- group takes part in a reversible reaction in which glucose 1,6-bisphosphate is an intermediate.

Enz-P + Glucose 6-phosphate $\longleftrightarrow$

Enz + Glucose 1,6-bisphosphate $\longleftrightarrow$

Enz-P + Glucose 1-phosphate

Next, glucose 1-phosphate reacts with uridine triphosphate (UTP) to form the active nucleotide **uridine diphosphate glucose (UDPGlc)*** (Fig 19–2).

The reaction between glucose 1-phosphate and uridine triphosphate is catalyzed by the enzyme **UDPGlc pyrophosphorylase.**

UTP + Glucose 1-phosphate $\longleftrightarrow$ UDPGlc + PP$_i$

The subsequent hydrolysis of inorganic pyrophosphate by **inorganic pyrophosphatase** pulls the reaction to the right of the equation.

By the action of the enzyme **glycogen synthase (or glucosyltransferase),** the C_1 of the activated glucose of UDPGlc forms a glycosidic bond with the C_4 of a terminal glucose residue of glycogen, liberating uridine diphosphate (UDP). A preexisting glycogen molecule, or "primer," must be present to initiate this reaction. The glycogen primer may in turn be formed on a protein backbone, which may be a process similar to the synthesis of other glycoproteins (see Chapter 54).

$$\text{UDPGlc} + \underset{\text{glycogen}}{(C_6)_n} \longrightarrow \text{UDP} + \underset{\text{glycogen}}{(C_6)_{n+1}}$$

The Mechanism of Branching

The addition of a glucose residue to a preexisting glycogen chain, or "primer," occurs at the nonreducing, outer end of the molecule so that the "branches" of the glycogen "tree" become elongated as successive **-1→4- linkages** occur (Fig 19–3). When the chain has

Table 19–1. Storage of carbohydrate in postabsorptive normal adult humans (70 kg).

Liver glycogen	4.0% =	72 g*
Muscle glycogen	0.7% =	245 g†
Extracellular glucose	0.1% =	10 g‡
		327 g

*Liver weight, 1800 g.
†Muscle mass, 35 kg.
‡Total volume, 10 L.

---

*Other nucleoside diphosphate sugar compounds are known, eg, UDPGal. In addition, the same sugar may be linked to different nucleotides. For example, glucose may be linked to uridine (as shown above) as well as to guanosine, thymidine, adenosine, or cytidine nucleotides.

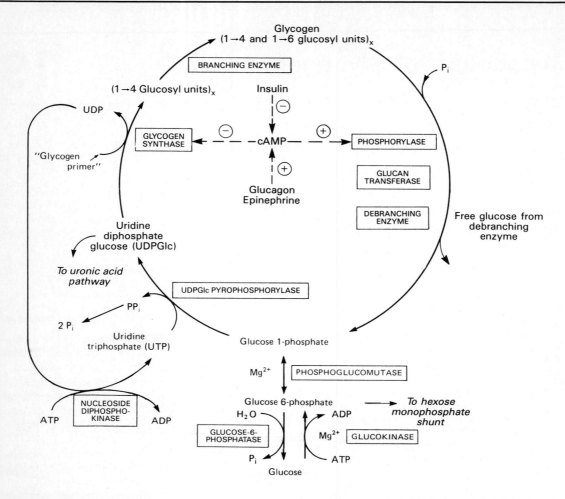

Figure 19–1. Pathway of glycogenesis and of glycogenolysis in the liver. Two high-energy phosphate bonds are used in the incorporation of 1 mol of glucose into glycogen. ⊕, Stimulation; ⊖, inhibition. Insulin decreases the level of cAMP only after it has been raised by glucagon or epinephrine; ie, it antagonizes their action.

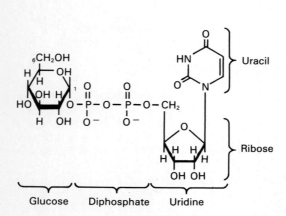

Figure 19–2. Uridine diphosphate glucose (UDPGlc).

been lengthened to a minimum of 11 glucose residues, a second enzyme, the **branching enzyme (amylo-[1→4]→[1 → 6]-transglucosidase)** transfers a part of the -1→4- chain (minimum length 6 glucose residues) to a neighboring chain to form a **-1→6- linkage,** thus establishing a **branch point** in the molecule. The branches grow by further additions of -1→4- glucosyl units and further branching.

The action of the branching enzyme has been studied in the living animal by feeding ^{14}C-labeled glucose and examining the liver glycogen at intervals thereafter (Fig 19–3).

GLYCOGENOLYSIS

Pathway of Degradation & Debranching
(See Fig 19–1.)

It is the step catalyzed by **phosphorylase** that is rate-limiting in glycogenolysis.

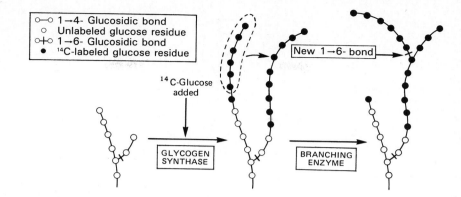

Figure 19–3. The biosynthesis of glycogen. The mechanism of branching as revealed by the addition of ¹⁴C-labeled glucose.

$$\underset{\text{glycogen}}{(C_6)_n} + P_i \longrightarrow \underset{\text{glycogen}}{(C_6)_{n-1}} + \text{Glucose 1-phosphate}$$

This enzyme is specific for the phosphorylytic breaking (phosphorolysis) of the -1→4- linkages of glycogen to yield glucose 1-phosphate. Glucosyl residues from the outermost chains of the glycogen molecule are removed until approximately 4 glucose residues remain on either side of a -1→6- branch (Fig 19–4). Another enzyme (α-[1→4]→α-[1→4] glucan transferase) transfers a trisaccharide unit from one branch to the other, exposing the -1→6- branch points. The **hydrolytic** splitting of the -1→6- linkages requires the action of a specific **debranching enzyme (amylo-[1→6]-glucosidase),** which appears to be a second ac-

tivity of the glucan transferase.* With the removal of the branch, further action by phosphorylase can proceed. The combined action of phosphorylase and these other enzymes leads to the complete breakdown of glycogen. The reaction catalyzed by phosphoglucomutase is reversible, so that glucose 6-phosphate can be formed from glucose 1-phosphate. In **liver** and **kidney** (but not in muscle), there is a specific enzyme, **glucose-6-phosphatase,** that removes phosphate from glucose 6-phosphate, enabling glucose to diffuse from the cell into the blood. This is the final step in hepatic glycogenolysis, which is reflected by a rise in the blood glucose.

CONTROL MECHANISMS OF GLYCOGENOLYSIS & GLYCOGENESIS

The principal enzymes controlling glycogen metabolism—glycogen phosphorylase and glycogen synthase—are regulated by a complex series of reactions involving both allosteric mechanisms (see p 87) and covalent modifications due to phosphorylation and dephosphorylation of enzyme protein (see p 90).

Activation & Inactivation of Phosphorylase (See Fig 19–5.)

In liver, the enzyme exists in both an active and an inactive form. Active phosphorylase (**phosphorylase a**) has one of its serine hydroxyl groups phosphorylated in an ester linkage. By the action of a specific

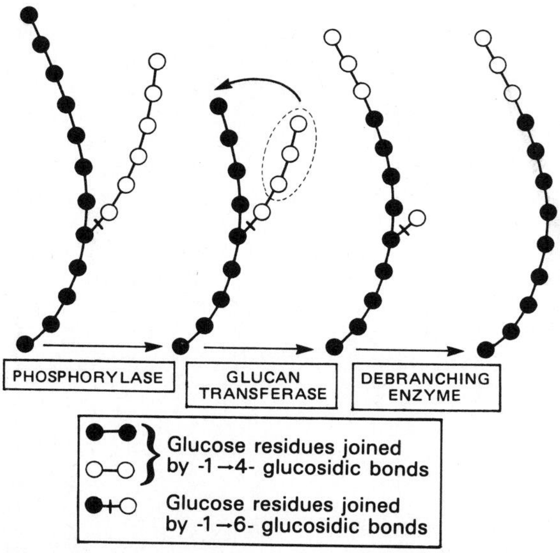

Figure 19–4. Steps in glycogenolysis.

*Because the -1→6- linkage is hydrolytically split, 1 mol of free glucose is produced rather than 1 mol of glucose 1-phosphate. In this way, it is possible for some rise in the blood glucose to take place even in the absence of glucose-6-phosphatase, as occurs in type I glycogen storage disease (von Gierke's disease; see below) after glucagon or epinephrine is administered.

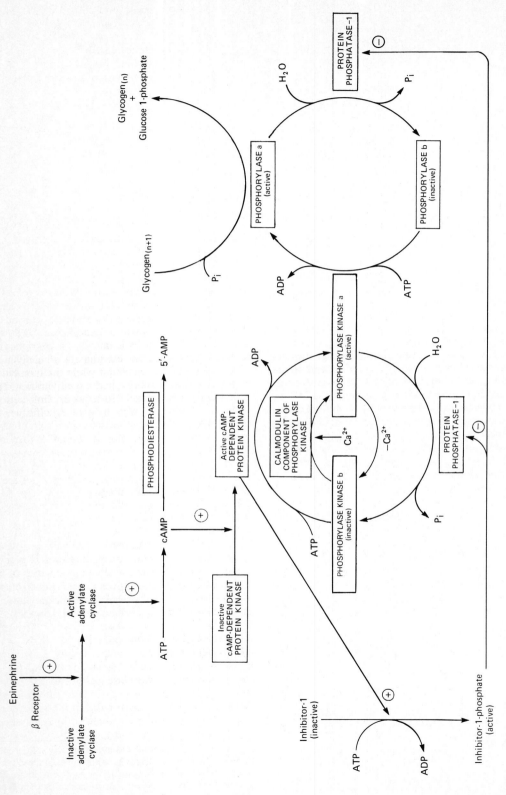

Figure 19–5. Control of phosphorylase in muscle (n = number of glucose residues). The sequence of reactions arranged as a cascade allows amplification of the hormonal signal at each step.

phosphatase (protein phosphatase-1), the enzyme is inactivated to phosphorylase b in a reaction that involves hydrolytic removal of the phosphate from the serine residue. Reactivation requires rephosphorylation with ATP and a specific enzyme, phosphorylase kinase.

Muscle phosphorylase is immunologically and genetically distinct from that of liver. It is present in 2 forms: phosphorylase a, which is phosphorylated and active in either the presence or absence of AMP (its allosteric modifier), and phosphorylase b, which is dephosphorylated and active only in the presence of AMP. Phosphorylase a is the normal physiologically active form of the enzyme. It is a dimer, each monomer containing 1 mol of pyridoxal phosphate.

Activation via cAMP

Phosphorylase in muscle is activated by epinephrine (Fig 19-5). However, this occurs not as a direct effect but rather by way of the action of cAMP (3',5'-cyclic adenylic acid; cyclic AMP) (see Fig 19-6 and p 475). cAMP is the intracellular intermediate compound or second messenger through which many hormones act. It is formed from ATP by an enzyme, adenylate cyclase, occurring in the inner surface of cell membranes. Adenylate cyclase is activated by hormones such as epinephrine and norepinephrine acting through β-adrenergic receptors on the cell membrane and additionally in liver by glucagon acting through an independent glucagon receptor. cAMP is destroyed by a phosphodiesterase, and it is the activity of this enzyme that maintains the normally low level of cAMP. Insulin has been reported to increase its activity in liver, thereby lowering the concentration of cAMP.

Increasing the concentration of cAMP activates an enzyme of rather wide specificity, cAMP-dependent protein kinase. This kinase catalyzes the phosphorylation by ATP of inactive phosphorylase kinase b to active phosphorylase kinase a, which in turn, by means of a further phosphorylation, activates phosphorylase b to phosphorylase a (Fig 19-5).

Inactive cAMP-dependent protein kinase comprises 2 pairs of subunits, each pair consisting of a regulatory subunit (R), which binds 2 mol of cAMP, and a catalytic subunit (C), which contains the active site. Combination with cAMP causes the R_2C_2 complex to dissociate, releasing active C monomers (see p 476).

$$R_2C_2 + 4cAMP \longleftrightarrow 2C + 2(R-cAMP_2)$$

Inactive enzyme   Active enzyme

Activation by Ca^{2+} & Synchronization With Muscle Contraction

Glycogenolysis increases in muscle several hundred-fold immediately after the onset of contraction. This involves the rapid activation of phosphorylase owing to activation of phosphorylase kinase by Ca^{2+}, the same signal that initiates contraction. Muscle phosphorylase kinase has 4 types of subunits, α, β, γ, and δ, in a structure represented as $(\alpha\beta\gamma\delta)_4$. The α and β subunits contain serine residues that are phosphorylated by cAMP-dependent protein kinase. The β subunit binds 4 Ca^{2+} and is identical to the Ca^{2+} binding protein calmodulin. The binding of Ca^{2+} activates the catalytic site of the γ subunit while the molecule remains in the dephosphorylated b configuration. However, the phosphorylated a form is only fully activated in the presence of Ca^{2+}. It is of significance that calmodulin is similar in structure to TpC, the Ca^{2+} binding protein in muscle. A second molecule of calmodulin or TpC can interact with the phosphorylase kinase, causing further activation. Thus, activation of muscle contraction and glycogenolysis are carried out by the same Ca^{2+} binding protein. Calmodulin is a protein that effects many of the actions of calcium in the cell (see p 479).

Glycogenolysis in Liver

Studies on liver have shown that α_1 receptors are the major mediators of catecholamine stimulation of glycogenolysis. This involves a cAMP-independent mobilization of Ca^{2+} from mitochondria into the cytosol, followed by the stimulation of a Ca^{2+}/calmodulin-sensitive phosphorylase kinase. Skeletal muscle phosphorylase is not affected by glucagon, although heart muscle is. Another important difference is that liver protein phosphatase-1 is inhibited by the active form of phosphorylase (see p 192).

Inactivation of Phosphorylase

Both phosphorylase a and phosphorylase kinase a are dephosphorylated and inactivated by protein phosphatase-1. Protein phosphatase-1 is inhibited by a protein called inhibitor-1, which is active only after it has been phosphorylated by cAMP-dependent protein kinase. Thus, cAMP controls both the activation and inactivation of phosphorylase (Fig 19-5).

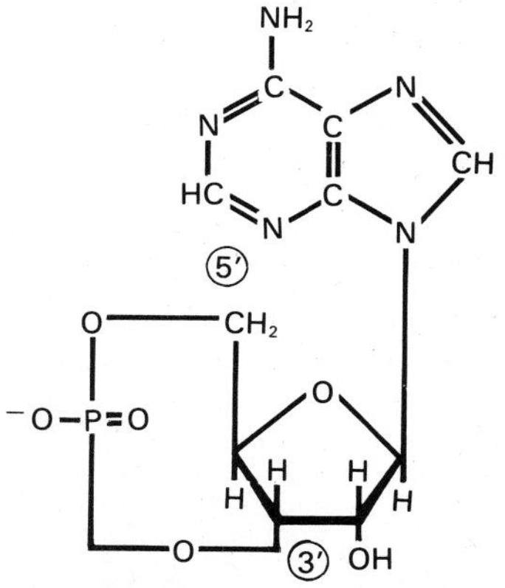

Figure 19-6. 3',5'-Adenylic acid (cyclic AMP; cAMP).

Activation & Inactivation of Glycogen Synthase
(See Fig 19–7.)

Like phosphorylase, glycogen synthase exists in either a phosphorylated or nonphosphorylated state. However, unlike phosphorylase, the active form is dephosphorylated (**glycogen synthase a**) and may be inactivated to **glycogen synthase b** by phosphorylation on 7 serine residues by no fewer than 6 different protein kinases. All 7 phosphorylation sites are contained on each of 4 identical subunits. Two of the protein kinases are Ca^{2+}/calmodulin-dependent (one of these is phosphorylase kinase). Another kinase is cAMP-dependent protein kinase, **which allows cAMP-mediated hormonal action to inhibit glycogen synthesis synchronously with the activation of glycogenolysis.** The remaining kinases are known as glycogen synthase kinase-3, -4, and -5.

Glucose 6-phosphate is an allosteric activator of glycogen synthase b, causing a decrease in K_m for UDP-glucose and allowing glycogen synthesis by the phosphorylated enzyme. Glycogen also exerts an inhi-

bition on its own formation, and **insulin** also stimulates glycogen synthesis in muscle by promoting dephosphorylation and activation of glycogen synthase b. Normally, dephosphorylation of glycogen synthase b is carried out by protein phosphatase-1, which is under the control of cAMP-dependent protein kinase (Fig 19–7).

Further aspects of the regulation of glycogen metabolism are discussed on p 192.

DISEASES OF GLYCOGEN STORAGE

The term "glycogen storage disease" is a generic one intended to describe a group of inherited disorders characterized by deposition of an abnormal type or quantity of glycogen in the tissues.

In **type I glycogenosis (von Gierke's disease)**, both the liver cells and the cells of the renal convoluted tubules are characteristically loaded with glycogen. However, these glycogen stores are unavailable, as

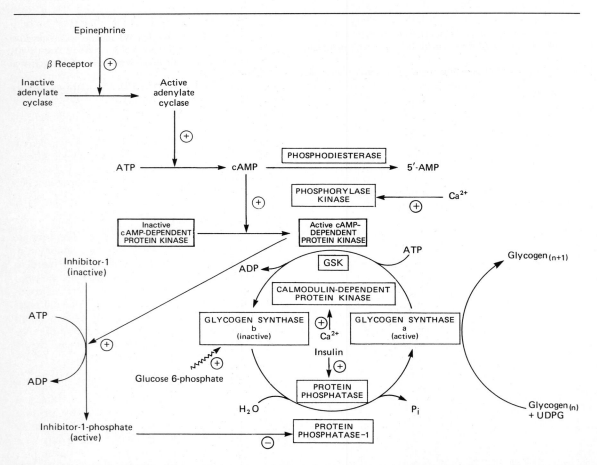

Figure 19–7. Control of glycogen synthase in muscle (n = number of glucose residues). The sequence of reactions arranged in a cascade causes amplification at each step, allowing only nanomole quantities of hormone to cause major changes in glycogen concentration. GSK, glycogen synthase kinase-3, -4, and -5; wavy arrow, allosteric activation.

evidenced by the occurrence of hypoglycemia and a lack of glucose release under stimulus by epinephrine or glucagon. Ketosis and hyperlipemia are also present in these patients, as would be characteristic of an organism deprived of carbohydrate. In liver, kidney, and intestinal tissue, **the activity of glucose-6-phosphatase is either extremely low or entirely absent.**

Type II (Pompe's disease) is fatal and is characterized by a deficiency of lysosomal α-1→4- and 1→6-glucosidase (acid maltase), whose function is to degrade glycogen, which otherwise accumulates in the lysosomes.

Type III (limit dextrinosis; Forbes', or Cori's, disease) is characterized by the absence of debranching enzyme, which causes the accumulation of a characteristic branched polysaccharide.

Type IV (amylopectinosis; Andersen's disease) is characterized by the absence of branching enzyme, with the result that a polysaccharide having few branch points accumulates. Death due to cardiac or liver failure usually occurs in the first year of life.

An absence of muscle phosphorylase (myophosphorylase) is the cause of **type V glycogenosis (myophosphorylase deficiency glycogenosis; McArdle's syndrome).** Patients with this disease exhibit a markedly diminished tolerance to exercise. Although their skeletal muscles have an abnormally high content of glycogen (2.5–4.1%), little or no lactate is detectable in their blood after exercise.

Also described among the glycogen storage diseases are phosphorylase deficiency in the liver **(type VI),** a deficiency of phosphofructokinase in the muscles and erythrocytes **(type VII; Tarui's disease),** and a glycogenosis in which liver phosphorylase kinase is deficient. Deficiencies of **adenylate kinase** and **cAMP-dependent protein kinase** have also been reported.

REFERENCES

Brown DH, Brown BI: Some inborn errors of carbohydrate metabolism. Page 391 in: *MTP International Review of Science.* Vol 5. Whelan WJ (editor). Butterworth, 1975.

Cohen P: *Control of Enzyme Activity,* 2nd ed. Chapman & Hall, 1983.

Cohen P: The role of protein phosphorylation in the hormonal control of enzyme activity. *Eur J Biochem* 1985;**151**:439.

Exton JH: Molecular mechanisms involved in α-adrenergic responses. *Mol Cell Endocrinol* 1981;**23**:233.

Hers HG: The control of glycogen metabolism in the liver. *Annu Rev Biochem* 1976;**45**:167.

Randle PJ, Steiner DF, Whelan WJ (editors): *Carbohydrate Metabolism and Its Disorders.* Vol 3. Academic Press, 1981.

Sperling O, de Vries A (editors): *Inborn Errors of Metabolism in Man.* Karger, 1978.

Stanbury JB et al (editors): *The Metabolic Basis of Inherited Disease,* 5th ed. McGraw-Hill, 1983.

20 Gluconeogenesis & the Pentose Phosphate Pathway

Peter A. Mayes, PhD, DSc

GLUCONEOGENESIS

INTRODUCTION

Gluconeogenesis includes all mechanisms and pathways responsible for converting noncarbohydrates to glucose or glycogen. The major substrates for gluconeogenesis are the glucogenic amino acids, lactate, glycerol, and (in ruminants) propionate. Liver and kidney are the major tissues involved, since they contain a full complement of the necessary enzymes.

BIOMEDICAL IMPORTANCE

Gluconeogenesis meets the needs of the body for glucose when carbohydrate is not available in sufficient amounts from the diet. A continual supply of glucose is necessary as a source of energy, especially for the nervous system and the erythrocytes. Below a critical blood glucose concentration, there is brain dysfunction, which under conditions of severe hypoglycemia can lead to coma and death. Glucose is also required in adipose tissue as a source of glyceride-glycerol, and it probably plays a role in maintaining the level of intermediates of the citric acid cycle in many tissues. It is clear that even under conditions where fat may be supplying most of the caloric requirement of the organism, **there is always a certain basal requirement for glucose.** In addition, glucose is the only fuel that will supply energy to skeletal muscle under anaerobic conditions. It is the precursor of milk sugar (lactose) in the mammary gland, and it is taken up actively by the fetus. In addition, gluconeogenic mechanisms are used to clear the products of the metabolism of other tissues from the blood, eg, lactate, produced by muscle and erythrocytes, and glycerol, which is continuously produced by adipose tissue. Propionate, the principal glucogenic fatty acid produced in the digestion of carbohydrates by ruminants, is a major substrate for gluconeogenesis in these species.

METABOLIC PATHWAYS INVOLVED IN GLUCONEOGENESIS
(See Fig 20–1.)

These pathways are modifications and adaptations of the glycolysis pathway and the citric acid cycle. Krebs pointed out that energy barriers obstruct a simple reversal of glycolysis (1) between pyruvate and phosphoenolpyruvate, (2) between fructose 1,6-bisphosphate and fructose 6-phosphate, (3) between glucose 6-phosphate and glucose, and (4) between glucose 1-phosphate and glycogen. These barriers are circumvented by special reactions.

(1) Present in mitochondria is an enzyme, **pyruvate carboxylase,** which in the presence of ATP, the B vitamin biotin, and CO_2 converts pyruvate to oxaloacetate. The function of the biotin is to bind CO_2 from bicarbonate onto the enzyme prior to the addition of the CO_2 to pyruvate (see below). In the extramitochondrial part of the cell is found a second enzyme, **phosphoenolpyruvate carboxykinase,** which catalyzes the conversion of oxaloacetate to phosphoenolpyruvate. High-energy phosphate in the form of GTP or ITP is required in this reaction, and CO_2 is liberated. Thus, with the help of these 2 enzymes and lactate dehydrogenase, lactate can be converted to phosphoenolpyruvate.

However, oxaloacetate does not diffuse readily from mitochondria. Alternative means are available to achieve the same end by converting oxaloacetate into compounds that can diffuse from the mitochondria, followed by their reconversion to oxaloacetate in the extramitochondrial portion of the cell. Such a compound is malate, whose formation from oxaloacetate within mitochondria and conversion back to oxaloacetate in the extramitochondrial compartment involves malate dehydrogenase.

(2) The conversion of fructose 1,6-bisphosphate to fructose 6-phosphate, necessary to achieve a reversal of glycolysis, is catalyzed by a specific enzyme, **fructose-1,6-bisphosphatase.** This is a key enzyme in the sense that its presence determines whether or not a tissue is capable of resynthesizing glycogen from pyruvate and triosephosphates. It is present in liver and kidney and has been demonstrated in striated muscle.

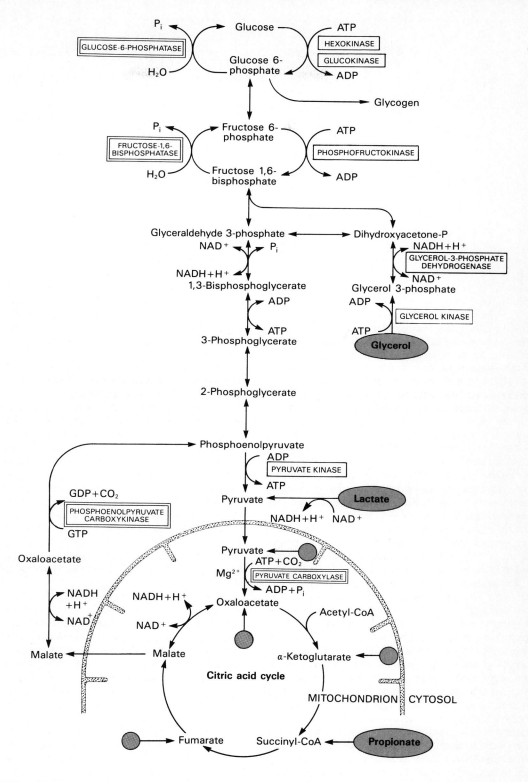

Figure 20–1. Major pathways of gluconeogenesis in the liver. Entry points of glucogenic amino acids after transamination are indicated by ●→ . (See also Fig 17–7.) The key gluconeogenic enzymes are shown thus ▭. The ATP required for gluconeogenesis is supplied by the oxidation of acetyl-CoA derived mainly from long-chain fatty acids or lactate (via pyruvate and pyruvate dehydrogenase). Propionate is of quantitative importance only in ruminants.

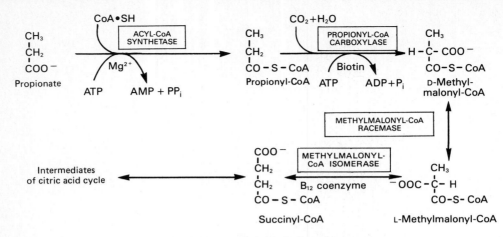

Figure 20–2. Metabolism of propionate.

It is held to be absent from heart muscle and smooth muscle.

(3) The conversion of glucose 6-phosphate to glucose is catalyzed by another specific phosphatase, **glucose-6-phosphatase.** It is present in liver and kidney but absent from muscle and adipose tissue. Its presence allows a tissue to add glucose to the blood.

(4) The breakdown of glycogen to glucose 1-phosphate is carried out by phosphorylase. The synthesis of glycogen involves an entirely different pathway through the formation of uridine diphosphate glucose and the activity of **glycogen synthase** (see Fig 19–1).

The relationships between these key enzymes of gluconeogenesis and the glycolysis pathway are shown in Fig 20–1. After transamination or deamination, glucogenic amino acids form either pyruvate or members of the citric acid cycle. Therefore, the reactions described above can account for the conversion of both glucogenic amino acids and lactate to glucose or glycogen. Thus, lactate forms pyruvate and enters the mitochondria before conversion to oxaloacetate and ultimate conversion to glucose.

Propionate, which is a major source of glucose in ruminants, enters the main gluconeogenic pathway via the citric acid cycle after conversion to succinyl-CoA. Propionate is first activated with ATP and CoA by an appropriate **acyl-CoA synthetase.** Propionyl-CoA, the product of this reaction, undergoes a CO_2 fixation reaction to form D-methylmalonyl-CoA, catalyzed by **propionyl-CoA carboxylase** (Fig 20–2). This reaction is analogous to the fixation of CO_2 in acetyl-CoA by acetyl-CoA carboxylase (see Chapter 23) in that it forms a malonyl derivative and requires **biotin** as a coenzyme. D-Methylmalonyl-CoA must be converted to its stereoisomer, L-methylmalonyl-CoA, by **methylmalonyl-CoA racemase** before its final isomerization to succinyl-CoA by the enzyme **methylmalonyl-CoA isomerase,** which requires **vitamin B_{12}** as a coenzyme. Vitamin B_{12} deficiency in humans and animals results in the excretion of large amounts of methylmalonate (**methylmalonic aciduria**).

Although the pathway to succinate is its main route of metabolism, propionate may also be used as the priming molecule for the synthesis—in adipose tissue and mammary gland—of fatty acids that have an odd number of carbon atoms in the molecule. **C_{15} and C_{17} fatty acids** are found particularly in the lipids of ruminants.

Glycerol is a product of the metabolism of adipose tissue, and only tissues that possess the activating enzyme, **glycerol kinase,** can utilize it. This enzyme, which requires ATP, is found in liver and kidney, among other tissues. Glycerol kinase catalyzes the conversion of glycerol to glycerol 3-phosphate. This pathway connects with the triosephosphate stages of the glycolysis pathway, because glycerol 3-phosphate may be oxidized to dihydroxyacetone phosphate by NAD^+ in the presence of **glycerol-3-phosphate dehydrogenase.** Liver and kidney are able to convert glycerol to blood glucose by making use of the above enzymes, some enzymes of glycolysis, and the specific enzymes of the gluconeogenic pathway, fructose-1,6-bisphosphatase and glucose-6-phosphatase (Fig 20–1).

Biotin

Biotin is a member of the B group of water-soluble vitamins. It is an imidazole derivative widely distributed in natural foods (Fig 20–3). A large portion of the human biotin requirement is probably **supplied from the intestinal bacteria.**

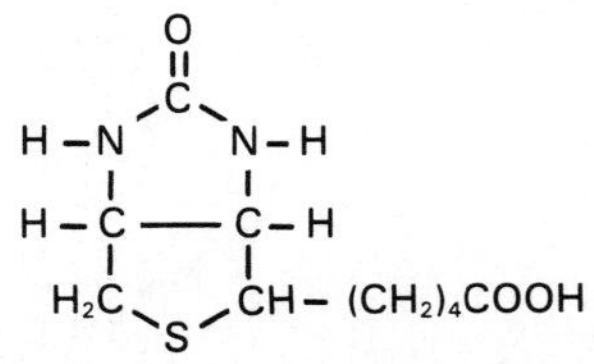

Figure 20–3. Biotin.

Table 20–1. Biotin-dependent enzymes in animals.

Enzyme	Role
Pyruvate carboxylase	First reaction in pathway that converts 3-carbon precursors to glucose (gluconeogenesis)
	Replenishes oxaloacetate for citric acid cycle
Acetyl-CoA carboxylase	Commits acetate units to fatty acid synthesis by forming malonyl-CoA
Propionyl-CoA carboxylase	Converts propionate to succinate, which can then enter citric acid cycle
β-Methylcrotonyl-CoA carboxylase	Catabolism of leucine and certain isoprenoid compounds

Biotin functions as a component of specific multisubunit enzymes (Table 20–1) that catalyze carboxylation reactions. It is attached to the apoenzyme by an amide linkage to the ϵ-amino group of a lysyl residue.

In the first step of the reaction of **pyruvate carboxylase,** a carboxylate ion is attached to the N^1 of the biotin, generating an activated intermediate, **carboxybiotin-enzyme** (Fig 20–4). This step requires HCO_3^-, ATP, Mg^{2+}, and acetyl-CoA (as an allosteric effector). The activated carboxyl group is then transferred from the carboxybiotin-enzyme intermediate to pyruvate to form oxaloacetate and the biotin-holoenzyme. The long flexible arm between the biotin and the enzyme probably enables this prosthetic group (biotin) to move from one active site of the multisubunit enzyme (eg, the phosphocarbonate-forming component) to the other site (eg, that possessing the pyruvate).

There appears to be a single enzyme responsible for attaching biotin to the specific lysyl residue of all the carboxylase apoenzymes. This enzyme is called holocarboxylase synthetase. In its absence, substrates of the biotin-dependent carboxylase enzyme accumulate and can be detected in urine. These metabolites include lactate, β-methylcrotonate, β-hydroxyisovalerate, and β-hydroxypropionate. Children with this enzyme deficiency exhibit dermatitis, retarded growth, alopecia, loss of muscular control, and, in some cases, immune deficiency diseases.

THE PENTOSE PHOSPHATE PATHWAY, OR HEXOSE MONOPHOSPHATE SHUNT

INTRODUCTION

The pentose phosphate pathway is an alternative route for the oxidation of glucose. It is a multicyclic process in which 3 molecules of glucose 6-phosphate give rise to 3 molecules of CO_2 and three 5-carbon residues. The latter are arranged to regenerate 2 molecules of glucose 6-phosphate and one molecule of the glycolytic intermediate, glyceraldehyde 3-phosphate. Since 2 molecules of glyceraldehyde 3-phosphate can regenerate glucose 6-phosphate, glucose may be completely oxidized by this pathway.

3 Glucose 6-phosphate + 6NADP⁺ ⟶

3CO₂ + 2 Glucose 6-phosphate + Glyceraldehyde 3-phosphate + 6NADPH + 6H⁺

BIOMEDICAL IMPORTANCE

The pentose phosphate cycle does not generate ATP but has 2 major functions: (1) The generation of NADPH for reductive syntheses such as fatty acid and

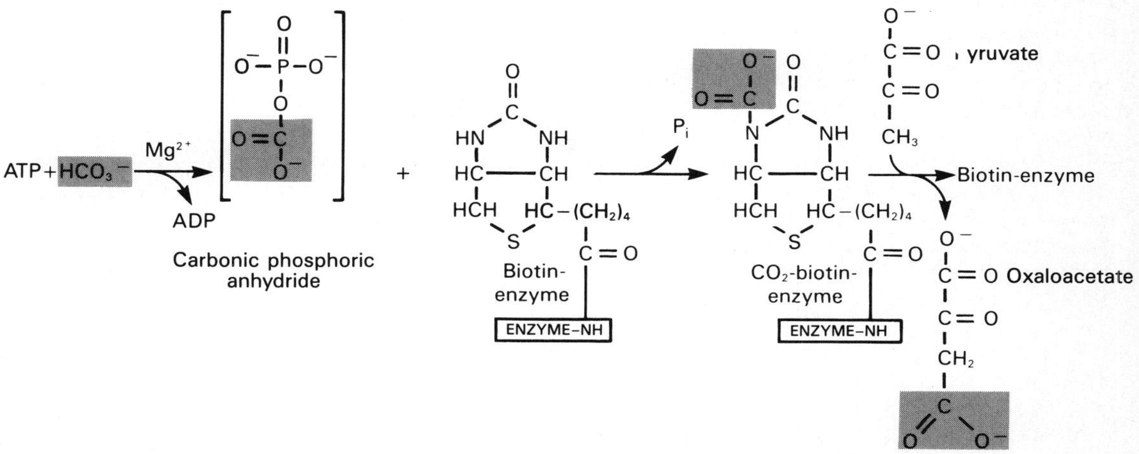

Figure 20–4. Formation of the CO_2-biotin enzyme complex and conversion of pyruvate to oxaloacetate by pyruvate carboxylase.

steroid biosynthesis. (2) The provision of ribose for nucleotide and nucleic acid biosynthesis. Deficiencies of certain enzymes of the pentose phosphate pathway are major causes of hemolysis of red blood cells. Such deficiencies result in one type of hemolytic anemia. The principal enzyme involved is glucose-6-phosphate dehydrogenase. As many as 100 million people in the world may have genetically determined low levels of glucose-6-phosphate dehydrogenase, owing to variant forms. The variant forms can be distinguished by electrophoresis and other techniques.

SEQUENCE OF REACTIONS

The enzymes of the pentose phosphate pathway are found in the **extramitochondrial** soluble portion of the cell, the cytosol. As in glycolysis, oxidation is achieved by dehydrogenation; but in the case of the pentose phosphate pathway, NADP and not NAD is used as a hydrogen acceptor.

The sequence of reactions of the pathway may be divided into 2 phases: oxidative and nonoxidative. In the first, glucose 6-phosphate undergoes dehydrogen-

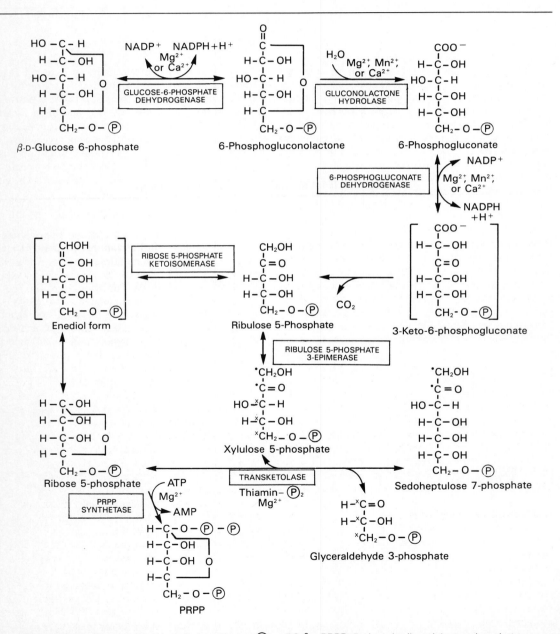

Figure 20−5. The pentose phosphate pathway. P, $-PO_3^{2-}$; PRPP, 5-phosphoribosyl-1-pyrophosphate.

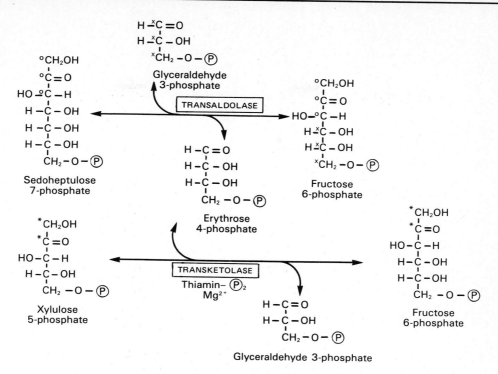

Figure 20–5 (cont'd). The pentose phosphate pathway.

ation and decarboxylation to give a pentose, ribulose 5-phosphate. In the second phase, ribulose 5-phosphate is converted back to glucose 6-phosphate by a series of reactions involving mainly 2 enzymes: **transketolase** and **transaldolase** (Fig 20–5).

The Oxidative Phase

Dehydrogenation of glucose 6-phosphate to 6-phosphogluconate occurs via the formation of 6-phosphogluconolactone catalyzed by **glucose-6-phosphate dehydrogenase,** an NADP-dependent enzyme. The hydrolysis of 6-phosphogluconolactone is accomplished by the enzyme **gluconolactone hydrolase.** A second oxidative step is catalyzed by **6-phosphogluconate dehydrogenase,** which also requires NADP⁺ as hydrogen acceptor. Decarboxylation follows with the formation of the ketopentose, ribulose 5-phosphate. The reaction probably takes place in 2 steps through the intermediate 3-keto-6-phosphogluconate.

The Nonoxidative Phase

Ribulose 5-phosphate now serves as substrate for 2 different enzymes. **Ribulose 5-phosphate 3-epimerase** alters the configuration about carbon 3, forming the epimer xylulose 5-phosphate, another ketopentose. **Ribose 5-phosphate ketoisomerase** converts ribulose 5-phosphate to the corresponding aldopentose, ribose 5-phosphate.

Transketolase transfers the 2-carbon unit comprising carbons 1 and 2 of a ketose to the aldehyde carbon of an aldose sugar. It therefore effects the conversion of a ketose sugar into an aldose with 2 carbons less, and simultaneously converts an aldose sugar into a ketose with 2 carbons more. The reaction requires a B vitamin, **thiamin,** as the coenzyme thiamin diphosphate in addition to Mg²⁺ ions. The 2-carbon moiety transferred is probably glycolaldehyde bound to thiamin diphosphate, ie, "active glycolaldehyde." Transketolase catalyzes the transfer of the 2-carbon unit from xylulose 5-phosphate to ribose 5-phosphate, producing the 7-carbon ketose sedoheptulose 7-phosphate and the aldose glyceraldehyde 3-phosphate. These 2 products then enter another reaction known as transaldolation. **Transaldolase** allows the transfer of a 3-carbon moiety, "active dihydroxyacetone" (carbons 1–3), from the ketose sedoheptulose 7-phosphate to the aldose glyceraldehyde 3-phosphate to form the ketose fructose 6-phosphate and the 4-carbon aldose erythrose 4-phosphate.

A further reaction takes place, again involving **transketolase,** in which xylulose 5-phosphate serves as a donor of "active glycolaldehyde." In this case the erythrose 4-phosphate formed above acts as acceptor, and the products of the reaction are fructose 6-phosphate and glyceraldehyde 3-phosphate.

In order to oxidize glucose completely to CO_2 via the pentose phosphate pathway, it is necessary that the enzymes are present in the tissue to convert glyceraldehyde 3-phosphate to glucose 6-phosphate. This involves enzymes of the glycolysis pathway working in

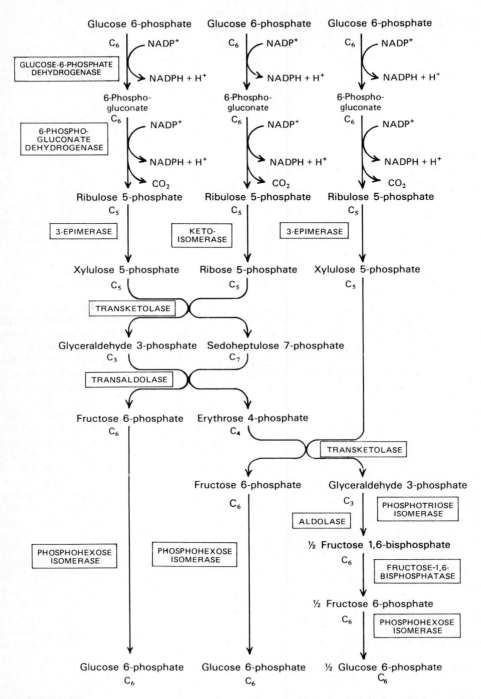

Figure 20–6. Flow chart of pentose phosphate pathway and its connections with the pathway of glycolysis.

a reverse direction and, in addition, the gluconeogenic enzyme **fructose-1,6-bisphosphatase.** A summary of the pathway is shown in Fig 20–6.

METABOLIC SIGNIFICANCE OF THE PENTOSE PHOSPHATE PATHWAY

Comparison With Glycolysis

The pentose phosphate pathway is markedly different from glycolysis. Oxidation occurs in the first reactions utilizing NADP rather than NAD, and CO_2, which is not produced at all in the glycolysis pathway, is a characteristic product. ATP is not generated.

Generation of Reducing Equivalents

Estimates of the activity of the pathway in various tissues give an indication of its metabolic significance. It is active in liver, adipose tissue, adrenal cortex, thyroid, erythrocytes, testis, and lactating mammary gland. It is not active in nonlactating mammary gland, and its activity is low in skeletal muscle. All of the tissues in which the pathway is active use NADPH in reductive syntheses, eg, synthesis of fatty acids, steroids, amino acids via glutamate dehydrogenase, or reduced glutathione in erythrocytes. It is probable that the presence of active lipogenesis or of a system which utilizes NADPH stimulates an active degradation of glucose via the pentose phosphate pathway by altering the NADP:NADPH ratio. The synthesis of glucose-6-phosphate dehydrogenase and 6-phosphogluconate dehydrogenase may also be induced during conditions associated with the "fed state."

Generation of Ribose

The pentose phosphate pathway provides ribose for nucleotide and nucleic acid synthesis (Fig 20–5). The source is the ribose 5-phosphate intermediate that reacts with ATP to form PRPP used in nucleotide biosynthesis (see p 344). Muscle tissue contains very small amounts of glucose-6-phosphate dehydrogenase and 6-phosphogluconate dehydrogenase. Nevertheless, skeletal muscle is capable of synthesizing ribose. This is probably accomplished by a reversal of the nonoxidative phase of the pentose phosphate pathway utilizing fructose 6-phosphate. **Thus, it is not necessary to have a completely functioning pentose phosphate pathway in order that a tissue may synthesize ribose.**

CLINICAL ASPECTS

The pentose phosphate pathway in the erythrocyte provides NADPH for the reduction of oxidized glutathione (G–S–S–G) to reduced glutathione (2G–SH), catalyzed by **glutathione reductase.** In turn, reduced glutathione removes H_2O_2 from the erythrocyte in a reaction catalyzed by **glutathione peroxidase.**

$$G-S-S-G + NADPH + H^+ \longrightarrow 2G-SH + NADP^+$$

GLUTHATHIONE REDUCTASE

$$2G-SH + H_2O_2 \longrightarrow G-S-S-G + 2H_2O$$

GLUTHATHIONE PEROXIDASE
Selenium

This reaction is important, since accumulation of H_2O_2 may decrease the life span of the erythrocyte by increasing the rate of oxidation of hemoglobin to methemoglobin. An inverse correlation has been found between the activity of **glucose-6-phosphate dehydrogenase** and the fragility of red cells (susceptibility to hemolysis). A mutation present in some populations causes a **deficiency in this enzyme** with consequent impairment of the generation of NADPH that is manifested as red cell hemolysis when the susceptible individual is subjected to oxidants, such as the antimalarial primaquine, aspirin, or sulfonamides, or when the susceptible individual has eaten fava beans (*Vicia fava*—favism).

Measurement of **transketolase** activity in blood reflects the degree of thiamin deficiency. The only condition in which its activity is raised is pernicious anemia.

REFERENCES

Krebs HA: Gluconeogenesis. *Proc R Soc Lond [Biol]* 1964; **159**:545.

Landau BR, Wood HG: The pentose cycle in animal tissues: Evidence for the classical and against the "L-type" pathway. *Trends Biochem Sci* 1983;**8**:292.

Williams JF: A critical examination of the evidence for the reactions of the pentose pathway in animal tissues. *Trends Biochem Sci* 1980;**5**:315.

Wood T: *The Pentose Phosphate Pathway.* Academic Press, 1985.

21

Metabolism of Important Hexoses

Peter A. Mayes, PhD, DSc

INTRODUCTION

Glucose, fructose, and galactose are quantitatively the most important hexoses absorbed from the gastrointestinal tract. They are derived from dietary starch, sucrose, and lactose, respectively. Specialized pathways have been developed, particularly in the liver, for the conversion of fructose and galactose to glucose.

BIOMEDICAL IMPORTANCE

The major metabolic routes for the utilization of glucose are glycolysis and the pentose phosphate pathway. Of minor quantitative importance but of major significance for the excretion of metabolites and foreign chemicals (xenobiotics) as glucuronides is the elaboration of glucuronic acid via the uronic acid pathway from glucose. A deficiency in the pathway leads to the condition of essential pentosuria. The total absence of one particular enzyme of the pathway in all primates and guinea pigs accounts for the fact that ascorbic acid (vitamin C) is required in the diet of humans but not most other mammals. Deficiencies in the enzymes of fructose and galactose metabolism lead to metabolic diseases such as essential fructosuria and the galactosemias. Fructose has been used for parenteral nutrition, but at high concentration it can cause depletion of adenine nucleotides in liver and hepatic necrosis.

THE URONIC ACID PATHWAY

Besides the major pathways of metabolism of glucose 6-phosphate that have been described, there exists a pathway for the conversion of glucose to glucuronic acid, ascorbic acid, and pentoses that is referred to as the uronic acid pathway. It is also an alternative oxidative pathway for glucose, but like the pentose phosphate pathway, it does not lead to the generation of ATP.

In the uronic acid pathway, glucuronic acid is formed from glucose by the reactions shown in Fig 21–1. Glucose 6-phosphate is converted to glucose 1-phosphate, which then reacts with uridine triphosphate (UTP) to form the active nucleotide, uridine diphosphate glucose (UDPGlc). This latter reaction is cata-

lyzed by the enzyme **UDPGlc pyrophosphorylase.** All of the steps up to this point are those previously indicated as in the pathway of glycogenesis in the liver. UDPGlc is oxidized at carbon 6 by a 2-step process to glucuronate. The product of the oxidation, which is catalyzed by an NAD-dependent **UDPGlc dehydrogenase,** is **UDP-glucuronate.**

UDP-glucuronate is the "active" form of glucuronate for reactions involving incorporation of glucuronic acid into proteoglycans or for reactions in which glucuronate is conjugated to such substrates as steroid hormones, certain drugs, or bilirubin (see Fig 33–13).

In an NADPH-dependent reaction, glucuronate is reduced to L-gulonate (Fig 21–1). This latter compound is the direct precursor of ascorbate in those animals capable of synthesizing this vitamin. In humans and other primates as well as in guinea pigs, ascorbic acid cannot be synthesized. Gulonate is oxidized to 3-keto-L-gulonate, which is then decarboxylated to the pentose, L-xylulose.

Xylulose is a constituent of the pentose phosphate pathway, but in the reactions shown in Fig 21–1, the L-isomer of xylulose is formed from ketogulonate. If the 2 pathways are to connect, it is necessary to convert L-xylulose to the D-isomer. This is accomplished by an NADPH-dependent reduction to xylitol, which is then oxidized in an NAD-dependent reaction to D-xylulose; this latter compound, after conversion to D-xylulose 5-phosphate, is further metabolized in the pentose phosphate pathway.

Clinical Aspects

In the rare hereditary disease **essential pentosuria,** considerable quantities of L-xylulose appear in the urine. It is now believed that this may be explained by the absence in pentosuric patients of the enzyme necessary to accomplish reduction of L-xylulose to xylitol. Parenteral administration of xylitol may lead to **oxalosis** involving calcium oxalate deposition in brain and kidneys. This results from the conversion of D-xylulose to oxalate via xylulose 1-phosphate, glycolaldehyde, and glycolate formation.

Various drugs markedly increase the rate at which glucose enters the uronic acid pathway. For example, administration of barbital or of chlorobutanol to rats results in a significant increase in the conversion of glucose to glucuronate, L-gulonate, and ascorbate. This effect on L-ascorbic acid biosynthesis is shown by

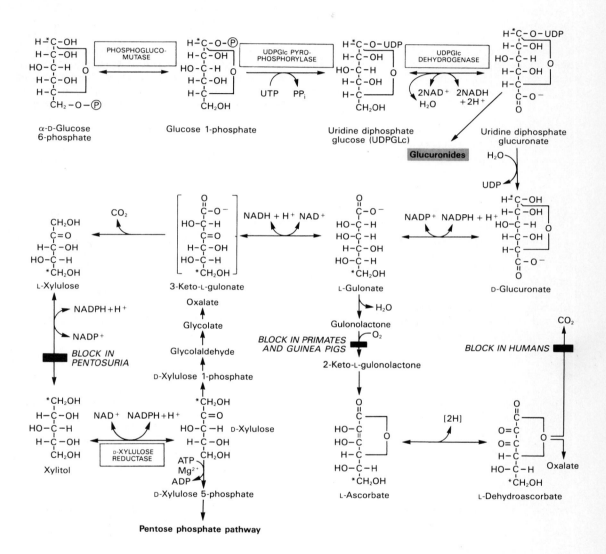

Figure 21–1. Uronic acid pathway. *Indicates the fate of carbon 1 of glucose; $\textcircled{P}$, $-PO_3^{2-}$.

many drugs, including various barbiturates, aminopy-rine, and antipyrine. It is of interest that these last 2 drugs have also been reported to increase the excretion of L-xylulose in pentosuric subjects.

METABOLISM OF FRUCTOSE

Fructose may be phosphorylated to form fructose 6-phosphate, catalyzed by the same enzyme, hexoki-nase, that accomplishes the phosphorylation of glu-cose (or mannose). (See Fig 21–2.) However, the affinity of the enzyme for fructose is very low com-pared with its affinity for glucose. It is unlikely, there-fore, that this is a major pathway for fructose utili-zation.

Another enzyme, **fructokinase,** is present in liver

and effects the transfer of phosphate from ATP to fruc-tose, forming fructose 1-phosphate. It has also been demonstrated in kidney and intestine. This enzyme will not phosphorylate glucose, and, unlike glucoki-nase, its activity is not affected by fasting or by in-sulin, which may explain why fructose disappears from the blood of diabetic patients at a normal rate. The K_m for fructose of the enzyme in liver is very low, indicating a very high affinity of the enzyme for its substrate. It seems probable that this is the major route for the phosphorylation of fructose. **Essential fructos-uria** results from a lack of hepatic fructokinase.

Fructose 1-phosphate is split into D-glyceraldehyde and dihydroxyacetone phosphate by **aldolase B,** an enzyme found in the liver. The enzyme also attacks fructose 1,6-bisphosphate. Absence of this enzyme leads to a **hereditary fructose intolerance.** D-Glycer-

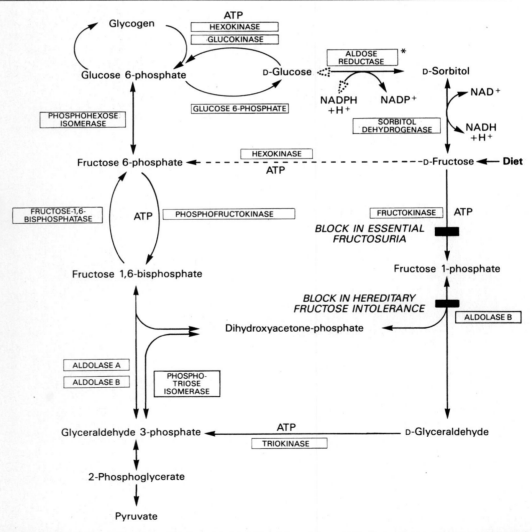

Figure 21–2. Metabolism of fructose. Aldolase A is found in all tissues except the liver, where only aldolase B is present. *Not found in liver.

......phate. The 2 triose phosphates, dihydroxyacetone phosphate and glyceraldehyde 3-phosphate, may be degraded via the glycolysis pathway or they may combine under the influence of aldolase and be converted to glucose. The latter is the fate of much of the fructose metabolized in the liver.

One consequence of hereditary fructose intolerance and of another condition due to **fructose-1,6-bisphosphatase deficiency** is a fructose-induced **hypoglycemia** despite the presence of high glycogen reserves. Apparently, the accumulation of fructose 1-phosphate and fructose 1,6-bisphosphate inhibits the activity of liver phosphorylase by allosteric mechanisms.

If the liver and intestines of an experimental animal are removed, the conversion of injected fructose to glucose does not take place and the animal succumbs to hypoglycemia unless glucose is administered. However, it is reported that the human kidney can convert fructose to glucose and lactate. In humans but not in the rat, a significant amount of the fructose resulting from the digestion of sucrose is converted to glucose in the intestinal wall prior to passage into the portal circulation.

Fructose is more rapidly glycolyzed by the liver than glucose. This is due to the fact that it bypasses the step in glucose metabolism catalyzed by **phosphofructokinase,** at which point metabolic control is exerted on the rate of catabolism of glucose. This allows fructose to flood the pathways in the liver leading to enhanced fatty acid synthesis, esterification of fatty acids, and VLDL secretion, which may raise serum triacylglycerols.

Free fructose is found in seminal plasma and is secreted in quantity into the fetal circulation of ungulates and whales, where it accumulates in the amniotic and allantoic fluids.

Sorbitol Metabolism

Both fructose and **sorbitol** are found in the human lens, where they increase in concentration in diabetes and may be involved in the pathogenesis of **diabetic cataract.** The **sorbitol (polyol) pathway** (not found in liver) is responsible for fructose formation from glucose (Fig 21–2) and increases in activity as the glucose concentration rises in diabetes. Glucose undergoes reduction by NADPH to sorbitol catalyzed by **aldose reductase,** followed by oxidation of sorbitol to fructose in the presence of NAD and **sorbitol dehydrogenase** (polyol dehydrogenase). Sorbitol does not diffuse through cell membranes easily and therefore accumulates. Aldose reductase is found in the placenta of the ewe and is responsible for the secretion of sorbitol into the fetal blood. The presence of sorbitol dehydrogenase in the liver, including the fetal liver, is responsible for the conversion of sorbitol into fructose. When sorbitol is administered intravenously, it is converted to fructose rather than to glucose, although if given by mouth, much escapes absorption from the gut and is fermented in the colon by bacteria to products such as acetate and H_2. Abdominal pain (**sorbitol intolerance**) may be caused by "sugar-free" sweeteners containing sorbitol.

METABOLISM OF GALACTOSE

Galactose is derived from intestinal hydrolysis of the disaccharide **lactose,** the sugar of milk. It is readily converted in the liver to glucose. The ability of the liver to accomplish this conversion may be used as a test of hepatic function in the **galactose tolerance test.** The pathway by which galactose is converted to glucose is shown in Fig 21–3.

In reaction 1, galactose is phosphorylated with the aid of **galactokinase,** using ATP as phosphate donor. The product, galactose 1-phosphate, reacts with **uridine diphosphate glucose** (UDPGlc) to form **uridine diphosphate galactose** (UDPGal) and glucose 1-phosphate. In this step (reaction 2), which is catalyzed by an enzyme called **galactose 1-phosphate uridyl transferase,** galactose is transferred to a position on UDPGlc, replacing glucose. The conversion of galactose to glucose takes place (reaction 3) in a reaction of the galactose-containing nucleotide that is catalyzed by an **epimerase.** The product is UDPGlc. Epimerization probably involves an oxidation and reduction at carbon 4 with NAD as coenzyme. Finally (reaction 4), glucose is liberated from UDPGlc as glucose 1-phosphate, probably after incorporation into glycogen followed by phosphorolysis.

Reaction 3 is freely reversible. In this manner, glucose can be converted to galactose, so that preformed galactose is not essential in the diet. Galactose is required in the body not only in the formation of lactose but also as a constituent of glycolipids (cerebrosides), proteoglycans, and glycoproteins.

In the synthesis of lactose in the mammary gland, glucose is converted to UDPGal by the enzymes described above. UDPGal condenses with glucose to yield lactose, catalyzed by **lactose synthase.**

Clinical Aspects

Inability to metabolize galactose occurs in the **galactosemias,** which may be caused by inherited defects in any of the 3 enzymes marked 1, 2, and 3 in Fig 21–3, although a deficiency in the **uridyl transferase** (2) is the best known. Galactose, which increases in concentration in the blood, is reduced by aldose reductase in the eye to the corresponding polyol (galactitol), which accumulates, causing cataract. The general condition is more severe if it is due to a defect in the uridyl transferase, since galactose 1-phosphate accumulates and depletes the liver of inorganic phosphate. Ultimately, liver failure and mental deterioration result.

In inherited **galactose 1-phosphate uridyl transferase deficiency** affecting the liver and red blood cells (reaction 2), the epimerase (reaction 3) is, however, present in adequate amounts, so that the galac-

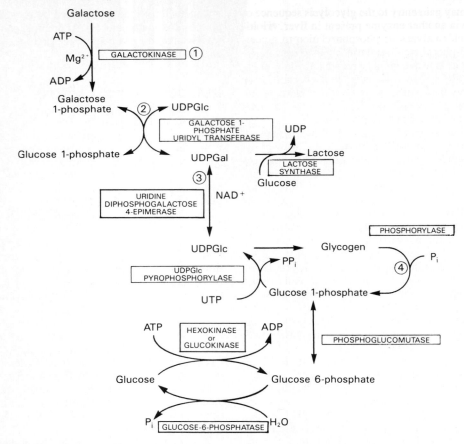

Figure 21–3. The pathway for conversion of galactose to glucose and for the synthesis of lactose.

tosemic individual can still form UDPGal from glucose. This explains how it is possible for normal growth and development of affected children to occur regardless of the galactose-free diets used to control the symptoms of the disease. Several different genetic defects have been described that cause reduced rather than total transferase deficiency. As the enzyme is normally present in excess, a reduction in activity to 50% or even less does not cause clinical manifestation of the disease, which manifests itself only in homozygotes. The epimerase has been found deficient in erythrocytes but present in liver and elsewhere, and this third condition appears to be symptom-free.

Metabolism of Amino Sugars (Hexosamines)
(See Fig 21–4.)

Amino sugars are important components of **glycoproteins** (see Chapter 54), of certain **glycosphingolipids** (eg, gangliosides) (see Chapter 15), and of **glycosaminoglycans** (see Chapter 54). The major amino sugars are **glucosamine, galactosamine,** and

mannosamine (these are all hexosamines) and the 9-carbon compound **sialic acid;** the principal sialic acid found in human tissues is N-acetylneuraminic acid (NeuAc). A summary of the interrelationships among the amino sugars is shown in Fig 21–4; the following are its important features: (1) **Glucosamine** is the major amino sugar. It is formed as glucosamine 6-phosphate from fructose 6-phosphate, using glutamine as the donor of the amino group. (2) The amino sugars occur mainly in the **N-acetylated** form. The acetyl donor is acetyl-CoA. (3) N-acetyl-mannosamine 6-phosphate is formed by epimerization of glucosamine 6-phosphate. (4) **NeuAc** is formed by the condensation of mannosamine 6-phosphate with phosphoenolpyruvate. (5) **Galactosamine** is formed by the epimerization of UDP-N-acetylglucosamine (UDPGlcNAc) to UDP-N-acetylgalactosamine (UDPGalNAc). (6) **Nucleotide sugars** are the forms in which amino sugars are used for the biosynthesis of glycoproteins and other complex compounds; the important amino sugar-containing nucleotide sugars are UDPGlcNAc, UDPGalNAc, and CMP-NeuAc.

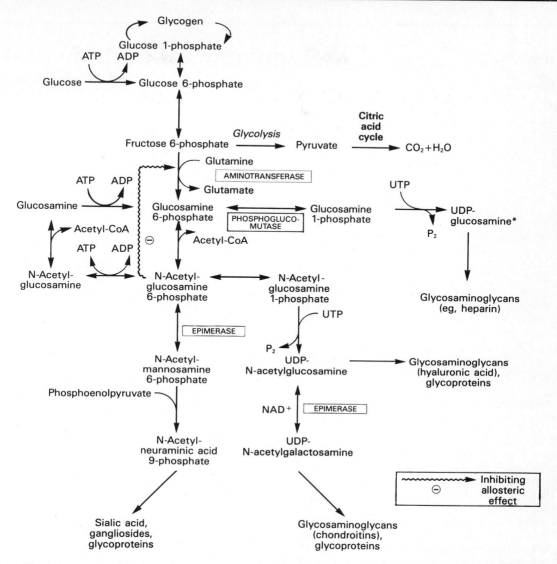

Figure 21–4. A summary of the interrelationships in metabolism of amino sugars. *Analogous to UDPGlc. Other purine or pyrimidine nucleotides may be similarly linked to sugars or amino sugars. Examples are thymidine diphosphate (TDP)-glucosamine and TDP-N-acetylglucosamine.

REFERENCES

Brown DH, Brown BI: Some inborn errors of carbohydrate metabolism. Page 391 in: *MTP International Review of Science*. Vol 5. Whelan WJ (editor). Butterworth, 1975.

Dickens F, Randle PJ, Whelan WJ (editors): *Carbohydrate Metabolism and Its Disorders*. 2 vols. Academic Press, 1968.

Huijing F: Textbook errors: Galactose metabolism and galactosemia. *Trends Biochem Sci* 1978;**3:**N129.

James HM et al: Models for the metabolic production of oxalate from xylitol in humans. *Aust J Exp Biol Med Sci* 1982; **60:**117.

Kador PF, Akagi Y, Kinoshita JH: The effects of aldose reductase and its inhibition on sugar cataract formation. *Metabolism* 1986;**35:**15.

Macdonald I, Vrana A (editors): *Metabolic Effects of Dietary Carbohydrates*. Karger, 1986.

Randle PJ, Steiner DF, Whelan WJ (editors): *Carbohydrate Metabolism and Its Disorders*. Vol 3. Academic Press, 1981.

Sperling O, de Vries A (editors): *Inborn Errors of Metabolism in Man*. Karger, 1978.

Stanbury JB et al (editors): *The Metabolic Basis of Inherited Disease,* 5th ed. McGraw-Hill, 1983.

22 Regulation of Carbohydrate Metabolism

Peter A. Mayes, PhD, DSc

INTRODUCTION

All metabolic pathways must be regulated if they are to coordinate to serve the needs of individual cells, organs, or the whole body. Regulation of the metabolic pathways that provide fuel molecules (eg, carbohydrate) is essential if the supply is to be maintained under the various nutritional, metabolic, and pathologic conditions encountered in vivo. The term "caloric homeostasis" has been given to this type of metabolic regulation.

BIOMEDICAL IMPORTANCE

Caloric homeostasis involves provision of the special fuel needs of each tissue, including making available alternative fuels. It also involves transport of various substrates about the body, together with mechanisms to control their concentration in the blood. These mechanisms ensure continuous supply of glucose between meals and during starvation. Many conditions—usually associated with enzyme deficiency—result in hypoglycemia (low blood glucose), and pathologic changes in endocrine status cause perturbations in carbohydrate metabolism, eg, insulin deficiency leading to diabetes mellitus and hyperglycemia.

GENERAL PRINCIPLES OF REGULATION OF METABOLIC PATHWAYS

Regulation of the overall flux along a metabolic pathway is often concerned with the control of only one or perhaps 2 key reactions in the pathway, catalyzed by **"regulatory enzymes."** The physicochemical factors that control the rate of an enzyme-catalyzed reaction, eg, substrate concentration (see Chapter 9), are of primary importance in the control of the overall rate of a metabolic pathway. However, temperature and pH, factors that can influence enzyme activity, are held constant in warm-blooded vertebrates and have little regulatory significance. (Note, however, the variation in pH in the gastrointestinal tract and its effect on digestion [see Chapter 53].)

Equilibrium & Nonequilibrium Reactions

In a reaction at equilibrium, the forward and reverse reactions take place at equal rates, and there is therefore no net flux in either direction. Many reactions in metabolic pathways are of this type, ie, "equilibrium reactions":

$$A \longleftrightarrow B \longleftrightarrow C \longleftrightarrow D$$

In vivo, under "steady-state" conditions, there would probably be a net flux from left to right due to continuous supply of A and continuous removal of D. Such a pathway could function, but there would be little scope for control of the flux via regulation of enzyme activity, since an increase in activity would only serve to speed up attainment of the equilibrium.

In practice, there are invariably one or more "nonequilibrium" type reactions in a metabolic pathway, where the reactants are present in concentrations that are far from equilibrium. In attempting to reach equilibrium, large losses of free energy occur as heat, which cannot be reutilized, making this type of reaction essentially nonreversible, eg,

$$A \longleftrightarrow B \xrightarrow{\text{Heat}} C \longleftrightarrow D$$
Nonequilibrium reaction

Such a pathway has both flow and direction but would exhaust itself if control were not exerted. The enzymes catalyzing nonequilibrium reactions are usually low in concentration and are subject to other controlling mechanisms. This is similar to the opening and shutting of a "one-way" valve, making it possible to control the net flow.

The Flux-Generating Reaction

A flux-generating reaction is the first reaction in the pathway that is saturated with substrate. It may be identified as a nonequilibrium reaction in which the K_m of the enzyme is considerably lower than the normal substrate concentration. The first reaction in glycolysis catalyzed by **hexokinase** (see Fig 22–2) is such a flux-generating step.

METABOLIC CONTROL OF AN ENZYME-CATALYZED REACTION

A hypothetical metabolic pathway, A,B,C,D, is shown in Fig 22–1, in which reactions A $\longleftrightarrow$ B and C $\longleftrightarrow$ D are equilibrium reactions and B $\longrightarrow$ C is a nonequilibrium reaction. The flux through such a pathway can be regulated by the availability of substrate A. This depends on its supply from the blood, which in turn depends on adequate food intake to the gut or on certain key reactions that maintain and release major substrates to the blood, eg, the flux-generating reactions catalyzed by phosphorylase in liver, which provides blood glucose, and hormone-sensitive lipase in adipose tissue, which supplies free fatty acids. It also depends on the ability of substrate A to permeate the cell membrane. The flux will also be determined by the efficiency of removal of the end product D and on the availability of cosubstrate or cofactors represented by X and Y.

Enzymes catalyzing nonequilibrium reactions are often allosteric proteins subject to the rapid action of "feed-back" or "feed-forward" control by **allosteric modifiers,** often in immediate response to the needs of the cell (see Chapter 9). Other control mechanisms depend on the action of hormones responding to the needs of the body as a whole. These act by several different mechanisms (see Chapter 43). One is **covalent modification** of the enzyme by **phosphorylation** and **dephosphorylation.** This is rapid and is often mediated through the formation of **cAMP,** which in turn causes the conversion of an inactive enzyme into an active enzyme. This change is brought about via the activity of a **cAMP-dependent protein kinase** that phosphorylates the enzyme or of specific **phosphatases** that dephosphorylate the enzyme. The active

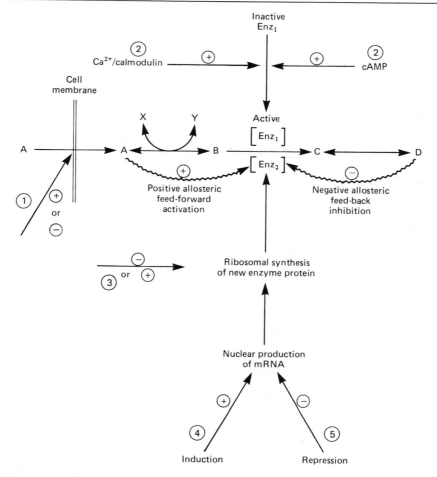

Figure 22–1. Mechanisms of control of an enzyme-catalyzed reaction. Circled numbers indicate possible sites of action of hormones. **1,** Alteration of membrane permeability; **2,** conversion of an inactive to an active enzyme; **3,** alteration of the rate of translation of mRNA at the ribosomal level; **4,** induction of new mRNA formation; and **5,** repression of mRNA formation.

form of the enzyme can be either the phosphorylated enzyme, as in enzymes catalyzing **degradative** pathways (eg, phosphorylase a), or the dephosphorylated enzyme, as in enzymes catalyzing **synthetic** processes (eg, glycogen synthase a).

Some regulatory enzymes can be phosphorylated without the mediation of cAMP and cAMP-dependent protein kinase. These enzymes respond to other metabolic signals such as the [ATP]/[ADP] ratio, eg, pyruvate dehydrogenase (Fig 22–3), or $Ca^{2+}/$ calmodulin-dependent protein kinase, eg, phosphorylase kinase (see Fig 19–5).

The synthesis of rate-controlling enzymes can be affected by hormones. Because this involves new protein synthesis, it is not a rapid change but is often a response to a change in nutritional state. Hormones can act as inducers or repressors of mRNA formation in the nucleus or as stimulators of the translation stage of protein synthesis at the ribosomal level (see Chapters 41 and 43).

Table 22–1. Regulatory and adaptive enzymes of the rat (mainly liver).

| | Activity In | | | | | |
	Carbo-hydrate Feeding	Starva-tion and Diabetes	Inducer	Repressor	Activator	Inhibitor
Enzymes of glycolysis and glycogenesis						
Hexokinase						*Glucose 6-phosphate
Glucokinase	↑	↓	Insulin			
Glycogen synthase system	↑	↓	Insulin		Insulin	Glucagon (cAMP), phosphorylase, glycogen
Phosphofructokinase-1	↑	↓	Insulin		*AMP, *fructose 6-P, *Pᵢ, *fructose 2,6-bisphosphate	*Citrate (fatty acids, ketone bodies), *ATP, glucagon (cAMP)
Pyruvate kinase	↑	↓	Insulin, fructose		*Fructose 1,6-bisphosphate	ATP, alanine, glucagon (cAMP), epinephrine
Pyruvate dehydrogenase	↑	↓			CoA, NAD, insulin†, ADP, pyruvate	Acetyl-CoA, NADH, ATP (fatty acids, ketone bodies)
Enzymes of gluconeogenesis						
Pyruvate carboxylase	↓	↑	Glucocorticoids, glucagon, epinephrine	Insulin	*Acetyl-CoA	*ADP
Phosphoenolpyruvate carboxykinase	↓	↑	Glucocorticoids, glucagon, epinephrine	Insulin	Glucagon?	
Fructose-1,6-bisphosphatase	↓	↑	Glucocorticoids, glucagon, epinephrine	Insulin	Glucagon (cAMP)	*Fructose 1,6-bisphosphate, *AMP, fructose 2,6-bisphosphate
Glucose-6-phosphatase	↓	↑	Glucocorticoids, glucagon, epinephrine	Insulin		
Enzymes of the pentose phosphate pathway and lipogenesis						
Glucose-6-phosphate dehydrogenase	↑	↓	Insulin			
6-Phosphogluconate dehydrogenase	↑	↓	Insulin			
"Malic enzyme"	↑	↓	Insulin			
ATP-citrate lyase	↑	↓	Insulin			ADP
Acetyl-CoA carboxylase	↑	↓	Insulin?		*Citrate, insulin	Long-chain acyl-CoA, cAMP, glucagon
Fatty acid synthase	↑	↓	Insulin?			

*Allosteric.
†In adipose tissue but not in liver.

REGULATION OF CARBOHYDRATE METABOLISM AT THE CELLULAR & ENZYMATIC LEVEL

Gross effects on metabolism of changes in nutritional state or in the endocrine balance of an animal may be studied by observing changes in the concentration of blood metabolites. By such techniques as catheterization, it is also possible to study effects on individual organs by measuring arteriovenous differences, etc. However, the changes that occur in the metabolic balance of the intact animal are due to shifts in the pattern of metabolism in individual tissues which are usually associated with changes in availability of metabolites or changes in activity of key enzymes.

Changes in availability of substrates are either directly or indirectly responsible for most changes in metabolism. Fluctuations in their blood concentrations due to changes in dietary availability may alter the rate of secretion of hormones that influence, in turn, the pattern of metabolism in metabolic pathways—often by affecting the activity of key enzymes which attempt to compensate for the original change in substrate availability. Three types of mechanisms can be identified as responsible for regulating the activity of enzymes concerned in carbohydrate metabolism and may be identified in Table 22–1: (1) changes in the rate of enzyme synthesis, (2) conversion of an inactive to an active enzyme by covalent modification, and (3) allosteric effects.

REGULATION OF GLYCOLYSIS, GLUCONEOGENESIS, & THE PENTOSE PHOSPHATE PATHWAY

Induction & Repression of Enzyme Synthesis

Some of the better-documented changes in enzyme activity that are considered to occur under various metabolic conditions are listed in Table 22–1. The information in this table applies mainly to the liver. The enzymes involved catalyze nonequilibrium reactions that may be regarded physiologically as "one-way" rather than balanced reactions. Often the effects are reinforced because the activity of the enzymes catalyzing the changes in the opposite direction vary reciprocally (Fig 22–2). It is of importance that the key enzymes involved in a metabolic pathway are all activated or depressed in a coordinated manner. Table 22–1 shows that this is clearly the case. The enzymes involved in the utilization of glucose (ie, those of glycolysis and lipogenesis) all become more active under the circumstance of a superfluity of glucose, and under these conditions the enzymes responsible for producing glucose by the pathway of gluconeogenesis are all low in activity. The secretion of insulin, which is responsive to the blood glucose concentration, controls the activity of the enzymes responsible for glycolysis and, together with the glucocorticoids, those responsi-

ble for gluconeogenesis. All of these effects, which can be explained on the basis of new enzyme synthesis, can be prevented by agents that block the synthesis of protein, such as puromycin and ethionine.

Both dehydrogenases of the pentose phosphate pathway can be classified as adaptive enzymes, since they increase in activity in the well-fed animal and when insulin is given to a diabetic animal. Activity is low in diabetes or fasting. "Malic enzyme" and ATP-citrate lyase behave similarly, indicating that these 2 enzymes are involved in lipogenesis rather than gluconeogenesis.

Covalent Modification

Pyruvate dehydrogenase may be regulated by phosphorylation involving an ATP-specific kinase that causes a decrease in activity, and by dephosphorylation by a phosphatase that causes an increase in activity of the dehydrogenase. The kinase is activated by increases in the [acetyl-CoA]/[CoA], [NADH]/[NAD$^+$], or [ATP]/[ADP] ratios. Thus, pyruvate dehydrogenase—and therefore glycolysis—**is inhibited under conditions of fatty acid oxidation,** which leads to increases in these ratios (Fig 22–3). In starvation, there is a decrease in the proportion of the enzyme in the active form, and an increase in activity occurs after administration of insulin in adipose tissue but not in the liver. **Glucagon inhibits glycolysis and stimulates gluconeogenesis** in the liver by increasing the concentration of cAMP, which in turn activates cAMP-dependent protein kinase, leading to the phosphorylation and inactivation of **pyruvate kinase.** Glucagon also affects the concentration of fructose 2,6-bisphosphate and therefore glycolysis and gluconeogenesis, as explained below.

Allosteric Modification

Several examples are available from carbohydrate metabolism to illustrate allosteric control of the activity of an enzyme. In gluconeogenesis, the synthesis of oxaloacetate from bicarbonate and pyruvate, which is catalyzed by the enzyme **pyruvate carboxylase,** requires the presence of acetyl-CoA as an allosteric activator. The addition of acetyl-CoA results in a change in the tertiary structure of the protein, lowering the K_m value for bicarbonate. This effect has important implications for the self-regulation of intermediary metabolism, for, as acetyl-CoA is formed from pyruvate, it automatically ensures the provision of oxaloacetate and its further oxidation in the citric acid cycle by activating pyruvate carboxylase. The activation of pyruvate carboxylase and the inhibition of pyruvate dehydrogenase by acetyl-CoA formed from the oxidation of fatty acids helps to explain the sparing action of fatty acid oxidation on the oxidation of pyruvate and the stimulation of gluconeogenesis in the liver. There is a reciprocal relationship between the regulation of pyruvate dehydrogenase and pyruvate carboxylase in both liver and kidney that alters the metabolic fate of pyruvate as the tissue changes from carbohydrate oxidation, via glycolysis, to gluconeogenesis (Fig 22–2).

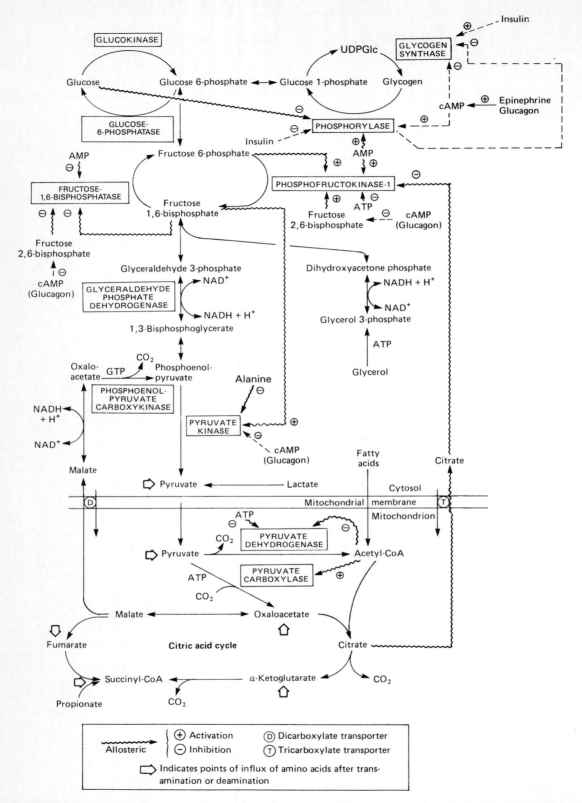

Figure 22–2. Key enzymes in the control of glycolysis, gluconeogenesis, and glycogen metabolism in liver. Indications of hormone action do not necessarily imply a direct action on the enzymes concerned. The effects of cAMP on phosphofructokinase-1 and on fructose-1,6-bisphosphatase are due to a combination of covalent and allosteric modifications (see Fig 22–4). High concentrations of alanine act as a "gluconeogenic signal" by inhibiting glycolysis at the pyruvate kinase step.

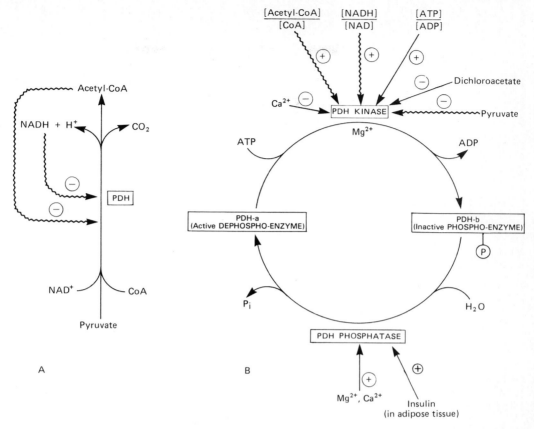

Figure 22–3. Regulation of pyruvate dehydrogenase (PDH). Arrows with wavy shafts indicate allosteric effects. *A:* Regulation by end product inhibition. *B:* Regulation by interconversion of active and inactive forms.

A major role of fatty acid oxidation in promoting gluconeogenesis is to supply ATP required in the pyruvate carboxylase and phosphoenolpyruvate carboxykinase reactions.

Another enzyme that is subject to feedback control is **phosphofructokinase (phosphofructokinase-1).** It occupies a key position in regulating glycolysis. Phosphofructokinase-1 is inhibited by citrate and by ATP and is activated by AMP. AMP acts as an indicator of the energy status of the cell. The presence of **adenylate kinase** in liver and many other tissues allows rapid equilibration of the reaction:

$$\text{ATP} + \text{AMP} \longleftrightarrow 2\,\text{ADP}$$

Thus, when ATP is used in energy-requiring processes resulting in formation of ADP, [AMP] rises. As [ATP] may be 50 times [AMP] at equilibrium, a small fractional decrease in [ATP] will cause a several-fold increase in [AMP]. Thus, a large change in [AMP] acts as a metabolic amplifier of a small change in [ATP]. This mechanism allows the activity of phosphofructokinase-1 to be **highly sensitive to even small changes in energy status of the cell** and to **control the quantity of carbohydrate undergoing glycolysis**

prior to its entry into the citric acid cycle. The increase in [AMP] can also explain why glycolysis is increased during anoxia when [ATP] decreases. Simultaneously, AMP activates phosphorylase, increasing glycogenolysis. The inhibition of phosphofructokinase-1 by citrate and ATP could be another explanation of the sparing action of fatty acid oxidation on glucose oxidation and also of the **Pasteur effect** whereby aerobic oxidation (via the citric acid cycle) inhibits the anaerobic degradation of glucose. A consequence of the inhibition of phosphofructokinase-1 is an accumulation of glucose 6-phosphate which, in turn, inhibits further uptake of glucose in extrahepatic tissues by allosteric inhibition of hexokinase.

Role of Fructose 2,6-Bisphosphate

The most potent positive allosteric effector of phosphofructokinase-1 and inhibitor of fructose-1,6-bisphosphatase in liver is **fructose 2,6-bisphosphate.** It relieves inhibition of phosphofructokinase-1 by ATP and increases affinity for fructose 6-phosphate. It inhibits fructose-1,6-bisphosphatase by increasing the K_m for fructose 1,6-bisphosphate. Its concentration is

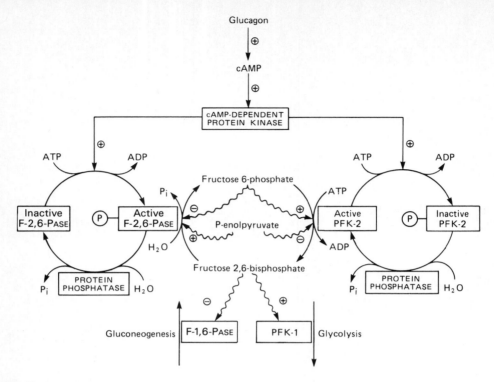

Figure 22–4. Control of glycolysis and gluconeogenesis in the liver by fructose 2,6-bisphosphate. PFK-1, 6-phospho-fructo-1-kinase; PFK-2, 6-phosphofructo-2-kinase; F-1,6-Pase, fructose-1,6-bisphosphatase; F-2,6-Pase, fructose-2,6-bisphosphatase. Arrows with wavy shafts indicate allosteric effects.

under both substrate and hormonal controls (Fig 22–4). Fructose 2,6-bisphosphate is formed by phosphorylation of fructose 6-phosphate by **phosphofructokinase-2.** The same enzyme protein is also responsible for its breakdown, since it contains **fructose-2,6-bisphosphatase** activity. This bifunctional enzyme is under the allosteric control of fructose 6-phosphate, which when raised in concentration owing to an abundance of glucose, stimulates the kinase and inhibits the phosphatase. On the other hand, when glucose is short, glucagon stimulates the production of cAMP, activating cAMP-dependent protein kinase, which in turn inactivates phosphofructokinase-2 and activates fructose-2,6-bisphosphatase by phosphorylation. Thus, **under a superfluity of glucose, fructose 2,6-bisphosphate increases in concentration, stimulating glycolysis by activating phosphofructokinase-1 and inhibiting fructose-1,6-bisphosphatase.** Under conditions of glucose shortage, gluconeogenesis is stimulated by glucagon by decreasing the concentration of fructose 2,6-bisphosphate, which in turn inhibits phosphofructokinase-1 and activates fructose-1,6-bisphosphatase. This mechanism also ensures that glucagon stimulation of glycogenolysis in liver results in glucose release rather than glycolysis.

REGULATION OF GLYCOGEN METABOLISM
(See Fig 22–5.)

Regulation of glycogen metabolism is effected by a balance in activities between **glycogen synthase** and **phosphorylase,** which are under substrate control (through allostery) as well as hormonal control. Not only is phosphorylase activated by a rise in concentration of cAMP (via phosphorylase kinase), but glycogen synthase is at the same time converted to the inactive form (see Chapter 18); both effects are mediated via **cAMP-dependent protein kinase.** Thus, inhibition of glycogenolysis enhances net glycogenesis, and inhibition of glycogenesis enhances net glycogenolysis. Of further significance in the regulation of glycogen metabolism is the finding that the dephosphorylation of phosphorylase a, phosphorylase kinase, and glycogen synthase b is accomplished by a single enzyme of wide specificity—protein phosphatase-1. In turn, protein phosphatase-1 is inhibited by cAMP-dependent protein kinase via inhibitor-1 (Fig 22–5). Thus, glycogenolysis can be terminated and glycogenesis can be stimulated synchronously, or vice versa, because both processes are keyed to the activity of

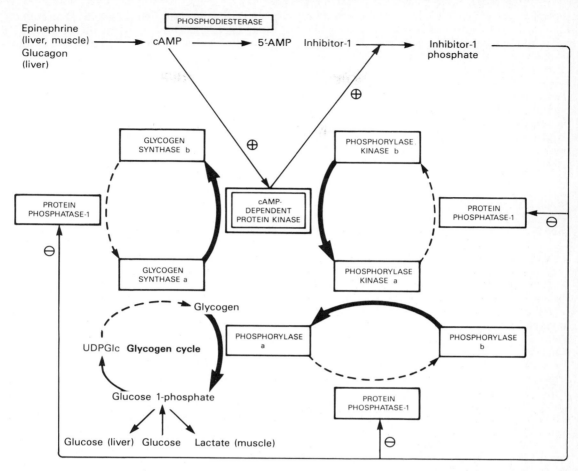

Figure 22–5. Control of glycogenolysis and glycogenesis by cAMP-dependent protein kinase. The reactions that lead to glycogenolysis as a result of an increase in cAMP concentrations are shown with bold arrows, and those that are inhibited are shown as broken arrows. The reverse occurs when cAMP concentrations decrease as a result of phosphodiesterase activity, leading to glycogenesis.

cAMP-dependent protein kinase. Both phosphorylase kinase and glycogen synthase may be reversibly phosphorylated in more than one site by separate kinases and phosphatases. These secondary phosphorylations modify the sensitivity of the primary sites to phosphorylation and dephosphorylation. Pyruvate dehydrogenase also shows evidence of **multisite phosphorylation.**

Liver. The major factor that controls glycogen metabolism in the liver is the concentration of **phosphorylase a.** Not only does this enzyme control the rate-limiting step in glycogenolysis, but it also inhibits the activity of protein phosphatase-1 and thereby controls glycogen synthesis (Fig 22–2). Inactivation of phosphorylase occurs as a result of allosteric inhibition by glucose as it rises in concentration after a meal. Activation is caused by 5′-AMP responding to depletion of ATP (see above). Catecholamines, including epinephrine, stimulate glycogenolysis by an additional mechanism not involving cAMP but via α_1-adrenergic

receptors. These mechanisms involve direct stimulation of phosphorylase b kinase by Ca^{2+} and calmodulin. cAMP-independent glycogenolysis is also caused by vasopressin, oxytocin, and angiotensin II acting through calcium or the phosphatidylinositol bisphosphate pathway. Administration of insulin causes an immediate inactivation of phosphorylase followed by activation of glycogen synthase. The effects of insulin require the presence of glucose.

Regulation of the branching and debranching enzymes does not occur.

Significance of Substrate (Futile) Cycles

It will be apparent that many of the control points in glycolysis and glycogen metabolism involve a cycle of phosphorylation and dephosphorylation, eg, glucokinase/glucose-6-phosphatase; phosphofructokinase-1/fructose-1,6-bisphosphatase; pyruvate kinase/pyruvate carboxylase/phosphoenolypyruvate carboxykinase;

glycogen synthase/phosphorylase. If these were allowed to cycle unchecked, they would amount to futile cycles whose net result was hydrolysis of ATP. That this does not occur is due to the various control mechanisms which ensure that one limb of the cycle is inhibited as the other is stimulated, according to the need of the tissue and of the body. However, there may be a physiologic advantage in allowing some cycling, as in the glucokinase/glucose-6-phosphatase cycle, where large changes in net flux of metabolites in either direction can occur, controlled by substrate concentrations only.

REGULATION OF THE CITRIC ACID CYCLE
(See Fig 18–3.)

In most tissues, where the primary function of the citric acid cycle is to provide energy, there is little doubt that **respiratory control** via the respiratory chain and oxidative phosphorylation is the overriding control on citric acid cycle activity. Thus, activity is immediately dependent on the supply of oxidized dehydrogenase cofactors (eg, NAD), which in turn is dependent on the availability of ADP and ultimately, therefore, on the rate of utilization of ATP. In addition to this overall or coarse control, the properties of some of the enzymes of the cycle indicate that control might also be exerted at the level of the cycle itself. In a tissue such as brain, which is largely dependent on carbohydrate to supply acetyl-CoA, control of the citric acid cycle may occur at the pyruvate dehydrogenase step. In the cycle proper, control may be exercised by allosteric inhibition of citrate synthase by ATP or long-chain fatty acyl-CoA. Allosteric activation of mitochondrial NAD-dependent isocitrate dehydrogenase by ADP is counteracted by ATP and NADH. The α-ketoglutarate dehydrogenase complex appears to be under control analogous to that of pyruvate dehydrogenase. Succinate dehydrogenase is inhibited by oxaloacetate, and the availability of oxaloacetate, as controlled by malate dehydrogenase, depends on the [NADH]/[NAD$^+$] ratio. Since the K_m for oxaloacetate of citrate synthase is of the same order of magnitude as the intramitochondrial concentration, it would appear that the concentration of oxaloacetate could play a part in controlling the rate of citrate formation. Which (if any) of these mechanisms operates in vivo has still to be resolved.

REGULATION OF THE BLOOD GLUCOSE

Sources of Blood Glucose
A. From Carbohydrates of the Diet: Most carbohydrates in the diet form glucose, galactose, or fructose upon digestion. These are transported to the liver via the **hepatic portal vein.** Galactose and fructose are readily converted to glucose in the liver (see Figs 21–2 and 21–3).

B. From Various Glucogenic Compounds That Undergo Gluconeogenesis: (Fig 22–2.) These compounds fall into 2 categories: (1) those which involve a direct net conversion to glucose without significant recycling, such as some **amino acids** and **propionate;** and (2) those which are the products of the partial metabolism of glucose in certain tissues and which are conveyed to the liver and kidney, where they are resynthesized to glucose. Thus, **lactate,** formed by the oxidation of glucose in skeletal muscle and by erythrocytes, is transported to the liver and kidney where it re-forms glucose, which again becomes available via the circulation for oxidation in the tissues. This process is known as the **Cori cycle** or lactic acid cycle (Fig 22–6). **Glycerol** for the triacylglycerols of adipose tissue is derived initially from the blood glucose, since free glycerol cannot be utilized readily for the synthesis of triacylglycerols in this tissue. Acylglycerols of adipose tissue are continuously undergoing hydrolysis to form free glycerol, which diffuses out of the tissue into the blood. It is converted back to glucose by gluconeogenetic mechanisms in the liver and kidney. Thus, a continuous cycle exists in which glucose is transported from the liver and kidney to adipose tissue and glycerol is returned from adipose tissue to be synthesized into glucose by the liver and kidney.

It has been noted that, of the amino acids transported from muscle to the liver during starvation, alanine predominates. This has led to the postulation of a **glucose-alanine cycle,** as shown in Fig 22–6, which has the effect of cycling glucose from liver to muscle and alanine from muscle to liver, effecting a net transfer of amino nitrogen from muscle to liver and of free energy from liver to muscle. The energy required for the hepatic synthesis of glucose from pyruvate is derived from the oxidation of fatty acids.

C. From Liver Glycogen by Glycogenolysis.

The Concentration of the Blood Glucose
In the postabsorptive state, the blood glucose concentration in humans varies between 80 and 100 mg/dL. After the ingestion of a carbohydrate meal, it may rise to 120–130 mg/dL. During fasting, the level falls to around 60–70 mg/dL. Under normal circumstances, the level is controlled within these limits. The blood glucose level in ruminants is considerably lower, being approximately 40 mg/dL in sheep and 60 mg/dL in cattle. These lower normal levels appear to be associated with the fact that ruminants ferment virtually all dietary carbohydrate to lower (volatile) fatty acids, and these largely replace glucose as the main metabolic fuel of the tissues in the fed condition.

Control of the Concentration of the Blood Glucose
The maintenance of stable levels of glucose in the blood is one of the most finely regulated of all homeostatic mechanisms and one in which the liver, the extrahepatic tissues, and several hormones play a part.

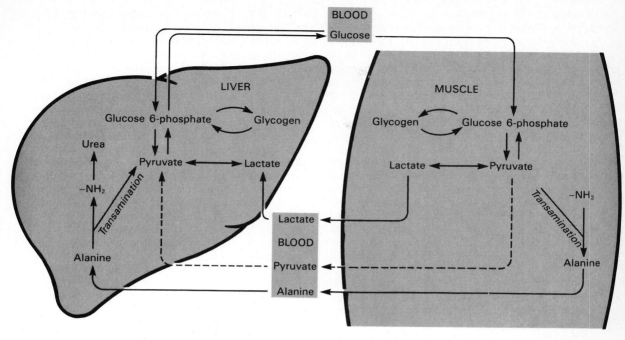

Figure 22–6. The lactic acid (Cori) cycle and glucose-alanine cycle.

Liver cells appear to be freely permeable to glucose, whereas cells of **extrahepatic tissues are relatively impermeable.** As a result, the passage through the cell membrane is the rate-limiting step in the uptake of glucose in extrahepatic tissues, and glucose is rapidly phosphorylated by hexokinase on entry into the cells. On the other hand, it is probable that the activity of certain enzymes and the concentration of key intermediates exert a much more direct effect on the uptake or output of glucose from liver. Nevertheless, the concentration of glucose in the blood is an important factor controlling the rate of uptake of glucose in both liver and extrahepatic tissues.

Role of glucokinase. It is to be noted that hexokinase is inhibited by glucose 6-phosphate, so that some feedback control may be exerted on glucose uptake in extrahepatic tissues dependent on hexokinase for glucose phosphorylation. The liver is not subject to this constraint because glucokinase is not affected by glucose 6-phosphate. Glucokinase, which has a higher K_m (lower affinity) for glucose than does hexokinase, increases in activity over the physiologic range of glucose concentrations (Fig 22–7) and seems to be specifically concerned with glucose uptake into the liver at the higher concentrations found in the hepatic portal vein after a carbohydrate meal. Its absence in ruminants, which have little glucose entering the portal circulation from the intestines, is compatible with this function.

At normal blood glucose concentrations (80–100 mg/dL), the liver appears to be a net producer of glucose. However, as the glucose level rises, the output of glucose ceases, so that at high levels there is a net uptake. In the rat, it has been estimated that the rate of uptake of glucose and the rate of output are equal at a hepatic portal vein blood glucose concentration of 150 mg/dL.

Role of insulin. In addition to the direct effects of hyperglycemia in enhancing the uptake of glucose into both the liver and peripheral tissues, the hormone insulin plays a central role in the regulation of the blood

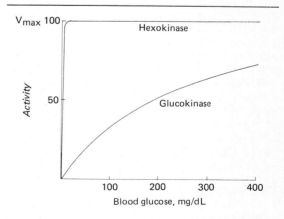

Figure 22–7. Variation in glucose phosphorylating activity of hexokinase and glucokinase with increase of blood glucose concentration. The K_m for glucose of hexokinase is 0.05 mmol/L (0.9 mg/dL) and of glucokinase is 10 mmol/L (180 mg/dL).

glucose concentration. It is produced by the B cells of the islets of Langerhans in the pancreas and is secreted into the blood as a direct response to hyperglycemia. Its concentration in the blood parallels that of the blood glucose, and its administration results in prompt **hypoglycemia.** Substances causing release of insulin include also amino acids, free fatty acids, ketone bodies, glucagon, secretin, and the drug tolbutamide. Epinephrine and norepinephrine block the release of insulin. Insulin has an immediate effect of increasing glucose uptake in tissues such as adipose tissue and muscle. This action is due to an enhancement of glucose transport through the cell membrane by recruitment of insulin transporters from the interior of the cell to the plasma membrane. In contrast, there is no direct effect of insulin on glucose penetration of hepatic cells; this finding agrees with the fact that glucose metabolism by liver cells is not rate-limited by their permeability to glucose. However, insulin does indirectly enhance uptake of glucose by the liver as a result of its actions on the enzymes controlling glycolysis and glycogenesis (see above).

The **anterior pituitary gland** secretes hormones that tend to elevate the blood glucose and therefore antagonize the action of insulin. These are growth hormone, ACTH (corticotropin), and possibly other "diabetogenic" principles. Growth hormone secretion is stimulated by hypoglycemia. Growth hormone decreases glucose uptake in certain tissues, eg, muscle. Some of this effect may not be direct, since it mobilizes free fatty acids from adipose tissue which themselves inhibit glucose utilization. Chronic administration of growth hormone leads to diabetes. By producing hyperglycemia, it stimulates secretion of insulin, eventually causing B cell exhaustion.

The **glucocorticoids** (11-oxysteroids) are secreted by the adrenal cortex and are important in carbohydrate metabolism. Administration of these steroids causes increased gluconeogenesis. This is as a result of increased protein catabolism in the tissues, increased hepatic uptake of amino acids, and increased activity of transaminases and other enzymes concerned with gluconeogenesis in the liver. In addition, glucocorticoids **inhibit the utilization of glucose in extrahepatic tissues.** In all these actions, glucocorticoids act in a manner **antagonistic to insulin.**

Epinephrine is secreted by the adrenal medulla as a result of stressful stimuli (fear, excitement, hemorrhage, hypoxia, hypoglycemia, etc) and leads to glycogenolysis in liver and muscle owing to stimulation of phosphorylase. In muscle, as a result of the absence of glucose-6-phosphatase, glycogenolysis ensues with the formation of lactate, whereas in liver, glucose is the main product leading to increase in blood glucose.

Glucagon is the hormone produced by the A cells of the islets of Langerhans of the pancreas. Its secretion is stimulated by hypoglycemia, and, when it reaches the liver (via the portal vein), it causes glycogenolysis by activating phosphorylase in a manner similar to epinephrine. Most of the endogenous

glucagon is cleared from the circulation by the liver. Unlike epinephrine, glucagon does not have an effect on muscle phosphorylase. Glucagon also enhances gluconeogenesis from amino acids and lactate. Both hepatic glycogenolysis and gluconeogenesis contribute to the **hyperglycemic** effect of glucagon.

Thyroid hormone should also be considered as affecting the blood glucose. There is experimental evidence that thyroxine has a diabetogenic action and that thyroidectomy inhibits the development of diabetes. It has also been noted that there is a complete absence of glycogen from the livers of thyrotoxic animals. In humans, the fasting blood glucose is elevated in hyperthyroid patients and decreased in hypothyroid patients. However, hyperthyroid patients apparently utilize glucose at a normal or increased rate, whereas hypothyroid patients have a decreased ability to utilize glucose. In addition, hypothyroid patients are much less sensitive to insulin than are normal or hyperthyroid individuals.

The Renal Threshold for Glucose—Glycosuria

When the blood glucose rises to relatively high levels, the kidney also exerts a regulatory effect. Glucose is continuously filtered by the glomeruli but is ordinarily returned completely to the blood by the reabsorptive system of the renal tubules. The reabsorption of glucose is linked to the provision of ATP in the tubular cells. The capacity of the tubular system to reabsorb glucose is limited to a rate of about 350 mg/min. When the blood levels of glucose are elevated, the glomeru-

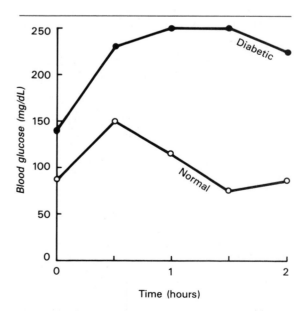

Figure 22–8. Glucose tolerance test. Blood glucose curves of a normal and a diabetic individual after oral administration of 50 g of glucose. Note the initial raised concentration in the diabetic. A criterion of normality is the return of the curve to the initial value within 2 hours.

lar filtrate may contain more glucose than can be reabsorbed; the excess passes into the urine to produce glycosuria. In normal individuals, glycosuria occurs when the venous blood glucose concentration exceeds 170–180 mg/dL. This is termed the renal threshold for glucose.

Glycosuria may be produced in experimental animals with **phlorhizin,** which inhibits the glucose reabsorptive system in the tubule. This is known as renal glycosuria. Glycosuria of renal origin may result from inherited defects in the kidney, or it may be acquired as a result of disease processes. The presence of glycosuria is frequently an indication of **diabetes mellitus.**

Glucose Tolerance

The ability of the body to utilize glucose may be ascertained by measuring its glucose tolerance. It is indicated by the nature of the blood glucose curve following the administration of a test amount of glucose (Fig 22–8). **Diabetes mellitus** ("sugar" diabetes) is characterized by decreased glucose tolerance due to decreased secretion of insulin. This is manifested by elevated blood glucose levels (hyperglycemia) and glycosuria and may be accompanied by changes in fat metabolism. Tolerance to glucose declines not only in diabetes but also in conditions where the liver is damaged, in some infections, in obesity, under the influence of some drugs, and sometimes in atherosclerosis. It would also be expected to occur in the presence of hyperactivity of the pituitary or adrenal cortex, because of the antagonism of the hormones of these endocrine glands to the action of insulin.

Insulin increases glucose tolerance. Injection of insulin lowers the content of the glucose in the blood and increases its utilization and its storage in the liver and muscle as glycogen. An excess of insulin may cause severe **hypoglycemia** resulting in convulsions and even in death unless glucose is administered promptly. In humans, hypoglycemic convulsions may occur when the blood glucose is lowered acutely to about 20 mg/dL. Increased tolerance to glucose is observed in pituitary or adrenocortical insufficiency, attributable to a decrease in the normal antagonism to insulin which results in a relative excess of that hormone.

REFERENCES

Cohen P: *Control of Enzyme Activity,* 2nd ed. Chapman & Hall, 1983.

Hers HG: The control of glycogen metabolism in the liver. *Annu Rev Biochem* 1976;**45**:167.

Hers HG, Hue L: Gluconeogenesis and related aspects of glycolysis. *Annu Rev Biochem* 1983;**52**:617.

Hers HG, Van Schaftingen E: Fructose 2-6-bisphosphate two years after its discovery. *Biochem J* 1982;**206**:1.

Hue L, Van de Werve G (editors): *Short-Term Regulation of Liver Metabolism.* Elsevier/North Holland, 1981.

Newsholme EA, Crabtree B: Flux-generating and regulatory steps in metabolic control. *Trends Biochem Sci* 1981;**6**:53.

Newsholme EA, Start C: *Regulation in Metabolism.* Wiley, 1973.

Storey KB: A re-evaluation of the Pasteur effect. *Mol Physiol* 1985;**8**:439.

23

Oxidation & Biosynthesis of Fatty Acids

Peter A. Mayes, PhD, DSc

INTRODUCTION

Fatty acids are both oxidized to acetyl-CoA and synthesized from acetyl-CoA. Although the starting material of one process is identical to the product of the other and the chemical stages involved are comparable, fatty acid biosynthesis is not the simple reverse of fatty acid oxidation but an entirely different process taking place in a separate compartment of the cell. Fatty acid oxidation takes place in mitochondria; each step involves acyl-CoA derivatives catalyzed by separate enzymes, utilizes NAD and FAD as coenzymes, and generates ATP. In contrast, fatty acid biosynthesis (lipogenesis) takes place in the cytosol, involves acyl derivatives continuously attached to a multienzyme complex, utilizes NADP as coenzyme, and requires both ATP and bicarbonate ion.

BIOMEDICAL IMPORTANCE

The separation of fatty acid oxidation from biosynthesis allows each process to be individually controlled and integrated with tissue requirements. Increased fatty acid oxidation is characteristic of starvation and of diabetes mellitus, leading to ketone body production by the liver (ketosis). Ketone bodies are acidic and when produced in excess over long periods, as in diabetes, cause ketoacidosis, which is ultimately fatal. Because gluconeogenesis is dependent upon fatty acid oxidation, any impairment in fatty acid oxidation leads to hypoglycemia. This occurs in various states of carnitine deficiency or deficiency of essential enzymes in fatty acid oxidation (eg, carnitine palmitoyltransferase) or inhibition of fatty acid oxidation by poisons (eg, hypoglycin). Fatty acid biosynthesis is of primary importance in disposing of excess carbohydrate intake, either for temporary storage for use between meals or on a longer-term basis. Fatty acids are stored in adipose tissue as triacylglycerols.

OXIDATION OF FATTY ACIDS

Free Fatty Acids

The term "free fatty acid" (FFA) refers to fatty acids that are in the unesterified state. Alternative nomenclature is UFA (unesterified fatty acids) or NEFA (nonesterified fatty acids). In plasma, FFA of longer-chain fatty acids are combined with **albumin,** and in the cell they are attached to a **fatty acid binding protein,** or Z-protein, so that in fact they are never really "free." Shorter-chain fatty acids are more water-soluble and exist as the un-ionized acid or as a fatty acid anion.

Activation of Fatty Acids

As in the metabolism of glucose, fatty acids must first be converted in a reaction with ATP to an active intermediate before they will react with the enzymes responsible for their further metabolism. This is the only step in the complete degradation of a fatty acid that requires energy from ATP. In the presence of ATP and coenzyme A, the enzyme **acyl-CoA synthetase (thiokinase)** catalyzes the conversion of a fatty acid (or free fatty acid) to an "active fatty acid" or acyl-CoA, accompanied by the expenditure of one high-energy phosphate bond.

$$\textbf{Fatty acid + ATP + CoA} \longrightarrow \textbf{Acyl-CoA + PP}_i\textbf{ + AMP}$$

ACYL-CoA SYNTHETASE

The presence of **inorganic pyrophosphatase** ensures that activation goes to completion by facilitating the loss of the additional high-energy phosphate bond of pyrophosphate. Thus, in effect, 2 high-energy phosphate bonds are expended during the activation of each fatty acid molecule.

$$\textbf{PP}_i\textbf{ + H}_2\textbf{O} \longrightarrow \textbf{2 P}_i$$

INORGANIC PYROPHOSPHATASE

Acyl-CoA synthetases are found in the endoplasmic reticulum and inside and on the outer membrane of mitochondria. Several acyl-CoA synthetases have been described, each specific for fatty acids of different chain length.

Role of Carnitine in Fatty Acid Oxidation

Carnitine (β-hydroxy-γ-trimethylammonium butyrate), $(CH_3)_3N^+-CH_2-CH(OH)-CH_2-COO^-$, is widely distributed and is particularly abundant in muscle. It is synthesized from lysine and methionine in liver and kidney. Activation of

$$\text{Acyl-CoA + Carnitine} \longleftrightarrow \text{Acylcarnitine + CoA}$$

| CARNITINE |
| PALMITOYLTRANSFERASE |

lower fatty acids, and their oxidation may occur within the mitochondria, independently of carnitine, but long-chain acyl-CoA (or FFA) **will not penetrate mitochondria** and become oxidized unless it forms acylcarnitines. An enzyme, **carnitine palmitoyltransferase I,** is associated with the outer side of the inner mitochondrial membrane and converts long-chain acyl groups to acylcarnitine, which are able to penetrate mitochondria and gain access to the β-oxidation system of enzymes. A possible mechanism to account for the action of carnitine in facilitating the oxidation of fatty acids by mitochondria is shown in Fig 23–1. In addition, another enzyme, **carnitine acetyltransfer-**

ase, is present within mitochondria and catalyzes the transfer of short-chain acyl groups between CoA and carnitine. The function of this enzyme is obscure, but it may facilitate transport of acetyl groups through the mitochondrial membrane.

$$\text{Acetyl-CoA + Carnitine} \longleftrightarrow \text{Acetylcarnitine + CoA}$$

| CARNITINE |
| ACETYLTRANSFERASE |

β-Oxidation of Fatty Acids

Overview (Fig 23–2). In β-oxidation, 2 carbons are cleaved at a time from acyl-CoA molecules, starting at the carboxyl end. The chain is broken between the α(2)- and β(3)-carbon atoms, hence the name β-oxidation. The 2-carbon units formed are acetyl-CoA; thus, palmitoyl-CoA forms 8 acetyl-CoA molecules.

Sequence of Reactions

Several enzymes, known collectively as "fatty acid oxidase," are found in the mitochondrial matrix adjacent to the respiratory chain (which is found in the inner membrane). These catalyze the oxidation of acyl-CoA to acetyl-CoA, the system being coupled with the phosphorylation of ADP to ATP (Fig 23–3).

After the penetration of the acyl moiety through the mitochondrial membrane via the carnitine transporter system and the re-formation of acyl-CoA, there follows the removal of 2 hydrogen atoms from the 2(α)- and 3(β)-carbon atoms, catalyzed by **acyl-CoA dehydrogenase.** This results in the formation of Δ^2-*trans*-enoyl-CoA. The coenzyme for the dehydrogenase is a flavoprotein, containing FAD as prosthetic group, whose reoxidation by the respiratory chain requires the mediation of another flavoprotein, termed **electron-transferring flavoprotein** (see p 103). Water is added to saturate the double bond and form 3-hydroxyacyl-CoA, catalyzed by the enzyme Δ^2**-enoyl-CoA hydratase.** The 3-hydroxy derivative undergoes further dehydrogenation on the 3-carbon (**3-hydroxyacyl-CoA dehydrogenase**) to form the corresponding 3-ketoacyl-CoA compound. In this case, NAD is the coenzyme involved in the dehydrogenation. Finally, 3-ketoacyl-CoA is split at the 2,3-position by **thiolase** (3-ketothiolase or acetyl-CoA acyltransferase), which catalyzes a thiolytic cleavage involving another molecule of CoA. The products of this reaction are acetyl-CoA and an acyl-CoA derivative containing 2 carbons less than the original acyl-CoA molecule that underwent oxidation. The acyl-CoA formed in the cleavage reaction reenters the oxidative pathway at reaction 2 (Fig 23–3). In this way, a long-chain fatty acid may be degraded completely to acetyl-CoA (C_2 units). As acetyl-CoA can be oxidized to CO_2 and water via the citric acid cycle (which is also found within the mitochondria), the complete oxidation of fatty acids is achieved.

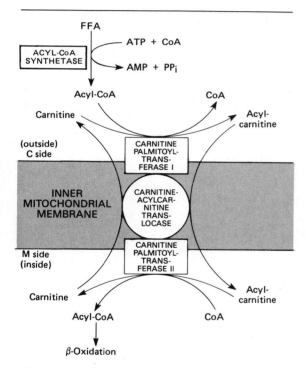

Figure 23–1. Role of carnitine in the transport of long-chain fatty acids through the inner mitochondrial membrane. Long-chain acyl-CoA cannot pass through the inner mitochondrial membrane, but its metabolic product, acylcarnitine, formed by the action of **carnitine palmitoyltransferase I,** can. **Carnitine-acylcarnitine translocase** acts as a membrane carnitine exchange transporter. Acylcarnitine is transported in, coupled with the transport out of one molecule of carnitine. The acylcarnitine then reacts with CoA, catalyzed by **carnitine palmitoyltransferase II,** attached to the inside of the inner membrane. Acyl-CoA is re-formed in the mitochondrial matrix, and carnitine is liberated.

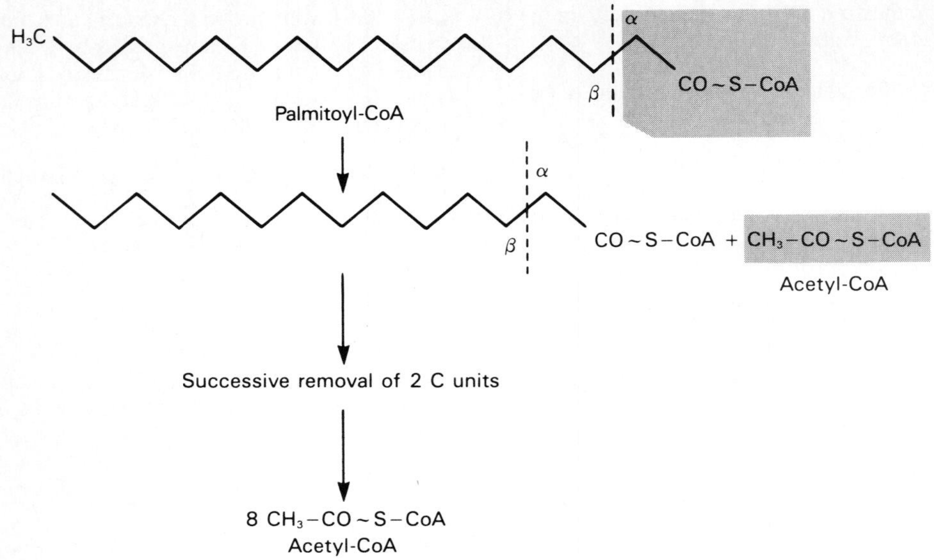

Figure 23–2. Overview of β-oxidation of fatty acids.

Oxidation of Fatty Acids With Odd Numbers of Carbon Atoms

Fatty acids with an odd number of carbon atoms are oxidized by the pathway of β-oxidation until a 3-carbon (propionyl-CoA) residue remains. This compound is converted to succinyl-CoA, a constituent of the citric acid cycle (see also Fig 20–2).

Energetics of Fatty Acid Oxidation

Transport in the respiratory chain of electrons from reduced flavoprotein and NAD will lead to the synthesis of 5 high-energy phosphate bonds (see Chapter 13) for each of the first 7 acetyl-CoA molecules formed by β-oxidation of palmitate ($7 \times 5 = 35$). A total of 8 mol of acetyl-CoA is formed, and each will give rise to 12 high-energy bonds on oxidation in the citric acid cycle, making $8 \times 12 = 96$ high-energy bonds derived from the acetyl-CoA formed from palmitate, minus 2 for the initial activation of the fatty acid, yielding a net gain of 129 high-energy bonds/mol, or $129 \times 30.5 = 3935$ kJ. As the free energy of combustion of palmitic acid is 9791 kJ/mol, the process captures as high-energy phosphate on the order of 40% of the total energy of combustion of the fatty acid.

Peroxisomal Fatty Acid Oxidation

A modified form of β-oxidation is found in peroxisomes and leads to the formation of acetyl-CoA and H_2O_2 (from the flavoprotein-linked dehydrogenase step). The system is not linked directly to phosphorylation and the generation of ATP but aids the oxidation of very long chain fatty acids (eg, C_{20}, C_{22}) and is induced by high-fat diets and hypolipidemic drugs such as clofibrate.

The enzymes in peroxisomes do not attack shorter-chain fatty acids; the β-oxidation sequence ends at octanoyl-CoA. Octanoyl and acetyl groups are subsequently removed from the peroxisomes in the forms of octanoyl and acetyl carnitine, and both are further oxidized in mitochondria.

α- & ω-Oxidation of Fatty Acids

Quantitatively, β-oxidation is the most important pathway for fatty acid oxidation. However, α-oxidation, ie, the removal of one carbon at a time from the carboxyl end of the molecule, has been detected in brain tissue. It does not require CoA intermediates and does not generate high-energy phosphates.

ω-Oxidation is normally a very minor pathway and is brought about by hydroxylase enzymes involving cytochrome P-450 in the endoplasmic reticulum (see p 103). The $-CH_3$ group is converted to a $-CH_2OH$ group that subsequently is oxidized to $-COOH$, thus forming a dicarboxylic acid. This is β-oxidized usually to adipic (C_6) and suberic (C_8) acids, which are excreted in the urine.

Clinical Aspects

Ketosis (see p 258) occurs whenever there are high rates of fatty acid oxidation in the liver, particularly when there is an associated deficiency of carbohydrate. This occurs under **high-fat diets,** and in **starvation, diabetes mellitus, ketosis of lactating cattle,** and **pregnancy toxemia** (ketosis) of sheep. The following conditions are all associated with impaired oxidation of fatty acids.

Carnitine deficiency can occur particularly in the newborn—and especially in preterm infants—owing to inadequate biosynthesis or renal leakage. Losses

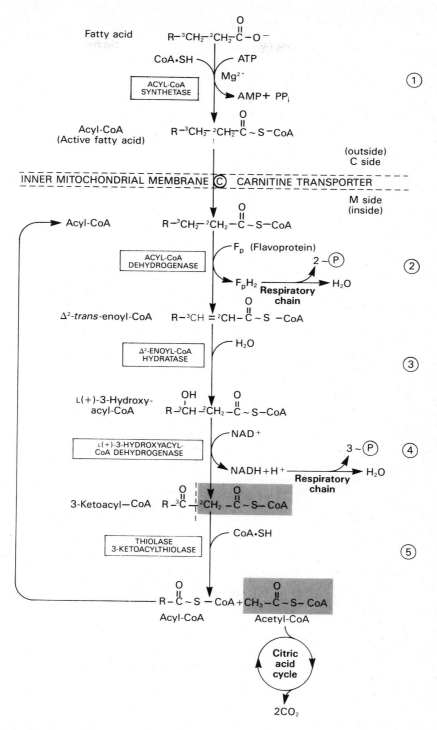

Figure 23-3. β-Oxidation of fatty acids. Long-chain acyl-CoA is cycled through reactions 2–5, acetyl-CoA being split off each cycle by thiolase (reaction 5). When the acyl radical is only 4 carbon atoms in length, 2 acetyl-CoA molecules are formed in reaction 5.

can also occur in hemodialysis; patients with organic aciduria have large losses of carnitine, which is excreted conjugated to the organic acids. This indicates a vitaminlike dietary requirement for carnitine in some individuals. Signs and symptoms of deficiency include episodic periods of hypoglycemia owing to reduced gluconeogenesis resulting from impaired fatty acid oxidation, impaired ketogenesis in the presence of raised plasma FFA, muscular weakness, and lipid accumulation. Treatment is by oral supplementation of carnitine. The symptoms are similar to Reye's syndrome, in which carnitine is adequate, but the cause of Reye's syndrome is unknown.

Hepatic carnitine palmitoyltransferase deficiency results in hypoglycemia and low plasma ketone bodies, whereas **muscle carnitine palmitoyltransferase deficiency** leads to impaired fatty acid oxidation that results in recurrent muscle weakness and myoglobinuria.

Jamaican vomiting sickness is caused by eating the unripe fruit of the akee tree, which contains a toxin, **hypoglycin,** that inactivates acyl-CoA dehydrogenase, inhibiting β-oxidation and causing hypoglycemia.

Dicarboxylic aciduria is characterized by the excretion of C_6-C_{10} ω-dicarboxylic acids and by nonketotic hypoglycemia. It is caused by a lack of mitochondrial medium-chain acyl-CoA dehydrogenase. This impairs β-oxidation but increases ω-oxidation of long- and medium-chain fatty acids, which are then shortened by β-oxidation to medium-chain dicarboxylic acids, which are excreted.

Refsum's disease is a rare neurologic disorder caused by accumulation of phytanic acid, formed from phytol, a constituent of chlorophyll found in plant foodstuffs. Phytanic acid contains a methyl group on carbon 3 that blocks β-oxidation. Normally, an initial α-oxidation removes the methyl group, but persons with Refsum's disease have an inherited defect in α-oxidation that allows accumulation of phytanic acid.

Zellweger's (cerebrohepatorenal) syndrome occurs in individuals with a rare inherited absence of peroxisomes in all tissues. They accumulate $C_{26}-C_{38}$ polyenoic acids in brain tissue owing to inability to oxidize fatty acids of long-chain length in peroxisomes.

Oxidation of Unsaturated Fatty Acids

The CoA esters of these acids are degraded by the enzymes normally responsible for β-oxidation until either a Δ^3-*cis*-acyl-CoA compound or a Δ^4-*cis*-acyl-CoA compound is formed, depending upon the position of the double bonds (Fig 23–4). The former compound is isomerized (Δ^3-*cis*$\rightarrow\Delta^2$-*trans*-**enoyl-CoA isomerase**) to the corresponding Δ^2-*trans*-CoA stage of β-oxidation for subsequent hydration and oxidation. Any Δ^4-*cis*-acyl-CoA either remaining, as in the case of linoleic acid (Fig 23–4), or entering the pathway at this point is converted by acyl-CoA dehydrogenase to Δ^2-*trans*-Δ^4-*cis*-dienoyl-CoA. This is converted to Δ^3-*trans*-enoyl-coA by an NADP-dependent

enzyme, **Δ^2-*trans*-Δ^4-*cis*-dienoyl-CoA reductase.** Δ^3-*cis*$\rightarrow\Delta^2$-*trans*-enoyl-CoA isomerase will also attack the Δ^3-*trans* double bond to produce Δ^2-*trans*-enoyl-CoA, an intermediate in β-oxidation.

Microsomal Peroxidation of Polyunsaturated Fatty Acids

NADPH-dependent peroxidation of unsaturated fatty acids is catalyzed by microsomal enzymes (see p 106). The antioxidants BHT (butylated hydroxytoluence) and α-tocopherol (vitamin E) inhibit microsomal lipid peroxidation.

BIOSYNTHESIS OF SATURATED FATTY ACIDS

Like many other degradative and synthetic processes (eg, glycogenolysis and glycogenesis), fatty acid synthesis was formerly considered to be merely the reversal of oxidation. However, it now seems clear that a mitochondrial system for fatty acid synthesis, involving some modification of the β-oxidation sequence, is responsible only for elongation of existing fatty acids of moderate chain length, whereas a radically different and highly active **extramitochondrial** system is responsible for the complete synthesis of palmitate from acetyl-CoA. An active system for **chain elongation** is also present in liver endoplasmic reticulum.

Extramitochondrial System for De Novo Synthesis of Fatty Acids (Lipogenesis)

This system has been found in the soluble (cytosol) fraction of many tissues, including liver, kidney, brain, lung, mammary gland, and adipose tissue. Its cofactor requirements include NADPH, ATP, Mn^{2+}, and HCO_3^- (as a source of CO_2). Acetyl-CoA is the substrate, and free palmitate is the end product. These characteristics contrast markedly with those of β-oxidation.

Production of Malonyl-CoA

Bicarbonate as a source of CO_2 is required in the initial reaction for the carboxylation of acetyl-CoA to **malonyl-CoA** in the presence of ATP and **acetyl-CoA carboxylase.** Acetyl-CoA carboxylase has a requirement for the vitamin **biotin** (Fig 23–5). The enzyme contains a variable number of identical subunits, each monomer containing biotin, biotin carboxylase, biotin carboxyl carrier protein, and transcarboxylase, as well as a regulatory allosteric site. It is therefore a **multienzyme protein.** The reaction takes place in 2 steps: (1) carboxylation of biotin (involving ATP; see Fig 20–4) and (2) transfer of the carboxyl to acetyl-CoA to form malonyl-CoA. Acetyl-CoA carboxylase is activated by citrate and inhibited by long-chain acyl-CoA. The activated form readily polymerizes into filaments containing 10–20 protomers.

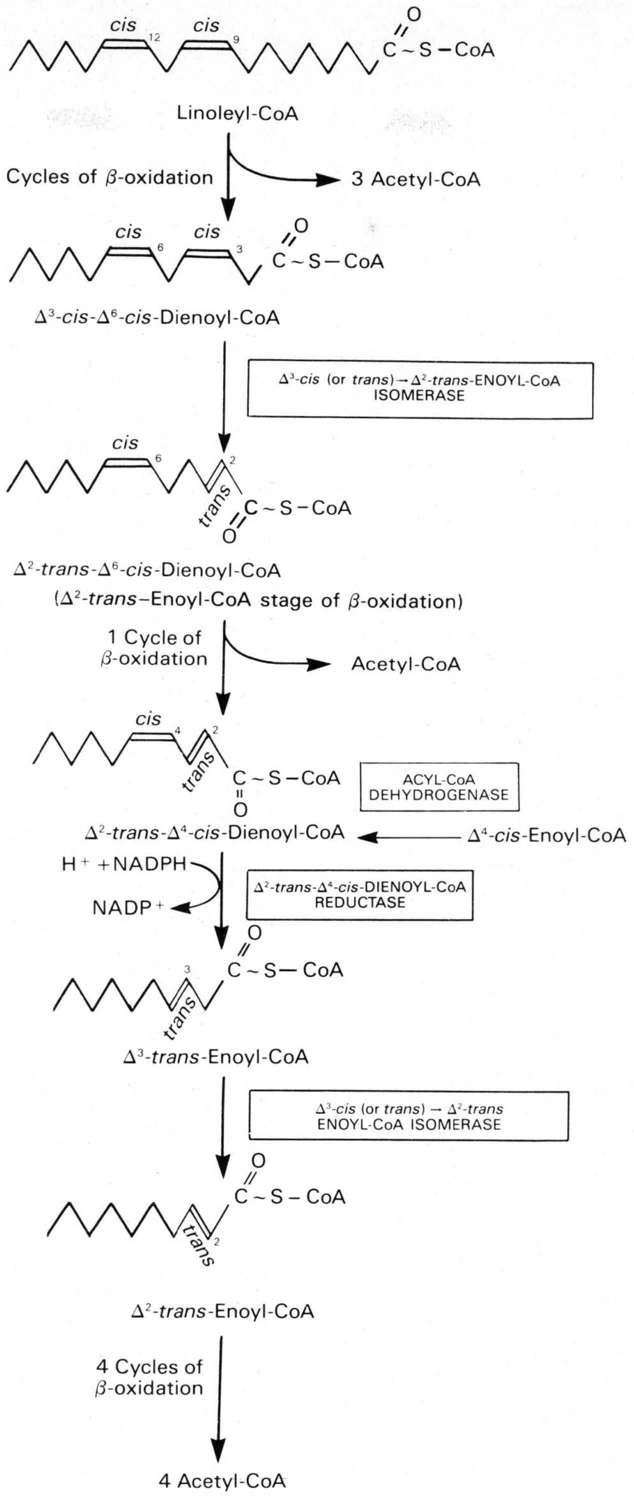

cis 12   *cis* 9 $\overset{O}{\underset{}{\|}}$ C ~ S —CoA

Linoleyl-CoA

3 Cycles of β-oxidation   ⟶   3 Acetyl-CoA

cis 6   *cis* 3 $\overset{O}{\underset{}{\|}}$ C ~ S—CoA

Δ^3-*cis*-Δ^6-*cis*-Dienoyl-CoA

Δ^3-*cis* (or *trans*) → Δ^2-*trans*-ENOYL-CoA
ISOMERASE

cis 6   2   *trans*   C ~ S—CoA
$\overset{}{\underset{\|}{O}}$

Δ^2-*trans*-Δ^6-*cis*-Dienoyl-CoA
(Δ^2-*trans*–Enoyl-CoA stage of β-oxidation)

1 Cycle of
β-oxidation   ⟶   Acetyl-CoA

cis 4   2   *trans*   C ~ S—CoA
$\overset{}{\underset{\|}{O}}$

ACYL-CoA
DEHYDROGENASE

Δ^2-*trans*-Δ^4-*cis*-Dienoyl-CoA ⟵ Δ^4-*cis*-Enoyl-CoA

H^+ +NADPH

NADP$^+$ ⟵

Δ^2-*trans*-Δ^4-*cis*-DIENOYL-CoA
REDUCTASE

$\overset{O}{\underset{}{\|}}$ C ~ S— CoA
3 *trans*

Δ^3-*trans*-Enoyl-CoA

Δ^3-*cis* (or *trans*) → Δ^2-*trans*
ENOYL-CoA ISOMERASE

$\overset{O}{\underset{}{\|}}$ C ~ S — CoA
2 *trans*

Δ^2-*trans*-Enoyl-CoA

4 Cycles of
β-oxidation

4 Acetyl-CoA

Figure 23–4. Sequence of reactions in the oxidation of unsaturated fatty acids, eg, linoleic acid. Δ^4-*cis*-fatty acids or fatty acids forming Δ^4-*cis*-enoyl-CoA enter the pathway at the position shown.

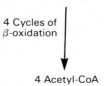

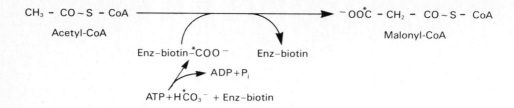

Figure 23–5. Biosynthesis of malonyl-CoA. Enz, acetyl-CoA carboxylase.

Fatty Acid Synthase Complex

There appear to be 2 types of **fatty acid synthase** systems found in the soluble portion of the cell. In bacteria, plants, and lower forms such as *Euglena,* the individual enzymes of the system are separate, and the acyl radicals are found in combination with a protein called the **acyl carrier protein (ACP).** However, in yeast, mammals, and birds, the synthase system is a multienzyme complex that may not be subdivided without loss of activity, and ACP is part of this complex. Both ACP of bacteria and the multienzyme complex contain the vitamin pantothenic acid in the form of 4'-phosphopantetheine (see Fig 17–6). In this system, ACP takes over the role of CoA.

The fatty acid synthase complex is a dimer (Fig 23–6). In animals, each monomer is identical, con-

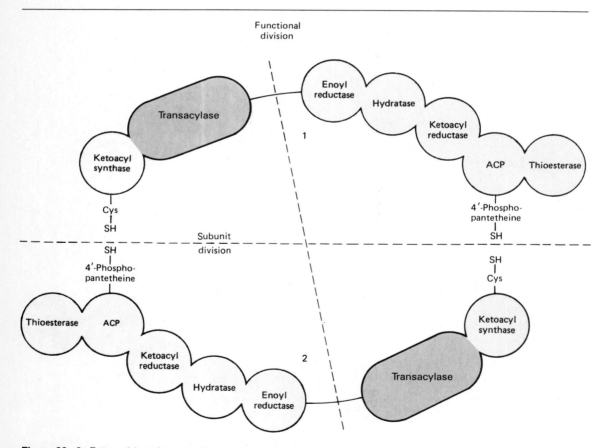

Figure 23–6. Fatty acid synthase multienzyme complex. The complex is a dimer of 2 identical polypeptide monomers, 1 and 2, each consisting of 6 separate enzyme activities and the acyl carrier protein (ACP). Cys–SH, cysteine thiol. The –SH of the 4'-phosphopantetheine of one monomer is in close proximity to the –SH of the cysteine residue of the ketoacyl synthase of the other monomer, suggesting a "head-to-tail" arrangement of the 2 monomers. The detailed sequence of the enzymes in each monomer is tentative (based on Tsukamoto). Though each monomer contains all the partial activities of the reaction sequence, the actual functional unit consists of one-half of a monomer interacting with the complementary half of the other. Thus, 2 acyl chains are produced simultaneously.

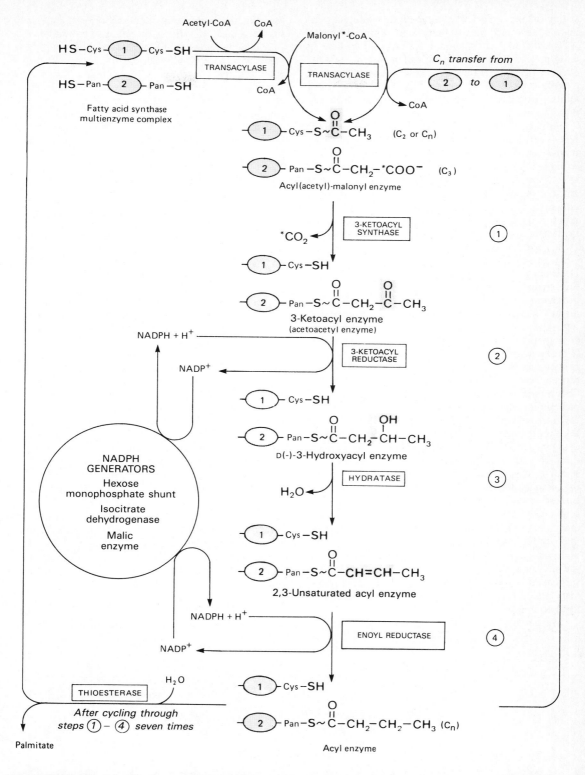

Figure 23-7. Biosynthesis of long-chain fatty acids. Details of how addition of a malonyl residue causes the acyl chain to grow by 2 carbon atoms. Cys-, cysteine residue; pan, 4′-phosphopantetheine. Details of the fatty acid synthase dimer are shown in Fig 23-6. ① and ② represent the individual monomers of fatty acid synthase. Two acyl chains are synthesized simultaneously on one dimer using each pair of Cys/pan –SH groups.

sisting of one remarkable polypeptide chain containing all the 6 enzymes of fatty acid synthase and an ACP with a 4′-phosphopantetheine –SH group. In close proximity is another thiol of a cysteine residue attached to **3-ketoacyl synthase (condensing enzyme)** of the other monomer (Fig 23–6). Since both thiols participate in the synthase activity, **only the dimer is active.**

Initially, a priming molecule of acetyl-CoA combines with the cysteine –SH group catalyzed by **transacylase** (Fig 23–7). Malonyl-CoA combines with the adjacent –SH on the 4′-phosphopantetheine of ACP of the other monomer, catalyzed by the same enzyme, **transacylase,** to form **acetyl (acyl)-malonyl enzyme.** The acetyl group attacks the methylene group of the malonyl residue, catalyzed by **3-ketoacyl synthase,** and liberates CO_2, forming 3-ketoacyl enzyme (acetoacetyl enzyme). This frees the cysteine –SH group, hitherto occupied by the acetyl group. Decarboxylation allows the reaction to go to completion, acting as a pulling force for the whole sequence of reactions. The 3-ketoacyl group is reduced, dehydrated, and reduced again to form the corresponding saturated acyl-S-enzyme. These reactions are analogous to those in β-oxidation, except that the 3-hydroxy acid is the D(−) isomer instead of the L(+) isomer and NADPH rather than NADH serves as hydrogen donor for both reductions. A new malonyl-CoA molecule combines with the –SH of 4′-phosphopantetheine, displacing the saturated acyl residue onto the free cysteine –SH group. The sequence of reactions is repeated 6 more times, a new malonyl residue being incorporated during each sequence, until a saturated 16-carbon acyl radical (palmityl) has been assembled. It is liberated from the enzyme complex by the activity of a sixth enzyme in the complex, **thioesterase** (deacylase). The free palmitate must be activated to acyl-CoA before it can proceed via any other metabolic pathway. Its usual fate is esterification into acylglycerols (Fig 23–8).

In mammary gland, there is a separate thioesterase specific for acyl residues of C_8, C_{10}, or C_{12}, which are subsequently found in milk lipids. In ruminant mammary gland, this enzyme is part of the fatty acid synthase complex.

There would appear to be 2 centers of activity in one dimer complex that function independently and simultaneously to form 2 molecules of palmitate.

The aggregation of all the enzymes of a particular pathway into one multienzyme functional unit offers great efficiency and freedom from interference by competing processes, thus achieving the effect of compartmentalization of the process within the cell, without the erection of permeability barriers.

The equation for the overall synthesis of palmitate from acetyl-CoA and malonyl-CoA is shown below:

$$CH_2CO \cdot S \cdot CoA + 7HOOC \cdot CH_2CO \cdot S \cdot CoA$$
$$+ 14NADPH + 14H^+ \longrightarrow$$

$$CH_3(CH_2)_{14}COOH + 7CO_2 + 6H_2O$$
$$+ 8CoA \cdot SH + 14NADP^+$$

The acetyl-CoA used as a primer forms carbon atoms 15 and 16 of palmitate. The addition of all the subsequent C_2 units is via malonyl-CoA formation. Butyryl-CoA may act as a primer molecule in mammalian liver and mammary gland. If propionyl-CoA acts as primer, long-chain fatty acids having an odd number of carbon atoms result. These are found particularly in ruminants, where propionate is formed by microbial action in the rumen.

Sources of reducing equivalents and acetyl-CoA. NADPH is involved as coenzyme in both the reduction of the 3-ketoacyl and of the 2,3-unsaturated acyl derivatives. The oxidative reactions of the pentose phosphate pathway are the chief source of the hydrogen required for the reductive synthesis of fatty acids. It is significant that tissues which possess an active pentose phosphate pathway are also the tissues specializing in active lipogenesis, ie, liver, adipose tissue, and the lactating mammary gland. Moreover, both metabolic pathways are found in the extramitochondrial region of the cell, so that there are no membranes or permeability barriers for the transfer of NADPH/NADP from one pathway to the other. Other sources of NADPH include the reaction that converts malate to pyruvate catalyzed by the **"malic enzyme"** (NADP malate dehydrogenase) (Fig 23–9) and the extramitochondrial **isocitrate dehydrogenase** reaction (probably not a substantial source).

Acetyl-CoA, the main building block for fatty acids, is formed from carbohydrate via the oxidation

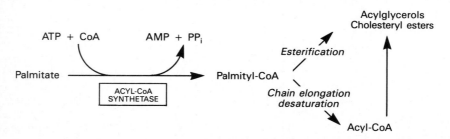

Figure 23–8. Fate of palmitate.

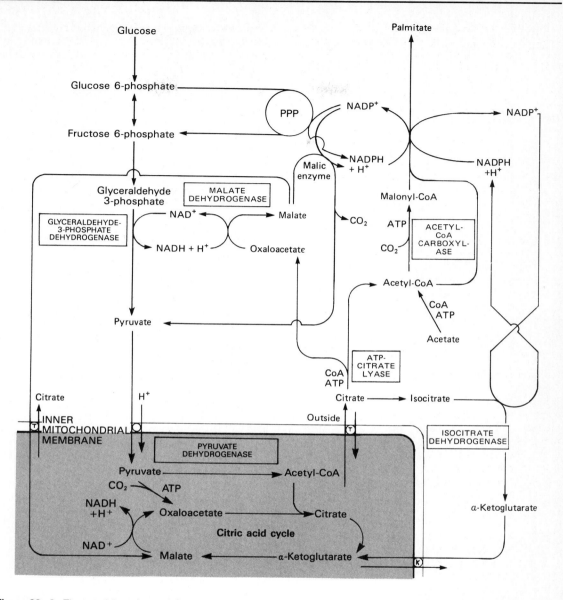

Figure 23–9. The provision of acetyl-CoA and NADPH for lipogenesis. PPP, pentose phosphate pathway; T, tricarboxylate transporter; K, α-ketoglutarate transporter.

of pyruvate within the mitochondria. However, acetyl-CoA does not diffuse readily into the extramitochondrial compartment, the principal site of fatty acid synthesis. The activity of the extramitochondrial **ATP-citrate lyase,** like the "malic enzyme," increases in activity in the well-fed state, closely paralleling the activity of the fatty acid synthesizing system. It is now believed that utilization of pyruvate for lipogenesis is by way of citrate. The pathway involves glycolysis followed by the oxidative decarboxylation of pyruvate to acetyl-CoA within the mitochondria and subsequent condensation with oxalocetate to form citrate, as part of the citric acid cycle. This is followed

by the **translocation of citrate** into the extramitochondrial compartment, where in the presence of CoA and ATP, it undergoes cleavage to acetyl-CoA and oxaloacetate catalyzed by ATP-citrate lyase. The acetyl-CoA is then available for malonyl-CoA formation and synthesis to palmitate (Fig 23–9). The oxaloacetate can form malate via NADH-linked malate dehydrogenase, followed by the generation of NADPH via the malic enzyme. In turn, the NADPH becomes available for lipogenesis. This pathway is a means of transferring reducing equivalents from extramitochondrial NADH to NADP. Alternatively, malate can be transported into the mitochondrion, where it is able to re-

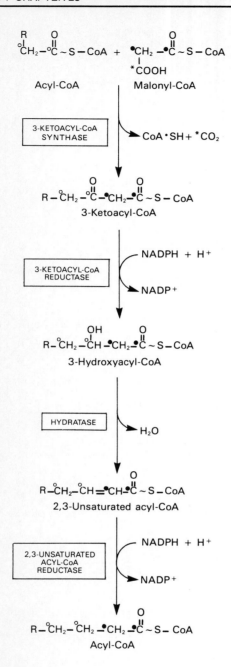

Figure 23–10. Microsomal system for chain elongation (elongase).

form oxaloacetate. It is to be noted that the citrate (tricarboxylate) transporter in the mitochondrial membrane requires malate to exchange with citrate (see Fig 13–16).

There is little ATP-citrate lyase or malic enzyme in ruminants, probably because in these species acetate (derived from the rumen) is the main source of acetyl-CoA. Since the acetate is activated to acetyl-CoA extramitochondrially, there is no necessity for it to enter mitochondria and form citrate prior to incorporation into long-chain fatty acids. Generation of NADPH via extramitochondrial isocitrate dehydrogenase is more important in these species because of the deficiency in malic enzyme.

Microsomal System for Chain Elongation of Fatty Acids (Elongase)

This is probably the main site for the elongation of existing long-chain fatty acid molecules. The pathway converts acyl-CoA compounds of fatty acids to acyl derivatives having 2 carbons more, using malonyl-CoA as acetyl donor and NADPH as reductant. Intermediates in the process are the CoA thioesters. The acyl groups that may act as a primer molecule include the saturated series from C_{10} upward, as well as unsaturated fatty acids. Fasting largely abolishes chain elongation. Elongation of stearyl-CoA in brain increases rapidly during myelination in order to provide C_{22} and C_{24} fatty acids that are present in sphingolipids (Fig 23–10).

REFERENCES

Boyer PD (editor): *The Enzymes,* 3rd ed. Vol 16 of *Lipid Enzymology.* Academic Press, 1983.

Debeer LJ, Mannaerts GP: The mitochondrial and peroxisomal pathways of fatty acid oxidation in rat liver. *Diabete Metab (Paris)* 1983;**9**:134.

Goodridge AG: Fatty acid synthesis in eukaryotes. Page 143 in: *Biochemistry of Lipids and Membranes.* Vance DE, Vance JE (editors). Benjamin/Cummings, 1985.

Gurr MI, James AT: *Lipid Biochemistry: An Introduction,* 3rd ed. Wiley, 1980.

Pande SV, Parvin R: Page 143 in: *Carnitine Biosynthesis, Metabolism, and Functions*. Frenkel RA, McGarry JD (editors). Academic Press, 1980.

Schulz H: Oxidation of fatty acids. Page 116 in: *Biochemistry of Lipids and Membranes*. Vance DE, Vance JE (editors). Benjamin/Cummings, 1985.

Singh N, Wakil SJ, Stoops JK: On the question of half- or full-site reactivity of animal fatty acid synthetase. *J Biol Chem* 1984;**259**:3605.

Tsukamoto Y et al: The architecture of the animal fatty acid synthetase complex. *J Biol Chem* 1983;**258**:15312.

Various authors: Disorders characterized by evidence of abnormal lipid metabolism. In: *The Metabolic Basis of Inherited Disease,* 5th ed. Stanbury JB et al (editors). McGraw-Hill, 1983.

24 Metabolism of Unsaturated Fatty Acids & Eicosanoids

Peter A. Mayes, PhD, DSc

INTRODUCTION

Compared with plants, animal tissues have limited ability in desaturating fatty acids. This necessitates dietary intake of certain polyunsaturated fatty acids derived ultimately from a plant source. These essential fatty acids give rise to eicosanoic (C_{20}) fatty acids, from which are derived families of compounds known as eicosanoids. These make up the prostaglandins, thromboxanes, and leukotrienes.

BIOMEDICAL IMPORTANCE

The content of unsaturated fatty acids in a fat determines its melting point and therefore its fluidity. Similarly, phospholipids of the cell membrane contain unsaturated fatty acids important in maintaining membrane fluidity. A high ratio of polyunsaturated fatty acids to saturated fatty acids (P:S ratio) in the diet is a major factor in lowering plasma cholesterol concentrations by dietary means and is considered to be beneficial in preventing coronary heart disease. The prostaglandins and thromboxanes are local hormones that are synthesized rapidly when required and act near their sites of synthesis. Nonsteroidal anti-inflammatory drugs, such as aspirin, act by inhibiting prostaglandin synthesis. The major physiologic roles played by prostaglandins are as modulators of adenyl cyclase activity, eg, (1) in controlling platelet aggregation, and (2) in inhibiting the effect of antidiuretic hormone in the kidney. Leukotrienes have muscle contractant and chemotactic properties, suggesting that they could be important in allergic reactions and inflammation. A mixture of leukotrienes have been identified as the slow-reacting substance of anaphylaxis (SRS-A). By varying the proportions of the different polyunsaturated fatty acids in the diet, it is possible to influence the type of eicosanoids synthesized, indicating that it might be possible to influence the course of disease by dietary means.

METABOLISM OF UNSATURATED FATTY ACIDS

Some long-chain unsaturated fatty acids of metabolic significance in mammals are shown in Fig 24–1.

Palmitoleic acid ($\omega7$, 16:1, Δ^9)

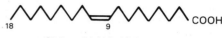

Oleic acid ($\omega9$, 18:1, Δ^9)

*Linoleic acid ($\omega6$, 18:2, $\Delta^{9,12}$)

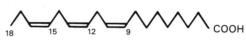

*α-Linolenic acid ($\omega3$, 18:3, $\Delta^{9,12,15}$)

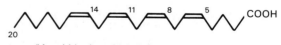

*Arachidonic acid ($\omega6$, 20:4, $\Delta^{5,8,11,14}$)

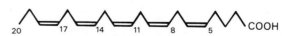

Eicosapentaenoic acid ($\omega3$, 20:5, $\Delta^{5,8,11,14,17}$)

Figure 24–1. Structure of some unsaturated fatty acids. Although the carbon atoms in the molecules are conventionally numbered, ie, numbered from the carboxyl terminal, the ω numbers (eg, $\omega7$ in palmitoleic acid) are calculated from the reverse end (the methyl terminal) of the molecules. The information in parentheses shows, for instance, that palmitoleic acid contains double bonds starting at the seventh carbon from the methyl terminal ($\omega7$), has 16 carbons and one double bond (16:1), and is unsaturated at the ninth carbon from the carboxyl terminal (Δ^9). Likewise, α-linolenic acid contains double bonds starting at the third carbon from the methyl terminal, has 18 carbons and 3 double bonds, and has these double bonds at the ninth, 12th, and 15th carbons from the carboxyl terminal. *Classified as "essential fatty acids."

(For a review of the nomenclature of fatty acids, see Chapter 15.)

Other C_{20}, C_{22}, and C_{24} polyenoic fatty acids may be detected in the tissues. These may be derived from linoleic and α-linolenic acids by chain elongation. It is to be noted that all double bonds present in naturally occurring unsaturated fatty acids of mammals are of the *cis* configuration.

Palmitoleic and oleic acids are not essential in the diet, because the tissues are capable of introducing one double bond into the corresponding saturated fatty acid. Experiments with labeled palmitate have demonstrated that the label enters freely into palmitoleic and oleic acids but is absent from linoleic, α-linolenic, and arachidonic acids. **Linoleic, α-linolenic,** and **arachidonic acids** are the only fatty acids known to be essential for the complete nutrition of many species of animals, including the human, and must therefore be supplied in the diet; as a consequence, they are known as the **nutritionally essential fatty acids.** Although linoleic acid cannot be synthesized and therefore must be supplied preformed in the diet, arachidonic acid can be formed from linoleic acid in most mammals (Fig 24–4). In animals, double bonds can be introduced at the Δ^4, Δ^5, Δ^6, and Δ^9 positions (counting from the carboxyl terminal; see Chapters 15 and 23) but never beyond the Δ^9 position. In contrast, plants are able to introduce new double bonds at the Δ^6, Δ^9, Δ^{12}, and Δ^{15} positions.

SYNTHESIS OF MONOUNSATURATED FATTY ACIDS BY Δ^9 DESATURASE
(See Fig 24–2.)

As far as the nonessential monounsaturated fatty acids are concerned, several tissues including the liver are considered to be responsible for their formation from saturated fatty acids. The first double bond introduced into a saturated fatty acid is nearly always in the Δ^9 position. An enzyme system, Δ^9 desaturase, in the endoplasmic reticulum will catalyze the conversion of palmitoyl-CoA or stearoyl-CoA to palmitoleyl-CoA or oleyl-CoA, respectively. Oxygen and either NADH or NADPH are necessary for the reaction. The enzymes appear to be those of a typical monooxygenase system involving cytochrome b_5 (hydroxylase).

SYNTHESIS OF POLYUNSATURATED FATTY ACIDS
(See Fig 24–3.)

Additional double bonds introduced into existing monounsaturated fatty acids are always separated from each other by a methylene group (methylene interrupted), except in bacteria. In animals, the additional double bonds are **all introduced between the existing double bond and the carboxyl group,** but in plants they may also be introduced between the existing double bond and the ω (methyl terminal) carbon.

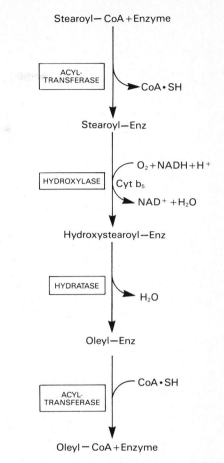

Figure 24–2. Microsomal Δ^9 desaturase system.

Thus, since animals have a Δ^9 desaturase, they are able to completely synthesize the $\omega9$ (oleic acid) series or family of unsaturated fatty acids by a combination of chain elongation and desaturation (Fig 24–3). However, since they are unable to synthesize either linoleic ($\omega6$) or α-linolenic ($\omega3$) acids, the required desaturases being absent, **these acids must be supplied in the diet to accomplish the synthesis of the other members of the $\omega6$ and $\omega3$ series of polyunsaturated fatty acids.** Linoleate may be converted to arachidonate (Fig 24–4). The pathway is first by dehydrogenation of the CoA ester through γ-linolenate followed by the addition of a 2-carbon unit via malonyl-CoA in the microsomal system for chain elongation, to give eicosatrienoate (dihomo γ-linolenate). The latter forms arachidonate by a further dehydrogenation. The dehydrogenating system is similar to that described above for saturated fatty acids. **The nutritional requirement for arachidonate may thus be dispensed with if there is adequate linoleate in the diet.**

The desaturation and chain elongation system is greatly diminished in the fasting state and in the absence of insulin.

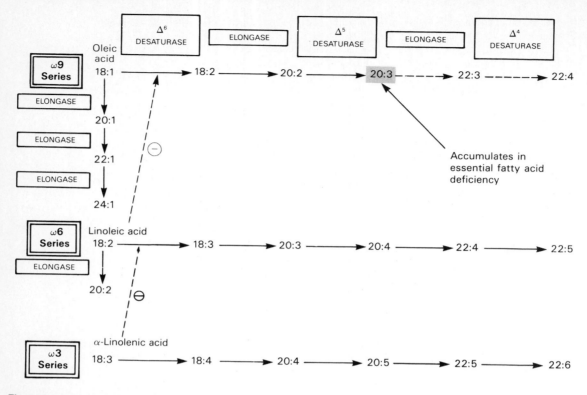

Figure 24–3. Biosynthesis of the ω9, ω6, and ω3 series of polyunsaturated fatty acids. Each step is catalyzed by the microsomal chain elongation or desaturase systems. ω9 Polyunsaturated fatty acids only become quantitatively significant when linoleic and α-linolenic acids are withheld from the diet. This is because each series competes for the same enzyme systems, and affinities decrease from the ω3 to ω9 series. ⊖, Inhibition.

THE ESSENTIAL FATTY ACIDS (EFA)

In 1928, Evans and Burr noticed that rats fed on a purified nonlipid diet to which vitamins A and D were added exhibited a reduced growth rate and a reproductive deficiency. Later work showed that the deficiency syndrome was cured by the addition of **linoleic, α-linolenic,** and **arachidonic acids** to the diet. Further diagnostic features of the syndrome include scaly skin, necrosis of the tail, and lesions in the urinary system, but the condition is not fatal. These fatty acids are found in high concentrations in various vegetable oils (see p 131) and in small amounts in animal carcasses.

The functions of the essential fatty acids appear to be various, though not well defined, apart from prostaglandin and leukotriene formation (see below). Essential fatty acids are found in the structural lipids of the cell, are concerned with the structural integrity of the mitochondrial membrane, and occur in high concentration in the reproductive organs. In many of their structural functions, essential fatty acids are present in phospholipids, mainly in the 2 position. **In essential fatty acid deficiency,** nonessential polyenoic acids of the ω9 series replace the essential fatty acids in phospholipids, other complex lipids, and membranes, particularly $\Delta^{5,8,11}$ eicosatrienoic acid. The triene:tetraene

(arachidonate) ratio in plasma lipids can be used to diagnose the extent of essential fatty acid deficiency.

Skin symptoms and impairment of lipid transport have been noted in human subjects ingesting a diet lacking in essential fatty acids. In adults subsisting on ordinary diets, no signs of essential fatty acid deficiencies have been reported. However, infants receiving formula diets low in fat developed skin symptoms that were cured by giving linoleate. Deficiencies attributable to a lack of essential fatty acids, including α-linolenic acid, also occur among patients maintained for long periods exclusively by intravenous nutrition low in essential fatty acids. Deficiency can be prevented by an essential fatty acid intake of 1–2% of the total caloric requirement.

Trans-Fatty Acids

Traces of *trans*-unsaturated fatty acids are found in ruminant fat, where they arise from the action of microorganisms in the rumen, but the presence of large amounts of *trans*-unsaturated fatty acids in partially hydrogenated vegetable oils (eg, margarine) raises the question of their safety as food additives. Their long-term effects in humans are not known, but up to 15% of tissue fatty acids have been found at autopsy to be in the *trans* configuration. To date, no serious effects have been substantiated. They are metabolized more like saturated than like the *cis*-unsaturated fatty acids.

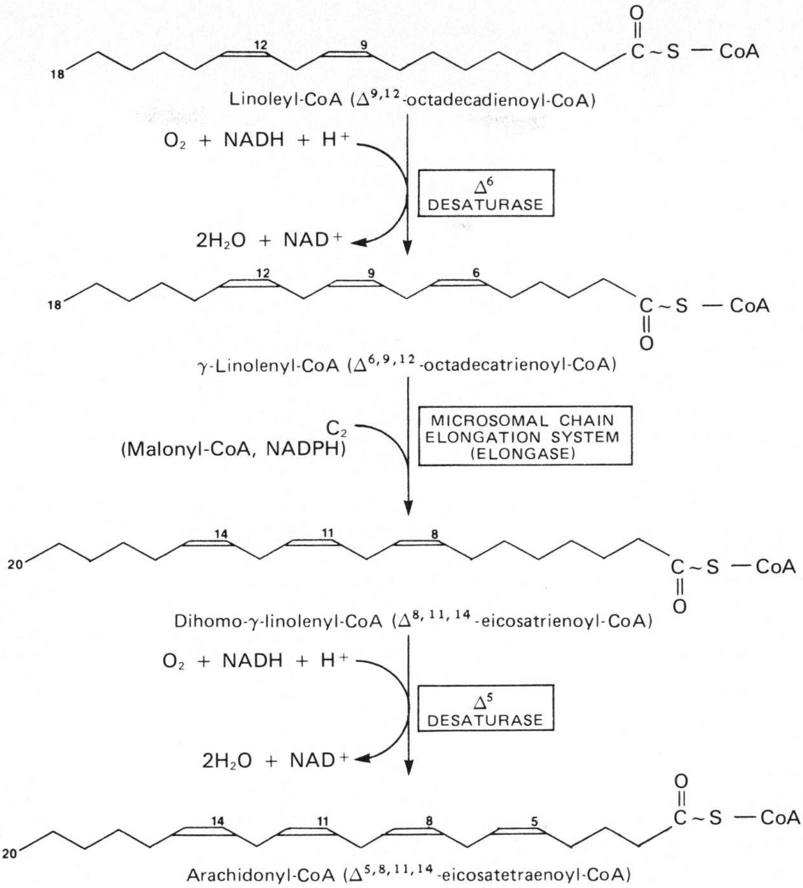

Figure 24–4. Conversion of linoleate to arachidonate. Cats cannot carry out this conversion owing to absence of Δ^6 desaturase and must be given arachidonate in their diet.

This may be due to their similar straight-chain conformation (see Chapter 15). *Trans*-polyunsaturated fatty acids do not possess essential fatty acid activity and may antagonize the metabolism of essential fatty acids and exacerbate essential fatty acid deficiency.

Clinical Aspects

Apart from essential fatty acid deficiency and changes in unsaturated fatty acid patterns in chronic malnutrition, abnormal metabolism of essential fatty acids, which may be connected with dietary insufficiency, has been noted in cystic fibrosis, acrodermatitis enteropathica, hepatorenal syndrome, Sjögren-Larsson syndrome, multisystem neuronal degeneration, Crohn's disease, cirrhosis and alcoholism, and Reye's syndrome. Diets with a high P:S (polyunsaturated:saturated fatty acid) ratio lower serum cholesterol levels, particularly in low-density lipoproteins. This is considered to be beneficial in view of the relationship between serum cholesterol level and coronary heart disease.

EICOSANOIDS

Arachidonate and some other C_{20} fatty acids with methylene-interrupted bonds give rise to **eicosanoids,** physiologically and pharmacologically active compounds known as **prostaglandins (PG), thromboxanes (TX),** and **leukotrienes (LT).** (See Chapter 15.)

Arachidonate, usually derived from the 2-position of phospholipids in the plasma membrane, as a result of phospholipase A_2 activity (see Fig 25–5), is the substrate for the synthesis of the PG_2, TX_2, and LT_4 compounds. The pathways of metabolism are divergent, the synthesis of the PG_2 and TX_2 series **(prostanoids)** competing with the synthesis of LT_4 for the arachidonate substrate. These 2 pathways are known as the **cyclooxygenase** and **lipoxygenase pathways,** respectively (Fig 24–5).

There are 3 groups of eicosanoids (each comprising PG, TX, and LT) that are synthesized from each of the essential fatty acids, respectively, **linoleate, arachidonate,** and **α-linolenate** (Fig 24–6).

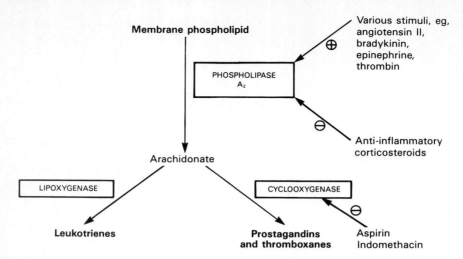

Figure 24–5. Conversion of arachidonic acid to prostaglandins and thromboxanes via the cyclooxygenase pathway and to leukotrienes via the lipoxygenase pathway. The figure indicates why steroids, which inhibit total eicosanoid production, are better anti-inflammatory agents than aspirinlike drugs, which only inhibit the cyclooxygenase pathway.

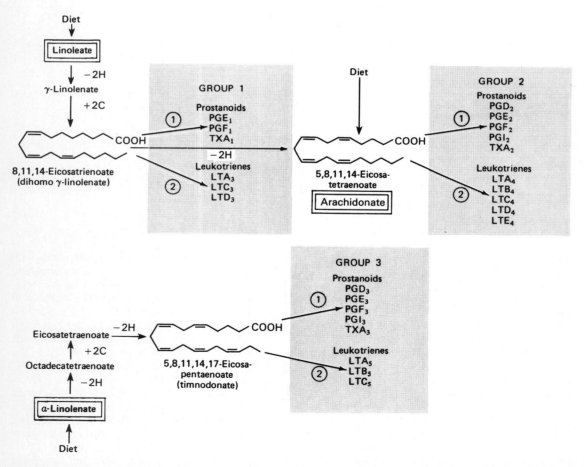

Figure 24–6. The 3 groups of eicosanoids and their biosynthetic origins. PG, prostaglandin; PGI, prostacyclin; TX, thromboxane; LT, leukotriene; **1**, cyclooxygenase pathway; **2**, lipoxygenase pathway. The subscript denotes the total number of double bonds in the molecule and the series to which the compound belongs.

PROSTANOIDS

Biosynthesis

Prostanoid synthesis (Fig 24–7) involves the consumption of 2 molecules of O_2 catalyzed by **prostaglandin endoperoxide synthase,** which possesses 2 separate enzyme activities, **cyclooxygenase** and **peroxidase.** The product of the cyclooxygenase pathway, an endoperoxide (PGH), is converted to prostaglandins D, E, and F as well as to the thromboxane (TXA₂) and prostacyclin (PGI₂). Each cell type produces only one type of prostanoid. Aspirin inhibits the cyclooxygenase, as does indomethacin.

Clinical Aspects

Thromboxanes are synthesized in platelets and upon release cause vasoconstriction and platelet aggregation. **Prostacyclins (PGI₂)** are produced by blood vessel walls and are potent inhibitors of platelet aggregation. Thus, thromboxanes and prostacyclins are antagonistic. The low incidence of heart disease, diminished platelet aggregation, and prolonged clotting times in Greenland Eskimos have been attributed to their high intake of fish oils containing 20:5 ω3 (EPA, or eicosapentaenoic acid), which gives rise to the series 3 prostaglandins (PG₃) and thromboxane TX₃ (Fig 24–6). PG₃ and TX₃ inhibit the release of arachidonate from phospholipids and the formation of PG₂ and TX₂. PGI₃ is as potent an antiaggregator of platelets as PGI₂, but TXA₃ is a weaker aggregator than TXA₂; thus, the balance of activity is shifted toward nonaggregation. In addition, the plasma concentrations of cholesterol, triacylglycerol, and low-density and very low density lipoproteins are all low in Eskimos, whereas the high-density lipoprotein concentration is raised—all factors considered to militate against atherosclerosis and myocardial infarction.

The prostaglandins are potent biologically active substances. As little as 1 ng/mL causes contraction of smooth muscle in animals. Potential therapeutic uses include prevention of conception, induction of labor at term, termination of pregnancy, prevention or allevia-

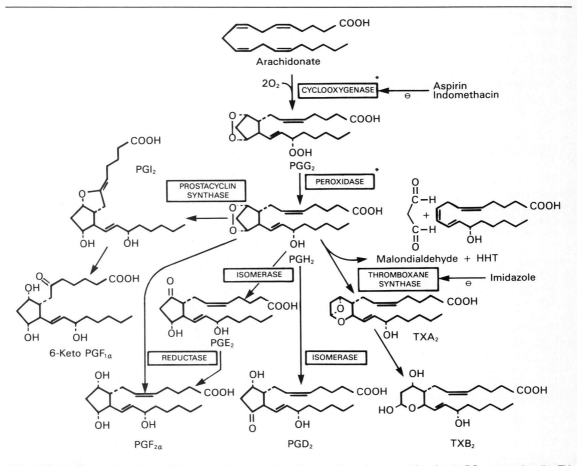

Figure 24–7. Conversion of arachidonic acid to prostaglandins and thromboxanes of series 2. PG, prostaglandin; TX, thromboxane; PGI, prostacyclin; HHT, hydroxyheptadecatrienoate. *Both of these activities are attributed to one enzyme—prostaglandin endoperoxide synthase. Similar conversions occur in series 1 and 3.

tion of gastric ulcers, control of inflammation and of blood pressure, and relief of asthma and nasal congestion.

Prostaglandins increase cAMP in platelets, thyroid, corpus luteum, fetal bone, adenohypophysis, and lung but lower cAMP in renal tubule cells and adipose tissue (see p 222).

Essential Fatty Acids & Prostaglandins

Although there is a marked correlation between essential fatty acid activity of various fatty acids and their ability to be converted to prostaglandins, it does not seem that essential fatty acids exert all of their physiologic effects via prostaglandin synthesis. The role of essential fatty acids in membrane formation is unrelated to prostaglandin formation. Prostaglandins do not relieve symptoms of essential fatty acid deficiency, and an essential fatty acid deficiency syndrome is not caused by chronic inhibition of prostaglandin synthesis.

Removal & Inactivation

"Switching off" of prostaglandin formation is partly achieved by a remarkable property of cyclooxygenase—that of self-catalyzed destruction; ie, it is a "suicide enzyme." The inactivation of prostaglandins, once formed, is rapid. The presence of the enzyme **15-hydroxyprostaglandin dehydrogenase** in most mammalian tissues is probably the principal cause. It has been shown that blocking the action of this enzyme with sulfasalazine or indomethacin can prolong the half-life of prostaglandins in the body.

LEUKOTRIENES

The leukotrienes are a family of conjugated trienes formed from eicosanoic acids in leukocytes, mastocytoma cells, platelets, and macrophages by the **lipoxygenase pathway,** in response to both immunologic and nonimmunologic stimuli. Three different lipoxygenases (dioxygenases) insert oxygen into the 5, 12, and 15 positions of arachidonic acid, giving rise to hydroperoxides (HPETE). Only **5-lipoxygenase** forms leukotrienes. The first formed is leukotriene A_4, which in turn is metabolized to either leukotriene B_4 or leukotriene C_4 (Fig 24–8). Leukotriene C_4 is formed by the addition of the peptide glutathione via a thioether bond. The subsequent removal of glutamate

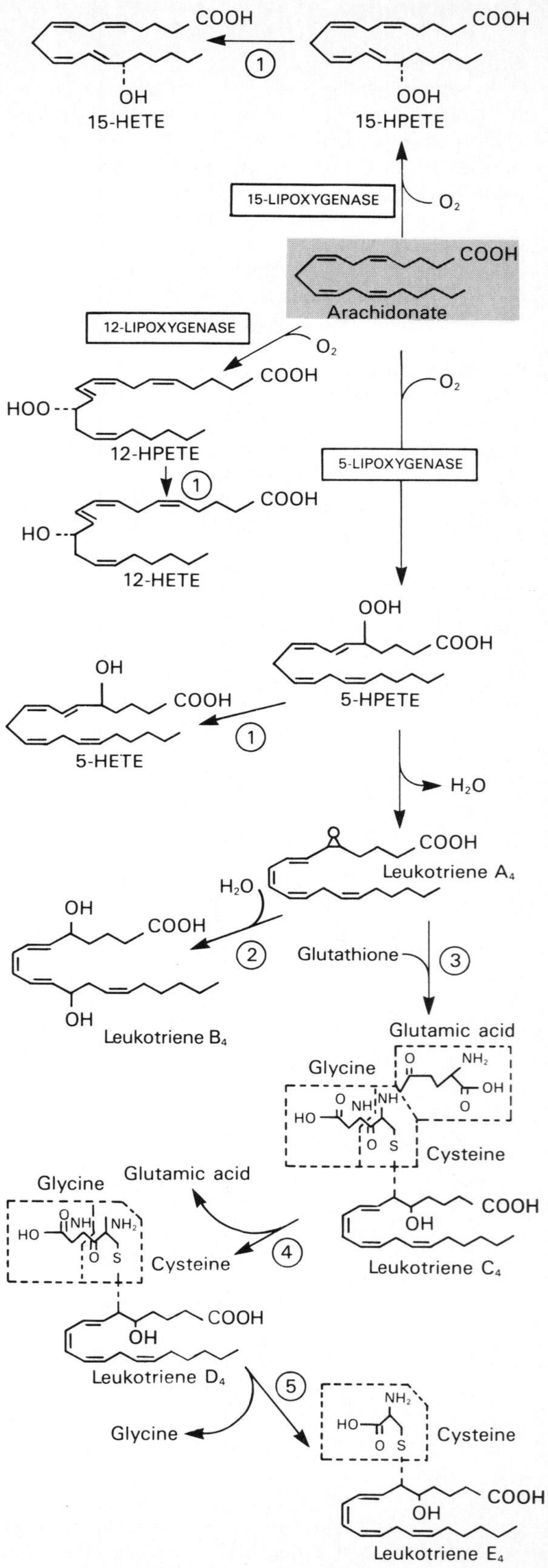

Figure 24–8. Conversion of arachidonic acid to leukotrienes of series 4 via the lipoxygenase pathway. HPETE, hydroperoxyeicosatetraenoate; HETE, hydroxyeicosatetraenoate. Some similar conversions occur in series 3 and 5 leukotrienes. 1, Peroxidase; 2, leukotriene A_4 epoxide hydrolase; 3, glutathione S-transferase; 4, γ-glutamyltransferase; 5, cysteinyl-glycine dipeptidase.

and glycine generates leukotriene D_4 and leukotriene E_4, sequentially.

Clinical Aspects

The slow-reacting substance of anaphylaxis (**SRS-A**) is a mixture of leukotrienes C_4, D_4, and E_4. This mixture of leukotrienes is 100–1000 times more potent than histamine or prostaglandins as a constrictor of the bronchial airway musculature. These leukotrienes together with leukotriene B_4 also cause vascular permeability and attraction and activation of leukocytes and seem to be important regulators in many diseases involving inflammatory or immediate hypersensitivity reactions, such as asthma.

REFERENCES

Hammarström S: Leukotrienes. *Annu Rev Biochem* 1983;**52**:355.

Holman RT: Control of polyunsaturated acids in tissue lipids. *J Am Coll Nutr* 1986;**5**:183.

Kinsella JE: Food components with potential therapeutic benefits: The n-3 polyunsaturated fatty acids of fish oils. *Food Technology* 1986;**40**:89.

Lewis RA, Austen KF: The biologically active leukotrienes: Biosynthesis, metabolism, receptors, functions, and pharmacology. *J Clin Invest* 1984;**73**:889.

Moncada S (editor): Prostacyclin, thromboxane and leukotrienes. *Br Med Bull* 1983;**39**:209.

Piper P: Formation and actions of leukotrienes. *Physiol Rev* 1984;**64**:744.

Rivers JPW, Frankel TL: Essential fatty acid deficiency. *Br Med Bull* 1981;**37**:59.

Smith WL, Borgeat P: The eicosanoids. In: *Biochemistry of Lipids and Membranes*. Vance DE, Vance JE (editors). Benjamin/Cummings, 1985.

25

Metabolism of Acylglycerols & Sphingolipids

Peter A. Mayes, PhD, DSc

INTRODUCTION

Acylglycerols in the form of triacylglycerols constitute the majority of lipids in the body. They are the major lipids in fat deposits and in food. In addition, acylglycerols, including phospholipids, are major components of the plasma and other membranes. Phospholipids take part in the metabolism of many lipids. Glycosphingolipids, which contain sphingosine and sugar residues as well as fatty acids, account for 5–10% of the lipids of the plasma membrane.

BIOMEDICAL IMPORTANCE

The role of triacylglycerol in lipid transport and storage and in various diseases such as obesity, diabetes, and hyperlipoproteinemia will be described in detail in subsequent chapters. Phosphoglycerols, phosphosphingolipids, and glycosphingolipids are all amphipathic lipids and consequently ideally suited as the main lipid constituents of the plasma membrane. Some phospholipids have specialized functions; eg, dipalmitoyl lecithin is a major component of lung surfactant, the lack of which in premature infants is responsible for respiratory distress syndrome of the newborn. Inositol phospholipids act as precursors of hormone second messengers, and platelet-activating factor is an alkylphospholipid. Glycosphingolipids, found in the outer leaflet of the plasma membrane with their oligosaccharide chains facing outward, form part of the glycocalyx of the cell surface and are considered to be important (1) in intercellular communication and contact; (2) as receptors for bacterial toxins (eg, the toxin that causes cholera); and (3) as ABO blood group substances. A dozen or so glycolipid storage diseases have been described (eg, Gaucher's disease, Tay-Sachs disease); these are due to deficiencies in glycolipid hydrolases present in lysosomes.

METABOLISM OF ACYLGLYCEROLS

Catabolism of Triacylglycerol

Triacylglycerols must be hydrolyzed by lipase to their constituent fatty acids and glycerol before further catabolism can proceed. Much of this hydrolysis occurs in adipose tissue with release of free fatty acids into the plasma, where they are found combined with serum albumin. This is followed by free fatty acid uptake into tissues and subsequent oxidation or reesterification. Many tissues (including liver, heart, kidney, muscle, lung, testis, brain, and adipose tissue) have the ability to oxidize long-chain fatty acids, although brain cannot extract them readily from the blood. The utilization of glycerol depends upon whether such tissues possess the necessary activating enzyme, **glycerol kinase** (Fig 25–1). The enzyme has been found in significant amounts in liver, kidney, intestine, brown adipose tissue, and lactating mammary gland.

Biosynthesis of Acylglycerols

Although reactions involving the hydrolysis of triacylglycerols by lipase can be reversed in the laboratory, this is not the mechanism by which acylglycerols are synthesized in tissues. Both glycerol and fatty acids must be activated by ATP before they become incorporated into acylglycerols. Glycerol kinase will catalyze the activation of glycerol to *sn*-glycerol 3-phosphate. If this enzyme is absent—or low in activity, as it is in muscle or adipose tissue—most of the glycerol 3-phosphate **must be derived from an intermediate of the glycolytic system, dihydroxyacetone phosphate,** which forms glycerol 3-phosphate by reduction with NADH catalyzed by **glycerol-3-phosphate dehydrogenase** (Fig 25–1).

A. Triacylglycerol: Fatty acids are activated to acyl-CoA by the enzyme **acyl-CoA synthetase,** utilizing ATP and CoA. Two molecules of acyl-CoA combine with glycerol 3-phosphate to form 1,2-diacylglycerol phosphate (phosphatidate). This takes place in 2 stages via lysophosphatidate, catalyzed first by **glycerol-3-phosphate acyltransferase** and then by **1-acylglycerol-3-phosphate acyltransferase** (lysophosphatidate acyltransferase). Phosphatidate is converted by **phosphatidate phosphohydrolase** to a 1,2-diacylglycerol. In intestinal mucosa, a **monoacylglycerol pathway** exists whereby monoacylglycerol is converted to 1,2-diacylglycerol as a result of the presence of **monoacylglycerol acyltransferase.** A further molecule of acyl-CoA is esterified with the diacylglycerol to form a triacylglycerol, catalyzed by **diacylglycerol acyltransferase.** Most of the activity of these enzymes resides in the endoplasmic reticulum

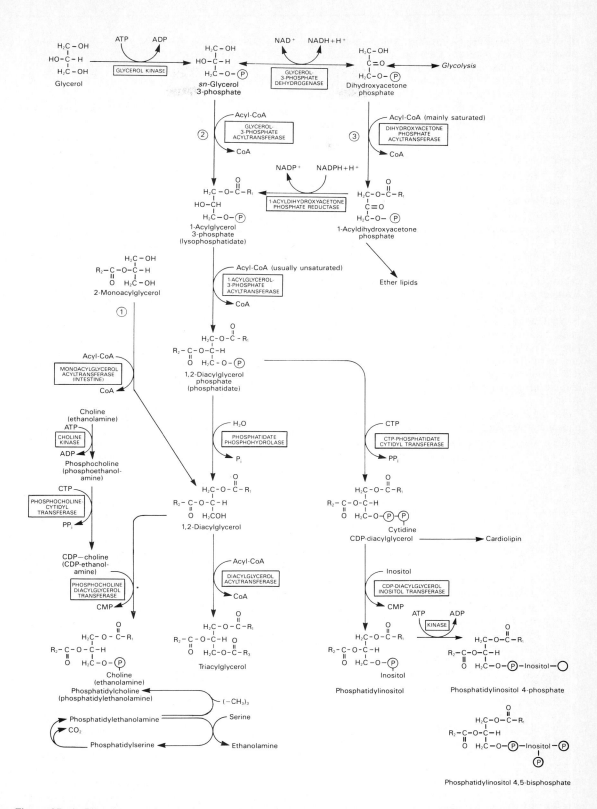

Figure 25–1. Biosynthesis of triacylglycerol and phospholipids. 1, Monoacylglycerol pathway; 2, glycerol phosphate pathway; 3, dihydroxyacetone phosphate pathway. *Phosphoethanolamine-diacylglycerol transferase is not present in liver.

of the cell, but some is found in mitochondria, eg, glycerol 3-phosphate acyltransferase. Phosphatidate phosphohydrolase activity is found mainly in the particle-free supernatant fraction but also is membrane-bound. Dihydroxyacetone phosphate may be acylated and converted to lysophosphatidate after reduction by NADPH. The quantitative significance of this pathway remains in dispute. The pathway appears to be more important in peroxisomes, where it is involved in **ether lipid synthesis.**

B. Phosphoglycerols: These phospholipids are synthesized either from phosphatidate, eg, phosphatidylinositol, or from 1,2-diacylglycerol, eg, phosphatidylcholine and phosphatidylethanolamine. In the synthesis of phosphatidylinositol, cytidine triphosphate (CTP) reacts with phosphatidate to form a cytidine-diphosphate-diacylglycerol (CDP-diacylglycerol). Finally, this compound reacts with inositol, catalyzed by the enzyme **CDP-diacylglycerol inositol transferase,** to form phosphatidylinositol (Fig 25–1). By successive phosphorylations, phosphatidylinositol is transformed first to phosphatidylinositol 4-phosphate and then to phosphatidylinositol 4,5-bisphosphate. The latter is broken down into diacylglycerol and inositol triphosphate by hormones that increase $[Ca^{2+}]$, eg, vasopressin. These 2 products act as second messengers in the action of the hormone (see Fig 44–5).

In the biosynthesis of phosphatidylcholine and phosphatidylethanolamine (lecithins and cephalins),

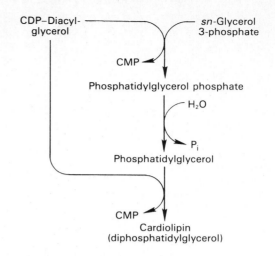

Figure 25–2. Biosynthesis of cardiolipin.

choline or ethanolamine must first be converted to "active choline" or "active ethanolamine," respectively. This is a 2-stage process involving, first, a reaction with ATP to form the corresponding monophosphate, followed by a further reaction with CTP to form either cytidine diphosphocholine (CDP-choline) or cytidine diphosphoethanolamine (CDP-ethanola-

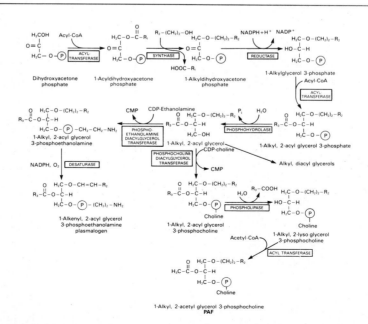

Figure 25–3. Biosynthesis of ether lipids, plasmalogens, and platelet-activating factor (PAF).

mine). In this form, choline or ethanolamine reacts with 1,2-diacylglycerol so that a phosphorylated base (either phosphocholine or phosphoethanolamine) is transferred to the diacylglycerol to form either phosphatidylcholine or phosphatidylethanolamine, respectively. The cytidyl transferase appears to be the regulatory enzyme of the phosphatidylcholine pathway.

Phosphatidylserine is formed from phosphatidylethanolamine directly by reaction with serine. Phosphatidylserine may re-form phosphatidylethanolamine by decarboxylation. An alternative pathway in liver enables phosphatidylethanolamine to give rise directly to phosphatidylcholine by progressive methylation of the ethanolamine residue utilizing S-adenosylmethionine as the methyl donor.

A phospholipid present in mitochondria is **cardiolipin** (diphosphatidylglycerol). It is formed from phosphatidylglycerol, which in turn is synthesized from CDP-diacylglycerol (Fig 25–1) and glycerol 3-phosphate according to the scheme shown in Fig 25–2.

Lung surfactant is a secretion with marked surface active properties that prevent the alveoli from collapsing. Surfactant activity is largely attributed to the presence of a phospholipid, **dipalmitoylphosphatidylcholine,** which is synthesized shortly before parturition in full-term infants. Deficiency of lung surfactant in the lungs of many preterm newborns gives rise to the respiratory distress syndrome.

C. Glycerol Ether Phospholipids and Plasmalogens: A plasmalogenic diacylglycerol is one in which the 1 (or 2) position has an alkenyl residue containing the vinyl ether aldehydogenic linkage ($-CH_2-O-CH=CH-R'$). Dihydroxyacetone phosphate is the precursor of the glycerol moiety (Fig 25–3). This compound combines with acyl-CoA to give 1-acyldihydroxyacetone phosphate. An exchange reaction takes place between the acyl group and a long-chain alcohol to give a 1-alkyldihydroxyacetone phosphate (containing the ether link) which in the presence of NADPH is converted to 1-alkylglycerol 3-phosphate. After further acylation in the 2 position, the resulting 1-alkyl, 2-acyl glycerol 3-phosphate (analogous to phosphatidate in Fig 25–1) is hydrolyzed to give the free glycerol derivative. Plasmalogens are formed by desaturation of the analogous 3-phosphoethanolamine derivative (Fig 25–3). Much of the phospholipid in mitochondria consists of plasmalogens. **Platelet activating factor (PAF)** is synthesized from the corresponding 3-phosphocholine derivative and has been identified as 1-alkyl-2-acetyl-*sn*-glycerol-3-phosphocholine. It is formed by many blood cells and other tissues and aggregates platelets at concentrations as low as 10^{-11} mol/L. It also has hypotensive properties.

Degradation & Turnover of Glycerophospholipids

Degradation of many complex molecules in tissues is complete, eg, proteins. Thus, a turnover time can be determined for such a molecule. Although phospholipids are actively degraded, each portion of the molecule turns over at a different rate; eg, the turnover time of the phosphate group is different from that of the 1-acyl group. This is due to the presence of enzymes that allow partial degradation followed by resynthesis (Fig 25–4). **Phospholipase A$_2$** catalyzes the hydrolysis of the ester bond in position 2 of glycerophospholipids to form a free fatty acid and lysophospholipid, which in turn may be reacylated by acyl-CoA in the presence of an acyltransferase. Alternatively, lysophospholipid (eg, lysolecithin) is attacked by **lysophospholipase** (phospholipase B), removing the remaining 1-acyl group and forming the corresponding glyceryl phosphoryl base, which in turn may be split by a hydrolase liberating glycerol 3-phosphate plus base. **Phospholipase A$_1$** attacks the ester bond in position 1 of phospholipids (Fig 25–5). **Phos-**

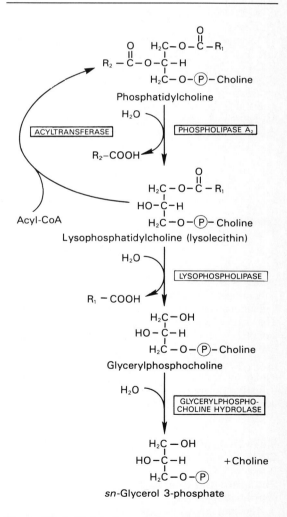

Figure 25–4. Metabolism of phosphatidylcholine (lecithin).

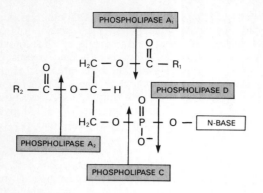

Figure 25–5. Sites of the hydrolytic activity of phospholipases on a phospholipid substrate.

pholipase **C** attacks the ester bond in position 3, liberating 1,2-diacylglycerol plus a phosphoryl base. It is one of the major toxins secreted by bacteria. **Phospholipase D** is an enzyme, described mainly in plants, that hydrolyzes the nitrogenous base from phospholipids.

Lysolecithin may be formed by an alternative route that involves **lecithin:cholesterol acyltransferase (LCAT).** This enzyme, found in plasma and synthesized in liver, catalyzes the transfer of a fatty acid residue from the 2 position of lecithin to cholesterol to form cholesteryl ester and is considered to be responsible for much of the cholesteryl ester in plasma lipoproteins. The consequences of **LCAT deficiency** are discussed on p 247.

> **LECITHIN: CHOLESTEROL ACYLTRANSFERASE**

Lecithin + Cholesterol $\longrightarrow$

Lysolecithin + Cholesteryl ester

Long-chain saturated fatty acids are found predominantly in the 1 position of phospholipids, whereas the polyunsaturated acids (eg, the precursors of prostaglandins) are incorporated more into the 2 position. The incorporation of fatty acids into lecithin occurs by complete synthesis of the phospholipid, by transacylation between cholesteryl ester and lysolecithin, and by direct acylation of lysolecithin by acyl-CoA. Thus, a continuous exchange of the fatty acids is possible, particularly with regard to **introducing essential fatty acids into phospholipid molecules.**

METABOLISM OF SPHINGOLIPIDS

Phosphosphingolipids (Sphingomyelins)

The sphingomyelins are phospholipids containing a fatty acid, phosphate, choline, and a complex amino alcohol, sphingosine. **No glycerol is present.**

Sphingosine (Fig 25–6) is synthesized in the endoplasmic reticulum. Following activation by combination with pyridoxal phosphate, the amino acid serine combines with palmitoyl-CoA to form 3-ketosphinganine after loss of CO_2. Sphingosine itself is formed after 2 reductive steps, one of which is known to utilize NADPH as H donor and the other to involve a flavoprotein enzyme, analogous to the acyl-CoA dehydrogenase step in β-oxidation.

Ceramide (N-acylsphingosine) is a component in sphingomyelin biosynthesis and is formed by a combination of either a free fatty acid or acyl-CoA and sphingosine (Fig 25–7). The acyl group is represented frequently by long-chain saturated or monoenoic acids. **Sphingomyelin** is formed when ceramide reacts with either CDP-choline or phosphatidylcholine; the former reaction is analogous to that employed in the biosynthesis of phosphatidylcholine (Fig 25–1).

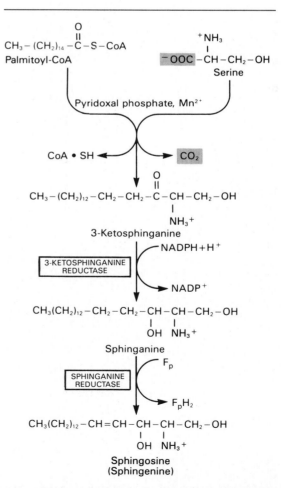

Figure 25–6. Biosynthesis of sphingosine. Fp, flavoprotein.

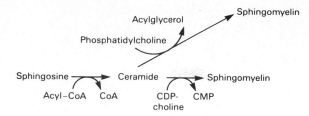

Figure 25–7. Biosynthesis of ceramide and sphingo-myelin.

GLYCOLIPIDS

Glycosphingolipids

The glycosphingolipids are glycolipids that contain the sphingosine-fatty acid combination **ceramide** (Fig 25–7) in combination with one or more sugar residues. Characteristically, C_{24} fatty acids occur in many glycosphingolipids, particularly those in brain (ligno-ceric, cerebronic, and nervonic acids). Lignoceric acid ($C_{23}H_{47}COOH$) is completely synthesized from acetyl-CoA. Cerebronic acid, the 2-hydroxy derivative of lignoceric acid, is formed from it. Nervonic acid ($C_{23}H_{45}COOH$), a monounsaturated acid, is formed by elongation of oleic acid.

The simplest glycosphingolipids are **galactosylceramide (GalCer)** and **glucosylceramide (GlcCer).** GalCer is a major lipid of myelin, whereas GlcCer is the major glycosphingolipid of extraneural tissues and a precursor of most of the more complex glycosphingolipids.

Biosynthesis of glycosphingolipids is catalyzed by an enzyme preparation obtained from young rat brain (Fig 25–8). **Uridine diphosphogalactose epimerase** utilizes uridine diphosphate glucose (UDPGlc) as substrate and accomplishes epimerization of the glucose moiety to galactose, thus forming uridine diphosphate galactose (UDPGal). The reaction in brain is similar to that described in Fig 21–3 for the liver and mammary gland. Galactosylceramide is formed in a reaction between ceramide and UDPGal. **Sulfogalactosylceramide** is formed after further reaction with 3′-phosphoadenosine-5′-phosphosulfate (PAPS; "active sulfate"). PAPS is also involved in the biosynthesis of the other sulfolipids, ie, the **sulfo(galacto)glycerolipids** and the **steroid sulfates.**

Gangliosides are synthesized from ceramide by the stepwise addition of the activated sugars (eg, UDPGlc and UDPGal) and a **sialic acid,** usually **N-acetylneuraminic acid** (Fig 25–9). A large number of gangliosides of increasing molecular weight may be formed. Most of the enzymes transferring sugars from nucleotide sugars (glycosyl transferases) are found in the Golgi apparatus.

Functions of glycosphingolipids. As constituents of the outer leaflet of plasma membranes, glycosphingolipids may be important in **intercellular communication and contact.** Some are antigens, eg, the Forssman antigen and ABO blood group substances. Similar oligosaccharide chains are found in glycoproteins in the plasma membrane. Certain gangliosides function as receptors for bacterial toxins (eg, for cholera toxin, which subsequently activates adenylate cyclase).

PHOSPHOLIPIDS & SPHINGOLIPIDS IN DISEASE (Lipidoses)

Certain diseases are characterized by abnormal quantities of these lipids in the tissues, often in the nervous system. They may be classified into 3 groups: (1) true demyelinating diseases, (2) sphingolipidoses, and (3) leukodystrophies.

In **multiple sclerosis,** which is a demyelinating disease, there is loss both of phospholipids (particularly ethanolamine plasmalogen) and of sphingolipids from white matter. Thus, the composition of white matter resembles that of gray matter. Cholesteryl esters may be found in white matter, although they are normally absent. The cerebrospinal fluid shows raised phospholipid levels.

The **sphingolipidoses** are a group of inherited diseases that are often manifested in childhood. These diseases are part of a larger group of lysosomal disorders (Neufeld, Lim, and Shapiro, 1975).

Lipid storage diseases exhibit several constant features: (1) In various tissues, there is an accumulation of complex lipids that have a portion of their structure in common—**ceramide.** (2) The rate of **synthesis** of the stored lipid is comparable to that in normal humans. (3) The enzymatic defect in each of these diseases is **a deficiency of a specific lysosomal hydrolytic enzyme necessary to break down the lipid.** (4) The extent to which the activity of the affected en-

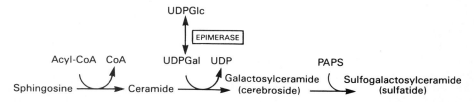

Figure 25 – 8. Biosynthesis of galactosylceramide and its sulfo- derivative. PAPS, "active sulfate," phosphoadenosine-5′-phosphosulfate.

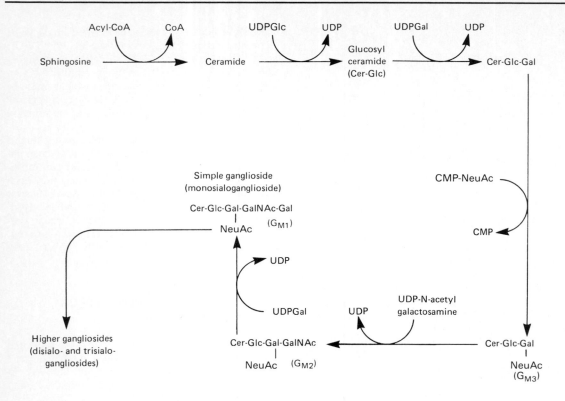

Figure 25–9. Biosynthesis of gangliosides. NeuAc, N-acetylneuraminic acid.

Table 25–1. Summary of the sphingolipidoses.

Disease	Enzyme Deficiency	Lipid Accumulating : Site of Deficient Enzymatic Reaction		Clinical Symptoms
Fucosidosis	α-Fucosidase	Cer−Glc−Gal−GalNAc−Gal÷Fuc H-Isoantigen		Cerebral degeneration, muscle spasticity, thick skin.
Generalized gangliosidosis	G$_{M1}$-β-galactosidase	Cer−Glc−Gal(NeuAc)−GalNAc÷Gal G$_{M1}$ Ganglioside		Mental retardation, liver enlargement, skeletal deformation.
Tay-Sachs disease	Hexosaminidase A	Cer−Glc−Gal(NeuAc)÷GalNAc G$_{M2}$ Ganglioside		Mental retardation, blindness, muscular weakness.
Tay-Sachs variant or Sandhoff's disease	Hexosaminidase A and B	Cer−Glc−Gal−Gal÷GalNAc Globoside plus G$_{M2}$ ganglioside		Same as Tay-Sachs but progressing more rapidly.
Fabry's disease	α-Galactosidase	Cer−Glc−Gal÷Gal Globotriaosylceramide		Skin rash, kidney failure (full symptoms only in males; X-linked recessive).
Ceramide lactoside lipidosis	Ceramide lactosidase (β-galactosidase)	Cer−Glc÷Gal Ceramide lactoside		Progressing brain damage, liver and spleen enlargement.
Metachromatic leukodystrophy	Arylsulfatase A	Cer−Gal÷OSO$_3$ 3-Sulfogalactosylceramide		Mental retardation and psychologic disturbances in adults; demyelination.
Krabbe's disease	β-Galactosidase	Cer÷Gal Galactosylceramide		Mental retardation; myelin almost absent.
Gaucher's disease	β-Glucosidase	Cer÷Glc Glucosylceramide		Enlarged liver and spleen, erosion of long bones, mental retardation in infants.
Niemann-Pick disease	Sphingomyelinase	Cer÷P−choline Sphingomyelin		Enlarged liver and spleen, mental retardation; fatal in early life.
Farber's disease	Ceramidase	Acyl÷Sphingosine Ceramide		Hoarseness, dermatitis, skeletal deformation, mental retardation; fatal in early life.

NeuAc, N-acetylneuraminic acid; Cer, ceramide; Glc, glucose; Gal, galactose; Fuc, fucose.

zyme is decreased is similar in all of the tissues of the affected individual. As a result of these unifying basic considerations, procedures for the diagnosis of patients with these disorders have been developed. It has also become possible to detect heterozygous carriers of the genetic abnormalities responsible for these diseases as well as to discover in the unborn fetus the fact that a sphingolipodystrophy is present. A summary of the more important lipidoses is shown in Table 25–1.

Multiple sulfatase deficiency results in accumulation of sulfogalactosylceramide, steroid sulfates, and proteoglycans, owing to a combined deficiency of arylsulfatases A, B, and C and steroid sulfatase (see p 604).

REFERENCES

Bell RM, Coleman RA: Enzymes of glycerolipid synthesis in eukaryotes. *Annu Rev Biochem* 1980;**49:**459.

Boyer PD (editor): *The Enzymes,* 3rd ed. Vol 16: *Lipid Enzymology.* Academic Press, 1983.

Brady RO: Sphingolipidoses. *Annu Rev Biochem* 1978;**47:**687.

Hanahan DJ: Platelet-activating factor. *Annu Rev Biochem* 1986;**55:**483.

Hawthorne JN, Ansell GB (editors): *Phospholipids.* Elsevier, 1982.

Neufeld EF, Lim TW, Shapiro LJ: Inherited disorders of lysosomal metabolism. *Annu Rev Biochem* 1975;**44:**357.

Various authors: Disorders characterized by evidence of abnormal lipid metabolism. In: *The Metabolic Basis of Inherited Disease,* 5th ed. Stanbury JB et al (editors). McGraw-Hill, 1983.

Various authors: Metabolism of triacylglycerols; phospholipid metabolism; ether-linked glycerolipids; sphingolipids. In: *Biochemistry of Lipids and Hormones.* Vance DE, Vance JE (editors). Benjamin/Cummings, 1985.

26

Lipid Transport & Storage

Peter A. Mayes, PhD, DSc

INTRODUCTION

Fat absorbed from the diet and lipids synthesized by the liver and adipose tissue must be transported between the various tissues and organs for utilization and storage. Since lipids are insoluble in water, the problem arises of how to transport them in an aqueous environment—the blood plasma. This is solved by associating nonpolar lipids (triacylglycerol and cholesteryl esters) with amphipathic lipids (phospholipids and cholesterol) and proteins to make water-miscible lipoprotein.

BIOMEDICAL IMPORTANCE

In a meal-eating omnivore such as the human, excess calories are ingested in the anabolic phase of the feeding cycle, followed by a period of negative caloric balance when the organism draws upon its carbohydrate and fat stores. Lipoproteins mediate this cycle by transporting lipids from the intestines as chylomicrons, and from the liver as very low density lipoproteins (VLDL), to most tissues for oxidation and to adipose tissue for storage. Lipid is mobilized from adipose tissue as free fatty acids (FFA) attached to serum albumin. Abnormalities of lipid metabolism occur at the sites of production or utilization of lipoproteins, causing various hypo- or hyperlipoproteinemias. The most common of these is diabetes mellitus, where insulin deficiency causes excessive mobilization of FFA and underutilization of chylomicrons and VLDL, leading to hypertriacylglycerolemia. Most other pathologic conditions affecting lipid transport primarily are due to inherited defects in synthesis of the apoprotein portion of the lipoprotein, of key enzymes, or of lipoprotein receptors. Some of these defects cause hypercholesterolemia and premature atherosclerosis. Excessive fat deposits constitute obesity, one form of which may be due to defective diet-induced thermogenesis in brown adipose tissue.

THE PLASMA LIPIDS & LIPOPROTEINS

Composition

Extraction of the plasma lipids with a suitable lipid solvent and subsequent separation of the extract into various classes of lipids show the presence of **triacyl-glycerols, phospholipids, cholesterol,** and **cholesteryl esters** and, in addition, the existence of a much smaller fraction of unesterified long-chain fatty acids (free fatty acids) that accounts for less than 5% of the total fatty acid present in the plasma. This latter fraction, the **free fatty acids (FFA),** is now known to be metabolically the most active of the plasma lipids. An analysis of blood plasma showing the major lipid classes is given in Table 26–1.

Pure fat is less dense than water; it follows that as the proportion of lipid to protein in lipoproteins increases, the density decreases (Table 26–2). Use is made of this property in separating the various lipoproteins in plasma by **ultracentrifugation.** The rate at which each lipoprotein floats through a solution of NaCl (specific gravity 1.063) may be expressed in Svedberg (Sf) units of flotation. One Sf unit is equal to 10^{-13} cm/s/dyne/g at 26 °C. The composition of the various lipoprotein fractions obtained by centrifugation is shown in Table 26–2. The various chemical classes of lipids are seen to occur in varying amounts in most of the lipoprotein fractions. Since the fractions represent the physiologic entities present in the plasma, mere chemical analysis of the plasma lipids (apart from FFA) yields little information on their physiology.

In addition to the use of techniques depending on their density, lipoproteins may be separated according

Table 26–1. Lipids of the blood plasma in humans.

Lipid	mg/dL	
	Mean	Range
Total lipid	570	360–820
Triacylglycerol	142	80–180*
Total phospholipid†	215	123–390
Phosphatidylcholine		50–200
Phosphatidylethanolamine		50–130
Sphingomyelins		15–35
Total cholesterol	200	107–320
Free cholesterol (nonesterified)	55	26–106
Free fatty acids (nonesterified)	12	6–16*

Total fatty acids (as stearic acid) range from 200 to 800 mg/dL: 45% are triacylglycerols, 35% phospholipids, 15% cholesteryl ester, and less than 5% free fatty acids.

*Varies with nutritional state.
†Analyzed as lipid phosphorus. Lipid phosphorus × 25 = phospholipid as phosphatidylcholine (4% phosphorus).

Table 26–2. Composition of the lipoproteins in plasma of humans.

Fraction	Source	Diameter (nm)	Density	Sf	Protein (%)	Total Lipid (%)	Triacyl-glycerol	Phospho-lipid	Cholesteryl Ester	Cholesterol (Free)	Free Fatty Acids
							Composition				
							Percentages of Total Lipid				
Chylomicrons	Intestine	100–1000	<0.96	>400	1–2	98–99	88	8	3	1	. . .
Very low density lipoproteins (VLDL)	Liver and intestine	30–90	0.96–1.006	20–400	7–10	90–93	56	20	15	8	1
Intermediate-density lipo-proteins (IDL)	VLDL and chylomi-crons	25–30	1.006–1.019	12–20	11	89	29	26	34	9	1
Low-density lipoproteins (LDL)		20–25	1.019–1.063	2–12	21	79	13	28	48	10	1
High-density lipoproteins HDL₂	Liver and intestine VLDL?	10–20	1.063–1.125		33	67	16	43	31	10	. . .
HDL₃	Chylomi-crons?	7.5–10	1.125–1.210		57	43	13	46	29	6	6
Albumin-FFA	Adipose tissue		>1.2810		99	1	0	0	0	0	100

FFA, free fatty acids. VHDL (very high density lipoprotein) is a minor fraction occurring at density 1.21–1.25.

to their electrophoretic properties (Fig 26–1) and may be identified more accurately by means of immunoelectrophoresis. Apart from FFA, **4 major groups of lipoproteins have been identified that are important physiologically and in clinical diagnosis.** These are (1) **chylomicrons,** derived from intestinal absorption of triacylglycerol; (2) **very low density lipoproteins** (VLDL, or pre-β-lipoproteins), derived from the liver for the export of triacylglycerol; (3) **low-density lipoproteins** (LDL, or β-lipoproteins), representing a final stage in the catabolism of VLDL; and (4) **high-density lipoproteins** (HDL, or α-lipoproteins), involved in VLDL and chylomicron metabolism and also in cholesterol metabolism. Triacylglycerol is the predominant lipid in chylomicrons and VLDL, whereas cholesterol and phospholipid are the predominant lipids in LDL and HDL, respectively (Table 26–2).

Structure

The protein moiety of lipoproteins is known as an **apolipoprotein** or **apoprotein,** constituting nearly 60% of some HDL and as little as 1% of chylomicrons.

A typical lipoprotein—such as a chylomicron or VLDL—consists of a **lipid core** of mainly **nonpolar triacylglycerol** and **cholesteryl ester** surrounded by a surface layer of more polar **phospholipid, cholesterol,** and **apoproteins.** Some apoproteins are integral and cannot be removed, whereas others are free to transfer to other lipoproteins (Fig 26–2).

The Apolipoproteins (Apoproteins)

The lipoproteins are characterized by the presence of one or more proteins or polypeptides known as apoproteins. According to the ABC nomenclature, the 2 major apoproteins of HDL are designated A-I and A-II. The main apoprotein of LDL is apoprotein B, which is found also in VLDL and chylomicrons. However, apo-B of chylomicrons (B-48) is smaller than apo-B of LDL or VLDL (B-100) and has a different amino acid composition. B-48 is synthesized in the intestine and B-100 in the liver. (In the rat, the liver appears to form B-48 in addition to B-100.) Apoproteins

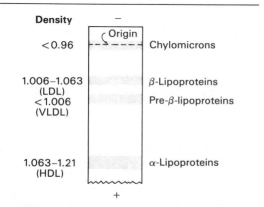

Figure 26–1. Separation of plasma lipoproteins by electrophoresis.

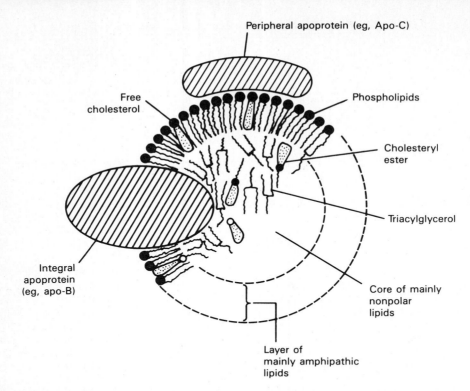

Figure 26–2. Generalized structure of a plasma lipoprotein. The similarities with the structure of the plasma membrane are to be noted. Recent work indicates that some cholesteryl ester and triacylglycerol are to be found in the surface layer and some free cholesterol in the core.

C-I, C-II, and C-III are smaller polypeptides freely transferable between several different lipoproteins (Table 26–3). Carbohydrates account for approximately 5% of apo-B and include mannose, galactose, fucose, glucose, glucosamine, and sialic acid. Thus, some lipoproteins are also glycoproteins. C-II is an important activator of extrahepatic lipoprotein lipase, involved in the clearance of triacylglycerol from the circulation. A-I in HDL is an activator of the plasma enzyme lecithin:cholesterol acyltransferase, which is responsible for most of the formation of plasma cholesteryl ester in humans.

Several apoproteins other than apo-A, -B, or -C have been found in plasma lipoproteins. One is the arginine-rich apoprotein E isolated from VLDL; it contains arginine to the extent of 10% of the total

Table 26–3. Apoproteins of human plasma lipoproteins.

Apoprotein	Lipoprotein	Molecular Weight	Additional Remarks
A-I	HDL, chylomicrons	28,300	Activator of lecithin:cholesterol acyltransferase (LCAT).
A-II	HDL, chylomicrons	17,400	Structure is 2 identical monomers joined by a disulfide bridge.
B-100	LDL, VLDL, IDL	350,000	Synthesized in liver.
B-48	Chylomicrons, chylomicron remnants	200,000	Synthesized in intestine.
C-I	VLDL, HDL	7000	Possible activator of LCAT.
C-II	VLDL, HDL, chylomicrons	8800	Activator of extrahepatic lipoprotein lipase.
C-III	VLDL, HDL, chylomicrons	8750	Several polymorphic forms depending on content of sialic acids.
D	Subfraction of HDL	32,500	Possibly identical to the cholesteryl ester transfer protein.
E (arginine-rich)	VLDL, HDL, chylomicrons, chylomicron remnants	34,000	Present in excess in the β-VLDL of patients with type III hyperlipoproteinemia. It is the sole apoprotein found in HDL$_C$ of diet-induced hypercholesterolemic animals.

amino acids and accounts for 5–10% of total VLDL apoproteins in normal subjects but is present in excess in the broad β-VLDL of patients with type III hyperlipoproteinemia.

METABOLISM OF THE PLASMA LIPOPROTEINS

FREE FATTY ACIDS (FFA)

The free fatty acids (nonesterified fatty acids, unesterified fatty acids) arise in the plasma from lipolysis of triacylglycerol in adipose tissue or as a result of the action of lipoprotein lipase during uptake of plasma triacylglycerols into tissues. They are found **in combination with serum albumin** in concentrations varying between 0.1 and 2 μeq/mL of plasma and comprise the long-chain fatty acids found in adipose tissue, ie, palmitic, stearic, oleic, palmitoleic, linoleic, and other polyunsaturated acids, and smaller quantities of other long-chain fatty acids. Binding sites on albumin of varying affinity for the fatty acids have been described. Low levels of free fatty acids are recorded in the fully fed condition, rising to about 0.5 μeq/mL in the postabsorptive and between 0.7 and 0.8 μeq/mL in the fully fasting state. In uncontrolled **diabetes mellitus,** the level may rise to as much as 2 μeq/mL. In meal eaters, the level falls just after eating and rises again prior to the next meal, whereas in such continuous feeders as ruminants—where there is uninterrupted influx of nutrient from the intestine—the free fatty acids remain at a low level.

The rate of removal of free fatty acids from the blood is extremely rapid. Estimates suggest that the oxidation of free fatty acids supplies about 25–50% of the energy requirements in fasting. The remainder of the uptake is esterified and, according to evidence using radioactive free fatty acids, eventually recycled. In starvation, the respiratory quotient (RQ) would indicate that considerably more fat is being oxidized than can be traced to the oxidation of free fatty acids. This difference is accounted for by the oxidation of esterified lipids from the circulation or of those present in tissues. The latter are thought to occur particularly in heart and skeletal muscle, where considerable stores of lipid are to be found in the muscle cells. **The free fatty acid turnover is related directly to free fatty acid concentration.** Thus, the rate of free fatty acid production in adipose tissue controls the free fatty acid concentration in plasma, which in turn determines the free fatty acid uptake by other tissues. The nutritional condition does not appear to have a great effect on the fractional uptake of free fatty acids by tissues. It does, however, alter the proportion of the uptake which is oxidized compared to the fraction which

is esterified, more being oxidized in the fasting than in the fed state.

The presence of a **fatty acid-binding protein,** or **Z-protein,** in the cytosol of many of the major tissues has been reported. The role of this protein in intracellular transport is thought to be similar to the role of serum albumin in extracellular transport of long-chain fatty acids.

FORMATION OF CHYLOMICRONS & VERY LOW DENSITY LIPOPROTEINS (VLDL)

By definition, **chylomicrons** are found in **chyle** formed only by the lymphatic system **draining the intestine.** Smaller and denser particles having the physical characteristics of VLDL are also to be found in chyle. However, their apoprotein composition resembles chylomicrons rather than VLDL, indicating that they should be regarded as small chylomicrons. Their formation occurs even in the fasting state, in which they are responsible for transporting 50% of lymphatic triacylglycerol and cholesterol; their lipids originate mainly from bile and intestinal secretions. On the other hand, chylomicron formation increases with the load of triacylglycerol absorbed. Most of the plasma **VLDL** are of hepatic origin. They are the vehicles of transport of **triacylglycerol from the liver to the extrahepatic tissues.**

There are many similarities in the mechanism of formation of chylomicrons by intestinal cells and of VLDL by hepatic parenchymal cells (Fig 26–3). Apoprotein B is synthesized by ribosomes in the rough endoplasmic reticulum and is incorporated into lipoproteins in the smooth endoplasmic reticulum, which is the main site of synthesis of triacylglycerol. Lipoproteins pass through the Golgi apparatus, where, it is thought, carbohydrate residues are added to the lipoprotein. The chylomicrons and VLDL are released from either the intestinal or hepatic cell by fusion of the secretory vacuole with the cell membrane (reverse pinocytosis). Chylomicrons pass into the spaces between the intestinal cells, eventually making their way into the lymphatic system (lacteals) draining the intestine. VLDL are secreted by hepatic parenchymal cells into the space of Disse and then into the hepatic sinusoids through fenestrae in the endothelial lining. The similarities between the 2 processes and the anatomic mechanisms are striking, for—apart from the mammary gland—the intestine and liver are the only tissues from which particulate lipid is secreted. The inability of particulate lipid of the size of chylomicrons and VLDL to pass through endothelial cells of the capillaries without prior hydrolysis is probably the reason dietary fat enters the circulation via the lymphatics (thoracic duct) and not via the hepatic portal system.

Although both chylomicrons and VLDL isolated from blood contain apoproteins C and E, the newly secreted or "nascent" lipoproteins contain little or none, and it would appear that the full complement of

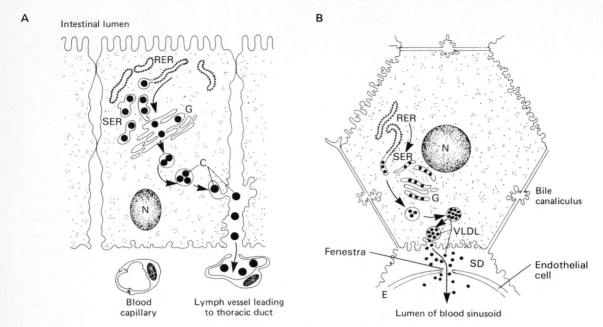

Figure 26–3. The formation and secretion of (**A**) chylomicrons by an intestinal cell and (**B**) very low density lipoproteins by a hepatic cell. RER, rough endoplasmic reticulum; SER, smooth endoplasmic reticulum; G, Golgi apparatus; N, nucleus; C, chylomicrons; VLDL, very low density lipoproteins; E, endothelium; SD, space of Disse, containing blood plasma. The figure is a diagrammatic representation of events that can be seen with electron microscopy.

apoprotein C and E polypeptides is taken up by transfer from HDL once the chylomicrons and VLDL have entered the circulation (Figs 26–4 and 26–5). A more detailed account of the factors controlling hepatic VLDL secretion is given below.

Apoprotein B is essential for chylomicron and VLDL formation. In **abetalipoproteinemia** (a rare disease), apoprotein B is not synthesized; lipoproteins containing this apoprotein are not formed, and lipid droplets accumulate in the intestine and liver.

CATABOLISM OF CHYLOMICRONS & VERY LOW DENSITY LIPOPROTEINS

The clearance of labeled chylomicrons from the blood is rapid, the half-time of disappearance being on the order of minutes in small animals (eg, rats) but longer in larger animals (eg, humans), in whom it is still under 1 hour. Larger particles are catabolized more quickly than smaller ones. When chylomicrons labeled in the triacylglycerol fatty acids are administered intravenously, some 80% of the label is found in adipose tissue, heart, and muscle and approximately 20% in the liver. As experiments with the perfused organ have shown that the **liver does not metabolize native chylomicrons or VLDL significantly,** the label in the liver must result secondarily from their metabolism in extrahepatic tissues.

Role of Lipoprotein Lipase

There is a significant correlation between the ability of a tissue to incorporate lipoprotein triacylglycerol fatty acids and the activity of the enzyme **lipoprotein lipase.** It is located on the **walls of blood capillaries,** anchored by proteoglycan chains of heparan sulfate, and has been found in extracts of heart, adipose tissue, spleen, lung, renal medulla, aorta, diaphragm, and lactating mammary gland. Normal blood does not contain appreciable quantities of the enzyme; however, following injection of **heparin,** lipoprotein lipase is released from its heparan sulfate binding into the circulation and is accompanied by the clearing of lipemia. A lipase is also released from the liver by large quantities of heparin (**heparin-releasable hepatic lipase),** but this enzyme has properties different from those of lipoprotein lipase and does not react readily with chylomicrons. Its function is obscure, but it has been implicated in the metabolism of HDL₂ by the liver (Fig 26–6) and in the metabolism of chylomicron and VLDL remnants (see below).

Both **phospholipids** and **apolipoprotein C-II** are required as cofactors for lipoprotein lipase activity. Apo C-II contains a specific phospholipid binding site through which it is attached to the lipoprotein. Thus, chylomicrons and VLDL provide the enzyme for their metabolism with both its substrate and cofactors. Hydrolysis takes place while the lipoproteins are attached to the enzyme on the endothelium. The triacylglycerol is hydrolyzed progressively through a diacylglycerol

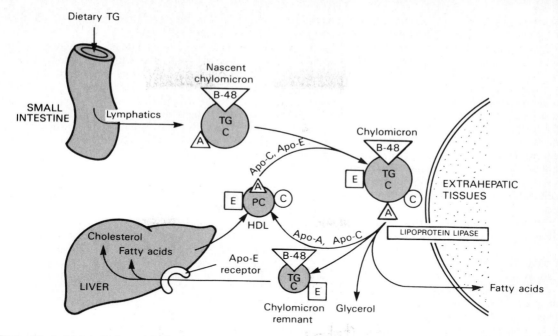

Figure 26–4. Metabolic fate of chylomicrons. Apo-A, apolipoprotein A; Apo-B, apolipoprotein B; Apo-C, apolipoprotein C; Apo-E, apolipoprotein E; HDL, high-density lipoprotein; TG, triacylglycerol; C, cholesterol and cholesteryl ester; P, phospholipid. Only the predominant lipids are shown.

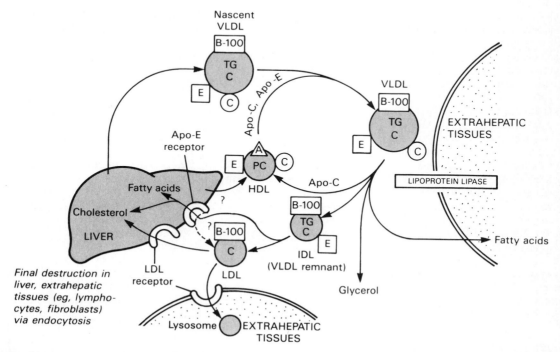

Figure 26–5. Metabolic fate of very low density lipoproteins (VLDL) and production of low-density lipoproteins (LDL). Apo-A, apolipoprotein A; Apo-B, apolipoprotein B; Apo-C, apolipoprotein C; Apo-E, apolipoprotein E; HDL, high-density lipoprotein; TG, triacylglycerol; IDL, intermediate-density lipoprotein; C, cholesterol and cholesteryl ester; P, phospholipid. Only the predominant lipids are shown. There is disagreement whether IDL is taken up by the liver via the Apo-E or LDL receptor.

to a monoacylglycerol that is finally hydrolyzed to free fatty acid plus glycerol. Some of the released free fatty acids return to the circulation, attached to albumin, but the bulk are transported into the tissue (Figs 26–4 and 26–5). Heart lipoprotein lipase has a low K_m for triacylglycerol, whereas the K_m of the enzyme in adipose tissue is 10 times greater. As the concentration of plasma triacylglycerol decreases in the transition from the fed to the starved condition, the heart enzyme remains saturated with substrate but the saturation of the enzyme in adipose tissue diminishes, thus **redirecting uptake from adipose tissue toward the heart.** A similar redirection occurs during lactation, in which adipose tissue activity diminishes and mammary gland activity increases, allowing uptake of lipoprotein triacylglycerol long-chain fatty acid for milk fat synthesis.

Formation of Remnant Lipoproteins

Reaction with lipoprotein lipase results in the loss of approximately 90% of the triacylglycerol of chylomicrons and in the loss of the apo-C (which returns to HDL) but not apo-E (which is retained). The resulting lipoprotein or **chylomicron remnant** is about half the diameter of the parent chylomicron and in terms of the percentage composition becomes relatively enriched in cholesterol and cholesteryl esters because of the loss of triacylglycerol. Similar changes occur to VLDL, with the formation of VLDL remnants or IDL (intermediate-density lipoprotein).

Role of the Liver

Chylomicron remnants are taken up by the liver, and the cholesteryl esters and triacylglycerols are hydrolyzed and metabolized. Uptake appears to be mediated by a **receptor specific for apo-E** (Fig 26–4). Up to 50% of VLDL remnants are taken up into the liver via this route; the rest form LDL.

Studies using apoprotein B-labeled VLDL have shown that VLDL is the precursor of IDL and IDL the precursor of LDL. Calculations indicate that only one or 2 molecules of apoprotein B-100 are present in each of these lipoprotein particles and these are conserved during the transformations. Each LDL particle is derived from only one VLDL particle (Fig 26–5). The role of the liver in this process is uncertain. In the rat, most of the apo-B from VLDL appears in the liver, and only a small percentage appears in LDL. This may be due to the fact that VLDL in the rat contains apo-B-48 as well as B-100. If the hepatic apo-E receptor is more specific for B-48 than for B-100, this would account for much of the hepatic removal of IDL in the rat and for the low production of LDL in this species.

METABOLISM OF LDL

Most LDL appears to be formed from VLDL, as described above, but there is evidence for some production directly by the liver. The half-time of disappearance from the circulation of apoprotein B in LDL is approximately 2½ days.

Studies on cultured human fibroblasts, lymphocytes, and arterial smooth muscle cells have shown the existence of specific binding sites, or receptors, for LDL (B-100 receptors), which are defective in **familial hypercholesterolemia.** Approximately 50% of LDL is degraded in extrahepatic tissues and 50% in the liver. A positive correlation exists between the incidence of **coronary atherosclerosis** and plasma concentration of LDL.

METABOLISM OF HDL (See Fig 26–6.)

HDL is synthesized and secreted from both liver and intestine. However, nascent HDL from intestine does not contain apoprotein C but only apoprotein A. Thus, apoprotein C is synthesized in the liver and is transferred to intestinal HDL when the latter enters the plasma. A major function of HDL is to act as a repository for apoproteins C and E that are required in the metabolism of chylomicrons and VLDL.

Nascent HDL consists of discoid phospholipid bilayers containing apoprotein and free cholesterol. These lipoproteins are similar to the particles found in the plasma of patients with a deficiency of the plasma enzyme **lecithin:cholesterol acyltransferase (LCAT)** and in the plasma of patients with obstructive jaundice. LCAT—and possibly the LCAT activator apoprotein A-I—bind to the disk. Catalysis by LCAT converts surface phospholipid and free cholesterol into cholesteryl esters and lysolecithin. The nonpolar cholesteryl esters move into the hydrophobic interior of the bilayer, whereas lysolecithin is transferred to plasma albumin. The reaction continues, generating a nonpolar core that pushes the bilayer apart until a spherical, pseudomicellar HDL is formed, covered by a surface film of polar lipids and apoproteins. The esterified cholesterol can be transferred from HDL to the lower density lipoproteins, eg, chylomicrons, VLDL, and LDL, by means of the **cholesteryl ester transfer protein** (apoprotein D), which is another protein component of HDL. Thus, the cholesteryl ester transfer protein allows cholesteryl ester of HDL to be transported to the liver via the remnants of chylomicrons and VLDL or via hepatic uptake of LDL. Thus, the LCAT system is involved in the removal of excess unesterified cholesterol from lipoproteins and from the tissues. The liver and possibly the intestines seem to be the final sites of degradation of HDL apoproteins.

An HDL cycle has been proposed to account for the transport of cholesterol from the tissues to the liver (see Fig 26–6 for details). This explains why **HDL_2 concentrations in the plasma vary reciprocally with the chylomicron and VLDL concentration and directly with the activity of lipoprotein lipase.** HDL (HDL_2) concentrations are **inversely related to the incidence of coronary atherosclerosis,** possibly because they reflect the efficiency of cholesterol-scav-

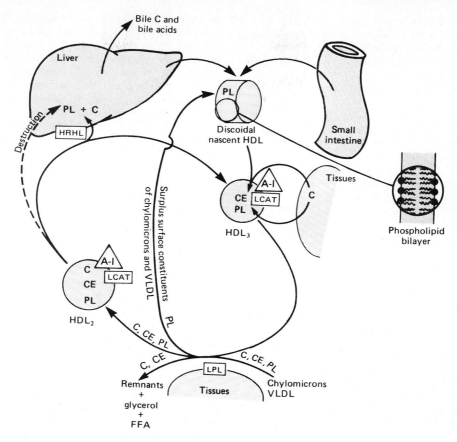

Figure 26–6. Metabolism of high-density lipoprotein (HDL). HRHL, heparin-releasable hepatic lipase; LCAT, lecithin:cholesterol acyltransferase; LPL, lipoprotein lipase; C, cholesterol; CE, cholesteryl ester; PL, phospholipid; FFA, free fatty acids; A-I, apoprotein A-I. The figure illustrates the role of the 3 enzymes HRHL, LCAT, and LPL in the postulated HDL cycle for the transport of cholesterol from the tissues to the liver. HDL$_2$, HDL$_3$—see Table 26–2. HRHL hydrolyzes phospholipid on the surface of HDL$_2$, releasing cholesterol for uptake into the liver.

enging from the tissues. HDL$_c$ is found in the blood of diet-induced hypercholesterolemic animals. It is rich in cholesterol, and its sole apoprotein is apo-E. It is taken up by the liver via the apo-E remnant receptor but also by LDL receptors. As a consequence, the latter are sometimes designated apo-B-100,E receptors. Atherosclerotic plaques contain scavenger cells that have taken up so much cholesterol that they are converted into cholesteryl ester-laden foam cells. Most of the cells arise from macrophages that ingest the more abnormal cholesterol-rich lipoproteins such as chemically modified LDL or β-VLDL (see p 250). Recent work (Brown and Goldstein, 1983) has shown that macrophages secrete both cholesterol (to a suitable receptor such as HDL) and apo-E. This apo-E, after suitable processing in the presence of LCAT, may be the source of cholesterol-rich HDL$_c$. Thus, HDL$_c$ could be an important component in the movement of cholesterol from the tissues to the liver ("reverse cholesterol transport").

It would therefore appear that all plasma lipoproteins are interrelated components of one or more metabolic cycles that together are responsible for the complex process of plasma lipid transport.

ROLE OF THE LIVER IN LIPID TRANSPORT & METABOLISM

Much of the lipid metabolism of the body was formerly thought to be the prerogative of the liver. The discovery that most tissues have the ability to oxidize fatty acids completely and the knowledge which has accumulated showing that adipose tissue is extremely active metabolically have tended to modify the former emphasis on the role of the liver. Nonetheless, the concept of a central and unique role for the liver in lipid metabolism is still an important one. The liver carries out the following major functions in lipid metabolism: (1) It facilitates the digestion and absorption of lipids by the production of bile, which contains cholesterol and bile salts synthesized within the liver. (2) The liver has active enzyme systems for synthesizing and oxidizing fatty acids and for synthesizing triacylglycerols, phospholipids, and cholesterol. (3) It synthesizes plasma lipoproteins. (4) It converts fatty acids to ketone bodies (ketogenesis). (5) It plays an integral part in the metabolism of plasma lipoproteins. Some of these processes have already been described.

TRIACYLGLYCEROL SYNTHESIS & THE FORMATION OF VLDL
(See Fig 26–7.)

Experiments involving a comparison between hepatectomized and intact animals have shown that the liver is the main source of plasma lipoproteins derived from endogenous sources. Hepatic triacylglycerols are

the immediate precursors of triacylglycerols contained in plasma VLDL. The fatty acids used in the synthesis of hepatic triacylglycerols are derived from 2 possible sources: (1) synthesis within the liver from acetyl-CoA derived mainly from carbohydrate and (2) uptake of free fatty acids from the circulation. The first source is predominant in the well-fed condition, when fatty acid synthesis is high and the level of circulating free fatty acids is low. As triacylglycerol does not normally accumulate in the liver under this condition, it must be inferred that it is transported from the liver in VLDL as rapidly as it is synthesized. On the other hand, during fasting, during the feeding of high-fat diets, or in diabetes mellitus, the level of circulating free fatty acids is raised and more is abstracted into the liver. Under these conditions, lipogenesis is inhibited and free fatty

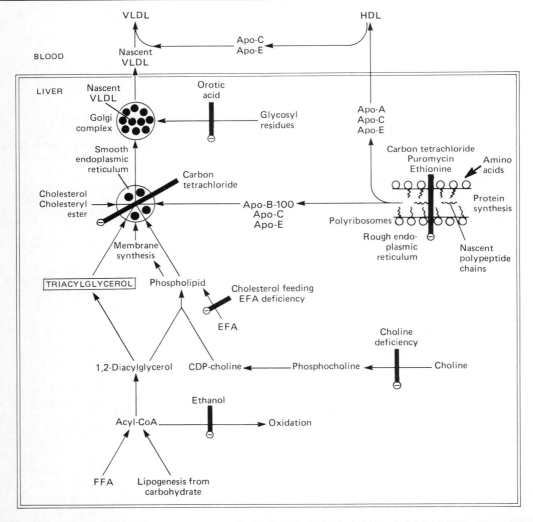

Figure 26–7. The synthesis of very low density lipoprotein (VLDL) and the possible loci of action of factors causing accumulation of triacylglycerol and a fatty liver. EFA, essential fatty acids; FFA, free fatty acids; HDL, high-density lipoproteins; Apo-A, apolipoprotein A; Apo-B, apolipoprotein B; Apo-C, apolipoprotein C; Apo-E, apolipoprotein E. The pathways indicated form a basis for events depicted in Fig 26–3B.

acids are the main source of triacylglycerol fatty acids in the liver and in VLDL. The enzyme mechanisms responsible for the synthesis of triacylglycerols and phospholipids have been described on pp 218 and 220. Factors that enhance both the synthesis of triacylglycerol and the secretion of VLDL by the liver include (1) the feeding of diets high in carbohydrate (particularly if they contain sucrose or fructose); (2) high levels of circulating free fatty acids; (3) ingestion of ethanol; and (4) the presence of high concentrations of insulin and low concentrations of glucagon.

Fatty Liver
(See Fig 26–7.)

For a variety of reasons, lipid—mainly as triacylglycerol—can accumulate in the liver. Extensive accumulation is regarded as a pathologic condition. When accumulation of lipid in the liver becomes chronic, fibrotic changes occur in the cells that progress to **cirrhosis** and impaired liver function.

Fatty livers fall into 2 main categories. (1) The First type is associated with **raised levels of plasma free fatty acids** resulting from mobilization of fat from adipose tissue or from the hydrolysis of lipoprotein or chylomicron triacylglycerol by lipoprotein lipase in extrahepatic tissues. Increasing amounts of free fatty acids are taken up by the liver and esterified. The production of plasma lipoprotein does not keep pace with the influx of free fatty acids, allowing triacylglycerol to accumulate, causing a fatty liver. The quantity of triacylglycerol present in the liver is significantly increased during **starvation** and the feeding of **high-fat diets.** In many instances (eg. in starvation), the ability to secrete VLDL is also impaired. In uncontrolled **diabetes mellitus, pregnancy toxemia of ewes,** and **ketosis in cattle,** fatty infiltration is sufficiently severe to cause visible pallor or fatty appearance and enlargement of the liver.

(2) The second type of fatty liver is usually due to a **metabolic block in the production of plasma lipoproteins,** thus allowing triacyglycerol to accumulate. Theoretically, the lesion may be due to (a) a block in lipoprotein apoprotein synthesis, (b) a block in the synthesis of the lipoprotein from lipid and apoprotein, (c) a failure in provision of phospholipids that are found in lipoproteins, or (d) a failure in the secretory mechanism itself.

One type of fatty liver that has been studied extensively in rats of due to a deficiency of **choline,** which has therefore been called a **lipotropic factor.** As choline may be synthesized using labile methyl groups donated by methionine in the process of **transmethylation** (see Chapters 31 and 32), the deficiency is basically due to a shortage of the type of methyl group donated by methionine. Several mechanisms have been suggested to explain the role of choline as a lipotropic agent, including its absence, causing an impairment in synthesis of lipoprotein phospholipids.

The antibiotic puromycin inhibits protein synthesis and causes a fatty liver and a marked reduction in concentration of VLDL in rats. Other substances that act similarly include ethionine (α-amino-γ-mercaptobutyric acid), carbon tetrachloride, chloroform, phosphorus, lead, and arsenic. Choline will not protect the organism against these agents but appears to aid in recovery. It is very likely that carbon tetrachloride also affects the secretory mechanism itself or the conjugation of the lipid with lipoprotein apoprotein. Its effect is not direct but depends on further transformation of the molecule. This probably involves formation of free radicals that disrupt lipid membranes in the endoplasmic reticulum by formation of lipid peroxides. Some protection against carbon tetrachloride-induced lipid peroxidation is provided by vitamin E-supplemented diets. The action of ethionine is thought to be due to a reduction in availability of ATP. This results when ethionine, replacing methionine in S-adenosylmethionine, traps available adenine and prevents synthesis of ATP. Orotic acid also causes fatty livers; as VLDL accumulate in the Golgi apparatus, it is considered that orotic acid interferes with glycosylation of the lipoprotein, thus inhibiting its release and accounting for the marked decrease in plasma lipoproteins containing apo-B.

A deficiency of vitamin E enhances the hepatic necrosis of the choline deficiency type of fatty liver. Added vitamin E or a source of selenium has a protective effect. In addition to protein deficiency, essential fatty acid and vitamin deficiencies (eg, pyridoxine and pantothenic acid) can cause fatty infiltration of the liver. A deficiency of essential fatty acids is thought to depress the synthesis of phospholipids; therefore, other substances such as cholesterol that compete for available essential fatty acids for esterification can also cause fatty livers.

Metabolism of Ethanol

Alcoholism also leads to fat accumulation in the liver, hyperlipidemia, and ultimately cirrhosis. The exact mechanism of action of alcohol in the long term is still uncertain. Whether or not extra free fatty acid mobilization plays some part in causing the accumulation of fat is not clear, but several studies have demonstrated elevated levels of free fatty acids in the rat after administration of a single intoxicating dose of ethanol. However, alcohol consumption over a long period leads to the accumulation of fatty acids in the liver that are derived from endogenous synthesis rather than from adipose tissue. There is no impairment of hepatic synthesis of protein after ethanol ingestion. There is good evidence of increased hepatic triacylglycerol synthesis, decreased fatty acid oxidation, and decreased citric acid cycle activity, caused by oxidation of ethanol in the hepatic cytosol by **alcohol dehydrogenase** leading to excess production of NADH.

$$CH_3-CH_2-OH + NAD^+ \xrightarrow{\boxed{\text{ALCOHOL DEHYDROGENASE}}} CH_3-CHO + NADH + H^+$$

Ethanol   Acetaldehyde

The NADH generated competes with reducing equivalents from other substrates for the respiratory chain, inhibiting their oxidation. The increased [NADH]/[NAD$^+$] ratio causes a shift to the left in the equilibrium malate $\rightleftharpoons$ oxaloacetate, which may reduce activity of the citric acid cycle. The net effect of inhibiting fatty acid oxidation is to cause increased esterification of fatty acids in triacylglycerol, which appears to be the cause of the fatty liver. Oxidation of ethanol leads to the formation of acetaldehyde, which is oxidized by **aldehyde dehydrogenase** in mitochondria, acetate being the end product. Other effects of alcohol may include increased lipogenesis and cholesterol synthesis from acetyl-CoA. The increased [NADH]/[NAD$^+$] ratio also causes an increased [lactate]/[pyruvate] ratio that results in hyperlactacidemia, which in turn decreases the capacity of the kidney to excrete uric acid. The latter is probably the cause of aggravation of gout by drinking alcohol. Although the major route for ethanol metabolism is via the alcohol dehydrogenase pathway, some metabolism takes place via a cytochrome P-450-dependent microsomal ethanol oxidizing system involving NADPH and O$_2$. This system increases in activity in **chronic alcoholism** and may account for the increased metabolic clearance in this condition as indicated by increased blood levels of both acetaldehyde and acetate.

$$CH_2-CH_2-OH + NADPH + H^+ + O_2 \longrightarrow$$
Ethanol
$$CH_3-CHO + NADP^+ + 2H_2O$$
Acetaldehyde

METABOLISM OF ADIPOSE TISSUE & MOBILIZATION OF FAT

The triacylglycerol stores in adipose tissue are continually undergoing lipolysis (hydrolysis) and reesterification (Fig 26–8). These 2 processes are not the forward and reverse phases of the same reaction. Rather, they are entirely different pathways involving different reactants and enzymes. Many of the nutritional, metabolic, and hormonal **factors that regulate the metabolism of adipose tissue act either upon the process of esterification or on lipolysis.** The resultant of these 2 processes determines the magnitude of the free fatty acid pool in adipose tissue, which in turn is the source and determinant of the level of free fatty acids circulating in the plasma. Since the level of plasma free fatty acids has most profound effects upon the metabolism of other tissues, particularly liver and muscle, **the factors operating in adipose tissue that regulate the outflow of free fatty acids exert an influence far beyond the tissue itself.**

METABOLIC PATHWAYS
(See Fig 26–8.)

Esterification & Lipolysis

In adipose tissue, triacylglycerol is synthesized from acyl-CoA and glycerol 3-phosphate according to the mechanism shown in Fig 25–1. Because the enzyme **glycerol kinase** is low in activity in adipose tissue, glycerol cannot be utilized to any great extent in the esterification of acyl-CoA. **For the provision of glycerol 3-phosphate, the tissue is dependent on a supply of glucose.**

Triacylglycerol undergoes hydrolysis by a **hormone-sensitive lipase** to form free fatty acids and glycerol. This lipase is distinct from lipoprotein lipase that catalyzes lipoprotein triacylglycerol hydrolysis prior to its uptake into extrahepatic tissues (see p 230). Since glycerol cannot be utilized readily in this tissue, it diffuses out into the plasma, from where it is utilized by such tissues as liver and kidney, which possess an active glycerol kinase. The free fatty acids formed by lipolysis can be reconverted in the tissue to acyl-CoA by **acyl-CoA synthetase** and reesterified with glycerol 3-phosphate to form triacylglycerol. Thus, there is a continuous cycle of lipolysis and reesterification within the tissue. However, when the rate of reesterification is not sufficient to match the rate of lipolysis, free fatty acids accumulate and diffuse into the plasma, where they bind to albumin and raise the concentration of plasma free fatty acids. These are a most important source of fuel for many tissues.

Glucose Metabolism

When the utilization of glucose by adipose tissue is increased, the free fatty acid outflow decreases. However, the release of glycerol continues, demonstrating that the effect of glucose is not mediated by reducing the rate of lipolysis. It is believed that the effect is due to the provision of glycerol 3-phosphate, which enhances esterification of free fatty acids via acyl-CoA.

Glucose can take several pathways in adipose tissue, including oxidation to CO$_2$ via the citric acid cycle, oxidation in the pentose phosphate pathway, conversion to long-chain fatty acids, and formation of acylglycerol via glycerol 3-phosphate. When glucose utilization is high, a larger proportion of the uptake is oxidized to CO$_2$ and converted to fatty acids. However, as total glucose utilization decreases, the greater proportion of the glucose is directed to the formation of glycerol 3-phosphate and acylglycerol, which helps to minimize the efflux of free fatty acids.

Uptake of Free Fatty Acids

There is more than one free fatty acid pool within adipose tissue. It has been shown that the free fatty acid pool (Fig 26–8, pool 1) formed by lipolysis of triacylglycerol is the same pool that supplies fatty acids for reesterification; also, it releases them into the external medium (plasma). Fatty acids taken up from the external medium as a result of the action of lipoprotein lipase on the triacylglycerol of chylomicrons and

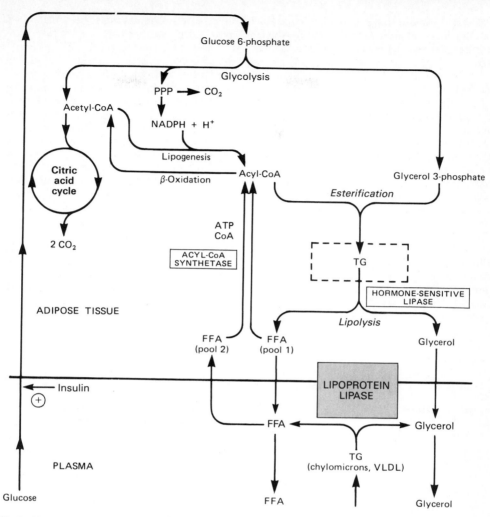

Figure 26–8. Metabolism of adipose tissue. Hormone-sensitive lipase is activated by ACTH, TSH, glucagon, epinephrine, norepinephrine, and vasopressin and inhibited by insulin, prostaglandin E_1, and nicotinic acid. Details of the formation of glycerol 3-phosphate from intermediates of glycolysis are shown in Fig 25–1. PPP, pentose phosphate pathway; TG, triacylglycerol; FFA, free fatty acids; VLDL, very low density lipoproteins.

VLDL do not label pool 1 before they are incorporated into triacylglycerol.

ROLE OF HORMONES IN FAT MOBILIZATION

Insulin

The rate of release of free fatty acids from adipose tissue is affected by many hormones that influence either the rate of esterification or the rate of lipolysis. Insulin inhibits the release of free fatty acids from adipose tissue, which is followed by a fall in circulating plasma free fatty acids. It enhances lipogenesis and the synthesis of acylglycerol and increases the oxidation of glucose to CO_2 via the pentose phosphate pathway. All of these effects are dependent on the presence of glucose and can be explained, to a large extent, on the basis of the ability of insulin to enhance the uptake of glucose into adipose cells. This is achieved by insulin causing the translocation of glucose transporters from the Golgi apparatus to the plasma membrane. Insulin has also been shown to increase the activity of **pyruvate dehydrogenase, acetyl-CoA carboxylase, and glycerol phosphate acyltransferase,** which would reinforce the effects arising from increased glucose uptake on the enhancement of fatty acid and acylglycerol synthesis. These 3 enzymes are now known to be regulated in a coordinate manner by covalent modification, ie, by phosphorylation-dephosphorylation mechanisms.

A principal action of insulin in adipose tissue is to inhibit the activity of the **hormone-sensitive lipase,** reducing the release not only of free fatty acids but of glycerol as well. Adipose tissue is much more sensitive to insulin than are many other tissues, **which points to adipose tissue as a major site of insulin action in vivo.**

Lipolytic Hormones
(See Fig 26–9.)

Other hormones accelerate the release of free fatty acids from adipose tissue and raise the plasma free fatty acid concentration by increasing the rate of lipolysis of the triacylglycerol stores. These include epinephrine, norepinephrine, glucagon, adrenocorticotropic hormone (ACTH), α- and β-melanocyte-stimulating hormones (MSH), thyroid-stimulating hormone (TSH), growth hormone (GH), and vasopressin. Many of these activate the hormone-sensitive lipase. For an optimal effect, most of these lipolytic processes require the presence of **glucocorticoids** and **thyroid hormones.** On their own, these particular hormones do not increase lipolysis markedly but act in a **facilitatory** or **permissive** capacity with respect to other lipolytic endocrine factors.

The hormones that act rapidly in promoting lipoly-

sis, ie, catecholamines, do so by stimulating the activity of **adenylate cyclase,** the enzyme that converts ATP to cAMP. The mechanism is analogous to that responsible for hormonal stimulation of glycogenolysis (see Chapter 19). It appears that cAMP, by stimulating **cAMP-dependent protein kinase,** converts inactive hormone-sensitive triacylglycerol lipase into active lipase. Lipolysis is controlled largely by the amount of cAMP present in the tissue. It follows that processes which destroy or preserve cAMP have an effect on lipolysis. cAMP is degraded to 5′-AMP by the enzyme **cyclic 3′,5′-nucleotide phosphodiesterase.** This enzyme is inhibited by methyl xanthines such as **caffeine** and **theophylline.** It is significant that the drinking of coffee, containing caffeine, causes marked and prolonged elevation of plasma FFA in humans.

Insulin antagonizes the effect of the lipolytic hormones. It is now considered that lipolysis may be more sensitive to changes in concentration of insulin than are glucose utilization and esterification. The antilipolytic effects of insulin, nicotinic acid, and prostaglandin E_1 may be accounted for by inhibition of the synthesis of cAMP, possibly at the adenylate cy-

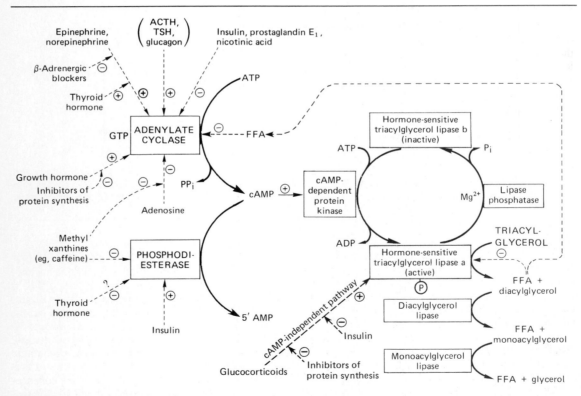

Figure 26–9. Control of adipose tissue lipolysis. TSH, thyroid-stimulating hormone; FFA, free fatty acids. Note the cascade sequence of reactions affording amplification at each step. The lipolytic stimulus is "switched off" by (1) removal of the stimulating hormone; (2) the action of lipase phosphatase; (3) the inhibition of the lipase and adenylate cyclase by high concentrations of FFA; (4) the inhibition of adenylate cyclase by adenosine; and (5) the removal of cAMP by the action of phosphodiesterase. ACTH, TSH, and glucagon may not activate adenylate cyclase in vivo, since the concentration of each hormone required in vitro is much higher than is found in the circulation. Positive ($\oplus$) and negative ($\ominus$) regulatory effects are represented by broken lines and substrate flow by solid lines.

clase site or by stimulating phosphodiesterase. Possible mechanisms for the action of thyroid hormones include an augmentation of the level of cAMP by facilitation of the passage of the stimulus from the receptor site on the outside of the cell membrane to the adenylate cyclase site on the inside of the membrane and an inhibition of phosphodiesterase activity. The effect of growth hormone in promoting lipolysis is slow. It is dependent on synthesis of proteins involved in the formation of cAMP. Glucocorticoids promote lipolysis via synthesis of new lipase protein by a cAMP-independent pathway, which may be inhibited by insulin. These findings help to explain the role of the pituitary gland and the adrenal cortex in enhancing fat mobilization.

The sympathetic nervous system, through liberation of norepinephrine in adipose tissue, plays a central role in the mobilization of free fatty acids by exerting a tonic influence even in the absence of augmented nervous activity. Thus, the increased lipolysis caused by many of the factors described above can be reduced or abolished by denervation of adipose tissue, by ganglionic blockade with hexamethonium, or by depleting norepinephrine stores with reserpine.

COMPARATIVE ASPECTS

Human adipose tissue may not be an important site of lipogenesis. This is indicated by the observation that there is not significant incorporation of label into long-chain fatty acids from labeled glucose or pyruvate; ATP-citrate lyase, a key enzyme in lipogenesis, does not appear to be present and has extremely low activity in liver. Other enzymes—eg, glucose-6-phosphate dehydrogenase and the malic enzyme—which in the rat undergo adaptive changes coincident with increased lipogenesis, do not undergo similar changes in human adipose tissue. Indeed, it has been suggested that in humans there is a "carbohydrate excess syndrome" due to a unique limitation in ability to dispose of excess carbohydrate by lipogenesis. In birds, lipogenesis is confined to the liver, where it is particularly important in providing lipids for egg formation.

Human adipose tissue is unresponsive to most of the lipolytic hormones apart from the catecholamines. Of further interest is the lack of lipolytic response to epinephrine in the rabbit, guinea pig, pig, and chicken; the pronounced lipolytic effect of glucagon in birds, together with an absence of any antilipolytic effect of insulin; and the lack of acylglycerol-glycerol synthesis from glucose in the pigeon. It would appear that, in the various species studied, a variety of mechanisms have been evolved for fine control of adipose tissue metabolism.

On consideration of the profound derangement of metabolism in diabetes mellitus (which is due mainly to increased release of free fatty acids from the depots) and the fact that insulin to a large extent corrects the condition, it must be concluded that **insulin plays a prominent role in the regulation of adipose tissue metabolism.**

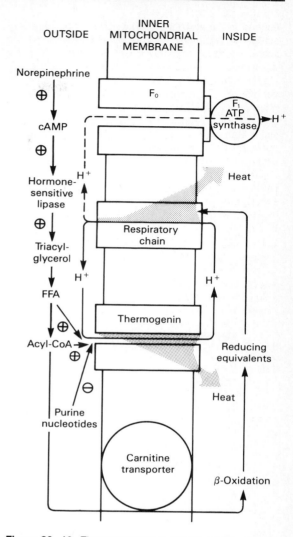

Figure 26–10. Thermogenesis in brown adipose tissue. Activity of the respiratory chain produces heat in addition to translocating protons (see p 110). These protons dissipate heat when returned to the inner mitochondrial compartment via thermogenin, instead of generating ATP when returning via the F_1 ATP synthase. The passage of H^+ via thermogenin is inhibited by purine nucleotides when brown adipose tissue is unstimulated. Under the influence of norepinephrine, the inhibition is removed by the production of free fatty acids (FFA) and acyl-CoA. Note the dual role of acyl-CoA in both facilitating the action of thermogenin and supplying reducing equivalents for the respiratory chain. Positive ($\oplus$) or negative ($\ominus$) regulatory effects.

ROLE OF BROWN ADIPOSE TISSUE IN THERMOGENESIS

Brown adipose tissue is involved in metabolism particularly at times when heat generation is necessary. Thus, the tissue is extremely active in some species in arousal from hibernation, in animals exposed to cold (nonshivering thermogenesis), and in heat production in the newborn animal. Though not a prominent tissue in humans, recently it has been shown to be active in normal individuals, where it appears to be responsible for **"diet-induced thermogenesis,"** which may account for how some persons can "eat and not get fat." It is noteworthy that brown adipose tissue is reduced or absent in obese persons. Brown adipose tissue is characterized by a well-developed blood supply and a high content of mitochondria and cytochromes but low activity of ATP synthase. Metabolic emphasis is placed on oxidation of both glucose and fatty acids.

Norepinephrine liberated from sympathetic nerve endings is important in increasing lipolysis in the tissue. Oxidation and phosphorylation are not coupled in mitochondria of this tissue, since dinitrophenol has no effect and there is no respiratory control by ADP. The phosphorylation that does occur is at the substrate level, eg, at the succinate thiokinase step and in glycolysis. Thus, **oxidation produces much heat, and little free energy is trapped in ATP.** In terms of the chemiosmotic theory, it would appear that the proton gradient, normally present across the inner mitochondrial membrane of coupled mitochondria, is continually dissipated in brown adipose tissue by a thermogenic protein, **thermogenin,** which acts as a proton conductance pathway through the membrane. This would explain the apparent lack of effect of uncouplers (Fig 26–10).

REFERENCES

Brown MS, Goldstein JL: Lipoprotein metabolism in the macrophage: Implications for cholesterol deposition in atherosclerosis. *Annu Rev Biochem* 1983;**52:**223.

Cryer A: Tissue lipoprotein lipase activity and its action in lipoprotein metabolism. *Int J Biochem* 1981;**13:**525.

Eisenberg S: Lipoproteins and lipoprotein metabolism. *Klin Wochenschr* 1983;**61:**119.

Fain JN: Hormonal regulation of lipid mobilization from adipose tissue. Page 119 in: *Biochemical Actions of Hormones.* Vol 7. Litwack G (editor). Academic Press, 1980.

Fielding CJ, Fielding PE: Metabolism of cholesterol and lipoproteins. Page 404 in: *Biochemistry of Lipids and Membranes.* Vance DE, Vance JE (editor). Benjamin/Cummings, 1985.

Himms-Hagen J: Brown adipose tissue metabolism and thermogenesis. *Annu Rev Nutr* 1985;**5:**69.

Krauss RM: Regulation of high density lipoprotein levels. *Med Clin North Am* 1982;**66:**403.

Lieber CS: Alcohol and the liver: Metabolism of ethanol, metabolic effects and pathogenesis of injury. *Acta Med Scand [Suppl]* 1985;**703:**11.

Sparks JD, Sparks CE: Apolipoprotein B and lipoprotein metabolism. *Adv Lipid Res* 1985;**21:**1.

Cholesterol Synthesis, Transport, & Excretion

27

Peter A. Mayes, PhD, DSc

INTRODUCTION

Cholesterol is present in tissues and in plasma lipoproteins either as free cholesterol or (combined with a long-chain fatty acid) as cholesteryl ester. It is synthesized in many tissues from acetyl-CoA and is ultimately eliminated from the body in the bile as cholesterol or bile salts. Cholesterol is the precursor of all other steroids in the body such as corticosteroids, sex hormones, bile acids, and vitamin D. It is typically a product of animal metabolism and therefore occurs in foods of animal origin such as egg yolk, meat, liver, and brain.

BIOMEDICAL IMPORTANCE

Cholesterol is an amphipathic lipid and as such is an essential structural component of membranes and of the outer layer of plasma lipoproteins. Additionally, lipoproteins transport free cholesterol in the circulation, where it readily equilibrates with cholesterol in other lipoproteins and in membranes. Cholesteryl ester is a storage form of cholesterol found in most tissues. It is transported as cargo in lipoproteins. LDL is the mediator of cholesterol and cholesteryl ester uptake into many tissues. Free cholesterol is removed from tissues by HDL and transported to the liver for conversion to bile acids. Cholesterol is a major constituent of gallstones. However, its chief role in pathologic processes is as a factor in the genesis of atherosclerosis of vital arteries, causing cerebrovascular, coronary, and peripheral vascular disease. Coronary atherosclerosis correlates with a high plasma LDL:HDL cholesterol ratio.

CHOLESTEROL BIOSYNTHESIS

Approximately half the cholesterol of the body arises by synthesis (about 500 mg/d), and the remainder is provided by the average diet. The liver accounts for approximately 50% of total synthesis, the gut for about 15%, and the skin for a large proportion of the remainder.

PATHWAY OF BIOSYNTHESIS

Virtually all tissues containing nucleated cells are capable of synthesizing cholesterol. The microsomal (endoplasmic reticulum) and cytosol fraction of the cell is responsible for cholesterol synthesis.

Acetyl-CoA is the source of all the carbon atoms in cholesterol. The manner of synthesis of this complex molecule has been the subject of investigation by many workers, with the result that it is possible at the present time to chart the origin of all parts of the cholesterol molecule (Figs 27–1, 27–2, and 27–3).

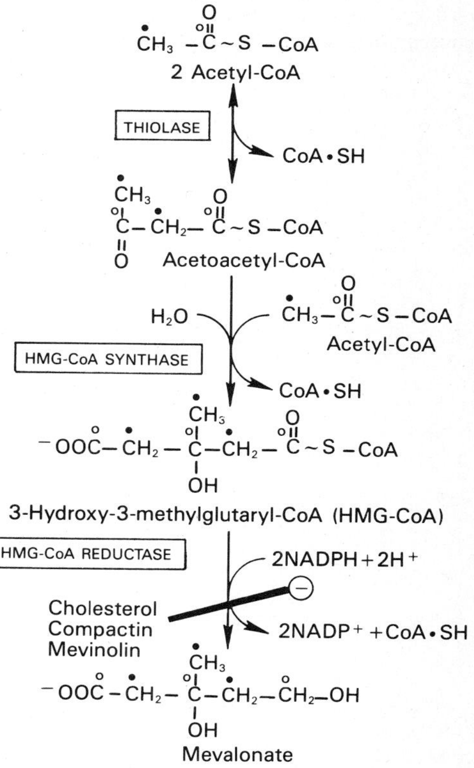

Figure 27–1. Biosynthesis of mevalonate. HMG, 3-hydroxy-3-methylglutaryl. HMG-CoA reductase is inhibited by cholesterol and the fungal metabolites compactin and mevinolin, which are competitive with HMG-CoA.

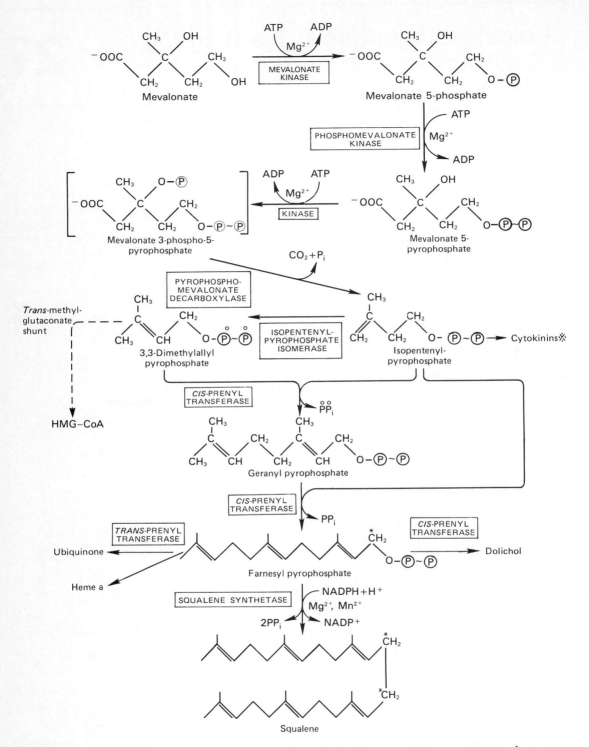

Figure 27–2. Biosynthesis of squalene, ubiquinone, and dolichol. HMG, 3-hydroxy-3-methylglutaryl; ☀, isopenten-yladenine, a component of tRNA. A farnesyl residue is present in heme a of cytochrome oxidase. The carbon marked * becomes C_{11} or C_{12} in squalene. Squalene synthetase is a microsomal enzyme; all other enzymes indicated are soluble cytosolic proteins.

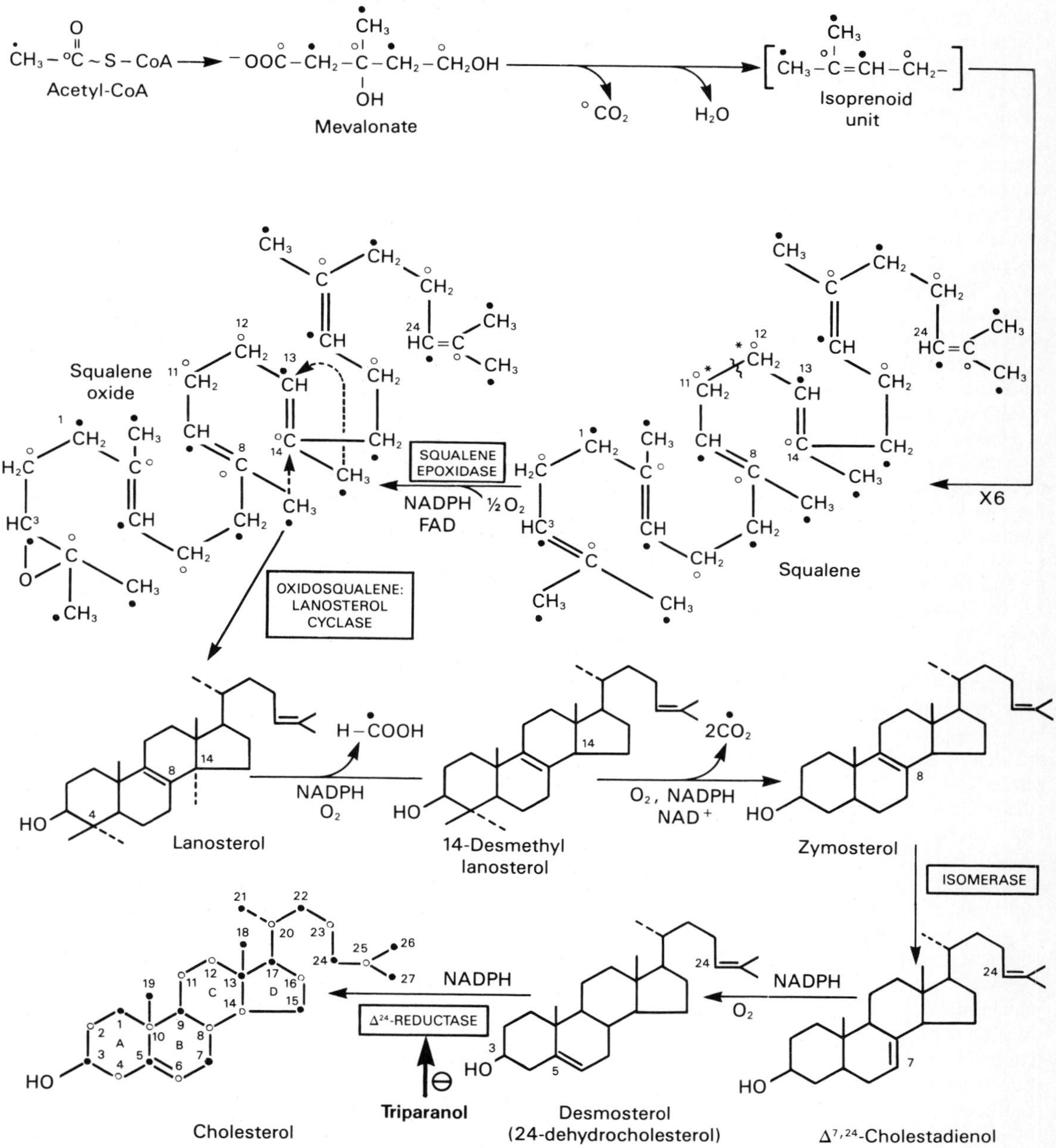

Figure 27–3. Biosynthesis of cholesterol. The numbered positions are those of the steroid nucleus. *Refers to labeling of squalene in Fig 27–2.

Synthesis takes place in several stages. (1) Mevalonate, a 6-carbon compound, is synthesized from acetyl-CoA (Fig 27–1). (2) Isoprenoid units are formed from mevalonate by loss of CO_2 (Fig 27–2). (3) Six isoprenoid units condense to form the intermediate, squalene. (4) Squalene cyclizes to give rise to the parent steroid, lanosterol. (5) Cholesterol is formed from lanosterol after several further steps, including the loss of 3 methyl groups (Fig 27–3).

(1) The pathway through HMG-CoA (3-hydroxy-3-methylglutaryl-CoA) follows the same sequence of reactions described in Chapter 28 for the synthesis in mitochondria of ketone bodies. However, since cholesterol synthesis is extramitochondrial, the 2 pathways are distinct. Initially, 2 molecules of acetyl-CoA condense to form acetoacetyl-CoA catalyzed by a cytosolic **thiolase** enzyme. Alternatively, in liver, acetoacetate made inside the mitochondrion in the pathway of ketogenesis (see Chapter 28) diffuses into the cytosol and may be activated to acetoacetyl-CoA by **acetoacetyl-CoA synthetase,** requiring ATP and CoA. Acetoacetyl-CoA condenses with a further molecule of acetyl-CoA catalyzed by **HMG-CoA synthase** to form HMG-CoA.

HMG-CoA is converted to **mevalonate** in a 2-stage reduction by NADPH catalyzed by **HMG-CoA reductase,** a microsomal enzyme considered to catalyze the rate-limiting step in the pathway of cholesterol synthesis (Fig 27–1).

(2) Mevalonate is phosphorylated by ATP to form several active phosphorylated intermediates (Fig 27–2). By means of a decarboxylation, the active isoprenoid unit, **isopentenylpyrophosphate,** is formed.

(3) The next stage involves the condensation of 3 molecules of isopentenylpyrophosphate to form **farnesyl pyrophosphate.** This occurs via an isomerization of isopentenylpyrophosphate involving a shift of the double bond to form **dimethylallyl pyrophosphate,** followed by condensation with another molecule of isopentenylpyrophosphate to form the 10-carbon intermediate, **geranyl pyrophosphate** (Fig 27–2). A further condensation with isopentenylpyrophosphate forms farnesyl pyrophosphate. Two molecules of farnesyl pyrophosphate condense at the pyrophosphate end in a reaction involving first an elimination of pyrophosphate to form pre-squalene pyrophosphate, followed by a reduction with NADPH with elimination of the remaining pyrophosphate radical. The resulting compound is **squalene.** An alternative pathway known as the "trans-methylglutaconate shunt" may be present. This pathway removes a significant proportion (20%) of the dimethylallyl pyrophosphate and returns it, via trans-3-methylglutaconate-CoA, to HMG-CoA. This pathway may have regulatory potential with respect to the overall rate of cholesterol synthesis.

(4) Squalene has a structure that closely resembles the steroid nucleus (Fig 27–3). Before ring closure occurs, squalene is converted to squalene 2,3-oxide by a mixed-function oxidase in the endoplasmic reticulum, **squalene epoxidase.** The methyl group on C_{14} is

transferred to C_{13} and that on C_8 to C_{14} as cyclization occurs, catalyzed by **oxidosqualene:lanosterol cyclase.**

(5) The last stage (Fig 27–3), the formation of cholesterol from **lanosterol,** takes place in the membranes of the endoplasmic reticulum and involves changes in the steroid nucleus and side chain. The methyl group on C_{14} is oxidized to CO_2 to form 14-desmethyl lanosterol. Likewise, 2 more methyl groups on C_4 are removed to produce zymosterol. $\Delta^{7,24}$-Cholestadienol is formed from zymosterol by the double bond between C_8 and C_9 moving to a position between C_8 and C_7. **Desmosterol** is formed at this point by a further shift in the double bond in ring B to take up a position between C_5 and C_6, as in cholesterol. Finally, cholesterol is produced when the double bond of the side chain is reduced, although this can occur at any stage of the overall conversion to cholesterol. The exact order in which the steps described actually take place is not known with certainty.

It is probable that the intermediates from squalene to cholesterol are attached to a special carrier protein known as the **squalene and sterol carrier protein.** This protein binds sterols and other insoluble lipids, allowing them to react in the aqueous phase of the cell. In addition, it seems likely that it is in the form of cholesterol-sterol carrier protein that cholesterol is converted to steroid hormones and bile acids and participates in the formation of membranes and of lipoproteins.

Synthesis of Other Isoprenoid Compounds

Farnesyl pyrophosphate is the branch point for the synthesis of the other polyisoprenoids, **dolichol** and **ubiquinone.** The polyisoprenyl alcohol dolichol is formed by the further addition of up to 16 isopentenylpyrophosphate residues, whereas the side chain of ubiquinone is formed by the addition of a further 3–7 isoprenoid units.

REGULATION OF CHOLESTEROL SYNTHESIS

Regulation of cholesterol synthesis is exerted near the beginning of the pathway. There is a marked decrease in the activity of HMG-CoA reductase in fasting rats, which explains the **reduced synthesis of cholesterol during fasting.** There is a feedback mechanism whereby HMG-CoA reductase in liver is inhibited by cholesterol. Since a direct inhibition of the enzyme by cholesterol cannot be demonstrated, cholesterol (or a metabolite, eg, oxygenated sterol) may act either by repression of the synthesis of new reductase or by inducing the synthesis of enzymes that degrade existing reductase. Cholesterol synthesis is also inhibited by LDL-cholesterol taken up via LDL receptors (apo-B-100 receptors). A **diurnal variation** occurs in both cholesterol synthesis and reductase activity. However, other work indicates more rapid ef-

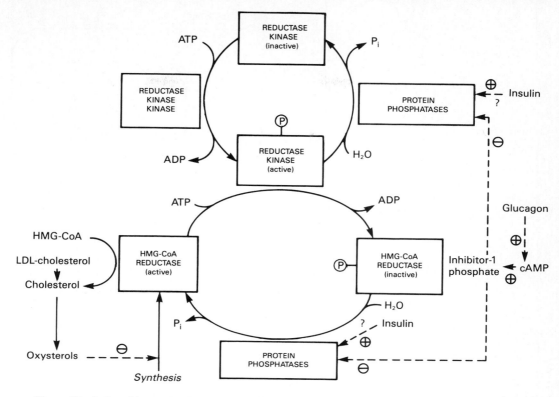

Figure 27–4. Possible mechanisms in the regulation of cholesterol synthesis by HMG-CoA reductase.

fects of cholesterol on reductase activity than can be explained solely by changes in the rate of protein synthesis. Administration of insulin or thyroid hormone increases HMG-CoA reductase activity, whereas glucagon or glucocorticoids decrease it. The enzyme exists in both active and inactive forms that may be reversibly modified by phosphorylation-dephosphorylation mechanisms, some of which may be cAMP-dependent and therefore responsive to glucagon (Fig 27–4).

The effect of variations in the amount of cholesterol in the diet on the endogenous production of cholesterol has been studied in rats. When there was only 0.05% cholesterol in the diet, 70–80% of the cholesterol of the liver, small intestine, and adrenal gland was synthesized within the body, whereas when there was 2% cholesterol in the diet, the endogenous production fell. However, endogenous production could not be completely suppressed by raising the dietary intake. It appears that it is only hepatic synthesis which is inhibited. Experiments with the perfused liver have demonstrated that cholesterol-rich chylomicron remnants (see p 232) inhibit sterol synthesis.

The ability to suppress cholesterol synthesis after feeding cholesterol varies among human subjects. However, attempts to lower plasma cholesterol in humans by reducing the amount of cholesterol in the diet are effective. A decrease of 100 mg in dietary choles-

terol causes a decrease of 5 mg cholesterol per 100 mL serum.

CHOLESTEROL TRANSPORT

CHOLESTEROL BALANCE IN TISSUES

At the tissue level, the following processes are considered to govern the cholesterol balance of cells (Fig 27–5).

Increase is due to (1) uptake of cholesterol-containing lipoproteins by receptors, eg, the LDL receptor; (2) uptake of cholesterol-containing lipoproteins by a non-receptor-mediated pathway; (3) uptake of free cholesterol from cholesterol-rich lipoproteins to the cell membrane; (4) cholesterol synthesis; and (5) hydrolysis of cholesteryl esters by the enzyme **cholesteryl ester hydrolase.**

Decrease is due to (1) efflux of cholesterol from the membrane to lipoproteins of low cholesterol potential, particularly to HDL₃ or nascent HDL, promoted by **LCAT** (lecithin:cholesterol acyltransferase); (2) esterification of cholesterol by **ACAT** (acyl-CoA:cho-

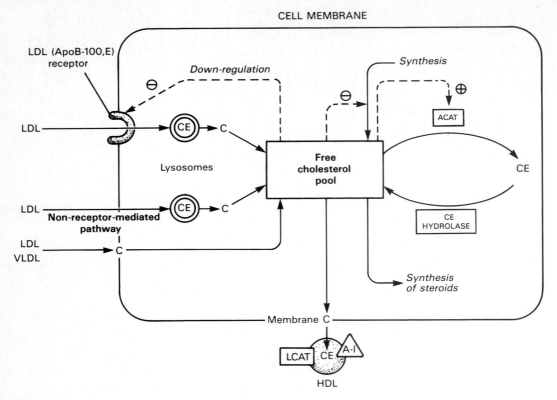

Figure 27–5. Factors affecting cholesterol balance at the cellular level. C, cholesterol; CE, cholesteryl ester; ACAT, acyl-CoA:cholesterol acyltransferase; LCAT, lecithin:cholesterol acyltransferase; A-I, apoprotein A-I; LDL, low-density lipoprotein; VLDL, very low density lipoprotein.

lesterol acyltransferase); and (3) utilization of cholesterol for synthesis of other steroids, such as hormones or bile acids in the liver.

The LDL Receptor (Apo-B-100,E Receptor)

In normal cells, the LDL receptor reacts with the ligand on LDL, apo-B-100, and the LDL is taken up intact by endocytosis. It is broken down in the lysosomes, which involves hydrolysis of cholesteryl ester followed by translocation of cholesterol into the cell. This influx of cholesterol inhibits HMG-CoA reductase and cholesterol synthesis and stimulates ACAT activity. It also appears that the number of LDL receptors on the cell surface is regulated by the cellular need for cholesterol for membranes and steroid hormone synthesis. Thus, influx of cholesterol down-regulates the number of LDL receptors (Fig 27–5).

The apo-B-100,E receptor is a "high-affinity" LDL receptor, which may be saturated under most circumstances. Other "low-affinity" LDL receptors also appear to be present.

CHOLESTEROL TRANSPORT BETWEEN TISSUES (See Fig 27–6.)

In humans, the total plasma cholesterol is about 200 mg/dL, rising with age, although there are wide variations between individuals. The greater part is found in the esterified form. It is transported in lipoproteins in the plasma, and the highest proportion of cholesterol is found in the LDL (β-lipoproteins). However, under conditions where the VLDL are quantitatively predominant, an increased proportion of the plasma cholesterol will reside in this fraction.

Dietary cholesterol takes several days to equilibrate with cholesterol in the plasma and several weeks to equilibrate with cholesterol of the tissues. The turnover of cholesterol in the liver is relatively fast compared with the half-life of the total body cholesterol, which is several weeks. Free cholesterol in plasma and liver equilibrates in a matter of hours.

Cholesteryl ester in the diet is hydrolyzed to free cholesterol, which mixes with dietary free cholesterol

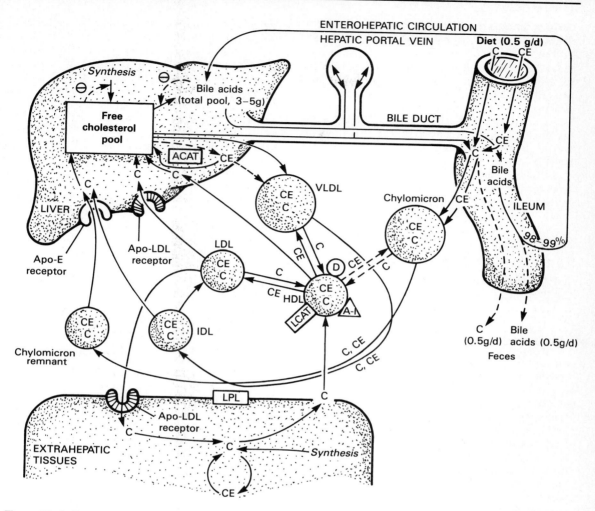

Figure 27–6. Transport of cholesterol between the tissues in humans. C, free cholesterol; CE, cholesteryl ester; VLDL, very low density lipoprotein; IDL, intermediate-density lipoprotein; LDL, low-density lipoprotein; HDL, high-density lipoprotein; ACAT, acyl-CoA:cholesterol acyltransferase; LCAT, lecithin:cholesterol acyltransferase; A-I, apoprotein A-I; D, apoprotein D; LPL, lipoprotein lipase.

and biliary cholesterol before absorption from the intestine in company with other lipids. It mixes with cholesterol synthesized in the intestines and is incorporated into chylomicrons. Of the cholesterol absorbed, 80–90% is esterified with long-chain fatty acids in the intestinal mucosa. The plant sterols (sitosterols) are poorly absorbed. When chylomicrons react with lipoprotein lipase to form chylomicron remnants, only about 5% of the cholesteryl ester is lost. The rest is taken up by the liver when the remnant reacts with the apo-E receptor and is hydrolyzed to free cholesterol. VLDL formed in the liver transport cholesterol into the plasma. However, in humans this is mainly free cholesterol because of low ACAT activity in liver. The cholesteryl ester in VLDL is derived mainly from the action of LCAT in plasma. Most of the cholesterol in VLDL is retained in the VLDL remnant that is taken up by the liver or converted to LDL,

which in turn is taken up by the LDL receptor in liver and extrahepatic tissues.

Role of LCAT

The activity of plasma LCAT is responsible for virtually all plasma cholesteryl ester in humans. (This is not so in other species such as the rat, where there is appreciable ACAT activity in liver, allowing significant export of cholesteryl ester in nascent VLDL.) LCAT activity is associated with a species of HDL containing apo-A-I. As cholesterol in HDL becomes esterified, it creates a concentration gradient and draws in cholesterol from tissues and from other lipoproteins (Fig 27–6). It becomes less dense, forming HDL$_2$, which is thought to deliver cholesterol to the liver (see Fig 26–6). Thus, HDL features prominently in **reverse cholesterol transport,** the process whereby tissue cholesterol is transported to the liver.

Role of Cholesteryl Ester Transfer Protein

This protein, found in plasma of the human but not the rat, is also associated with HDL and seems identical to apo-D. It facilitates transfer of cholesteryl ester from HDL to VLDL, LDL, and to a lesser extent, chylomicrons. It therefore relieves product inhibition of LCAT activity in HDL. Thus, in humans, much of the cholesteryl ester formed by LCAT in HDL finds its way to the liver via VLDL remnants (IDL) or LDL (see Fig 26–6).

Ultimately, all cholesterol destined for excretion from the body must enter the liver and be excreted in the bile, either as cholesterol or as bile salts.

CHOLESTEROL EXCRETION & FORMATION OF BILE ACIDS (SALTS)

About 1 g of cholesterol is eliminated from the body per day. Approximately half is excreted in the feces after conversion to bile acids. The remainder is excreted as neutral steroids. Much of the cholesterol secreted in the bile is reabsorbed, and it is believed that at least some of the cholesterol that serves as precursor for the fecal sterols is derived from the intestinal mucosa. **Coprostanol** is the principal sterol in the feces; it is formed from cholesterol in the lower intestine by the bacterial flora therein. A large proportion of the biliary excretion of bile salts is reabsorbed into the portal circulation, taken up by the liver, and reexcreted in the bile. This is known as the **enterohepatic circulation.** The bile salts not reabsorbed, or their derivatives, are excreted in the feces. Bile salts undergo changes brought about by intestinal bacteria to form secondary bile acids.

BIOSYNTHESIS OF BILE ACIDS

The **primary bile acids** are synthesized in the liver from cholesterol by several intermediate steps. **Cholic acid** is the bile acid found in the largest amount in the bile itself. Both cholic acid and **chenodeoxycholic acid** are formed from a common precursor, itself derived from cholesterol (Fig 27–7).

The 7α-hydroxylation of cholesterol is the first committed step in the biosynthesis of bile acids, and it is this reaction that is rate-limiting in the pathway for synthesis of the acids. The reaction is catalyzed by **7α-hydroxylase,** a microsomal enzyme. It requires oxygen, NADPH, and cytochrome P-450 and appears to be a typical monooxygenase, as are subsequent hydroxylation steps. Vitamin C deficiency interferes with bile acid formation at the 7α-hydroxylation step

and leads to cholesterol accumulation and atherosclerosis in scorbutic guinea pigs.

The pathway of bile acid biosynthesis divides early into one subpathway leading to **cholic acid,** characterized by an extra α-OH group on position 12, and another pathway leading to **chenodeoxycholic acid.** Apart from this difference, both pathways involve similar hydroxylation reactions and shortening of the side chain (Fig 27-7) to give the typical bile acid structures of α-OH groups on positions 3 and 7 and full saturation of the steroid nucleus.

The bile acids normally enter the bile as glycine or taurine conjugates. The newly synthesized primary bile acids are considered to exist within the liver cell as esters of CoA, ie, cholyl- or chenodeoxycholyl-CoA (Fig 27–7). The CoA derivatives are formed with the aid of an activating enzyme occurring in the microsomes of the liver. A second enzyme catalyzes conjugation of the CoA derivatives with glycine or taurine to form glycocholic or glycochenodeoxycholic and taurocholic or taurochenodeoxycholic acids. These are the primary bile acids. In humans, the ratio of the glycine to the taurine conjugates is normally 3:1.

Since bile contains significant quantities of sodium and potassium and the pH is alkaline, it is assumed that the bile acids and their conjugates are actually in a salt form—hence the term "bile salts."

A portion of the primary bile acids in the intestine may be subjected to some further changes by the activity of the intestinal bacteria. These include deconjugation and 7α-dehydroxylation, which produce the **secondary bile acids,** deoxycholic acid from cholic acid, and lithocholic acid from chenodeoxycholic acid (Fig 27–7).

THE ENTEROHEPATIC CIRCULATION

Although fat digestion products, including cholesterol, are absorbed in the first 100 cm of small intestine, the primary and secondary bile acids are absorbed almost exclusively in the ileum, returning to the liver by way of the portal circulation about 98–99% of the bile acids secreted into the intestine. This is known as the enterohepatic circulation. However, lithocholic acid, because of its insolubility, is not reabsorbed to any significant extent.

A small fraction of the bile salts—perhaps only as little as 500 mg/d—escapes absorption and is therefore eliminated in the feces. Even though this is a very small amount, it nonetheless represents a major pathway for the elimination of cholesterol. The enterohepatic circulation of the bile salts is so efficient that each day the relatively small pool of bile acids (about 3–5 g) can be cycled through the intestine 6–10 times with only a small amount lost in the feces; ie, approximately 1–2% per pass through the enterohepatic circulation. However, **each day, an amount of bile acid equivalent to that lost in the feces is synthesized**

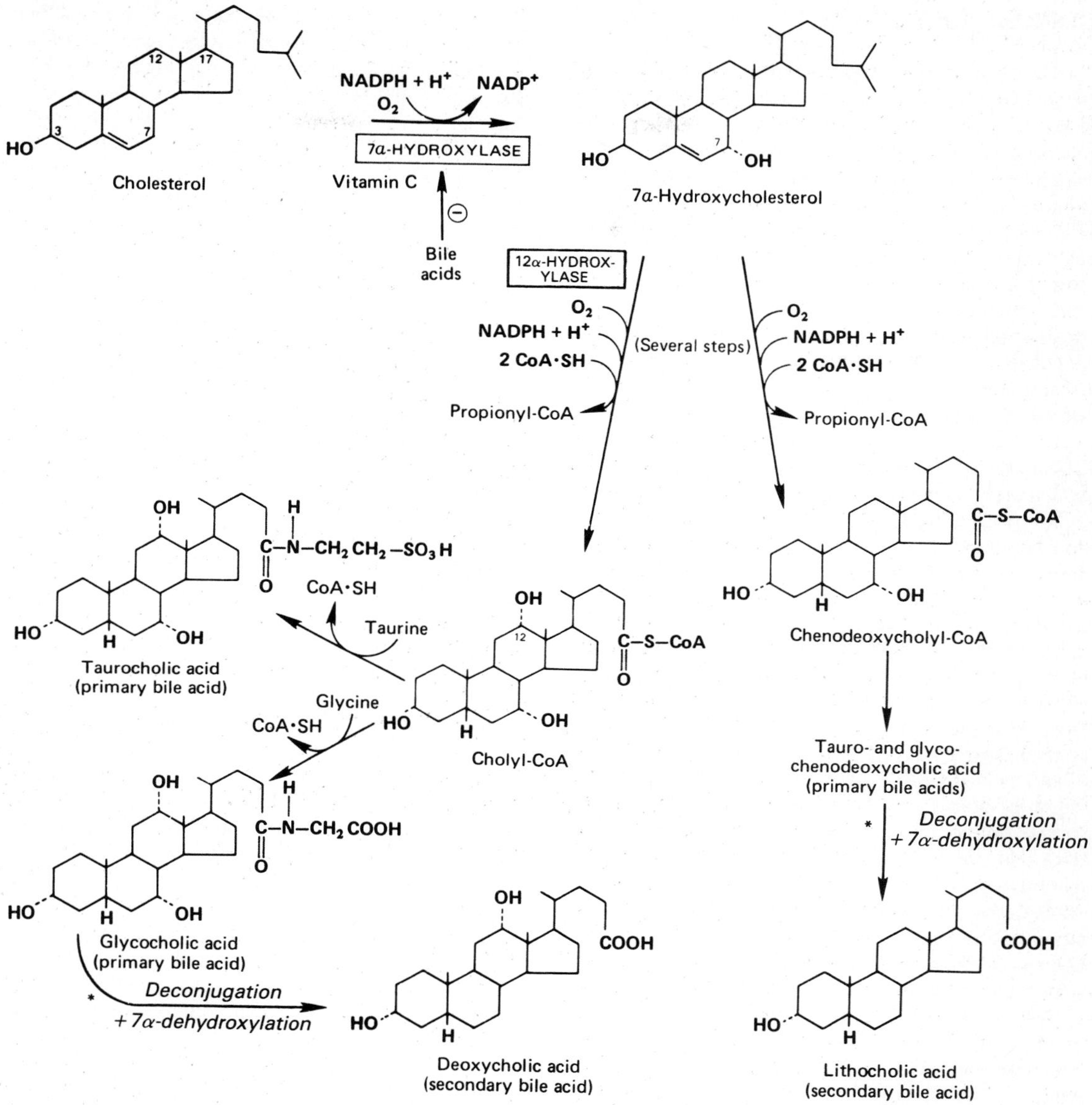

Figure 27–7. Biosynthesis and degradation of bile acids. *Catalyzed by microbial enzymes.

from cholesterol by the liver, so that a pool of bile acids of constant size is maintained. This is accomplished by a system of feedback control.

Regulation of Bile Acid Synthesis

The principal rate-limiting step in the biosynthesis of bile acids is at the **7α-hydroxylase reaction,** and in the biosynthesis of cholesterol it is at the HMG-CoA reductase step (Fig 27–1). The activities of these 2 enzymes often change in parallel, and consequently it is difficult to ascertain whether inhibition of bile acid synthesis takes place primarily at the HMG-CoA reductase step or at the 7α-hydroxylase reaction. Both enzymes undergo similar diurnal variation in activity, but whether cholesterol directly stimulates 7α-hydroxylase activity is not clear. Bile acids certainly exert feedback inhibition on 7α-hydroxylase, but it does not seem to be a direct allosteric mechanism. In this regard, the return of bile salts to the liver via the enterohepatic circulation is an important control that, if interrupted, leads to activation of 7α-hydroxylase. However, recent work indicates that 7α-hydroxylase (as well as HMG-CoA reductase) can be controlled by covalent phosphorylation-dephosphorylation. In contrast to HMG-CoA reductase, it is the phosphorylated form that results in increased activity of 7α-hydroxylase.

CLINICAL ASPECTS

Clinically, **hypercholesterolemia** may be treated by interrupting the enterohepatic circulation of bile acids. It is reported that significant reductions of plasma cholesterol can be effected by this procedure, which can be accomplished by the use of cholestyramine resin (Questran) or surgically by the ileal exclusion operations. Both procedures cause a block in the reabsorption of bile acids. Then, because of release from feedback regulation normally exerted by bile acids, the conversion of cholesterol to bile acids is greatly enhanced in an effort to maintain the pool of bile acids. LDL receptors in the liver are up-regulated, causing increased uptake of LDL with consequent lowering of plasma cholesterol.

Cholesterol, Atherosclerosis, & Coronary Heart Disease

Many investigators have demonstrated a correlation between raised serum lipid levels and the incidence of coronary heart disease and atherosclerosis in humans. Of the serum lipids, cholesterol has been the one most often singled out as being chiefly concerned in the relationship. However, other parameters—such as serum triacylglycerol concentration—show similar correlations. Patients with arterial disease can have any one of the following abnormalities: (1) elevated concentrations of VLDL with normal concentrations of LDL; (2) elevated LDL with normal VLDL; (3) elevation of both lipoprotein fractions. There is also an inverse relationship between HDL (HDL_2) concentrations and coronary heart disease, and some consider that the most predictive relationship is the **LDL:HDL cholesterol ratio.** This relationship is explainable in terms of the proposed roles of LDL in transporting cholesterol to the tissues and of HDL acting as the scavenger of cholesterol.

Atherosclerosis is characterized by the deposition of cholesterol and cholesteryl ester of lipoproteins containing apo-B-100 in the connective tissue of the arterial walls. Diseases in which prolonged elevated levels of VLDL, IDL, or LDL occur in the blood (eg, diabetes mellitus, lipid nephrosis, hypothyroidism, and other conditions of hyperlipidemia) are often accompanied by premature or more severe atherosclerosis.

Experiments on the induction of atherosclerosis in animals indicate a wide species variation in susceptibility. The rabbit, pig, monkey, and humans are species in which atherosclerosis can be induced by feeding cholesterol. The rat, dog, and cat are resistant. Thyroidectomy or treatment with thiouracil drugs will allow induction of atherosclerosis in the dog and rat. Low blood cholesterol is a characteristic of hyperthyroidism.

Hereditary factors play the greatest role in determining individual blood cholesterol concentrations, but of the dietary and environmental factors that lower blood cholesterol, the substitution in the diet of **polyunsaturated fatty acids** for some of the saturated fatty acids has been the most intensely studied. Naturally occurring oils that contain a high proportion of linoleic acid are beneficial in lowering plasma cholesterol and include peanut, cottonseed, corn, and soybean oil, whereas butterfat, beef fat, and coconut oil, containing a high proportion of saturated fatty acids, raise the level. Sucrose and fructose have a greater effect in raising blood lipids, particularly triacylglycerols, than do other carbohydrates.

The reason for the cholesterol-lowering effect of polyunsaturated fatty acids is still not clear. However, several hypotheses have been advanced to explain the effect, including the stimulation of cholesterol excretion into the intestine and the stimulation of the oxidation of cholesterol to bile acids. It is possible that cholesteryl esters of polyunsaturated fatty acids are more rapidly metabolized by the liver and other tissues, which might enhance their rate of turnover and excretion. There is other evidence that the effect is largely due to a shift in distribution of cholesterol from the plasma into the tissues because of increased catabolic rate of LDL. Saturated fatty acids cause the formation of smaller VLDL particles that contain relatively more cholesterol, and they are utilized by extrahepatic tissues at a slower rate than are larger particles. All of these tendencies may be regarded as atherogenic.

Additional factors considered to play a part in coronary heart disease include high blood pressure, smoking, obesity, lack of exercise, and drinking soft as opposed to hard water. Elevation of plasma free fatty acids will also lead to increased VLDL secretion by the liver, involving extra triacylglycerol and cholesterol output into the circulation. Factors leading to higher or fluctuating levels of free fatty acids include emotional stress, nicotine from cigarette smoking, coffee drinking, and partaking of a few large meals rather than more continuous feeding. Premenopausal women appear to be protected against many of these deleterious factors, possibly because they have higher concentrations of HDL than do men and postmenopausal women.

Hypolipidemic Drugs

When dietary measures fail to achieve reduced serum lipid levels, the use of hypolipidemic drugs may be resorted to. Several drugs are known to block the formation of cholesterol at various stages in the biosynthetic pathway. Many of these drugs have harmful effects, but the fungal inhibitors of HMG-CoA reductase, **compactin** and **mevinolin,** reduce LDL cholesterol levels with few adverse effects. **Sitosterol** is a hypocholesterolemic agent that acts by blocking the absorption of cholesterol in the gastrointestinal tract. Resins such as **colestipol** and **cholestyramine** (Questran) prevent the reabsorption of bile salts by combining with them, thereby increasing their fecal loss. **Neomycin** also inhibits reabsorption of bile salts. **Clofibrate** and **gemfibrozil** exert at least part of their hypolipidemic effect by diverting the hepatic inflow of free fatty acids from the pathways of esterification into those of oxidation, thus decreasing the secretion of triacylglycerol and cholesterol containing VLDL by the liver. In addition, they facilitate hydrolysis of VLDL triacylglycerols by lipoprotein lipase. **Probucol** appears to increase LDL catabolism via receptor-independent pathways. **Nicotinic acid** reduces the flux of FFA by inhibiting adipose tissue lipolysis, thereby inhibiting VLDL production by the liver.

Disorders of the Plasma Lipoproteins (Dyslipoproteinemias)

A few individuals in the population exhibit inherited defects in their lipoproteins, leading to the primary condition of either **hypo-** or **hyperlipoproteinemia.** Many others having defects such as diabetes mellitus, hypothyroidism, and atherosclerosis show abnormal lipoprotein patterns that are very similar to one or another of the primary inherited conditions. Virtually all of these primary conditions are due to a defect at one or another stage in the course of lipoprotein formation, transport, or destruction (see Figs 26–4, 26–5, and 27–6). Not all of the abnormalities are harmful.

A. Hypolipoproteinemia:

1. Abetalipoproteinemia–This is a rare inherited disease characterized by absence of β-lipoprotein (LDL) in plasma. The blood lipids are present in low concentrations—especially acylglycerols, which are virtually absent, since **no chylomicrons or VLDL are formed.** Both the intestine and the liver accumulate acylglycerols. Abetalipoproteinemia is due to a defect in apoprotein B synthesis.

2. Familial hypobetalipoproteinemia–In hypobetalipoproteinemia, LDL concentration is between 10 and 50% of normal, but chylomicron formation occurs. It must be concluded that apo-B is essential for triacylglycerol transport. Most individuals are healthy and long-lived.

3. Familial alpha-lipoprotein deficiency (Tangier disease)–In the homozygous individual, there is near absence of plasma HDL and accumulation of cholesteryl esters in the tissues. There is no impairment of chylomicron formation or secretion of VLDL by the liver. However, on electrophoresis, there is no pre-β-lipoprotein, but a broad β-band is found containing the endogenous triacylglycerol. This is because the normal pre-β-band contains other apoproteins normally provided by HDL. Patients tend to develop hypertriacylglycerolemia as a result of the absence of apo-C-II, which normally activates lipoprotein lipase.

B. Hyperlipoproteinemia:

1. Familial lipoprotein lipase deficiency (type I)–This condition is characterized by very slow clearing of chylomicrons from the circulation, leading to abnormally raised levels of chylomicrons. VLDL may be raised, but there is a decrease in LDL and HDL. Thus, the condition is fat-induced. It may be corrected by reducing the quantity of fat and increasing the proportion of complex carbohydrate in the diet.

A variation of this disease is caused by a deficiency in apo-C-II, required as a cofactor for lipoprotein lipase.

2. Familial hypercholesterolemia (type II)–Patients are characterized by hyperbetalipoproteinemia (LDL), which is associated with increased plasma total cholesterol. There may also be a tendency for the VLDL to be elevated in type IIb. Therefore, the patient may have somewhat elevated triacylglycerol levels but the plasma—as is not true in the other types of hyperlipoproteinemia—remains clear. Lipid deposition in the tissue (eg, xanthomas, atheromas) is common. A type II pattern may also arise as a secondary result of hypothyroidism. The disease appears to be associated with reduced rates of clearance of LDL from the circulation due to **defective LDL receptors** and is associated with an increased incidence of atherosclerosis. Reduction of dietary cholesterol and saturated fats may be of use in treatment. A disease producing hypercholesterolemia but due to a different cause is **Wolman's disease** (cholesteryl ester storage disease). This is due to a deficiency of cholesteryl ester hydrolase in lysosomes of cells such as fibroblasts that normally metabolize LDL.

3. Familial type III hyperlipoproteinemia (broad beta disease, remnant removal disease, familial dysbetalipoproteinemia)–This condition is characterized by an increase in both chylomicron and VLDL remnants; these are lipoproteins of density less than 1.019 but appear as a broad β-band on electrophoresis (β-VLDL). They cause hypercholesterolemia and hypertriacylglycerolemia. Xanthomas and atherosclerosis of both peripheral and coronary arteries are present. Treatment by weight reduction and diets containing complex carbohydrates, unsaturated fats, and little cholesterol is recommended. The disease is due to a **deficiency in remnant metabolism by the liver caused by an abnormality in apo-E,** which is normally present in 3 isoforms, E2, E3, and E4. Patients with type III hyperlipoproteinemia possess only E2, which does not react with the E receptor.

4. Familial hypertriacylglycerolemia (type IV)–This condition is characterized by high levels of endogenously produced triacylglycerol (VLDL). Cholesterol levels rise in proportion to the hypertriacylglycerolemia, and glucose intolerance is frequently present. Both LDL and HDL are subnormal in quantity. This lipoprotein pattern is also commonly associated with coronary heart disease, type II non-insulin-dependent diabetes mellitus, obesity, and many other conditions, including alcoholism and the taking of progestational hormones. Treatment of primary type IV hyperlipoproteinemia is by weight reduction; replacement of soluble diet carbohydrate with complex carbohydrate, unsaturated fat, low-cholesterol diets; and also hypolipidemic agents.

5. Familial type V hyperlipoproteinemia–The lipoprotein pattern is complex, since both chylomicrons and VLDL are elevated, causing both triacylglycerolemia and cholesterolemia. Concentrations of LDL and HDL are low. Xanthomas are frequently present, but the incidence of atherosclerosis is apparently not striking. Glucose tolerance is abnormal and frequently associated with obesity and diabetes. The reason for the condition, which is familial, is not clear. Treatment has consisted of weight reduction followed by a diet not too high in either carbohydrate or fat.

It has been suggested that a further cause of hypolipoproteinemia is overproduction of apo-B, which can influence plasma concentrations of VLDL and LDL.

6. Familial hyperalphalipoproteinemia–This is a rare condition associated with increased concentrations of HDL apparently beneficial to health.

C. Familial Lecithin:Cholesterol Acyltransferase (LCAT) Deficiency: In affected subjects, the plasma concentration of cholesteryl esters and lysolecithin is low, whereas the concentration of cholesterol and lecithin is raised. The plasma tends to be turbid. Abnormalities are also found in the lipoproteins. One HDL fraction contains disk-shaped structures in stacks or rouleaux that are clearly nascent HDL unable to take up cholesterol owing to the absence of LCAT. Also present as an abnormal LDL subfraction is lipoprotein-X, otherwise found only in patients with **cholestasis.** VLDL are also abnormal, migrating as β-lipoproteins upon electrophoresis (β-VLDL). Patients with **parenchymal liver disease** also show a decrease of LCAT activity and abnormalities in the serum lipids and lipoproteins.

REFERENCES

Brown MS, Goldstein JL: Lipoprotein metabolism in the macrophage: Implications for cholesterol deposition in atherosclerosis. *Annu Rev Biochem* 1983;**52**:223.

Connor WE, Connor SL: The dietary treatment of hyperlipidemia. *Med Clin North Am* 1982;**66**:485.

Eisenberg S: Lipoproteins and lipoprotein metabolism. *Klin Wochenschr* 1983;**61**:119.

Fears R, Sabine JR (editors): *Cholesterol 7α-Hydroxylase (7α-Monooxygenase)*. CRC Press, 1986.

Fielding CJ, Fielding PE: Metabolism of cholesterol and lipoproteins. Page 404 in: *Biochemistry of Lipids and Membranes*. Vance DE, Vance JE (editors). Benjamin/Cummings, 1985.

Frohlich J, McLeod R, Hon K: Lecithin:cholesterol acyltransferase (LCAT). *Clin Biochem* 1982;**15**:269.

Kane JB, Havel RJ: Treatment of hypercholesterolemia. *Annu Rev Med* 1986;**37**:427.

Mahley RW, Innerarity TL: Lipoprotein receptors and cholesterol homeostasis. *Biochim Biophys Acta* 1983;**737**:197.

Regulation of Lipid Metabolism & Tissue Fuels

28

Peter A. Mayes, PhD, DSc

INTRODUCTION

Lipids play many structural and metabolic roles, but it is as the provider of a large proportion of dietary calories (~40% of Western diets) that they have their greatest impact on metabolism and health. The regulation of this fuel influx and the manner in which it is integrated with other tissue fuels are of central interest, since they impinge on many other metabolic processes.

BIOMEDICAL IMPORTANCE

Under positive caloric balance, a significant proportion of the food energy intake is stored as either glycogen or fat. However, in many tissues, even under fed conditions, fatty acids are oxidized in preference to glucose, but particularly under conditions of caloric deficit or starvation. The purpose is to spare glucose for those tissues (eg, brain and erythrocytes) that require it under all conditions. Thus, regulatory mechanisms, often hormone-mediated, ensure a supply of suitable fuel for all tissues, at all times, from the fully fed to the totally starved state. Breakdown of these mechanisms occurs owing to hormone imbalance (eg, insulin deficiency in diabetes mellitus), to metabolic imbalance due to heavy lactation (eg, ketosis of cattle), or to high metabolic demands in pregnancy (eg, pregnancy toxemia in sheep). All of these conditions are pathologic aberrations of the starvation syndrome. This is a complication of many medical situations when appetite diminishes.

REGULATION OF FATTY ACID BIOSYNTHESIS (Lipogenesis)

PHYSIOLOGIC FACTORS REGULATING LIPOGENESIS

Many animals, including humans, take their food as spaced meals and must therefore store much of the energy of their diet for use between meals. The process of lipogenesis is concerned with the conversion of glucose and intermediates such as pyruvate, lactate, and acetyl-CoA to fat, which constitutes the anabolic phase of this cycle. The **nutritional state** of the organism and tissues is the **main factor controlling the rate of lipogenesis.** Thus, the rate is high in the well-fed animal whose diet contains a high proportion of carbohydrate. It is depressed under conditions of restricted caloric intake, on a high-fat diet, or when there is a deficiency of insulin, as in diabetes mellitus. All of these conditions are associated with increased concentrations of plasma free fatty acids. The regulation of the mobilization of free fatty acids from adipose tissue is described in Chapter 26.

There is an **inverse relationship between hepatic lipogenesis and the concentration of serum free fatty acids** (Fig 28–1). The greatest inhibition of lipogenesis occurs over the range of free fatty acids (0.3–0.8 μmol/mL of plasma) through which the plasma free fatty acids increase during transition from the fed to the starved state. Fat in the diet also causes depression of lipogenesis in the liver, and when there is more than 10% of fat in the diet, there is little conversion of dietary carbohydrate to fat. Lipogenesis is higher when sucrose is fed instead of glucose because

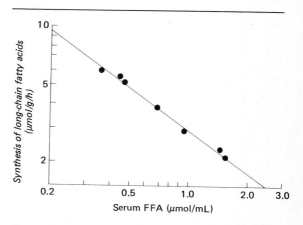

Figure 28–1. Direct inhibition of hepatic lipogenesis by free fatty acids. Lipogenesis was determined from the incorporation of ^3H_2O into long-chain fatty acids in the perfused rat liver. FFA, free fatty acids. (Experiments from the author's laboratory with DL Topping.)

fructose bypasses the phosphofructokinase control point in glycolysis (see Fig 21–2). Because of the close association between the activities of the pentose phosphate pathway on the one hand and of the lipogenic pathway on the other, it was considered that the block in lipogenesis in fasting was due to lack of NADPH generation from the pathway. However, subsequent work in which an NADPH-generating system was added to a liver homogenate from fasting rats failed to promote fatty acid synthesis.

MOLECULAR FACTORS REGULATING LIPOGENESIS

Long-chain fatty acid synthesis is controlled in the short term by allosteric and covalent modification of enzymes and in the long term by changes in rates of synthesis and degradation of enzymes.

Regulation of Acetyl-CoA Carboxylase

At present it is recognized that the rate-limiting reaction in the lipogenic pathway is at the **acetyl-CoA carboxylase step** (see Fig 23–5). Acetyl-CoA carboxylase is activated by **citrate** but inhibited by long-chain acyl-CoA molecules; this is an example of metabolic negative feedback inhibition by a product of a reaction sequence. Thus, if acyl-CoA accumulates because it is not esterified quickly enough, it will automatically reduce the synthesis of new fatty acid. Likewise, if acyl-CoA accumulates as a result of increased lipolysis or an influx of free fatty acids into the tissue, this will also inhibit synthesis of new fatty acid. Acyl-CoA may also inhibit the mitochondrial **tricarboxylate transporter,** thus preventing egress of citrate from the mitochondria into the cytosol.

Regulation of Pyruvate Dehydrogenase

There is also an inverse relationship between free fatty acids and the proportion of active to inactive pyruvate dehydrogenase, which would regulate the availability of acetyl-CoA for lipogenesis. Acyl-CoA causes an inhibition of pyruvate dehydrogenase by inhibiting the ATP-ADP exchange transporter of the inner mitochondrial membrane, which leads to increased intramitochondrial [ATP]/[ADP] ratios and therefore to conversion of active to inactive pyruvate dehydrogenase (see Fig 22–3). Also, oxidation of fatty acids due to increased levels of free fatty acids may increase the ratio of [acetyl-CoA]/[CoA] and [NADH]/NAD$^+$] in mitochondria, inhibiting pyruvate dehydrogenase.

Hormonal Regulation

Insulin stimulates lipogenesis by several possible mechanisms. It increases the transport of glucose into the cell (eg, in adipose tissue) and thereby increases the availability of both pyruvate for fatty acid synthesis and glycerol 3-phosphate for esterification of the fatty acids. Insulin converts the inactive form of pyruvate dehydrogenase to the active form in adipose tissue but not in liver. In addition, insulin activates acetyl-CoA carboxylase, possibly by activation of a protein phosphatase. Also, insulin, by its ability to depress the level of intracellular cAMP, inhibits lipolysis and thereby reduces the concentration of long-chain acyl-CoA, an inhibitor of lipogenesis. **Glucagon** and **epinephrine** inhibit acetyl-CoA carboxylase, and therefore lipogenesis, by increasing **cAMP,** allowing cAMP-dependent protein kinase to inactivate the enzyme by phosphorylation. In addition, catecholamines inhibit the enzyme through α-adrenergic receptors and a Ca^{2+}/calmodulin-dependent protein kinase. In ruminants, **acetate**—not glucose—is the starting material for lipogenesis. It follows that, in these species, many of the control mechanisms involving mitochondria are bypassed and thus do not apply.

Enzyme Adaptation

Both the **fatty acid synthase** complex and **acetyl-CoA carboxylase** are adaptive enzymes, increasing in total amount in the fed state and decreasing in fasting, feeding of fat, and diabetes. **Insulin** is an important hormone causing induction of enzyme biosynthesis. These effects on lipogenesis take several days to become fully manifested and augment the direct and immediate effect of free fatty acids and hormones such as insulin and glucagon.

REGULATION OF FATTY ACID OXIDATION

KETOGENESIS

Under certain metabolic conditions associated with a high rate of fatty acid oxidation, the liver produces considerable quantities of **acetoacetate** and D(−)-3-**hydroxybutyrate** (β-hydroxybutyrate) that pass by diffusion into the blood. Acetoacetate continually undergoes spontaneous decarboxylation to yield **acetone.** These 3 substances are collectively known as the **ketone bodies** (also called acetone bodies or [incorrectly*] "ketones") (Fig 28–2). Acetoacetate and 3-hydroxybutyrate are interconverted by the mitochondrial enzyme D(−)-3-hydroxybutyrate **dehydrogenase;** the equilibrium is controlled by the mitochondrial ratio of [NAD$^+$] to [NADH], ie, the **redox state.** The ratio [3-hydroxybutyrate]/[acetoacetate] in blood varies between 1:1 and 10:1.

---

*The term "blood ketones" should not be used because 3-hydroxybutyrate is not a ketone and there are many ketones in blood that are not ketone bodies, eg, pyruvate, fructose.

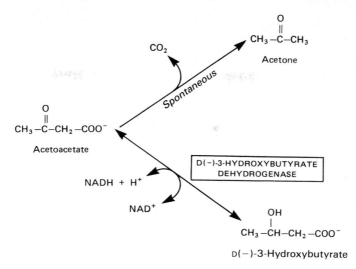

Figure 28–2. Interrelationships of the ketone bodies. D(−)-3-Hydroxybutyrate dehydrogenase is a mitochondrial enzyme.

The concentration of total ketone bodies in the blood of well-fed mammals does not normally exceed 1 mg/dL (as acetone equivalents). It is somewhat higher than this in ruminants, owing to 3-hydroxybutyrate formation from butyric acid (a product of ruminal fermentation) in the rumen wall. Loss via the urine is usually less than 1 mg/24 h in humans. Higher than normal quantities present in the blood or urine constitute **ketonemia** (hyperketonemia) or **ketonuria,** respectively. The overall condition is called **ketosis.** Acetoacetic and 3-hydroxybutyric acids are both moderately strong acids and are buffered when present in blood or the tissues. However, their continual excretion in quantity entails some loss of buffer cation (in spite of ammonia production by the kidney) that progressively depletes the alkali reserve, causing **ketoacidosis.** This may be fatal in uncontrolled **diabetes mellitus.**

The simplest form of ketosis occurs in **starvation** and involves depletion of available carbohydrate coupled with mobilization of free fatty acids. No other condition in which ketosis occurs seems to differ qualitatively from this general pattern of metabolism, but quantitatively it may be exaggerated to produce the pathologic states found in diabetes mellitus, pregnancy toxemia in sheep, and ketosis in lactating cattle. Nonpathologic forms of ketosis are found under conditions of high-fat feeding and after severe exercise in the postabsorptive state.

In vivo, the liver appears to be the only organ in nonruminants to add significant quantities of ketone bodies to the blood. Extrahepatic tissues utilize them as respiratory substrates. Extrahepatic sources of ketone bodies, as in fed ruminants, do not contribute significantly to the occurrence of ketosis in these species.

The net flow of ketone bodies from the liver to the extrahepatic tissues results from an active enzymatic mechanism in the liver for the production of ketone bodies coupled with very low activity of enzymes responsible for their utilization. The reverse situation occurs in extrahepatic tissues (Fig 28–3).

The Pathway of Ketogenesis in Liver

Enzymes responsible for ketone body formation are associated mainly with the mitochondria. Originally it was thought that only one molecule of acetoacetate was formed from the terminal 4 carbons of a fatty acid upon oxidation. Later, to explain both the production of more than one equivalent of acetoacetate from a long-chain fatty acid and the formation of ketone bodies from acetic acid, it was proposed that C_2 units formed in β-oxidation condensed with one another to form acetoacetate. This may occur by a reversal of the **thiolase** reaction whereby 2 molecules of acetyl-CoA condense to form acetoacetyl-CoA. Thus, acetoacetyl-CoA, which is the starting material for ketogenesis, arises either directly during the course of β-oxidation or as a result of the condensation of acetyl-CoA (Fig 28–4). Two pathways for the formation of acetoacetate from acetoacetyl-CoA have been proposed. The first is by simple deacylation. The second pathway (Fig 28–5) involves the condensation of acetoacetyl-CoA with another molecule of acetyl-CoA to form 3-hydroxy-3-methylglutaryl-CoA (HMG-CoA), catalyzed by **3-hydroxy-3-methylglutaryl-CoA synthase.** The presence of another enzyme in the mitochondria, **3-hydroxy-3-methylglutaryl-CoA lyase,** causes acetyl-CoA to split off from the HMG-CoA, leaving free acetoacetate. The carbon atoms split off in the acetyl-CoA molecule are derived from

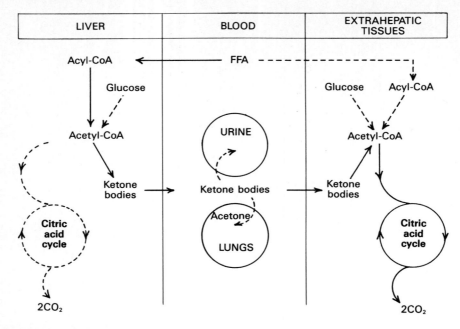

Figure 28–3. Formation, utilization, and excretion of ketone bodies. (The main pathway is indicated by the solid arrows.)

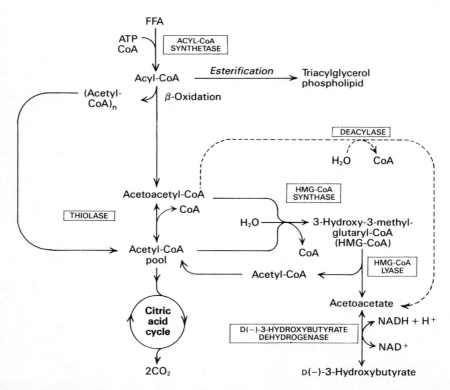

Figure 28–4. Pathways of ketogenesis in the liver. FFA, free fatty acids; HMG, 3-hydroxy-3-methylglutaryl.

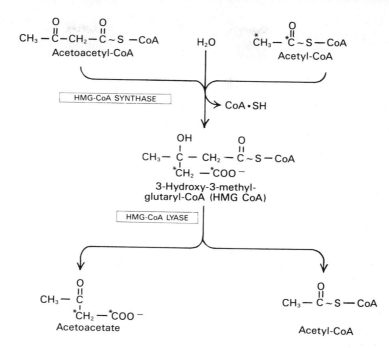

Figure 28–5. Formation of acetoacetate through intermediate production of HMG-CoA.

the original acetoacetyl-CoA molecule (Fig 28–5). **Both of these enzymes must be present in mitochondria for ketogenesis to take place.** This occurs solely in liver and rumen epithelium.

Present opinion favors the HMG-CoA pathway as the major route of ketone body formation. Although there is a marked increase in activity of HMG-CoA lyase in fasting, evidence does not suggest that this enzyme is rate-limiting in ketogenesis.

Acetoacetate may be converted to D(−)-3-hydroxybutyrate by D(−)-**3-hydroxybutyrate dehydrogenase,** which is present in many tissues, including the liver. D(−)-3-Hydroxybutyrate is quantitatively the predominant ketone body present in the blood and urine in ketosis.

Utilization of Ketone Bodies in Extrahepatic Tissues

While the liver is equipped with an active enzymatic mechanism for the production of acetoacetate from acetoacetyl-CoA, acetoacetate once formed cannot be reactivated directly in the liver except in the cytosol, where it is a precursor in cholesterol synthesis, a much less active pathway. This accounts for the net production of ketone bodies by the liver.

Two reactions take place in extrahepatic tissues. These activate acetoacetate to acetoacetyl-CoA. One mechanism involves succinyl-CoA and the enzyme **succinyl-CoA-acetoacetate-CoA transferase.** Acetoacetate reacts with succinyl-CoA, the CoA being transferred to form acetoacetyl-CoA and succinate.

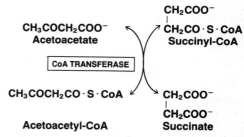

The other reaction involves the activation of acetoacetate with ATP in the presence of CoA catalyzed by **acetoacetyl-CoA synthetase.**

$$\underset{\text{Acetoacetate}}{CH_3COCH_2COO^-} + ATP + CoA \cdot SH \xrightarrow{\boxed{\begin{array}{c}\text{ACETOACETYL-CoA}\\\text{SYNTHETASE}\end{array}}}$$

$$\underset{\text{Acetoacetyl-CoA}}{CH_3COOH_2CO \cdot S \cdot CoA} + AMP + PP_i$$

D(−)-3-Hydroxybutyrate may be activated directly in extrahepatic tissues by a synthetase; however, conversion to acetoacetate with D(−)-3-hydroxybutyrate dehydrogenase and NAD⁺, followed by activation to acetoacetyl-CoA, is the more important route leading to its further metabolism. The acetoacetyl-CoA formed by these reactions is split to acetyl-CoA by thiolase and oxidized in the citric acid cycle as shown in Fig 28–4.

Ketone bodies are oxidized in extrahepatic tissues proportionately to their concentration in the blood. They are also oxidized in preference to glucose and to FFA. If the blood level is raised, oxidation of ketone bodies increases until, at a concentration of approximately 70 mg/dL, they saturate the oxidative machinery. When this occurs, a large proportion of the oxygen consumption of the animal may be accounted for by the oxidation of ketone bodies.

Most of the evidence suggests that **ketonemia is due to increased production of ketone bodies** by the liver rather than to a deficiency in their utilization by extrahepatic tissues. However, the results of experiments on depancreatized rats support the possibility that ketosis in the severe diabetic may be enhanced by a reduced ability to catabolize ketone bodies.

In moderate ketonemia, the loss of ketone bodies via the urine is only a few percent of the total ketone body production and utilization. Since there are renal threshold-like effects (there is not a true threshold) that vary between species and individuals, measurement of the ketonemia, not the ketonuria, is the preferred method of assessing the severity of ketosis.

While acetoacetate and D(−)-3-hydroxybutyrate are readily oxidized by extrahepatic tissues, acetone is difficult to oxidize in vivo.

Regulation of Ketogenesis

Three crucial steps can be identified in which control may be exercised in regulating ketogenesis: (1) Ketosis does not occur in vivo unless there is a concomitant rise in the level of circulating free fatty acids that arise from lipolysis of triacylglycerol in adipose tissue. Fatty acids are the precursors of ketone bodies in the liver. The liver, both in fed and in fasting conditions, has the ability to extract about 30% or more of the free fatty acids passing through it, so that at high concentrations of free fatty acids the flux passing into the liver is substantial. Therefore, the factors regulating mobilization of free fatty acids from adipose tissue are important in controlling ketogenesis (Fig 28–6). (2) One of 2 fates awaits the free fatty acids upon uptake and after they are activated to acyl-CoA: They are **esterified** mainly to triacylglycerol and phospholipid, or they are **β-oxidized** to acetyl-CoA. (3) In turn, acetyl-CoA is oxidized in the citric acid cycle, or it enters the pathway of ketogenesis to form ketone bodies.

Among several possible factors, the capacity for esterification as an antiketogenic factor depends on the availability of precursors in the liver to supply sufficient glycerol 3-phosphate. However, the availability of glycerol 3-phosphate does not limit esterification in fasting perfused livers. Whether the availability of glycerol 3-phosphate in the liver is ever rate-limiting on esterification is not clear; neither is there critical information on whether the in vivo activities of the enzymes involved in esterification are rate-limiting. It does not seem that they are, since neither free fatty acids nor any intermediates in their pathway of esterification to triacylglycerol (see Fig 25–1) ever accumulate in the liver. Phosphatidate phosphohydro-

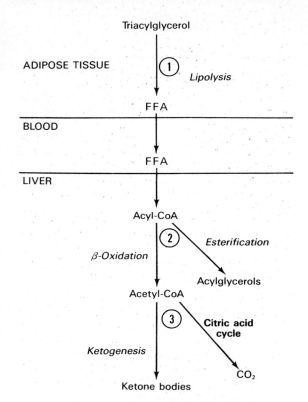

Figure 28–6. Regulation of ketogenesis. 1–3, three crucial steps in the pathway of metabolism of free fatty acids (FFA) that determine the magnitude of ketogenesis.

lase increases in activity in livers in which extra triacylglycerol synthesis is taking place.

By use of the perfused liver, it has been shown that livers from fed rats esterify considerably more ^{14}C-free fatty acids than livers from fasted rats, and the balance not esterified in the livers from fasted rats is oxidized to either $^{14}CO_2$ or ^{14}C-ketone bodies. These results may be explained by the fact that **carnitine palmitoyltransferase I** activity in the mitochondrial membrane regulates the entry of long-chain acyl groups into mitochondria prior to β-oxidation (see Fig 23–1). Its activity is low in the fed state, when fatty acid oxidation is depressed, and high in fasting, when fatty acid oxidation increases. McGarry and Foster (1980) have shown that **malonyl-CoA,** the initial intermediate in fatty acid biosynthesis (see Fig 23–5), which increases in concentration in the fed state, inhibits this enzyme, switching off β-oxidation. Thus, in the fed condition there is active lipogenesis and high [malonyl-CoA], which inhibits carnitine palmitoyltransferase I (Fig 28–7). Free fatty acids entering the liver cell in low concentrations are nearly all esterified to acylglycerols and transported out of the liver in VLDL. However, as the concentration of free fatty acids increases with the onset of starvation, acetyl-CoA carboxylase is inhibited and [malonyl-CoA] decreases, releasing the inhibition of carnitine palmitoyl-

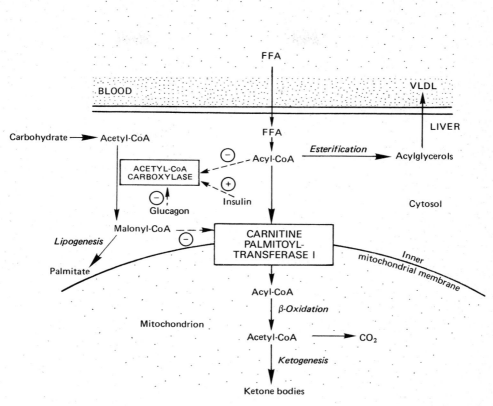

Figure 28–7. Regulation of long-chain fatty acid oxidation in the liver. FFA, free fatty acids; VLDL, very low density lipoprotein. Positive (⊕) and negative (⊖) regulatory effects are represented by broken lines and substrate flow by solid lines.

transferase and allowing more acyl-CoA to be oxidized. These events are reinforced in starvation by the **[insulin]/[glucagon] ratio,** which decreases, causing increased lipolysis in adipose tissue, the release of free fatty acids, and inhibition of acetyl-CoA carboxylase in the liver.

As the level of serum free fatty acids is raised, proportionately more free fatty acid is converted to ketone bodies and less is oxidized via the citric acid cycle to CO_2. The partition of acetyl-CoA between the ketogenic pathway and the pathway of oxidation to CO_2 is so regulated that the total free energy trapped in ATP which results from the oxidation of free fatty acids remains constant. **Complete oxidation of 1 mol of palmitate involves a net production of 129 mol of ATP** via β-oxidation and CO_2 production in the citric acid cycle (see Chapter 23), whereas **only 33 mol of ATP is produced when acetoacetate is the end product and only 21 mol when 3-hydroxybutyrate is the end product.** Thus, ketogenesis may be regarded as a mechanism that allows the liver to oxidize large quantities of fatty acids within an apparently tightly coupled system of oxidative phosphorylation, without increasing its total energy expenditure.

Several other hypotheses have been advanced to account for the diversion of fatty acid oxidation from CO_2 formation to ketogenesis. Theoretically, a fall in concentration of oxaloacetate, particularly within the mitochondria, could impair the ability of the citric acid cycle to metabolize acetyl-CoA. This fall may occur because of an increase in the [NADH]/[NAD$^+$] ratio caused by increased β-oxidation. Krebs has suggested that since oxaloacetate is also on the main pathway of gluconeogenesis, enhanced gluconeogenesis leading to a fall in the level of oxaloacetate may be the cause of the severe forms of ketosis found in diabetes and the ketosis of cattle. Utter and Keech have shown that pyruvate carboxylase, which catalyzes the conversion of pyruvate to oxaloacetate, is activated by acetyl-CoA. Consequently, when there are significant amounts of acetyl-CoA, there should be sufficient oxaloacetate to initiate the condensing reaction of the citric acid cycle.

In summary, ketosis arises as a result of a deficiency in available carbohydrate. This has the following actions in fostering ketogenesis (Figs 28–6 and 28–7). (1) It causes an imbalance between esterification and lipolysis in adipose tissue, with consequent release of free fatty acids into the circulation. Free fatty acids are the principal substrates for ketone body formation in the liver, and therefore all factors, metabolic or endocrine, affecting the release of free fatty acids from adipose tissue influence ketogenesis. (2) Upon entry of free fatty acids into the liver, the balance between their esterification and oxidation is governed by carnitine palmitoyltransferase I, whose activity is increased indirectly by the concentration of free fatty acids and the hormonal state of the liver. (3) As more fatty acid is oxidized, more forms ketone bodies and less forms CO_2, regulated in such a manner that the total ATP production remains constant.

Ketosis in Vivo

The ketosis that occurs in starvation and fat feeding is relatively mild compared with the condition encountered in uncontrolled diabetes mellitus, pregnancy toxemia of ewes, or ketosis of lactating cattle. The main reason appears to be that in the severe conditions carbohydrate is even less available to the tissues than in the mild conditions. Thus, in the milder forms of diabetes mellitus, in fat feeding, and in chronic starvation, glycogen is present in the liver in variable amounts, and free fatty acid levels are lower, which probably accounts for the less severe ketosis associated with these conditions.

In ketosis of ruminants, there is a severe drain of glucose from the blood owing to excessive fetal demands or the demands of heavy lactation (Fig 28–8). Extreme hypoglycemia results, coupled with negligible amounts of glycogen in the liver. Ketosis in these conditions tends to be severe. As hypoglycemia develops, the secretion of insulin diminishes, allowing not only less glucose utilization but also enhancement of lipolysis in adipose tissue.

In diabetes mellitus, the lack (or relative lack) of insulin probably affects adipose tissue more than any other tissue, because of its extreme sensitivity to this hormone. As a result, free fatty acids are released in quantities that give rise to plasma free fatty acid levels more than twice those in fasting normal subjects. Many changes also occur in the activity of enzymes within the liver, and these changes enhance both the rate of gluconeogenesis and the rate of transfer of glucose to the blood despite high levels of circulating glucose.

INTERCONVERSION OF MAJOR FOODSTUFFS
(See Fig 28–9.)

That animals may be fattened on a predominantly carbohydrate diet demonstrates the ease of conversion of carbohydrate into fat. A most significant reaction in this respect is the conversion of pyruvate to acetyl-CoA, as acetyl-CoA is the starting material for the synthesis of long-chain fatty acids. However, **the pyruvate dehydrogenase reaction is essentially nonreversible,** which prevents the direct conversion of acetyl-CoA, formed from the oxidation of fatty acids,

to pyruvate. There cannot be a net conversion of acetyl-CoA to oxaloacetate via the citric acid cycle, since one molecule of oxaloacetate is required to condense with acetyl-CoA and only one molecule of oxaloacetate is regenerated. For similar reasons, **there cannot be a net conversion of fatty acids having an even number of carbon atoms** (which form acetyl-CoA) **to glucose or glycogen.**

Only the terminal 3-carbon portion of a fatty acid having an odd number of carbon atoms is glycogenic, as this portion of the molecule will form **propionate** upon oxidation. Nevertheless, it is possible for labeled carbon atoms of fatty acids to be found ultimately in glycogen after traversing the citric acid cycle: This is because oxaloacetate is an intermediate both in the citric acid cycle and in the pathway of gluconeogenesis. Many of the carbon skeletons of the nonessential amino acids can be produced from carbohydrate via the citric acid cycle and transamination. By reversal of these processes, glycogenic amino acids yield carbon skeletons that are either members or precursors of the citric acid cycle. They are therefore readily converted by gluconeogenic pathways to glucose and glycogen. The ketogenic amino acids give rise to acetoacetate, which will in turn be metabolized as ketone bodies, forming acetyl-CoA in extrahepatic tissues.

For the same reasons that it is not possible for a net conversion of fatty acids to carbohydrate to occur, it is not possible for a net conversion of fatty acids to glucogenic amino acids to take place. Neither is it possible to reverse the pathways of breakdown of ketogenic amino acids, all of which fall into the category of "essential amino acids." Conversion of the carbon skeletons of glucogenic amino acids to fatty acids is possible, either by formation of pyruvate and acetyl-CoA or by reversal of nonmitochondrial reactions of the citric acid cycle from α-ketoglutarate to citrate followed by the action of ATP-citrate lyase to give acetyl-CoA (see Chapter 23). However, under most natural conditions, eg, starvation, a net breakdown of protein and amino acids is usually accompanied by a net breakdown of fat. The net conversion of amino acids to fat is therefore not a significant process except possibly in animals receiving a high-protein diet.

THE ECONOMICS OF CARBOHYDRATE & LIPID METABOLISM IN THE WHOLE BODY

The Necessity for Glucose

Many of the details of the interplay between carbohydrate and lipid metabolism in various tissues have been described. The conversion of glucose to fat is a process that occurs readily under conditions of optimal nutritional intake. With the exception of the glycerol moiety, fat (triacylglycerol) cannot give rise to a net

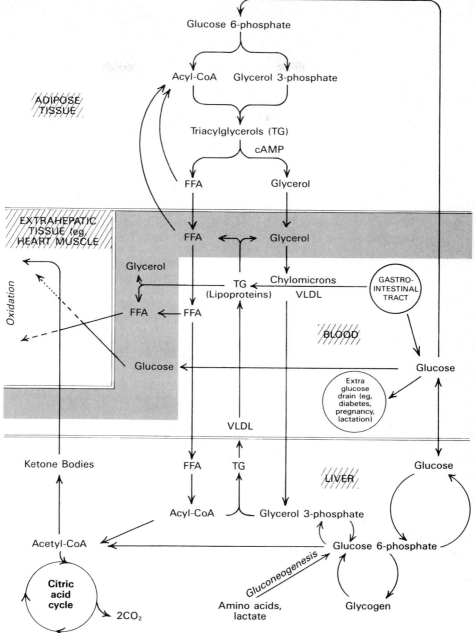

Figure 28–8. Metabolic interrelationships between adipose tissue, the liver, and extrahepatic tissues. Stippled area, lipoprotein lipase region of capillary wall; FFA, free fatty acids, VLDL, very low density lipoproteins.

formation of glucose, because of the irreversible nature of the oxidative decarboxylation of pyruvate to acetyl-CoA (see above). Certain tissues and cell types, including the central nervous system and the erythrocytes, are much more dependent upon a continual supply of glucose than others. A minimal supply of glucose is probably necessary in extrahepatic tissues to maintain the integrity of the citric acid cycle. In addi-

tion, glucose appears to be the main source of glycerol 3-phosphate in tissues devoid of glycerol kinase. There is a minimal and obligatory rate of glucose oxidation. Large quantities of glucose are also required for fetal nutrition and the synthesis of milk. Certain mechanisms, in addition to gluconeogenesis, safeguard essential supplies of glucose in times of shortage **by allowing other substrates to spare its oxidation.**

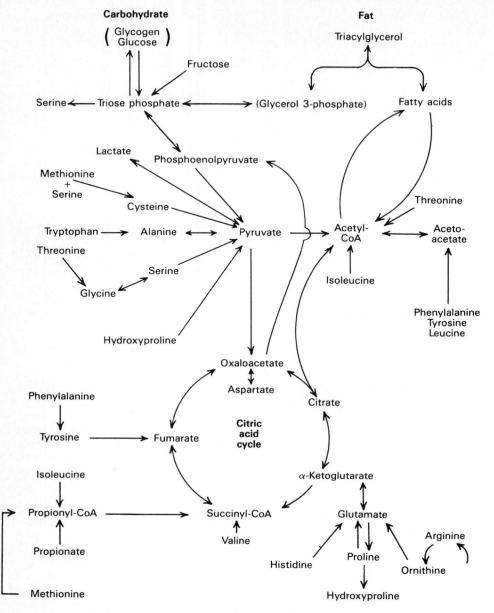

Figure 28-9. Interconversion of the major foodstuffs.

Preferential Utilization of Ketone Bodies & Free Fatty Acids

Ketone bodies and free fatty acids spare the oxidation of glucose in muscle by impairing its entry into the cell, its phosphorylation to glucose 6-phosphate, the phosphofructokinase reaction, and the oxidative decarboxylation of pyruvate. Oxidation of free fatty acids and ketone bodies causes an increase in the intracellular concentration of citrate that in turn inhibits phosphofructokinase. These observations and others, which demonstrated that acetoacetate was oxidized in the perfused heart preferentially to free fatty acids, justify the conclusion that under conditions of carbohydrate shortage available **fuels are oxidized in the following order of preference: (1) ketone bodies** (and probably other short-chain fatty acids, eg, acetate), (2) **free fatty acids,** and (3) **glucose.** This does not imply that any particular fuel is oxidized to the total exclusion of any other (Fig 28–8). However, these mechanisms are more important in tissues having a high capacity for aerobic oxidation of fatty acids, eg, heart and slow-twitch muscle, than in tissues with a low capacity, eg, fast-twitch muscle.

The combination of the effects of free fatty acids in

sparing glucose utilization in muscle and heart and the effect of the spared glucose in inhibiting free fatty acid mobilization in adipose tissue has been called the "glucose-fatty acid cycle."

STARVATION

In animals fed high-carbohydrate diets, fatty acid oxidation is spared. As the animal passes from the fed to the fasting condition, glucose availability becomes less, and liver glycogen is drawn upon in an attempt to maintain the blood glucose. The concentration of insulin in the blood decreases, and glucagon increases. As glucose utilization diminishes in adipose tissue and the inhibitory effect of insulin on lipolysis becomes less, fat is mobilized as free fatty acids and glycerol. The free fatty acids are transported to nonadipose tissues, where they are either oxidized or esterified. Glycerol joins the carbohydrate pool after activation to glycerol 3-phosphate, mainly in the liver and kidney. During this transition phase from the fully fed to the fully fasting state, endogenous glucose production (from amino acids and glycerol) does not keep pace with its utilization and oxidation, since the liver glycogen stores become depleted and blood glucose tends to fall. Thus, fat is mobilized at an ever-increasing rate, but in several hours the plasma free fatty acids and blood glucose stabilize at the fasting level (0.7–0.8 μmol/mL and 60–70 mg/dL, respectively). At this point, it must be presumed that in the whole animal the supply of glucose balances the obligatory demands for glucose utilization and oxidation. This is achieved by the increased oxidation of free fatty acids and ketone bodies, sparing the nonobligatory oxidation of glucose. This fine balance is disturbed in conditions that demand more glucose or in which glucose utilization is impaired and which therefore lead to further mobilization of fat. The provision of carbohydrate by adipose tissue, in the form of **glycerol,** is an important function, for it is only this source of carbohydrate together with that provided by **gluconeogenesis from protein** that can supply the fasting organism with the glucose needed for those processes which must utilize glucose. In prolonged starvation in humans, **gluconeogenesis from protein is diminished** owing to reduced release of amino acids, particularly alanine, from muscle. This coincides with adaptation of the brain to **replace approximately half of the glucose oxidized with ketone bodies.**

A feedback mechanism for controlling free fatty acid output from adipose tissue in starvation may operate as a result of the action of ketone bodies and free fatty acids to directly stimulate the pancreas to produce insulin. Under most conditions free fatty acids are mobilized in excess of oxidative requirements, since a large proportion are esterified, even during fasting. As the liver takes up and esterifies a considerable proportion of the free fatty acid output, it plays a regulatory role in removing excess free fatty acids from the circulation. When carbohydrate supplies are adequate, most of the influx is esterified and ultimately retransported from the liver as VLDL to be utilized by other tissues. However, in the face of an increased influx of free fatty acids, an alternative route, **ketogenesis,** is available that enables the liver to continue to retransport much of the influx of free fatty acids in a form which is readily utilized by extrahepatic tissues under all nutritional conditions.

Most of these principles are depicted in Fig 28–8. It will be noted that there is a carbohydrate cycle involving release of glycerol from adipose tissue and its conversion in the liver to glucose, followed by its transport back to adipose tissue to complete the cycle. The other cycle, a lipid cycle, involves release of free fatty acid by adipose tissue, its transport to and esterification in the liver, and retransport as VLDL back to adipose tissue.

REFERENCES

Cohen P: *Control of Enzyme Activity,* 2nd ed. Chapman & Hall, 1983.

Hue L, Van de Werve G (editors): *Short-Term Regulation of Liver Metabolism.* Elsevier/North Holland, 1981.

Laker ME, Mayes PA: Investigations into the direct effects of insulin on hepatic ketogenesis, lipoprotein secretion and pyruvate dehydrogenase activity. *Biochim Biophys Acta* 1984; **795:**4.

Mayes PA, Laker ME: Regulation of ketogenesis in the liver. *Biochem Soc Trans* 1981;**9:**339.

McGarry JD, Foster DW: Regulation of hepatic fatty acid oxidation and ketone body production. *Annu Rev Biochem* 1980; **49:**395.

Siess EA, Kientsch-Engel RI, Wieland OH: Concentration of free oxaloacetate in the mitochondrial compartment of isolated liver cells. *Biochem J* 1984;**218:**171.

Wakil SJ, Stoops JK, Joshi VC: Fatty acid synthesis and its regulation. *Annu Rev Biochem* 1983;**52:**537.

Zorzano A et al: Effects of starvation and exercise on concentrations of citrate, hexose phosphates and glycogen in skeletal muscle and heart: Evidence for selective operation of the glucose-fatty acid cycle. *Biochem J* 1985;**232:**585.

Section III.
Metabolism of Proteins & Amino Acids

Biosynthesis of Amino Acids

29

Victor W. Rodwell, PhD

INTRODUCTION

We often refer to nutritionally essential amino acids as "essential" or "indispensable" and to nutritionally nonessential amino acids as "nonessential" or "dispensable" (Table 29–1). While in a nutritional context these terms are correct, they obscure the biologically essential nature of all 20 amino acids. It might be argued that the nutritionally nonessential amino acids are more important to the cell than the nutritionally essential ones, since organisms (eg, humans) have evolved that lack the ability to manufacture the latter but not the former group.

The existence of nutritional requirements suggests that dependence on an external supply of a required intermediate can be of greater survival value than the ability to biosynthesize it. If a specific intermediate is present in the food, an organism that can synthesize it is reproducing and transferring to future generations genetic information of negative survival value. The survival value is negative rather than nil because ATP and nutrients are used to synthesize unnecessary DNA. The number of enzymes required by prokaryotic cells to synthesize the nutritionally essential amino acids is large relative to the number of enzymes required to synthesize the nutritionally nonessential amino acids (Table 29–2). This suggests that there is a survival advantage in retaining the ability to manufacture "easy" amino acids while losing the ability to make "difficult" amino acids.

BIOMEDICAL IMPORTANCE

Medical implications of the material in this chapter relate to amino acid deficiency states that can result if any of the nutritionally essential amino acids are omitted from the diet or are present in inadequate amounts. Since certain grains are relatively poor sources of tryptophan and lysine, in regions where the diet relies heavily on these grains for total protein and is unsupplemented by protein sources such as milk, fish, or

Table 29–1. Amino acid requirements of humans.

Nutritionally Essential	Nutritionally Nonessential
Arginine*	Alanine
Histidine*	Asparagine
Isoleucine	Aspartate
Leucine	Cysteine
Lysine	Glutamate
Methionine	Glutamine
Phenylalanine	Glycine
Threonine	Hydroxyproline†
Tryptophan	Hydroxylysine†
Valine	Proline
	Serine
	Tyrosine

*"Nutritionally semiessential." Synthesized at rates inadequate to support growth of children.
†Not necessary for protein synthesis but formed during posttranslational processing of collagen.

Table 29–2. Enzymes required for the synthesis of amino acids from amphibolic intermediates.

Number of Enzymes Required to Synthesize:			
Nutritionally Essential		**Nutritionally Nonessential**	
Arg*	7	Ala	1
His	6	Asp	1
Thr	6	Asn†	1
Met	5 (4 shared)	Glu	1
Lys	8	Gln*	1
Ile	8 (6 shared)	Hyl¶	1
Val	1 (7 shared)	Hyp#	1
Leu	3 (7 shared)	Pro*	3
Phe	10	Ser	3
Trp	5 (8 shared)	Gly‡	1
	59	Cys§	2
		Tyr‖	1
			17

*From Glu. †From Asp. ¶From Lys. #From Pro. ‡From Ser. §From Ser plus S^{2-}. ‖From Phe.

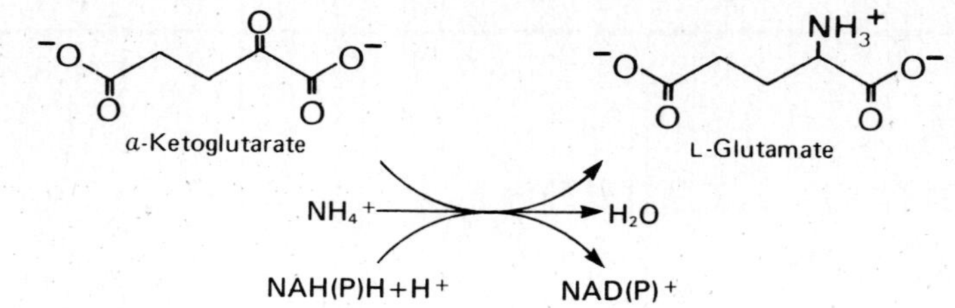

Figure 29–1. The glutamate dehydrogenase reaction. Reductive amination of α-ketoglutarate by NH_4^+ proceeds at the expense of NAD(P)H.

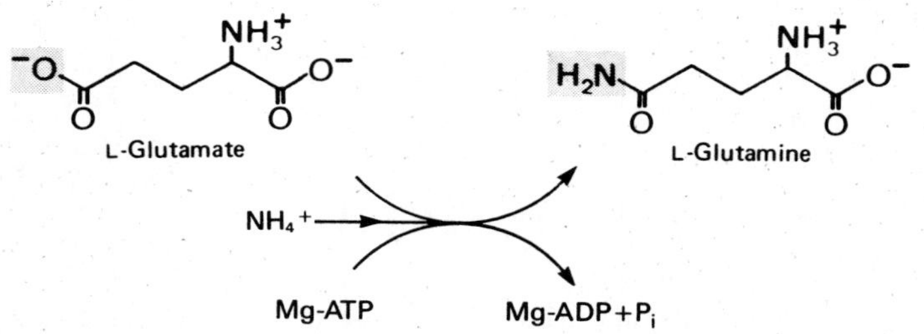

Figure 29–2. The glutamine synthetase reaction.

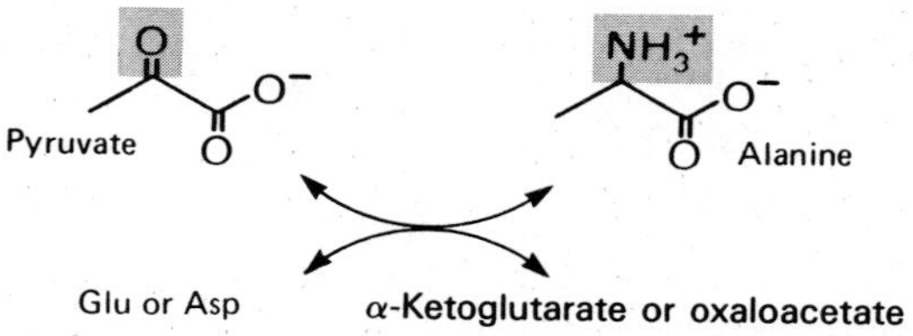

Figure 29–3. Formation of alanine by transamination of pyruvate. The amino donor may be glutamate or aspartate. The other product thus is α-ketoglutarate or oxaloacetate.

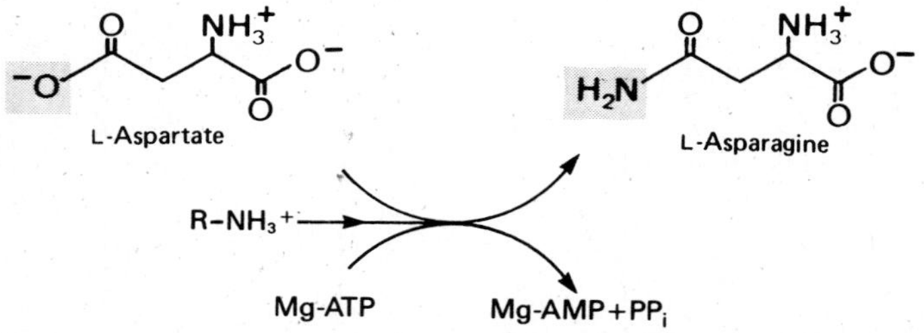

Figure 29–4. The asparagine synthetase reaction. Note similarities to and differences from the glutamine synthetase reaction (Fig 29–2). The nature of the amino donor ($R–NH_3^+$) differs depending on the life form considered.

meat, dramatic deficiency states may be observed. Kwashiorkor and marasmus are endemic in certain regions of West Africa. Kwashiorkor results when a child is weaned onto a starchy diet poor in protein. In marasmus, both caloric intake and specific amino acids are deficient.

BIOSYNTHESIS OF NUTRITIONALLY NONESSENTIAL AMINO ACIDS

Of the 12 nutritionally nonessential amino acids (Table 29–1), 9 are formed from amphibolic intermediates. The remaining 3 (Cys, Tyr, Hyl) are formed from nutritionally essential amino acids.

Glutamate dehydrogenase, glutamine synthetase, and transaminases occupy central positions in amino acid biosynthesis. Their combined effect is to catalyze transformation of inorganic ammonium ion into the organic α-amino nitrogen of various amino acids.

Glutamate

Reductive amination of α-ketoglutarate is catalyzed by glutamate dehydrogenase (Fig 29–1). In addition to forming L-glutamate from the amphibolic intermediate α-ketoglutarate, this reaction constitutes a key first step in the biosynthesis of many additional amino acids.

Glutamine

Biosynthesis of glutamine from glutamate is catalyzed by glutamine synthetase (Fig 29–2). The reaction exhibits both similarities to and differences from the glutamate dehydrogenase reaction. Both "fix" inorganic nitrogen—one into amino and the other into amide linkage. Both reactions are coupled to highly exergonic reactions—for glutamate dehydrogenase the oxidation of NAD(P)H and for glutamine synthetase the hydrolysis of ATP.

Alanine & Aspartate

Transamination of pyruvate forms L-alanine, and transamination of oxaloacetate forms L-aspartate (Fig 29–3). Transfer of the α-amino group of glutamate to these amphibolic intermediates illustrates the ability of a transaminase to channel ammonium ion, via glutamate, into the α-amino nitrogen of amino acids.

Asparagine

Formation of asparagine from aspartate, catalyzed by asparagine synthetase (Fig 29–4), resembles glutamine synthesis (Fig 29–2). However, since the mammalian enzyme uses glutamine rather than ammonium ion as the nitrogen source, mammalian asparagine synthetase does not "fix" inorganic nitrogen. By contrast, bacterial asparagine synthetases do use ammonium ion and hence do "fix" nitrogen. As for other reactions in which PP_i is formed, hydrolysis of PP_i to P_i by pyrophosphatase ensures that the reaction is strongly favored energetically.

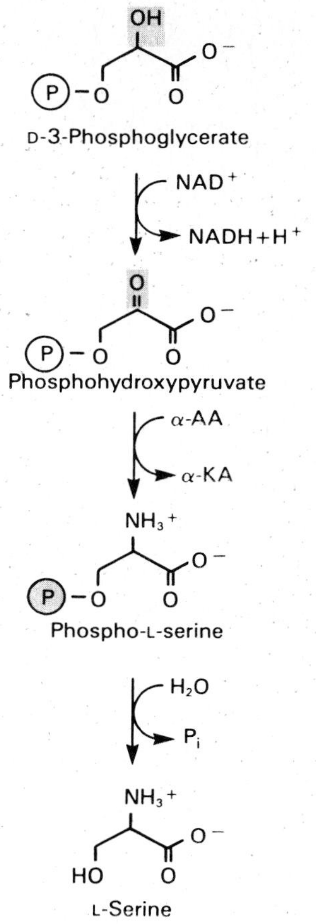

Figure 29–5. Serine biosynthesis. α-AA, α-amino acids; α-KA, α-keto acids.

Serine

Serine is formed from the glycolytic intermediate D-3-phosphoglycerate (Fig 29–5). The α-hydroxyl group is reduced to an oxo group by NAD$^+$, then transaminated, forming phosphoserine. This is then dephosphorylated, forming serine.

Glycine

Synthesis of glycine in mammalian tissues can occur in several ways. Liver cytosol contains glycine transaminases that catalyze the synthesis of glycine from glyoxylate and glutamate or alanine. Unlike most transaminase reactions, this strongly favors glycine synthesis. Two additional important mammalian routes for glycine formation are from choline (Fig 29–6) and from serine via the serine hydroxymethyl-transferase reaction (Fig 29–7).

Proline

In mammals and some other life forms, proline is formed from glutamate by reversal of the reactions of proline catabolism (Fig 29–8).

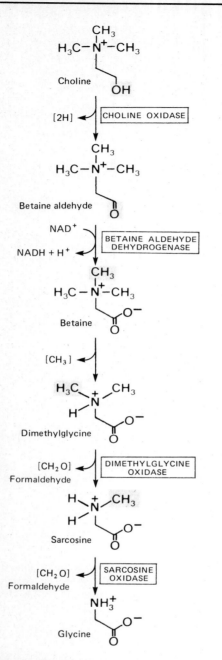

Figure 29-6. Formation of glycine from choline.

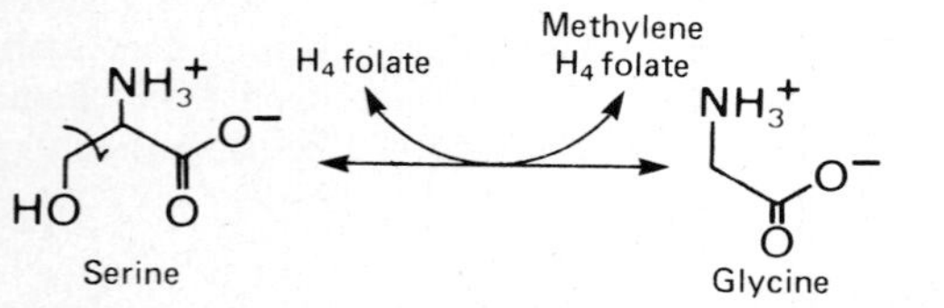

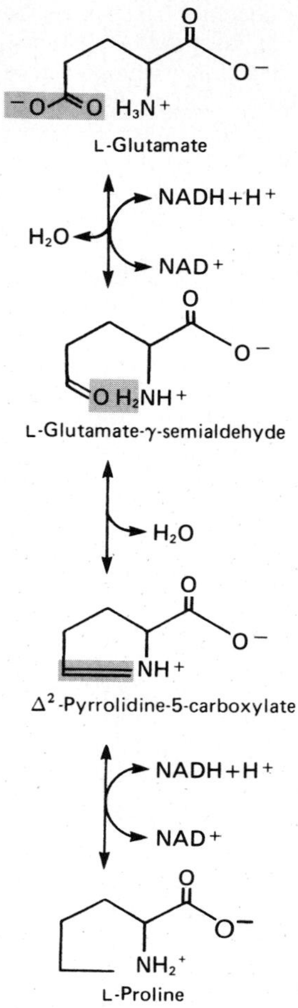

Figure 29-8. Biosynthesis of proline from glutamate by reversal of the reactions of proline catabolism.

Figure 29-7. The serine hydroxymethyltransferase reaction. The reaction is freely reversible. H₄folate, tetrahydrofolate.

Hydroxyproline

Since proline serves as a precursor of hydroxyproline, proline and hydroxyproline belong to the glutamate family of amino acids. Although both 3- and 4-hydroxyprolines occur in mammalian tissues, what follows refers solely to *trans*-4-hydroxyproline.

Hydroxyproline, like hydroxylysine, is almost exclusively associated with collagen, the most abundant protein of mammalian tissues. Collagen contains about one-third glycine and one-third proline and hydroxyproline. Hydroxyproline, which accounts for many of the amino acid residues of collagen, stabilizes the collagen triple helix to digestion by proteases. Unlike the hydroxyl groups of hydroxylysine, which serve as sites for attachment of galactosyl and glucosyl residues, the hydroxyl groups of collagen hydroxyproline are unsubstituted.

A unique feature of hydroxyproline and hydroxylysine metabolism is that the preformed amino acids, as they may occur in ingested food protein, are not incorporated into collagen. There is no tRNA capable of

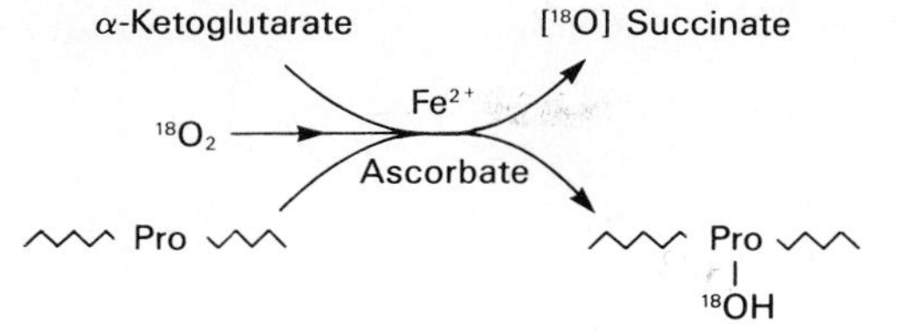

Figure 29–9. The prolyl hydroxylase reaction. The substrate is a proline-rich peptide. During the course of the reaction, molecular oxygen is incorporated into both succinate and proline (shown by the use of heavy oxygen, $^{18}O_2$).

accepting hydroxyproline or hydroxylysine and inserting them into an elongating polypeptide chain. Dietary proline is, however, a precursor of collagen hydroxyproline, and dietary lysine a precursor of collagen hydroxylysine. Hydroxylation of proline or lysine is catalyzed by prolyl hydroxylase or by lysyl hydroxylase, enzymes associated with the microsomal fraction of many tissues (skin, liver, lung, heart, skeletal muscle,

and granulating wounds). These enzymes are peptidyl hydroxylases, since hydroxylation only occurs subsequent to incorporation of proline or lysine into polypeptide linkage (see Chapter 55).

Both hydroxylases are mixed function oxygenases that require, in addition to substrate, molecular O_2, ascorbate, Fe^{2+}, and α-ketoglutarate. Prolyl hydroxylase has been more extensively studied, but lysyl hydroxylase appears to be an entirely analogous enzyme. For every mole of proline hydroxylated, 1 mol of α-ketoglutarate is decarboxylated to succinate. During this process, one atom of molecular O_2 is incorporated into proline and one into succinate (Fig 29–9).

Cysteine

Cysteine, while not itself nutritionally essential, is formed from methionine (nutritionally essential) and serine (nutritionally nonessential). Methionine is first converted to homocysteine via S-adenosylmethionine and S-adenosylhomocysteine (see Chapter 31). Conversion of homocysteine and serine to cysteine and homoserine is shown in Fig 29–10.

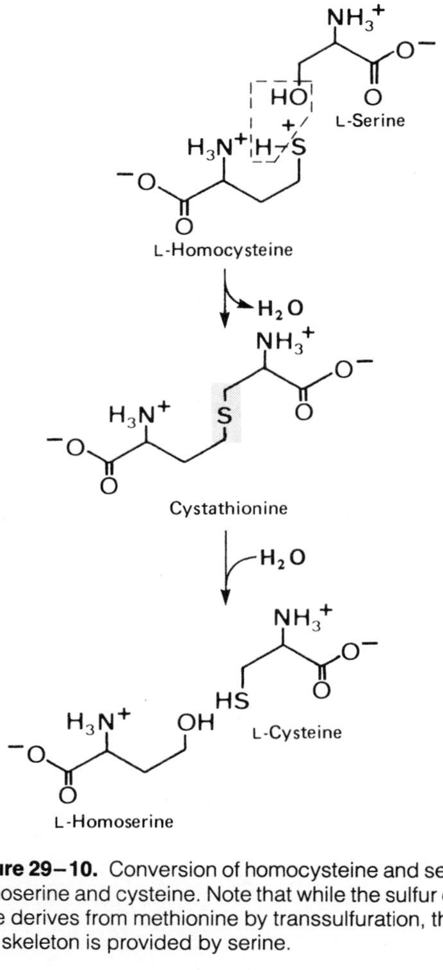

Tyrosine

Tyrosine is formed from phenylalanine by the reaction catalyzed by phenylalanine hydroxylase (Fig 29–11). Thus, whereas phenylalanine is a nutrition-

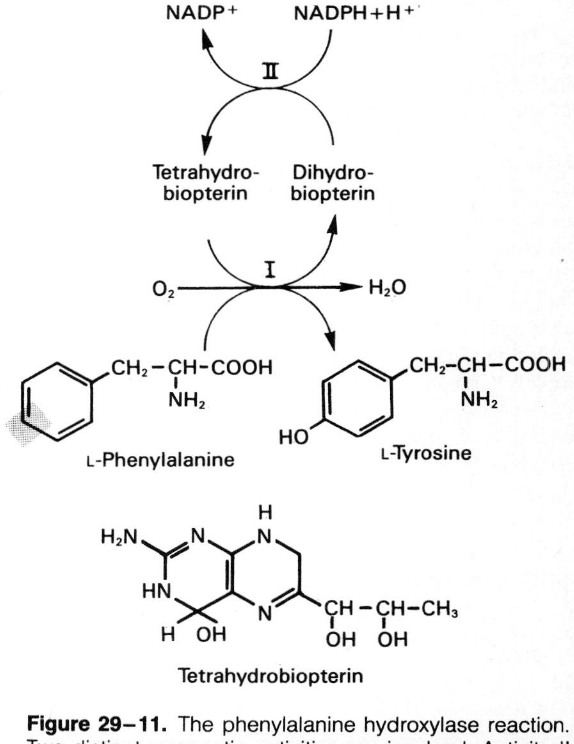

Figure 29–10. Conversion of homocysteine and serine to homoserine and cysteine. Note that while the sulfur of cysteine derives from methionine by transsulfuration, the carbon skeleton is provided by serine.

Figure 29–11. The phenylalanine hydroxylase reaction. Two distinct enzymatic activities are involved. Activity II catalyzes reduction of dihydrobiopterin by NADPH, and activity I the reduction of O_2 to H_2O and of phenylalanine to tyrosine. This reaction is associated with several defects of phenylalanine metabolism discussed in Chapter 31.

ally essential amino acid, tyrosine is not—provided the diet contains adequate quantities of phenylalanine. The reaction is not reversible, so tyrosine cannot replace the nutritional requirement for phenylalanine. The **phenylalanine hydroxylase complex** is a mixed-function oxygenase present in mammalian liver but absent from other tissues. The reaction involves incorporation of one atom of molecular oxygen into the para position of phenylalanine while the other atom is reduced, forming water (Fig 29–11). The reducing power, supplied ultimately by NADPH, is immediately provided as **tetrahydrobiopterin,** a pteridine resembling folic acid.

Hydroxylysine

5-Hydroxylysine (α,ϵ-diamino-δ-hydroxycaproate) is present in collagen but absent from most other mammalian proteins. Collagen hydroxylysine arises directly from dietary lysine, not dietary hydroxylysine. Before lysine is hydroxylated, it must first be incorporated into peptide linkage. Hydroxylation of the lysyl peptide is then catalyzed by lysyl hydroxylase, a mixed-function oxidase analogous to prolyl hydroxylase (Fig 29–9).

BIOSYNTHESIS OF NUTRITIONALLY ESSENTIAL AMINO ACIDS

Biosynthesis of nutritionally essential amino acids by bacteria from glutamate, aspartate, or other amphibolic intermediates is outlined below. These reactions do not occur in mammalian tissues.

Arginine

Arginine, a nutritionally essential amino acid for growing humans, can be synthesized by rats but not in quantities sufficient to permit normal growth. Microorganisms biosynthesize arginine from glutamate, via N-acetylated intermediates. One intermediate, N-acetylglutamate-γ-semialdehyde, is also a precursor of proline in bacteria. In humans and other animals, however, proline is formed from glutamate.

Aspartate is the precursor of a family of amino acids that includes lysine, methionine, threonine, and isoleucine (see Fig 10–6). The regulatory implications of this relationship in bacteria are discussed in Chapter 10.

Methionine & Threonine

Following conversion of aspartate-β-semialdehyde to homoserine, the pathways for methionine and threonine biosynthesis diverge. The interconversion of homoserine and methionine is discussed in Chapter 31.

Lysine

Bacteria form lysine from aspartate-β-semialdehyde via condensation with pyruvate. The dihydropicolinate formed serves, in addition, a role in spore formation in certain spore-forming bacteria, and the diaminopimelate performs a role in bacterial cell wall synthesis.

Lysine biosynthesis in yeast starts from α-ketoglutarate and acetyl-CoA and utilizes a series of reactions analogous to those of the citric acid cycle but catalyzed by a set of enzymes with slightly different substrate specificities.

Leucine, Valine, & Isoleucine

While leucine, valine, and isoleucine are all nutritionally essential amino acids for humans and other higher animals, mammalian tissues do contain transaminases that reversibly catalyze interconversion of all 3 amino acids with their corresponding α-keto acids (see Chapter 31). This explains the ability of the appropriate keto acids to replace their amino acids in the diet.

Histidine

Histidine, like arginine, is nutritionally semiessential. Adult humans and adult rats have been maintained in nitrogen balance for short periods in the absence of histidine. The growing animal does, however, require histidine in the diet. If studies were to be carried on for longer periods, it is probable that a requirement for histidine in adult human subjects would also be elicited.

Biosynthesis starts with 5-phosphoribosyl-1-pyrophosphate (PRPP), which condenses with ATP, forming N'-(5-phosphoribosyl)-ATP. This reaction thus closely resembles the initial reaction of purine biosynthesis.

REFERENCES

Burnstein P: The biosynthesis of collagen. *Annu Rev Biochem* 1974;**43**:567.

Calvo JM, Fink GR: Regulation of biosynthetic pathways in bacteria. *Annu Rev Biochem* 1971;**40**:943.

Cardinale GJ, Udenfriend S: Prolyl hydroxylase. *Adv Enzymol* 1974;**41**:245.

Rosenberg LE, Scriver CR: Disorders of amino acid metabolism. Chapter 11 in: *Metabolic Control and Disease*. Bondy PK, Rosenberg LE (editors). Saunders, 1980.

Tyler B: Regulation of the assimilation of nitrogen compounds. *Annu Rev Biochem* 1978;**47**:1127.

Umbarger HE: Amino acid biosynthesis and its regulation. *Annu Rev Biochem* 1978;**47**:533.

Catabolism of Amino Acid Nitrogen

30

Victor W. Rodwell, PhD

INTRODUCTION

We shall consider how nitrogen is removed from amino acids and converted to urea and the medical problems that arise when there are defects in these reactions.

BIOMEDICAL IMPORTANCE

Ammonia, derived mainly from the α-amino nitrogen of amino acids, is potentially toxic to humans. The mechanisms by which ammonia causes toxicity are not fully understood. The body disposes of ammonia by converting it to the nontoxic compound urea. Normal operation of the metabolic pathway that converts ammonia to urea—the urea cycle—is essential for maintenance of health. In conditions in which liver function is seriously compromised—eg, in individuals with massive cirrhosis (where normal liver cells are replaced by fibroblasts and collagen) or severe hepatitis—ammonia accumulates in the blood and results in clinical signs and symptoms. In those few infants born with a deficiency in the activity of one of the enzymes of the urea cycle, proper treatment requires an understanding of the biochemistry of the formation of urea.

OVERALL VIEW

In a healthy human adult, normal protein turnover amounts to 1–2% of total body protein per day. This protein turnover results predominantly from degradation of muscle protein to amino acids. However, approximately 75–80% of the released amino acids are reutilized for new protein synthesis. The remainder are metabolized to nitrogenous waste and glucose, ketones, and/or carbon dioxide (Fig 30–1). The net daily loss of protein amounts to 30–40 g. Since approximately 16% of the atomic mass of proteins is nitrogen, 5–7 g of nitrogen is lost per day. To maintain a healthy steady state, the average adult requires 30–60 g of protein or the equivalent in amino acids per day, but the **quality is important.** Protein quality here refers to the concentration of essential amino acids in a food relative to their concentrations in protein molecules being synthesized. Regardless of their source, amino acids that are not immediately incorpo-

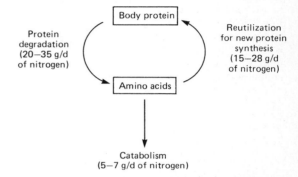

Figure 30–1. Quantitative relationships for protein and amino acid turnover.

rated into new protein are rapidly degraded; ie, **excess amino acids are not stored.** Consumption of excess amino acids thus is wasteful, since this surplus is catabolized to form energy, a function that carbohydrates and lipids can serve at a lower cost. This lower cost reflects the high energy requirement for fixing nitrogen.

Several terms describing the overall state of nitrogen metabolism are in wide nutritional and medical use. **Nitrogen balance** refers to the difference between the total nitrogen intake of a human (or other organism) and its total nitrogen loss. When more nitrogen is taken in than is excreted, the individual is said to be in **positive nitrogen balance.** Important examples of this are growth and pregnancy; both the healthy growing infant and the healthy pregnant woman are in positive nitrogen balance. Healthy adult humans are in **nitrogen equilibrium,** with intake of nitrogen matching output in feces and urine. In the state of **negative nitrogen balance,** output of nitrogen exceeds intake. Important examples of this are individuals receiving inadequate dietary sources of nitrogen (eg, in kwashiorkor), patients suffering from advanced cancer, and patients in certain postsurgical states.

Amino Acid Catabolism

Amino acids in excess of needs for protein biosynthesis cannot be stored, nor are they excreted as such. Amino groups of surplus amino acids are removed by transamination or oxidative deamination, and the carbon skeletons are converted to amphibolic intermediates. Some organisms (fishes) excrete free ammonia as the end product of nitrogen catabolism and are referred to as **ammonotelic.** Some organisms (birds and amphibians) excrete uric acid instead and are referred to as **uricotelic,** and some organisms (mammals) excrete urea and are referred to as **ureotelic.**

Ammonia is toxic to the central nervous system by mechanisms that are not fully understood but likely involve the reversal of glutamate dehydrogenase (discussed below) and consequent depletion of α-ketoglutarate. **Uric acid** and its salts are highly **insoluble** and precipitate in tissues and fluids when their concentrations exceed several milligrams per deciliter. Therefore, neither of these end products of nitrogen metabolism is well tolerated by higher organisms. Accordingly, humans and other mammals convert their nitrogenous waste to the **highly soluble, nontoxic compound urea.**

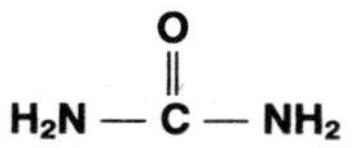

$$H_2N - \overset{\overset{\displaystyle O}{\|}}{C} - NH_2$$

Urea

The biosynthesis of urea will be divided for discussion into 4 stages: (1) transamination, (2) oxidative deamination, (3) ammonia transport, and (4) reactions of the urea cycle. Fig 30–2 relates these areas to overall catabolism of amino acid nitrogen. Although each stage also plays a role in amino acid biosynthesis (see Chapter 29), what follows is discussed from the viewpoint of amino acid catabolism.

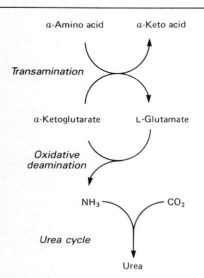

Figure 30–2. Overall flow of nitrogen in amino acid catabolism. Although the reactions shown are reversible, they are represented as being unidirectional to emphasize the direction of metabolic flow in mammalian amino acid catabolism.

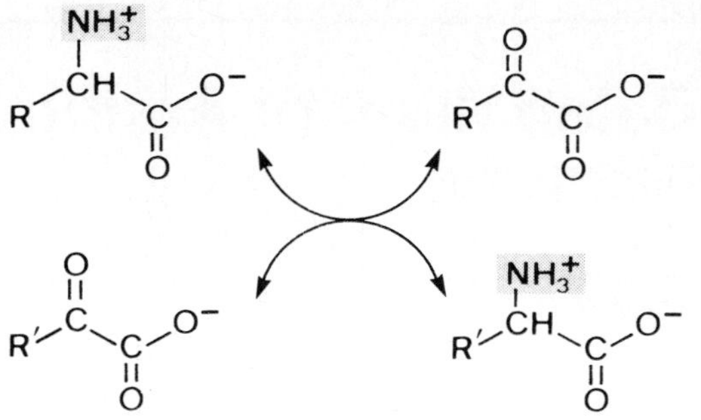

Figure 30–3. Transamination. The reaction is shown for 2 α-amino and 2 α-keto acids. Non-α-amino or carbonyl groups also participate in transamination, although this is relatively uncommon. The reaction is freely reversible with an equilibrium constant of about 1.

TRANSAMINATION

Transamination, catalyzed by enzymes termed **transaminases** or **aminotransferases,** interconverts a pair of amino acids and a pair of keto acids. These generally are α-amino and α-keto acids (Fig 30–3).

Pyridoxal phosphate forms an essential part of the active site of transaminases and of many other enzymes with amino acid substrates. In all pyridoxal phosphate-dependent reactions of amino acids, the initial step is formation of an enzyme-bound Schiff base intermediate. This intermediate, stabilized by interaction with a cationic region of the active site, can be rearranged in ways that include release of a keto acid with formation of enzyme-bound pyridoxamine phosphate. The bound amino form of the coenzyme can then form an analogous Schiff base intermediate with a keto acid. During transamination, bound coenzyme thus serves as a carrier of amino groups. Since the equilibrium constant for most transaminase reactions is close to unity, transamination is a freely reversible process. This permits transaminases to function in both amino acid catabolism and biosynthesis.

Two transaminases, alanine-pyruvate transaminase (**alanine transaminase**) and glutamate-α-keto-

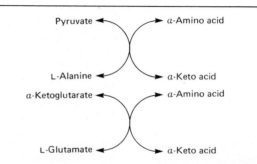

Figure 30–4. Alanine (*top*) and glutamate (*bottom*) transaminases.

glutarate transaminase (**glutamate transaminase**), present in most mammalian tissues, catalyze transfer of amino groups from most amino acids to form alanine (from pyruvate) or glutamate (from α-ketoglutarate) (Fig 30–4).

Each transaminase is specific for the specified pair of amino and keto acids as one pair of substrates but nonspecific for the other pair, which may be any of a wide variety of amino acids and their corresponding keto acids. Since alanine is also a substrate for glutamate transaminase, all of the amino nitrogen from amino acids that can undergo transamination can be concentrated in glutamate. This is important, because **L-glutamate is the only amino acid in mammalian tissues that undergoes oxidative deamination** at an appreciable rate. The formation of ammonia from α-amino groups thus occurs mainly via conversion to the α-amino nitrogen of L-glutamate.

Most (but not all) amino acids are substrates for transamination. Exceptions include lysine, threonine, and the cyclic imino acids, proline and hydroxyproline. Transamination is not restricted to α-amino groups. The δ-amino group of ornithine (but not the ϵ-amino group of lysine) is readily transaminated, forming glutamate-γ-semialdehyde (see Fig 31–3). Serum levels of transaminases are elevated in some disease states (see the Appendix).

OXIDATIVE DEAMINATION

Oxidative conversion of many amino acids to their corresponding α-keto acids occurs in mammalian liver and kidney. Although most of the activity toward L-α-amino acids is due to the coupled action of transaminases plus **L-glutamate dehydrogenase**, both L- and D-amino acid oxidase activities occur in mammalian liver and kidney tissue and are widely distributed in

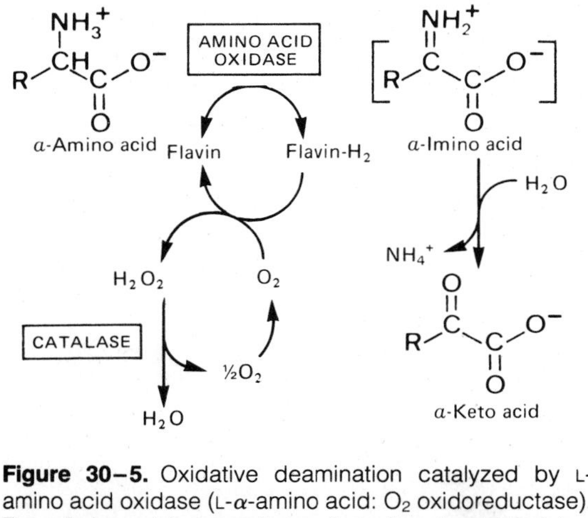

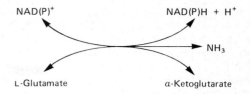

Figure 30–6. The L-glutamate dehydrogenase reaction. $NAD(P)^+$ means that either NAD^+ or $NADP^+$ can serve as cosubstrate. The reaction is reversible, but the equilibrium constant favors glutamate formation.

other animals and microorganisms. However, the physiologic function of L- and D-amino acid oxidase of mammalian tissue is not known.

Amino acid oxidases are **auto-oxidizable flavoproteins;** ie, the reduced FMN or FAD is reoxidized directly by molecular oxygen, forming hydrogen peroxide (H_2O_2) without participation of cytochromes or other electron carriers (Fig 30–5). The toxic product H_2O_2 is then split to O_2 and H_2O by **catalase,** which occurs widely in tissues, especially liver. Although the amino acid oxidase reactions are reversible, if catalase is absent the α-keto acid product is nonenzymatically decarboxylated by H_2O_2, forming a carboxylic acid with one less carbon atom. It is doubtful, however, whether this decarboxylation occurs to any great extent in intact human tissues.

In the amino acid oxidase reactions (Fig 30–5), the amino acid is first dehydrogenated by the flavoprotein of the oxidase, forming an α-imino acid. This spontaneously adds water, then decomposes to the corresponding α-keto acid with loss of the α-imino nitrogen as ammonium ion.

Mammalian L-amino acid oxidase, an FMN-flavoprotein, is restricted to kidney and liver tissue. Its activity is quite low, and it is essentially without activity toward glycine or the L-isomers of dicarboxylic or β-hydroxy-α-amino acids. It thus is not likely that this enzyme fulfills a major role in mammalian amino acid catabolism.

Mammalian D-amino acid oxidase, an FAD-flavoprotein of broad substrate specificity, occurs in the liver and kidney tissue of most mammals. D-Asparagine and D-glutamine are not oxidized, and glycine and the D-isomers of the acidic and basic amino acids are poor substrates. The physiologic significance of this enzyme in mammals is not known.

L-Glutamate Dehydrogenase

The amino groups of most amino acids ultimately are transferred to α-ketoglutarate by transamination, forming L-glutamate (Fig 30–2). Release of this nitrogen as ammonia is catalyzed by **L-glutamate dehydrogenase,** an enzyme of high activity widely distributed in mammalian tissues (Fig 30–6). Liver glutamate dehydrogenase is a regulated enzyme whose activity is affected by allosteric modifiers such as ATP, GTP, and NADH, which inhibit the enzyme,

Figure 30–5. Oxidative deamination catalyzed by L-amino acid oxidase (L-α-amino acid: O_2 oxidoreductase). The α-imino acid, shown in brackets, is not a stable intermediate.

and ADP, which activates the enzyme. Certain hormones appear also to influence glutamate dehydrogenase activity in vitro.

Glutamate dehydrogenase uses either NAD^+ or $NADP^+$ as cosubstrate. The reaction is reversible and functions in both amino acid catabolism and biosynthesis. It therefore functions not only to funnel nitrogen from glutamate to urea (catabolism) but also to catalyze **amination** of α-ketoglutarate by free ammonia (see Chapter 29).

FORMATION OF AMMONIA*

In addition to ammonia formed in the tissues, a considerable quantity is produced by intestinal bacteria from dietary protein and from urea present in fluids secreted into the gastrointestinal tract. This ammonia is absorbed from the intestine into the portal venous blood, which characteristically contains higher levels of ammonia than does systemic blood. Under normal circumstances, the liver promptly removes the ammonia from the portal blood, so that blood leaving the liver (and indeed all of the peripheral blood) is virtually ammonia-free. This is essential, since even minute quantities of ammonia are toxic to the central nervous system. The symptoms of **ammonia intoxication** include a peculiar flapping tremor, slurring of speech, blurring of vision, and, in severe cases, coma and death. These symptoms resemble those of the syndrome of hepatic coma which occurs when blood and, presumably, brain ammonia levels are elevated. Ammonia intoxication is assumed to be a factor in the etiology of hepatic coma. Therefore, treatment includes measures designed to reduce blood ammonia levels.

With severely impaired hepatic function or development of collateral communications between the portal and systemic veins (as may occur in cirrhosis), portal blood may bypass the liver. Ammonia may thus rise to toxic levels in the systemic blood. Surgically produced shunting procedures (Eck fistula, or other forms of portacaval shunts) are also conducive to ammonia intoxication, particularly after ingestion of protein or after gastrointestinal hemorrhage, which provides blood proteins to colonic bacteria.

The ammonia content of the blood in renal veins exceeds that in renal arteries, indicating that the kidneys produce ammonia and add it to the blood. However, the excretion into the urine of the ammonia produced by renal tubular cells constitutes a more significant aspect of renal ammonia metabolism. Ammonia production, an important renal tubular mechanism for regulation of acid-base balance and conservation of cations, is markedly increased in metabolic acidosis and depressed in alkalosis. This ammonia is derived, not from urea, but from intracellular amino

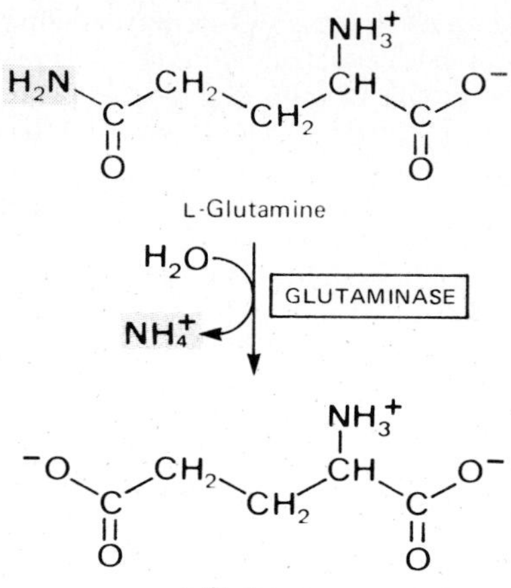

Figure 30–7. The glutaminase reaction proceeds essentially irreversibly in the direction of glutamate and NH_4^+ formation. Note that the amide nitrogen, not the α-amino nitrogen, is removed.

acids, particularly glutamine. Ammonia release is catalyzed by renal **glutaminase** (Fig 30–7).

TRANSPORT OF AMMONIA

Although ammonia may be excreted as ammonium salts—particularly in metabolic acidosis—the vast majority is excreted as urea, the principal nitrogenous component of urine. Ammonia, constantly produced in the tissues but present only in traces in peripheral blood (10–20 μg/dL), is rapidly removed from the circulation by the liver and converted to glutamate, to glutamine, or to urea.

Removal of ammonia via **glutamate dehydrogenase** was mentioned above. Formation of glutamine is catalyzed by **glutamine synthetase** (Fig 30–8), a mitochondrial enzyme present in highest quantities in renal tissue. Synthesis of the amide bond of glutamine is accomplished at the expense of hydrolysis of one equivalent of ATP to ADP and P_i. The reaction is thus strongly favored in the direction of glutamine synthesis (see also Chapter 29).

Liberation of the amide nitrogen of glutamine as ammonia occurs by hydrolytic removal of ammonia catalyzed by **glutaminase** (Fig 30–7). The glutaminase reaction, unlike the glutamine synthetase reaction, does not involve adenine nucleotides, strongly favors glutamate formation, and does not function in glutamine synthesis. Glutamine synthetase and glutaminase thus catalyze interconversion of free ammonium ion and glutamine (Fig 30–9) in a manner reminiscent of the interconversion of glucose and glucose 6-phosphate by glucokinase and glucose-6-phosphatase (see Chapter 17).

A reaction analogous to that catalyzed by glutaminase is catalyzed by **L-asparaginase** of animal, plant,

*Present at physiologic pH almost exclusively as ammonium ion, NH_4^+.

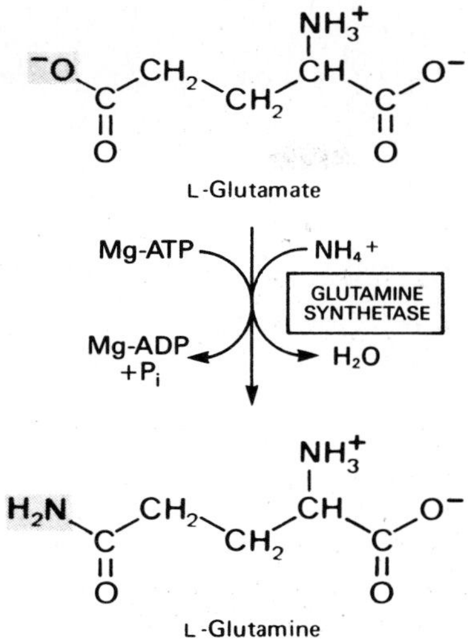

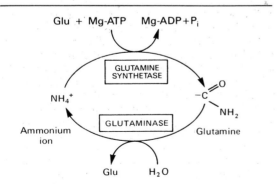

Figure 30-8. The glutamine synthetase reaction. The reaction strongly favors glutamine synthesis.

and microbial tissue. Asparaginase and glutaminase have both been investigated as antitumor agents, since certain tumors exhibit abnormally high requirements for glutamine and asparagine.

Whereas in brain the major mechanism for removal of ammonia is glutamine formation, in the liver the most important pathway is urea formation. Brain tissue can form urea, although this does not play a significant role in ammonia removal. Formation of glutamine in the brain must be preceded by synthesis of glutamate in the brain, because the supply of blood glutamate is inadequate in the presence of high levels of blood ammonia. The immediate precursor of glutamate is α-ketoglutarate. Thus, formation of glutamine from ammonia would rapidly deplete citric acid cycle

Figure 30-9. Interconversion of ammonia and of glutamine catalyzed by glutamine synthetase and glutaminase. Both reactions are strongly favored in the directions indicated by the arrows. Glutaminase thus serves solely for glutamine deamidation and glutamine synthetase solely for synthesis of glutamine from glutamate. Glu, glutamate.

intermediates unless they could be replaced by CO_2 fixation with conversion of pyruvate to oxaloacetate (see Chapter 17). A significant fixation of CO_2 into amino acids does indeed occur in the brain, presumably by way of the citric acid cycle, and after infusion of ammonia more oxaloacetate is diverted to the synthesis of glutamine (rather than to aspartate) via α-ketoglutarate.

INTERORGAN AMINO ACID EXCHANGE IN THE POSTABSORPTIVE STATE

The maintenance of steady-state concentrations of circulating plasma amino acids between meals (eg, after fasting overnight following supper) depends upon the net balance between release from endogenous protein stores and utilization by various tissues. Muscle accounts for the generation of greater than 50% of the total body pool of free amino acids, while liver is the site of the urea cycle enzymes necessary for disposal of nitrogenous waste. Thus, muscle and liver play a major role in determining the circulating levels and the turnover of amino acids.

Muscle

Alanine and glutamine account for more than 50% of the total α-amino acid nitrogen **released** from muscle tissue. In contrast to the significant output of α-amino acids, muscle consistently takes up small quantities of serine, cysteine, and glutamate from the circulation.

Liver & Gut

The liver and gut (the splanchnic tissues) consistently **take up** from the plasma large quantities of alanine and glutamine, the predominant amino acids released by muscle. The liver is the primary site of uptake of alanine, and the gut is the site of utilization of glutamine. In the gut, the majority of the amino groups of glutamine are released from that tissue as alanine or free ammonia. Serine is extracted by these splanchnic tissues as well as by muscle.

Kidney

The kidney is the major source of release of serine; in addition, it releases small but significant quantities of alanine. The kidney takes up glutamine, proline, and glycine from the circulation.

There thus is a fairly close correspondence between the output of most amino acids from peripheral muscle and their uptake by the splanchnic tissues.

Brain

The uptake of valine by the brain exceeds that of all other amino acids, and the capacity of the rat brain to oxidize the branched-chain amino acids (leucine, isoleucine, and valine) is at least 4-fold greater than that of muscle and liver. Although in the postabsorptive state significant quantities of these branched-chain

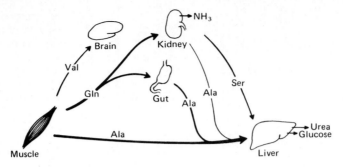

Figure 30–10. Interorgan amino acid exchange in normal postabsorptive humans. The key role of alanine in amino acid output from muscle and gut and uptake by the liver is shown. (Reproduced, with permission, from Felig P: Amino acid metabolism in man. *Annu Rev Biochem* 1975;**44**:937. Copyright © 1975 by Annual Reviews, Inc.)

amino acids are released from muscle, they are not extracted by liver, and thus it is likely that the brain is the primary site for their utilization.

Fig 30–10 summarizes the postabsorptive state. Free amino acids, particularly alanine and glutamine, are released from muscle into the circulation. Alanine, which appears to be the vehicle of nitrogen transport in the plasma, is extracted primarily by the liver. Glutamine is extracted by the gut and the kidney, both of which convert a significant portion to alanine. Glutamine also serves as a source of ammonia for excretion by the kidney. The kidney provides a major source of serine for uptake by peripheral tissues, including liver and muscle. Branched-chain amino acids, particularly valine, are released by muscle and taken up predominantly by brain.

Alanine serves as a key protein-derived glucose precursor, ie, a key **gluconeogenic amino acid** (Fig 30–11). In the liver, the rate of glucose synthesis from alanine and serine is far higher than that observed from all other amino acids. The capacity of the liver for gluconeogenesis from alanine is enormous; it does not reach saturation until the alanine concentration is 9 mmol/L, some 20–30 times its physiologic level. The predominance of alanine in the outflow of α-amino acids from muscle reflects its **synthesis in muscle** by transamination of pyruvate.

INTERORGAN AMINO ACID EXCHANGE IN THE FED STATE

After a protein-rich meal is ingested, the splanchnic tissues release amino acids, predominantly the branched-chain amino acids (Fig 30–12). Valine, isoleucine, and leucine account for at least 60% of the total amino acids entering the systemic circulation (as opposed to the portal circulation), even though they make up only 20% of the total amino acids in a lean protein meal. Concomitant with the release of amino acids from the splanchnic tissues following a meal, the peripheral muscles extract amino acids, predominantly the branched-chain amino acids. Amino acid uptake by muscle tissue in the first postprandial hour consists of at least 50% branched-chain amino acids. After 2–3 hours, 90–100% of the amino acids taken up by peripheral tissues are branched-chain amino acids. The branched-chain amino acids, which are also oxidized in muscle in response to feeding, probably serve as the major donors of amino groups for the transamination of pyruvate to alanine.

Thus, the branched-chain amino acids have a special role in nitrogen metabolism, both in the fasting state, when they provide the brain with an energy source, and after feeding, when they are extracted pre-

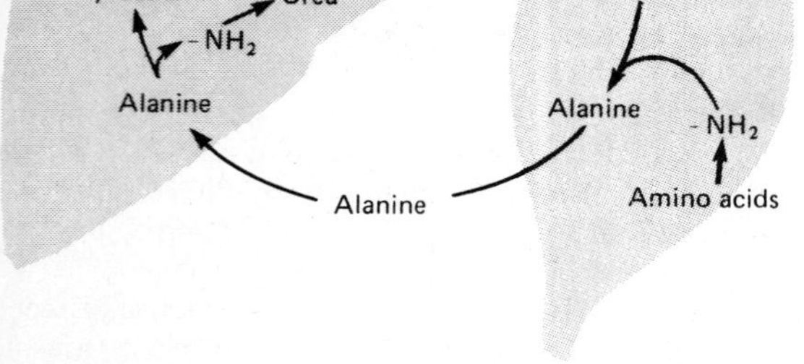

Figure 30–11. The glucose-alanine cycle. Alanine is synthesized in muscle by transamination of glucose-derived pyruvate, released into the bloodstream, and taken up by the liver. In the liver, the carbon skeleton of alanine is reconverted to glucose and released into the bloodstream, where it is available for uptake by muscle and resynthesis of alanine. (Reproduced, with permission, from Felig P: Amino acid metabolism in man. *Annu Rev Biochem* 1975; **44**:938. Copyright © 1975 by Annual Reviews, Inc.)

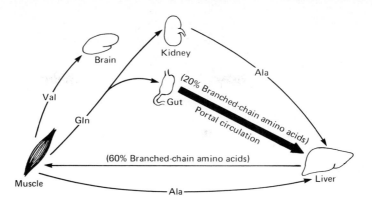

Figure 30–12. Summary of amino acid exchange between organs immediately after feeding.

dominantly by muscles, having been spared by the liver. In muscle, they seem to provide an important source of energy as well as of nitrogen.

BIOSYNTHESIS OF UREA

Overall View

A moderately active man consuming about 300 g of carbohydrate, 100 g of fat, and 100 g of protein daily must excrete about 16.5 g of nitrogen daily. Ninety-five percent is eliminated by the kidneys and the remaining 5% in the feces. The **major pathway of nitrogen excretion in humans is as urea** synthesized in the liver, released into the blood, and cleared by the kidney. In humans eating an occidental diet, urea constitutes 80–90% of the nitrogen excreted.

Reactions of the Urea Cycle

The reactions and intermediates in biosynthesis of 1 mol of urea from 1 mol each of ammonium ion, of carbon dioxide (activated with Mg^{2+} and ATP), and of the α-amino nitrogen of aspartate are shown in Fig 30–13. The overall process requires 3 mol of ATP (2 of which are converted to ADP + P_i and 1 to AMP + PP_i) and the successive participation of 5 enzymes catalyzing the numbered reactions of Fig 30–13. Of the 6 amino acids involved in urea synthesis, one (N-acetylglutamate) functions as an enzyme activator rather than as an intermediate. The remaining 5—aspartate, arginine, ornithine, citrulline, and argininosuccinate—all function as carriers of atoms which ultimately become urea. Two (aspartate and arginine) occur in proteins, while the remaining 3 (ornithine, citrulline, and argininosuccinate) do not. The major metabolic role of these latter 3 amino acids in mammals is urea synthesis. Note that urea formation is in part a **cyclical process.** The ornithine used in reaction 2 is regenerated in reaction 5. There is thus no net loss or gain of ornithine, citrulline, argininosuccinate, or arginine during urea synthesis; however, ammonium ion, CO_2, ATP, and aspartate are consumed.

Reaction 1: Synthesis of carbamoyl phosphate. Condensation of 1 mol each of ammonium ion, carbon dioxide, and phosphate (derived from ATP) to form carbamoyl phosphate is catalyzed by **carbamoyl phosphate synthase,** an enzyme present in liver **mitochondria** of all ureotelic organisms, including humans. The 2 mol of ATP hydrolyzed during this reaction provide the driving force for synthesis of 2 covalent bonds—the amide bond and the mixed carboxylic acid-phosphoric acid anhydride bond of carbamoyl phosphate. In addition to Mg^{2+}, a dicarboxylic acid, preferably N-acetylglutamate, is required. Its presence brings about a profound conformational change in the structure of carbamoyl phosphate synthase that exposes certain sulfhydryl groups, conceals others, and affects the affinity of the enzyme for ATP.

Reaction 2: Synthesis of citrulline. Transfer of a carbamoyl moiety from carbamoyl phosphate to ornithine, forming citrulline + P_i, is catalyzed by **L-ornithine transcarbamoylase** of liver mitochondria. The reaction is highly specific for ornithine, and the equilibrium strongly favors citrulline synthesis.

Reaction 3: Synthesis of argininosuccinate. In the argininosuccinate synthase reaction, aspartate and citrulline are linked together via the amino group of aspartate. The reaction requires ATP, and the equilibrium strongly favors argininosuccinate synthesis.

Reaction 4: Cleavage of argininosuccinate to arginine and fumarate. Reversible cleavage of argininosuccinate to arginine plus fumarate is catalyzed by **argininosuccinase,** an enzyme of mammalian liver and kidney tissues. The reaction proceeds via a *trans* elimination mechanism. The fumarate formed may be converted to oxaloacetate via the fumarase and malate dehydrogenase reactions and then transaminated to regenerate aspartate.

Reaction 5: Cleavage of arginine to ornithine and urea. This reaction completes the urea cycle and regenerates ornithine, a substrate for reaction 2. Hydrolytic cleavage of the guanidino group of arginine is catalyzed by **arginase,** present in the livers of all ureotelic organisms. Smaller quantities of arginase also occur in renal tissue, brain, mammary gland, testicular

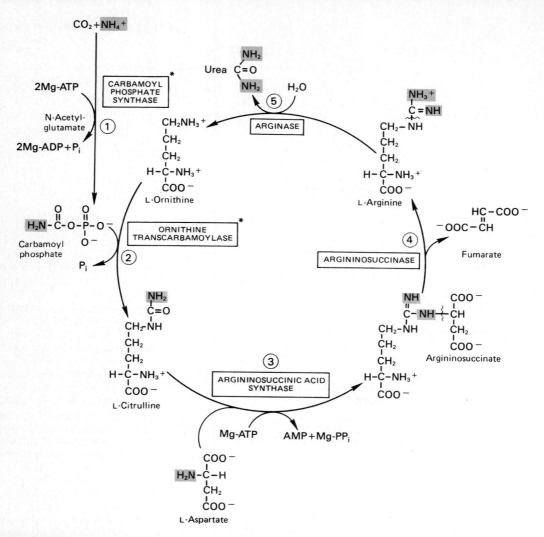

Figure 30-13. Reactions and intermediates of urea biosynthesis. The amines contributing to the formation of urea are shaded. *Mitochondrial enzymes.

tissue, and skin. Mammalian liver arginase is activated by Co^{2+} or Mn^{2+}. Ornithine and lysine are potent inhibitors competitive with arginine.

REGULATION OF UREA SYNTHESIS

Functional Interaction Between Glutamate Dehydrogenase & Carbamoyl Phosphate Synthase

Carbamoyl phosphate synthase acts with mitochondrial glutamate dehydrogenase to channel nitrogen from glutamate (and hence from all amino acids; see Fig 30-2) into carbamoyl phosphate and thus into urea. While the equilibrium constant of the glutamate dehydrogenase reaction favors glutamate rather than ammonia formation, removal of ammonia by car-

bamoyl phosphate synthase and oxidation of α-keto-glutarate by citric acid cycle enzymes in the mitochondrion serve to favor glutamate catabolism. This effect is enhanced by ATP, which, in addition to being a substrate for carbamoyl phosphate synthesis, stimulates glutamate dehydrogenase activity unidirectionally, favoring ammonia formation.

METABOLIC DISORDERS OF THE UREA CYCLE

Metabolic disorders associated with a deficiency of each of the 5 enzymes of hepatic urea synthesis (Fig 30-13) are known. The rate-limiting reactions of urea synthesis appear to be catalyzed by carbamoyl phosphate synthase (reaction 1), ornithine transcarbamoylase (reaction 2), and arginase (reaction 5). Since the urea cycle converts ammonia to the nontoxic compound urea, all disorders of urea synthesis cause ammonia intoxication. This intoxication is more severe when the metabolic block occurs at reaction 1 or 2, since some covalent linking of ammonia to carbon has already occurred if citrulline can be synthesized. Clinical symptoms common to all urea cycle disorders include vomiting in infancy, avoidance of high-protein foods, intermittent ataxia, irritability, lethargy, and mental retardation.

The clinical features and the treatment of all 5 of the disorders discussed below are similar. Significant improvement is noted on a low-protein diet, and much of the brain damage may thus be prevented. Food intake should be in frequent small meals to avoid sudden increases in blood ammonia levels.

Hyperammonemia Type I

One case of **carbamoyl phosphate synthase** deficiency (reaction 1, Fig 30–13) has been reported. This probably is a familial disorder.

Hyperammonemia Type II

Numerous patients have been shown to suffer from a deficiency of **ornithine transcarbamoylase** (reaction 2, Fig 30–13). This disease is X chromosome-linked. The mothers also exhibited hyperammonemia and an aversion to high-protein foods. The only consistent clinical finding was an elevation of glutamine in blood, cerebrospinal fluid, and urine. This probably reflects enhanced synthesis of glutamine by glutamine synthase (Fig 30–8) consequent to elevated tissue levels of ammonia.

Citrullinemia

This rare disorder probably is recessively inherited. Large quantities (1–2 g/d) of citrulline are excreted in the urine, and both plasma and cerebrospinal fluid citrulline levels are markedly elevated. In one patient, complete absence of **argininosuccinate synthase** activity (reaction 3, Fig 30–13) was noted. In another, a less profound modification of this enzyme had occurred. The K_m for citrulline for the synthase from cultured fibroblasts from this patient was 25 times normal. This suggests a mutation causing a significant but not "lethal" modification of the catalytic site of the synthase.

Citrulline (and argininosuccinate; see below) may serve as a carrier of waste nitrogen, since it contains nitrogen intended for urea synthesis. Feeding arginine enhances the excretion of citrulline in these patients. Similarly, feeding benzoate diverts ammonium nitrogen to hippurate via glycine (see Fig 31–2).

Argininosuccinicaciduria

This rare recessive inherited disease, characterized by elevated levels of argininosuccinate in the blood, cerebrospinal fluid, and urine, frequently is associated with the occurrence of friable, tufted hair (trichorrhexis nodosa). While both early- and late-onset types are known, the disease is always manifest by age 2 and usually terminates fatally early in life.

Argininosuccinicaciduria reflects the absence of **argininosuccinase** (reaction 4, Fig 30–13). Cultured skin fibroblasts from normal subjects contain this enzyme whereas those from patients with argininosuccinicacidemia do not. Argininosuccinase is also absent from brain, liver, kidney, and erythrocytes of patients with this disease. While the diagnosis is readily made by 2-dimensional paper chromatography of the urine, additional abnormal spots appear in urine on standing owing to the tendency of argininosuccinate to form cyclic anhydrides. Confirmatory diagnosis is by measurement of erythrocyte levels of argininosuccinase. This test can be performed on umbilical cord blood for early detection. Since argininosuccinase is present in amniotic fluid cells, diagnosis by amniocentesis is also possible. For reasons discussed relevant to citrullinemia, feeding arginine and benzoate promotes nitrogen waste excretion in these patients also.

Hyperargininemia

This defect in urea synthesis is characterized by elevated blood and cerebrospinal fluid arginine levels, low erythrocyte levels of **arginase** (reaction 5, Fig 30–13), and a urinary amino acid pattern resembling that of lysine-cystinuria. Possibly this pattern reflects competition by arginine with lysine and cystine for reabsorption in the renal tubule. In patients, a low-protein diet resulted in lowering of plasma ammonia levels and disappearance of the urinary lysine-cystinuria pattern.

REFERENCES

Adams E, Frank L: Metabolism of proline and the hydroxyprolines. *Annu Rev Biochem* 1980;**49**:1005.

Batshaw ML et al: Treatment of inborn errors of urea synthesis: Activation of alternative pathways of waste nitrogen synthesis and excretion. *N Engl J Med* 1982;**306**:1387.

Felig P: Amino acid metabolism in man. *Annu Rev Biochem* 1975;**44**:933.

Msall M et al: Neurologic outcome in children with inborn errors of urea synthesis: Outcome of urea-cycle enzymopathies. *N Engl J Med* 1984;**310**:1500.

Nyhan WL: *Heritable Disorders of Amino Acid Metabolism: Patterns of Clinical Expression and Genetic Variation.* Wiley, 1974.

Ratner S: Enzymes of arginine and urea synthesis. *Adv Enzymol* 1973;**39**:1.

Ratner S: A long view of nitrogen metabolism. *Annu Rev Biochem* 1977;**46**:1.

Rosenberg LE, Scriver CR: Disorders of amino acid metabolism. Chapter 11 in: *Metabolic Control and Disease.* Bondy PK, Rosenberg LE (editors). Saunders, 1980.

Stanbury JB et al: *The Metabolic Basis of Inherited Disease*, 5th ed. McGraw-Hill, 1983.

Torchinsky YM: Transamination; Its discovery, biological and clinical aspects (1937–1987). *Trends Biochem Sci* 1987; **12:**115.

Tyler B: Regulation of the assimilation of nitrogen compounds. *Annu Rev Biochem* 1978;**47:**1127.

Wellner D, Meister A: A survey of inborn errors of amino acid metabolism and transport in man. *Annu Rev Biochem* 1981; **50:**911.

Catabolism of the Carbon Skeletons of Amino Acids

31

Victor W. Rodwell, PhD

INTRODUCTION

This section deals with conversions of the carbon skeletons of common L-amino acids to amphibolic intermediates and with the metabolic diseases or "inborn errors of metabolism" associated with these catabolic pathways.

BIOMEDICAL IMPORTANCE

Historically, certain metabolic disorders of amino acid metabolism in humans played key roles in elucidation of the pathways by which amino acids are metabolized in normal human subjects. Most of these diseases are rare and thus are unlikely to be encountered by most practicing physicians. These disorders nevertheless present a formidable challenge to the psychiatrist, pediatrician, genetic counselor, or biochemist. They are detected most frequently at infancy, often are fatal at an early age, and often result in irreversible brain damage if untreated. Early detection and rapid initiation of appropriate treatment, if available, is essential. Since several of the enzymes concerned are detectable in cultures of amniotic fluid cells, prenatal diagnosis of these disorders by amniocentesis is possible. While current treatment consists primarily of feeding diets low in the amino acids whose catabolism is impaired, more effective treatment may someday be available. For example, circulating the patient's blood through a column containing the missing enzyme in an immobilized state may "replace" the deficient or defective enzyme in question. Alternatively, recombinant DNA technology may eventually provide a means of correcting genetic defects by "gene therapy."

These metabolic disorders, which result from genetic mutations, cause production of proteins with modified primary structures. Depending on the nature of the primary change, other orders of protein structure may also be affected. While some changes in the primary structures of enzymes may have little or no effect, others may profoundly modify the 3-dimensional structure of catalytic or regulatory sites (see Chapters 6 and 7). The modified or mutant enzyme may possess altered catalytic efficiency (low V_{max} or high K_m) or altered ability to bind an allosteric regulator of its catalytic activity. In principle, a wide variety of mutations may cause the same clinical disease. For example, any mutation that causes a substantial loss of the catalytic activity of argininosuccinase (see Fig 30–13) will cause the metabolic disorder known as argininosuccinicacidemia. It is extremely unlikely, however, that all cases of argininosuccinicacidemia represent the same alteration in primary structure of argininosuccinase. In this sense they are, therefore, distinct molecular diseases. Some known disorders of amino acid metabolism are discussed in this chapter. For further examples, the reader should consult major reference works that specialize in this subject, eg, Stanbury et al, 1983.

We first consider the pathways by which the carbon skeletons of the L-α-amino acids are converted to amphibolic intermediates. Subsequently, and within each section, we discuss certain representative metabolic defects of these pathways in human subjects.

CONVERSION OF CARBON SKELETONS OF COMMON L-α-AMINO ACIDS TO AMPHIBOLIC INTERMEDIATES

That the carbon skeletons of the common amino acids are converted to amphibolic intermediates was evident from nutritional studies carried out in the period 1920–1940. These data, reinforced and confirmed by studies using isotopically labeled amino

Table 31–1. Fates of the carbon skeletons of the common L-α-amino acids.

Converted to amphibolic intermediates forming:			
Glycogen ("Glycogenic")		Fat ("Ketogenic")	Glycogen and Fat ("Glycogenic" and "Ketogenic")
Ala	Hyp	Leu	Ile
Arg	Met		Lys
Asp	Pro		Phe
Cys	Ser		Trp
Glu	Thr		Tyr
Gly	Val		
His			

acids conducted from 1940 to 1950, supported the concept of the interconvertibility of fat, carbohydrate, and protein carbons and established that each amino acid is convertible either to carbohydrate (13 amino acids), fat (one amino acid), or both (5 amino acids) (Table 31–1). Although at the time a detailed explanation of these interconversions was not possible, it was established that they indeed occur. How they occur is outlined in Fig 31–1.

In what follows, individual amino acids are grouped for discussion on the basis of the first amphibolic intermediates formed as end products of their catabolism. An early step in amino acid catabolism—frequently the first reaction—involves removal of the α nitrogen. This usually (but not always, eg, proline, hydroxyproline, lysine) involves transamination. Once removed, the nitrogen enters the general metabolic pool. Depending upon demand, it may then be reutilized for anabolic processes (eg, protein synthesis) or, if in excess, converted to urea and excreted (see Chapter 30). The nitrogen-free carbon skeleton that remains is, in most instances, merely an oxidized hydrocarbon and can no longer be specifically

identified as an amino acid derivative. As such, it is degraded to amphibolic intermediates by reactions similar to those by which other oxidized hydrocarbons (eg, linear and branched fatty acids) are catabolized. Analogies to other areas of metabolism, particularly of fatty acids (see Chapter 24), are particularly striking. For example, the carbon skeletons derived from the branched-chain amino acids leucine, isoleucine, and valine are degraded by reactions analogous to those for catabolism of fatty acids.

AMINO ACIDS FORMING OXALOACETATE

Asparagine & Aspartate

All 4 carbons of asparagine and of aspartate are converted to oxaloacetate via asparaginase and a transaminase (Fig 31–2, top).

No known metabolic defect is associated with this short catabolic pathway, possibly because a defect in the transaminase might have grave consequences incompatible with life. Transaminases fulfill central an-

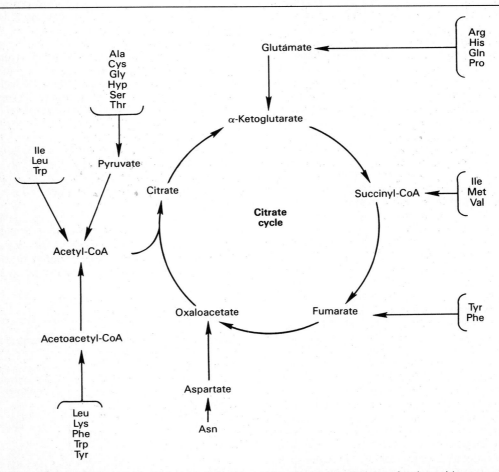

Figure 31–1. Amphibolic intermediates formed from the carbon skeleton of amino acids.

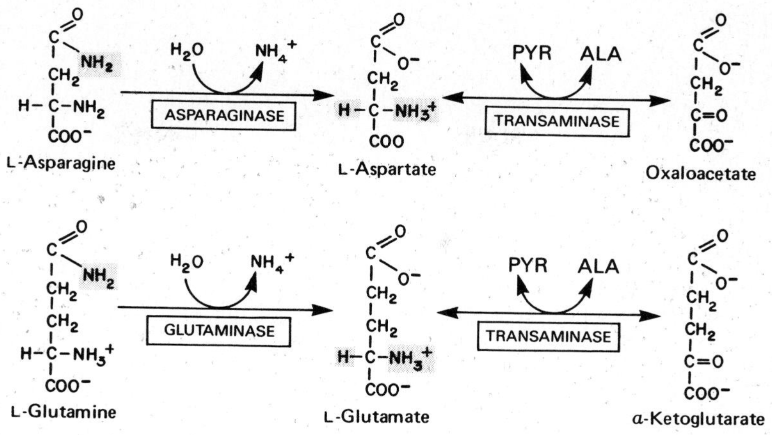

Figure 31–2. Catabolism of L-asparagine (*top*) and of L-glutamine (*bottom*) to amphibolic intermediates. PYR, pyruvic acid; ALA, L-alanine. In this and subsequent figures, shading on functional groups highlights portions of the molecules undergoing chemical change.

abolic as well as catabolic functions in the metabolism of several different amino acids (see Chapter 29 and below).

AMINO ACIDS FORMING α-KETOGLUTARATE

Glutamine & Glutamate

Catabolism of glutamine and of glutamate proceeds like that of asparagine and aspartate but with formation of α-ketoglutarate, the methylene homolog of oxaloacetate (Fig 31–2, bottom). While both glutamate and aspartate are substrates for the same transaminase, deamidation of asparagine and glutamine is catalyzed by distinct enzymes. A dual specificity glutaminase-asparaginase exists in some bacteria.

Possibly for the reasons alluded to above for asparagine and aspartate, there are no known metabolic defects of the glutamine-glutamate catabolic pathway.

Proline

All 5 carbons of L-proline form α-ketoglutarate (Fig 31–3, left). Proline is oxidized to a dehydroproline which, on addition of water, forms glutamate-γ-semialdehyde. This is then oxidized to glutamate and transaminated to α-ketoglutarate.

Metabolic disorders of proline catabolism. Two genetically distinct hyperprolinemias have been described. Both type I and type II hyperprolinemias are inherited, apparently as autosomal recessive traits. Despite the occurrence of mental retardation in half of the known cases, both type I and type II hyperprolinemia are believed to be harmless.

A. Hyperprolinemia Type I: The site of the metabolic block in hyperprolinemia type I is proline dehydrogenase (Fig 31–3). In contrast to hyperprolinemia type II, there is no associated impairment of *hydroxy*proline catabolism. An animal model for type

I hyperprolinemia, the Pro/Re mouse, has only 10% of normal hepatic proline dehydrogenase activity. Type I heterozygotes exhibit only a mild hyperprolinemia.

B. Hyperprolinemia Type II: The extent of the hyperprolinemia in the type II condition exceeds that seen in type I patients. The urine contains Δ^1-pyrroline-3-hydroxy-5-carboxylate. The site of the metabolic block in hyperprolinemia type II is the dehydrogenase that catalyzes the oxidation of glutamate-γ-semialdehyde to glutamate (Fig 31–3). Since the same dehydrogenase functions in hydroxyproline catabolism to oxidize γ-hydroxy-L-glutamate-γ-semialdehyde to erythro-γ-hydroxy-L-glutamate (Fig 31–12), both proline and hydroxyproline catabolism are affected. Unlike type I heterozygotes, type II heterozygotes exhibit no hyperprolinemia. This suggests that the shared dehydrogenase plays different roles in regulation of free proline and hydroxyproline pools.

Arginine

While arginine and histidine also form α-ketoglutarate, one carbon and either 2 (histidine) or 3 (arginine) nitrogens must first be removed from these 6-carbon amino acids. With arginine, this requires but a single step: hydrolytic removal of the guanidino group catalyzed by arginase. The product, ornithine, then undergoes transamination of the δ-amino group, forming glutamate-γ-semialdehyde, which is converted to α-ketoglutarate as described above for proline (Fig 31–3).

Hyperargininemia, a metabolic disorder of arginine catabolism in which the affected enzyme is liver arginase, is discussed in Chapter 30 in conjunction with metabolic disorders of enzymes of the urea cycle.

Histidine

For histidine, removal of the extra carbon and nitrogens requires 4 reactions (Fig 31–4). Deamination

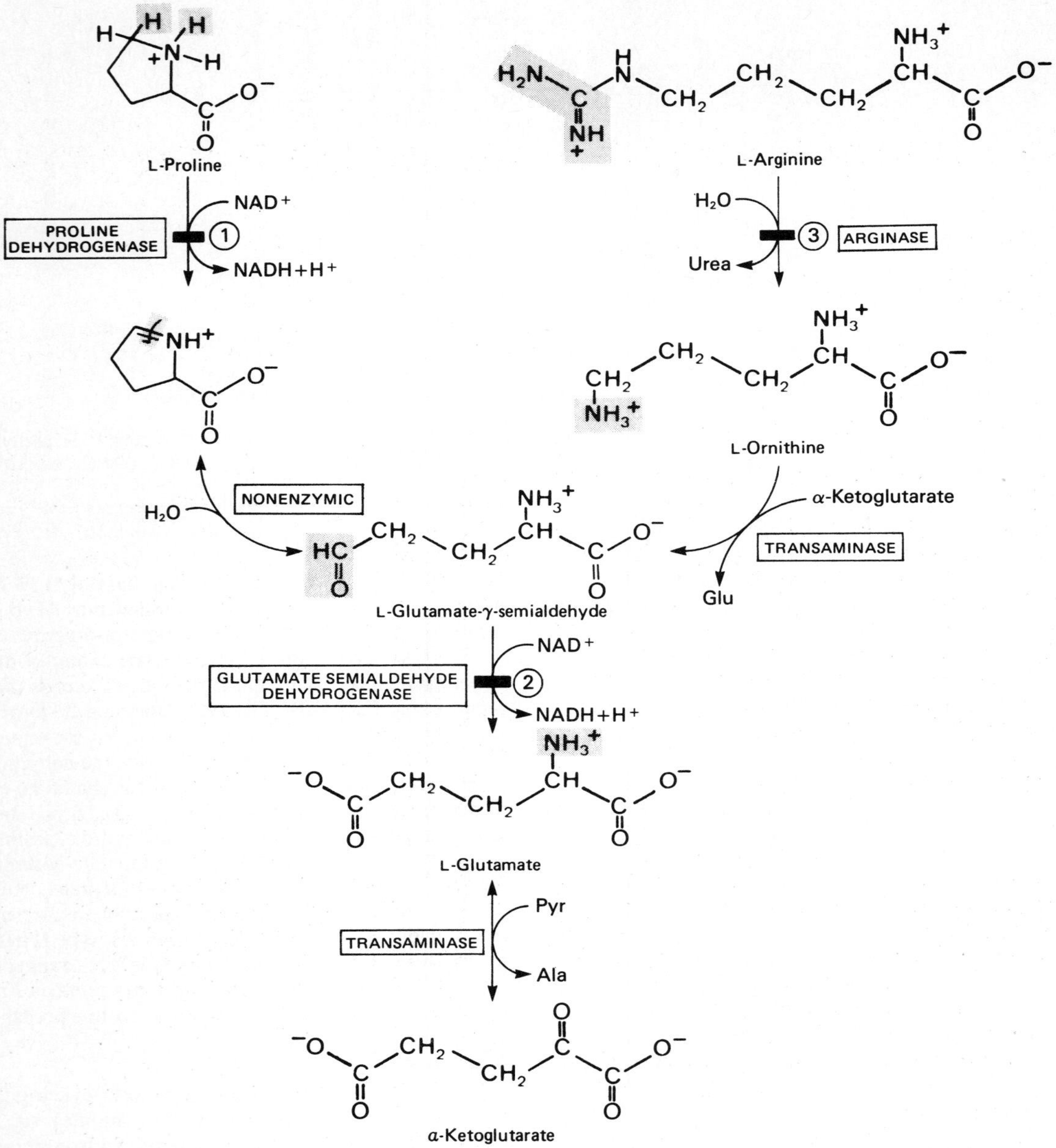

Figure 31–3. Catabolism of L-proline (*left*) and of L-arginine (*right*) to α-ketoglutarate. The circled numerals represent the sites of the metabolic defects in 1, type I hyperprolinemia; 2, type II hyperprolinemia; and 3, hyperargininemia (see Chapter 30).

of histidine produces urocanate. Conversion of urocanate to 4-imidazolone-5-propionate, catalyzed by urocanase, involves both addition of H_2O and an internal oxidation-reduction. Although 4-imidazolone-5-propionate may undergo additional fates, conversion to α-ketoglutarate involves hydrolysis to N-formiminoglutamate followed by transfer of the formimino group on the α carbon to tetrahydrofolate, forming N^5-formiminotetrahydrofolate. In patients with folic acid deficiency, this last reaction is partially or totally blocked and N-formiminoglutamate (Figlu) is excreted in the urine. This forms the basis for a test for folic acid deficiency in which N-formiminoglutamate is detected in the urine following a large dose of histidine.

Metabolic disorders of histidine catabolism. Histidinemia, a metabolic disorder of histidine catabolism in which the affected enzyme is histidase (Fig 31–4), is inherited as an autosomal recessive trait. Over half of patients with histidinemia are men-

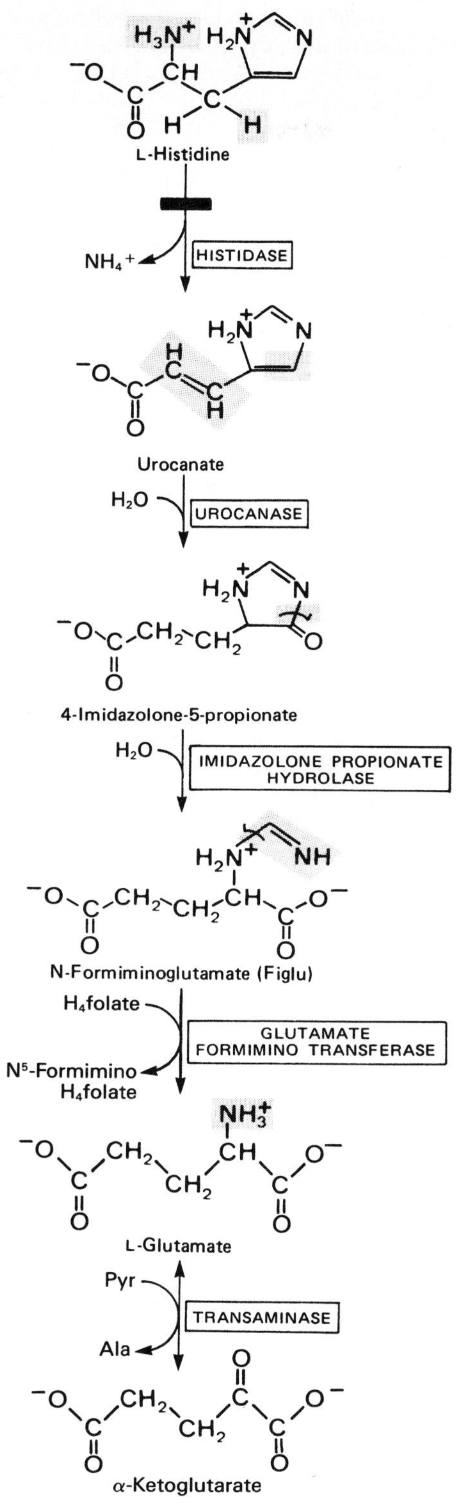

Figure 31–4. Catabolism of L-histidine to α-ketoglutarate. H₄folate, tetrahydrofolate. The reaction catalyzed by histidase represents the site of the probable metabolic defect in histidinemia.

tally retarded and exhibit a characteristic speech defect.

In addition to increased levels of histidine in blood and urine, there is also increased excretion of imidazolepyruvate (which in a color test with ferric chloride may be mistaken for phenylpyruvate, so that a mistaken diagnosis of phenylketonuria could be made). The metabolic defect in histidinemia is inadequate activity of liver histidase, which impairs conversion of histidine to urocanate (Fig 31–4). An alternative route of histidine metabolism, transamination to imidazolepyruvate, is then favored, and the excess imidazolepyruvate is excreted in the urine. Imidazoleacetate and imidazolelactate, the reduction products of imidazolepyruvate, have also been detected in the urine of histidinemic patients.

Since the quantity of histidine in normal urine is relatively large, it is readily detected. A conspicuous increase in histidine excretion is a characteristic finding in normal pregnancy but does not occur in gestational hypertensive disorders. The normally increased excretion of histidine during pregnancy apparently does not result from a metabolic defect in histidine metabolism. The phenomenon may be explained largely on the basis of the changes in renal function characteristic of normal pregnancy as well as gestational hypertensive disorders. Furthermore, alterations in amino acid excretion during pregnancy are not confined to histidine.

AMINO ACIDS FORMING PYRUVATE

Conversion of the carbon skeletons of alanine, cysteine, cystine, glycine, threonine, and serine to pyruvate is summarized diagrammatically below. All of the carbons of glycine, alanine, cysteine, and serine—but only 2 of the carbons of threonine—form pyruvate. Pyruvate may then be converted to acetyl-CoA.

L-Threonine
↓
Glycine
↓
L-Serine   **L-Cystine**
↓   ↓
L-Alanine → Pyruvate ← L-Cysteine
↓
Acetyl-CoA

Glycine

Amphibolic intermediates formed from glycine include pyruvate, CO_2, and N^5,N^{10}-methylenetetrahydrofolate. Formation of pyruvate from glycine can occur by conversion to serine, catalyzed by serine hydroxymethyltransferase (Fig 31–5), followed by the serine dehydratase reaction (Fig 31–7; see also Serine, below).

The major pathway for glycine catabolism in vertebrates involves conversion to CO_2, NH_4^+, and N^5,N^{10}-methylenetetrahydrofolate catalyzed by the glycine

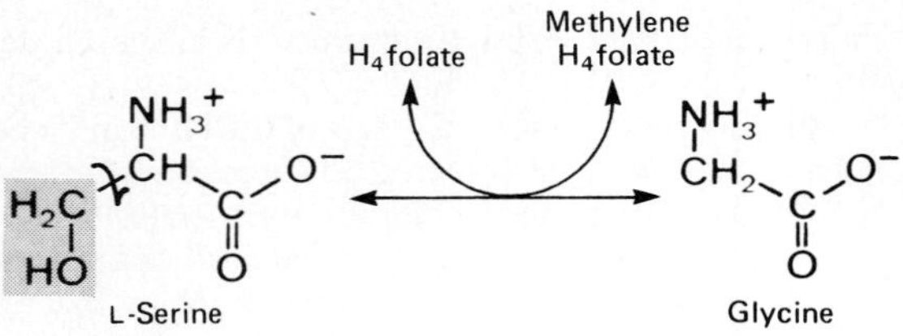

Figure 31−5. The freely reversible serine hydroxymethyltransferase reaction. H₄folate, tetrahydrofolate.

synthase complex. This reversible reaction (Fig 31−6) resembles conversion of pyruvate to acetyl-CoA by enzymes of the pyruvate dehydrogenase complex. Both complexes comprise macromolecular aggregates in liver mitochondria. The reactions of glycine cleavage occur in liver tissue of most vertebrates, including humans, other mammals, birds, and reptiles.

The reactions of the glycine cleavage system probably constitute the major route not only for glycine but also for serine catabolism in humans and many other vertebrates (see also Serine, below).

Metabolic disorders of glycine catabolism. Discussed below are 2 disorders of glycine metabolism.

A. Glycinuria: Glycinuria, described in only one family, is characterized by excess urinary excretion of glycine (glycinuria) in association with a tendency to formation of oxalate renal stones, although the amount of oxalate excreted in the urine is normal. Glycinuria appears to be inherited as a dominant, possibly X-linked, trait. The plasma glycine levels are normal, while the urinary excretion of glycine ranges from 600 to 1000 mg/d. Consequently, glycinuria is attributed to a defect in renal tubular reabsorption of glycine.

B. Primary Hyperoxaluria: Primary hyperoxaluria is characterized by continuous high urinary excretion of oxalate unrelated to dietary intake of oxalate. The history of the disease is that of progressive bilateral calcium oxalate urolithiasis, nephrocalcinosis, and recurrent infection of the urinary tract. Death occurs in childhood or early adult life from renal failure or hypertension. The excess oxalate is apparently of endogenous origin, possibly from glycine, which may be deaminated to form glyoxylate, a precursor of oxalate. The metabolic defect is considered

to be a disorder of glyoxylate metabolism associated with failure to convert glyoxylate to formate or to glycine by transamination. As a result, the excess glyoxylate is oxidized to oxalate. Glycine transaminase deficiency, together with some impairment of oxidation of glyoxylate to formate, may be the biochemical explanation for the inherited metabolic disease primary hyperoxaluria.

Alanine

Transamination of L-alanine (Fig 31−7) forms pyruvate, which may then be decarboxylated to acetyl-CoA.

Possibly for the reasons advanced under glutamate and aspartate catabolism, there is no known metabolic defect of α-alanine catabolism.

Serine

Conversion of serine to pyruvate by serine dehydratase, a pyridoxal phosphate protein, involves both elimination of water and hydrolytic loss of ammonia from an amino acid intermediate (Fig 31−7). Rat and guinea pig liver is rich in serine dehydratase. Whereas in these species conversion of serine to pyruvate by serine dehydratase is of considerable physiologic significance, in humans and many other vertebrates, serine is degraded primarily to glycine and N^5,N^{10}-methylenetetrahydrofolate. The initial reaction is catalyzed by serine hydroxymethyltransferase (Fig 31−5). Further catabolism of serine then merges with glycine catabolism (Fig 31−6).

Cystine

Like carbon and nitrogen, sulfur is continuously recycled through the biosphere through the combined metabolic activities of prokaryotic and eukaryotic organisms. Mammals, which play no role in sulfur assimilation, participate in this cycle by catabolism of organic sulfur compounds to inorganic sulfur compounds. For example, human subjects excrete approximately 20−30 mmol of sulfur per day, at least 80% of which is inorganic sulfate.

The major catabolic fate of cystine in mammals is conversion to cysteine, principally by the reaction catalyzed by cystine reductase (Fig 31−8). From this point, catabolism of cystine merges with that of cysteine (discussed below).

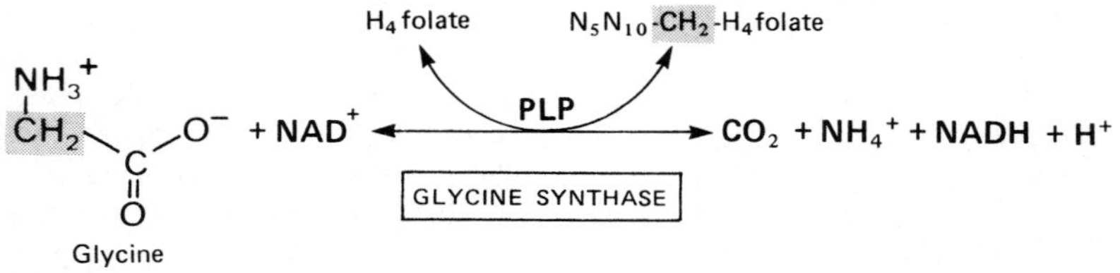

Figure 31−6. The reversible cleavage of glycine by the mitochondrial glycine synthase complex. PLP, pyridoxal phosphate.

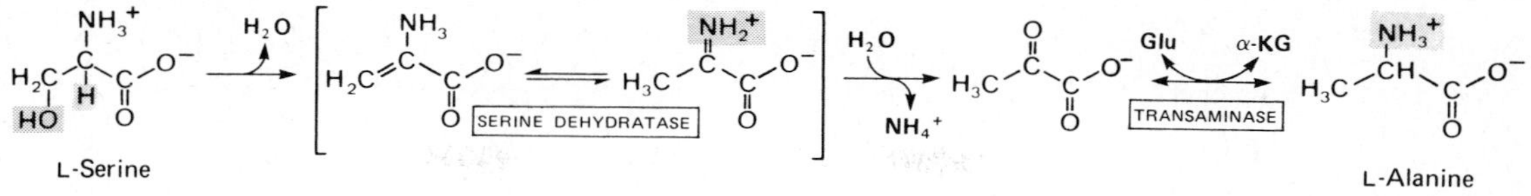

Figure 31–7. Conversion of alanine and serine to pyruvate. Both the alanine transaminase and serine dehydratase reactions require pyridoxal phosphate as coenzyme. The serine dehydratase reaction proceeds via elimination of H_2O from serine, forming an unsaturated amino acid. This rearranges to an α-imino acid that is spontaneously hydrolyzed to pyruvate plus ammonia. There is thus no net gain or loss of water during the serine dehydratase reaction. Glu, glutamate; α-KG, α-ketoglutarate.

Cysteine

Cysteine is catabolized in mammals via 2 principal catabolic pathways: the direct oxidative (cysteine sulfinate) pathway and the transamination (3-mercaptopyruvate) pathway (Fig 31–9). A third pathway involving cysteine desulfhydrase, present in bacteria, was formerly thought also to be functional in mammals. However, since cysteine desulfhydrase activity has never been detected in mammalian tissues, catabolism of cysteine by this route is unlikely.

A. The Direct Oxidative Pathway of Cysteine Catabolism: Conversion of cysteine to cysteine sulfinate (Fig 31–9) is catalyzed by cysteine dioxygenase, an enzyme that requires Fe^{2+} and NAD(P)H. Further catabolism of cysteine sulfinate probably involves its transamination to β-sulfinylpyruvate. However, while transaminases present in mammalian tissues accept cysteine sulfinate as the amino donor, it is not clear whether these are distinct from classic glutamate-aspartate transaminase. In addition, the presumed product, β-sulfinylpyruvate, has yet to be isolated as a catabolite of cysteine sulfinate. Conversion of the putative intermediate β-sulfinylpyruvate to pyruvate and sulfite may not be enzyme-catalyzed. Desulfination is extremely rapid even in the absence of enzymatic catalysis, for transamination of cysteine sulfinate forms stoichiometric quantities of sulfite.

B. The Transaminase (3-Mercaptopyruvate) Pathway of Cysteine Catabolism: Reversible transamination of cysteine to 3-mercaptopyruvate (thiolpyruvate) is catalyzed by specific cysteine transami-

nases or by glutamate or asparagine transaminases of mammalian liver and kidney (Fig 31–9). 3-Mercaptopyruvate may then be reduced in a reaction catalyzed by L-lactate dehydrogenase. The product, 3-mercaptolactate, is a normal constituent of human urine in the form of its mixed disulfide with cysteine and is excreted in increased amounts in the urine of patients with mercaptolactate-cysteine disulfiduria. Alternatively, 3-mercaptopyruvate undergoes desulfuration, forming pyruvate and H_2S (Fig 31–9).

Metabolic disorders of sulfur-containing amino acids. Table 31–2 summarizes known defects of sulfur-containing amino acid catabolism. Several disorders are discussed below.

A. Cystinuria (Cystine-Lysinuria): In this inherited metabolic disease, urinary excretion of cystine is 20–30 times normal. Excretion of lysine, arginine, and ornithine is also markedly increased. Cystinuria is considered to be due to a renal transport defect. The greatly increased excretion of lysine, arginine, and ornithine as well as cystine in urine of cystinuric patients suggests a defect in the renal reabsorptive mechanisms for these 4 amino acids. It is possible that a single reabsorptive site is involved. The term "cystinuria" is therefore a misnomer. Cystine-lysinuria may now be the preferred descriptive term for this disease.

Because cystine is relatively insoluble, in cystinuric patients it may precipitate in the kidney tubules and form cystine calculi. Were it not for this possibility, cystinuria would be an entirely benign anomaly and probably would escape recognition in many cases.

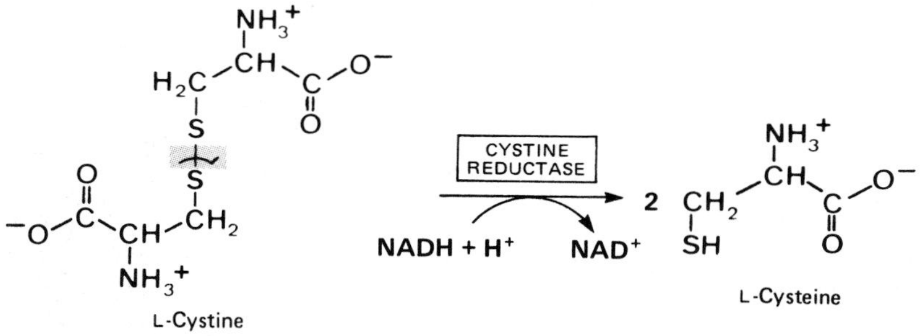

Figure 31–8. The cystine reductase reaction.

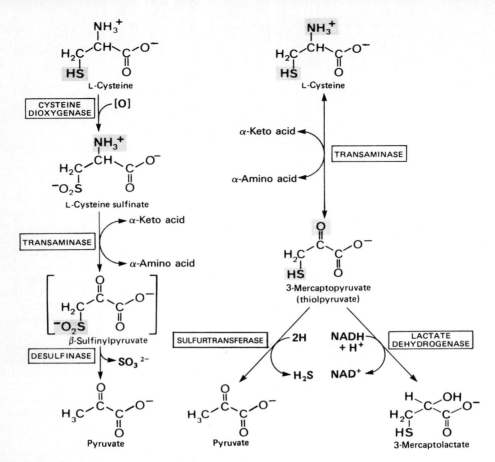

Figure 31–9. Catabolism of L-cysteine via the direct oxidative (cysteine sulfinate) pathway (*left*) and by the transamination (3-mercaptopyruvate) pathway (*right*). β-Sulfinylpyruvate is bracketed, since it is a putative intermediate. Oxidation of the sulfite produced in the last reaction of the direct oxidative pathway is catalyzed by sulfite oxidase. α-KA, α-keto acid, α-AA, α-amino acid.

The mixed disulfide of L-cysteine and L-homocysteine (Fig 31–10) has also been detected in the urine of cystinuric patients. This compound is somewhat more soluble than cystine. To the extent that it may be formed at the expense of cystine, it reduces the tendency to formation of cystine crystals and calculi in the urine.

B. Cystinosis (Cystine Storage Disease): In cystinosis, which is also inherited, cystine crystals are deposited in many tissues and organs (particularly the reticuloendothelial system) throughout the body. It is usually accompanied by a generalized aminoaciduria in which all amino acids are considerably increased in the urine. Various other renal functions are also seriously impaired, and affected patients usually die at an early age with all of the manifestations of acute renal failure. Recent evidence implicates impaired lysosomal function as the primary defect.

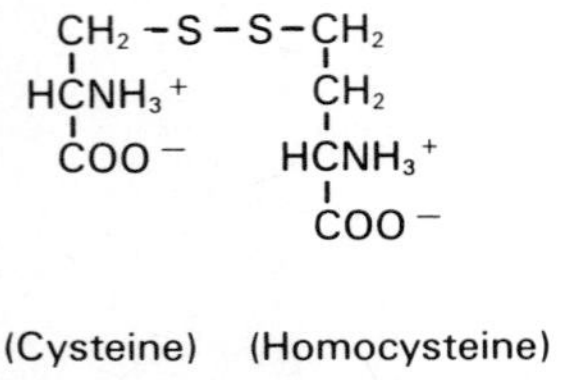

C. Homocystinurias: The incidence of these heritable defects of methionine catabolism is estimated at 1:160,000 births. Homocystine (up to 300 mg/d), together with S-adenosylmethionine in some cases, is excreted in the urine, and plasma methionine levels are elevated. At least 4 known metabolic defects give rise to homocystinuria (Table 31–2). In homocystinuria type I, associated clinical findings include the occurrence of thromboses, osteoporosis, dislocated lenses in the eyes, and frequently mental retardation. Two forms of this disease are known: a vitamin B₆-responsive form and a vitamin B₆-unresponsive form. Feeding a diet low in methionine and high in

$$CH_2-S-S-CH_2$$
$$HCNH_3^+ \qquad CH_2$$
$$COO^- \qquad HCNH_3^+$$
$$\qquad\qquad COO^-$$

(Cysteine)   (Homocysteine)

Figure 31–10. Mixed disulfide of cysteine and homocysteine.

Table 31–2. Inborn errors of sulfur-containing amino acid metabolism.

Name	Defect	Reference
Homocystinuria I	Cystathionine β-synthase	Fig 29–10, reaction 1
Homocystinuria II	N^5, N^{10}-methylenetetrahydrofolate reductase	
Homocystinuria III	Low N^5-methyltetrahydrofolate-homocysteine transmethylase owing to inability to synthesize methylcobalamin	
Homocystinuria IV	Low N^5-methyltetrahydrofolate-homocysteine transmethylase owing to defective intestinal absorption of cobalamin	
Hypermethioninemia	Liver methionine adenosyltransferase*	Fig 31–22
Cystathioninuria	Cystathionase	Fig 29–10, reaction 2
Sulfituria (sulfocysteinuria)	Sulfite oxidase	Fig 29–8, legend
Cystinosis	Defect in lysosomal function	
3-Mercaptopyruvate-cysteine disulfiduria	3-Mercaptopyruvate sulfurtransferase	Fig 31–9
Methionine malabsorption syndrome	Inability to absorb methionine from gut	

*May also occur in cystathioninuria, tyrosinemia, and fructose intolerance.

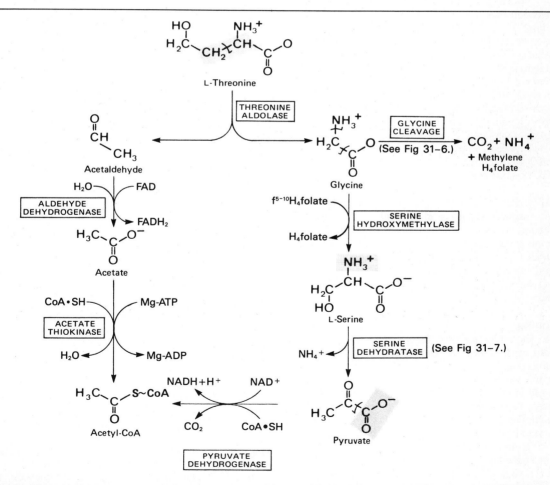

Figure 31–11. Conversion of threonine and glycine to serine, pyruvate, and acetyl-CoA. $f^{5-10} \cdot H_4$folate, formyl [5–10] tetrahydrofolic acid.

cystine effectively prevents pathologic changes if initiated early in life. Other types of homocystinuria reflect defects in the remethylation cycle (Table 31–2).

Threonine

Threonine is cleaved to acetaldehyde and glycine by **threonine aldolase.** Acetaldehyde then forms acetyl-CoA (Fig 31–11). Catabolism of glycine is discussed above.

Hydroxyproline

4-Hydroxy-L-proline is converted to pyruvate and glyoxylate (Fig 31–12). A mitochondrial dehydrogenase catalyzes conversion of hydroxyproline to L-Δ^1-pyrroline-3-hydroxy-5-carboxylate. This is in nonenzymatic equilibrium with γ-hydroxy-L-glutamate-γ-semialdehyde, formed by addition of water. The semialdehyde is oxidized to the corresponding carboxylic acid, erythro-γ-hydroxy-L-glutamate, and transaminated to α-keto-γ-hydroxyglutarate. An aldol type cleavage then forms glyoxylate plus pyruvate.

Metabolic disorders of hydroxyproline catabolism. Hyperhydroxyprolinemia is a metabolic disorder characterized by high plasma levels of 4-hydroxyproline. The site of the metabolic defect in this autosomal recessive trait is 4-hydroxyproline dehydrogenase (Fig 31–12). In contrast to type II hyperprolinemia, there is no accompanying impairment of proline catabolism, since the affected enzyme functions solely in hydroxyproline catabolism. The condition has no effect on collagen metabolism and, like the hyperprolinemias, appears to be harmless.

AMINO ACIDS FORMING ACETYL-COENZYME A

All amino acids forming pyruvate (alanine, cysteine, cystine, glycine, hydroxyproline, serine, and threonine) are convertible to acetyl-CoA. In addition, 5 amino acids form acetyl-CoA without first forming pyruvate. These include the aromatic amino acids phenylalanine, tyrosine, and tryptophan, the basic amino acid lysine, and the neutral branched-chain amino acid leucine.

Tyrosine

A. Overall Reaction Sequence: Five sequential enzymatic reactions convert tyrosine to fumarate and to acetoacetate (Fig 31–13): (1) transamination to p-hydroxyphenylpyruvate; (2) simultaneous oxidation and migration of the 3-carbon side chain and decarboxylation, forming homogentisate; (3) oxidation of homogentisate to maleylacetoacetate; (4) isomerization of maleylacetoacetate to fumarylacetoacetate; and (5) hydrolysis of fumarylacetoacetate to fumarate and acetoacetate. Acetoacetate may then undergo thiolytic cleavage to acetate plus acetyl-CoA.

Several intermediates of tyrosine metabolism were discovered during studies of the human genetic dis-

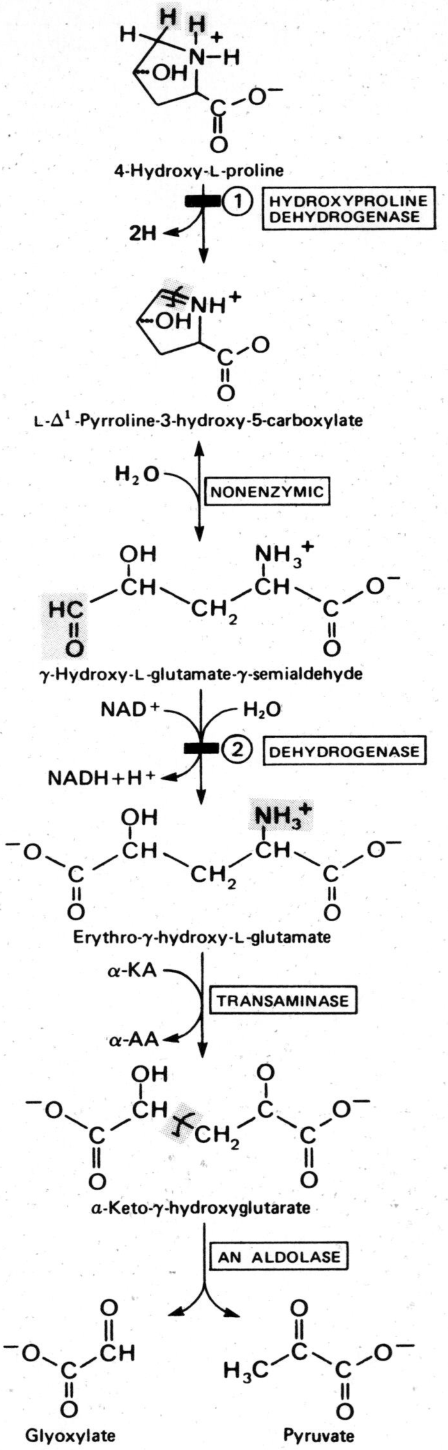

Figure 31–12. Intermediates in L-hydroxyproline catabolism in mammalian tissues. α-KA, α-keto acid; α-AA, α-amino acid. The circled numerals represent the sites of the probable metabolic defects in **1**, hyperhydroxyprolinemia; and **2**, type II hyperprolinemia.

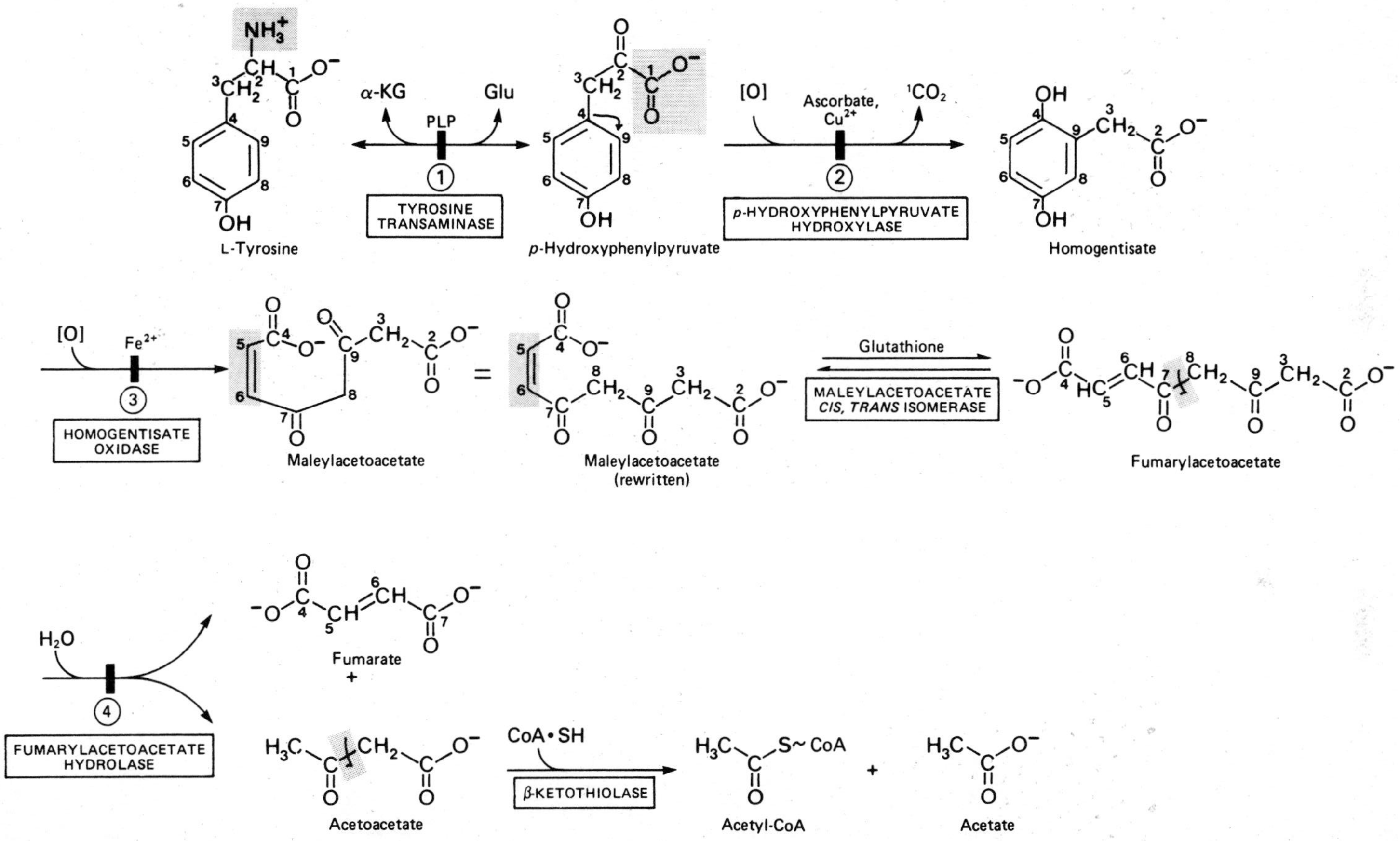

Figure 31–13. Intermediates in tyrosine catabolism. With the exception of β-ketothiolase, reactions are discussed in the text. Carbon atoms of intermediates are numbered to assist readers in determining the ultimate fate of each carbon (see also Fig 31–15). α-KG, α-ketoglutarate; Glu, glutamate; PLP, pyridoxal phosphate. The circled numerals represent the probable sites of the metabolic defects in **1**, type II tyrosinemia; **2**, neonatal tyrosinemia; **3**, alkaptonuria; and **4**, type I tyrosinemia, or tyrosinosis.

ease alkaptonuria. Patients with alkaptonuria excrete homogentisate in the urine, and much useful information was obtained by feeding suspected precursors of homogentisate to these patients.

B. Transamination of Tyrosine: Transamination of tyrosine to *p*-hydroxyphenylpyruvate is catalyzed by **tyrosine-α-ketoglutarate transaminase,** an inducible enzyme of mammalian liver.

C. Oxidation of *p*-Hydroxyphenylpyruvate to Homogentisate: Although the reaction (Fig 31–13) appears to involve hydroxylation of *p*-hydroxyphenylpyruvate in the ortho position accompanied by oxidative loss of the carboxyl carbon, it actually involves migration of the side chain. Ring hydroxylation and side chain migration occur in a concerted manner. ***p*-Hydroxyphenylpyruvate hydroxylase** is a copper metalloprotein similar to **tyrosinase.** Although other reducing agents can replace ascorbate as a cofactor for this reaction in vitro, scorbutic patients excrete incompletely oxidized products of tyrosine metabolism.

D. Conversion of Homogentisate to Fumarate and Acetoacetate: The benzene ring of homogentisate is ruptured, forming maleylacetoacetate in an oxidative reaction catalyzed by **homogentisate oxidase,** an iron metalloprotein of mammalian liver.

Conversion of maleylacetoacetate to fumarylacetoacetate, a *cis* to *trans* isomerization about the double bond, is catalyzed by **maleylacetoacetate *cis, trans* isomerase,** an –SH enzyme of mammalian liver. Hydrolysis of fumarylacetoacetate by **fumarylacetoacetate hydrolase** forms fumarate and acetoacetate. Acetoacetate can then be converted to acetyl-CoA plus acetate by the β-ketothiolase reaction (see Chapter 23).

Metabolic disorders of tyrosine catabolism. Several metabolic disorders are characterized by tyrosinemia, tyrosinuria, and phenolaciduria.

A. Tyrosinemia Type I (Tyrosinosis): Tyrosinosis is characterized by accumulation of metabolites that adversely affect the activities of several enzymes and transport systems. The pathophysiology of this disorder thus is complex. The proposed metabolic defect is in fumarylacetoacetate hydrolase (Fig 31–13) and possibly in maleylacetoacetate hydrolase as well.

Both acute and chronic forms of tyrosinosis are known. In acute tyrosinosis, infants exhibit diarrhea, vomiting, a "cabbagelike" odor, and failure to thrive. Death from liver failure in untreated acute tyrosinosis ensues within 6–8 months. In chronic tyrosinemia, similar but milder symptoms lead to death by the age of 10 years. Plasma tyrosine levels are elevated (6–12 mg/dL), as are those of additional amino acids, notably methionine. Treatment involves a diet low in tyrosine and phenylalanine and, on occasion, low in methionine also.

B. Tyrosinemia Type II (Richner-Hanhart Syndrome): The probable site of the metabolic defect in tyrosinemia type II is hepatic tyrosine transaminase (Fig 31–13). Clinical findings include elevated plasma tyrosine levels (4–5 mg/dL), characteristic eye and skin lesions, and moderate mental retardation. Self-mutilation and disturbances of fine coordination have also been reported. Tyrosine is the only amino acid whose urinary concentration is elevated. However, renal clearance and reabsorption of tyrosine fall within normal limits. Metabolites excreted in the urine include *p*-hydroxyphenylpyruvate, *p*-hydroxyphenyllactate, *p*-hydroxyphenylacetate, N-acetyltyrosine, and tyramine (Fig 31–14).

C. Neonatal Tyrosinemia: The disorder is thought to result from a relative deficiency of *p*-hydroxyphenylpyruvate hydroxylase (Fig 31–13). Blood levels of tyrosine and phenylalanine are elevated, as are urinary levels of tyrosine, *p*-hydroxyphenylacetate, N-acetyltyrosine, and tyramine. Therapy involves feeding a diet low in protein.

D. Alkaptonuria: This inherited metabolic disorder, noted in medical literature as early as the 16th century, was characterized in 1859. The disease is of considerable historic interest, because it formed the basis for Garrod's ideas concerning heritable metabolic disorders. Its most striking clinical manifestation is the occurrence of dark urine on standing in air. Late in the disease, there occur generalized pigmentation of connective tissues (ochronosis) and a form of arthritis. The metabolic defect is lack of **homogentisate oxidase** (Fig 31–13). The substrate, homogentisate, is excreted in the urine, where it is oxidized in air to a brownish-black pigment. Over 600 cases have been reported; the estimated incidence of alkaptonuria is 2–5 per million live births.

Alkaptonuria is inherited as an autosomal recessive trait. At present, no diagnostic procedure for the detection of heterozygotes is available. While the precise mechanism of the ochronosis is not known, it is believed to involve oxidation of homogentisate by polyphenol oxidase, forming benzoquinoneacetate, which polymerizes and binds to connective tissue macromolecules.

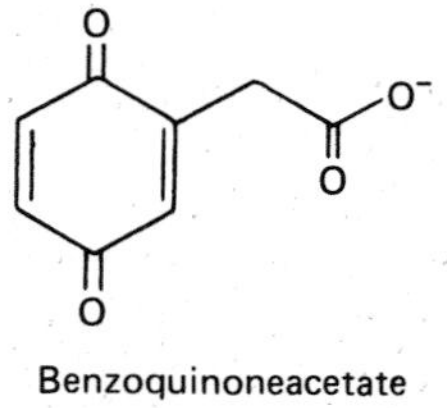

Benzoquinoneacetate

Phenylalanine

Phenylalanine is first converted to tyrosine by phenylalanine hydroxylase (see Fig 29–12). The labeling pattern in the amphibolic products fumarate and acetoacetate (Fig 31–15) thus is identical to that for tyrosine (Fig 31–13).

Metabolic disorders of phenylalanine catabolism. Major metabolic disorders associated with impaired ability to convert phenylalanine to tyrosine (see Fig 29–12) may be classified into 3 broad groups: defects in phenylalanine hydroxylase (hyperphenylalaninemia type I, or classic phenylketonuria), defects in

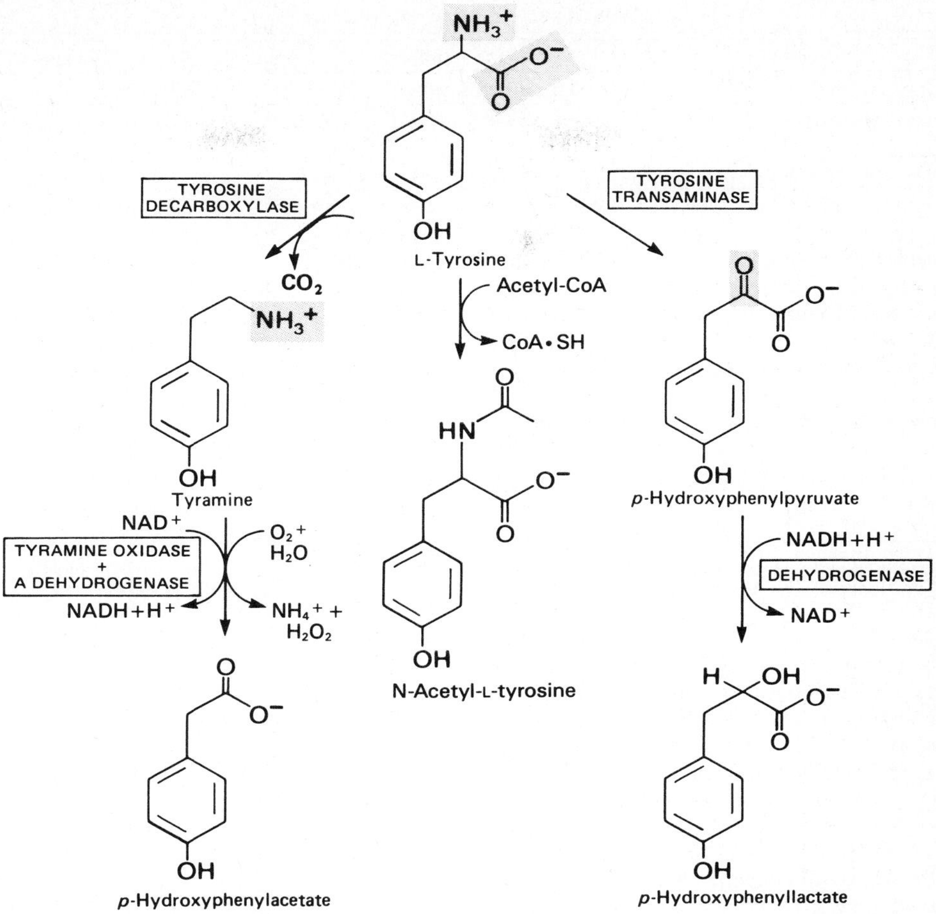

Figure 31–14. Alternative catabolites of tyrosine. *p*-Hydroxyphenylacetaldehyde is formed as an intermediate during oxidation of tyramine to *p*-hydroxyphenylacetate.

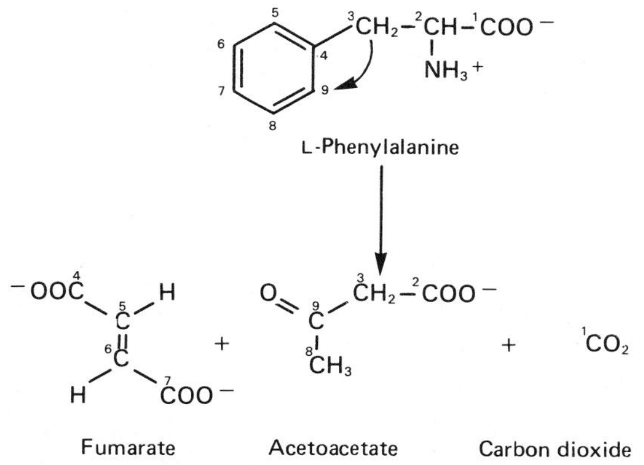

Figure 31–15. Ultimate catabolic fate of each carbon atom of phenylalanine. Pattern of isotopic labeling in the ultimate catabolites of phenylalanine (and tyrosine).

dihydrobiopterin reductase (hyperphenylalaninemia types II and III), and defects in dihydrobiopterin biosynthesis (hyperphenylalaninemia types IV and V). Additional types have, however, been identified (Table 31–3).

The major consequence of untreated **hyperphenylalaninemia type I (classic phenylketonuria; PKU)** is the mental retardation that results in IQ below 70 in late childhood. Additional clinical signs include seizures, psychoses, eczema, and a "mousy" odor. However, if diagnosis and initiation of appropriate treatment are prompt, these symptoms may be avoided. Because of the availability of animal models and because prompt dietary intervention can ameliorate the otherwise inevitable mental retardation, PKU has served as a model for study of the mental retardation associated with metabolic diseases. In classic PKU, a heritable disorder with a frequency of about 1:10,000 live births, levels of component I of liver phenylalanine hydroxylase (see Fig 29–12) average approximately 25% of normal, and the hydroxylase is insensitive to regulation by phenylalanine.

Table 31–3. Hyperphenylalaninemias.*

Type	Condition	Defect	Treatment
I	Phenylketonuria	Phe hydroxylase absent	Low Phe diet
II	Persistent hyperphenylalaninemia	Decreased Phe hydroxylase	None, or temporary dietary therapy
III	Transient mild hyperphenylalaninemia	Maturational delay of hydroxylase	Same as type II
IV	Dihydropteridine reductase deficiency	Deficient or absent dihydropteridine reductase	Dopa, 5-hydroxytryptophan, carbidopa
V	Abnormal dihydrobiopterin function	Dihydrobiopterin synthesis defect	Dopa, 5-hydroxytryptophan, carbidopa
VI	Persistent hyperphenylalaninemia and tyrosinemia	? Catabolism tyrosine	Reduced Phe intake
VII	Transient neonatal tyrosinemia	p-Hydroxyphenylpyruvic oxidase inhibition	Vitamin C
VIII	Hereditary tyrosinemia	Deficiency: 1. p-Hydroxyphenylpyruvate deoxygenase 2. Cytoplasmic tyrosine aminotransferase 3. Fumarylacetoacetate	Low Tyr diet Low Tyr diet plus glutathione injections

*Modified and reproduced, with permission, from Tourian A, Sidbury JB: Phenylketonuria and hyperphenylalaninemia. Page 273 in: Stanbury JB et al (editors): *The Metabolic Basis of Inherited Disease,* 5th ed. McGraw-Hill, 1983.

The patient is unable to convert phenylalanine to tyrosine, and, as a result, alternative catabolites of phenylalanine are produced (Fig 31–16). These include phenylpyruvic acid, the product of deamination of phenylalanine; phenyllactic acid, the reduction product of phenylpyruvic acid; and phenylacetic acid, produced by decarboxylation and oxidation of phenylpyruvic acid. Much of the phenylacetate is conjugated in the liver with glutamine and excreted in the urine as the conjugate, phenylacetylglutamine. Table 31–4 illustrates the chemical pattern in the blood and urine of a phenylketonuric patient. The presence in urine of the keto acid phenylpyruvate gives the disease its name—phenylketonuria.

In the absence of a normal catabolic pathway for phenylalanine, several reactions of otherwise minor quantitative importance in normal liver assume a major catabolic role. In phenylketonurics, phenylpyruvate, phenyllactate, phenylacetate, and its glutamine conjugate phenacetylglutamine are formed and occur in the blood and urine (Fig 31–16). Although phenylpyruvate, present in the urine of most phenylketonuric patients, can be detected by a simple biochemical spot test, definitive diagnosis requires determination of elevated plasma phenylalanine levels.

Further deterioration of mental performance of phenylketonuric children can be prevented if they are maintained on a diet containing very low levels of phenylalanine. The diet can be terminated at 6 years of age, when high concentrations of phenylalanine and its derivatives no longer are injurious to the brain.

Plasma phenylalanine may be measured by an automated micro method that requires as little as 20 μL of blood. However, abnormally high blood phenylalanine levels may not occur in phenylketonuric infants until the third or fourth day of life, because of their initially low intake of dietary protein. Furthermore,

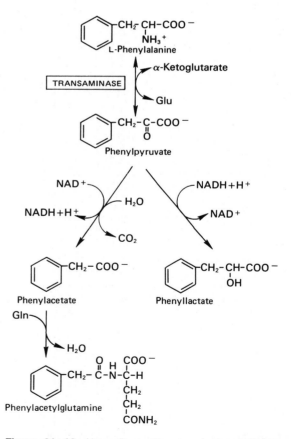

Figure 31–16. Alternative pathways of phenylalanine catabolism of particular importance in phenylketonuria. The reactions shown also occur in the liver tissue of normal individuals but are of minor significance if a functional phenylalanine hydroxylase is present. Glu, glutamate; Gln, glutamine.

Table 31-4. Metabolites of phenylalanine accumulating in the plasma and urine of phenylketonuric patients.

Metabolite	Plasma (mg/dL)		Urine (mg/dL)	
	Normal	Phenylketonuric	Normal	Phenylketonuric
Phenylalanine	1-2	15-63	30	300-1000
Phenylpyruvate		0.3-1.8		300-2000
Phenyllactate				290-550
Phenylacetate				Increased
Phenylacetylglutamine			200-300	2400

false-positive results may occur in premature infants owing to delayed maturation of the enzymes required for phenylalanine catabolism. A useful but less reliable screening test depends on detecting elevated urinary levels of phenylpyruvate with ferric chloride.

Administration of phenylalanine to a phenylketonuric subject should result in prolonged elevation of the level of this amino acid in the blood, indicating diminished tolerance to phenylalanine. However, abnormally low tolerance to injected phenylalanine and a high fasting level of phenylalanine are also characteristic of the parents of phenylketonurics. The defective gene responsible for phenylketonuria thus can be detected biochemically in the phenotypically normal, heterozygous parents.

Lysine

Lysine provides an exception to the rule that the first step in catabolism of an amino acid is removal of its α-amino group by transamination. In mammalian tissues, neither the α- nor ϵ-nitrogen atoms of L-lysine undergo transamination. Mammals convert the intact carbon skeleton of L-lysine to α-aminoadipate and α-ketoadipate (Fig 31-17). L-Lysine was formerly thought to be degraded via pipecolic acid, a cyclic imino acid. However, while liver degrades D-lysine via pipecolate, L-lysine is degraded via saccharopine (Fig 31-18), an intermediate in lysine biosynthesis by fungi.

L-Lysine first condenses with α-ketoglutarate, splitting out water and forming a Schiff base. This is reduced to saccharopine by a dehydrogenase and then oxidized by a second dehydrogenase. Addition of water forms L-glutamate and L-α-aminoadipate-δ-semialdehyde. The net effect of this reaction sequence is equivalent to removal of the ϵ nitrogen of lysine by transamination. One mole each of L-lysine and of α-ketoglutarate are converted to α-aminoadipate-δ-

semialdehyde and glutamate. However, NAD$^+$ and NADH are specifically required as cofactors, even though no net oxidation or reduction occurs.

Further catabolism of α-aminoadipate involves transamination to α-ketoadipate, probably followed by oxidative decarboxylation to glutaryl-CoA. While lysine is both glycogenic and ketogenic, the nature of the subsequent catabolites of glutaryl-CoA in mammalian systems is not known.

Metabolic disorders of lysine catabolism. Two rare metabolic abnormalities of lysine catabolism have been described. Both result from defects in enzymes that catabolize lysine to acetoacetyl-CoA, and in both instances the primary defect appears to involve impaired conversion of L-lysine and α-ketoglutarate to saccharopine (Fig 31-18).

A. Periodic Hyperlysinemia With Associated Hyperammonemia: In periodic hyperlysinemia, ingestion of normal levels of protein triggers hyperlysinemia. Secondary to the hyperlysinemia, hyperammonemia results from competitive inhibition of liver arginase activity (see Fig 30-13) by elevated levels of tissue lysine. Fluid therapy and restriction of dietary lysine intake relieve both the hyperammonemia and its clinical manifestations. Conversely, administration of a lysine load precipitates severe crises and coma. No information is available concerning the genetic basis of this disorder.

B. Persistent Hyperlysinemia Without Hyperammonemia: Clinical and biochemical findings have varied widely in the 12 reported cases of persistent hyperlysinemia. Some but not all patients are mentally retarded. There is no associated hyperammonemia, even in response to a lysine load. Lysine catabolites may or may not accumulate in biologic fluids. Persistent hyperlysinemia is believed to be inherited as an autosomal recessive trait. In addition to impaired conversion of lysine and α-ketoglutarate to

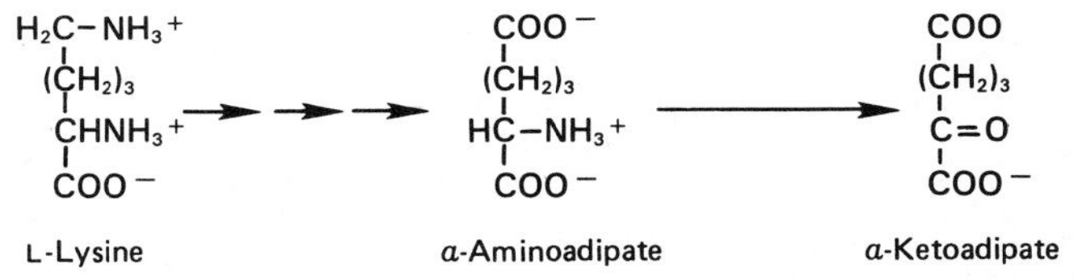

Figure 31-17. Conversion of L-lysine to α-aminoadipate and α-ketoadipate. Multiple arrows represent multiple reactions.

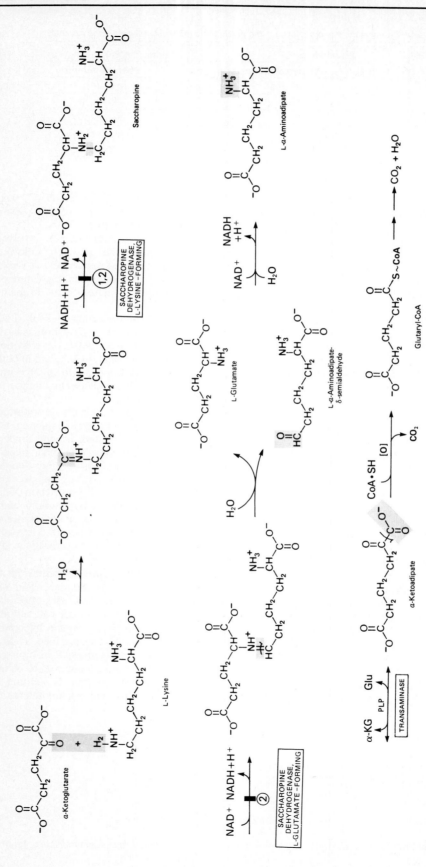

Figure 31–18. Catabolism of L-lysine. (α-KG, α-ketoglutarate; Glu, glutamate; PLP, pyridoxal phosphate.) The circled numerals indicate the probable sites of the metabolic defects in **1**, periodic hyperlysinemia with associated hyperammonemia; and **2**, persistent hyperlysinemia without associated hyperammonemia.

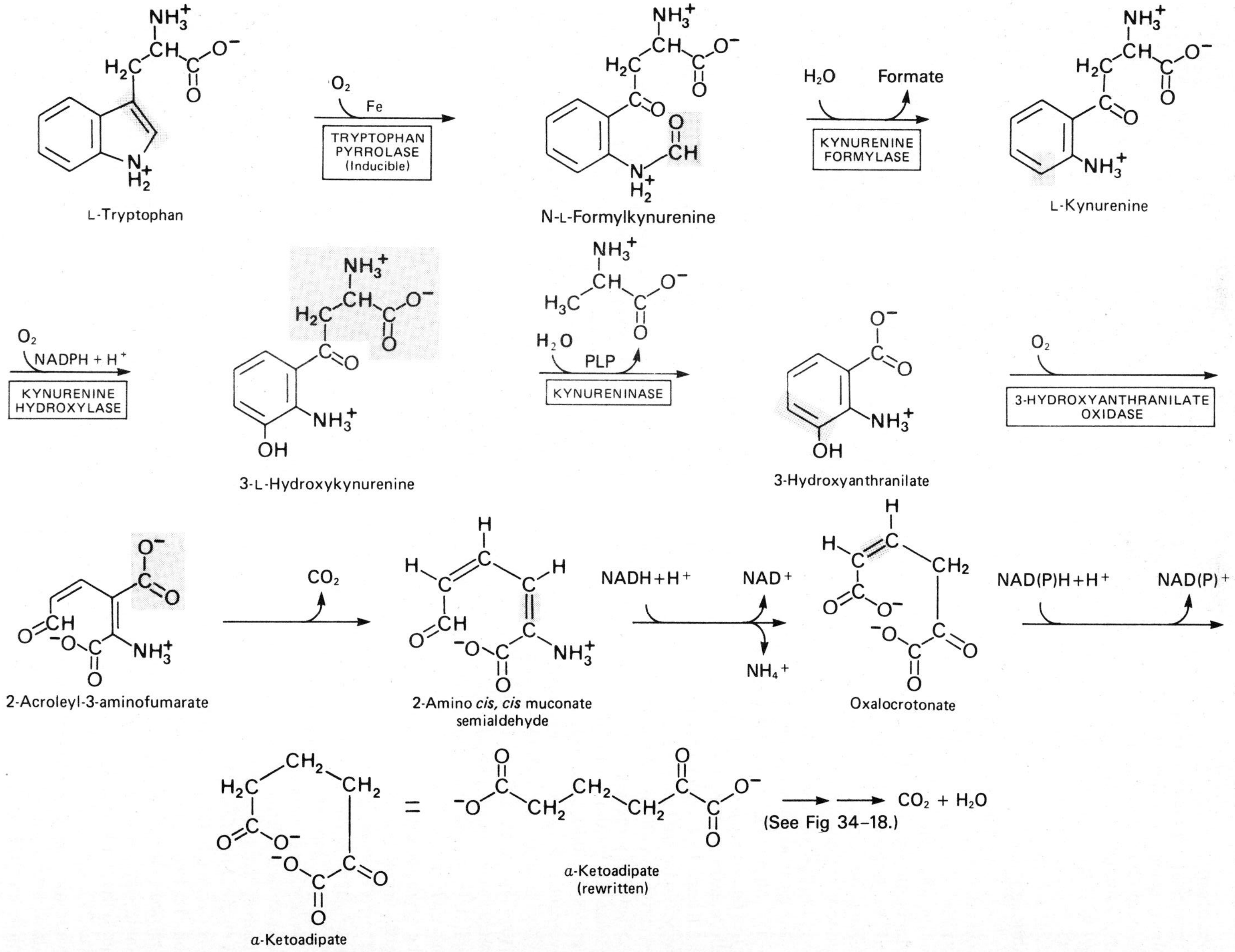

Figure 31–19. Catabolism of L-tryptophan. PLP, pyridoxal phosphate.

saccharopine, some patients appear to have an additional deficiency in the conversion of saccharopine to L-glutamate and α-aminoadipate-δ-semialdehyde (Fig 31–18).

Tryptophan

Tryptophan, notable for its variety of important metabolic reactions and products, was among the first amino acids shown to be nutritionally essential. *Neurospora* mutants, the bacterium *Pseudomonas,* and isolation of tryptophan metabolites from urine have proved invaluable aids in unraveling the details of tryptophan metabolism.

Although a large portion of the isotope of administered ^{14}C-L-tryptophan is incorporated into proteins, a considerable fraction appears in the urine as various catabolites. The carbon atoms both of the side chain and of the aromatic ring may be completely degraded to amphibolic intermediates via the **kynurenine-anthranilate pathway** (Fig 31–19), important both for tryptophan degradation and for conversion of tryptophan to **nicotinamide.**

Tryptophan oxygenase (tryptophan pyrrolase) catalyzes cleavage of the indole ring with incorporation of 2 atoms of molecular oxygen, forming N-formylkynurenine. The oxygenase, an iron porphyrin metalloprotein, is inducible in liver by adrenal corticosteroids and by tryptophan. A considerable portion of newly synthesized enzyme is in a latent form that requires activation. Tryptophan also stabilizes the oxygenase toward proteolytic degradation. Tryptophan oxygenase is feedback-inhibited by nicotinic acid derivatives, including NADPH.

Hydrolytic removal of the formyl group of N-formylkynurenine is catalyzed by **kynurenine formylase** of mammalian liver. Hydrolysis in $H_2{}^{18}O_2$ incorporates one equivalent of ^{18}O into the formate formed. The enzyme catalyzes similar reactions with various arylformylamines.

The reaction catalyzed by kynurenine formylase produces **kynurenine** (Fig 31–19). This may be deaminated by transamination of the amino group of the side chain to ketoglutarate. The resulting keto derivative, 2-amino-3-hydroxybenzoyl pyruvate, loses water, and spontaneous ring closure forms **kynurenic acid.** This compound, a by-product of kynurenine, is not formed in the main pathway of tryptophan breakdown (Fig 31–19).

Further metabolism of kynurenine involves conversion to **3-hydroxykynurenine,** which is converted to **3-hydroxyanthranilate.** Hydroxylation requires molecular oxygen in an NADPH-dependent reaction similar to that for hydroxylation of phenylalanine (see Chapter 29).

Kynurenine and hydroxykynurenine are converted to hydroxyanthranilate by **kynureninase,** a pyridoxal phosphate enzyme. A deficiency of vitamin B_6 results in partial failure to catabolize these kynurenine derivatives, which thus reach extrahepatic tissues where they are converted to **xanthurenate** (Fig 31–20). This abnormal metabolite occurs in the urine of humans,

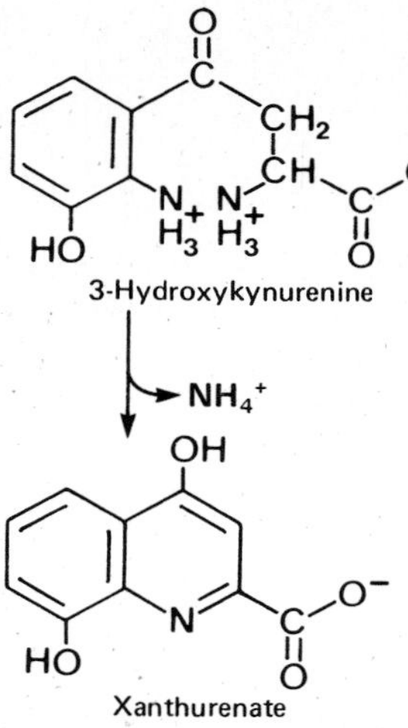

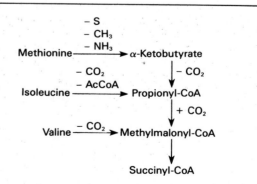

Figure 31–20. Formation of xanthurenate in vitamin B_6 deficiency. Conversion of the tryptophan metabolite 3-hydroxykynurenine to 3-hydroxyanthranilate is impaired (see Fig 31–19). A large portion is therefore converted to xanthurenate.

monkeys, and rats when dietary vitamin B_6 is inadequate. Feeding excess tryptophan induces excretion of xanthurenate in vitamin B_6 deficiency.

In many animals, conversion of tryptophan to nicotinic acid makes a supply of the vitamin in the diet unnecessary. In the rat, rabbit, dog, and pig, tryptophan can completely replace the vitamin in the diet; in humans and other animals, tryptophan increases the urinary excretion of nicotinic acid derivatives (eg, N-methylnicotinamide). In vitamin B_6 deficiency, synthesis of NAD^+ and $NADP^+$ may be impaired, a result of inadequate conversion of tryptophan to nicotinic acid for pyridine nucleotide synthesis. If an adequate supplement of nicotinic acid is supplied, pyridine nucleotide synthesis proceeds normally even in the absence of vitamin B_6.

Figure 31–21. Overall catabolism of methionine, isoleucine, and valine to succinyl-CoA. AcCoA, acetyl-CoA.

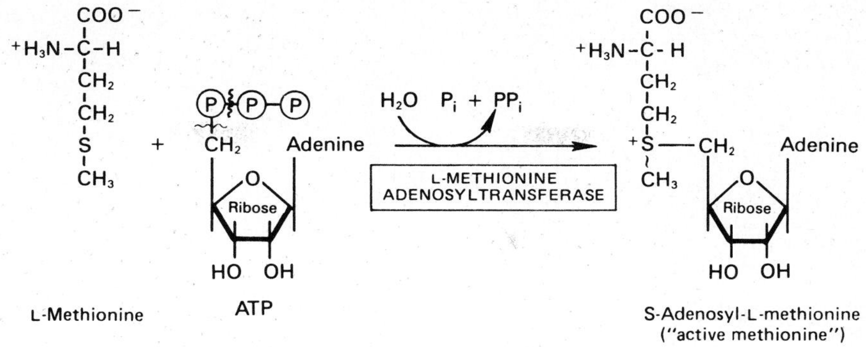

Figure 31–22. Formation of S-adenosylmethionine. The ~CH_3 represents the high transfer potential of the CH_3 of "active methionine."

Metabolic disorders of tryptophan catabolism. Hartnup disease, a hereditary abnormality in metabolism of tryptophan, is characterized by a pellagralike skin rash, intermittent cerebellar ataxia, and mental deterioration. The urine of patients with Hartnup disease contains greatly increased amounts of indoleacetate (α-N[indole-3-acetyl]glutamine) and tryptophan.

AMINO ACIDS FORMING SUCCINYL-COENZYME A

Overall Reactions

While succinyl-CoA is the amphibolic end product for catabolism of methionine, isoleucine, and valine, only portions of the skeletons are converted (Fig 31–21). Four-fifths of the carbons of valine, three-fifths of those of methionine, and half of those of isoleucine form succinyl-CoA. The carboxyl carbons of all 3 form CO_2. The terminal 2 carbons of isoleucine form acetyl-CoA, and the S-methyl group of methionine is removed as such.

What follows relates only to conversion of methionine and isoleucine to propionyl-CoA and of valine to methylmalonyl-CoA. The reactions leading from propionyl-CoA through methylmalonyl-CoA to succinyl-CoA are discussed in Chapter 23 in connection with catabolism of propionate and of fatty acids containing an odd number of carbon atoms.

Methionine

L-Methionine condenses with ATP, forming S-adenosylmethionine, or "active methionine" (Fig 31–22). The activated S-methyl group may transfer to

various acceptor compounds.* Removal of the methyl group forms S-adenosylhomocysteine. Hydrolysis of the S–C bond yields L-homocysteine plus adenosine. Homocysteine then condenses with serine, forming cystathionine (Fig 31–23). Hydrolytic cleavage of cystathionine forms L-homoserine plus cysteine, so that the net effect is conversion of homocysteine to homoserine and of serine to cysteine. These 2 reactions are therefore also involved in biosynthesis of cysteine from serine (see Chapter 29). Homoserine is converted to α-ketobutyrate by homoserine deaminase (Fig 31–24). Conversion of α-ketobutyrate to propionyl-CoA then occurs in the usual manner for oxidative decarboxylation of α-keto acids (eg, pyruvate, α-ketoglutarate) to form acyl-CoA derivatives.

Metabolic disorders of methionine catabolism. See Table 31–2.

Leucine, Valine, & Isoleucine

As might be suspected from their structural similarities, catabolism of L-leucine, L-valine, and L-isoleucine initially involves the same reactions. This common pathway then diverges, and each amino acid skeleton follows a unique pathway to amphibolic intermediates (Figs 31–25 and 31–26). The nature of these amphibolic end products determines whether an amino acid is glycogenic (valine), ketogenic (leucine), or both (isoleucine). **Many of the reactions involved are analogous to reactions of straight- and branched-chain fatty acid catabolism.** Because of the similarities noted in Fig 31–26, it is convenient to discuss initial reactions in catabolism of all 3 amino acids together. In what follows, reaction numbers refer to reactions of Figs 31–26 through 31–29.

A. Transamination: Reversible transamination (reaction 1) of all 3 branched L-α-amino acids in mammalian tissues probably involves a single transaminase. Reversibility of this reaction accounts for the ability of the corresponding α-keto acids to replace the L-α-amino acids in the diet if other adequate sources of nitrogen are available.

---

*Compounds whose methyl groups derive from S-adenosylmethionine include betaines, choline, creatine, epinephrine, melatonin, sarcosine, N-methylated amino acids, nucleotides, and many plant alkaloids.

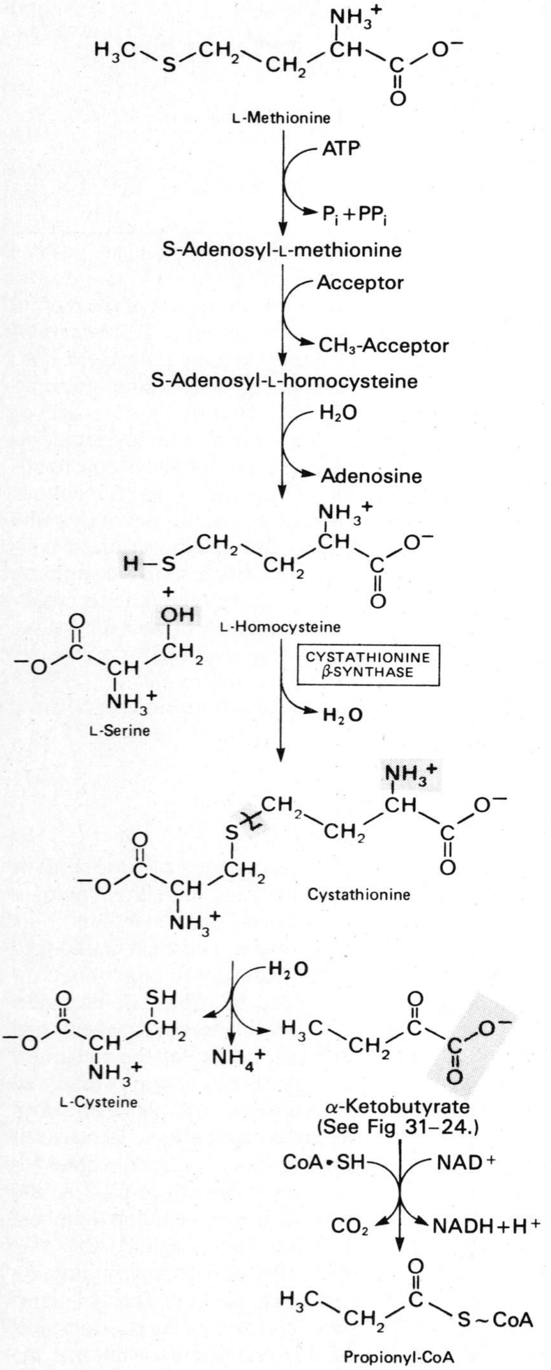

Figure 31–23. Conversion of methionine to propionyl-CoA.

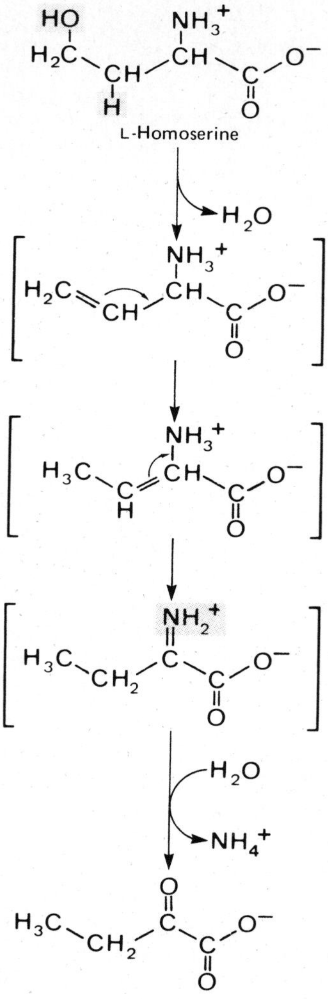

Figure 31–24. Conversion of L-homoserine to α-ketobutyrate, catalyzed by homoserine deaminase.

B. Oxidative Decarboxylation to Acyl-CoA Thioesters: This reaction (reaction 2) is analogous to oxidation of pyruvate to acetyl-CoA and CO_2 by pyruvate dehydrogenase and of α-ketoglutarate to CO_2 and succinyl-CoA by α-ketoglutarate dehydrogenase (see Chapter 18). The mammalian branched-chain α-keto acid dehydrogenase is an intramitochondrial multienzyme complex that catalyzes oxidative decarboxylation of α-ketoisocaproate (from leucine), α-keto-β-methylvalerate (from isoleucine), and α-ketoisovalerate (from valine).

The α-keto acid dehydrogenase complex subunits are analogous to those of pyruvate dehydrogenase. The catalytic activities are an α-keto acid decarboxylase, a transacylase, and a dihydrolipoyl dehydrogenase. As for pyruvate dehydrogenase, the complex is inactivated when phosphorylated by ATP in a reaction catalyzed by a protein kinase. A Ca^{2+}-independent

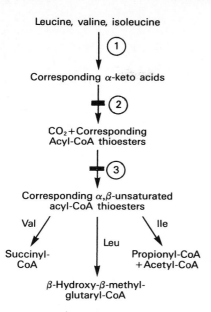

Figure 31–25. Catabolism of the branched-chain amino acids in mammals. Reactions 1–3 are common to all 3 amino acids; thereafter, the pathways diverge. Double lines intersecting arrows mark sites of metabolic blocks in 2 rare human diseases: at **2**, maple syrup urine disease, a defect in catabolism of all 3 amino acids; and at **3**, isovaleric acidemia, a defect of leucine catabolism.

phosphoprotein phosphatase catalyzes its dephosphorylation and accompanying reactivation. The phosphorylation state thus can regulate the catabolism of the branched-chain amino acids. The protein kinase is inhibited by ADP, the branched-chain α-keto acid products, the hypolipidemic agents clofibrate and dichloroacetate, and coenzyme A thioesters (eg, acetoacetyl-CoA). Of the branched-chain α-keto acids, α-ketoisocaproate (α-ketoleucine) is the most potent inhibitor.

C. Dehydrogenation to α,β-Unsaturated Acyl-CoA Thioesters: This reaction (reaction 3) is analogous to dehydrogenation of straight-chain acyl-CoA thioesters in fatty acid catabolism. It is not known whether a single enzyme catalyzes dehydrogenation of all 3 branched acyl-CoA thioesters. Indirect evidence suggesting that at least 2 enzymes are required derives from **isovaleric acidemia** wherein, following ingestion of protein-rich foods, isovalerate accumulates in the blood. An increase in other branched α-keto acids does not occur. Isovalerate is formed by deacylation of isovaleryl-CoA, the substrate for the above dehydrogenase.

Reactions Specific to Leucine Catabolism (See Fig 31–27.)

Reaction 4L: Carboxylation of β-methyl-crotonyl-CoA. A key observation leading to explanation of the ketogenic action of leucine was the discov-

ery that 1 mol of CO_2 was "fixed" (ie, covalently bound) per mole of isopropyl groups (from the terminal isopropyl group of leucine) converted to acetoacetate. This CO_2 fixation (reaction 4L, Fig 31–27) requires biotinyl-CO_2, formed from enzyme-bound biotin and CO_2 at the expense of ATP. This reaction forms β-methylglutaconyl-CoA as an intermediate.

Reaction 5L: Hydration of β-methylglutaconyl-CoA. The reaction product, β-hydroxy-β-methylglutaryl-CoA, is a precursor not only of ketone bodies (reaction 6L, Fig 31–27) but also of mevalonate, and hence of cholesterol and other polyisoprenoids (see Chapter 28).

Reaction 6L: Cleavage of β-hydroxy-β-methylglutaryl-CoA. Cleavage of β-hydroxy-β-methylglutaryl-CoA to acetyl-CoA and acetoacetate occurs in mammalian liver, kidney, and heart mitochondria. It explains the strongly ketogenic effect of leucine, since not only is 1 mol of acetoacetate formed per mole of leucine catabolized but another ½ mole of ketone bodies may be formed indirectly from the remaining product, acetyl-CoA (see Chapter 28).

Reaction Specific to Valine Catabolism (See Fig 31–28.)

Reaction 4V: Hydration of methylacrylyl-CoA. This reaction, which occurs nonenzymatically at a relatively rapid rate, is catalyzed by crotonase, a hydrolase of broad specificity for L-β-hydroxyacyl-CoA thioesters having 4–9 carbon atoms.

Reaction 5V: Deacylation of β-hydroxyisobutyryl-CoA. Since the CoA thioester is not a substrate for the subsequent reaction (reaction 6V, Fig 31–28), it must first be deacylated to β-hydroxyisobutyrate (reaction 5V, Fig 31–28). This is catalyzed by a deacylase, present in many animal tissues, whose only other substrate is β-hydroxypropionyl-CoA.

Reaction 6V: Oxidation of β-hydroxyisobutyrate. Mammalian tissues catalyze the NAD^+-dependent oxidation of the primary alcohol group of β-hydroxyisobutyrate to an aldehyde (reaction 6V, Fig 31–28), forming methylmalonate semialdehyde. The reaction is readily reversible.

Reaction 7V: Fate of methylmalonate semialdehyde. Two fates are possible for methylmalonate semialdehyde in mammalian tissues: transamination to β-aminoisobutyrate (reaction 7V, Fig 31–28) and conversion to succinyl-CoA (reactions 8V through 10V, Fig 31–28). Transamination to α-aminoisobutyrate, a normal urinary amino acid, is catalyzed by various mammalian tissues including kidney. The second major fate involves oxidation to methylmalonate, acylation to methylmalonyl-CoA, and isomerization to succinyl-CoA (reactions 8V through 10V, Fig 31–28). Isomerization (reaction 10V, Fig 31–28) requires adenosylcobalamin coenzyme and is catalyzed by methylmalonyl-CoA mutase. This reaction is important not only for valine catabolism but also for that of propionyl-CoA, a catabolite of isoleucine (Fig 31–29). In cobalamin (vitamin B_{12}) deficiency, mu-

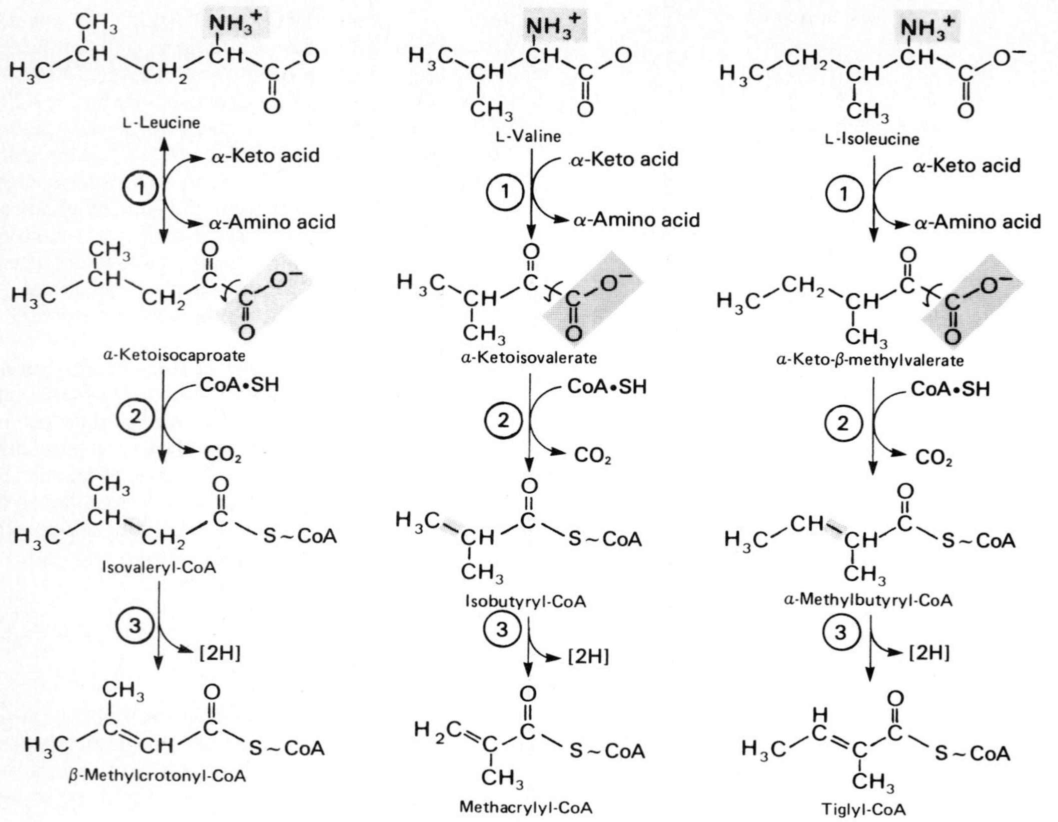

L-Leucine

α-Keto acid

α-Amino acid

CH₃
H₃C—CH—CH₂—C—C—O⁻
α-Ketoisocaproate

CoA·SH

CO₂

CH₃
H₃C—CH—CH₂—C—S~CoA
Isovaleryl-CoA

[2H]

CH₃
H₃C—C=CH—C—S~CoA
β-Methylcrotonyl-CoA

L-Valine

α-Keto acid

α-Amino acid

H₃C—CH—C—C—O⁻
CH₃
α-Ketoisovalerate

CoA·SH

CO₂

H₃C—CH—C—S~CoA
CH₃
Isobutyryl-CoA

[2H]

H₂C=C—C—S~CoA
CH₃
Methacrylyl-CoA

L-Isoleucine

α-Keto acid

α-Amino acid

H₃C—CH₂—CH—C—C—O⁻
CH₃
α-Keto-β-methylvalerate

CoA·SH

CO₂

H₃C—CH—CH—C—S~CoA
CH₃
α-Methylbutyryl-CoA

[2H]

H₃C—C=C—C—S~CoA
H   CH₃
Tiglyl-CoA

Figure 31–26. The analogous first 3 reactions in the catabolism of leucine, valine, and isoleucine. Note also the analogy in reactions 2 and 3 to the catabolism of fatty acids. This latter analogy continues, as shown in subsequent figures. α-KA, α-keto acid; α-AA, α-amino acid.

tase activity is impaired. This produces a "dietary metabolic defect" in ruminants that utilize propionate (from fermentation in the rumen) as an energy source. The purified mutase from sheep liver contains 2 mol of deoxyadenosyl-B_{12} per mole. Rearrangement to succinyl-CoA occurs via an intramolecular shift of the CoA-carboxyl group. Although the overall reaction resembles isomerization of threo-β-methylaspartate to glutamate, the reaction mechanisms appear to differ.

Reactions Specific to Isoleucine Catabolism
(See Fig 31–29.)

As with valine and leucine, the first data concerning isoleucine catabolism came from dietary studies in intact animals that identified isoleucine as glycogenic and weakly ketogenic. Glycogen synthesis from isoleucine was confirmed using D_2O. Use of ^{14}C-labeled intermediates and liver slice preparations revealed that the isoleucine skeleton was cleaved, forming acetyl-CoA and propionyl-CoA (Fig 31–29).

Reaction 4I: Hydration of tiglyl-CoA. This reaction, like the analogous reaction in valine catabolism (reaction 4V, Fig 31–28), is catalyzed by mammalian crotonase.

Reaction 5I: Dehydrogenation of α-methyl-β-hydroxybutyryl-CoA. This reaction is analogous to reaction 5V of valine catabolism (Fig 31–28). In valine catabolism, it will be recalled, the hydroxylated acyl-CoA thioester is first deacylated and then oxidized.

Reaction 6I: Thiolysis of α-methylacetoacetyl-CoA. Thiolytic cleavage of the covalent bond linking carbons 2 and 3 of α-methylacetoacetyl-CoA resembles thiolysis of acetoacetyl-CoA to 2 mol of acetyl-CoA catalyzed by β-ketothiolase. The products, acetyl-CoA (ketogenic) and propionyl-CoA (glycogenic), account for the ketogenic and glycogenic properties of isoleucine.

Metabolic Defects of Branched-Chain Amino Acid Catabolism (Leucine, Valine, Isoleucine)

Four defects in branched-chain amino acid catabolism are known. Of these, maple syrup urine disease has been most extensively studied. Over 50 cases have been reported. The incidence of the disease has been estimated at 5–10 per million live births. Hypervalinemia, intermittent branched-chain ketonuria,

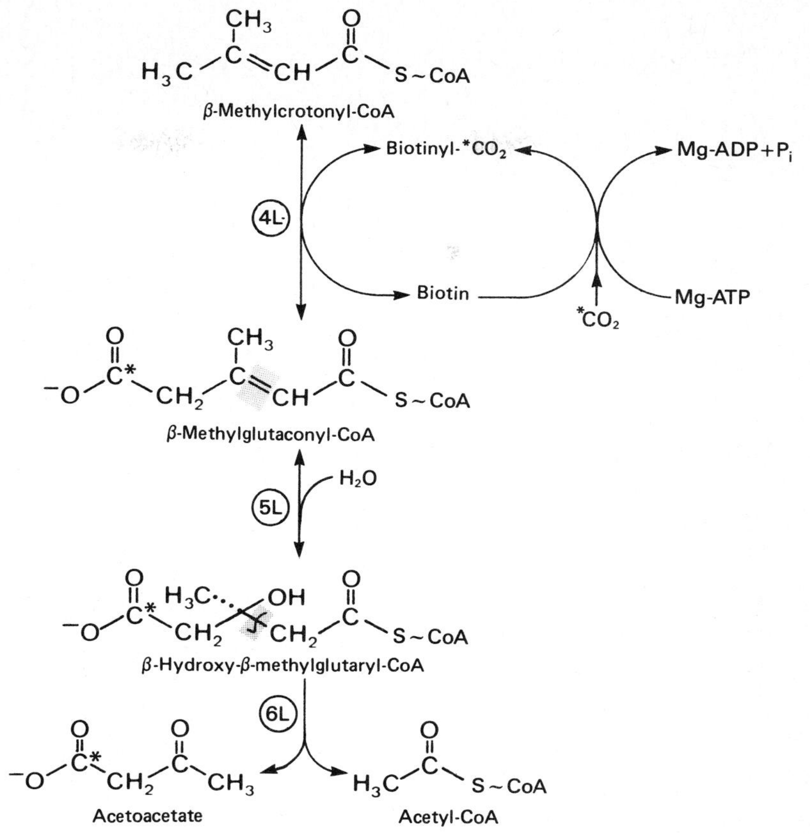

Figure 31–27. Subsequent catabolism of the β-methylcrotonyl-CoA formed from L-leucine (see Fig 31–26). *Carbon atoms derived from CO_2. For structure of biotinyl-CO_2, see Fig 20–4.

and isovaleric acidemia have been reported in fewer than 5 children.

A. Hypervalinemia: This metabolic disease, characterized by elevated plasma levels of valine (but not of leucine or isoleucine), reflects the inability to transaminate valine to α-ketoisovalerate (reaction 1, Fig 31–26). However, transamination of leucine and isoleucine (reaction 1, Fig 31–26) is unimpaired.

B. Maple Syrup Urine Disease: The most striking feature of this hereditary disease is the characteristic odor of the urine, which resembles that of maple syrup or burnt sugar. Plasma and urinary levels of the branched-chain amino acids leucine, isoleucine, and valine and their corresponding α-keto acids are greatly elevated. For this reason, the disease has also been termed **branched-chain ketonuria.** Smaller quantities of branched-chain α-hydroxy acids, formed by reduction of the α-keto acids, also are present in the urine.

Characteristic signs of the disease are evident by the end of the first week of extrauterine life. In addition to the biochemical abnormalities described above, the infant is difficult to feed and may vomit. The patient may also exhibit lethargy. Diagnosis prior to 1 week of age is possible only by enzymatic analysis.

Extensive brain damage occurs in surviving children. Without treatment, death usually occurs by the end of the first year of life.

The biochemical defect is the absence or greatly reduced activity of the α-keto acid decarboxylase that catalyzes conversion of all 3 branched-chain α-keto acids to CO_2 plus acyl-CoA thioesters (reaction 2, Fig 31–26). This was established by enzymatic analysis of leukocytes and of cultured skin fibroblasts from afflicted children. The mechanism of toxicity is unknown.

Patients are placed on a diet in which protein is replaced by a mixture of purified amino acids from which leucine, isoleucine, and valine are omitted. When plasma levels of these amino acids fall within the normal range, they are restored to the diet in the form of milk and other foods in amounts adequate to supply—but not to exceed—the requirements for branched-chain amino acids. There is no indication when, if ever, dietary restrictions may be eased. When treatment was initiated in the first week of life, considerable success was achieved in mitigating the dire consequences of the disease.

C. Intermittent Branched-Chain Ketonuria: This disease, a variant of maple syrup urine disease,

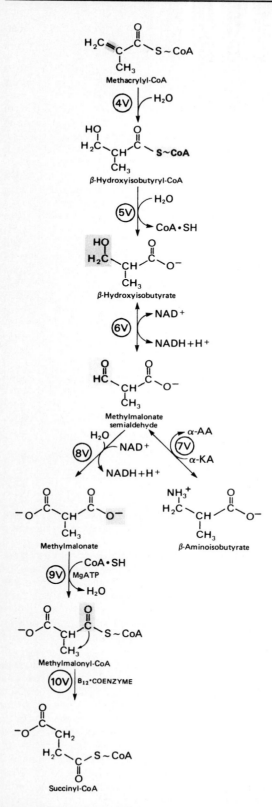

Figure 31–28. Subsequent catabolism of the methacryl-yl-CoA formed from L-valine (see Fig 31–26). α-KA, α-keto acid; α-AA, α-amino acid.

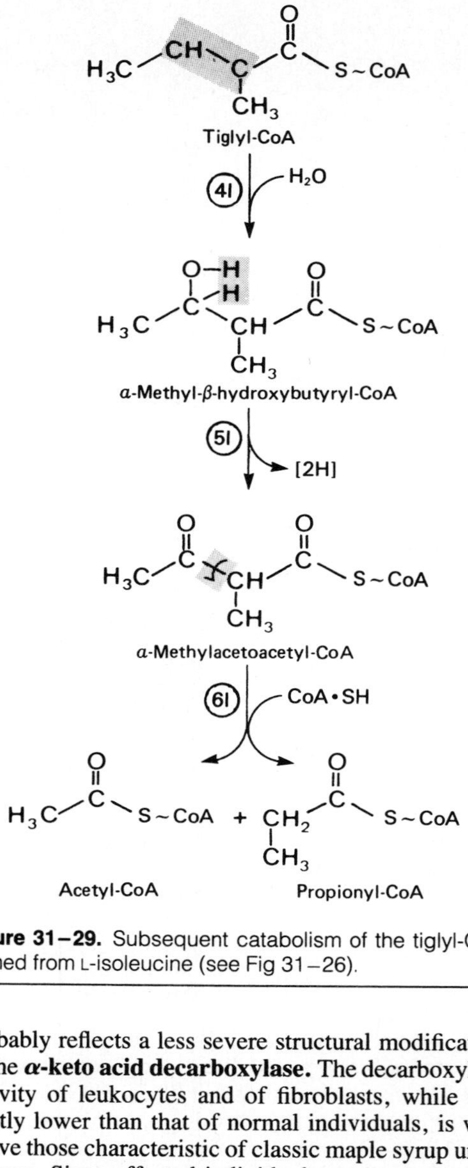

Figure 31–29. Subsequent catabolism of the tiglyl-CoA formed from L-isoleucine (see Fig 31–26).

probably reflects a less severe structural modification of the **α-keto acid decarboxylase.** The decarboxylase activity of leukocytes and of fibroblasts, while distinctly lower than that of normal individuals, is well above those characteristic of classic maple syrup urine disease. Since affected individuals appear to possess an impaired but nevertheless distinct capability for catabolism of leucine, valine, and isoleucine, it is perhaps understandable that the typical symptoms of maple syrup urine disease occur later in life and only intermittently. The prognosis for successful use of dietary therapy would appear to be far more favorable in these individuals.

Taken together, maple syrup urine disease and intermittent branched-chain ketonuria appear to illustrate the situation described in the introduction to this section—mutations causing different changes in the primary structure of the same enzyme. It is probable that a spectrum of activities ranging from frank disease through intermittent manifestations to normal values in fact occurs in individual subjects.

D. Isovaleric Acidemia: Relevant findings include a persistent "cheesy" odor of the breath and body fluids, vomiting, acidosis, and coma precipitated by excessive ingestion of protein or by an episode of infectious disease. Mild mental retardation was associated with the 3 known cases. The impaired enzyme is **isovaleryl-CoA dehydrogenase** (reaction 3, Fig 31–26). Isovaleryl-CoA thus accumulates, is hydrolyzed to isovalerate, and is excreted in the urine and sweat.

Additional Metabolic Defects Related to Amino Acid Catabolism (Propionate, Methylmalonate, & Vitamin B₁₂)

Propionyl-CoA (Fig 31–21) is formed from isoleucine (Figs 31–26 and 31–29) and methionine (Fig 31–23), as well as from the side chain of cholesterol and from fatty acids with odd numbers of carbon atoms. The conversion of propionyl-CoA to amphibolic intermediates involves biotin-dependent carboxylation to methylmalonyl-CoA. Methylmalonyl-CoA also is formed directly (ie, without prior formation of propionyl-CoA) from valine (Figs 31–21 and 31–28, reaction 9V). A vitamin B_{12} coenzyme-dependent isomerization converts methylmalonyl-CoA to succinyl-CoA, a citric acid cycle intermediate, which is oxidized to CO_2 and water.

Shortly after the discovery that 5'-deoxyadenosyl-cobalamin is a cofactor for the isomerization of methylmalonyl-CoA to succinyl-CoA, patients with acquired vitamin B_{12} deficiency were observed to excrete large quantities of methylmalonate in their urine. This methylmalonic aciduria disappeared when sufficient vitamin B_{12} was administered.

A. Propionic Acidemia: Propionyl-CoA carboxylase deficiency is characterized by high serum propionate levels and by defective catabolism of propionate by leukocytes. Treatment involves feeding a low-protein diet and measures to counteract metabolic acidosis.

B. Methylmalonic Aciduria: Two forms of methylmalonic aciduria are known. One responds to parenteral administration of physiologic doses of vitamin B_{12}; the other does not. A patient with this latter condition responded favorably to massive (pharmacologic) doses (1 g/d) of vitamin B_{12}. When fibroblasts from this patient were cultured in a medium containing vitamin B_{12} (25 pg/mL) and were oxidized with ^{14}C-propionate, they were found to oxidize poorly. The cultured cells contained only about 10% as much 5'-deoxyadenosylcobalamin as did control cells. When the concentration of vitamin B_{12} in the medium was increased 10,000-fold, the rate of propionate oxidation and the intracellular concentration of 5'-deoxyadenosylcobalamin both approached normal. No defect in binding the coenzyme to the mutase apoenzyme was observed. The defect in the latter form of methylmalonic aciduria thus appears to be the inability to form 5'-deoxyadenosylcobalamin from normal levels of the vitamin.

The above selection of inherited diseases of amino acid catabolism is generally confined to the better studied diseases now known. For a more extensive recent review of this subject, see Wellner and Meister, 1981.

REFERENCES

Bremer HJ et al: *Amino Acid Metabolism: Clinical Chemistry and Diagnosis*. Urban & Schwarzenberg, 1981.

Cooper AJL: Biochemistry of the sulfur-containing amino acids. *Annu Rev Biochem* 1983;**52**:187.

Felig P: Amino acid metabolism in man. *Annu Rev Biochem* 1975;**44**:933.

Frimter GW: Aminoacidurias due to disorders of metabolism. (2 parts.) *N Engl J Med* 1973;**289**:835, 895.

Paxton R, Harris RA: Isolation of rabbit liver branched chain α-ketoacid dehydrogenase and regulation by phosphorylation. *J Biol Chem* 1982;**257**:14433.

Paxton R, Harris RA: Regulation of branched-chain α-ketoacid

dehydrogenase kinase. *Arch Biochem Biophys* 1984;**231**:48.

Rosenberg LE, Scriver CR: Disorders of amino acid metabolism. Chapter 11 in: *Metabolic Control and Disease*. Bondy PK, Rosenberg LE (editors). Saunders, 1980.

Schwarz V: *A Clinical Companion to Biochemical Studies*. Freeman, 1978.

Stanbury JB et al (editors): *The Metabolic Basis of Inherited Disease*, 5th ed. McGraw-Hill, 1983.

Wellner D, Meister A: A survey of inborn errors of amino acid metabolism and transport in man. *Annu Rev Biochem* 1981;**50**:911.

32

Conversion of Amino Acids to Specialized Products

Victor W. Rodwell, PhD

INTRODUCTION

Amino acids, the primary source of nitrogen for animals, serve as precursors for other nitrogenous compounds. Since most of these products are not amino acids per se, the discussion merges with metabolic pathways discussed elsewhere in this book. Physiologically important products derived from amino acids include heme, purines, pyrimidines, hormones, and neurotransmitters including biologically active peptides. In addition, many proteins contain amino acids that have been modified for a specific function, eg,

calcium binding or cross-linking, and thus the amino acid residues in those proteins serve as precursors for these modified residues. Finally, there are small peptides or peptidelike molecules not synthesized on ribosomes that carry out specific functions in cells.

BIOMEDICAL IMPORTANCE

Specific compounds of great medical interest formed from amino acids include histamine. This biologically active amine is formed by the decarboxyl-

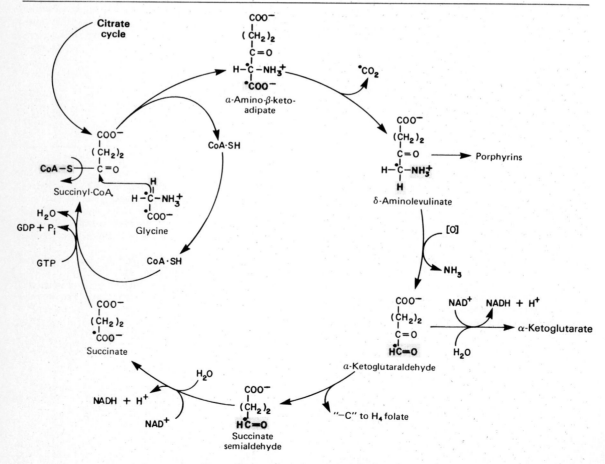

Figure 32–1. The succinate-glycine cycle.

ation of histidine and plays a central role in many allergic reactions in the human body. Specific neurotransmitters derived from amino acids include γ-aminobutyrate from glutamate; 5-hydroxytryptamine (serotonin) from tryptophan; and dopamine, norepinephrine, and epinephrine from tyrosine. Understanding how the brain works depends in part on a thorough knowledge of various aspects of neurotransmitters. In addition, many drugs used to treat neurologic and psychiatric conditions affect the metabolism of the neurotransmitters mentioned above.

GLYCINE

Synthesis of Heme

The α-carbon and the nitrogen atom of glycine are used for synthesis of the porphyrin moiety of hemoglobin (see Chapter 33). The pyrrole nitrogen is derived from glycine nitrogen and an adjoining carbon from the α carbon of glycine. The α carbon is also the source of the methylene bridge atoms linking the pyrrole rings.

In the **"succinate-glycine cycle"** (Fig 32–1), succinyl-CoA condenses on the α-carbon atom of glycine to form α-amino-β-ketoadipate. The citric acid cycle provides succinyl-CoA. α-Amino-β-ketoadipate is decarboxylated to δ-aminolevulinate, a precursor for porphyrin synthesis. Succinate and α-ketoglutarate (α-KG), which may return to the citric acid cycle, are also formed.

Metabolic disorders of heme metabolism are discussed in Chapter 33.

Synthesis of Purines

The entire glycine molecule is utilized to form positions 4, 5, and 7 of the purine skeleton. (see Chapter 35).

Formation of Glycine Conjugates

Glycine conjugates with cholic acid, forming glycocholic acid. With benzoate, it forms hippurate (Fig 32–2). The quantitative ability of liver to convert a measured dose of benzoate to hippurate formerly was used as a test of liver function.

Synthesis of Creatine

The sarcosine (N-methylglycine) component of creatine is derived from glycine and S-adenosylmethionine.

α-ALANINE

Alanine, together with glycine, makes up a considerable fraction of the amino nitrogen in human plasma. Alanine is also a major component of bacterial cell walls, partly as the D-isomer: 39–50% in *Streptococcus faecalis;* 67% in *Staphylococcus aureus.*

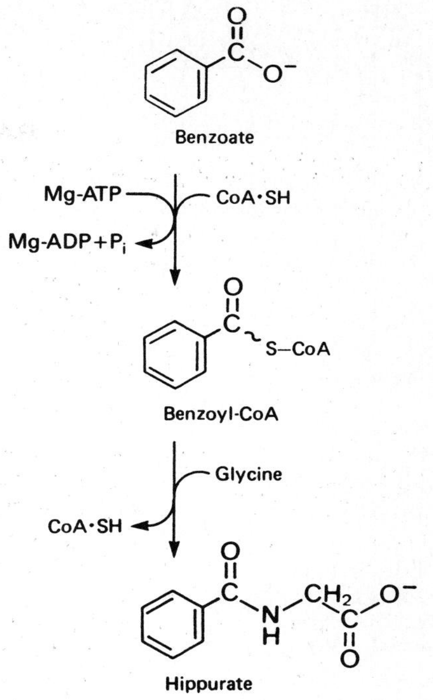

Figure 32–2. Hippurate biosynthesis.

β-ALANINE

Little free β-alanine is present in tissues. Considerably more is present as β-alanyl dipeptides (see below) and as coenzyme A (see Fig 17–6).

Biosynthesis

While microorganisms form β-alanine by α-decarboxylation of aspartate, mammalian tissue β-alanine arises principally from catabolism of uracil (see Fig 35–17), carnosine, and anserine (Fig 32–3).

Catabolism

Catabolism of β-alanine in mammals involves transamination to malonate semialdehyde, which is then oxidized to acetate and thence to CO_2.

Hyper-β-alaninemia

In this rare metabolic disorder, free β-alanine levels are elevated in body fluids (plasma, cerebrospinal fluid, urine) and tissues (brain, kidney, liver, and skeletal muscle). Taurine and β-aminoisobutyrate levels also are elevated.

β-ALANYL & RELATED DIPEPTIDES

The major fraction of β-alanine in humans is present in the skeletal muscle dipeptide carnosine (Fig 32–3). The closely related β-alanyl dipeptide anser-

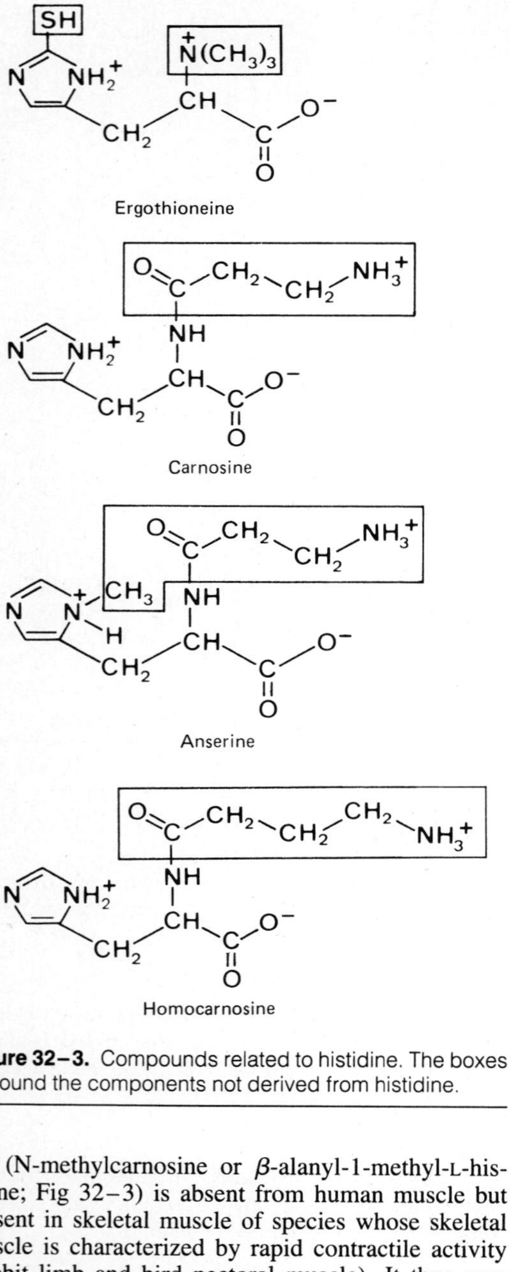

Figure 32–3. Compounds related to histidine. The boxes surround the components not derived from histidine.

ine (N-methylcarnosine or β-alanyl-1-methyl-L-histidine; Fig 32–3) is absent from human muscle but present in skeletal muscle of species whose skeletal muscle is characterized by rapid contractile activity (rabbit limb and bird pectoral muscle). It thus may fulfill physiologic functions distinct from carnosine.

The physiologic functions of β-alanyl-imidazole dipeptides are incompletely understood. They may serve to buffer the pH of anaerobically contracting skeletal muscle. Carnosine and anserine both activate myosin ATPase activity in vitro. Both dipeptides also chelate copper and enhance copper uptake. They thus may participate in pathologic processes in Wilson's disease (see Chapter 7).

Biosynthesis

Carnosine is formed from β-alanine and L-histidine in an ATP-requiring reaction catalyzed by carnosine synthetase:

ATP + L-Histidine + β-Alanine $\longrightarrow$

AMP + PP$_i$ + Carnosine

Anserine is formed from carnosine (the methyl group donor is S-adenosylmethionine) in a reaction catalyzed by carnosine N-methyltransferase:

S-Adenosylmethionine + Carnosine $\longrightarrow$

S-Adenosylhomocysteine + Anserine

The overall reaction involves enzyme-bound β-alanyl-adenylate.

Uptake of β-Alanyl Dipeptides

Kidney tissue and intestinal enterocytes take up carnosine and β-alanine by membrane carriers that discriminate one substrate from the other and both from other dipeptides. This ability to differentiate β-alanine from carnosine uptake has been applied to the definition of Hartnup disease, a heritable disorder in the transport of certain neutral α-amino acids (see Chapter 31). In patients with Hartnup disease, the blood histidine response is normal if carnosine is fed but is attenuated following administration of L-histidine.

Catabolism

Carnosine is hydrolyzed to β-alanine and L-histidine by the serum zinc metalloenzyme carnosinase (carnosine hydrolase). Two forms of carnosinase are present in serum.

Homocarnosine

The physiologic function of homocarnosine (γ-aminobutyryl-L-histidine; Fig 32–3), a central nervous system dipeptide closely related structurally and metabolically to carnosine, is not known. This dipeptide of γ-aminobutyrate and L-histidine is present in human brain tissue, where its concentration varies with the region examined. Biosynthesis of homocarnosine in human brain appears to be catalyzed by carnosine synthetase. However, serum carnosinase does not hydrolyze homocarnosine.

Serum Carnosinase Deficiency

This presumably autosomal recessive, heritable disorder is characterized by persistent carnosinuria and occasionally also by carnosinemia. Carnosinuria persists even if carnosine is excluded from the diet.

Homocarnosinosis

Levels of homocarnosine are elevated in cerebrospinal fluid and in brain but not in plasma or urine. A single case has been reported.

SERINE

Much of the serine in phosphoproteins is present as O-phosphoserine.

Serine is involved in synthesis of sphingosine. (See Chapter 25.)

Serine participates in purine and pyrimidine synthesis. The β carbon is a source of the methyl groups of thymine (and of choline) and of the carbon in positions 2 and 8 of the purine nucleus. (See Chapter 35.)

THREONINE

Since threonine does not participate in transamination, the D-isomer and the α-keto acid are not utilized by mammals. Threonine is present in certain proteins as O-phosphothreonine.

METHIONINE

Methionine as a methyl group donor is discussed in Chapter 31. In the form of S-adenosylmethionine, it is the principal source of methyl groups in the body. In addition to direct utilization, the methyl group is also oxidized. The methyl carbon may be used to produce the one-carbon moiety that conjugates with glycine in synthesis of serine. As S-adenosylmethionine, methionine serves as precursor to the 1,3-diaminopropane portions of the polyamines spermine and spermidine (see Ornithine, below, and Fig 32–6).

CYSTEINE

Urinary sulfate arises almost entirely from oxidation of L-cysteine. The sulfur of methionine (as homocysteine) is transferred to serine (see Fig 29–10) and thus contributes to the urinary sulfate indirectly (ie, via cysteine). L-Cysteine serves as a precursor of the thioethanolamine portion of coenzyme A. Cysteine is also a precursor of the taurine that conjugates with bile acids, forming taurocholic acid and other products.

HISTIDINE

Histamine is derived from histidine by decarboxylation, a reaction catalyzed in mammalian tissues by an **aromatic L-amino acid decarboxylase.** This enzyme also catalyzes decarboxylation of dopa, 5-hydroxytryptophan, phenylalanine, tyrosine, and tryptophan (see below). The decarboxylase is inhibited by α-methyl amino acids in vitro and in vivo that thus have clinical application as antihypertensive agents. In addition to the aromatic amino acid decarboxylase, a different enzyme, **histidine decarboxylase,** present in most cells, catalyzes decarboxylation of histidine.

Histidine compounds present in the body include **ergothioneine,** in red blood cells and liver; carnosine; and anserine (Fig 32–3). 1-Methylhistidine in human urine probably is derived from anserine. 3-Methylhistidine, identified in human urine in amounts of about 50 mg/dL, is unusually low in the urine of patients with Wilson's disease.

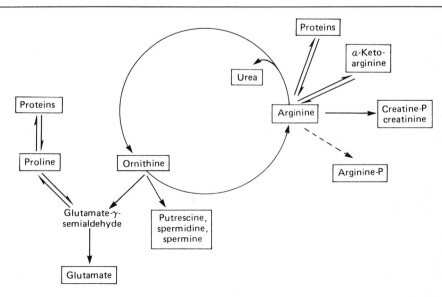

Figure 32–4. Arginine, ornithine, and proline metabolism. Reactions with solid arrows all occur in mammalian tissues. Putrescine and spermine synthesis occurs in both mammals and bacteria. Arginine phosphate occurs in invertebrate muscle, where it functions as a phosphagen analogous to creatine phosphate in mammalian tissues.

ARGININE

Arginine serves as a formamidine donor for creatine synthesis in primates (Fig 32–11) and for streptomycin synthesis in *Streptomyces*. Other fates include conversion, via ornithine, to putrescine, spermine, and spermidine (Fig 32–4) and synthesis of arginine phosphate (functionally analogous to creatine phosphate) in invertebrate muscle.

ORNITHINE

In addition to its role in urea biosynthesis (see Chapter 30), ornithine (in conjunction with methionine) serves as a precursor of the ubiquitous mammalian (and bacterial) polyamines spermidine and spermine (Fig 32–5). Normal humans biosynthesize approximately 0.5 mmol of spermine per day. Pharmacologic doses of polyamines are hypothermic and hypotensive.

Spermidine and spermine are implicated in diverse physiologic processes that share as a common thread a close relationship to cell proliferation and growth. They are growth factors for cultured mammalian and bacterial cells and have been implicated in the stabilization of intact cells, subcellular organelles, and membranes. As a consequence of their multiple positive charges, polyamines associate readily with polyanions such as DNA and RNAs and have been implicated in such fundamental processes as stimulation of DNA and RNA biosynthesis, DNA stabilization, and packaging of DNA in bacteriophages. Polyamines also exert diverse effects on protein synthesis and act as inhibitors of enzymes that include protein kinases.

While it is not presently possible to explain (in precise mechanistic terms) the mode of action of polyamines on any specific metabolic process, the essential nature of polyamines in mammalian metabolism is convincingly documented by experiments of the following type. The initial reaction in polyamine biosynthesis is catalyzed by ornithine decarboxylase (Fig 32–6). Addition to cultured mammalian cells of inhibitors of ornithine decarboxylase activity (eg, α-methylornithine or difluoromethylornithine) triggers overproduction of ornithine decarboxylase. This suggests an essential physiologic role for this enzyme, whose only known function is polyamine biosynthesis.

Biosynthesis of Polyamines

Fig 32–6 summarizes the pathway of polyamine biosynthesis in mammalian tissues. Note that the putrescine portion of spermidine and spermine derives from L-ornithine (a urea cycle intermediate; see Chapter 30) and the diaminopropane portion from L-methionine via intermediate formation of S-adenosylmethionine. Ornithine decarboxylase and S-adenosylmethionine decarboxylase both are inducible enzymes with short half-lives. Spermine and spermidine synthases are, by contrast, neither inducible nor unusually labile enzymes.

Of the enzymes of mammalian polyamine biosynthesis, 2 (ornithine decarboxylase and S-adenosylmethionine decarboxylase) are of interest with respect to both their regulation and their potential for enzyme-directed chemotherapy. The half-life of ornithine decarboxylase (approximately 10 minutes) is shorter than that of any other known mammalian enzyme, and its activity responds rapidly and dramatically to many

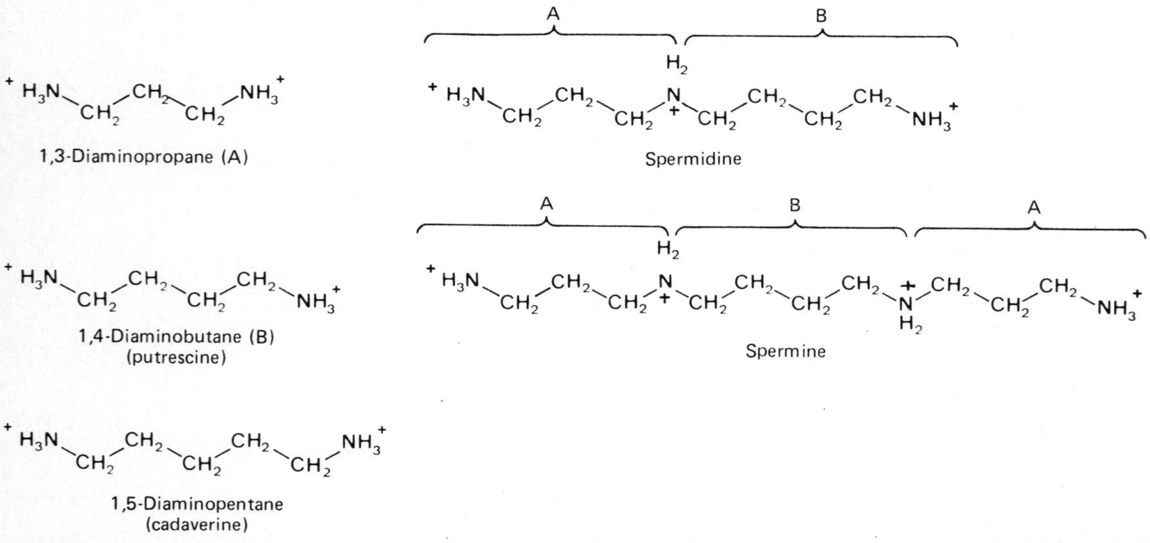

Figure 32–5. Structures of the natural polyamines. Note that spermidine and spermine are polymers of diaminopropane (A) and diaminobutane (B). Diaminopentane (cadaverine) also occurs in mammalian tissues.

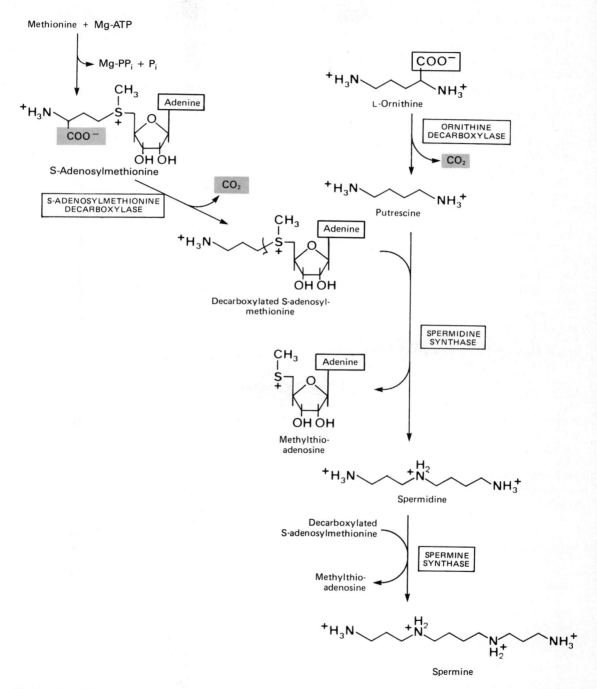

Figure 32–6. Intermediates and enzymes that participate in the biosynthesis of spermidine and spermine. Methylene groups are abbreviated to facilitate visualization of the overall process.

stimuli. Increases in ornithine decarboxylase activity of 10- to 200-fold rapidly follow administration to cultured mammalian cells of growth hormone, corticosteroids, testosterone, or epidermal growth factor. Polyamines added to cultured cells induce synthesis of a protein antizyme that binds to ornithine decarboxylase and inhibits its activity. The activity of ornithine decarboxylase thus appears also to be controlled by a protein-protein interaction reminiscent of the regulation of trypsin activity by protein trypsin inhibitors. Difluoromethylornithine, a "suicide inhibitor" of ornithine decarboxylase, has been used both to isolate mutant cell lines that overproduce ornithine decarboxylase and to inhibit cell replication by enzyme-directed chemotherapy.

S-Adenosylmethionine decarboxylase, the only known eukaryotic enzyme that contains bound pyruvate as an essential cofactor (decarboxylases normally contain pyridoxal phosphate, which is absent from S-adenosylmethionine decarboxylase), has a short half-life (1–2 hours) and responds to promoters of cell growth in a manner qualitatively similar to ornithine decarboxylase. Both the rapidity and the extent of the response are, however, less dramatic than for ornithine decarboxylase. The activity of S-adenosylmethionine decarboxylase (Fig 32–6) is inhibited by decarboxylated S-adenosylmethionine and activated by putrescine.

Catabolism of Polyamines

Fig 32–7 summarizes the catabolism of polyamines in mammalian tissues. The enzyme polyamine oxidase, present in liver peroxisomes, oxidizes spermine to spermidine and subsequently oxidizes spermidine to putrescine. Both diaminopropane moieties are converted to β-aminopropionaldehyde. Subsequently, putrescine is partially oxidized to NH_4^+ and CO_2 by mechanisms that remain to be elucidated. However, major portions of putrescine and spermidine are excreted in urine as conjugates, principally as acetyl derivatives.

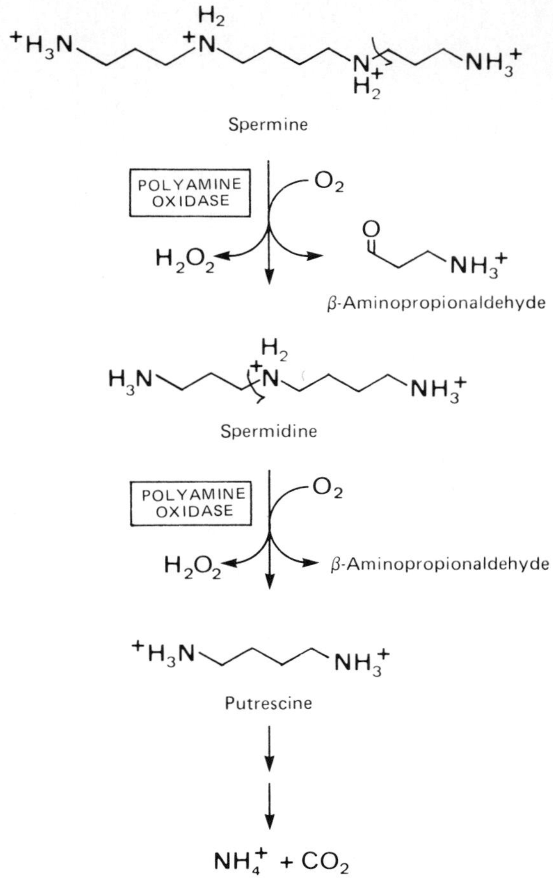

TRYPTOPHAN

Serotonin

A secondary pathway for the metabolism of tryptophan involves hydroxylation to 5-hydroxytryptophan. Oxidation of tryptophan to the hydroxy derivative is analogous to conversion of phenylalanine to tyrosine (Fig 29–11), and liver phenylalanine hydroxylase also catalyzes hydroxylation of tryptophan. Decarboxylation of 5-hydroxytryptophan forms **5-hydroxytryptamine (serotonin)** (reaction 1, Fig 32–8) a potent vasoconstrictor and stimulator of smooth muscle contraction.

The 5-hydroxytryptophan decarboxylase that forms serotonin from hydroxytryptophan is present in the kidney (hog and guinea pig), liver, and stomach. However, the widely distributed aromatic L-amino

Figure 32–7. Catabolism of polyamines. Structures are abbreviated to facilitate presentation.

acid decarboxylase will also catalyze decarboxylation of 5-hydroxytryptophan.

Most serotonin is metabolized by oxidative deamination to 5-hydroxyindoleacetate. The enzyme that catalyzes this reaction is **monoamine oxidase** (reaction 2, Fig 32–8). Inhibitors of this enzyme include iproniazid. It is hypothesized that the psychic stimulation that follows the administration of this drug is attributable to its ability to prolong the stimulating action of serotonin through inhibition of monoamine oxidase. In normal human urine, 2–8 mg of 5-hydroxyindoleacetate is excreted per day.

Greatly increased production of serotonin occurs in malignant **carcinoid** (argentaffinoma), a disease characterized by widespread serotonin-producing tumor cells in the argentaffin tissue of the abdominal cavity. Carcinoid has been considered an abnormality in tryptophan metabolism in which a much greater proportion of tryptophan than normal is metabolized by way of hydroxyindole. One percent of tryptophan is normally converted to serotonin, but in the carcinoid patient as much as 60% may follow this pathway. This metabolic diversion markedly reduces production of

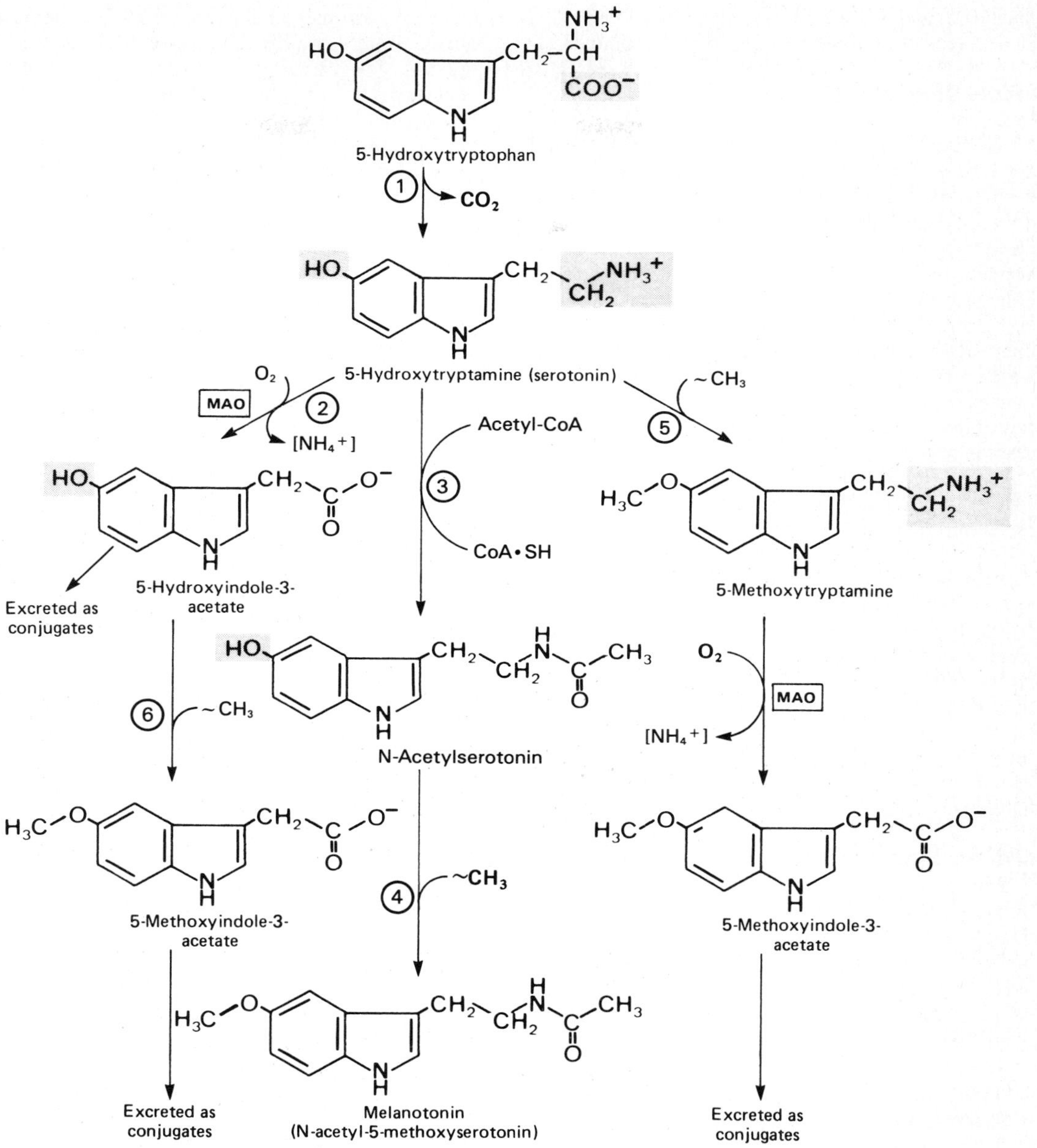

Figure 32–8. Biosynthesis and metabolism of melatonin. [NH_4^+], by transamination; MAO, monoamine oxidase. The numbered reactions are referred to in the text.

nicotinic acid from tryptophan; consequently, symptoms of pellagra as well as negative nitrogen balance may occur. Other metabolites of serotonin identified in the urine of patients with carcinoid include 5-hydroxyindoleaceturate (the glycine conjugate of 5-hydroxyindoleacetate) and N-acetylserotonin conjugated with glucuronic acid.

Melatonin

Melatonin is derived from serotonin by N-acetylation (reaction 3, Fig 32–8) followed by methylation of the 5-hydroxy group (reaction 4, Fig 32–8). Methylation is localized in pineal body tissue. In addition to methylation of N-acetylserotonin, direct methylation of serotonin (reaction 5, Fig 32–8) and of 5-hydroxyindoleacetate (reaction 6, Fig 32–8), the serotonin metabolite, also occurs.

Serotonin and 5-methoxytryptamine are metabolized to the corresponding acids by monoamine oxidase. Circulating melatonin is taken up by all tissues, including brain, but is rapidly metabolized by hydroxylation at position 6 followed by conjugation with sulfate (70%) and with glucuronic acid (6%). A portion is also converted to nonindolic reacting compounds.

Indole Derivatives in Urine

Tryptophan may be converted to several indole derivatives (Fig 32–8). The end products of these conversions that appear in the urine are principally 5-hydroxyindoleacetate, the major end product of the hydroxytryptophan-to-serotonin pathway, and indole-3-acetate, from decarboxylation and oxidation of indolepyruvate, the keto acid of tryptophan.

Mammalian kidney and liver and bacteria from human feces decarboxylate tryptophan to tryptamine, which can then be oxidized to indole-3-acetate. Patients with phenylketonuria excrete increased quantities of indoleacetate (and indolelactate, formed by reduction of indolepyruvate).

MELANINS

Eumelanins are insoluble, heterogeneous, high-molecular-weight, black to brown heteropolymers of 5,6-dihydroxyindole and several of its biosynthetic precursors. **Pheomelanins** are yellow to reddish-brown polymers but, while also of high molecular weight, are soluble in dilute alkali. The low-molecular-weight **trichochromes** are related to pheomelanins (both are derived from cysteine and dopaquinone). Pheomelanins and trichochromes are present primarily in hair and feathers.

Melanin Biosynthesis

Discussion of melanin biosynthesis is complicated by the protracted, branched, and incomplete biosynthetic pathway, the complex chemical structures of melanin heteropolymers, and the insolubility of these polymers, all of which hinder their structural determination. Melanins are synthesized in **melanosomes**—membrane-bound particles within melanocytes, which are cells of neural crest origin. The developing eumelanin polymer is thought to entrap free radicals and to undergo partial degradation by H_2O_2 generated during the auto-oxidative process. Pheomelanins and eumelanins then complex with proteins of the melanosomal matrix, forming **melanoprotein.**

Fig 32–9 summarizes the known intermediates and reactions of eumelanin and pheomelanin biosynthesis. The initial reaction is catalyzed by tyrosinase, a copper-dependent enzyme. **The tyrosinase reaction is defective in tyrosinase-negative oculocutaneous albinism** (see below).

Metabolic Defects of Melanin Biosynthesis

As might be anticipated from the number of reactions involved in melanin biosynthesis, mutations in many genes can give rise to metabolic defects in melanin biosynthesis.

The term "albinism" encompasses a spectrum of clinical syndromes characterized by **hypomelanosis** arising from heritable defects in the pigment cells (melanocytes) of the eye and skin. Several useful rodent models of albinism are known.

Clinical signs common to all 10 human forms of **oculocutaneous albinism** include decreased pigmentation of the skin and eye. All 10 forms can be differentiated on the basis of their clinical, biochemical, ultrastructural, and genetic characteristics. All but the dominant form are inherited as autosomal recessive traits.

Tyrosinase-negative albinos completely lack visual pigment. Hair bulbs from these patients fail to convert added tyrosine to pigment in vitro, and their melanocytes contain unpigmented melanosomes. **Tyrosinase-positive albinos** have some visible pigment, although this may not be evident in white infants. Hair color ranges from white-yellow to light tan, and lightly pigmented nevi may be present. Hair bulb melanocytes may contain lightly pigmented melanosomes, which convert tyrosine to black eumelanin in vitro.

Ocular albinism occurs both as an autosomal recessive and as an X-linked trait. The melanocytes of X-linked and heterozygous (but not autosomal recessive) ocular albinos contain macromelanosomes. The retinas of females heterozygous for X-linked ocular albinism (Nettleship variety) exhibit a mosaic pattern of pigment distribution due to random X-chromosome inactivation. The precise metabolic defects leading to hypomelanosis in ocular albinism are unknown.

Oculocutaneous albinoidism is inherited as an autosomal recessive trait. With rare exceptions, patients lack associated nystagmus, photophobia, and decreased visual acuity.

TYROSINE

Tyrosine is a precursor of **epinephrine** and **norepinephrine,** which are formed in cells of neural origin. Although dopa is an intermediate in the formation of both melanin in melanocytes and norepinephrine in neuronal cells, different enzymes carry out the tyrosine hydroxylation reactions in the different cell types. **Tyrosine hydroxylase,** an enzyme that is not copper-dependent but utilizes tetrahydrobiopterin as much as does phenylalanine hydroxylase, forms dopa in the neuronal and adrenal cells on the pathway to norepinephrine and epinephrine production (Fig 32–10). **Dopa decarboxylase,** a pyridoxal phosphate-dependent enzyme, forms dopamine. The latter is subjected to further hydroxylation by dopamine β-oxidase, a copper-dependent enzyme that seems to utilize vitamin C to generate norepinephrine. In the **adrenal medulla,** the enzyme phenylethanolamine-N-methyltransferase utilizes S-adenosylmethionine to methylate the primary amine of norepinephrine to form **epinephrine** (Fig 32–10).

Tyrosine is also a precursor of the thyroid hor-

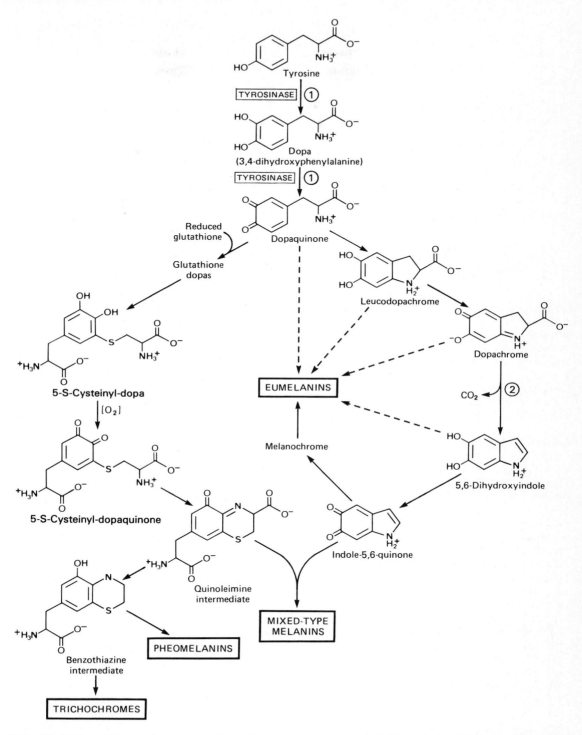

Figure 32–9. Known intermediates and reactions of eumelanin and pheomelanin biosynthesis. Melanin polymers contain both eumelanin and pheomelanin in varying proportions. Dotted arrows indicate that intermediates contribute toward synthesis of eumelanins in varying proportions. Circled numerals indicate probable regulated reactions of the biosynthetic pathway. Reaction 1, that catalyzed by tyrosinase, is defective in tyrosinase-negative oculocutaneous albinism.

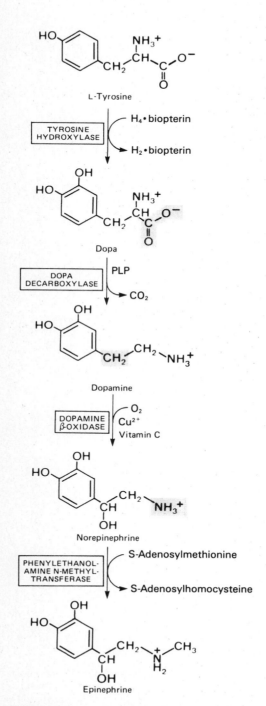

Figure 32–10. Conversion of tyrosine to epinephrine and norepinephrine in neuronal and adrenal cells. PLP, pyridoxal phosphate.

mones triiodothyronine and thyroxine (see Chapter 46).

Tyrosine is excreted in urine both free and as a sulfate, but most phenolic compounds are conjugated with sulfate when present in the urine.

CREATINE & CREATININE

Creatine is present in muscle, brain, and blood, both as phosphocreatine and in the free state. Traces of creatine are also normally present in urine. Creatinine, the anhydride of creatine, is formed largely in muscle by irreversible nonenzymatic dehydration of creatine phosphate (Fig 32–11).

The 24-hour excretion of creatinine in the urine of a given subject is remarkably constant from day to day and proportionate to muscle mass.

For synthesis of creatine, 3 amino acids—**glycine, arginine,** and **methionine**—are directly involved. The first reaction is transamidination from arginine to glycine to form guanidoacetate (glycocyamine). This occurs in the kidney but not in the liver or in heart muscle. Synthesis of creatine is completed by methylation of glycocyamine by "active methionine" in the liver.

γ-AMINOBUTYRATE

Biosynthesis

γ-Aminobutyrate (GABA) is formed by decarboxylation of L-glutamate, a reaction catalyzed by the pyridoxal phosphate-dependent enzyme L-glutamate decarboxylase (Fig 32–12). This decarboxylase is present in the tissues of the central nervous system, principally in the gray matter. While decarboxylation of L-glutamate represents the major route of γ-aminobutyrate biosynthesis, 2 reaction sequences convert putrescine (Fig 32–5) to γ-aminobutyrate. One involves deamination by diamine oxidase; the other utilizes N-acetylated intermediates. The relative importance of these 3 routes of γ-aminobutyrate biosynthesis varies among tissues and with developmental stage. For example, the polyamine precursor ornithine (Fig 32–6) is efficiently converted to γ-aminobutyrate in embryonic chick retinal tissue and in adult brain nerve terminals.

Catabolism

Catabolism of γ-aminobutyrate (Fig 32–12) involves transamination, catalyzed by γ-aminobutyrate transaminase, to succinate semialdehyde. Succinate semialdehyde may undergo reduction to γ-hydroxybutyrate, a reaction catalyzed by L-lactate dehydrogenase, or oxidation to the citric acid cycle intermediate succinate and thence to CO_2 and H_2O.

Aminobutyric Acidemia

γ-Aminobutyrate, in common with the anions of other ω-amino acids, is poorly transported across plasma cell membranes. The urinary levels of γ-

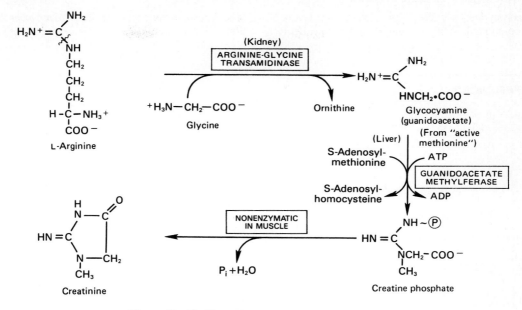

Figure 32–11. Biosynthesis of creatine and creatinine.

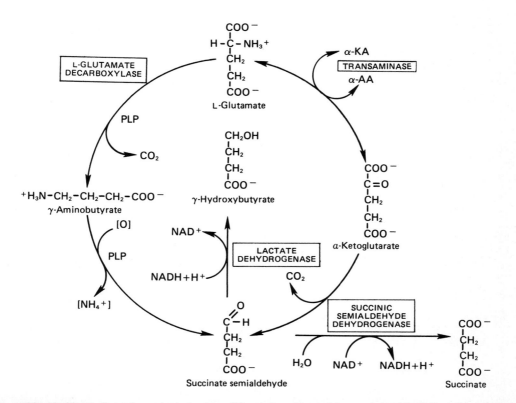

Figure 32–12. Metabolism of γ-aminobutyrate. α-KA, α-keto acids; α-AA, α-amino acids; PLP, pyridoxal phosphate.

aminobutyrate vary directly with the serum levels of this compound. While the precise biochemical defect is not known, it may result from impaired transamination of γ-aminobutyrate to succinate semialdehyde.

REFERENCES

Stanbury JB et al (editors): *The Metabolic Basis of Inherited Disease*, 5th ed. McGraw-Hill, 1983.

Tabor CW, Tabor H: Polyamines. *Annu Rev Biochem* 1984;**53:**749.

Porphyrins & Bile Pigments

33

Robert K. Murray, MD, PhD

INTRODUCTION

The biochemistry of the porphyrins and of the bile pigments is presented in this chapter. These topics are closely related, because heme is synthesized from porphyrins and iron, and the products of its degradation are the bile pigments and iron.

BIOMEDICAL IMPORTANCE

Knowledge of the biochemistry of the porphyrins and of heme is basic to understanding the varied functions of hemoproteins (involvement in oxygen transport, electron transport, drug metabolism, etc) in the body. The porphyrias are a group of diseases due to abnormalities in the pathway of biosynthesis of the various porphyrins. They are uncommon, but medical practitioners must be aware of them, and dermatologists, hepatologists, and psychiatrists will encounter patients with these conditions. A more common clinical condition is jaundice, due to elevation of bilirubin in the plasma. This elevation is due to overproduction of bilirubin or to failure of its excretion and is seen in numerous diseases, ranging from viral hepatitis to cancer of the pancreas.

PORPHYRINS

Porphyrins are cyclic compounds formed by the linkage of 4 pyrrole rings through methenyl bridges (Fig 33–1). A characteristic property of the porphyrins is the formation of complexes with metal ions bound to the nitrogen atom of the pyrrole rings. Examples are the iron porphyrins such as **heme** of hemoglobin and the magnesium-containing porphyrin **chlorophyll,** the photosynthetic pigment of plants.

In nature, the metalloporphyrins are conjugated to proteins to form many compounds important in biologic processes. These include the following:

A. Hemoglobins: Iron porphyrins attached to the protein, globin. These conjugated proteins possess the ability to combine reversibly with oxygen. They serve as the transport mechanism for oxygen within the blood (see Chapter 6). The structure of heme is shown in Fig 6–2.

B. Erythrocruorins: Iron porphyrinoproteins occuring in the blood and in the tissue fluids of some invertebrates. They correspond in function to hemoglobin.

C. Myoglobins: Respiratory pigments that occur in the muscle cells of vertebrates and invertebrates. An example is the myoglobin obtained from the heart muscle of the horse and crystallized by Theorell in 1934. A myoglobin molecule is similar to a subunit of hemoglobin.

D. Cytochromes: Compounds that act as electron transfer agents in oxidation-reduction reactions. An important example is **cytochrome c,** which has a

Figure 33–1. The porphin molecule. Rings are labeled I, II, III, IV. Substituent positions on rings are labeled 1, 2, 3, 4, 5, 6, 7, 8. Methenyl bridges are labeled α, β, γ, δ.

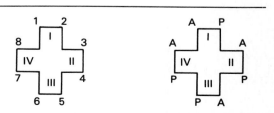

Figure 33–2. Uroporphyrin III.

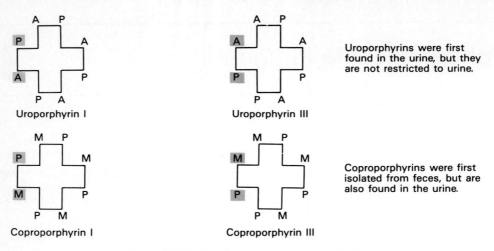

Uroporphyrin I

Uroporphyrin III

Uroporphyrins were first found in the urine, but they are not restricted to urine.

Coproporphyrin I

Coproporphyrin III

Coproporphyrins were first isolated from feces, but are also found in the urine.

Figure 33–3. Uroporphyrins and coproporphyrins.

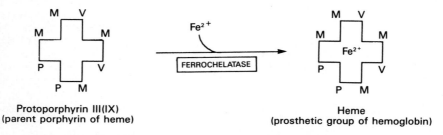

Protoporphyrin III(IX)
(parent porphyrin of heme)

Heme
(prosthetic group of hemoglobin)

Figure 33–4. Addition of iron to protoporphyrin to form heme.

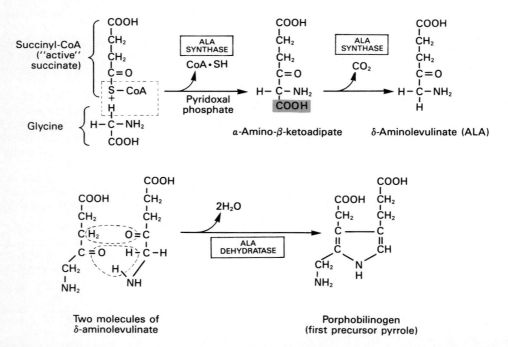

Two molecules of δ-aminolevulinate

Porphobilinogen
(first precursor pyrrole)

Figure 33–5. Biosynthesis of porphobilinogen. ALA synthase occurs in the mitochondria, whereas ALA dehydratase is present in the cytosol.

molecular weight of about 13,000 and contains 1 gram-atom of iron per mole.

E. Catalases: Iron porphyrin enzymes, several of which have been obtained in crystalline form. In plants, catalase activity is minimal, but the iron porphyrin enzyme peroxidase performs similar functions.

F. Tryptophan Pyrrolase: This enzyme catalyzes the oxidation of tryptophan to formyl kynurenine. It is an iron porphyrinoprotein.

Structure of Porphyrins

The porphyrins found in nature are compounds in which various side chains are substituted for the 8 hydrogen atoms numbered in the porphin nucleus shown in Fig 33–1. As a simple means of showing these substitutions, Fischer proposed a shorthand formula in which the methenyl bridges are omitted and each pyrrole ring is shown as a bracket with the 8 substituent positions numbered as shown (Fig 33–2). Various porphyrins are represented in Figs 33–2 through 33–4 (A [acetate] = $-CH_2COOH$; P [propionate] = $-CH_2CH_2COOH$; M [methyl] = $-CH_3$; V [vinyl] = $-CH{=}CH_2$).

The arrangement of the A and P substituents in the uroporphyrin shown in Fig 33–2 is asymmetric (in ring IV, the expected order of the acetate and propionate substituents is reversed). A porphyrin with this type of **asymmetric substitution** is classified as a type III porphyrin. A porphyrin with a completely symmetric arrangement of the substituents is classified as a type I porphyrin. Only types I and III are found in nature, and the **type III series** is by far the more abundant (Fig 33–3).

The compounds shown in Fig 33–4 are both type III porphyrins (ie, the methyl groups are asymmetrically distributed, as in type III coproporphyrin). However, they are sometimes identified as belonging to series IX, because they were designated ninth in a series of isomers postulated by Hans Fischer, the pioneer worker in the field of porphyrin chemistry.

Biosynthesis of Porphyrins

Both chlorophyll, the photosynthetic pigment of plants, and heme, the iron protoporphyrin of hemoglobin in animals, are synthesized in living cells by a common pathway. The 2 starting materials are "active succinate," the **coenzyme A derivative of succinic acid,** derived from the citric acid cycle in mitochondria, and the amino acid **glycine.** Pyridoxal phosphate is also necessary in this reaction to "activate" glycine. It is probable that pyridoxal reacts with glycine to form a Schiff base, whereby the α carbon of glycine can be combined with the carbonyl carbon of succinate. The product of the condensation reaction between succinyl-CoA and glycine is α-amino-β-ketoadipic acid, which is rapidly decarboxylated to form δ-aminolevulinate (ALA) (Fig 33–5). This step is catalyzed by the enzyme **ALA synthase.** This appears to be the **rate-controlling** enzyme in porphyrin biosynthesis in mammalian liver. Synthesis of aminolevulinic acid

occurs in the **mitochondria.** In the cytosol, 2 molecules of ALA are condensed by the enzyme **ALA dehydratase** to form 2 molecules of water and one of **porphobilinogen** (Fig 33–5). ALA dehydratase is a Zn-containing enzyme and is sensitive to inhibition by lead.

The formation of a tetrapyrrole, ie, a porphyrin, occurs by condensation of 4 monopyrroles derived from porphobilinogen (Fig 33–6). In each instance, the amino carbon (originally derived from the α carbon of glycine) serves as the source of the methylene (alpha, beta, gamma, delta) carbons that connect each pyrrole in the tetrapyrrole structure. Although the conversion of porphobilinogen to a porphyrin can be accomplished simply by heating under acid conditions, such as in an acid urine, this conversion is catalyzed in the tissues by specific enzymes.

It has been pointed out that only types I and III porphyrins occur in nature, and it may be assumed that the type III isomers are the more abundant, since the biologically important porphyrins such as heme and the cytochromes are type III isomers.

At present, the detailed steps leading to the formation of the uroporphyrinogens from condensation of porphobilinogens remain obscure. The formation from porphobilinogen of uroporphyrinogen III, the obligatory intermediate in heme biosynthesis, is catalyzed by a complex interaction of 2 enzymes. **Uroporphyrinogen-I synthase** (also called porphobilinogen deaminase) condenses porphobilinogen to uroporphyrinogen I in vitro (Fig 33–6). However, when a second enzyme, **uroporphyrinogen-III cosynthase,** is present, interaction between these 2 enzymes results in the formation of uroporphyrinogen III rather than the symmetric isomer uroporphyrinogen I (Fig 33–6). Under normal conditions, the uroporphyrinogen formed is almost exclusively the III isomer, but in certain of the porphyrias (discussed below) the type I isomers of porphyrinogens are also formed in excess.

Note that both of these uroporphyrinogens have the pyrrole rings connected by **methylene** bridges, which do not form a conjugated ring system. Thus, these compounds (as are all porphyrinogens) are **colorless.** However, the porphyrinogens are readily auto-oxidized to their respective porphyrins. These oxidations are catalyzed by light and by the porphyrins that are formed.

Uroporphyrinogen III is converted to coproporphyrinogen III by decarboxylation of all of the acetate (A) groups, which changes them to methyl (M) substituents. The reaction is catalyzed by **uroporphyrinogen decarboxylase,** which is also capable of converting uroporphyrinogen I to coproporphyrinogen I (Fig 33–7). Coproporphyrinogen III then enters the mitochondria, where it is converted to **protoporphyrinogen III** and then to **protoporphyrin III.** Several steps seem to be involved in this conversion. The mitochondrial enzyme **coproporphyrinogen oxidase** catalyzes the decarboxylation and oxidation of 2 propionic side chains to form protoporphyrinogen. This enzyme is able to act only on type III coproporphyrin-

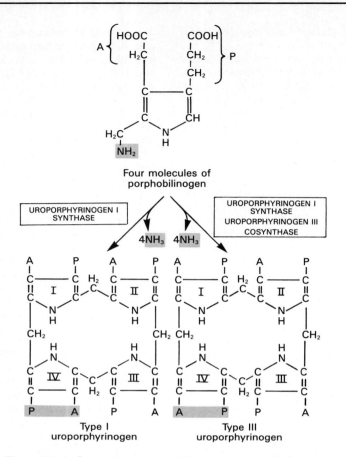

Four molecules of
porphobilinogen

Type I
uroporphyrinogen

Type III
uroporphyrinogen

Figure 33–6. Conversion of porphobilinogen to uroporphyrinogens.

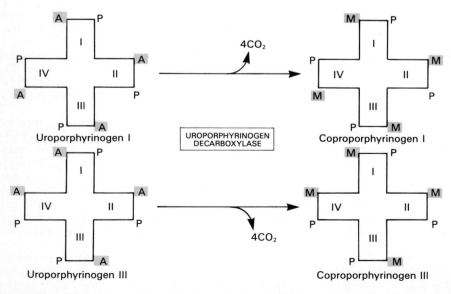

Uroporphyrinogen I

Coproporphyrinogen I

Uroporphyrinogen III

Coproporphyrinogen III

Figure 33–7. Decarboxylation of uroporphyrinogens to coproporphyrinogens in cytosol. A, acetyl; M, methyl; P, propyl.

ogen, which would explain why a type I protopor-phyrin has not been identified in natural materials. The oxidation of protoporphyrinogen to protoporphyrin is catalyzed by another mitochondrial enzyme, **proto-porphyrinogen oxidase.** In mammalian liver, the conversion of coproporphyrinogen to protoporphyrin requires molecular oxygen.

Formation of Heme

The final step in heme synthesis involves the incor-poration of ferrous iron into protoporphyrin in a reac-tion catalyzed by **heme synthase** or **ferrochelatase,** another mitochondrial enzyme (Fig 33–4). This reac-tion occurs readily in the absence of enzymes, but it is noted to be much more rapid in the presence of tissue preparations, presumably because of the tissue contri-bution of enzymes active in catalyzing this iron chela-tion.

A summary of the steps in the biosynthesis of the porphyrin derivatives from porphobilinogen is given in Fig 33–8. Heme biosynthesis occurs in most mam-malian tissues with the exception of mature erythro-cytes, which do not contain mitochondria.

The porphyrinogens that have been described above are colorless, containing 6 extra hydrogen atoms as compared to the corresponding **colored por-phyrins.** It is now apparent that these reduced por-

phyrins (the porphyrinogens) and not the correspond-ing porphyrins are the actual intermediates in the biosynthesis of protoporphyrin and of heme.

Regulation of Heme Biosynthesis

The rate-limiting reaction for the synthesis of heme occurs at the condensation of succinyl-CoA and glycine to form ALA (Fig 33–5), a reaction catalyzed by the enzyme aminolevulinic acid synthase (ALA synthase). The levels of ALA synthase activity in nor-mal tissues capable of synthesizing heme are significantly lower than those of the other enzymes of the heme synthetic pathway. However, ALA synthase is a regulated enzyme. It appears that heme, probably acting through an aporepressor molecule, acts as a negative regulator of the accumulation of ALA syn-thase. This repression and derepression mechanism is depicted diagrammatically in Fig 33–9. It is possible that there is also significant feedback inhibition at this step, but the major regulatory effect of heme appears to be one in which the rate of accumulation of ALA synthase increases greatly in the absence of heme and is diminished in its presence. The rate of ALA syn-thase turnover is normally rapid (half-life is about 1 hour) in mammalian liver, not a surprising property for an enzyme catalyzing a rate-limiting reaction.

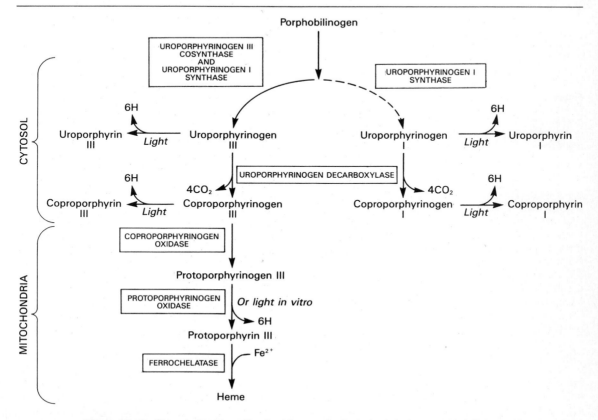

Figure 33–8. Steps in the biosynthesis of the porphyrin derivatives from porphobilinogen.

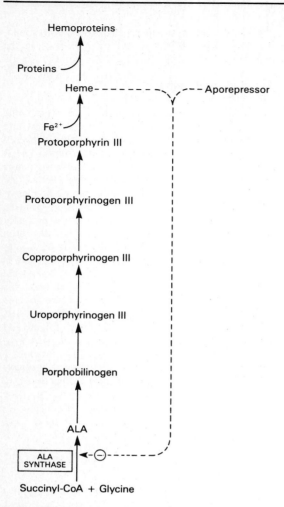

Figure 33–9. Regulation of heme synthesis at the level of ALA synthase by a repression-derepression mechanism mediated by heme and its hypothetical aporepressor. The dotted lines indicate the negative ($\ominus$) regulation by repression.

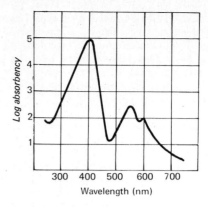

Figure 33–10. Absorption spectrum of hematoporphyrin (0.01% solution in 5% HCl).

the drug-mediated derepression of ALA synthase in vivo. The administration of hematin in vivo can prevent the drug-mediated derepression of ALA synthase, as well as that of other hemoproteins in liver. In erythropoietic tissues, hypoxia increases ALA synthase activity without having a demonstrable effect on ALA synthase activity in liver.

The importance of these regulatory mechanisms is discussed below along with the diseases classified among the porphyrias.

Chemistry of Porphyrins

Because of the presence of tertiary nitrogens in the 2 pyrrolene rings contained in each porphyrin, these compounds act as weak bases. Those which possess a carboxyl group on one or more side chains act also as acids. Their isoelectric points range from pH 3.0 to 4.5, and within this pH range the porphyrins may easily be precipitated from an aqueous solution.

The various porphyrinogens are colorless, whereas the various **porphyrins are all colored.** In the study of porphyrins or porphyrin derivatives, the characteristic absorption spectrum that each exhibits, in both the visible and the ultraviolet regions of the spectrum, is of great value. An example is the absorption curve for a solution of porphyrin in 5% hydrochloric acid (Fig 33–10). Note the sharp absorption band near 400 nm. This is a distinguishing feature of the porphin ring and is characteristic of all porphyrins regardless of the side chains present. This band is termed the **Soret band,** after its discoverer. Hematoporphyrin in acid solution has, in addition to the Soret band, 2 weaker absorption bands with maxima at 550 and 592 nm.

When porphyrins dissolved in strong mineral acids or in organic solvents are illuminated by ultraviolet light, they emit a strong red fluorescence. This **fluorescence** is so characteristic that it is frequently used to detect small amounts of free porphyrins. The double bonds in the porphyrins are responsible for the characteristic absorption and fluorescence of these compounds, and, as previously noted, the reduction (by addition of hydrogen) of the methenyl ($-HC=$)

Many compounds of diverse structures, including presently used insecticides, carcinogens, and pharmaceuticals, when administered to humans, can result in a marked increase in hepatic ALA synthase. Most of these drugs are metabolized by a system in the liver that utilizes a specific hemoprotein, cytochrome P-450. During the process of metabolizing these drugs, the utilization of heme by cytochrome P-450 is greatly increased, which in turn diminishes the intracellular heme concentration. This latter event effects a derepression of ALA synthase with a corresponding increased rate of heme synthesis to meet the needs of the cells.

Several other factors affect the induction of ALA synthase in the liver. Glucose can prevent the induction of ALA synthase; iron in chelated form exerts a synergistic effect on the induction of hepatic ALA synthase; and steroids play at least a permissive role in bridges to methylene ($-CH_2-$) leads to the formation of colorless compounds termed **porphyrinogens.**

When a porphyrin combines with a metal, its absorption in the visible spectrum becomes changed. This is exemplified by protoporphyrin, the iron-free precursor of heme. In alkaline solution, protoporphyrin shows several sharp absorption bands (at 645, 591, and 540 nm), whereas heme has a broad band with a plateau extending from 540 to 580 nm.

Tests for Porphyrins

The presence of coproporphyrins or of uroporphyrins is of clinical interest, since these 2 types of compounds are excreted in increased amounts in the porphyrias. Coproporphyrins I and III are soluble in glacial acetic acid-ether mixtures, from which they may then be extracted by hydrochloric acid. Uroporphyrins, on the other hand, are not soluble in acetic acid-ether mixtures but are partially soluble in ethyl acetate, from which they may be extracted by hydrochloric acid. In the HCl solution, ultraviolet illumination gives a characteristic red fluorescence. A spectrophotometer may then be used to demonstrate the characteristic absorption bands.

The upper limits of normal excretory values of porphyrins and porphyrin precursors are given in Table 33–1. In healthy subjects, the total urinary coproporphyrin averages about 67 μg/24 h; the type I isomer comprises on the average 14 μg/24 h and type III 53 μg/24 h. An alteration in the normal ratio of the excretion of types I and III coproporphyrins may be of value in detection of certain types of diseases of the liver.

During the synthesis of heme from ALA, there is an increase in the hydrophobic qualities of the various intermediate compounds. The acetyl carboxyl groups on uroporphyrinogen are removed when it is converted to coproporphyrinogen, and 2 of the propyl groups are decarboxylated in the course of the conversion of coproporphyrinogen to protoporphyrinogen. The relative distributions in the urine and feces of the intermediates of heme biosynthesis reflect this increasing hydrophobic quality. Thus, the more polar uroporphyrinogen will be excreted to a greater extent in urine than in feces, whereas the more hydrophobic copro-

Table 33–1. Upper limits of normal excretory values and concentrations of porphyrins and porphyrin precursors.*

	Urine (μg/24 h)	Feces (μg/g dry wt)	Erythrocytes (μg/dL cells)
ALA	4000	—	—
Porphobilinogen	1500	—	—
Uroporphyrin	50	5	trace
Coproporphyrin	300	50	3
Protoporphyrin	—	120	80

*Modified and reproduced, with permission, from Meyer UA, Schmid R: The porphyrias. In: *The Metabolic Basis of Inherited Disease,* 4th ed. Stanbury JB, Wyngaarden JB, Fredrickson DS (editors). McGraw-Hill, 1978.

porphyrinogen and protoporphyrinogen will increasingly distribute themselves in the bile and ultimately the feces rather than in the aqueous urine.

THE PORPHYRIAS

The porphyrias constitute a heterogeneous group of diseases, all of which exhibit increased excretion of porphyrins or porphyrin precursors. Some forms of porphyria are **inherited,** whereas others are **acquired.** Several different classifications of the porphyrias have been proposed. It is convenient to divide the inherited porphyrias into 3 general groups—the **erythropoietic** porphyrias, the **hepatic** porphyrias, and those with both erythropoietic and hepatic abnormalities (Table 33–2). In most types of inherited porphyria, the defect is present in all tissues, but for reasons that are not clear, the metabolic abnormalities are expressed preferentially in one or another tissue type. There follows a brief description of the biochemical abnormalities characteristic of the porphyrias.

The pattern of excretion of porphyrin and porphyrin precursors is characteristic for each type of porphyria. In Fig 33–11 these patterns and their relationships to the heme synthetic pathway are depicted.

Intermittent acute porphyria (IAP) is an autosomal, dominantly inherited disease in humans that usually is not expressed before puberty. It results from an

Table 33–2. Classification of human porphyrias.*

Condition	Mode of Inheritance	Demonstrated or Suspected Enzyme Defect	Predominant Site(s) of Metabolic Expression
Congenital erythropoietic porphyria	Autosomal recessive	Uroporphyrinogen I synthase and/or uroporphyrinogen III cosynthase	Erythroid cells
Hepatic porphyrias			
Intermittent acute porphyria	Autosomal dominant	Uroporphyrinogen I synthase	Liver
Hereditary coproporphyria	Autosomal dominant	Coproporphyrinogen oxidase	Liver
Variegate porphyria	Autosomal dominant	Protoporphyrinogen oxidase	Liver
Porphyria cutanea tarda	Autosomal dominant (?)	Uroporphyrinogen decarboxylase	Liver
Toxic porphyria	Acquired	Variable	Liver
Protoporphyria	Autosomal dominant	Ferrochelatase	Erythroid cells and liver (?)

*Reproduced, with permission, from Meyer UA, Schmid R: The porphyrias. In: *The Metabolic Basis of Inherited Disease,* 4th ed. Stanbury JB, Wyngaarden JB, Fredrickson DS (editors). McGraw-Hill, 1978.

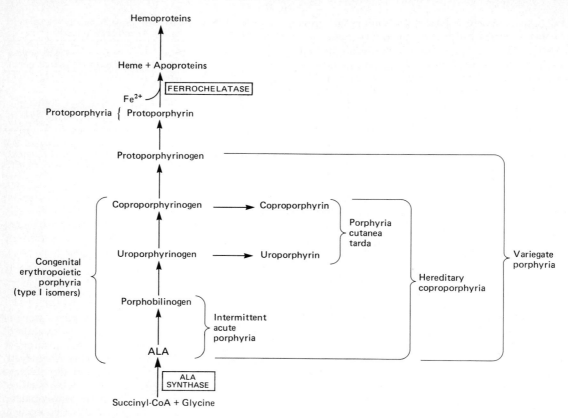

Figure 33–11. Patterns of urinary porphyrin and porphyrin precursor excretion in the porphyrias in relation to the pathway of heme biosynthesis. Intermediates of the pathway excessively excreted during the acute phase of each of the porphyrias are within the respective brackets. ALA, δ-aminolevulinic acid. (Modified from Kaufman L, Marver HS: Biochemical defects in two types of human hepatic porphyria. *N Engl J Med* 1970;**283**:954.)

inherited partial deficiency of **uroporphyrinogen I synthase.** Individuals with this disease are heterozygous for a defective structural gene for uroporphyrinogen I synthase, with the result that only 50% of the normal specific catalytic activity of that enzyme is present within their cells. Patients with IAP excrete massive quantities of **porphobilinogen** and **ALA** in the urine. Both of these compounds are **colorless,** but porphobilinogen upon exposure to light and air polymerizes spontaneously but slowly to form 2 colored compounds: porphobilin and porphyrin. These cause a **darkening of the urine upon standing** in light and air.

Both porphobilinogen and ALA are present in the plasma and spinal fluid of these patients, particularly during acute exacerbations. Drugs and steroid hormones that require metabolism by heme-containing proteins such as cytochrome P-450 can precipitate acute exacerbations. Apparently, because of the increased consumption of heme proteins necessitated by the metabolism of these porphyria-inducing compounds, there is a **derepression** of ALA synthase activity brought about by diminished intracellular heme concentration. The increased ALA synthase activity

combined with the partial block in the uroporphyrinogen I synthase results in a massive accumulation of ALA and porphobilinogen. The acute attacks of abdominal pain, vomiting, constipation, cardiovascular abnormalities, and neuropsychiatric signs and symptoms correlate with the increased production of ALA and porphobilinogen in these patients. However, experimental data suggest that depletion of heme diminishes the activity of tryptophan pyrrolase and thereby leads to the accumulation of the neuroactive substances tryptophan and 5-hydroxytryptamine.

Patients with IAP **do not have abnormal sensitivity to light,** as do patients with other types of hepatic porphyrias. This is understandable when it is considered that these patients do not accumulate porphyrins or porphyrinogens, because the inherited metabolic defect occurs in the heme synthetic pathway *prior to* the formation of the first porphyrinogen (uroporphyrinogen).

As mentioned above, the metabolic defect in IAP occurs in other cells, including erythrocytes, cultured fibroblasts, and cultured amniotic fluid cells. Although the enzymatic deficiency is ubiquitous, the increased ALA synthase activity responsible for the overproduction of ALA and porphobilinogen is **predominantly a hepatic phenomenon.** This probably is because the liver is the organ in which the inducing agents are metabolized. IAP is one of the rare examples of a disease phenotype being expressed in a heterozygote in whom the known enzyme deficiency is only 50%.

As might be predicted from the proposed mechanism of regulation of ALA synthase by a repression-derepression system, the **infusion of hematin** into patients with IAP can ameliorate the induction of ALA synthase and thereby ameliorate the clinical signs and symptoms.

Congenital erythropoietic porphyria is an even rarer congenital disease having an autosomal recessive mode of inheritance. The molecular nature of the defect in congenital erythropoietic porphyria is not clearly defined, but there is a definite imbalance between the relative activities of uroporphyrinogen III cosynthase and uroporphyrinogen I synthase. The formation of uroporphyrinogen I greatly exceeds that of uroporphyrinogen III, the normal isomer on the pathway to heme synthesis. Although the genetic defect is present in all cells, it is for an unknown reason expressed predominantly in erythropoietic tissue. Patients with congenital erythropoietic porphyria excrete large quantities of the **type I isomers** of both uroporphyrinogen and coproporphyrinogen, which in the urine are spontaneously oxidized to uroporphyrin I and coproporphyrin I, both fluorescent, red pigments. There is reported to be a small increase in uroporphyrin III, but the ratio of the type I to type III isomer approaches 100:1. Circulating erythrocytes contain high concentrations of uroporphyrin I, although the highest concentration of this porphyrin is present in bone marrow cells and not in the hepatocytes.

Apparently because of the decreased formation of the true precursor of heme, uroporphyrinogen III, and thus a relative deficiency of heme, ALA synthase is induced in the erythropoietic tissues of patients with congenital erythropoietic porphyria. This induction of ALA synthase promotes the massive overproduction of the type I porphyrinogens. Concomitant with the increase in ALA synthase and overproduction of the type I porphyrinogens is an increased production and excretion of porphobilinogen and ALA. Thus, from the biochemical abnormalities one can predict the existence of clinical symptoms comparable to those of IAP but with the addition of **cutaneous photosensitivity,** because of the absorption spectrum of the porphyrin compounds that are formed in abnormal quantities in this disorder. These patients also exhibit a prominent increased cutaneous fragility and hemolysis.

Hereditary coproporphyria is an autosomal dominant disorder due to partial deficiency of **coproporphyrinogen oxidase,** the mitochondrial enzyme responsible for the conversion of coproporphyrinogen III to protoporphyrinogen IX. Coproporphyrinogen III is excreted in excessive quantities in feces, but, because of its solubility in water, it is excreted also in large quantities in urine. As is true also of uroporphyrinogen, in the presence of air and light, coproporphyrinogen is rapidly oxidized to coproporphyrin, a red pigment.

The limited capacity to produce heme in this disease—particularly under conditions of stress—will result in derepression of ALA synthase. This leads to the overproduction of ALA and porphobilinogen and the other intermediates in the heme synthetic pathway proximal to the inherited block. Accordingly, patients with hereditary coproporphyria exhibit the signs and symptoms associated with the excess ALA and porphobilinogen, such as those present in IAP, along with some photosensitivity due to the presence of excessive coproporphyrinogens and uroporphyrinogens. Again, the infusion of hematin can effect at least a partial repression of ALA synthase and amelioration of the signs and symptoms secondary to overproduction of intermediates in heme synthesis.

Variegate porphyria, or protocoproporphyria hereditaria, is an autosomal dominant disorder in which there is a partial block in the enzymatic conversion of protoporphyrinogen to heme. Two enzymes, protoporphyrinogen oxidase and ferrochelatase, both located in the **mitochondria,** seem to be normally responsible for this conversion. Patients with variegate porphyria have only half the normal level of **protoporphyrinogen oxidase** in their cultured skin fibroblast cells. Patients with variegate porphyria also exhibit a relative heme deficiency under stressful conditions, and the hepatic ALA synthase is derepressed. As discussed above, this increased activity of ALA synthase leads to overproduction of all of the intermediates in the heme synthetic pathway proximal to the block. Accordingly, patients with variegate porphyria excrete excessive quantities of ALA, porphobilinogen, uroporphyrin, and coproporphyrin in their urines, and uroporphyrin, coproporphyrin, and protoporphyrin in their feces. Thus, their urines are pigmented and fluoresce, and they exhibit cutaneous photosensitivity—the latter indistinguishable from that observed in porphyria cutanea tarda, discussed below.

As with the other porphyria syndromes, excretion of the accumulated intermediates of the heme pathway may be normal or only slightly elevated under nonstressful conditions but increases greatly when the demand for heme is increased—particularly in the liver, for reasons explained above. The plasma of patients frequently exhibits a remarkable red fluorescence which seems to correlate with the high concentration of coproporphyrinogen in that fluid. Erythrocyte porphyrin levels remain normal in this disease.

Porphyria cutanea tarda is probably the most common form of porphyria. It is usually associated with some form of hepatic injury, particularly alcohol or iron overload. The nature of the metabolic defect has not been well defined, but it is perhaps attributable to a partial deficiency of **uroporphyrinogen decarboxylase.** The defect appears to be transmitted as an autosomal dominant disorder, but the penetrance of the disease is variable, in most cases being dependent upon the existence of some form of hepatic injury. Predictably, the urine contains increased quantities of uroporphyrins of both type I and type III, but elevated urinary excretion of ALA and porphobilinogen occurs only rarely. Although the urine may occasionally contain sufficient porphyrins to produce a pinkish color, upon acidification it frequently exhibits a pink fluorescence under ultraviolet light.

The liver contains large quantities of porphyrins, so that it fluoresces intensely, whereas the erythrocytes and cells of the bone marrow do not. In porphyria cutanea tarda, the major clinical manifestation is **cutaneous photosensitivity.** The lack of increased ALA synthase activity and corresponding lack of excess porphobilinogen and ALA in the urine of these patients correlate positively with the lack of the acute manifestations typical of IAP.

Protoporphyria, or erythropoietic protoporphyria, appears to result from a dominantly inherited partial deficiency of **ferrochelatase** activity in the mitochondria of all tissues, and it is associated clinically with acute urticaria on exposure to sunlight. The erythrocytes, plasma, and feces contain increased quantities of protoporphyrin IX, and the reticulocytes (young erythrocytes) and skin obtained by biopsy frequently exhibit red fluorescence.

The liver probably also contributes to the overproduction of protoporphyrin IX, but there is no increased urinary excretion of porphyrin precursors or porphyrins.

Acquired (toxic) porphyria can result from exposure to toxic compounds such as hexachlorobenzene, lead, and other salts of heavy metals, as well as drugs such as griseofulvin. Heavy metals inhibit several enzymes in the heme synthetic pathway, including ALA dehydratase, uroporphyrinogen synthase, and ferrochelatase.

CATABOLISM OF HEME; FORMATION OF BILE PIGMENTS

Under physiologic conditions in the human adult, $1-2 \times 10^8$ erythrocytes are destroyed per hour. Thus, in 1 day, a 70-kg human turns over approximately 6 g of hemoglobin. When hemoglobin is destroyed in the body, the protein portion, globin, may be reutilized either as such or in the form of its constituent amino acids, and the iron of heme enters the iron pool, also for reuse. However, the iron-free porphyrin portion of heme is degraded, mainly in the reticuloendothelial cells of the liver, spleen, and bone marrow.

The catabolism of heme from all of the heme proteins appears to be carried out in the microsomal fractions of the reticuloendothelial cells by a complex enzyme system called **heme oxygenase.** By the time the heme of heme proteins reaches the heme oxygenase system, the iron has usually been oxidized to the ferric form, constituting **hemin,** and may be loosely bound to albumin as methemalbumin. The heme oxygenase system is substrate-inducible. It is located in close proximity to the microsomal electron transport system. As depicted in Fig 33–12, the hemin is reduced with NADPH, and, with the aid of more NADPH, oxygen is added to the α-methenyl bridge between pyrroles I and II of the porphyrin. The ferrous iron is again oxidized to the ferric form. With the further addition of oxygen, **ferric ion** is released, **carbon monoxide** is produced, and an equimolar quantity of **biliverdin IX**-α results from the splitting of the tetrapyrrole ring. The heme itself participates in this reaction as a catalyst.

In birds and amphibia, the green biliverdin IX-α is excreted; in mammals, a soluble enzyme called **biliverdin reductase** reduces the methenyl bridge between pyrrole III and pyrrole IV to a methylene group to produce **bilirubin IX**-α, a yellow pigment (Fig 33–12).

It is estimated that 1 g of hemoglobin yields 35 mg of bilirubin. The daily bilirubin formation in human adults is approximately 250–350 mg.

The chemical conversion of heme to bilirubin by the reticuloendothelial cells can be observed in vivo as the purple color of the heme in a hematoma is slowly converted to the yellow pigment of bilirubin.

The further metabolism of bilirubin occurs primarily in the liver. It can be divided into 3 processes: (1) uptake of bilirubin by liver parenchymal cells, (2) conjugation of bilirubin in the smooth endoplasmic reticulum, and (3) secretion of conjugated bilirubin into the bile. Each of these processes will be considered separately.

Uptake of Bilirubin by the Liver

Bilirubin is only sparingly soluble in plasma and water, but in the plasma it is protein-bound, specifically to albumin. Each molecule of albumin appears to have one high-affinity site and one low-affinity site for bilirubin. In 100 mL of plasma, approximately 25 mg of bilirubin can be **tightly bound to albumin** at its high-affinity site. Bilirubin in excess of this quantity can be bound only loosely and thus can easily be detached and diffused into tissues. A number of compounds such as antibiotics and other drugs compete with bilirubin for the high-affinity binding site on albumin. Thus, these compounds can displace bilirubin from albumin and have significant clinical effects.

In the liver, the bilirubin seems to be removed from the albumin and taken up at the sinusoidal surface of the hepatocytes by a carrier-mediated saturable system. This facilitated transport system has a very large capacity, so that even under pathologic conditions the system does not appear to be rate-limiting in the metabolism of bilirubin.

Since this facilitated transport system allows the equilibration of bilirubin across the sinusoidal membrane of the hepatocyte, the net uptake of bilirubin will be dependent upon the removal of bilirubin by subsequent metabolic pathways.

Conjugation of Bilirubin

By adding polar groups to bilirubin, the liver con-

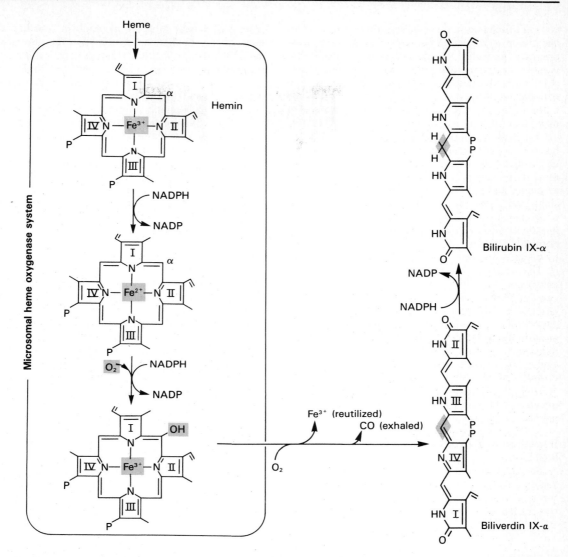

Figure 33–12. Schematic representation of the microsomal heme oxygenase system. (Modified from Schmid R, Mc-Donough AF in: *The Porphyrins.* Dolphin D [editor]. Academic Press, 1978.)

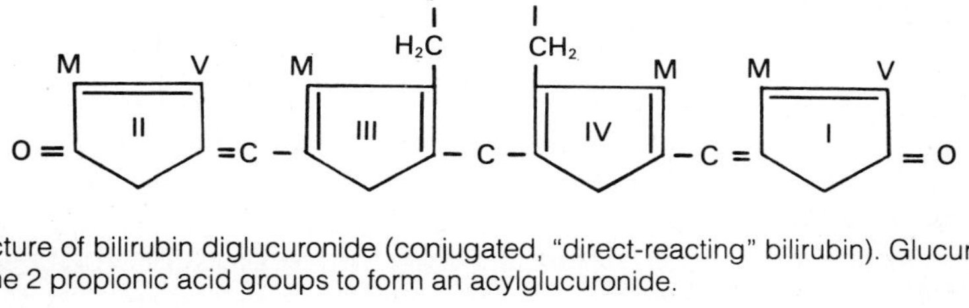

Figure 33–13. Structure of bilirubin diglucuronide (conjugated, "direct-reacting" bilirubin). Glucuronic acid is attached via ester linkage to the 2 propionic acid groups to form an acylglucuronide.

verts bilirubin to a water-soluble form that can subsequently be secreted into the bile. This process of **increasing the water solubility** or polarity of bilirubin is achieved by conjugation. It is a process carried out, at least initially, in the smooth endoplasmic reticulum with the aid of a specific set of enzymes. Most of the bilirubin excreted in the bile of mammals is in the form of a **bilirubin diglucuronide** (Fig 33–13). The formation of the intermediate monoglucuronide of bilirubin is catalyzed by uridine diphosphate glucuronate glucuronyltransferase **(UDP-glucuronyltransferase),** an enzyme that exists in the smooth endoplasmic reticulum and is probably composed of more than a single entity. The reaction is depicted in Fig 33–14. It occurs chiefly in the liver but also in the kidney and the intestinal mucosa. When bilirubin conjugates exist abnormally in human serum, they are predominantly monoglucuronides.

The formation of the diglucuronide of bilirubin may occur in the bile canalicular region of the hepatocyte membrane by a similar UDP-glucuronyltransferase (Fig 33–14) or by an enzyme-catalyzed dismutation of 2 mol of bilirubin monoglucuronide to 1 mol of bilirubin diglucuronide and 1 mol of unconjugated bilirubin (Fig 33–14). More will be said about this conjugation system in the discussion of the inherited disorders of bilirubin conjugation.

UDP-glucuronyltransferase activity can be **induced** by a number of clinically useful drugs, including phenobarbital.

Secretion of Bilirubin into Bile

Secretion of conjugated bilirubin into the bile occurs against a large concentration gradient and must be carried out by an active transport mechanism. The **active transport** is probably **rate-limiting** for the entire process of hepatic bilirubin metabolism. The hepatic transport of conjugated bilirubin into the bile is inducible by those same drugs that are capable of inducing the conjugation of bilirubin. Thus, the conjugation and excretion systems for bilirubin behave as a coordinated functional unit.

Under physiologic conditions, essentially all (> 97%) of the bilirubin secreted into the bile is conjugated. Only after phototherapy can significant quantities of unconjugated bilirubin be found in bile.

In the liver, there are multiple systems for secreting naturally occurring and pharmaceutical compounds into the bile after their metabolism. Some of these secreting systems are shared by the bilirubin diglucuronides, but others seem to operate independently.

METABOLISM OF BILIRUBIN IN THE INTESTINE

As the conjugated bilirubin reaches the terminal ileum and the large intestine, the glucuronides are removed by specific bacterial enzymes (β-**glucuronidases**), and the pigment is subsequently reduced by the fecal flora to a group of colorless tetrapyrrolic compounds called **urobilinogens** (Fig 33–15). In the terminal ileum and large intestine, a small fraction of the urobilinogens is reabsorbed and reexcreted through the liver to constitute the **intrahepatic urobilinogen cycle.** Under abnormal conditions, particularly when excessive bile pigment is formed or liver disease interferes with this intrahepatic cycle, urobilinogen may also be excreted in the urine.

Normally, most of the **colorless urobilinogens** formed in the colon by the fecal flora are oxidized there to urobilins (colored compounds) and are excreted in the feces (Fig 33–15). Darkening of feces

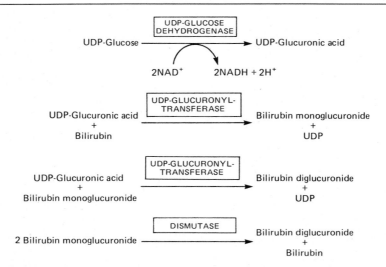

Figure 33–14. Conjugation of bilirubin with glucuronic acid. The glucuronate donor, UDP-glucuronic acid, is formed from UDP-glucose as depicted.

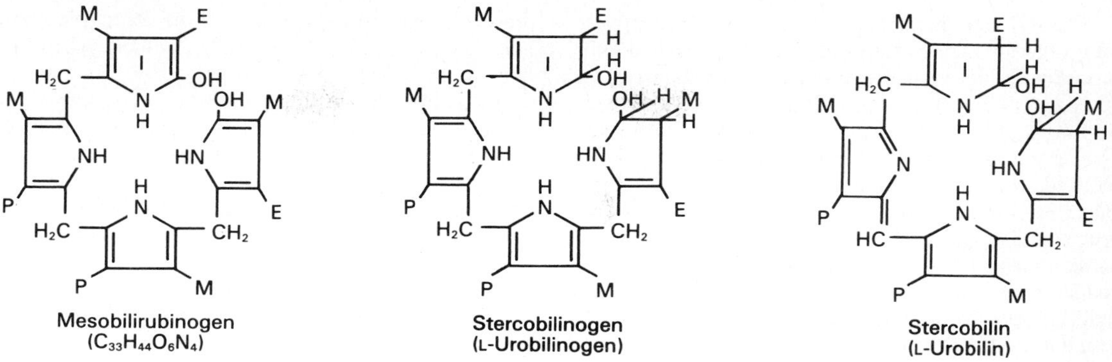

Figure 33–15. Structure of some bile pigments.

Mesobilirubinogen ($C_{33}H_{44}O_6N_4$)

Stercobilinogen (L-Urobilinogen)

Stercobilin (L-Urobilin)

upon standing in air is due to the oxidation of residual urobilinogens to urobilins.

HYPERBILIRUBINEMIA

When bilirubin in the blood exceeds 1 mg/dL (17.1 μmol/L), hyperbilirubinemia exists. Hyperbilirubinemia may be due to the **production of more bilirubin** than the normal liver can excrete, or it may result from the **failure of a damaged liver to excrete bilirubin** produced in normal amounts. In the absence of hepatic damage, obstruction to the excretory ducts of the liver—by preventing the excretion of bilirubin—will also cause hyperbilirubinemia. In all of these situations, bilirubin accumulates in the blood, and when it reaches a certain concentration, it diffuses into the tissues, which then become yellow. The condition is called **jaundice** or **icterus.**

In clinical studies of jaundice, measurement of bilirubin in the serum is of great value. A method for quantitatively assaying the bilirubin content of the serum was first devised by Van den Bergh by application of Ehrlich's test for bilirubin in urine. The Ehrlich reaction is based on the coupling of diazotized sulfanilic acid (Ehrlich's diazo reagent) and bilirubin to produce a reddish-purple azo compound. In the original procedure as described by Ehrlich, methanol was used to provide a solution in which both bilirubin and the diazo reagent were soluble. Van den Bergh inadvertently omitted the methanol on an occasion when assay of bile pigment in human bile was being attempted. To his surprise, normal development of the color occurred "directly." This form of bilirubin that would react without the addition of methanol was thus termed **"direct-reacting."** It was then found that this same direct reaction would also occur in serum from cases of jaundice due to biliary obstruction. However, it was still necessary to add methanol to detect bilirubin in normal serum or that which was present in excess in serum from cases of hemolytic jaundice where no evidence of obstruction was to be found. To that form of bilirubin which could be measured only after the addition of methanol, the term **"indirect-reacting"** was applied.

It has now been demonstrated that the **indirect** bilirubin is **"free" (unconjugated) bilirubin** en route to the liver from the reticuloendothelial tissues where the bilirubin was originally produced by the breakdown of heme porphyrins. Since this bilirubin is not water-soluble, it requires methanol to initiate coupling with the diazo reagent. In the liver, the free bilirubin becomes conjugated with glucuronic acid, and the conjugate, bilirubin glucuronide, can then be excreted into the bile. Furthermore, conjugated bilirubin, being water-soluble, can react directly with the diazo reagent, so that the **"direct bilirubin"** of Van den Bergh is actually a **bilirubin conjugate** (bilirubin glucuronide).

Depending on the type of bilirubin present in plasma, ie, unconjugated bilirubin or conjugated bilirubin, the hyperbilirubinemia may be classified as **retention** hyperbilirubinemia or **regurgitation** hyperbilirubinemia, respectively.

Only unconjugated bilirubin can cross the blood-brain barrier into the central nervous system; thus, encephalopathy due to hyperbilirubinemia (kernicterus) can occur only in connection with retention bilirubin or unconjugated hyperbilirubinemia. On the other hand, only conjugated bilirubin can appear in urine. Accordingly, **choluric jaundice** occurs only in regurgitation hyperbilirubinemia, and **acholuric jaundice** occurs only in the presence of an excess of unconjugated bilirubin.

Unconjugated Hyperbilirubinemia

Even in the event of extensive hemolysis, unconjugated hyperbilirubinemia is usually only slight (< 4 mg/dL; < 68.4 μmol/L), because of the liver's large capacity for handling bilirubin. However, if the handling of bilirubin is defective owing to either an acquired defect or an inherited abnormality, unconjugated hyperbilirubinemia may occur.

The most common cause of unconjugated hyperbilirubinemia is the transient neonatal **"physiologic jaundice."** This hyperbilirubinemia results from an accelerated hemolysis and an immature hepatic system for the uptake, conjugation, and secretion of bilirubin. Not only is the **UDP-glucuronyltransferase** activity reduced, but there probably is reduced synthesis of the substrate for that enzyme, UDP-glucuronic acid. Since the increased bilirubin is unconjugated, it is capable of penetrating the blood-brain barrier when its concentration in plasma exceeds that which can be tightly bound by albumin (20–25 mg/dL). This can result in a hyperbilirubinemic toxic encephalopathy, or **kernicterus.** Because of the recognized inducibility of this bilirubin metabolizing system, phenobarbital has been administered to jaundiced neonates and is effective in this disorder. In addition, exposure to visible light (by a mechanism that is not understood) can promote the hepatic excretion of unconjugated bilirubin by converting some of the bilirubin to other derivatives such as maleimide fragments and geometric isomers that are excreted in the bile.

A. Crigler-Najjar Syndrome, Type I; Congenital Nonhemolytic Jaundice: Type I Crigler-Najjar syndrome, a rare autosomal recessive disorder of humans, is due to a primary metabolic defect in the conjugation of bilirubin. It is characterized by severe congenital jaundice due to the **inherited absence of bilirubin UDP-glucuronyltransferase activity** in hepatic tissues. The disease is usually fatal within the first 15 months of life, but a few teenagers have been reported who did not develop difficulties until puberty. These children have been treated with phototherapy with some reduction in plasma bilirubin levels. Phenobarbital and other drugs that induce the bilirubin metabolizing systems in normal liver have no effect on the formation of bilirubin glucuronides in patients with type I Crigler-Najjar syndrome. Serum bilirubin usually exceeds 20 mg/dL when untreated.

B. Crigler-Najjar Syndrome, Type II: This rare inherited disorder seems to result from a milder defect in the bilirubin conjugating system and has a more benign course. The serum bilirubin concentrations usually do not exceed 20 mg/dL, but all of the bilirubin accumulated is of the unconjugated type. Surprisingly, the bile in these patients does contain bilirubin monoglucuronide, and it has been proposed that the genetic defect **may involve the hepatic UDP-glucuronyltransferase that adds the second glucuronyl group** to bilirubin monoglucuronide.

It has been demonstrated that patients with this syndrome can respond to treatment with large doses of phenobarbital. In these patients, the drug-mediated reduction of hyperbilirubinemia seems to result from induction of the entire bilirubin metabolizing system and not simply a stimulation of bilirubin conjugation.

In several instances, presumed heterozygotes for this disorder exhibited a mild unconjugated hyperbilirubinemia indistinguishable from that seen in Gilbert's disease (see below). Thus, it is possible that the autosomal recessive type II Crigler-Najjar syndrome is the homozygous state of the defect present in heterozygous form in the mild chronic hyperbilirubinemia of Gilbert.

C. Gilbert's Disease: Gilbert's disease is a heterogeneous group of disorders, many of which are now recognized to be due to a compensated hemolysis associated with unconjugated hyperbilirubinemia. There also appears to be a defect in the hepatic clearance of bilirubin, possibly due to a **defect in the uptake of bilirubin** by the liver parenchymal cells. However, **bilirubin UDP-glucuronyltransferase activities** in the livers of those patients studied with this disease were found to be reduced.

In general, the benign disorders collectively termed Gilbert's disease seem to be transmitted in an autosomal dominant manner.

D. Toxic Hyperbilirubinemia: Unconjugated hyperbilirubinemia can result from toxin-induced liver dysfunction such as that caused by chloroform, arsphenamines, carbon tetrachloride, acetaminophen, hepatitis virus, cirrhosis, and *Amanita* mushroom poisoning. Although most of these acquired disorders are due to hepatic parenchymal cell damage, there is frequently a component of obstruction of the biliary tree within the liver that results in the presence of some conjugated hyperbilirubinemia.

Conjugated Hyperbilirubinemia

Because conjugated bilirubin is water-soluble, it is detectable in the urine of most patients with conjugated hyperbilirubinemia; thus, they are frequently said to have choluric jaundice.

A. Chronic Idiopathic Jaundice (Dubin-Johnson Syndrome): This autosomal recessive disorder consists of conjugated hyperbilirubinemia in childhood or during adult life. The hyperbilirubinemia is apparently caused by a **defect in the hepatic secretion of conjugated bilirubin into the bile.** However, this secretory defect of conjugated compounds is not restricted to bilirubin but also involves secretion of conjugated estrogens and test compounds such as the dye sulfobromophthalein. In fact, the secretory defect of conjugated sulfobromophthalein results in its reflux into the plasma, leading to a secondary rise in the plasma concentration of this test dye, a phenomenon that is pathognomonic for Dubin-Johnson syndrome. When test compounds such as indocyanine green and rose bengal, which do not require conjugation for excretion, are used, such secondary rises in plasma concentration do not appear in these patients. Thus, the defect appears to be in the secretory process that normally deals only with conjugated compounds including conjugated bilirubins. Characteristically, in patients with Dubin-Johnson syndrome, the hepatocytes in the centrilobular area contain an abnormal pigment that has not been identified.

B. Biliary Tree Obstruction: Conjugated hyperbilirubinemia also results from blockage of the hepatic or common bile ducts. The bile pigment is believed to pass from the blood into the liver cells as usual but fails to be excreted. As a consequence of

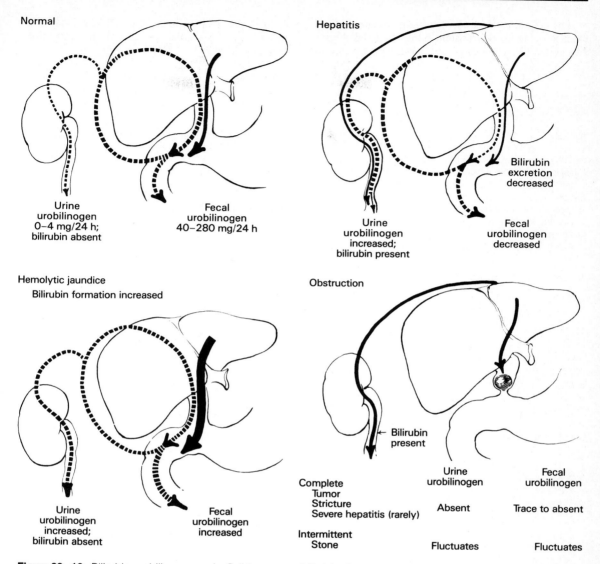

Figure 33–16. Bilirubin-urobilinogen cycle. Solid arrows = bilirubin glucuronide; dotted arrows = urobilinogen. (Reproduced, with permission, from Krupp MA et al: *Physician's Handbook,* 21st ed. Lange, 1985.)

this, the conjugated bilirubin is absorbed into the hepatic veins and lymphatics.

The term **cholestatic jaundice** may be used to include all forms of extrahepatic obstructive jaundice in addition to some forms of parenchymal jaundice characterized by conjugated hyperbilirubinemia.

Urine Urobilinogen

Normally, there are mere traces of urobilinogen in the urine (average, 0.64 mg; maximum normal, 4 mg [in 24 hours]). In complete obstruction of the bile duct, no urobilinogen is found in the urine, since bilirubin has no access to the intestine where it can be converted to urobilinogen. In this case, the presence of bilirubin in the urine without urobilinogen suggests

obstructive jaundice, either intrahepatic or posthepatic.

In **hemolytic jaundice,** the increased production of bilirubin leads to increased production of urobilinogen, which appears in the urine in large amounts. Bilirubin is not usually found in the urine in hemolytic jaundice, so that the combination of increased urobilinogen and absence of bilirubin is suggestive of hemolytic jaundice. Increased blood destruction from any cause (eg, pernicious anemia) will, of course, also bring about an increase in urine urobilinogen. Furthermore, infection of the biliary passages may increase the urobilinogen in the absence of any reduction in liver function, because of the reducing activity of the infecting bacteria.

The diagrams in Fig 33–16 summarize the events characterizing the handling of bilirubin and urobilinogen by the liver, intestine, and kidney under normal circumstances and in the presence of hemolytic jaundice, hepatitis, or jaundice associated with obstruction of the bile duct.

REFERENCES

Battersby AR et al: Biosynthesis of the pigments of life: Formation of the macrocycle. *Nature* 1980;**285**:17.

Berk PD et al: Disorders of bilirubin metabolism. In: *Metabolic Control and Disease,* 8th ed. Bondy PK, Rosenberg LE (editors). Saunders, 1980.

Kappas A, Sassa S, Anderson KE: The porphyrias. In: *The Metabolic Basis of Inherited Disease,* 5th ed. Stanbury JB et al (editors). McGraw-Hill, 1983.

Lemberg R, Legge JW: *Hematin Compounds and Bile Pigments.* Interscience, 1949.

Schmid R, McDonough AF: Formation and metabolism of bile pigments in vivo. In: *The Porphyrins.* Dolphin D (editor). Academic Press, 1978.

Tschudy DP, Lamon JM: Porphyrin metabolism and the porphyrias. In: *Metabolic Control and Disease,* 8th ed. Bondy PK, Rosenberg LE (editors). Saunders, 1980.

Watson CJ: Gold from dross: The first century of the urobilinoids. *Ann Intern Med* 1969;**70**:839.

Section IV.
Structure, Function, & Replication of Informational Macromolecules

Nucleotides

34

Victor W. Rodwell, PhD

INTRODUCTION

The nucleotides participate in a wide variety of biochemical processes. Perhaps the best known role of the purine and pyrimidine nucleotides is to serve as the monomeric precursors of RNA and DNA. However, the **purine** ribonucleotides serve also as the ubiquitous high-energy source, ATP; as regulatory signals (cyclic AMP [cAMP] and cyclic GMP [cGMP]); and as components of the coenzymes FAD, NAD, and NADP and of the methyl group donor S-adenosylmethionine. The **pyrimidine** nucleotides, in addition to providing monomeric precursors for nucleic acid synthesis, also serve as high-energy intermediates, such as UDP-glucose and UDP-galactose in carbohydrate metabolism and CDP-acylglycerol in lipid synthesis.

BIOMEDICAL IMPORTANCE

The heterocyclic bases purine and pyrimidine are the parent molecules of nucleosides and nucleotides. Nucleotides are ubiquitous in living cells, where they perform numerous key functions. Examples include incorporation, as their ribose (RNA) or deoxyribose (DNA) monophosphates, into nucleic acids, energy transduction (ATP), parts of coenzymes (AMP), acceptors for oxidative phosphorylation (ADP), allosteric regulators of enzyme activity, and "second messengers" (cAMP, cGMP). Synthetic analogs of naturally occurring nucleotides find application in cancer chemotherapy as enzyme inhibitors and can replace the naturally occurring nucleotides in nucleic acids. Therapeutic attempts to inhibit the growth of cancer cells or certain viruses have often employed administration of analogs of bases, nucleosides, or nucleotides that inhibit the synthesis of either DNA or RNA. Such compounds include 5-fluorouracil, 5'-iodo-2'-deoxyuridine, 6-thioguanine, 6-mercaptopu-

rine, 6-azauridine, and arabinosyl cytosine, as described below. Allopurinol, a purine analog, is widely used in the treatment of gout.

STRUCTURES OF PURINE & PYRIMIDINE BASES

Purine and pyrimidine bases that occur in the nucleotides are derived by substitution on the ring structures of the parent substances, purine or pyrimidine (Fig 34–1). The positions on the rings are numbered according to the international system. Note that the direction of the numbering of the purine ring is different from that of the pyrimidine ring but that the number 5 carbon is the same in both heterocyclic compounds. Because of their π electron clouds, both the purine and pyrimidine bases are planar molecules, the significance of which is discussed in Chapter 37.

Major Bases

The 3 major pyrimidine bases present in the nucleotides of both prokaryotes and eukaryotes are **cytosine, thymine,** and **uracil** (Fig 34–2). The **purine bases**

Purine

Pyrimidine

Figure 34–1. Structures of purine and pyrimidine with the positions of the elements numbered according to the international system.

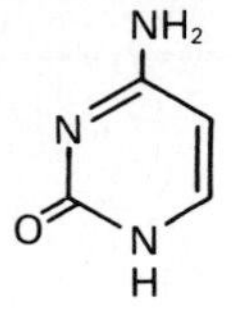

Cytosine
(2-oxy-4-aminopyrimidine)

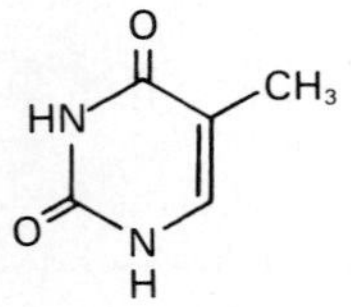

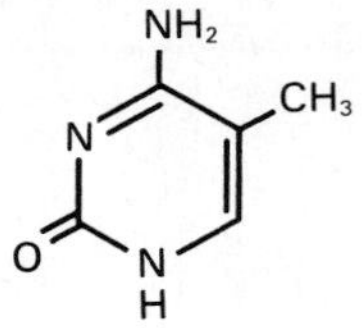

5-Methylcytosine

5-Hydroxymethylcytosine

Figure 34–4. The structures of 2 uncommon naturally occurring pyrimidine bases.

Thymine
(2,4-dioxy-5-methylpyrimidine)

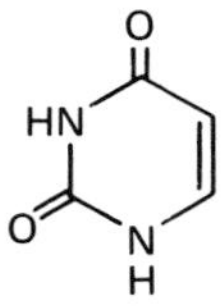

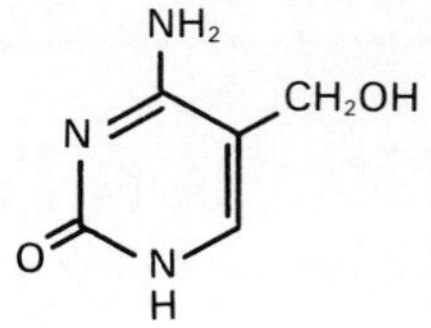

Uracil
(2,4-dioxypyrimidine)

Figure 34–2. The 3 major pyrimidine bases found in nucleotides.

adenine and **guanine** are the 2 major purines found in living organisms. Two other purine bases, **hypoxanthine** and **xanthine,** occur as intermediates in the metabolism of adenine and guanine (Fig 34–3). In humans, a completely oxidized purine base, **uric acid,** is formed as the end product of purine catabolism (see Chapter 35).

In natural materials, numerous minor (ie, unusual) bases occur in addition to the 5 major bases described above. Some of these unusual substituted bases are present only in the nucleic acids of bacteria and viruses, but many are also found in the DNA and transfer RNAs of both prokaryotes and eukaryotes. For ex-

ample, both bacterial and human DNA contain significant quantities of 5-methylcytosine; bacteriophages contain 5-hydroxymethylcytosine (Fig 34–4). Unusual bases present in the messenger RNA molecules of mammalian cells include N^6-methyladenine, N^6,N^6-dimethyladenine, and N^7-methylguanine (Fig 34–5). A uracil modified at the N_3 position by the attachment of an (α-amino, α-carboxyl)-propyl group has also been detected in bacteria. The functions of these substituted purine and pyrimidine nucleotide bases are not fully understood.

In plants, a series of purine bases containing methyl substituents occurs (Fig 34–6). Many have pharmacologic properties. Examples are coffee, which contains caffeine (1,3,7-trimethylxanthine); tea, which contains theophylline (1,3-dimethylxanthine); and cocoa, which contains theobromine (3,7-dimethylxanthine). The biologic properties of these compounds are described in Chapter 35 in the discussion of the metabolism of cyclic nucleotides.

PHYSICOCHEMICAL PROPERTIES OF PURINE & PYRIMIDINE BASES

Tautomerism

Because of keto-enol tautomerism, these aromatic molecules can exist in a lactim or lactam form (Fig 34–7); the latter is by far the predominant tautomer of guanine or thymine under physiologic conditions. (The importance of the lactim versus the lactam form becomes apparent in the discussions on base pairing and mutagenesis in Chapters 38 and 40.)

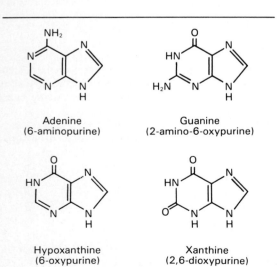

Adenine
(6-aminopurine)

Guanine
(2-amino-6-oxypurine)

Hypoxanthine
(6-oxypurine)

Xanthine
(2,6-dioxypurine)

Figure 34–3. The major purine bases present in nucleotides.

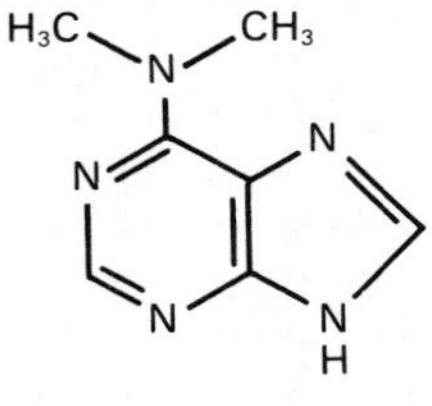

N^6,N^6-Dimethyladenine

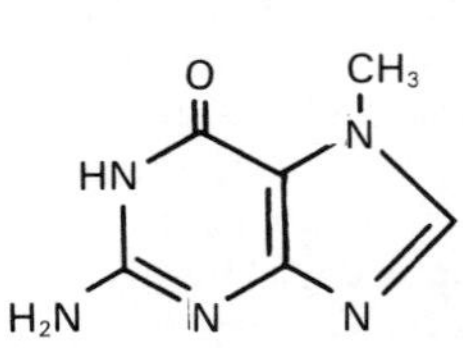

N^7-Methylguanine

Figure 34–5. The structures of 2 uncommon naturally occurring purine bases.

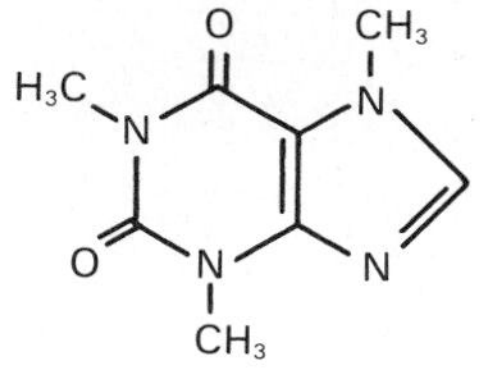

Caffeine (1,3,7-trimethyl-xanthine)

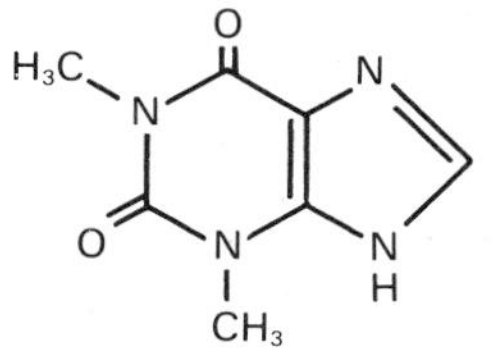

Theophylline (1,3-dimethyl-xanthine)

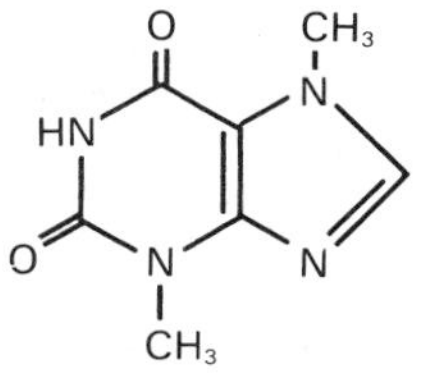

Theobromine (3,7-dimethyl-xanthine)

Figure 34–6. The structures of some methyl xanthines commonly occurring in foodstuffs.

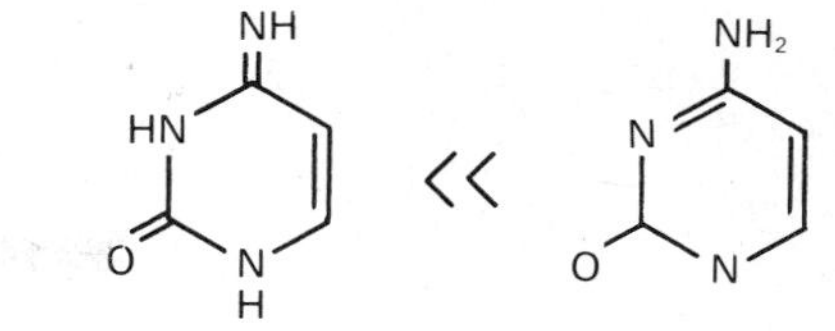

Cytosine (lactam) Cytosine (lactim)

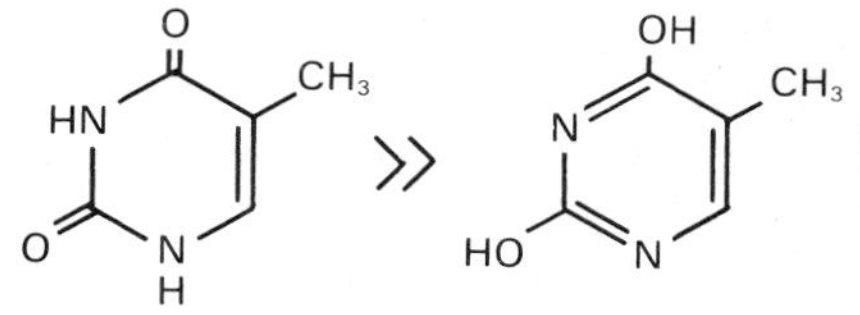

Thymine (lactam) Thymine (lactim)

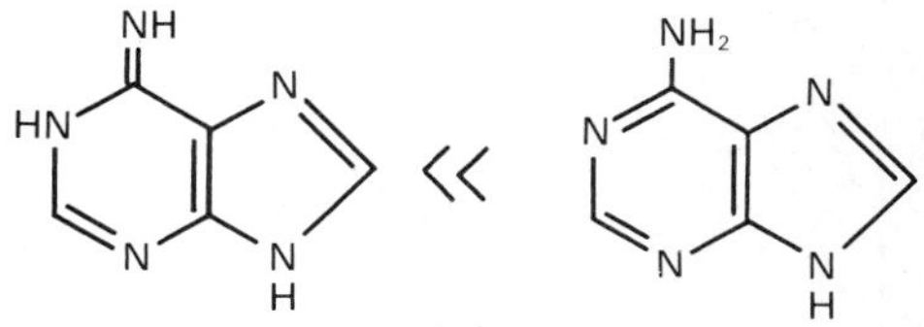

Adenine (lactam) Adenine (lactim)

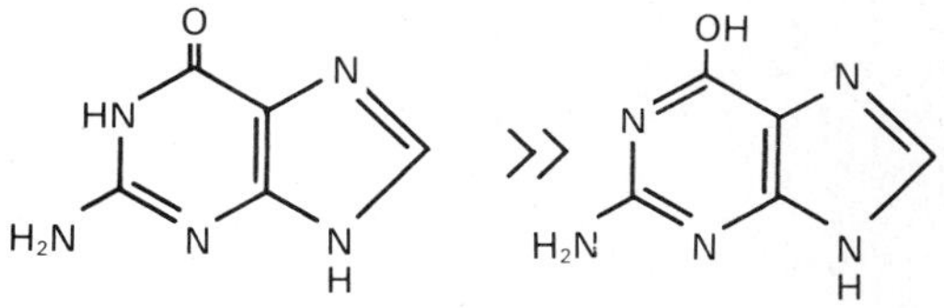

Guanine (lactam) Guanine (lactim)

Figure 34–7. The structures of the tautomers of cytosine, thymine, adenine, and guanine with the predominant forms indicated.

Solubility

At neutral pH, guanine is the least soluble of the bases, followed in this respect by xanthine. Although uric acid as urate is relatively soluble at a neutral pH, it is highly insoluble in solutions with a lower pH, such as urine. Guanine is not a normal constituent of human urine, but xanthine and uric acid do occur in human urine. These latter 2 purines frequently occur as constituents of urinary tract stones.

NUCLEOSIDES & NUCLEOTIDES

The free bases are much less abundant in nature than are their nucleosides and nucleotides. A **nucleoside** (Fig 34–8) is composed of a purine or a pyrimidine base to which a sugar (usually either D-ribose or 2-deoxyribose) is attached in β-linkage at N_9 or N_1, respectively. Thus, the adenine ribonucleoside **adenosine** consists of adenine with D-ribose attached at the N_9. **Guanosine** consists of guanine with D-ribose attached at N_9. Cytidine is cytosine with ribose attached at its N_1 position. Uridine consists of ribose attached at the N_1 position of uracil.

The 2′-deoxyribonucleosides consist of 2-deoxyribose attached to purine or pyrimidine bases at the positions described above. Attachment of the ribose or 2-deoxyribose to the ring structures is through an N-glycosidic bond, which is relatively acid-labile. Although, theoretically, free rotation of the sugar moiety and the purine or pyrimidine ring structure occurs about this N-glycosidic bond, steric hindrance in fact hinders free rotation. In the naturally occurring nucleosides, the **anti** conformation is strongly favored over the **syn** form (Fig 34–9). As is discussed in Chapter 37, the anti form is necessary for the proper positioning of the complementary purine and pyrimidine bases in the double-stranded B form of deoxyribonucleic acid. (Because of the conventional representation of the D-ribose, in most figures of this and other chapters, the purine and pyrimidine nucleosides and nucleotides are shown in the less favored syn conformation.)

Nucleotides are nucleosides phosphorylated on one or more of the hydroxyl groups of the sugar (ribose or deoxyribose) (Fig 34–10). Thus, adenosine monophosphate (AMP or adenylate) is adenine + ribose + phosphate. 2′-Deoxyadenosine monophosphate (dAMP or deoxyadenylate) consists of adenine + 2-deoxyribose + phosphate. The only sugar commonly found attached to uracil is ribose, and that commonly found attached to thymine is 2-deoxyribose. There-

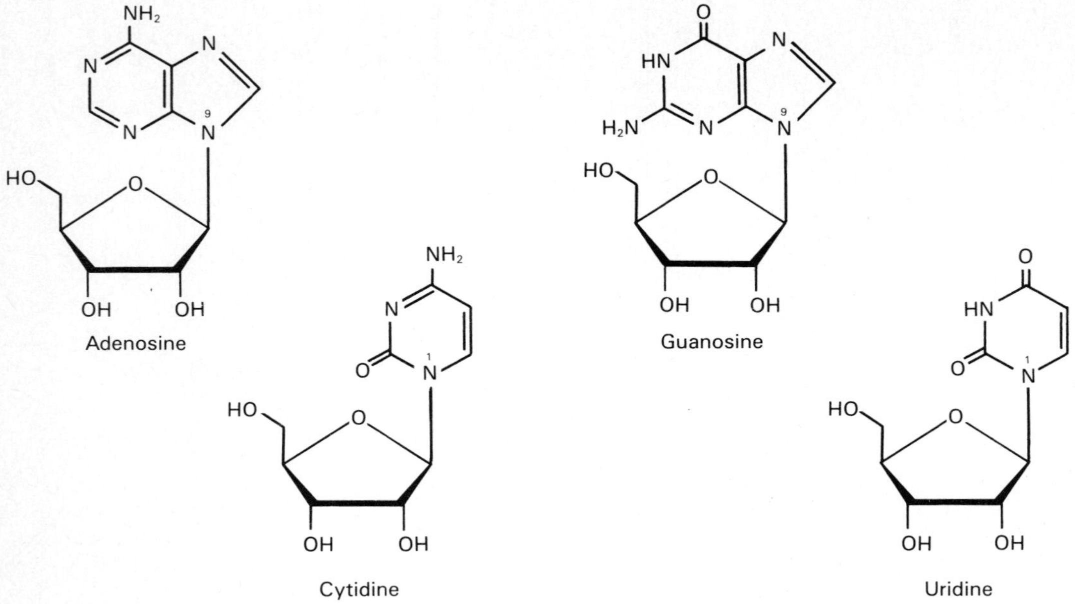

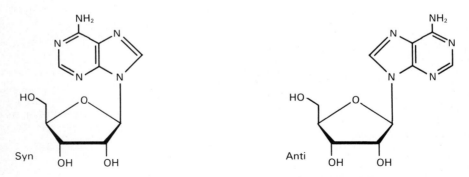

Figure 34–8. Structures of ribonucleosides.

Figure 34–9. The structures of the **syn** and **anti** configurations of adenosine.

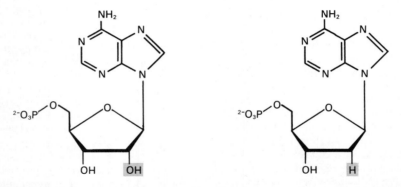

Figure 34–10. The structures of adenylic acid (AMP) (*left*) and 2'-deoxyadenylic acid (dAMP) (*right*).

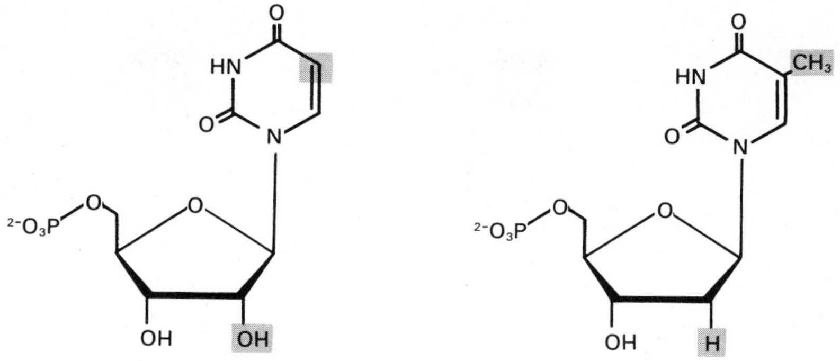

Figure 34–11. The structures of uridylic acid (UMP) (*left*) and thymidylic acid (TMP) (*right*).

fore, thymidylic acid (TMP) is thymine + 2-deoxyribose + phosphate, and uridylic acid (UMP) is uracil + ribose + phosphate (Fig 34–11). DNA is a polymer of thymidylic acid, 2'-deoxycytidylic acid, 2'-deoxyadenylic acid, and 2'-deoxyguanylic acid. RNA is a polymer containing uridylate, cytidylate, adenylate, and guanylate.

There are exceptions to the above structures of nucleotides. For example, in tRNA the ribose is occasionally attached to uracil at the 5 position, establishing a carbon-to-carbon linkage instead of the usual nitrogen-to-carbon linkage. This unusual compound is called pseudouridine (Ψ). tRNA molecules contain another unusual nucleotide structure, ie, thymine attached to ribose monophosphate. This compound is formed subsequent to the synthesis of the tRNA by methylation of the UMP residue by S-adenosylmethionine (see below). Pseudouridylic acid (Ψ MP) is similarly rearranged from uridylic acid after the tRNA molecule has been synthesized.

Nomenclature of Nucleosides & Nucleotides

The position of the phosphate in the nucleotide is indicated by a numeral. For example, adenosine with

the phosphate attached to carbon 3 of the sugar ribose would be designated adenosine 3'-monophosphate. The prime mark after the numeral is required to differentiate the numbered position on the sugar moiety from the numbered position on a purine or pyrimidine base; the latter would not be followed by the prime mark. A nucleotide of 2'-deoxyadenosine with the phosphate moiety attached to the carbon 5 position of the sugar would be designated 2'-deoxyadenosine-5'-monophosphate (Fig 34–12).

The abbreviations A, G, C, T, and U may be used to designate the nucleosides that contain adenine, guanine, cytosine, thymine, or uracil, respectively. The prefix d is added if the sugar of the nucleoside is 2'-deoxyribose. When the nucleoside occurs in the free form as a mononucleotide (ie, not a component of nucleic acid polynucleotide), the abbreviation MP (monophosphate) may be added to the abbreviation designating the nucleoside. For example, guanosine containing 2'-deoxyribose would be designated dG (deoxyguanosine) and the corresponding monophosphate with the phosphate esterified to the carbon 3 of

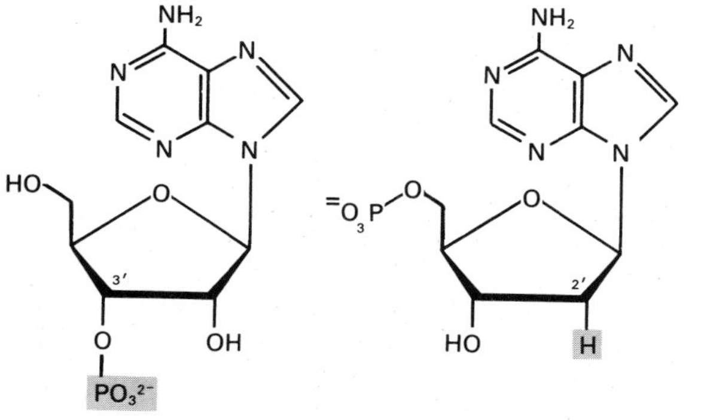

Figure 34–12. The structures of adenosine 3'-monophosphate (*left*) and 2'-deoxyadenosine-5'-monophosphate (*right*).

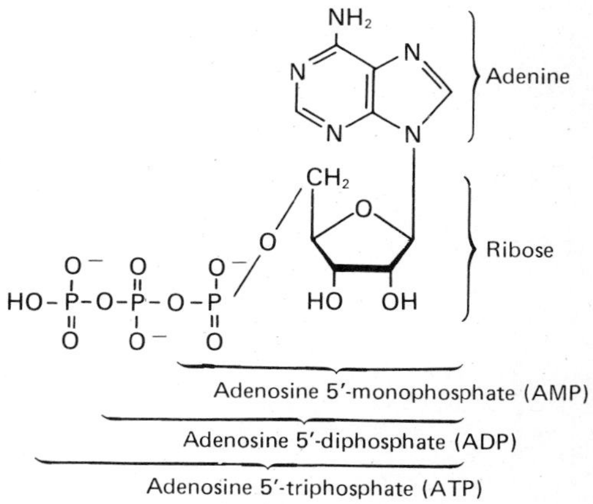

Figure 34–13. The structure of ATP and the structures of the corresponding diphosphate and monophosphate forms.

the deoxyribose moiety designated dG-3'-MP. Generally, when the phosphate is esterified to the carbon 5 of the ribose or deoxyribose moiety, the prefixed primed number (5') is omitted. For example, guanosine 5'-monophosphate would be abbreviated GMP, while the 5'-monophosphate of 2'-deoxyguanosine would be designated dGMP. When 2 or 3 phosphates are attached to the sugar moiety in the acid anhydride form, the abbreviations DP (diphosphate) and TP (triphosphate) are added to the abbreviations for the corresponding purine or pyrimidine nucleoside. Thus, adenosine triphosphate with 3 phosphate residues attached to the 5' carbon of the adenosine would be abbreviated ATP. The structure of ATP is shown in Fig 34–13 along with its corresponding diphosphate and monophosphate forms. Because the phosphates are in the acid anhydride form—a low-entropy situation—the phosphates are said to be high-energy ones, ie, high potential energy. The hydrolysis of 1 mol of ATP to ADP releases about 7 kcal of potential energy.

NATURALLY OCCURRING NUCLEOTIDES

Free nucleotides also perform important functions in tissues (see below).

Adenosine Derivatives

ADP and **ATP** are substrates and products, respectively, for oxidative phosphorylation (see Chapter 13), and ATP serves as the major intracellular transducer of free energy (see Chapter 11). The mean intracellular

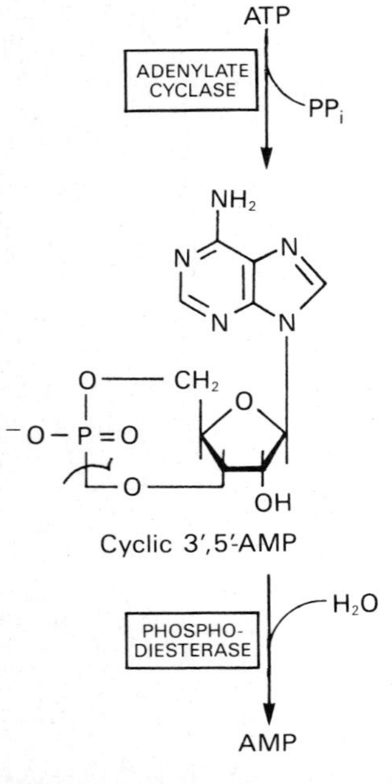

Figure 34–14. Formation of cAMP from ATP and destruction of cAMP by phosphodiesterase.

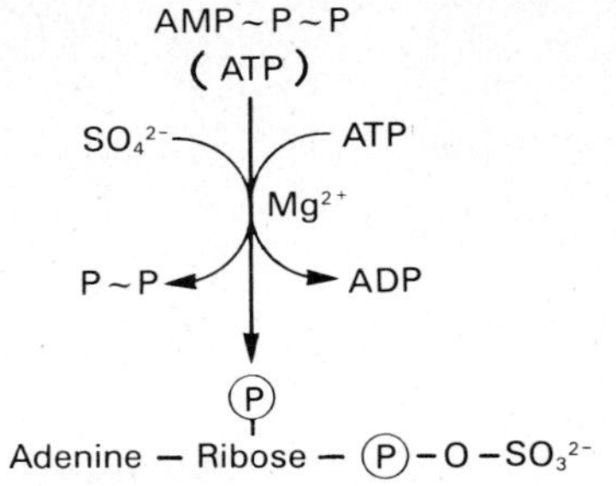

Figure 34–15. Formation of adenosine 3'-phosphate-5'-phosphosulfate.

concentration of ATP, the most abundant free nucleotide in mammalian cells, is about 1 mmol/L.

Cyclic AMP (3',5'-adenosine monophosphate; cAMP), a mediator of diverse extracellular signals in mammalian cells, is formed from ATP in a reaction catalyzed by **adenylate cyclase** (Fig 34–14). The activity of adenylate cyclase is regulated by complex interactions, many of which involve hormone receptors (see Chapter 43). Intracellular concentrations of cAMP (about 1 μmol/L) are about 3 orders of magnitude below those of ATP. Hydrolysis of cAMP to 5'-AMP is catalyzed by **cAMP phosphodiesterase** (Fig 34–14).

The incorporation of sulfate into ester linkages in compounds such as sulfated proteoglycans (see Chapter 42) requires the preliminary "activation" of the sulfate molecule. Sulfate is "activated" by reacting with ATP to form adenosine 3'-phosphate-5'-phosphosulfate (PAPS) in the reaction shown in Fig 34–15. The active sulfate moiety is also required as the substrate for sulfate conjugation reactions.

S-Adenosylmethionine (Fig 34–16) serves as a form of "active" methionine. S-Adenosylmethionine serves widely as a methyl donor in many diverse methylation reactions and as a source of propylamine for the synthesis of polyamines.

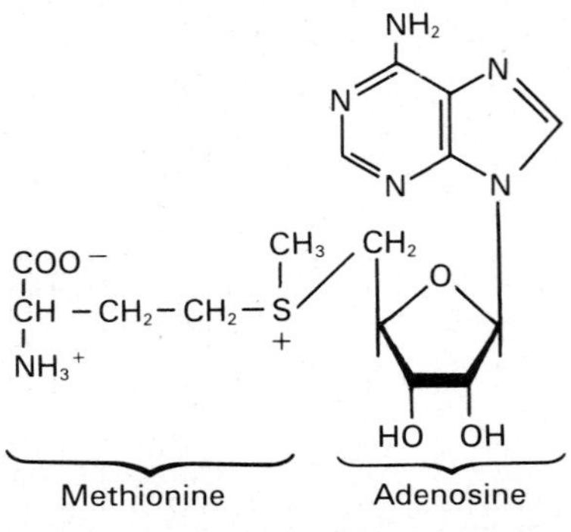

Figure 34–16. S-Adenosylmethionine.

Guanosine Derivatives

Guanosine nucleotides, particularly guanosine diphosphate and guanosine triphosphate, serve in several energy-requiring systems. These are analogs of ADP and ATP, respectively. For example, the oxidation of α-ketoglutaric acid to succinyl-CoA in the tricarboxylic acid cycle involves oxidative phosphorylation with transfer of phosphate to GDP to form GTP. GTP is required for the activation of adenylate cyclase by some hormones and serves both as an allosteric regulator and as an energy source for protein synthesis on polyribosomes. It therefore has an important role in the maintenance of the internal milieu.

Cyclic GMP (3′,5′-guanosine monophosphate; cGMP [Fig 34–17]) appears also to be an intracellular signal of extracellular events. In some cases, cGMP acts antagonistically to cAMP. cGMP is formed from GTP by **guanylate cyclase,** an enzyme similar in many ways to adenylate cyclase. Guanylate cyclase, like adenylate cyclase, appears to be regulated by a variety of effectors, including hormones. Like cAMP, cGMP is also catabolized by a phosphodiesterase to its 5′-monophosphate.

Hypoxanthine Derivatives

Hypoxanthine ribonucleotide, usually called inosinic acid (IMP, or inosinate in the salt form), is a precursor of all purine ribonucleotides synthesized de novo. Inosinate can also be formed by the deamination of AMP, a reaction which occurs particularly in muscle as part of the purine nucleotide cycle (Fig 34–18). Inosinate, derived from AMP, when reconverted to AMP results in the net production of ammonia from aspartate. Removal of the phosphate group from IMP forms the nucleoside inosine (hypoxanthine riboside), an intermediate in another cycle referred to as the purine salvage cycle (see Chapter 35).

Inosine diphosphate (IDP) and inosine triphosphate (ITP), analogs of ADP and ATP in which the purine nucleoside derivative is inosine rather than adenosine, occasionally participate in phosphorylation reactions.

Uracil Derivatives

Uridine nucleotide derivatives are important coenzymes in reactions involving the metabolism of

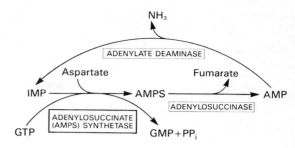

Figure 34–18. The purine nucleotide cycle.

hexoses and the polymerization of sugars to form starch and the oligosaccharide moieties of glycoproteins and proteoglycans. (See Chapter 54.) In these reactions, the substrates are uridine diphospho-sugars. For example, uridine diphosphate glucose (UDPGlc) is the precursor of glycogen. Another uridine nucleotide coenzyme, uridine diphosphoglucuronic acid (UDPGlcUA), serves as the "active" glucuronide for conjugation reactions such as the formation of bilirubin glucuronide (see Chapter 33).

Uracil also participates in the formation of high-energy phosphate compounds analogous to ATP, GTP, or ITP. Uridine triphosphate (UTP) is utilized, for example, in the reactions involving conversion of galactose to glucose in which the UDPGlc and UDPGal also are formed. UTP is the precursor for the polymerization of uridine nucleotides into RNA.

Cytosine Derivatives

Cytidine (cytosine-ribose) may form the high-energy phosphate compounds cytidine diphosphate (CDP) and cytidine triphosphate (CTP). The latter serves also as the precursor for the polymerization of CMP into nucleic acids. CTP is required for the biosynthesis of some phosphoglycerides in animal tissue. Reactions involving ceramide and CDP-choline are responsible for the formation of sphingomyelin and other substituted sphingosines. Cyclic nucleotide derivatives of cytidine, analogous to those of adenosine and guanosine, have been described.

Coenzyme Nucleotides

The functional moieties of many vitamins are coenzyme nucleotides with structures analogous to purine and pyrimidine nucleotides (Table 34–1).

SYNTHETIC NUCLEOTIDE ANALOGS

Synthetic analogs of nucleobases, nucleosides, and nucleotides are widely used in the medical sciences and clinical medicine. In the past, most of these uses have depended upon the role of nucleotides as components of nucleic acids for cellular growth and division. For a cell to divide, its nucleic acids must be replicated. This requires that the precursors of nucleic

Figure 34–17. Cyclic 3′,5′-guanosine monophosphate (cyclic GMP; cGMP).

Table 34-1. Many coenzymes and related compounds are derivatives of adenosine monophosphate.

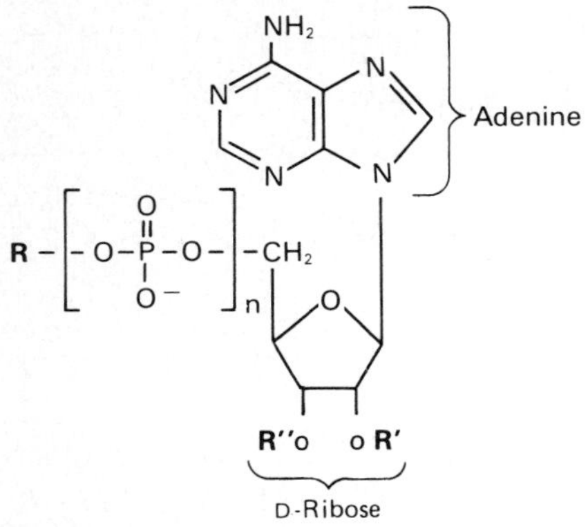

Coenzyme	R	R'	R''	n
Active methionine	Methionine*	H	H	0
Amino acid adenylates	Amino acid	H	H	1
Active sulfate	SO_3^{2-}	H	PO_3^{2-}	1
3',5'-Cyclic AMP	H	H	PO_3^{2-}	
NAD†	†	H	H	2
NADP†	†	PO_3^{2-}	H	2
FAD	†	H	H	2
CoA·SH	†	H	PO_3^{2-}	2

*Replaces phosphate group.
†R is a B-vitamin derivative.

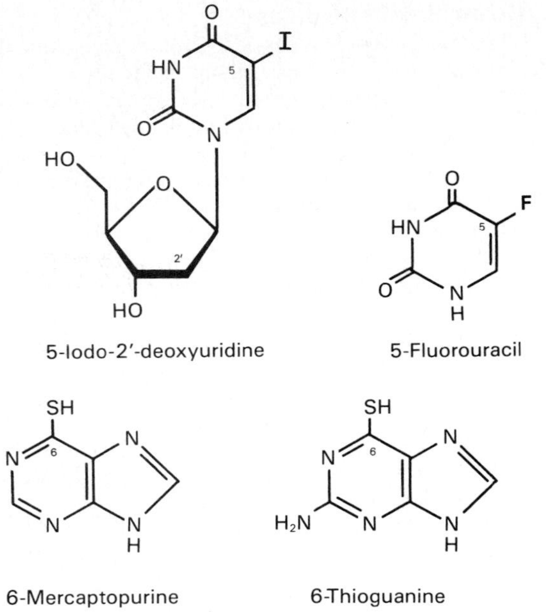

5-Iodo-2'-deoxyuridine   5-Fluorouracil

6-Mercaptopurine   6-Thioguanine

Figure 34-19. The structures of 2 synthetic pyrimidine analogs (*above*) and 2 synthetic purine analogs (*below*).

acids—the normal purine and pyrimidine deoxyribonucleotides—be readily available. One of the most important components of the oncologist's pharmacopeia is the group of synthetic analogs of purine and pyrimidine nucleobases and nucleosides.

The pharmacologic approach has been to use an analog in which either the heterocyclic ring structure or the sugar moiety has been altered in such a way as to induce toxic effects when the analog becomes incorporated into various cellular constituents. Many of these effects result from inhibition by the drug of specific enzyme activities necessary for nucleic acid synthesis or from the incorporation of metabolites of the drug into the nucleic acids where they alter the base pairing essential to accurate transfer of information. Examples of these would be the 5-fluoro or 5-iodo derivatives of uracil or deoxyuridine, which serve as thymine or thymidine analogs, respectively (Fig 34-19). Both 6-thioguanine and 6-mercaptopurine, in which naturally occurring hydroxyl groups are replaced with thiol groups at the 6 position, are widely used clinically. The analogs in which the purine or pyrimidine ring contains extra nitrogen atoms, such as 5- or 6-azauridine or azacytidine and 8-azaguanine (Fig 34-20), also have been tested clinically.

The purine analog 4-hydroxypyrazolopyrimidine (allopurinol) is widely marketed as an inhibitor of de novo purine biosynthesis and of xanthine oxidase. It is used for the treatment of hyperuricemia and gout. Nucleosides containing arabinose rather than ribose as the sugar moieties, eg, cytarabine (arabinosyl cytosine, Ara-C), are used in the chemotherapy of cancer and viral infections (Fig 34-21).

Azathioprine, which is catabolized to 6-mercaptopurine, is useful in organ transplantation as a suppressor of events involved in immunologic rejection. A series of nucleoside analogs with antiviral activities has been studied for several years; one, 5-iododeoxyuridine (see above), is effective in the local treatment of herpetic keratitis, an infection of the cornea by herpesvirus.

Numerous analogs of purine and pyrimidine ribonucleotides have been synthesized so as to generate

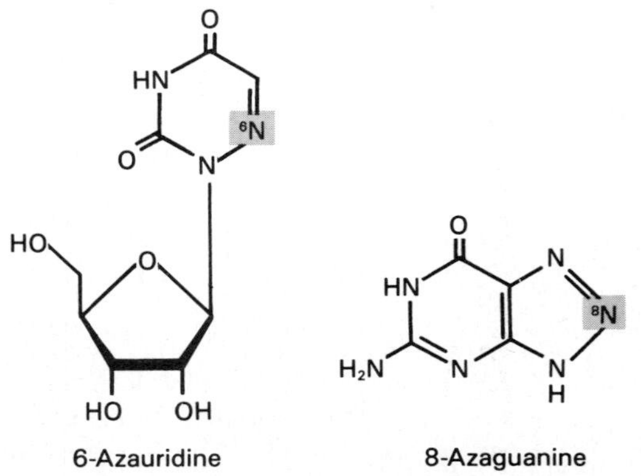

6-Azauridine   8-Azaguanine

Figure 34-20. The structures of 6-azauridine (*left*) and 8-azaguanine (*right*).

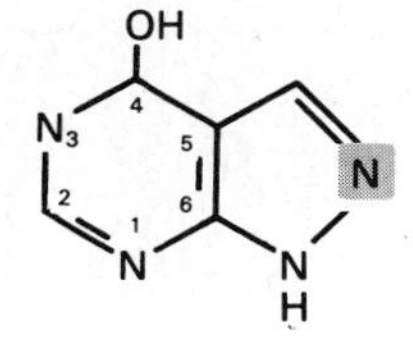

Allopurinol
(lactim)

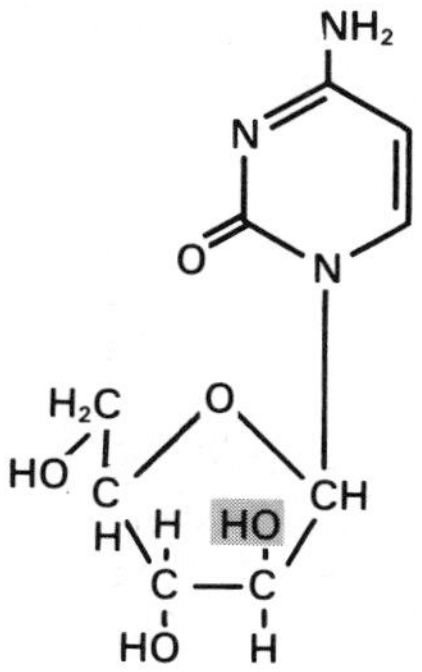

Arabinosyl
cytosine

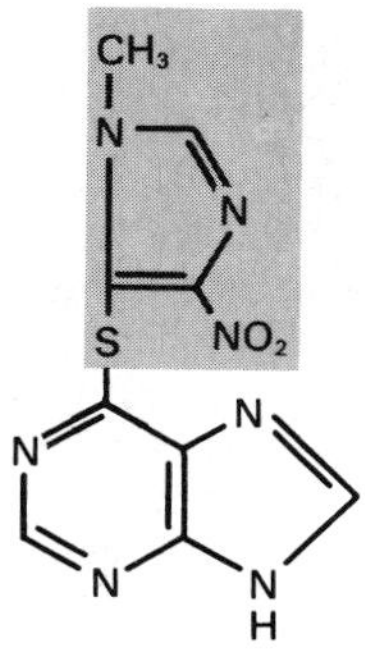

Azathioprine

Figure 34–21. The structures of 4-hydroxypyrazolopyrimidine (allopurinol), arabinosyl cytosine (cytarabine), and azathioprine.

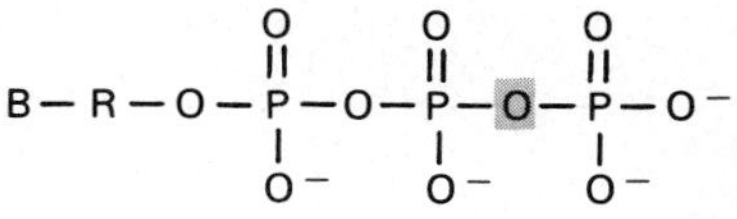

Parent nucleoside triphosphate

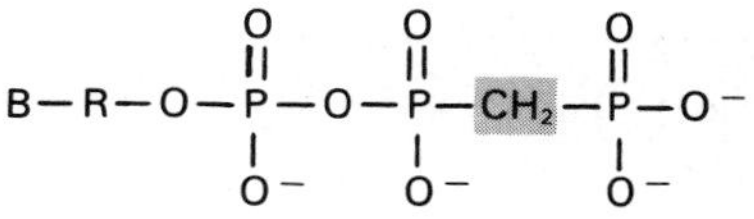

β,γ-Methylene derivative

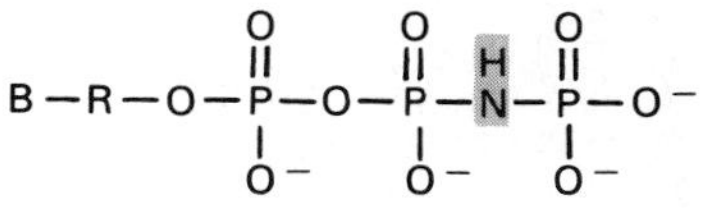

β,γ-Imino derivative

Figure 34–22. Synthetic derivatives of nucleoside triphosphates incapable of undergoing hydrolytic release of the terminal phosphate group. B, a purine or pyrimidine base; R, ribose or deoxyribose. Shown are the parent (hydrolyzable) nucleoside triphosphate (*top*) and the unhydrolyzable β,γ-methylene (*center*) and β,γ-imino derivatives (*bottom*).

nonhydrolyzable di- or triphosphates for use in vitro. These analogs allow the investigator to determine whether given biochemical effects of nucleoside di- or triphosphates require hydrolysis or whether their effects are mediated by occupying specific nucleotide binding sites on enzymes or regulatory proteins. Fig 34–22 depicts 2 such analogs of guanosine triphosphate.

REFERENCES

Henderson JF, Paterson ARP: *Nucleotide Metabolism: An Introduction*. Academic Press, 1973.

Michelson AM: *The Chemistry of Nucleosides and Nucleotides*. Academic Press, 1963.

Prusoff WH, Ward DC: Nucleoside analogs with antiviral activity. *Biochem Pharmacol* 1976;**25:**1233.

35

Metabolism of Purine & Pyrimidine Nucleotides

Victor W. Rodwell, PhD

INTRODUCTION

In this chapter, the metabolism of the purines and pyrimidines and their nucleosides and nucleotides will be discussed.

BIOMEDICAL IMPORTANCE

Neither nucleotides nor their parent purine and pyrimidine bases in the diet are incorporated into human tissue nucleic acids or into purine or pyrimidine coenzymes such as ATP or NAD. Even when a diet rich in nucleoproteins is ingested, human subjects form the constituents of tissue nucleic acids from amphibolic intermediates. This de novo synthesis permits purine and pyrimidine analogs with potential as anticancer drugs to be incorporated into DNA. The rates of synthesis of purine and pyrimidine oxy- and deoxyribonucleotides are subject to precise regulation. Mechanisms have evolved to ensure production of these compounds in quantities and at times appropriate to meet varying physiologic demand. In addition to de novo synthesis, these include "salvage" pathways for reutilization of purine or pyrimidine bases released by degradation of nucleic acids in vivo. Human diseases that involve abnormalities in purine or pyrimidine metabolism include gout, Lesch-Nyhan syndrome, Reye's syndrome, adenosine deaminase deficiency, and purine nucleoside phosphorylase deficiency.

DIGESTION

Mammals and most lower vertebrates are "prototrophic" for purines and pyrimidines; ie, they synthesize purine and pyrimidine nucleotides de novo. Although mammals consume significant quantities of nucleic acids and nucleotides in their food, their survival is not dependent upon the absorption of these compounds or their breakdown products.

Most dietary nucleic acids are ingested in the form of nucleoproteins from which the nucleic acids are liberated in the intestinal tract by the action of proteolytic enzymes. The pancreatic juice contains ribonucleases and deoxyribonucleases that degrade nucleic acids into nucleotides. Intestinal polynucleotidases or phosphoesterases supplement the action of the pancreatic nucleases to produce mononucleotides from the nucleic acids. The mononucleotides are subsequently hydrolyzed to nucleosides by nucleotidases and phosphatases, and the nucleosides are either absorbed directly or further degraded by intestinal phosphorylase to purine or pyrimidine bases. The bases may be oxidized; eg, guanine may be converted to xanthine and then to uric acid, or adenosine may be converted to inosine, to hypoxanthine, and then to uric acid (Fig 35–1). Uric acid can be absorbed across the intestinal mucosa and excreted in the urine as uric acid per se. In humans, the majority of purines in ingested nucleic acids are **directly converted to uric acid** without having previously been incorporated into the nucleic acids of the ingesting organism. Free pyrimidine orally administered to rats is mostly catabolized and excreted without having entered the nucleic acids of the ingesting organism. Thus, few or no dietary nucleic acids serve as a direct precursor of tissue nucleic acids.

Different results are obtained when purines or pyrimidines are administered parenterally as nucleosides or nucleotides. Injected thymidine may be incorporated into DNA unaltered. This is the basis of a valuable technique for labeling newly produced DNA in a great variety of biologic materials both in vivo and in vitro. For these purposes, ^{3}H-thymidine—ie, thymidine containing tritium (^{3}H), the radioactive isotope of hydrogen—is used.

PURINES

Biosynthesis of Purine Nucleotides

In humans and other mammals, purine nucleotides are synthesized to meet the needs of the organism for the monomeric precursors of nucleic acids and for those other functions described in Chapter 34. In some organisms (birds, amphibians, and reptiles), the synthesis of purine nucleotides has an additional function, which is to serve as the chemical vehicle to excrete nitrogen waste products as uric acid. Such organisms are referred to as **uricotelic,** whereas those organisms which dispose of nitrogenous waste products in the form of urea, as humans do, are referred to as **ureotelic.** Because the uricotelic organisms must dispose

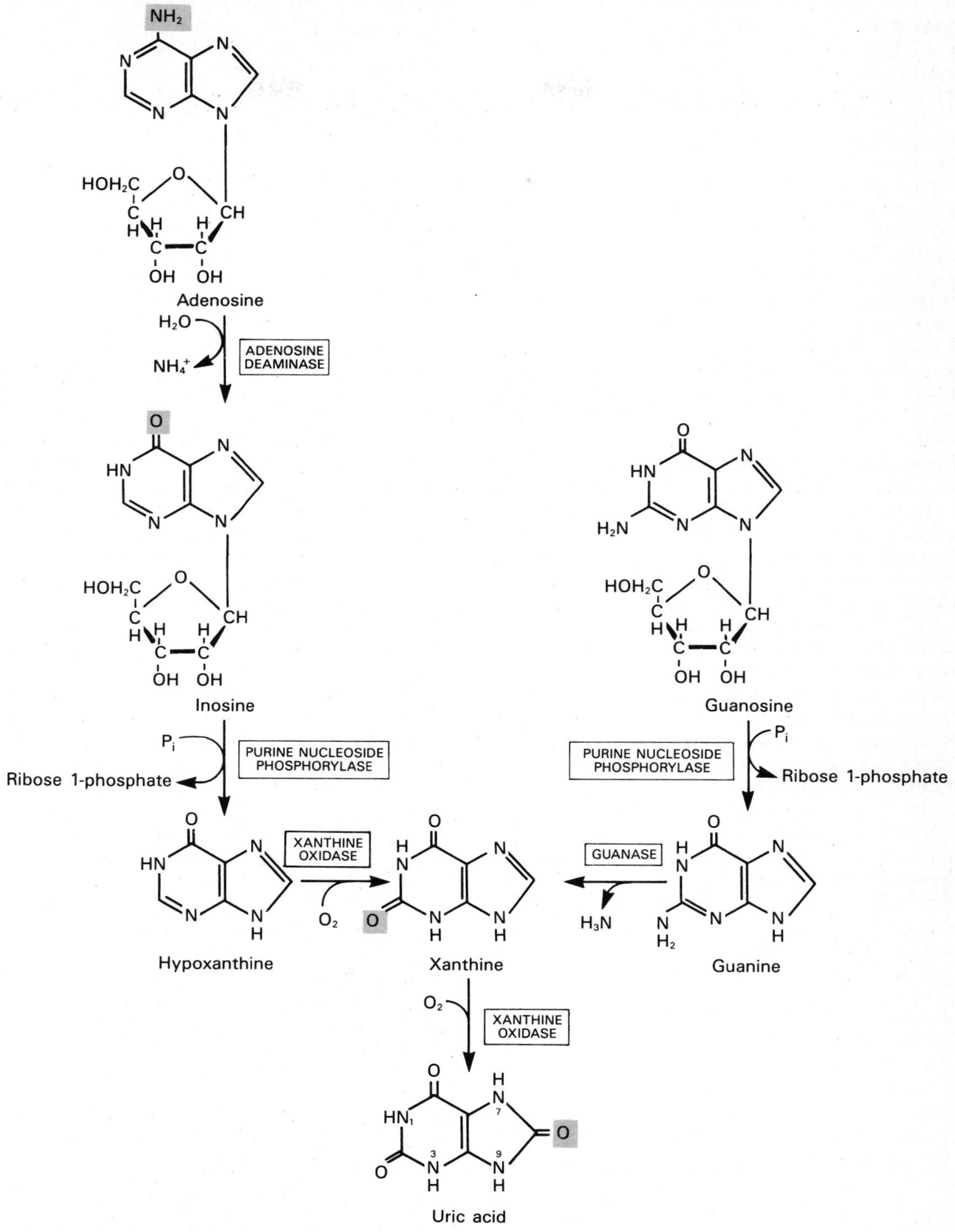

Figure 35–1. Generation of uric acid from purine nucleosides by way of the purine bases hypoxanthine, xanthine, and guanine. Purine deoxyribonucleosides are degraded by the same pathway and enzymes, all of which exist in the mucosa of the mammalian gastrointestinal tract.

of their nitrogenous wastes in the form of uric acid, they synthesize purine nucleotides at a relatively greater rate than do ureotelic organisms. However, the steps involved in de novo purine nucleotide synthesis in mammals (ureotelic) are analogous to those in birds (uricotelic).

Information on the sources of the various atoms of the purine base obtained by tracer studies in birds, rats, and humans is presented in Fig 35–2.

Fig 35–3 summarizes the biosynthetic pathway for the synthesis of purine nucleotides. The first step (reaction 1, Fig 35–3) in the synthesis of purine nucleotides is the formation of 5-phosphoribosyl-1-pyrophosphate (PRPP). The conversion of ribose 5-phosphate and ATP to AMP + PRPP (Fig 35–3) is not, however, unique to the synthesis of purine nucleotides. PRPP also serves as a precursor of the pyrimidine nucleotides and is required for the synthesis of NAD and NADP, 2 coenzymes derived from niacin.

PRPP then reacts (reaction 2, Fig 35–3) with glutamine in a reaction catalyzed by **phosphoribosylpyrophosphate amidotransferase** to form 5-phosphoribosylamine. The reaction is accompanied by the displacement of pyrophosphate and the formation of glutamate. Although other mechanisms have been proposed for the synthesis of 5-phosphoribosylamine, the physiologically important reaction in mammalian tissues is that catalyzed by the amidotransferase.

5-Phosphoribosylamine then reacts (reaction 3) with glycine to produce glycinamide ribosylphosphate (glycinamide ribotide [GAR]). The amido group from glutamine contributes what will become the 9 N of the purine ring, while the glycine contributes carbons 4 and 5 and the 7 N. The enzyme catalyzing reaction 3 is **glycinamide kinosynthetase.**

The N_7 of glycinamide ribosylphosphate is then formylated (reaction 4) by N^5,N^{10}-methenyltetrahydrofolate and the enzyme **glycinamide ribosylphosphate formyltransferase.** The C_1 moiety becomes the C_8 of the purine base.

In reaction 5, again with glutamine as the amide donor, amidation occurs at the C_4 of the formylglycin-

amide ribosylphosphate, catalyzed by **formylglycinamidine ribosylphosphate synthetase.** The amide N becomes position 3 in the purine.

Closure of the imidazole ring, catalyzed by **aminoimidazole ribosylphosphate synthetase,** forms aminoimidazole ribosylphosphate (reaction 6).

The synthesis progresses (reaction 7) to aminoimidazole carboxylate ribosylphosphate by addition of a carbonyl group, the source of which is respiratory CO_2.

The source of the nitrogen in the 1 position is the α-amino group of aspartate (reaction 8), the remaining portion of which is indicated as the succinyl moiety of aminoimidazole succinyl carboxamide ribosylphosphate (SAICAR).

In reaction 9, the succinyl group of SAICAR is removed as fumarate. Aminoimidazole carboxamide ribosylphosphate, which remains, is then formylated (reaction 10) by N^{10}-formyltetrahydrofolate ($f^{10} \cdot H_4$folate) to form amidoimidazole carboxamide ribosylphosphate in a reaction catalyzed by the appropriate **formyltransferase.** The newly added carbon, which, like the C_8 of the purine base, is derived from the C_1 pool via the tetrahydrofolate carrier, will be C_2 of the purine nucleus.

Ring closure now occurs (reaction 11) via **IMP cyclohydrolase,** and the first purine nucleotide, **inosinic acid (inosine monophosphate, IMP),** is thus formed.

Importance of Folate Metabolism

Two one-carbon moieties are added to the purine ring at positions 8 and 2 by N^5,N^{10}-methenyltetrahydrofolate and N^{10}-formyltetrahydrofolate, respectively. The latter is derived from the former. The N^5,N^{10}-methenyltetrahydrofolate is derived from the NADP-dependent dehydrogenation of N^5,N^{10}-methylenetetrahydrofolate. The N^5,N^{10}-methylenetetrahydrofolate can donate a one-carbon moiety to numerous acceptors, but once N^5,N^{10}-methenyltetrahydrofolate is formed, the one-carbon group is committed to transfer only into purines, whence it is donated either directly or after conversion to N^{10}-formyltetrahydrofolate. Thus, any inhibition of the formation of these tetrahydrofolate compounds will have a detrimental effect upon the de novo synthesis of purines.

Conversion of IMP to AMP & GMP

As shown in Fig 35–4, adenine nucleotides (reactions 12 and 13) and guanine nucleotides (reactions 14 and 15) are derived from **inosine monophosphate (IMP)** by amination and by oxidation and amination, respectively. The amination of IMP is accomplished through the formation of an intermediate compound in which aspartate is attached to inosinic acid to form adenylosuccinate. This reaction resembles reaction 9, in which the nitrogen at position 1 of the purine nucleus was added by way of the α nitrogen of aspartate. Formation of adenylosuccinate, catalyzed by **adenylosuccinate synthase,** requires GTP, which provides a

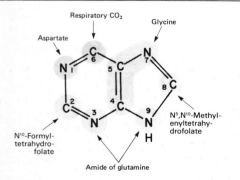

Figure 35–2. The sources of the nitrogen and carbon atoms of the purine ring.

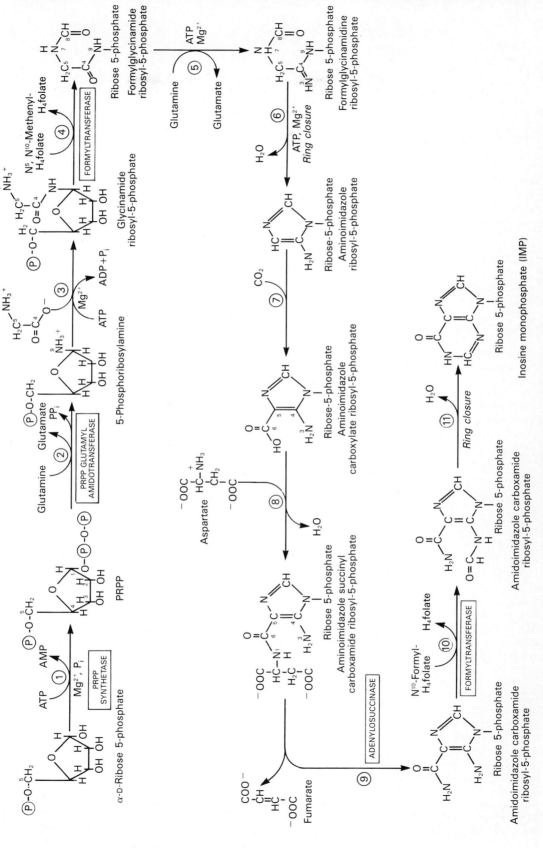

Figure 35–3. The pathway of de novo purine biosynthesis from ribose 5-phosphate and ATP. (See text for explanation.) $\textcircled{P}$, PO_3^{2-} or PO_2^-.

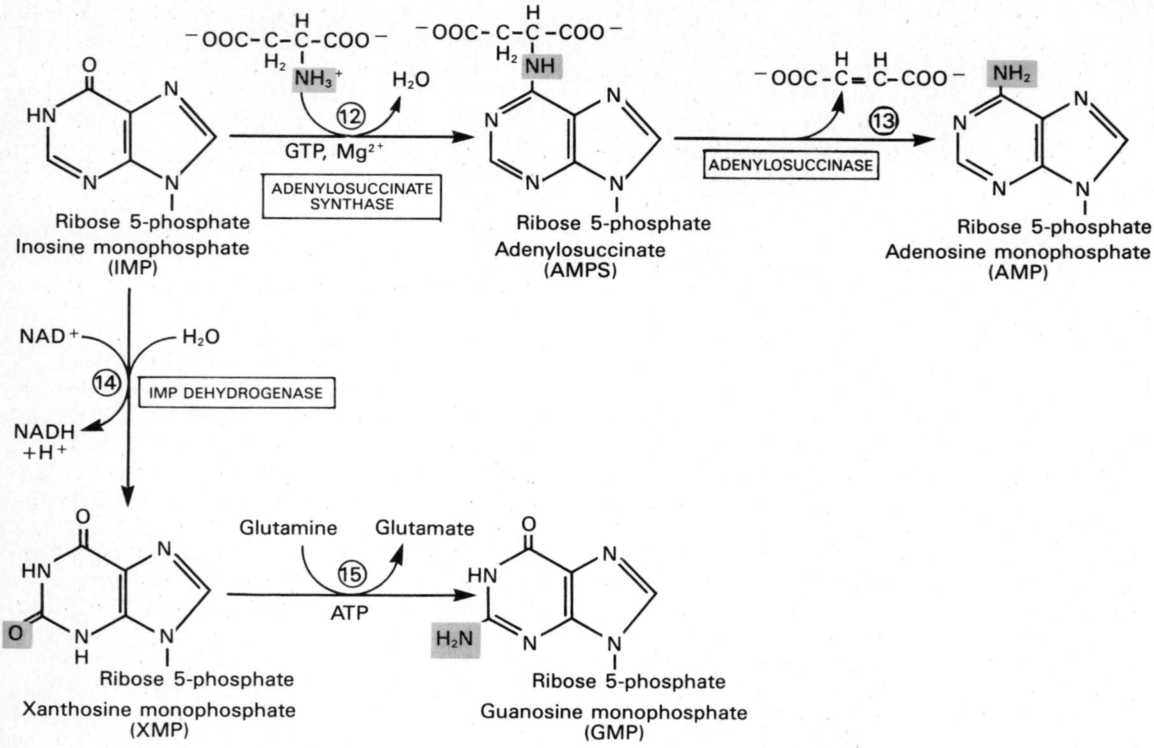

Figure 35–4. The conversions of IMP to AMP and GMP. (See text for explanation.)

potential regulatory mechanism. The splitting off, as fumarate, of the remaining portion of aspartate from adenylosuccinate produces the final product, adenylic acid (adenosine monophosphate, AMP). Cleavage of fumarate from adenylosuccinate is catalyzed by **adenylosuccinase,** which is also responsible for the cleavage of fumarate from the succinyl of aminoimidazole succinyl carboxamide ribosylphosphate (reaction 9).

Also in 2 steps, IMP is converted to guanosine monophosphate (GMP). The first reaction in this sequence (reaction 14) is an oxidation utilizing NAD and water to form xanthosine monophosphate (XMP). XMP is then aminated by the amido group of glutamine in a reaction that requires ATP, somewhat analogous to the requirement of GTP for the conversion of IMP to AMP (reaction 15).

Inhibitors of Purine Biosynthesis

Several antimetabolites that are glutamine analogs are effective inhibitors of purine biosynthesis. **Azaserine** (O-diazoacetyl-L-serine) is an antagonist to glutamine, particularly at reaction 5. **Diazonorleucine** ([6-diazo-5-oxo]-L-norleucine) blocks reaction 2 in purine synthesis, and **6-mercaptopurine,** among its other actions, inhibits reactions 13 and 14 in the synthesis of AMP and GMP, respectively. **Mycophenolic acid** inhibits reaction 14.

Formation of Purine Nucleoside Di- & Triphosphates

Conversion of AMP and GMP to their respective nucleoside diphosphates and nucleoside triphosphates occurs in 2 successive steps (Fig 35–5). The successive transfers of phosphate groups from ATP are catalyzed by **nucleoside monophosphate kinase** and **nucleoside diphosphate kinase,** respectively. The enzyme that phosphorylates adenylate is also called **myokinase.**

Formation of Purine Deoxyribonucleotides

Synthesis of purine and pyrimidine deoxyribonucleotides occurs by **direct reduction at the 2′ carbon** in the ribose moiety of the corresponding nucleotide, not by synthesis of the entire nucleotide utilizing a 2′-deoxy analog of PRPP. Reduction at the 2′ carbon occurs only after the purine and pyrimidine nucleotides

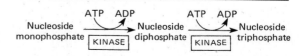

Figure 35–5. The reactions responsible for the conversion of nucleoside monophosphates to nucleoside diphosphates and nucleoside triphosphates.

have been converted to their respective nucleoside diphosphates. In some bacteria, cobalamin (vitamin B_{12}) is required for this reductive process, although it is not required for the same reaction in mammals. Reduction of ribonucleoside diphosphates to deoxyribonucleoside diphosphates (Fig 35–6) is catalyzed by **ribonucleotide reductase** and requires **thioredoxin** (a protein cofactor), **thioredoxin reductase** (a flavoprotein), and NADPH as a cofactor. The immediate electron donor to the nucleotide is thioredoxin that has been reduced by NADPH. The reversible oxidation-reduction of thioredoxin is catalyzed by **thioredoxin reductase**. Reduction of the ribonucleoside diphosphate by reduced thioredoxin is catalyzed by ribonucleotide reductase (Fig 35–6). This complex enzyme system is present in cells only when they are actively synthesizing DNA and dividing.

Tissue Specificity of Purine Biosynthesis

Not all human tissues catalyze the de novo synthesis of purine nucleotides. Erythrocytes and polymorphonuclear leukocytes are incapable of synthesizing 5-phosphoribosylamine and therefore are dependent upon exogenous purines for the formation of purine nucleotides. Peripheral lymphocytes do possess some ability to synthesize purines de novo. Mammalian brain appears to have a reduced content of PRPP amidotransferase; indeed, it has been suggested that the human brain is dependent upon exogenous purines for the formation of purine nucleotides. Mammalian liver is a major site of purine nucleotide synthesis and provides purines in the form of bases or nucleosides to be salvaged and utilized by those tissues incapable of synthesizing purines de novo.

Purine Salvage Pathways

Salvage of preformed purines can occur by 2 general mechanisms. The quantitatively more important mechanism is the **phosphoribosylation of the free purine bases** by enzymes that require PRPP as the ribose phosphate donor. The second general mechanism is the **phosphorylation of purine nucleosides** on their 5'-hydroxyl groups.

A. Of Purine Bases: Two human tissue enzymes phosphoribosylate purine bases. One, **adenine phosphoribosyltransferase** (Fig 35–7), phosphoribosylates adenine with PRPP to generate AMP. The second, **hypoxanthine-guanine phosphoribosyltransferase** (Fig 35–8), phosphoribosylates hypoxanthine and guanine with PRPP to yield IMP and GMP, respectively. As discussed below, the latter pathway is more active than the formation of AMP from adenine.

B. Of Purine Ribonucleosides: Salvage of purine ribonucleosides to purine ribonucleotides in humans is carried out by **adenosine kinase** (Fig 35–9). Adenosine kinase also phosphorylates 2'-deoxyadenosine but demonstrates minimal ability to phosphorylate guanosine, inosine, or their 2'-deoxy derivatives to their respective ribonucleotides. Deoxycytidine kinase, in addition to phosphorylating deoxycytidine, can phosphorylate 2'-deoxyadenosine and 2'-deoxyguanosine to deoxy AMP and deoxy GMP, respectively.

In addition there is in humans a cycle (Fig 35–10) in which IMP and GMP, as well as their respective deoxyribonucleotides, are converted to their respective nucleosides (inosine, deoxyinosine, guanosine, and deoxyguanosine) by a **purine 5'-nucleotidase**. These purine ribonucleosides and 2'-deoxynucleosides are

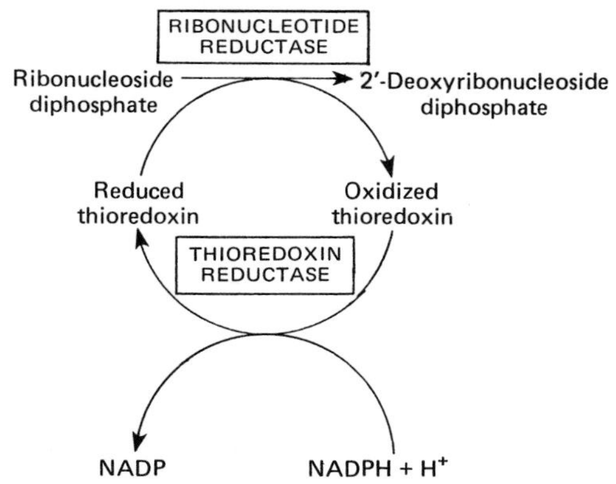

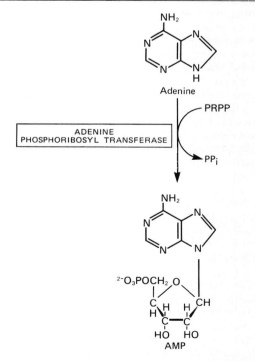

Figure 35–7. Phosphoribosylation of adenine catalyzed by adenine phosphoribosyltransferase.

Figure 35–6. Reduction of ribonucleoside diphosphates to 2'-deoxyribonucleoside diphosphates.

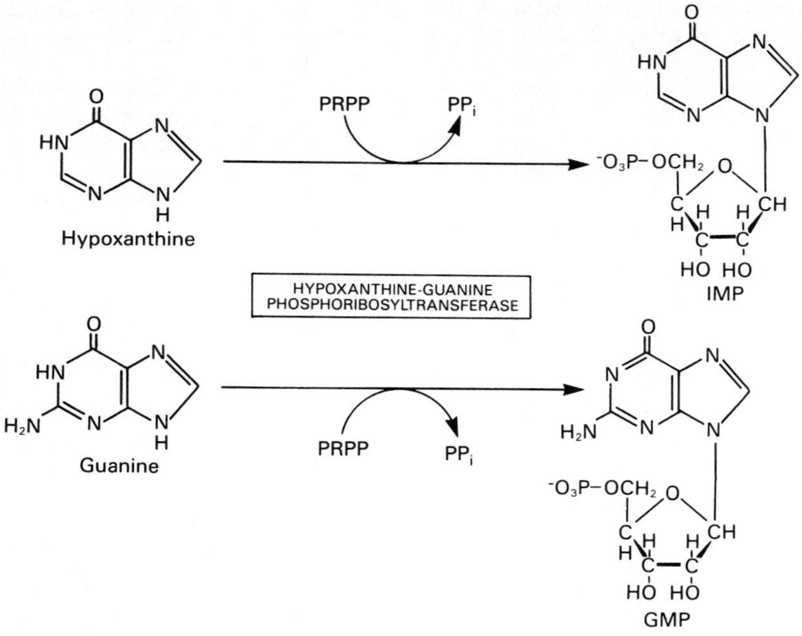

Figure 35–8. Phosphoribosylation of hypoxanthine and guanine to form IMP and GMP, respectively. Both reactions are catalyzed by hypoxanthine-guanine phosphoribosyltransferase.

converted to hypoxanthine or guanine by **purine nucleoside phosphorylase,** producing ribose 1-phosphate or 2′-deoxyribose 1-phosphate as phosphorolysis products. The hypoxanthine and guanine can then again be phosphoribosylated by PRPP to IMP and GMP to complete the cycle. The functions of this purine salvage cycle are unknown, but it is clear that, in the human organism as a whole, the consumption of PRPP by this salvage cycle is greater than the consumption of PRPP for the synthesis of purine nucleotides de novo.

A lateral pathway of this cycle involves the conversion of IMP to AMP (reactions 12 and 13, Fig 35–4), with subsequent conversion of AMP to adenosine. The latter reaction probably is catalyzed by the same purine 5′-nucleotidase that hydrolyzes IMP to inosine.

The adenosine so produced is then either salvaged directly back to AMP via **adenosine kinase** or is converted to inosine by **adenosine deaminase.** Quantitatively, the function of this "inosine loop" is less important than the previously described cycle. Qualitatively, the action of adenosine deaminase is important, particularly for the immune system, as described in the discussion of inherited disorders of purine metabolism.

Regulation of Purine Biosynthesis

The de novo synthesis of IMP consumes the equivalent of 6 high-energy phosphodiester bonds (by ATP hydrolysis) along with the other required precursors, glycine, glutamine, methenyltetrahydrofolate, and as-

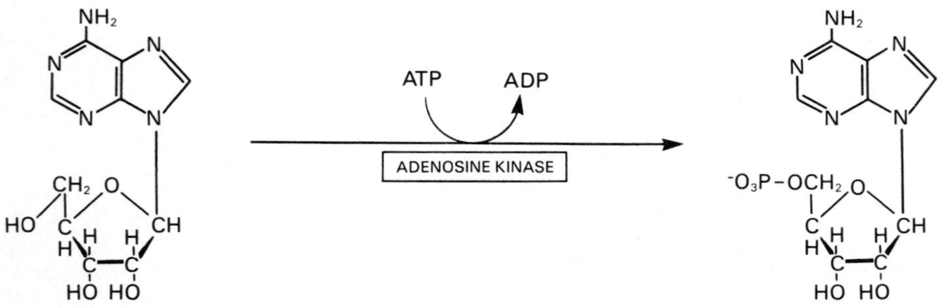

Figure 35–9. Phosphorylation of adenosine to AMP by adenosine kinase.

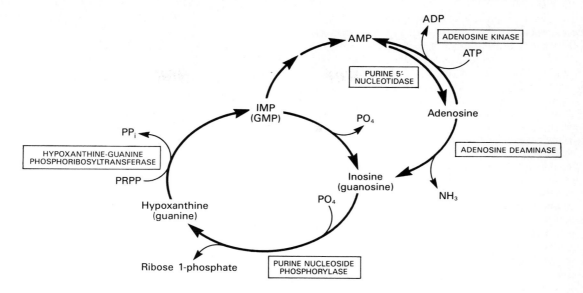

Figure 35–10. The purine salvage cycles involving the interconversion of AMP, IMP, and—to a lesser extent—GMP to their respective ribonucleosides and their eventual reconversion to purine ribonucleotides. Deoxyadenosine, deoxyinosine, and deoxyguanosine share the same pathways except that deoxyadenosine and deoxyguanosine can be directly phosphorylated to deoxy AMP and deoxy GMP, respectively.

partate. Thus, it is important for the conservation of energy and nutrients that the cell economically regulate its rate of de novo purine biosynthesis. **The most important regulator of de novo purine biosynthesis is the intracellular concentration of PRPP.** As with so many other intracellular compounds, PRPP concentration depends upon its rates of synthesis, utilization, and degradation. The rate of synthesis of PRPP is dependent upon (1) the availability of its substrates, particularly ribose 5-phosphate; and (2) the catalytic activity of PRPP synthetase, which is dependent upon the intracellular phosphate concentration as well as the concentrations of the purine and pyrimidine ribonucleotides acting as allosteric regulators (Fig 35–11). The rate of utilization of PRPP is dependent to a large extent on its consumption by the salvage pathway that phosphoribosylates hypoxanthine and guanine to their respective ribonucleotides. To a lesser extent, utilization is dependent upon the rate of de novo purine synthesis. This conclusion stems from the observation that, in males with inherited deficiencies of hypoxanthine-guanine phosphoribosyltransferase, the levels of PRPP in their erythrocytes and cultured fibroblasts are elevated severalfold.

The first enzyme uniquely committed to de novo purine synthesis, **PRPP amidotransferase, demonstrates in vitro a sensitivity to feedback inhibition by purine nucleotides, particularly adenosine monophosphate and guanosine monophosphate.** These feedback inhibitors of the amidotransferase are competitive with the substrate PRPP, and thus, again, PRPP plays a major role in the regulation of de novo purine synthesis. Numerous indirect experiments sug-

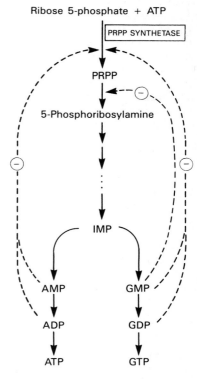

Figure 35–11. Control of the rate of de novo purine synthesis. Solid lines represent chemical flow, and broken lines represent feedback inhibition ($\ominus$) by end products of the pathway.

gest that the regulation of de novo purine synthesis by amidotransferase is physiologically less important than that by PRPP synthetase.

The conversion of IMP to GMP or to AMP is regulated by 2 mechanisms (Fig 35–12). **AMP feedback regulates its own synthesis at the level of adenylosuccinate synthetase; GMP regulates its own synthesis by feedback inhibition of IMP dehydrogenase.** Furthermore, the conversion of IMP to adenylosuccinate en route to AMP requires the presence of GTP. The conversion of xanthinylate to GMP requires ATP. Thus, **there is significant cross-regulation between the divergent pathways in the metabolism of IMP. This regulation prevents the synthesis of one purine nucleotide when there is a deficiency of the other.** Hypoxanthine-guanine phosphoribosyltransferase, which converts hypoxanthine and guanine to IMP and GMP, respectively, is quite sensitive to product inhibition by these same nucleotides.

The reduction of ribonucleoside diphosphates to deoxyribonucleoside diphosphates is subject to complex regulation. This process (Fig 35–13) provides for balanced production of deoxyribonucleotides for the synthesis of DNA.

Catabolism of Purines

In humans, the ultimate catabolite (end product) of purines is uric acid. On the basis of observations in humans with inherited enzyme deficiencies, it appears that over 99% of the uric acid is derived from substrates of purine nucleoside phosphorylase, a component of the purine salvage pathway. The purine products of purine nucleoside phosphorylase, guanine and hypoxanthine, are converted to uric acid by way of xanthine in reactions catalyzed by the enzymes **guanase** and **xanthine oxidase,** respectively (Fig 35–1), of liver, small intestine, and kidney.

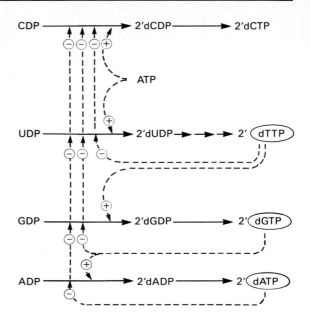

Figure 35–13. Regulation of the reduction of purine and pyrimidine ribonucleotides to their respective 2′-deoxyribonucleotides. Solid lines represent chemical flow, and broken lines represent negative ($\ominus$) or positive ($\oplus$) feedback regulation.

Xanthine oxidase is an important site for pharmacologic intervention in patients with hyperuricemia and gout. In lower primates and other mammals, but not in humans, **uricase** hydrolyzes uric acid to allantoin (Fig 35–14), a highly water-soluble end product. Amphibians, birds, and reptiles, which lack uricase, excrete uric acid and guanine as the end products of both purine metabolism and nitrogen (protein) metabolism.

Birds, amphibians, and reptiles have evolved a uricotelic system to regain water of hydration from uric acid after it precipitates out. If they were to use urea as the end product of nitrogen metabolism, the water of hydration could not be regained, since urea is water-soluble up to 10 mol/L, a concentration far higher than any kidney can attain.

Uric Acid Metabolism in Human Subjects (Gout)

The metabolism of uric acid in humans has been studied by the use of isotopically labeled uric acid as well as its precursors, glycine and formate. Single doses of N^{15} uric acid were injected intravenously into normal human subjects and patients suffering from gout, a disease characterized by increased accumulation of uric acid and sodium urate. The dilution of the injected labeled isotope was used to calculate the quantity of total uric acid equilibrating with body water, a quantity referred to as the **miscible urate pool.** The mass of the rapidly miscible pool of uric acid in 25 normal male adult subjects averaged 1200 mg with a

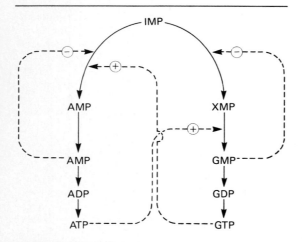

Figure 35–12. Regulation of the interconversion of IMP to adenosine nucleotides and guanosine nucleotides. Solid lines represent chemical flow, and broken lines represent both positive ($\oplus$) and negative ($\ominus$) feedback regulation.

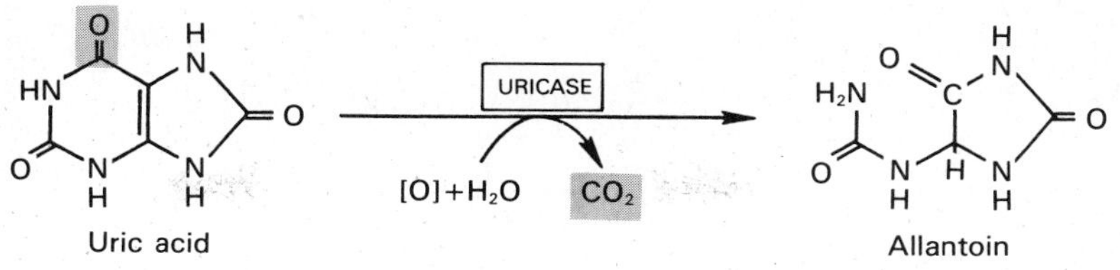

Figure 35–14. Conversion of uric acid to allantoin.

range of 866–1578 mg. In 3 normal female subjects, the pool ranged from 541 to 687 mg. In gouty subjects, the miscible urate pool was much larger, generally ranging from 2000 to 4000 mg in patients without tophi, ie, deposits of sodium urate in soft tissues. However, in severe tophaceous gout, the pool was as high as 31,000 mg. The turnover of the miscible pool of total uric acid in normal persons is approximately 600 mg/24 h. Of the lost uric acid, 18–20% is degraded to CO_2 and ammonia and excreted in the feces, where it can be further metabolized by intestinal flora. Some uric acid is excreted in the bile and thus is subject to degradation by intestinal flora. However, in humans, the breakdown of uric acid to CO_2 and NH_3 is independent of intestinal bacteria.

In humans, urate appears to play a role beyond that of an end product of purine metabolism. Urate by itself can serve as an antioxidant, undergoing **nonenzymatic** conversion to allantoin. It has been proposed that this naturally occurring antioxidant, urate, has in primates replaced ascorbate, which primates have lost the ability to synthesize. Thus, it may well be that during evolution the loss of uricase provided selective advantage to those organisms which had previously lost the ability to reduce gulonolactone to ascorbate.

Sodium urate is freely filtered by the mammalian glomerulus, is extensively reabsorbed and partially secreted in the proximal tubule, is further secreted in the loop of Henle, and perhaps is again partially reabsorbed in the distal convoluted tubule. The net excretion of total uric acid in normal men is 400–600 mg/24 h. Many pharmacologic and naturally occurring compounds influence the renal absorption and secretion of sodium urate. Aspirin in high doses competitively inhibits urate excretion as well as reabsorption.

PYRIMIDINES

Biosynthesis of Pyrimidines

Although the pyrimidine nucleus is simpler and its synthetic pathway briefer than that of the purine structure, the 2 share several common precursors. **PRPP, glutamine, CO_2,** and **aspartate** are required for the synthesis of all pyrimidine and purine nucleotides. For the thymidine nucleotides and for all purine nucleotides, **tetrahydrofolate** derivatives are also necessary. There is one striking difference between the synthesis of pyrimidine nucleotides and that of purine nucleotides, namely, that the synthesis of the purine nucleotides commences with ribose phosphate as an integral part of the earliest precursor molecule, whereas the pyrimidine base is formed and **attachment of the ribose phosphate moiety delayed until the later steps** of the pathway.

Synthesis of the pyrimidine ring (Fig 35–15) commences with the formation of **carbamoyl phosphate** from glutamine, ATP, and CO_2 in a reaction catalyzed by the carbamoyl phosphate synthase in the **cytosol** (reaction 1). The carbamoyl phosphate synthase enzyme responsible for the early steps in urea synthesis resides in the mitochondria.

The first step uniquely committed to the biosynthesis of pyrimidines is the formation of carbamoyl aspartate by the condensation of carbamoyl phosphate and aspartate, a reaction catalyzed by the enzyme **aspartate transcarbamoylase** (reaction 2).

A ring structure can then be formed from carbamoyl aspartate by loss of H_2O catalyzed by the enzyme **dihydroorotase** (reaction 3).

In a subsequent dehydrogenation step catalyzed by **dihydroorotate dehydrogenase** and utilizing NAD as a cofactor, **orotic acid** is formed (reaction 4).

In reaction 5, a ribose phosphate moiety is added to orotic acid to form **orotidylate (orotidine monophosphate, OMP).** This reaction is catalyzed by **orotate phosphoribosyltransferase,** an enzyme analogous to the hypoxanthine-guanine phosphoribosyltransferase and the adenine phosphoribosyltransferase involved in the phosphoribosylation of preformed purine rings.

The first true pyrimidine ribonucleotide is formed by the decarboxylation of orotidylate to form **uridylate (uridine monophosphate, UMP)** (reaction 6). Thus, only at the penultimate step in the formation of UMP is the heterocyclic ring phosphoribosylated.

Dihydroorotate dehydrogenase is **mitochondrial;** all the other enzymes in the de novo pyrimidine nucleotide pathway are in the **cytosol.**

By mechanisms analogous to those described for the further phosphorylation of the purine nucleoside monophosphates, the pyrimidine nucleoside monophosphates are converted to their diphosphate and triphosphate derivatives (reactions 7–12). UTP is aminated to CTP by glutamine and ATP (reaction 9). The reduction of the pyrimidine nucleoside diphosphates to the respective 2'-deoxynucleoside diphosphates (reaction 10) occurs by a mechanism also analogous to

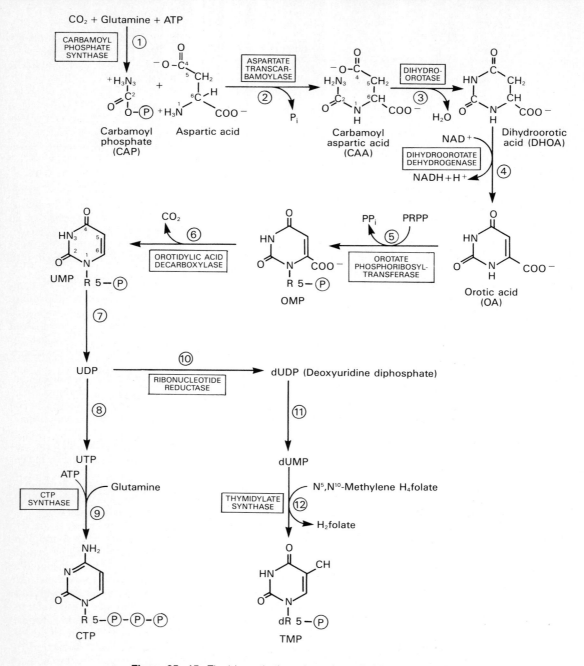

Figure 35–15. The biosynthetic pathway for pyrimidine nucleotides.

that described for the purine nucleotides (Figs 35–6 and 35–13).

The formation of **thymidylate (thymidine monophosphate, TMP)** (reaction 12) is the one reaction in pyrimidine nucleotide biosynthesis that requires a **tetrahydrofolate** donor of a single carbon compound. The 2′-deoxy UMP is methylated by **thymidylate synthase,** which utilizes as a methyl donor N^5,N^{10}-methylenetetrahydrofolate. The methyl-ene group of N^5,N^{10}-methylenetetrahydrofolate, which is added as a methyl group to the C_5 of deoxy UMP, must be reduced in the process of its donation. While the methylene is reduced to a methyl group, the tetrahydrofolate carrier is oxidized to dihydrofolate, and the net redox state of the reaction is thus unchanged. The methylation of deoxy UMP to TMP results in an overall reduction of the hydroxymethyl group from serine to a methyl group with the simultaneous oxidation of tetrahydrofolate to dihydrofolate. In order to continue to use the folate carrier, the cell must reduce dihydrofolate to tetrahydrofolate, a reaction carried out by the enzyme dihydrofolate reductase. Thus, dividing cells that by necessity are generating TMP and dihydrofolate are especially sensitive to inhibitors of dihydrofolate reductase. An example of such an inhibitor is methotrexate (amethopterin), a widely used anticancer drug.

Pyrimidine Ribonucleoside Salvage Pathways

Mammalian cells lack an efficient means of salvaging free pyrimidine bases. However, they do possess salvage pathways for converting the pyrimidine **ribonucleosides** uridine and cytidine and the 2'-deoxyribonucleosides thymidine and deoxycytidine to their respective nucleotides (Fig 35–16). 2'-Deoxycytidine is phosphorylated by deoxycytidine kinase, an enzyme that can also phosphorylate deoxyguanosine and deoxyadenosine.

The enzyme required for de novo pyrimidine biosynthesis, **orotate phosphoribosyltransferase,** is capable of salvaging orotic acid to OMP, but in a strict sense orotic acid is not considered a complete pyrimidine base. The orotate phosphoribosyltransferase cannot use normal pyrimidine bases as substrates, although it is capable of converting allopurinol (4-hydroxypyrazolopyrimidine) to a nucleotide in which the ribosyl phosphate is attached to the N_1 of the pyrimidine ring of that drug. The anticancer drug 5-fluorouracil is also phosphoribosylated by orotate phosphoribosyltransferase.

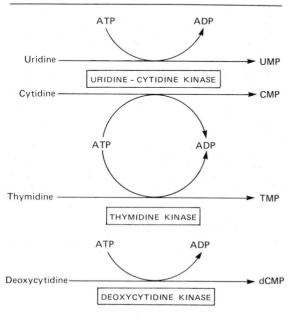

Figure 35–16. The pyrimidine nucleoside kinase reactions responsible for formation of the respective pyrimidine nucleoside monophosphates.

Catabolism of Pyrimidines

The catabolism of pyrimidines, which occurs mainly in the liver, produces highly soluble end products (Fig 35–17). This contrasts with the production of the sparingly soluble uric acid and sodium urate by purine catabolism. The release of respiratory CO_2 from the ureido carbon (C_2) of the pyrimidine nucleus represents a major pathway for the catabolism of uracil, cytosine, and thymine. β-Alanine and β-aminoisobutyrate are the major end products of cytosine, uracil, and thymine catabolism, respectively.

Thymine is the precursor of β-aminoisobutyrate in laboratory animals and in humans. The excretion of β-aminoisobutyrate is increased in leukemia as well as after the body has been subjected to x-irradiation. This is undoubtedly a reflection of increased destruction of cells and their DNA. A familial occurrence of an abnormally high excretion of β-aminoisobutyrate has also been observed in otherwise normal individuals. This genetic trait is traceable to a recessively expressed gene. High excretors result only when the trait is homozygous. Approximately 25% of tested persons of Chinese or Japanese ancestry consistently excreted large amounts of β-aminoisobutyrate. Although little is known about the mechanisms whereby β-aminoisobutyrate is degraded in humans, an enzyme that catalyzes the reversible transamination reaction has been identified in pig kidney. The β-aminoisobutyrate is converted to methylmalonic semialdehyde and thence to propionate, which in turn proceeds to succinate.

The initial steps in the degradation of pyrimidine nucleotides, including the removal of the sugar phosphate moiety by hydrolysis of the N-glycosidic bond, are similar to reversing the latter part of the synthetic pathway. For pseudouridine, which is formed in situ in tRNA by a rearrangement reaction, there is no mechanism to catalyze the hydrolysis or phosphorolysis of this unusual nucleoside to its respective pyrimidine base, uracil. Consequently, pseudouridine is excreted unchanged in the urine of normal persons.

Regulation of Pyrimidine Biosynthesis

The pathway of pyrimidine nucleotide biosynthesis is regulated by 2 general mechanisms. The first 2 enzymes in the pathway are sensitive to **allosteric regulation,** while the first 3 enzymes are regulated by an apparently coordinate **repression and derepression,** as are the last 2 enzymes of the pathway. **Carbamoyl phosphate synthase** is inhibited by UTP and purine nucleotides but activated by PRPP (Fig 35–18). **Aspartate transcarbamoylase** is particularly sensitive to inhibition by CTP. The allosteric properties of the aspartate transcarbamoylase in microorganisms have been the subject of extensive and now classic studies in allostery.

On a molar basis, the rate of pyrimidine biosynthesis parallels that of purine biosynthesis, demonstrating a coordinate control of purine and pyrimidine nucleotide synthesis. PRPP synthetase, an enzyme that forms

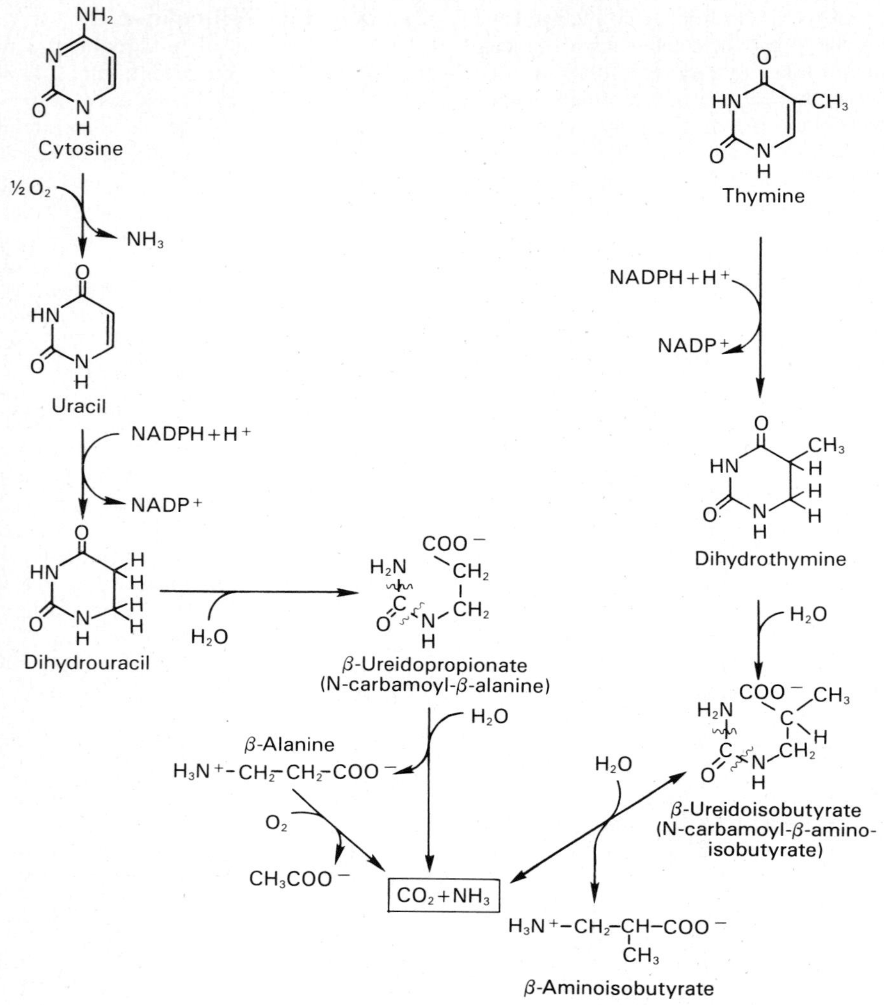

Figure 35–17. Catabolism of pyrimidines.

a necessary precursor for both purine nucleotide and pyrimidine nucleotide biosynthesis, is subject to feedback inhibition by both purine and pyrimidine nucleotides. Furthermore, carbamoyl phosphate synthase is sensitive to feedback inhibition by both purine and pyrimidine nucleotides and activation by PRPP. Thus, there are several sites at which there is significant cross-regulation between purine and pyrimidine nucleotide synthesis.

CLINICAL DISORDERS OF PURINE METABOLISM
(See Table 35–1.)

Hyperuricemia & Gout

The predominant form of uric acid is determined by the pH of its milieu (eg, blood, urine, cerebrospinal fluid). The **pK of the N_9 proton is 5.75,** and the pK of the N_1 proton is 10.3. Thus, under physiologic conditions—ie, at the usual pH of physiologic fluids—only uric acid and its monosodium salt, sodium urate, are found. In a fluid where the pH is less than 5.75, the predominant molecular species will be uric acid. In a fluid at pH 5.75, the concentration of sodium urate will equal that of uric acid. At a pH greater than 5.75, sodium urate will predominate in the solution.

The miscible urate pool in the body is reflected by the sodium urate concentration in the serum. When this level exceeds the solubility of sodium urate in serum **(hyperuricemia),** crystals of sodium urate may precipitate. The solubility of sodium urate in serum at 37 °C is 7 mg/dL. Crystals of sodium urate can collect and deposit in soft tissues, particularly in or about joints. These urate deposits are referred to as **tophi.** Accumulation of sodium urate crystals in the tissues,

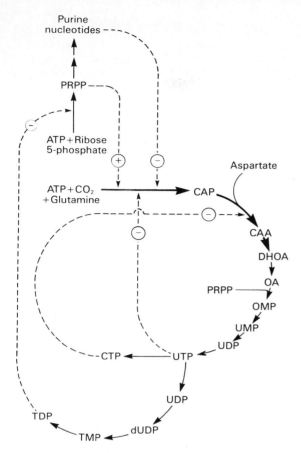

Figure 35–18. Control of pyrimidine nucleotide synthesis. Solid lines represent chemical flow, and broken lines represent positive ($\oplus$) and negative ($\ominus$) feedback regulation.

including phagocytosis of the crystals by polymorphonuclear leukocytes in joint spaces, can lead to an acute inflammatory reaction called **acute gouty arthritis.** The chronic inflammatory changes induced by the deposition of sodium urate tophi can generate **chronic gouty arthritis,** resulting in joint destruction.

In aqueous solutions, uric acid—the protonated form of urate—is only one-seventeenth as soluble as sodium urate. Urine at pH 5 becomes saturated with urates at 15 mg/dL. Because the pH of urine of normal persons generally is below the pK of uric acid (ie, 5.75), the predominant form of urate in urine is as uric acid, the highly insoluble form. Accordingly, if the urine is alkalinized to pH 7, it will accommodate 150–200 mg of urates per deciliter.

Uric acid becomes the predominant form once the urine is acidified to a pH of less than 5.75, a process that occurs in the distal tubule and collecting ducts of the kidney. If crystals of this end product of purine catabolism are formed in the urinary system, they will be sodium urate at any site proximal to the site of acidification of urine; at any site distal to the acidification, uric acid crystals will be formed. Therefore, most stones of the urinary collecting system are uric acid. As noted above, the precipitation of uric acid stones can be prevented to a considerable extent by alkalinization of the urine to ensure that sodium urate, the more soluble form, will predominate.

The needle-shaped sodium urate crystals are intensely **negatively birefringent** (optically anisotropic); thus, when viewed through a polarizing microscope, they can be distinguished from other types of crystals. If the synovial or joint fluid of a patient shows polymorphonuclear leukocytes containing crystals whose color is yellow when viewed with their long axis parallel to the plane of polarized light and blue when perpendicular to the plane of light, then sodium urate crystals are present. The diagnosis is gout. It should be noted, however, that calcium pyrophosphate crystals, which are found in synovial fluid, are positively birefringent and can be responsible for a syndrome referred to as "pseudogout."

The classification of the disorders of purine metabolism includes those exhibiting **hyperuricemia,** those exhibiting **hypouricemia,** and the immunodeficiency diseases. As shown in Table 35–2, individuals with hyperuricemia can be divided into 2 groups: those with normal urate excretion rates and those excreting excessive quantities of total urates.

Lesch-Nyhan Syndrome & Von Gierke's Disease

Some individuals with urate overexcretion (> 600 mg of uric acid per 24 hours) can be categorized as having secondary hyperuricemia. They have other disease processes such as cancer or psoriasis that lead to enhanced tissue turnover.

Finally, there are persons with identifiable enzyme defects, including abnormalities of PRPP synthetase (feedback-resistant and enhanced enzyme activities), the HGPRTase (hypoxanthine-guanine phosphoribosyltransferase) deficiencies (both the complete [Lesch-Nyhan syndrome] and incomplete deficiencies), and **glucose-6-phosphatase deficiency** (von Gierke's disease). There exists also a group of patients exhibiting idiopathic overproduction hyperuricemia, which will certainly be regarded as a heterogeneous group of diseases once the molecular bases for their metabolic defects are recognized.

Lesch-Nyhan syndrome (complete HGPRTase deficiency) is an inherited X-linked recessive disorder characterized by cerebral palsy with choreoathetosis and spasticity, a bizarre syndrome of self-mutilation, and severe overproduction hyperuricemia. There is usually an associated uric acid lithiasis. The mothers of affected children are heterozygous and mosaic for the HGPRTase deficiency and frequently exhibit overproduction hyperuricemia but without any neurologic manifestations. There also exist male patients with partial deficiencies of HGPRTase attributable to a different mutation of the same gene. These males have

Table 35–1. Inherited disorders of purine metabolism and their associated enzyme abnormalities.

Clinical Disorder	Defective Enzyme	Nature of the Defect	Characteristics of Clinical Disorder	Inheritance Pattern
Gout	PRPP synthetase	Superactive (increased V_{max})	Purine overproduction and overexcretion	X-linked recessive
Gout	PRPP synthetase	Resistance to feedback inhibition	Purine overproduction and overexcretion	X-linked recessive
Gout	PRPP synthetase	Low K_m for ribose 5-phosphate	Purine overproduction and overexcretion	Probably X-linked recessive
Gout	HGPRTase*	Partial deficiency	Purine overproduction and overexcretion	X-linked recessive
Lesch-Nyhan syndrome	HGPRTase*	Complete deficiency	Purine overproduction and overexcretion; cerebral palsy and self-mutilation.	X-linked recessive
Immune deficiency	Adenosine deaminase	Severe deficiency	Combined (T cell and B cell) immunodeficiency, deoxyadenosinuria	Autosomal recessive
Immune deficiency	Purine nucleoside phosphorylase	Severe deficiency	T cell deficiency, inosinuria, deoxyinosinuria, guanosinuria, deoxyguanosinuria, hypouricemia	Autosomal recessive
Renal lithiasis	Adenine phosphoribosyltransferase	Complete deficiency	2,8-Dihydroxyadenine renal lithiasis	Autosomal recessive
Xanthinuria	Xanthine oxidase	Complete deficiency	Xanthine renal lithiasis, hypouricemia	Autosomal recessive

*HGPRTase = hypoxanthine-guanine phosphoribosyltransferase.

severe overproduction hyperuricemia but usually are without significant neurologic signs and symptoms.

Purine overproduction by patients deficient in hypoxanthine-guanine phosphoribosyltransferase is related to the increased intracellular concentrations of PRPP. Increased PRPP levels seem to result from the sparing of PRPP by the deficient salvage pathway. The biochemical basis for the neurologic disorder in Lesch-Nyhan syndrome is unknown.

The basis of purine overproduction and hyperuricemia in **von Gierke's disease** is purportedly secondary to the enhanced activity of the hexose monophosphate shunt and thus enhanced generation of ribose 5-phosphate, from which PRPP is synthesized. However, patients with glucose-6-phosphatase deficiency also have chronic lactic acidosis and thus have elevated renal thresholds for secretion of urate contributing to the accumulation of total body urates.

All of the known enzyme defects (except the glucose-6-phosphatase deficiency, which has not been tested) are associated with increased intracellular concentrations of PRPP, and the theoretic basis for purine

Table 35–2. Classification of patients with hyperuricemia.

I. Normal excretion of urate; renal disorder responsible for elevated serum urate.
II. Excessive excretion of urate because of overproduction.
  A. Secondary to other disease, eg, cancer, psoriasis.
  B. Known enzyme defects responsible for overproduction.
    1. PRPP synthetase abnormalities.
    2. Hypoxanthine-guanine phosphoribosyltransferase deficiencies.
    3. Glucose-6-phosphatase deficiencies.
  C. Unrecognized defects.

overproduction in the glucose-6-phosphatase deficiency is probably similar. Thus, it seems likely that many more disorders of overproduction hyperuricemia will eventually be found to be associated with increased accumulation of intracellular PRPP.

Other Purine Disorders

Hypouricemia is due either to enhanced excretion or to decreased production of urate and uric acid. Dalmatian dogs, although possessing uricase activity, as do all dogs, are not capable of reabsorbing completely the filtered uric acid in their kidneys. They excrete urate and uric acid in amounts that are excessive in respect to their serum urate levels. Similar defects have been discovered in humans with hypouricemia.

Deficiency of the enzyme xanthine oxidase, due either to an inherited genetic defect or to severe liver damage, results in hypouricemia and increased excretion of the oxypurines, hypoxanthine and xanthine. In severe xanthine oxidase deficiencies, patients frequently exhibit **xanthinuria** and xanthine lithiasis.

Two immunodeficiency diseases associated with deficiencies of purine metabolizing enzymes have been described in recent years. **Adenosine deaminase deficiency** is associated with a severe combined immunodeficiency disease in which both thymus-derived lymphocytes (T cells) and bone marrow-derived lymphocytes (B cells) are sparse and dysfunctional. **Purine nucleoside phosphorylase deficiency** is associated with a severe thymus-derived lymphocyte deficiency with apparently normal B cell function, a milder form of immunodeficiency. Both of these immunodeficiency diseases are inherited as autosomal recessive disorders. The metabolic bases for the im-

Table 35–3. Inherited disorders of pyrimidine metabolism and their associated enzyme abnormalities.

Clinical Disorder	Defective Enzyme	Nature of the Defect	Characteristics of Clinical Disorder	Inheritance Pattern
β-Aminoisobutyric aciduria	Transaminase	Deficiency	No symptoms; frequent in Orientals.	Autosomal recessive
Orotic aciduria, type I	Orotate phosphoribosyl-transferase and oroti-dylate decarboxylase	Deficiencies	Orotic acid crystalluria, failure to thrive, and megaloblastic anemia. (?) Immune deficiency. Remission with oral uridine.	Autosomal recessive
Orotic aciduria, type II	Orotidylate decarbox-ylase	Deficiency	Orotidinuria and orotic aciduria, meg-aloblastic anemia. Remission with oral uridine.	Autosomal recessive
Ornithine transcarbam-oylase deficiency	Ornithine transcar-bamoylase	Deficiency	Protein intolerance, hepatic encepha-lopathy, and mild orotic aciduria.	X-linked recessive.

mune dysfunctions seem to involve the intracellular accumulation of the triphosphates of the deoxyribonu-cleoside substrates of purine nucleoside phosphoryl-ase and adenosine deaminase, respectively. These toxic deoxynucleoside triphosphates, deoxy GTP and deoxy ATP, are capable of allosterically inhibiting ri-bonucleotide reductase and thereby depleting cells, such as T cells, of the precursors of DNA synthesis, particularly deoxy CTP.

Purine deficiency states are rare in humans. These are limited to circumstances attributable primarily to deficiencies of folic acid and perhaps of vitamin B_{12} when the latter results in a secondary deficiency of fo-late derivatives.

CLINICAL DISORDERS OF PYRIMIDINE METABOLISM (See Table 35–3.)

As described above, the end products of pyrimidine metabolism, unlike those of purine metabolism, are **highly water-soluble compounds** such as CO_2, am-monia, β-alanine, and propionate. Thus, in circum-stances where pyrimidine overproduction occurs, clin-ically detectable abnormalities are rarely evident. In cases of hyperuricemia associated with severe PRPP overproduction, there is concomitant overproduction of pyrimidine nucleotides with increased excretion of compounds such as β-alanine. Because of the require-ment for N^5,N^{10}-methylenetetrahydrofolate for thymi-dylate synthesis, disorders of folate and vitamin B_{12} metabolism result in deficiencies of TMP (in the case of vitamin B_{12} deficiency, by an indirect mechanism).

β-Aminoisobutyric aciduria is an autosomal, re-cessively inherited disorder prevalent among Orien-tals. It is not associated with any pathologic state. It has been discussed above in connection with pyrim-idine catabolism.

Two types of primary **hereditary orotic aciduria** have been reported. The more common type (type I), although still rare, is that in which both orotate phos-phoribosyltransferase and orotidylate (OMP) decar-boxylase are missing in all cell types tested (Fig 35–19). The patients are pyrimidine auxotrophs.

They are readily treated with uridine. As infants, these patients exhibit failure to thrive, megaloblastic ane-mias, and orange crystalluria (orotic acid). Unless treated with a source of pyrimidine nucleosides, they succumb to infections. The second type of hereditary orotic aciduria (type II) is due to a deficiency only of OMP decarboxylase (Fig 35–19). In patients with type I orotic aciduria, orotic acid is the major abnor-mal excretory product. In the one patient with type II, orotidine is the major excretory product, although some orotic acid is also excreted. In the erythrocytes of patients with type I orotic aciduria, the specific cata-lytic activities of aspartate transcarbamoylase and di-hydroorotase were found to be greatly increased but returned to normal upon treatment of the patient with oral uridine. These observations suggest that one or more end products of the pathway are normally re-sponsible for the maintenance of these enzyme activi-ties at a regulated level. In a deficient state when the cells are deprived of the end products of this pathway, there is a derepression, probably a coordinate one, of at least those 2 enzymes.

The enzymology of the de novo pyrimidine path-way has suggested that there is a common protein molecule providing the carbamoyl phosphate syn-thase, aspartate transcarbamoylase, and dihydrooro-tase catalytic activities, and another accounting for both orotate phosphoribosyltransferase and OMP de-carboxylase activities.

Increased excretion of orotic acid, uracil, and uridine has been described in patients deficient in or-nithine transcarbamoylase, a liver mitochondrial en-zyme responsible for an early step in urea and arginine biosynthesis. In these patients there is apparently mi-tochondrial carbamoyl phosphate accumulation in re-sponse to the enzyme deficiency. The mitochondrial carbamoyl phosphate diffuses into the cytosol to be utilized as a substrate for de novo pyrimidine nucleo-tide synthesis. The excess production of orotic acid is then manifested as orotic aciduria, which usually oc-curs in a mild degree and appears without crystal for-mation but increases upon the ingestion of foodstuffs such as meat that contain large amounts of nitrogen.

At least 2 drugs can result in orotic aciduria. Allo-purinol, 4-hydroxypyrazolopyrimidine, is a purine

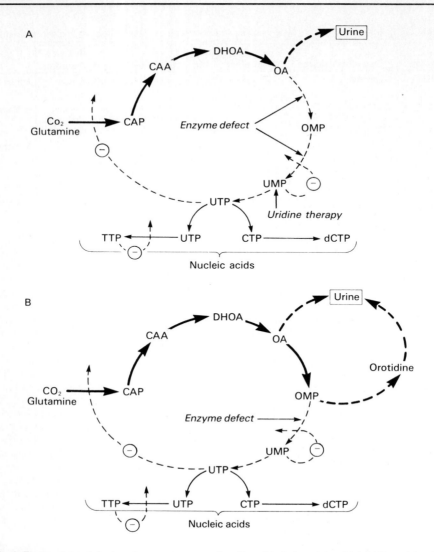

Figure 35–19. *A:* The enzyme defect and consequences of orotic aciduria type I, in which both orotate phosphoribosyl-transferase and orotidylic decarboxylase are deficient. *B:* The defect and consequences of orotic aciduria type II, in which orotidylic decarboxylase is deficient. The broken lines in which a negative sign is inserted represent feedback inhibition that exists under normal conditions. In type I orotic aciduria, orotic acid is spilled in the urine, whereas in type II both orotic acid and orotidine appear in the urine. The abbreviations used are defined in Fig 35–15. (Redrawn and reproduced, with permission, from Smith LH Jr: Pyrimidine metabolism in man. *N Engl J Med* 1973;288:764.)

analog that directly inhibits xanthine oxidase; it is widely used in the treatment of certain patients with gout. It can be phosphoribosylated by orotate phosphoribosyltransferase, thereby competitively inhibiting the phosphoribosylation of orotic acid. Furthermore, the unusual nucleotide formed inhibits orotidylate decarboxylase, producing orotic aciduria and orotidinuria. In humans, at least, the pyrimidine pathway appears to readjust itself to this inhibition so that the organism is only transiently starved for pyrimidine nucleotides during the early stages of treatment.

6-Azauridine, after conversion to 6-azauridylate, is a competitive inhibitor of OMP decarboxylase, inducing high rates of excretion of orotic acid and orotidine as a result.

In specific liver mitochondrial failure, such as in **Reye's syndrome,** there is a secondary orotic aciduria. It is probably secondary to the inability of the mitochondria to utilize carbamoyl phosphate, which then, as in the inherited deficiency of ornithine transcarbamoylase, causes overproduction of orotic acid and a resultant orotic aciduria.

REFERENCES

Ames BN et al: Uric acid provides an antioxidant defense in humans against oxidant- and radical-caused aging and cancer: A hypothesis. *Proc Natl Acad Sci USA* 1981;**78:**6858.

Henderson JF: *Regulation of Purine Biosynthesis.* Monograph No. 170. American Chemical Society, 1972.

Henderson JF, Paterson ARP: *Nucleotide Metabolism: An Introduction.* Academic Press, 1973.

Jones M: Pyrimidine nucleotide biosynthesis in animal cells. *Annu Rev Biochem* 1980;**49:**253.

Kempe TD et al: Stable mutants of mammalian cells that overproduce the first three enzymes of pyrimidine nucleotide biosynthesis. *Cell* 1976;**9:**541.

Martin DW Jr, Gelfand EW: Biochemistry of diseases of immunodevelopment. *Annu Rev Biochem* 1981;**50:**845.

Smith LH Jr: Pyrimidine metabolism in man. *N Engl J Med* 1973;**288:**764.

Stanbury JB et al (editors): *The Metabolic Basis of Inherited Disease,* 5th ed. McGraw-Hill, 1983.

Thelander L, Reichard P: Reduction of ribonucleotides. *Annu Rev Biochem* 1979;**48:**133.

Wyngaarden JB, Kelley WN: *Gout and Hyperuricemia.* Grune & Stratton, 1976.

36

Recombinant DNA Technology

Daryl K. Granner, MD

INTRODUCTION

Recombinant DNA technology, often referred to as **genetic engineering,** has revolutionized biology and is having an ever-increasing impact on clinical medicine. Much has been learned about human genetic disease from pedigree analysis and study of affected proteins, but in many cases where the specific genetic defect is unknown, these approaches cannot be used. The new technology circumvents these limitations by going directly to the DNA molecule for information.

This chapter is aimed at clarifying this rather complex topic. It presents the basic concepts of recombinant DNA technology, its applications to clinical medicine, and a glossary. In order for the chapter to be complete in itself, some repetition of subjects discussed in other chapters will be found.

BIOMEDICAL IMPORTANCE

Understanding recombinant DNA technology is important for several reasons. (1) The information explosion occurring in this area is truly staggering. To understand and keep up with this field, one must have an appreciation of the fundamental concepts involved. (2) There is now a rational approach to understanding the molecular basis of a number of diseases (eg, familial hypercholesterolemia, sickle cell disease, the thalassemias, cystic fibrosis, Huntington's chorea). (3) Using recombinant DNA technology, human proteins can be produced in abundance for therapy (eg, insulin, growth hormone, plasminogen activator). (4) Proteins for vaccines (eg, hepatitis B) and for diagnostic tests (eg, AIDS test) can be obtained. (5) Recombinant DNA technology is used to diagnose existing diseases and predict the risk of developing a given disease. (6) Gene therapy for sickle cell disease, the thalassemias, adenosine deaminase deficiency, and other diseases may be devised. Gene therapy has already been accomplished in mice, in whom hereditary hypogonadism has been corrected by transgenic injection of the gonadotropin-releasing hormone gene into the fertilized ovum.

BASIC FEATURES OF DNA

DNA Structure

DNA is a complex biopolymer that is organized as a double helix. The fundamental organizational element is the sequence of purine (adenine [A] or guanine [G]) and pyrimidine (cytosine [C] or thymine [T]) bases. These bases are attached to the C-1' position of the sugar deoxyribose, and the bases are linked together through joining of the sugar moieties at their 3' and 5' positions via a phosphodiester bond (see Fig 37–1). The alternating deoxyribose and phosphate groups form the backbone of the double helix (see Fig 37–2). These 3'–5' linkages also define the orientation of a given strand of the DNA molecule, and since the 2 strands run in opposite directions, they are said to be antiparallel.

Base pairing is one of the most fundamental concepts of DNA structure and function. Adenine and thymine always pair, by hydrogen bonding, as do guanine and cytosine (see Fig 37–3). These base pairs are said to be **complementary,** and the guanine content of a fragment of double-stranded DNA will always equal its cytosine content; likewise the thymine and adenine contents are equal. Base pairing and hydrophobic base-stacking interactions hold the 2 DNA strands together. These interactions can be reduced by heating the DNA to denature it. The laws of base pairing predict that 2 complementary DNA strands will reanneal exactly in register upon renaturation, as happens when the temperature of the solution is slowly reduced to normal. Indeed, the degree of base-pair matching (or mismatching) can be estimated from the temperature required for denaturation-renaturation. Segments of DNA with high degrees of base-pair matching require more energy input (heat) to accomplish denaturation, or, to put it another way, a closely matched segment will withstand more heat before the strands separate. This reaction is used to determine whether there are significant differences between 2 DNA sequences, and it underlies the concept of **hybridization,** which is fundamental to the processes described below.

There are about 3×10^9 base pairs (bp) in each human haploid genome. If an average gene length is 3×10^3 bp (3 kilobases [kb]), the genome could consist of 10^6 genes, assuming that there is no overlap and that transcription proceeds in only one direction. It is thought that there are only about 10^5 genes in the human and that only 10% of the DNA codes for proteins. The function of the remaining 90% of the human genome has not yet been defined.

The double-helical DNA is packaged into a more compact structure by a number of proteins, most notably the basic proteins called **histones.** This condensation may serve a regulatory role and certainly has a practical purpose. The DNA present within the nucleus of a cell, if simply extended, would be about a meter long. The chromosomal proteins compact this long length of DNA so that it can be packaged into a nucleus with a volume of a few cubic microns.

Gene Organization

In general, prokaryotic genes consist of a small regulatory region (100–500 bp) and a large protein-coding segment (500–10,000 bp). Several genes are often controlled by a single regulatory unit. Most mammalian genes are more complicated, in that the coding regions are interrupted by noncoding regions that are eliminated when the primary RNA transcript is processed into mature **messenger RNA (mRNA).** The **coding regions** (those regions that appear in the mature RNA species) are called **exons,** and the **noncoding regions,** which interpose or intervene between the exons, are called **introns** (Fig 36–1). Introns are always removed from precursor RNA before transport into the cytoplasm occurs. The process by which introns are removed from precursor RNA and by which exons are ligated together is called **RNA splicing.** Incorrect processing of the primary transcript into the mature mRNA can result in disease in humans (see below); this underscores the importance of these post-

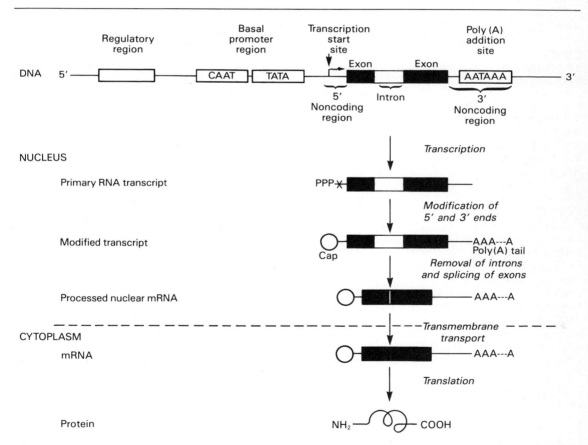

Figure 36–1. Organization of a eukaryotic transcription unit and the pathway of eukaryotic gene expression. Eukaryotic genes have structural and regulatory regions. The structural region consists of the coding DNA and 5′ and 3′ noncoding DNA sequences. The coding regions are divided into 2 parts: (1) exons, which eventually become mature RNA, and (2) introns, which are processed out of the primary transcript. The structural region is bounded at its 5′ end by the transcription initiation site and at its 3′ end by the polyadenylate addition or termination site. The promoter region, which contains specific DNA sequences that interact with various protein factors to regulate transcription, is discussed in detail in Chapters 39 and 41. The primary transcript has a special structure, a cap, at the 5′ end and a stretch of A's at the 3′ end. This transcript is processed to remove the introns, and the mature mRNA is then transported to the cytoplasm, where it is translated into protein.

transcriptional processing steps. Regulatory regions for specific eukaryotic genes are usually located in the DNA that flanks the transcription initiation site at its 5' end (**5' flanking-sequence DNA**). Occasionally, such sequences are found within the gene itself or in the region that flanks the 3' end of the gene. In mammalian cells, each gene has its own regulatory region. Many eukaryotic genes (and some viruses that replicate in mammalian cells) have special regions, called **enhancers**, that increase the rate of transcription. Some genes also have DNA sequences, known as **silencers**, that diminish transcription. Mammalian genes are obviously complicated, multicomponent structures.

Gene Transcription

Information generally flows from DNA to mRNA to protein, as illustrated in Fig 36–1 and discussed in more detail in Chapter 41. This is a rigidly controlled process involving a number of complex steps, each of which no doubt is regulated by one or more enzymes or factors; faulty function at any of these steps can cause disease.

CONCEPTS USED IN RECOMBINANT DNA TECHNOLOGY

Isolation and manipulation of DNA, including end-to-end joining of sequences from very different sources to make chimeric molecules (eg, molecules containing both human and bacterial DNA sequences in a sequence-independent fashion), is the essence of recombinant DNA research. This involves several unique techniques and reagents.

Restriction Enzymes

Certain **endonucleases**, enzymes that cut DNA at specific DNA sequences within the molecule (as opposed to **exonucleases**, which digest from the ends of DNA molecules), are a key tool in recombinant DNA research. These enzymes were originally called restriction enzymes because their presence in a given bacterium restricted the growth of certain bacterial viruses called bacteriophages. Restriction enzymes

cut DNA into short pieces in a sequence-specific manner, in contrast to most other enzymatic, chemical, or physical methods, which break DNA randomly. These defensive enzymes (over 200 have been discovered) protect the host bacterial DNA from DNA from foreign organisms (primarily infective phages). However, they are only present in cells that also have a companion enzyme that methylates the host DNA, rendering it an unsuitable substrate for digestion by the restriction enzyme. Thus, **site-specific DNA methylases** and restriction enzymes always exist in pairs in a bacterium.

Restriction enzymes are named after the bacterium from which they are isolated (Table 36–1), eg, Eco RI from *Escherichia coli*, Bam HI from *Bacillus amyloliquefaciens*. The first 3 letters in the restriction enzyme name consist of the first letter of the genus (E) and the first 2 letters of the species (co). These may be followed by a strain designation (R) and a roman numeral (I) to indicate the order of discovery (eg, Eco RI, Eco RII). Each enzyme recognizes and cleaves a specific double-stranded DNA sequence that is 4–7 bp long. These DNA cuts result in **blunt ends** (Hpa I) or overlapping (**sticky**) **ends** (Bam HI) (Fig 36–2), depending on the mechanism used by the enzyme. Sticky ends are particularly useful in constructing hybrid or chimeric DNA molecules (see below). If the nucleotides are distributed randomly in a given DNA molecule, one can calculate how frequently a given enzyme would cut a length of DNA. For each position in the DNA molecule there are 4 possibilities (A, C, G, or T); therefore, a restriction enzyme that recognizes a 4-bp sequence will cut, on average, once every 256 bp (4^4), whereas another enzyme that recognizes a 6-bp sequence will cut once every 4096 bp (4^6). A given piece of DNA will have a characteristic linear array of sites for the various enzymes; hence, a **restriction map** can be constructed. When DNA is digested with a given enzyme, the ends of all the fragments will have the same DNA sequence. The fragments produced can be isolated by electrophoresis on agarose or polyacrylamide (see Blot Transfer, below); this is an essential step in cloning and a major use of these enzymes.

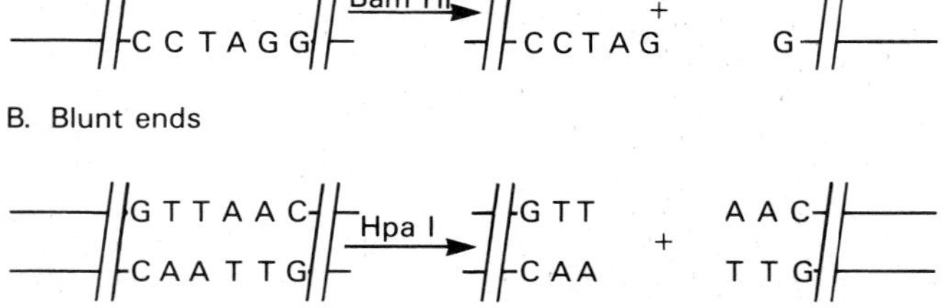

A. Sticky, or staggered, ends

B. Blunt ends

Figure 36–2. Results of restriction endonuclease digestion. Digestion with a restriction endonuclease can result in the formation of DNA fragments with sticky, or cohesive, ends (**A**) or blunt ends (**B**). This is an important consideration in devising cloning strategies.

Table 36–1. Selected restriction endonucleases and their sequence specificities.*

Endonuclease	Sequence Cleaved	Bacterial Source
Bam HI	G G A T C C C C T A G G	*Bacillus amyloliquefaciens* H
Bgl II	A G A T C T T C T A G A	*Bacillus globigii*
Eco RI	G A A T T C C T T A A G	*Escherichia coli* RY13
Eco RII	C C T G G G G A C C	*Escherichia coli* R245
Hind III	A A G C T T T T C G A A	*Haemophilus influenzae* R_d
Hha I	G C G C C G C G	*Haemophilus haemolyticus*
Hpa I	G T T A A C C A A T T G	*Haemophilus parainfluenzae*
Mst II	C C T N A G G G G A N T C C	*Microcoleus* strain
Pst I	C T G C A G G A C G T C	*Providencia stuartii* 164
Taq I	T C G A A G C T	*Thermus aquaticus* YTI

*A, adenine; C, cytosine; G, guanine; T, thymine. Arrows show the site of cleavage; depending on the site, sticky ends (Bam HI) or blunt ends (Hpa I) may result. The length of the recognition sequence can be 4 bp (Taq I), 5 bp (Eco RII), 6 bp (Eco RI), or 7 bp (Mst II). By convention, these are written in the 5′ to 3′ direction for the upper strand of each recognition sequence, and the lower strand is shown with the opposite (ie, 3′ to 5′) polarity. Note that most recognition sequences are palindromes (ie, the sequence reads the same in opposite directions on the 2 strands). A residue designated N means that any nucleotide is permitted.

Other Enzymes

A number of other enzymes that act on DNA and RNA are an important part of recombinant DNA technology. Many of these are referred to in this and subsequent chapters (Table 36–2).

Preparation of Chimeric DNA Molecules

Sticky-end ligation is technically easy, but some special techniques are often required to overcome problems inherent in this approach. Sticky ends of a vector may reconnect with themselves, with no net gain of DNA. Sticky ends of fragments can also an-

neal, so that tandem heterogeneous inserts form. Also, sticky-end sites may not be available or in a convenient position. To circumvent these problems, an enzyme that generates blunt ends is used, and new ends are added using the enzyme terminal transferase. If poly d(G) is added to the 3′ ends of the vector and poly d(C) is added to the 3′ ends of the foreign DNA, the 2 molecules can only anneal to each other, thus circumventing the problems listed above. This procedure, called **homopolymer tailing,** also generates an Sma I restriction site, and so it is easy to retrieve the fragment. Sometimes, synthetic oligonucleotide linkers with a convenient restriction enzyme sequence are li-

Table 36–2. Enzymes used in recombinant DNA research.*

Enzyme	Reaction	Primary Use
Alkaline phosphatase	Dephosphorylates 5′ ends of RNA and DNA.	Removal of 5′–PO₄ groups prior to kinase labeling to prevent self-ligation.
BAL 31 nuclease	Degrades both the 3′ and 5′ ends of DNA.	Progressive shortening of DNA molecules.
DNA ligase	Catalyzes bonds between DNA molecules.	Joining of DNA molecules.
DNA polymerase I	Synthesizes double-stranded DNA from single-stranded DNA.	Synthesis of double-stranded cDNA; nick translation.
DNase I	Under appropriate conditions, produces single-stranded nicks in DNA.	Nick translation; mapping of hypersensitive sites.
Exonuclease III	Removes nucleotides from 3′ ends of DNA.	DNA sequencing; mapping of DNA-protein interactions.
λ Exonuclease	Removes nucleotides from 5′ ends of DNA.	DNA sequencing.
Polynucleotide kinase	Transfers terminal phosphate (γ position) from ATP to 5′–OH groups of DNA or RNA.	^{32}P labeling of DNA or RNA.
Reverse transcriptase	Synthesizes DNA from RNA template.	Synthesis of cDNA from mRNA; RNA (5′ end) mapping studies.
SI nuclease	Degrades single-stranded DNA.	Removal of "hairpin" in synthesis of cDNA; RNA mapping studies (both 5′ and 3′ ends).
Terminal transferase	Adds nucleotides to the 3′ ends of DNA.	Homopolymer tailing.

*Adapted and reproduced, with permission, from Emery AEH: Page 41 in: *An Introduction to Recombinant DNA*. Wiley, 1984.

gated to the blunt-ended DNA. Direct blunt-end ligation is accomplished using the enzyme bacteriophage T4 DNA ligase. This technique, though more difficult than sticky-end ligation, has the advantage of joining together any pairs of ends. The disadvantages are that there is no control over the orientation of insertion or the number of molecules annealed together, and there is no easy way of retrieving the insert.

Cloning

A **clone** is a large population of identical molecules, bacteria, or cells that arise from a common ancestor. Cloning allows for the production of a large number of identical DNA molecules, which can then be characterized or used for other purposes. This technique is based on the fact that chimeric or hybrid DNA molecules can be constructed in **cloning vectors,** typically bacterial plasmids, phages, or cosmids, which then continue to replicate in a host cell under their own control systems. In this way, the chimeric DNA is amplified. The general procedure is illustrated in Fig 36–3.

Bacterial **plasmids** are small, circular duplex DNA molecules whose natural function is to confer antibiotic resistance to the host cell. Plasmids have several properties that make them extremely useful as cloning vectors. They exist as single or multiple copies within the bacterium and replicate independently from the bacterial DNA. The complete DNA sequence of many plasmids is known; hence, the precise location of restriction enzyme cleavage sites for inserting the foreign DNA is available. Plasmids are smaller than the host chromosome and are therefore easily separated from the latter, and the desired DNA is readily removed by cutting the plasmid with the enzyme specific for the restriction site into which the original piece of DNA was inserted.

Phages usually have linear DNA molecules into which foreign DNA can be inserted at several restriction enzyme sites. The chimeric DNA is collected after the phage proceeds through its lytic cycle and produces mature, infective phage particles. A major advantage of phage vectors is that while plasmids accept DNA pieces about 6–10 kb long, phages can accept DNA fragments 10–20 kb long, a limitation imposed by the amount of DNA that can be packed into the phage head.

Even larger fragments of DNA can be cloned in **cosmids,** which combine the best features of plasmids and phages. Cosmids are plasmids that contain the DNA sequences, so-called **cos sites,** required for packaging lambda DNA into the phage particle. These vectors grow in the plasmid form in bacteria, but since much of the unnecessary lambda DNA has been removed, more chimeric DNA can be packaged into the particle head. It is not unusual for cosmids to carry inserts of chimeric DNA that are 35–50 kb long. A comparison of these vectors is shown in Table 36–3.

Insertion of DNA into a functional region of the vector will interfere with the action of this region, and so care must be taken not to interrupt an essential function of the vector. This concept can be exploited, however, to provide a selection technique. The common plasmid vector **pBR322** has both **tetracycline** (tet) and **ampicillin** (amp) **resistance genes.** A single Pst I site within the amp-resistance gene is commonly used

Table 36–3. Common cloning vectors.

Vector	DNA Insert Size
Plasmid pBR322	0.01–10 kb
Lambda charon 4A	10–20 kb
Cosmids	35–50 kb

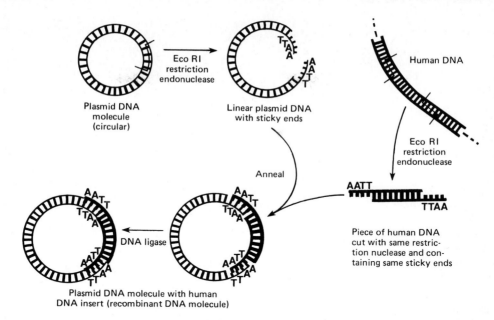

Figure 36–3. Use of restriction nucleases to make new recombinant or chimeric DNA molecules. When inserted back into a bacterial cell (by the process called transformation), the plasmid DNA replicates not only itself but also the physically linked new DNA insert. Since recombining the sticky ends, as indicated, regenerates the same DNA sequence recognized by the original restriction enzyme, the cloned DNA insert can be cleanly cut back out of the recombinant plasmid circle with this endonuclease. If a mixture of all of the DNA pieces created by treatment of total human DNA with a single restriction nuclease is used as the source of human DNA, a million or so different types of recombinant DNA molecules can be obtained, each pure in its own bacterial clone. (Modified and reproduced, with permission, from Cohen SN: The manipulation of genes. *Sci Am* [July] 1975;**233**:34.)

as the insertion site for a piece of foreign DNA. In addition to having sticky ends (Table 36–1 and Fig 36–2), the DNA inserted at this site disrupts the amp-resistance gene and makes the bacterium carrying this plasmid amp-sensitive (Fig 36–4). Thus, the parental plasmid, which provides resistance to both antibiotics, can be readily separated from the chimeric plasmid, which is resistant only to tetracycline. Additional confirmation that insertion has taken place comes from sizing the plasmid DNA obtained from the putative recombinant on an agarose gel, since the chimeric DNA molecule is now larger than the host vector DNA.

Libraries & Library Construction

The combination of restriction enzymes and various cloning vectors allows the entire genome of an organism to be packaged into a vector. A collection of these different recombinant clones is called a library. A **genomic library** is prepared from the total DNA of a cell line or tissue. A **cDNA library** represents the population of mRNAs in a tissue. Genomic libraries are prepared by performing partial digestion of total DNA with a restriction enzyme that cuts DNA frequently (eg, Sau IIIA). The idea is to generate rather large fragments, so that most genes will be left intact. Phage vectors are preferred for these libraries because they accept large pieces of DNA (up to 20 kb). The goal is to achieve a complete library. The number of

fragments required to attain this objective is inversely related to fragment size and directly related to genome size (Table 36–4). A human library that contains 10^6 recombinant fragments of large size has a 99% probability of being complete. Thus, the chances of finding any single-copy gene are excellent.

cDNA libraries are prepared by first isolating all the mRNAs in a tissue and then copying these molecules into double-stranded DNA, using (sequentially) the enzymes reverse transcriptase and DNA polymerase. For technical reasons, full-length cDNA copies are seldom obtained, and so smaller DNA fragments are cloned. Plasmids are often the favored vectors for cDNA libraries because they are much more convenient to work with than are phages or cosmids, although various lambda phage vectors have special advantages for cDNA cloning (see below).

A vector in which the protein coded by the gene introduced by recombinant DNA technology is actually synthesized is known as an **expression vector.** Such vectors are now commonly used to detect specific cDNA molecules in libraries and to produce proteins by genetic engineering techniques. These vectors are specially constructed to contain very active inducible promoters, proper in-phase translation initiation codons, both transcription and translation termination signals, and appropriate protein processing signals, if needed. Some expression vectors even contain genes

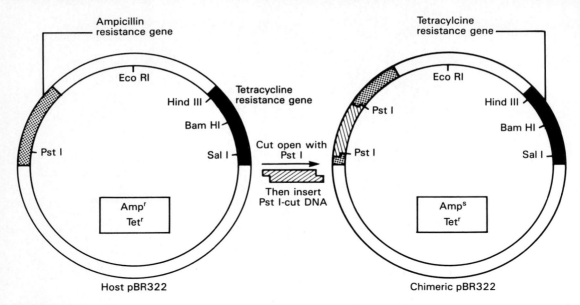

Figure 36–4. A method of screening recombinants for inserted DNA fragments. Using the plasmid pBR322, a piece of DNA is inserted into the unique Pst I site. This insertion disrupts the gene coding for a protein that provides ampicillin resistance to the host bacterium. Hence, the chimeric plasmid will no longer survive when plated on a substrate medium that contains this antibiotic. The differential sensitivity to tetracycline and ampicillin can therefore be used to distinguish clones of plasmid that contain an insert.

Table 36–4. The composition of complete genomic libraries.*

Source	Complete Genomic Library
E coli	1500 fragments
Yeast	4500 fragments
Drosophila	50,000 fragments
Mammals	800,000 fragments

*The number of random fragments (unique clones) that a library should have to ensure that any single gene is represented is inversely related to the average fragment size used to construct the library and directly related to the number of genes in the organism. The numbers given above represent the number of fragments (independent clones) necessary to achieve a 99% probability of finding a given DNA sequence in a recombinant DNA library with an average insert size of 2×10^4 nucleotides. The differences represent the variations in genomic complexity between the creatures.

The number of clones necessary is calculated from the following formula:

$$N = \frac{\ln(1 - P)}{\ln(1 - f)}$$

where P is the probability desired and f is the fraction of the total genome in a single clone. In the case of the mammalian genomic library cited above, given the presence of 3×10^9 nucleotides in the haploid genome, the equation is as follows:

$$N = \frac{\ln(1 - 0.99)}{\ln\left(1 - \left[\dfrac{2 \times 10^4}{3 \times 10^9}\right]\right)}$$

The advantage of having a library composed of large DNA inserts is immediately apparent if this equation is solved using an average fragment size of 5×10^3 nucleotides rather than 2×10^4.

that code for protease inhibitors, so that the final yield of product is enhanced. The vector λgt11 is popular for library construction because it accepts large cDNA molecules that are replicated and translated into proteins; thus, λgt11 recombinant libraries can be screened with either cDNA or antibody probes.

Probes

A variety of molecules can be used to "probe" libraries in search of a specific gene or cDNA molecule or to define and quantitate DNA or RNA separated by electrophoresis through various gels. Probes are generally pieces of DNA or RNA labeled with a ^{32}P-containing nucleotide. The probe must recognize a complementary sequence to be effective. A cDNA synthesized from a specific mRNA can be used to screen either a cDNA library for a longer cDNA or a genomic library for a complementary sequence in the coding region of a gene. A popular technique for finding specific genes entails taking a short amino acid sequence and, using the codon usage for that species (see Chapter 40), making an oligonucleotide probe that will detect the corresponding DNA fragment in a genomic library. If the sequences match exactly, probes 15–20 nucleotides long will hybridize. cDNA probes are used to detect DNA fragments on Southern blot transfers and to detect and quantitate RNA on Northern blot transfers.

Blotting & Hybridization Techniques

Visualization of a specific DNA or RNA fragment

among the many thousands of "contaminating" molecules requires the convergence of a number of techniques, which are collectively termed **blot transfer.** Fig 36–5 illustrates the **Southern** (DNA), **Northern** (RNA), and **Western** (protein) **blot transfer** procedures. (The first is named for the person who devised the technique, and the other names began as laboratory jargon but are now accepted terms.) These procedures are useful in determining how many copies of a gene are in a given tissue or whether there are any gross alterations in a gene (deletions, insertions, or rearrangements). Occasionally, if a specific base is changed and a restriction site is altered, these procedures can detect a point mutation. The Northern and Western blot transfer techniques are used to size and quantitate specific RNA and protein molecules, respectively.

Colony or **plaque hybridization** is the method by which specific clones are identified and purified. Bacteria are grown on colonies on an agar plate and overlaid with a nitrocellulose filter paper. Cells from each colony stick to the filter and are permanently fixed thereto by heat, which with NaOH treatment also lyses the cells and denatures the DNA so that it will hybridize with the probe. A radioactive probe is added to the filter, and after washing, the hybrid complex is lo-

calized by exposing the filter to x-ray film. By matching the spot on the autoradiograph to a colony, the latter can be picked from the plate. A similar strategy is used to identify fragments in phage libraries. Successive rounds of this procedure result in a clonal isolate (bacterial colony) or individual phage plaque.

All of the hybridization procedures discussed in this section depend on the specific base-pairing properties of complementary nucleic acid strands described above. Perfect matches hybridize readily and withstand high temperatures in the hybridization and washing reactions. These complexes also form in the presence of low salt concentrations. Less than perfect matches do not tolerate these **stringent conditions** (ie, elevated temperatures and low salt concentrations); thus, hybridization either never occurs or is disrupted during the washing step. Gene families, in which there is some degree of homology, can be detected by varying the stringency of the hybridization and washing steps. Cross-species comparisons of a given gene can also be made using this approach.

DNA Sequencing

The segments of specific DNA molecules obtained by recombinant DNA technology can be analyzed for

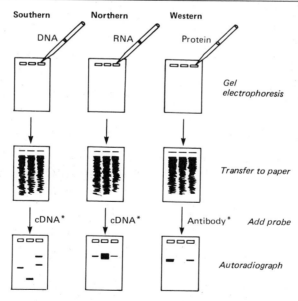

Figure 35–5. The blot transfer procedure. In a Southern, or DNA, blot transfer, DNA isolated from a cell line or tissue is digested with one or more restriction enzymes. This mixture is pipetted into a well in an agarose or polyacrylamide gel and exposed to a direct electrical current. DNA, being negatively charged, migrates toward the cathode; the smaller fragments move the most rapidly. After a suitable time, the DNA is denatured by exposure to mild alkali and transferred to nitrocellulose paper, in an exact replica of the pattern on the gel, by the blotting technique devised by Southern. The DNA is annealed to the paper by exposure to heat, and the paper is then exposed to the labeled cDNA probe, which hybridizes to complementary fragments on the filter. After thorough washing, the paper is exposed to x-ray film, which is developed to reveal several specific bands corresponding to the DNA fragment that recognized the sequences in the cDNA probe. The RNA, or Northern, blot is conceptually similar. RNA is subjected to electrophoresis before blot transfer. This requires some different steps from those of DNA transfer, primarily to ensure that the RNA remains intact, and is generally somewhat more difficult. In the protein, or Western, blot, proteins are electrophoresed and transferred to nitrocellulose and then probed with a specific antibody or other probe molecule.

their nucleotide sequence (Fig 36–6). This method depends upon having a large number of identical DNA molecules. This requirement can be satisfied by cloning the fragment of interest, using the techniques described above. The method shown is that of Maxam and Gilbert and employs chemical methods to cleave the DNA molecules where they contain the specific nucleotides. A second, enzymatic method (Sanger's) employs specific deoxynucleotide analogs that terminate DNA strand synthesis at specific nucleotides as the strand is synthesized on purified template nucleic acid.

SOME PRACTICAL APPLICATIONS OF RECOMBINANT DNA TECHNOLOGY

The isolation of a specific gene from an entire genome requires a technique that will discriminate one part in a million. The identification of a regulatory region that may be only 10 bp in length requires a sensitivity of one part in 3×10^8; a disease such as sickle cell anemia is caused by a single base change, or one part in 3×10^9. Recombinant DNA technology is powerful enough to accomplish all of these things.

Gene Mapping

This technique is used to localize specific genes to distinct chromosomes and, thus, to define a map of the human genome. This is already yielding useful information in the definition of human disease. Somatic cell hybridization and in situ hybridization are 2 techniques used to accomplish this. In **in situ hybridization,** the simpler and more direct procedure, a radioactive probe is added to a metaphase spread of chromosomes on a glass slide. The exact area of hybridization is localized by layering photographic emulsion over the slide and, after exposure, lining up the grains with some histologic identification of the chromosome. This often places the gene at a location on a given band or region on the chromosome. Some

of the human genes localized using this technique are listed in Table 36–5.

This table represents only a sampling, since more than 100 genes have been mapped. This human map will become more complete in ensuing years, and there is presently talk of sequencing the entire human genome. The following conclusions can already be drawn: (1) Genes that code for proteins with similar functions can be located on separate chromosomes (α- and β-globin). (2) Genes that form part of a family can also be on separate chromosomes (growth hormone and prolactin). (3) The genes involved in many hereditary disorders known to be due to specific protein deficiencies, including X chromosome-linked conditions, are indeed located at specific sites. Of most interest, perhaps, is the fact that because of the availability of defined and cloned restriction fragments, the chromosomal location for many disorders for which the protein deficiency is unknown is being defined, eg, Huntington's chorea, chromosome 4; cystic fibrosis, chromosome 7; adult polycystic kidney disease, chromosome 16; and Duchenne-type muscular dystrophy, chromosome X. Once the defect is localized to a region of DNA that has the characteristic structure of a gene (Fig 36–1), a synthetic gene can be constructed and expressed in an appropriate vector and its function can be assessed, or the putative peptide, deduced from the open reading frame in the coding region, can be synthesized. Antibodies directed against this peptide can be used to assess whether this peptide is expressed in normal persons and whether it is absent in those with the genetic syndrome.

Protein Production

A practical goal of recombinant DNA research is the production of materials for biomedical application. This technology has 2 distinct merits: (1) It can supply large amounts of material that could not be obtained by conventional purification methods (eg, interferon, plasminogen activating factor). (2) It can provide human material (eg, insulin, growth hormone). The ad-

Table 36–5. Localization of human genes.*

Gene	Chromosome	Disease
Insulin	11p15	
Prolactin	6p23-q12	
Growth hormone	17q21-qter	Growth hormone deficiency
α-Globin	16p12-pter	α-Thalassemia
β-Globin	11p12	β-Thalassemia, sickle cell
Adenosine deaminase	20q13-qter	Adenosine deaminase deficiency
Phenylalanine hydroxylase	12q24	Phenylketonuria
Hypoxanthine-guanine phosphoribosyltransferase	Xq26-q27	Lesch-Nyhan syndrome
DNA segment G8	4p	Huntington's chorea

*This table indicates the chromosomal location of several genes and the diseases associated with deficient or abnormal production of the gene products. The chromosome involved is indicated by the first (underlined) number or letter. The other numbers and letters refer to precise localizations, as defined in McKusick VA: *Mendelian Inheritance in Man,* 6th ed. Johns Hopkins Univ Press, 1983.

Step 1

Isolate the population ($\sim 10^{12}$) of identical DNA molecules. (Identical molecules will, of course, have identical termini, nucleotide sequence, and length.) DNA molecular cloning and restriction endonuclease digestion clearly provide the most effective means for obtaining the molecules.

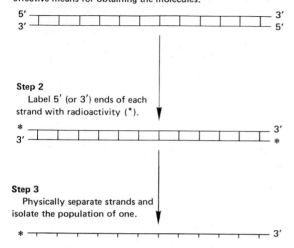

Step 2

Label 5′ (or 3′) ends of each strand with radioactivity (*).

Step 3

Physically separate strands and isolate the population of one.

Step 4

Divide into 4 tubes. To each tube is added a different specific chemical reagent that will destroy specifically one or 2 of the 4 bases (A, T, C, G) at the sites where they occur in the DNA strand and thereby break the strand at that site. The destruction must be controlled so that it is incomplete and only some of the strands are broken at each of the sites where a given base exists.

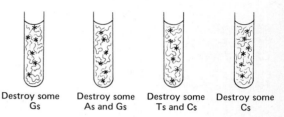

| Destroy some Gs | Destroy some As and Gs | Destroy some Ts and Cs | Destroy some Cs |

This will generate in each tube a mixture of radioactively labeled strand-fragments (and many unlabeled fragments) of different lengths. The lengths of the labeled fragments will reflect the number of nucleotides between the labeled (*) end and the specific nucleobases (which were destroyed) where they appear in the single-stranded molecule.

Step 5

The components of each mixture of strand-fragments are then separated by size (length) using polyacrylamide gel slab electrophoresis. The shorter fragments move more rapidly than the longer fragments. The slab gel is autoradiographed, and the bands of the labeled strand-fragment components of each mixture produce an image on the x-ray film.

Sequence of original strand:
$*$ — A — G — T — C — T — T — G — G — A — G — C — T — 3′

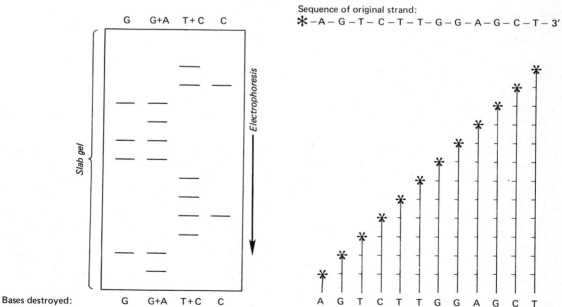

The ladderlike arrays represent from bottom to top all of the successively longer fragments of the original DNA strand. Knowing which specific one or 2 nucleotides were chemically destroyed to produce each mixture of fragments, one can determine the sequence of nucleotides from the labeled end toward the unlabeled end by reading up the gel. The base-pairing rules of Watson and Crick (A—T, G—C) dictate the sequence of the other (complementary) strand.

Figure 36–6. Sequencing of DNA by the method devised by Maxam and Gilbert.

vantages in both cases are obvious. Although the primary aim is to supply products, generally proteins, for treatment (insulin) and diagnosis (AIDS test) of human and other animal diseases and for disease prevention (hepatitis B vaccine), there are other real and potential commercial applications, especially in agriculture. An example of the latter is the attempt to engineer plants that are more resistant to drought or temperature extremes or more efficient at fixing nitrogen.

RECOMBINANT DNA TECHNOLOGY IN MOLECULAR ANALYSIS OF DISEASE

Normal Gene Variations

There is a normal variation of DNA sequence just as there is with more obvious aspects of human structure. Variations of DNA sequence, **polymorphisms,** occur approximately once in every 500 nucleotides, or about 10^7 times per genome. There are, no doubt, deletions and insertions of DNA as well as single-base substitutions. In healthy people, these alterations obviously occur in noncoding regions of DNA or at sites that cause no change in function of the encoded protein. This polymorphism of DNA structure can be associated with certain diseases and can be used to search for the specific gene involved, as is illustrated below.

Gene Variations Causing Disease

Classical genetics taught that most genetic diseases

Table 36–6. Structural alterations of the β-globin gene.

Alteration	Function Affected	Disease
Point mutations	Protein folding Transcriptional control Frameshift and nonsense mutations RNA processing	Sickle cell disease β-Thalassemia β-Thalassemia β-Thalassemia
Deletion	mRNA production	β^0-Thalassemia Hemoglobin Lepore
Rearrangement	mRNA production	β-Thalassemia type III

were due to point mutations that resulted in an impaired protein. This may still be true, but if on reading the initial sections of this chapter one predicted that genetic disease could result from derangement of any of the steps illustrated in Fig 36–1, one would have made a proper assessment.

This point is nicely illustrated by an examination of the β-globin gene. This gene is located in a cluster on chromosome 11 (Fig 36–7), and an expanded version of the gene is illustrated in Fig 36–8. Defective production of β-globin results in a variety of diseases and is due to many different lesions in and around the β-globin gene (Table 36–6).

Point Mutations

The classic example is **sickle cell disease,** which is caused by mutation of a single base out of the 3×10^9 in the genome, a T-to-A DNA substitution, which in turn results in an A-to-U change in the mRNA corresponding to the sixth codon of the β-globin gene (see

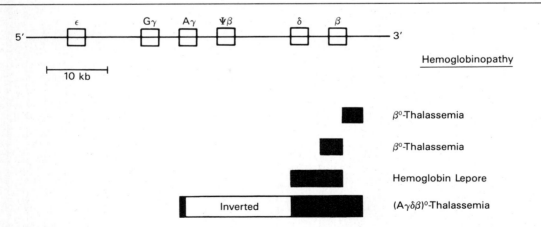

Figure 36–7. Schematic representation of the β-globin gene cluster and some genetic disorders. The β-globin gene is located on chromosome 11 in close association with the 2 γ-globin genes and the δ-globin gene. The β-gene family is arranged in the order 5'-ξ-Gγ-Aγ-$\psi\beta$-δ-β-3'. The ξ locus is expressed in early embryonic life ($\alpha_2\xi_2$). The γ genes are expressed in fetal life, making fetal hemoglobin (HbF, $\alpha_2\gamma_2$). Adult hemoglobin consists of HbA ($\alpha_2\beta_2$) or HbA$_2$ ($\alpha_2\delta_2$). The $\psi\beta$ is a pseudogene that has sequence homology with β but contains mutations that prevent its expression. Deletions (solid bar) of the β locus cause β-thalassemia (deficiency or absence [β^0] of β-globin). A deletion of δ and β causes hemoglobin Lepore (only hemoglobin α is present). An inversion (A$\gamma\delta\beta$)^0 in this region (open bar) disrupts gene function and also results in thalassemia (type III). Each type of thalassemia tends to be found in a certain group of people, eg, the (A$\gamma\delta\beta$)^0 deletion inversion occurs in persons from India. Many more deletions in this region have been mapped, and each causes some type of thalassemia.

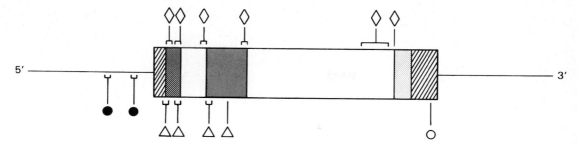

Figure 36–8. Mutations in the β-globin gene causing β-thalassemia. The β-globin gene is shown in the 5′ to 3′ orientation. The cross-hatched areas indicate the 5′ and 3′ nontranslated regions. Reading from the 5′ to 3′ direction, the shaded areas are exons 1–3 and the clear spaces are introns 1 and 2. Mutations that affect transcription control (●) are located in the 5′ flanking-region DNA. Examples of nonsense mutations (Δ), RNA processing mutations (◇), and RNA cleavage mutations (○) have been identified and are indicated. In some regions, many mutations have been found. These are indicated by a bracket (⌐⊥⌐).

Fig 6–20). The altered codon specifies a different amino acid (valine rather than glutamic acid), and this causes a structural abnormality of the β-globin molecule. Other point mutations in and around the β-globin gene result in decreased or, in some instances, no production of β-globin; β-thalassemia is the result of these mutations. (The thalassemias are characterized by defects in the synthesis of hemoglobin subunits, and so β-thalassemia results when there is insufficient production of β-globin.) Fig 36–8 illustrates that point mutations affecting each of the many processes involved in generating a normal mRNA (and therefore a normal protein) have been implicated as a cause of β-thalassemia.

Deletions, Insertions, & Rearrangements of DNA

Studies of bacteria, viruses, yeasts, and fruit flies show that pieces of DNA can move from one place to another within a genome. The deletion of a critical piece of DNA, the rearrangement of DNA within a gene, or the insertion of a piece of DNA within a coding or regulatory region can all cause changes in gene expression resulting in disease. Again, a molecular analysis of β-thalassemia produces numerous examples of these processes, particularly deletions, as a cause of disease (Fig 36–7). The globin gene clusters seem particularly prone to this lesion. Deletions in the α-globin cluster, located on chromosome 16, cause α-thalassemia. There is a strong ethnic association for many of these deletions, so that northern Europeans, Filipinos, blacks, and Mediterranean peoples have different lesions all resulting in the absence of hemoglobin A and α-thalassemia.

A similar analysis could be made for a number of other diseases. Point mutations are usually defined by sequencing the gene in question, though occasionally, if the mutation destroys or creates a restriction enzyme site, the technique of restriction fragment analysis can be used to pinpoint the lesion. Deletions or insertions of DNA larger than 50 bp can often be detected by the Southern blotting procedure.

Pedigree Analysis

Sickle cell disease again provides an excellent example of how recombinant DNA technology can be applied to the study of human disease. The substitution of T for A in the coding strand of DNA in the β-globin gene changes the sequence in the region that corresponds to the sixth codon from

$$\downarrow$$
G G A C Ⓣ C C   **coding strand**
C C T G A G G   **noncoding strand**
$$\uparrow$$

to

G G A C Ⓐ C C   **coding strand**
C C T G T G G   **noncoding strand**

and destroys a recognition site for the restriction enzyme Mst II (CCTNAGG; denoted by the small vertical arrows; Table 36–1). Other Mst II sites 5′ and 3′ from this site (Fig 36–9) are not affected and so will be cut. Therefore, incubation of DNA from normal (AA), heterozygous (AS), and homozygous (SS) individuals results in 3 different patterns on Southern blot transfer (Fig 36–9). This illustrates how a DNA pedigree can be established using the principles discussed in this chapter. Pedigree analysis has been applied to a number of genetic diseases and is most useful in those caused by deletions and insertions or the rarer instances in which a restriction endonuclease cleavage site is affected, as in the example cited in this paragraph.

Prenatal Diagnosis

If the genetic lesion is understood and a specific probe is available, prenatal diagnosis is possible. DNA from cells collected from as little as 10 mL of amniotic fluid (or by chorionic villus biopsy) can be analyzed by Southern blot transfer. A fetus with the restriction pattern AA in Fig 36–9 does not have sickle cell disease, nor is it a carrier. A fetus with the SS pattern will develop the disease. Probes are now available for this type of analysis of many genetic diseases.

A. Mst II restriction sites around and in the β-globin gene

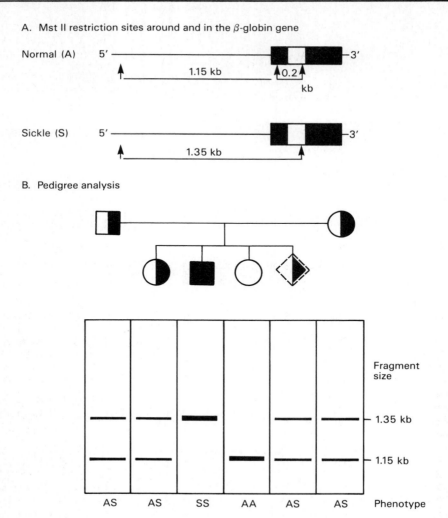

B. Pedigree analysis

Figure 36–9. Pedigree analysis of sickle cell disease. The top part of the figure (**A**) shows the first part of the β-globin gene and the Mst II restriction enzyme sites (↑) in the normal (A) and sickle cell (S) β-globin genes. Digestion with the restriction enzyme Mst II results in DNA fragments 1.15 kb and 0.2 kb long in normal individuals. The T-to-A change in individuals with sickle cell disease abolishes one of the 3 Mst II sites around the β-globin gene; hence, a single restriction fragment 1.35 kb in length is generated in response to Mst II. This size difference is easily detected on a Southern blot (**B**). (The 0.2-kb fragment would run off the gel in this illustration.) Pedigree analysis shows 3 possibilities: AA = normal (0); AS = heterozygous (◑,◨); SS = homozygous (■). This approach allows for prenatal diagnosis of sickle cell disease or trait (◈).

Restriction Fragment Length Polymorphism (RFLP)

The differences in DNA sequence cited above can result in variations of restriction sites and thus in the length of restriction fragments. Inherited differences in the pattern of restriction (eg, a DNA variation occurring in more than 1% of the general population) is known as restriction fragment length polymorphism, or RFLP. RFLPs result from single-base changes (eg, sickle cell disease) or from deletions or insertions of DNA into a restriction fragment (eg, the thalassemias) and are proving to be a useful diagnostic tool. They have been found at known gene loci and in sequences that have no known function; thus, RFLPs may disrupt

the function of the gene or may have no biologic consequences.

RFLPs are inherited, and they segregate in a mendelian fashion. A major use of RFLPs (nearly 350 are now known) is in the definition of inherited diseases in which the functional deficit is unknown. RFLPs can be used to establish linkage groups, which in turn, by the process of **chromosome walking,** will eventually define the disease locus. In chromosome walking (Fig 36–10), a fragment representing one end of a long piece of DNA is used to isolate another that overlaps but extends the first. The direction of extension is determined by restriction mapping, and the procedure is repeated sequentially until the desired se-

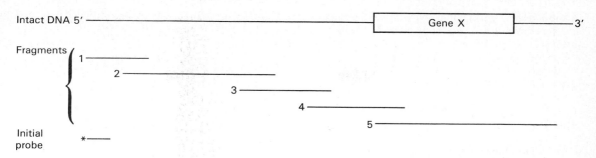

Figure 36–10. The technique of chromosome walking. Gene X is to be isolated from a large piece of DNA. The exact location of this gene is not known, but a probe (*——) directed against a fragment of DNA (shown at the 5′ end in this representation) is available, as is a library containing a series of overlapping DNA fragments. For the sake of simplicity, only 5 of these are shown. The initial probe will hybridize only with clones containing fragment 1, which can then be isolated and used as a probe to detect fragment 2. This procedure is repeated until fragment 4 hybridizes with fragment 5, which contains the entire sequence of gene X.

quence is obtained. The X chromosome-linked disorders are particularly amenable to this approach, since only a single allele is expressed. Hence, 20% of the defined RFLPs are on the X chromosome, and a reasonably complete linkage map of this chromosome exists. X-linked disorders such as Duchenne type muscular dystrophy will be defined using RFLPs. Likewise, the defect in Huntington's chorea has been localized to the terminal region of the short arm of chromosome 4, and the defect that causes polycystic kidney disease is linked to the α-globin locus on chromosome 16.

Gene Therapy

Diseases caused by deficiency of a gene product (Table 36–5) are amenable to replacement therapy. The strategy is to clone a gene (eg, the gene that codes for adenosine deaminase) into a vector that will readily be taken up and incorporated into the genome of a host cell. Bone marrow precursor cells are being investigated for this purpose because they presumably will resettle in the marrow and replicate there. The introduced gene would begin to direct the expression of its protein product, and this would correct the deficiency in the host cell.

This somatic cell gene replacement would obviously not be passed on to offspring. Other strategies to alter germ cell lines have been devised but have been tested only in experimental animals. A certain percentage of genes injected into a fertilized mouse ovum will be incorporated into the genome and found in both somatic and germ cells. These **transgenic animals** are proving to be useful for analysis of tissue-specific effects on gene expression and effects of overproduction of gene products (eg, those from the growth hormone gene or oncogenes) and in discovering genes involved in development, a process that heretofore has been difficult to study. The transgenic approach has recently been used to correct a genetic deficiency in mice. Fertilized ova obtained from mice with genetic hypogonadism (see Chapter 55) were injected with DNA containing the coding sequence for the gonadotropin-releasing hormone (GnRH) precursor protein. This gene was expressed and regulated normally in the hypothalamus of a certain number of the resultant mice, and these animals were in all respects normal. Their offspring also showed no evidence of GnRH deficiency. This is, therefore, evidence of somatic cell expression of the transgene and of its maintenance in germ cells.

GLOSSARY

Autoradiography: The detection of radioactive molecules (eg, DNA, RNA, protein) by visualization of their effects on photographic film.

Bacteriophage: A virus that infects a bacterium.

Blunt-ended DNA: Two strands of a DNA duplex having ends that are flush with each other.

cDNA: A single-stranded DNA molecule that is complementary to an mRNA molecule and is synthesized from it by the action of reverse transcriptase.

Chimeric molecule: A molecule (eg, DNA, RNA, protein) containing sequences derived from 2 different species.

Clone: A large number of cells or molecules that are identical with a single parental cell or molecule.

Cosmid: A plasmid into which the DNA sequences from bacteriophage lambda that are necessary for the packaging of DNA (cos sites) have been inserted; this permits the plasmid DNA to be packaged in vitro.

Endonuclease: An enzyme that cleaves internal bonds in DNA or RNA.

Exon: The sequence of a gene that is represented (expressed) as mRNA.

Exonuclease: An enzyme that cleaves nucleotides from either the 3' or 5' ends of DNA or RNA.

Hairpin: A double-helical stretch formed by base pairing between neighboring complementary sequences of a single strand of DNA or RNA.

Hybridization: The specific reassociation of complementary strands of nucleic acids (DNA with DNA, DNA with RNA, or RNA with RNA).

Insert: An additional length of base pairs in DNA, generally introduced by the techniques of recombinant DNA technology.

Intron: The sequence of a gene that is transcribed but excised before translation.

Library: A collection of cloned fragments that represents the entire genome. Libraries may be either genomic DNA (in which both introns and exons are represented) or cDNA (in which only exons are represented).

Ligation: The enzyme-catalyzed joining in phosphodiester linkage of 2 stretches of DNA or RNA into one; the respective enzymes are DNA and RNA ligases.

Nick translation: A technique for labeling DNA based on the ability of the DNA polymerase from *E coli* to degrade a strand of DNA that has been nicked and then to resynthesize the strand; if a radioactive nucleoside triphosphate is employed, the rebuilt strand becomes labeled and can be used as a radioactive probe.

Northern blot: A method for transferring RNA from an agarose gel to a nitrocellulose filter, on which the RNA can be detected by a suitable probe.

Palindrome: A sequence of duplex DNA that is the same when the 2 strands are read in opposite directions.

Plasmid: A small, extrachromosomal, circular molecule of DNA that replicates independently of the host DNA.

Probe: A molecule used to detect the presence of a specific fragment of DNA or RNA in, for instance, a bacterial colony that is formed from a genetic library or during analysis by blot transfer techniques; common probes are cDNA molecules, synthetic oligodeoxynucleotides of defined sequence, or antibodies to specific proteins.

Pseudogene: An inactive segment of DNA arising by mutation of a parental active gene.

Recombinant DNA: The altered DNA that results from the insertion of a sequence of deoxynucleotides not previously present into an existing molecule of DNA by enzymatic or chemical means.

Restriction enzyme: An endodeoxynuclease that causes cleavage of both strands of DNA at highly specific sites dictated by the base sequence.

Reverse transcription: RNA-directed synthesis of DNA, catalyzed by reverse transcriptase.

Signal: The end product observed when a specific sequence of DNA or RNA is detected by autoradiography or some other method. Hybridization with a complementary radioactive polynucleo-tide (eg, by Southern or Northern blotting) is commonly used to generate the signal.

Southern blot: A method for transferring DNA from an agarose gel to a nitrocellulose filter, on whch the DNA can be detected by a suitable probe (eg, complementary DNA or RNA).

Splicing: The removal of introns from RNA accompanied by the joining of its exons.

Sticky-ended DNA: Complementary single strands of DNA that protrude from opposite ends of a DNA duplex or from the ends of different duplex molecules (see also Blunt-ended DNA, above).

Tandem: Used to describe multiple copies of the same sequence (eg, DNA) that lie adjacent to one another.

Terminal transferase: An enzyme that adds nucleotides of one type (eg, deoxyadenonucleotidyl residues) to the 3' end of DNA strands.

Transcription: DNA-directed synthesis of RNA.

Transgenic: Describing the introduction of new DNA into germ cells by its injection into the nucleus of the ovum.

Translation: Synthesis of protein using mRNA as template.

Vector: A plasmid or bacteriophage into which foreign DNA can be introduced for the purposes of cloning.

Western blot: A method for transferring protein to a nitrocellulose filter, on which the protein can be detected by a suitable probe (eg, an antibody).

REFERENCES

Beaudet AL: Bibliography of cloned human and other selected DNAs. *Am J Hum Genet* 1985;**37**:386.

DNA in medicine. *Lancet* 1984;**2**:853, 908, 966, 1022, 1086, 1138, 1194, 1257, 1329, 1380, 1440.

Gusella JF: Recombinant DNA techniques in the diagnosis and treatment of inherited disorders. *J Clin Invest* 1986;**77**:1723.

Kan YW et al: Pages 275–283 in: *Thalassemia: Recent Advances in Detection and Treatment.* Cao A, Carcassi U, Rowley P (editors). AR Liss, 1982.

Lewin B: *Genes II.* Wiley, 1985.

Maniatis T, Fritsch EF, Sambrook J: *Molecular Cloning,* 2nd ed. Cold Spring Harbor Laboratory, 1983.

Martin JB, Gusella JF: Huntington's disease: Pathogenesis and management. *N Engl J Med* 1986;**315**:1267.

Maxam AM: Sequencing the DNA of recombinant chromosomes. *Fed Proc* 1980;**39**:2830.

Orkin SH et al: Improved detection of the sickle mutation by DNA analysis: Application to prenatal diagnosis. *N Engl J Med* 1982;**307**:32.

Watson JD, Tooze J, Kurtz DT: *Recombinant DNA: A Short Course.* Freeman, 1983.

Weatherall DJ: *The New Genetics and Clinical Practice,* 2nd ed. Oxford Univ Press, 1986.

Yuan R: Structure and mechanism of multifunctional restriction endonucleases. *Annu Rev Biochem* 1981;**50**:285.

Nucleic Acid Structure & Function 37

Daryl K. Granner, MD

INTRODUCTION

The discovery that genetic information is coded along the length of a polymeric molecule composed of only 4 types of monomeric units is one of the major scientific achievements of this century. This polymeric molecule, **DNA, is the chemical basis of heredity** and is organized into **genes,** the fundamental units of genetic information. Genes control the synthesis of various types of RNA, most of which are involved in protein synthesis. Genes do not function autonomously; their replication and function are controlled, in still vaguely understood ways, by feedback loops in which gene products themselves play a critical role. Knowledge of the structure and function of nucleic acids is essential in understanding genetics and provides the basis for future research.

BIOMEDICAL IMPORTANCE

The chemical basis of heredity and of genetic diseases is found in the structure of DNA. The basic information pathway (ie, DNA directs the synthesis of RNA, which in turn regulates protein synthesis) has been elucidated. This knowledge is being used to define normal cellular physiology and the pathophysiology of disease at the molecular level.

DNA

The demonstration that DNA contained the genetic information was first made in 1944 in a series of experiments by Avery, MacLeod, and McCarty, who showed that the genetic determination of the character (type) of the capsule of a specific pneumococcus could be transmitted to another of a distinctly different capsular type by introducing purified DNA from the former coccus into the latter. These authors referred to the agent (DNA) accomplishing the change as "transforming factor." Subsequently, this type of genetic manipulation has become commonplace. Similar experiments have recently been performed utilizing yeast, cultured mammalian cells, and insect and rodent embryos as recipients, and cloned DNA as the donor of genetic information.

Chemical Nature of DNA

The chemical nature of the monomeric units of DNA—**deoxyadenylate, deoxyguanylate, deoxycytidylate,** and **thymidylate**—is described in Chapter 34. These monomeric units of DNA are held in polymeric form by $3',5',$-phosphodiester bridges constituting a single strand, as depicted in Fig 37–1. The informational content of DNA (the genetic code) resides in the sequence in which these monomers—purine and pyrimidine deoxyribonucleotides—are ordered. The polymer as depicted possesses a polarity; one end has a $5'$-hydroxyl or phosphate terminus while the other has a $3'$-phosphate or hydroxyl moiety. The importance of this polarity will become evident. Since the genetic information resides in the order of the monomeric units within the polymers, there must exist a mechanism of reproducing or replicating this specific information with a high degree of fidelity. That requirement, together with x-ray diffraction data from the DNA molecule and the observation of Chargaff that in DNA molecules the concentration of deoxyadenosine (A) nucleotides equals that of thymidine (T) nucleotides ($A = T$), while the concentration of deoxyguanosine (G) nucleotides equals that of deoxycytidine (C) nucleotides ($G = C$), led Watson, Crick, and Wilkins to propose in the early 1950s a model of a double-stranded DNA molecule. A model of the B form of DNA is depicted in Fig 37–2. The 2 strands of this right-handed, double-stranded molecule are held together by **hydrogen bonds** between the purine and pyrimidine bases of the respective linear molecules. The pairings between the purine and pyrimidine nucleotides on the opposite strands are very specific and are dependent upon hydrogen bonding of **A with T, and G with C** (Fig 37–3).

In the double-stranded molecule, restrictions imposed by the rotation about the phosphodiester bond, the favored **anti** configuration of the glycosidic bond (see Fig 34–9), and the predominant tautomers (see Fig 34–4) of the 4 bases (A, G, T, and C) allow A to pair only with T, and G only with C, as depicted in Fig 37–3. This base-pairing restriction explains the earlier observation that in a double-stranded DNA molecule the content of A equals that of T and the content of G equals that of C. The 2 strands of the double-helical molecule, each of which possesses a polarity, are **antiparallel;** ie, one strand runs in the $5'$ to $3'$ direction and the other in the $3'$ to $5'$ direction. This is analogous to 2 parallel streets, each running one way but carrying

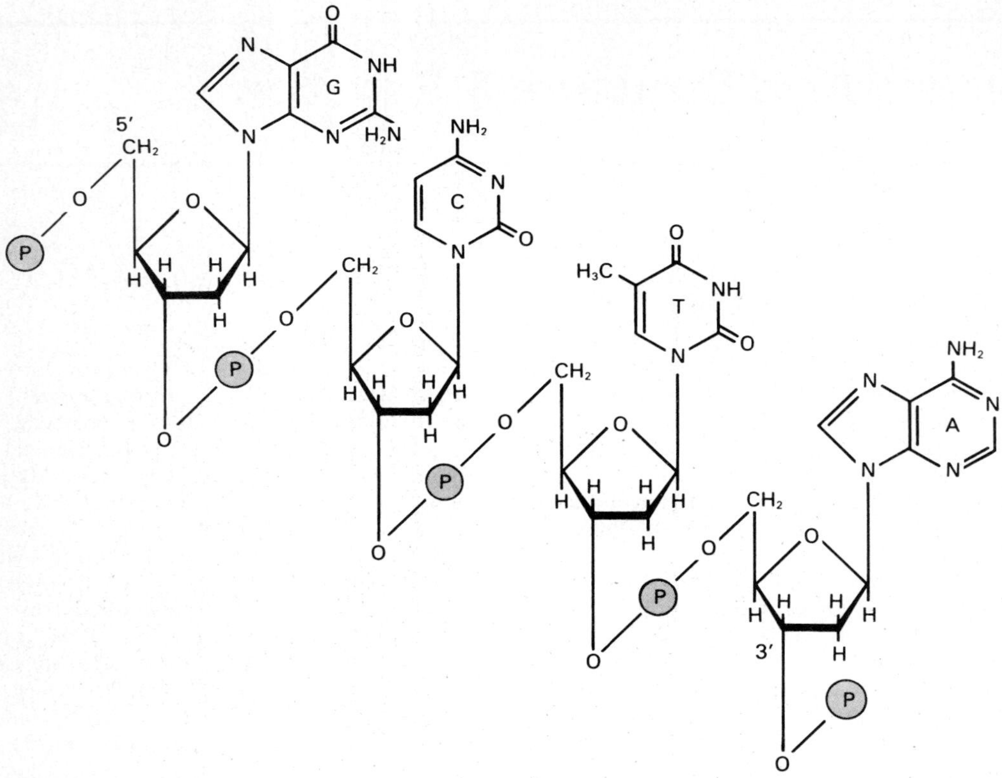

Figure 37–1. A segment of a structure of DNA molecule in which the purine and pyrimidine bases adenine (A), thymine (T), cytosine (C), and guanine (G) are held together by a phosphodiester backbone between 2'-deoxyribosyl moieties attached to the nucleobases by an N-glycosidic bond. Note that the backbone has a polarity (ie, a direction).

traffic in opposite directions. In the double-stranded DNA molecules, since the information resides in the sequence of nucleotides on one strand, the **coding strand (template),** the opposite strand is considered the **noncoding strand.**

As depicted in Fig 37–3, 3 hydrogen bonds hold the deoxyguanosine nucleotide to the deoxycytidine nucleotide, whereas the other pair, the A-T pair, is held together by 2 hydrogen bonds. Thus, the G-C bond is stronger by approximately 50%. Because of this added strength and also because of stacking interactions, regions of DNA that are rich in G-C bonds are much more resistant to denaturation, or "melting," than A-T-rich regions.

Structures of DNA

DNA exists in several double-helical structures. To date 6 forms (A to E and Z) have been described, but most of these have been found only under rigidly controlled experimental conditions. These forms are distinguished by (1) the number of base pairs that occupy each turn of the helix, (2) the pitch or angle between each base pair, (3) the helical diameter of the molecule, and (4) the handedness (right or left) of the double helix (Table 37–1). Some of these forms interconvert if salt and hydration conditions are manipulated.

It is possible that interconversion might happen in vivo.

The **B form,** the overwhelmingly dominant form of DNA under physiologic conditions (low salt, high degree of hydration), has a pitch of 3.4 nm per turn (Fig 37–2). Within a single turn, 10 base pairs (bp) exist, each planar base being stacked to resemble 2 winding stacks of coins side by side. The 2 stacks are held together by hydrogen bonding at each level between the 2 coins on opposite stacks and by 2 ribbons wound in a right-hand turn about the 2 stacks and representing the phosphodiester backbone.

A variation of this basic structure is seen in the A form, which is favored by an environment slightly less hydrous and richer in Na^+ or K^+ ions. This right-handed structure is bulkier than the B form, has more base pairs per turn, and resembles the structure of double-stranded RNA and DNA-RNA hybrid duplexes. The C-E forms, also right-handed, are seen in very special experimental circumstances and are not thought to exist in vivo.

Z-DNA forms a left-handed double helix in which the phosphodiester backbone zigzags along the molecule; hence, the name Z-DNA. Z-DNA is the least twisted (12 bp per turn) and skinniest DNA helix known to exist, and it has only one groove (see be-

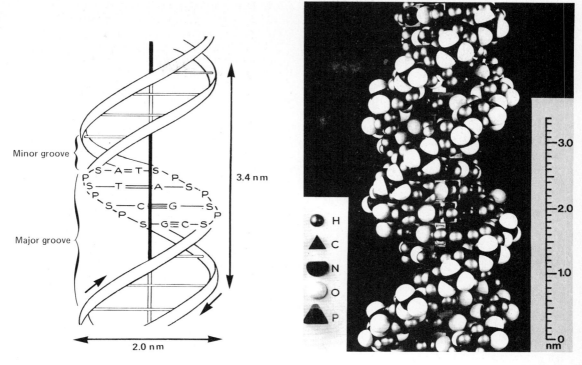

Figure 37–2. The Watson and Crick model of the double-helical structure of the B form of DNA. *Left:* Diagrammatic representation of structure (modified). (A, adenine; C, cytosine; G, guanine; T, thymine; P, phosphate; S, sugar [deoxyribose].) *Right:* Space-filling model of DNA structure. (Photograph from James D. Watson, *Molecular Biology of the Gene,* 3rd ed. Copyright © 1976, 1970, 1965, by W.A. Benjamin, Inc., Menlo Park, Calif.)

low). Z-DNA occurs in repeated sequences of alternating purine and pyrimidine deoxynucleotides (GC or AC) but also requires one or more stabilizing influences. These stabilizing influences include (1) the presence of high-salt or specific cations such as spermine or spermidine, (2) a high degree of negative supercoiling of the DNA (see Chapter 38), (3) the binding of Z-DNA-specific proteins, and (4) the methylation of the 5-carbon of some of the deoxycytidine nucleotides in the alternating sequence.

Z-DNA could exert regulatory effects both proximal and distal to the site of its existence. For instance, some proteins that bind in the minor or major groove of B-form DNA could probably not bind to the Z form. In addition, the reversion of Z form to a B form of DNA, an event that might occur as a consequence of loss of methyl groups from 5-methyldeoxycytidine, would likely result in torsional differences of DNA distal to the actual site of Z-DNA. As discussed below, torsional winding and unwinding are thought to affect gene activity, as is the methylation of deoxycytidine.

The existence of Z-DNA in *Drosophila* (fruit fly) chromosomes has been demonstrated utilizing antibodies that recognize and bind specifically to Z-DNA. Human DNA contains potential Z-DNA-forming regions dispersed throughout the genome, and the stabilizing influences may also exist.

Denaturation (Melting) of DNA

The double-stranded structure of DNA can be melted in solution by increasing the temperature or decreasing the salt concentration. Not only do the 2 stacks of bases pull apart but the bases themselves unstack while still connected in the polymer by the phosphodiester backbone. Concomitant with this denaturation of the DNA molecule is an increase in the optical absorbance of the purine and pyrimidine bases—a phenomenon referred to as **hyperchromicity** of denaturation. Because of the stacking of the bases and the hydrogen bonding between the stacks, the double-stranded DNA molecule exhibits properties of a fiber and in solution is a viscous material that loses its viscosity upon denaturation.

The strands of a given molecule of DNA separate over a temperature range. The midpoint is called the **melting temperature,** or T_m. The T_m is influenced by the base composition of the DNA and by the salt concentration of the solution. DNA rich in G-C pairs, which have 3 hydrogen bonds, melts at a higher temperature than that rich in A-T pairs, which have 2 hydrogen bonds. A 10-fold increase of monovalent cation concentration increases the T_m by 16.6 °C. Formamide, which is commonly used in recombinant DNA experiments, destabilizes hydrogen bonding between bases, thereby lowering the T_m. This allows the

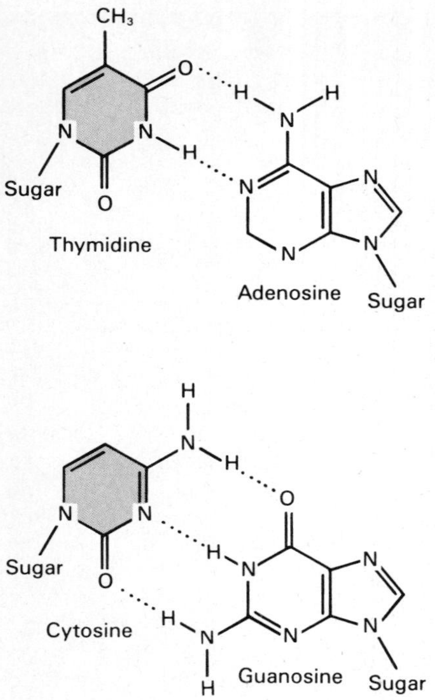

Thymidine

Adenosine Sugar

Cytosine

Guanosine Sugar

Figure 37-3. Base pairing between deoxyadenosine and thymidine involves the formation of 2 hydrogen bonds. Three such bonds form between deoxycytidine and deoxyguanosine. The broken lines represent hydrogen bonds. In DNA, the sugar moiety is 2-deoxyribose, whereas in RNA it is D-ribose.

strands of DNA or DNA-RNA hybrids to be separated at much lower temperatures and minimizes the strand breakage that occurs at high temperatures.

Grooves in the DNA Molecule

Careful examination of the model depicted in Fig 37-2 reveals a **major groove** and a **minor groove** winding along the molecule parallel to the phosphodiester backbones. In these grooves, proteins can interact specifically with exposed atoms of the nucleotides and thereby recognize and bind to specific nucleotide sequences without disrupting the base pairing of the double-helical DNA molecule. As discussed in Chapters 39 and 41, regulatory proteins can control the expression of specific genes via such interactions.

Relaxed & Supercoiled DNA

In some organisms such as bacteria, bacteriophages, and many DNA-containing animal viruses, the 2 ends of the DNA molecules are joined to create a **closed circle** with no terminus. This of course does not destroy the polarity of the molecules, but it eliminates all free 3' and 5' hydroxyl and phosphoryl groups. Closed circles exist in relaxed or supercoiled forms. Supercoils are introduced when a closed circle is twisted around its own axis or when a linear piece of duplex DNA, whose ends are fixed, is twisted. This

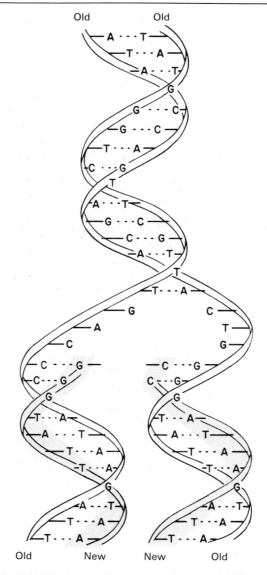

Figure 37-4. The double-stranded structure of DNA and the template function of each old strand on which a new complementary strand (shaded) is synthesized. (From James D. Watson, *Molecular Biology of the Gene,* 3rd ed. Copyright © 1976, 1970, 1965, by W.A. Benjamin, Inc., Menlo Park, Calif.)

Table 37-1. Features of some structures of DNA.

Type	Handedness	Base Pairs per Turn	Pitch per Base Pair	Helical Diameter
A	Right	11	0.256 nm	2.3 nm
B	Right	10	0.338 nm	1.9 nm
Z	Left	12	0.371 nm	1.8 nm

energy-requiring process puts the molecule under stress, and the greater the number of supercoils, the greater the stress or torsion (test this with a rubber band). **Negative supercoils** are formed when the molecule is twisted in the direction opposite from the clockwise turns of the right-handed double helix found in B-DNA. Such DNA is said to be **underwound.** The energy required to achieve this state is, in a sense, stored in the supercoils. The transition to another form that requires energy is thereby facilitated by the underwinding. One such transition is strand separation, which is a prerequisite of DNA replication and transcription. Supercoiled DNA is therefore a preferred form in biologic systems. Enzymes that catalyze topologic changes of DNA are called **topoisomerases.** The best characterized is **bacterial gyrase,** which induces negative supercoiling in DNA.

Function of DNA

The genetic information stored in the nucleotide sequence of DNA serves 2 purposes. It is the source of information for the synthesis of all protein molecules of the cell and organism, and it provides the information inherited by daughter cells or offspring. Both of these functions require that the DNA molecule serve as a template—in the first case for the **transcription** of the information into RNA and in the second case for the **replication** of the information into daughter DNA molecules.

The complementarity of the Watson and Crick dou-

ble-stranded model of DNA strongly suggests that **replication of the DNA molecule occurs in a semiconservative manner.** Thus, when each strand of the double-stranded parental DNA molecule separates from its complement during replication, each serves as a template on which a new complementary strand is synthesized (Fig 37–4). The 2 newly formed double-stranded daughter DNA molecules, each containing one strand (but complementary rather than identical) from the parent double-stranded DNA molecule, are then sorted between the 2 daughter cells (Fig 37–5). Each daughter cell contains DNA molecules with information identical to that which the parent possessed; yet in each daughter cell the DNA molecule of the parent cell has been only semiconserved.

The semiconservative nature of DNA replication in the bacterium *Escherichia coli* was unequivocally demonstrated by Meselson and Stahl in a classic experiment using the heavy isotope of nitrogen and centrifugal equilibrium techniques. The DNA of *E coli* is chemically identical to that of humans, although the sequences of nucleotides are, of course, different, and the human cell contains about 1000 times more DNA per cell than does the bacterium. Furthermore, the chemistry of replication of DNA in prokaryotes such as *E coli* appears to be identical to that in eukaryotes, including humans, even though the enzymes carrying out the reactions of DNA synthesis and replication are different. Thus, any observations on the chemical nature or chemical reactions of nucleic acids of pro-

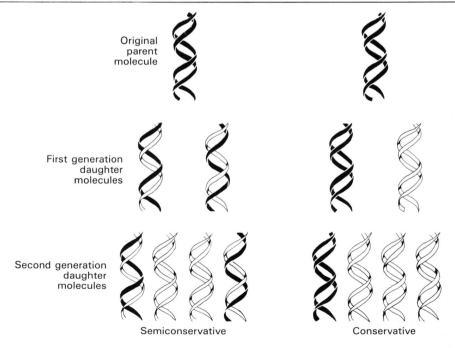

Figure 37–5. The expected distributions of parental DNA strands during semiconservative and conservative replication. The parental strands are solid, and the newly synthesized strands are open. (Redrawn and reproduced, with permission, from Lehninger AL: *Biochemistry,* 2nd ed. Worth, 1975.)

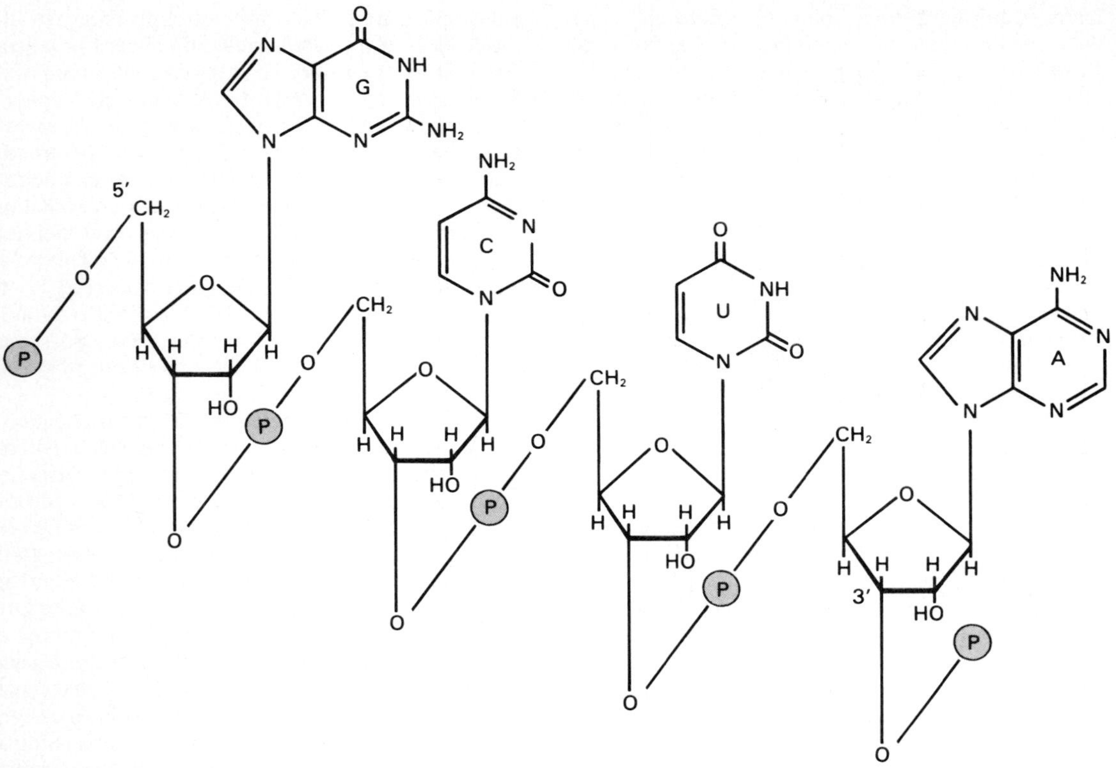

Figure 37–6. A segment of a ribonucleic acid (RNA) molecule in which the purine and pyrimidine bases—adenine (A), uracil (U), cytosine (C), and guanine (G)—are held together by phosphodiester bonds between ribosyl moieties attached to the nucleobases by N-glycosidic bonds. Note that the polymer has a polarity as indicated by the labeled 3'- and 5'- attached phosphates.

karyotes are very likely applicable to eukaryotic organisms. Indeed, the Meselson and Stahl type of experiment has now been performed in mammalian cells and has yielded results comparable to those obtained with *E coli*.

RNA

Chemical Nature of RNA

Ribonucleic acid (RNA) is a polymer of purine and pyrimidine ribonucleotides linked together by 3',5'-phosphodiester bridges analogous to those in DNA (Fig 37–6). Although sharing many features with DNA, RNA possesses several specific differences:

(1) In RNA, the sugar moiety to which the phosphates and purine and pyrimidine bases are attached is ribose rather than the 2'-deoxyribose of DNA.

(2) The pyrimidine components of RNA differ from those of DNA. Although RNA contains the ribonucleotides of adenine, guanine, and cytosine, it does not possess thymine except in the rare case mentioned below. Instead of thymine, **RNA contains the ribonucleotide of uracil.**

(3) RNA exists as a single stand, whereas DNA

exists as a double-stranded helical molecule. However, given the proper complementary base sequence with opposite polarity, the single strand of RNA, as demonstrated in Fig 37–7, is capable of folding back on itself like a hairpin and thus acquiring double-stranded characteristics.

(4) Since the RNA molecule is a single strand complementary to only one of the 2 strands of a gene, its **guanine content does *not* necessarily equal its cytosine content,** nor does its adenine content necessarily equal its uracil content.

(5) RNA can be hydrolyzed by alkali to 2',3' cyclic diesters of the mononucleotides. A necessary intermediate in this hydrolysis is the 2',3',5'-triester, an intermediate that cannot be formed in alkali-treated DNA because of the absence of a 2'-hydroxyl group. The alkali lability of RNA is useful both diagnostically and analytically.

Information within the single strand of RNA is contained in its sequence ("primary structure") of purine and pyrimidine nucleotides within the polymer. The sequence is complementary to the coding strand of the gene from which it was transcribed. Because of this complementarity, an RNA molecule will bind specifically via the base-pairing rules to its coding DNA strand; it will not bind ("hybridize") with the

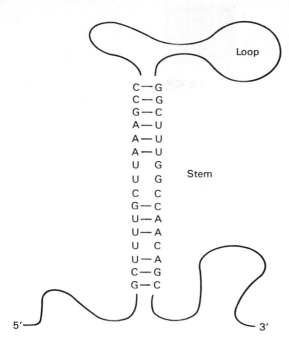

Figure 37–7. Diagrammatic representation of the secondary structure of a single-stranded RNA molecule in which a stem loop, or "hairpin," has been formed and is dependent upon the intramolecular base pairing.

other (noncoding) strand of its gene. The sequence of the RNA molecule (except for U replacing T) is the same as that of the noncoding strand of the gene (Fig 37–8).

Biologic Function of RNA

Nearly all of the several species of RNA are involved in some aspect of protein synthesis. Those cytoplasmic RNA molecules that serve as templates for protein synthesis are designated **messenger RNA,** or **mRNA.** Many other cytoplasmic RNA molecules (**ribosomal RNA,** or **rRNA**) have structural roles wherein they contribute to the formation of ribosomes (the organellar machinery for protein synthesis) or serve as adapter molecules (**tRNA**) for the translation of RNA information into specific sequences of polymerized amino acids.

Much of the RNA synthesized from DNA templates in eukaryotic cells, including mammalian cells, is **degraded within the nucleus,** and it never serves as either a structural or an informational entity within the cellular cytoplasm.

In cultured human cells there are **small nuclear RNA (snRNA)** species which are not directly involved in protein synthesis but which may have roles in RNA processing and the cellular architecture. These relatively small molecules vary in size from 90 to about 300 nucleotides (Table 37–3).

The genetic material for some animal and plant viruses is RNA rather than DNA. Although some RNA viruses do not ever have their information transcribed into a DNA molecule, many animal RNA viruses—specifically the retroviruses—are transcribed by an **RNA-dependent DNA polymerase, reverse transcriptase,** to produce a double-stranded DNA copy of their RNA genome. In many cases, the resulting double-stranded DNA transcript is integrated into the host genome and subsequently serves as a template for gene expression and from which new viral RNA genomes can be transcribed.

Structural Organization of RNA

In all prokaryotic and eukaryotic organisms, 3 main classes of RNA molecules exist: messenger RNA (mRNA), transfer RNA (tRNA), and ribosomal RNA (rRNA). Each class differs from the others by size, function, and general stability.

The **messenger RNA (mRNA)** class is the most heterogeneous in size and stability. All of the members of the class function as messengers conveying the information in a gene to the protein-synthesizing machinery, where each serves as a template on which a specific sequence of amino acids is polymerized to form a specific protein molecule, the ultimate gene product (Fig 37–9).

Messenger RNAs, particularly in eukaryotes, have some unique chemical characteristics. The 5' terminus of mRNA is "capped" by a 7-methylguanosine triphosphate that is linked to an adjacent 2'-O-methyl ribonucleoside at its 5'-hydroxyl through the 3 phosphates (Fig 37–10). The mRNA molecules frequently contain internal 6-methyladenylates and other 2'-O-ribose methylated nucleotides. Although the function of this capping of mRNAs is not completely understood, the cap is probably involved in the recognition of

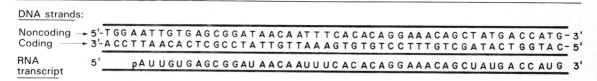

DNA strands:

Noncoding → 5'-T G G A A T T G T G A G C G G A T A A C A A T T T C A C A C A G G A A A C A G C T A T G A C C A T G- 3'
Coding ——→ 3'-A C C T T A A C A C T C G C C T A T T G T T A A A G T G T G T C C T T T G T C G A T A C T G G T A C- 5'

RNA   5'
transcript   pA U U G U G A G C G G A U A A C A A U U U C A C A C A G G A A A C A G C U A U G A C C A U G   3'

Figure 37–8. The relationship between the sequences of an RNA transcript and its gene, in which the coding and noncoding strands are shown with their polarities. The RNA transcript with a 5' to 3' polarity is complementary to the coding strand with its 3' to 5' polarity. Note that the sequence in the RNA transcript and its polarity is the same as that in the noncoding strand, except that the U of the transcript replaces the T of the gene.

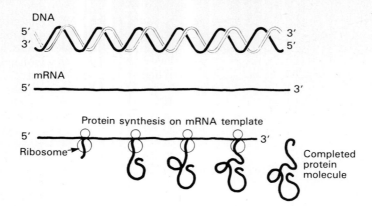

Figure 37–9. The expression of genetic information in DNA into the form of an mRNA transcript. This is subsequently translated by ribosomes into a specific protein molecule.

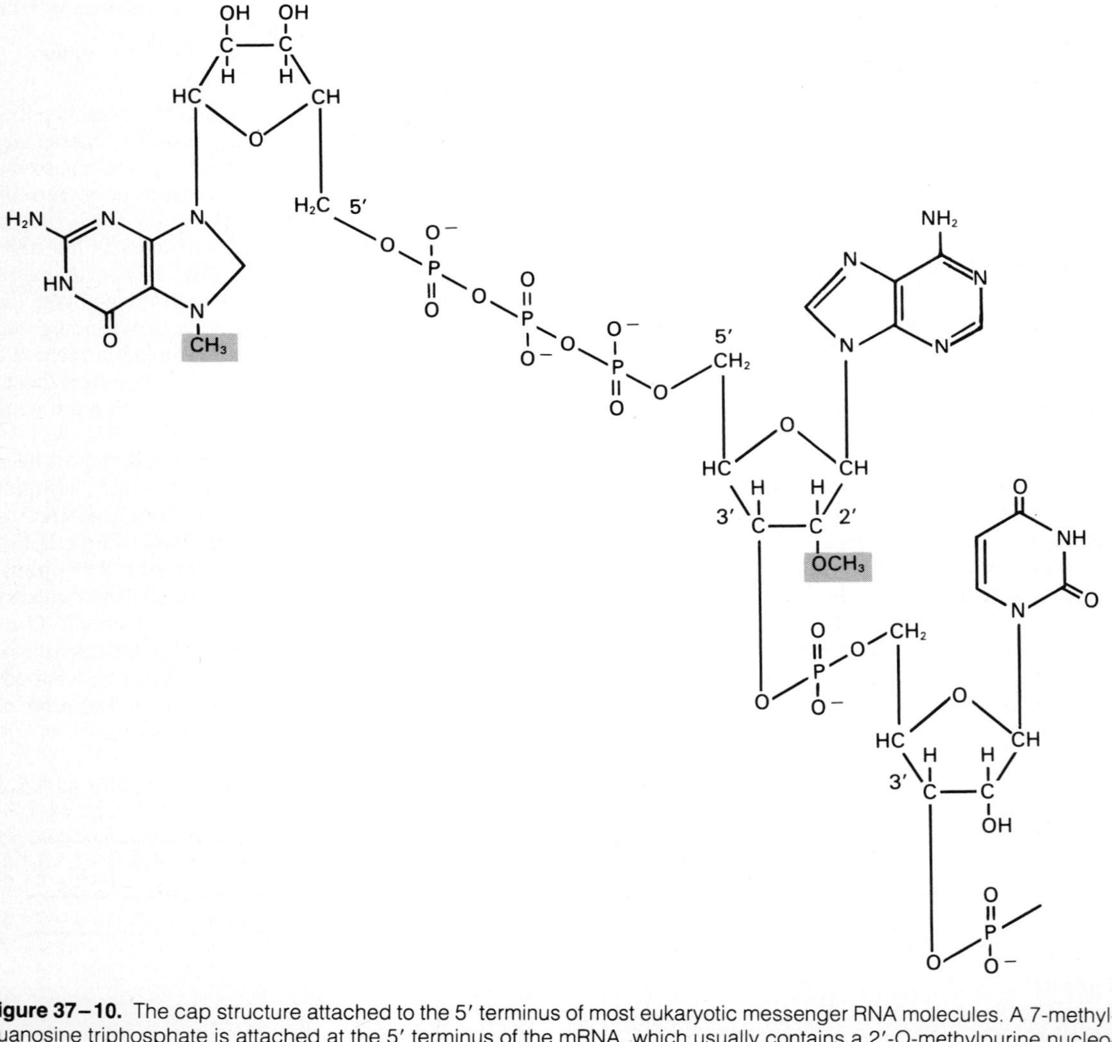

Figure 37–10. The cap structure attached to the 5′ terminus of most eukaryotic messenger RNA molecules. A 7-methylguanosine triphosphate is attached at the 5′ terminus of the mRNA, which usually contains a 2′-O-methylpurine nucleotide.

mRNA by the translating machinery. The protein-synthesizing machinery begins translating the mRNA into proteins at the 5' or capped terminus. The other end of most mRNA molecules, the 3'-hydroxyl terminus, has attached a polymer of adenylate residues 20–250 nucleotides in length. The specific function of the **poly(A) "tail"** at the 3'-hydroxyl terminus of mRNAs is not fully understood, but it seems that it maintains the intracellular stability of the specific mRNA. Some mRNAs, including those for some histones, do not contain poly(A). The poly(A) tail, because it will form a base pair with oligodeoxythymidine polymers attached to a solid substrate like cellulose, can be used to separate mRNA from other species of RNA.

In **mammalian cells,** including cells of humans, the mRNA molecules present in the cytoplasm are not the RNA products immediately synthesized from the DNA template but must be formed by processing from a precursor molecule before entering the cytoplasm. Thus, in mammalian nuclei, the immediate products of gene transcription constitute a fourth class of RNA molecules. These nuclear RNA molecules are very heterogeneous in size and are quite large. The **heterogeneous nuclear RNA (hnRNA)** molecules may have a MW in excess of 10^7, whereas the MW of mRNA molecules is generally less than 2×10^6. As is discussed in Chapter 39, hnRNA molecules are processed to generate the mRNA molecules which then enter the cytoplasm to serve as templates for protein synthesis.

Transfer RNA (tRNA) molecules consist of approximately 75 nucleotides and thus have a molecular weight of 25,000. They also are generated by nuclear processing of a precursor molecule (see Chapter 39). The tRNA molecules serve as adapters for the translation of the information in the sequence of nucleotides of the mRNA into specific amino acids. There are at least 20 species of tRNA molecules in every cell, at least one (and often several) corresponding to each of the 20 amino acids required for protein synthesis. Although each specific tRNA differs from the others in its sequence of nucleotides, the tRNA molecules as a class have many features in common. The primary structure—ie, the nucleotide sequence—of all tRNA molecules allows extensive folding and intrastrand complementarity to generate a secondary structure that appears like a **cloverleaf** (Fig 37–11).

All tRNA molecules contain 4 main arms. The **acceptor arm** consists of a base-paired stem that terminates in the sequence CCA (5' to 3'). It is through an ester bond to the 3'-hydroxyl group of the adenosyl moiety that the carboxyl groups of amino acids are attached. The other arms have base-paired stems and unpaired loops (Fig 37–7). The **anticodon arm** at the end of a base-paired stem recognizes the triplet nucleotide or codon (discussed in Chapter 40) of the template mRNA. The **D arm** is named for the presence of the base dihydrouridine, and the **TψC arm** for the sequence T, pseudouridine, and C. The **extra arm** is the most variable feature of tRNA, and it provides a basis for classification. Class 1 tRNAs (about 75% of all

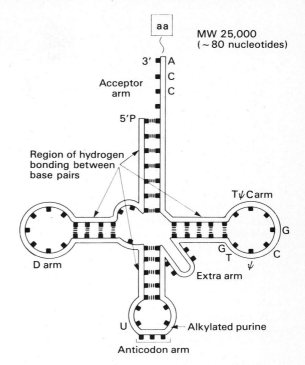

Figure 37–11. Typical aminoacyl tRNA in which the amino acid (aa) is attached to the 3' CCA terminus. The anticodon, TψC, and dihydrouracil (DHU) arms are indicated, as are the positions of the intramolecular hydrogen bonding between these base pairs. (From James D. Watson, *Molecular Biology of the Gene,* 3rd ed. Copyright © 1976, 1970, 1965, by W.A. Benjamin, Inc., Menlo Park, Calif.)

tRNAs) have an extra arm that is 3–5 bp long. Class 2 tRNAs have an extra arm 13–21 bp long and often have a stem-loop structure.

The secondary structure of tRNA molecules is maintained by the base pairing in these arms, and this is a consistent feature. The acceptor arm has 7 bp, the TψC and anticodon arms 5 bp, and the D arm 3 (or 4) bp.

Although tRNAs are quite stable in prokaryotes, they are somewhat less stable in eukaryotes. The opposite is true for mRNAs, which are quite unstable in prokaryotes but generally stable in eukaryotic organisms.

Ribosomal RNA. A ribosome is a cytoplasmic nucleoprotein structure that acts as the machinery for the synthesis of proteins from the mRNA templates. On the ribosomes, the mRNA and tRNA molecules interact to translate into a specific protein molecule information transcribed from the gene.

The components of the mammalian ribosome, which has a molecular weight of about 4.2×10^6 and a sedimentation velocity of 80S (Svedburg units), are shown in Table 37–2. The mammalian ribosome contains 2 major nucleoprotein subunits, a large one of

Table 37–2. Components of mammalian ribosomes.*

Component	Mass (mw)	Protein Number	Protein Mass	RNA Size	RNA Mass	RNA Bases
40S subunit	1.4×10^6	~35	7×10^5	18S	7×10^5	1900
60S subunit	2.8×10^6	~50	1×10^6	5S	35,000	120
				5.8S	45,000	160
				28S	1.6×10^6	4700

*The ribosomal subunits are defined according to their sedimentation velocity in Svedburg units (40S or 60S). This table illustrates the total mass (MW) of each. The number of unique proteins and their total mass (MW) and the RNA components of each subunit in size (Svedberg units), mass, and base composition are listed.

MW 2.8×10^6 (60S) and a small subunit of 1.4×10^6 MW (40S). The **60S subunit** contains a **5S ribosomal RNA** (rRNA), a **5.8S rRNA,** and a **28S rRNA;** there are also probably more than 50 specific polypeptides. The smaller, or **40S, subunit** contains a single **18S rRNA** and approximately 30 polypeptide chains. All of the ribosomal RNA molecules except the 5S rRNA are processed from a single 45S precursor RNA molecule in the nucleolus (see Chapter 40). The 5S rRNA apparently has its own precursor that is independently transcribed. The highly methylated ribosomal RNA molecules are packaged in the nucleolus with the specific ribosomal proteins. In the cytoplasm, the ribosomes remain quite stable and capable of many translations. The functions of the ribosomal RNA molecules in the ribosomal particle are not fully understood, but they are necessary for ribosomal assembly and seem to play key roles in the binding of mRNA to ribosomes and its translation.

Small stable RNA. A large number of discrete, highly conserved, and small stable RNA species are found in eukaryotic cells. The majority of these molecules exist as ribonucleoproteins and are distributed in the nucleus, in the cytoplasm, or in both. They range in size from 90 to 300 nucleotides and are present in 100,000–1,000,000 copies per cell.

Small nuclear ribonucleoprotein particles, often called **snurps,** may be significantly involved in gene regulation. The U7 snurp appears to be involved in the production of correct 3′ ends of histone mRNA. The U4 and U6 RNAs may be required for poly(A) processing, and the U1 snurp is implicated in intron removal and mRNA processing (see Chapter 39).

Table 37–3 summarizes some characteristics of small stable RNAs.

Table 37–3. Some of the species of small stable RNAs found in mammalian cells.

Name	Length (nucleotides)	Molecules per Cell	Localization
U1	165	1×10^6	Nucleoplasm/hnRNA
U2	188	5×10^5	Nucleoplasm
U3	216	3×10^5	Nucleolus
U4	139	1×10^5	Nucleoplasm
U5	118	2×10^5	Nucleoplasm
U6	106	3×10^5	Perichromatin granules
4.5S	91–95	3×10^5	Nucleus and cytoplasm
7S	280	5×10^5	Nucleus and cytoplasm
7-2	290	1×10^5	Nucleus and cytoplasm
7-3	300	2×10^5	Nucleus

REFERENCES

Darnell J et al: *Molecular Cell Biology.* Scientific American Books, 1986.

Hunt T: *DNA Makes RNA Makes Protein.* Elsevier, 1983.

Lewin B: *Genes,* 2nd ed. Wiley, 1985.

Rich A et al: The chemistry and biology of left-handed Z-DNA. *Annu Rev Biochem* 1984;**53**:847.

Turner P: Controlling roles for snurps. *Nature* 1985;**316**:105.

Watson JD: *The Double Helix.* Atheneum, 1968.

Watson JD, Crick FHC: Molecular structure of nucleic acids. *Nature* 1953;**171**:737.

Zieve GW: Two groups of small stable RNAs. *Cell* 1981; **25**:296.

DNA Organization & Replication

Daryl K. Granner, MD

INTRODUCTION

The DNA in prokaryotic organisms is not combined with proteins other than those involved in DNA replication or transcription. Much of the DNA in eukaryotic organisms is covered with a variety of proteins. These proteins and DNA form a complex structure, chromatin, that allows for a variety of configurations of the DNA molecule and types of control unique to the eukaryotic organism.

The genetic information in the DNA of a chromosome can be transmitted by exact replication, or it can be exchanged by a number of processes, including crossing-over, recombination, transposition, and conversion. These provide a means of ensuring adaptability and diversity for the organism but can also result in disease.

DNA replication, a highly complex and ordered process, follows the 5' to 3' polarity typical of the synthesis of RNA and protein described in other chapters. Replication of the DNA in a chromosome begins at multiple sites and proceeds in both directions simultaneously. A number of enzymes are required for the synthesis and repair of DNA, both of which follow the rules of Watson-Crick base pairing.

BIOMEDICAL IMPORTANCE

Mutations, due to the faulty replication, movement, or repair of DNA, occur with a frequency of about one in every 10^6 cell divisions. An abnormal gene product can be the result of mutations that occur in coding or regulatory-region DNA. A mutation in a germ cell will be transmitted to offspring (so-called vertical transmission of hereditary disease). A number of factors, including viruses, chemicals, ultraviolet light, and ionizing radiation, increase the rate of mutation. Mutations often affect somatic cells and so are passed on to successive generations of cells within an organism. It is becoming apparent that a number of diseases, and perhaps most cancers, are due to horizontal transmission of induced mutations.

CHROMATIN

Chromatin is the chromosomal material extracted from nuclei of cells of eukaryotic organisms.* Chromatin consists of very long double-stranded **DNA molecules** and a nearly equal mass of rather small basic proteins termed **histones** as well as a smaller amount of **nonhistone proteins** (most of which are acidic and larger than histones) and a small quantity of **RNA.** Electron microscopic studies of chromatin have demonstrated dense spherical particles called **nucleosomes,** which are approximately 10 nm in diameter and connected by DNA filaments (Fig 38–1).

Histones & Nucleosomes

The **histones** are somewhat heterogeneous, consisting of a series of closely related **basic proteins.** Among the histones, **H1** histones are the least tightly bound to chromatin and are, therefore, easily removed with a salt solution, after which chromatin becomes soluble. The isolated core nucleosomes contain 4 classes of histones: **H2A, H2B, H3,** and **H4.** The structures of slightly lysine-rich histones—H2A and H2B—appear to have been significantly conserved between species, while the structures of arginine-rich histones—H3 and H4—have been highly conserved between species. This severe conservation implies that the function of histones is identical in all eukaryotes and that the entire molecule is involved quite specifically in carrying out this function. The C-terminal two-thirds of the molecules have a usual amino acid composition, while their N-terminal thirds contain most of the basic amino acids. These 4 core histones are subject to 5 types of **covalent modifications:** acetylation, methylation, phosphorylation, ADP-ribosylation, and covalent linkage (H2A only) to ubiquitin, the nuclear protein. These histone modifications likely play some role in chromatin structure and function, but little is currently understood.

When removed from chromatin, the histones interact with each other in very specific ways. **H3 and H4 aggregate to form a tetramer** containing 2 molecules

---

*So far as possible, the discussion of this chapter and of Chapters 39, 40, and 41 will pertain to mammalian organisms, which are, of course, among the higher eukaryotes. At times it will be necessary to refer to observations made in prokaryotic organisms such as bacteria and viruses, but when such occurs it will be acknowledged as being information that must be extrapolated to mammalian organisms. The division of the material presented in Chapters 37–41 is somewhat arbitrary and should not be taken to mean that the processes described are not fully integrated and interdependent.

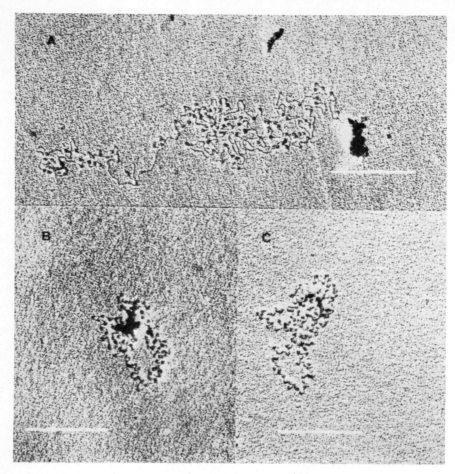

Figure 38–1. Electron micrograph of nucleosomes attached by strands of nucleic acid. (White bar represents 2.5 μm.) (Reproduced, with permission, from Oudet P, Gross-Bellard M, Chambon P: Electron microscopic and biochemical evidence that chromatin structure is a repeating unit. *Cell* 1975;4:281.)

of each (H3$_2$-H4$_2$), while **H2A and H2B form dimers** (H2A-H2B) and higher oligomeric complexes ([H2A-H2B]$_n$). The tetrameric H3-H4 does not associate with the H2A-H2B dimer or oligomer, and H1 does not associate directly with any of the other histones in solution.

However, when the H3$_2$-H4$_2$ tetramer and H2A-H2B dimers are mixed with purified, double-stranded DNA, the same x-ray diffraction pattern is formed as that observed in freshly isolated chromatin. Electron microscopic studies confirm the existence of reconstituted **nucleosomes.** Furthermore, the reconstitution of nucleosomes from DNA and histones H2A, H2B, H3, and H4 is independent of the organismal or cellular origin of the various components. The histone H1 and the nonhistone proteins are *not* necessary for the reconstitution of the nucleosome core.

In the nucleosome, the DNA is supercoiled in a left-handed helix over the surface of the disk-shaped histone octamer consisting of a central H3-H4 tetramer (H3$_2$-H4$_2$) and two H2A-H2B dimers (Fig

38–2). The core histones interact with the DNA on the inside of the supercoil without protruding.

The H3$_2$-H4$_2$ itself can confer nucleosomelike properties on DNA and thus has a central role in the formation of the nucleosome. The addition of two H2A-H2B dimers stabilizes the primary particle and binds firmly 2 additional half-turns of DNA previously bound only loosely to the H3$_2$-H4$_2$. Thus, **1.75 super-helical turns of DNA** are wrapped around the surface of the histone octamer, **protecting 146 base pairs of DNA** and forming the **nucleosome core** (Fig 38–2). As the DNA wraps around the surface of the histone octamer to form the nucleosome, it comes in contact with the histones in the order

H2A–H2B–H4–H3–H3–H4–H2B–H2A

Histone H1 binds to the DNA, where it enters and leaves the nucleosome core to seal a **2-turn, 166-base-pair DNA superhelix** generating the **nucleosome** (Fig 38–2).

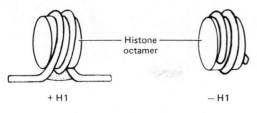

Figure 38–3. Proposed structure of the 10-nm fibril of chromatin made up of disk-shaped nucleosomes. The positions of the H1 histone-spacer regions are undefined.

+H1

166 base pairs protected
(2 superhelical turns) and
exposed (unprotected)
linker DNA

−H1

146 base pairs protected
(1.75 superhelical turns)

Figure 38–2. Model for the structure of the nucleosome (*left*) and nucleosome core (*right*), in which DNA is wrapped around the surface of a flat protein cylinder consisting of 2 each of histones H2A, H2B, H3, and H4, Histone H1 (shaded area) expands the number of base pairs protected. (Reproduced, with permission, from Laskey RA, Earnshaw WC: Nucleosome assembly. *Nature* 1980;**286**:763.)

The assembly of nucleosomes is probably mediated by the anionic nuclear protein **nucleoplasmin.** Histones, which are strongly cationic, can bind nonspecifically to the strongly anionic DNA by forming salt bridges. Clearly, such a nonspecific interaction of histones and DNA would be detrimental to nucleosome formation and chromatin function. Nucleoplasmin is an anionic pentameric protein that binds neither to DNA nor to chromatin, but it can interact reversibly with a histone octamer in such a way that the histones no longer adhere nonspecifically to negatively charged surfaces such as DNA. It seems that nucleoplasmin thereby maintains in the nucleus an ionic environment conducive to the specific interaction of histones and DNA and assembly nucleosomes. As the nucleosome is assembled, nucleoplasmin must be released from the histones. Nucleosomes appear to exhibit preference for certain regions on specific DNA molecules, but the basis for this nonrandom distribution, termed **phasing,** is unknown, although it is probably related to the relative physical flexibility of certain nucleotide sequences that are able to accommodate the regions of kinking within the supercoil.

The super-packing of nucleosomes in nuclei is seemingly dependent upon the interaction of the H1 histones with the double-stranded DNA connecting the nucleosomes. The topology of the interaction of the double-stranded DNA with the H1 histones to form the internucleosome spacer regions is not well delineated.

Electron microscopy of chromatin reveals 2 higher orders of structure—the 10-nm fibril and the 25- to 30-nm chromatin fiber—beyond that of the nucleosome itself. The disklike nucleosome structure has a 10-nm diameter and a height of 5 nm. The **10-nm fibril** seems to consist of nucleosomes arranged with their edges touching and their flat faces parallel with the fibril axis (Fig 38–3). The 10-nm fibril is probably further supercoiled with 6–7 nucleosomes per turn to form the

30-nm chromatin fiber (Fig 38–4). Each turn of the supercoil would be relatively flat, and the faces of the nucleosomes of successive turns would be nearly parallel to each other. H1 histones appear to stabilize the 30-nm fiber, but their position and that of the variable length spacer DNA are not clear. It is probable that nucleosomes can form a variety of packed structures. In order to form a mitotic chromosome, the 30-nm fiber must be compacted in length another 100-fold (see below).

In **interphase chromosomes,** chromatin fibers appear to be organized into 30,000–100,000 base-pair **loops or domains** anchored in a scaffolding (or supporting matrix) within the nucleus. Within these domains, some DNA sequences may be located nonrandomly. It has been suggested that each looped domain of chromatin may correspond to a separate genetic function, containing both coding and noncoding regions of the gene.

Active Chromatin

Generally, every cell of an individual metazoan organism contains the same genetic information in the form of the same DNA sequences. Thus, the differences between different cell types within an organism must be explained by differential expression of the common genetic information. Chromatin containing active genes (ie, transcriptionally active chromatin) has been shown to differ in several ways from that of

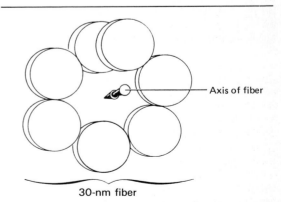

Figure 38–4. Proposed structure of the 30-nm chromatin fiber consisting of superhelixes of 10-nm fibrils of nucleosomes. The axis of the 30-nm fiber is perpendicular to the plane of the page.

nonactive regions. The nucleosome structure of active chromatin appears to be altered or even absent in highly active regions. DNA in active chromatin contains large regions (about 100,000 bases long) that are **sensitive to digestion by a nuclease** such as DNase I. The sensitivity to DNase I of chromatin regions being actively transcribed reflects only a potential for transcription rather than transcription itself and in several systems can be correlated with a relative lack of 5-methyldeoxycytidine in the DNA.

Within the large regions of active chromatin there exist shorter stretches of 100–300 nucleotides that exhibit an even greater (another 10-fold) sensitivity to DNase I. These **hypersensitive sites** probably result from a structural conformation that favors access of the nuclease to the DNA. These regions are generally located immediately upstream from the active gene and may be effected by the presence of so-called enhancer elements (see Chapters 39 and 41). In many cases, it seems that if a gene is capable of being transcribed, it must have a DNase hypersensitive site in

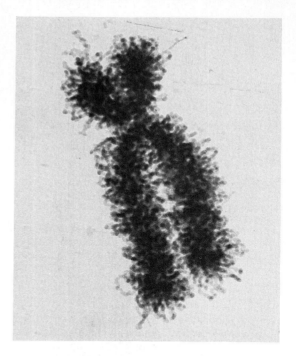

Figure 38–6. The 2 sister chromatids of human chromosome 12. × 27,850. (Reproduced, with permission, from DuPraw EJ: *DNA and Chromosomes*. Holt, Rinehart, & Winston, 1970.)

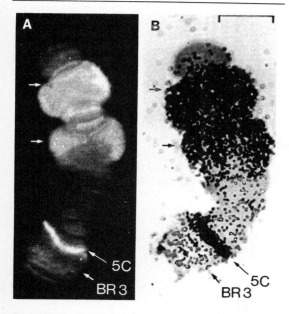

Figure 38–5. The correlation between RNA polymerase II activity and RNA synthesis. A number of genes are activated when *Chironomus tentans* larvae are subjected to heat shock (39 °C for 30 minutes). *A:* Distribution of RNA polymerase B (also called type II) in isolated chromosome IV from the salivary gland. The enzyme was detected by immunofluorescence using an antibody directed against the polymerase. The 5C and BR3 are specific bands of chromosome IV, and the arrows indicate puffs. *B:* Autoradiogram of a chromosome IV that was incubated in ³H-uridine to label the RNA. Note the correspondence of the immunofluorescence and presence of the radioactive RNA (dots). Bar = 7 μm. (Reproduced, with permission, from Sass H: RNA polymerase B in polytene chromosomes. *Cell* 1982:**28**:274. Copyright © 1982 by the Massachusetts Institute of Technology.)

the chromatin immediately upstream. Perhaps the hypersensitive sites provide proteins involved in gene transcription with the necessary **access to the coding strand.** Hypersensitive sites often provide the first clue about the presence and location of a transcription control element.

Transcriptionally inactive chromatin is densely packed during interphase as observed by electron microscopic studies and is referred to as **heterochromatin;** transcriptionally active chromatin stains less densely and is referred to as **euchromatin.** Generally, euchromatin is replicated earlier in the mammalian cell cycle (see below) than is heterochromatin.

There are 2 types of heterochromatin, constitutive heterochromatin and facultative heterochromatin. **Constitutive heterochromatin** is always condensed and, thus, inactive. Constitutive heterochromatin is found in the regions near the chromosomal centromere and at chromosomal ends (telomeres). **Facultative heterochromatin** is at times condensed, but at other times it is actively transcribed and, thus, uncondensed and appearing as euchromatin. Of the 2 members of the X chromosome pair in mammalian females, one X chromosome is almost completely inactive transcriptionally and is heterochromatic. However, the heterochromatic X chromosome decondenses during gametogenesis and becomes transcriptionally active during early embryogenesis; thus, it is facultative heterochromatin.

Certain cells of insects, eg, *Chironomus,* contain giant chromosomes that have been replicated for 10 cycles without separation of daughter chromatids. These copies of DNA line up side by side in precise register and produce a banded chromosome containing regions of condensed chromatin and lighter bands of more extended chromatin. Transcriptionally active regions of these **polytene chromosomes** are especially decondensed into **"puffs,"** which can be shown to contain the enzymes responsible for transcription and to be the sites of RNA synthesis (Fig 38–5).

Chromosomes

At metaphase, mammalian **chromosomes** possess a 2-fold symmetry, with identical **sister chromatids** connected at a **centromere,** the relative position of which is characteristic for a given chromosome (Fig 38–6). Each sister chromatid contains one double-stranded DNA molecule. During interphase, the packing of the DNA molecule is less dense than it is in the condensed chromosome during the metaphase. Metaphase chromosomes are transcriptionally **inactive.**

The human haploid genome consists of 3.5×10^9 base pairs or pairs of nucleotides and about 1.7×10^7 nucleosomes. Thus, each of the 23 chromatids in the human haploid genome would contain on the average 1.5×10^8 nucleotides in one double-stranded DNA molecule. The length of each DNA molecule must be **compressed about 8000-fold** to generate the structure of a condensed metaphase chromosome! In metaphase chromosomes, the 25- to 30-nm chromatin fibers are

Table 38–1. The packing ratios of each of the orders of DNA structure.

Chromatin Form	Packing Ratio
Bare double-helix DNA	~1.0
~2 turns of DNA on nucleosome	2.5
10-nm fibril of nucleosomes	5
25- to 30-nm chromatin fiber of superhelical nucleosomes	30
Condensed metaphase chromosome of loops	8000

also folded into a series of **looped domains,** the proximal portions of which are anchored to a nonhistone proteinaceous scaffolding. The packing ratios of each of the orders of DNA structure are summarized in Table 38–1.

The packaging of nucleoproteins within chromatids is not random, as evidenced by the characteristic patterns observed when chromosomes are stained with specific dyes such as quinacrine or Giemsa's stain (Fig 38–7).

From individual to individual within a single species, the pattern of staining (banding) of the entire chromosome complement is highly reproducible; nonetheless, it differs significantly from other species, even those closely related. Thus, the packaging of the nucleoproteins in chromosomes of higher eukaryotes must in some way be dependent upon species-specific characteristics of the DNA molecules.

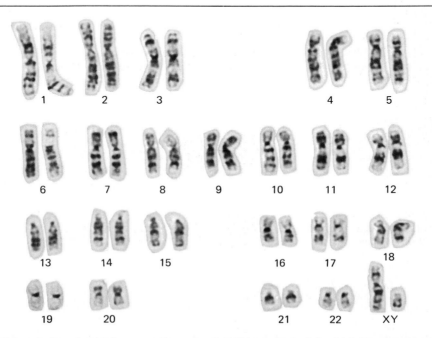

Figure 38–7. A human karyotype (of a man with a normal 46 XY constitution), in which the chromosomes have been stained by the Giemsa method and aligned according to the Paris Convention. (Courtesy of H Lawce and F Conte.)

GENETIC ORGANIZATION OF THE MAMMALIAN GENOME

The haploid genome of each human cell consists of 3.5×10^9 base pairs of DNA, subdivided into 23 chromosomes. The entire haploid genome contains sufficient DNA to code for nearly 1.5 million pairs of genes. However, studies of mutation rates and of the complexities of the genomes of higher organisms strongly suggest that humans have only about 100,000 essential proteins. This implies that most of the DNA is noncoding; ie, its information is never translated into an amino acid sequence of a protein molecule. Certainly, some of the excess DNA sequences serve to regulate the expression of genes during development, differentiation, and adaptation to the environment. Some excess clearly makes up the intervening sequences that split the coding regions of genes, but most of the excess appears to be composed of many families of repeated sequences for which no functions have been clearly defined.

The DNA in a eukaryotic genome can be divided into different "sequence classes." These are **unique-sequence, or nonrepetitive, DNA** and **repetitive-sequence DNA.** Unique-sequence DNA generally includes the single copy genes that code for proteins. The repetitive DNA includes sequences that vary in copy number from 2 to as many as 10^7 copies per cell.

Unique-Sequence, or Nonrepetitive, DNA

More than half the DNA in eukaryotic organisms is in unique or nonrepetitive sequences. This (and the distribution of repetitive sequence DNA) is based on a variety of DNA-RNA hybridization techniques, not on direct measurement, and as such is considered an estimation. Similar techniques are used to estimate the number of active genes in a population of unique-sequence DNA. In yeast, a lower eukaryote, about 4000 genes are expressed. In typical tissues in a higher eukaryote (eg, mammalian liver and kidney), between 10,000 and 15,000 genes are expressed. Different genes are expressed in each tissue, of course, and how this is accomplished is one of the major unanswered questions in biology.

Intervening Sequences in Coding Regions

The coding regions of DNA, the transcripts of which ultimately appear in the cytoplasm as single mRNA molecules, **are interrupted in the genome by large intervening sequences of noncoding DNA.** Accordingly, the primary transcripts of DNA—hnRNA—contain noncoding intervening sequences of RNA that must be removed in a process which also joins together the appropriate coding segments to form the mature mRNA. Most coding sequences for a single mRNA are interrupted in the genome (and thus in the primary transcript) by at least one—and as many as 50 in some cases—noncoding intervening sequences (**introns**). In most cases the **introns** are much longer than the continuous coding regions (**exons**). The processing of the primary transcript, which involves removal of introns and splicing of adjacent exons, is described in detail in Chapter 39.

The function of the intervening sequences, or introns, is not clear. They may serve to separate functional domains (exons) of coding information in a form that permits genetic rearrangement by recombination to occur more rapidly than if all coding regions for a given genetic function were contiguous. Such an enhanced rate of genetic rearrangement of functional domains might allow more rapid evolution of biologic function.

Repetitive Sequences in DNA

Repetitive-sequence DNA can be broadly classified as moderately repetitive or as highly repetitive. In human DNA, at least 20–30% of the genome consists of repetitive sequences.

The **highly repetitive sequences** consist of 5–500 base-pair lengths repeated many times in tandem. These sequences are usually **clustered** in centromeres and telomeres of the chromosome and are present in about 1–10 million copies per haploid genome. These sequences are transcriptionally inactive and may play a structural role in the chromosome.

The **moderately repetitive sequences,** which are defined as being present at less than 10^6 copies per haploid genome, are not clustered but are interspersed with unique sequences and can be categorized as being long or short in length. The **long interspersed repeats** are 5000–7000 base pairs in length and are present at 1000–100,000 copies per haploid genome. These long interspersed repeats are flanked on either side by 300–600 base-pair **direct repeats** (Fig 38–8) that closely resemble the **long terminal repeats** (LTR) at the ends of integrated retrovirus DNAs. In many cases, these long interspersed repeats are transcribed by **RNA polymerase II** and contain caps indistinguishable from those on mRNA.

The **short interspersed repeats** are families of related but individually distinct members that consist of a few to several hundred nucleotide pairs. The short

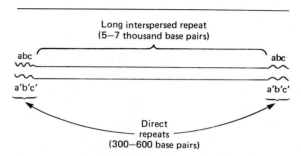

Figure 38–8. Representation of a long interspersed repeated sequence with its short direct repeat sequences (abc) and complements (a′b′c′) indicated at the termini.

interspersed repeats are actively transcribed either as integral components of introns or by DNA-dependent RNA polymerase III (see Chapter 39) as discrete elements. Of the short interspersed repeats in the human genome, one family, the **Alu family,** is present in about 500,000 copies per haploid genome and accounts for at least 3–6% of the human genome. Members of the human Alu family and their closely related analogs in other animals are transcribed as integral components of hnRNA or as discrete RNA molecules, including the well-studied 4.5S RNA and 7S RNA. These particular family members are highly conserved within a species as well as between mammalian species. The structures of the short interspersed repeats, including the members of the Alu family, resemble the retroviral long terminal repeat itself and are likely to be mobile elements, capable of jumping into and out of various sites within the genome (see below).

ALTERATION & REARRANGEMENT OF GENETIC MATERIAL

An alteration in the sequence of purine and pyrimidine bases in a gene due to a change, a removal, or an insertion of one or more bases may result in an altered gene product that in most instances ultimately is a protein. Such alteration in the genetic material results in a **mutation,** the consequences of which are discussed in detail in Chapter 40.

Chromosomal Recombination

Prokaryotic and eukaryotic organisms are capable of exchanging genetic information between similar or homologous chromosomes. The exchange or **recombination** event occurs primarily during meiosis in mammalian cells and requires alignment of homologous chromosomes, an alignment that almost always occurs with great exactness. A process of crossing-over occurs as shown in Fig 38–9. This usually results in an equal and reciprocal exchange of genetic information between homologous chromosomes. If the homologous chromosomes possess different alleles of the same genes, the crossover may produce noticeable and heritable genetic linkage differences. In the rare case where the alignment of homologous chromosomes is not exact, the crossing-over or recombination event may result in an unequal exchange of information. One chromosome may receive less genetic material and thus a deletion, while the other partner of the chromosome pair receives more genetic material and thus an insertion or duplication (Fig 38–9). Unequal crossing-over does occur in humans, as evidenced by the existence of hemoglobins designated Lepore and anti-Lepore. **Unequal crossover** affects tandem arrays of repeated DNAs whether they are related globin genes, as in Fig 38–10, or more abundant repetitive DNA. The unequal crossover through slippage in the pairing can result in expansion or contraction in the copy number of the repeat family and may

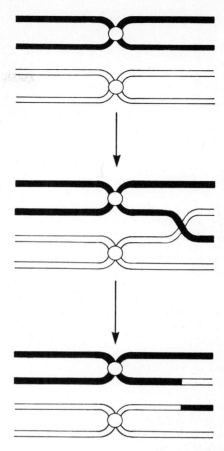

Figure 38–9. The process of crossing-over between homologous chromosomes to generate recombinant chromosomes.

contribute to the expansion and fixation of variant members throughout the array.

Chromosomal Integration

Some bacterial viruses (bacteriophages) are capable of recombining with the DNA of a bacterial host in such a way that the genetic information of the bacteriophage is incorporated in a linear fashion into the genetic information of the host. This integration, which is a form of recombination, occurs by the mechanism simplified in Fig 38–11. The backbone of the circular bacteriophage genome is broken, as is that of the DNA molecule of the host; the appropriate ends are resealed with the proper polarity. The bacteriophage DNA is figuratively straightened out ("linearized") as it is integrated into the bacterial DNA molecule—frequently a closed circle as well. The site at which the bacteriophage genome integrates or recombines with the bacterial genome is chosen by one of 2 mechanisms. If the bacteriophage contains a DNA sequence **homologous** to a sequence in the host DNA molecule, then a recombination event analogous to that occurring between homologous chromosomes can occur. However, some

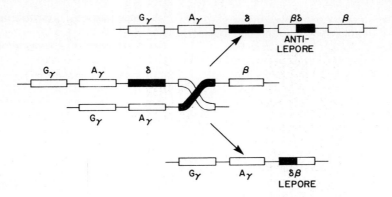

Figure 38–10. The process of unequal crossover in the region of the mammalian genome that harbors the structural genes for hemoglobin and the generation of the unequal recombinant products hemoglobin delta-beta Lepore and beta-delta anti-Lepore. The examples given show the locations of the crossover regions between amino acid residues. (Redrawn and reproduced, with permission, from Clegg JB, Weatherall DJ: $\beta°$ Thalassemia: Time for a reappraisal? *Lancet* 1974;2:133.)

bacteriophages synthesize proteins that bind specific sites on bacterial chromosomes with a nonhomologous site specifically of the bacteriophage DNA molecule. Integration occurs at the site and is said to be **"site-specific."**

Many animal viruses, particularly the oncogenic viruses—either directly or, in the case of RNA viruses, their DNA transcripts—can be integrated into chromosomes of the mammalian cell. The integration of the animal virus DNA into the animal genome generally is not "site-specific."

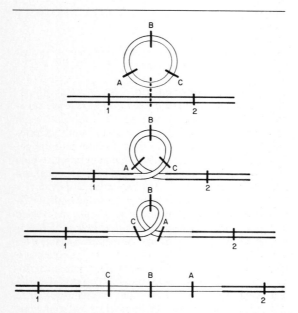

Figure 38–11. The integration of a circular genome (with genes A, B, and C) into the DNA molecule of a host (with genes 1 and 2) and the consequent ordering of the genes.

Transposition

In eukaryotic cells, small DNA elements that clearly are not viruses are capable of transposing themselves in and out of the host genome in ways that affect the function of neighboring DNA sequences. These mobile elements, sometimes called "jumping DNA," can carry flanking regions of DNA and, therefore, profoundly affect evolution. As mentioned above, the Alu family of moderately repeated DNA sequences has structural characteristics similar to the termini of retroviruses, which would account for the ability of the latter to move into and out of the mammalian genome.

Direct evidence for the transposition of other small DNA elements into the human genome has been provided by the discovery of "processed genes" for immunoglobulin molecules, α-globin molecules, and several others. These **processed genes** consist of DNA sequences identical or nearly identical to those of the messenger RNA for the appropriate gene product. That is, the 5' nontranscribed region, the coding region without intron representation, and the 3' poly(A) tail are all present contiguously. This particular DNA sequence arrangement must have resulted from the reverse transcription of an appropriately processed messenger RNA molecule from which the intron regions had been removed and the poly(A) tail added. The only recognized mechanism that this reverse transcript could have used to integrate into the genome would have been a transposition event. In fact, these "processed genes" have short terminal repeats at each end, as do known transposed sequences in lower organisms. Some of the processed genes have been randomly altered through evolution so that they now contain nonsense codons that preclude their expression (see Chapter 40). Thus, they are referred to as **"pseudogenes."**

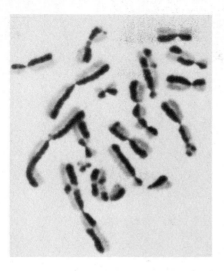

Figure 38–12. Sister chromatid exchanges between human chromosomes. These are detectable by Giemsa staining of the chromosomes of cells replicated for 2 cycles in the presence of bromodeoxyuridine. (Courtesy of S Wolff and J Bodycote.)

Gene Conversion

Besides unequal crossover and transposition, a third mechanism can effect rapid changes in the genetic material. Similar sequences on homologous or nonhomologous chromosomes may occasionally pair up and eliminate any mismatched sequences between them. This may lead to the accidental fixation of one variant or another throughout a family of repeated sequences and thereby homogenize the sequences of the members of repetitive DNA families. This latter process is referred to as gene conversion.

In diploid eukaryotic organisms such as humans, after cells progress through the S phase they contain a tetraploid content of DNA. This is in the form of sister chromatids of chromosome pairs. Each of these sister chromatids contains identical genetic information, since each is a product of the semiconservative replication of the original parent DNA molecule of that chromosome. Crossing-over occurs between these genetically identical sister chromatids. Of course, these **sister chromatid exchanges** (Fig 38–12) have no genetic consequence so long as the exchange is the result of an equal crossover.

In mammalian cells, some interesting gene rearrangements occur normally during development and differentiation. For example, in mice the V_L and C_L genes for a single immunoglobulin molecule (see Chapter 41) are widely separated in the germ line DNA. In the DNA of a differentiated immunoglobulin-producing (plasma) cell, the same V_L and C_L genes have been moved physically closer together in the genome, and into the same transcription unit. However, even then, this rearrangement of DNA during

differentiation does not bring the V_L and C_L genes into contiguity in the DNA. Instead, the DNA contains an interspersed or interruption sequence of about 1200 base pairs at or near the junction of the V and C regions. The interspersed sequence is transcribed into RNA along with the V_L and C_L genes, and the interspersed information is removed from the RNA during its nuclear processing (see Chapters 39 and 41).

DNA SYNTHESIS & REPLICATION

The primary function of DNA replication is understood to be the provision of progeny with the genetic information possessed by the parent. Thus, the replication of DNA must be complete and carried out with **high fidelity to maintain genetic stability** within the organism and the species. The process of DNA replication is complex and involves many cellular functions and several verification procedures to ensure fidelity in replication. The first enzymologic observations on DNA replication were made in *Escherichia coli* by Arthur Kornberg, who described in that organism the existence of an enzyme now called DNA polymerase I. This enzyme has multiple catalytic activities, a complex structure, and a requirement for the triphosphates of the 4 deoxyribonucleosides of adenine, guanine, cytosine, and thymine. The polymerization reaction catalyzed by DNA polymerase I of *E coli* has served as a prototype for all DNA polymerases of both prokaryotes and eukaryotes, even though it is now recognized that the major role of this polymerase is to ensure fidelity and to repair rather than to replicate DNA.

Initiation of DNA Synthesis

The initiation of DNA synthesis (Fig 38–13) requires priming by a **short length of RNA,** about 10–200 nucleotides long. This priming process involves the nucleophilic attack by the 3'-hydroxyl group of the RNA primer to the alpha phosphate of the deoxynucleoside triphosphate with the splitting off of pyrophosphate. The 3'-hydroxyl group of the recently attached deoxyribonucleoside monophosphate is then free to carry out a nucleophilic attack on the next entering deoxyribonucleoside triphosphate, again at its alpha phosphate moiety, with the splitting off of pyrophosphate. Of course, the selection of the proper deoxyribonucleotide whose terminal 3'-hydroxyl group is to be attacked is dependent upon **proper pairing with the other (template) strand** of the DNA molecule according to the rules proposed originally by Watson and Crick (Fig 38–14). When an adenine deoxyribonucleoside monophosphoryl moiety is in the template position, a thymidine triphosphate will enter and its alpha phosphate will be attacked by the 3'-hydroxyl group of the deoxyribonucleoside monophosphoryl most recently added to the polymer. By this stepwise process, the template dictates which deoxyri-

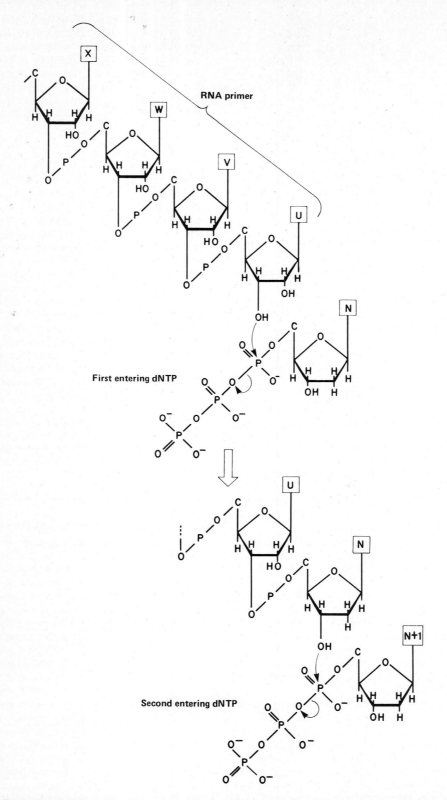

Figure 38–13. The initiation of DNA synthesis upon a primer of RNA and the subsequent attachment of the second deoxyribonucleoside triphosphate.

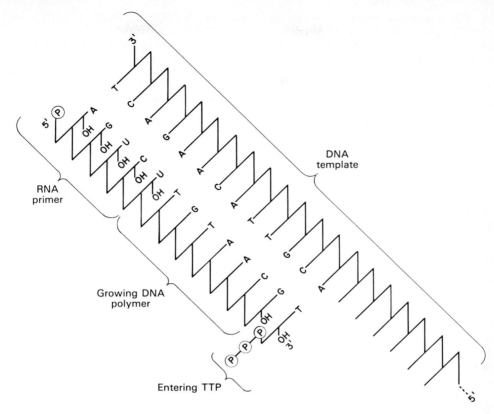

Figure 38–14. The RNA-primed synthesis of DNA demonstrating the template function of the complementary strand of parental DNA.

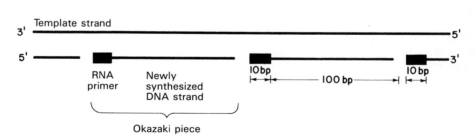

Figure 38–15. The discontinuous polymerization of deoxyribonucleotides and formation of Okazaki pieces.

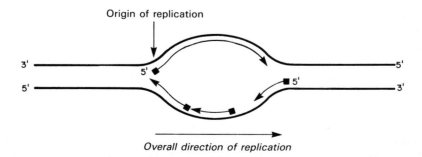

Figure 38–16. The process of semidiscontinuous, simultaneous replication of both strands of double-stranded DNA.

bonucleoside triphosphate is complementary and by hydrogen bonding holds it in place while the 3′-hydroxyl group of the growing strand attacks and incorporates the new nucleotide into the polymer. These fragments of DNA attached to an RNA initiator component, discovered by Okazaki, are referred to as **Okazaki pieces** (Fig 38–15). In mammals, after many Okazaki pieces are generated, the replication complex begins to remove the RNA primers, to fill in the gaps left by their removal with the proper base-paired deoxynucleotide, and then to seal the fragments of newly synthesized DNA by enzymes referred to as **DNA ligases.**

Polarity of Replication

As has already been noted, DNA molecules are double-stranded and the 2 strands are antiparallel, ie, running in opposite directions. The replication of DNA in prokaryotes and eukaryotes occurs on **both strands simultaneously.** However, an enzyme capable of polymerizing DNA in the 3′ to 5′ direction does not exist in any organism, so that both of the newly replicated DNA strands cannot grow in the same direction simultaneously. Nevertheless, the same enzyme appears to replicate both strands at the same time. The single enzyme replicates **one strand ("leading strand") in a continuous manner in the 5′ to 3′ direction,** with the same overall forward direction. It replicates the **other strand ("lagging strand") discontinuously** while polymerizing the nucleotides in short spurts of 150–250 nucleotides again in the 5′ to 3′ direction, but at the same time it faces toward the back end of the preceding RNA primer rather than toward the unreplicated portion. This process of **semidiscontinuous DNA synthesis** is shown diagrammatically in Fig 38–16.

In the mammalian nuclear genome, most of the RNA primers are eventually removed as part of the replication process, whereas after replication of the mitochondrial genome the small piece of RNA remains as an integral part of the closed circular DNA structure.

Enzymes Involved in DNA Polymerization & Repair

In mammalian cells, there is one class of DNA polymerase enzymes, called **polymerase alpha,** which is present in the nucleus and responsible for chromosome replication. One polymerase alpha molecule is capable of polymerizing about 100 nucleotides per second, a rate 10-fold less than the rate of polymerization of deoxynucleotides by the bacterial DNA polymerase. This reduced rate may result from the interference by nucleosomes. It is not known how DNA polymerase negotiates nucleosomes. However, after DNA replication, the assembling nucleosomes are distributed randomly to either daughter strand. Newly synthesized core histones in the octameric form would then attach to the other strand as the replication fork progresses.

A lower-molecular-weight polymerase, **polymerase beta,** is also present in mammalian nuclei but is not responsible for the usual DNA replication. It may function in DNA repair (see below). Mitochondrial DNA polymerase, **polymerase gamma,** is responsible for replication of the mitochondrial genome, another DNA molecule that exists in circular form.

The entire mammalian genome replicates in 9 hours, the average period required for formation of a tetraploid genome from a diploid genome in a replicating cell. This requires the presence of **multiple origins** of DNA replication that occur in clusters of up to 100 of these replication units. Replication occurs in **both directions** along the chromosome, and both strands are replicated simultaneously. This replication process generates "replication bubbles" (Fig 38–17).

The multiple sites that serve as origins for DNA replication in eukaryotes are poorly defined except in a few animal viruses and in yeast. However, it is clear that initiation is regulated both spatially and temporally, since clusters of adjacent sites initiate synchronously. There are suggestions that functional domains of chromatin replicate as intact units, implying that the origins of replication are specifically located with respect to transcription units.

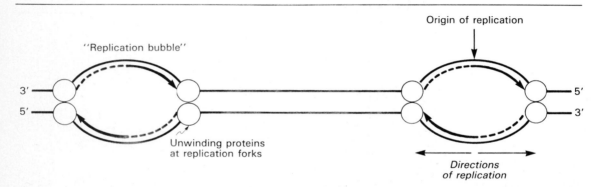

Figure 38–17. The generation of "replication bubbles" during the process of DNA synthesis. The bidirectional replication and the proposed positions of unwinding proteins at the replication forks are depicted.

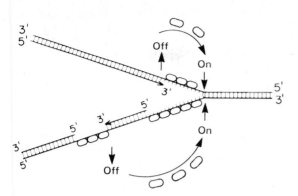

Figure 38–18. Hypothetical scheme for single-strand binding protein action at a replicating fork. The protein is recycled after binding single-stranded regions of the template and facilitating replication. (Courtesy of B Alberts.)

During the replication of DNA, there must be a separation of the 2 strands to allow each to serve as a template by hydrogen bonding its nucleotide bases to the incoming deoxynucleoside triphosphate. The separation of the DNA double helix is promoted by specific protein molecules that **stabilize the single-stranded structure** as the replication fork progresses. These stabilizing proteins bind stoichiometrically to the single strands without interfering with the abilities of the nucleotides to serve as templates (Fig 38–18). In addition to separating the 2 strands of the double helix, there must be an **unwinding** of the molecule (once every 10 nucleotide pairs) to allow the rewinding of the newly synthesized daughter strands. Given the time during which DNA replication must occur in prokaryotes, it can be calculated that the molecule must unwind at approximately 400,000 turns per second, which is clearly an impossible feat. Thus, there must be multiple "swivels" interspersed in the DNA molecules of all organisms. The swivel function is

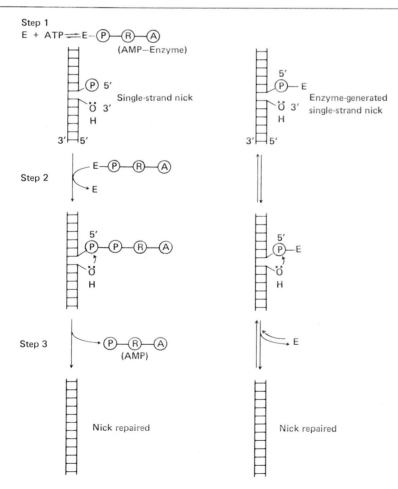

Figure 38–19. Comparison of 2 types of nick-sealing reactions on DNA. The series of reactions at left is catalyzed by DNA ligase; that at right by DNA topoisomerase I. (Slightly modified and reproduced, with permission, from Lehninger AL: *Biochemistry,* 2nd ed. Worth, 1975.)

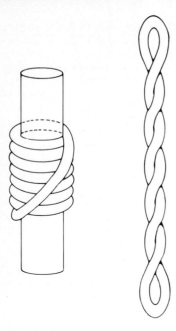

Figure 38–20. Supercoiling of DNA. A left-handed toroidal (selenoidal) supercoil, at left, will convert to a right-handed interwound supercoil, at right, when the cylindric core is removed. Such a transition is analogous to that which occurs when nucleosomes are disrupted by the high salt extraction of histones from chromatin.

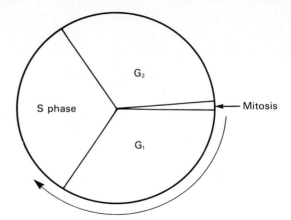

Figure 38–21. Mammalian cell cycle. The DNA synthetic phase (S phase) is separated from mitosis by gap 1 (G_1) and gap 2 (G_2). (Arrow indicates direction of cell progression.)

provided by specific enzymes that introduce **"nicks" in one strand of the unwinding double helix,** thereby allowing the unwinding process to proceed. The nicks are quickly resealed without requiring energy input, because of the formation of a high-energy covalent bond between the nicked phosphodiester backbone and the nicking-sealing enzyme. This process is depicted diagrammatically in Fig 38–19 and there compared to the ATP-dependent resealing carried out by the DNA ligases. The nicking-resealing enzymes are called DNA **topoisomerases** and are also capable of unwinding supercoiled DNA. Supercoiled DNA is a higher-ordered structure occurring in circular (or extraordinarily long) DNA molecules wrapped around a core, as depicted in Fig 38–20.

There exists in one species of animal viruses (retroviruses) a class of enzymes capable of synthesizing a single-stranded and then a double-stranded DNA molecule from a single-stranded RNA template. This polymerase, RNA-dependent DNA polymerase or **"reverse transcriptase,"** first synthesizes a DNA-RNA hybrid molecule utilizing the RNA genome as a template. A specific enzyme, RNase H, degrades the RNA strand, and the remaining DNA strand in turn serves as a template to form a double-stranded DNA molecule containing the information originally present in the RNA genome of the animal virus.

Regulation of DNA Synthesis

In animal cells, including human cells, the replication of the DNA genome occurs only at a specified time during the life span of the cell. This period is referred to as the synthetic or S phase. This is usually temporally separated from the mitotic phase by nonsynthetic periods referred to as gap 1 (G_1) and gap 2 (G_2), occurring before and after the S phase, respectively (Fig 38–21). The cell regulates its DNA synthesis grossly by allowing it to occur only at specific times and mostly in cells preparing to divide by a mitotic process. The regulation of the entry of a cell into an S phase may involve cyclic purine nucleotides and perhaps the substrates for DNA synthesis, but the mechanisms are unknown. Many of the cancer-causing viruses (oncoviruses) are capable of alleviating or disrupting the apparent restriction that normally controls the entry of mammalian cells from G_1 into the S phase. Again, this mechanism is currently unknown but probably involves the phosphorylation of specific host protein molecules.

During the S phase, mammalian cells contain greater quantities of DNA polymerase alpha than during the nonsynthetic phases of the cell cycle. Furthermore, those enzymes responsible for the formation of the substrates for DNA synthesis, ie, deoxyribonucleoside triphosphates, are also increased in activity, and their activity will diminish following the synthetic phase until the reappearance of the signal for renewed DNA synthesis. During S phase, the nuclear DNA is **completely replicated once and only once.** It seems that once chromatin has been replicated, it is marked so as to prevent its further replication until it again passes through mitosis. It has been suggested that DNA methylation may serve as such a covalent marker.

In general, a given pair of chromosomes will replicate simultaneously and within a fixed portion of the S phase upon every replication. On a chromosome, clusters of replication units replicate coordinately. The nature of the signals that regulate DNA synthesis at these levels is unknown, but the regulation does appear to be an intrinsic property of each individual chromosome.

Degradation & Repair of DNA

The maintenance of the integrity of the information in DNA molecules is of utmost importance to the survival of a particular organism as well as to survival of the species. Thus, it might be concluded that surviving species must have evolved mechanisms for repairing DNA damage incurred as a result of either replication errors or environmental insults. It has been estimated that DNA replication and environmentally induced DNA damage result in an average of about 6 nucleotide changes per year in the human germ line. Presumably, at least that number of nucleotide changes or mutations must occur per year in the somatic cells as well.

As described in Chapter 37, the major responsibility for the fidelity of replication resides in the specific pairing of nucleotide bases. Proper pairing is dependent upon the presence of the favored tautomers of the purine and pyrimidine nucleotides (see Fig 34−7), but the equilibrium wherein one tautomer is more stable than another is only about 10^4 or 10^5 in favor of that with the greater stability. Although this is not sufficiently favorable to ensure the high fidelity that is necessary, the favoring of the preferred tautomers, and thus of the proper base pairing, could be ensured by monitoring the base pairing twice. Such double monitoring does appear to occur in both bacterial and mammalian systems: once at the time of insertion of the deoxyribonucleoside triphosphates, and later by a follow-up, energy-requiring mechanism which removes all improper bases that may occur in the newly formed strand. This double monitoring does not permit errors of mispairing due to the presence of the unfavored tautomers to occur more frequently than once every 10^8−10^{10} base pairs. The molecule responsible for this monitoring mechanism in *E coli* is the built-in $3' \rightarrow 5'$ exonuclease activity of DNA polymerase, but mammalian DNA polymerases do not clearly possess such a nuclease proofreading function.

Damage to DNA by environmental, physical, and chemical agents may be classified into 4 types (Table 38−2). The damaged regions of DNA may be **repaired, replaced** by recombination, or **retained.** Retention leads to mutations and, potentially, cell death. Repair and replacement exploit the redundancy of information inherent in the double helical DNA structure. The defective region in one strand can be returned to its original form by relying on the complementary information stored in the unaffected strand.

The key to all of the repair or recombinational processes is the initial **recognition of the defect** and either repairing it during the recognition step or marking it for future attention. The **depurination** of DNA, which happens spontaneously owing to the thermal

Table 38−2. Types of damage to DNA.

I. **Single-base alteration**
   A. Depurination
   B. Deamination of cytosine to uracil
   C. Deamination of adenine to hypoxanthine
   D. Alkylation of base
   E. Insertion or deletion of nucleotide
   F. Base-analog incorporation
II. **Two-base alteration**
   A. UV light-induced thymine-thymine dimer
   B. Bifunctional alkylating agent cross-linkage
III. **Chain breaks**
   A. Ionizing radiation
   B. Radioactive disintegration of backbone element
IV. **Cross-linkage**
   A. Between bases in same or opposite strands
   B. Between DNA and protein molecules (eg, histones)

lability of the purine N-glycosidic bond, occurs at a rate of 5000−10,000/cell/d at 37 °C. Specific enzymes recognize a depurinated site and replace the appropriate purine directly, without interruption of the phosphodiester backbone.

Both cytosine and adenine bases in DNA spontaneously **deaminate** to form uracil and hypoxanthine, respectively. Since neither uracil nor hypoxanthine normally exists in DNA, it is not surprising that specific N-glycosylases can recognize these abnormal bases and remove the base itself from the DNA. This removal marks the site of the defect and allows an apurinic or apyrimidinic endonuclease to incise the appropriate backbone near the defect. Subsequently, the sequential actions of an exonuclease, a repair DNA polymerase, and a ligase return the DNA to its original state (Fig 38−22). This series of events is called **excision-repair.** By a similar series of steps involving initially the recognition of the defect, alkylated bases and base analogs can be removed from DNA and the DNA returned to its original informational content.

The repair of insertions or deletions of nucleotides normally occurs by recombinational mechanisms either with or without replication.

Ultraviolet light induces the formation of pyrimidine-pyrimidine dimers, predominantly the dimerization of 2 juxtaposed thymines in the same strand (Fig 38−23). There are apparently 2 mechanisms for removing or repairing these thymine-thymine dimers. One is excision-repair analogous to that described above. The second mechanism involves visible light photoactivation of a specific enzyme that directly reverses the dimer formation in situ.

Single-strand breaks induced by ionizing radiation can be repaired by direct ligation or by recombination. The mechanisms responsible for the repair of cross-linkages between bases on opposite strands of the DNA double helix or between the DNA and protein molecules are poorly understood.

In general, damage caused by ionizing radiation and by alkylation of bases is repaired in short patches of excision and resynthesis. Ultraviolet light damage and strand cross-linkages are repaired by long patches of excision and resynthesis. In mammalian cells, re-

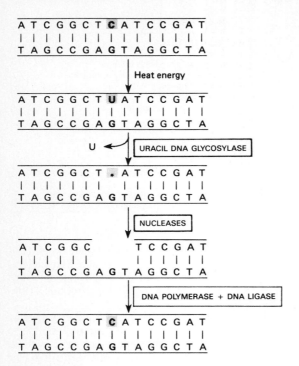

Figure 38–22. The enzyme uracil DNA glycosylase removes the uracil created by spontaneous deamination of cytosine in the DNA. (Courtesy of B Alberts.)

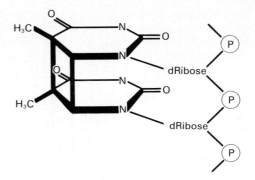

Figure 38–23. A thymine-thymine dimer formed via a cyclobutane moiety between juxtaposed thymine residues of DNA.

pair replication can be observed as **unscheduled DNA synthesis,** ie, incorporation of DNA precursors (radioactive thymidine) into DNA when a cell is not in S phase.

Associated with the increased excision-repair activity in response to DNA damaging agents, mammalian cells exhibit increased activity of the enzyme **poly(ADP-ribose) polymerase.** This enzyme uses the coenzyme NAD^+ to ADP-ribosylate chromatin proteins. It adds mostly mono(ADP-ribose), but to some extent homopolymeric chains of ADP-ribose are added. It is not evident what function poly(ADP-ribose) polymerase or its product, $(ADP-ribose)_n$, has in the excision-repair process. There is a temporal relationship between the increased repair activity and the increased specific enzyme. Furthermore, inhibition of the enzyme by specific inhibitors prevents the rejoining of broken DNA strands. The increased activity of poly(ADP-ribose) polymerase appears to be a response to DNA fragmentation in the nucleus. This fragmentation might be induced primarily by physical agents such as x-ray, or secondarily by the incision mechanism responding to other chemical or physical agents such as ultraviolet light or alkylating agents.

The activity of the polymerase is sufficiently great to cause a depletion of intracellular NAD^+ following environmentally induced DNA damage.

Xeroderma pigmentosum is an autosomal recessive genetic disease. The clinical syndrome includes marked sensitivity to sunlight (ultraviolet) with subsequent formation of multiple skin cancers and premature death. The inherited defect seems to involve the repair of damaged DNA. Cells cultured from patients with xeroderma pigmentosum exhibit low activity for the photoactivated thymine dimer cleavage process. However, the involved DNA repair processes in this disease are quite complex; there are at least 7 genetic complementation groups.

In cells from most if not all complementation groups of xeroderma pigmentosum, there is an abnormal temporal or quantitative response of the poly(ADP-ribose) polymerase to ultraviolet light exposure. It seems that the abnormal response in at least one complementation group is due to the inability to incise the DNA strand at the site of damage, since the addition of deoxyribonuclease to permeabilized defective cells is followed by a normal or nearly normal increase in poly(ADP-ribose) polymerase activity.

In patients with **ataxia-telangiectasia,** an autosomal recessive disease in humans resulting in the development of cerebellar ataxia and lymphoreticular neoplasms, there appears to exist an increased sensitivity to damage by x-ray. Patients with **Fanconi's anemia,** an autosomal recessive anemia characterized also by an increased frequency of cancer and by chromosomal instability, probably have defective repair of crosslinking damage. All 3 of these clinical syndromes are associated with increased frequency of cancer. It is likely that other human diseases resulting from disordered DNA repair capabilities will be found in the future.

REFERENCES

Bauer WR et al: Supercoiled DNA. *Sci Am* (July) 1980; **243**:118.

Cantor CR: DNA choreography. *Cell* 1981;**25**:293.

Igo-Kemenes T, Hörz W, Zachau HG: Chromatin. *Annu Rev Biochem* 1982;**51**:89.

Jelinek WR, Schmid CW: Repetitive sequences in eukaryotic DNA and their expression. *Annu Rev Biochem* 1982;**51**:813.

Jongstra J et al: Induction of altered chromatin structures by simian virus 40 enhancer and promoter elements. *Nature* 1984;**307**:708.

Kornberg A: *DNA Replication*. Freeman, 1980.

Lindahl T: DNA repair enzymes. *Annu Rev Biochem* 1982; **51**:61.

Loeb LA, Kunkel TA: Fidelity of DNA synthesis. *Annu Rev Biochem* 1982;**51**:429.

McGhee JD, Felsenfeld G: Nucleosome structure. *Annu Rev Biochem* 1980;**49**:1115.

Nossal NG: Prokaryotic DNA replication systems. *Annu Rev Biochem* 1983;**52**:581.

39

RNA Synthesis & Processing

Daryl K. Granner, MD

INTRODUCTION & BIOMEDICAL IMPORTANCE

The synthesis of an RNA transcript from DNA is a very complex process involving the enzyme RNA polymerase and a number of associated proteins. The general steps required to synthesize the primary transcript are initiation, elongation, and termination. Most is known about initiation. A number of DNA regions (generally located upstream from the initiation site) and protein factors that bind to these sequences to regulate the initiation of transcription have been identified. This process is best understood in prokaryotes and viruses, but considerable progress has been made in deciphering mammalian cell transcription in recent years. It is important to understand the basic principles of RNA synthesis, for modulation of this process results in altered rates of protein synthesis and hence a variety of metabolic changes, and this is how organisms adapt to changes of environment.

The RNA molecules synthesized in mammalian cells are often very different from those made in prokaryotic organisms, particularly the mRNA-encoding transcripts. Prokaryotic mRNA can be translated as it is being synthesized, whereas in mammalian cells most RNAs are made as precursor molecules that have to be processed into mature, active RNA. Erroneous processing and splicing of mRNA transcripts are a cause of disease, eg, certain types of thalassemia (see Chapter 36).

RNA SYNTHESIS

The process of synthesizing RNA from a DNA template has been characterized best in prokaryotes. Although in mammalian cells the regulation of RNA synthesis and the processing of the RNA transcripts are different from that in prokaryotes, the process of RNA synthesis per se is quite similar in these 2 classes of organisms. Therefore, the description of RNA synthesis in prokaryotes will be applicable to eukaryotes even though the enzymes involved and the regulatory signals are different.

The sequence of ribonucleotides in an RNA molecule is complementary to the sequence of deoxyribonucleotides in one strand of the double-stranded DNA molecule (see Fig 37–8). The strand that is transcribed into an RNA molecule is referred to as the **coding strand** of the DNA. The other DNA strand is fre-

![Diagram showing template strands of linked genes A, B, C, and D]

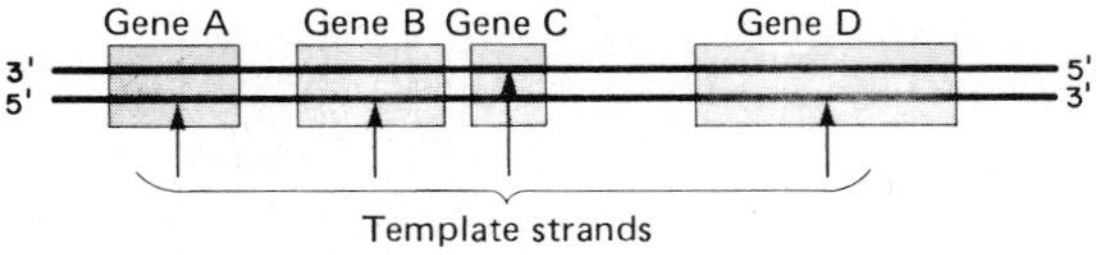

Figure 39–1. Template strands of the linked genes. These are not necessarily the same strand of the DNA double helix.

quently referred to as the **noncoding strand** of that gene. In the case of a double-stranded DNA molecule containing many genes, the coding strand for each gene will not necessarily be the same strand of the DNA double helix (Fig 39–1). Thus, a given strand of a double-stranded DNA molecule will serve as the coding strand for some genes and the noncoding strand of other genes. Note that the nucleotide sequence of an RNA transcript will be the same (except for U replacing T) as that of the noncoding strand.

DNA-dependent RNA polymerase is the enzyme responsible for the polymerization of ribonucleotides into a sequence complementary to the coding strand of the gene (Fig 39–2). The enzyme attaches at a specific site, the **promoter,** on the coding strand. This is followed by initiation of RNA synthesis at the starting point, and the process continues until a termination se-

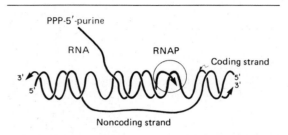

Figure 39–2. The RNA polymerase (RNAP)-catalyzed polymerization of ribonucleotides into an RNA sequence complementary to the coding strand of the gene. (From James D. Watson, *Molecular Biology of the Gene,* 3rd ed. Copyright © 1976, 1970, 1965, by W.A. Benjamin, Inc., Menlo Park, Calif.)

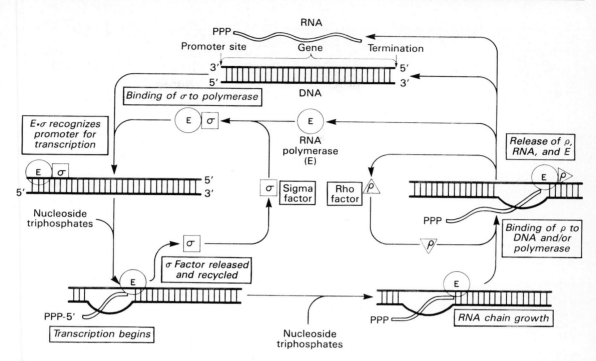

Figure 39–3. The process of RNA synthesis. It begins at the upper left-hand portion of the figure with the binding of sigma factor to the core polymerase to form a complex that can recognize the promoter for transcription. The process is completed as the RNA polymerase is released from the gene, and all of the catalytic components are free to recycle. E, enzyme. (From James D. Watson, *Molecular Biology of the Gene*, 3rd ed. Copyright © 1976, 1970, 1965, by W.A. Benjamin, Inc., Menlo Park, Calif.)

quence is reached. A **transcription unit** is defined as that region of DNA that extends between the promoter and the terminator. The RNA product, which is synthesized in the 5′ to 3′ direction, is the **primary transcript.** In prokaryotes, this can represent the product of several genes; in mammalian cells, it usually represents the product of a single gene. The 5′ termini of the primary RNA transcript and the mature cytoplasmic RNA are identical. Thus, the **start point of transcription corresponds to the 5′ nucleotide of the mRNA.** The primary transcripts generated by RNA polymerase II are promptly capped by 7-methylguanosine triphosphate (see Fig 37–10)—caps that persist and eventually appear on the 5′ end of mature cytoplasmic mRNA. These caps are presumably necessary for both the subsequent processing of the primary transcript to mRNA as described below and for its subsequent translation.

The DNA-dependent RNA polymerase of the bacterium *Escherichia coli* exists as a core molecule composed of 4 subunits; 2 of these are identical to each other (the α subunits), and 2 are similar in size to each other but not identical (the β subunit and β' subunit). The core RNA polymerase utilizes a specific protein factor (the sigma [σ] factor) that assists the core enzyme to attach more tightly to the specific deoxynucleotide sequence of the promoter region. Bacteria contain multiple σ factors, each of which acts as a reg-

ulatory protein that modifies the **promoter recognition specificity** of the RNA polymerase. The appearance of different σ factors can be correlated temporally with various programs of gene expresson in prokaryotic systems, such as bacteriophage development, sporulation, and the response to heat shock.

The process of RNA synthesis, depicted in Fig 39–3, involves first the binding of the RNA holopolymerase molecule to the template at the promoter site. Initiation of formation of the RNA molecule at its 5′ end then follows with the release of the σ factor, while the elongation of the RNA molecule from the 5′ to its 3′ end continues **antiparallel** to its template. The enzyme polymerizes the ribonucleotides in a specific sequence that is dictated by the coding strand and interpreted by Watson-Crick base-pairing rules. Pyrophosphate is released in the polymerization reaction. In both prokaryotes and eukaryotes, a purine ribonucleotide is usually the first to be polymerized into the RNA molecule.

As the elongation complex containing the core RNA polymerase progresses along the DNA molecule, **DNA unwinding** must occur in order to provide access for the appropriate base pairing to the nucleotides of the coding strand. The extent of DNA unwinding is constant throughout transcription and has been estimated to be about 17 base pairs per polymerase molecule. Thus, it appears that the size of the un-

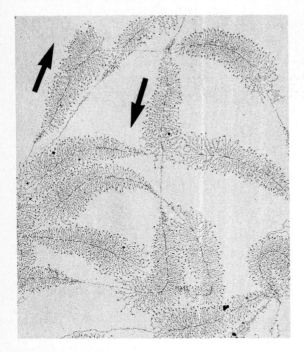

Figure 39–4. Electron photomicrograph of multiple copies of ribosomal RNA genes being transcribed in an amphibian cell. × 6000. Note that the length of the transcripts increases as the RNA polymerase molecules have progressed along the individual ribosomal RNA genes. Thus, the proximal end of the transcribed gene has short transcripts attached to it, while much longer transcripts are attached to the distal end of the gene. The arrows indicate the direction (5' to 3') of transcription. (Reproduced, with permission, from Miller OL Jr, Beatty BR: Portrait of a gene. *J Cell Physiol* 1969;**74[Suppl 1]:**225.)

Table 39–1. Nomenclature and localization of animal DNA-dependent RNA polymerases.

Class of Enzyme	Sensitivity to α-Amanitin	Products	Principal Localization
I (A)	Insensitive	rRNA	Nucleolar
II (B)	Sensitive to low concentration (10^{-8} to 10^{-9} mol/L)	hnRNA (mRNA)	Nucleoplasmic
III (C)	Sensitive to high concentration	tRNA and 5S RNA	Nucleoplasmic

Mammalian cells possess several DNA-dependent RNA polymerases, the properties of which are described in Table 39–1. Each of these DNA-dependent RNA polymerases seems to be responsible for the transcription of different sets of genes. The sizes of the RNA polymerases for the 3 major classes of eukaryotic RNA range from MW 500,000 to 600,000. All of these enzymes share the basic subunit structural organization of bacterial RNA polymerase. They all have 2 large subunits and a number of smaller subunits. Recent DNA cloning and sequencing work indicates that eukaryotic RNA polymerases have extensive amino acid homologies with prokaryotic RNA polymerases. The functions of each of the subunits are not yet understood. Many could have regulatory functions, such as serving to assist the polymerase in the recognition of specific sequences like promoters and termination signals.

One toxin from the mushroom *Amanita phalloides,* α-amanitin, is a specific inhibitor of the eukaryotic nucleoplasmic DNA-dependent RNA polymerase (RNA polymerase II) and as such has proved to be a powerful research tool (Table 39–1).

Transcription Signals

The DNA sequence analysis of specific genes obtained by recombinant DNA technology has allowed the recognition of a number of sequences important in gene transcription. From the large number of bacterial genes studied it is possible to construct consensus models of transcription promoter and termination signals. Bacterial **promoters** are approximately 40 nucleotide pairs (4 turns of the DNA double helix) in length, a region sufficiently small to be covered by an *E coli* RNA holopolymerase molecule. In this consensus promoter region are 2 short, conserved sequence elements. Approximately 35 base pairs upstream of the transcription start site there is a consensus sequence of 8 nucleotide pairs, shown in Fig 39–5. More proximal to the transcription start site, about 10 nucleotides upstream, is a 6-nucleotide-pair AT-rich sequence. The latter sequence will have a low melting temperature because of its deficiency of GC nucleotide pairs. Thus, the TATA or **Pribnow box** is thought to ease the dissociation between the coding and noncoding strands so that RNA polymerase bound to the promoter region can have access to the nucleotide sequence of its immediately downstream coding strand.

wound DNA region is dictated by the polymerase and is independent of the DNA sequence in the complex. This suggests that RNA polymerase has associated with it an "unwindase" activity that opens the DNA helix. The fact that the DNA double helix must unwind and the strands part at least transiently for transcription implies some disruption of the nucleosome structure of eukaryotic cells.

Termination of the synthesis of the RNA molecule is signaled by a sequence in the coding strand of the DNA molecule, a signal that is recognized by a termination protein, the rho (ρ) factor. After termination of synthesis of the RNA molecule, the core enzyme separates from the DNA template. With the assistance of another σ factor, the core enzyme then recognizes a promoter at which the synthesis of a new RNA molecule commences. More than one RNA polymerase molecule may transcribe the same coding strand of a gene simultaneously, but the process is phased and spaced in such a way that at any one moment each is transcribing a different portion of the DNA sequence. An electron micrograph of RNA synthesis is shown in Fig 39–4.

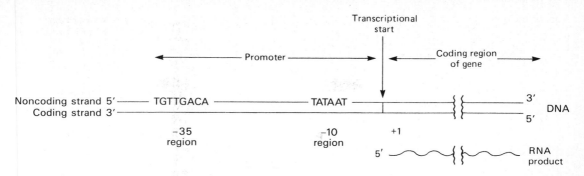

Figure 39–5. Bacterial promoters share 2 regions of highly conserved nucleotide sequence. These are centered 35 and 10 base pairs upstream from the start site of transcription, indicated as +1.

Rho-dependent transcription **termination signals** in *E coli* also appear to have a distinct consensus sequence, as shown in Fig 39–6. The conserved consensus sequence, which also extends about 40 nucleotide pairs in length, can be seen to contain a hyphenated or interrupted inverted repeat, followed by a series of AT base pairs. As transcription proceeds through the hyphenated, inverted repeat, the generated transcript can form the intramolecular hairpin structure, also depicted in Fig 39–6. Transcription continues into the AT region, and with the aid of a termination protein factor called rho (ρ) the RNA polymerase stops and dissociates, releasing the primary transcript.

The transcription signals in mammalian cells are, not unexpectedly, more complex. An extensive analysis of eukaryotic transcription signals has been conducted utilizing recombinant DNA techniques. It is clear that the signals in DNA which control transcription are of several types. Two types of sequence elements are promoter-proximal. One of these defines **where** transcription is to commence along the DNA,

and the other determines **how frequently** this event is to occur. In the thymidine kinase gene of herpes simplex, which utilizes the transcription system of its mammalian host for gene expression, there is a unique transcription start site, and accurate transcription from this start site depends upon a nucleotide sequence 32–16 nucleotides upstream from the start site. This region has the sequence of **TATAAAAG** and bears remarkable homology to the functionally related **Pribnow box** (TATAAT) located about 10 base pairs upstream from prokaryotic mRNA start points. The RNA polymerase II probably binds to DNA in the region of the TATA box and then commences transcription of the template strand about 32 nucleotides downstream at a T, which is surrounded by purines (Fig 39–7). Therefore, the TATA box seems to provide the "where" signal.

Two segments of nucleotides farther upstream from the start site form a single functional element that is responsible for determining how frequently this transcription event occurs. Mutations in either of these

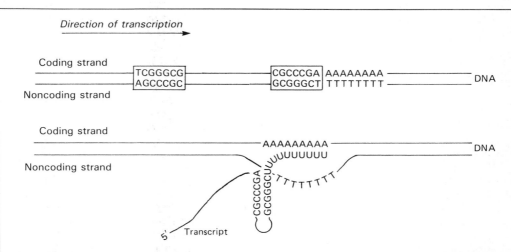

Figure 39–6. The bacterial transcriptional termination signal in the gene contains an inverted, hyphenated repeat followed by a stretch of AT base pairs (above) that upon being transcribed can generate the secondary structure in the RNA transcript shown below.

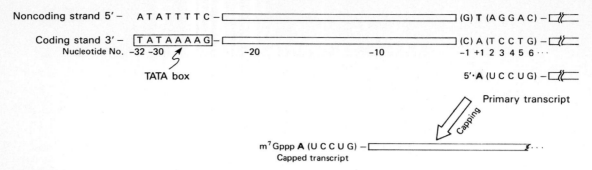

Figure 39–7. Transcription of the thymidine kinase gene. The DNA-dependent RNA polymerase II binds to the complement of the TATA box and commences transcription of the coding strand about 32 nucleotides downstream at a T that is surrounded by purines. The primary transcript is rapidly capped at the first nucleotide (5′-purine).

regions, located between −61 to −47 and −105 to −80 bp upstream from the start site of the thymidine kinase gene, reduce the frequency of transcriptional starts 10- to 20-fold. These basal promoter elements that confer fidelity and frequency of initiation have rigid requirements for position and orientation. Single base changes have dramatic effects on function; spacing with respect to the start site is critical, and they generally do not work if the 5′ to 3′ orientation is reversed (Fig 39–8).

A third class of sequences increase or decrease the basal rate of transcription of eukaryotic genes. These elements are called **enhancers** or **silencers,** depending on which effect they have. These elements can be found in a variety of locations upstream or downstream from the transcription start site. In contrast to basal promoter elements, they can exert their effects when located hundreds or thousands of bases away from *cis*-linked transcription units, and their function is orientation-independent.

Finally, another class of regulatory elements convey adaptive regulation to some genes. Hormone regulatory elements (for steroids, T_3, TRH, cAMP, prolactin, etc) are the prototype of this class (see Chapter 44). Other processes that require gene regulation, such as the response to heat shock, metals (Cd^{2+} and Zn^{2+}), and some toxic chemicals (eg, dioxin) are mediated through specific regulatory elements. Tissue-specific expression of genes, eg, the albumin gene in liver, is also mediated by specific DNA sequences. Some of these adapter structures function like enhancers or silencers in some cases (the glucocorticoid regulatory element acting as an enhancer is an example).

A common feature of all the elements, basal and regulatory, is that the interaction of specific proteins with the DNA sequences is involved. A number of these factors have been identified (Table 39–2), and a great deal of research is directed at analyzing how these protein-DNA interactions influence gene transcription.

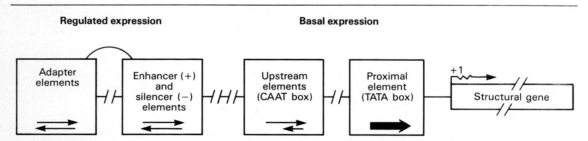

Figure 39–8. Schematic diagram showing the transcription control regions in a typical eukaryotic gene. A functional gene can be divided into its structural and regulatory regions, as defined by the transcription start site (arrow). The regulatory region consists of 2 elements responsible for ensuring basal expression. The proximal component, generally the TATA box, directs RNA polymerase to the correct site (fidelity). Another component, the upstream element, specifies the frequency of initiation. The best-studied of these is the CAAT box, but several other elements may be used in various genes. Regulated expression consists of elements that enhance or silence expression and of others that mediate the response to various signals, including hormones, heat shock, metals, and chemicals. Tissue-specific expression also involves specific sequences of this sort. It is possible that these 2 regulatory regions overlap in function (the connecting line). The orientation-dependence of all the elements is indicated by the arrows within the boxes. For example, the proximal elements must be in the 5′ to 3′ orientation. The upstream elements work best in the 5′ to 3′ orientation, but some of them can be reversed. The broken lines indicate that some elements are not fixed with respect to the transcription start site. Indeed, the elements responsible for regulated expression can be located downstream from the start site.

Table 39–2. Some of the transcription control elements and the factors that bind to them which are found in genes transcribed by RNA polymerase II.

Element	Factor
TATA box	TATA binding protein
CAAT box	CAAT binding protein
$TGG(N)_{6-7}GCCAA$	NF-1 (nuclear factor 1)
GGGCCG	SP-1
ATTTCGAT	NF-A (immunoglobulin gene nuclear factor A)
Enhancers	Many
CNNGAANNTTCNNG	Heat shock transcription factor
TGTTCT (glucocorticoid regulatory element; GRE)	Glucocorticoid receptor
GGCCACGTGACC	USF (upstream factor) _or_ MLTF (major late promoter transcription factor) } from adenovirus

The signals for the termination of transcription by eukaryotic RNA polymerase II are very poorly understood. However, it appears that the termination signals exist far downstream of the coding sequence of eukaryotic genes. For example, the transcription termination signal for mouse β-globin occurs at several positions 1000–2000 bases beyond the site at which the poly(A) tail will eventually be added. Little is known about the termination process, or whether specific termination factors similar to the bacterial ρ factor are involved. The mRNA 3′ terminus is generated posttranscriptionally, and it appears to involve 2 steps. After RNA polymerase II has traversed the region of the transcription unit encoding the 3′ end of the transcript, an RNA endonuclease cleaves the primary transcript at a position about 15 bases 3′ to a consensus sequence of **AAUAAA** that seems to serve in eukaryotic transcripts as a cleavage signal. Finally, this newly formed 3′ terminus is polyadenylated in the nucleoplasm, as described below.

DNA-dependent RNA polymerase III, which transcribes the tRNA genes and the small stable RNA genes (ssRNA; see Chapter 37), recognizes a promoter that is **internal to the gene** to be expressed, rather than upstream of the transcription starting point. In the case of eukaryotic tRNA genes, 2 internal, separated blocks (A and B) of sequences exist that act as an intragenic promoter. The sequences within the A and B blocks exist in the mature tRNA molecule in regions that are highly conserved and participate in the formation of the DHU loop and the TψC loop, respectively (Fig 37–11). By manipulating the structure of tRNA genes, it has been shown that for promoter function the optimal distance between the A and B blocks is 30–40 base pairs and that the transcription start point occurs between 10 and 16 base pairs upstream from the A block. For the 5S RNA gene, which is also transcribed by RNA polymerase III, there appears to be a specific transcription protein factor which, once bound to the intragenic promoter for that gene, probably interacts with an RNA polymerase III molecule to position its catalytic sites on the transcription start point of the DNA.

PROCESSING OF RNA MOLECULES

In prokaryotic organisms the primary transcripts of mRNA-encoding genes begin to serve as translation templates even before their transcription has been completed. Thus, transcription and translation are coupled. Consequently, prokaryotic mRNAs are subjected to little modification and processing prior to carrying out their intended function in protein synthesis. Prokaryotic rRNA and tRNA molecules are transcribed in units considerably longer than the ultimate molecule. In fact, many of the tRNA transcription units contain more than one molecule. Thus, in prokaryotes the processing of these rRNA and tRNA precursor molecules is required for the generation of the ultimate functional molecules.

Nearly all eukaryotic RNA primary transcripts undergo extensive processing between the time they are synthesized and the time at which they serve their ultimate function, whether it be as mRNA or as a structural molecule such as rRNA, 5S RNA, or tRNA. The processing occurs primarily within the nucleus. The processing includes **capping, nucleolytic and ligation reactions, terminal additions** of nucleotides, and **nucleoside modifications.** However, it is clear that, for mammalian cells, 50–75% of the nuclear RNA, including those with capped 5′ termini, do not contribute to the cytoplasmic mRNA. This nuclear RNA loss is significantly greater than can be reasonably accounted for by the loss of intervening sequences alone (see below). Thus, the exact function of the seemingly excessive transcripts in the nucleus of a mammalian cell is unknown.

As a result of advances in techniques for mapping DNA molecules by restriction endonucleases and DNA sequencing, it is now apparent that interspersed within the amino acid-coding portions (**exons**) of many genes are long sequences of DNA that do not contribute to the genetic information ultimately translated into the amino acid sequence of a protein molecule (see Chapter 38). These **intervening sequences (introns)** exist within most but not all genes of higher eukaryotes. The primary transcripts of the

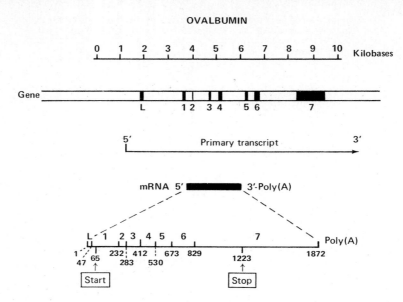

Figure 39–9. The arrangement of noncoding intervening sequences within the gene for chicken ovalbumin. Those informational segments eventually appearing in the mature mRNA are numbered and shown solid. The primary transcript commences upstream of the untranslated L exon and extends beyond the untranslated region of exon 7. The expanded depiction of the mature mRNA includes the corresponding numbers of the exons along the top and the numbers of nucleotides along the bottom. The positions of the translation start and stop codons are indicated.

structural genes contain the transcripts of the interspersed sequences. However, the intron RNA sequences are cleaved out of the transcript, and the exons of the transcript are appropriately spliced together in the nucleus before the resulting mRNA molecule appears in the cytoplasm for translation (Fig 39–9).

The precise mechanisms whereby the introns are removed from the primary transcript in the nucleus, the exons are ligated to form the mRNA molecule, and the mRNA molecule is transported to the cytoplasm are not known. However, as described below, recent experiments have provided us with considerable insight. Although the sequences of nucleotides in the introns of the various eukaryotic transcripts, and even those within a single transcript, are very heterogeneous, there is a consensus sequence at each of the 2 exon-intron (splice) junctions (Fig 39–10). This **consensus sequence at the splice junction** is not sufficiently unique to allow a specific nuclease to cleave only at exon-intron junctions. The very **abundant small nuclear RNA, U1 RNA,** possesses a ribonucleotide sequence with **complementarity to the consensus sequence across the potential splice sites** (Fig

39–11). The U1 RNA molecules are also found associated with specific protein molecules in eukaryotic nuclei. These RNA-protein complexes preferentially bind to the 5′ and 3′ splice junction sequences in RNA, and antibodies to the U1-protein complexes inhibit intron excision in vitro.

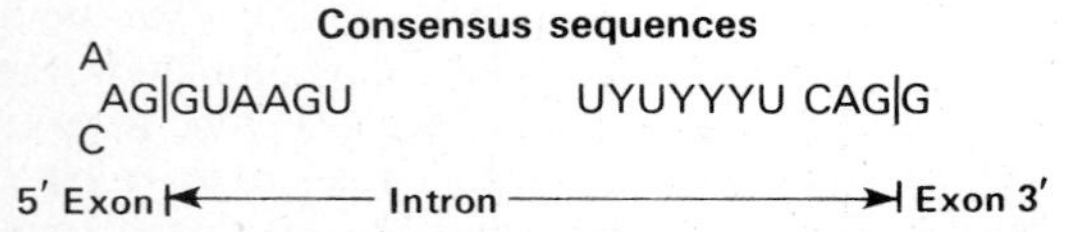

Interestingly, patients with lupus erythematosus, an autoimmune disease, have antibodies to some of these specific nuclear U1-protein molecules. The relationship to the autoimmune disease per se is unknown.

Recently, it has been discovered that during the process of removing the intron sequence from pre-mRNA, there is formed an unusual RNA molecule resembling a lariat. It appears that the 5′ end of the intervening sequence is joined via a 2′–5′ phosphodiester linkage to an adenylate residue 28–37 nucleotides upstream from the 3′ end of the intervening sequence. This process and structure are diagrammed in Fig 39–12.

It seems that the mystery of the relationship between hnRNA and the corresponding mature mRNA in eukaryotic cells is solved. The hnRNA molecules are the primary transcripts plus their early processed products, which, after the addition of caps and poly(A) tails and removal of the portion corresponding to the introns, are transported to the cytoplasm as mature mRNA molecules.

The processing of hnRNA molecules is a potential site for regulation of gene expression. In fact, it has been demonstrated that alternative patterns of RNA splicing are subject to developmental control. For example, the cytoplasmic mRNAs for α-amylase in rat

---

Consensus sequences

A
 AG|GUAAGU   UYUYYYU CAG|G
C

5′ Exon |◄———— Intron ————►| Exon 3′

Figure 39–10. Consensus sequences at splice junctions.

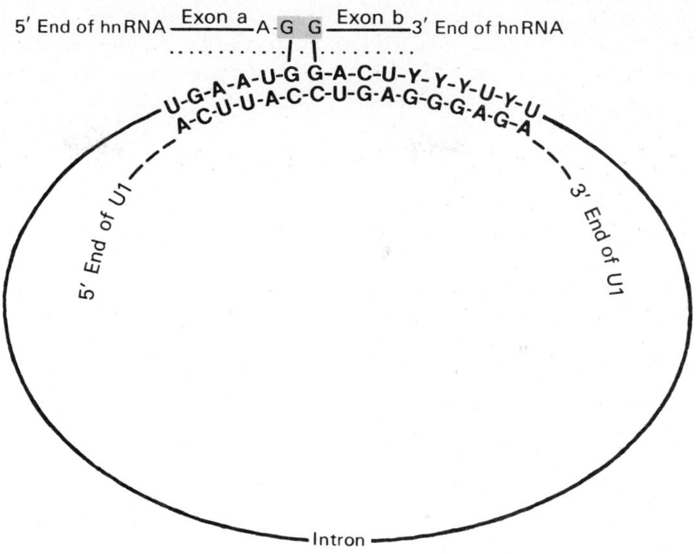

5' End of hnRNA ____Exon a___ A-G G ___Exon b___ 3' End of hnRNA

U-G-A-A-U-G G-A-C-U-Y-Y-Y-U-Y-U
A-C-U-U-A-C-C-U-G-A-G-G-G-A-G-A

5' End of U1

3' End of U1

—Intron—

Figure 39–11. Proposed mechanism for establishing the splice site for intron removal from the hnRNA. The 5' end of U1 snRNA forms a base-pair complex with the distal end of the consensus splice sequence at the 3' end of exon a, and the other end of U1 forms a base pair with the consensus site of exon b. The structure below the dotted line is excised, and the molecule is spliced at the G residues in the shaded circle.

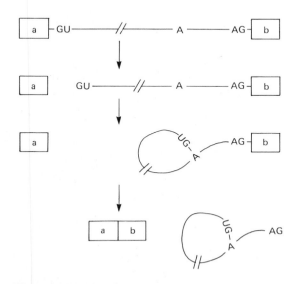

Figure 39–12. Proposed pathway for pre-mRNA splicing. Cleavage at the 5' site is followed by lariat formation and subsequent release of the lariat by cleavage from exon b. The intron is represented by the line and exons a and b by boxes. These reactions all occur with the snRNA and pre-mRNA tightly and specifically bound as part of a ribonucleoprotein structure called the "splicosome."

salivary gland and in rat liver differ in their 5' nucleotide sequence, while the remainder of the mRNA genes containing the coding region and poly(A) addition sites are identical. Further analysis has revealed that although the primary transcripts are huge and overlapping, different splice sites are used to join 2 different cap and leader sequences to the same mRNA "body." In addition, alternative patterns of RNA splicing are used to generate 2 different immunoglobulin heavy-chain mRNAs—one that codes for a membrane-bound heavy chain protein and another that codes for a secreted heavy chain protein (see Chapter 41). Thus, RNA splicing is often necessary for the generation of messenger RNA molecules, and the splicing function also provides a site for differential regulation of gene expression.

At least one form of β-thalassemia, a disease in which the β-globin gene of hemoglobin is severely underexpressed, appears to result from a nucleotide change at an exon-intron junction, precluding removal of the intron and therefore leading to diminished or absent synthesis of the β chain.

Messenger RNA (mRNA)

As mentioned above, mammalian mRNA molecules contain a capped structure at their 5' terminus, and most have a poly(A) tail at the 3' terminus. The cap structures are added in the nucleus prior to transport of the mRNA molecule to the cytoplasm. The poly(A) tails (when present) appear to be added either in the nucleus or in the cytoplasm. The secondary methylations of mRNA molecules, those on the 2'-hydroxy groups and the N_6 of adenylate residues, occur after the mRNA molecule has appeared in the cytoplasm. They also occur in the nucleus and may play a positive role in RNA splicing. The 5' cap of the RNA transcript appears to be required for the formation of the ribonucleoprotein complex necessary for the splicing reactions and may be involved in mRNA transport and translation initiation.

The function of the poly(A) tail is unknown. In any event, the presence or absence of the poly(A) tail does not determine whether a precursor molecule in the nucleus appears in the cytoplasm, because all poly(A)-tailed hnRNA molecules do not contribute to cytoplasmic mRNA, nor do all cytoplasmic mRNA molecules contain poly(A) tails. Cytoplasmic processes in mammalian cells can both add and remove adenylate residues from the poly(A) tails.

The turnover of poly(A)-containing mRNA in cultured mammalian cells is a first-order process with a half-time approximately equal to the doubling time of the cell culture. The kinetics of the degradation of histone mRNA that does not contain a poly(A) tail appears to be a zero-order process in which there is an

age-dependent decay with a lifetime of approximately 6 hours. It is not clear whether this difference is related to the presence or absence of poly(A) or to some intrinsic property of these mRNA molecules.

The size of the cytoplasmic mRNA molecules even after the poly(A) tail is removed is still considerably greater than the size required to code for the specific protein for which it is template, often by a factor of 2 or 3. The extra nucleotides occur in untranslated regions both 5' and 3' to the coding region; the longest sequences are usually at the 3' end. The exact function of these sequences is unknown, but they have been implicated in RNA processing, transport, degradation, and translation.

Transfer RNA (tRNA)

The tRNA molecules, as described in Chapters 37 and 40, serve as adapter molecules for the translation of mRNA into protein sequences. The tRNAs contain many peculiar bases; some are simply methylated derivatives and some possess rearranged glycosidic bonds. The tRNA molecules are transcribed in both

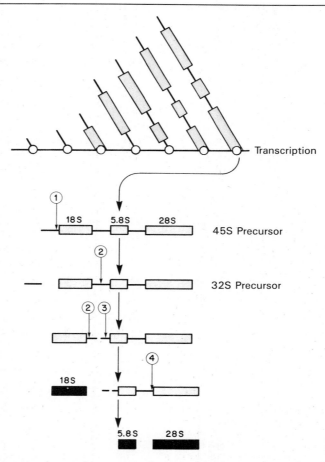

Figure 39–13. Diagrammatic representation of the processing of ribosomal RNA from precursor RNA molecules. The final products are indicated by the solid bars. (Reproduced, with permission, from Perry RP: Processing of RNA. *Annu Rev Biochem* 1976;45:605.)

prokaryotes and eukaryotes as large precursors, frequently containing the sequence for more than one tRNA, which are then subjected to **nucleolytic processing** and reduced in size by a specific class of ribonucleases. In addition, the genes of some tRNA molecules contain—very near the portion corresponding to the anticodon loop—a single intron 10–40 nucleotides long. These introns in the tRNA genes are transcribed; thus, the processing of the precursor transcripts of many tRNA molecules must include removal of the introns and proper splicing of the anticodon region to generate an active adapter molecule for protein synthesis. The nucleolytic processing of tRNA precursors is apparently not determined directly by nucleotide sequence but recognizes 3-dimensional structure and thereby processes only molecules capable of folding into functionally competent products.

The further modification of the tRNA molecules includes nucleotide **alkylations** and the **attachment of the characteristic C · C · A terminus** at the 3′ end of the molecule. This C · C · A terminus is the point of attachment for the specific amino acid that is to enter into the polymerization reaction of protein synthesis. The methylation of mammalian tRNA precursors probably occurs in the nucleus, whereas the cleavage and attachment of C · C · A are cytoplasmic functions, since the termini turn over more rapidly than do the tRNA molecules themselves. Enzymes within the cytoplasm of mammalian cells are required for the attachment of amino acids to the C · C · A residues.

Ribosomal RNA (rRNA)

In mammalian cells, the 2 major rRNA molecules and one minor rRNA molecule are transcribed as part of a single large precursor molecule (Fig 39–13). The precursor is subsequently processed in the **nucleolus** to provide the RNA for the ribosome subunits for the cytoplasm. The rRNA genes are located in the nucleoli of mammalian cells. Hundreds of copies of these genes are present in every cell. The rRNA genes are transcribed as units, each of which encodes (5′ to 3′) an 18S, a 5.8S, and a 28S ribosomal RNA. The primary transcript is a 45S molecule that is highly methylated in the nucleolus. In the **45S precursor,** the eventual 28S segment contains 65 ribose-methyl groups and 5 base-methyl groups. Only those portions of the precursor that eventually become rRNA molecules are methylated. The 45S precursor is nucle-

olytically processed, but the processing signals are clearly distinct from those in hnRNA. Therefore, the nucleolytic processing likely is mediated by a mechanism distinct from that responsible for processing hnRNA to mRNA.

Nearly half of the original primary transcript is degraded, as shown in Fig 39–13. During the processing of rRNA, further methylation occurs, and eventually, in the nucleoli, the 28S chains self-assemble with ribosomal proteins to form the larger 60S subunit. The 5.8S rRNA molecule also formed from the 45S precursor RNA in the nucleolus becomes an integral part of the larger ribosomal subunit. The smaller ribosomal subunits (40S) are formed by the association of constituent peptides with the 18S rRNA molecule.

NUCLEASES

Enzymes capable of degrading nucleic acids have been recognized for many years. These can be classified in several ways. Those which exhibit specificity for deoxyribonucleic acid are referred to as **deoxyribonucleases.** Those which specifically hydrolyze ribonucleic acids are **ribonucleases.** Within both of these classes are enzymes capable of cleaving internal phosphodiester bonds to produce either 3′-hydroxyl and 5′-phosphoryl termini or 5′-hydroxyl and 3′-phosphoryl termini. These are referred to as **endonucleases.** Some are capable of hydrolyzing both strands of a **double-stranded** molecule, whereas others can only cleave **single strands** of nucleic acids. Some nucleases can hydrolyze only unpaired single strands, while others are capable of hydrolyzing single strands participating in the formation of a double-stranded molecule. There exist classes of endonucleases that recognize specific sequences in DNA; the majority of these are the **restriction endonucleases,** which have in recent years become important tools in molecular genetics and medical sciences.

A list of some currently recognized restriction endonucleases is presented in Table 36–1.

Some nucleases are capable of hydrolyzing a nucleotide only when it is present at a terminus of a molecule; these are referred to as **exonucleases.** Exonucleases may act in one direction (3′ → 5′ or 5′ → 3′) only. In bacteria, a 3′ → 5′ exonuclease is an integral part of the DNA replication machinery and there serves to edit the most recently added deoxynucleotide for base-pairing errors.

REFERENCES

Breathnach R, Chambon P: Organization and expression of eucaryotic split genes coding for proteins. *Annu Rev Biochem* 1981;**50**:349.

Busch H et al: SnRNAs, SnRNPs, and RNA processing. *Annu Rev Biochem* 1982;**51**:617.

Chambon P: Eukaryotic nuclear RNA polymerases. *Annu Rev Biochem* 1975;**44**:613.

Cordin J et al: Promoter sequences of eukaryotic protein-coding

genes. *Science* 1980;**209**:1406.

Nevins JR: The pathway of eukaryotic mRNA formation. *Annu Rev Biochem* 1983;**52**:441.

Ruskin B et al: Excision of an intact intron as a novel lariat structure during pre-mRNA splicing in vitro. *Cell* 1984;**38**:317.

Sharp PA: On the origin of RNA splicing and introns. *Cell* 1985;**42**:397.

Protein Synthesis & the Genetic Code

Daryl K. Granner, MD

INTRODUCTION

The language of life consists of a 4-letter alphabet: A, G, T, and C. These letters correspond to the nucleotides found in DNA. They are organized into 3-letter code words called **codons,** and the collection of these codons comprises the **genetic code.** A linear array of codons (a **gene**) specifies the synthesis of various RNA molecules, most of which are involved in some aspect of protein synthesis. Protein synthesis occurs in 3 major steps: initiation, elongation, and termination. This process resembles DNA replication and transcription in its general features and in the fact that it, too, follows a 5' to 3' polarity.

BIOMEDICAL IMPORTANCE

It was impossible to understand protein synthesis, or to explain mutations, before the genetic code was elucidated. The genetic code provides a foundation for explaining the way in which protein defects may cause genetic disease and for the diagnosis and perhaps treatment of these disorders.

INFORMATION FLOW

The genetic information within the nucleotide sequence of DNA is transcribed in the nucleus into the specific nucleotide sequence of an RNA molecule. The sequence of nucleotides in the RNA transcript is complementary to the nucleotide sequence of the coding strand of its gene in accordance with the base-pairing rules.

In prokaryotes there is a linear correspondence between the gene, the **messenger RNA (mRNA)** transcribed from the gene, and the polypeptide product. The situation is more complicated in higher eukaryotic cells, in which the primary transcript, **heterogeneous nuclear RNA (hnRNA),** is much larger than the mature mRNA. The large hnRNA contains coding regions **(exons)** that will form the mature mRNA and long intervening sequences **(introns)** that separate the exons. The hnRNA is processed within the nucleus, and the introns, which often make up much more of the hnRNA than the exons, are removed. Exons are spliced to form mature mRNA, which is transported to the cytoplasm, where it is translated into protein.

The cell must possess the machinery necessary to accurately and efficiently translate information from the nucleotide sequence of an mRNA into the sequence of amino acids of the corresponding specific protein. Clarification of our understanding of this process, which is termed **translation,** awaited deciphering of the genetic code. It was realized early that mRNA molecules in themselves have no affinity for amino acids and, therefore, that the translation of the information in the mRNA nucleotide sequence into the amino acid sequence of a protein requires an intermediate adapter molecule. This **adapter molecule** must recognize a specific nucleotide sequence on the one hand as well as a specific amino acid on the other. With such an adapter molecule, the cell can direct a specific amino acid into the proper sequential position of a protein as dictated by the nucleotide sequence of the specific mRNA. In fact, the functional groups of the amino acids do not themselves actually come into contact with the mRNA template.

CODONS IN PROTEIN SYNTHESIS

In the nucleotide sequence of the mRNA molecule, codons exist for each amino acid. The adapter molecules that translate the codons into the amino acid sequence of a protein are the **transfer RNA (tRNA)** molecules. The **ribosome** is the cellular component on which these various functional entities interact to assemble the protein molecule. Many of these subcellular units (ribosomes) can assemble to translate simultaneously a single mRNA molecule and, in so doing, form a **polyribosome.** The **rough endoplasmic reticulum** is a compartment of membrane-attached polyribosomes that provides for the synthesis of integral membrane proteins and proteins to be exported. Polyribosomal structures also exist free in the cytoplasm, where they synthesize proteins that remain within the cell.

Twenty different amino acids are required for the synthesis of the cellular complement of proteins; thus, there must be at least 20 distinct codons that comprise the genetic code. Since there are only 4 different nucleotides in mRNA, each codon must consist of more than a single purine or pyrimidine nucleotide. Codons consisting of 2 nucleotides each could provide for only 16 (4^2) specific codons, whereas codons of 3 nucleotides could provide 64 (4^3) specific codons.

As a result of the initial observations of Matthaei and Nirenberg, it is now known that each codon consists of a sequence of 3 nucleotides; ie, it is a triplet code. The deciphering of the genetic code was carried out largely in the laboratory of Nirenberg. It depended heavily on the chemical synthesis of nucleotide polymers, particularly triplets, by Khorana.

Three codons do not code for specific amino acids; these have been termed **nonsense codons.** At least 2 of these nonsense codons are utilized in the cell as **termination signals;** they specify where the polymerization of amino acids into a protein molecule is to stop. The remaining 61 codons code for 20 amino acids. Thus, there must be "**degeneracy**" in the genetic code; ie, multiple codons must decode the same amino acid. An examination of the genetic code in Table 40–1 reveals that the 64 possible codons may be arranged in **16 families,** a family of codons being those which have the **same first 2 bases.** Each family occupies a single column between the horizontal lines. For example, the codons CCN, where N can be U, C, A, or G, define a family located in the second column of the second box from the top. In some families, all 4 codons code for the same amino acid, as do members of the CC family described immediately above. These are referred to as **unmixed** families. Eight of the 16 families of codons are unmixed (Table 40–1). Those families of codons which code for more than one amino acid are said to be **mixed** families. In 6 of the mixed families, codons with pyrimidines (U or C) at the third position code for one amino acid, while members with purines (A or G)

at the third position code for another amino acid or chain termination signal (Table 40–1). The 2 remaining families—the UG family and the AU family—do not exhibit either pattern and are unique. Thus, in general, the third nucleotide in a codon is less important than the other 2 in determining the specific amino acid to be incorporated, and this accounts for most of the degeneracy of the code. However, for any specific codon only a single amino acid is indicated; the genetic code is **unambiguous**—ie, given a specific codon, only a single amino acid is indicated. **The distinction between ambiguity and degeneracy is an important concept to be emphasized.**

The unambiguous but degenerate code can be explained in molecular terms. The recognition of specific codons in the mRNA by the tRNA adapter molecules is dependent upon their **anticodon region** and specific base-pairing rules. Each tRNA molecule contains a specific sequence, complementary to a codon, which is termed its anticodon. For a given codon in the mRNA, only a single species of tRNA molecule possesses the proper anticodon. Since each tRNA molecule can be charged with only one specific amino acid, each codon therefore specifies only one amino acid. However, some tRNA molecules can utilize the anticodon to recognize more than one codon. As can be seen from the example of unmixed families, the nucleotide in the anticodon that recognizes the third (3'-) base of the codon could be less discriminating (mixed families) or nondiscriminating (unmixed families) and still manage to insert the proper amino acid when called for. This reduced stringency between the third base of the codon and the complementary nucleotide in the anticodon is referred to as **wobble. Therefore, given a specific codon, only a specific amino acid will be incorporated—although, given a specific amino acid, more than one codon may call for it.**

As discussed below, the reading of the genetic code during the process of protein synthesis does not involve any overlap of codons. **Thus, the genetic code is nonoverlapping.** Furthermore, once the reading is commenced at a specific codon, there is **no punctuation** between codons, and the message is read in a continuing sequence of nucleotide triplets until a nonsense codon is reached.

Until recently, the genetic code was thought to be universal. It has now been shown that the set of tRNA molecules in mitochondria (which contain their own separate and distinct set of translation machinery) from lower and higher eukaryotes, including humans, reads 4 codons differently from the tRNA molecules in the cytoplasm of even the same cells. As noted in Table 40–1, the codon AUA is read as Met, and UGA codes for Trp in mammalian mitochondria. These 2 codons reside in the 2 codon families noted to be unique: the UG family and the AU family. Apparently, in order to minimize the number of tRNA molecules necessary to translate the genetic code, mitochondria have managed to convert the UG family and the AU family to simple mixed types. In addition,

Table 40–1. The genetic code (codon assignments in messenger RNA).*

First Nucleotide	Second Nucleotide				Third Nucleotide
	U	C	A	G	
U	Phe	Ser	Tyr	Cys	U
	Phe	Ser	Tyr	Cys	C
	Leu	Ser	Term	Term†	A
	Leu	Ser	Term	Trp	G
C	Leu	Pro	His	Arg	U
	Leu	Pro	His	Arg	C
	Leu	Pro	Gln	Arg	A
	Leu	Pro	Gln	Arg	G
A	Ile	Thr	Asn	Ser	U
	Ile	Thr	Asn	Ser	C
	Ile†	Thr	Lys	Arg†	A
	Met	Thr	Lys	Arg†	G
G	Val	Ala	Asp	Gly	U
	Val	Ala	Asp	Gly	C
	Val	Ala	Glu	Gly	A
	Val	Ala	Glu	Gly	G

*The terms first, second, and third nucleotide refer to the individual nucleotides of a triplet codon. U, uridine nucleotide; C, cytosine nucleotide; A, adenine nucleotide; G, guanine nucleotide; Met, chain initiator codon; Term, chain terminator codon. Aug, which codes for Met, serves as the initiator codon in mammalian cells. (Abbreviations of amino acids are explained in Chapter 3.)

†In mammalian mitochondria, AUA codes for Met and UGA for Trp, and AGA and AGG serve as chain terminators.

the codons AGA and AGG are read as stop or chain terminator codons rather than as Arg. As a result, mitochondria require only 22 tRNA molecules to read their genetic code, whereas the cytoplasmic translation system possesses a full complement of 31 tRNA species. These exceptions noted, **the genetic code is universal.** The frequency of use of each amino acid codon varies considerably between species and in different tissues within a species. Tables of codon usage are becoming more accurate as more genes are sequenced. This is of considerable importance because investigators often need to deduce mRNA structure from the primary sequence of a portion of protein in order to synthesize an oligonucleotide probe and initiate a recombinant DNA cloning project.

TRANSFER RNA IN PROTEIN SYNTHESIS

There exists at least one species of transfer RNA (tRNA) for each of the 20 amino acids. All of the tRNA molecules have extraordinarily similar functions and 3-dimensional structures. The adapter function of the tRNA molecules requires the charging of each specific tRNA with its specific amino acid. Since there is no affinity of nucleic acids for specific functional groups of amino acids, this recognition must be carried out by a protein molecule capable of recognizing both a specific tRNA molecule and a specific amino acid. At least 20 specific enzymes are required for these specific recognition functions and for the proper attachment of the 20 amino acids to specific tRNA molecules. The process of recognition and attachment (charging) is carried out in 2 steps by one enzyme for each of the 20 amino acids. These enzymes are termed **aminoacyl-tRNA synthetases.** They form an activated intermediate of aminoacyl-AMP-enzyme complex as depicted in Fig 40–1. The specific aminoacyl-AMP-enzyme complex then recognizes a specific tRNA to which it attaches the aminoacyl moiety at the 3′-hydroxyl adenosine terminus. The amino acid remains attached to its specific tRNA in an ester linkage until it is polymerized at a specific position in

the fabrication of a polypeptide precursor of a protein molecule.

The regions of the tRNA molecule referred to in Chapter 37 (and illustrated in Fig 37–11) now become important. The thymidine-pseudouridine-cytidine (TψC) arm is involved in binding of the aminoacyl-tRNA to the ribosomal surface at the site of protein synthesis. The D arm is one of the sites important for the proper recognition of a given tRNA species by its proper aminoacyl-tRNA synthetase. The acceptor arm, located at the 3′-hydroxyl adenosyl terminus, is the site of attachment of the specific amino acid.

The anticodon region consists of 7 nucleotides, and it recognizes the 3-letter codon in mRNA (Fig 40–2). The sequence read from the 3′ to 5′ direction in that anticodon loop consists of a variable base · modified purine · X · Y · Z · pyrimidine · pyrimidine-5′. Note that this direction of reading the anticodon is 3′ to 5′, whereas the genetic code in Table 40–1 is read 5′ to 3′, since the codon and the anticodon loop of the mRNA and tRNA molecules, respectively, are **antiparallel** in their complementarity.

The degeneracy of the genetic code resides mostly in the last nucleotide of the codon triplet, suggesting that the base pairing between this last nucleotide and the corresponding nucleotide of the anticodon is not strict. As referred to above, this is called **wobble;** the pairing of the codon and anticodon can "wobble" at this specific nucleotide-to-nucleotide pairing site. For example, the 2 codons for arginine, A · G · A and A · G · G, can bind to the same anticodon having a uracil at its 5′ end. Similarly, 3 codons for glycine, G · G · U, G · G · C, and G · G · A, can form a base pair from one anticodon, C · C · I. I is an inosine nucleotide, another of the peculiar bases appearing in tRNA molecules.

The codon recognition by a tRNA molecule does not depend upon the amino acid that is attached at its 3′-hydroxyl terminus. This has been ingeniously demonstrated by charging a tRNA specific for cysteine (tRNA$_{cys}$) with radioactively labeled cysteine. By chemical means, the cysteinyl residue was then altered to generate a tRNA molecule specific for cysteine but charged instead with alanine. The chemical transfor-

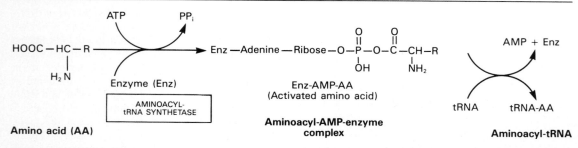

Figure 40–1. Formation of aminoacyl-tRNA. A 2-step reaction, involving the enzyme aminoacyl-tRNA synthetase, results in the formation of aminoacyl-tRNA. The first reaction involves the formation of an AMP-amino acid-enzyme complex. This activated amino acid is next transferred to the corresponding tRNA molecule. The AMP and enzyme are released, and the latter can be reutilized.

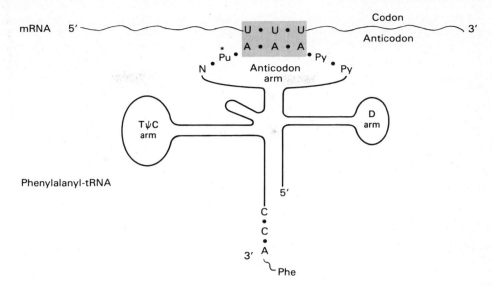

Figure 40–2. Recognition of the codon by the anticodon. One of the codons for phenylalanine is U · U · U. A tRNA charged with phenylalanine (Phe) has the complementary sequence A · A · A; hence, it forms a base pair complex with the codon. The anticodon region typically consists of a sequence of 7 nucleotides: variable (N), modified purine (Pu*), X, Y, Z, and 2 pyrimidines (Py) in the 3' to 5' direction.

mation of the cysteinyl to the alanyl moiety did not alter the anticodon portion of the cysteine-specific tRNA molecule. When this alanyl-tRNA$_{cys}$ was used in the translation of a hemoglobin mRNA, a radioactive alanine was incorporated at what was normally a cysteine site in the hemoglobin protein molecule. The experiment demonstrated that the aminoacyl derivative of an aminoacyl-tRNA molecule does not play a role in the codon recognition. As already noted, the aminoacyl moiety never comes in contact with the template mRNA containing the codons.

MUTATIONS

A mutation is a change in the nucleotide sequence of a gene. Although the initial change may not occur in the coding strand of the double-stranded DNA molecule for that gene, after replication, daughter DNA molecules with mutations in the coding strand will segregate and appear in the population of organisms.

Base-Substitution Mutations

Single base changes (**point mutations**) may be **transitions** or **transversions.** In the former, a given pyrimidine is changed to the other pyrimidine or a given purine is changed to the other purine. Transversions are changes from a purine to either of the 2 pyrimidines or the change of a pyrimidine into either of the 2 purines, as shown in Fig 40–3.

If the nucleotide sequence of the gene containing the mutation is transcribed into an RNA molecule, then the RNA molecule will possess a complementary base change at this corresponding locus.

Single base changes in the mRNA molecules may have one of several effects when translated into protein:

(1) There may be **no detectable effect** because of the degeneracy of the code. This would be more likely if the changed base in the mRNA molecule were to fall on the third nucleotide of a codon. Because of wobble, the translation of a codon is least sensitive to a change at the third position.

(2) A **missense** effect will occur when a different amino acid is incorporated at the corresponding site in the protein molecule. This mistaken amino acid, or missense, depending upon its location in the specific protein, might be **acceptable, partially acceptable,** or **unacceptable** to the function of that protein molecule. From a careful examination of the genetic code, one can conclude that most single base changes would result in the replacement of one amino acid by another with rather similar functional groups. This is an effective mechanism to avoid drastic change in the physical properties of a protein molecule. If an acceptable missense effect occurs, the resulting protein molecule may not be distinguishable from the normal one. A

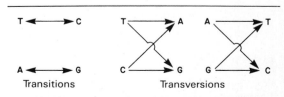

Figure 40–3. Diagrammatic representation of transition mutations and transversion mutations.

partially acceptable missense will result in a protein molecule with partial but abnormal function. If an unacceptable missense effect occurs, then the protein molecule will not be capable of functioning in its assigned role.

(3) A **nonsense** codon may appear that would then result in the **premature termination** of amino acid incorporation into a peptide chain and the production of only a fragment of the intended protein molecule. The probability is high that a prematurely terminated protein molecule or peptide fragment would not function in its assigned role.

Globin Gene Mutations

Much information is available on the amino acid sequences of the normal and abnormal human hemoglobins (see Chapter 6). The hemoglobin molecule can be used to demonstrate the effects of single base changes in the hemoglobin structural gene and to illustrate several principles of the genetic code that have been discussed above.

Some mutations have no apparent effect. The lack of effect of a single base change would be demonstrable only by sequencing the nucleotides in the mRNA molecules or structural genes for hemoglobin from a large number of humans with normal hemoglobin molecules. However, it can be deduced that the codon for valine at position 67 of the β chain of hemoglobin is not identical in all persons possessing the normal β chain of hemoglobin. Hemoglobin Milwaukee has at position 67 a glutamic acid; hemoglobin Bristol contains aspartic acid at position 67. In order to account for the amino acid change by the change of a single nucleotide residue in the codon for amino acid 67, one must infer that the mRNA encoding hemoglobin Bristol possessed a $G \cdot U \cdot U$ or $G \cdot U \cdot C$ codon

prior to a later change to $G \cdot A \cdot U$ or $G \cdot A \cdot C$, both codons for aspartic acid (Fig 40-4). However, the mRNA encoding hemoglobin Milwaukee would have to possess at position 67 a codon $G \cdot U \cdot A$ or $G \cdot U \cdot G$ in order that a single nucleotide change could provide for the appearance of the glutamic acid codons $G \cdot A \cdot A$ or $G \cdot A \cdot G$. Hemoglobin Sydney, which contains an alanine at position 67, could have arisen by the change of a single nucleotide in any of the 4 codons for valine ($G \cdot U \cdot U$, $G \cdot U \cdot C$, $G \cdot U \cdot A$, or $G \cdot U \cdot G$) to the alanine codons ($G \cdot C \cdot U$, $G \cdot C \cdot C$, $G \cdot C \cdot A$, or $G \cdot C \cdot G$, respectively).

Missense Mutations

A. Acceptable Missense Mutations: An example of an acceptable missense mutation (Fig 40-5, top) in the structural gene for the β chain of hemoglobin could be detected by the presence of an electrophoretically altered hemoglobin in the red cells of an apparently healthy individual. Hemoglobin Hikari has been found in at least 2 families of Japanese people. This hemoglobin has asparagine substituted for lysine at the 61 position in the β chain. The corresponding transversion might be either $A \cdot A \cdot A$ or $A \cdot A \cdot G$ changed to either $A \cdot A \cdot U$ or $A \cdot A \cdot C$. The replacement of the specific lysine with asparagine apparently does not alter the normal function of the β chain in these individuals.

B. Partially Acceptable Missense Mutations: A partially acceptable missense mutation (Fig 40-5, center) is best exemplified by **hemoglobin S,** sickle hemoglobin, in which the normal amino acid in position 6 of the β chain, glutamic acid, has been replaced by valine. The corresponding single nucleotide change within the codon would be $G \cdot A \cdot A$ or $G \cdot A \cdot G$ of glutamic acid to $G \cdot U \cdot A$ or $G \cdot U \cdot G$ of

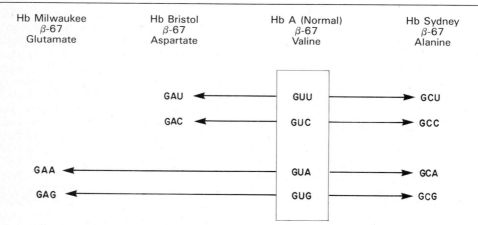

Hb Milwaukee β-67 Glutamate	Hb Bristol β-67 Aspartate	Hb A (Normal) β-67 Valine	Hb Sydney β-67 Alanine
	GAU	GUU	GCU
	GAC	GUC	GCC
GAA		GUA	GCA
GAG		GUG	GCG

Figure 40-4. The normal valine at position 67 of the β chain of hemoglobin A can be coded for by one of the 4 codons shown in the box. In abnormal hemoglobin Milwaukee, the amino acid at position 67 of the β chain contains glutamate, coded for by $G \cdot A \cdot A$ or $G \cdot A \cdot G$, either one of which could have resulted from a single-step transversion from the valine codons $G \cdot U \cdot A$ or $G \cdot U \cdot G$. Similarly, the alanine present at position 67 of the β chain of hemoglobin Sydney could have resulted from a single-step transition from any one of the 4 valine codons. However, the aspartate residue at position 67 of hemoglobin Bristol could have resulted from a single-step transversion only from the $G \cdot U \cdot U$ or $G \cdot U \cdot C$ valine codons.

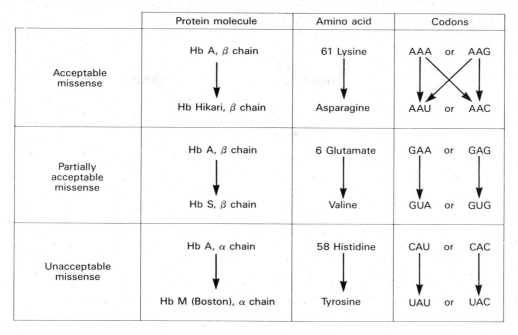

	Protein molecule	Amino acid	Codons
Acceptable missense	Hb A, β chain ↓ Hb Hikari, β chain	61 Lysine ↓ Asparagine	AAA or AAG ↓ (crossed) ↓ AAU or AAC
Partially acceptable missense	Hb A, β chain ↓ Hb S, β chain	6 Glutamate ↓ Valine	GAA or GAG ↓ ↓ GUA or GUG
Unacceptable missense	Hb A, α chain ↓ Hb M (Boston), α chain	58 Histidine ↓ Tyrosine	CAU or CAC ↓ ↓ UAU or UAC

Figure 40–5. Examples of 3 types of missense mutations resulting in abnormal hemoglobin chains. The amino acid alterations and possible alterations in the respective codons are indicated. The hemoglobin Hikari β-chain mutation has apparently normal physiologic properties but is electrophoretically altered. Hemoglobin S has a β-chain mutation and partial function; hemoglobin S binds oxygen but precipitates when deoxygenated. Hemoglobin M Boston, an α-chain mutation, permits the oxidation of the heme ferrous iron to the ferric state and thus will not bind oxygen at all.

valine. Clearly, this missense mutation hinders normal function and results in sickle cell anemia when the mutant gene is present in the homozygous state. The glutamate-to-valine change may be considered to be partially acceptable because hemoglobin S does bind and release oxygen, although abnormally.

C. Unacceptable Missense Mutations: An unacceptable missense mutation (Fig 40–5, bottom) in a hemoglobin gene generates a nonfunctioning hemoglobin molecule. For example, the hemoglobin M mutations generate molecules that allow the Fe^{2+} of the heme moiety to be oxidized to Fe^{3+}, producing methemoglobin. Methemoglobin cannot transport oxygen. (See Chapter 6.)

Frame Shift Mutations

Frame shift mutations result from the deletion or insertion of nucleotides in the gene and thus generate altered nucleotide sequences of mRNA molecules. The deletion of a single nucleotide from the coding strand of a gene results in an altered reading frame in the mRNA. The machinery translating the mRNA does not recognize that a base was missing, since there is no punctuation in the reading of codons. Thus, a major alteration in the sequence of polymerized amino acids, as depicted in example 1, Fig 40–6, results. Altering the reading frame results in a garbled translation of the mRNA distal to the single nucleotide deletion. Not only is the sequence of amino acids distal to this deletion garbled, but reading of the message can also result

in the appearance of a nonsense codon and thus the production of a polypeptide both garbled and prematurely terminated near its carboxyl terminus (example 3, Fig 40–6).

If 3 nucleotides or a multiple of 3 were deleted from a gene, the corresponding messenger when translated would provide a protein from which was missing the corresponding number of amino acids (example 2, Fig 40–6). Because the reading frame is a triplet, the reading phase would not be disturbed for those codons distal to the deletion. If, however, deletion of one or 2 nucleotides occurs just prior to or within the normal termination codon (nonsense codon), the reading of the normal termination signal is disturbed. Such a deletion might result in reading through a termination signal until another nonsense codon was encountered (example 1, Fig 40–6). Excellent examples of this phenomenon are described in discussions of hemoglobinopathies.

Insertions of one or 2 or nonmultiples of 3 nucleotides into a gene result in an mRNA in which the reading frame is distorted upon translation, and the same effects that occur with deletions are reflected in the mRNA translation. This may be **garbled amino acid sequences** distal to the insertion, and the generation of a **nonsense codon** at or distal to the insertion, or perhaps **reading through** the normal termination codon. Following a deletion in a gene, an insertion (or vice versa) can reestablish the proper reading frame (example 4, Fig 40–6). The corresponding mRNA, when

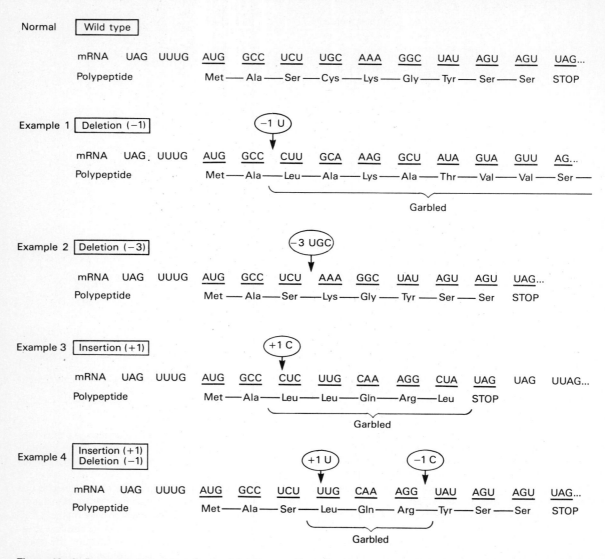

Figure 40-6. Demonstration of the effects of deletions and insertions in a gene on the sequence of the mRNA transcript and of the polypeptide chain translated therefrom. The arrows indicate the sites of deletions or insertions, and the numbers in the circles indicate the number of nucleotide residues deleted or inserted.

translated, would contain a garbled amino acid sequence between the insertion and deletion. Beyond the reestablishment of the reading frame, the amino acid sequence would be correct. One can imagine that different combinations of deletions, of insertions, or of deletions and insertions would result in formation of a protein wherein a portion is abnormal, but this portion is surrounded by the normal amino acid sequences. Such phenomena have been demonstrated convincingly in the bacteriophage T4, a finding which contributed significantly to evidence that the reading frame is a triplet.

Suppressor tRNA Molecules

The above discussion of the altered protein prod-

ucts of gene mutations is based on the presence of normally functioning tRNA molecules. However, in prokaryotic and lower eukaryotic organisms, **abnormally functioning tRNA molecules** have been discovered that are themselves the results of mutations. Some of these abnormal tRNA molecules are capable of suppressing the effects of mutations in distant structural genes. These suppressor tRNA molecules, usually formed as the result of alterations in their anticodon regions, are capable of suppressing missense mutations, nonsense mutations, and frame shift mutations. However, since the suppressor tRNA molecules are not capable of distinguishing between a normal codon and one resulting from a gene mutation, their presence in a cell usually results in decreased viability.

protein, has a single domain but also binds the operator DNA more tightly as a **dimer** (Fig 41–7D). Obviously, the cro protein's single domain mediates both operator binding and dimerization.

In a lysogenic bacterium, ie, a bacterium containing a lambda prophage, the lambda repressor dimer binds **preferentially to O_R1** but in so doing, by a cooperative interaction, **enhances** the binding of another repressor dimer to O_R2 (Fig 41–8). The affinity of repressor for O_R3 is the least of the 3 operator subregions. The binding of repressor to O_R1 has 2 major effects. The occupation of O_R1 by repressor **blocks the binding of RNA polymerase to the rightward promoter** and thereby prevents the expression of the cro gene. Secondly, as mentioned above, repressor dimer bound to O_R1 enhances the binding of repressor dimer to O_R2. The binding of repressor to O_R2 has the important added effect of **enhancing the binding of RNA polymerase to the leftward promoter** that overlaps O_R2 and thereby enhances the transcription and subsequent expression of the repressor gene. This enhancement of transcription is apparently mediated through direct protein-protein interactions between promoter-bound RNA polymerase and O_R2-bound repressor. Thus, the lambda repressor is both a **negative regulator,** by preventing transcription of the cro gene, and a **positive regulator,** by enhancing the transcription of its own gene, the repressor gene. This dual effect of repressor is responsible for the stable state of the dormant lambda bacteriophage; not only does the repressor prevent the expression of the genes necessary for lysis, but it also promotes the expression of itself to stabilize this state of differentiation. In the event that the repressor protein concentration becomes very high, repressor can bind to O_R3 and by so doing diminish the transcription of the repressor gene from the leftward promoter, until the repressor concentration drops and repressor dissociates itself from O_R3.

When a DNA-damaging signal, such as ultraviolet light, strikes the lysogenic host bacterium, fragments of single-stranded DNA are generated that activate a specific **protease** coded by a bacterial gene and referred to as **recA** (Fig 41–8). The activated recA protease hydrolyzes the portion of the repressor protein that connects the amino-terminal and carboxy-terminal domains of that molecule. Such cleavage of the repressor domains causes the **repressor dimers to dissociate,** which in turn causes a **dissociation of the repressor molecules from O_R2** and eventually from O_R1. The effects of removal of repressor from O_R1 and O_R2 are predictable. RNA polymerase immediately has access to the rightward promoter and commences transcribing the **cro gene,** and the enhancement effect of the repressor at O_R2 on leftward transcription is lost (Fig 41–8).

The cro protein translated from the newly transcribed cro gene also binds to the operator region as dimers, but its order of preference is the opposite of that of repressor (Fig 41–8). That is, **cro binds most tightly to O_R3,** but there is no cooperative effect of cro at O_R3 on the binding of cro to O_R2. At increasingly

higher concentrations of cro, the protein will bind to O_R2 and eventually to O_R1.

The occupancy of O_R3 by cro immediately turns off the transcription from the leftward promoter and, hence, **prevents any further expression of the repressor gene.** Thereby the switch is completely effected: the cro gene is now expressed, and the repressor gene is fully turned off. This event is irreversible, and the expression of other lambda genes commences as part of the lytic cycle. When cro repressor concentration becomes quite high, it will eventually occupy O_R1 and in so doing turn down the expression of its own gene, a process that is necessary in order to effect the final stages of the lytic cycle.

The 3-dimensional structure of the cro protein and that of the lambda repressor protein have been determined by x-ray crystallography, and models for their binding and effecting the above-described molecular and genetic events have been proposed and tested. **To date, this system provides the best understanding of the molecular events involved in gene regulation.**

Attenuation of Transcription

Bacterial operons responsible for the biosynthesis of amino acids often have their expression modulated by a **transcriptional termination** process, termed attenuation, which is independent of promoter-operator regulation. This modulation occurs in response to various factors in the cell, particularly those responsible for the translation of codons specifying the amino acid products of the particular synthetic pathway. Regulation by the attenuation process involves translation, ribosome stalling, and shifts between alternative **RNA secondary structures** that form the transcriptional terminator or prevent the formation of the transcriptional terminator. In *E coli* and other bacteria, those operons involved in the biosynthesis of tryptophan, phenylalanine, histidine, threonine, leucine, isoleucine, and valine are all subject to modulation by attenuation.

Attenuation in the tryptophan (Trp) operon of *E coli* has been studied extensively, and that system will be described. The topology of the Trp operon is shown in Fig 41–9. Regulation by the Trp promoter-operator system is analogous to that described above for the lac operon. Repression of the Trp operon can reduce transcription about 70-fold, but mutants **lacking** a functional repressor system are still **capable of responding to tryptophan starvation** by enhancing 8- to 10-fold their rate of synthesis of Trp mRNA. From analyses of other types of mutants in *E coli,* it became apparent that the attenuation process involved the **translation of tryptophan codons** rather than simply a recognition of the concentration of free tryptophan in the cell. It was soon established that within the TrpL region (Fig 41–9), premature transcriptional termination could occur prior to the transcription of the distal Trp genes (TrpEDCBA) in the operon. This transcriptional termination occurred when tryptophan codons within the TrpL region were translated at a normal rate. The premature termination (attenuation) gener-

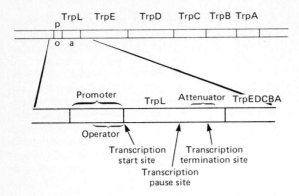

Figure 41–9. The regulatory and structural gene regions of the Trp operon of *E coli*. Transcription initiation is controlled at a promoter-operator. Transcription termination is regulated at an attenuator in the transcribed 162-base-pair leader region, TrpL. All RNA polymerase molecules transcribing the operon pause at the transcription pause site before proceeding further. (Reproduced, with permission, from Yanofsky C: Attenuation in the control of expression of bacterial operons. *Nature* 1981;**289**:751.)

ated a **leader transcript 140 nucleotides long** rather than a transcript of the entire polycistronic operon, as is necessary to generate the enzymes of the tryptophan synthetic pathway. Thus, the attenuation process is capable of detecting a **tryptophan deficiency** by a process involving translation of tryptophan codons and then **communicating this information to the transcribing RNA polymerase** in order that it transcribe beyond the attenuator into the structural genes of the operon. Recombinant DNA technology allowed the construction of specific mutants within this attenuator region and the isolation of pure DNA molecules for

nucleotide sequence analysis. The combination of genetics and recombinant DNA technology allowed the following picture to emerge.

At the Trp promoter, an RNA polymerase molecule that has escaped repressor control begins transcription of the operon and proceeds into the TrpL region, the sequence of which is shown in Fig 41–10, to nucleotide number 90 where it **pauses**. During this pause, a **ribosome attaches** at its own binding site centered on the AUG start codon at nucleotides 27–29 and commences translation of the 14-amino-acid **leader peptide**. At nucleotide 54 in this transcript appear 2 sequential **Trp codons** which, of course, require tryptophan-charged tRNA^Trp if the ribosome is going to proceed beyond these codons. It should be noted that tryptophan is a rare amino acid and the occurrence of 2 sequential tryptophans in a peptide is extremely rare. Thus, the translation of this leader peptide provides a means of sensing the availability of appropriately charged tryptophan-tRNA^Trp molecules and therefore intracellular tryptophan levels. When the RNA polymerase molecule resumes transcription beyond its pause site at nucleotide 90, the ribosome translating the leader will have proceeded to the translation stop codon centered at nucleotide 70 if adequate Trp-tRNA^Trp is available, or the ribosome will have **stalled** at the 2 Trp codons earlier in the peptide. The **position of the ribosome** on the leader transcript will determine which of **2 alternative RNA secondary structures** will be formed in the transcript generated by the RNA polymerase.

The nucleotide sequence of the **leader transcript** is such that by the base-pairing rules (A:U, G:C) **2 hairpin loops** can be formed between regions designated 1 and 2 and between regions 3 and 4 (Fig 41–11). The hairpin loop between regions 3 and 4 generates a transcriptional termination signal that causes the RNA polymerase to stop transcribing be-

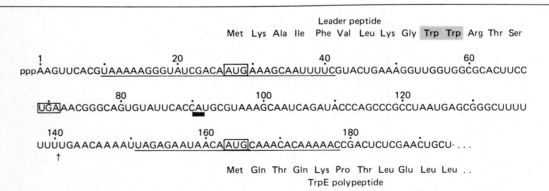

Figure 41–10. The nucleotide sequence of the 5' end of Trp mRNA. The nonterminated transcript is presented. When transcription is terminated at the attenuator, a 140-nucleotide transcript is produced. Its 3' terminus is marked by an arrow. The 3' terminus of the pause transcript, at nucleotide 90, is underlined by a heavy bar. The two AUG-centered ribosome binding sites in this transcript segment are underlined. The boxed AUGs are where translation starts and the boxed UGA where it stops. The predicted amino acid sequence of the Trp leader peptide and the beginning of the TrpE protein are shown. (Reproduced, with permission, from Yanofsky C: Attenuation in the control of expression of bacterial operons. *Nature* 1981;**289**:751.)

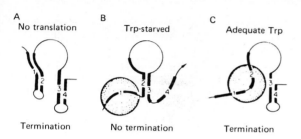

Figure 41–11. Model for attenuation in the *E coli* Trp operon. With excess tryptophan available, the ribosome (dotted circle) translating the newly transcribed leader RNA will synthesize the complete leader peptide. The ribosome will mask regions 1 and 2 of the RNA and will prevent the formation of stem and loop 1:2 or 2:3. Stem and loop 3:4 will be free to form, causing the RNA polymerase molecule (not shown) transcribing the leader region to terminate transcription. During tryptophan starvation, charged tRNA^{Trp} will be the limiting factor, and the ribosome will stall at the adjacent Trp codons in the leader peptide-coding region. Because only region 1 is masked, stem and loop 2:3 will be free to form and will exclude the formation of stem and loop 3:4 (required for termination). RNA polymerase will therefore continue transcription into the structural genes. Under conditions in which the leader peptide is not translated, stem and loop 1:2 can form, as regions 1 and 2 are synthesized. Formation of 1:2 will prevent the formation of 2:3 and will thereby permit the formation of stem and loop 3:4, causing transcription termination. (Reproduced, with permission, from Oxender D, Zurawski G, Yanofsky C: Attenuation in the *Escherichia coli* tryptophan operon: Role of RNA secondary structure involving the tryptophan codon region. *Proc Natl Acad Sci USA* 1979;**76**:5524.)

Phe, His, Leu, Thr, and Ilv leader peptides

PheA: Met-Lys-His-Ile-Pro-*PHE-PHE-PHE*-Ala-*PHE-PHE-PHE*-Thr-*PHE*-Pro

His: Met-Thr-Arg-Val-Gln-Phe-Lys-*HIS-HIS-HIS-HIS-HIS-HIS-HIS*-Pro-Asp

Leu: Met-Ser-His-Ile-Val-Arg-Phe-Thr-Gly-*LEU-LEU-LEU-LEU*-Asn-Ala-Phe-Ile-Val-Arg-Gly-Arg-Pro-Val-Gly-Gly-Ile-Gln-His

Thr: Met-Lys-Arg-*Ile*-Ser-*THR-THR-Ile-THR-THR-THR-Ile-THR-Ile-THR-THR*-Gly-Asn-Gly-Ala-Gly

Ilv: Met-Thr-Ala-*LEU-LEU*-Arg-*VAL-ILE*-Ser-*LEU-VAL-VAL-ILE*-Ser-*VAL-VAL-VAL-ILE-ILE-ILE*-Pro-Pro-Cys-Gly-Ala-Ala-Leu-Gly-Arg-Gly-Lys-Ala

Figure 41–12. The predicted amino acid sequences of the leader peptides of the PheA, His, Leu, Thr, and Ilv (isoleucine, leucine, valine) operons of *E coli* or *S typhimurium*. Amino acids that regulate the respective operons are in italicized capitals and are underlined. (Reproduced, with permission, from Yanofsky C: Attenuation in the control of expression of bacterial operons. *Nature* 1981;**289**:751.)

yond nucleotide 140 and to release a **140-nucleotide-long prematurely terminated transcript.**

However, the nucleotide sequences are such that **regions 2 and 3** are also capable of base pairing, generating an **alternative secondary structure** that **excludes the formation of the termination signal** by regions 3 and 4. When the ribosome stalls at the 2 Trp codons in the leader sequence (Fig 41–11B), **region 1 is protected** and regions 2 and 3 can generate the hairpin loop that excludes the formation of a premature termination signal. This allows the RNA polymerase to transcribe beyond nucleotide 140 and thereby to generate the polycistronic messenger RNA for the synthesis of the enzymes of the tryptophan synthetic pathway.

In the event that there is adequate Trp-tRNA^{Trp} present and the ribosome proceeds through the Trp codons in the leader to the UGA translation stop signal in region 2, both **regions 1 and 2 are protected** by the ribosome. Therefore, regions 3 and 4 can base-pair to generate the premature **termination signal** beyond which RNA polymerase will not transcribe. Accordingly, a **140-nucleotide terminated transcript** is formed rather than the polycistronic message necessary for the enzymes of the tryptophan operon. The formation of these **2 mutually exclusive secondary**

structures between regions 2 and 3 or between regions 3 and 4 transmits to the RNA polymerase information concerning the ability of the cell to translate tryptophan codons.

The predicted amino acid sequences of the leader peptides of several other operons in *E coli* or *Salmonella typhimurium* are shown in Fig 41–12 and reveal the remarkable abundance of codons calling for the amino acid that is the end product of the pathway genetically dictated by the respective operons.

The phenomenon of attenuation has recently been described in mammalian cells. The mechanism is unknown in these cases, but it must be very different from that just described, since transcription and translation occur in separate cellular compartments in eukaryotic organisms.

REGULATION IN EUKARYOTES

The nuclear membrane of eukaryotic cells physically segregates gene transcription from translation, since ribosomes exist only in the cytoplasm. There are many more steps, especially in RNA processing, involved in the expression of eukaryotic genes than of prokaryotic genes, and these steps provide sites for regulatory influences that cannot exist in prokaryotes. These RNA processing steps in eukaryotes include capping of the 5′ end of the primary transcript, addition of a polyadenylate tail to the 3′ end of transcripts, and excision of intron regions to generate spliced exons in the mature mRNA molecule. To date, the analyses of eukaryotic gene expression have provided evidence that regulation occurs at the level of **transcription, nuclear RNA processing,** and **mRNA sta-**

bility. In addition, **gene amplification** and **rearrangement** have been shown to occur and to influence gene expression.

Owing to the advent of recombinant DNA technology, much progress has been made in recent years in the understanding of eukaryotic gene expression. However, because most eukaryotic organisms contain so much more genetic information than do prokaryotes and the manipulation of their genes is so much more limited, molecular aspects of eukaryotic gene regulation are less well understood. This section briefly describes a few different types of eukaryotic gene regulation.

Gene Amplification During Development

During early development of metazoans, there is an abrupt increase in the need for specific molecules such as ribosomal RNAs and messenger RNA molecules for proteins that make up such organs as the eggshell. One way to increase the rate at which such molecules can be formed is to increase the number of genes available for transcription of these specific molecules. Among the repetitive DNA sequences are hundreds of copies of ribosomal RNA genes and tRNA genes. These genes preexist repetitively in the genomic material of the gametes and, thus, are transmitted in high copy number from generation to generation. In some specific organisms such as the fruit fly *(Drosophila)*, there occurs during oogenesis an amplification of a few preexisting genes, such as those for the chorion (eggshell) proteins. Subsequently, these amplified genes, presumably generated by a process of repeated initiations during DNA synthesis (compounded replication bubbles) provide multiple sites for gene transcription (Figs 38–16 and 41–13).

In recent years, it has been possible to promote the amplification of specific genetic regions in cultured mammalian cells. In some cases, a several thousand-fold increase in the copy number of specific genes can be achieved over a period of time involving increasing doses of selective drugs. In fact, it has been demonstrated in patients receiving methotrexate for treatment of cancer that malignant cells can develop **drug resistance** by increasing the number of genes for dihydrofolate reductase, the target of methotrexate. Gene amplification events such as these occur spontaneously in vivo, ie, in the absence of exogenously supplied selective agents, and these unscheduled extra rounds of replication can become "frozen" in the genome under appropriate selective pressures.

Immunoglobulin Gene Rearrangement

Some of the most interesting and perplexing questions raised by biologists in recent decades concern the genetic and molecular basis of antibody diversity (see Chapter 55). In addition, advances in immunology have made it apparent that as cells of the humoral immunity system differentiate, they produce antibodies with the same specificity but different effector func-

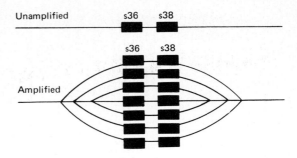

Figure 41–13. Schematic representation of the amplification of chorion protein genes s36 and s38. (Reproduced, with permission, from Chisholm R: Gene amplification during development. *Trends Biochem Sci* 1982; 7:161.)

tions. Within the last several years, many laboratories have contributed greatly to the understanding of the genetic basis of antibody diversity and regulation of the expression of immunoglobulin genes during development and differentiation.

As described in Chapter 39, the coding segments responsible for the generation of specific protein molecules are frequently not contiguous in the mammalian genome. The coding segments for the variable and the constant domains of the immunoglobulin (antibody) light chain were the first recognized to be separated in the genome. As described in more detail in Chapter 55, immunoglobulin molecules are composed of 2 types of polypeptide chains, light (L) and heavy (H) chains (see Fig 55–3). The L and H chains are each divided into N-terminal variable (V) and carboxy-terminal constant (C) regions. The V regions are responsible for the recognition of antigens (foreign molecules) and the constant regions for effector functions that determine how the antibody molecule will dispense with the antigen.

There are 3 unlinked families of genes responsible for immunoglobulin molecule structure. Two families are responsible for the light chains (λ and κ chains) and one family for heavy chains.

Each **light chain** is encoded by 3 distinct segments: the variable (V_L), the joining (J_L), and the constant (C_L) segments. The mammalian haploid genome contains over 500 V_L segments, five or six J_L segments, and perhaps ten or twenty C_L segments. During the differentiation of a lymphoid B cell, a V_L segment is brought from a distant site on the same chromosome to a position closer to the region of the genome containing the J_L and C_L segments. This **DNA rearrangement** then allows the V_L, J_L, and C_L segments to be transcribed as a single mRNA precursor and subsequently processed to generate the mRNA for a specific antibody light chain. By rearrangement of the various V_L, J_L, and C_L segments in the genome, the immune system can generate an immensely diverse library of antigen-specific immunoglobulin molecules. This DNA rearrangement is referred to as **V-J joining** of the light chain.

The **heavy chain** is encoded by 4 gene segments: the V_H, the D (diversity), the J_H, and the C_H DNA segments. The variable region of the heavy chain is generated by joining the V_H with a D and a J_H segment. The resulting V_H-D-J_H DNA region is in turn linked to a C_H gene, of which there are 8. These C_H genes (C_μ, C_δ, $C_\gamma 3$, $C_\gamma 1$, $C_\gamma 2b$, $C_\gamma 2a$, C_α, and C_ϵ) determine the immunoglobulin class or subclass—IgM, IgG, IgA, etc—of the immunoglobulin molecule (see Chapter 55). An example of the rearrangements and processing events that result in the formation of a $C_\gamma 2b$ heavy chain gene is shown in Fig 41–14.

During its differentiation, a B cell that secretes antibody to a **specific antigen** will secrete antibodies of different classes having the same antigen specificity but different biologic roles. The different classes of immunoglobulins contain the same light chains and V_H regions but different C_H regions. Thus, a single B cell and its clonal derivatives can undergo "class switching." Class switching is the result of a second type of

DNA rearrangement occurring during differentiation of the immune system.

Developmentally and temporally, the **V-J joining** for light chain expression and the **V-D-J joining** for the heavy chain expression **precede the class-switching DNA rearrangement.**

Class Switching

During the ontogeny of an immunoglobulin-secreting B cell and its clonal derivatives, including the terminally differentiated plasma cell, the sequence of immunoglobulin production and secretion commences with IgM and subsequently switches to IgA or IgG, etc. In the germ line genome, the J_H segments are next to the C_μ genes; thus, once the V_H-D-J_H rearrangement has occurred, no further DNA rearrangement is necessary to allow transcription of an mRNA precursor for a μ chain. However, as differentiation proceeds and immunoglobulin production switches from IgM to IgA, the V-D-J region of the parent B cell must be rear-

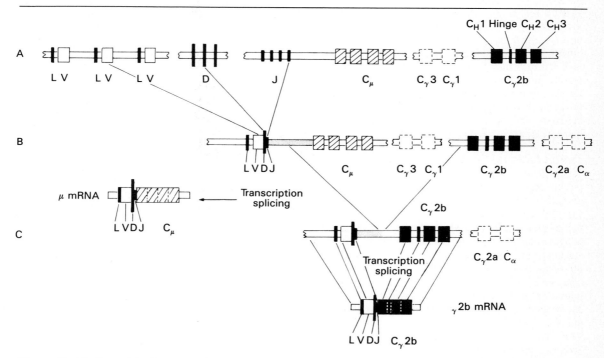

Figure 41–14. Recombination events leading to a complete immunoglobulin heavy chain ($\gamma 2b$) gene. **A:** The germ line DNA before rearrangement. There is a cluster of at least 50 genes (each with a short leader sequence L) coding for part of the variable (V) region, a cluster of D gene segments coding for most of the third hypervariable region, and some distance away there are four J segments that complete the V region coding sequence. The J segments lie about 8000 bases from the C_μ gene that lies at the start of a cluster containing all the C region genes. The C region gene sequences are interrupted by noncoding sequences to give a series of exons that coincide with the domains and the hinge region in the C region amino acid sequence. **B:** In the first translocation event, one of each of the V, D, and J segments are recombined to give a complete μ chain transcription unit. The transcript is a copy of the gene as shown, but the noncoding sequences (introns) are removed by splicing events that lead to a continuous coding sequence in the μ mRNA. **C:** A second translocation event, the heavy-chain switch, deletes the C_μ, $C_\gamma 3$, and $C_\gamma 1$ gene segments and places the V-D-J segment and part of the J-C_μ intron near the $C_\gamma 2b$ gene. Following transcription, the introns are removed by splicing, leading to a continuous coding sequence in the $\gamma 2b$ mRNA. (Reproduced, with permission, from Molgaard HV: Assembly of immunoglobulin heavy chain genes. *Nature* 1980;**286**:659.)

ranged with a C_α gene to permit the transcription of an mRNA precursor for an α chain containing the same antigen-specific variable region.

The physical order of the 8 closely linked C_H genes is C_μ, C_δ, $C_\gamma 3$, $C_\gamma 1$, $C_\gamma 2b$, $C_\gamma 2a$, C_α, and C_ϵ. The **temporal order of the class switching is unidirectional** within this physical order, from left to right. In most cases studied to date, rearrangement of the C_H genes seems to involve deletion of those C_H genes 5' to (left of) the C_H gene joined to the V-D-J region.

An example of the recombination or rearrangement events leading to the complete $\gamma 2b$ gene is shown in Fig 41–14. The sequence first involves rearrangement of the V-D-J segments and subsequently the appropriate deletion or rearrangement of the C_H genes. The C region gene sequences coincide with the domains in the hinge region described in Chapter 55. The intervening sequences or introns that are transcribed and appear in the primary transcript are removed by RNA splicing events described in Chapter 39.

The information encoded in the genome can clear be increased by this combinatorial joining of gene segments. This mechanism would not only increase the variable region diversity but would also effectively allow rearranged (useful or advantageous) information to be retained and perpetuated as a cell line alters its effector function during differentiation.

This seemingly complex DNA rearrangement during development and differentiation could be rather simply regulated by the appropriate induction and repression of specific joining proteins that recognize the highly conserved sequences flanking the coding sequences to be rearranged.

Transcriptional Control

In Chapter 39, a promoter is described as that region of a gene to which RNA polymerase attaches in order to commence transcription of a gene at a specific site. In general (as illustrated in Fig 39–12), promoters dictate precisely **where** the RNA polymerase is to start transcription, but the issue of **when** (ie, how frequently) to commence transcription is a more complex issue that is less well understood. As described in Chapter 39, 2 separate DNA elements and associated binding proteins mediate the where and when. In the extreme case of a zero rate of transcription, the where and when issues become indistinguishable and irrelevant, since transcription is not starting at all. Thus, the "where" signal is a potential signal and is meaningless if the more dominant "when" signal is "not now." As discussed in Chapter 38, there appear to be quite large regions of chromatin that are transcriptionally inactive, either constitutively or facultatively, while other areas of chromatin are potentially active chromatin. Additionally, as described in Chapter 38, there is evidence that the methylation of deoxycytidine residues in DNA may effect gross changes in chromatin so as to preclude its active transcription. For example, in mouse liver only the unmethylated ribosomal genes can be expressed, and there is evidence that many animal viruses are not transcribed when their DNA is

methylated. However, it is *not* possible to generalize that methylated DNA is transcriptionally inactive, that all inactive chromatin is methylated, or that active DNA is not methylated.

The Role of Enhancers

In addition to gross changes in chromatin affecting transcriptional activity, there is increasing evidence that there are signals in DNA which facilitate or enhance initiation at the promoter. For example, in simian virus 40 (SV40) there exists about 200 bp upstream from the promoter of the early genes a region of 2 identical, tandem 72-base-pair lengths that can greatly increase the expression of genes in vivo. These so-called **enhancer elements** differ from the promoter in 2 remarkable ways. They can exert their positive influence on transcription even when separated by thousands of base pairs from a promoter, and they work when oriented in either direction. Enhancers are promiscuous; they can stimulate any promoter in the vicinity. The SV40 enhancer element can exert an influence on, for example, the transcription of β-globin by increasing its transcription 200-fold in cells containing both the enhancer and the β-globin gene on the same plasmid. The enhancer element does not seem to be producing a product that in turn acts on the promoter, since it is active only when it exists within the same DNA molecule as (ie, *cis* to) the promoter. Enhancer binding proteins are now being isolated, and these should help elucidate how these elements work. Enhancer elements do appear to convey nuclease hypersensitivity to those regions where they reside (see Chapter 38).

Many genes have now been recognized to harbor enhancer elements in various locations relative to their coding regions. In addition to being able to enhance gene transcription, some of these enhancer elements clearly possess the ability to do so in a tissue-specific manner. Thus, the enhancer element associated with the immunoglobulin genes between the J and C regions enhances the expression of those genes preferentially in lymphoid cells. Enhancer elements associated with the genes for pancreatic enzymes are capable of enhancing even unrelated but physically linked genes preferentially in the pancreatic cells of mice into which the specifically engineered gene constructions were introduced microsurgically at the single-cell embryo stage. Tissue-specific gene expression may therefore be mediated by enhancers or enhancerlike elements.

Other Regulatory Elements

By ligating regions of DNA suspected of harboring regulatory sequences to various reporter genes (the **fusion** or **chimeric gene approach,** described in Chapter 36), one can determine which regions in the vicinity of structural genes have an influence on their expression. In many cases, it has been demonstrated that regions 5' to the transcriptional start site exert profound influence on how often the RNA polymerase commences transcription. For example, **metallothio-**

nein is a heavy metal-binding protein that contains many cysteine residues and exists in most organs of mammals. When an organism or its cultured cells are exposed to metal ions such as zinc or cadmium, there is an enhanced rate of transcription of the metallothionein gene and a subsequent increase in the metallothionein protein to bind the potentially toxic heavy metal. By the use of recombinant DNA technology, it has been possible to isolate the DNA region a few hundred base pairs proximal to the transcriptional start site of the metallothionein gene. Another structural gene, such as that for thymidine kinase, can then be ligated to this **"metallothionein regulatory element"** and the synthetic construct introduced to cultured cells, a small number of which will integrate the DNA into its own genome. When those cells are exposed to heavy metals, the metallothionein promoter region effects an induction of thymidine kinase. By using progressively shorter pieces of DNA, one can pinpoint the location of the element. A similar experiment has recently been conducted in mice. The metallothionein promoter region was ligated to the structural gene for thymidine kinase or to the structural gene for growth hormone. The engineered genetic constructions were introduced into the male pronuclei of single-cell mouse embryos and the embryos placed into the uterus of a surrogate mother to develop. Offspring have been generated under these conditions, and in some the addition of zinc ions to their drinking water will effect an increase in liver thymidine kinase or growth hormone. In the latter case, these transgenic animals have responded to the high levels of growth hormone by becoming twice as large as their normal litter mates.

Glucocorticoids are one class of steroid hormones that regulate gene expression (see Chapter 44). Once glucocorticoids enter a mammalian cell, they bind to a steroid-specific receptor molecule that undergoes a conformational change in the cytoplasm and enters the nucleus. The glucocorticoid-receptor complex in the nucleus binds to a specific receptor recognition site on DNA a few hundred base pairs 5′ upstream from the transcription start site for steroid-responsive genes, eg, mouse mammary tumor virus. The occupancy of this receptor recognition site on DNA appears to influence the efficiency of utilization of the promoter by RNA polymerase and thereby influence the expression of the steroid-responsive gene. Again by recombinant DNA technology, the DNA region that binds the glucocorticoid-receptor complex can be molecularly cloned and ligated to other unrelated structural genes. When the new constructions are introduced into cultured cells and integrated into their genome, the structural genes come under the regulatory influences of glucocorticoids added to the culture medium and, thus, are converted to steroid-inducible genes. By whittling away at the ends with nucleases and introducing changes (mutations) into the apparent regulatory region of the cloned DNA, molecular biologists can identify the specific DNA sequence to which the glucocorticoid-receptor complex binds. In this case, it seems that the binding of the steroid-recep-

tor complex to the specific DNA sequence converts it to an active enhancer element. In the near future, we should understand at the molecular level in eukaryotes precisely how gene expression is regulated, for example, by steroid hormones.

RNA Processing as a Control Mechanism

In addition to regulating gene expression by affecting the efficiency of promoter utilization, eukaryotic cells can utilize **alternative RNA processing** to control gene expression. There are 2 general types of RNA processing control, **process versus discard** decisions and **differential processing.** Relative to the former, it is apparent that when primary transcripts contain introns, those intron sequences must be removed before the transcript can mature to a messenger RNA and appear in the cytoplasm for expression. There are **many more primary transcripts in the nucleus than are represented as messenger RNA molecules in the cytoplasm.** Thus, there must exist regulatory decisions as to which transcripts will ultimately be expressed and which will be discarded. There is no information available as to the mechanisms involved in such processes or even direct evidence that such decisions can be changed during differentiation or development or in response to environmental influences.

The direct evidence for differential processing of primary transcripts is derived from studies of the regulation of immunoglobulin synthesis.

Differential RNA Processing

As described above, the first immunoglobulin synthesized by differentiating B cells is IgM. However, the IgM of the earliest B cell is not extruded all the way through the membrane to be secreted. Instead, the C-terminal region of the μ chain remains trapped in the membrane as an integral protein (see Chapter 42). The μ chain of **secreted IgM** is referred to as μ_s, and the μ chain of the **membrane-associated IgM** is called μ_m. The μ_s and μ_m chains of the same B cell or B cell lineage contain identical amino acid sequences up to the very C-terminal region of the $C_\mu 4$ domain (see Fig 55–3). The μ_s chain has a 20-amino-acid **hydrophilic** C-terminal segment after the $C_\mu 4$ domain, while the μ_m chain has a C-terminal segment containing 38 **hydrophobic** amino acids followed by -Lys-Val-Lys. This hydrophobic amino acid sequence can thus embed itself in the membrane bilayer up to the charged Lys residues (see Chapter 42). Accordingly, μ_s and μ_m must be translated from different mRNA molecules.

These different mRNA molecules have been isolated and their nucleotide sequences determined indirectly by DNA sequencing. The μ_m mRNA consists of 2700 bases and the μ_s mRNA 2400 bases. By isolating and sequencing the genomic region coding for these μ mRNA molecules, it has been possible to demonstrate that they are both derived from a **common mRNA precursor** molecule as a result of **alternative RNA**

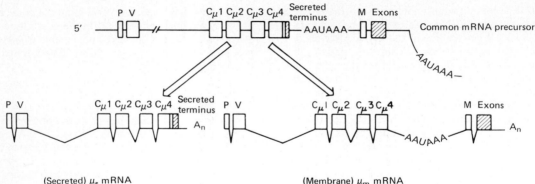

Figure 41–15. Splicing patterns deduced for μ_m and μ_s mRNAs. The μ_m and μ_s mRNAs are identical 5′ to $C_\mu 4$. Raised boxes indicate exons. 3′ Untranslated sequences are crosshatched. P refers to the signal peptide exon and V to the rearranged V_H exon. Bent lines indicate RNA splicing between exons. The signals for alternative poly(A) track (A_n)) additions (AAUAAA) are indicated in the transcripts. (Reproduced, with permission, from Early P et al: Two mRNAs can be produced from a single immunoglobulin μ gene by alternative RNA processing pathways. *Cell* 1980;20:313.)

processing pathways in the nucleus. Fig 41–15 depicts the 2 splicing patterns deduced for the μ_m and μ_s mRNA molecules transcribed from a single μ gene.

The common mRNA precursor contains **2 poly(A) addition sites** (see Chapter 39), one between the $C_\mu 4$ exon and the M exons and the other 3′ to the M exons. Depending upon which potential poly(A) site is endonucleolytically cleaved in preparation for the addition of the poly(A) track, 2 μ heavy-chain mRNAs with different 3′ region sequences can be formed, one for the μ_m chain and one for the μ_s chain. Thus, it may well be that the site chosen for poly(A) addition dictates which exons in the primary transcript shall be spliced.

Messenger RNA Stability

The stability of messenger RNA molecules in the cytoplasm can clearly affect the level of gene expression in a positive or negative direction. Stabilization of mRNA, given a fixed rate of transcription, would lead to increased accumulation, and vice versa. Little is known about the mechanisms involved in mRNA degradation, but a few examples of control of this process have been reported. Estradiol prolongs the half-life of vitellogenin mRNA from a few hours to more than 200 hours. This, coupled with the fact that estrogens enhance the rate of transcription of this gene by 4- to 6-fold, results in a tremendous increase of vitellogenin mRNA. The half-life of the mRNA for the milk protein casein is also greatly increased by exposing mammary cells to the hormone prolactin.

Differential mRNA Translation

There are specific examples of organisms or cultured cells that appear to differentially translate mature mRNA molecules, the translation efficiencies of which cannot be distinguished using in vitro translation systems. This suggests that in these systems there exist molecules capable of recognizing specific mature mRNA molecules and affecting the rates at which they are translated relative to other messenger molecules.

Table 41–2 summarizes the frequency of various types of control of eukaryotic gene expression.

Table 41–2. Summary of frequency of various types of control.*

	Examples Proved or Strongly Suggested	Possible
Nuclear		
Transcriptional		
Initiation	Many, >100	
Termination		
Premature (attenuation)	1	+
readthrough		
RNA processing		
Poly(A) choice	~3	
Splicing choice	1	
Process versus discard		+
Cytoplasmic		
mRNA stability	~5 Specific, many general	
mRNA translation efficiency	~10 Specific, many general	

For those cases of control where only one or a few cases are proved, it is anticipated that these numbers will increase. It is not possible at present to even guess at the true frequency with which various mechanisms will be found ultimately to be used.

*Reproduced, with permission, from Darnell JE Jr: Variety in the level of gene control in eukaryotic cells. *Nature* 1982; **297**:365. Copyright © 1982 by Macmillan Journals Ltd.

REFERENCES

Compere SJ, Palmiter RD: DNA methylation controls the inducibility of the mouse metallothionein-I gene in lymphoid cells. *Cell* 1981;**25:**233.

Darnell JE: Variety in the level of gene control in eukaryotic cells. *Nature* 1982;**297:**365.

Jacob F, Monod J: Genetic regulatory mechanisms in protein synthesis. *J Mol Biol* 1961;**3:**318.

McKnight S, Tjian R: Transcriptional selectivity of viral genes in mammalian cells. *Cell* 1986;**46:**795.

Ptashne M, Johnson AD, Pabo CO: A genetic switch in a bacterial virus. *Sci Am* (Nov) 1982;**247:**128.

Shimizu A, Honjo T: Immunoglobulin class switching. *Cell* 1984;**36:**801.

Swift GH et al: Tissue-specific expression of the rat pancreatic elastase I gene in transgenic mice. *Cell* 1984;**38:**639.

Wu R, Bahl CP, Narang SA: Lactose operator-repressor interaction. *Curr Top Cell Regul* 1978;**13:**137.

Yamamoto K: Steroid receptor regulated transcription of specific gene and gene networks. *Annu Rev Genet* 1985; **19:**209.

Yanofsky C: Attenuation in the control of expression of bacterial operons. *Nature* 1981;**289:**751.

Section V.
Biochemistry of Extracellular & Intracellular Communication

Membranes: Structure, Assembly, & Function

42

Daryl K. Granner, MD

INTRODUCTION

Membranes are highly viscous yet plastic structures that surround all living cells. Plasma membranes form closed compartments around cellular protoplasm to separate one cell from another and thus permit cellular individuality. The plasma membrane has selective permeabilities and acts as a barrier, thereby maintaining differences in composition between the inside and the outside of the cell. The selective permeabilities are provided by channels and pumps for ions and substrates and by specific receptors for signals, eg, hormones. Plasma membranes also exchange material with the extracellular environment by exocytosis and endocytosis, and there are special areas of membrane structures—the gap junctions—through which adjacent cells exchange material.

Membranes also form specialized compartments within the cell. Such intracellular membranes form many of the morphologically distinguishable structures (organelles), eg, mitochondria, endoplasmic reticulum, sarcoplasmic reticulum, Golgi complexes, secretory granules, lysosomes, and the nuclear membrane. Membranes localize enzymes, function as integral elements in excitation-response coupling, and provide sites of energy transduction, such as in photosynthesis and oxidative phosphorylation.

BIOMEDICAL IMPORTANCE

Gross alterations of membrane structure can affect water balance and ion flux and therefore every process within the cell. Specific deficiencies or alterations of certain membrane components lead to a variety of diseases. Examples include the lysosomal absence of acid maltase, causing type II glycogen storage disease; the lack of an iodide transporter, causing congenital goiter (see Fig 46–2); and defective endocytosis of low-density lipoproteins, resulting in accelerated hypercholesterolemia and coronary artery disease. Normal cellular function obviously begins with normal membranes.

MAINTENANCE OF A NORMAL ENVIRONMENT

The maintenance of a normal environment around and in a cell is a fundamental requirement. Life originated in an aqueous environment; enzyme reactions, cellular and subcellular processes, and so forth have therefore evolved to work in this milieu. Since most mammals live in a gaseous environment, how is the aqueous state maintained? Membranes accomplish this by internalizing and compartmentalizing body water.

The Internal Water Compartments

Water makes up about 56% of the lean body mass of the human body and is distributed in 2 large compartments.

A. Intracellular Fluid (ICF): This compartment constitutes two-thirds of total water and provides the environment for the cell to (1) make, store, and utilize energy; (2) repair itself; (3) replicate; and (4) perform special functions.

B. Extracellular Fluid (ECF): This compartment contains about one-third of total water and is distributed between the plasma and interstitial compartments. The extracellular fluid is a delivery system. It brings to the cells nutrients (eg, glucose, fatty acids, amino acids), oxygen, various ions and trace minerals, and a variety of regulatory molecules (hormones) that coordinate the functions of widely separated cells. Extracellular fluid removes CO_2, waste products, and toxic or detoxified materials from the immediate cellular environment.

445

Table 42–1. Comparison of the mean concentration of various substances outside and inside a mammalian cell.

Substance	Extracellular Fluid	Intracellular Fluid
Na^+	140 mmol/L	10 mmol/L
K^+	4 mmol/L	140 mmol/L
Ca^{2+} (free)	2.5 mmol/L	0.1 μmol/L
Mg^{2+}	1.5 mmol/L	30 mmol/L
Cl^-	100 mmol/L	4 mmol/L
HCO_3^-	27 mmol/L	10 mmol/L
PO_4^{3-}	2 mmol/L	60 mmol/L
Glucose	5.5 mmol/L	0–1 mmol/L
Protein	2 g/dL	16 g/dL

Composition of Cells & Extracellular Fluid

The ionic composition of intracellular fluid differs greatly from that of extracellular fluid, as illustrated in Table 42–1. The internal environment is rich in K^+ and Mg^{2+}, and phosphate is its major anion, whereas extracellular fluid is characterized by high Na^+ and Ca^{2+} content and Cl^- as the major anion. Note also that glucose is higher in extracellular fluid than in the cell, whereas the opposite is true for proteins. Why is there such a difference? It is thought that the primordial sea in which life originated was rich in K^+ and Mg^{2+}. It therefore follows that enzyme reactions and other biologic processes evolved to function best in that environment, hence the high concentration of these ions within cells. Cells were faced with strong selection pressure as the sea gradually changed to a composition rich in Na^+ and Ca^{2+}. Vast changes would have been required for evolution of a completely new set of biochemical and physiologic machinery; instead, as it happened, cells developed barriers—membranes with associated "pumps"—to maintain the internal microenvironment.

THE STRUCTURE OF MEMBRANES

Membranes are complex structures composed of lipids, proteins, and carbohydrates. Different membranes within the cell and between cells have different compositions, as reflected in the ratio of protein to lipid (Fig 42–1). The difference is not surprising given the very different functions of membranes. Membranes are asymmetric sheetlike enclosed structures with an inside and an outside surface. These sheetlike structures are noncovalent assemblies that are thermodynamically stable and metabolically active. Specific protein molecules are anchored in membranes, where they carry out specific functions of the organelle, the cell, or the organism.

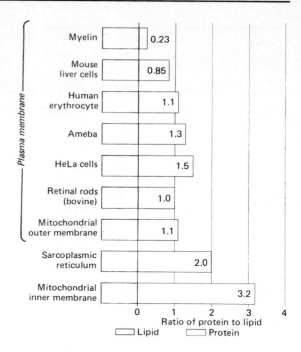

Figure 42–1. Ratio of protein to lipid in different membranes. Proteins equal or exceed the quantity of lipid in nearly all membranes. The outstanding exception is myelin, an electrical insulator found on many nerve fibers. (Reproduced, with permission, from Singer SJ: Architecture and topography of biologic membranes. Chapter 4 in: *Cell Membranes: Biochemistry, Cell Biology & Pathology.* Weissman G, Claiborne R [editors]. HP Publishing Co., 1975.)

LIPID COMPOSITION

The major lipids in mammalian membranes are phospholipids, glycosphingolipids, and cholesterol.

Phospholipids

Of the 2 major phospholipid groups present in membranes, **phosphoglycerides are the more common** and consist of a glycerol backbone to which are attached 2 fatty acids in ester linkage and a phosphorylated alcohol (Fig 42–2). The fatty acid constituents are usually even-numbered carbon molecules, most commonly containing 14 or 16 carbons. They are unbranched and can be saturated or unsaturated. The simplest phosphoglyceride is phosphatidic acid, which is 1,2-diacylglycerol 3-phosphate, a key intermediate in the formation of all other phospholipids (see Chapter 25). In other phospholipids, the 3-phosphate is esterified to an alcohol such as ethanolamine, choline, serine, glycerol, or inositol (see pp 134 and 135).

The second class of phospholipids is composed of **sphingomyelins,** which contain a sphingosine backbone rather than glycerol. A fatty acid is attached by an amide linkage to the amino group of sphingosine.

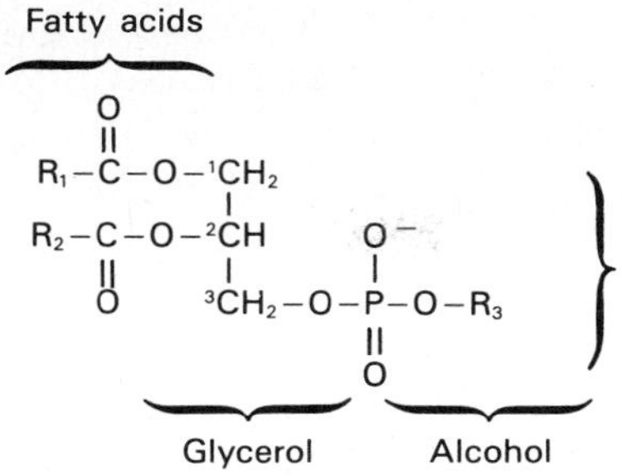

Fatty acids

$$R_1-\overset{O}{\overset{\|}{C}}-O-{}^1CH_2$$
$$R_2-\overset{}{\underset{\|}{C}}-O-{}^2CH \qquad \overset{O^-}{\underset{}{}}$$
$$\overset{}{\underset{O}{}} \qquad {}^3CH_2-O-\overset{}{\underset{\|}{P}}-O-R_3$$
$$\overset{}{\underset{}{O}}$$

Glycerol   Alcohol

Figure 42–2. A phosphoglyceride showing the fatty acids (R_1 and R_2), glycerol, and phosphorylated alcohol components. In phosphatidic acid, R_3 is hydrogen.

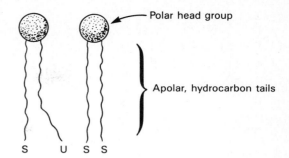

Polar head group

Apolar, hydrocarbon tails

S   U   S   S

Figure 42–3. Diagrammatic representation of a phospholipid or other membrane lipid. The polar head group is hydrophilic, and the hydrocarbon tails are hydrophobic or lipophilic. The fatty acids in the tails are saturated (S) or unsaturated (U); the former are usually attached to carbon 1 of glycerol and the latter to carbon 2.

The primary hydroxyl group of sphingosine is esterified to phosphorylcholine (see p 135). Sphingomyelins, as the name implies, are prominent in myelin sheaths.

Glycosphingolipids

The **glycosphingolipids are sugar-containing lipids** such as cerebrosides and gangliosides and are also derived from sphingosine. The **cerebrosides** and **gangliosides** differ from sphingomyelin in the moiety attached to the primary hydroxyl group of sphingosine. In sphingomyelin, a phosphorylcholine is attached to the alcohol group. A cerebroside contains a single hexose moiety, glucose or galactose, at that site (see p 135). A ganglioside contains a chain of 3 or more sugars, at least one of which is a sialic acid, attached to the primary alcohol of sphingosine.

Sterols

The **most common sterol in membranes is cholesterol,** which exists almost exclusively in the plasma membranes of mammalian cells but can also be found in lesser quantity in mitochondria, Golgi complexes, and nuclear membranes. Cholesterol is generally more abundant toward the outside of the plasma membrane. Cholesterol intercalates among the phospholipids of the membrane, with its hydroxyl group at the aqueous interface and the remainder of the molecule within the leaflet. At temperatures above the transition temperature (see under Fluid Mosaic Model, below), its rigid sterol ring interacts with the acyl chains of the phospholipids, limits their movement, and thus decreases membrane fluidity. On the other hand, when the temperature approaches the transition temperature, the interaction of cholesterol with the acyl chains interferes with their alignment with each other; this phenomenon lowers the temperature at which the fluid → gel transition occurs, thus assisting in keeping the membrane fluid at lower temperatures.

Amphipathic Nature of Membranes

All major lipids in membranes contain both hydrophobic and hydrophilic regions and are therefore termed **amphipathic.** Membranes themselves are thus amphipathic. If the hydrophobic regions were separated from the rest of the molecule, it would be insoluble in water but soluble in oil. Conversely, if the hydrophilic region were separated from the rest of the molecule, it would be insoluble in oil but soluble in water. The amphipathic membrane lipids have a polar head group and nonpolar tails, as represented in Fig 42–3. Saturated fatty acids make straight tails, whereas unsaturated fatty acids, which generally exist in the *cis* form in membranes, make kinked tails. As more kinks are inserted in the tails, the membrane becomes less tightly packed and therefore more fluid. Detergents are amphipathic molecules that are important in biochemistry and in the houshold. The molecular structure of a detergent is not unlike that of a phospholipid.

ORGANIZATION OF MEMBRANE LIPIDS

The amphipathic character of phospholipids suggests that the 2 regions of the molecule have incompatible solubilities; however, in a solvent such as water, phospholipids organize themselves into a form that thermodynamically satisfies both regions. A micelle (Fig 42–4) is such a structure; the hydrophobic regions are shielded from water, while the hydrophilic polar groups are immersed in the aqueous environment.

The Lipid Bilayer

As recognized nearly 60 years ago by Gorter and Grendel, a bimolecular layer, or bilayer, can also satisfy the thermodynamic requirements of amphipathic molecules in an aqueous environment. A bilayer exists as a sheet in which the hydrophobic regions of the phospholipids are protected from the aqueous environment, while the hydrophilic regions are immersed in water (Fig 42–5). Only the ends or edges of the bilayer sheet are exposed to an unfavorable environ-

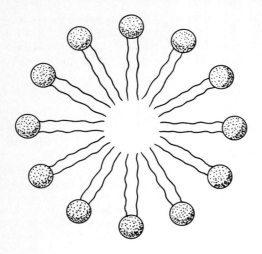

Figure 42–4. Diagrammatic cross section of a micelle. The polar head groups are bathed in water, whereas the hydrophobic hydrocarbon tails are surrounded by other hydrocarbons and thereby protected from water. Micelles are spherical structures.

ment, but even these exposed edges can be eliminated by folding the sheet back upon itself to form an enclosed vesicle with no edges. The closed bilayer provides one of the essential properties of membranes. It is impermeable to most water-soluble molecules, since they would be insoluble in the hydrophobic core of the bilayer.

Two questions immediately arise. First, **how many biologic materials are lipid-soluble and can therefore readily enter the cell?** Gases such as oxygen, CO_2, and nitrogen—small molecules with little interaction with solvents—readily diffuse through the hydrophobic regions of the membrane. Lipid-derived molecules, eg, steroid hormones, readily traverse the bilayer. Organic nonelectrolyte molecules exhibit diffusion rates that are dependent upon their oil-water partition coefficients (Fig 42–6); the greater the lipid solubility of a molecule, the greater its diffusion rate across the membrane.

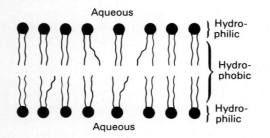

Figure 42–5. Diagram of a section of a bilayer membrane formed from phospholipid molecules. (Slightly modified and reproduced, with permission, from Stryer L: *Biochemistry*, 2nd ed. Freeman, 1981.)

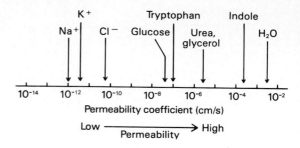

Figure 42–6. Permeability coefficients of water, some ions, and other small molecules in lipid bilayer membranes. Molecules that move rapidly through a given membrane are said to have a high permeability coefficient. (Slightly modified and reproduced, with permission, from Stryer L: *Biochemistry*, 2nd ed. Freeman, 1981.)

The second question concerns molecules that are not lipid-soluble. **How are the transmembrane concentration gradients for non-lipid-soluble molecules maintained?** The answer is that membranes contain proteins, and proteins are also amphipathic molecules that can be inserted into the correspondingly amphipathic lipid bilayer. Proteins form channels for the movement of ions and small molecules and serve as transporters for larger molecules that otherwise could not pass the bilayer. These processes are described below.

MEMBRANE PROTEINS

Proteins Associated With the Bilayer

Membrane phospholipids act as a solvent for membrane proteins, creating an environment in which the latter can function. Of the 20 amino acids contributing to the primary structure of proteins, the functional groups attached to the α carbon are strongly hydrophobic in 6, weakly hydrophobic in a few, and hydrophilic in the remainder. As described in Chapter 5, the α-helical structure of proteins minimizes the hydrophilic character of the peptide bonds themselves. Thus, proteins can be amphipathic and form an integral part of the membrane by having hydrophilic regions protruding at the inside and outside faces of the membrane but connected by a hydrophobic region traversing the hydrophobic core of the bilayer. In fact, those portions of membrane proteins that traverse membranes do contain substantial numbers of hydrophobic amino acids and a high α-helical or β-pleated sheet content.

The number of different proteins in a membrane varies from 6 to 8 in the sarcoplasmic reticulum to over 100 in the plasma membrane. The proteins consist of enzymes, transport proteins, structural proteins, antigens (eg, for histocompatibility), and receptors for various molecules. Because every membrane possesses a different complement of proteins, there is no such thing as a typical membrane structure. The enzymatic properties of several different membranes are shown in Table 42–2.

Membranes and their components are dynamic structures. The lipids and proteins in membranes turn over, just as they do in other compartments of the cell. Different lipids have different turnover rates, and the turnover rates of individual species of membrane proteins may vary widely. The membrane itself can turn over even more rapidly than any of its constituents. This is discussed in more detail in the section on endocytosis.

Asymmetric Nature of Membranes

Asymmetry is an important feature of membranes that can be partially attributed to the irregular distribution of proteins within the membranes. An inside-outside asymmetry is also provided by the external location of the carbohydrates attached to membrane proteins. In addition, specific enzymes are located exclusively on the outside or inside of membranes, as in the mitochondrial and plasma membranes.

There are regional asymmetries in membranes. Some, such as occur at the villous border of mucosal cells, are almost macroscopically visible. Others, such as those at gap junctions, tight junctions, and synapses, occupy much smaller regions of the membrane and generate correspondingly smaller local asymmetries.

There is also inside-outside (transverse) asymmetry of the phospholipids. The choline-containing phospholipids (phosphatidylcholine and sphingomyelin) are located mainly in the outer molecular layer; the aminophospholipids (phosphatidylserine and ethanolamine) are preferentially located in the inner layer. Cholesterol is generally present in larger amounts on the outside than on the inside. Obviously, if this asymmetry is to exist at all, there must be limited transverse mobility (flip-flop) of the membrane phospholipids. In fact, phospholipids in synthetic bilayers exhibit an extraordinarily slow rate of flip-flop; the half-life of the asymmetry can be measured in days

or weeks. However, when certain membrane proteins such as the erythrocyte protein glycophorin are inserted artificially into synthetic bilayers, the frequency of phospholipid flip-flop may increase as much as 100-fold.

The mechanisms involved in the establishment of lipid asymmetry are not understood. The enzymes involved in the synthesis of phospholipids are located on the cytoplasmic side of microsomal membrane vesicles. It has therefore been postulated that translocases exist which transfer certain phospholipids from the inner leaflet to the outer. Also, specific proteins that preferentially bind individual phospholipids may be present in the 2 leaflets, leading to the asymmetric distribution of these lipid molecules.

Integral & Peripheral Membrane Proteins

Most membrane proteins are integral components of the membrane (they interact with the phospholipids), and in fact all of those which have been adequately studied span the entire 5- to 10-nm transverse distance of the bilayer. These integral proteins are usually globular and are themselves amphipathic. They consist of 2 hydrophilic ends separated by an intervening hydrophobic region that traverses the hydrophobic core of the bilayer. As the structures of integral membrane proteins are being elucidated, it is apparent that some (notably transporter molecules) may span the bilayer many times, as illustrated in Fig 42–7.

Integral proteins are asymmetrically distributed across the membrane bilayer (Fig 42–8). If a membrane containing an asymmetrically distributed inte-

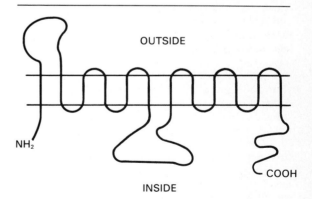

Figure 42–7. Proposed model for the human glucose transporter. The transporter is thought to span the membrane 12 times. These spanning regions may form amphipathic α helices with amide and hydroxyl side chains that could aid in glucose binding or form a channel through which glucose moves. The amino and carboxyl termini are both on the cytoplasmic surface. (Adapted, with permission, from Mueckler M et al: Sequence and structure of a human glucose transporter. *Science.* 1985;229:941. Copyright © 1985 by the American Association for the Advancement of Science.)

Table 42–2. Enzymatic markers of different membranes.*

Membrane	Enzyme
Plasma	5' nucleotidase Adenylate cyclase Na$^+$/K$^+$-ATPase
Endoplasmic reticulum	Glucose-6-phosphatase
Golgi complex	Galactosyltransferase
Inner mitochondrial membrane	ATP synthase

*Membranes contain many proteins, some of which have enzymatic activity. Some of these enzymes are located only in certain membranes and can therefore be used as markers to follow the purification of these membranes.

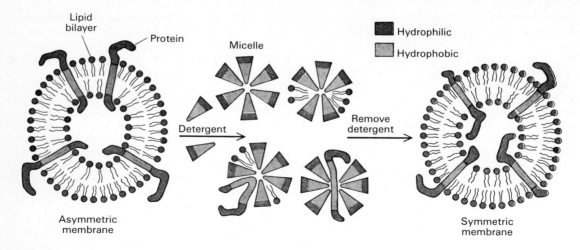

Figure 42–8. Self-assembly of a membrane preserves its basic structure but not its asymmetry. A membrane can be disrupted by a high concentration of a detergent, which is an amphipathic molecule that forms the small droplets called micelles. The detergent dissolves the components of the membrane by enveloping the hydrophobic portions of both lipids and proteins in micelles, where they are protected from contact with water. If the detergent is then removed, the lipids spontaneously form a new bilayer, incorporating the integral proteins in it. The proteins, however, generally assume random orientations. Experiments such as this have shown that all membranes in the cell cannot be self-assembled; instead, at least some integral proteins must be inserted in a membrane that already exists and has a defined sidedness. (Reproduced, with permission, from Lodish HF, Rothman JE: The assembly of cell membranes. *Sci Am* [Jan] 1979;**240**:43.)

gral protein is dissolved in detergent and the detergent is then slowly removed, the phospholipids and the integral proteins will self-assemble, but the latter will lose their specific inside-outside orientation in the membrane. Thus, at least some proteins must be given their asymmetric orientation in the membrane at the time of their insertion in the lipid bilayer. The hydrophilic external region of an amphipathic protein, which is clearly synthesized inside the cell, must traverse the hydrophobic core of the membrane and eventually be found outside the membrane. The molecular mechanisms of membrane assembly are discussed below.

Peripheral proteins do not interact directly with the phospholipids in the bilayer but are instead by definition weakly bound to the hydrophilic regions of specific integral proteins. For example, ankyrin, a peripheral protein, is bound to the integral protein "band III" of erythrocyte membrane. Spectrin, a cytoskeletal structure within the erythrocyte, is in turn bound to ankyrin and thereby plays an important role in the maintenance of the biconcave shape of the erythrocyte. The immunoglobulin molecules on the plasma membranes of lymphocytes are integral membrane proteins and can be released by the shedding of small fragments of the membrane. Many hormone receptor molecules are integral proteins, and the specific polypeptide hormones that bind to these receptor molecules may therefore be considered peripheral proteins. Peripheral proteins, such as peptide hormones, may even organize the distribution of integral proteins, such as their receptors, within the plane of the bilayer (see below).

ARTIFICIAL MEMBRANES

Artificial membrane systems can be prepared by appropriate techniques. These systems generally consist of mixtures of one or more phospholipids of natural or synthetic origin that can be treated (eg, by using mild sonication) to form spherical vesicles in which the lipids form a bilayer. Such vesicles, surrounded by a lipid bilayer, are termed **liposomes.**

Some of the advantages and uses of artificial membrane systems in the study of membrane function follow.

(1) The lipid content of the membranes can be varied, allowing systematic examination of the effects of varying lipid composition on certain functions. For instance, vesicles can be made that are composed solely of phosphatidylcholine or, alternatively, of known mixtures of different phospholipids, glycolipids, and cholesterol. The fatty acid moieties of the lipids used can also be varied by employing synthetic lipids of known composition to permit systematic examination of the effects of fatty acid composition on certain membrane functions (eg, transport).

(2) Purified membrane proteins or enzymes can be incorporated into these vesicles in order to assess what factors (eg, specific lipids or ancillary proteins) the proteins require to reconstitute their function. Investigations of purified proteins, eg, the Ca^{2+}/ATPase of the sarcoplasmic reticulum, have in certain cases suggested that only a single protein and a single lipid are required to reconstitute an ion pump.

(3) The environment of these systems can be rigidly controlled and systematically varied (eg, ion

concentrations). The systems can also be exposed to known ligands if, for example, the liposomes contain specific receptor proteins.

(4) When liposomes are formed, they can be made to entrap certain compounds inside themselves, eg, drugs and isolated genes. There is interest in using liposomes to distribute drugs to certain tissues, and if components (eg, antibodies to certain cell surface molecules) could be incorporated into liposomes so that they would be targeted to specific tissues or tumors, the therapeutic impact could be considerable. DNA entrapped inside liposomes appears to be less sensitive to attack by nucleases; this approach may prove useful in attempts at gene therapy.

THE FLUID MOSAIC MODEL OF MEMBRANE STRUCTURE

Functional membranes are 2-dimensional solutions of globular integral proteins dispersed in a fluid phospholipid matrix. This fluid mosaic model of membrane structure was proposed in 1972 by Singer and Nicolson (Fig 42–9). Early evidence for the model was the rapid and random redistribution of species-specific integral proteins in the plasma membrane of an interspecies hybrid cell formed by the artificially induced fusion of 2 different parent cells. It has subsequently been demonstrated that phospholipids also undergo rapid redistribution in the plane of the membrane. This diffusion within the plane of the membrane, termed translational diffusion, can be quite rapid for a phospholipid; in fact, within the plane of the membrane, one molecule of phospholipid can move several micrometers per second.

The phase changes, and thus the fluidity of membranes, are highly dependent upon the lipid composition of the membrane. In a lipid bilayer, the hydrophobic chains of the fatty acids can be highly aligned or ordered to provide a rather stiff structure. As the temperature increases, the hydrophobic side chains undergo a transition from the ordered state to a disordered one, taking on a more liquidlike or fluid arrangement. The temperature at which the structure undergoes the transition from ordered to disordered is called the transition temperature. The longer and more saturated fatty acid chains exhibit higher transition temperatures, ie, higher temperatures are required to increase the fluidity of the structure. Unsaturated bonds that exist in the *cis* configuration tend to increase the fluidity of a bilayer by decreasing the compactness of the side chain packing without diminishing hydrophobicity (Fig 42–3). The phospholipids of cellular membranes generally contain at least one unsaturated fatty acid with at least one *cis* double bond.

Cholesterol also acts as a moderator molecule in membranes, producing intermediate states of fluidity. If the acyl side chains exist in a disordered phase, cholesterol will have a condensing effect; if the acyl side chains are ordered or in a crystalline phase,

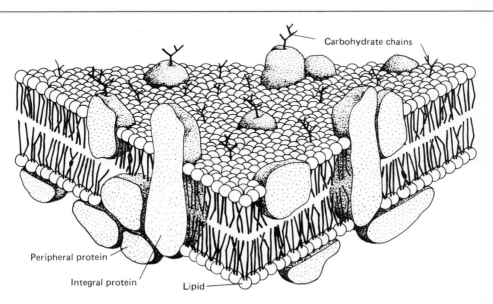

Carbohydrate chains

Peripheral protein

Integral protein

Lipid

Figure 42–9. The fluid mosaic model of membrane structure. The membrane consists of a bimolecular lipid layer with proteins inserted in it or bound to the cytoplasmic surface. Integral membrane proteins are firmly embedded in the lipid layers. Some of these proteins completely span the bilayer and are called transmembrane proteins, while others are embedded in either the outer or inner leaflet of the lipid bilayer. Loosely bound to the inner surface of the membrane are the peripheral proteins. Many of the proteins and lipids have externally exposed oligosaccharide chains. (Reproduced, with permission, from Junqueira LC, Carneiro J, Long JA: *Basic Histology,* 5th ed. Appleton & Lange, 1986.)

cholesterol will induce disorder. At high cholesterol:phospholipid ratios, transition temperatures are abolished altogether.

The fluidity of a membrane significantly affects its functions. As membrane fluidity increases, so does its permeability to water and other small hydrophilic molecules. The lateral mobility of integral proteins increases as the fluidity of the membrane increases. If the active site of an integral protein involved in some given function resides exclusively in its hydrophilic regions, changing lipid fluidity will probably have little effect on the activity of the protein; however, if the protein is involved in a transport function in which transport components span the membrane, lipid phase effects may significantly alter the transport rate. The insulin receptor is an excellent example of altered function with changes in fluidity (see Chapter 51). As the concentration of unsaturated fatty acids in the membrane is increased (by growing cultured cells in a medium rich in such molecules), fluidity increases. This alters the receptor so that it binds more insulin.

A state of fluidity and thus of translational mobility in a membrane may be confined to certain regions of membranes under certain conditions. For example, protein-protein interaction may take place within the plane of the membrane, such that the integral proteins form a rigid matrix, in contrast to the more usual situation, where the lipid acts as the matrix. Such regions of rigid protein matrix can exist side by side in the same membrane with the usual lipid matrix. Gap junctions, tight junctions, and bacteriorhodopsin-containing regions of the purple membranes of halobacteria are clear examples of such side-by-side coexistence of different matrices.

Some of the protein-protein interactions taking place within the plane of the membrane may be mediated by interconnecting peripheral proteins, such as cross-linking antibodies or lectins that are known to patch or cap on membrane surfaces. Thus, peripheral proteins, by their specific attachments, may restrict the mobility of integral proteins within the membrane.

MEMBRANE ASSEMBLY

As mentioned above, the enzymes responsible for the synthesis of phospholipids reside on the cytoplasmic surface of the vesicles of endoplasmic reticulum. As phospholipids are synthesized at that site, they probably self-assemble into thermodynamically stable biomolecular layers, thereby expanding the sheet of the vesicle. The lipid vesicles originating in the endoplasmic reticulum appear to migrate to the Golgi apparatus, which in turn eventually fuses with the plasma membrane. The Golgi apparatus and the endoplasmic reticulum vesicles exhibit transverse asymmetries of both lipid and protein, and these asymmetries are maintained during fusion with the plasma

membrane. The inside of the vesicle after fusion becomes the outside of the plasma membrane, and the cytoplasmic side of the vesicles remains the cytoplasmic side of the membrane (Fig 42–10). Since the transverse asymmetry of the membranes already exists in the vesicles of the endoplasmic reticulum well before they are fused to the plasma membrane, the major problem of membrane assembly becomes understanding how the integral proteins are inserted asymmetrically into the lipid bilayer of the endoplasmic reticulum.

Integral and secreted proteins frequently are synthesized with a **hydrophobic N-terminal leader se-**

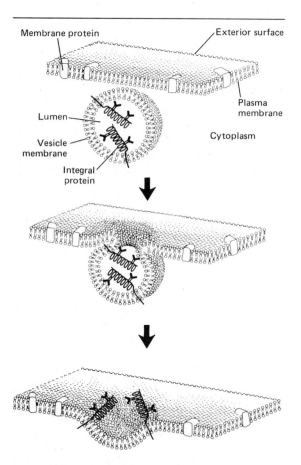

Figure 42–10. Fusion of a vesicle with the plasma membrane preserves the orientation of any integral proteins embedded in the vesicle bilayer. Initially, the N terminus of the protein faces the lumen, or inner cavity, of such a vesicle. After fusion, the N terminus is on the exterior surface of the plasma membrane. That the orientation of the protein has not been reversed can be perceived by noting that the other end of the molecule, the C terminus, is always immersed in the cytoplasm. The lumen of a vesicle and the outside of the cell are topologically equivalent. (Redrawn and modified, with permission, from Lodish HF, Rothman JE: The assembly of cell membranes. *Sci Am* [Jan] 1979;**240**:43.)

quence of 15–30 amino acids. Rarely, the hydrophobic leader sequence may be internal. The N-terminal leader sequence is usually removed from the protein during or after its integration into the membrane, producing the mature secreted or membrane protein.

There is strong evidence that the leader sequence is involved in the process of protein insertion. Mutant proteins containing altered leader sequences in which a hydrophobic amino acid is replaced by a hydrophilic one are not inserted into membranes. Nonmembrane proteins, to which leader sequences have been attached by genetic engineering, can be inserted into membranes or even secreted.

The Signal Hypothesis

Proteins are inserted into membranes in a variety of ways (Fig 42–11); the precise mechanism involved, in many cases, remains to be elucidated. Two models have been proposed to describe the integration of proteins into membrane; the signal hypothesis and the membrane trigger hypothesis. The signal hypothesis proposes that the protein is inserted into the membrane simultaneously with the translation of its mRNA on polyribosomes, so-called cotranslational insertion. As the leader sequence of the protein emerges from the ribosome, it is recognized by a **signal recognition particle (SRP)** that blocks further translation after about

70 amino acids have been polymerized, 40 buried in the large ribosomal complex and 30 exposed (Fig 42–12). The SRP contains 6 proteins and has associated with it a 7S RNA that is closely related to the "Alu family" of highly repeated DNA sequences (see Chapter 38). The SRP-imposed block is not released until the SRP-leader sequence-ribosome complex has bound to the so-called docking protein (a receptor for the SRP) on the endoplasmic reticulum. Cotranslational insertion of the protein into the endoplasmic reticulum then commences at that site. The process of elongation of the remaining portion of the protein molecule drives the protein chain across the lipid bilayer as the ribosomes remain attached to the endoplasmic reticulum. Thus, the rough (or ribosome-studded) endoplasmic reticulum is formed. Ribosomes remain attached to the endoplasmic reticulum during synthesis of the membrane protein but are released and dissociated into their respective subunits as the protein is completed. The leader sequence is cleaved off and carbohydrate attached as the early synthesized portion of the protein enters the lumen of the endoplasmic reticulum.

Integral membrane proteins do not completely cross the membrane and are probably prevented from doing so by a hydrophilic C-terminal anchor region. However, secreted proteins completely traverse the

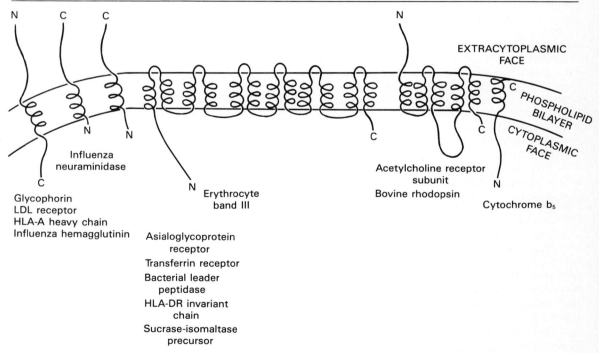

Figure 42–11. Variations in the way proteins are inserted into membranes. This schematic representation, which illustrates a number of possible orientations, shows the portion of the peptide within the membrane as α helices and the other portions as lines. N, NH$_2$ terminus; C, COOH terminus. (Adapted, with permission, from Wickner WT, Lodish HF: Multiple mechanisms of protein insertion into and across membranes. *Science* 1985;230:400. Copyright © 1985 by the American Association for the Advancement of Science.)

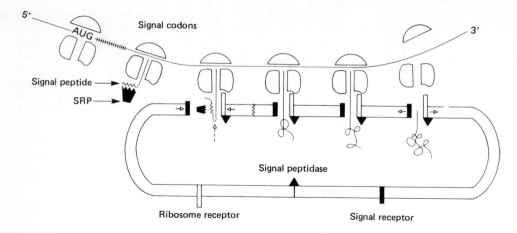

Figure 42–12. Diagram of the signal hypothesis for the transport of secreted proteins across the endoplasmic reticulum membrane. The ribosomes synthesizing a protein move along the messenger RNA specifying the amino acid sequence of the protein. (The messenger is represented by the line between 5′ and 3′.) The codon AUG marks the start of the message for the protein; the hatched lines that follow AUG represent the codons for the signal sequence. As the protein grows out from the larger ribosomal subunit, the signal sequence is exposed and bound by the signal recognition particle (SRP). Translation is blocked until the complex binds to the "docking protein" (represented by the solid bar) on the endoplasmic reticulum membrane. There is also a receptor (open bar) for the ribosome itself. The interaction of the ribosome and growing peptide chain with the endoplasmic reticulum membrane results in the opening of a pore through which the protein is transported to the interior space of the endoplasmic reticulum. During transport, the signal sequence of most proteins is removed by an enzyme called the signal peptidase. The completed protein is eventually released by the ribosome, which then separates into its 2 components, the large and small ribosomal subunits. The protein ends up inside the endoplasmic reticulum. (Slightly modified and reproduced, with permission, from: Newly made proteins zip through the cell. *Science* 1980;**207**:154. Copyright © 1980 by the American Association for the Advancement of Science.)

membrane bilayer and are discharged into the lumen of the endoplasmic reticulum. By the time they reach the interior of that vesicle, carbohydrate moieties have already been attached. Subsequently, the secretory proteins are found in the lumen of the Golgi apparatus, where their carbohydrate chains are modified prior to their being directed to specific intracellular organelles or cellular membranes or being secreted. Some proteins traverse one membrane and subsequently become anchored in a second, juxtaposed membrane, such as the mitochondrial inner membrane.

The Membrane Trigger Hypothesis

The membrane trigger hypothesis emphasizes the role of the leader sequence in altering the folding pathway of the protein itself. The leader sequence is said to promote an alternative folding of the usually hydrophobic integral protein, so that it can remain soluble in the aqueous environment of the cytoplasm where it is synthesized. The membrane lipid bilayer is said to trigger the refolding of the protein into a conformation that favors its insertion into a particular bilayer. Thus, the protein is triggered to self-assemble into the membrane in a manner that establishes the necessary transverse asymmetry. Once the protein is inserted or integrated, the leader sequence is cleaved off. The trigger hypothesis does not require specific ribosome-mem-

Table 42–3. A comparison of 2 models for membrane assembly.*

Stage of Synthesis	Signal Hypothesis	Membrane Trigger Hypothesis
Site of initiation	Soluble polysomes	Soluble polysomes
Role of the leader peptide	Recognized by the protein transport channel	To alter the folding pathway
Association of the new protein with membrane	When leader peptide is complete *Place:* protein transport channel	During or after protein synthesis *Place:* receptor protein or lipid portions of the bilayer
Specific ribosome associations	With the protein transport channel	None
Catalysis for assembly	A specific pore	The effect of the leader peptide on conformation
Driving force for assembly	Polypeptide chain elongation	Protein-protein and protein-lipid association: self-assembly
Removal of leader peptide	During polypeptide extrusion	During or after polypeptide assembly into bilayer
Final orientation	C terminus in, N terminus out	Specified by the primary sequence

*Modified and reproduced, with permission, from Wickner W: The assembly of proteins into biological membranes: The membrane trigger hypothesis. *Annu Rev Biochem* 1979;**48**:23.

brane interactions, but this still does not mean that protein synthesis on membranes cannot occur.

The major characteristics of the signal hypothesis and the trigger hypothesis are compared in Table 42–3. The signal and trigger mechanisms probably operate in the same cell. Several problems of membrane biosynthesis remain unsolved, since some membrane proteins and secreted proteins are synthesized on membrane-bound polysomes, whereas others are formed on free cytoplasmic polysomes. Certain proteins will not enter the so-called assembly pathway to be secreted or inserted unless they interact with the membrane bilayer early in the process of their own synthesis on ribosomes. Other proteins, such as the mitochondrial cytochrome b_5, can self-assemble or integrate in an oriented fashion into membranes after their synthesis is completed but still require the presence of a normal leader sequence during their biosynthesis. Some single-peptide chains or proteins such as bacteriorhodopsin can span a membrane back and forth several times, a phenomenon not easily attributed to the signal hypothesis.

SPECIALIZED FUNCTIONS OF MEMBRANES

If the plasma membrane is relatively impermeable, how do most molecules enter a cell? How is selectivity of this movement established? Answers to such questions are important in understanding how cells adjust to a constantly changing extracellular environment. Metazoan organisms also must have means of communicating between adjacent and distant cells, so that complex biologic processes can be coordinated. These signals must arrive at and be transmitted by the membrane, or they must be generated as a consequence of some interaction with the membrane. Some of the major mechanisms used to accomplish these different objectives are listed in Table 42–4.

Table 42–4. Transfer of material and information across membranes.

Cross-membrane movement of small molecules
  Diffusion (passive and facilitated)
  Active transport
Cross-membrane movement of large molecules
  Endocytosis
  Exocytosis
Signal transmission across membranes
  Cell surface receptors
    1. Signal transduction (eg, glucagon → cAMP)
    2. Signal internalization (coupled with endocytosis, eg, the LDL receptor)
  Movement to intracellular receptors (steroid hormones; a form of diffusion)
Intercellular contact and communication

CROSS-MEMBRANE MOVEMENT OF SMALL MOLECULES

Molecules can passively traverse the bilayer down electrochemical gradients by simple or facilitated diffusion. This spontaneous movement toward equilibrium contrasts with active transport, which requires energy because it constitutes movement against an electrochemical gradient. Fig 42–13 provides a schematic representation of these mechanisms.

Passive Diffusion

As described above, some solutes such as gases can enter the cell by diffusing down an electrochemical gradient across the membrane and do not require metabolic energy. The simple diffusion of a solute across the membrane is limited by the thermal agitation of that specific molecule, by the concentration gradient across the membrane, and by the solubility of that solute (the permeability coefficient, Fig 42–6) in the hydrophobic core of the membrane bilayer. Solubility is inversely proportionate to the number of hydrogen bonds that must be broken in order for a solute in the external aqueous phase to become incorporated in the hydrophobic bilayer. Electrolytes, poorly soluble in lipid, do not form hydrogen bonds with water, but they do acquire a shell of water from hydration by electrostatic interaction. The size of the shell is directly proportionate to the charge density of the electrolyte. Electrolytes with a large charge density have a larger shell of hydration and thus a slower diffusion rate. Na^+, for example, has a higher charge density than K^+. Hydrated Na^+ is therefore larger than hydrated K^+; hence, the latter tends to move more easily through the membrane.

In natural membranes, as opposed to synthetic membrane bilayers, there are **transmembrane channels,** porelike structures composed of proteins that constitute selective ion-conductive pathways. Cation-conductive channels have an average diameter of about 5–8 nm and are negatively charged within the channel. The permeability of a channel depends upon the size, extent of hydration, and extent of charge density on the ion. Specific channels for Na^+, K^+, and Ca^{2+} have been identified.

The membranes of nerve cells contain well-studied ion channels that are responsible for the action potentials generated across the membrane. The activity of some of these channels is controlled by neurotransmitters; hence, channel activity can be regulated. **One ion can regulate the activity of the channel of another ion.** For example, a decrease of Ca^{2+} concentration in the extracellular fluid increases membrane permeability and increases the diffusion of Na^+. This depolarizes the membrane and triggers nerve discharge. This may explain the numbness, tingling, and muscle cramps symptomatic of a low level of serum Ca^{2+}.

Channels are open transiently and thus are gated. Gates can be controlled by opening or closing. In **ligand-gated channels,** a specific molecule binds to a receptor and opens the channel. **Voltage-gated chan-**

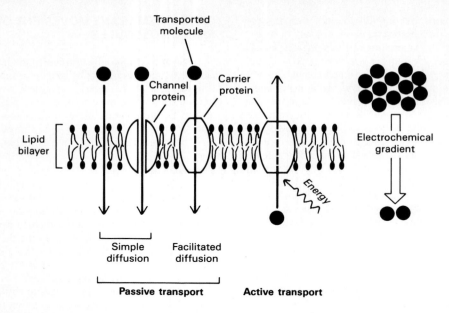

Figure 42–13. Many small uncharged molecules pass freely through the lipid bilayer. Charged molecules, larger uncharged molecules, and some small uncharged molecules are transferred through channels or pores or by specific carrier proteins. Passive transport is always down an electrochemical gradient, toward equilibrium. Active transport is against an electrochemical gradient and requires an input of energy, whereas passive transport does not. (Redrawn and reproduced, with permission, from Alberts B et al: *Molecular Biology of the Cell.* Garland, 1983.)

nels open (or close) in response to a change in membrane potential.

Some microbes synthesize small organic molecules, **ionophores,** that function as shuttles for the movement of ions across membranes. These ionophores contain hydrophilic centers that bind specific ions and are surrounded by peripheral hydrophobic regions; this arrangement allows the molecules to dissolve effectively in the membrane and diffuse transversely therein. Others, like the well-studied polypeptide gramicidin, form channels. Microbial toxins such as diphtheria toxin and activated serum complement components can produce large pores in cellular membranes and thereby provide macromolecules with direct access to the internal milieu.

In summary, net diffusion of a substance depends upon the following: (1) Its concentration gradient across the membrane. Solutes move from high to low concentration. (2) The electrical potential across the membrane. Solutes move toward the solution that has the opposite charge. The inside of the cell usually has a negative charge. (3) The permeability coefficient of the substance for the membrane. (4) The hydrostatic pressure gradient across the membrane. Increased pressure will increase the rate and force of the collision between the molecules and the membrane. (5) Temperature. Increased temperature will increase particle motion and thus increase the frequency of collisions between external particles and the membrane.

Facilitated Diffusion & Active Transport

Transport systems can be described in a functional sense according to the number of molecules moved and the direction of movement (Fig 42–14) or according to whether movement is toward or away from equilibrium. A **uniport** system moves one type of molecule bidirectionally. In **cotransport** systems, the transfer of one solute depends upon the stoichiometric simultaneous or sequential transfer of another solute. A **symport** moves these solutes in the same direction. Examples are the proton-sugar transporter in bacteria and the Na^+-sugar transporters (glucose, mannose, galactose, xylose, and arabinose) and the Na^+-amino acid transporters in mammalian cells. **Antiport** systems move 2 molecules in opposite directions (eg, Na^+ in and Ca^{2+} out).

Molecules that cannot pass freely through the lipid bilayer membrane by themselves do so in association with carrier proteins. This involves 2 processes, facilitated diffusion and active transport, and highly specific transport systems.

Facilitated diffusion and active transport share many features. Both appear to involve carrier proteins, and both show specificity for ions, sugars, and amino acids. Mutations in bacteria and mammalian cells (including some that result in human disease) have supported these conclusions. Facilitated diffusion and active transport resemble a substrate-enzyme

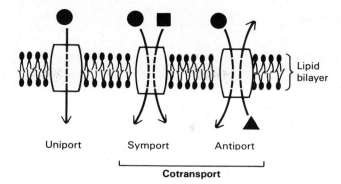

Figure 42–14. Schematic representation of types of transport systems. Transporters can be classified with regard to the direction of movement and whether one or more unique molecules are moved. (Redrawn and reproduced, with permission, from Alberts B et al: *Molecular Biology of the Cell.* Garland, 1983.)

reaction except that no covalent interaction occurs. These points of resemblance are as follows: (1) There is a specific binding site for the solute. (2) The carrier is saturable, so it has maximum rate of transport (V_{max}; see Fig 42–15). (3) There is a binding constant (K_m) for the solute, and so the whole system has a K_m (Fig 42–15). (4) Structurally similar competitive inhibitors block transport.

Major differences are the following: (1) Facilitated diffusion can operate bidirectionally, whereas active transport is usually unidirectional. (2) Active transport always occurs against an electrical or chemical gradient, and so it requires energy.

A. Facilitated Diffusion: Some specific solutes diffuse down electrochemical gradients across membranes more rapidly than might be expected from their size, charge, or partition coefficients. This facilitated diffusion exhibits properties distinct from those of simple diffusion. The rate of facilitated diffusion, a uniport system, can be saturated; ie, the number of sites involved in diffusion of the specific solutes appears finite. Many facilitated diffusion systems are stereospecific but, like simple diffusion, require no metabolic energy.

As described earlier, the inside-outside asymmetry of membrane proteins is stable, and mobility of proteins across (rather than in) the membrane is rare; therefore, transverse mobility of specific carrier proteins is not likely to account for facilitated diffusion processes except those for microbial ionophores, described earlier.

A "ping-pong" mechanism (Fig 42–16) explains facilitated diffusion. In this model, the carrier protein exists in 2 principal conformations. In the "pong" state it is exposed to high concentrations of solute, and molecules of the solute bind to specific sites on the carrier protein. Transport occurs when a conformational change exposes the carrier to a lower concentration of solute ("ping" state). This process is completely reversible, and net flux across the membrane depends upon the concentration gradient. The rate at which solutes enter a cell by facilitated diffusion is determined by the following factors: (1) The concentration gradient across the membrane. (2) The amount of carrier available (this is a key control step). (3) The rapidity of the solute-carrier interaction. (4) The rapidity of the conformational change for both the loaded and the unloaded carrier.

Hormones regulate facilitated diffusion by changing the number of transporters available. Insulin increases glucose transport in fat and muscle by recruiting transporters from an intracellular reservoir (see Fig 51–13). Insulin also enhances amino acid transport in liver and other tissues. One of the coordinated actions of glucocorticoid hormones is to enhance transport of amino acids into liver, where the amino acids then serve as a substrate for gluconeogenesis. Growth hormone increases amino acid transport in all cells, and estrogens do this in the uterus. There are at least 5 dif-

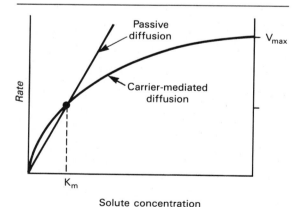

Figure 42–15. A comparison of the kinetics of carrier-mediated (facilitated) diffusion with passive diffusion. The rate of movement in the latter is directly proportionate to solute concentration, whereas the process is saturable when carriers are involved. V_{max}, maximal rate. The concentration at half-maximal velocity is equal to the binding constant (K_m) of the carrier for the solute.

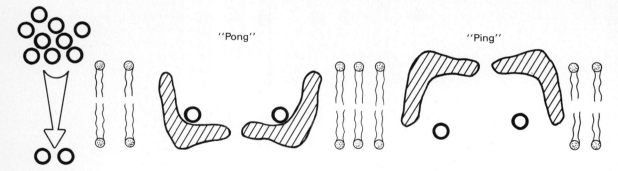

Figure 42–16. The "ping-pong" model of facilitated diffusion. A protein carrier (shaded structure) in the lipid bilayer associates with a solute in high concentration on one side of the membrane. A conformational change ensues ("pong" to "ping"), and the solute is discharged on the side favoring the new equilibrium. The empty carrier then reverts to the original conformation ("ping" to "pong") to complete the cycle.

ferent carrier systems for amino acids in animal cells. Each is specific for a group of closely related amino acids, and most operate as Na^+-symport systems (Fig 42–13).

B. Active Transport: The process of active transport differs from diffusion in that molecules are transported away from thermodynamic equilibrium; hence, energy is required. This energy can come from the hydrolysis of ATP, from electron movement, or from light. The maintenance of electrochemical gradients in biologic systems is so important that it consumes perhaps 30–40% of the total energy expenditure in a cell.

In general, cells maintain a low intracellular Na^+ concentration and a high intracellular K^+ concentration (Table 42–1), along with a net negative electrical potential inside. The pump that maintains these gradients is an ATPase that is activated by Na^+ and K^+ (Fig 42–17). The ATPase is an integral membrane protein and requires phospholipids for activity. The ATPase has catalytic centers for both ATP and Na^+ on the cytoplasmic side of the membrane, but the K^+ binding site is located on the extracellular side of the membrane. Ouabain (or digitalis) inhibits this ATPase by binding to the extracellular domain. Inhibition of the ATPase by ouabain can be antagonized by extracellular K^+.

Transmission of Nerve Impulses

The membrane forming the surface of neuronal cells maintains an asymmetry of inside-outside voltage (electrical potential) and is electrically excitable. When appropriately stimulated by a chemical signal mediated by a specific synaptic membrane receptor (see Transmission of Biochemical Signals, below), gates in the membrane are opened to allow the rapid influx of Na^+ or Ca^{2+} (with or without the efflux of K^+), so that the voltage difference rapidly collapses and that segment of the membrane is depolarized. However, as a result of the action of the ion pumps in the membrane, the gradient is quickly restored.

When large areas of the membrane are depolarized in this manner, the electrochemical disturbance propagates in wavelike form down the membrane, generating a nerve impulse. Myelin sheets, formed by Schwann cells, wrap around nerve fibers and provide an electrical insulator that surrounds most of the nerve and greatly speeds up the propagation of the wave (signal) by allowing ions to flow in and out of the membrane only where the membrane is free of the insulation. The myelin membrane is composed of phospholipids, including sphingomyelin, cholesterol, proteins, and glycosphingolipids. There are relatively few integral and peripheral proteins associated with the myelin membrane; those present appear to hold together multiple membrane bilayers to form the hydrophobic, insulating structure that is impermeable to ions and water. Certain diseases, eg, multiple sclerosis and the Guillain-Barré syndrome, are characterized by demyelination and impaired nerve conduction.

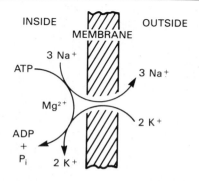

Figure 42–17. Stoichiometry of the Na^+/K^+-ATPase pump. This pump moves 3 Na^+ ions from inside the cell to the outside and brings 2 K^+ ions from the outside to the inside for every molecule of ATP hydrolyzed to ADP by the membrane-associated ATPase. Ouabain and other cardiac glycosides inhibit this pump by acting on the extracellular surface of the membrane. (Courtesy of R Post.)

Transport of Glucose

A discussion of the transport of glucose summarizes many of the points made in this chapter. Glucose must enter cells as the first step in energy utilization. The exception is liver, wherein no specific carrier process has been identified. In hepatic cells, glucose enters by simple diffusion down a concentration gradient that is always steep because of the rapid intracellular conversion of glucose to glucose 6-phosphate. In other cells, adipocytes and muscle in particular, glucose enters by a specific transport system that is regulated by insulin (see Fig 51–13). Changes in transport are primarily due to alterations of V_{max} (presumably from more or fewer active transporters) but changes in K_m may also be involved. The structure of the red blood cell glucose transporter has been deduced from the cDNA sequence (Fig 42–7). cDNA-transfected cells express this protein and insert it into membranes in a functional state; future studies with site-directed mutagenesis should reveal how the protein actually works.

Glucose transport involves different aspects of the principles of transport discussed above. Glucose and Na^+ bind to different sites on the glucose transporter. Na^+ moves into the cell down its electrochemical gradient and "drags" glucose with it (Fig 42–18). Therefore, the greater the Na^+ gradient, the more glucose enters, and if Na^+ in extracellular fluid is low, glucose transport stops. To maintain a steep Na^+ gradient, this Na^+-glucose symport is dependent on gradients generated by an Na^+/K^+ pump that maintains a low intracellular Na^+ concentration. Similar mechanisms are used to transport other sugars as well as amino acids.

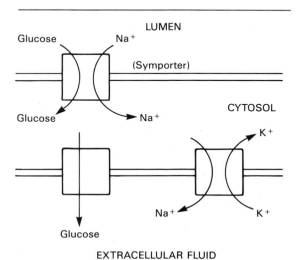

Figure 42–18. The transcellular movement of glucose in an intestinal cell. Glucose follows Na^+ across the luminal epithelial membrane. The Na^+ gradient that drives this symport is established by Na^+/K^+ exchange, which occurs at the basal membrane facing the extracellular fluid compartment. Glucose at high concentration within the cell moves "downhill" into the extracellular fluid by facilitated diffusion (a uniport mechanism).

The transcellular movement of sugars involves one additional component, a uniport that allows the glucose accumulated within the cell to move across a different surface toward a new equilibrium; this occurs in intestinal and renal cells, for example.

CROSS-MEMBRANE MOVEMENT OF MACROMOLECULES

Cells also transport macromolecules across the plasma membrane. **Endocytosis** is the process by which cells take up large molecules. Some of these (eg, polysaccharides, proteins, and polynucleotides) can be sources of nutritional elements. Endocytosis provides a mechanism for regulating the content of certain membrane components, hormone receptors being a case in point. Endocytosis can be used to learn more about how cells function. DNA from one cell type can be used to transfect a different cell and alter the latter's function or phenotype. A specific gene is often employed in these experiments, and this provides a unique way to study and analyze the regulation of that gene. DNA transfection depends upon endocytosis; endocytosis is responsible for the entry of DNA into the cell. Such experiments commonly use calcium phosphate, since Ca^{2+} stimulates endocytosis and precipitates DNA, which makes the DNA a better object for endocytosis. Cells also release macromolecules by **exocytosis.** Endocytosis and exocytosis both involve vesicle formation with or from the plasma membrane.

Endocytosis

All eukaryotic cells are continuously ingesting parts of their plasma membranes. Endocytotic vesicles are generated when segments of the plasma membrane invaginate, enclosing a minute volume of extracellular fluid and its contents. The vesicle then pinches off as the fusion of plasma membranes seals the neck of the vesicle at the original site of invagination (Fig 42–19). This vesicle fuses with other membrane structures and thus achieves the transport of its contents to other cellular compartments or even back to the cell exterior. Most endocytotic vesicles fuse with **primary lysosomes** to form secondary lysosomes, which contain hydrolytic enzymes and are therefore specialized organelles for intracellular disposal. The macromolecular contents are digested to yield amino acids, simple sugars, and nucleotides, and they diffuse out of the vesicles to be reused in the cytoplasm. Endocytosis requires (1) energy, usually from the hydrolysis of ATP; (2) Ca^{2+} in extracellular fluid; and (3) contractile elements in the cell (probably the microfilament system).

There are 2 general types of endocytosis. **Phagocytosis** occurs only in specialized cells such as macrophages and granulocytes. Phagocytosis involves the ingestion of large particles such as viruses, bacteria, cells, or debris. Macrophages are extremely active in this regard and may ingest 25% of their volume per hour. In so doing, a macrophage may internalize 3%

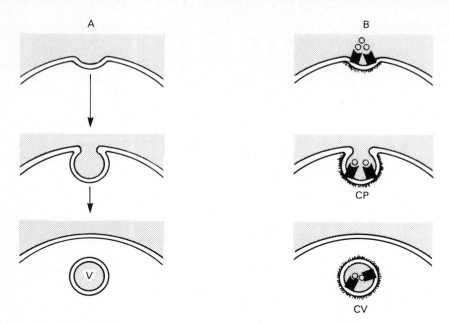

Figure 42–19. Two types of endocytosis. An endocytotic vesicle (V) forms as a result of invagination of a portion of the plasma membrane. Fluid-phase endocytosis (*A*) is random and nondirected. Receptor-mediated endocytosis (*B*) is selective and occurs in coated pits (CP) lined with the protein clathrin (the fuzzy material). Targeting is provided by receptors (■) specific for a variety of molecules. This results in the formation of a coated vesicle (CV).

of its plasma membrane each minute or the entire membrane every 30 minutes.

Pinocytosis is a property of all cells and leads to the cellular uptake of fluid and fluid contents. There are 2 types. **Fluid-phase pinocytosis** is a nonselective process in which the uptake of a solute by formation of small vesicles is simply proportionate to its concentration in the surrounding extracellular fluid. The formation of these vesicles is an extemely active process. Fibroblasts, for example, internalize their plasma membrane at about one-third the rate of macrophages. This process occurs more rapidly than membranes are made. The surface area and volume of a cell do not change much, so membranes must be replaced by exocytosis or by being recycled as fast as they are removed by endocytosis.

The other type of pinocytosis, **absorptive pinocytosis,** is a receptor-mediated selective process primarily responsible for the uptake of macromolecules for which there are a finite number of binding sites on the plasma membrane. These high-affinity receptors permit the selective concentration of ligands from the medium, minimize the uptake of fluid or soluble unbound macromolecules, and markedly increase the rate at which specific molecules enter the cell. The vesicles formed during absorptive pinocytosis are derived from invaginations (pits) that are coated on the cytoplasmic side with a filamentous material. In many systems, **clathrin** is the filamentous material; it is probably a peripheral membrane protein. **Coated pits** may constitute as much as 2% of the surface of some cells.

For example, the low-density lipoprotein (LDL) molecule and its receptor (see Chapter 26) are internal-

ized by means of coated pits containing the LDL receptor. These endocytotic vesicles containing LDL and its receptor fuse to lysosomes in the cell. The receptor is released and recycled back to the cell surface membrane, but the apoprotein of LDL is degraded and the cholesteryl esters metabolized. Synthesis of the LDL receptor is regulated by secondary or tertiary consequences of pinocytosis, eg, by metabolic products, such as cholesterol, released during the degradation of LDL. Disorders of the LDL receptor and its internalization are medically important and are discussed in Chapter 26.

Other macromolecules, including several hormones, are subject to **adsorptive pinocytosis** and form **receptosomes,** vesicles that avoid lysosomes and deliver their contents to other intracellular sites, such as the Golgi system.

Adsorptive pinocytosis of extracellular glycoproteins requires that the glycoproteins carry specific carbohydrate recognition signals. These recognition signals are bound by membrane receptor molecules, which play a role analogous to that of the LDL receptor. A galactosyl receptor on the surface of hepatocytes is instrumental in the adsorptive pinocytosis of asialoglycoproteins from the circulation. Acid hydrolases taken up by adsorptive pinocytosis in fibroblasts are recognized by their mannose 6-phosphate moieties. Interestingly, the mannose 6-phosphate moiety also seems to play an important role in the intracellular targeting of the acid hydrolases to the lysosomes of the cells in which they are synthesized (see Chapter 54).

There is a dark side to receptor-mediated endocytosis, for viruses that cause diseases like hepatitis (af-

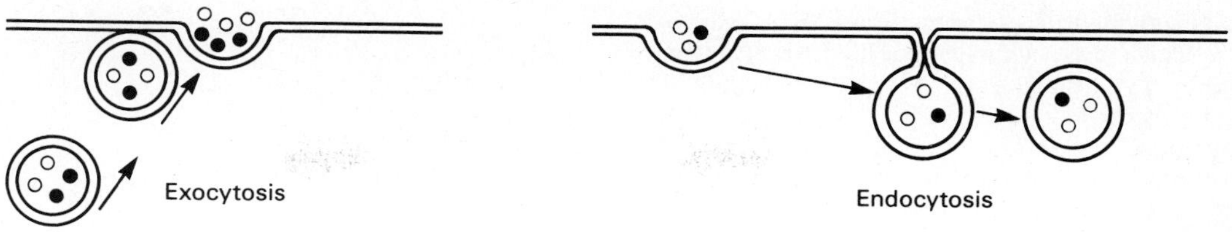

Figure 42–20. A comparison of the mechanisms of endocytosis and exocytosis. Exocytosis involves the contact of 2 inside surface (cytoplasmic side) monolayers, whereas endocytosis results from the contact of 2 outer surface monolayers.

fecting liver cells), poliomyelitis (affecting motor neurons), and AIDS (affecting T cells) initiate their damage by this mechanism. Iron toxicity also begins with excessive uptake due to endocytosis.

Exocytosis

Most cells also release macromolecules to the exterior by exocytosis. This process is also involved in membrane remodeling when the components synthesized in the Golgi apparatus are carried in vesicles to the plasma membrane. The signal for exocytosis is often a hormone which, when it binds to a cell-surface receptor, induces a local and transient change in Ca^{2+} concentration. Ca^{2+} triggers exocytosis. Fig 42–20 provides a comparison of the mechanisms of exocytosis and endocytosis.

Molecules released by exocytosis fall into 3 categories: (1) They can attach to the cell surface and become peripheral proteins, eg, antigens. (2) They can become part of the extracellular matrix, eg, collagen and glycosaminoglycans. (3) They can enter extracellular fluid and signal other cells. Insulin, parathyroid hormone, and the catecholamines are all packaged in granules and processed within cells, to be released upon appropriate stimulation (see Chapters 47, 49, and 51).

SIGNAL TRANSMISSION ACROSS MEMBRANES

Transmission of Biochemical Signals

Specific biochemical signals such as neurotransmitters, hormones, and immunoglobulins bind to specific receptors (integral proteins) exposed to the outside of cellular membranes and transmit information through these membranes to the cytoplasm. For example, the β-adrenergic receptor, which stereospecifically binds catecholamines, is asymmetrically located on the outer aspect of plasma membranes of target cells. The binding of the catecholamine on the outside stimulates the catalytic activity of adenylate cyclase, asymmetrically located on the inside of the membrane. Adenylate cyclase generates cAMP from ATP (see Chapter 44). Thus, the information that a specific catecholamine is present on the outside is transmitted to the inside of the cell, where a second messenger, cAMP, takes up the role of conveying the information further. The receptor-coupled adenylate cyclase system, which has stimulatory and inhibitory components, mediates the response of many hormones, and is discussed in detail in Chapter 44.

Recently, another type of signal transmission system has been discovered in mammalian cells. In this signaling system, **inositol triphosphate** (Fig 42–21) acts as the second messenger, and its intracellular concentration is regulated by extracellular signals via a transmembrane receptor coupling. On the surface of most mammalian cells are specific receptors for a series of protein growth factors such as insulin, epidermal growth factor, and platelet-derived growth factor. When such a receptor is occupied by its respective effector molecule, a kinase activity that is an integral component of the transmembrane receptor molecule is activated on the cytoplasmic side of the membrane. This kinase is capable of phosphorylating phosphatidylinositol to phosphatidylinositol 4-phosphate and the latter compound to phosphatidylinositol 4,5-bisphosphate. Interestingly, several oncogenes, the ex-

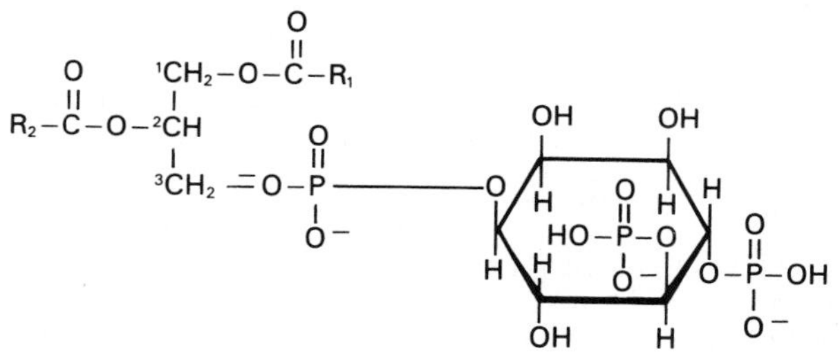

Figure 42–21. The structure of phosphatidylinositol 4,5-bisphosphate.

pression of which can effect malignant transformation of a cell, also possess kinase activities that can generate these polyphosphatidylinositides (see Chapter 57).

Other receptors on cell surfaces, such as those for acetylcholine, antidiuretic hormone, and α_1-type catecholamines, are, when occupied by their respective ligands, potent activators of phospholipase C. The latter catalyzes the hydrolysis of phosphatidylinositol 4,5-bisphosphate to inositol triphosphate and 1,2-diacylglycerol. The diacylglycerol is itself capable of activating protein kinase C, the activity of which is also dependent upon free ionic calcium. On the other hand, the inositol triphosphate is an effective releaser of calcium from intracellular storage sites such as the sarcoplasmic reticulum and mitochondria. Thus, the hydrolysis of phosphatidylinositol 4,5-bisphosphate leads to activation of protein kinase C and promotes an increase of cytoplasmic calcium ion. This activates the Na^+/K^+ pump transporter, and there is a net loss of protons from inside the cell, hence an increase of the internal pH. This results in cellular proliferation and other specific responses. In this signaling system, it appears that calcium and 1,2-diacylglycerol are tertiary messengers of the signal. Interestingly, some oncogenes appear to affect the signaling system process indirectly, in this case by possessing phosphatidylinositol kinase activity, which increases the accumulation of polyphosphatidylinositides that in turn serve as precursors for secondary and tertiary messengers. It is likely that other complex systems for transmitting information to the interior of the cell will be discovered. The role that this transmembrane signaling system plays in hormone action is discussed in Chapter 44.

INTERCELLULAR CONTACT & COMMUNICATION

There are many areas of intercellular contact in a metazoan organism. This necessitates contact between the plasma membranes of the individual cells. Cells have developed specialized regions on their mem-

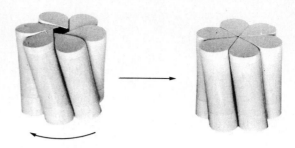

Figure 42–22. Simple model of the connexon, depicting the transition from the "open" to the "closed" configuration. It is proposed that the closure on the cytoplasmic face (uppermost) is achieved by the subunits sliding against one another, decreasing their inclination and hence rotating, in a clockwise sense, at the base. The darker shading on the side of the model indicates the portion that would be embedded in the membrane. The radial displacement of each subunit at the cytoplasmic end would be about 0.6 nm, given the observed inclination change of 5 degrees distributed over its 7.5-nm length. (Reproduced, with permission, from Unwin PNT, Zampighi G: Structure of the junction between communicating cells. *Nature* 1980; **283**:545.)

branes for intercellular communication in close proximity. **Gap junctions** mediate and regulate the passage of ions and small molecules through a narrow hydrophilic pore connecting the cytoplasm of adjacent cells. These pores are composed of subunits called **connexons** that have been studied by x-ray crystallography. As diagrammed in Fig 42–22, connexons consist of 6 protein subunits that span the membrane and connect with the analogous structures on the adjacent cells. Each subunit is apparently rigid, and in response to specific chemical stimuli, the subunits rearrange themselves relative to one another (compare the behavior of hemoglobin on oxygenation, Fig 6–12) to provide a tangential central opening about 2 nm in diameter. It is through this central opening that ions and small molecules can pass from one cell to another in a regulated fashion.

REFERENCES

Blobel G et al: Translocation of proteins across membranes: The signal hypothesis and beyond. *Symp Soc Exp Biol* 1979;**33**:9.

Dautry-Varsat A, Lodish HF: How receptors bring proteins and particles into cells. *Sci Am* (May) 1984;**250**:52.

Goldstein J et al: Receptor-mediated endocytosis. *Annu Rev Cell Biol* 1985;**1**:1.

Houslay MD, Stanley KK: *Dynamics of Biological Membranes.* Wiley, 1982.

Mueckler M et al: Sequence and structure of a human glucose transporter. *Science* 1985;**229**:941.

Sabatini DD et al: Mechanisms for the incorporation of proteins in membranes and organelles. *J Cell Biol* 1982;**92**:1.

Singer SJ, Nicolson GL: The fluid mosaic model of the structure of cell membranes. *Science* 1972;**175**:720.

Stahl P, Schwartz AL: Receptor-mediated endocytosis. *J Clin Invest* 1986;**77**:657.

Stein WD: *Transport and Diffusion Across Cell Membranes.* Academic Press, 1986.

Unwin N, Henderson R: The structures of proteins in biological membranes. *Sci Am* (Feb) 1984;**250**:78.

Vance DE, Vance JE (editors): *Biochemistry of Lipids and Membranes.* Benjamin/Cummings, 1985.

Walter P, Gilmore R, Blobel G: Protein translocation across the endoplasmic reticulum. *Cell* 1984;**38**:5.

Wickner WT, Lodish HF: Multiple mechanisms of protein insertion into and across membranes. *Science* 1985;**230**:400.

Characteristics of Hormone Systems

43

Daryl K. Granner, MD

INTRODUCTION

The objective of this chapter is to impart an appreciation for the varied nature of the endocrine system and to define several of the fundamental concepts that will reappear in subsequent chapters.

BIOMEDICAL IMPORTANCE

The advances made in understanding the general features of the endocrine system—such as why some glands are located in proximity to others, how hormones are produced, and the concepts of target cells, feedback control, and receptors—have direct clinical application. We are now able to provide some precise definitions of the causes of endocrine diseases. Most of these involve the identification of specific hormone receptor defects, but other types of causes will surely be found.

GENERAL FEATURES

A distinguishing characteristic of multicellular organisms is the presence of differentiated tissues that perform the specialized functions necessary for the survival of the organism. Mechanisms are required for intercellular communication to ensure the coordination of the responses necessary for adjusting to a constantly changing external and internal environment. Two general systems have evolved to serve these functions. These are the **nervous system,** often viewed as conducting signals or messages through a fixed, structural system; and the **endocrine system,** in which various **hormones** secreted by specific glands are transported as mobile messages to act on adjacent and distant tissue.

It is now clear that there is an exquisite convergence of these regulatory systems. Neural regulation of the endocrine system is very important; eg, epinephrine is produced and secreted by postganglionic cells in the adrenal medulla, and vasopressin is synthesized in the hypothalamus and transported by axons to the posterior pituitary, from which it is released. Likewise, many **neurotransmitters** (catecholamines, dopamine, acetylcholine, etc) are similar to hormones with regard to their synthesis, release, transport, and mechanism of action. In fact, catecholamines are neurotransmitters in one tissue and hormones in other tissues. The recent demonstration that certain metabolites of adrenal steroid hormones are barbituratelike modulators of the γ-aminobutyrate (GABA) receptor in brain is another example of this interaction. Finally, many hormones such as insulin, ACTH, vasoactive intestinal polypeptide (VIP), somatostatin, thyrotropin-releasing hormone (TRH), and cholecystokinin have recently been found in brain. It remains to be established whether all of these molecules are synthesized in brain and whether they act there as neuromodulators or as neurotransmitters. Since specific receptors for many of these hormones are found in brain, the possibility exists that these molecules act in brain.

The word hormone is derived from a Greek term meaning "to arouse to activity." By the classic definition, a hormone is a substance that is synthesized in one tissue and transported by the circulatory system to act on another organ. This original description is too restrictive; it is now appreciated that hormones act on adjacent cells in a given tissue (**paracrine function**) as well as on the cells in which they are synthesized (**autocrine function**).

DIVERSITY OF THE ENDOCRINE SYSTEM

One of the most remarkable features of the endocrine system is that it provides an organism with a number of different ways for solving problems. The purpose of this section is to present a very brief discussion of selected examples that highlight this diversity.

Derivation & Location of Endocrine Glands

The endocrine glands are mostly derived from epithelial cells. Notable exceptions include the connective tissue origin of the testosterone-producing Leydig cells in the testis and the estrogen-producing granulosa cells in the ovary, and the neuronal origin of the secretory cells in the neurohypophysis. The **neural crest** has been suggested as the embryologic origin of a number of endocrine cell types. If true, this would provide a rational link between the central nervous system and the endocrine system. Since neural crest tissue can appear in any organ, this could explain why some hormones appear to be made in brain and in tissues predominantly derived from the midgut and foregut. It may also explain the **ectopic hormone syndromes,** in which there is production of hormones by the "wrong tissue" (eg, production of parathyroid hormone [PTH] and ACTH by malignant cells in the case of lung cancer). These syndromes generally involve a rather restricted number of peptide hormones but a large and apparently diverse number of tissues. Although commonly thought to represent the activation of silent genes within a given cell, these syndromes could represent the activation of silent cells of common embryologic ancestry within a tissue. Another curious example is afforded by the **multiple endocrine neoplasia (MEN) syndromes,** in which there is a peculiar familial clustering of neoplasia of several endocrine glands. The production of excessive amounts of peptide or catecholamine hormones, often with one tissue making several, is a feature of these syndromes.

Hormone-producing cells are not randomly distributed; they are present in various tissues for specific reasons. Locally high concentrations of some hormones (ie, values in excess of plasma hormone levels) are often required for specific biologic processes. For example, a level of testosterone higher than that available in the plasma is required for spermatogenesis; thus, the testosterone-secreting Leydig cells and the seminiferous tubules are in juxtaposition. A very high concentration of estrogen is required for corpus luteum formation; hence, there is close proximity of this structure and the granulosa cells. A major action of insulin and glucagon is to regulate hepatic production of glucose; thus, there is close association of the pancreatic islets with the hepatic portal circulation. Cortisol, which is required in high concentration in the adrenal medulla for the induction of phenylethanolamine-N-methyltransferase (a rate-limiting enzyme in catecholamine biosynthesis), reaches this site by a portal vascular system that originates in the adrenal cortex. There is an intimate association of the hypothalamus and anterior pituitary, so that high concentrations of the very labile hypothalamic releasing hormones can easily reach the pituitary target via another special portal vascular system. Finally, there is a unique anatomic relationship between the various cells of the pancreatic islets in which locally high gradients of somatostatin, pancreatic polypeptide, glucagon, and insulin interact to regulate the secretion of one another.

Hormone Biosynthesis & Modification

There is great diversity in the chemical nature of and in the biosynthetic and postsynthetic mechanisms employed in the generation of active hormones. Hormones are derived from lipid precursors, from modifications of the amino acid tyrosine, and from the synthesis of proteins (simple and complex peptides and carbohydrate-containing glycoproteins).

Hormones may be synthesized and secreted in final form; examples include aldosterone, hydrocortisone, triiodothyronine (T_3), estradiol, and the catecholamines. Others must be modified within the cell before they are secreted or before they have full biologic activity. Examples include insulin, which is synthesized as proinsulin, the prototype of **precursor proteins,** and parathyroid hormone (PTH), which has at least 2 precursor peptides (a prepro segment) that must be removed to achieve full biologic activity. A description of precursor proteins, their synthesis, and intracellular processing into the final product can be found in Chapter 42. Pro-opiomelanocortin (POMC), a 285-amino-acid peptide that is the product of a single gene, represents an even more complicated case. POMC is cleaved to form ACTH, β-lipotropin, β-endorphin, α-MSH, and β-MSH, and the parent or precursor molecule may contain sequences of peptide hormones as yet unidentified. The **processing** of the precursor molecule is tissue-specific (see Chapter 45).

Perhaps the most exaggerated example of a large precursor of a hormone is thyroglobulin. This is a large protein (MW 660,000) found in the lumen of the thyroid follicle. Among the 5000 amino acids of the thyroglobulin molecule, there are 120 tyrosyl residues, some of which are iodinated in the process of thyroid hormone biosynthesis (see Chapter 46). The entire thyroglobulin molecule must be degraded to release the few tetraiodothyronine (T_4) and T_3 molecules present.

Some hormones are converted into more active molecules in peripheral tissues. This can occur in target tissues, as is the case in conversion of T_4 to T_3 in liver and pituitary and in conversion of testosterone to dihydrotestosterone in secondary sex tissues. **Peripheral conversion** can also occur in nontarget tissues; dehydroepiandrosterone is synthesized in the adrenal and converted to androstenedione in the liver. This latter molecule can then be converted to testosterone or to estrone and estradiol in fat cells, liver, or skin. Combined target and nontarget tissue peripheral conversion of an inactive molecule to an active hormone occurs. This is illustrated by the conversion of vitamin D_3 (from skin) to 25-hydroxycholecalciferol in liver, with subsequent conversion to 1,25-dihydroxycholecalciferol in kidney (see Chapter 47).

Hormones that are secreted from very different tissues and have different cell specificity may have structural similarities. The glycoprotein hormones from the pituitary and placenta (TSH, LH, FSH, and hCG) are heterodimers consisting of α and β subunits in which the α subunits are identical.

TARGET GLAND CONCEPT

There are about 200 types of differentiated cells in humans. Only a few of these produce hormones, but the 75 trillion cells in a human probably are all targets for one or more of the approximately 50 known hormones. A hormone may have one target tissue or it may affect a number of tissues. A target tissue was classically defined as having a unique biochemical or physiologic response to a hormone. For example, the thyroid is a specific target gland of TSH; TSH increases the number and size of the thyroid acinar cells and enhances all of the enzymatic steps involved in thyroid hormone biosynthesis. In contrast, insulin affects many tissues. Insulin enhances glucose uptake and oxidation in muscle, lipogenesis in fat, amino acid transport in liver and lymphocytes, and protein synthesis in liver and muscle, to name a few effects. More recently, with the delineation of specific cell surface and intracellular hormone receptors, the definition of a target has been expanded to include any tissue in which the hormone can be demonstrated to bind to its receptor, whether or not a classic biochemical or physiologic response has been determined (eg, insulin binding to endothelial cells). This definition is also incomplete, but it has heuristic merit, since it recognizes that not all actions of hormones have been elucidated.

Several factors determine the overall response of a target tissue to a hormone. The local concentration of a hormone around the target tissue depends upon (1) the rate of synthesis and secretion of the hormone; (2) the proximity of target and source; (3) the association-dissociation constants of the hormone with specific carrier proteins in the plasma, if such exist; (4) the rate of conversion of an inactive or suboptimally active form of the hormone into the active form; and (5) the rate of clearance of the hormone from blood by degradation or excretion, primarily accomplished by the liver and kidneys. The actual response then depends upon (1) the relative activity and state of occupancy, or both, of the specific hormone receptors on the plasma membrane or within the cytoplasm or nucleus; and (2) the postreceptor sensitization-desensitization of the cell. Alterations of any of these processes can result in a change of the hormonal activity on a given target tissue and must be considered as an addition to the classic feedback loops.

FEEDBACK CONTROL CONCEPT

Physiologic hormone levels in blood are maintained by a variety of homeostatic mechanisms that entail precise signaling between the hormone-secreting gland and the target tissue, and this often involves one or more intermediate glands. **Negative feedback control** is commonly employed, especially by the hypothalamic-pituitary target gland systems. An example is illustrated in Fig 43–1. A hypothalamic releasing hormone stimulates the synthesis and release of an anterior pituitary hormone, which in turn stimulates the production of the target organ hormone. High levels of the latter inhibit the system by decreasing hypothalamic hormone synthesis and action, while low levels result in the system being activated at the level of the hypothalamus. A unique feature of this particular axis is that the pituitary hormone may itself blunt the system by **"short-loop" feedback inhibition** of its own

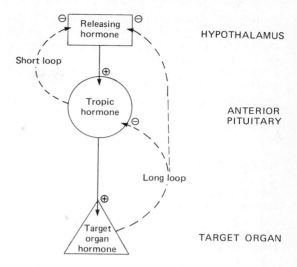

Figure 43–1. Example of the negative feedback control system used to regulate the function of the thyroid, adrenal, ovary, and testes.

synthesis. This tonic system provides for exquisite control of the plasma hormone level and also illustrates that there are several hormones and several target tissues within one endocrine system. Such control loops are recognized for the adrenal, thyroid, testicular, and ovarian systems.

In other instances, negative feedback regulation is accomplished through various metabolites or substrates whose plasma concentration is changed as a result of a hormone acting on a target cell. For example, an increased plasma glucose concentration (hyperglycemia) provokes a measured release of insulin, which enhances glucose uptake and utilization by a number of tissues, thereby decreasing the plasma glucose to a normal level and, in turn, diminishing insulin release. Under certain pathologic conditions, the insulin response may be excessive, in which case hypoglycemia ensues. The physiologic response to this life-threatening condition is the release of catecholamines, growth hormone, glucagon, ACTH, vasopressin, and angiotensin II, all of which act to increase the plasma glucose concentration. Thus, a complex network has evolved to regulate a critical metabolite (in this example, glucose) which is required for brain function.

In other cases, hormones exert **positive feedback control.** For example, estrogen and progesterone are required for the acute burst of LH secretion that results in ovulation and follicular luteinization and the further production of these steroid hormones. In many instances, the feedback loops have not been established, usually because the end products of the action of the hormone are not known.

A number of **pathophysiologic events,** including shock, trauma, hypoglycemia, pain, and stress, affect the hypothalamic-pituitary target gland axis through higher brain centers. These stimuli have profound effects on catecholamine and growth hormone metabolism and on the function of the adrenal cortex, thyroid, and gonads, but the precise components of these circuits are still poorly defined.

Endocrine and metabolic diseases result from the disruption of these normal feedback control mechanisms, and diagnostic perturbations of these systems (eg, the metyrapone test), are used to distinguish normal from pathologic conditions.

HORMONE RECEPTORS

General Features of Receptors

One of the major challenges faced in making the hormone-based communication system work is depicted in Fig 43–2. Hormones are present at very low concentrations in the extracellular fluid, generally in the range of $10^{-15}-10^{-9}$ mol/L. This is a much lower concentration than that of the many structurally similar molecules (sterols, amino acids, peptides, proteins) and other molecules that circulate at concentrations in the $10^{-5}-10^{-3}$ mol/L range. Target cells, therefore, must distinguish not only between different hormones present in small amounts but also between a given hormone and the 10^6- to 10^9-fold excess of other molecules. This high degree of discrimination is provided by cell-associated recognition molecules called receptors. Hormones initiate their biologic effects by binding to specific receptors, and since any effective control system also must provide a means of stopping a response, hormone-induced actions generally terminate when the effector dissociates from the receptor.

A target cell is defined by its ability to bind selectively a given hormone via such a receptor, an interaction that is often quantitated using radioactive ligands that mimic hormone binding. Several features of this interaction are important: (1) the radioactivity must not alter the biologic activity of the ligand; (2) the binding should be specific, ie, displaceable by unlabeled agonist or antagonist; (3) binding should be saturable; and (4) binding should occur within the concentration range of the expected biologic response.

Recognition & Coupling Domains of Receptors

All receptors, whether for polypeptides or steroids, have at least 2 functional domains. A recognition domain binds the hormone, and a second region generates a signal that couples hormone recognition to some intracellular function. The binding of hormone by receptor implies that some region of the hormone molecule has a conformation that is complementary to a region of the receptor molecule. The degree of similarity, or fit, determines the tightness of the association; this is measured as the affinity of binding (K). If the native hormone has a relative K value of 1, other natural molecules range between 0 and 1. In absolute terms, this actually spans a binding affinity range of more than a trillion. Ligands with a relative K value of >1 for some receptors have been synthesized and are used to study various aspects of receptor biology.

Coupling (signal transduction) occurs in 2 general ways. Polypeptide and protein hormones and the catecholamines bind to receptors located in the plasma membrane and thereby generate a signal that regulates various intracellular functions, often by changing the

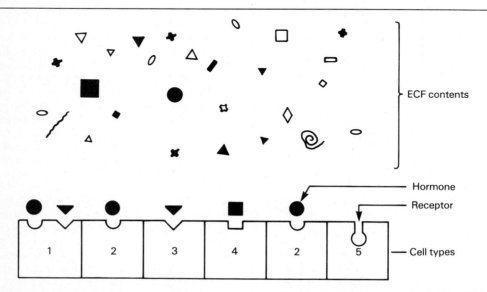

Figure 43–2. The specificity and selectivity of hormone receptors. Many different molecules circulate in the extracellular fluid (ECF), but only a few are recognized by hormone receptors. Receptors must select these molecules from high concentrations of the other molecules. There can be one class of receptor on each cell, or a given cell can have receptors for several hormones.

activity of an enzyme. Steroid and thyroid hormones interact with intracellular receptors, and this complex provides the signal (see below).

The amino acid sequences of these 2 domains in many polypeptide hormone receptors have now been identified. Hormone analogs with specific amino acid substitutions were used to change binding and to alter the biologic activity of the hormone. Steroid hormone receptors also have at least 2 functional domains: one site binds the hormone and the other binds to specific DNA regions. A number of these receptors are currently being characterized by recombinant DNA techniques, and structural analysis shows that these domains are highly homologous.

The dual functions of binding and coupling ultimately define a receptor, and it is the coupling of hormone binding to signal transduction, so-called **receptor-effector coupling,** that provides the first step in the amplification of the hormonal response. This dual purpose also distinguishes the target cell receptor from the plasma carrier proteins that bind hormone but do not generate a signal.

Comparison of Receptors & Transport Proteins

It is important to distinguish the binding of hormones to receptors from the association hormones have with various transport (carrier) proteins. Table 43–1 illustrates several comparative features. Thousands of receptor molecules per cell bind the ligand, and there is high binding affinity and specificity. Receptors are capable of recognizing and selecting specific molecules against a concentration gradient of 10^6 or 10^7, and this binding is saturable at physiologic concentrations of the hormone. Receptor-hormone interactions exhibit a salt, temperature, and pH dependency that is characteristic for each hormone. Binding

Table 43–1. Comparison of hormone receptors with proteins that transport hormones in the plasma.

Feature	Receptors	Plasma Transport Proteins
Concentration	Very low (thousands/cell)	Very high (billions/μL)
Binding affinity	Very high (10^{-11}–10^{-9} mol/L)	Low (10^{-7}–10^{-5} mol/L)
Binding specificity	High	Low
Saturability at physiologic concentrations	Yes	No
Reversibility	Yes	Yes
Signal transduction	Yes	No

is by **hydrophobic** and **electrostatic** mechanisms and thus is readily reversible except in special cases.

The hydrophobic steroid and thyroid hormones circulate bound to specific transport proteins. Transport proteins are much more abundant than the intracellular receptor proteins, but they bind hormones with lower affinity and lower specificity. Transport proteins provide a circulating reservoir of hormones, which when bound cannot be metabolized or excreted. Only the unbound, or free, hormone has biologic activity. Peptide and protein hormones do not have plasma transport proteins and thus have a much shorter plasma half-life than do the steroid hormones (seconds and minutes as opposed to hours).

Relationship Between Receptor Occupancy & Biologic Effect

The concentrations of hormone required for occupancy of the receptor and for elicitation of a specific biologic response are often very similar (Fig 43–3A).

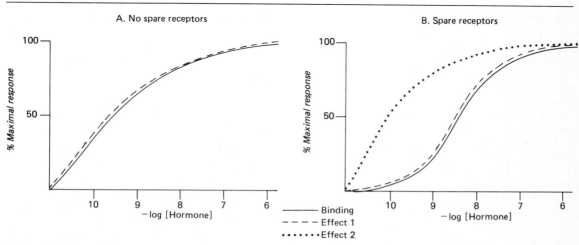

Figure 43–3. A comparison of hormone binding with biologic effect in the absence (*A*) or presence (*B*, effect 2) of spare receptors. In some cases, one biologic effect may be tightly coupled to binding in a tissue, whereas another effect shows the spare receptor phenomenon (compare effects 1 and 2 in *B*).

This is especially true for steroid hormones, but some peptide hormones also exhibit the same characteristics. This is remarkable, considering the many steps that must occur between hormone binding and complex responses, such as enzyme induction, cell lysis, and amino acid transport. In other instances, there is a marked dissociation of these 2 processes, so that a maximal biologic effect occurs when only a small percentage of the receptors are occupied (Fig 43–3B, effect 2). Those receptors not involved in the elicitation of the response are said to be **spare receptors.**

Spare receptors are observed in the response of several polypeptide hormones and are thought to provide a means of increasing the sensitivity of a target cell to activation by low concentrations of hormone and to provide a reservoir of receptors. The concept of spare receptors is an operational one and may depend on which aspect of the response is examined and which tissue is involved. For example, there is excellent agreement between LH binding and cAMP production in granulosa cells (there generally are no spare receptors when any hormone activates adenylate cyclase), but steroidogenesis in these cells, which is cAMP-dependent, occurs when fewer than 1% of the receptors are occupied (compare effect 1 with effect 2, Fig 43–3B). Transcription of the phosphoenolpyruvate carboxykinase gene is repressed when far fewer than 1% of liver cell insulin receptors are occupied, whereas in thymocytes there is a high correlation between insulin binding and amino acid transport. Other examples of the dissociation of receptor binding and biologic effects include the effects of catecholamines on muscle contraction, lipolysis, and ion transport. The presumption is that these end responses reflect a cascade or multiplier effect of the hormone. Variable sensitivities of different responses even within the same cell have been noted. Successively greater degrees of occupancy of the adipose cell insulin receptor increase (in sequence) lipolysis, glucose oxidation, amino acid transport, and protein synthesis.

Regulation of Receptors

The number of receptors on or in a cell is in a dynamic state and can be regulated physiologically or be influenced by diseases or therapeutic measures. Most is known about the plasma membrane receptors. Both the receptor concentration and the affinity of hormone binding can be regulated in plasma membrane. These changes can be acute and can significantly affect hormone responsiveness of the cell. For instance, cells exposed to β-adrenergic agonists for minutes to hours no longer activate adenylate cyclase in response to the addition of more agonist, and the biologic response is lost. This **desensitization** occurs by 2 mechanisms. The first involves a loss of receptors from the plasma membrane. This **down-regulation** involves the internal sequestration of receptors, thereby segregating them from the other components of the response system including the regulatory and catalytic subunits of adenylate cyclase (see Chapter 44). Removal of the agonist results in the return of receptors to the cell surface and restoration of hormonal sensitivity. A second form of desensitization of the β-adrenergic system involves the covalent modification of receptor by phosphorylation. This cAMP-dependent process entails no change in receptor number and no translocation. Reconstitution experiments show that the phosphorylated receptor is unable to activate cyclase, so that the activation and hormone binding functions are uncoupled. Other examples of physiologic adaptation accomplished through down-regulation of receptor number by the homologous hormone include insulin, glucagon, TRH, growth hormone, LH, FSH, and catecholamines. A few hormones, such as angiotensin II and prolactin, **"up-regulate"** their receptors. These changes in receptor number can occur rapidly (minutes to hours) and are probably an important means of regulating biologic responses.

How the loss of receptors affects the biologic response elicited at a given hormone concentration depends on whether or not there are spare receptors. Fig 43–4 illustrates the effects of a 5-fold loss of receptors on the concentration-response curve in both conditions. In condition A (with no spare receptors), the maximal response obtained is 20% that of control; hence, the effect is on the "V_{max}." In condition B (with spare receptors), the maximal response is obtained but at 5 times the originally effective hormone concentration, analogous to a "K_m" effect.

Structure of Receptors

Most is known about the acetylcholine receptor, which was easy to purify, since it exists in relatively large amounts in the electric organ of the eel *Torpedo californica*. The acetylcholine receptor consists of 4 subunits in the configuration α_2, β, γ, δ. The two α subunits bind acetylcholine; the technique of site-directed mutagenesis has been used to show which regions of this subunit are involved in the formation of the transmembrane ion channel, which performs the major function of the acetylcholine receptor.

Other receptors are present in very small amounts; thus, purification and characterization have been difficult. Recombinant DNA techniques can now provide the requisite amounts of material for such studies, and so this has become an active area of investigation. The insulin receptor is a heterotetramer ($\alpha_2\beta_2$) linked by multiple disulfide bonds, in which the extramembrane α subunit binds insulin and the membrane-spanning β subunit transduces the signal, presumably through the tyrosine kinase component of the cytoplasmic portion of this polypeptide. The receptors for insulinlike growth factor I (IGF-I), epidermal growth factor (EGF), and low-density lipoprotein (LDL) are generally similar to the insulin receptor (see Fig 51–16). Other polypeptide hormone receptors are less well characterized but are assumed to have a fundamental protein component because of their sensitivity to a variety of peptidases and proteolytic enzymes. Hormone binding appears to require disulfide bonds, phospholipid, and carbohydrate moieties in many instances.

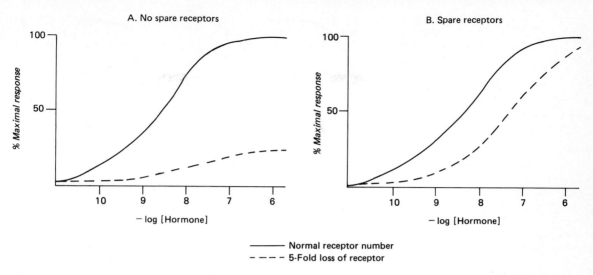

Figure 43–4. Effect of a 5-fold loss of receptors on a biologic effect in a system that lacks (*A*) or has (*B*) spare receptors.

Steroid hormone receptors also are proteins. The function of this group of molecules has been defined in recent years, and we now are beginning to learn about their structure. The glucocorticoid receptor is a good example (Fig 43–2). This molecule has 3 functional domains: (1) a hormone-binding region in the carboxy-terminal portion; (2) an adjacent DNA-binding region; and (3) a specifier region, located in the amino-terminal half of the molecule, which is necessary for high-affinity binding to the proper region of DNA (and which contains most of the antigenic sites in the molecule). These functional regions are being confirmed by analyses of receptor synthesized in cDNA-driven expression systems. It appears that the various types of steroid receptors share this general structure and that there is a great deal of amino acid sequence homology in these regions in the different receptors. There is also a curious homology between this class of receptors and the v-*erb* A oncogene.

The Agonist-Antagonist Concept

Molecules can be divided into 4 groups with respect to their ability to elicit a given hormone receptor-mediated response. These classes are agonists, partial agonists, antagonists, and inactive agents. This classification has been elaborated in great detail for glucocorticoid hormone action (see Chapter 48).

Agonists elicit the maximal response, although different concentrations may be required (see Fig 43–5 and example A of Fig 43–6). In Fig 43–5, 1, 2, and 3 could be porcine insulin, porcine proinsulin, and guinea pig insulin, respectively. In all systems tested, these insulins have the same rank order of potency, yet each elicits a maximal response if present in sufficient concentration. Likewise, 1, 2, and 3 could be dexamethasone, cortisol, and corticosterone (see Table 48–4). Partial agonists evoke an incomplete response even when very large concentrations of the hormone are employed, as illustrated in example B of Fig 43–6). Antagonists generally have no effect themselves, but they competitively inhibit the action of agonists or partial agonists (see examples A + C or B + C in Fig 43–6). A large group of structurally similar compounds elicit no effect at all and have no effect on the action of the agonists or antagonists. These are classified as inactive agents and are represented as example D in Fig 43–6.

Partial agonists also compete with agonists for binding to (and activation of) the receptor, in which case they become partial antagonists. The extent of the inhibition of agonist activity caused by partial or complete antagonists depends on the relative concentration of the various steroids. Generally much higher concentrations of the antagonist are required to inhibit an agonist than are necessary for the latter to exert its maximal effect. Such concentrations are rarely, if ever, achieved in vivo; thus, this phenomenon is generally employed for studies of the mechanism of action of glucocorticoid hormones in vitro.

The coupling function of the glucocorticoid receptor—namely, that in addition to binding to the receptor a ligand also must facilitate a change in this molecule so that it can bind to DNA—was first suggested in studies that employed the steroids listed in Table 48–4. The hypothesis suggested is that (1) agonists bind to and fully activate the receptor and elicit the maximal biologic response; (2) partial agonists fully occupy the receptor but afford incomplete activation and thus a partial response; and (3) antagonists fully occupy the receptor, but this complex is unable to bind to DNA, and so it elicits no intrinsic response but does inhibit the action of agonists.

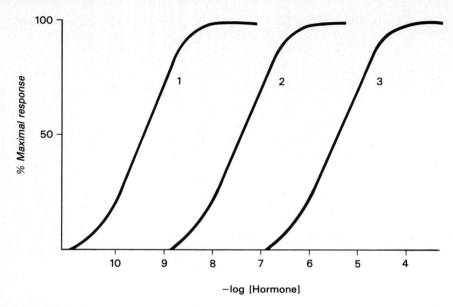

Figure 43–5. Within a class of hormones, different molecules may have different potency. The same response is obtained, but different concentrations of hormones are required.

Hormone Receptors & Disease

The realization that **certain diseases involve abnormalities of receptor function** followed the basic elucidation of the role that receptors play in hormone action. Three general categories of receptor-related defects are shown in Table 43–2. In the first category, antibodies directed against a specific hormone receptor are responsible for the disease. These antibodies of the IgG class can block hormone binding (acanthosis nigricans with insulin resistance; asthma), mimic hormone binding (Graves' disease), or enhance receptor turnover (myasthenia gravis).

In the second category, no hormone binding to receptor can be detected. Whether receptors are absent in these diseases or present but defective cannot be de-

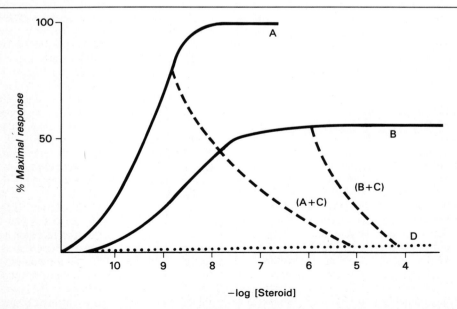

Figure 43–6. Hormones can be classified as agonists (A), partial agonists (B), antagonists (A + C and B + C), or inactive agents (D).

Table 43–2. Hormone receptors and disease.

Disease	Receptor	Problem
Graves' disease (hyperthyroidism)	TSH	Antibody stimulates TSH receptor
Acanthosis nigricans with insulin resistance	Insulin	Antibody blocks insulin binding to receptor
Myasthenia gravis	Acetylcholine	Antibody enhances turnover of the acetylcholine receptor
Asthma	β-Adrenergic receptor	Antibody blocks β-adrenergic binding
Congenital nephrogenic diabetes insipidus	ADH	Receptor deficiency
Testicular feminization syndrome	Androgen	Receptor deficiency
Pseudohypoparathyroidism	PTH	Receptor deficiency
Vitamin D-resistant rickets type II	Calcitriol	Receptor deficiency
Obesity	Insulin	Hormone binding decreased
Diabetes mellitus type II (non-insulin-dependent diabetes mellitus [NIDDM])	Insulin	Hormone binding decreased

termined as yet, since the assays for most receptors depend upon hormone binding.

The third category consists of diseases involving abnormal receptor regulation. Patients with obesity and those with type II diabetes mellitus and obesity often manifest glucose intolerance and insulin resistance in spite of elevated plasma insulin levels. These individuals have fewer insulin receptors (owing to down-regulation) on target cells such as fat, liver, and muscle. With weight reduction, the plasma insulin level decreases, the number of receptors increases, insulin sensitivity improves, and glucose intolerance is reduced. Recent studies of the molecular basis of cancer strongly suggest that abnormalities of growth factor receptor-effector coupling can account for the uncontrolled growth of malignant cells. These examples are illustrative of the many hormone receptor-mediated diseases.

REFERENCES

Ginsberg BH: Synthesis and regulation of receptors for polypeptide hormones. Pages 59–97 in: *Biological Regulation and Development*. Vol 3B. Yamamoto K (editor). Plenum Press, 1985.

Granner DK, Lee F: The multiple endocrine neoplasia syndromes. Chapter 76 in: *Comprehensive Textbook of Oncology*. Moosa AR, Robson MC, Schimpff SC (editors). Williams & Wilkins, 1984.

Mishina M et al: Expression of functional acetylcholine receptor from cloned cDNAs. *Nature* 1984;**307:**604.

Roth J, Taylor SI: Receptors for peptide hormones: Alterations in disease of humans. *Annu Rev Physiol* 1982;**44:**639.

Roth J et al: The evolutionary origins of hormones, neurotransmitters, and other extracellular messengers. *N Engl J Med* 1982;**306:**523.

Sibley DR, Lefkowitz RJ: Molecular mechanisms of receptor desensitization using the β-adrenergic receptor-coupled adenylate cyclase system as a model. *Nature* 1985;**317:**124.

44

Hormone Action

Daryl K. Granner, MD

INTRODUCTION

Hormone action at the cellular level begins with the association of the hormone and its specific receptor. Hormones can be classified by the location of the receptor and by the nature of the signal or second messenger used to mediate hormone action within the cell. A number of these second messengers have been defined, but for several hormones the intracellular signal has not been discovered. Considerable progress has been made in elucidating how hormones work intracellularly, particularly in regard to the regulation of expression of specific genes.

BIOMEDICAL IMPORTANCE

The rational diagnosis and therapy of a disease depend upon understanding the pathophysiology involved and the ability to quantitate it. Diseases of the endocrine system, which are generally due to excessive or deficient production of hormones, are an excellent example of the application of basic principles to clinical medicine. Knowing the general aspects of hormone action and understanding the physiologic and biochemical effects of the individual hormones enable one to recognize endocrine disease syndromes that result from hormone imbalance and to apply effective therapy.

CLASSIFICATION OF HORMONES

Hormones can be classified according to chemical composition, solubility properties, location of receptors, and nature of the signal used to mediate hormone action within the cell. A classification based on the last 2 properties is illustrated in Table 44–1, and general features of each group are illustrated in Table 44–2.

The hormones in group I are lipophilic and, with the exception of T_3 and T_4, are derived from cholesterol. After secretion, these hormones associate with transport proteins, a process that circumvents the problem of solubility while prolonging the plasma half-life. The free hormone readily traverses the plasma membrane of all cells and encounters receptors in either the cytosol or nucleus of target cells. The ligand-receptor complex is assumed to be the intracellular messenger in this group.

The second major group consists of water-soluble hormones that bind to the plasma membrane of the target cell. Hormones that bind to the surface of cells communicate with intracellular metabolic processes through intermediary molecules, so-called **second messengers** (the hormone itself is the first messenger), which are generated as a consequence of the ligand-receptor interaction. The second-messenger concept arose from Sutherland's observation that epinephrine binds to the plasma membrane of pigeon erythrocytes and increases intracellular cAMP. This was followed by a series of experiments in which cAMP was found to mediate the metabolic effects of many hormones. Hormones that clearly employ this mechanism are shown in group II.A. Several hormones, many of which were previously thought to affect cAMP, appear to use calcium or metabolites of complex phosphoinositides (or both) as the intracellular signal. These are shown in group II.B. The intracellular messenger has not been definitively identified for group II.C, a large and very interesting class of hormones. Several candidates have been proposed as mediators of the actions of insulin—eg, cAMP, cGMP, H_2O_2, calcium, several small peptides, a phospholipid, insulin itself, and the insulin receptor—but none of these fulfill the necessary criteria as yet. It would not be surprising if a number of mediators or different mechanisms are involved in the action of this group of hormones. A few hormones fit into more than one category, and assignments change with new information.

MECHANISM OF ACTION OF GROUP I HORMONES

The general features of the action of this group of hormones are illustrated in Fig 44–1. These lipophilic molecules diffuse through the plasma membrane of all cells but only encounter their specific, high-affinity receptor in target cells. The hormone-receptor complex next undergoes a temperature- and salt-dependent **"activation" reaction** that results in size, conformation, and surface charge changes that render it able to bind to chromatin. Whether this association and "activation" process occurs in the cytoplasm or nucleus is debatable but not crucial to understanding the whole process. The hormone-receptor complex binds to specific regions of DNA and activates or inactivates specific genes. By selectively affecting gene transcription and the production of the respective mRNAs, the

472

Table 44–1. Classification of hormones by mechanism of action.

Group I. Hormones that bind to intracellular receptors

Estrogens	Calcitriol (1,25[OH]$_2$-D$_3$)
Glucocorticoids	Androgens
Mineralocorticoids	Thyroid hormones (T$_3$ and T$_4$)
Progestins	

Group II. Hormones that bind to cell surface receptors

A. The second messenger is cAMP:

Adrenocorticotropic hormone (ACTH)	Parathyroid hormone (PTH)
Angiotensin II	Opioids
Antidiuretic hormone (ADH)	Acetylcholine
Follicle-stimulating hormone (FSH)	Glucagon
Human chorionic gonadotropin (hCG)	α_2-Adrenergic catecholamines
Lipotropin (LPH)	Corticotropin-releasing hormone (CRH)
Luteinizing hormone (LH)	Calcitonin
Melanocyte-stimulating hormone (MSH)	Somatostatin
Thyroid-stimulating hormone (TSH)	β-Adrenergic catecholamines

B. The second messenger is calcium or phosphatidylinositides (or both):

α_1-Adrenergic catecholamines	Acetylcholine (muscarinic)
Cholecystokinin	
Gastrin	
Substance P	
Thyrotropin-releasing hormone (TRH)	Gonadotropin-releasing hormone (GnRH)
Vasopressin	Angiotensin II

C. The intracellular messenger is unknown:

Chorionic somatomammotropin (CS)	
Growth hormone (GH)	Oxytocin
Insulin	Nerve growth factor (NGF)
Insulinlike growth factors (IGF-I, IGF-II)	Epidermal growth factor (EGF)
	Fibroblast growth factor (FGF)
Prolactin (PRL)	Platelet-derived growth factor

amounts of specific proteins are changed and metabolic processes are influenced. The effect of each of these hormones is quite specific; generally, the hormone affects less than 1% of the proteins or mRNA in a target cell. This discussion has concentrated on nuclear actions of steroid and thyroid hormones because these are quite well defined. Direct actions in the cytoplasm and upon various organelles and membranes have also been described.

An effect of estrogens and glucocorticoids on mRNA degradation has been demonstrated, and it is known that glucocorticoids affect posttranslational processing of some proteins. However, most evidence suggests that these hormones exert their predominant effect on gene transcription. Although the biochemistry of gene transcription in mammalian cells is not well understood, a general model of the structural requirements for steroid and thyroid regulation of gene transcription can be drawn (Fig 44–2). These genes must be in regions of "open" or transcriptionally active chromatin (depicted as the bubble in Fig 44–1), as defined by their susceptibility to digestion by the enzyme DNase I. The genes studied to date have at least 2 separate regulatory elements (control sites) in the DNA sequence immediately 5′ of the transcription initiation site (Fig 44–2). The first of these, the **promoter element (PE),** is generic, since it is present in some form or other in all genes. This element specifies the site of RNA polymerase II attachment to DNA and therefore the accuracy of transcript initiation (see Chapter 41).

Table 44–2. General features of hormone classes.

	Group I	Group II
Types	Steroids, iodothyronines, calcitriol	Polypeptides, proteins, glycoproteins, catecholamines
Solubility	Lipophilic	Hydrophilic
Transport proteins	Yes	No
Plasma half-life	Long (hours to days)	Short (minutes)
Receptor	Intracellular	Plasma membrane
Mediator	Receptor-hormone complex	cAMP, Ca^{2+}, metabolites of complex phosphoinositides, others

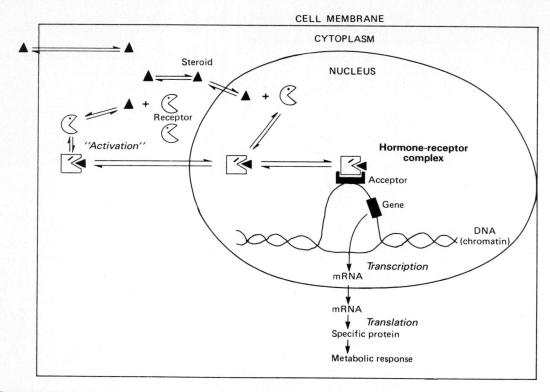

Figure 44–1. The steroid binds to an intracellular receptor and causes a conformational change of the latter. This complex then binds to a specific region on chromatin, which results in the activation of a restricted number of genes. Thyroid hormones bind to a receptor that is part of the chromatin complex. In other respects, the action of these hormones appears to be similar.

A second element, the **hormone responsive element (HRE),** has been identified in many genes regulated by steroid hormones. This is located slightly farther 5' than the PE and may consist of several discrete elements. The HRE presumably modulates the frequency of transcript initiation and is less dependent on position and orientation; in these respects, it resembles the transcription **enhancer elements** found in other genes (see Chapter 41). Generally, the HRE is found within a few hundred nucleotides upstream of the transcription initiation site, but the precise location of the HRE varies from gene to gene. In some instances, it is located within the gene.

The identification of an HRE requires that it bind the hormone-receptor complex more avidly than does surrounding DNA or DNA from another source. In the cases just cited, such specific binding has been demonstrated. The HRE must also confer hormone respon-

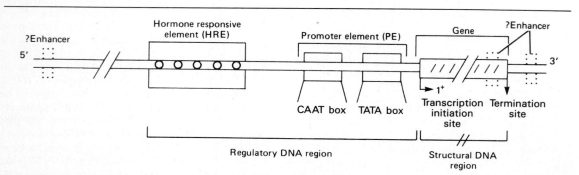

Figure 44–2. Structural requirements for steroid regulation of gene transcription.

siveness. Putative regulatory sequence DNA can be ligated to reporter genes to assess this point. Usually, these **"fusion genes"** contain reporter genes not ordinarily influenced by the hormone, and often these genes are not normally expressed in the tissue being tested. Commonly used reporter genes are globin, thymidine kinase, and bacterial chloramphenicol acetyltransferase. The "fusion gene" is transfected into a target cell, and if the hormone now regulates the transcription of the reporter gene, one has functionally defined an HRE. Position, orientation, and base substitution effects can be precisely defined using this technique. Exactly how the hormone-receptor interaction with the HRE affects transcription is an area of active investigation. Transcript initiation is a probable control site, but effects on elongation and termination might also occur. Control sites farther 5′ from the initiation site, or 3′ downstream, either within or beyond the gene, have been proposed. Finally, *trans*-acting control mechanisms (eg, from another chromosome) may also be operative.

MECHANISM OF ACTION OF GROUP II (PEPTIDE) HORMONES

The largest number of hormones are water-soluble, have no transport proteins (and therefore a short plasma half-life), and initiate a response by binding to a receptor located in the plasma membrane (Tables 44–1 and 44–2). The mechanism of action of this group of hormones can best be discussed in terms of their intracellular messengers.

1. cAMP AS THE SECOND MESSENGER

cAMP (cyclic AMP, 3′,5′-adenylic acid; see p 169), a ubiquitous nucleotide derived from ATP through the action of the enzyme adenylate cyclase, plays a crucial role in the action of a number of hormones. The intracellular level of cAMP is increased or decreased by various hormones (Table 44–3), and this

Table 44–3. Subclassification of group II.A hormones.

Hormones That Stimulate Adenylate Cyclase (H_s)	Hormones That Inhibit Adenylate Cyclase (H_i)
ACTH	Acetylcholine
ADH	α_2-Adrenergics
β-Adrenergics	Angiotensin II
Calcitonin	Opioids
CRH	Somatostatin
FSH	
Glucagon	
hCG	
LH	
LPH	
MSH	
PTH	
TSH	

effect varies from tissue to tissue. Epinephrine causes large increases of cAMP in muscle and relatively small changes in liver. The opposite is true of glucagon. Tissues that respond to several hormones of this group do so through unique receptors converging upon a single adenylate cyclase molecule. The best example of this is the adipose cell, in which epinephrine, ACTH, TSH, glucagon, MSH, and vasopressin (ADH) stimulate adenylate cyclase and increase cAMP. Combinations of maximally effective concentrations are not additive, and treatments that destroy one receptor have no effect on the cellular response to other hormones.

Adenylate Cyclase System

The components of this system in mammalian cells are illustrated in Fig 44–3. The interaction of the hormone with its receptor results in the activation or inactivation of adenylate cyclase. This process is mediated by at least 2 GTP-dependent regulatory proteins, designated G_s (stimulatory) and G_i (inhibitory) (sometimes referred to as N_s and N_i), each of which is composed of 3 subunits, α, β, and γ. Adenylate cyclase, located on the inner surface of the plasma membrane, catalyzes the formation of cAMP from ATP in the presence of magnesium (see Fig 34–14).

What was originally conceived of as a single protein with 2 functional domains is now viewed as a system of extraordinary complexity. Over the past 15 years, a number of studies have established the biochemical uniqueness of the hormone receptor and GTP regulatory and catalytic domains of the adenylate cyclase complex, a current model of which is illustrated in Fig 44–3. This model explains how different peptide hormones can either stimulate (s) or inhibit (i) the production of cAMP (Table 44–3).

Two parallel systems, a stimulatory (s) one and an inhibitory (i) one, converge upon a single catalytic molecule (C). Each consists of a receptor, R_s or R_i, and regulatory complex, G_s and G_i. G_s and G_i are each trimers composed of α, β, and γ subunits. The β and γ subunits in G_s appear to be identical to their respective counterparts in G_i. Because the α subunit in G_s differs from that in G_i, the proteins are designated α_s (MW 45,000) and α_i (MW 41,000). The binding of a hormone to R_s or R_i results in a receptor-mediated activation of G, which entails Mg^{2+}-dependent binding of GTP by α and the concomitant dissociation of β and γ from α.

$$\alpha\beta\gamma \xrightarrow[\text{GTPase}]{\text{GTP}} \alpha \cdot GTP + \beta\gamma$$

The α_s has intrinsic GTPase activity, and the active form, $\alpha_s \cdot GTP$, is inactivated upon hydrolysis of the GTP to GDP, and the trimeric G_s complex is reformed. Cholera toxin, known to be an irreversible **activator** of cyclase, causes ADP ribosylation of α_s and in so doing inactivates the GTPase; therefore, α_s is frozen in the active form. The α_i also has a GTPase activity; however, GDP does not freely dissociate from $\alpha_i \cdot GDP$. The α_i is reactivated by an exchange of GTP

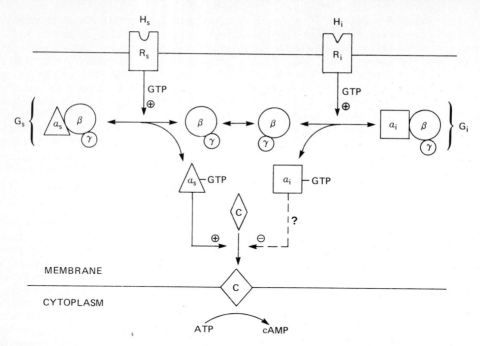

Figure 44–3. The hormone-receptor signal is transmitted through a stimulatory (s) or inhibitory (i) regulatory complex (G_s or G_i) to either stimulate (s) or inhibit (i) the activity of adenylate cyclase (C). Adenylate cyclase catalyzes the formation of cAMP from ATP. (Modified, with permission, from Gilman AG: G proteins and dual control of adenylate cyclase. *Cell* 1984;**36**:577. Copyright © the Massachusetts Institute of Technology.)

for GDP. Pertussis toxin irreversibly activates adenylate cyclase by promoting the ADP ribosylation of α_i, which prevents the α_i subunit from being activated. NaF, another irreversible activator of cyclase, presumably acts on the α_s or α_i subunit, because it affects G_s and G_i similarly. The exact role of each of the α, β, and γ subunits has not been defined. Two possibilities have been tested. The α_s and α_i could interact noncompetitively with C, causing opposite effects; the net effect would depend on the balance of active α_s and active α_i. Unfortunately, active α_i has little inhibitory effect on C in isolated systems. The second possibility, therefore, is that the β subunit of G_i inhibits α_s. In this model, α_i would be an anti-inhibitor of adenylate cyclase by binding β and, as such, would have no direct effect.

Many of the components of the cyclase system have been purified, including the catalytic subunit, and it is now apparent that there is a family of G proteins. Transducin, the protein that is important in coupling light to photoactivation in the retina, is closely related to the G protein of adenylate cyclase, as are the products of the *ras* oncogenes. A unique protein, G_0, has been isolated from brain. Others, not yet well-characterized, may be involved in calcium and potassium flux or phosphoinositide metabolism.

The importance of these components is underscored by an "experiment of nature." Pseudohypoparathyroidism is a syndrome characterized by hypo-

calcemia and hyperphosphatemia, the biochemical hallmarks of hypoparathyroidism, and by a number of congenital defects. Affected individuals do not have defective parathyroid function; in fact, they secrete large amounts of biologically active PTH. Some have target organ resistance on the basis of a postreceptor defect. They are partially deficient in G protein (probably only the α_s subunit) and thus fail to couple binding to adenylate cyclase stimulation. Others appear to generate cAMP in response to PTH but fail to respond to the cAMP signal. The observation that patients with pseudohypoparathyroidism often show evidence of defective responses to other hormones, including TSH, glucagon, and β-adrenergic agents, is not surprising.

Protein Kinase

In **prokaryotic cells,** cAMP binds to a specific protein, called **catabolite regulatory protein (CRP),** which binds directly to DNA and influences gene expression. The analogy of this to steroid hormone action described above is apparent. In **eukaryotic cells,** cAMP binds to a protein kinase that is a heterotetrameric molecule consisting of 2 **regulatory** subunits (R) and 2 **catalytic** subunits (C). cAMP binding results in the following reaction:

$$4 \text{ cAMP} + R_2C_2 \rightleftharpoons R_2 \cdot (4 \text{ cAMP}) + 2 \text{ C}$$

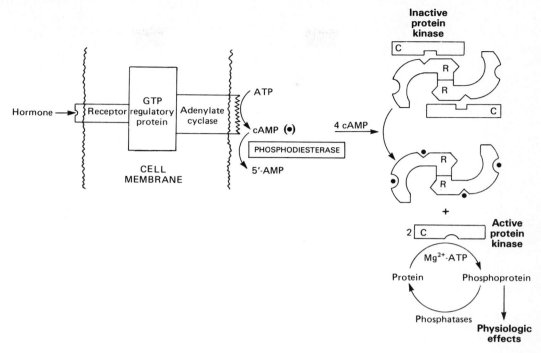

Figure 44−4. Hormonal regulation of cellular processes through cAMP-dependent protein kinases. (Courtesy of J Corbin.)

The R_2C_2 complex has no enzymatic activity, but the binding of cAMP by R dissociates R from C, thereby activating the latter (Fig 44−4). The active C subunit catalyzes the transfer of the γ phosphate of ATP (Mg^{2+}) to a serine or threonine residue in a variety of proteins. The consensus phosphorylation sites are -Arg-Arg-X-Ser- and -Lys-Arg-X-X-Ser-, where X can be any amino acid.

Protein kinase activities were originally described as being **cAMP-dependent** or **cAMP-independent.** Table 44−4 shows that this too has become considerably more complex, as protein phosphorylation is now recognized as being an important regulatory mechanism. The kinases listed are all unique molecules and show considerable variability with respect to subunit composition, molecular weight, autophosphorylation, K_m for ATP, and substrate specificity.

The cAMP-dependent protein kinases I and II have been studied in greatest detail. These kinases share a common C subunit and have different R subunits. They were originally classified by type on the basis of their different surface charges and hence differential elution from ion exchange chromatography columns (type I is less acidic and elutes at a lower salt concentration than does type II). Most tissues have both forms, but there is a marked species-to-species and tissue-to-tissue variation in the distribution of the 2 isozymes. Recent studies suggest that types I and II respond differently to various combinations of cAMP

Table 44−4. Purified protein kinases.

Hormone-responsive
  Calcium/calmodulin-dependent kinase
  Calcium/phospholipid-dependent kinase
  cAMP-dependent kinase I and II
  cGMP-dependent kinase
  Epidermal growth factor-dependent tyrosine kinase
  Insect cyclic nucleotide-dependent kinase
  Insulin-dependent tyrosine kinase
  Myosin light chain kinase
  Phosphorylase kinase
  Pyruvate dehydrogenase kinase
Not known to be hormone-responsive
  Casein kinase II
  dsRNA-dependent eIF-2α kinase
  Hemin-dependent eIF-2α kinase
  Protease-activated kinase I
  Rhodopsin kinase
  Viral tyrosine kinases I, II, and III
May be hormone-responsive
  Casein kinase I
  Protease-activated kinase II

analogs, and this approach may be useful in defining which isozyme mediates a specific biologic response. There is also some evidence that hormone stimulation selectively enhances either type I or type II kinase activity.

Several other protein kinases are involved in hormone action. The role some of these play is illustrated in Fig 44−5 and in the later parts of this chapter.

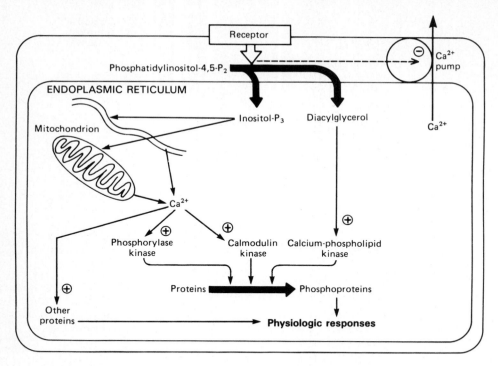

Figure 44–5. The regulation of hormone action by Ca^{2+}. (Courtesy of JH Exton.)

The epidermal growth factor and insulin-dependent kinases are unique in that the enzymatic activity resides within the hormone receptor and depends upon ligand-receptor binding for activation (see Chapter 51). Another distinguishing feature is that these kinases preferentially phosphorylate tyrosine residues, and tyrosine phosphorylation is infrequent in mammalian cells. The role these receptor-associated kinases play in hormone action is not clear, but it is possible that the hormone initiates a phosphorylation cascade and that one or more products of the cascade are the intracellular messenger.

Phosphoproteins

The effects of cAMP in eukaryotic cells are all thought to be mediated by protein phosphorylation-dephosphorylation. The control of any of the effects of cAMP, including such diverse processes as steroidogenesis, secretion, ion transport, carbohydrate and fat metabolism, enzyme induction, gene regulation, and cell growth and replication, could be conferred by a specific protein kinase, a specific phosphatase, or by specific substrates for phosphorylation. In some instances a phosphoprotein that is a known participant in a metabolic pathway has been identified; however, in most processes cited above, the phosphoproteins involved have not been identified. These substrates may help define a target tissue and certainly are involved in defining the extent of the response within a given cell. Many proteins can be phosphorylated, including ca-

sein, histones, and protamine; such phosphorylation may be epiphenomena, although they are useful for assaying protein kinase activity. Until recently, the only actions of cAMP that had been defined were actions that occurred outside the nucleus. Effects of cAMP on the transcription of several genes have been described. Whether these nuclear actions of cAMP occur by protein phosphorylation or by a CRP-like mechanism is not known.

Phosphodiesterases

Actions caused by hormones that increase cAMP concentration can be terminated in a number of ways, including the hydrolysis of cAMP by phosphodiesterases. The presence of these hydrolytic enzymes ensures a rapid turnover of the signal (cAMP) and hence a rapid termination of the biologic process once the hormonal stimulus is removed. cAMP phosphodiesterases exist in low and high K_m forms and are themselves subject to regulation by hormones as well as by intracellular messengers such as calcium, probably acting through calmodulin. Inhibitors of phosphodiesterase, most notably methylated xanthine derivatives such as caffeine, increase intracellular cAMP and mimic or prolong the actions of hormones.

Phosphoprotein Phosphatases

Another means of controlling hormone action is the regulation of the protein dephosphorylation reaction. The phosphoprotein phosphatases are themselves subject to regulation by phosphorylation-dephosphorylation and by a variety of other substances. Most is known about the role of phosphatase in regulation of glycogen metabolism in muscle. In this tissue, 2 types of phosphoprotein phosphatases have been described. Type I preferentially dephosphorylates the β subunit of phosphorylase kinase, whereas type II dephosphorylates the α subunit. Two heat-stable protein inhibitors regulate type I phosphatase activity. Inhibitor-1 is phosphorylated and activated by cAMP-dependent protein kinase, and inhibitor-2, which may be a subunit of the inactive phosphatase, is also phosphorylated, possibly by glycogen synthase kinase-3. Phosphorylation of inhibitor-2 also results in activation of the phosphatase. Certain phosphatases may attack specific residues; eg, there may be a phosphatase that removes phosphate from tyrosine residues.

Extracellular cAMP

Some cAMP leaves cells and can be readily detected in extracellular fluids. The action of glucagon on liver and vasopressin or PTH on kidney is reflected in elevated levels of cAMP in plasma and urine, respectively; this has led to diagnostic tests of target organ responsiveness. Extracellular cAMP has little if any biologic activity in mammals, but it is an extremely important intercellular messenger in lower eukaryotes and prokaryotes.

Guanylate Cyclase; cGMP; cGMP-Dependent Protein Kinase

Cyclic GMP is made from GTP by the enzyme guanylate cyclase, which exists in soluble and membrane-bound forms. Each of these isozymes has unique kinetic, physiochemical, and antigenic properties. For some time, cGMP was thought to be the functional counterpart of cAMP. It now appears that cGMP has its unique place in hormone action. The atriopeptins, a family of peptides produced in cardiac atrial tissues, cause natriuresis, diuresis, vasodilation, and inhibition of aldosterone secretion. These peptides (eg, atrial natriuretic factor) bind to and activate the membrane-bound form of guanylate cyclase. This results in an increase of cGMP of as much as 50-fold in some cases, which is thought to mediate these effects. Other evidence links cGMP to vasodilation. A series of compounds, including nitroprusside, nitroglycerin, sodium nitrite, and sodium azide, all cause smooth muscle relaxation and are potent vasodilators. These agents increase cGMP by activating the soluble form of guanylate cyclase, and inhibitors of cGMP phosphodiesterase enhance and prolong these responses. The increased cGMP activates cGMP-dependent protein kinase, which in turn phosphorylates a number of smooth muscle proteins, including the myosin light chain. Presumably, this is involved in relaxation of smooth muscle and vasodilation.

2. HORMONES THAT ACT THROUGH CALCIUM OR PHOSPHOINOSITIDES

Ionized calcium is an important regulator of a variety of cellular processes including muscle contraction, stimulus-secretion coupling, the blood clotting cascade, enzyme activity, and membrane excitability. It is also an intracellular messenger of hormone action.

Calcium Metabolism

The extracellular calcium (Ca^{2+}) concentration is about 5 mmol/L and is very rigidly controlled (see Chapter 47). The intracellular concentration of this free ion is much lower, $0.1-10$ μmol/L, and the concentration associated with intracellular organelles such as mitochondria and endoplasmic reticula is in the range of $1-20$ μmol/L. In spite of this 5000- to 10,000-fold concentration gradient and a favorable transmembrane electrical gradient, Ca^{2+} is restrained from entering the cell. There are 3 ways of changing cytosolic Ca^{2+}. Certain hormones (class IIB) enhance membrane permeability to Ca^{2+} and thereby increase Ca^{2+} influx. This is probably accomplished by an Na^+/Ca^{2+} exchange mechanism that has a high capacity but low affinity for Ca^{2+}. There also is a $Ca^{2+}/2H^+$-ATPase-dependent pump that extrudes Ca^{2+} in exchange for H^+. This has a high affinity for Ca^{2+} but a low capacity and is probably responsible for fine-tuning cytosolic Ca^{2+}. Finally, Ca^{2+} can be mobilized (or deposited) from (or into) the mitochondrial and endoplasmic reticulum pools.

Two observations led to the current understanding of how Ca^{2+} serves as an intracellular messenger of hormone action. First was the ability to quantitate the rapid changes of intracellular Ca^{2+} concentration that are implicit in a role for Ca^{2+} as an intracellular messenger. Such evidence was provided by a variety of techniques, including the use of Quin 2 or Fura 2, fluorescent Ca^{2+} chelators. Rapid changes of Ca^{2+} in the submicromolar range can be quantitated using these compounds. The second important observation linking Ca^{2+} to hormone action involved the definition of the intracellular targets of Ca^{2+} action. The discovery of a Ca^{2+}-dependent regulator of phosphodiesterase activity provided the basis for understanding how Ca^{2+} and cAMP interact within cells.

Calmodulin

The calcium-dependent regulatory protein is now referred to as calmodulin, a 17,000-MW protein that is homologous to the muscle protein troponin C in structure and function. Calmodulin has four Ca^{2+} binding sites, and full occupancy of these sites leads to a marked conformational change, so that most of the molecule assumes an alpha-helical structure. This conformational change is presumably linked to calmodulin's ability to activate or inactivate enzymes. The interaction of Ca^{2+} with calmodulin (with the resultant change of activity of the latter) is conceptually similar to the binding of cAMP to protein kinase and the subsequent activation of this molecule. Calmodulin is often one of numerous subunits of complex proteins and is particularly involved in regulating various kinases and enzymes of cyclic nucleotide generation and degradation. A partial list of the enzymes regulated directly or indirectly by Ca^{2+}, probably through calmodulin, is given in Table 44–5.

In addition to its effects on enzymes and ion transport, Ca^{2+}/calmodulin regulates the activity of many structural elements in cells. These include the actin-myosin complex of smooth muscle, which is under β-adrenergic control, and various microfilament-mediated processes in noncontractile cells including cell motility, conformation changes, mitosis, granule release, and endocytosis.

Calcium as a Mediator of Hormone Action

A role for ionized calcium in hormone action is suggested by the observations that the effect of many hormones (1) is blunted in Ca^{2+}-free media or when intracellular calcium is depleted; (2) can be mimicked by agents that increase cytosolic Ca^{2+}, such as the Ca^{2+} ionophore A23187; and (3) influences cellular calcium flux. These processes have been studied in some detail in pituitary, smooth muscle, platelets, and salivary gland, but most is probably known about how vasopressin and α-adrenergic catecholamines regulate glycogen metabolism in liver. This is shown schematically in Figs 19–5 and 19–7.

Addition of α_1 agonists or vasopressin to isolated hepatocytes results in a 3-fold increase of cytosolic Ca^{2+} (from 0.2 to 0.6 μmol/L) within a few seconds. This change precedes and equals the increase in phosphorylase a activity, and the hormone concentrations required for both processes are comparable. This effect on Ca^{2+} is inhibited by α_1 antagonists, and removal of the hormone results in a prompt decline of both cytosolic Ca^{2+} and phosphorylase a. The initial source of the Ca^{2+} appears to be the intracellular organelle reservoirs, which seem to be sufficient for the early effects of the hormones. More prolonged action appears to require enhanced influx or inhibition of Ca^{2+} efflux through the Ca^{2+} pump. The latter may depend upon concomitant increases of cAMP.

Phosphorylase activation results from the conversion of phosphorylase b to phosphorylase a through the action of the enzyme phosphorylase b kinase. This enzyme contains calmodulin as its δ subunit, and its

Table 44–5. Enzymes regulated by calcium/calmodulin.

Adenylate cyclase	Glycogen synthase
Ca^{2+}-dependent protein kinase	Guanylate cyclase
	Myosin kinase
Ca^{2+}/Mg^{2+}-ATPase	NAD kinase
Ca^{2+}/phospholipid-dependent protein kinase	Phospholipase A_2
	Phosphorylase kinase
Cyclic nucleotide phosphodiesterase	Pyruvate carboxylase
	Pyruvate dehydrogenase
Glycerol-3-phosphate dehydrogenase	Pyruvate kinase

activity is increased through a Ca^{2+} concentration range of 0.1–1 μmol/L, the range through which hormones increase Ca^{2+} in liver. The link between Ca^{2+} and phosphorylase activation is definite.

A number of critical metabolic enzymes are regulated by Ca^{2+}, phosphorylation, or both, including glycogen synthase, pyruvate kinase, pyruvate carboxylase, glycerol-3-phosphate dehydrogenase, and pyruvate dehydrogenase. It is uncertain whether calmodulin is directly involved or whether the newly discovered Ca^{2+}/calmodulin-dependent or Ca^{2+}/phospholipid-dependent protein kinases are responsible.

Role of Phosphoinositide Metabolism in Ca^{2+}-Dependent Hormone Action

Some signal must provide communication between the hormone receptor on the plasma membrane and the intracellular Ca^{2+} reservoirs. The best candidates appear to be products of phosphoinositide metabolism. Phosphatidylinositol-4,5-bisphosphate is hydrolyzed to **myo-inositol 1,4,5-triphosphate** and **diacylglycerol** through the action of phospholipase C (Fig 44–5). This reaction occurs within seconds after the addition of either vasopressin or epinephrine to hepatocytes. Myo-inositol-P_3, at 0.1–0.4 μmol/L, releases Ca^{2+} from a variety of membrane and organelle preparations with appropriately rapid kinetics. Attempts to mimic hormone action using this compound, an essential step in establishing this relationship, have met with mixed success, perhaps because it is difficult to get the compound into cells and because it is rapidly hydrolyzed intracellularly. Another product of phosphoinositide hydrolysis, 1,2-diacylglycerol, activates a Ca^{2+}-phospholipid-dependent protein kinase by decreasing the K_m of the enzyme for Ca^{2+}. The role this process plays in the action of Ca^{2+}-dependent hormones is under investigation.

Steroidogenic agents, including ACTH and cAMP in the adrenal cortex; angiotensin II, K^+, serotonin, ACTH, and dibutyryl cAMP in the zona glomerulosa of the adrenal; LH in the ovary; and LH and cAMP in the Leydig cells of the testes, have been associated with increased amounts of phosphatidic acid, phosphoinositol, and polyphosphoinositides in the respective target tissues.

Several other examples can be cited. The addition of TRH to pituitary cells is followed, within 5–10 seconds, by a marked increase of inositol degradation by phospholipase C. The intracellular levels of inositol diphosphate and triphosphate increase markedly, and this results in mobilization of intracellular calcium. The calcium-dependent protein kinase is activated, which in turn phosphorylates several proteins, one of which is presumably involved in TSH release. Calcium also appears to be the intracellular mediator of GnRH action on LH release. This reaction probably also involves calmodulin.

The roles that Ca^{2+} and polyphosphoinositide breakdown products might play in hormone action are presented in Fig 44–5. In this scheme, the polyphosphoinositide products are the second messengers and Ca^{2+} is actually a tertiary messenger. Not shown is the possible involvement of a G protein in linking the membrane events to Ca^{2+} release. It is likely that this particular complex networking of the intracellular messengers is not unique.

3. HORMONES FOR WHICH THE INTRACELLULAR MESSENGER IS UNKNOWN

A large number of important hormones have no identified intracellular messenger. It is curious that these hormones cluster into 2 groups. One group consists of insulin, the insulinlike growth factors (IGF-I and IGF-II), and a variety of other growth factors, all of which may share a common ancestor. The other major group consists of proteins from the growth hormone gene family (growth hormone, prolactin, chorionic somatomammotropin), which clearly are related to one another (see Chapter 45). There is some overlap between these 2 groups, since many of the actions of growth hormone appear to be mediated by IGF-I. Oxytocin appears to stand alone.

Much effort has been directed toward finding the intracellular mediator of insulin action. A variety of candidates including cAMP, cGMP, H_2O_2, Ca^{2+}, and insulin itself have been proposed. Various "mediator" substances of peptide or phospholipid derivation have been found in tissue extracts, but to date these have not been purified or characterized. The recent observation that the insulin receptor has intrinsic tyrosine kinase activity has spurred interest in finding a phosphorylation cascade that might explain the actions of this hormone. This is not an isolated observation, since the epidermal growth factor receptor is also a tyrosine kinase; indeed, this observation led to the studies of the insulin receptor. Finally, the platelet-derived growth factor is also a tyrosine kinase that closely resembles v-*sis* and c-*sis,* specific virus-associated and cell-associated oncogene products. Stimulation of platelet-derived growth factor target cells (fibroblasts, glial cells, smooth muscle cells) results in the production of several gene products that appear to be involved in the replication of those cells.

It is probable that entirely different mechanisms of intracellular signaling are employed by this large group of hormones. The traditional messengers certainly do not seem to be involved.

REFERENCES

Anderson JE: The effect of steroid hormones on gene transcription. In: *Biological Regulation and Development.* Goldberger RF, Yamamoto KR (editors). Vol 3B: Hormone Action. Plenum Press, 1985.

Blackmore PF, Exton JH: Mechanisms involved in the actions of calcium dependent hormones. In: *Biochemical Action of the Hormones.* Vol 12. Litwack G (editor). Academic Press, 1985.

Catt KJ, Dufau ML: Hormone action: Control of target-cell function by peptide, thyroid, and steroid hormones. Pages 61–105 in: *Endocrinology and Metabolism.* Felig P et al (editors). McGraw-Hill, 1981.

Codina J et al: Mechanism in the vectorial receptor-adenylate cyclase signal transduction. *Adv Cyclic Nucleotide Res* 1984;**17:**111.

Enhancers and eukaryotic gene expression. In: *Current Communications in Molecular Biology.* Gluzman Y, Shenk T (editors). Cold Spring Harbor Press, 1983.

Gilman A: G proteins and dual control of adenylate cyclase. *Cell* 1984;**36:**577.

Means AR, Chafouleas JG: Calmodulin in endocrine cells. *Annu Rev Physiol* 1982;**44:**667.

Niall HD: The evolution of peptide hormones. *Annu Rev Physiol* 1982;**44:**615.

O'Malley BW: Steroid hormone action in eukaryotic cells. *J Clin Invest* 1984;**74:**307.

Rasmussen H: The calcium messenger system. (2 parts.) *N Engl J Med* 1986;**314:**1094, 1164.

Pituitary & Hypothalamic Hormones

Daryl K. Granner, MD

INTRODUCTION

The anterior pituitary, under control of hypothalamic hormones, secretes a number of hormones (trophic hormones) that regulate the growth and function of other endocrine glands or influence metabolic reactions in other target tissues. The posterior pituitary produces hormones that regulate water balance and milk ejection from the lactating mammary gland.

BIOMEDICAL IMPORTANCE

The loss of anterior pituitary function (panhypopituitarism) results in atrophy of the thyroid, adrenal cortex, and gonads. Secondary effects due to the absence of the hormones secreted by these targets glands affect most body organs and tissues and many general processes such as protein, fat, carbohydrate, and fluid and electrolyte metabolism. The loss of posterior pituitary function results in diabetes insipidus, the inability to concentrate the urine.

HYPOTHALAMIC HORMONES

The release (and in some cases production) of each of the pituitary hormones listed in Table 45–1 is under tonic control by at least one hypothalamic hormone. The hypothalamic hormones are released from the hypothalamic nerve fiber endings around the capillaries of the hypothalamic-hypophyseal system in the pituitary stalk and reach the anterior lobe through the special portal system that connects the hypothalamus and the anterior lobe. The structures of several hypothalamic hormones are illustrated in Table 45–2.

The hypothalamic hormones are released in a pulsatile manner, and isolated anterior pituitary target cells respond better to pulsatile administration of these hormones than to continuous exposure. The release of LH and FSH is controlled by the concentration of one releasing hormone, GnRH; this in turn is primarily a function of circulating levels of gonadal hormones that reach the hypothalamus (see the feedback loop in Fig 43–1). The release of ACTH is primarily controlled by CRH, but a number of other hormones, including ADH, catecholamines, VIP, and angiotensin II, may

Acronyms Used in This Chapter

ACTH	Adrenocorticotropic hormone
ADH	Antidiuretic hormone
CG	Chorionic gonadotropin
CLIP	Corticotropinlike intermediate lobe peptide
CRH	Corticotropin-releasing hormone
CS	Chorionic somatomammotropin; placental lactogen
FSH	Follicle-stimulating hormone
GAP	GnRH-associated peptide
GH	Growth hormone
GHRH or GRH	Growth hormone-releasing hormone
GHRIH	Growth hormone release-inhibiting hormone; somatostatin
GnRH	Gonadotropin-releasing hormone
IGF	Insulinlike growth factor
LH	Luteinizing hormone
LPH	Lipotropin
MSH	Melanocyte-stimulating hormone
NGF	Nerve growth factor
POMC	Pro-opiomelanocortin peptide family
PRIH or PIH	Prolactin release-inhibiting hormone
PRL	Prolactin
SRIH	Somatotropin release-inhibiting hormone
T_3	Triiodothyronine
T_4	Thyroxine; tetraiodothyronine
TRH	Thyrotropin-releasing hormone
TSH	Thyroid-stimulating hormone
VIP	Vasoactive intestinal polypeptide

be involved. CRH release is influenced by cortisol, a glucocorticoid hormone secreted by the adrenal. TSH release is primarily affected by TRH, which in turn is regulated by the thyroid hormones T_3 and T_4; but TSH release is also inhibited by somatostatin (see Fig 46–4). Growth hormone release and production are under tonic control by both stimulating and inhibiting hypothalamic hormones. In addition, a peripheral feedback loop is involved in GH regulation. IGF-I (somatomedin C), which mediates some of the effects of

Table 45–1. Hypothalamic-hypophyseal hormones.

Hypothalamic Hormone	Acronym	Pituitary Hormone Affected*
Corticotropin-releasing hormone	CRH	ACTH (LPH, MSH, endorphins)
Thyrotropin-releasing hormone	TRH	TSH (PRL)
Gonadotropin-releasing hormone	GnRH (LHRH, FSHRH)	LH, FSH
Growth hormone-releasing hormone	GHRH or GRH	GH
Growth hormone release-inhibiting hormone; somatostatin; somatotropin release-inhibiting hormone	GHRIH or SRIH	GH (TSH, FSH, ACTH)
Prolactin release-inhibiting hormones; dopamine and GAP	PRIH or PIH	PRL

*The hypothalamic hormone has a secondary or lesser effect on the hormones in parentheses.

GH, stimulates the release of somatostatin (GHRIH) while inhibiting the release of GHRH (Fig 45–5). The regulation of PRL synthesis and secretion is primarily under tonic inhibition by hypothalamic agents. It is unique because of the combined neural (nipple stimulation) and neurotransmitter/neurohormone link. Dopamine (Table 45–2) inhibits PRL synthesis (by inhibiting transcription of the PRL gene) and release, but evidence existed that this did not account for overall PRL inhibition. Recently, a 56-amino-acid neuropeptide was discovered that has both GnRH and PRIH activities—thus the name GnRH-associated peptide (GAP). GAP, the general and specific structures of which are shown in Fig 45–1, is a potent inhibitor of PRL release and may be the elusive PRIH peptide. GAP may explain the curious link between GnRH and PRL secretion that is particularly obvious in some species.

Many of the hypothalamic hormones, in particular TRH, CRH, and somatostatin, are found in other portions of the nervous system and in a variety of peripheral tissues. The concentration of somatostatin is higher in the pancreas than in the hypothalamus. In the pancreas, it is produced by the D cells of the islets of Langerhans and probably regulates glucagon and insulin secretion. Somatostatin is also one of the more than 40 peptides produced by neurons of the central and peripheral nervous systems.

Although cAMP was originally thought to mediate the action of releasing hormones on the adenohypophysis, recent studies with GnRH and TRH suggest that a calcium-phospholipid mechanism, similar to that described above, is involved (see Fig 44–5). Whether the releasing hormones also affect the synthesis of the corresponding pituitary hormone has been argued, but recently GHRH has been shown to stimulate the rate of transcription of the GH gene, and TRH has a similar effect on the prolactin gene.

ANTERIOR PITUITARY HORMONES

The anterior pituitary produces a large number of hormones that stimulate a variety of physiologic and biochemical processes in target tissues. In addition, the placenta produces hormones that are closely re-

Table 45–2. Structures of hypothalamic releasing hormones.

Hormone	Structure
TRH	(pyro)Glu-His-Pro-NH$_2$
Somatostatin	Ala-Gly-Cys-Lys-Asn-Phe-Phe-Trp-Lys-Thr-Phe-Thr-Ser-Cys-NH$_2$ (with S—S disulfide bridge between the two Cys residues)
GnRH	(pyro)Glu-His-Trp-Ser-Tyr-Gly-Leu-Arg-Pro-Gly-NH$_2$
PRIH	HO—(benzene ring)—CH$_2$CH$_2$NH$_2$ with HO at adjacent position; GnRH-associated peptide (GAP)
Ovine CRH	Ser-Gln-Glu-Pro-Pro-Ile-Ser-Leu-Asp-Leu-Thr-Phe-His-Leu-Leu-Arg-Glu-Val-Leu-Glu-Met-Thr-Lys-Ala-Asp-Gln-Leu-Ala-Gln-Gln-Ala-His-Ser-Asn-Arg-Lys-Leu-Leu-Asp-Ile-Ala-NH$_2$
Human GHRH	Tyr-Ala-Asp-Ala-Ile-Phe-Thr-Asn-Ser-Tyr-Arg-Lys-Val-Leu-Gly-Gln-Leu-Ser-Ala-Arg-Lys-Leu-Leu-Gln-Asp-Ile-Met-Ser-Arg-Gln-Gln-Gly-Glu-Ser-Asn-Gln-Glu-Arg-Gly-Ala-Arg-Ala-Arg-Leu-NH$_2$

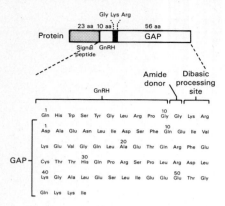

Figure 45–1. Structure and encoded amino acid sequence of human placental cDNA for prepro-GnRH. Schematic representation of the encoded protein identifies 3 domains: signal peptide, GnRH, and GAP, with the respective sizes in amino acid (aa) residues. Amino acid sequences of GnRH and GAP with an enzymatic processing site separating the 2 moieties are shown below. Numbers refer to the respective positions within GnRH (1–10) or GAP (1–56). (Reproduced, with permission, from Nikolics K et al: A prolactin-inhibiting factor with the precursor for human gonadotropin-releasing hormone. *Nature* 1986; **316**:511. Copyright © 1986 by Macmillan Journals Ltd.)

lated to some anterior lobe hormones. These hormones have traditionally been discussed individually, but recent studies dealing with the mechanism of synthesis and with the intracellular mediators of action (see Table 44–1) allow one to classify these hormones into 3 categories: (1) the growth hormone-prolactin-chorionic somatomammotropin group, (2) the glycoprotein hormone group, and (3) the pro-opiomelanocortin peptide family.

1. THE GROWTH HORMONE-PROLACTIN-CHORIONIC SOMATOMAMMOTROPIN GROUP

Growth hormone (GH), prolactin (PRL), and chorionic somatomammotropin (CS; placental lactogen) are a family of protein hormones having considerable sequence homology. GH, CS, and PRL range in size

from 190 to 199 amino acids in different species. Each has a single tryptophan residue (locus 85 in GH and CS; locus 91 in PRL), and each has 2 homologous disulfide bonds. The amino acid homology between hGH and hCS is 85%, whereas that between hGH and hPRL is 35%. In view of this homology, it is not surprising that these 3 hormones share common antigenic determinants and that all have growth-promoting and lactogenic activity. The hormones are produced in a tissue-specific manner, with GH and PRL produced in the anterior pituitary and CS in the syncytiotrophoblast cells of the placenta. Each appears to be under different regulation (see below).

On the basis of these striking similarities, it was postulated several years ago that these hormones may have arisen by duplication of an ancestral gene. Recombinant DNA technology has revealed that there are multiple genes for GH and CS in primates and humans; that the single PRL gene, while encoding a very similar protein, is 5 times as large as those for GH and CS; that hCS is a variant of hGH; and that the GH-CS group in humans is located on chromosome 17, while PRL in humans is found on chromosome 6. There is marked evolutionary divergence of these genes. Rat and bovine tissues have a single copy of GH and PRL per haploid genome, and humans have a single PRL gene. Humans have one functional GH gene (GH-N) and a variant (GH-V), two CS genes that are expressed (CS-A and CS-B), and one CS gene that is not expressed (CS-L). Several simian species have at least 4 of the genes from the GH-CS family. The coding sequence of all of these genes is organized into 5 exons interrupted by 4 introns (Fig 45–2). The genes are highly homologous in the 5′ flanking regions and the coding sequence areas (~93% homology in the latter) and diverge in the 3′ flanking regions. The splice junctions are highly conserved, even though the introns in the PRL gene are much longer.

The human GH-CS gene family is located on a linkage group in region q22–24 on the long arm of chromosome 17. Fig 45–3 indicates the relative positions of each of these genes in a 5′ to 3′ orientation. The genes are all transcribed in the 5′ to 3′ direction, and GH-N is separated from CS-B by about 45 kb.

The GH-N coding sequence matches the amino acid sequence for circulating GH, and the gene is DNase I-sensitive, signifying its location in a region of **"active chromatin."** The GH-V gene, if expressed, would encode for a protein with 13 amino acid differ-

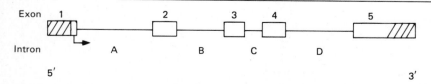

Figure 45–2. Schematic representation, drawn to scale, of the structure of the human growth hormone gene. The gene is about 45 kb in length and consists of 5 exons and 4 introns. Cross-hatching represents noncoding regions in exons 1 and 5. Arrow indicates direction of transcription.

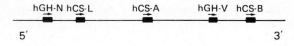

Figure 45–3. Location and orientation of the human GH-CS gene family on chromosome 17. Relative positions of hGH and hCS genes are shown in a 5′ to 3′ orientation. Arrows indicate direction of transcription.

ences. This gene is DNase I-resistant, and thus it may not be active. The GH-V gene is present in patients who lack the GH-N gene (inherited GH deficiency), but since these persons have complete GH deficiency, the GH-V gene either is silent or is producing an inactive GH molecule. The first possibility is most likely, because these individuals form antibodies in response to exogenous GH, an indication that this molecule has not previously been seen by the immune system.

The CS-A and CS-B genes are expressed in placenta; CS-L is silent.

Growth Hormone (GH)

A. Synthesis and Structure: Growth hormone is synthesized in **somatotropes,** a subclass of the pituitary acidophilic cells; somatotropes are the most abundant cell in the gland. The concentration of GH in the pituitary is 5–15 mg/g, which is much higher than the μg/g quantities of other pituitary hormones. Growth hormone is a single polypeptide, with a molecular weight of about 22,000 in all mammalian species. The 191-amino-acid sequence of human growth hormone is shown in Fig 45–4. Although there is a high degree of sequence homology between various mammalian growth hormones, only human growth hormone or that of other higher primates is active in humans.

B. Regulation of Secretion and Synthesis: GH secretion is influenced by a variety of stimuli (eg, sleep, stress) and, like that of many of the pituitary hormones, is episodic and pulsatile. Plasma GH levels may change as much as 10-fold within a few minutes. One of the largest increases occurs shortly after the onset of sleep, lending support to the adage "If you don't get your sleep, you won't grow." Other stimuli include stress (pain, cold, apprehension, surgery), exercise, severe hypoglycemia or fasting, a protein meal, and the amino acid arginine. The stress responses may be mediated by catecholamines acting through the hypothalamus. These and many other effectors may be a consequence of the major physiologic action of GH to spare glucose. In stress, hypoglycemia, sleep, or fasting, GH promotes lipolysis (delivery of fatty acids to cells) and amino acid entry into cells (potential gluconeogenic substrates), thus sparing glucose for brain metabolism. The intracellular glucose (or metabolite) level in the GH-regulating region of the ventromedial nucleus of the hypothalamus may be key.

A variety of other agents influence GH release, including estrogens, dopamine, α-adrenergic agents,

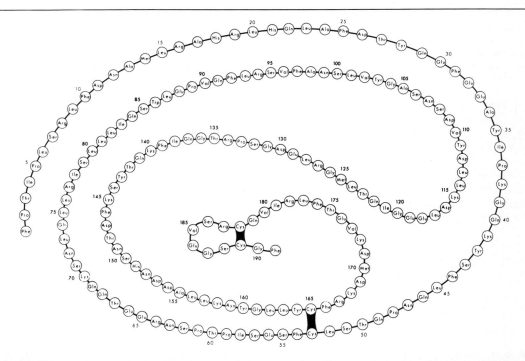

Figure 45–4. Structure of human growth hormone. The numbers identify the amino acid residues, starting from the N terminus.

serotonin, opiate polypeptides, gut hormones, and glucagon. All of these appear to converge in the ventromedial nucleus of the hypothalamus, where GH release is regulated by the feedback system illustrated in Fig 45–5. In this system, the **short loop** involves the positive regulator GHRH and the negative regulator GHRIH (somatostatin); the **peripheral loop** involves insulinlike growth factor I (IGF-I; also known as somatomedin C and sulfation factor).

The growth-promoting actions of GH are presumably mediated by IGF-I, which is produced in the liver. IGF-I regulates GH secretion by inhibiting the release of GHRH from the hypothalamus and stimulating the release of somatostatin. Short-loop feedback inhibition is provided by GH itself, which inhibits the release of GHRH. GHRH is produced in the median eminence and, in addition to its effects on release, has recently been shown to stimulate transcription of the GH gene. Some effects of GHRH are duplicated by dopamine, which also increases GH production.

Negative control of GH release is provided by **somatostatin.** Somatostatin also inhibits the release of glucagon, insulin, TSH, FSH, ACTH, and many gut hormones but has no effect on PRL. This tetradecapeptide has a disulfide bridge but is active in either the linear or cyclic form (Table 45–2). Somatostatin is synthesized as part of an 11,500-MW prohormone that has biologic activity, as does a 28-amino-acid precursor. Secretion of somatostatin is increased by such substances as Ca^{2+}, Na^+, thyroid hormones, cAMP, and vasoactive intestinal peptide (VIP) and is decreased by atropine, acetylcholine, and GABA. The

effect of GABA is reversed by picrotoxin and the benzodiazepine class of drugs. Exactly how these agents affect hypothalamic release of somatostatin and GH remains to be elucidated. Since somatostatin is produced in a variety of tissues, there may be many different types of control of its synthesis and release.

Somatostatin appears to inhibit GH by inhibiting calcium mobilization. Whether this occurs by changing Ca^{2+} influx or by stabilizing intracellular stores is uncertain. It also inhibits K^+ efflux, which in turn might decrease Ca^{2+} influx.

C. Physiologic and Biochemical Actions: GH is essential for postnatal growth and for normal carbohydrate, lipid, nitrogen, and mineral metabolism. As mentioned above, the growth-related effects are primarily mediated by **IGF-I,** a member of the insulinlike gene family. This was originally known as "sulfation factor" because of its ability to enhance the incorporation of sulfate into cartilage. It next was known as somatomedin C. Structurally, it is similar to proinsulin (see Chapter 51 and Fig 51–8). Another closely related peptide found in human plasma, **IGF-II,** has activity similar or identical to what is often referred to in the rat as multiplication-stimulating activity (MSA). IGF-I and IGF-II both bind to membrane receptors; however, they can be differentiated on the basis of specific radioimmunoassays. IGF-I has 70 amino acids, and IGF-II has 67. Plasma levels of IGF-II are twice those of IGF-I, but it is IGF-I that correlates most directly with GH effects. Individuals who lack sufficient IGF-I but have IGF-II (GH-deficient dwarfs and pygmies; see Table 45–3) fail to grow normally.

1. Protein synthesis–GH increases the transport of amino acids into muscle cells and also increases protein synthesis by a mechanism separate from the transport effect. Animals treated with GH show positive nitrogen balance, reflecting a generalized increase in protein synthesis and a decrease in plasma and urinary levels of amino acids and urea. This is accompanied by increased synthesis of RNA and DNA in some tissues. In these respects, GH actions resemble some of the actions of insulin.

2. Carbohydrate metabolism–GH generally antagonizes the effects of insulin. Hyperglycemia after growth hormone administration is the combined result of decreased peripheral utilization of glucose and increased hepatic production via gluconeogenesis. In

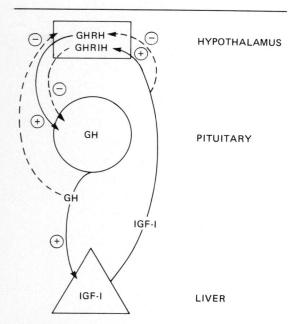

Figure 45–5. Feedback system regulating growth hormone release. Dashed lines show inhibitory effects; solid lines show stimulatory effects. See text for description.

Table 45–3. Relationship of GH, IGF-I, and IGF-II to dwarfism.

	Plasma Levels			Response to GH Stimulation
	GH	IGF-I	IGF-II	
GH-deficient dwarfs	Low	Low	Low to normal	Yes
Pygmies	Normal	Low	Normal	No
Laron type dwarfs	High	Low	Low	No

liver, GH increases liver glycogen, probably from activation of gluconeogenesis from amino acids. Impairment of glycolysis may occur at several steps, and GH may also inhibit the transport of glucose. Whether the latter is a direct effect on transport or a result of the inhibition of glycolysis has not yet been established. The mobilization of fatty acids from triacylglycerol stores may also contribute to the inhibition of glycolysis in muscle. Prolonged administration of GH may result in diabetes mellitus.

3. Lipid metabolism–GH promotes the release of free fatty acids and glycerol when incubated in vitro with adipose tissue. The in vivo administration of GH causes a rapid (30–60 minutes) increase of circulating free fatty acids and increased oxidation of fatty acids in the liver. Under conditions of insulin deficiency (eg, diabetes), increased ketogenesis may occur. These effects and those on carbohydrate metabolism probably are not mediated by IGF-I.

4. Mineral metabolism–GH, or more likely IGF-I, promotes a positive calcium, magnesium, and phosphate balance and causes the retention of Na^+, K^+, and Cl^-. The first effect probably relates to the action of GH in bone, where it promotes growth of long bones at the epiphyseal plates in growing children and appositional or acral growth in adults. In children, GH also increases formation of cartilage.

5. Prolactinlike effects–GH binds to lactogenic receptors and thus has many of the properties of prolactin, such as stimulation of the mammary glands, lactogenesis, and stimulation of the pigeon crop sac.

D. Pathophysiology: Deficient amounts of GH, whether from panhypopituitarism or isolated GH deficiency, are most serious in infancy because affected infants fail to grow properly. The other metabolic effects are less troublesome. Several types of dwarfism help illustrate the importance of the various steps in GH action (Table 45–3). **GH-deficient dwarfs** respond normally to exogenous GH. Two types of target organ resistance have been described. **Laron type dwarfs** have excessive amounts of GH-N, but they lack hepatic GH receptors. **Pygmies** apparently have a post-GH receptor defect, and this may be limited to the action GH exerts through IGF-I.

GH excess, usually from an acidophilic tumor, causes **gigantism** if it occurs before the epiphyseal plates close, since there is accelerated growth of the long bones. **Acromegaly** results from excessive release of GH that begins after epiphyseal closure and the cessation of long bone growth. Acral bone growth causes the characteristic facial changes (protruding jaw, enlarged nose) and enlargement of the hands, feet, and skull. Other findings include enlarged viscera, thickening of the skin, and a variety of metabolic problems, including diabetes mellitus.

A knowledge of GH regulation allows one to understand the clinical tests used to confirm these diagnoses. GH-deficient patients fail to increase GH levels in response to induced hypoglycemia or administration of arginine or levodopa. Patients with increased GH from a tumor (gigantism or acromegaly) fail to suppress GH levels in response to glucose administration.

Prolactin (PRL: Lactogenic Hormone, Mammotropin, Luteotropic Hormone)

A. Synthesis and Structure: PRL is a protein hormone with a molecular weight of about 23,000; its primary structure is illustrated in Fig 45–6. It is secreted by **lactotropes,** which are acidophilic cells in the anterior pituitary. The number of these cells and their size increase dramatically during pregnancy. The similarities between the structures and functions of PRL, GH, and CS are noted above.

B. Regulation of Secretion: An early and important observation about the control of PRL secretion was that—as is not true of other pituitary hormones—PRL secretion increased when the gland was removed from the sella turcica or when the pituitary stalk was completely transected. PRL thus seemed to be under tonic inhibition due to a prolactin release-inhibiting hormone (PRIH), which may, in fact, be dopamine. Pituitary cells have dopamine receptors, and dopamine decreases PRL release and inhibits PRL gene transcription, possibly mediated by decreasing cAMP levels. Levodopa, the dopamine precursor used in clinical testing, inhibits PRL release while stimulating GH release. GAP (Fig 45–2) is also involved in inhibiting PRL release. The presence of a positive releasing factor (prolactin-releasing hormone; PRH) is less well established. PRL levels rise during late pregnancy and lactation. The physiologic stimulus for PRL release is nipple stimulation, but stress, sleep, and sexual intercourse also promote release.

C. Physiologic and Biochemical Actions: PRL is involved in the initiation and maintenance of lactation in mammals. Physiologic levels act only upon breast tissue primed by female sex hormones, but excessive levels can trigger breast development in ovariectomized females or in males. In rodents, PRL is capable of maintaining the corpus luteum—hence the name **luteotropic hormone.** Related molecules appear to be responsible for the adaptation of saltwater fish to fresh water, for the molting of reptiles, and for crop-sac milk production in birds. The intracellular mediator of PRL action is unknown. A peptide has been proposed, but this has not been verified.

D. Pathophysiology: Tumors of prolactin-secreting cells cause **amenorrhea** (cessation of menses) and **galactorrhea** (breast discharge) in women. Excessive PRL has been associated with **gynecomastia** (breast enlargement) and **impotence** in men.

Chorionic Somatomammotropin (CS; Placental Lactogen)

The final member of the GH-PRL-CS family has no definite function in humans. In bioassays, CS has lactogenic and luteotropic activity and metabolic effects that are qualitatively similar to those of growth hormone, including inhibition of glucose uptake, stimulation of free fatty acid and glycerol release, enhance-

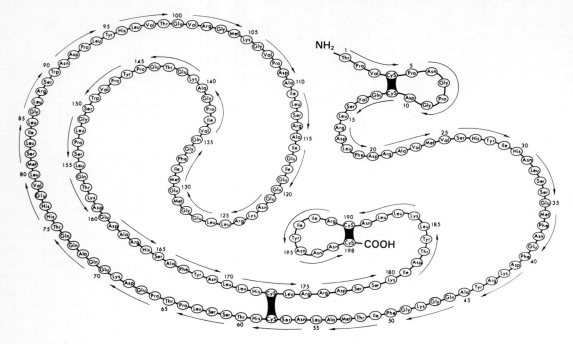

Figure 45–6. Structure of ovine prolactin.

ment of nitrogen and calcium retention (despite increased urinary calcium excretion), and reduction in the urinary excretion of phosphorus and potassium. It may provide growth-supporting activity to the developing fetus, but pregnancies in which the fetus and placenta lack all the genes in the GH-CS group except GH-N and CS-L result in infants with normal in utero development and normal growth in the neonatal period. Since the human CS-L gene is not expressed, there is no possible source of CS in such individuals.

2. THE GLYCOPROTEIN HORMONE GROUP

The most complex protein hormones yet discovered are the pituitary and placental glycoproteins: **thyroid-stimulating hormone (TSH), luteinizing hormone (LH), follicle-stimulating hormone (FSH),** and **chorionic gonadotropin (CG).** These hormones affect diverse biologic processes and yet have remarkable structural similarities. This class of hormones is found in all mammals. Hormones with similar action are found in lower forms, and molecules with TSH- and hCG-like activities have been found in bacteria. These molecules, like other peptide and protein hormones, interact with cell surface receptors and activate adenylate cyclase; thus, they employ cAMP as their intracellular messenger.

Each of these hormones consists of 2 subunits, α and β, joined by noncovalent bonding. The α subunits

are identical for all of these hormones within a species, and there is considerable interspecies homology. The specific biologic activity is determined by the β subunit, which also is highly conserved between hormones but to a lesser extent than that noted for the α subunit. The β subunit is not active by itself, and receptor recognition involves the interaction of regions of both subunits. Interhormone and interspecies hybrid molecules are fully active; eg, $TSH_\alpha LH_\beta$ = LH activity, and $hTSH_\alpha mTSH_\beta$ = mouse TSH activity. Thus, interspecies differences between α and β do not affect subunit association or the biologic function domain on β. Each subunit is synthesized from unique mRNAs from separate genes. It is thought that all hormones in this class evolved from a common ancestral gene that resulted in 2 molecules, α and β, and that the latter evolved further to provide the separate hormones.

A great deal is known about the structure of these molecules. For example, the carboxy-terminal pentapeptide of α is essential for receptor binding but not for $\alpha\beta$ association. The feature that distinguishes hormones in the glycoprotein group from hormones in other groups is their glycosylation. In each glycoprotein hormone, the α subunit contains 2 complex asparagine-linked oligosaccharides, and the β subunit has either one or 2. The glycosylation may be necessary for $\alpha\beta$ interaction. The α subunit has five S—S bridges, and the β moiety has 6.

Free α subunits are found in the pituitary and placenta. This finding and the observation that α and β

are translated from separate mRNAs support the concept that the syntheses of α and β are under separate control and that β is limiting for the production of the complete hormone. All are synthesized as preprohormones and are subject to posttranslational processing within the cell to yield the glycosylated proteins.

The Gonadotropins
(FSH, LH, & hCG)

These hormones are responsible for gametogenesis and steroidogenesis in the gonads. Each is a glycoprotein with a molecular weight of about 25,000.

A. Follicle-Stimulating Hormone (FSH): FSH binds to specific receptors on the plasma membranes of its target cells, the **follicular cells** in the ovary and the **Sertoli cells** in the testis. This results in activation of adenylate cyclase and increased cAMP production. FSH promotes follicular growth, prepares the follicle for the ovulation-inducing action of LH, and enhances the LH-induced release of estrogen. In the male, it binds to the Sertoli cells, where it induces the synthesis of an **androgen-binding protein** that appears to be involved in transporting testosterone to the seminiferous tubule and epididymis, a mechanism important for achieving the high local level of testosterone required for spermatogenesis. FSH stimulates seminiferous tubule and testicular growth and is important in initiating spermatogenesis. In the absence of FSH, the testes are atrophic and sperm production is absent. FSH also stimulates estradiol production in isolated Sertoli cells. The role of this in male physiology is unclear. Plasma FSH concentrations increase through puberty from the low levels of infancy. The appearance of the pulsatile release of FSH and LH, particularly during sleep, may signal the onset of puberty. In the female, there is marked cycling of levels, with peak values 10-fold or more over basal level at or slightly before the time of ovulation.

B. Luteinizing Hormone (LH): LH binds to specific plasma membrane receptors and stimulates the production of progesterone by **corpus luteum cells** and of testosterone by the **Leydig cells.** The intracellular signal of LH action is cAMP. This nucleotide mimics the actions of LH, which include enhanced conversion of acetate to squalene (the precursor for cholesterol synthesis) and enhanced conversion of cholesterol to 2α-hydroxycholesterol, a necessary step in the formation of progesterone and testosterone. There is tight coupling between the binding of LH and the production of cAMP, but steroidogenesis occurs when very small increases of cAMP have occurred. There thus are spare receptors in this response (see Fig 43–3). Prolonged exposure to LH results in desensitization, perhaps owing to down-regulation of LH receptors.

An estradiol-dependent LH spike in midcycle triggers human ovulation, and LH is required for maintenance of the corpus luteum, which is the transformed follicle that begins to make progesterone as well as estradiol. This function of LH is later assumed by the placental hormone hCG if fertilization and implantation occur. For the first 6–8 weeks, the corpus luteum maintains the pregnancy, and then the placenta makes sufficient progesterone to continue the pregnancy, but production of hCG continues throughout.

In males, LH stimulates testosterone production, which with FSH promotes spermatogenesis. Systemic actions include the development of secondary sex characteristics and development and maintenance of accessory sex organs including the prostate, vasa deferentia, and seminal vesicles.

LH can stimulate interstitial cells in ovarian nongerminal tissues to produce a number of androgens and androgen precursors, including androstenedione, dehydroepiandrosterone, and testosterone. Patients with **polycystic ovaries** (Stein-Leventhal syndrome) have elevated LH levels, increased androgen production, poor fertility, and increased body and facial hair growth. This syndrome is presumed to be secondary to overactivity of the ovarian struma.

C. Human Chorionic Gonadotropin (hCG): hCG is a glycoprotein synthesized in the **syncytiotrophoblast cells** of the placenta. It has the $\alpha\beta$ dimer structure characteristic of this class of hormones and most closely resembles LH. It increases in blood and urine shortly after implantation (see above); hence, its detection is the basis of many pregnancy tests.

D. Regulation of LH and FSH Release: LH and FSH secretion are regulated by a classic negative feedback loop regulated by the gonadal steroid hormones. Sex hormones given over a prolonged period inhibit the secretion of LH and FSH. Castration or physiologic atrophy of the ovary at menopause results in hypersecretion of both LH and FSH. Positive feedback control may also be involved, since estradiol (progesterone or 20α-hydroxyprogesterone in some species) is responsible for, or permits, the ovulatory burst of LH release. LH and FSH are released episodically; this is especially notable during puberty. There are great fluctuations in mean plasma levels as well, with mid-menstrual cycle peaks of both FSH and LH.

The secretion of LH and FSH is regulated by a single hypothalamic releasing factor generally referred to as **gonadotropin-releasing hormone (GnRH).** GnRH is a decapeptide whose N-terminal amino acid, pyroglutamate, is a cyclized derivative of glutamate (Table 45–2). GnRH release is inhibited by the target organ hormones testosterone and estradiol and by endorphin. GnRH acts directly on the anterior pituitary to effect gonadotropin release through a calcium-phospholipid-dependent mechanism. Although separate releasing factors have not been found, plasma FSH and LH do not always change concordantly. Men in whom spermatogenesis fails to proceed beyond the secondary spermatocyte stage have elevated FSH levels. This finding and other observations led to the hypothesis that a testicular factor, termed **inhibin,** inhibits FSH release. Inhibin has now been purified, and its physiologic role is being defined. Various GnRH analogs are being tested for the contrary purposes of promoting fertility or contraception.

Thyroid-Stimulating Hormone (TSH)

A. Structure and Mechanism of Action: TSH is a glycoprotein of $\alpha\beta$ dimer structure with a molecular weight of about 30,000. Like other hormones of this class, TSH binds to plasma membrane receptors and activates adenylate cyclase. The consequent increase of cAMP is responsible for the action of TSH in thyroid hormone biosynthesis. Its relationship to the trophic effects of TSH on the thyroid is less certain.

TSH has several acute effects on thyroid function. These occur in minutes and involve increases of all phases of T_3 and T_4 biosynthesis, including iodide concentration, organification, coupling, and thyroglobulin hydrolysis. TSH also has several chronic effects on the thyroid. These require several days and include increases in the synthesis of proteins, phospholipids, and nucleic acids and in the size and number of thyroid cells. Long-term metabolic effects of TSH are due to the production and action of the thyroid hormones.

B. Regulation of TSH Release: TSH release is governed by a negative feedback system that includes the target gland hormones T_3 and T_4 and the hypothalamic hormone thyrotropin-releasing hormone (TRH). The details of this system are illustrated in Fig 46–4.

TRH is a neutral tripeptide consisting of pyroglutamic acid, histidine, and prolinamide (Table 45–2). The TRH peptide shows no species specificity, and synthetic 3-methylation of histidine increases its activity 8-fold. TRH increases TSH release and cAMP levels within 1 minute, but its action appears to be more closely related to a Ca^{2+}-phospholipid-dependent mechanism, as is the case for GnRH. Prolonged exposure of cells to TRH also results in desensitization.

TRH, like somatostatin, is found in many extrahypothalamic locations, where it may serve as a neurotransmitter. TRH is used to distinguish between hypothyroidism of pituitary versus hypothalamic origin, since a person with the latter increases TSH in response to exogenous TRH whereas a person with the former does not. Because of its rapid plasma disappearance time ($t_{1/2} \sim 4$ minutes), TRH is not useful for long-term therapy.

3. THE PRO-OPIOMELANOCORTIN (POMC) PEPTIDE FAMILY

The POMC family consists of peptides that act as hormones (ACTH, LPH, MSH) and others that may serve as neurotransmitters or neuromodulators (endorphins). POMC is synthesized as a precursor molecule of $\sim$ 285 amino acids and is processed differently in various regions of the pituitary.

Distribution, Processing, & Functions of the POMC Gene Products

The POMC gene is expressed in the anterior and intermediate lobes of the pituitary. The most conserved sequences between species are within the N-terminal fragment, the ACTH region, and the β-endorphin region. POMC or related products are found in several other vertebrate tissues, including the brain, placenta, gastrointestinal tract, reproductive tract, lung, and lymphocytes. This is presumably due to gene expression in these tissues (rather than to absorption from plasma) but has only been proved for brain, placenta, and testes. Related peptides have also been found in many invertebrate species.

The POMC protein is processed differently in the anterior lobe than in the intermediate lobe. The intermediate lobe is rudimentary in adult humans, but it is active in human fetuses and in pregnant women during late gestation and is also active in many animal species. Processing of the POMC protein in the peripheral tissues (gut, placenta, male reproductive tract) resembles that in the intermediate lobe. There are 3 basic peptide groups: (1) ACTH, which can give rise to α-MSH and corticotropinlike intermediate lobe peptide (CLIP); (2) β-lipotropin (β-LPH), which can yield γ-LPH, β-MSH, and β-endorphin (and thus α- and γ-endorphins); and (3) a large N-terminal peptide, which generates γ-MSH. The diversity of these products is due to the many dibasic amino acid clusters that are potential cleavage sites for trypsinlike enzymes. Each of the peptides mentioned is preceded by Lys-Arg, Arg-Lys, Arg-Arg, or Lys-Lys residues. The prehormone segment is cleaved, and modification by glycosylation, acetylation, and phosphorylation occurs after translation. The next cleavage, in both anterior and intermediate lobes, is between ACTH and β-LPH, resulting in an N-terminal peptide with ACTH and a β-LPH segment (Fig 45–7). $ACTH_{1-39}$ is subsequently cleaved from the N-terminal peptide, and in the anterior lobe essentially no further cleavages occur. In the intermediate lobe, $ACTH_{1-39}$ is cleaved into α-MSH (residues 1–13) and CLIP (18–39); β-LPH (42–134) is converted to γ-LPH (42–101) and β-endorphin (104–134). β-MSH (84–101) is derived from γ-LPH.

There are extensive additional modifications of these peptides. Much of the N-terminal peptide and $ACTH_{1-39}$ in the anterior pituitary is glycosylated. α-MSH is found predominantly in an N-acetylated and carboxy-terminal amidated form; deacetylated α-MSH is much less active. β-Endorphin is rapidly acetylated in the intermediate lobe; acetylated β-endorphin, in contrast to α-MSH, is 1000 times less active than the unmodified form. β-Endorphin may therefore be inactive in the pituitary. In the hypothalamus, these molecules are not acetylated and presumably are active. β-Endorphin is also trimmed at the C-terminal end to form α- and γ-endorphin (Fig 45–7). These form the 3 major endorphins in the rodent intermediate lobe. The large N-terminal fragment is probably also extensively cleaved, but while γ-MSH has been found in rat and bovine pituitaries, less is known about this fragment. This structural information has come largely from studies of the rodent pituitary, but the general scheme is thought to apply to other species.

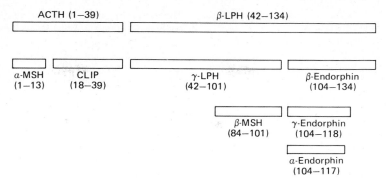

Figure 45–7. Products of pro-opiomelanocortin (POMC) cleavage. MSH, melanocyte-stimulating hormone; CLIP, corticotropinlike intermediate lobe peptide; LPH, lipotropin.

Precise functions for most of the POMC peptides have not been established. Postulated actions are listed in Table 45–4.

Regulation of POMC Production

About 5% of the cells of the anterior pituitary synthesize POMC, whereas all of the intermediate lobe cells synthesize this peptide. The regulation of synthesis and secretion is very different in the 2 areas.

A. Anterior Lobe: Corticotropin-releasing hormone (CRH) is the major factor controlling POMC release from the anterior pituitary. It works through a cAMP-mediated system in which Ca^{2+} is required. The stimulatory effects of CRH on POMC secretion are directly prevented by glucocorticoid hormones. In addition, glucocorticoids may act on the hypothalamus by inhibiting CRH production, CRH release, or both. Adrenalectomy (which causes decreased glucocorticoid and increased CRH levels) results in a 20-fold increase of POMC gene transcription. Through the concurrent administration of a glucocorticoid, however, the increase of POMC gene transcription can be inhibited, presumably through a receptor-mediated process. The inhibition of ACTH secretion by glucocorticoids occurs more rapidly than the effect on POMC gene transcription, so that these effects are presumably mediated independently. Minor effects on anterior lobe POMC (ACTH) secretion include direct stimulation by vasopressin and α-adrenergic agents, indirect stimulation (via the central nervous system) by serotonin and acetylcholine, and inhibition by GABA. Dopamine has no effect.

B. Intermediate Lobe: This lobe is poorly vascularized and is not reached by the hypothalamic-hypophyseal portal system; thus, it is unaffected by CRH. This tissue also has no glucocorticoid receptors, and thus glucocorticoids do not regulate POMC. The lobe is heavily innervated by dopaminergic fibers, and it also has serotoninergic and catecholaminergic nerve endings. Dopamine agonists (ergocryptine) decrease and antagonists (haloperidol) increase POMC mRNA and the release of POMC peptides. The time and magnitude of these changes are consistent with production-secretion coupling. These agents have no effect in the anterior lobe. POMC release is stimulated by serotonin and β-adrenergic agents in the intermediate lobe.

C. Other Tissues: Little is known about the regulation of POMC in other tissues. Hypophysectomy, adrenalectomy, CRH, and glucocorticoids do not affect POMC in these tissues. Chronic stress (eg, in immobilization) increases plasma ACTH and decreases pituitary ACTH, while brain POMC is unchanged. In contrast, acute stress decreases hypothalamic β-endorphin. Estrogens may promote β-endorphin release from the hypothalamus.

Action & Regulation of Specific Peptides

A. Adrenocorticotropic Hormone (ACTH):

1. Structure and mechanism of action– ACTH, a single-chain polypeptide consisting of 39 amino acids (Fig 45–8), regulates the growth and function of the adrenal cortex. The 24 N-terminal amino acids are required for full biologic activity and are invariant between species, whereas the 16 C-ter-

Table 45–4. Postulated functions of POMC peptides.*

Peptide	Function
ACTH	Adrenal growth and steroid production†
α-MSH	Melanin dispersion in amphibians† Learning and sexual behavior Growth and function of the testicular Sertoli cells
β-LPH	Lipolysis and fatty acid mobilization
β-Endorphin	Analgesia† Behavior (feeding, emotion, learning) Temperature and blood pressure regulation Contraction of reproductive tract muscles
N-terminal fragment	Potentiation of ACTH action on steroidogenesis

*Adapted, with permission, from Krieger DT: The multiple faces of pro-opiomelanocortin, a prototype precursor molecule. *Clin Res* 1983;**31**:342.
†Established functions.

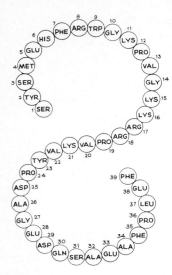

Figure 45–8. Structure of human ACTH.

minal amino acids are quite variable. A synthetic ACTH$_{1-24}$ analog is widely used in diagnostic testing.

ACTH increases the synthesis and release of adrenal steroids by enhancing the conversion of cholesterol to pregnenolone. This step entails the conversion from a C$_{27}$ to a C$_{21}$ steroid by removal of a 6-carbon side chain. Since pregnenolone is the precursor of all adrenal steroids (see Fig 48–3), prolonged ACTH stimulation results in excessive production of glucocorticoids, mineralocorticoids, and dehydroepiandrosterone (an androgen precursor). However, the contribution of ACTH to the last 2 classes of steroids is minimal under physiologic conditions. ACTH increases adrenal cortical growth (the trophic effect) by enhancing protein and RNA synthesis.

ACTH, like other peptide hormones, binds to a plasma membrane receptor. Within a few seconds of this interaction, intracellular cAMP levels increase markedly. cAMP analogs mimic the action of ACTH, but calcium also is involved.

ACTH activates adenylate cyclase in adipose cells and results in a cAMP-mediated activation of lipase and increased lipolysis. ACTH also stimulates insulin release from the pancreas, but these extra-adrenal effects are small and require supraphysiologic concentrations of the hormone.

2. Regulation–The production of ACTH from the POMC precursor protein and the regulation of synthesis and secretion of the latter have been discussed above. Primary regulation is accomplished through a negative feedback loop that involves glucocorticoids and CRH. Excessive levels of ACTH may also inhibit CRH production through a "short loop." The central nervous system is prominently involved in the regulation of ACTH production and secretion. This is accomplished through a variety of neurotransmitters including norepinephrine, serotonin, and acetylcholine.

The ACTH response, which stimulates the glucocorticoids necessary for adaptation to stresses such as hypoglycemia, surgery, physical or emotional trauma, cold, and pyrogens, is probably mediated by these neurotransmitters.

3. Pathophysiology–Excessive production of ACTH by the pituitary or by ectopic production from a tumor results in **Cushing's syndrome.** The weak MSH-like activity of ACTH or associated release of β- or α-MSH results in hyperpigmentation. The metabolic manifestations are due to excessive production of adrenal steroids and include (1) negative nitrogen, potassium, and phosphorus balance; (2) sodium retention, which can result in hypertension, edema, or both; (3) glucose intolerance or overt diabetes mellitus; (4) increased plasma fatty acids; and (5) decreased circulating eosinophils and lymphocytes, with increased polymorphonuclear leukocytes. Patients with Cushing's syndrome may have muscle atrophy and a peculiar redistribution of fat, ie, truncal obesity. Loss of ACTH owing to tumor, infection, or infarction of the pituitary results in an opposite constellation of findings.

B. β-Lipotropin (β-LPH): This peptide consists of the carboxy-terminal 91 amino acids of POMC (Fig 45–7). β-LPH contains the sequences of β-MSH, γ-LPH, Met-enkephalin, and β-endorphin. Of these, β-LPH, γ-LPH, and β-endorphin have been found in human pituitary but β-MSH has not been detected. β-LPH is found only in the pituitary, since it is rapidly converted to γ-LPH and β-endorphin in other tissues. β-LPH contains a 7-amino-acid sequence (β-LPH$_{47-53}$) that is identical to ACTH$_{4-10}$ (Fig 45–9). β-LPH causes lipolysis and fatty acid mobilization, but its physiologic role is minimal. It probably serves only as the precursor for β-endorphin.

C. Endorphins: β-Endorphin consists of the carboxy-terminal 31 amino acids of β-LPH (Fig 45–7). The α- and γ-endorphins are modifications of β-endorphin from which 15 and 14 amino acids, respectively, are removed from the C-terminal end. These peptides are found in the pituitary, but they are acetylated there (see above) and probably are inactive. In other sites (eg, central nervous system neurons), they are not modified and hence probably serve as neu-

---

-Tyr-Ser-Met-Glu-His-Phe-Arg-Trp-Gly-Lys-Pro-

ACTH$_{2-12}$

-Tyr-Lys-Met-Glu-His-Phe-Arg-Trp-Gly-Ser-Pro-

β-LPH$_{45-55}$

Figure 45–9. A comparison of the amino acid sequences of portions of the ACTH and β-LPH molecules. Underlined residues indicate the differences. The entire ACTH molecule consists of 39 amino acids, and β-LPH has 91 amino acids.

rotransmitters or neuromodulators. Endorphins bind to the same central nervous system receptors as do the morphine opiates and may play a role in endogenous control of pain perception. They have higher analgesic potencies (18–30 times on a molar basis) than morphine. The sequence for enkephalin is present in POMC, but it is not preceded by dibasic amino acids and presumably is not cleaved or expressed.

D. Melanocyte-Stimulating Hormone (MSH):
MSH stimulates **melanogenesis** in some species by causing the dispersion of intracellular melanin granules, resulting in darkening of the skin. Three different MSH molecules, α, β, and γ, are contained within the POMC molecule, and 2 of these, α and β, are secreted in some nonhuman species. In humans, the actual circulating MSH activity is contained within the larger molecules γ- or β-LPH. α-MSH contains an amino acid sequence that is identical to $ACTH_{1-13}$, but it has an acetylated N terminus. α-MSH (and CLIP) are generally found in animals that have a well-developed intermediate lobe. These peptides are not found in postnatal humans.

Patients with insufficient production of glucocorticoids (**Addison's disease**) have hyperpigmentation associated with increased plasma MSH activity. This could be due to ACTH but is more likely the result of concomitant secretion of β- and γ-LPH, with their associated MSH activity.

POSTERIOR PITUITARY HORMONES

The posterior pituitary contains 2 active hormones, vasopressin and oxytocin. **Vasopressin,** originally named because of its ability to increase blood pressure when administered in pharmacologic amounts, is more appropriately called **antidiuretic hormone (ADH)** because its most important physiologic action is to promote reabsorption of water from the distal renal tubules. **Oxytocin** is also named for an effect of questionable physiologic significance, the acceleration of birth by stimulation of uterine smooth muscle contraction. Its probable physiologic role is to promote milk ejection from the mammary gland.

Both hormones are produced in the hypothalamus and transported by axoplasmic flow to nerve endings in the posterior pituitary where, upon appropriate stimulation, the hormones are released into the circulation. The probable reason for this arrangement is to escape the blood-brain barrier. ADH is primarily synthesized in the **supraoptic nucleus** and oxytocin in the **paraventricular nucleus.** Each is transported through axons in association with specific carrier proteins called **neurophysins.** Neurophysins I and II are synthesized with oxytocin and ADH, respectively, each as a part of a single protein (sometimes referred to as propressophysin) from a single gene. Neurophysins I and II are unique proteins with molecular weights of 19,000 and 21,000, respectively. ADH and oxytocin are secreted separately into the bloodstream along

with their appropriate neurophysins. They circulate unbound to proteins and have very short plasma half-lives, on the order of 2–4 minutes. The structures for ADH and oxytocin are shown below.

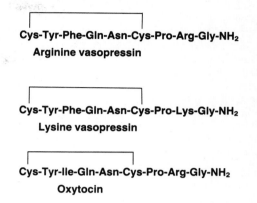

Cys-Tyr-Phe-Gln-Asn-Cys-Pro-Arg-Gly-NH₂
Arginine vasopressin

Cys-Tyr-Phe-Gln-Asn-Cys-Pro-Lys-Gly-NH₂
Lysine vasopressin

Cys-Tyr-Ile-Gln-Asn-Cys-Pro-Arg-Gly-NH₂
Oxytocin

Each is a nonapeptide containing cysteine molecules at positions 1 and 6 linked by an S–S bridge. Most animals have arginine vasopressin; however, the hormone in pigs and related species has a lysine substituted in position 8. Because of the close structural similarity, it is not surprising that ADH and oxytocin each exhibit some of the effects of the other molecule. These peptides are primarily metabolized in the liver, although renal excretion of ADH accounts for a significant part of its loss from blood.

1. OXYTOCIN

Regulation of Secretion

The neural impulses that result from stimulation of the nipples are the primary stimulus for oxytocin release. Vaginal and uterine distention are secondary stimuli. PRL is released by many of the stimuli that release oxytocin, and a fragment of oxytocin has been proposed as prolactin-releasing factor. Estrogen stimulates the production of oxytocin and of neurophysin I, and progesterone inhibits the production of these compounds.

Mechanism of Action

The mechanism of action of oxytocin is unknown. It causes contraction of uterine smooth muscle and thus is used in pharmacologic amounts to induce labor in humans. Interestingly, pregnant animals in which the hypothalamic-hypophyseal tract has been destroyed do not necessarily have trouble delivering their young. The most likely physiologic function of oxytocin is to stimulate contraction of myoepithelial cells surrounding the mammary alveoli. This promotes the movement of milk into the alveolar duct system and allows for milk ejection. Membrane receptors for oxytocin are found in both uterine and mammary tissues. These receptors are increased in number by estrogens and decreased by progesterone. The concomitant rise in estrogens and fall in progesterone occurring immediately before parturition probably explains the onset of lactation prior to delivery. Progesterone derivatives are commonly used to inhibit postpartum lactation in humans. Oxytocin and neurophysin I appear to be produced in the ovary, wherein oxytocin may inhibit steroidogenesis.

The chemical groups important for oxytocin action include the primary amino group of the N-terminal cysteine; the phenolic group of tyrosine; the 3 carboxamide groups of asparagine, glutamine, and glycinamide; and the disulfide (S−S) linkage. By deleting or substituting these groups, numerous analogs of oxytocin have been produced. For example, deletion of the free primary amino group of the terminal half cysteine residue (position 1) results in desamino oxytocin, which has 4−5 times the antidiuretic activity of oxytocin.

2. ANTIDIURETIC HORMONE (ADH; VASOPRESSIN)

Regulation of Secretion

The neural impulses that trigger ADH release are activated by a number of different stimuli. Increased osmolality of plasma is the primary physiologic stimulus. This is mediated by **osmoreceptors** located in the hypothalamus and by **baroreceptors** located in the heart and other regions of the vascular system. Hemodilution (decreased osmolality) has the opposite effect. Other stimuli include emotional and physical stress and pharmacologic agents including acetylcholine, nicotine, and morphine. Most of these effects involve increased synthesis of ADH and neurophysin II, since the depletion of stored hormone is not associated with this action. Epinephrine and agents that expand plasma volume inhibit ADH secretion, as does ethanol.

Mechanism of Action

The most important physiologic target cells of ADH in mammals are those of the distal convoluted tubules and collecting structures of the kidney. These ducts pass through the renal medulla, in which the extracellular solute pool has an osmolality gradient up to 4 times that of plasma. These cells are relatively impermeable to water, so that in the absence of ADH, the urine is not concentrated and may be excreted in amounts exceeding 20 L/d. ADH increases the permeability of the cells to water and permits osmotic equilibration of the collecting tubule urine with the hypertonic interstitium, resulting in urine volumes in the range of 0.5−1 L/d. ADH receptors exist on the mucosal (urinary) membrane of these epithelial cells. This receptor is linked to adenylate cyclase, and cAMP is thought to mediate the effects of ADH in the renal tubule. This physiologic action is the basis of the name "antidiuretic hormone." cAMP and inhibitors of phosphodiesterase activity mimic the actions of ADH. In vivo, an elevated level of calcium in the medium bathing the mucosal surface of the tubular cells inhibits the action of ADH on water movement, apparently by inhibiting the action of adenylate cyclase, since it does not diminish the action of cAMP per se. This may account, in part, for the excessive volumes of urine that are characteristic of patients with hypercalcemia.

Pathophysiology

Abnormalities of ADH secretion or action lead to **diabetes insipidus,** which is characterized by the excretion of large volumes of dilute urine. Primary diabetes insipidus, an insufficient amount of the hormone, is usually due to destruction of the hypothalamic-hypophyseal tract from a basal skull fracture, tumor, or infection, but it can be hereditary. In **hereditary nephrogenic diabetes insipidus,** ADH is secreted normally but the target cell is incapable of responding, presumably because of a receptor defect (see Table 43−2). This hereditary lesion is distinguished from **acquired nephrogenic diabetes insipidus,** which most often is due to the pharmacologic administration of lithium for manic-depressive illness. The **inappropriate secretion of ADH** occurs in association with ectopic production by a variety of tumors (usually tumors of the lung) but can also occur in conjunction with diseases of the brain, pulmonary infections, or hypothyroidism. It is called inappropriate secretion because ADH is produced at a normal or increased rate in the presence of hypo-osmolality, thus causing a persistent and progressive dilutional hyponatremia with excretion of hypertonic urine.

REFERENCES

Anterior Pituitary Hormones

Douglass J, Civelli O, Herbert E: Polyprotein gene expression: Generation of diversity of neuroendocrine peptides. *Annu Rev Biochem* 1984;**53**:665.

Frantz AG: Prolactin. *N Engl J Med* 1978;**298**:201.

Krieger DT: The multiple faces of pro-opiomelanocortin, a prototype precursor molecule. *Clin Res* 1983;**3**:342.

Krulich L: Central neurotransmitters and the secretion of prolactin, GH, LH, and TSH. *Annu Rev Physiol* 1979; **41**:603.

Nikolics K et al: A prolactin-inhibiting factor with the precursor for human gonadotropin-releasing hormone. *Nature* 1986;**316**:511.

Pierce JG, Parsons TF: Glycoprotein hormones: Structure and function. *Annu Rev Biochem* 1981;**50**:465.

Seeburg P: The human growth hormone gene family: Structure and evolution of the chromosomal locus. *Nucleic Acids Res* 1983;**11**:3939.

Posterior Pituitary Hormones

Chord IT: The posterior pituitary gland. *Clin Endocrinol* 1975;**4:**89.

Robertson GL: Regulation of vasopressin function in health and disease. *Recent Prog Horm Res* 1977;**33:**333.

Hypothalamic Hormones

Imura H et al: Effect of CNS peptides on hypothalamic regulation of pituitary secretion. *Adv Biochem Psychopharmacol* 1981;**28:**557.

Labrie F et al: Mechanism of action of hypothalamic hormones in the adenohypophysis. *Annu Rev Physiol* 1979; **41:**555.

Reichlin S: Systems for the study of regulation of neuropeptide secretion. In: *Neurosecretion and Brain Peptides: Implications for Brain Function and Neurological Disease.* Martin JB, Reichlin S, Bick KL (editors). Raven Press, 1981.

Thyroid Hormones

Daryl K. Granner, MD

INTRODUCTION

The thyroid gland produces 2 iodoamino acid hormones, **3,5,3'-triiodothyronine (T_3)** and **3,5,3',5'-tetraiodothyronine (T_4, thyroxine),** which have long been recognized for their importance in regulating general metabolism, development, and tissue differentiation. These hormones, whose structures are illustrated in Fig 46–1, regulate gene expression using mechanisms similar to those employed by steroid hormones.

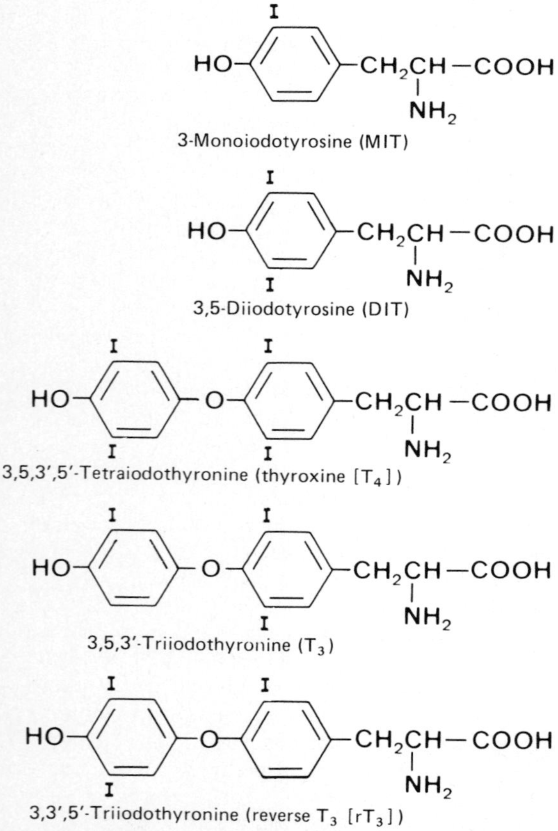

<table>
<tr><td colspan="2">Acronyms Used in This Chapter</td></tr>
<tr><td>DIT</td><td>Diiodotyrosine</td></tr>
<tr><td>GH</td><td>Growth hormone</td></tr>
<tr><td>IGF</td><td>Insulinlike growth factor</td></tr>
<tr><td>MIT</td><td>Monoiodotyrosine</td></tr>
<tr><td>SRIH</td><td>Somatostatin</td></tr>
<tr><td>T_3</td><td>Triiodothyronine</td></tr>
<tr><td>T_4</td><td>Thyroxine; tetraiodothyronine</td></tr>
<tr><td>TBG</td><td>Thyroxine-binding globulin</td></tr>
<tr><td>TBPA</td><td>Thyroxine-binding prealbumin</td></tr>
<tr><td>TRH</td><td>Thyrotropin-stimulating hormone</td></tr>
<tr><td>TSH</td><td>Thyroid-stimulating hormone</td></tr>
<tr><td>TSI</td><td>Thyroid-stimulating IgG</td></tr>
</table>

BIOMEDICAL IMPORTANCE

Diseases of the thyroid are among the most common afflictions involving the endocrine system. Diagnosis and therapy are firmly based on the principles of thyroid hormone physiology and biochemistry. The availability of radioisotopes of iodine has greatly aided in the elucidation of these principles. Radioactive iodine, because it localizes in the gland, is widely used in the diagnosis and treatment of thyroid disorders. Radioiodine has a dangerous aspect as well, since excessive exposure, such as from nuclear fallout, is a major risk factor for thyroid cancer. This is especially true in infants and adolescents, whose thyroid cells are still actively dividing.

THYROID HORMONE BIOSYNTHESIS

Thyroid hormones are unique in that they require the trace element **iodine** for biologic activity. In most parts of the world, iodine is a scarce component of soil, and hence there is little in food. A complex mechanism has evolved to acquire and retain this crucial element and to convert it into a form suitable for incorporation into organic compounds. At the same time, the thyroid must synthesize thyronine, and this synthesis takes place in thyroglobulin. These processes will be discussed separately, although they occur concurrently.

Figure 46–1. Structure of thyroid hormones and related compounds.

1. THYROGLOBULIN METABOLISM

Biosynthesis

Thyroglobulin is the precursor of T_4 and T_3. It is a large, iodinated, glycosylated protein with a molecular weight of 660,000. Carbohydrate accounts for 8–10% of the weight of thyroglobulin and iodide for about 0.2–1%, depending upon the iodine content in the diet. Thyroglobulin is composed of 2 subunits. It contains 115 tyrosine residues, each of which is a potential site of iodination. About 70% of the iodide in thyroglobulin exists in the inactive precursors, **monoiodotyrosine (MIT)** and **diiodotyrosine (DIT),** while 30% is in the **iodothyronyl** residues, T_4 and T_3. When iodine supplies are sufficient, the T_4:T_3 ratio is about 7:1. In **iodine deficiency,** this ratio decreases, as does the DIT:MIT ratio. The reason for synthesizing a molecule of 5000 amino acids to generate a few molecules of a modified diamino acid seems to be that the conformation of this large structure is required for tyrosyl coupling or iodide organification. Thyroglobulin is synthesized in the basal portion of the cell, moves to the lumen, where it is stored in the extracellular **colloid,** and reenters the cell and moves in an apical to basal direction during its hydrolysis into the active T_3 and T_4 hormones.

Amino acids for thyroglobulin synthesis, including tyrosine, enter the cell through the basal membrane and are incorporated into nascent thyroglobulin subunits by polyribosomes attached to the endoplasmic reticulum. The addition of carbohydrate starts in the cisternae of the rough endoplasmic reticulum but continues in the Golgi apparatus. Each molecule contains over 20 carbohydrate chains, which may be short or long and simple or branched. Packaging, including polymerization, continues in the Golgi vesicles, which migrate toward the apical membrane of the cell. Thyroglobulin is secreted by exocytosis into the follicular lumen. All of these steps are enhanced by TSH, and this hormone (or cAMP) also enhances transcription of the thyroglobulin gene.

Hydrolysis

Thyroglobulin is a storage form of T_3 and T_4 in the colloid; a several weeks' supply of these hormones exists in the normal thyroid. Within minutes after the stimulation of the thyroid by TSH (or cAMP), there is a marked increase of microvilli on the apical membrane. This microtubule-dependent process entraps thyroglobulin, and subsequent pinocytosis brings it back into the follicular cell. These phagosomes fuse with lysosomes to form **phagolysosomes** in which various acid proteases and peptidases hydrolyze the thyroglobulin into amino acids, including the iodothyronines. T_4 and T_3 are discharged from the basal portion of the cell, perhaps by a facilitated process, into the blood. The T_4:T_3 ratio in this blood is lower than that in thyroglobulin, so that some selective deiodination of T_4 must occur in the thyroid. About 50 μg of thyroid hormone iodide is secreted each day. With an average uptake of iodide (25–30% of the iodide in-

gested), the daily iodide requirement is between 150 and 200 μg.

As mentioned above, most of the iodide in thyroglobulin is not in iodothyronine; about 70% is in the inactive compounds MIT and DIT. These amino acids are released when thyroglobulin is hydrolyzed and the iodide is scavenged by an environmentally conscious enzyme, **deiodinase.** This NADPH-dependent enzyme is also present in kidney and liver. The iodide removed from MIT and DIT constitutes an important pool within the thyroid, as distinguished from that I^- which enters from blood. Under steady-state conditions, the amount of iodide that enters the thyroid matches the amount that leaves. If one-third of the iodide in thyroglobulin leaves (as T_4 and T_3), it follows that two-thirds of the iodide available for biosynthesis comes from the deiodination of MIT and DIT within the thyroid.

2. IODIDE METABOLISM

Iodide metabolism involves a number of discrete steps, as illustrated in Fig 46–2.

Concentration of Iodide (I^-)

The thyroid, along with several other epithelial tissues including mammary gland, chorion, salivary gland, and stomach, is able to concentrate I^- against a strong electrochemical gradient. This is an energy-dependent process and is linked to the ATPase-dependent Na^+/K^+ pump. The activity of the **thyroidal I^- pump** can be isolated from subsequent steps in hormone biosynthesis by inhibiting organification of I^- with drugs of the thiourea class (Fig 46–3). The ratio of iodide in thyroid to iodide in serum (T:S ratio) is a reflection of the activity of this pump or concentrating mechanism. This activity is primarily controlled by TSH and ranges from 500 in animals chronically stimulated with TSH to 5 or less in hypophysectomized animals. The T:S ratio in humans on a normal iodine diet is about 25.

A very small amount of iodide also enters the thyroid by diffusion. Any intracellular I^- that is not incorporated into MIT or DIT (generally $< 10\%$) is free to leave by this mechanism.

The transport mechanism is inhibited by 2 classes of molecules. The first group consists of perchlorate (ClO_4^-), perrhenate (ReO_4^-), and pertechnetate (TcO_4^-), all anions with a similar partial specific volume to I^-. These anions compete with I^- for its carrier and are concentrated by the thyroid. A radioisotope of TcO_4^- is commonly used to study iodide transport in humans. The linear anion thiocyanate (SCN^-), an example of the second class, is a competitive inhibitor of I^- transport but is not concentrated by the thyroid. These inhibitors of I^- transport unmask the rapid diffusion of exchangeable I^- from the thyroid and are used to diagnose organification deficiencies. After the acute administration of a blocking concentration of a transport inhibitor, the amount of accumulated I^-

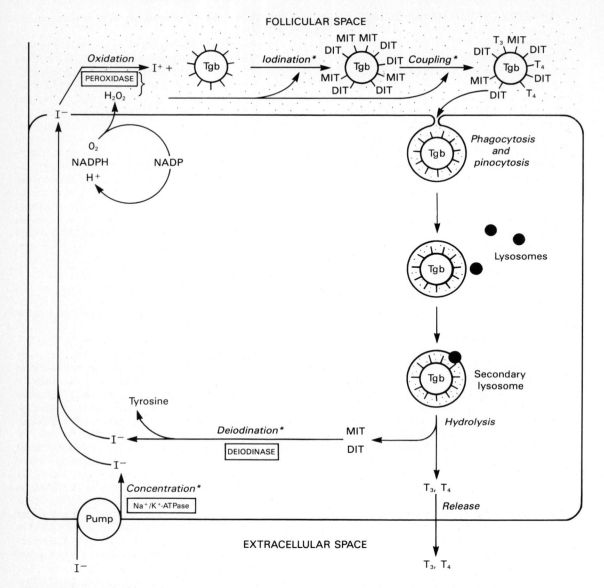

Figure 46–2. Model of iodide metabolism in the thyroid follicle. A follicular cell is shown facing the follicular lumen (stippled area) and the extracellular space (at bottom). Iodide enters the thyroid by a pump and by passive diffusion. Thyroid hormone synthesis occurs in the follicular space through a series of reactions, many of which are peroxidase-mediated. Thyroid hormones are released from thyroglobulin by hydrolysis. Tgb, thyroglobulin; MIT, monoiodotyrosine; DIT, diiodotyrosine; T_3, triiodothyronine; T_4, tetraiodothyronine. Asterisks indicate steps or processes that are inherited enzyme deficiencies which cause congenital goiter and often result in hypothyroidism.

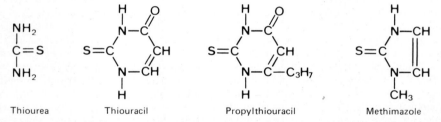

Figure 46–3. Thiourea class of antithyroid drugs.

(usually measured as the isotope ^{131}I) that leaves the thyroid is directly related to the unbound, or nonorganified, fraction. Individuals with incomplete organification will "discharge" more ^{131}I than will normal persons in response to ClO_4^- .

Oxidation of I⁻

The thyroid is the only tissue that can oxidize I⁻ to a higher valence state, an obligatory step in I⁻ organification and thyroid hormone biosynthesis. This step involves a heme-containing peroxidase and occurs at the luminal surface of the follicular cell.

Thyroperoxidase, a tetrameric protein with a molecular weight of 60,000, requires hydrogen peroxide as an oxidizing agent. The H_2O_2 is produced by an NADPH-dependent enzyme resembling cytochrome c reductase. A number of compounds inhibit I⁻ oxidation and therefore its subsequent incorporation into MIT and DIT. The most important of these clinically are the thiourea drugs, some of which are shown in Fig 46–3. They are known as **antithyroid drugs** because of their ability to inhibit thyroid hormone biosynthesis at this step.

Iodination of Tyrosine

Oxidized iodide reacts with the tyrosyl residues in thyroglobulin in a reaction that probably also involves thyroperoxidase. The 3 position of the aromatic ring is iodinated first and then the 5 position to form MIT and DIT, respectively. This reaction, sometimes called organification, occurs within seconds in luminal thyroglobulin. Once iodination occurs, the iodine does not readily leave the thyroid. Free tyrosine can be iodinated, but it is not incorporated into proteins, since no tRNA recognizes iodinated tyrosine.

Coupling of Iodotyrosyls

The coupling of two DIT molecules to form T_4 or of an MIT and DIT to form T_3 occurs within the thyroglobulin molecule, although the addition of a free MIT or DIT to a bound DIT has not been conclusively excluded. A separate coupling enzyme has not been found, and since this is an oxidative process, it is assumed that the same thyroperoxidase catalyzes this reaction by stimulating free radical formation of iodotyrosine. This hypothesis is supported by the observation that the same drugs which inhibit I⁻ oxidation also inhibit coupling. The formed thyroid hormones remain as integral parts of thyroglobulin until the latter is degraded, as described above. Thyroglobulin hydrolysis is stimulated by TSH but is inhibited by I⁻; this latter effect is occasionally exploited by using KI to treat hyperthyroidism.

TRANSPORT & METABOLISM OF THYROID HORMONES

One-half to two-thirds of T_4 and T_3 in the body is extrathyroidal, and most of this circulates in bound form, ie, bound to 2 specific binding proteins, **thyrox-**

Table 46–1. Comparison of T_4 and T_3 in plasma.

Total Hormone (μg/dL)	Free Hormone			$t_{1/2}$ in Blood (days)
	Percent of Total	ng/dL	Molarity	
T_4 8	0.03	~2.24	3.0×10^{-11}	6.5
T_3 0.15	0.3	~0.4	~0.6×10^{-11}	1.5

ine-binding globulin (TBG) and **thyroxine-binding prealbumin (TBPA).** TBG, a glycoprotein of 50,000 MW, is quantitatively the more important. It binds T_4 and T_3 with 100 times the affinity of TBPA and has the capacity to bind 20 μg/dL of plasma. Under normal circumstances, TBG binds, noncovalently, nearly all of the T_4 and T_3 in plasma (Table 46–1). The small, unbound (free) fraction is responsible for the biologic activity. In spite of the great difference in total amount, the free fraction of T_3 approximates that of T_4, but the plasma half-life of T_4 is 4–5 times that of T_3.

TBG is also subject to regulation, an important consideration in diagnostic testing of thyroid function, since most assays of T_4 or T_3 measure the total amount in plasma rather than the free hormone. TBG is produced in liver, and its synthesis is increased by estrogens (pregnancy and birth control pills). Decreased production of TBG occurs following androgen or glucocorticoid therapy and in certain liver diseases. Inherited increases or decreases of TBG also occur. All of these conditions result in changes of total T_4 and T_3 without a change of the free level. Phenytoin and salicylates compete with T_3 and T_4 for binding to TBG. This decreases the total level of hormone without changing the free fraction and must be considered when interpreting diagnostic tests.

Extrathyroidal deiodination converts T_4 to T_3. Since T_3 binds to the thyroid receptor in target cells with 10 times the affinity of T_4, T_3 is thought to be the preponderant metabolically active form of the molecule. About 80% of circulating T_4 is converted to T_3 or reverse T_3 (rT_3) in the periphery, and this conversion accounts for most of the production of T_3. Reverse T_3 is a very weak agonist that is made in relatively larger amounts in chronic disease, in carbohydrate starvation, and in the fetus. Propylthiouracil and propranolol decrease the conversion of T_4 to T_3.

Other forms of thyroid hormone metabolism include total deiodination and inactivation by deamination or decarboxylation. Hepatic glucuronidation and sulfation result in a more hydrophilic molecule that is excreted into bile, reabsorbed in the gut, deiodinated in the kidney, and excreted as the glucuronide conjugate in the urine.

REGULATION OF THYROID HORMONE SYNTHESIS & RELEASE

The primary components of the negative feedback loop consist of T_4, T_3, TSH, and TRH (Fig 46–4). T_4

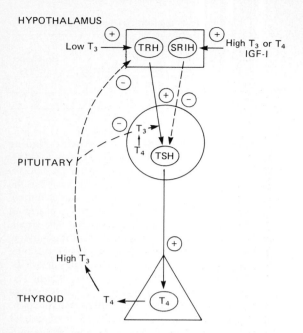

HYPOTHALAMUS

PITUITARY

THYROID

Figure 46–4. Feedback regulation of thyroid hormone biosynthesis. Solid lines and ⊕ symbols indicate stimulatory pathways, and dashed lines and ⊖ symbols indicate inhibitory pathways. IGF-I, insulinlike growth factor I; SRIH, somatostatin; TRH, thyrotropin-stimulating hormone; TSH, thyroid-stimulating hormone.

and T_3 cause feedback inhibition of their own synthesis. T_3 may be the actual feedback mediator, since T_4 is converted to T_3 in the pituitary. Feedback at this level inhibits TSH release. T_3 (or possibly T_4) may also inhibit the release and production of TRH by the hypothalamus. The stimulus for increased release of TRH and TSH is a decrease of free thyroid hormone in the blood. Even if thyroid hormone biosynthesis is completely blocked, as with antithyroid drug therapy, enhanced release of TRH and TSH does not occur immediately. The thyroid contains several weeks' supply of preformed hormone, and there are also substantial extrathyroidal stores (in liver and bound to TBG) that first must be depleted. In addition, a thyroid autoregulatory mechanism helps compensate when iodine deficiency threatens to decrease hormone biosynthesis.

There is an interesting entwining of the thyroid and growth hormone feedback loops that accounts for the regulatory mechanisms shown in Fig 46–4. T_3 and T_4 enhance the release of somatostatin (SRIH) from the hypothalamus, and this peptide inhibits TSH release from the pituitary. Somatostatin is involved in another way. Levels of somatostatin increase in response to increased plasma IGF-I, which in turn is stimulated by GH (see Chapter 45 and Fig 45–5). Children treated with GH for short stature occasionally develop hypothyroidism, presumably because the increase in IGF-I stimulates somatostatin release, which in turn shuts off TSH release.

MECHANISM OF ACTION OF THYROID HORMONES

Thyroid hormones bind to specific high affinity receptors in the target cell nucleus; T_3 binds with approximately 10 times the affinity of T_4. The question as to whether all biologic activity of the thyroid hormones is mediated by T_3 is thus moot; both T_3 and T_4 are active. A comparison of various thyroid hormone analogs shows a high correlation between binding affinity and ability to elicit a biologic response. Thyroid hormones bind to low-affinity sites in cytoplasm, but this is apparently not the same protein as the nuclear receptor. The cytoplasmic binding may serve to keep thyroid hormones "in the neighborhood." Plasma membrane binding of T_3 has been described; the role this plays in hormone transport is uncertain.

The general metabolic function of thyroid hormones is to increase oxygen consumption. This effect is seen in all organs except the brain, reticuloendothelial system, and gonads. This observation directed attention to the mitochondria, where T_4 was shown to cause morphologic changes and to uncouple oxidative phosphorylation. These observed effects require massive amounts of T_4 and almost certainly are not physiologic events. Thyroid hormones induce mitochondrial α-glycerophosphate dehydrogenase, and this may be related to the effects on O_2 consumption.

Edelman and coworkers hypothesize that much of the energy utilized by a cell is for driving the **Na^+/K^+-ATPase pump.** Thyroid hormones enhance the function of this pump by increasing the number of pump units. Since all cells have this pump and virtually all cells respond to thyroid hormones, this increased utilization of ATP and the associated increase of oxygen consumption via oxidative phosphorylation could be the basic mechanism of thyroid hormone action.

Thyroid hormones, like steroids, induce proteins through an enhanced gene transcription mechanism (see Fig 44–1). Although few specific examples are known, this is presumed to be the mechanism by which T_3 enhances general protein synthesis and causes positive nitrogen balance. Here again there is a curious association between the 2 classes of hormones related to growth, the thyroid hormones and growth hormone itself. T_3 and glucocorticoids enhance transcription of the GH gene, so that more GH is produced. This explains a classic observation in which the pituitaries of T_3-deficient animals were found to lack GH, and it may account for some of the general anabolic effects of T_3. Very high concentrations of T_3 inhibit protein synthesis and cause negative nitrogen balance.

Thyroid hormones are known to be important modulators of developmental processes. This is most apparent in amphibian metamorphosis. Thyroid hormones are required for the conversion of a tadpole into a frog, a process that involves resorption of the tail, limb-bud proliferation, conversion from fetal to adult hemoglobin, stimulation of urea cycle enzymes (carbamoyl phosphate synthase) so that urea is excreted rather than ammonia, and epidermal changes. Thyroid hormones are required for normal development in humans. Intrauterine or neonatal hypothyroidism results in **cretinism,** a condition characterized by multiple congenital defects and severe, irreversible mental retardation.

PATHOPHYSIOLOGY

Goiter

Any enlargement of the thyroid is referred to as a goiter. Simple goiter represents an attempt to compensate for decreased thyroid hormone production; thus, in all of these situations, elevated TSH is the common denominator. Causes include iodide deficiency; iodide excess, when an autoregulatory mechanism fails; and a variety of rare inherited metabolic defects that illustrate the importance of various steps in thyroid hormone biosynthesis. These defects include (1) I^- transport defect; (2) iodination defect; (3) coupling defect; (4) deiodinase deficiency; and (5) production of abnormal iodinated proteins. Partial deficiencies of these functions may cause simple goiter in adults. Any of these causes of simple goiter can, when severe, cause hypothyroidism. Simple goiter is treated with exogenous thyroid hormone. Supplementation or restriction of iodide intake is appropriate for specific types of goiter.

Hypothyroidism

Insufficient amounts of free T_3 or T_4 result in the clinical condition known as hypothyroidism. This is usually due to thyroid failure but can be due to disease of the pituitary or hypothalamus. In hypothyroidism, the basal metabolic rate is decreased, as are other processes dependent upon thyroid hormones. Prominent features include slow heart rate, diastolic hypertension, sluggish behavior, sleepiness, constipation, sensitivity to cold, dry skin and hair, and a sallow complexion. Other features depend upon the age at onset. Cretinism is discussed above. Hypothyroidism later in childhood results in short stature but no mental retardation. The various kinds of hypothyroidism are treated with exogenous thyroid hormone replacement.

Hyperthyroidism

Hyperthyroidism, or **thyrotoxicosis,** is due to the excessive production of thyroid hormone. There are many causes, but most cases in the USA are due to **Graves' disease,** which results from the production of **thyroid-stimulating IgG (TSI)** that activates the TSH receptor (see Table 43–2). This causes a diffuse enlargement of the thyroid and excessive, uncontrolled production of T_3 and T_4, since the production of TSI is not under feedback control. Findings are multisystemic and include rapid heart rate, widened pulse pressure, nervousness, inability to sleep, weight loss in spite of increased appetite, weakness, excessive sweating, sensitivity to heat, and red, moist skin. The hyperthyroidism of Graves' disease is treated by blocking hormone production with an antithyroid drug, by ablating the gland with a radioactive isotope of iodide (such as ^{131}I), or by a combination of these 2 methods. Occasionally, the gland is removed surgically.

REFERENCES

Chopra IJ et al: Pathways of metabolism of thyroid hormones. *Recent Prog Horm Res* 1978;**34:**531.

Jackson IMD: Thyrotropin-releasing hormone. *N Engl J Med* 1982;**306:**145.

Larsen PR: Thyroid-pituitary interaction: Feedback regulation of thyrotropin secretion by thyroid hormones. *N Engl J Med* 1982;**396:**23.

Lo GS et al: Dependence of renal (Na^+ and K^+)-ATPase activity on thyroid status. *J Biol Chem* 1976;**251:**7826.

Oppenheimer JH: Thyroid hormone action at the nuclear level. *Ann Intern Med* 1985;**102:**374.

Robins J et al: Thyroxine transport proteins of plasma: Molecular properties and biosynthesis. *Recent Prog Horm Res* 1978; **34:**477.

47

Hormones That Regulate Calcium Metabolism

Daryl K. Granner, MD

INTRODUCTION

Calcium ion regulates a number of important physiologic and biochemical processes including **neuromuscular excitability, blood coagulation, secretory processes, membrane integrity and plasma membrane transport, enzyme reactions,** the **release of hormones and neurotransmitters,** and the **intracellular action** of a number of hormones. In addition, the proper extracellular fluid (ECF) and periosteal concentrations of Ca^{2+} and PO_4^{3-} are required for **bone mineralization.** To ensure that these processes operate normally, the plasma Ca^{2+} concentration is maintained within *very* narrow limits. The purpose of this chapter is to explain how this is accomplished.

BIOMEDICAL IMPORTANCE

Deviations of the ionized calcium from the normal range cause many disorders and can be life-threatening. As many as 3% of hospitalized patients may have disorders of calcium homeostasis.

GENERAL FEATURES

There is approximately 1 kg of calcium in the human body. Ninety-nine percent of this is located in bone where, with phosphate, it forms the **hydroxyapatite crystals** that provide the inorganic and structural component of the skeleton. Bone is a dynamic tissue, and it undergoes constant remodeling as stresses change; in the steady-state condition, there is a balance between new bone formation and bone resorption. Most of the calcium in bone is not freely exchangeable with **extracellular fluid (ECF) calcium.** Thus, in addition to its mechanical role, bone serves as a large reservoir of calcium. About 1% of skeletal Ca^{2+} is in a freely exchangeable pool and this, with another 1% of the total found in the periosteal space, constitutes the **miscible pool of Ca^{2+}.** The hormones discussed in this chapter regulate the amount of calcium in the ECF by influencing the transport of calcium across the membrane that separates the ECF space from the periosteal fluid space. This transport is

primarily stimulated by parathyroid hormone (PTH), but calcitriol is also involved. Calcitonin may counteract these effects.

Plasma calcium exists in 3 forms: (1) **complexed** with organic acids; (2) **protein-bound;** and (3) **ionized.** About 6% of total calcium is complexed with citrate, phosphate, and other anions. The remainder is divided nearly equally between a protein-bound form (bound primarily to albumin) and an ionized (unbound) form. The ionized calcium (Ca^{2+}), which is maintained at concentrations between 1.1 and 1.3 mmol/L in most mammals, birds, and freshwater fish, is the biologically active fraction. The organism has very little tolerance for significant deviation from this normal range. If the ionized calcium level falls, the animal becomes increasingly hyperexcitable and may develop tetanic convulsions. A marked elevation of plasma calcium may result in death owing to muscle paralysis and coma.

Calcium ion and the counter-ion, phosphate, exist at or near their solubility product in plasma; hence, protein binding may protect against precipitation and **ectopic calcification.** An alteration of the plasma protein concentration (primarily albumin, but globulins also bind calcium) results in parallel changes in total plasma calcium. For example, hypoalbuminemia results in a decrease of total plasma calcium of approximately 0.8 mg/dL for each g/dL of albumin decrease. The converse is noted when the plasma albumin is increased. The association of calcium with plasma proteins is pH-dependent; acidosis favors the ionized form, whereas alkalosis enhances binding and causes a concomitant decrease in Ca^{2+}. The latter probably accounts for the numbness and tingling associated with the **hyperventilation syndrome,** which causes acute respiratory alkalosis.

CALCIUM HOMEOSTASIS

The primordial sea primarily contained K^+ and Mg^{2+}, and so proteins evolved with the ability to function best in that ionic milieu. With time, the composition of the sea changed, and Na^+ and Ca^{2+} became relatively more abundant. Thus, in order to preserve intracellular protein function in the changing environment, it became necessary to limit the intracellular concentrations of sodium and calcium while conserving the potassium and magnesium environment. This was achieved by the development of membrane-associated pumps for sodium and calcium which, in the case of Ca^{2+}, maintain about a 1000-fold concentration gradient between cytosol and extracellular fluid. Na^+ and Ca^{2+} now constitute the major extracellular ionic environment of multicellular animals. Hormones and other effectors cause rapid, transient changes in the flux of calcium ion across the cellular plasma membrane and from one intracellular compartment to another. Thus, calcium ion serves as an intracellular mediator in a variety of metabolic reactions (see Chapter 44).

The movement from an aquatic environment rich in Ca^{2+} to a terrestrial environment relatively deficient in this element necessitated the development of an intricate homeostatic mechanism to extract calcium from dietary sources and to ensure against marked changes of the Ca^{2+} concentration in ECF. This mechanism involves the actions of 3 hormones—**parathyroid hormone (PTH), calcitriol** [1,25(OH)$_2$-D$_3$], and **calcitonin (CT)**—acting on 3 organs—bone, kidney, and intestine. The parathyroid glands increase the secretion of PTH when the level of ionized calcium in plasma falls below the lower limit of the normal range (< 1.1 mmol/L). PTH stimulates the movement of calcium and phosphate from bone to blood and acts upon the kidney to increase calcium resorption and phosphate excretion.

A second important action of PTH on the kidney is to stimulate the formation of 1,25(OH)$_2$-D$_3$. This compound, currently referred to as calcitriol, is the active form of what previously was called vitamin D. Calcitriol acts upon the intestine to increase calcium absorption and probably plays a permissive role in the actions of PTH on bone and kidney. The concerted actions of these agents are to raise the level of Ca^{2+} in ECF while maintaining or decreasing the concentration of phosphate. There is a feedback inhibition of PTH secretion when extracellular Ca^{2+} reaches the normal level. An increase of Ca^{2+} also inhibits the formation of calcitriol (in part by lowering PTH) and increases the production of inactive metabolites of this compound. This results in decreased intestinal calcium absorption and decreased PTH action on bone and kidney. In some animals, an elevation of extracellular Ca^{2+} stimulates the secretion of calcitonin (CT) from C cells of the thyroid or the ultimobranchial bodies. The role of CT in normal calcium homeostasis is obscure in humans, but in some in vitro test systems CT appears to inhibit bone resorption.

HORMONES INVOLVED IN CALCIUM HOMEOSTASIS

1. PARATHYROID HORMONE (PTH)

Structure

PTH is an 84-amino-acid single-chain peptide (MW 9500) that contains no carbohydrate or other covalently bound molecules (Fig 47–1). Full biologic activity resides in the N-terminal third of the molecule; PTH$_{1-34}$ has full biologic activity. The region 25–34 is primarily responsible for receptor binding.

PTH is synthesized as a 115-amino-acid precursor molecule (Fig 47–1). The immediate precursor of PTH is **proPTH,** which differs from the native hormone by having an N-terminal highly basic hexapeptide extension whose function is obscure. The primary gene product and the immediate precursor for proPTH is **preproPTH.** This differs from proPTH by having an additional 25-amino-acid N-terminal extension that, in common with the other leader or signal sequences characteristic of secreted proteins, is hydrophobic. The complete structure of preproPTH and the sequences of proPTH and PTH are illustrated in Fig 47–1.

PTH was the first preprohormone identified, and the sequence of events involved in the conversion of this to PTH is shown in schematic form in Fig 47–2. PreproPTH is transferred to the cisternal space of the endoplasmic reticulum while the molecule is still being translated from PTH mRNA by the ribosomes.

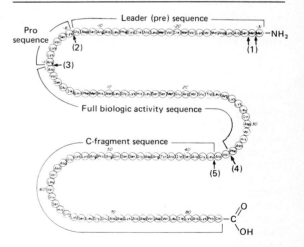

Figure 47–1. Structure of bovine preproparathyroid hormone. Arrows indicate sites cleaved by processing enzymes in the parathyroid gland (1–5) and in the liver after secretion of the hormone (4–5). The biologically active region of the molecule is flanked by sequence not required for activity on target receptors. (Slightly modified and reproduced, with permission, from Habener JF: Recent advances in parathyroid hormone research. *Clin Biochem* 1981;14:223.)

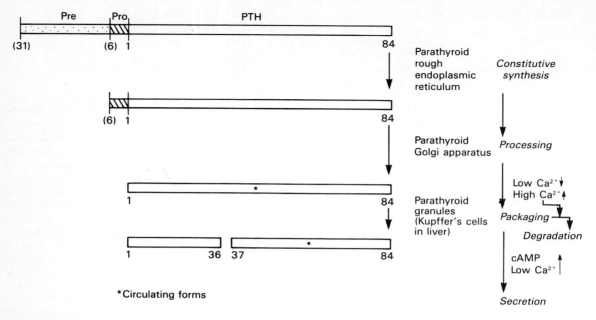

Figure 47–2. The precursors and cleavage products of PTH and the location of these steps in the parathyroid gland and liver. The numbers in parentheses indicate the number of amino acids in the pre (31) and pro (6) fragments.

During this transfer, the 25-amino-acid prepeptide (signal or leader peptide) is removed to yield proPTH. ProPTH is then transported to the Golgi apparatus, where an enzyme removes the pro-extension to yield the mature PTH molecule. The PTH released from the Golgi apparatus in secretory vesicles has 3 possible fates: (1) transport into a storage pool; (2) degradation; or (3) immediate secretion.

Role in Mineral Homeostasis

A. Calcium Homeostasis: The central role of PTH in calcium metabolism is underscored by the observation that the first evolutionary appearance of this hormone was in animals attempting to adapt to a terrestrial existence. The physiologic maintenance of calcium balance depends on the long-term effects of PTH acting on intestinal absorption through the formation of calcitriol. If in the face of prolonged dietary Ca^{2+} deficiency intestinal calcium absorption is inadequate, a complex regulatory system involving PTH is brought into play. PTH restores normal ECF calcium concentration by acting directly on bone and kidney and by acting indirectly on the intestinal mucosa (through stimulation of synthesis of calcitriol). PTH (1) increases the rate of dissolution of bone, including both organic and inorganic phases, which moves Ca^{2+} into ECF; (2) reduces the renal clearance or excretion of calcium, hence increasing the ECF concentration of this cation; and (3) increases the efficiency of calcium absorption from the intestine by promoting the synthesis of calcitriol. The most rapid changes occur through the action on the kidney, but the largest effect is from bone. Therefore, although PTH prevents hypocal-

cemia in the face of dietary calcium deficiency, it does so at the expense of bone substance.

B. Phosphate Homeostasis: The usual counter-ion for Ca^{2+} is phosphate, and the hydroxyapatite crystal in bone consists of calcium phosphate. Phosphate is released with calcium from bone whenever PTH increases dissolution of the mineral matrix. PTH increases renal phosphate clearance; thus, the net effect of PTH on bone and kidney is to increase the ECF calcium concentration and decrease the ECF phosphate concentration. Importantly, this prevents the development of a supersaturated concentration of calcium and phosphate in plasma.

Biochemistry

A. Regulation of Synthesis: The rate of synthesis and degradation of proPTH is unaffected by the ambient Ca^{2+} concentration, even though the rate of formation and secretion of PTH is always markedly enhanced at low Ca^{2+} concentrations. Indeed, 80–90% of the proPTH synthesized cannot be accounted for as intact PTH in cells or in the incubation medium of experimental systems. This led to the conclusion that most of the proPTH synthesized is quickly degraded. It was later discovered that this rate of degradation decreases when Ca^{2+} concentrations are low and increases when Ca^{2+} concentrations are high. This indicates that calcium affects PTH production through control of degradation and not synthesis. The constitutive synthesis of proPTH is reflected in PTH mRNA levels, which also do not change in spite of wide fluctuations of extracellular Ca^{2+}. It appears that the only way that the organism can enhance PTH synthesis is to increase the size and number of PTH-producing chief cells in the parathyroid glands.

B. Regulation of Metabolism: The degradation of PTH begins about 20 minutes after proPTH is synthesized, is initially unaffected by the Ca^{2+} concentration, and occurs after the hormone is in secretory vesicles. Newly formed PTH can either be secreted immediately or be placed in storage vesicles for subsequent secretion. Degradation occurs as soon as the secretory vesicle begins to enter the storage compartment.

Very specific fragments of PTH are generated during its proteolytic digestion (Figs 47–1 and 47–2), and large amounts of carboxy-terminal fragments of PTH are found in the circulation. These molecules, with molecular weights of about 7000, consist of PTH_{37-84} and lesser amounts of PTH_{34-84}. Most of the newly synthesized PTH is degraded. About 2 mol of the C-terminal fragments is secreted for each mole of intact PTH; hence, the bulk of circulating PTH consists of the C-terminal molecules. No biologic function for the C-terminal fragment of PTH has been defined, but it may prolong the half-life of the hormone in the circulation. A number of proteolytic enzymes, including **cathepsins B and D,** have been identified in parathyroid tissue. Cathepsin B cleaves PTH into 2 fragments, PTH_{1-36} and PTH_{37-84}. PTH_{37-84} is not further degraded; however, PTH_{1-36} is rapidly and progressively cleaved into di- and tripeptides. ProPTH has never been found in circulation, and little (if any) PTH_{1-34} escapes from the gland. PreproPTH was identified by deciphering the coding sequence of the PTH gene.

Most of the proteolysis of PTH occurs within the gland; however, there are a number of studies which confirm that PTH, once secreted, is proteolytically degraded in other tissues. The exact contribution of extraglandular proteolysis has not been defined, nor is it clear whether the proteolytic enzymes in the 2 sites are similar or whether the patterns and products of cleavage are identical.

The liver and kidneys are involved in peripheral metabolism of secreted PTH. After hepatectomy, no 34–84 or 37–84 fragments are detected, indicating that the liver is the principal organ involved in the generation of these fragments. The role of the kidneys may be to remove and excrete these fragments. The principal site of **peripheral proteolysis** appears to be the **Kupffer cells** lining the intrasinusoidal passages of the liver. The endopeptidase responsible for the initial cleavage into the amino- and carboxy-terminal fragments is located on the surface of these macrophagelike cells, which are in intimate contact with plasma. This enzyme, also a cathepsin B, cleaves PTH between residues 36 and 37; as in the parathyroid, the resulting carboxy-terminal fragment continues to circulate, whereas the amino-terminal fragment is rapidly degraded.

C. Regulation of Secretion: PTH secretion is inversely related to the ambient concentration of ionized calcium and magnesium, as is the circulating level of immunoreactive PTH. Fig 47–3 illustrates that serum PTH declines in a rectilinear fashion in relation to serum calcium levels between 4 mg/dL and 10.5 mg/dL. The presence of biologically active PTH when the serum calcium level is 10.5 mg/dL or greater is an indication of **hyperparathyroidism.**

There is a linear relationship between PTH release and the parathyroid intracellular level of cAMP. The intracellular Ca^{2+} level may be involved in this process, since there is an inverse relationship between the intracellular concentrations of calcium and cAMP. Calcium may exert this effect through its known action on phosphodiesterase (via Ca^{2+}/calmodulin-dependent protein kinase) or through a similar mechanism by inhibiting adenylate cyclase. Phosphate has no effect on PTH secretion.

Parathyroid glands have relatively few storage granules and contain enough hormone to maintain maximal secretion for only 1.5 hours. This is in contrast to the pancreatic islets, which contain insulin stores sufficient for several days, and to the thyroid, which contains hormone stores adequate for several weeks. PTH must therefore be continually synthesized and secreted.

Mechanism of Action

A. The PTH Receptor: PTH binds to a single membrane receptor protein of approximately 70,000 MW. This receptor appears to be identical in bone and kidney, and it is not found in nontarget cells. The hormone-receptor interaction initiates a typical cascade: activation of adenylate cyclase → increased intracellular cAMP → increased intracellular calcium → phos-

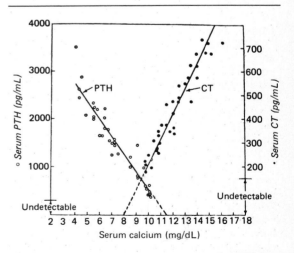

Figure 47–3. Concentration of calcitonin (CT) and parathyroid hormone (PTH) as a function of serum calcium level. (Modified and reproduced, with permission, from Arnaud CD, Littledike T, Tsao HS: Calcium homeostasis and the simultaneous measurement of calcitonin and parathyroid hormone in the pig. Pages 95–101 in: *Calcitonin: Proceedings of the Second International Symposium.* Taylor S [editor]. Heinemann, 1969.)

phorylation of specific intracellular proteins by kinases → activation of the intracellular enzymes or proteins that finally mediate the biologic actions of the hormone. The PTH response system, like that for many other peptide and protein hormones, is subject to **down-regulation** of receptor number and to "**desensitization,**" which may involve a post-cAMP mechanism.

B. Effects of PTH on Bone: PTH has multiple effects on bone and apparently influences several types of bone cells. The net effect is bone destruction, with the concomitant release of calcium, phosphate, and organic matrix elements, including collagen breakdown products. The cell responsible for this may be the **osteoclast,** which certainly can destroy bone when stimulated chronically with PTH, or the **osteocyte,** which also can resorb bone. PTH may stimulate the differentiation of precursor cells into bone-resorbing cells. At low concentrations, perhaps within the physiologic range, PTH has an anabolic effect and is responsible for bone remodeling. At these concentrations, **osteoblasts** increase in number, there is an increase in alkaline phosphatase activity that reflects new bone formation, and the incorporation of radioactive sulfate into cartilage is promoted. Calcitriol may play a permissive role in the action of PTH on bone.

Ca^{2+} may be the intracellular messenger for PTH. The earliest action of PTH is to decrease pericellular Ca^{2+} while increasing intracellular Ca^{2+}. Increased intracellular calcium promotes bone cell RNA synthesis and the release of enzymes associated with bone resorption. These processes appear to be mediated through the binding of calcium to calmodulin. In the absence of extracellular calcium, PTH still increases cAMP but no longer stimulates bone resorption. Thus, a major requirement for PTH stimulation of bone resorption may be a paradoxic increase of the uptake of ionized calcium into the bone-resorbing cells.

C. Effects of PTH on Kidney: PTH has many effects on the kidney; it influences the transport of several ions and regulates the synthesis of calcitriol. Under normal conditions, Ca^{2+} resorption exceeds 90% of the filtered load, and PTH increases Ca^{2+} resorption to greater than 98%. Phosphate resorption varies between 75% and 90%, since it is more dependent on dietary intake and other factors, but PTH inhibits phosphate resorption whatever the basal rate. PTH also inhibits sodium, potassium, and bicarbonate transport. The effects of PTH on calcitriol metabolism (see below) may occur at the same cellular sites as its effects on mineral resorption.

An infusion of PTH results in a prompt increase of intracellular cAMP in renal cells and in the prompt excretion of cAMP in urine. This response precedes the characteristic phosphaturia and is presumed to be responsible for the latter. The PTH-stimulated adenylate cyclase is in the basolateral portion of renal cortical tubular cells and is distinct from the renal adenylate cyclase that is stimulated by calcitonin, catecholamines, and ADH. The intracellular receptor proteins for cAMP (presumably protein kinases) are found in the brush border of these cells on the luminal surface of the tubule. The cAMP generated by PTH stimulation therefore migrates from the basolateral aspect of the cell to the luminal surface, where it facilitates ion transport.

Calcium may also be involved in the renal mechanism of action of PTH. Indeed, the first physiologic effect of PTH administration is to decrease extracellular fluid Ca^{2+} while intracellular Ca^{2+} increases. These changes, however, occur after the changes in cAMP, so that the association of Ca^{2+} flux to PTH action is not so clear in the kidney.

D. Effects of PTH on Intestinal Mucosa: PTH probably has no direct effect on Ca^{2+} transport across the intestinal mucosa, but it is critically involved in the biosynthesis of calcitriol (see below) and certainly has an important indirect effect on the intestine.

Pathophysiology

Insufficient amounts of PTH result in **hypoparathyroidism.** The biochemical hallmarks of this condition are decreased serum ionized calcium and elevated serum phosphate levels. Symptoms include neuromuscular irritability which, when mild, causes muscle cramps and **tetany.** Severe, acute hypocalcemia results in tetanic paralysis of the respiratory muscles, laryngospasm, severe convulsions, and death. Long-standing hypocalcemia results in cutaneous changes, cataracts, and calcification of the basal ganglia of the brain. The usual cause of hypoparathyroidism is accidental removal or damage of the glands during neck surgery (secondary hypoparathyroidism), but the disorder occasionally results from **autoimmune destruction** of the glands (primary hypoparathyroidism).

Pseudohypoparathyroidism is discussed in Chapter 44. Biologically active PTH is produced in this inherited disorder, but there is end-organ resistance to its effects. The biochemical consequences are the same, however. There are usually associated developmental anomalies including short stature, short metacarpal or metatarsal bones, and mental retardation. There are several types of pseudohypoparathyroidism, and they have been attributed to (1) partial deficiency of the G_s adenylate cyclase regulatory protein, and (2) a defective step beyond the formation of cAMP.

Hyperparathyroidism, the excessive production of PTH, is usually due to the presence of a functioning **parathyroid adenoma** but can be due to **parathyroid hyperplasia** or to **ectopic production** of PTH in a malignant tumor. The biochemical hallmarks of hyperparathyroidism are elevated serum ionized calcium and PTH and depressed serum phosphate levels. In long-standing hyperparathyroidism, findings include extensive resorption of bone and a variety of renal effects, including kidney stones, nephrocalcinosis, frequent urinary tract infections, and (in severe cases) decreased renal function. **Secondary hyperparathyroidism,** characterized by hyperplasia of the glands and hypersecretion of PTH, may be seen in patients with progressive renal failure. Hyperparathyroidism in these patients is presumably due to the decreased conversion of 25OH-D$_3$ to 1,25(OH)$_2$-D$_3$ in the diseased renal parenchyma, which results in inefficient calcium absorption in the gut and the secondary release of PTH in a compensatory attempt to maintain normal ECF calcium levels.

2. CALCITRIOL (1,25[OH]$_2$-D$_3$)

General Role in Calcium Homeostasis

A. Historical Perspective: Rickets, a childhood disorder characterized by deficient mineralization of the skeleton and severe, crippling bone deformities, was epidemic in North America and Western Europe early in this century. Results of a series of studies suggested that rickets was due to a dietary deficiency. After the discovery that rickets could be prevented by ingestion of cod-liver oil and that the active ingredient in this agent was not vitamin A, the preventive factor was termed fat-soluble **vitamin D.** About the same time, it was found that ultraviolet light, either artificial or from sunlight, would also prevent the disorder. It was subsequently determined that

there was an adult equivalent to rickets. **Osteomalacia,** in which there is a failure to mineralize bone, also responded to vitamin D. Clues to further developments resulted from the observation that patients with liver or kidney disease did not respond normally to vitamin D. For the last 50 years, efforts to elucidate the structure of vitamin D and to define its mechanism of action have proceeded, greatly accelerated during the last 10 years.

B. Homeostatic Role: The principal biologic role of calcitriol is **to stimulate intestinal absorption of calcium and phosphate.** Calcitriol is the only hormone which can promote this translocation of calcium against the concentration gradient that exists across the intestinal cell membrane. Since the production of calcitriol is tightly regulated (Fig 47–4), a fine mechanism exists for controlling ECF Ca^{2+} in spite of marked fluctuations of the calcium content of food. This ensures a proper concentration of calcium and phosphate for deposition, as hydroxyapatite crystals, onto the collagen fibrils in bone. In vitamin D deficiency (calcitriol deficiency), new bone formation slows and bone remodeling is also impaired. These processes are primarily regulated by PTH acting on bone cells, but small concentrations of calcitriol are

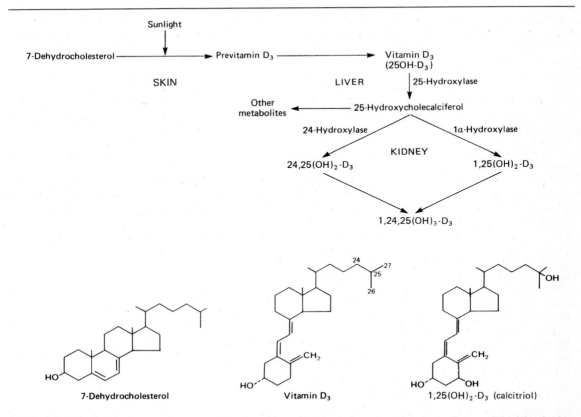

Figure 47–4. Formation and hydroxylation of vitamin D$_3$. 25-Hydroxylation takes place in the liver, and the other hydroxylations occur in the kidneys. 25,26(OH)$_2$-D$_3$ and 1,25,26(OH)$_3$-D$_3$ are probably formed as well. The formulas of 7-dehydrocholesterol, vitamin D$_3$, and 1,25(OH)$_2$-D$_3$ (calcitriol) are also shown. (Reproduced, with permission, from Ganong WF: *Review of Medical Physiology,* 13th ed. Appleton & Lange, 1987.)

also required. Calcitriol may also augment the actions of PTH on renal calcium reabsorption.

Biochemistry

A. Biosynthesis: Calcitriol is a hormone in every respect. It is produced by a complex series of enzymatic reactions that involve the plasma transport of precursor molecules to a number of different tissues (Fig 47–4). The active molecule, calcitriol, is transported to other organs where it activates biologic processes in a manner similar to that employed by the steroid hormones.

1. Skin—Small amounts of **vitamin D** occur in food (fish-liver oil, egg yolk), but most of the vitamin D available for calcitriol synthesis is produced in the malpighian layer of the epidermis from 7-dehydrocholesterol in an ultraviolet light-mediated, nonenzymatic **photolysis reaction.** The extent of this conversion is directly related to the intensity of the exposure and inversely related to the extent of pigmentation in the skin. There is an age-related loss of 7-dehydrocholesterol in the epidermis that may be related to the negative calcium balance associated with old age.

2. Liver—A specific transport protein called the **D-binding protein** binds vitamin D_3 and its metabolites and moves D_3 from the skin or intestine to the liver where it undergoes 25-hydroxylation, the first obligatory reaction in the production of calcitriol. 25-Hydroxylation occurs in the endoplasmic reticulum in a reaction that requires magnesium, NADPH, molecular oxygen, and an uncharacterized cytoplasmic factor. Two enzymes, an NADPH-dependent cytochrome P-450 reductase and a cytochrome P-450, are involved. This reaction is not regulated, and it also occurs with low efficiency in kidney and intestine. The $25OH$-D_3 enters the circulation, where it is the major form of vitamin D found in plasma, and is transported to the kidney by the D-binding protein.

3. Kidney—$25OH$-D_3 is a weak agonist and must be modified by hydroxylation at position C_1 for full biologic activity. This is accomplished in mitochondria of the renal proximal convoluted tubule in a complex, 3-component monooxygenase reaction that requires NADPH, Mg^{2+}, molecular oxygen, and at least 3 enzymes: (1) a flavoprotein, renal ferredoxin reductase; (2) an iron sulfur protein, renal ferredoxin; and (3) cytochrome P-450. This system produces $1,25(OH)_2$-D_3, which is the most potent naturally occurring metabolite of vitamin D.

4. Other tissues—The placenta has a 1α-hydroxylase that appears to be an important extrarenal source of calcitriol. Enzyme activity is found in a variety of other tissues, including bone; however, the physiologic significance of this appears to be minimal, since very little calcitriol is found in nonpregnant, nephrectomized animals.

B. Regulation of Metabolism and Synthesis: Like other steroid hormones, calcitriol is subject to tight feedback regulation (Fig 47–4 and Table 47–1). Low-calcium diets and hypocalcemia result in marked increases of 1α-hydroxylase activity in intact animals.

Table 47–1. Regulation of renal 1α-hydroxylase

Primary Regulators	Secondary Regulators
Hypocalcemia (↑)	Estrogens
PTH (↑)	Androgens
Hypophosphatemia (↑)	Progesterone
Calcitriol (↑)	Insulin
	Growth hormone
	Prolactin
	Thyroid hormone

This effect requires PTH, which is also released in response to hypocalcemia. The action of PTH is as yet unexplained, but it stimulates 1α-hydroxylase activity in both vitamin D-deficient and vitamin D-treated animals. Low-phosphorus diets and hypophosphatemia also induce 1α-hydroxylase activity, but this appears to be a weaker stimulus than that provided by hypocalcemia.

Calcitriol is an important regulator of its own production. High levels of calcitriol inhibit renal 1α-hydroxylase and stimulate the formation of a 24-hydroxylase that leads to the formation of $24,25(OH)_2$-D_3, an apparently inactive by-product. Estrogens, progestins, and androgens cause marked increases of 1α-hydroxylase in ovulating birds. The role that these hormones, along with insulin, growth hormone, and prolactin, play in mammals is uncertain.

The basic sterol molecule can be modified by **alternative metabolic pathways,** ie, by hydroxylation at positions 1, 23, 24, 25, and 26 and by the formation of a number of lactones. Over 20 metabolites have been found; none has unequivocally been shown to have biologic activity.

Mechanism of Action

Calcitriol acts at the cellular level in a manner similar to other steroid hormones (Fig 47–5). Studies using radioactive calcitriol revealed localization in the nuclei of intestinal villus and crypt cells, osteoblasts, and distal renal tubular cells. There also is nuclear accumulation of this hormone in cells not previously suspected of being targets, including cells in the malpighian layer of the skin; pancreatic islet cells; some brain cells; some cells in the pituitary, ovary, testis, placenta, uterus, mammary gland, and thymus; and myeloid precursors. Calcitriol binding has also been noted in parathyroid cells, which leads to the intriguing possibility that it might be involved in PTH metabolism.

A. The Calcitriol Receptor: An intestinal cell protein with a molecular weight of 90,000–100,000 binds calcitriol with high affinity and low capacity. This binding is saturable, specific, and reversible. This protein, which meets the basic criteria for being a receptor, has been found in many of the tissues listed above.

When examined under physiologic salt conditions, most of the unoccupied receptor is found in the nucleus in association with chromatin. This resembles the cases of receptors for T_3 and progesterone, if not

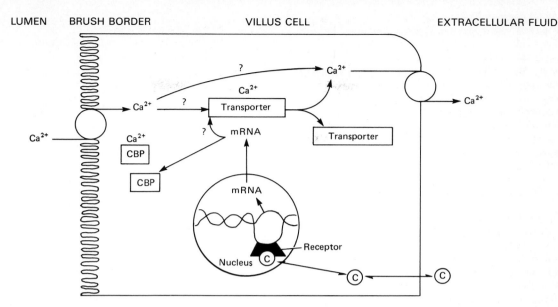

Figure 47–5. Calcitriol (C) acts like other steroid hormones. It induces gene products that promote the movement of calcium from the intestinal lumen to the extracellular fluid. CBP, calcium-binding protein.

for all steroid hormones. It is not certain whether the association of the calcitriol-receptor complex with chromatin requires an activation step, as do typical steroid hormone-receptor complexes.

B. Calcitriol-Dependent Gene Products: It has been known for several years that the response of intestinal transport to calcitriol requires RNA and protein synthesis. The observation of binding of the calcitriol receptor to chromatin in the nucleus suggests that calcitriol stimulates gene transcription and the formation of specific mRNAs. One such example, the induction of an mRNA that codes for a calcium-binding protein (CBP), has been reported.

There are several cytosolic proteins that bind Ca^{2+} with high affinity. One group, comprised of several proteins of different molecular weight, antigenicity, and tissue location (intestine, skin, and bone), is calcitriol-dependent. Of these, intestinal CBP has been studied most intensively. No CBP is found in the intestine of vitamin D-deficient rats, and the concentration of CBP is highly correlated with the extent of nuclear localization of calcitriol.

C. Effects of Calcitriol on Intestinal Mucosa: The transfer of Ca^{2+} or PO_4^{3-} across the intestinal mucosa requires (1) uptake across the brush border and microvillar membrane; (2) transport across the mucosal cell membrane; and (3) efflux across the basal lateral membrane into the ECF. It is clear that calcitriol enhances one or more of these steps, but the precise mechanism has not been established. CBP was thought to be actively involved until it was observed that Ca^{2+} translocation occurs within 1–2 hours after administration of calcitriol, well before CBP increases in response to calcitriol. CBP may bind Ca^{2+} and protect the mucosa cell against the large fluxes of Ca^{2+} coincident with the transport process. Several investiga-

tors are searching for other proteins that may be involved in Ca^{2+} transport, whereas others suggest that the process, particularly the early increase of Ca^{2+} flux, may be mediated by a membrane change. Metabolites of polyphosphoinositides have been implicated.

D. Effects of Calcitriol on Other Tissues: Much less is known about the action of calcitriol in other tissues. Nuclear receptors are present in bone cells, and the increase in ECF Ca^{2+} following treatment with calcitriol requires concomitant RNA and protein synthesis. The gene products presumably induced by calcitriol are unknown, as is the mechanism by which PTH and calcitriol interact in bone cells.

An interesting role of calcitriol in cellular differentiation is suggested by studies demonstrating that this hormone promotes the conversion of promyelocytic leukemia cells into macrophages. Since osteoclasts are thought to be related to or derived from macrophages, calcitriol may be involved in promoting the differentiation of bone cells.

Pathophysiology

Rickets is a childhood disorder characterized by low plasma calcium and phosphorus levels and by poorly mineralized bone with associated skeletal deformities. Rickets is most commonly due to **vitamin D deficiency.** There are 2 types of **vitamin D-dependent rickets. Type I** is an inherited autosomal recessive trait characterized by a defect in the conversion of $25OH\text{-}D_3$ to calcitriol. **Type II** is an autosomal recessive disorder in which there is apparent absence of the calcitriol receptor.

Vitamin D deficiency in the adult results in **osteomalacia.** Calcium and phosphorus absorption are decreased, as are the ECF levels of these ions. Consequently mineralization of osteoid to form bone is impaired, and such undermineralized bone is structurally weak.

When substantial renal parenchyma is lost or diseased, the formation of calcitriol is reduced and calcium absorption decreases. When hypocalcemia ensues, there is a compensatory increase of PTH, which acts on bone in an attempt to increase ECF Ca^{2+}. The associated extensive bone turnover, structural changes, and symptoms are known as **renal osteodystrophy.** Early treatment with vitamin D will blunt this process.

3. CALCITONIN (CT)

Origin & Structure

Calcitonin (CT) is a 32-amino-acid peptide (Fig 47–6) secreted by the parafollicular C cells of the human thyroid (less commonly, the parathyroid or thymus) or by similar cells located in the ultimobranchial gland of other species. These cells originate in the neural crest and are biochemically related to cells in a variety of other endocrine glands.

The entire CT molecule, including the 7-member N-terminal loop, formed by a Cys–Cys bridge, is required for biologic activity. There is tremendous interspecies variation of the amino acid sequence of CT (human and porcine CT share only 14 of 32 amino acids), but in spite of these differences there is cross-species bioactivity. The most potent naturally occurring CT is isolated from salmon.

Regulation of Secretion

CT and PTH secretion are inversely related (Fig 47–3), and both are controlled by the ECF ionized calcium (and probably magnesium) level. CT secretion increases linearly when calcium concentrations are between 9.5 and 15 mg/dL. Glucagon and pentagastrin are potent CT secretagogues, and the latter is used as a provocative test for medullary thyroid cancer, a malignant tumor of the parafollicular C cells.

Mechanism of Action

CT has a history unmatched by any other hormone. Within a 7-year span (1962–1968), CT was discovered, isolated, sequenced, and synthesized, yet its role in human physiology is still uncertain. Removal of the thyroid in animals does not result in hypercalcemia, and the injection of CT into healthy adults has little calcium-lowering effect.

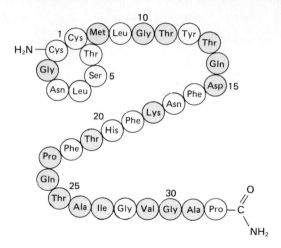

Figure 47–6. Structure of human calcitonin.

In test systems, the primary target of CT is bone, where it decreases matrix resorption and therefore decreases the release of calcium and phosphate. In this action, CT is independent of PTH; it increases bone cAMP, presumably in cells other than those affected by PTH.

CT also has significant effects on phosphate metabolism. It causes phosphate to enter bone cells and the periosteal fluid while reducing the movement of calcium from bone to blood plasma. This entry of phosphate may be accompanied by calcium, since the hypocalcemic effects of CT depend upon phosphate. This action and the apparent ability of CT to inhibit osteoclast-mediated bone resorption probably account for its effectiveness in treating the hypercalcemia of cancer.

Pathophysiology

No clinical manifestations of CT deficiency have been described. CT excess occurs in **medullary thyroid carcinoma (MTC),** which can be sporadic or familial. CT levels in MTC are often thousands of times greater than normal, yet hypocalcemia is a very rare occurrence in MTC. Although the biologic importance of these elevated CT levels is obscure, the observation has important diagnostic implications. Measurement of plasma CT, often coupled to a calcium or pentagastrin challenge, allows early diagnosis of this potentially fatal disease at a time when it is curable.

REFERENCES

Cohn DV, Elting J: Biosynthesis, processing, and secretion of parathormone and secretory protein-1. *Recent Prog Horm Res* 1983;**39**:181.

Copp CH: Parathyroids, calcitonin and control of plasma calcium. *Recent Prog Horm Res* 1964;**20**:59.

DeLuca HF, Schnoes HK: Vitamin D: Recent advances. *Annu Rev Biochem* 1983;**52**:411.

Norman AW, Roth J, Orci L: The vitamin D endocrine system: Steroid metabolism, hormone receptors, and biological response (calcium binding). *Endocr Rev* 1982;**3**:331.

Potts JT Jr, Kronenberg HM, Rosenblatt M: Parathyroid hormone: Chemistry, biosynthesis and mode of action. *Adv Protein Chem* 1982;**35**:323.

Rosenblatt M: Pre-proparathyroid hormone, proparathyroid hormone, and parathyroid hormone. *Clin Orthop* (Oct) 1982; **170**:260.

Talmadge RV, VanderWiel CJ, Matthews JL: Calcitonin and phosphate. *Mol Cell Endocrinol* 1981;**24**:235.

Hormones of the Adrenal Cortex

48

Daryl K. Granner, MD

INTRODUCTION

The adrenal cortex synthesizes dozens of different steroid molecules, but only a few of these have biologic activity. These sort into 3 classes of hormones: glucocorticoids, mineralocorticoids, and androgens. These hormones initiate their actions by combining with specific intracellular receptors, and this complex binds to specific regions of DNA to regulate gene expression. This results in altered rates of synthesis of a small number of proteins, which, in turn, affect a variety of metabolic processes, eg, gluconeogenesis and Na^+ and K^+ balance.

BIOMEDICAL IMPORTANCE

The hormones of the adrenal cortex, particularly the glucocorticoids, are an essential component of adaptation to severe stress. The mineralocorticoids are required for normal Na^+ and K^+ balance. Synthetic analogs of both classes are used therapeutically. In particular, many glucocorticoid analogs are potent anti-inflammatory agents. Excessive or deficient plasma levels of any of these 3 classes of hormones, whether due to disease or therapeutic use, result in serious, sometimes life-threatening, complications. A series of inherited enzyme deficiencies help define the steps involved in steroidogenesis and illustrate the capacity of the adrenal cortex to alter the relative rates of production of these different hormones.

ZONES OF THE ADRENAL CORTEX

The adult cortex has 3 distinct layers or zones. The subcapsular area is called the **zona glomerulosa** and is associated with the production of mineralocorticoids. Next is the **zona fasciculata,** which, with the **zona reticularis,** produces glucocorticoids and androgens.

HORMONES

Some 50 steroids have been isolated and crystallized from adrenal tissue. Most of these are intermediates; only a small number are secreted in significant

Acronyms Used in This Chapter	
ACTH	Adrenocorticotropic hormone
ADH	Antidiuretic hormone
CBG	Corticosteroid-binding globulin
CRH	Corticotropin-releasing hormone
DHEA	Dehydroepiandrosterone
DOC	Deoxycorticosterone
GH	Growth hormone
3β-OHSD	3β-Hydroxysteroid dehydrogenase
PEPCK	Phosphoenolpyruvate carboxykinase
POMC	Pro-opiomelanocortin
PTH	Parathyroid hormone
TBG	Thyroid-binding globulin

amounts; and few have significant hormonal activity. The adrenal cortex makes 3 general classes of steroid hormones, which are grouped according to their dominant action. There is an overlap of biologic activity, since all natural glucocorticoids have mineralocorticoid activity and vice versa. The **glucocorticoids** are 21-carbon steroids with many actions, the most important of which is to promote gluconeogenesis. **Cortisol** is the predominant glucocorticoid in humans, and it is made in the zona fasciculata. **Corticosterone,** made in the zonae fasciculata and glomerulosa, is less abundant in humans, but it is the dominant glucocorticoid in rodents. **Mineralocorticoids** are also 21-carbon steroids. The primary action of these hormones is to promote retention of Na^+ and excretion of K^+ and H^+, particularly in the kidney. **Aldosterone** is the most potent hormone in this class, and it is made exclusively in the zona glomerulosa. The zonae fasciculata and reticularis of the adrenal cortex also produce significant amounts of the androgen precursor **dehydroepiandrosterone** and of the weak androgen **androstenedione.** These steroids are converted into more potent androgens in extra-adrenal tissues and become pathologic sources of androgens when specific steroidogenic enzymes are deficient. Estrogens are not made in the normal adrenal in significant amounts, but in certain cancers of the adrenal they may be produced, and androgens of adrenal origin are important precursors of estrogen (converted by peripheral aromatization) in postmenopausal women.

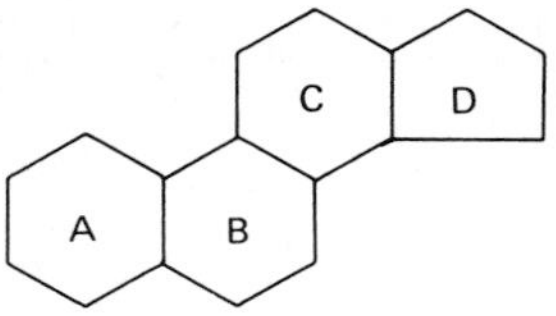

Cyclopentanoperhydrophenanthrene nucleus

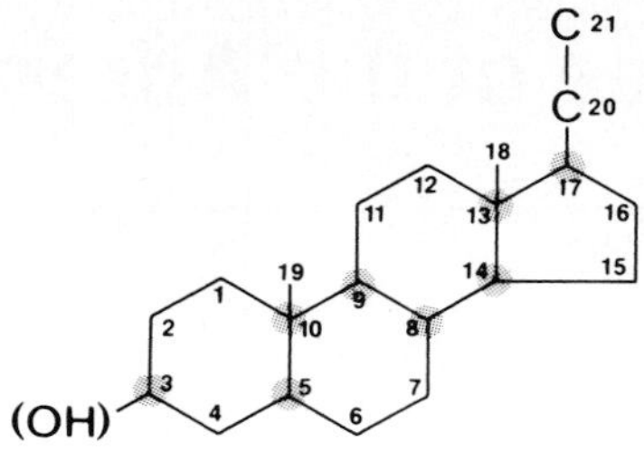

Numbering of the carbon atoms. Asymmetric carbons are shaded.

Figure 48–1. Structural features of steroid molecules.

NOMENCLATURE & CHEMISTRY OF STEROIDS

All steroid hormones have in common the 17-carbon **cyclopentanoperhydrophenanthrene** structure with the 4 rings labeled A–D (Fig 48–1). Additional carbons can be added at positions 10 and 13 or as a side chain attached to C_{17}. Steroid hormones and their precursors and metabolites differ in number and type of substituted groups, number and location of double bonds, and stereochemical configuration. A precise nomenclature for designating these chemical formulations has been devised. The asymmetric carbon atoms (shaded on the C_{21} molecule in Fig 48–1) allow for **stereoisomerism.** The angular methyl groups (C_{19} and C_{18}) at positions 10 and 13 project in front of the ring system and serve as the point of reference. Nuclear substitutions in the same plane as these groups are designated *cis* or "β" and are represented in drawings by solid lines. Substitutions that project behind the plane of the ring system are designated *trans* or "α" and are represented as a dashed line. Double bonds are referred to by the number of the preceding carbon (eg, Δ^3-Δ^4). The steroid hormones are named according to whether they have one angular methyl group (estrane, 18 carbons), 2 angular methyl groups (androstane, 19 carbons), or 2 angular groups plus a 2-carbon side chain at C_{17} (pregnane, 21 carbons). This information (Fig 48–2), together with the glossary provided in Table 48–1, should allow one to understand the chemical names of the natural and synthetic hormones listed in Table 48–2.

BIOSYNTHESIS OF ADRENAL STEROID HORMONES

Steroid Precursors & General Enzymatic Steps

The adrenal steroid hormones are synthesized from cholesterol that is mostly derived from the plasma, but a small portion is synthesized in situ from acetyl-CoA via mevalonate and squalene. Much of the cholesterol in the adrenal is esterified and stored in cytoplasmic lipid droplets. Upon stimulation of the adrenal by ACTH (or cAMP), an esterase is activated, and the free cholesterol formed is transported into the mitochondrion, where a **cytochrome P-450 side chain cleavage enzyme** (P-450_{scc}) converts cholesterol to pregnenolone. Cleavage of the side chain involves sequential hydroxylations, first at C_{22} and then at C_{20}, followed by side chain cleavage (removal of the 6-carbon fragment isocaproaldehyde) to give the 21-carbon steroid (Fig 48–2). An ACTH-dependent protein may bind and activate cholesterol or P-450_{scc}. Aminoglutethimide is a very efficient inhibitor of P-450_{scc} and of steroid biosynthesis.

All mammalian steroid hormones are formed from cholesterol via pregnenolone through a series of reactions that occur in either the mitochondria or endoplasmic reticulum of the adrenal cell. **Hydroxylases** that require molecular oxygen and NADPH are essential,

Table 48–1. Nomenclature of steroids.

Prefix	Suffix	Chemical Nature
Hydroxy-	-ol	Alcohols
Dihydroxy-	-diol	
Oxo-	-one	Ketones (eg, -dione = 2 keto groups)
Cis-		Arrangement of 2 groups in same plane as C_{19}
Trans-		Arrangement of 2 groups in opposing plane to C_{19}
α-		A group *trans* to the 19-methyl
β-		A group *cis* to the 19-methyl
Deoxy-		Lacking a hydroxy group
Iso- or epi-		Isomerism at a C–C, C–OH, or C–H bond, eg, androsterone (5α) versus isoandrosterone (5β)
Dehydro-		Removal of 2 hydrogen atoms to form a double bond
Dihydro-		Addition of 2 hydrogen atoms to a double bond
Allo-		*Trans* configuration of the A and B rings

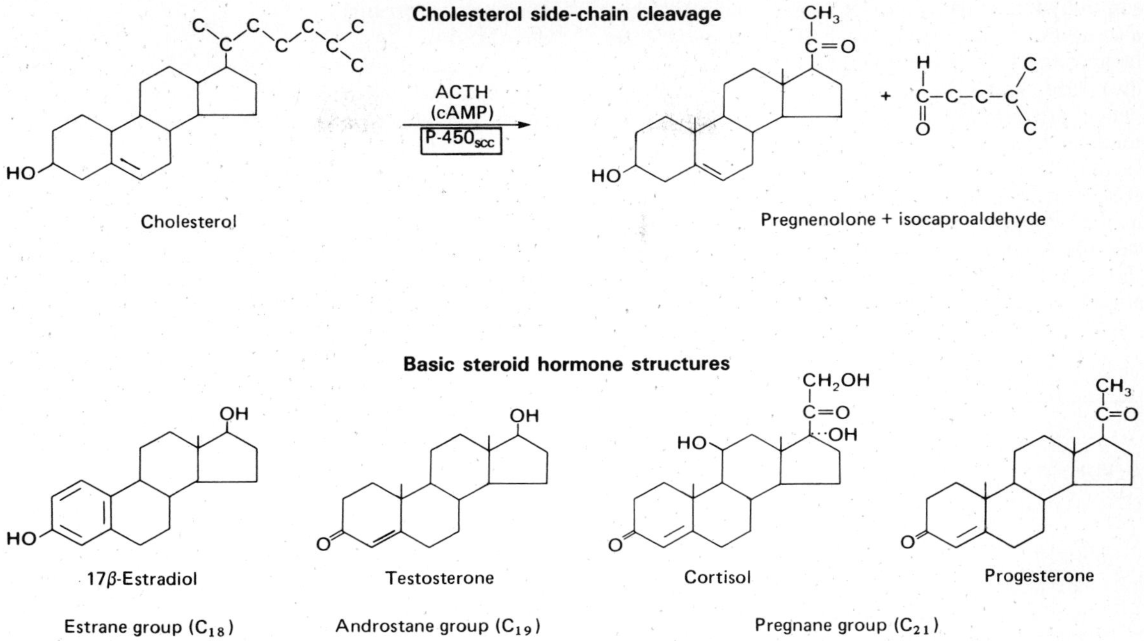

Cholesterol side-chain cleavage

Figure 48–2. Cholesterol side-chain cleavage and basic steroid hormone structures.

Basic steroid hormone structures

17β-Estradiol

Estrane group (C_{18})

Testosterone

Androstane group (C_{19})

Cortisol

Pregnane group (C_{21})

Progesterone

Table 48–2. Trivial and chemical names of some steroids.

Trivial Name	Chemical Name
Aldosterone	11β,21-Dihydroxy-3,20-dioxo-4-pregnen-18-al
Androstenedione	4-Androstene-3,17-dione
Cholesterol	5-Cholesten-3β-ol
Corticosterone (compound B)	11β,21-Dihydroxy-4-pregnene-3,20-dione
Cortisol (compound F)	11β,17α,21-Trihydroxy-4-pregnene-3,20-dione
Cortisone (compound E)	17α,21-Dihydroxy-4-pregnene-3,11,20-trione
Dehydroepiandrosterone (DHEA)	3β-Hydroxy-5-androsten-17-one
11-Deoxycorticosterone (DOC)	21-Hydroxy-4-pregnene-3,20-dione
11-Deoxycortisol (compound S)	17,21-Dihydroxy-4-pregnene-3,20-dione
Dexamethasone	9α-Fluoro-16α-methyl-11β,17α,21-trihydroxypregna-1,4-diene-3,20-dione
Estradiol	1,3,5,(10)-Estratriene-3,17β-diol
Estriol	1,3,5,(10)-Estratriene-3,16α,17β-triol
Estrone	3-Hydroxy-1,3,5(10)-estratriene-3-ol-17-one
Etiocholanolone	3α-Hydroxy-5β-androstan-17-one
9α-Fluorocortisol	9α-Fluoro-11β,17α,21-trihydroxypregn-4-ene,3,20-dione
Prednisolone	11β,17α,21-Trihydroxypregna-1,4-diene-3,20-dione
Prednisone	17α,21-Dihydroxypregna-1,4-diene-3,11,20-trione
Pregnanediol	5β-Pregnane-3α,20α-diol
Pregnanetriol	5β-Pregnane-3α,17α,20α-triol
Pregnenolone	3β-Hydroxy-5-pregnen-20-one
Progesterone	4-Pregnene-3,20-dione
Testosterone	17β-Hydroxy-4-androsten-3-one
Triamcinolone	9α-Fluoro-11β,16α,17α,21-tetrahydroxypregna-1,4-diene-3,20-dione

and **dehydrogenases,** an **isomerase,** and a **lyase** are also necessary for certain steps. There is some cellular specificity in steroidogenesis. For instance, 18-hydroxylase and 18-hydroxysteroid dehydrogenase, which are required for aldosterone synthesis, are found only in glomerulosa cells, so that the biosynthesis of this mineralocorticoid is confined to this region. A schematic representation of the pathways involved in the synthesis of the 3 major classes of adrenal steroids is presented in Fig 48–3. The enzymes are shown in the rectangular boxes, and the modifications at each step are shaded.

Mineralocorticoid Synthesis

Synthesis of aldosterone follows the mineralocorticoid pathway and occurs in the zona glomerulosa. Pregnenolone is converted to progesterone by the action of 2 smooth endoplasmic reticulum enzymes, **3β-hydroxysteroid dehydrogenase (3β-OHSD)** and **$\Delta^{5,4}$ isomerase.** Progesterone is hydroxylated at the C_{21} position to form 11-deoxycorticosterone (DOC), which is an active (Na^+-retaining) mineralocorticoid. The next hydroxylation, at C_{11}, produces corticosterone, which has glucocorticoid activity and is a weak

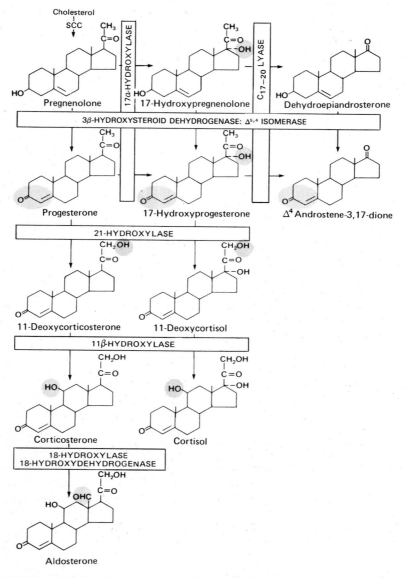

Figure 48–3. Pathways involved in the synthesis of the 3 major classes of adrenal steroids. Enzymes are shown in the rectangular boxes, and the modifications at each step are shaded. (Slightly modified and reproduced, with permission, from Harding BW. Page 1135 in: *Endocrinology.* Vol 2. DeGroot LJ [editor]. Grune & Stratton, 1979.)

mineralocorticoid (it has less than 5% of the potency of aldosterone). In some species (eg, rodents), it is the most potent glucocorticoid. C_{21} hydroxylation is necessary for both mineralocorticoid and glucocorticoid activity, but most steroids with a C_{17} hydroxyl group have more glucocorticoid and less mineralocorticoid action. In the zona glomerulosa, which does not have the smooth endoplasmic reticulum enzyme 17α-hydroxylase, a mitochondrial 18-hydroxylase is present. The **18-hydroxylase** acts on corticosterone to form 18-hydroxycorticosterone, which is changed to aldosterone by the conversion of the 18-alcohol to an aldehyde. This unique distribution of enzymes and the special regulation of the zona glomerulosa (see below) have led some investigators to suggest that, in addition to the adrenal being 2 glands, the adrenal cortex is actually 2 separate organs.

Glucocorticoid Synthesis

Cortisol synthesis requires 3 hydroxylases that act sequentially on the C_{17}, C_{21}, and C_{11} positions. The first 2 reactions are rapid, while C_{11} hydroxylation is relatively slow. If the C_{21} position is hydroxylated first, the action of 17α-hydroxylase is impeded and the mineralocorticoid pathway is followed (forming corticosterone or aldosterone, depending on the cell type). 17α-Hydroxylase is a smooth endoplasmic reticulum enzyme that acts upon either progesterone or, more commonly, pregnenolone. 17α-Hydroxyprogesterone is hydroxylated at C_{21} to form 11-deoxycortisol, which is then hydroxylated at C_{11} to form cortisol, the most potent natural glucocorticoid hormone in humans. The **21-hydroxylase** is a smooth endoplasmic reticulum enzyme, whereas the **11β-hydroxylase** is a mitochondrial enzyme. Steroidogenesis thus involves the repeated shuttling of substrates into and out of the mitochondria of the fasciculata and reticularis cells (Fig 48–4).

Androgen Synthesis

The major androgen or androgen precursor produced by the adrenal cortex is **dehydroepiandrosterone (DHEA).** Most 17-hydroxypregnenolone follows the glucocorticoid pathway, but a small fraction is subjected to oxidative fission and removal of the 2-carbon side chain through the action of 17,20-lyase. This enzyme is found in the adrenals and gonads and acts exclusively on 17α-hydroxy-containing molecules. Adrenal androgen production increases markedly if glucocorticoid biosynthesis is impeded by the lack of one of the hydroxylases (see adrenogenital syndrome, below). Most DHEA is rapidly modified by the addition of sulfate, about half of which occurs in the adrenal and the rest in the liver. DHEA sulfate is inactive, but removal of the sulfate results in reactivation. DHEA is really a prohormone, since the actions of 3β-OHSD and $\Delta^{5,4}$ isomerase convert the weak androgen DHEA into the more potent **androstenedione.** Small amounts of androstenedione are also formed in the adrenal by the action of the lyase on 17α-hydroxyprogesterone. Reduction of androstenedione at the C_{17}

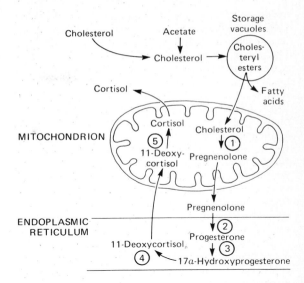

Figure 48–4. Subcellular compartmentalization of glucocorticoid biosynthesis. Adrenal steroidogenesis involves the shuttling of precursors between mitochondria and the endoplasmic reticulum. The enzymes involved are (1) C_{20-22} lyase, (2) 3β-hydroxysteroid dehydrogenase and $\Delta^{5,4}$ isomerase, (3) 17α-hydroxylase, (4) 21-hydroxylase, and (5) 11β-hydroxylase. (Slightly modified and reproduced, with permission, from Harding BW. Page 1135 in: *Endocrinology.* Vol 2. DeGroot LJ [editor]. Grune & Stratton, 1979.)

position results in the formation of **testosterone,** the most potent adrenal androgen. Small amounts of testosterone are produced in the adrenal by this mechanism, but most of this conversion occurs in other tissues.

Small amounts of other steroids can be isolated from adrenal venous blood, including 11-deoxycorticosterone, progesterone, pregnenolone, 17α-hydroxyprogesterone, and a very small amount of estradiol (from the aromatization of testosterone). None of these amounts are important in relation to production from other glands, however.

SECRETION, TRANSPORT, & METABOLISM OF ADRENAL STEROID HORMONES

Secretion

There is little, if any, storage of steroid hormones within the adrenal (or gonad) cell, since these hormones are released into the plasma when they are made. Cortisol release occurs with a periodicity that is regulated by the **diurnal rhythm** of ACTH release.

Plasma Transport

A. Glucocorticoids: Cortisol circulates in plasma in protein-bound and free forms. The main plasma binding protein is an α-globulin called **trans-**

cortin or **corticosteroid-binding globulin (CBG).** CBG is produced in the liver, and its synthesis, like that of thyroid-binding globulin (TBG), is increased by estrogens. CBG binds most of the hormone when plasma cortisol levels are within the normal range; much smaller amounts of cortisol are bound to albumin. The avidity of binding helps determine the biologic half-lives of various glucocorticoids. Cortisol binds tightly to CBG and has a $t_{1/2}$ of 1.5–2 hours, while corticosterone, which binds less tightly, has a $t_{1/2}$ of less than 1 hour. Binding to CBG is not restricted to glucocorticoids. Deoxycorticosterone and progesterone interact with CBG with sufficient affinity to compete for cortisol binding. The unbound, or free, fraction constitutes about 8% of the total plasma cortisol and represents the biologically active fraction of cortisol.

B. Mineralocorticoids: Aldosterone, the most potent natural mineralocorticoid, does not have a specific plasma transport protein, but it forms a very weak association with albumin. Corticosterone and 11-deoxycorticosterone, other steroids with mineralocorticoid effects, bind to CBG. These observations are important in understanding the mechanism of action of aldosterone (see below).

Metabolism & Excretion

A. Glucocorticoids: Cortisol and its metabolites constitute about 80% of the 17-hydroxycorticoids in plasma; the other 20% consist of cortisone and 11-deoxycortisol. About half of the cortisol (as well as cortisone and 11-deoxycortisol) circulates in the form of the reduced dihydro- and tetrahydro- metabolites that are produced from reduction of the A ring double bond by NADPH-requiring hydrogenases and from reduction of the 3-ketone group by a reversible dehydrogenase reaction. Substantial amounts of all of these compounds are also modified by conjugation at the C_3 position with glucuronide or, to a lesser extent, with sulfate. These modifications occur primarily in the liver and make the lipophilic steroid molecule water-soluble and excretable. In humans, most of the conjugated steroids that enter the intestine by biliary excretion are reabsorbed by the enterohepatic circulation. About 70% of the conjugated steroids are excreted in the urine, 20% leave in feces, and the rest exit through the skin.

B. Mineralocorticoids: Aldosterone is very rapidly cleared from the plasma by the liver, no doubt because it lacks a plasma carrier protein. The liver forms tetrahydroaldosterone 3-glucuronide, which is excreted in the urine.

C. Androgens: Androgens are excreted as 17-keto compounds including DHEA (sulfate) as well as androstenedione and its metabolites. Testosterone, secreted in small amounts by the adrenal, is not a 17-keto compound, but the liver converts about 50% of testosterone to androsterone and etiocholanolone, which are 17-keto compounds.

REGULATION OF SYNTHESIS OF ADRENAL STEROID HORMONES

Glucocorticoid Hormones

The secretion of cortisol is dependent on ACTH, which in turn is regulated by corticotropin-releasing hormone (CRH). These hormones are linked by a classic negative feedback loop (Fig 48–5). Excessive levels of free cortisol inhibit CRH release. Abnormally low levels of free cortisol activate the system by enhancing CRH release from the hypothalamus. This 41-amino-acid peptide stimulates the production and release of ACTH (from the precursor pro-opiomelanocortin [POMC] molecule as described in Chapter 45). In the adrenal cortex, ACTH enhances the rate of cholesterol side chain cleavage, the rate-limiting step in steroidogenesis. This completes one side of the negative feedback loop. The restoration of a normal free cortisol level results in reduced release of CRH from the hypothalamus, decreased release of ACTH from the pituitary, and diminished production of cortisol, thereby completing the other side of the loop. This intricate mechanism provides for rapid adjustments of the circulating level of cortisol.

ACTH release (and cortisol secretion) is controlled by neural input from a number of sites within the ner-

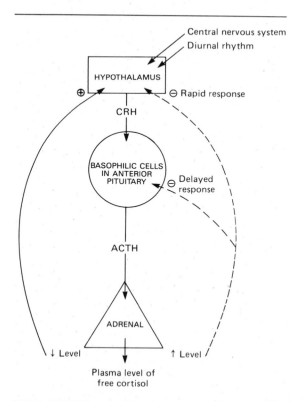

Figure 48–5. Feedback regulation of cortisol biosynthesis. Solid arrows = stimulatory pathways. Dashed arrows = inhibitory pathways.

vous system. There is an endogenous rhythm that controls the release of CRH and therefore ACTH. This circadian cycle is normally set to provide for an increase of plasma cortisol shortly after the onset of sleep. The plasma cortisol level gradually increases during the sleeping hours, peaks shortly after waking, gradually falls over the next several hours, and reaches a nadir in late afternoon and early evening. This general pattern is due to a series of episodic, pulsatile bursts of cortisol release, which are preceded by bursts of ACTH release (see Chapter 45). This is a complex cycle, and it is affected by the period of light exposure and feeding-fasting cycles as well as the sleep-wake cycle. Loss of the diurnal variation in steroid secretion is often seen in association with diseases of the adrenal-pituitary system, some types of depressive illness, and "jet lag."

Cortisol secretion is also affected by physical and emotional stress, apprehension, fear, anxiety, and pain. These responses can override both the negative feedback system and the diurnal rhythm.

Mineralocorticoid Hormones

The production of aldosterone by the glomerulosa cells is regulated in a completely different manner. The primary regulators are the renin-angiotensin system and potassium. Sodium, ACTH, and neural mechanisms are also involved.

A. The Renin-Angiotensin System: This system is involved in the regulation of blood pressure and electrolyte metabolism. The primary hormone in these processes is **angiotensin II,** an octapeptide made from **angiotensinogen** (Fig 48–6). Angiotensinogen, an

Table 48–3. Factors that influence renin release.

Stimulators	Inhibitors
Decreased blood pressure	Increased blood pressure
Change from supine to erect posture	Change from erect to supine posture
Salt depletion	Salt loading
β-Adrenergic agents	β-Adrenergic antagonists
Prostaglandins	Prostaglandin inhibitors
	Potassium
	Vasopressin
	Angiotensin II

α_2-globulin made in liver, is the substrate for renin, an enzyme produced in the **juxtaglomerular cells** of the renal afferent arteriole. The position of these cells makes them particularly sensitive to blood pressure changes, and many of the physiologic regulators of renin release act through renal **baroreceptors** (Table 48–3). The juxtaglomerular cells are also sensitive to changes of Na^+ and Cl^- concentration in the renal tubular fluid; therefore, any combination of factors that decreases fluid volume (dehydration, decreased blood pressure, fluid or blood loss) or decreases NaCl concentration stimulates renin release. Renal sympathetic nerves that terminate in the juxtaglomerular cells mediate the central nervous system and postural effects on renin release independent of the baroreceptor and salt effects, a mechanism that involves the β-adrenergic receptor.

Renin acts upon the substrate angiotensinogen to produce the decapeptide **angiotensin I.** The synthesis of angiotensinogen in liver is enhanced by glucocorticoids and estrogens. Hypertension associated with these hormones may be due in part to increased plasma

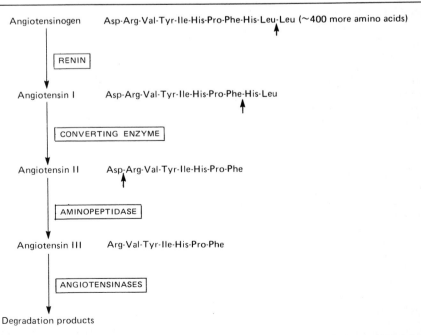

Figure 48–6. Formation and metabolism of angiotensins. Small arrows indicate cleavage sites.

levels of angiotensinogen. Since this protein circulates at about the K_m for its interaction with renin, small changes could markedly affect the generation of angiotensin II.

Angiotensin-converting enzyme, a glycoprotein found in lung, endothelial cells, and plasma, removes 2 carboxy-terminal amino acids from the decapeptide angiotensin I to form angiotensin II in a step that is not thought to be rate-limiting. Various nonapeptide analogs of angiotensin I inhibit converting enzyme and are used to treat **renin-dependent hypertension.** Converting enzyme also degrades **bradykinin,** a potent vasodilator; thus, this enzyme increases blood pressure in 2 distinct ways.

Angiotensin II increases blood pressure by causing vasoconstriction of the arteriole and is the most potent vasoactive substance known. It inhibits renin release from the juxtaglomerular cells and is a potent stimulator of aldosterone production. Although angiotensin II stimulates the adrenal directly, it has no effect on cortisol production.

In some species, angiotensin II is converted to the des-Asp[1] heptapeptide **angiotensin III** (Fig 48–6), an equally potent stimulator of aldosterone production. In humans, the plasma level of angiotensin II is 4 times greater than that of angiotensin III, so most effects are exerted by the octapeptide. Angiotensins II and III are rapidly inactivated by **angiotensinases.**

Angiotensin II binds to specific glomerulosa cell receptors. The concentration of these receptors is "up-regulated" by potassium and the hormone itself and "down-regulated" by a low potassium concentration; hence, this ion plays a central role in the action of angiotensin II on the adrenal. The hormone-receptor interaction does not activate adenylate cyclase, and cAMP does not appear to mediate the action of this hormone. The actions of angiotensin II, which are to stimulate the conversion of cholesterol to pregnenolone and of corticosterone to 18-hydroxycorticosterone and aldosterone, may involve changes in the concentration of intracellular calcium and of phospholipid metabolites by mechanisms similar to those described in Chapter 44. **Prostaglandin biosynthesis** may also be involved, since prostaglandins E_1 and E_2 stimulate aldosterone release while $F_{1\alpha}$ and $F_{2\alpha}$ are inhibitory, a pattern typical of prostaglandin-mediated responses. Indomethacin, an inhibitor of prostaglandin biosynthesis, inhibits basal and angiotensin II-stimulated aldosterone release.

B. Potassium: Aldosterone secretion is sensitive to changes in plasma potassium level; an increase as small as 0.1 meq/L stimulates production, whereas a similar decrease reduces aldosterone production and secretion. The effects of K^+ are independent of Na^+ and the plasma level of angiotensin II. Prolonged hyperkalemia results in hypertophy of the zona glomerulosa and increased sensitivity of glomerulosa cells to the ion. K^+ affects the same enzymatic steps as does angiotensin II, although the mechanism involved is obscure. Like angiotensin II, K^+ does not affect the biosynthesis of cortisol.

C. ACTH: An acute decrease in the level of ACTH (eg, following hypophysectomy or glucocorticoid suppression) has little effect on aldosterone production in humans, but long-term ACTH deficiency may reduce the effect other regulators (angiotensin II, Na^+, K^+) have on aldosterone. ACTH appears to be more important in aldosterone production in other species (eg, rats), and it does stimulate cAMP and the early steps of steroidogenesis in isolated glomerulosa cells.

D. Sodium: Na^+ deficiency increases and Na^+ loading decreases aldosterone production, but these effects are largely mediated through the renin-angiotensin system. Na^+ can influence aldosterone synthesis directly, but this regulation requires high Na^+ concentrations and the effects are small and transient.

METABOLIC EFFECTS OF ADRENAL STEROID HORMONES

Loss of adrenal cortical function results in death unless replacement therapy is instituted. In humans, treatment of adrenal insufficiency with mineralocorticoids is generally not sufficient; glucocorticoids seem to be more critical in this regard. Rats, in contrast, do quite well with mineralocorticoid replacement. Excessive or deficient plasma levels of either of these classes of hormones, whether due to disease or therapeutic use, cause a number of serious complications directly related to their metabolic actions. Only selected biochemical and physiologic effects are discussed in this chapter.

Glucocorticoid Hormones
A. Intermediary Metabolism:
1. Gluconeogenesis–Glucocorticoid hormones are named for their ability to promote glucose production. This is accomplished through a concerted action that involves a number of tissues, several complex biochemical pathways, and both catabolic and anabolic actions.

Glucocorticoids promote hepatic glucose production (1) by increasing the rate of gluconeogenesis; (2) by releasing amino acids, the gluconeogenic substrate, from peripheral tissues such as muscle and lymphoid cells through catabolic actions; and (3) by "permitting" other hormones to stimulate key metabolic processes, including gluconeogenesis, at maximal efficiency. Glucocorticoids are active in fasted or insulin-deficient animals and are required for the maximal action of other hormones in fed animals. Glucocorticoid hormones also inhibit glucose uptake and utilization by extrahepatic tissues, so that the net effect is an elevated plasma glucose level. In normal animals, this is counteracted by the release of insulin, which has effects opposite to those of glucocorticoids. These counterbalancing effects usually result in a normal blood glucose level, but the insulin-deficient animal develops hyperglycemia in response to glucocorticoids. Conversely, the glucocorticoid-deficient animal has decreased glucose production and decreased glycogen reserves and is particularly sensitive to insulin.

Glucocorticoid hormones enhance gluconeogenesis by increasing the amount (and activity) of several key hepatic enzymes. The induction of several enzymes that catalyze rate-limiting steps in amino acid degradation (alanine aminotransferase, tryptophan oxygenase, and tyrosine aminotransferase) has been studied in detail. Although these studies provide examples of how glucocorticoids regulate gene transcription, these enzymes probably have modest effects on gluconeogenesis. The rate-limiting enzyme in gluconeogenesis is **phosphoenolpyruvate carboxykinase (PEPCK)** (see Fig 17–7). The synthesis of PEPCK is increased by glucagon (acting through cAMP) and to a lesser extent by glucocorticoids. These hormones in combination result in an additive response. Insulin inhibits the synthesis of PEPCK and overrides the action of the 2 inducers. All of these effects are exerted at the level of gene transcription.

2. Glycogen synthesis–Glucocorticoids increase glycogen deposition in the liver of fasted or fed animals (an effect that led to the development of an assay for glucocorticoid potency) by promoting the conversion of glycogen synthase from its inactive to active (b to a) form, possibly by activating a phosphatase that enhances this conversion.

3. Lipid metabolism–Excessive amounts of glucocorticoids promote lipolysis in some areas of the body (extremities) and lipogenesis in others (face and trunk). It is not clear whether the lipogenic action is a direct steroid effect or one mediated by the increased plasma insulin that accompanies glucocorticoid excess. In any case, there seems to be some tissue specificity, because not all areas show increased fat deposition or lipolysis.

Plasma free fatty acid levels increase in humans given glucocorticoids. This effect may be due in part to a direct effect on lipolysis, since these hormones do release fatty acids from isolated hepatocytes. Glucocorticoids decrease glucose uptake and utilization in adipose tissue and thus cause a decrease in production of glycerol. Since glycerol is required for the reesterification of fatty acids, this decrease results in the release of fatty acids into the plasma. The net increase of plasma free fatty acids, coupled with their enhanced conversion to ketones, favors the development of ketosis, particularly in insulin-deficient animals. Such responses are important, but the most significant action of the glucocorticoids in lipid metabolism derives from their ability to augment the lipolytic actions of the catecholamines and growth hormone. This "permissive effect" of the glucocorticoids is discussed below.

4. Protein and nucleic acid metabolism–Glucocorticoids generally have anabolic effects on protein and RNA metabolism in liver and catabolic effects at other sites including muscle, lymphoid tissues, adipose tissue, skin, and bone. This pattern is in keeping with the general action of these hormones, which is to provide the optimal milieu for gluconeogenesis. The anabolic effects are reasonably well described; they involve the stimulation of specific gene products, which in turn results in corresponding increases in the rate of synthesis of specific proteins. The catabolic effects have not been extensively characterized at the molecular level.

B. Effects of Glucocorticoids on Host Defense Mechanisms:

1. Immune response–High concentrations of glucocorticoids suppress the host immune response. These steroids kill lymphocytes and cause involution of lymphoid tissue, effects that are dependent upon species and cell type. For example, mouse lymphocytes are much more sensitive to this killing effect than are human lymphocytes, and precursor cells seem to be spared in all species. Glucocorticoids affect the proliferative response of lymphocytes to antigens and, to a lesser extent, to mitogens. Steroids may also affect several other steps in the immune response, including the processing of antigen by macrophages, antibody production by B lymphocytes, suppressor and helper T lymphocyte functions, and antibody metabolism. Most of these effects require glucocorticoids in supraphysiologic levels, eg, in doses used to treat autoimmune diseases or suppress transplant rejection. The role of physiologic levels of these hormones in modulating immune function is not clear.

2. Anti-inflammatory response–The ability of glucocorticoids to suppress the inflammatory response is well known and provides the basis for the major therapeutic use of this class of hormones. In rodents, glucocorticoids decrease the number of circulating lymphocytes, monocytes, and eosinophils, presumably by cell lysis. In humans, this change is not due to cell death but rather to a shift of the cells from the vascular compartment to sites such as bone marrow, lymphoid tissue, and spleen. These hormones enhance the release of polymorphonuclear leukocytes from bone marrow, thus increasing the number of these cells in circulation. Glucocorticoids also inhibit the accumulation of leukocytes at the site of inflammation and cause substances involved in the inflammatory response (eg, kinins, plasminogen-activating factor, prostaglandins, and histamine) to be released from the leukocytes. Glucocorticoids inhibit fibroblast proliferation at the site of an inflammatory response and can inhibit some fibroblast functions, such as the production of collagen and fibronectin. The combination of these effects accounts for the poor wound healing, increased susceptibility to infection, and decreased inflammatory response that are characteristically seen in patients with glucocorticoid excess.

C. Effects of Glucocorticoids on Other Functions:

1. Cardiovascular function–Glucocorticoids are necessary for maintenance of normal blood pressure and cardiac output. These responses may not reflect direct physiologic effects but may represent examples of how glucocorticoids are required for the maximal action of other hormones, the catecholamines in this case (the "permissive effect" of glucocorticoids).

2. Fluid and electrolyte metabolism – Humans with glucocorticoid deficiency are unable to excrete water normally. This may be related to ADH secretion. Glucocorticoids have been observed to inhibit ADH release; thus, in the absence of glucocorticoids, ADH levels may increase and contribute to water retention. The glomerular filtration rate decreases in glucocorticoid deficiency, and this can result in decreased free water clearance.

Glucocorticoids mimic the action of mineralocorticoids in that they increase angiotensinogen, which in turn increases angiotensin II; thus, they increase blood pressure, promote sodium retention, and cause potassium excretion. Some portion of the effects of glucocorticoids on electrolyte metabolism are due to the intrinsic mineralocorticoid activity of these molecules.

3. Growth and development of connective tissue, muscle, and bone – High concentrations of glucocorticoids are catabolic. Fibroblast growth and replication are inhibited, as is formation of collagen and fibronectin. The skin substratum is therefore weakened, leading to development of the thin skin, easy bruising, and poor wound healing characteristic of glucocorticoid excess.

Muscle is the major source of amino acids, the gluconeogenic substrates; hence, it is a primary target of glucocorticoid action. Protein, RNA, and DNA synthesis are inhibited by glucocorticoids, and the rate of degradation of RNA and protein is enhanced. Severe muscle atrophy and weakness are typical findings in patients with prolonged exposure to excessive amounts of corticosteroids.

Glucocorticoids inhibit bone cell replication and function (collagen deposition) and augment the effect of PTH on bone. The net result of prolonged exposure to these hormones is decreased bone mass (**osteoporosis**).

D. Role of Glucocorticoids in the "Stress Response": The **"fight or flight"** response is discussed in Chapter 49, with the adrenal medulla, since it is a complex entwining of a number of physiologic responses primarily mediated by catecholamine hormones. Many components of this response require glucocorticoids for maximal activity.

Glucocorticoids are more directly involved in the response to the acute stress occasioned by surgery, trauma, or infection. Cortisol secretion increases severalfold in such circumstances, and a deficient response markedly reduces the chances for survival. Mineralocorticoid replacement helps in such situations, but glucocorticoid therapy is required for optimal results.

Mineralocorticoid Hormones

Mineralocorticoid hormones act in the kidney to stimulate active Na^+ transport by the distal convoluted tubules and collecting tubules, the net result being Na^+ retention. These hormones also promote the secretion of K^+, H^+, and NH_4^+ by the kidney and affect ion transport in other epithelial tissues including sweat glands, intestinal mucosa, and salivary glands. Aldosterone is 30–50 times more potent than 11-deoxycorticosterone (DOC) and 1000 times more potent than cortisol or corticosterone. As the most potent naturally occurring mineralocorticoid, aldosterone accounts for most of this action in humans. Cortisol, although far less potent, has a much higher production rate and thus has a significant effect on Na^+ retention and K^+ excretion. Since the amount of DOC produced is very small, it is much less important in this regard.

The regulation of cation transport by aldosterone occurs without changes in renal blood flow or glomerular filtration rate, and thus it appears to be a direct effect. RNA and protein synthesis are required for the action of aldosterone, which appears to involve the production of specific gene products (see below).

CLASSIFICATION & MECHANISM OF ACTION OF ADRENAL STEROID HORMONES

Glucocorticoid Hormones

A. Classes of Glucocorticoid Hormones: Glucocorticoid hormones initiate their action in a target cell by interacting with a specific receptor. This interaction results in "activation" of the receptor, a step thought to be necessary for DNA binding. There is generally a high correlation between the association of a steroid with receptor and the elicitation of a given biologic response. This correlation holds true for a wide range of activities, so that a steroid with one-tenth the binding affinity evokes a correspondingly decreased biologic effect at a given steroid concentration. There are generally no "spare receptors" involved in steroid hormone action.

The biologic effect of a steroid depends upon both its ability to bind to the receptor and the concentration of free hormone in the plasma. Cortisol, corticosterone, and aldosterone all bind with high affinity to the glucocorticoid receptor, but in physiologic circumstances cortisol is the dominant glucocorticoid because of its much greater plasma concentration. Corticosterone is an important glucocorticoid in certain pathologic conditions (17α-hydroxylase deficiency), but aldosterone never reaches a concentration in plasma sufficient to exert glucocorticoid effects.

A comparison of the ability of a number of steroids to mediate a well-known glucocorticoid effect, the induction of the liver enzyme tyrosine aminotransferase, revealed that these hormones can be divided into 4 classes; **agonists, partial agonists, antagonists,** and **inactive steroids** (Table 48–4).

B. The Glucocorticoid Receptor: A number of biochemical, immunologic, and genetic studies have led to the formulation of the glucocorticoid receptor illustrated in Fig 48–7. The amino-terminal half contains most of the antigenic sites and has a region that

Table 48–4. Classification of steroids according to their glucocorticoid action.

Agonists
Dexamethasone
Cortisol
Corticosterone
Aldosterone
Partial agonists
11β-Hydroxyprogesterone
21-Deoxycortisol
17α-Hydroxyprogesterone
Progesterone
Antagonists
Testosterone
17β-Estradiol
19-Nortestosterone
Cortisone
Inactive steroids
11α-Hydroxyprogesterone
Androstenedione
11α,17α-Methyltestosterone
Tetrahydrocortisol

modulates promoter function. The carboxy-terminal half contains the DNA- and hormone-binding sites. The DNA-binding domain is closer to the center of the molecule, while the hormone-binding domain is near the carboxyl terminus. The carboxyl half of the receptor is homologous in several regions to the oncogene protein v-*erb*-A, to the DNA binding region of TFIII A, and to a similar region in the estrogen and progesterone receptors. The sequence of the receptor has been deduced by analysis of cDNA molecules. The amino acid sequence in the DNA binding region reveals 2 regions with an abundance of Cys-Lys-Arg residues. By comparing these regions with other known DNA-binding proteins (eg, TFIII A), it is possible to fold the protein into the "fingered" structure (with zinc in its center) that is speculated to insert into a turn of DNA. The human glucocorticoid receptor exists in 2 forms, α and β, of 777 and 742 amino acids, respectively.

C. General Features of Hormone Action: The general features of glucocorticoid hormone action are described in Chapter 44 and illustrated in Fig 44–1. Numerous examples support the concept that this class of hormones affects specific cellular processes by influencing the amount of critical proteins, usually en-

zymes, within the cell. Glucocorticoids accomplish this by regulating the rate of transcription of specific genes in the target cell. This process requires that the steroid-receptor complex bind to specific regions of DNA in the vicinity of the transcription initiation site and that such regions confer specificity to the response. How this interaction actually enhances or inhibits transcription, how tissue specificity is accomplished, and how a given gene can be stimulated in one tissue and inhibited in another are a few of the important questions that remain unanswered.

A brief description of how glucocorticoid hormones affect transcription of mouse mammary tumor virus DNA provides a good illustration of what is known about steroid hormone action. The mammary tumor virus system has been useful because the steroid effect is rapid and large and the molecular biology of the virus has been extensively studied. The glucocorticoid hormone receptor complex binds with high selectivity and specificity to a region, the glucocorticoid regulatory element, a few hundred base pairs upstream from the transcription initiation site. Within the glucocorticoid regulatory element are sequences closely related to the consensus sequence $AGA\frac{A}{T}CAG\frac{A}{T}$ that has been found in the regulatory elements of several glucocorticoid-regulated genes. The receptor-charged glucocorticoid regulatory element enhances transcription initiation of the mouse mammary tumor virus genome and will also activate heterologous promoters. This cis-acting element also works when moved to different regions upstream and downstream, and it works in either the forward or backward orientation. In these respects, the glucocorticoid regulatory element qualifies as a transcription enhancer. These features have also been demonstrated in several other glucocorticoid-regulated genes.

Control of the rate of gene transcription appears to be the major action of the glucocorticoid hormones, but it is not the sole mechanism employed. The ability to measure specific processes has revealed that these hormones also regulate the processing and transport of nuclear transcripts (eg, α_1-acid glycoprotein), the rate of degradation of specific mRNAs (eg, growth hormone and phosphoenolpyruvate-carboxykinase), and posttranslational processing (various mammary tumor

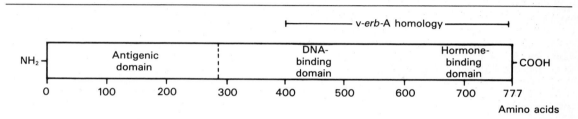

Figure 48–7. Schematic representation of the human glucocorticoid receptor. The human glucocorticoid receptor exists in 2 forms that are 742 and 777 amino acids long, differing at their carboxyl termini. The latter form is shown here. The receptor can be functionally divided into antigenic, DNA-binding, and hormone-binding domains. A region of extensive homology to the oncogene v-*erb*-A is shown.

virus proteins). It would appear that these and other classes of steroid hormones can act at any level of the "information flow" from DNA to protein and that the relative importance of each varies from system to system.

Mineralocorticoid Hormones

A. General Features of Hormone Action: The general features of aldosterone action are similar to those of other steroid hormones (Fig 48–8). Target cells contain specific receptors that bind aldosterone. The hormone-receptor complex binds to chromatin and alters the rate of transcription of specific genes. Although specific gene products have not been isolated, protein and RNA synthesis are known to be required for aldosterone action, and it is presumed that specific proteins are involved in mediating the effects of aldosterone on ion transport.

B. Aldosterone Receptor Binding: Receptors that bind aldosterone with high affinity ($K_d \sim 1$ nmol/L) are found in the cytoplasm and nuclei of target cells. The total binding capacity in the cytoplasm exceeds that in the nucleus by a factor of 80–100; however, the specificity and affinity of nuclear binding are much greater than those of the total cytosol binding activity. In vitro studies have shown that there are actually 3 types of cytosol receptors. Types I and II both bind aldosterone with high affinity, while type III has a low affinity for aldosterone. Type I is a mineralocorticoid receptor, whereas type II appears to be a glucocorticoid receptor that coincidentally binds aldosterone. The type I receptor avidly binds aldosterone but also binds DOC and corticosterone very well. Given the fact that the plasma level of aldosterone is much lower than that of either of the other 2 steroids, one might suppose that these would preferentially occupy the type I sites and that aldosterone would exert little effect. Recall that DOC and corticosterone are avidly bound to corticosteroid-binding globulin, the plasma glucocorticoid transport protein, while aldosterone has no specific carrier protein. Consequently, the effective "free" concentration of aldosterone in plasma is greater than that of either corticosterone or DOC. Aldosterone therefore is readily able to enter cells, and this ensures a competitive advantage for aldosterone with respect to occupying the type I receptor in vivo.

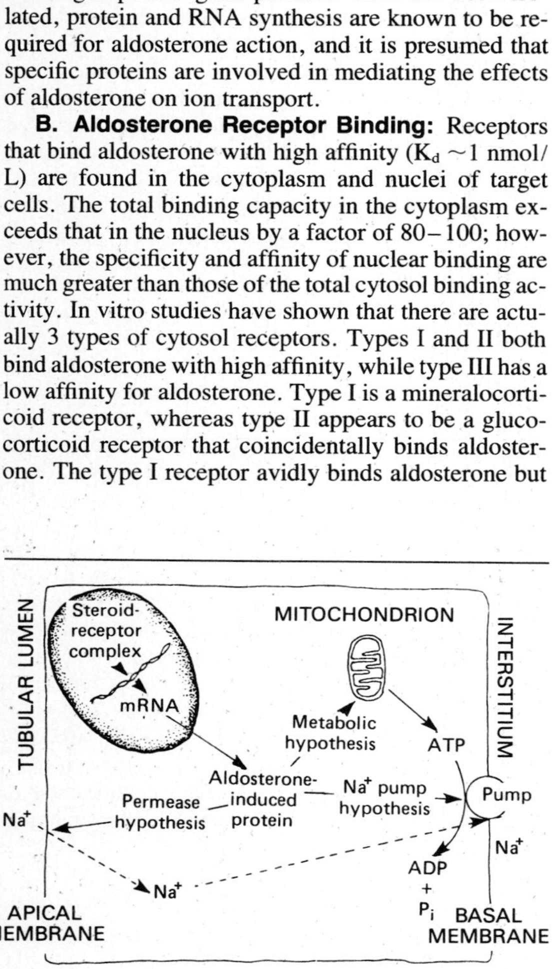

C. Actions of Aldosterone on Ion Transport: The molecular mechanisms of aldosterone action on Na^+ transport have not been elucidated, but several studies point to the general model illustrated in Fig 48–8. Na^+ from the luminal fluid bathing the apical surface of the renal cell enters passively through Na^+ channels. Na^+ is then transported into the interstitial fluid through the serosal side of the cell by the Na^+/K^+-dependent ATPase pump. ATP provides the energy required for this active process.

Aldosterone increases the number of apical membrane Na^+ channels, and this presumably increases intracellular Na^+. Aldosterone also increases the activity of several mitochondrial enzymes, and this could result in the generation of the ATP required to drive the serosal membrane Na^+/K^+ pump. The NADH:NAD ratio increases as a result of aldosterone action, as do the activities of several mitochondrial enzymes including citrate synthase. The increased activity of citrate synthase involves a true induction (perhaps mediated by the gene transcription effects alluded to above), and the temporal increase of this protein correlates highly with the effect of aldosterone on Na^+ transport. Aldosterone has not been shown to have an effect on the Na^+ pump itself; therefore, it appears that the hormone increases the intracellular concentration of Na^+ and creates the energy source required for removal of this ion through the serosal pump. Other mechanisms, involving different aldosterone-regulated proteins, may be involved in the handling of K^+ and H^+.

Figure 48–8. Mechanism of action of aldosterone. The steroid induces the formation of one or more proteins that in turn increase the permeability of the apical (luminal) membrane to Na^+, increase the active transport of Na^+ out of the cell across the basal and lateral membranes to the interstitium, or increase the energy available to the pump. (Modified from Edelman IS: Candidate mediators in the action of aldosterone on Na^+ transport. In: *Membrane Transport Processes.* Vol I. Hoffman JF [editor]. Raven Press, 1978.)

PATHOPHYSIOLOGY OF THE ADRENAL CORTEX

Disorders Involving Glucocorticoid Hormones

Primary adrenal insufficiency (**Addison's disease**) results in hypoglycemia, extreme sensitivity to insulin, intolerance to stress, anorexia, weight loss, nausea, and severe weakness. Patients with Addison's disease have low blood pressure, decreased glomerular filtration rate, and decreased ability to excrete a water load. They often have a history of salt craving. Plasma Na^+ levels are low, K^+ levels are high, and blood lymphocyte and eosinophil counts are increased. Such patients often show increased pigmentation of skin and mucous membranes because of the exaggerated compensatory secretion of ACTH and associated products of the POMC gene. **Secondary adrenal insufficiency** is due to a deficiency of ACTH resulting from tumor, infarction, or infection. This results in a similar metabolic syndrome without hyperpigmentation.

Glucocorticoid excess, commonly called **Cushing's syndrome,** is usually due to the pharmacologic use of steroids, but it may result from an ACTH-secreting pituitary adenoma, from adrenal adenomas or carcinomas, or from the ectopic production of ACTH by a neoplasm. Patients with Cushing's syndrome typically lose the diurnal pattern of ACTH/cortisol secretion. They have hyperglycemia or glucose intolerance (or both) because of accelerated gluconeogenesis. Related to this are severe protein catabolic effects, which result in thinning of the skin, muscle wasting, osteoporosis, extensive lymphoid tissue involution, and generally a negative nitrogen balance. There is a peculiar redistribution of fat, with truncal obesity and the typical "buffalo hump." Resistance to infections and inflammatory responses is impaired, as is wound healing. Several findings, including hypernatremia, hypokalemia, alkalosis, edema, and hypertension, are due to the mineralocorticoid actions of cortisol.

Disorders Involving Mineralocorticoid Hormones

Small adenomas of the glomerulosa cells result in **primary aldosteronism (Conn's syndrome),** the classic manifestations of which include hypertension, hypokalemia, hypernatremia, and alkalosis. Patients with primary aldosteronism do not have evidence of glucocorticoid hormone excess, and plasma renin and angiotensin II levels are suppressed.

Renal artery stenosis, with the attendant decrease in perfusion pressure, can lead to hyperplasia and hyperfunction of the juxtaglomerular cells and cause elevated levels of renin and angiotensin II. This action results in **secondary aldosteronism,** which resembles the primary form, except for the elevated renin and angiotensin II levels.

Congenital Adrenal Hyperplasia

Insufficient amounts of steroidogenic enzymes result in the deficiency of end products, the accumulation of intermediates, and the exaggerated production of steroids from alternative pathways. A common feature of most of these syndromes, which develop in utero, is deficient cortisol production with ACTH overproduction and adrenal hyperplasia—hence, the term **congenital adrenal hyperplasia.** The overproduction of adrenal androgens is another common feature. This hormone excess results in increased body growth, virilization, and ambiguous external genitalia—hence, the alternative designation **adrenogenital syndrome.** The virilizing aspect of the pathophysiology of congenital adrenal hyperplasia can be understood by referring to the discussion of sexual differentiation in Chapter 49. Other symptoms depend on whether there is excessive or deficient production of aldosterone with hypertension or salt wasting, respectively.

Two types of **21-hydroxylase deficiency** (partial, or simple virilizing, and complete, or salt wasting) account for more than 90% of cases of congenital adrenal hyperplasia, and most of the rest are due to **11β-hydroxylase deficiency.** Only a few cases of other deficiencies (3β-hydroxysteroid dehydrogenase, 17α-hydroxylase, cholesterol desmolase, 18-hydroxylase, and 18-dehydrogenase) have been described. The **18-hydroxylase and -dehydrogenase deficiencies** affect only aldosterone biosynthesis and so do not cause adrenal hyperplasia. The **cholesterol desmolase deficiency** prevents any steroid biosynthesis and so is usually incompatible with extrauterine life.

REFERENCES

Baxter JD, Forsham PH: Tissue effects of glucocorticoids. *Am J Med* 1972;**53**:573.

Chandler VL et al: DNA sequences bound specifically by glucocorticoid receptor in vitro render a heterologous promoter responsive in vivo. *Cell* 1983;**33**:489.

Gill GN: Mechanism of ACTH action. *Metabolism* 1972; **21**:571.

Granner DK: The role of glucocorticoid hormones as biological amplifiers. In: *Glucocorticoid Hormone Action.* Baxter JD, Rousseau GG (editors). Springer-Verlag, 1979.

Morris DJ: The metabolism and mechanism of action of aldosterone. *Endocr Rev* 1981;**2**:234.

Samuels HH, Tomkins GM: Relation of steroid structure to enzyme induction in hepatoma tissue culture cells. *J Mol Biol* 1970;**52**:57.

Weinberger C et al: Domain structure of human glucocorticoid receptor and its relationship to the v-*erb*-A oncogene product. *Nature* 1985;**318**:670.

Yamamoto KR, Alberts BM: Steroid receptors: Elements for modulation of eukaryotic transcription. *Annu Rev Biochem* 1976;**45**:721.

49 Hormones of the Adrenal Medulla

Daryl K. Granner, MD

INTRODUCTION

The **sympathoadrenal** system consists of the **parasympathetic** nervous system with its cholinergic pre- and postganglionic nerves, the **sympathetic** nervous system with cholinergic preganglionic and adrenergic postganglionic nerves, and the **adrenal medulla.** The latter is actually an extension of the sympathetic nervous system, since preganglionic fibers from the splanchnic nerve terminate in the adrenal medulla, where they innervate the chromaffin cells that produce the catecholamine hormones **dopamine, norepinephrine,** and **epinephrine.** The adrenal medulla is thus a specialized ganglion without axonal extensions. Its chromaffin cells synthesize, store, and release products that act on distant sites, so that it also functions as an endocrine organ—a perfect illustration of the enmeshing of the nervous and endocrine systems alluded to in Chapter 44.

BIOMEDICAL IMPORTANCE

The hormones of the sympathoadrenal system, while not necessary for life, are required for adaptation to acute and chronic stress. Epinephrine, norepinephrine, and dopamine are the major elements in the **"fight or flight" response** (as occurs, for example, when one encounters a bear in a blueberry patch). The response to such a threat involves an acute, integrated adjustment of many complex processes in the organs vital to the response (brain, muscles, cardiopulmonary system, and liver) at the expense of other organs that are less immediately involved (skin, gastrointestinal system, and lymphoid tissue). The processes affected are listed in Table 49–1. In this integrated metabolic response, epinephrine (1) rapidly provides fatty acids as the primary fuel for muscle action; (2) mobilizes glucose as an energy source for the brain by increasing glycogenolysis and gluconeogenesis in the liver and by decreasing glucose uptake in the muscle and other organs; and (3) decreases insulin release, which also prevents glucose from being taken up by peripheral tissues and thus preserves it for the central nervous system. Catecholamines do not facilitate the "fight or flight" response alone but are aided by the glucocorticoids, growth hormone, vasopressin, angiotensin II, and glucagon.

Acronyms Used in This Chapter

COMT	Catechol-O-methyltransferase
DBH	Dopamine β-hydroxylase
MAO	Monoamine oxidase
PNMT	Phenylethanolamine-N-methyltransferase
VMA	Vanillylmandelic acid

BIOSYNTHESIS OF CATECHOLAMINES

The catecholamine hormones dopamine, norepinephrine, and epinephrine are 3,4-dihydroxy derivatives of phenylethylamine. These amines are synthesized in the chromaffin cells of the adrenal medulla, so named because they contain granules that develop a red-brown color when exposed to potassium dichromate. Collections of these cells are also found in the heart, liver, kidney, gonads, adrenergic neurons of the

Table 49–1. Physiologic responses in the "fight or flight" response.

Organ	Process or Result
Brain	Increased blood flow Increased glucose metabolism
Cardiovascular system	Increased rate and force of contraction Peripheral vasoconstriction
Pulmonary system	Increased oxygen supply Bronchodilatation Increased ventilation
Muscle	Increased glycogenolysis Increased contraction
Liver	Increased glucose production ↑ Gluconeogenesis ↑ Glycogenolysis ↓ Glycogen synthesis
Adipose tissue	Increased lipolysis ↑ Fatty acids and glycerol
Skin Skeleton Gastrointestinal and genitourinary tracts Lymphoid tissue	Decreased blood flow Decreased glucose uptake and utilization Decreased protein synthesis Increased proteolysis

postganglionic sympathetic system, and central nervous system.

The major product of the adrenal medulla is epinephrine. This compound constitutes about 80% of the catecholamines in the medulla, and it is not made in extramedullary tissue. In contrast, most of the norepinephrine present in organs innervated by sympathetic nerves is made in situ (~80% of the total), and most of the rest is made in other nerve endings and reaches the target sites via the circulation. Epinephrine and norepinephrine may be produced and stored in different cells in the adrenal medulla and other chromaffin tissues.

The conversion of tyrosine to epinephrine requires 4 sequential steps: (1) ring hydroxylation; (2) decarboxylation; (3) side chain hydroxylation; and (4) N-methylation. The biosynthetic pathway and the enzymes involved are illustrated in Fig 49–1, and a schematic representation is illustrated in Fig 49–2.

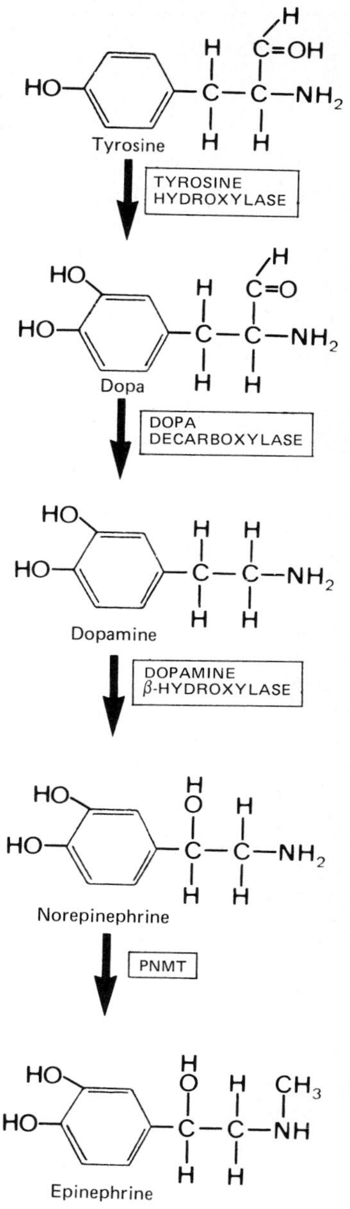

Tyrosine Hydroxylase

Tyrosine is the immediate precursor of catecholamines, and tyrosine hydroxylase is the rate-limiting enzyme in catecholamine biosynthesis. Tyrosine hydroxylase is found in both soluble and particle-bound forms; it functions as an oxidoreductase, with tetrahydropteridine as a cofactor, to convert L-tyrosine to L-dihydroxyphenylalanine (L-dopa). As the rate-limiting enzyme, tyrosine hydroxylase is regulated in a variety of ways. The most important mechanism involves feedback inhibition by the catecholamines, which compete with the enzyme for the pteridine cofactor by forming a Schiff base with the latter. Tyrosine hydroxylase is also competitively inhibited by a series of tyrosine derivatives, including α-methyltyrosine. This compound is occasionally used to treat catecholamine excess in pheochromocytoma, but other agents are more effective and have fewer side effects. A third group of compounds inhibit tyrosine hydroxylase by chelating iron and thus removing available cofactor. An example is α,α'-dipyridyl.

Catecholamines cannot cross the blood-brain barrier; hence, in the brain they must be synthesized locally. In certain central nervous system diseases, eg, Parkinson's disease, there is a local deficiency of dopamine synthesis. L-Dopa, the precursor of dopamine, readily crosses the blood-brain barrier and so is an important agent in the treatment of Parkinson's disease.

Dopa Decarboxylase

In contrast to tyrosine hydroxylase, which is found only in tissues that synthesize catecholamines, dopa decarboxylase is found in all tissues. This soluble enzyme requires pyridoxal phosphate for the conversion of L-dopa to 3,4-dihydroxyphenylethylamine (dopamine). Compounds that resemble L-dopa, such as α-methyldopa, are competitive inhibitors of this reaction. Halogenated compounds form a Schiff base with L-dopa and also inhibit the decarboxylase reaction.

α-Methyldopa and other related compounds, such as 3-hydroxytyramine (from tyramine), α-methylty-

Figure 49–1. Biosynthesis of catecholamines. PNMT, phenylethanolamine-N-methyltransferase. (Modified and reproduced, with permission, from Goldfien A: The adrenal medulla. In: *Basic & Clinical Endocrinology*, 2nd ed. Greenspan FS, Forsham PH [editors]. Appleton & Lange, 1986.)

rosine, and metaraminol, are effective in treating some kinds of hypertension. The antihypertensive action of these metabolites is thought to result from their ability to stimulate the α-adrenergic receptor (see below) of the corticobulbar system in the central nervous system. This results in reduced peripheral sympathetic nerve discharge and decreased blood pressure.

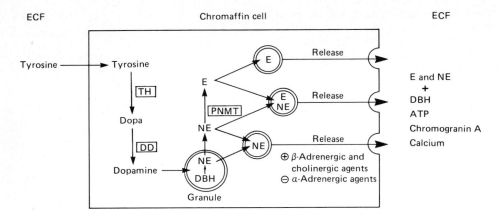

Figure 49–2. Schematic representation of catecholamine biosynthesis. TH, tyrosine hydroxylase; DD, dopa decarboxylase; PNMT, phenylethanolamine-N-methyltransferase; DBH, dopamine β-hydroxylase; ATP, adenosine triphosphate. The biosynthesis of catecholamines occurs within the cytoplasm and in various granules of the adrenal medullary cell. Some granules contain epinephrine (E), others have norepinephrine (NE), while still others have both hormones. Upon stimulation all the contents of the granules are released into the extracellular fluid (ECF).

Dopamine β-Hydroxylase

Dopamine β-hydroxylase (DBH) is a mixed-function oxidase that catalyzes the conversion of dopamine to norepinephrine. DBH uses ascorbate as an electron donor, copper at the active site, and fumarate as modulator. DBH is in the particulate fraction of the medullary cells, probably in the secretion granule; thus, the conversion of dopamine to norepinephrine occurs in this organelle. DBH is released from the adrenal medulla or nerve endings with norepinephrine, but (unlike norepinephrine) it cannot reenter nerve terminals via the reuptake mechanism.

Phenylethanolamine-N-Methyltransferase

The soluble enzyme phenylethanolamine-N-methyltransferase (PNMT) catalyzes the N-methylation of norepinephrine to form epinephrine in the epinephrine-forming cells of the adrenal medulla. Since PNMT is soluble, it is assumed that norepinephrine-to-epinephrine conversion occurs in the cytoplasm. The synthesis of PNMT is induced by glucocorticoid hormones that reach the medulla via the intra-adrenal portal system. This system provides for a 100-fold steroid concentration gradient over systemic arterial blood, and this high intra-adrenal concentration appears to be necessary for the induction of PNMT.

STORAGE & RELEASE OF CATECHOLAMINES

Storage

The adrenal medulla contains the **chromaffin granules**—organelles capable of the biosynthesis, uptake, storage, and secretion of catecholamines. These granules contain a number of substances in addition to the catecholamines, including ATP-Mg^{2+}, Ca^{2+},

DBH, and the protein chromogranin A. Catecholamines enter the granule via an ATP-dependent transport mechanism and bind this nucleotide in a 4:1 ratio (hormone:ATP). Norepinephrine is stored in these granules but can exit to be N-methylated; the epinephrine formed then enters a new population of granules.

Release

Neural stimulation of the adrenal medulla results in the fusion of the membranes of the storage granules with the plasma membrane, and this leads to the exocytotic release of norepinephrine and epinephrine. This is a calcium-dependent process and, like most exocytotic events, is stimulated by cholinergic and β-adrenergic agents and inhibited by α-adrenergic agents (Fig 49–2). Catecholamines and ATP are released in proportion to their intragranular ratio, as are the other contents including DBH, calcium, and chromogranin A.

Neuronal reuptake of catecholamines is an important mechanism for conserving these hormones and for quickly terminating hormonal or neurotransmitter activity. The adrenal medulla, unlike the sympathetic nerves, does not have a mechanism for the reuptake and storage of discharged catecholamines. The epinephrine discharged from the adrenal goes to the liver and skeletal muscle but then is rapidly metabolized. Very little adrenal norepinephrine reaches distal tissues. Catecholamines circulate in plasma in a loose association with albumin. They have an extremely short biologic half-life (10–30 seconds).

METABOLISM OF CATECHOLAMINES

Very little epinephrine (< 5%) is excreted in the urine. Catecholamines are rapidly metabolized by catechol-O-methyltransferase and monoamine oxidase to form the inactive O-methylated and deaminated metabolites (Fig 49–3). Most catecholamines are substrates for both of these enzymes, and these reactions can occur in any sequence.

Catechol-O-methyltransferase (COMT) is a cytosol enzyme found in many tissues. It catalyzes the addition of a methyl group, usually at the 3 position (meta) on the benzene ring, to a variety of catecholamines. The reaction requires a divalent cation, and S-adenosylmethionine is the methyl donor. The result of this reaction, depending on the substrate, is the production of homovanillic acid, normetanephrine, and metanephrine.

Monoamine oxidase (MAO) is an oxidoreductase that deaminates monoamines. It is located in many tissues, but it occurs in highest concentrations in the liver, stomach, kidney, and intestine. At least 2 isozymes of MAO have been described. MAO-A is found in neural tissue and deaminates serotonin, epinephrine, and norepinephrine, while MAO-B is found in extraneural tissues and is most active against 2-phenylethylamine and benzylamine. Dopamine and tyramine are metabolized by both forms. Much research effort is directed at correlating affective disorders with increases or decreases of the activity of these isozymes. MAO inhibitors have been used to treat hypertension and depression, but serious reactions with foods or drugs that contain sympathomimetic amines limit their usefulness.

O-Methoxylated derivatives are further modified by conjugation with glucuronic or sulfuric acid.

A bewildering number of metabolites of cate-cholamines are formed. Two classes of these have diagnostic significance, since they are found in readily measurable amounts in urine. **Metanephrines** represent the methoxy derivatives of epinephrine and norepinephrine, while the O-methylated deaminated product of epinephrine and norepinephrine is **3-methoxy-4-hydroxymandelic acid** (also called **vanillylmandelic acid [VMA]**) (Fig 49–3). The concentration of metanephrines or VMA in urine is elevated in more than 95% of patients with pheochromocytoma. These tests have excellent diagnostic precision, particularly when coupled with a measurement of plasma or urine catecholamines.

REGULATION OF SYNTHESIS OF CATECHOLAMINES

Stimulation of the splanchnic nerve, which supplies the preganglionic fibers to the adrenal medulla, results in the exocytotic release of catecholamines, the granule carrier protein, and DBH. Such stimulation is controlled by the hypothalamus and brain stem, but the exact feedback loop has not been described.

Nerve stimulation also results in increased synthesis of catecholamines. Norepinephrine synthesis increases after acute stress, but the amount of tyrosine hydroxylase is unchanged even though tyrosine hydroxylase activity increases. Tyrosine hydroxylase is a substrate for cAMP-dependent protein kinase, and so this activation may involve phosphorylation. Prolonged stress accompanied by chronic sympathetic nerve activity results in an induction (increased

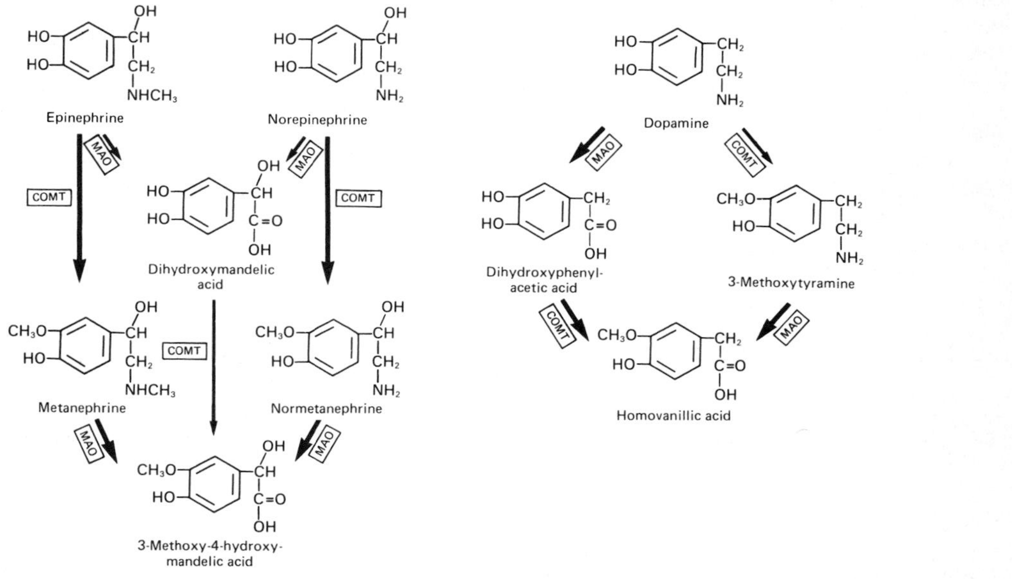

Figure 49–3. Metabolism of catecholamines by catechol-O-methyltransferase (COMT) and monoamine oxidase (MAO). (Reproduced, with permission, from Goldfien A: The adrenal medulla. In: *Basic & Clinical Endocrinology*, 2nd ed. Greenspan FS, Forsham PH [editors]. Appleton & Lange, 1986.)

amount) of tyrosine hydroxylase. A similar induction of DBH has also been reported. The induction of these enzymes of the catecholamine biosynthetic pathway is a means of adapting to physiologic stress and depends upon neural (tyrosine hydroxylase and DBH induction) and endocrine (PNMT induction) factors.

CLASSIFICATION & MECHANISM OF ACTION OF CATECHOLAMINES

Classification

The mechanism of action of the catecholamines has attracted the attention of investigators for nearly a century. Indeed, many of the general concepts of receptor biology and hormone action can be traced to these early studies.

The catecholamines act through 2 major classes of receptors. These are designated α-adrenergic and β-adrenergic, and each consists of 2 subclasses, ie, alpha$_1$, alpha$_2$, beta$_1$, and beta$_2$. This classification is based on the relative order of binding of various agonists and antagonists. Epinephrine binds to and activates both alpha and beta receptors, so that its action in a tissue which has both depends on the relative affinity of these receptors for the hormone. Norepinephrine at physiologic concentrations primarily binds to alpha receptors.

The β-Adrenergic Receptor

Molecular cloning of the gene and cDNA for the mammalian β-adrenergic receptor revealed some surprising features. First, the gene has no introns and thus joins the histone and interferon genes as the only mammalian genes to lack these structures. Second, the β-adrenergic receptor is closely homologous to rhodopsin (in 3 peptide regions, at least), the protein that initiates the process which converts light into the visual response. Further similarities are discussed below.

Mechanism of Action

Three of these adrenergic receptor subgroups are coupled to the adenylate cyclase system. Hormones that bind to the β_1 and β_2 receptors activate adenylate cyclase, whereas hormones that bind to α_2 receptors inhibit this enzyme (see Fig 44–3 and Table 44–3). Catecholamine binding induces the coupling of the receptor to a G protein that then binds GTP. This either stimulates (G_s) or inhibits (G_i) adenylate cyclase, thus stimulating or inhibiting the synthesis of cAMP. The response terminates when the α subunit-associated GTPase hydrolyzes the GTP (see Fig 44–2). α_1 Receptors are coupled to processes that alter intracellular calcium concentrations or modify phosphatidylinositide metabolism (or both). (See Chapter 44.) A separate G protein complex may be involved in this response.

There is functional similarity between the catecholamine receptor and the visual response system. The stimulation of rhodopsin by light couples it to transducin, a G protein complex whose α subunit also binds GTP. The activated G protein in turn activates a phosphodiesterase that hydrolyzes cGMP. This results in the closure of ion channels on the rod cell membrane and produces the visual response. The response terminates when the α subunit-associated GTPase hydrolyzes the bound GTP. A partial list of the biochemical and physiologic effects mediated by each of these receptors is provided in Table 49–2.

Activation of phosphoproteins by cAMP-dependent protein kinase (see Fig 44–4) accounts for many of the biochemical effects of epinephrine. In muscle, and to a lesser extent in liver, epinephrine stimulates glycogenolysis by activating a protein kinase that, in turn, activates the phosphorylase cascade (see Fig 19–7). Conversely, phosphorylation of glycogen synthase decreases glycogen synthesis. In heart muscle, epinephrine increases cardiac output by increasing the force **(inotropic effect)** and rate **(chronotropic effect)** of contraction, which are also related to increased cAMP.

In adipose tissue, epinephrine increases cAMP, which converts the hormone-sensitive lipase into the active (phosphorylated) form. This enzyme promotes lipolysis and release of fatty acids into the circulation. These fatty acids serve as an energy source in muscle and can activate gluconeogenesis in liver.

Table 49–2. Actions mediated through various adrenergic receptors.

Alpha$_1$	Alpha$_2$	Beta$_1$	Beta$_2$
Increased glycogenolysis	Smooth muscle relaxation	Stimulation of lipolysis	Increased hepatic gluconeogenesis
Smooth muscle contraction	Gastrointestinal tract	Myocardial contraction	Increased hepatic glycogenolysis
Blood vessels	Smooth muscle contraction	Increased rate	Increased muscle glycogenolysis
Genitourinary tract	Some vascular beds	Increased force	Increased release of:
	Inhibition of:		Insulin
	Lipolysis		Glucagon
	Renin release		Renin
	Platelet aggregation		Smooth muscle relaxation
	Insulin secretion		Bronchi
			Blood vessels
			Genitourinary tract
			Gastrointestinal tract

PATHOPHYSIOLOGY OF THE ADRENAL MEDULLA

Pheochromocytomas are tumors of the adrenal medulla that are usually not detected unless they produce and secrete enough epinephrine or norepineph-rine to cause a severe hypertension syndrome. The ratio of norepinephrine to epinephrine is often increased in pheochromocytoma. This may account for differences in clinical presentation, since norepinephrine is thought to be primarily responsible for hypertension and epinephrine for hypermetabolism.

REFERENCES

Benovic JL et al: β-Adrenergic receptor kinase: Identification of a novel protein kinase that phosphorylates the agonist-occupied form of the receptor. *Proc Natl Acad Sci USA* 1986;**83**:2797.

Cryer PE: Diseases of the adrenal medulla and sympathetic nervous system. Pages 511–550 in: *Endocrinology and Metabolism*. Felig P et al (editors). McGraw-Hill, 1981.

Dixon RAF et al: Cloning of the gene and cDNA for mammalian β-adrenergic receptor and homology with rhodopsin. *Nature* 1986;**321**:75.

Exton JH: Mechanisms involved in α-adrenergic phenomena: Role of calcium ion in actions of catecholamines in liver and other tissues. *Am J Physiol* 1980;**238**:E3.

Kaupp UB: Mechanism of photoreception in vertebrate vision. *Trends Biochem Sci* 1986;**11**:43.

Perlman FL, Chalfie M: Catecholamine release from the adrenal medulla. *Clin Endocrinol Metab* 1977;**6**:551.

Sibley DR, Lefkowitz RJ: Molecular mechanisms of receptor desensitization using the β-adrenergic receptor-coupled adenylate cyclase system as a model. *Nature* 1985;**317**:124.

Stiles GL, Caron MG, Lefkowitz RK: The β-adrenergic receptor: Biochemical mechanisms of physiological regulation. *Physiol Rev* 1984;**64**:661.

Young JB, Landsberg L: Catecholamines and intermediary metabolism. *Clin Endocrinol Metab* 1977;**6**:599.

Hormones of the Gonads

Daryl K. Granner, MD

INTRODUCTION

The gonads are bifunctional organs that produce germ cells and the sex hormones. These 2 functions are closely approximated, for high local concentrations of the sex hormones are required for germ cell development. The ovaries produce ova and the steroid hormones estrogen and progesterone; the testes produce spermatozoa and testosterone. As in the adrenal, a number of steroids are produced, but only a few are active as hormones. The production of these hormones is tightly regulated through a feedback loop that involves the pituitary and the hypothalamus. The gonadal hormones act by a nuclear mechanism similar to that employed by the adrenal steroid hormones.

BIOMEDICAL IMPORTANCE

Proper functioning of the gonads, which depends upon structural and hormonal integrity, is crucial for reproduction and, hence, survival of the species. Conversely, an understanding of the endocrine physiology and biochemistry of the reproductive process is the basis of many approaches to contraception. The gonadal hormones have other important actions; for example, they are anabolic and thus required for maintenance of metabolism in skin, bone, and muscle.

HORMONES OF THE TESTES

The testes are bifunctional organs that produce **testosterone** (the male sex hormone) and **spermatozoa** (the male germ cells). These functions are carried out by 3 specialized cell types; (1) the **spermatogonia** and more differentiated germ cells, which are located in the seminiferous tubules; (2) the **Leydig cells** (also called interstitial cells), which are scattered in the connective tissue between the coiled seminiferous tubules and which produce testosterone in response to LH; and (3) the **Sertoli cells,** which form the basement membrane of the seminiferous tubules and provide the environment necessary for germ cell differentiation and maturation, including the secretion of androgen-binding protein (ABP) in response to FSH. Spermatogenesis is stimulated by FSH and LH from the pituitary. It

Acronyms Used in This Chapter	
ACTH	Adrenocorticotropic hormone
ABP	Androgen-binding protein
CBG	Corticosteroid-binding globulin
DHEA	Dehydroepiandrosterone
DHT	Dihydrotestosterone
E_2	Estradiol
FSH	Follicle-stimulating hormone
GnRH	Gonadotropin-releasing hormone
hCG	Human chorionic gonadotropin
hCS	Human chorionic somatomammotropin
LH	Luteinizing hormone
MIF	Müllerian inhibiting factor
3β-OHSD	3β-Hydroxysteroid dehydrogenase
17β-OHSD	17β-Hydroxysteroid dehydrogenase
PL	Placental lactogen
SHBG	Sex hormone-binding globulin
TBG	Thyroid-binding globulin
TEBG	Testosterone-estrogen-binding globulin

requires an environment conducive to germ cell differentiation and a concentration of testosterone in excess of that found in the systemic circulation—requirements that are met by local secretion of ABP and testosterone by the Sertoli and Leydig cells, respectively.

BIOSYNTHESIS & METABOLISM OF TESTICULAR HORMONES

Synthesis

Testicular androgens are synthesized in the interstitial tissue by the Leydig cells: these cells contain virtually all of the 3β-hydroxysteroid dehydrogenase found in the testis, and this enzyme catalyzes the key step in testosterone biosynthesis.

A. Biosynthetic Pathways:

1. Testosterone–The immediate precursor of the gonadal steroids, as with the adrenal steroids, is cholesterol. The rate-limiting step, as in the adrenal, is cholesterol side chain cleavage. The conversion of cholesterol to pregnenolone is identical in adrenal, ovary, and testis. In the latter 2 tissues, however, the reaction is promoted by LH rather than ACTH.

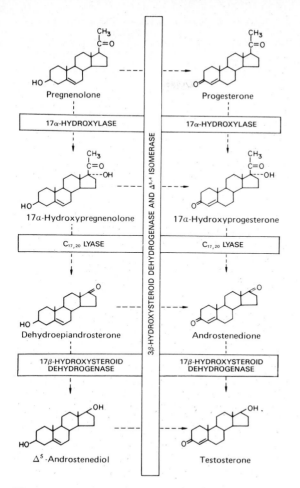

Figure 50–1. Pathways of testosterone biosynthesis. The pathway on the left side of the figure is called the Δ^5 or dehydroepiandrosterone pathway; the pathway on the right side is called the Δ^4 or progesterone pathway.

The conversion of pregnenolone to testosterone requires the action of 5 enzymes: (1) **3β-hydroxysteroid dehydrogenase (3β-OHSD);** (2) **$\Delta^{5,4}$ isomerase;** (3) **17α-hydroxylase;** (4) **C$_{17-20}$ lyase;** and (5) **17β-hydroxysteroid dehydrogenase (17β-OHSD).** This sequence, referred to as the progesterone (or Δ^4) pathway, is shown on the right side of Fig 50–1. Pregnenolone can also be converted to testosterone by the dehydroepiandrosterone (or Δ^5) pathway, which is illustrated on the left side of Fig 50–1. The Δ^4 route appears to be preferred in human testes; however, since these are seldom available for study, most information about these pathways comes from studies in other animals. There may be significant species differences.

The 5 enzymes are localized in the microsomal fraction in rat testes, and **there is a close functional association between the activities of 3β-OHSD and $\Delta^{5,4}$ isomerase and between those of 17α-hydroxylase and C$_{17-20}$ lyase.** These enzyme pairs are shown in the general reaction sequence in Fig 50–1 and in the schematic representation of androgen biosynthesis in the testicular microsomal membrane in Fig 50–2. The latter shows how the various substrates for testosterone biosynthesis can enter the microsomal compartment and how they might proceed via the Δ^4 pathway from one reaction to the next. Since there are 4 potential substrates for what appears to be a single 3β-OHSD, multiple alternative pathways exist; thus, the route taken probably depends on the substrate concentration in the vicinity of the various enzymes. Partitioning in the microsomal membrane may provide these gradients.

2. Other testicular hormones–Dihydrotestosterone (DHT) is formed from testosterone by the reduction of the A ring through the action of the enzyme **5α-reductase.** Human testes secrete about 50–100 μg of DHT per day, but most DHT is derived from peripheral conversion (see below).

The testes also make small but significant amounts

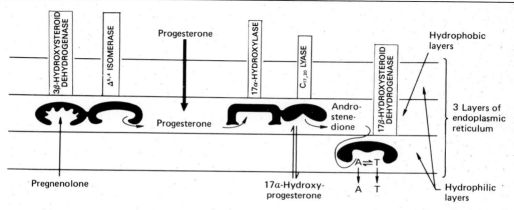

Figure 50–2. Schematic representation of androgen biosynthesis in the testicular microsomal membrane. The membrane is shown as horizontal, which may be the case in the cell; in microsomal preparations, however, it forms vesicles. A, androstenedione; T, testosterone. (Reproduced, with permission, from DeGroot LJ: *Endocrinology.* Vol 3. Grune & Stratton, 1979.)

of 17β-estradiol (E_2), the female sex hormone, but most of the E_2 produced by the male is derived from peripheral aromatization of testosterone and androstenedione. The Leydig cells, the Sertoli cells, and the seminiferous tubules are thought to be involved in E_2 production. The role of E_2 in the male has not been determined, but it may contribute to FSH regulation. Abnormally high plasma levels of E_2 and changes in the free E_2:testosterone ratio have been associated with pubertal or postpubertal gynecomastia (male breast enlargement), particularly in older individuals and in patients with chronic liver disease or hyperthyroidism.

B. Age-Related Changes in Testicular Hormone Production: Testosterone is the dominant hormone in the fetal and neonatal rat, but the testes make only androsterone soon after birth. The ability to produce testosterone is restored at puberty and continues throughout life. Similar observations have been made in other species, and these age-related changes may also occur in humans.

Secretion & Transport

A number of steroids are present in testicular venous blood, but **testosterone is the major steroid secreted by the adult testes.** The secretion rate is about 5 mg/d in normal adult men. There is no apparent regulation of the secretion of testicular steroids; like other steroid hormones, testosterone seems to be released as it is produced.

Most mammals, humans included, have a plasma β-globulin that binds testosterone with specificity, relatively high affinity, and limited capacity (Table 50–1). This protein, usually called **sex hormone-binding globulin (SHBG) or testosterone-estrogen-binding globulin (TEBG),** is produced in the liver. Its production is increased by estrogens (women have twice the serum concentration of SHBG as men), certain types of liver disease, and hyperthyroidism; it is decreased by androgens, advancing age, and hypothyroidism. Many of these conditions also affect the production of CBG (see Chapter 48) and TBG (see Chapter 46). Since SHBG and albumin bind 97–99% of circulating testosterone, only a small fraction of the hormone in circulation is in the free (biologically active) form. The primary function of SHBG may be to restrict the free concentration of testosterone in the serum. Testosterone binds to SHBG with higher affinity than does estradiol (Table 50–2). Therefore, a

Table 50–1. Hormone binding to sex hormone-binding globulin (SHBG).

Steroids Bound	Steroids Not Bound
Testosterone	Conjugated androgens
17β-Estradiol	17α-Testosterone
Dihydrotestosterone	Dehydroisoandrosterone
Other 17β-hydroxysteroids	Cortisol
Estrone	Progesterone

change in the level of SHBG causes a greater change in the free testosterone level than in the free estradiol level. An increase of SHBG may contribute to the increased free E_2:testosterone ratio noted in aging, cirrhosis, and hyperthyroidism and hence contribute to the attendant signs and symptoms of "estrogenization" alluded to above.

Peripheral Metabolism & Excretion

A. Metabolic Pathways: Testosterone is metabolized by 2 pathways. One involves oxidation at the 17-position, and the other involves reduction of the A ring double bond and the 3-ketone. Metabolism via the first pathway occurs in many tissues, including liver, and produces 17-ketosteroids that are generally inactive or less active than the parent compound. Metabolism via the second pathway, which is less efficient, occurs primarily in target tissues and produces the potent metabolite DHT as well as estradiol and androstanediol. Etiocholanolone and androsterone are 5β-reduced products of androgens.

B. Metabolites of Testosterone: The most significant metabolic product of testosterone is DHT, since in many tissues, including seminal vesicles, prostate, external genitalia, and some areas of the skin, this is the active form of the hormone. The plasma content of DHT in the adult male is about one-tenth that of testosterone, and approximately 400 μg of DHT is produced daily, as compared to about 5 mg of testosterone. The reaction is catalyzed by 5α-reductase, an NADPH-dependent enzyme. (See below.)

Testosterone can thus be considered a prohormone, since it is converted into a much more potent compound (dihydrotestosterone) and since most of this conversion occurs outside the testes. A small percentage of testosterone is also converted into estradiol by aromatization, a reaction that is especially important in the brain, where these hormones help determine the

Testosterone → [5α-REDUCTASE] [NADPH] → Dihydrotestosterone (DHT)

sexual behavior of the animal. Androstanediol, another potent androgen, is also produced from testosterone.

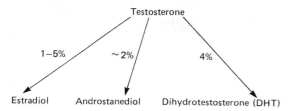

The major 17-ketosteroid metabolites, **androsterone** and **etiocholanolone,** are conjugated with glucuronide and sulfate in the liver to make water-soluble, excretable compounds. Quantification of urinary 17-ketosteroids was employed as an assay for androgen activity, but this is now recognized to be a poor reflection of the in vivo status of this class of hormones.

REGULATION OF TESTICULAR FUNCTION

Regulation of Steroidogenesis

Testicular function is regulated by **LH** and **FSH** (Fig 50–3). LH stimulates steroidogenesis and testosterone production by binding to receptors on the plasma membrane of the Leydig cells (an analogous LH receptor is found on cells of the corpus luteum) and activating adenylate cyclase, thus increasing intracellular cAMP. This action enhances the rate of cholesterol side-chain cleavage. It is not clear whether this cleavage results from activation or induction of the enzyme or from enhanced transport of cholesterol to the enzyme; the induction of a protein that facilitates this process has been proposed. The similarity between this action of LH and that of ACTH on the adrenal is apparent. Although LH (or hCG) has been shown to induce other steroidogenic enzymes, including 3β-OHSD, C_{17-20} lyase, and 5α-reductase, the primary effect appears to occur somewhere in the course of cholesterol conversion to pregnenolone.

The interaction of testosterone with its receptor probably provides for feedback control of gonadotropin release (Figs 50–3 and 50–4). This feedback appears to be accomplished at the hypothalamus through inhibition of GnRH release, GnRH production, or both, although inhibition of the action of GnRH on the LH-producing cells in the anterior pituitary cannot be excluded.

GnRH is normally released in a pulsatile fashion, and maximal production of the gonadotropins (LH and FSH) in experimental systems occurs only when this pulsatile pattern of GnRH release is simulated. Prolonged exposure to constantly elevated levels of GnRH appears to result in target cell desensitization and a profound suppression of LH and FSH release; thus, long-acting GnRH analogs are being tested as contraceptive agents.

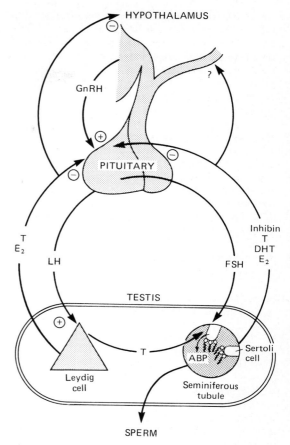

Figure 50–3. Feedback regulation of testicular function. GnRH, gonadotropin-releasing hormone; LH, luteinizing hormone; FSH, follicle-stimulating hormone; T, testosterone; DHT, dihydrotestosterone; ABP, androgen-binding protein; E_2, 17β-estradiol; $\oplus$, positive influence; $\ominus$, negative influence. (Reproduced, with permission, from Braunstein GD: The testes. In: *Basic & Clinical Endocrinology*, 2nd ed. Greenspan FG, Forsham PH [editors]. Appleton & Lange, 1986.)

Prior to puberty, serum testosterone levels are very low. An early event in this developmental process is an increase in sleep-associated, pulsatile bursts of LH and FSH release. When serum testosterone levels rise, growth and development of the secondary sex organs ensue. The signal that initiates this process has not been deciphered, but the prevailing hypothesis is that the hypothalamic centers responsible for GnRH production become less sensitive to feedback inhibition by gonadal steroid hormones.

Regulation of Spermatogenesis

This complex subject will be discussed only in terms of the feedback regulation of FSH secretion. **FSH binds to the Sertoli cells and promotes the synthesis of androgen-binding protein (ABP). ABP** is a glycoprotein that binds testosterone, but it is distinct from SHBG and the intracellular androgen receptor. ABP is secreted into the lumen of the seminiferous tubule, and in this process testosterone produced by the Leydig cells is transported in very high concentration to the site of spermatogenesis (Fig 50–3). This appears to be a critical step, since normal systemic levels of testosterone, such as might be achieved by replacement therapy, do not support spermatogenesis.

Although only one GnRH has been identified, **there is evidence for selective release of FSH and LH.** In men with selective destruction of seminiferous tubules or with arrested spermatogenesis due to various causes, there are normal LH and testosterone levels but elevated FSH levels. A substance from the seminiferous tubules or the Sertoli cells that regulates FSH release, **"inhibin,"** has been isolated. Part of this feedback effect on FSH release may be mediated by estradiol, which is made in small amounts by the Leydig and Sertoli cells. Production of estradiol in these cells is stimulated by FSH, and this steroid does result in feedback inhibition of FSH under certain test conditions. These possible mechanisms of FSH regulation are illustrated in Fig 50–3.

PHYSIOLOGIC ACTION OF TESTICULAR HORMONES

The androgens, principally testosterone and DHT, are involved in (1) sexual differentiation, (2) spermatogenesis, (3) development of secondary sexual organs and ornamental structures, (4) anabolic metabolism and gene regulation, and (5) male-pattern behavior (Fig 50–4). It is therefore difficult to define target and nontarget tissues, since so many are involved. In a more specific sense, such target tissues must also be defined according to whether they are affected by testosterone or DHT. The classic target cells for DHT (and those which coincidentally have the highest 5α-reductase activity) are the prostate, seminal vesicles, external genitalia, and genital skin. Targets for testosterone include the embryonic wolffian structures, spermatogonia, muscles, bone, kidney, and brain. The specific androgen involved in regulating the many other processes mentioned above has not been determined.

Sexual Differentiation

Sexual differentiation is a major action of the androgens, and since it represents a unique opportunity to discuss how specific hormones influence tissue differentiation, it is discussed separately below.

Spermatogenesis

The requirement for testosterone in spermatogenesis is unquestioned, yet its precise role in this process has not been defined. Testosterone may affect a relatively late stage, since it does not initiate spermatogenesis in the immature testis, nor does it restore sperm production in hypophysectomized animals in which the process has been allowed to regress.

Maturation of Secondary Sex Tissues

The increased production of testosterone at puberty results in the maturation of secondary sex tissues. The phallus, prostate, and seminal vesicles enlarge. The texture of skin and hair changes from the soft, smooth, infantile variety to a coarser form, beard growth commences, vocal cords thicken, and sebaceous glands enlarge. Castration causes most of these processes to revert toward the prepubertal form.

Anabolic Metabolism

Rapid growth of the musculoskeletal system is associated with puberty. Cartilage in the epiphyseal plates enlarges, and new bone formation ensues. This increase in height is accompanied by a marked in-

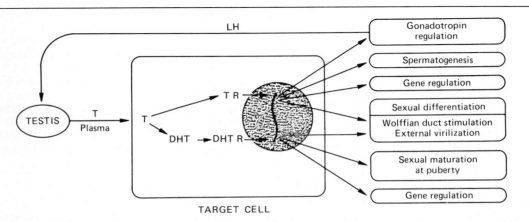

Figure 50–4. Mechanism of androgen action. LH, luteinizing hormone; T, testosterone; DHT, dihydrotestosterone; R, androgen receptor. (Modified and reproduced, with permission, from Wilson JD et al: The endocrine control of male phenotypic development. *Aust J Biol Sci* 1983;**36**:101.)

crease in skeletal muscle mass. These effects are difficult to analyze on a molecular level, but some clues exist. Androgens appear to induce the synthesis of protein in target tissues by the now well-accepted mechanism employed by other steroid hormones in which the hormone-receptor complex recognizes specific regions of chromatin and selectively activates certain genes. (See Chapter 44.) This appears to be true for ABP, aldolase, and the enzymes involved in polyamine biosynthesis. Similar mechanisms may be involved in androgen-mediated muscle hypertrophy, but as yet the specific proteins (genes) involved have not been identified.

MECHANISM OF ACTION OF TESTICULAR HORMONES

The current concept of androgen action is shown in Fig 50–4. Free testosterone enters cells through the plasma membrane by either passive or facilitated diffusion. Target cells retain testosterone, presumably because the hormone associates with a specific intracellular receptor. Although there is considerable tissue-to-tissue variability, most of the retained hormone is found in the cell nucleus. The cytoplasm of many (but not all) target cells contains the enzyme 5α-reductase, which converts testosterone to DHT. Whether there are distinct receptors for testosterone and DHT has been a point of controversy, but the consensus is that while there is but a single class of receptors, the affinity of the receptor for DHT exceeds that for testosterone. Single gene mutations in mice result in loss of binding of both testosterone and DHT to the receptor in various tissues, suggesting that a single protein is involved. The affinity difference, coupled with the ability of a target tissue to form DHT from testosterone, may determine whether the testosterone-receptor complex or the DHT-receptor complex is active.

Nuclear localization of the testosterone/DHT-receptor complex is a prerequisite for androgen action. Binding of the receptor-steroid complex to chromatin may involve a prior activation step, and some specificity or acceptor function is certainly supplied by chromatin. The nature of the nuclear acceptor for the active receptor-steroid complex has not been elucidated.

In keeping with other steroid (and some peptide) hormones, **the testosterone/DHT receptor presumably activates specific genes,** and the protein products of these mediate many (if not all) of the effects of the hormone. Testosterone stimulates protein synthesis in male accessory organs, an effect that is usually associated with increased accumulation of total cellular RNA, including mRNA, tRNA, and rRNA. A more specific example involves the effect of testosterone on the synthesis of ABP. The hormone increases the rate of transcription of the ABP gene, which results in an increased amount of the mRNA that codes for this protein. Another well-studied example is α_{2u} globulin, the major protein excreted in the urine of male rats. It is synthesized by the liver of mature (> 40-day-old) male rats but is not synthesized by female rats or castrated male rats unless they are given testosterone. Estrogen inhibits the production of α_{2u} globulin, and maximal rates of production require the concerted action of hormones including GH, thyroid hormones, insulin, and glucocorticoids. The rate of synthesis of α_{2u} globulin is directly related to the amount of the cognate mRNA, which in turn is related to the rate of transcription of the α_{2u} globulin gene.

The kidney is a major target tissue for androgens. These hormones cause a general enlargement of the kidney and induce the synthesis of a number of enzymes in various species.

Other more complex actions of androgens, such as their effects on muscle hypertrophy and glandular secretion, could be exerted through a different mechanism. **Androgens also stimulate the replication of cells in some target tissues,** an effect that is poorly understood. Testosterone or DHT, in combination with E_2, appears to be implicated in the extensive and uncontrolled division of prostate cells that results in **benign prostatic hypertrophy,** a condition that afflicts as many as 75% of men over the age of 60 years.

PATHOPHYSIOLOGY OF THE MALE REPRODUCTIVE SYSTEM

The lack of tetosterone synthesis is called **hypogonadism.** If this occurs before puberty, secondary sex characteristics fail to develop, and if it occurs in adults, many of these features regress. **Primary hypogonadism** is due to processes that affect the testes directly and cause testicular failure, whereas **secondary hypogonadism** is due to defective secretion of the gonadotropins. Isolated genetic deficiencies help to establish the importance of specific steps in the biosynthesis and action of androgens. Fig 50–5 represents the pathway involved in androgen action from testosterone biosynthesis through postreceptor actions of testosterone and DHT. At least 5 distinct genetic defects in testosterone biosynthesis have been described; the 5α-reductase deficiency is known; there are a number of instances in which either no testosterone/DHT receptor is detected or the receptor is abnormal in some manner; and there are a number of cases in which all measurable entities, including the receptor, are normal, but the patients (always genetic males) have variable degrees of feminization. The extent of the abnormality of differentiation is related to the severity of the deficit. Persons who completely lack a biosynthetic enzyme appear to be phenotypic females but have an XY (male) genotype, while the mildest cases may have only an abnormally located penile urethra. Genetic males who completely lack functioning receptors have testes and produce testosterone but have complete feminization of the external genitalia (the so-called **testicular feminization syndrome).** It is interesting that no comparable deficiencies in estrogen synthesis or action have been identified.

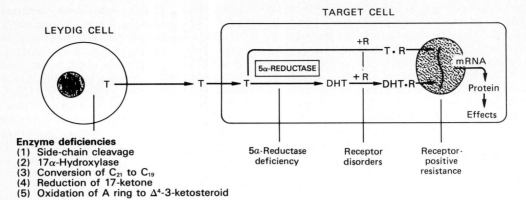

Figure 50–5. Steps involved in androgen resistance. Four stages at which mutations have been identified are shown. T, testosterone; DHT, dihydrotestosterone; R, androgen receptor. (Modified and reproduced, with permission, from Wilson JD et al: The endocrine control of male phenotypic development. *Aust J Biol Sci* 1983;**36**:101.)

HORMONES OF THE OVARIES

The ovaries are bifunctional organs that produce **estrogens** and **progestins** (the female sex hormones) and **ova** (the female germ cells). The most active naturally occurring hormones of these classes are 17β-estradiol (E_2) and progesterone.

BIOSYNTHESIS & METABOLISM OF OVARIAN HORMONES

Synthesis

The estrogens are a family of hormones synthesized in ovarian and extraovarian tissues. 17β-Estradiol is the primary estrogen of ovarian origin. In some species, estrone is more abundant but is of extraovarian origin. In pregnancy, relatively more estriol is produced, but this too is extraovarian. The general pathway and the subcellular localization of the enzymes involved in the early steps of estradiol synthesis are the same as are found in the adrenal glands and testes. Features unique to the ovary are illustrated in Fig 50–6.

Estrogens are formed by the aromatization of androgens in a complex process that involves 3 hydroxylation steps, each of which requires O_2 and NADPH. The aromatase enzyme complex is thought to include a P-450 mixed-function oxidase. Estradiol is formed if the substrate of this enzyme complex is testosterone, whereas estrone results from the aromatization of androstenedione (this generally occurs in extraovarian tissues).

The cellular source of the various ovarian steroids has been difficult to unravel, but it now appears that 2 cell types are involved. Theca cells are the major sources of 17α-hydroxyprogesterone and of androstenedione (the principal androgen produced by the ovary), and granulosa cells are the major source of estradiol. Progesterone is produced and secreted by the corpus luteum, which also makes some estradiol.

Significant amounts of estrogens are produced by the peripheral aromatization of androgens. In human males, the peripheral aromatization of testosterone to estradiol (E_2) accounts for 80% of the production rate of the latter. In females, adrenal androgens are important substrates, since as much as 50% of the E_2 produced during pregnancy comes from the aromatization of DHEA sulfate, and the conversion of androstenedione to estrone is the major source of estrogens in postmenopausal women. Aromatase activity is present in adipose cells and also in liver, skin, and other tissues. Increased activity of this enzyme may contribute to the "estrogenization" that characterizes diseases like cirrhosis of the liver, hyperthyroidism, aging, and obesity.

17β-Estradiol

Progesterone

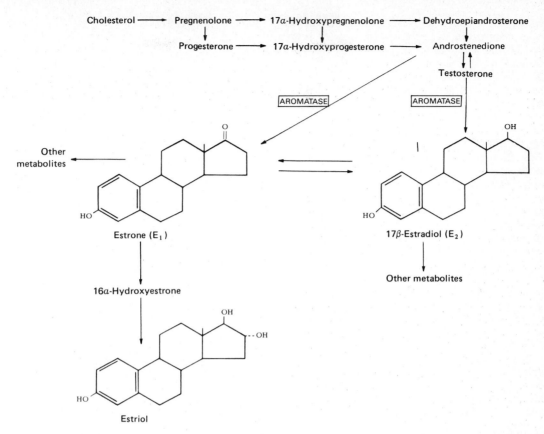

Cholesterol ⟶ Pregnenolone ⟶ 17α-Hydroxypregnenolone ⟶ Dehydroepiandrosterone

Progesterone ⟶ 17α-Hydroxyprogesterone ⟶ Androstenedione

Testosterone

AROMATASE   AROMATASE

Other metabolites

Estrone (E₁)

17β-Estradiol (E₂)

16α-Hydroxyestrone

Other metabolites

Estriol

Figure 50–6. Biosynthesis of estrogens. (Slightly modified and reproduced, with permission, from Ganong WF: *Review of Medical Physiology,* 13th ed. Appleton & Lange, 1987.)

Catechol estrogens are the major metabolites of estrogen in all mammalian species. They result from hydroxylation at position 2 of the aromatic ring. The catechol estrogens have weak estrogenic activity but are potent agents in the central nervous system, where they are also found.

Secretion & Transport

The rate of secretion of ovarian steroids varies considerably during the menstrual (or estrous) cycle and is directly related to rate of production in the ovary. There is no storage of these compounds; they are secreted when they are produced.

Estrogens and progestins, like other steroids, are bound in varying degrees to plasma transport proteins. Estrogens are bound to SHBG and progestins to CBG. SHBG binds estradiol about 5 times less avidly than it binds testosterone or DHT, but progesterone and cortisol have little affinity for this protein (Table 50–2). In contrast, progesterone and cortisol bind with nearly equal affinity to CBG, which in turn has little avidity for estradiol and even less for testosterone, DHT, or estrone.

These plasma transport proteins play no apparent role in the mechanism of action of these hormones at the cellular level, and (as with other steroids) probably only the free hormone has biologic activity. The binding proteins do provide a circulating reservoir of hormone, and because of the relatively large binding capacity, they buffer against sudden changes in the plasma level. The metabolic clearance rates of these steroids are inversely related to the affinity of their binding to SHBG; hence, estrone is cleared more rapidly than estradiol, which in turn is cleared more

Table 50–2. Approximate affinities of steroids for serum-binding proteins.*

	SHBG†	CBG†
Estradiol	5	>10
Estrone	>10	>100
Androstenedione	. . .	. . .
Testosterone	2	>100
Dihydrotestosterone	1	>100
Progesterone	>100	2
Cortisol	>100	3

*Adapted from Siiteri PK, Febres F: Ovarian hormone synthesis, circulation and mechanisms of action. Page 1401 in: *Endocrinology.* Vol 3. DeGroot LJ (editor). Grune & Stratton, 1979.
†Affinity expressed as K_d of molar quantity × 10^9.

rapidly than testosterone or DHT. In this regard, the conjugated derivatives of these hormones (see below) are not bound by either SHBG or CBG. SHBG has an additional role in that, given the difference in binding affinity of estradiol and testosterone or DHT, it can affect the amount of these sex hormones available for interaction with target tissues. The factors that regulate the production of SHBG are discussed above.

Metabolism & Excretion

A. Estrogens: The liver converts estradiol and estrone to estriol by the pathways shown in Fig 50–6. Estradiol, estrone, and estriol are substrates for hepatic enzymes that add glucuronide or sulfate moieties. Activity of these conjugating enzymes varies among species. Rodents have such active metabolizing enzyme systems, particularly hydroxylating systems, that estrogens are almost completely metabolized by the liver and thus are essentially without activity when given orally. These enzyme systems are less active in primates, so that oral estrogens are more effective. The conjugated steroids are water-soluble and do not bind to transport proteins; thus, they are excreted readily in the bile and feces and to a lesser extent in the urine.

B. Progestins: Because the liver actively metabolizes progesterone to several compounds, **progesterone is ineffective when given orally.** Sodium pregnanediol-20-glucuronide is the major progestin metabolite found in human urine (Fig 50–7). Certain synthetic steroids, eg, derivatives of 17α-hydroxyprogesterone and 17α-alkyl-substituted 19-nortestosterone compounds, have progestational activity and avoid hepatic metabolism. Thus, they are widely used in oral contraceptives.

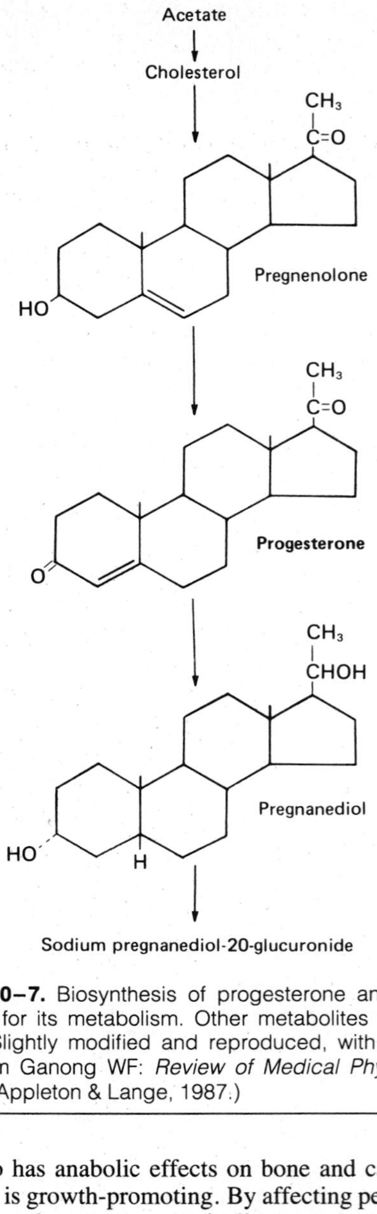

REGULATION & PHYSIOLOGIC ACTIONS OF OVARIAN HORMONES

Maturation & Maintenance of the Female Reproductive System

The major function of the ovarian hormones is to prepare the structural determinants of the female reproductive system (see below) for reproduction by (1) maturing the primordial germ cells; (2) developing the tissues that will allow for implantation of the blastocyst; (3) providing the "hormonal timing" for ovulation; (4) establishing, with placental hormones, the milieu required for the maintenance of pregnancy; and (5) providing the hormonal influences for parturition and lactation.

Estrogens stimulate the development of tissues involved in reproduction. In general, these hormones stimulate the size and number of cells by increasing the rate of synthesis of protein, rRNA, tRNA, mRNA, and DNA. Under estrogen stimulation, the vaginal epithelium proliferates and differentiates; the uterine endometrium proliferates and the glands hypertrophy and elongate; the myometrium develops an intrinsic, rhythmic motility; and breast ducts proliferate. Estra-

Figure 50–7. Biosynthesis of progesterone and major pathway for its metabolism. Other metabolites are also found. (Slightly modified and reproduced, with permission, from Ganong WF: *Review of Medical Physiology*, 13th ed. Appleton & Lange, 1987.)

diol also has anabolic effects on bone and cartilage, and so it is growth-promoting. By affecting peripheral blood vessels, estrogens typically cause vasodilation and heat dissipation.

Progestins generally require the previous or concurrent presence of estrogens; thus, the 2 classes of hormones often act synergistically, although they can be antagonists. Progestins reduce the proliferative activity of the estrogens on the vaginal epithelium and convert the uterine epithelium from proliferative to secretory (increased size and function of secretory glands and increased glycogen content), thus preparing the uterine epithelium for implantation of the fertilized ovum. Progestins enhance development of the acinar portions of breast glands after estrogens have stimulated ductal development. Progestins decrease peripheral blood flow, thereby decreasing heat loss, so that body temperature tends to increase during the luteal phase of the menstrual cycle, when these steroids are produced. This temperature spike, usually ~0.5 °C, is used as an indicator of ovulation.

The number of oogonia in the human fetal ovary reaches a maximum of 6–7 million at about the fifth month of gestation. This decreases to about 2 million by birth and is further diminished to 100,000–200,000 by the onset of menarche. Some 400–500 of these develop into mature oocytes; the rest are gradually lost through a process that is not understood, although ovarian androgens have been implicated. Follicular maturation begins in infancy, and the ovaries gradually enlarge in prepubertal years owing to increased volume of the follicles because of the growth of granulosa cells, to the accumulation of tissue from atretic follicles, and to the increased mass of medullary stromal tissue with the interstitial and theca cells that will produce the steroids.

The concentration of sex hormones is low in childhood, although exogenous gonadotropins increase production; hence, the immature ovary has the capacity to synthesize estrogen. It is thought that these low levels of sex steroids inhibit gonadotropin production in prepubertal girls and that at puberty the hypothalamic-pituitary system becomes less sensitive to suppression. At puberty, the pulsatile release of GnRH begins; LH causes a dramatic increase of ovarian hormone production; FSH, the main stimulus for estrogen secretion, stimulates a follicle to ripen, and ovulation ensues.

The Menstrual Cycle

Hormones determine the frequency of ovulation and receptivity to mating. Monestrous species ovulate and mate once a year, whereas polyestrous species repeat this cycle several times a year. Primates have menstrual cycles, with shedding of the endometrium at the end of each cycle, and mating behavior is not tightly coupled to ovulation. The human menstrual cycle results from a complex interaction between the hypothalamus, pituitary, and ovary. **The cycle normally varies between 25 and 35 days in length (average, 28 days).** It can be divided into a **follicular phase,** a **luteal phase,** and **menstruation** (Fig 50–8).

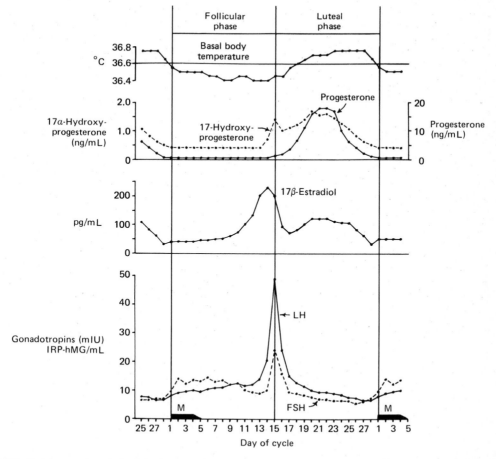

Figure 50–8. Hormonal and physiologic changes during a typical human menstrual cycle. M, menstruation; IRP-hMG, international reference standard for gonadotropins. (Reproduced, with permission, from Midgley AR in: *Human Reproduction.* Hafez ESE, Evans TN [editors]. Harper & Row, 1973.)

A. Follicular Phase: For reasons that are not clear, a particular follicle begins to enlarge under the general influence of FSH. E_2 levels are low during the first week of the follicular phase, but they begin to rise progressively as the follicle enlarges. E_2 reaches its maximal level 24 hours before the LH (FSH) peak and sensitizes the pituitary to GnRH. LH is released either in response to this high level of E_2 in a "positive feedback" manner or in response to a sudden decline of E_2 from this high level. Continual administration of high doses of estrogen (as in oral contraceptives) suppresses LH and FSH release and inhibits the action of GnRH on the pituitary. Progesterone levels are very low during the follicular phase. The LH peak heralds the end of the follicular phase and precedes ovulation by 16–18 hours.

B. Luteal Phase: After ovulation, the granulosa cells of the ruptured follicle luteinize and form the corpus luteum, a structure that soon begins to produce progesterone and some estradiol. Estradiol peaks about midway through the luteal phase and then declines to a very low level. **The major hormone of the luteal portion of the cycle is progesterone,** which (as noted above) is required for preparation and maintenance of the secretory endometrium that provides early nourishment for the implanted blastocyst. **LH is required for the early maintenance of the corpus luteum,** and the pituitary supplies it for about 10 days. If implantation occurs (day 22–24 of the average cycle), this LH function is assumed by chorionic gonadotropin (hCG), a placental hormone that is very similar to LH and is made by the cytotrophoblastic cells of the implanted early embryo (see Chapter 45). hCG supports progesterone synthesis by the corpus luteum until the placenta begins making large amounts of this steroid. In the absence of implantation (and hCG), the corpus luteum regresses and menstruation ensues; after the endometrium is shed, a new cycle commences. The luteal phase is always 14 ± 2 days in length. Variations in cycle length are almost always due to an altered follicular phase.

Pregnancy & Placental Hormones

The implanted blastocyst forms the trophoblast, which is subsequently organized into the placenta. The placenta provides the nutritional connection between the embryo and the maternal circulation and produces a number of hormones.

A. Human Chorionic Gonadotropin (hCG): The primary function of the glycoprotein hormone hCG is to support the corpus luteum until the placenta produces amounts of progesterone sufficient to support the pregnancy. hCG can be detected within a few days of implantation, and this provides the basis of early diagnostic tests for pregnancy. Peak hCG levels are reached in the middle of the first trimester, after which there is a gradual decline throughout the remainder of pregnancy. Changes in hCG and other hormone levels in pregnancy are illustrated in Fig 50–9.

B. Progestins: The corpus luteum is the major source of progesterone for the first 6–8 weeks of the

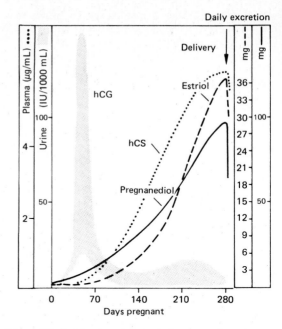

Figure 50–9. Hormone levels during normal pregnancy. hCG, human chorionic gonadotropin; hCS, human chorionic somatomammotropin. (Data from various authors.) (Reproduced, with permission, from Ganong WF: *Review of Medical Physiology,* 13th ed. Appleton & Lange, 1987.)

pregnancy, and then the placenta assumes this function. The corpus luteum continues to function, but late in pregnancy the placenta makes 30–40 times more progesterone than does the corpus luteum. The placenta cannot synthesize cholesterol and so depends upon a maternal supply.

C. Estrogens: Plasma concentrations of estradiol, estrone, and estriol gradually increase throughout pregnancy. **Estriol is produced in the largest amount, and its formation reflects a number of fetoplacental functions.** The fetal adrenal produces DHEA and DHEA sulfate, which are converted to 16α-hydroxy derivatives by the fetal liver. These are converted to estriol by the placenta; travel via the placental circulation to the maternal liver, where they are conjugated to glucuronides; and then are excreted in the urine (Fig 50–10). The measurement of urinary estriol levels is used to document the function of a number of maternal-fetal processes.

Another interesting exchange of substrates is required for fetal cortisol production. The fetal adrenal lacks the familiar 3β-hydroxysteroid dehydrogenase $\Delta^{5,4}$ isomerase complex and hence depends upon the placenta for the progesterone required for cortisol synthesis. The pregnenolone required for DHEA synthesis also comes from the placenta (Fig 50–10).

D. Placental Lactogens: The placenta makes a hormone called placental lactogen (PL). PL is also called chorionic somatomammotropin or placental growth hormone because it has biologic properties of prolactin and growth hormone. The genetic relation-

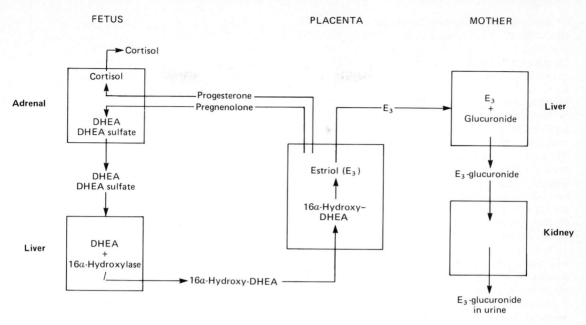

FETUS   PLACENTA   MOTHER

Figure 50–10. Steroid metabolism by the fetal-maternal unit. DHEA, dehydroepiandrosterone.

ship of these hormones is discussed in Chapter 45. The physiologic function of PL is uncertain, since women who lack this hormone appear to have normal pregnancies and deliver normal babies.

Parturition

Pregnancy lasts a predetermined number of days for each species, but the factors responsible for its termination are unknown. Hormonal influences are suspected but unproved. Estrogens and progestins are candidates, since they affect uterine contractility, and there is evidence that catecholamines are involved in induction of labor. Since oxytocin stimulates uterine contractility, it is used to facilitate delivery, but it will not initiate labor unless the pregnancy is at term. There are 100 times more oxytocin receptors in the uterus at term than there are at the onset of pregnancy. The increase correlates with the increased amount of estrogens at term, which increases the number of oxytocin receptors (see Chapter 45). Once labor begins, the cervix dilates, initiating a neural reflex that stimulates oxytocin release and hence further uterine contraction. Mechanical factors, such as the amount of stretch or force applied to the muscle, may be important. A sudden and dramatic change in the hormonal milieu of both the mother and newborn occurs with parturition, and plasma levels of progesterone (measured as pregnanediol) and estriol decline rapidly after the placenta is delivered (Fig 50–9).

Mammary Gland Development & Lactation

The differentiation and function of the mammary gland are regulated by the concerted action of several hormones. The female sex hormones initiate this process, since estrogens are responsible for ductal growth and progestins stimulate alveolar proliferation. Some growth of glandular tissue occurs during puberty, along with deposition of adipose tissue, but extensive development occurs during pregnancy when glandular tissue is exposed to high concentrations of estradiol and progesterone. Complete differentiation, studied mostly in rat mammary gland explants, requires the additional action of prolactin, a glucocorticoid, insulin or a growth peptide, and an unidentified serum factor. Of these hormones, only the concentration of prolactin changes dramatically in pregnancy; it increases from < 2 ng/dL to > 200 ng/dL in late pregnancy. The effects of these hormones on the synthesis of various milk proteins, including lactalbumin, lactoglobulin, and casein, have been studied in detail. These hormones increase the rate of synthesis of these proteins by increasing amounts of the specific mRNAs, and in the case of casein at least, this is due to an increase in gene transcription. Interestingly, gene transcription is not increased unless cortisol, prolactin, and insulin are all added to the explant cultures.

Progesterone, required for alveolar differentiation, inhibits milk production and secretion in late pregnancy. Lactation commences when levels of this hormone decrease abruptly after delivery. Prolactin levels also fall rapidly postpartum but are stimulated with each episode of suckling (see Chapter 45), thereby ensuring continual lactation. Lactation gradually decreases if suckling is not allowed and can be rapidly terminated by administration of a large parenteral dose of an androgen before suckling is allowed.

Suckling also results in the release of oxytocin from

the posterior pituitary. Oxytocin stimulates contraction of the myoepithelial cells that surround the alveolar ducts, thus expelling milk from the gland. The regulation of oxytocin synthesis and secretion is discussed in Chapter 45.

Menopause

Women in the western hemisphere cease having regular menstrual cycles at about age 53, coincident with loss of all follicles and ovarian function. There is no alternative source of progesterone, but substantial amounts of a weak estrogen, estrone, are produced by the peripheral aromatization of the adrenal steroid androstenedione (Fig 50–6). The levels of estrone are not sufficient to suppress pituitary gonadotropin levels; thus, marked increases of LH and FSH are characteristic of the postmenopausal years. Postmenopausal women are particularly prone to 2 problems associated with tissue catabolism. Estrone is not always able to prevent the **atrophy of secondary sex tissues,** particularly the epithelium of the lower urinary tract and vagina. **Osteoporosis** is a major health problem in older individuals, and women with the most severe decrease in bone mass have lower than normal estrone levels.

Synthetic Agonists & Antagonists

A. Estrogens: Several synthetic compounds have estrogenic activity and one or more favorable pharmacologic features. Most modifications are designed to retard hepatic metabolism, so that the compounds can be given orally. One of the first developed was diethylstilbestrol. Other examples of modified steroids include 17α-ethinyl estradiol and mestranol, which are used in oral contraceptives.

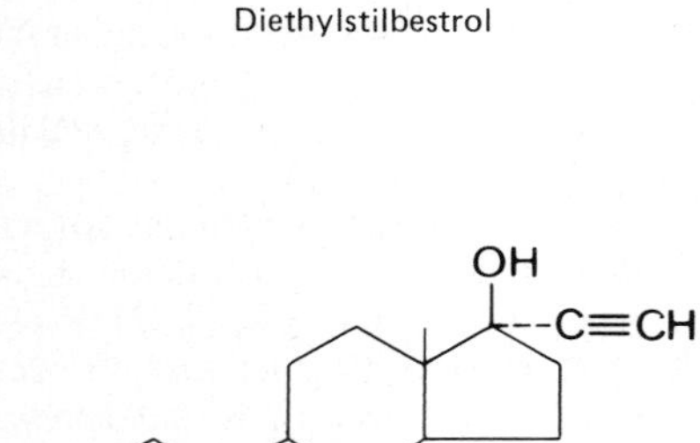

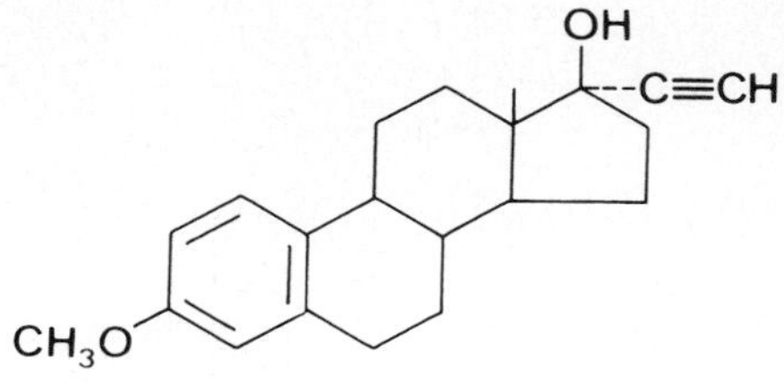

Mestranol

Numerous compounds with antiestrogenic activity have been synthesized, and several of these have clinical applications. Most of these antagonists act by competing with estradiol for its intracellular receptor (see below). Clomiphene citrate (Clomid) has a particular affinity for the estrogen receptor in the hypothalamus. Clomiphene was originally designed as an antifertility drug, but it has the opposite effect. Clomiphene competes with estradiol for hypothalamic receptor sites; thus, GnRH release is not restrained and excessive amounts of LH and FSH are released. Multiple follicles often mature simultaneously in response to clomiphene, and multiple pregnancies can ensue. Nafoxidine, a nonsteroidal compound, and tamoxifen combine with the estrogen receptor to form very stable complexes with chromatin; hence, the receptor cannot recycle and these agents inhibit the action of estradiol for prolonged periods. These antagonists are used in the treatment of estrogen-dependent breast cancer.

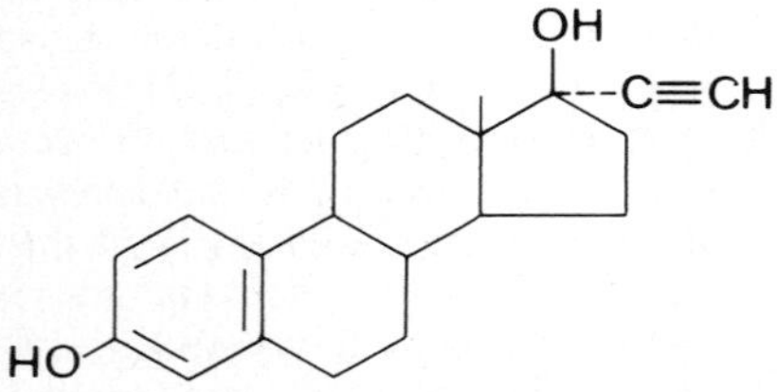

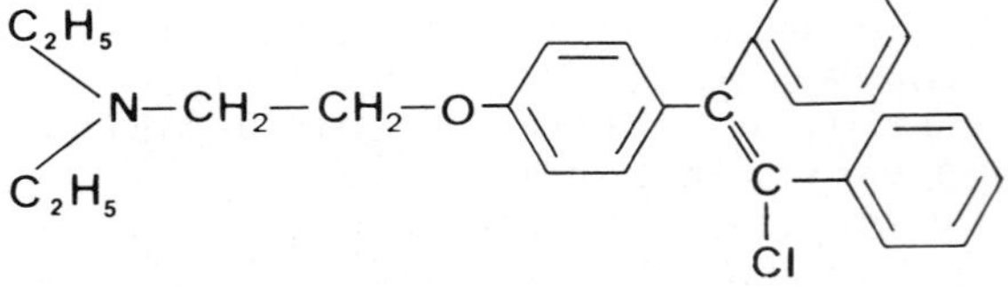

Diethylstilbestrol

Clomiphene citrate

B. Progestins: It has been difficult to synthesize compounds that have progestin activity but no estrogenic or androgenic action. The 17α-alkyl-substituted 19-nortestosterone derivatives (eg, norethindrone) have minimal androgenic activity in most women and are used in oral contraceptives. Another potent progestin is medroxyprogesterone acetate (Provera). Medroxyprogesterone inhibits ovulation for several months when given as an intramuscular depot injection, but it is more frequently used for treating well-differentiated endometrial carcinoma. It is believed to affect the replication of normal and malignant endometrial cells by forming a very stable complex with the progesterone receptor and thereby inhibiting action of the natural hormone.

17α-Ethinyl estradiol

Medroxyprogesterone acetate

Norethindrone

MECHANISM OF ACTION OF OVARIAN HORMONES

Estrogens and progestins exert major effects through their ability to combine with intracellular receptors which then bind to specific regions of chromatin or DNA (or both) to effect changes in the rate of transcription of specific genes. Much information has been learned from the analysis of how estradiol and progesterone stimulate transcription of the avian egg-white protein genes, especially ovalbumin and conalbumin. The determination of exactly how these hormones activate gene transcription is under intense investigation.

The Estrogen Receptor

The sequences of the human and chicken estrogen receptor proteins have been deduced from sequence analysis of full-length cDNA molecules. Each has 3 highly conserved regions, the second and third of which represent the DNA- and hormone-binding domains. The hormone-binding domain is very hydrophobic, and the DNA-binding domain is rich in cysteine and basic amino acid residues. These 2 domains of the estrogen receptor are conserved in the glucocorticoid, progesterone, and calcitriol receptors and in the product of the v-*erb*-A gene. All of the DNA-binding proteins may have arisen from a primordial gene.

Special Features

Some points, which may have application to the mechanism of action of other hormones, bear noting: (1) There is considerable cross talk between the sex hormone receptors. Progesterone binds to the androgen receptor and thus is a weak androgen; some androgens bind to the estrogen receptor and mimic the action of the latter in the uterus. (2) Estrogens increase the concentration of both the estrogen and the progesterone receptor. (3) Progesterone appears to enhance the rate of turnover of its receptor. (4) So-called weak estrogens, such as estriol, act as potent estrogens when given frequently.

Receptor-Independent Actions of Sex Hormones

Whether all actions of steroid hormones are mediated by receptors has been debated since their discovery. **Recent evidence suggests that some actions may be receptor-independent and probably do not involve nuclear effects.**

A. Estrogens: Estrogen may act independently of its receptor. Various estrogens stimulate uterine blood flow with a relative potency that does not match their binding to the receptor, nor does this effect require RNA synthesis. Direct effects on histamine release and prostaglandin production have been proposed to explain this effect, which, although not so well defined as the membrane effect of progesterone (see below), is an important physiologic action of estrogens.

B. Progestins: The interaction of progesterone with the membrane of amphibian oocytes results in a hormone- and time-dependent increase of intracellular calcium, which in turn stimulates the synthesis of specific cytoplasmic proteins that appear to be involved in oocyte maturation; these proteins, when transferred to other immature oocytes, stimulate maturation. This effect is observed in enucleated oocytes (hence, gene transcription is not involved), and the effect is not obtained if the progesterone is injected directly into the egg. This is probably not a unique example. The membrane anesthetic and anti-inflammatory actions of progesterone in mammalian cells may be other examples of receptor-independent actions of this hormone.

PATHOPHYSIOLOGY OF THE FEMALE REPRODUCTIVE SYSTEM

A discussion of all of the disorders that affect the female reproductive system is beyond the scope of this chapter, but a few illustrative disorders follow. **Primary hypogonadism** is due to processes that directly involve the ovaries and thus cause ovarian deficiency (decreased ovulation, decreased hormone production, or both), whereas **secondary hypogonadism** is due to the loss of pituitary gonadotropin function. **Gonadal dysgenesis (Turner's syndrome)** is a relatively frequent genetic disorder in which individuals have an XO karyotype, female internal and external genitalia, several developmental abnormalities, and delayed puberty.

Several syndromes are related to abnormal amounts of hormones. The most frequent is **polycystic ovary syndrome** (Stein-Leventhal syndrome), in which overproduction of androgens causes hirsutism, obesity, irregular menses, and impaired fertility. The rare **Leydig cell and arrhenoblastoma tumors** produce testosterone; **granulosa-theca cell tumors** produce estrogens; and **intraovarian adrenal rests** produce cortisol. Persistent trophoblastic tissue results in the benign **hydatidiform mole** or a malignant transformation of this—**choriocarcinoma;** both of these produce enormous quantities of hCG. The radioimmunoassay of hCG is a diagnostic test for these dangerous conditions and can also be used to monitor efficacy of therapy.

GONADAL HORMONES & SEXUAL DIFFERENTIATION

Sexual differentiation involves a series of sequential, ordered processes that can be described by the paradigm **chromosomal sex → gonadal sex → phenotypic sex.** This progression provides an excellent example of how hormones are involved in the differentiation and development of tissues, and an analysis of defects at several steps emphasizes the importance of various processes in overall androgen function.

CHROMOSOMAL SEX

Chromosomal sex, the first phase of sexual differentiation, is established at fertilization and is the only immutable portion of the paradigm. **A heterogametic distribution of sex chromosomes (XY) dictates the male genotype, whereas a homogametic complement (XX) specifies the female genotype.** In patients with ambiguous external genitalia or in whom a dissociation between phenotypic and genotypic sex is suspected, this fact is established by a Barr body analysis of buccal mucosal cells, fibroblasts, or leukocytes. The **Barr body** is an area of condensed chromatin that represents an inactivated X chromosome. The number of Barr bodies per cell is directly related to the number of X chromosomes. X chromosome number -1 = Barr body number; so that XY = 0, XX = 1, XXY = 1, XXX = 2. There is no known hormonal influence on chromosomal sex.

GONADAL SEX

During the initial stages of gestation, the development of male and female embryos is indistinguishable. In humans, the germ cells begin to migrate from the yolk sac to the genital ridges between the 35th and 50th days of gestation. This process results in the formation of the **primitive gonad.** The primitive gonad is ambiguous and consists of the primordial germ cells, connective or interstitial tissue, and a covering epithelial layer. At about day 56, hormones become involved, and gonadal differentiation begins. The exact initiating mechanism is not clear, but one fact is certain. **In the absence of a positive event, ie, the differentiation of the primitive gonad into a testis, all embryos develop into phenotypic females.** A male-specific cell surface antigen, the so-called **H-Y antigen,** has been associated with differentiation of the primitive gonad into a testis (Fig 50–11).

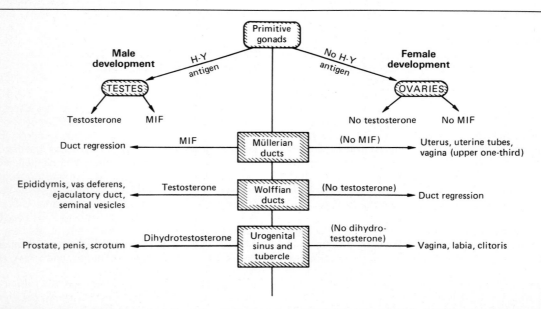

Figure 50–11. Hormonal involvement in sexual differentiation. MIF, müllerian inhibiting factor. (Modified and reproduced, with permission, from Fox Sl. Page 631 in: *Human Physiology.* William C. Brown, 1984.)

If the genetic sex is male, the Leydig cells appear in the connective tissue, testosterone synthesis begins, and development of the male reproductive tract starts. **In the absence of the Y chromosome effect, the female system develops** with a slight lag in time and is essentially completed by the 90th day of gestation. This sequence, in which the structural and functional development of the testis precedes development of the male phenotype and in which the ovary and female phenotype occur in the absence of testicular differentiation, is common to all mammals and is shown in schematic form in Fig 50–11. This scheme was first elucidated in rabbits, but the extrapolation to humans and other species has been subsequently validated.

PHENOTYPIC SEX

Internal Genital Structures

Male sex hormones are directly involved in the differentiation of the **primordial genital duct system** (wolffian and müllerian) and of the anlage for the external genitalia. The male internal genital tract develops from the primitive **wolffian duct system,** and the female tract is derived from the **müllerian duct system** (Fig 50–11).

The distinction between wolffian and müllerian development depends on the production of a testicular factor called **müllerian inhibiting factor (MIF).** MIF, a glycoprotein of MW ~ 70,000, is formed in the spermatogenic tubules and constitutes the first endocrine function of the testis. The mechanism by which MIF suppresses müllerian duct formation is not known, but it is clear that this is an active process unrelated to subsequent actions of testosterone.

External Genital Structures

The nature of the external genital structures, which develop from a common anlage, is determined by the presence or absence of another testicular hormone, testosterone, and its derivative, dihydrotestosterone (DHT).

Testosterone synthesis immediately precedes the initiation of fetal virilization. In the male rabbit embryo, testosterone synthesis begins between day 17 and day 17.5 and is associated with marked increases of cholesterol side-chain cleavage enzyme and 3β-hydroxysteroid dehydrogenase/$\Delta^{5,4}$ isomerase complex—key enzymes in testosterone biosynthesis; the other steroidogenic enzymes are always present in the primitive gonad. Estrogen synthesis begins in the ovary at exactly the same time. This coincidence suggests that perhaps female differentiation is not entirely passive, since estrogen synthesis by the immature gonad could play a role in stimulating the replication of the primordial germ cells or their differentiation into oogonia. The switch that determines the onset of steroidogenesis has not been determined, nor is it clear whether at early stages this process is regulated by other hormones. Later in embryogenesis, as in postnatal life, LH regulates steroidogenesis by controlling the rate of side-chain cleavage of cholesterol.

Until the discovery of DHT, it was assumed that the development of the male reproductive tract depended entirely on testosterone. That both DHT and testosterone are required for this process came from studies of embryonic tissues in which it was determined that, just before the onset of virilization, 5α-reductase activity was highest in the anlage that specified the prostate and external genitalia, whereas this enzyme was undetectable in wolffian tissue at that time. This biochemical observation received genetic support when patients lacking 5α-reductase activity were identified. Such individuals are genetic males, have normal wolffian structures, and have an external female phenotype, except that the vagina is incompletely developed.

REFERENCES

General

Huggins C: Two principles in endocrine therapy of cancer. Hormone deprival and hormone interference. *Cancer Res* 1965;**25:**1163.

Ohno S, Geller LN, Lai EVY: *Tfm* mutation and masculinization versus feminization of the mouse central nervous system. *Cell* 1974;**3:**235.

O'Malley BW: Steroid hormone action in eucaryotic cells. *J Clin Invest* 1984;**74:**307.

Testicular Hormones

Hall PF: Gonadotropic regulation of testicular function. Pages 1511–1519 in: *The Androgens of the Testis.* Eik-Nes KB (editor). Dekker, 1970.

Hall PF: Testicular hormones: Synthesis and control. Pages 1511–1520 in: *Endocrinology.* Vol 3. DeGroot LJ (editor). Grune & Stratton, 1979.

Longcope C, Kato T, Horton R: Conversion of blood androgens to estrogen in normal men and women. *J Clin Invest* 1969;**48:**2191.

Mainwaring WIP: *The Mechanism of Action of Androgens.* Springer, 1977.

Samuels LT, Matsumoto K: Localization of enzymes involved in testosterone biosynthesis by the mouse testis. *Endocrinology* 1974;**94:**55.

Wilson J: Metabolism of testicular androgens. Chap 25, pp 491–508, in: *Handbook of Endocrinology.* Section 7: *Endocrinology.* Vol 5: *Male Reproductive System.* Hamilton DW, Greep RO (editors). American Physiological Society, Washington DC, 1975.

Ovarian Hormones

Channing CP, Coudert SP: The role of granulosa cells and follicular fluid in estrogen secretion by the monkey ovary in vivo. *Endocrinology* 1976;**98:**590.

Channing CP, Tsafriri A: Mechanism of action of luteinizing hormone and follicle-stimulating hormone on the ovary in vitro. *Metabolism* 1977;**26:**413.

Green S et al: Human estrogen receptor DNA: Sequence, expression and homology to v-*erb*-A. *Nature* 1986; **320:**134.

Jensen EV, Jacobson HI: Basic guides to the mechanism of estrogen action. *Recent Prog Horm Res* 1962;**18:**387.

Siiteri PK, Febres F: Ovarian hormone synthesis, circulation and mechanisms of action. Pages 1401–1417 in: *Endocrinology*. Vol 3. DeGroot LJ (editor). Grune & Statton, 1979.

Toft D, Gorski J: A receptor molecule for estrogens. *Proc Natl Acad Sci USA* 1966;**55:**1574.

Sexual Differentiation

Bullock LP, Barden CW, Ohno S: The androgen insensitive mouse: Absence of intranuclear androgen retention in the kidney. *Biochem Biophys Res Commun* 1971;**44:**1537.

Jost A et al: Studies in sex differentiation in mammals. *Recent Prog Horm Res* 1973;**29:**1.

Ohno S: Major regulatory genes for mammalian sexual development. *Cell* 1976;**7:**315.

Ohno S: The role of H-Y antigen in primary sex determination. *JAMA* 1978;**239:**217.

Wilson JD, Walker JD: The conversion of testosterone to 5α-androsten-17β-ol-3-one (dihydrotestosterone) by skin slices of man. *J Clin Invest* 1969;**48:**371.

Wilson JD et al: The endocrine control of male phenotypic development. *Aust J Biol Sci* 1983;**36:**101.

Hormones of the Pancreas

51

Daryl K. Granner, MD

INTRODUCTION

The pancreas is 2 very different organs contained within one structure. The acinar portion has an **exocrine** function, secreting into the duodenal lumen the enzymes and ions used for the digestive process. The **endocrine** portion consists of the islets of Langerhans. The 1–2 million islets of the human pancreas make up 1–2% of its weight and are collections of the several different cell types listed in Table 51–1.

The pancreatic islets secrete at least 4 hormones: insulin, glucagon, somatostatin, and pancreatic polypeptide. The hormones are released into the pancreatic vein, which empties into the portal vein—a convenient arrangement, since the liver is the primary site of action of insulin and glucagon. These 2 hormones are chiefly involved in regulating carbohydrate metabolism but affect many other processes. Somatostatin, first identified in the hypothalamus as the hormone that inhibits growth hormone secretion, is present in higher concentration in the pancreatic islets than in the hypothalamus and is involved in the local regulation of insulin and glucagon secretion. Pancreatic polypeptide affects gastrointestinal secretion.

BIOMEDICAL IMPORTANCE

Insulin has been the model peptide hormone in many ways, being the first purified, crystallized, and synthesized by chemical and molecular biologic techniques. Studies of its biosynthesis led to the important concept of the propeptide. Insulin has important medical implications. Five percent of the population in developed countries has diabetes mellitus, and an equal number are liable to develop this disease. Diabetes mellitus is due to insufficient action of insulin, owing either to its absence or to resistance to its action. Glucagon, acting unopposed, aggravates this condition.

Table 51–1. Cell types in the islets of Langerhans.

Cell Type	Relative Abundance	Hormone Produced
A (or α)	~25%	Glucagon
B (or β)	~70%	Insulin
D (or δ)	<5%	Somatostatin
F	Trace	Pancreatic polypeptide

Acronyms Used in This Chapter

ACTH	Adrenocorticotropic hormone
EGF	Epidermal growth factor
FGF	Fibroblast growth factor
GIP	Gastric inhibitory polypeptide
IGF	Insulinlike growth factor
LDL	Low-density lipoproteins
IDDM	Insulin-dependent diabetes mellitus
NIDDM	Non-insulin-dependent diabetes mellitus
PDGF	Platelet-derived growth factor
PEPCK	Phosphoenolpyruvate carboxykinase
PGF$_{2\alpha}$	Prostaglandin F$_{2\alpha}$
PP	Pancreatic polypeptide
TSH	Thyroid-stimulating hormone
VLDL	Very low density lipoproteins

INSULIN

Historical Perspective

Langerhans identified the islets in the 1860s but did not understand their function—nor did von Mering and Minkowski, who demonstrated in 1889 that pancreatectomy produced diabetes. The link between the islets and diabetes was suggested by de Mayer in 1909 and by Sharpey-Schaffer in 1917, but it was Banting and Best who proved this association in 1921. These investigators used acid-ethanol to extract from the tissue an islet cell factor that had potent hypoglycemic activity. The factor was named insulin, and it was quickly learned that bovine and porcine islets contained insulin that was active in humans. Within a year, insulin was in widespread use for the treatment of diabetes and proved to be lifesaving.

Having large quantities of bovine or porcine insulin to study had an equally dramatic effect on biomedical research. Insulin was the first protein proved to have hormonal action, the first protein crystallized (Abel, 1926), the first protein sequenced (Sanger et al, 1955), the first protein synthesized by chemical techniques (Du et al; Zahn; Katsoyanis; ~1964), the first protein shown to be synthesized as a larger precursor molecule (Steiner et al, 1967), and the first protein prepared for commercial use by recombinant DNA technology. In spite of this impressive list of "firsts," less is known about how insulin works at the molecular level than about how most other hormones work at that level.

Chemistry

Insulin is a polypeptide consisting of 2 chains, A and B, linked by 2 interchain disulfide bridges that connect A7 to B7 and A20 to B19. A third intrachain disulfide bridge connects residues 6 and 11 of the A chain. The location of these 3 disulfide bridges is invariant, and the A and B chains have 21 and 30 amino acids, respectively, in most species. The covalent structure of human insulin (MW 5734) is illustrated in Fig 51–1, and a comparison of the amino acid substitutions found in a variety of species is presented in Table 51–2. Substitutions occur at many positions within either chain without affecting bioactivity and are particularly common in positions 8, 9, and 10 of the A chain. Thus, this region is not crucial for bioactivity. **Several positions and regions are highly conserved,** however, including (1) the positions of the 3 disulfide bonds, (2) the hydrophobic residues in the C-terminal region of the B chain, and (3) the N- and C-terminal regions of the A chain. Chemical modification or substitution of specific amino acids in these regions has allowed investigators to formulate a composite active region (Fig 51–2). The C-terminal hydrophobic region of the B chain is also involved in the dimerization of insulin.

Table 51–2 reveals the close similarity between human, porcine, and bovine insulins. Porcine insulin differs by a single amino acid, an alanine for threonine substitution at B30, while bovine insulin has this modification plus the substitutions of alanine for threonine at A8 and valine for isoleucine at A10. These modifications result in no appreciable change in biologic activity and very little antigenic difference. Although all patients given heterologous insulin develop low titers of circulating antibodies against the molecule, few develop clinically significant titers. Porcine and bovine insulins were standard therapy for diabetes mellitus until human insulin was produced by recombinant DNA technology. Despite a wide variation in primary structure, biologic activity is about 25–30 IU/mg dry weight for all insulins.

Table 51–2. Variations in the structure of insulin in mammalian species.*

Species	Variations From Human Amino Acid Sequence		
	A-Chain Position 8 9 10		B-Chain Position 30
Human	Thr-Ser-Ile		Thr
Pig, dog, sperm whale	Thr-Ser-Ile		Ala
Rabbit	Thr-Ser-Ile		Ser
Cattle, goat	Ala-Ser-Val		Ala
Sheep	Ala-Gly-Val		Ala
Horse	Thr-Gly-Ile		Ala
Sei whale	Ala-Ser-Thr		Ala

*Modified and reproduced, with permission, from Ganong WF: *Review of Medical Physiology,* 13th ed. Appleton & Lange, 1987.

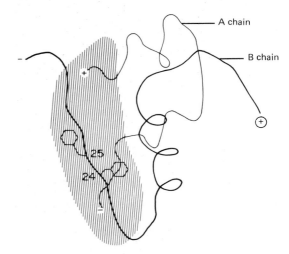

Figure 51–2. Region of the insulin molecule required for biologic activity. Diagrammatic structure of insulin as determined by x-ray crystallography. The shaded area illustrates the portion of insulin that is thought to be most important in conferring biologic activity to the hormone. The Phe residues B24 and B25 are the sites of mutations that affect insulin bioactivity. The N termini of the insulin A and B chains are indicated by ⊕, whereas the C termini are indicated by ⊖. (Redrawn and reproduced, with permission, from Tager HS: Abnormal products of the human insulin gene. *Diabetes* 1984;33:693.)

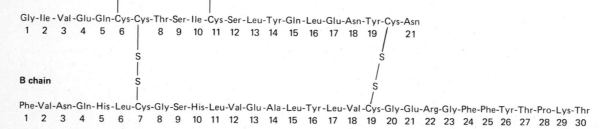

Figure 51–1. Covalent structure of human insulin. (Reproduced, with permission, from Ganong WF: *Review of Medical Physiology,* 13th ed. Appleton & Lange, 1987.)

Insulin forms very interesting complex structures. Zinc is present in high concentration in the B cell and forms complexes with insulin and proinsulin. Insulins from all vertebrate species form isologous dimers through hydrogen bonding between the peptide groups of the B24 and B26 residues of 2 monomers, and at high concentrations these are organized as hexamers, each with 2 atoms of zinc. This higher-order structure made studies of the crystalline structure of insulin feasible. Insulin is probably in the monomeric form at physiologic concentrations.

Biosynthesis

A. Precursors of Insulin: Insulin is synthesized as a **preprohormone** (MW ~11,500) and is the prototype for peptides that are processed from larger precursor molecules. The sequence and subcellular location of these biosynthetic events are depicted in Fig 51–3. The hydrophobic 23-amino-acid pre-, or leader, sequence directs the molecule into the cisternae of the endoplasmic reticulum and then is removed. This results in the 9000-MW proinsulin molecule that provides the conformation necessary for forming the proper disulfide bridges. As shown in Fig 51–4, the arrangement of proinsulin, starting from the amino terminus, is B chain—connecting (C) peptide—A chain. The proinsulin molecule undergoes a series of site-specific peptide cleavages that result in the formation of equimolar amounts of mature insulin and C peptide. These enzymatic cleavages, summarized in Fig 51–5, start with a protease with trypsinlike activity that cleaves at the carboxy-terminal end of 2 sequential basic amino acids at 2 separate sites (Arg 31–Arg 32↓, and Lys 64–Arg 65↓ in human proinsulin) within the C peptide. A second enzyme with carboxypeptidase B-like activity then removes the C-terminal basic amino acids from what will become insulin and the C peptide. The model illustrated in Fig 51–5 accounts for all of the intermediates found in the B cell secretion granule and can be duplicated by adding trypsin and chymotrypsin B to proinsulin.

B. Precursors of Other Islet Cell Hormones: The synthesis of other islet cell hormones also requires posttranslational enzymatic processing of higher-molecular-weight precursor molecules. Diagrammatic structures of pancreatic polypeptide, glucagon, and somatostatin are compared with that of insulin in Fig 51–6. Several combinations of endoproteolytic (trypsinlike) and exoproteolytic (carboxypeptidase B-like) cleavages are involved, since the hormone sequence may occur at the carboxyl terminus of the precursor (somatostatin), at the amino terminus (pancreatic polypeptide), at both ends (insulin), or in the middle (glucagon).

C. Subcellular Localization of Insulin Synthesis and Granule Formation: Insulin synthesis and the packaging of the hormone into secretion granules proceeds in orderly fashion (Fig 51–7). Proinsulin is synthesized by ribosomes on the rough endoplasmic reticulum, and the enzymatic removal of the leader peptide (pre- segment), disulfide bond forma-

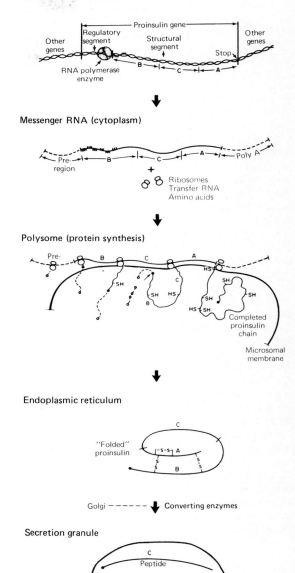

Figure 51–3. Biosynthesis of insulin via a short-lived preproinsulin. The letters A, B, and C identify the A and B chains of insulin and the connecting (C) peptide. A 23-amino-acid leader sequence transcribed by a segment of mRNA next to the portion that transcribes the B chain (dashed lines) is formed and then split off, possibly before the formation of the rest of the proinsulin molecule is completed. (Reproduced, with permission, from Steiner DF: Errors in insulin biosynthesis. *N Engl J Med* 1976; **294**:952.)

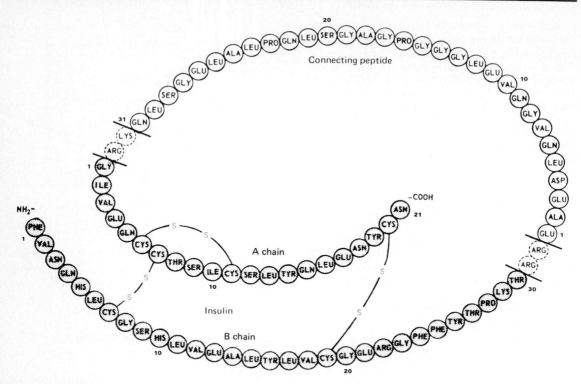

Figure 51–4. Structure of human proinsulin. Insulin and C-peptide molecules are connected at 2 sites by dipeptide links. (Slightly modified and reproduced, with permission, from Karam JH, Salber PR, Forsham PH: Pancreatic hormones and diabetes mellitus. In: *Basic & Clinical Endocrinology*, 2nd ed. Greenspan FS, Forsham PH [editors]. Appleton & Lange, 1986.)

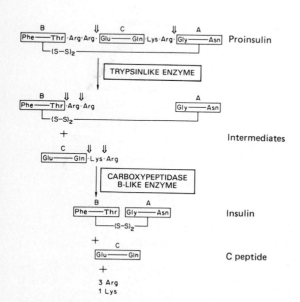

Figure 51–5. Stages in the cleavage of human proinsulin by the combined action of trypsinlike and carboxypeptidase B-like proteases. ⇓, cleavage sites. (Redrawn and reproduced, with permission, from Steiner DF, Tager HS. Page 927 in: *Endocrinology*. Vol 2. DeGroot LJ [editor]. Grune & Stratton, 1979.)

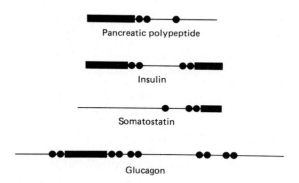

Figure 51–6. Diagrammatic structures of the precursors for the 4 major endocrine cell products of the pancreatic islet. Portions of the precursors that correspond to the named hormones are indicated by heavy bars, whereas portions that correspond to peptide extensions are indicated by lines; dibasic amino acid residues (arginine or lysine) corresponding to precursor conversion sites are shown as filled circles. Note that proinsulin has been drawn in an extended form that does not show disulfide bonds; the structure of proinsulin illustrated has the sequence B chain—C peptide—A chain. (Redrawn and reproduced, with permission, from Tager HS: Abnormal products of the human insulin gene. *Diabetes* 1984; 33:693.)

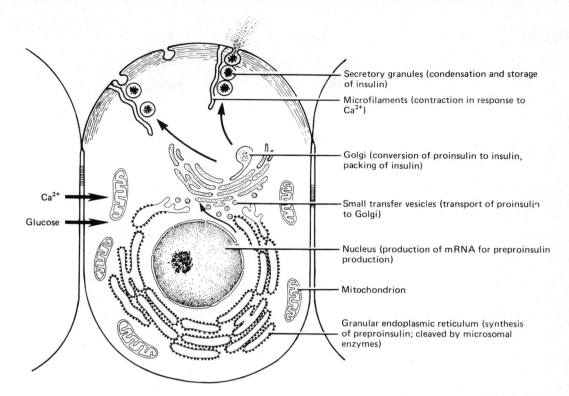

Figure 51–7. Structural components of the pancreatic B cell involved in glucose-induced biosynthesis and release. Schematic representation of secretory granular alignment on microfilament "tracks" that contract in response to calcium. (Based on data presented by Orci L: A portrait of the pancreatic B cell. *Diabetologia* 1974;10:163.) (Modified and reproduced, with permission, from Junqueira LC, Carneiro J, Long JA: *Basic Histology*, 5th ed. Appleton & Lange, 1986.)

tion, and folding (Fig 51–3) occur in the cisternae of this organelle. The proinsulin molecule is transported to the Golgi apparatus wherein proteolysis and packaging into the secretory granules begin. Granules continue to mature as they traverse the cytoplasm toward the plasma membrane. Proinsulin and insulin both combine with zinc to form hexamers, but since about 95% of the proinsulin is converted to insulin, it is the crystals of the latter that confer morphologic distinctness to the granules. Equimolar amounts of C peptide are present within these granules, but these molecules do not form a crystalline structure. Upon appropriate stimulation (see below), the mature granules fuse with the plasma membrane and discharge their contents into the extracellular fluid by **emiocytosis.**

D. Properties of Proinsulin and C Peptide: Proinsulins vary in length from 78 to 86 amino acids, with the variation occurring in the length of the C-peptide region. Proinsulin has the same solubility and isoelectric point as insulin; it also forms hexamers with zinc crystals, and it reacts strongly with insulin antisera. **Proinsulin has less than 5% of the bioactivity of insulin,** indicating that most of the active site of the latter is occluded in the precursor molecule. Some proinsulin is released with insulin and in certain conditions (islet cell tumors) in larger than usual

amounts. Since the plasma half-life of proinsulin is significantly longer than that of insulin and since proinsulin is strongly cross-reactive with insulin antisera, a radioimmunoassay for "insulin" may occasionally overestimate the bioactivity of "insulin" in plasma.

The C peptide has no known biologic activity. It is a distinct molecule from an antigenic standpoint. Thus, C-peptide immunoassays can distinguish insulin secreted endogenously from insulin administered exogenously and can quantitate the former when anti-insulin antibodies preclude the direct measurement of insulin. The C peptides of different species have a high rate of amino acid substitution, an observation which underscores the statement that this fragment probably has no biologic activity.

E. Precursors of Insulin-Related Peptides: The structural arrangement of the precursor molecule is not unique to insulin; very closely related peptide hormones (relaxin and the insulinlike growth factors) show the same arrangement (Fig 51–8). All of these hormones have highly homologous B- and A-chain regions at the amino and carboxyl termini of a precursor molecule, and these are joined by a connecting segment. In the relaxin and insulin precursor peptides, this connecting segment is bound on both ends by 2 basic amino acids. After the B and A chains are joined by

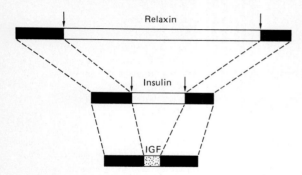

Figure 51–8. Diagrammatic structures of precursors for insulin-related peptides. Homologous regions of relaxin, insulin, and insulinlike growth factor are shown as solid bars. Amino acid sequences connecting B-chain and A-chain sequences in the precursors for relaxin and insulin are shown as open bars; these sequences are removed during processing of the precursors to their corresponding 2-chained products (vertical arrows). The amino acid sequence of insulinlike growth factor that corresponds to these connecting peptides, but which is not removed by proteolytic processing events, is shown as a stippled bar; insulinlike growth factor is a single-chain peptide hormone. (Redrawn and reproduced, with permission, from Tager HS: Abnormal products of the human insulin gene. *Diabetes* 1984;**33**:693.)

disulfide bonds, this piece is removed by endoproteolytic action, and these molecules are converted to 2-chain peptide hormones. The insulinlike growth factors, though highly homologous to insulin and relaxin in primary structure, lack these dibasic cleavage sites and thus remain single-chain peptide hormones.

F. The Human Insulin Gene: The human insulin gene (Fig 51–9) is located on the short arm of chromosome 11. Most mammals express a single insulin gene that is organized like the human gene, but rats and mice have 2 nonallelic genes. Each codes for a unique proinsulin that is processed into 2 distinct, active insulin molecules. The synthesis of human insulin in bacterial expression systems, using recombinant

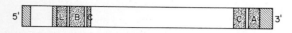

Figure 51–9. Diagrammatic structure of the human insulin gene. Areas with diagonal stripes correspond to untranslated regions of the corresponding mRNA; open regions correspond to intervening sequences; and stippled regions correspond to coding sequences. L, B, C, and A identify coding sequences for the leader (or signal) peptide, the insulin B chain, the C peptide, and the insulin A chain, respectively. Note that the coding sequence for the C peptide is split by an intervening sequence. The diagrammatic structure is drawn to scale. (Redrawn and reproduced, with permission, from Tager HS: Abnormal products of the human insulin gene. *Diabetes* 1984;**33:**693.)

DNA technology, affords an excellent source of this hormone for diabetic patients.

G. Abnormal Human Insulin Gene Products: Knowledge of the structure of the insulin gene and of the insulin molecule makes it possible to detect abnormal gene products, and this in turn provides additional information about the function of this hormone. Three cases of gene mutation have been documented, and the molecular basis of each defect has been defined. In one case, serine replaced phenylalanine at B24, owing to a single-base mutation. In another case, leucine replaced phenylalanine at B25, again because of a single-base mutation. One proinsulin processing mutation has been discovered. In this case, there was faulty cleavage at the 3′ terminus of the C peptide that abuts the A-chain peptide. The Lys-Arg at this site (Fig 51–5) was converted to Lys-X, so that trypsinlike cleavage could not occur. These cases were discovered because the mutations affected the active site of the insulin molecule; hence, the individuals had (1) hyperinsulinemia, detected by radioimmunoassay; (2) no evidence of insulin resistance; (3) decreased biologic activity of the circulating insulin; and (4) a normal response to exogenous insulin. At least 4 other single nucleotide substitutions have been identified in "normal" individuals, but since these occurred in intervening sequences or noncoding regions, they resulted in no functional impairment of the insulin molecule.

Regulation of Insulin Secretion

The human pancreas secretes 40–50 units of insulin daily, which represents about 15–20% of the hormone stored in the gland. Insulin secretion is an energy-requiring process that involves the microtubule-microfilament system in the B cells of the islets. A number of mediators have been implicated in insulin release.

A. Glucose: An increase in plasma glucose concentration is the most important physiologic regulator of insulin secretion. The threshold concentration for secretion is the fasting plasma glucose level (80–100 mg/dL), and the maximal response is obtained at glucose levels between 300 and 500 mg/dL. **Insulin secretion in response to glucose is biphasic** (Fig 51–10). There is an immediate, or first-phase, response that begins within 1 minute and lasts for 5–10 minutes. This is followed by a more gradual, prolonged second phase that terminates soon after the glucose stimulus is removed. It is postulated that these 2 phases reflect the existence of 2 intracellular compartments, or pools, of insulin. The **absolute** plasma glucose concentration is not the sole determinant of insulin secretion; the B cell responds to the **rate of change** of plasma glucose concentrations as well.

Glucose is much more effective in releasing insulin when given orally than intravenously; hence, various gastrointestinal hormones including secretin, cholecystokinin, glucagon, and gastrin have been implicated in insulin release, but gastric inhibitory polypep-

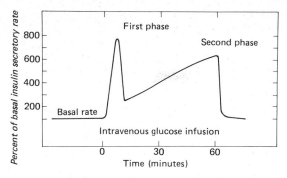

Figure 51-10. The biphasic pattern of insulin release in response to an increased plasma glucose concentration.

tide (GIP) is now thought to play the major role in this process.

Two different mechanisms have been proposed to explain how glucose regulates insulin secretion. One hypothesis suggests that glucose combines with a receptor, possibly located on the B cell membrane, that activates the release mechanism. The second hypothesis suggests that intracellular metabolites or the rate of metabolite flux through a pathway such as the pentose-phosphate shunt, the citric acid cycle, or the glycolytic pathway, is involved. There is experimental evidence to support both positions.

B. Hormonal Factors: Numerous hormones affect insulin release. α-Adrenergic agonists, principally epinephrine, inhibit insulin release even when this process has been stimulated by glucose. β-Adrenergic agonists stimulate insulin release, probably by increasing intracellular cAMP (see below). This is the probable mechanism by which GIP increases insulin release and may explain the actions of high concentrations of TSH, ACTH, gastrin, secretin, cholecystokinin, and enteroglucagon.

Chronic exposure to excessive levels of growth hormone, cortisol, placental lactogen, estrogens, and progestins also increases insulin secretion. It is therefore not surprising that insulin secretion increases markedly during the later stages of pregnancy.

C. Pharmacologic Agents: Many drugs stimulate insulin secretion, but the **sulfonylurea compounds** are used most frequently for therapy in humans. Drugs such as tolbutamide stimulate insulin release by a mechanism different from that employed by glucose and have achieved widespread use in the treatment of type II (non-insulin-dependent) diabetes mellitus.

$$H_3C-\langle\bigcirc\rangle-SO_2-NH-\underset{\underset{O}{\|}}{C}-NH-(CH_2)_3-CH_3$$

Tolbutamide

D. Intracellular Mediators of Secretion: Oxygen consumption and ATP utilization increase as glucose stimulates insulin release. This is associated with a K^+-induced depolarization of the membrane that results in the rapid entry of Ca^{2+} via a voltage-dependent channel. The fusion of insulin-containing secretory granules with the plasma membrane, and thus insulin secretion, is calcium-dependent. Metabolites of phosphatidylinositol (see Chapter 44) have been linked to glucose-stimulated insulin secretion.

cAMP is associated with insulin secretion, and this nucleotide potentiates the actions of glucose and amino acids. cAMP may release Ca^{2+} from intracellular organelles or it may activate a kinase that phosphorylates a component of the microfilament-microtubule system, thereby rendering this structure Ca^{2+}-sensitive and contractile. Replacement of extracellular Na^+ with another monovalent cation blunts the effects of glucose and other secretagogues; Na^+ may regulate the intracellular concentration of Ca^{2+} through a cotransport system.

Metabolism of Insulin

Unlike the insulinlike growth factors, **insulin has no plasma carrier protein;** thus, its plasma half-life is less than 3–5 minutes under normal conditions. The major organs involved in insulin metabolism are the liver, kidneys, and placenta; about 50% of insulin is removed in a single pass through the liver. **Mechanisms involving 2 enzyme systems are responsible for the metabolism of insulin.** The first involves an insulin-specific **protease** found in many tissues but in highest concentration in those listed above. This protease has been purified from skeletal muscle and is known to be sulfhydryl-dependent and active at physiologic pH. The second mechanism involves hepatic **glutathione-insulin transhydrogenase.** This enzyme reduces the disulfide bonds, and then the individual A and B chains are rapidly degraded. It is not clear which of these mechanisms is most active under physiologic conditions, nor is it clear whether either process is regulated.

Physiologic Effects of Insulin

The central role of insulin in carbohydrate, lipid, and protein metabolism can be best appreciated by examining the consequences of insulin deficiency in humans. The cardinal manifestation of **diabetes mellitus** is **hyperglycemia,** which results from (1) decreased entry of glucose into cells, (2) decreased utilization of glucose by various tissues, and (3) increased production of glucose (gluconeogenesis) by the liver. Each of these is discussed in more detail below.

Polyuria, polydipsia, and weight loss in spite of adequate caloric intake are the major symptoms of insulin deficiency. How is this explained? The plasma glucose level rarely exceeds 120 mg/dL in normal humans, but much higher levels are routinely found in patients with deficient insulin action. After a certain plasma glucose level is attained (generally > 180 mg/dL in humans), the maximum level of renal tubular reabsorption of glucose is exceeded, and sugar is excreted in the urine (glycosuria). The urine volume is increased owing to osmotic diuresis and coincident obligatory water loss (polyuria), and this in turn leads to dehydration (hyperosmolarity), increased thirst, and excessive drinking (polydipsia). Glycosuria causes a substantial loss of calories (4.1 kcal for every gram of glucose excreted); this loss, when coupled with the loss of muscle and adipose tissue, results in severe weight loss in spite of increased appetite (polyphagia) and normal or increased caloric intake.

Protein synthesis decreases in the absence of insulin, partly because the transport of amino acids into muscle is diminished (the amino acids serve as gluconeogenic substrates). Thus, insulin-deficient persons are in negative nitrogen balance. The antilipolytic action of insulin is lost, as is its lipogenic effect; hence, plasma fatty acid levels rise. When the capacity of the liver to oxidize fatty acids to CO_2 is exceeded, **β-hydroxybutyric acid and acetoacetic acid accumulate (ketosis).** The organism initially compensates for the accumulation of these organic acids by increasing respiratory losses of CO_2, but if unchecked by the administration of insulin, severe **metabolic acidosis** supervenes and the patient dies of **diabetic coma.** The pathophysiology of insulin deficiency is summarized in Fig 51–11.

A. Effects on Membrane Transport: The intracellular free glucose concentration is very low compared with the extracellular concentration. Most studies suggest that **the rate of glucose transport across the plasma membrane of muscle and adipose cells determines the rate of phosphorylation of glucose and its further metabolism.** D-Glucose and other sugars with a similar configuration at the $C_1–C_3$ positions (galactose, D-xylose, and L-arabinose) enter cells by **carrier-mediated facilitated diffusion,** a process enhanced in many cells by insulin (Fig 51–12). This involves a V_{max} effect (increased number of transporters) rather than a K_m effect (increased affinity of binding). Data suggest that in adipose cells this is accomplished by recruiting **glucose transporters** from an inactive pool in the Golgi fraction and then moving them to an active site in the plasma membrane. This

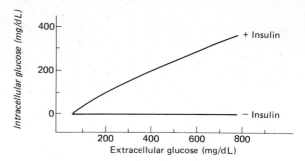

Figure 51–12. Entry of glucose into muscle cells.

transporter translocation is temperature- and energy-dependent and is protein synthesis-independent (Fig 51–13).

The hepatic cell represents a notable exception to this scheme. Insulin does not promote the facilitated diffusion of glucose into hepatocytes, but it indirectly enhances net inward flux by converting intracellular glucose to glucose 6-phosphate through the action of glucokinase, an enzyme induced by insulin. This rapid phosphorylation keeps the free glucose con-

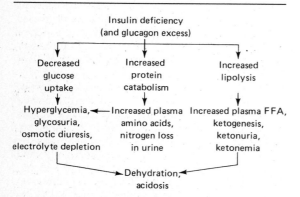

Figure 51–11. Pathophysiology of insulin deficiency. (Courtesy of RJ Havel.)

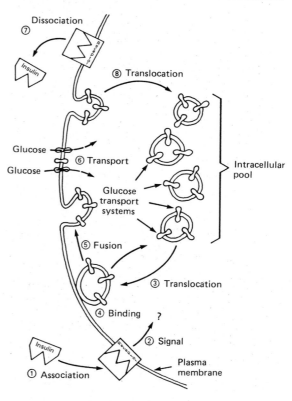

Figure 51–13. Translocation of glucose transporters by insulin. (Reproduced, with permission, from Karnieli E et al: Insulin-stimulated translocation of glucose transport systems in the isolated rat adipose cell. *J Biol Chem* 1981;**256**:4772. Courtesy of S Cushman.)

centration very low in the hepatocyte, thus favoring entry by simple diffusion down a concentration gradient.

Insulin also promotes the entry of amino acids into cells, particularly in muscle, and enhances the movement of K^+, Ca^{2+}, nucleosides, and inorganic phosphate. These effects are independent of the action of insulin on glucose entry.

B. Effects on Glucose Utilization: Insulin influences the intracellular utilization of glucose in a number of ways, as illustrated below.

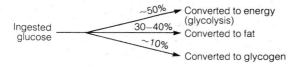

In a normal person, about half of the glucose ingested is converted to energy through the glycolytic pathway and about half is stored as fat or glycogen. Glycolysis decreases in the absence of insulin, and the anabolic processes of glycogenesis and lipogenesis are impeded. Indeed, only 5% of an ingested glucose load is converted to fat in an insulin-deficient diabetic.

Insulin increases hepatic glycolysis by increasing the activity and amount of several key enzymes including glucokinase, phosphofructokinase, and pyruvate kinase. Enhanced glycolysis increases glucose utilization and thus indirectly decreases glucose release into plasma. Insulin also decreases the activity of glucose-6-phosphatase, an enzyme found in liver but not in muscle. Since glucose 6-phosphate cannot exit from the plasma membrane, this action of insulin results in the retention of glucose within the liver cell.

Insulin stimulates lipogenesis in adipose tissue by (1) providing the acetyl-CoA and NADPH required for fatty acid synthesis; (2) maintaining a normal level of the enzyme acetyl-CoA carboxylase, which catalyzes the conversion of acetyl-CoA to malonyl-CoA; and (3) providing the glycerol involved in triacylglycerol synthesis. In insulin deficiency, all of these are decreased; thus, lipogenesis decreases. Another reason for the decreased lipogenesis in insulin deficiency is that fatty acids, released in large amounts by several hormones when unopposed by insulin, feedback-inhibit their own synthesis by inhibiting acetyl-CoA carboxylase. The net effect of insulin on fat is therefore anabolic.

The final action of insulin on glucose utilization involves another anabolic process. In liver and muscle, insulin stimulates the conversion of glucose to glucose 6-phosphate, which then undergoes isomerization to glucose 1-phosphate and is incorporated into glycogen by the enzyme glycogen synthase, the activity of which is stimulated by insulin. This action is indirect and dual in nature. Insulin decreases intracellular cAMP levels by activating a phosphodiesterase. Since cAMP-dependent phosphorylation inactivates glycogen synthase, low levels of this nucleotide allow the

enzyme to stay in the active form. Insulin also activates a phosphatase that dephosphorylates glycogen synthase, thereby resulting in the activation of this enzyme. Finally, insulin inhibits phosphorylase by a mechanism involving cAMP and phosphatase as described above, and this decreases glucose liberation from glycogen. The net effect of insulin on glycogen metabolism is also anabolic.

C. Effects on Glucose Production (Gluconeogenesis): The actions of insulin on glucose transport, glycolysis, and glycogenesis occur within seconds or minutes, since they primarily involve the activation or inactivation of enzymes by phosphorylation or dephosphorylation. A more long-term effect on plasma glucose involves **the inhibition of gluconeogenesis by insulin.** The formation of glucose from noncarbohydrate precursors involves a series of enzymatic steps, many of which are stimulated by glucagon (acting through cAMP), by glucocorticoid hormones, and to a lesser extent by α- and β-adrenergic agents, angiotensin II, and vasopressin. Insulin inhibits these same steps. The key gluconeogenic enzyme in the liver is phosphoenolpyruvate carboxykinase (PEPCK), which converts oxaloacetate to phosphoenolpyruvate. Recent studies (see below) show that insulin decreases the amount of this enzyme by selectively inhibiting transcription of the gene that codes for PEPCK mRNA.

D. Effects on Glucose Metabolism: The net action of all of the above effects of insulin is to **decrease the blood glucose level.** In this action, insulin stands alone against an array of hormones that attempt to counteract this effect. This no doubt represents one of the organism's most important defense mechanisms, since prolonged hypoglycemia poses a potentially lethal threat to the brain and must be avoided.

E. Effects on Lipid Metabolism: The lipogenic actions of insulin were discussed in the context of glucose utilization. Insulin also is a potent inhibitor of lipolysis in liver and adipose tissue and thus has an indirect anabolic effect. This is partly due to the ability of insulin to decrease tissue cAMP levels (which are increased in these tissues by the lipolytic hormones glucagon and epinephrine) but also to the fact that insulin inhibits hormone-sensitive lipase activity. This inhibition is presumably due to the activation of a phosphatase that dephosphorylates and thereby inactivates the lipase or cAMP-dependent protein kinase. Insulin therefore decreases circulating free fatty acids. This contributes to the action of insulin on carbohydrate metabolism, since fatty acids inhibit glycolysis at several steps and stimulate gluconeogenesis. This illustrates the point that one cannot discuss metabolic regulation in the context of a single hormone or metabolite. Regulation is a complex process in which the flux through a given pathway is the result of the interplay of a number of hormones and metabolites.

In patients with insulin deficiency, lipase activity increases, resulting in enhanced lipolysis and increased concentration of free fatty acids in plasma and liver. Glucagon levels also increase in these patients, and this enhances the release of free fatty acids. (Glucagon opposes most of the actions of insulin, and the metabolic state in the diabetic is a reflection of the relative levels of glucagon and insulin.) A portion of the free fatty acids is metabolized to acetyl-CoA (the reverse of lipogenesis) and then to CO_2 and H_2O via the citric acid cycle. In patients with insulin deficiency, the capacity of this process is rapidly exceeded and the acetyl-CoA is converted to acetoacetyl-CoA and then to acetoacetic and β-hydroxybutyric acids. Insulin reverses this pathway.

Insulin apparently affects the formation or clearance of VLDL and LDL, since levels of these particles, and consequently the level of cholesterol, are often elevated in poorly controlled diabetics. Accelerated atherosclerosis, a serious problem in many diabetics, is attributed to this metabolic defect.

The actions of insulin can be inferred by inspecting Fig 51–14, which depicts the flux through several critical pathways in the absence of the hormone.

F. Effects on Protein Metabolism: Insulin generally has an anabolic effect on protein metabolism in that it stimulates protein synthesis and retards protein degradation. Insulin stimulates the uptake of type A neutral amino acids into muscle, an effect that is not linked to glucose uptake or to subsequent incorporation of the amino acids into protein. The effects of insulin on general protein synthesis in skeletal and cardiac muscle and in liver are thought to be exerted at the level of mRNA translation.

In recent years, insulin has been shown to influence the synthesis of specific proteins by effecting changes in the corresponding mRNAs. This action of insulin, which may ultimately explain many of the effects the hormone has on the activity or amount of specific proteins, is discussed in more detail below.

G. Effects on Cell Replication: Insulin stimulates the proliferation of a number of cells in culture, and it may also be involved in the regulation of growth in vivo. Cultured fibroblasts are the most frequently

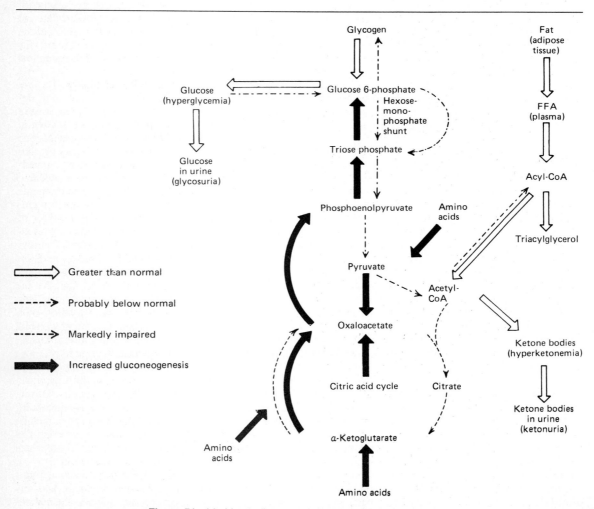

Figure 51–14. Metabolic consequences of insulin deficiency.

used cells in studies of growth control. In such cells, insulin potentiates the ability of fibroblast growth factor (FGF), platelet-derived growth factor (PDGF), epidermal growth factor (EGF), tumor-promoting phorbol esters, prostaglandin $F_{2\alpha}$ ($PGF_{2\alpha}$), vasopressin, and cAMP analogs to stimulate cell cycle progression of cells arrested in the G_1 phase of the cycle by serum deprivation.

The temporal requirements for the various growth factors have led to the concept that there are 2 classes. One class, including PDGF, FGF, $PGF_{2\alpha}$, and phorbol esters, is thought to cause a biochemical change early in G phase that, when established, removes the requirement for the growth factor and makes the cells **"competent"** for replication. The second class, including insulin, allows the cells to **"progress"** to and through S phase and must be present at all times. This model describes events in 3T3 fibroblasts, and its universality is not established. It also is not clear whether insulin acting through the insulin receptor is involved or whether insulin mediates this "progression" through an insulinlike growth factor (IGF) receptor, especially since IGF-I is also a "progression" factor.

Insulin supports the growth and replication of many cells of epithelial origin, including hepatocytes, hepatoma cells, adrenocortical tumor cells, and mammary carcinoma cells. Insulin stimulates replication at very low concentrations of the hormone, apparently through the insulin receptor, and often in the absence of other peptide growth factors. Indeed, insulin is a critical ingredient in all "defined" tissue culture media, so that a role in growth and replication is certain.

The biochemical actions of insulin in cell replication have not been established but have been linked to its known anabolic actions. Thus, effects on the uptake of glucose, phosphate, type A neutral amino acids, and cations have been implicated. The hormone might affect replication through its established ability to activate or inactivate enzymes by controlling the rate and extent of protein phosphorylation or by regulating enzyme synthesis.

An exciting new area of research involves the investigation of tyrosine kinase activity. The insulin receptor, along with receptors for many other growth-promoting peptides including those of PDGF and EGF, has tyrosine kinase activity. Interestingly, at least 10 oncogene products, many of which are suspected to be involved in stimulating malignant cell replication, are also tyrosine kinases. Mammalian cells contain analogs of these oncogenes (**proto-oncogenes**), which may be involved in the replication of normal cells. Support for the theory that they are involved comes from recent observations that the expression of at least 2 proto-oncogene products, c-*fos* and c-*myc*, increases following addition of serum to growth-arrested cells. PDGF has also been shown to stimulate the production of specific mRNAs. Whether insulin acts in a similar way remains to be established.

Mechanism of Action of Insulin

A. The Insulin Receptor: Insulin action begins when the hormone binds to a specific glycoprotein receptor on the surface of the target cell. The diverse actions of the hormone (Fig 51–15) can occur within seconds or minutes (transport, protein phosphorylation, enzyme activation and inhibition, RNA synthesis) or after a few hours (protein and DNA synthesis and cell growth).

The insulin receptor has been studied in great detail using biochemical and recombinant DNA techniques. It is a heterodimer consisting of 2 subunits, designated α and β, in the configuration α_2-β_2, linked by disulfide bonds (Fig 51–15). Both subunits are extensively glycosylated, and removal of sialic acid and galactose decreases insulin binding and insulin action. Each of these glycoprotein subunits has a unique structure and function. The α subunit (MW 135,000) is entirely extracellular, and it binds insulin, probably via a cysteine-rich domain. The β subunit (MW 95,000) is a transmembrane protein that performs the second major function of a receptor (see Chapter 44), ie, signal transduction. The cytoplasmic portion of the β subunit has tyrosine kinase activity and an autophosphorylation site. Both of these are thought to be involved in signal transduction and insulin action (see below). The striking similarity between 3 receptors with very different functions is illustrated in Fig 51–16. Indeed, several regions of the β subunit have sequence homology with the EGF receptor.

The insulin receptor is constantly being synthesized and degraded, and its half-life is 7–12 hours. The receptor is synthesized as a single-chain peptide in the rough endoplasmic reticulum and is rapidly glycosylated in the Golgi region. The precursor of the human insulin receptor has 1382 amino acids, has a molecular weight of 190,000 and is cleaved to form the mature α and β subunits. The human insulin receptor gene is located on chromosome 19.

Insulin receptors are found on most mammalian cells, in concentrations of up to 20,000 per cell, and often on cells not typically thought of as being insulin targets. Insulin has a well-known set of effects on metabolic processes but also is involved in growth and replication of cells (see above) as well as in fetal organogenesis and differentiation and in tissue repair and regeneration. The structure of the insulin receptor and the ability of different insulins to bind to receptors and elicit biologic responses are virtually identical in all cells and all species. Thus, porcine insulin is always 10–20 times more effective than porcine proinsulin, which in turn is 10–20 times more effective than guinea pig insulin, even in the guinea pig. The insulin receptor has apparently been highly conserved, more so than even insulin itself.

When insulin binds to the receptor, several events occur. (1) There is a conformational change of the receptor; (2) the receptors cross-link and form microaggregates, patches, or caps; (3) the receptor is internalized; and (4) some signal is generated. The significance of the conformational change is unknown, and

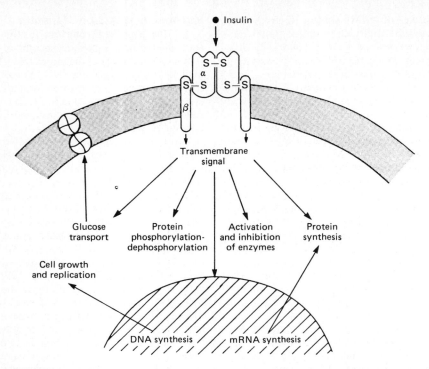

Figure 51–15. Relationship of the insulin receptor to insulin action. (Courtesy of CR Kahn.)

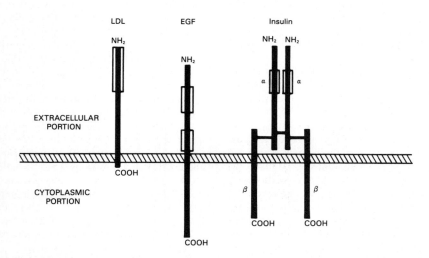

Figure 51–16. Schematic representation of the structure of the LDL, EGF, and insulin receptors. The amino terminus of each is in the extracellular portion of the molecule. The boxes represent cysteine-rich regions, which are thought to be involved in ligand binding. Each receptor has a short domain (~25 amino acids) that traverses the plasma membrane (the hatched line), and there is an intracellular domain of variable length. The EGF and insulin receptors have tyrosine kinase activity associated with the cytoplasmic domain and also have autophosphorylation sites in this region. The insulin receptor is a heterotetramer connected by disulfide bridges (vertical bars).

internalization probably represents a means of controlling receptor concentration and turnover. In conditions in which plasma insulin levels are high, eg, obesity or acromegaly, the number of insulin receptors is decreased and target tissues become less sensitive to insulin. This "down-regulation" results from the loss of receptors by internalization, the process whereby insulin-receptor complexes enter the cell through endocytosis in clathrin-coated vesicles (see Chapter 41). Down-regulation explains part of the insulin resistance in obesity and type II diabetes mellitus.

B. Intracellular Mediators: Although the mechanism of insulin action has been under investigation for 60 years, certain critical points, such as the nature of the intracellular signal, remain obscure. Insulin is not unique in this respect—the intracellular messenger has not been identified for a large number of hormones (see Table 44–1). A variety of different molecules have been proposed as the intracellular second messenger or mediator. These include insulin itself, calcium, cyclic nucleotides (cAMP, cGMP), H_2O_2, membrane-derived peptides, membrane phospholipids, monovalent cations, and tyrosine kinase (the insulin receptor). None has withstood rigorous testing.

Current interest centers on the observation that the insulin receptor is itself an insulin-sensitive enzyme, since it undergoes autophosphorylation in response to insulin binding. This function is conferred by the β subunit, which acts as a protein kinase in transferring the γ-phosphate of ATP to a tyrosine residue in the β subunit. Insulin increases the V_{max} of this enzymatic reaction, and divalent cations, particularly Mn^{2+}, decrease the K_m for ATP.

Tyrosine phosphorylation is unusual in mammalian cells (phosphotyrosine accounts for only 0.03% of the phosphoamino acid content of normal cells), and it is perhaps more than coincidental that the EGF, PDGF, and IGF-I receptors also have tyrosine kinase activity. Tyrosine kinase activity is thought to be an essential factor in the action of a number of viral oncogene products; the relationship of this and of cellular oncogene analogs with similar properties in malignant and normal cell growth was discussed above. As structures are elucidated, a high degree of homology between receptors and oncogenes is becoming apparent, eg, between the EGF receptor and *erb*-B, between the PDGF receptor and v-*sis,* and between the insulin receptor and v-*ros.*

Tyrosine kinase activity has not been proved to be involved in transduction of the insulin-receptor signal, but it could accomplish this by phosphorylating a specific protein that initiates insulin action, by initiating a phosphorylation-dephosphorylation cascade, by changing some property of the cell membrane, or by generating a membrane-related product, eg, a phospholipid.

C. Protein Phosphorylation-Dephosphorylation: Many of the metabolic effects of insulin, particularly those which occur rapidly, are mediated by influencing protein phosphorylation and dephosphorylation reactions that in turn alter the enzymatic activ-

ity of the protein. A list of enzymes affected in this way is presented in Table 51–3. In some instances, insulin decreases intracellular cAMP levels (by activating a cAMP-phosphodiesterase), thereby decreasing the activity state of cAMP-dependent protein kinase; examples of this action include glycogen synthase and phosphorylase. In other instances, this action is independent of cAMP and is exerted by activating other protein kinases (as is the case with the insulin receptor, tyrosine kinase); by inhibiting other protein kinases (see Table 44–4); or, more commonly, by stimulating the activity of phosphoprotein phosphatases. Dephosphorylation increases the activity of a number of key enzymes (Table 51–3). These covalent modifications allow for almost immediate changes in the activity of enzymes.

D. Effects on mRNA Translation: Insulin is known to affect the activity or amount of at least 50 proteins in a variety of tissues, and many of these effects involve covalent modification. A role for insulin in the translation of mRNA has been proposed, largely based on studies of ribosomal protein S6, a component of the 40S ribosomal subunit. Such a mechanism could account for the general effect insulin has on protein synthesis in liver, skeletal muscle, and cardiac muscle.

E. Effects on Gene Expression: The actions of insulin discussed heretofore all occur at the plasma

Table 51–3. Enzymes whose degree of phosphorylation and activity are altered by insulin.*

Enzyme	Change in Activity	Possible Mechanism
cAMP metabolism		
Phosphodiesterase (low K_m)	Increase	Phosphorylation
Protein kinase (cAMP-dependent)	Decrease	Association of R and C subunits
Glycogen metabolism		
Glycogen synthase	Increase	Dephosphorylation
Phosphorylase kinase	Decrease	Dephosphorylation
Phosphorylase	Decrease	Dephosphorylation
Glycolysis and gluconeogenesis		
Pyruvate dehydrogenase	Increase	Dephosphorylation
Pyruvate kinase	Increase	Dephosphorylation
6-Phosphofructo-2-kinase	Increase	Dephosphorylation
Fructose-2,6-bisphosphatase	Decrease	
Lipid metabolism		
Acetyl-CoA carboxylase	Increase	Phosphorylation
HMG-CoA reductase	Increase	Dephosphorylation
Triacylglycerol lipase	Decrease	Dephosphorylation
Other		
Tyrosine kinase (the insulin receptor)	?	Phosphorylation

*Modified and reproduced, with permission, from Denton RM et al: A partial view of the mechanism of insulin action. *Diabetologia* 1981;**21**:347.

membrane level or in the cytoplasm. In addition, insulin affects specific nuclear processes, presumably through its intracellular mediator. The enzyme **phosphoenolpyruvate carboxykinase (PEPCK)** catalyzes a rate-limiting step in gluconeogenesis. The synthesis of PEPCK is decreased by insulin; hence, gluconeogenesis decreases. Recent studies show that the rate of transcription of the PEPCK gene is selectively decreased within minutes after the addition of insulin to cultured hepatoma cells (Fig 51–17). The decrease in transcription accounts for the decreased amount of the primary transcript and of mature mRNA^{PEPCK}, which in turn is directly related to the decreased rate of PEPCK synthesis. This effect occurs at physiologic levels of insulin (10^{-12} to 10^{-9} mol/L), is mediated through the insulin receptor, and appears to be due to a decreased rate of mRNA^{PEPCK} transcript initiation.

Although studies of PEPCK regulation provided the first example of an effect of insulin on gene transcription, this case is no longer unique. Indeed, it appears that regulation of mRNA synthesis is a major action of insulin. A number of specific mRNAs are affected by insulin (Table 51–4), and a number of mRNAs in liver, adipose tissue, skeletal muscle, and cardiac muscle, as yet unidentified, are also affected by the hormone. In several instances, including those of ovalbumin, albumin, and casein, the hormone is known to affect gene transcription.

This effect of insulin involves enzymes retained in the cells, secreted enzymes and proteins, proteins involved in the reproductive process, and structural

Table 51–4. Messenger RNAs regulated by insulin.

Intracellular enzymes
  Tyrosine aminotransferase*
  Phosphoenolpyruvate carboxykinase*
  Fatty acid synthase
  Pyruvate kinase*
  Glycerol-3-phosphate dehydrogenase*
  Glyceraldehyde-1-dehydrogenase*
  Glucokinase
Secreted proteins and enzymes
  Albumin*
  Amylase
  α_{2u} Globulin
  Growth hormone*
Proteins involved in reproduction
  Ovalbumin*
  Casein*
Structural proteins
  δ-Crystallin
Other proteins
  Liver (p33, etc)
  Adipose tissue
  Cardiac muscle
  Skeletal muscle

*Insulin regulates the rate of specific mRNA transcription.

proteins (Table 51–4). A number of organs or tissues are involved, and the effect occurs in many species. The regulation of specific mRNA transcription by insulin is now well established, and as a means of modulating enzyme activity, it may soon rival phosphorylation-dephosphorylation in importance. The effect of insulin on gene transcription may also explain its effect on embryogenesis, differentiation, and growth and replication of cells.

Pathophysiology

Insulin deficiency or resistance to the action of insulin results in **diabetes mellitus.** About 90% of persons with diabetes have **non-insulin dependent (type II) diabetes mellitus (NIDDM).** Such patients are usually obese, have elevated plasma insulin levels, and have down-regulated insulin receptors. The other 10% have **insulin-dependent (type I) diabetes mellitus (IDDM).** The metabolic derangements discussed earlier most typically apply to the type I diabetic.

Certain rare conditions illustrate essential features about insulin action. A few individuals produce antibodies directed against their insulin receptors. These antibodies prevent insulin from binding to the receptor, so that such persons develop a syndrome of severe insulin resistance (see Table 43–2). Tumors of B cell origin cause hyperinsulinism and a syndrome characterized by severe hypoglycemia. The role of insulin (or perhaps of IGF-I or IGF-II) in organogenesis and development is illustrated by the rare cases of **leprechaunism.** This syndrome is characterized by low birth weight, decreased muscle mass, decreased subcutaneous fat, elfin facies, insulin resistance with markedly elevated plasma levels of biologically active insulin, and early death. Several individuals with leprechaunism have been shown to lack insulin receptors or to have defective receptors.

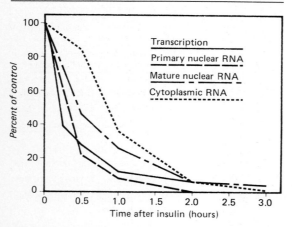

Figure 51–17. Effect of insulin on specific gene transcription. The addition of insulin to H4IIE hepatoma cells results in a rapid decrease in the rate of transcription of the PEPCK gene. This is followed by decreases in the amounts of primary transcript in the nucleus and mature mRNA^{PEPCK}. The rate of synthesis of PEPCK protein declines after the amount of cytoplasmic mRNA^{PEPCK} decreases. (Reproduced, with permission, from Sasaki K et al: Multihormonal regulation of phosphoenolpyruvate carboxykinase gene transcription. *J Biol Chem* 1984; **259:**15242.)

INSULINLIKE GROWTH FACTORS

The insulinlike growth factors (IGF-I and IGF-II) are not pancreatic hormones but are related to insulin in structure and function. It is difficult to separate the effects of insulin on cell growth and replication from similar actions exerted by IGF-I and IGF-II. Indeed, insulin and the IGFs may interact in this process. The structural similarity of these proteins was alluded to above and in Fig 51–8. A more detailed comparison is presented in Table 51–5. IGF-I and IGF-II are single-chain polypeptides of 70 and 67 amino acids, respectively. There is 62% homology between IGF-I and IGF-II, and these 2 hormones are identical with insulin in 50% of their residues. These molecules have unique antigenic sites and are regulated in different ways (Table 51–5). Insulin is the more potent metabolic hormone, whereas the IGFs are more potent in stimulating growth. Each hormone has a unique receptor. The IGF-I receptor, like the insulin receptor, is a heterodimer of α_2-β_2 structure and is a tyrosine kinase. The IGF-II receptor, in contrast, is a single-chain polypeptide of MW 260,000 and is not a tyrosine kinase.

There is some cross talk between these hormones and their receptors, and this probably accounts for the mixed biologic activity of these hormones (Table 51–6). In general, the growth-promoting effects of these hormones correlate best with their affinity for binding to the IGF-I or IGF-II receptor.

GLUCAGON

The early commercial preparations of insulin increased the plasma glucose level before lowering it,

Table 51–6. Binding of insulin, IGF-I, and IGF-II to various receptors.

Hormone	Receptor		
	Insulin	**IGF-I**	**IGF-II**
Insulin	High	Low	Negligible
IGF-I	Moderate	High	Moderate
IGF-II	Negligible	Low	High

owing to the presence of a contaminating peptide, glucagon, which was the second pancreatic islet cell hormone discovered.

Chemistry

Glucagon is a single-chain polypeptide (MW 3485) consisting of 29 amino acids (Fig 51–18). Glucagon contains no cysteine residues and thus has no disulfide bonds. Glucagon shares some immunologic and physiologic properties with enteroglucagon, a peptide extracted from the duodenal mucosa, and 14 of the 27 amino acid residues of secretin are identical to those of glucagon (Table 52–5).

Biosynthesis & Metabolism

Glucagon is synthesized mainly in the A cells of the pancreatic islets, although significant amounts may come from other sites in the gastrointestinal tract. Glucagon is synthesized as a much larger (MW ~9000) proglucagon precursor. Molecules larger than this have been detected, but whether they represent glucagon precursors or closely related peptides is unclear. Only 30–40% of the immunoreactive "glucagon" in plasma is pancreatic glucagon; the rest consists of biologically inactive larger molecules.

Glucagon circulates in plasma in the free form. Since it does not associate with a transport protein, its plasma half-life is short (~5 minutes). Glucagon is inactivated by the liver, which has an enzyme that removes the first 2 amino acids from the N-terminal end by cleaving between Ser 2 and Gln 3. Since the liver is the first stop for glucagon after it is secreted and since

Table 51–5. Comparison of insulin and the insulinlike growth factors. (Courtesy of CR Kahn.)

	Insulin	**IGF-I**	**IGF-II**
Other names	. . .	Somatomedin C	Multiplication-stimulating activity (MSA)
Number of amino acids	51	70	67
Source	Pancreatic B cells	Liver and other tissues	Diverse tissues
Level regulated by	Glucose	Growth hormone, nutritional status	Unknown
Plasma levels	0.3–2 ng/mL	ng/mL range	ng/mL range
Plasma binding protein	No	Yes	Yes
Major physiologic role	Control of metabolism	Skeletal and cartilage growth	Unknown; perhaps a role in embryonic development

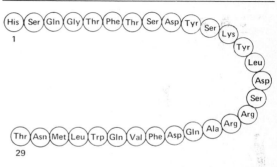

Figure 51–18. Amino acid sequence of glucagon. (Reproduced, with permission, from Katzung BG [editor]: *Basic & Clinical Pharmacology*, 3rd ed. Appleton & Lange, 1987.)

the liver rapidly inactivates the hormone, the level of glucagon in the portal vein is much higher than that in the peripheral circulation.

Regulation of Secretion

The secretion of glucagon is inhibited by glucose, an action that emphasizes the opposing metabolic roles of glucagon and insulin. It is not clear whether glucose directly inhibits glucagon secretion or whether this is mediated through the actions of insulin or IGF-I, since both of these islet cell hormones directly inhibit glucagon release. Many other substances, including amino acids, fatty acids and ketones, gastrointestinal tract hormones, and neurotransmitters, affect glucagon secretion.

Physiologic Effects

In general, the actions of glucagon oppose those of insulin. Whereas insulin promotes energy storage by stimulating glycogenesis, lipogenesis, and protein synthesis, glucagon causes the rapid mobilization of potential energy sources into glucose by stimulating glycogenolysis and into fatty acids by stimulating lipolysis. Glucagon is also the most potent gluconeogenic hormone, and it is ketogenic.

The liver is the primary target of glucagon action. Glucagon binds to specific receptors in the hepatic cell plasma membrane, and this activates adenylate cyclase. The cAMP generated activates phosphorylase, which enhances the rate of glycogen degradation while inhibiting glycogen synthase and thus glycogen formation (see Chapter 44). There is hormone and tissue specificity in this effect, since glucagon has no effect on glycogenolysis in muscle, whereas epinephrine is active in both muscle and liver.

The elevated cAMP level stimulates the conversion of amino acids to glucose by inducing a number of enzymes involved in the gluconeogenic pathway. Principal among these is PEPCK. Glucagon, through cAMP, increases the rate of transcription of mRNA from the PEPCK gene, and this stimulates the synthesis of more PEPCK. This is the opposite of the effect of insulin, which decreases PEPCK gene transcription. Other examples are illustrated in Table 51–7. The net action of glucagon in the liver is increased glucose production; since much of this glucose exits the liver, the plasma glucose concentration increases in response to glucagon.

Glucagon is a potent lipolytic agent; it increases adipose cell cAMP levels, and this activates the hor-

Table 51–7. Enzymes induced or repressed by insulin or glucagon.*

Enzymes induced by a high insulin:glucagon ratio and repressed by a low insulin:glucagon ratio
 Glucokinase
 Citrate cleavage enzyme
 Acetyl-CoA carboxylase
 HMG-CoA reductase
 Pyruvate kinase
 6-Phosphofructo-1-kinase
 6-Phosphofructo-2-kinase/fructose-2,6-bisphosphatase
Enzymes induced by a low insulin:glucagon ratio and repressed by a high insulin:glucagon ratio
 Glucose-6-phosphatase
 Phosphoenolpyruvate carboxykinase (PEPCK)
 Fructose-1,6-bisphosphatase

*Slightly modified and reproduced, with permission, from Karam JH, Salber PR, Forsham PH: Pancreatic hormones and diabetes mellitus. In: *Basic & Clinical Endocrinology,* 2nd ed. Greenspan FS, Forsham PH (editors). Appleton & Lange, 1986.

mone-sensitive lipase. The increased fatty acids can be metabolized for energy or converted to the ketone bodies acetoacetate and β-hydroxybutyrate. This is an important aspect of metabolism in the diabetic, since glucagon levels are always increased in insulin deficiency.

SOMATOSTATIN

Somatostatin, so-named because it was first isolated from the hypothalamus as the factor that inhibited growth hormone secretion, is a cyclic peptide synthesized as a large somatostatin prohormone (MW ~11,500) in the D cells of the pancreatic islets. The rate of transcription of the prosomatostatin gene is markedly enhanced by cAMP. The prohormone is first processed into a 28-amino-acid peptide and finally into a molecule that has a molecular weight of 1640 and contains 14 amino acids (Fig 51–19). All forms have biologic activity.

In addition to its presence in the hypothalamus and pancreatic islets, somatostatin is found in many gastrointestinal tissues, where it is thought to regulate a variety of functions, and in multiple sites in the central nervous system, where it may be a neurotransmitter.

Somatostatin inhibits the release of the other islet cell hormones through a paracrine action. In pharmacologic amounts, somatostatin significantly blunts the

Figure 51–19. Amino acid sequence of somatostatin. (Reproduced, with permission, from Karam JH, Salber PR, Forsham PH: Pancreatic hormones and diabetes mellitus. In: *Basic & Clinical Endocrinology,* 2nd ed. Greenspan FS, Forsham PH [editors]. Appleton & Lange, 1986.)

ketosis associated with acute insulin deficiency. This is apparently due to its ability to inhibit the glucagon release that accompanies insulinopenia. It also decreases the delivery of nutrients from the gastrointestinal tract into the circulation, because it (1) prolongs gastric emptying, (2) decreases gastrin secretion and therefore gastric acid production, (3) decreases pancreatic exocrine (digestive enzyme) secretion, (4) decreases splanchnic blood flow, and (5) slows sugar absorption. Little is known about the biochemical and molecular actions of this hormone.

PANCREATIC POLYPEPTIDE

Pancreatic polypeptide (PP), a 36-amino-acid peptide (MW ~4200), is a recently discovered product of the pancreatic F cells. Its secretion in humans is increased by a protein meal, fasting, exercise, and acute hypoglycemia and is decreased by somatostatin and intravenous glucose. The function of pancreatic polypeptide is unknown, but effects on hepatic glycogen levels and gastrointestinal secretion have been suggested.

REFERENCES

Chance RE, Ellis RM, Bromer WW: Porcine proinsulin: Characterization and amino acid sequence. *Science* 1968; **161:**165.

Cohen P: The role of protein phosphorylation in neural and hormonal control of cellular activity. *Nature* 1982;**296:**613.

Docherty K, Steiner DF: Post-translational proteolysis in polypeptide hormone biosynthesis. *Annu Rev Physiol* 1982; **44:**625.

Granner DK, Andreone T: Insulin modulation of gene expression. In: *Diabetes and Metabolism Reviews.* Vol 1. DeFronzo R (editor). Wiley, 1985.

Kahn CR: The molecular mechanism of insulin action. *Annu Rev Med* 1985; **36:**429.

Kono T: Action of insulin on glucose transport and cAMP phosphodiesterase in fat cells: Involvement of two distinct molecular mechanisms. *Recent Prog Horm Res* 1983;**30:**519.

Straus DS: Growth-stimulatory actions of insulin in vitro and in vivo. *Endocr Rev* 1984;**5:**356.

Tager HS: Abnormal products of the human insulin gene. *Diabetes* 1984;**33:**693.

Ullrich A et al: Human insulin receptor and its relationship to the tyrosine kinase family of oncogenes. *Nature* 1985;**313:**756.

Unger RH, Orci L: Glucagon and the A cell. (2 parts.) *N Engl J Med* 1981;**304:**1518,1575.

52

Gastrointestinal Hormones

Daryl K. Granner, MD

INTRODUCTION

The gastrointestinal tract secretes many hormones, perhaps more than any other single organ. The purpose of the gastrointestinal tract is to propel foodstuffs to sites of digestion, to provide the proper milieu (enzymes, pH, salt, etc) for the digestive process, to move the digested products across the intestinal mucosa through the mucosal cells and into the extracellular space, to move those products to distant cells via the circulation, and to expel waste products. The gastrointestinal hormones assist in all of these functions.

BIOMEDICAL IMPORTANCE

Disease syndromes due to excessive production of several of these hormones have been described. Signs and symptoms often involve many organ systems, and accurate diagnosis can be difficult unless the physician is aware of these syndromes. The gastrointestinal hormones are of interest also because of their close link to neuropeptides.

HISTORICAL PERSPECTIVE

The discipline of endocrinology began with the discovery of a gastrointestinal hormone. In 1902, Bayliss and Starling instilled hydrochloric acid into a denervated loop of a dog's jejunum and showed that this increased the secretion of fluid from the pancreas. Intravenous injection of HCl did not mimic this effect, but the intravenous injection of an extract of the jejunal mucosa did. These investigators postulated that "secretin," released from the mucosa of the upper intestine in response to a stimulus, moved to the pancreas through the circulation, where it exerted its effect. Bayliss and Starling were the first to use the word "hormone," and secretin was the first hormone whose function was identified.

Although the activity of secretin was identified in 1902, it took 60 years before its chemical identity was proved. In the meantime, many "younger" hormones were discovered, sequenced, and synthesized, often in a matter of a few years (eg, calcitonin; see Chapter 47). The reasons for this 60-year time span are now apparent; families of closely related gastrointestinal peptides have overlapping chemical structures and bio-

Acronyms Used In This Chapter	
BLI	Bombesinlike immunoreactivity
CCK	Cholecystokinin
GIP	Gastric inhibitory polypeptide
GRP	Gastrin-releasing peptide
PP	Pancreatic polypeptide
VIP	Vasoactive intestinal polypeptide

logic functions, and most of these peptides exist in multiple forms. Isolation techniques developed only recently are able to differentiate among them.

FEATURES OF GASTROINTESTINAL HORMONES

More than a dozen peptides with unique actions have been isolated from gastrointestinal tissues (Table 52–1). These peptides of the gastrointestinal hormone system differ in many respects from those of more typical hormone systems, and some of these differences are discussed below.

A. Diversity of Actions: Many of the gastrointestinal peptides fit the classic definition of a hormone, as discussed in Chapter 43. Examples include gastrin, secretin, gastric inhibitory polypeptide (GIP), and possibly cholecystokinin (CCK), motilin, pancreatic polypeptide (PP), and enteroglucagon (Table 52–1). Other gastrointestinal peptides are thought to have **paracrine** actions (see Chapter 43) or to act in a **neurocrine** fashion (as local neurotransmitters or neuromodulators). This is based on the observation that although these substances are found in high concentration in neurons or in various cells in the gastrointestinal tract, they either are not found in the circulation under normal conditions or have such short plasma half-lives that they would not be effective. Peptides with neurocrine action include vasoactive intestinal peptide (VIP), somatostatin, substance P, the enkephalins, bombesinlike peptides, and neurotensin (Table 52–1). Many of these substances are thought to have paracrine actions in vivo because they affect various cells when added to tissue or organ cultures.

B. Location of Gastrointestinal Peptide-Producing Cells: A unique aspect of the gastrointestinal

Table 52–1. Gastrointestinal hormones.*

	Mechanism of Action†			
	E	N	P	Major Action
Gastrin	+	(+)	–	Gastric acid and pepsin secretion.
Cholecystokinin (CCK)	(+)	(+)	–	Pancreatic amylase secretion.
Secretin	+	–	–	Pancreatic bicarbonate secretion.
Gastric inhibitory polypeptide (GIP)	+	–	–	Enhances glucose-mediated insulin release. Inhibits gastric acid secretion.
Vasoactive intestinal polypeptide (VIP)	–	+	(–)	Smooth muscle relaxation. Stimulates pancreatic bicarbonate secretion.
Motilin	(+)	–	–	Initiates interdigestive intestinal motility.
Somatostatin	–	+	(+)	Numerous inhibitory effects.
Pancreatic polypeptide (PP)	(+)	–	(+)	Inhibits pancreatic bicarbonate and protein secretion.
Enkephalins	–	+	(+)	Opiatelike actions.
Substance P	–	+	(+)	Physiologic actions uncertain.
Bombesinlike immunoreactivity (BLI)	–	+	(+)	Stimulates release of gastrin and CCK.
Neurotensin	–	+	(+)	Physiologic actions unknown.
Enteroglucagon	(+)	(+)	(+)	Physiologic actions unknown.

*Slightly modified and reproduced, with permission, from Deveney CW, Way LW: Regulatory peptides of the gut. In: *Basic & Clinical Endocrinology,* 2nd ed. Greenspan FS, Forsham PH (editors). Appleton & Lange, 1986.
†E, endocrine; N, neurocrine; P, paracrine; ( ), suggested but not proved; +, yes; –, no.

Table 52–2. Distribution of gastrointestinal hormones.*

	Endocrine Cell†	Localization	Localized in Gut Nerves
Gastrin	G	Gastric antrum, duodenum	(?)
CCK	I	Duodenum, jejunum	Yes
Secretin	S	Duodenum, jejunum	No
GIP	K	Small bowel	No
VIP	D_1	Pancreas	Yes
Motilin	EC_2	Small bowel	No
Substance P	EC_1	Entire gastrointestinal tract	Yes
Neurotensin	N	Ileum	(?)
Somatostatin	D	Stomach, duodenum, pancreas	Yes
Enkephalins	. . .	Stomach, duodenum, gallbladder	Yes
BLI	P	Stomach, duodenum	Yes
PP	D_2F	Pancreas	No
Enteroglucagon	A	Pancreas	No
	L	Small intestine	

*Slightly modified and reproduced, with permission, from Deveney CW, Way LW: Regulatory peptides of the gut. In: *Basic & Clinical Endocrinology,* 2nd ed. Greenspan FS, Forsham PH (editors). Appleton & Lange, 1986.
†Endocrine cells identified with a specific hormone are identified by a letter. EC, enterochromaffin cell. Note that several peptides are found both in nerves and in endocrine cells. VIP has been found only in nerves. The cells containing enkephalins have yet to be named.

endocrine system is that the cells are scattered throughout the gastrointestinal tract rather than collected in discrete organs as in more typical endocrine glands. The distribution of the gastrointestinal hormones is outlined in Table 52–2, which also includes the cellular nomenclature.

Since many of the gastrointestinal peptides are found in the nerves in gastrointestinal tissues, it is not surprising that most of them are also present in the central nervous system (Table 52–3). Synthesis of the peptides by central nervous system tissue has often been difficult to prove, but new techniques of molecular biology should establish whether genes coding for these substances are active. The function of these peptides in the central and peripheral nervous systems is under investigation.

C. Precursors and Multiple Forms: Of the major gastrointestinal hormones, only secretin exists in a single form (Table 52–4). The presence of multiple forms of gastrointestinal peptides in gastrointestinal tissues and in the circulation impeded the definition of the number and nature of these molecules. The concept of precursor molecules helped clarify this issue; much of tissue heterogeneity is due to this feature. Other help came from the construction of synthetic molecules that could be prepared free of contaminat-

Table 52–3. Peptides found in gut and central nervous system.*

Isolated from both brain and gut
    Substance P
    Neurotensin
    Secretin
    Somatostatin
    CCK
Isolated from either brain or gut: immunoreactivity found in the other organ
    VIP
    PP and motilin
    Enkephalins and endorphins
    BLI
    Insulin
    Glucagon

*Slightly modified and reproduced, with permission, from Deveney CW, Way LW: Regulatory peptides of the gut. In: *Basic & Clinical Endocrinology,* 2nd ed. Greenspan FS, Forsham PH (editors). Appleton & Lange, 1986.

Table 52–4. Multiple forms of gastrointestinal hormones.

Hormone	In Tissue	In Plasma
Secretin	27 Amino acids (S27)	S27
Gastrin	Large precursor	
	34 Amino acids (G34)	G34
	17 Amino acids (G17)	G17
	14 Amino acids (G14)	G14
Cholecystokinin	Large precursor	
	39 Amino acids (CCK39)	
	33 Amino acids (CCK33)	
	12 Amino acids (CCK12)	CCK12
	8 Amino acids (CCK8)	CCK8
	4 Amino acids (CCK4)	
GIP	Large precursor	Large form
	43 Amino acids (GIP43)	GIP43
Somatostatin	11,500-MW prohormone	
	28 Amino acids	
	14 Amino acids	

ing peptides and then used to define the function of specific peptides.

D. Overlapping Structure and Function of Gastrointestinal Peptides: The amino acid sequences of gastrointestinal peptides have been determined (Table 52–5). Many of these hormones can be placed in one of 2 families based on sequence and functional similarity. These are the **gastrin family,** which consists of gastrin and CCK, and the **secretin family,** which includes secretin, glucagon, GIP, VIP, and glicentin (which has glucagonlike immunoreactivity but is a distinct peptide). The neurocrine peptides neurotensin, bombesinlike peptides, substance P, and somatostatin bear no structural similarity to any other gastrointestinal peptide. A final general characteristic of this last group of molecules is that they have very short plasma half-lives and may play no physiologic role in plasma.

Table 52–5. Amino acid sequences of gastrointestinal peptides.*†

1 CCK	2 Gastrin	3 GIP	4 Glucagon	5 Secretin	6 VIP	7 Motilin	8 Substance P	9 Bombesin	10 Somatostatin
Tyr		Tyr	His	-	-	Phe	Arg	(pyro)Glu	Ala
Ile		Ala	Ser	-	-	Val	Pro	Gln	Gly
Gln		Glu	Gln	Asp	-	Pro	Lys	Arg	Cys
Gln		Gly	-	-	Ala	Ile	Pro	Leu	Lys
Ala		Thr	-	-	Val	Phe	Gln	Gly	Asn S
→Arg	(pyro)Glu	Phe	-	-	-	Thr	Gln	Asn	Phe
Lys	Leu	Ile	Thr	-	-	Tyr	Phe	Gln	Phe
Ala	Gly	Ser	-	-	Asp	Gly	Phe	Trp	-
Pro	-	Asp	-	Glu	Asn	Glu	Gly	Ala	Lys
Ser	Gln	Tyr	-	Leu	Tyr	Leu	-	Val	Thr
Gly	-	Ser	-	-	Thr	Gln	Met-NH₂	Gly	Phe
Arg	His	Ile	Lys	Arg	-	-		His	Thr S
Val	Pro	Ala	Tyr	Leu	-	Met		Leu	Ser
Ser	-	Met	Leu	Arg	-	Gln		Met-NH₂	Cys
Met	Leu	Asp	-	-	Lys	Glu			
Ile	Val	Lys	Ser	-	Gln	Lys			
Lys	Ala	Ile	Arg	Ala	Met	Glu			
Asn	Asp	Arg	-	-	Ala	Arg			
Leu	Pro	Gln	Ala	Leu	Val	Asn			
Gln	Ser	Gln	-	-	Lys	-			
Ser	Lys	Asp	-	Arg	Lys	Gly			
Leu	→Lys	Phe	-	Leu	Tyr	Gln			
Asp	Gln	Val	-	Leu	Leu				
Pro	Gly	Asn	Gln	-	Asn				
Ser	→Pro	Trp	-	Gly	Ser				
His	Trp	Leu	-	-	Ile				
→Arg	Leu	-	Met	Val-NH₂	Leu				
Ile	Glu	Ala	Asp		Asn-NH₂				
Ser	Glu	Gln	Thr						
Asp	Glu	Gln							
→Arg	Glu	Lys							
Asp	Glu	Gly							
Tys	Ala	Lys							
Met	Tys	Lys							
→Gly	→	Ser							
Trp	-	Asp							
Met	-	Trp							
Asp	-	Lys							
Phe-NH₂	-	His							
		Asn							
		Ile							
		Thr							
		Gln							

*Slightly modified and reproduced, with permission, from Grossman MI: The gastrointestinal hormones: An overview. In: *Endocrinology.* James VHT (editor). Excerpta Medica, 1977.

†Tys, tyrosine sulfate; -, same as preceding column; →, point of cleavage to form a smaller variant.

E. Mechanism of Action: Studies of the mechanism of action of the gastrointestinal peptide hormones have lagged behind those of other hormones, no doubt because most attention to date has been directed toward cataloging the various molecules and establishing their physiologic action. A notable exception to this statement involves the regulation of secretion of enzymes by the pancreatic acinar cell.

Six different classes of receptors on pancreatic acinar cells have been identified (Fig 52–1). These are for (1) muscarinic cholinergic agents, (2) the gastrin-CCK family, (3) bombesin and related peptides, (4) the physalaemin-substance P family, (5) secretin and VIP, and (6) cholera toxin.

Fig 52–1 illustrates that these peptide-receptor complexes activate 2 distinct intracellular mechanisms. One involves the mobilization of intracellular calcium stores, and the other involves the activation of adenylate cyclase and the generation of cAMP. These mechanisms do not cross over; ie, gastrin does not alter cAMP levels, nor does secretin affect the intracellular Ca^{2+} level. The 2 systems converge at some point, however, since combinations of secretagogues that act by these different mechanisms have a synergistic effect on enzyme secretion.

The peptides that cause Ca^{2+} mobilization in the pancreatic acinar cells also affect the metabolism of phosphatidylinositol and enhance its conversion to diacylglycerol and various inositol phosphates. These changes precede those of Ca^{2+} mobilization and thus may be involved in the primary response. These effects are associated with depolarization of the acinar cell, which may be involved in amylase secretion. The molecular basis of the cAMP-mediated secretion has not been elucidated. The convergence of the actions of cAMP and Ca^{2+} and phospholipids on amylase secretion is similar in many respects to that of others discussed in Chapter 44.

THE SECRETIN FAMILY

Secretin

Secretin, a 27-amino-acid peptide synthesized and released by the S cells in the duodenum and proximal jejunum in response to duodenal acidification, contains 4 arginine residues and one histidine and is basic. Its structure is identical to that of glucagon in 14 of 27 amino acids and quite similar to the structures of GIP and VIP (Table 52–5). The entire 27-amino-acid structure is required for stimulation of bicarbonate and water secretion from the pancreas.

Gastric Inhibitory Polypeptide

Gastric inhibitory polypeptide (GIP) is a 43-amino-acid peptide released from the duodenal and jejunal mucosa in response to glucose. Although GIP inhibits gastric motility and secretion, its major action appears to be the stimulation of insulin release. GIP is probably the physiologic B cell-stimulating hormone of the gastrointestinal tract. This action requires concomitant hyperglycemia.

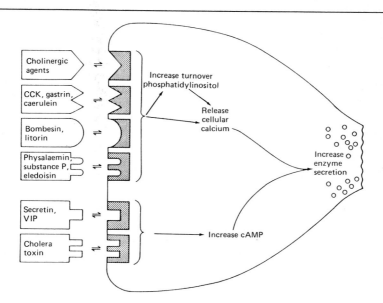

Figure 52–1. Mechanism of action of secretagogues on enzyme secretion from pancreatic acinar cells. There are 4 classes of receptors for secretagogues that can cause mobilization of cellular calcium and 2 classes of receptors for secretagogues that can cause activation of adenylate cyclase and increase cellular generation of cAMP. The interaction of these 2 pathways is described in the text. (Slightly modified and reproduced, with permission, from Gardner JD, Jensen RT: Gastrointestinal peptides: The basis of action at the cellular level. *Recent Prog Horm Res* 1983;**39:**211.)

Vasoactive Intestinal Polypeptide

Vasoactive intestinal polypeptide (VIP) is a basic 28-amino-acid peptide whose physiologic role has not been defined. It is present in the nerves of the submucosal plexus, the myenteric plexus, and blood vessels; it may be involved in gut motility, sphincter relaxation, and blood flow. At high concentration, VIP stimulates secretion by the pancreas and small intestine. Tumors that produce this peptide, **VIPomas,** cause a syndrome of watery diarrhea, hypokalemia, and achlorhydria.

Glucagon

Glucagon is the final member of this family; its chemical nature and actions were discussed above. It is made by gastric and duodenal A cells as well as by pancreatic A cells.

THE GASTRIN-CHOLECYSTOKININ FAMILY

Gastrin is produced by G cells located in the antral gastric mucosa and to a lesser extent in the duodenal mucosa. Gastrin exhibits more heterogeneity in size (number of forms) than any other gastrointestinal hormone (Table 52–4); in addition, each of these forms of gastrin has a sulfated and nonsulfated form (at the single Tyr residue; see explanatory note in Table 52–5). The carboxy-terminal 14 amino acids of G34, G17, and G14 are identical. G34 is more abundant in the circulation than G17, probably because its plasma half-life (15 minutes) is 5–7 times that of G17. The latter is thought to be the main stimulus for gastric acid secretion, which is under negative feedback control, since acidification of the antral region of the stomach decreases gastrin release. Gastrin also stimulates pepsin secretion and results in hypertrophy of the gastric mucosa. The carboxyl end of gastrin is responsible for biologic activity; the carboxy-terminal pentapeptide has the full range of physiologic action of G17 but is one-tenth as potent on a weight basis.

The homology between gastrin and CCK is particularly prominent in the C-terminal region, since the last 5 amino acids in these molecules are identical (Table 52–5). Both have a sulfated tyrosine residue, but this residue is not essential for gastrin activity (stimulation of gastric acid and pepsin secretion), whereas a 7-amino-acid C-terminal fragment with a sulfated tyrosine is required for maximal CCK activity (stimulation of pancreatic enzyme secretion and contraction of the gallbladder). This subtle change makes a large physiologic difference.

Gastrin-secreting tumors, **gastrinomas,** cause excessive gastric acid production and intractable peptic ulcer disease. The C-terminal pentapeptide, **pentagastrin,** causes calcitonin release and is used as a provocative test in the diagnosis of medullary thyroid cancer.

CCK exists in at least 5 molecular forms (Table 52–4), of which CCK8 appears to be the most abundant and potent form in the circulation. CCK is made in I cells in the mucosa of the duodenum and proximal jejunum and is released in response to peptides, amino acids, long-chain fatty acids, calcium, and acid. Its physiologic actions are to cause gallbladder contraction and to stimulate pancreatic enzyme secretion. Like all gastrointestinal peptides, it has numerous secondary actions, an intriguing one of which is to cause satiety (CCK8 is the form found in the brain).

OTHER GASTROINTESTINAL PEPTIDES

Several other peptides have been isolated from the gastrointestinal tract and appear to be involved in the regulation of digestion—probably by paracrine or neurocrine mechanisms, since appreciable concentrations are not found in the circulation. It should be emphasized that no certain physiologic role has been assigned to any of these molecules.

Substance P was the first peptide found in both the gut and the brain. The 5 carboxy-terminal amino acids of this 11-amino-acid peptide are required for its action, which appears to be stimulation of smooth muscle contraction in the intestine. **Bombesin** is found in frog skin, but a similar peptide, often called **gastrin-releasing peptide (GRP),** has been isolated from endocrine cells in the gut, gut neurons, and brain. The amino acids at positions 5–14 of bombesin are identical to those at positions 18–27 of gastrin-releasing peptide except at one residue. Bombesin stimulates gastric and pancreatic secretion and increases motility of the gallbladder and intestine. This peptide may have a growth-promoting autocrine action in small cell carcinoma of the lung. **Motilin,** a 22-amino-acid peptide, is made in the intestinal mucosa. It stimulates acid and pepsin secretion by gastric mucosa and is a stimulator of intestinal smooth muscle contraction. **Somatostatin** is produced in gastric D cells and inhibits (by paracrine action) the release of gastrin, secretin, CCK, motilin, and GIP. **Glucagon** is made in the gastric mucosa A cells and is thought to contribute to the metabolic action of pancreatic glucagon. Other peptides with glucagonlike immunoreactivity (GLI) have been isolated from the L cells of the ileum and colon. The major component of GLI is a large 100-amino-acid peptide called **glicentin,** which contains the exact sequence of pancreatic glucagon. Glicentin may mimic the actions of glucagon.

Neurotensin (a 13-amino-acid peptide), **Met-** and **Leu-enkephalins,** and **serotonin** are found in intestinal cells and may be active in these tissues. Some 40 peptides have been found in neural tissues, and it is likely that more gastrointestinal peptides will be discovered.

REFERENCES

Bloom SR, Polak JM: *Gut Hormones,* 2nd ed. Churchill Livingstone, 1981.

Boden G: Gastrointestinal hormones. Pages 1175-1190 in: *Endocrinology and Metabolism.* Felig P et al (editors). McGraw-Hill, 1981.

Buchanan KD: Gastrointestinal hormones: General concepts. *Clin Endocrinol Metab* 1979;**8:**249.

Chen WY, Gutierrez JG: The endocrine control of gastrointestinal function. *Adv Intern Med* 1978;**23:**61.

Gardner JD, Jensen RT: Gastrointestinal peptides: The basis of action at the cellular level. *Recent Prog Horm Res* 1983; **39:**211.

Walsh JH: Gastrointestinal hormones and peptides. In: *Physiology of the Gastrointestinal Tract.* Johnson LR (editor). Raven Press, 1981.

Section VI.
Special Topics

Nutrition, Digestion, & Absorption

53

Peter A. Mayes, PhD, DSc

INTRODUCTION

The science of nutrition seeks to define the qualitative and quantitative requirements of the diet necessary to maintain good health. Metabolic biochemistry provides much of the understanding of modern concepts of nutrition and has been discussed in earlier chapters of this text. Digestion and absorption of food constitute the essential link between nutrition and metabolism.

BIOMEDICAL IMPORTANCE

Overt nutritional deficiency is rare in more affluent populations, although some degree of nutritional deficiency may be present among the poor or the elderly and among groups with specialized nutritional requirements, eg, growing children, pregnant or lactating women, ill and convalescing patients, alcoholics, or individuals on restricted diets from choice or necessity. In more deprived populations, overt deficiencies are more widespread, eg, deficiency of protein **(kwashiorkor),** of vitamins (vitamin A in **xerophthalmia**), of minerals (iron, giving rise to **anemia**), and of energy **(starvation).** Although **obesity** has always been associated with dietary excess, the concept of excess intake of particular nutrients and their association with the prevalence of certain diseases in developed societies is gaining recognition, eg, atherosclerosis and coronary heart disease, diabetes, cancer of the breast and colon, cerebrovascular disease and strokes, and cirrhosis of the liver. Malabsorption of nutrients or defects in digestive enzymes also lead to pathologic conditions; for example, malabsorption of vitamin B_{12} and folate causes **anemia;** defects in calcium, magnesium, and vitamin D absorption lead to **tetany** and **osteoporosis;** and a general **malabsorption syndrome** includes these and other defects. Lactase deficiency gives rise to **milk intoler-** ance, and defects in absorption of neutral amino acids are involved in **Hartnup disease.**

NUTRITION

Table 53–1 summarizes nutritional requirements.

ENERGY REQUIREMENTS

Provision of Energy

The mammalian body requires nutrients sufficient to provide free energy to manufacture the daily requirement of high-energy phosphate (mainly ATP) and reducing equivalents (2H) needed to power all body functions (see Fig 16–1). Nutrients are provided by carbohydrate, fat, and protein in widely varying proportions among different human populations. Consumption of alcohol can also provide a significant proportion of energy intake.

A constant body weight under conditions of unaltered energy requirement indicates that there is enough energy in the diet.

The amount of energy available in the major food sources is indicated in Table 53–2. The large energy content per gram of fat compared to that of protein or carbohydrate and the relatively high energy content of alcohol are 2 notable facts. The recommended energy intake for average humans is given in Table 53–3.

Expenditure of Energy

Under conditions of **energy equilibrium** (calorie balance), energy intake must equal energy expenditure. Energy expenditure varies widely in different conditions and may be measured by placing an animal inside an insulated chamber and measuring the energy output represented by heat loss and excretory prod-

Table 53–1. Essential nutritional requirements.

	Humans	Selected Differences in Other Species
Amino acids	Histidine,[1] isoleucine, leucine, lysine, methionine (cysteine[3]), phenylalanine (tyrosine[3]), threonine, tryptophan, valine	Arginine[2] required by growing rats. Glycine required in chicks and taurine in cats. Most amino acids not essential in ruminants; requirement spared in other herbivores with substantial population of microorganisms in the gut.
Fatty acids	Linoleic acid (arachidonic acid[3]), α-linolenic acid[4]	Arachidonic acid is a specific requirement in cats.
Vitamins Water-soluble	Ascorbic acid (C), biotin,[5] cobalamin (B$_{12}$), folic acid, niacin, pantothenic acid, pyridoxine (B$_6$), riboflavin (B$_2$), thiamin (B$_1$)	Most mammals can synthesize ascorbic acid, but it is essential in the diet of primates, guinea pigs, and Indian fruit bats. Water-soluble vitamins are not essential in ruminants; requirements are spared in other herbivores with substantial populations of microorganisms in the gut.
Fat-soluble	Vitamin A, D[6], E, K[5]	Most species can utilize β-carotene as a source of vitamin A (retinol); must be supplied as retinol in cats.
Minerals Macrominerals	Calcium, chloride, magnesium, phosphorus, potassium, sodium	
Microminerals (trace elements)	Chromium, copper, iodine, iron, manganese, molybdenum, selenium, zinc	Silicon, vanadium, nickel, arsenic, fluoride, and tin have been shown to be essential in various species and may be required in humans. Cobalt is required for synthesis of cobalamin by ruminal microorganisms.
Fiber	Required for optimal health	
Water	The most critical component of the diet	
Energy	Utilization of carbohydrates, fats, and protein in variable proportions	

[1]Required in infants and probably in children and in adults.
[2]May be partly essential in infants.
[3]Cysteine, tyrosine, and arachidonic acid spare the requirement for methionine, phenylalanine, and linoleic acid, respectively.
[4]Workers disagree whether α-linolenic acid is essential in the human diet.
[5]Synthesized by intestinal microorganisms; therefore, dietary requirement uncertain.
[6]Exposure of the skin to sunlight reduces dietary requirement.

ucts. It is usually more convenient to measure **oxygen consumption,** since under most conditions 1 L of O_2 consumed accounts for approximately 4.83 kcal (20 kJ) of energy expended.

The energy expended by an individual depends on the following factors:

(1) The **basal metabolic rate** is the energy expenditure necessary to maintain basic physiologic functions under standardized conditions; the subject should be at rest, awake, and in a warm environment, and measurements should be taken at least 12 hours after the last meal. The basal metabolic rate is proportionate to lean body weight and to surface area. It is higher in males than females, in young children, and in people with **fever** and **hyperthyroidism.** It is lower in **hypothyroidism** and in **starvation.**

(2) The **thermogenic effect** (specific dynamic action) of food is equivalent to about 5–10% of total energy expenditure and is attributed to the added energy expenditure due to digestion and to any stimulation of metabolism caused by the influx of new substrate.

(3) **Physical activity** is the largest variable affect-

Table 53–2. Heats of combustion and energy available from the major food sources.*

	Energy kcal/g (kJ/g)		
	Heat of Combustion (Bomb Calorimeter)	Human Oxidation	Standard Conversion Factors†
Protein	5.4 (22.6)	4.1 (17.2)‡	4 (17)
Fat	9.3 (38.9)	9.3 (38.9)	9 (38)
Carbohydrate	4.1 (17.2)	4.1 (17.2)	4 (17)
Ethanol	7.1 (29.7)	7.1 (29.7)	7 (29)

*Adapted from Davidson S et al: *Human Nutrition and Dietetics,* 7th ed. Churchill Livingstone, 1979.
†Conversion factors are obtained by rounding off heats of combustion and correcting for estimates of absorption efficiency.
‡Protein oxidation corrected for loss of amino groups excreted in urine.

Table 53–3. Recommended energy intake for men and women.**

	Age	Weight		Energy Needs		
Category	(years)	(kg)	(lb)	(kcal)		(MJ)
				Mean	Range	
Men	23–50	70	154	2700	2300–3100	11.3
Women	23–50	55	120	2000	1600–2400	8.4
Pregnant				+300		
Lactating				+500		

*Data from *Recommended Dietary Allowances,* 9th ed. Food and Nutrition Board, National Research Council—National Academy of Sciences, 1980.

ing energy expenditure; the range is over 10-fold between resting and maximum athletic activity.

(4) When **environmental temperature** is low, it causes increased energy expenditure owing to shivering and to nonshivering thermogenesis in animals having brown fat (see p 239). At temperatures above blood heat, extra energy is expended in cooling.

AMINO ACID NITROGEN & SPECIFIC AMINO ACID REQUIREMENTS

Protein normally provides the body's requirement for amino acid nitrogen and amino acids themselves. All dietary protein is digested and enters the circulation as individual amino acids. The body requires 20 amino acids to synthesize specific proteins and other nitrogen-containing compounds such as purines, pyrimidines, and heme. Humans cannot synthesize 9 of these amino acids, the **essential amino acids,** which must therefore be supplied in the diet (Table 53–1). Two other amino acids, cysteine and tyrosine, may be formed from the essential amino acids methionine and phenylalanine, respectively. If insufficient methionine and phenylalanine are present in the diet, cysteine and tyrosine become essential amino acids. Conversely, if adequate cysteine and tyrosine are present in the diet, they spare the requirement for methionine and phenylalanine. As long as sufficient amounts of essential amino acids are present in the diet, the remaining 9 amino acids required for protein synthesis and other purposes can be formed through transamination reactions (see p 272).

Nitrogen Balance
(See also Chapter 30.)

When its diet is such that an adult animal is in a state of metabolic equilibrium, dietary protein is re-

quired to replace the essential amino acids and amino acid nitrogen lost during metabolic turnover. Nitrogen is lost in the urine, feces, saliva, desquamated skin, hair, and nails. The daily requirements for total protein and essential amino acids in humans are set forth in Table 53–4. When these requirements are calculated on the basis of body weight, the extra growth needs of infants and children are clearly evident. Pregnancy, lactation, tissue repair after injury, recovery from illness, and increased physical activity are other conditions requiring more dietary protein. For most situations, a diet in which 12% of the energy is supplied as protein is adequate.

The efficiency with which dietary protein is used determines the total quantity of protein required. This quantity is affected by 3 major factors: **protein quality, energy intake,** and **physical activity.**

A. Protein Quality: The quality of protein is measured by comparing the proportions of essential amino acids in a food with the proportions required for good nutrition. The closer the 2 numbers are, the higher the protein quality. Egg and milk proteins are **high-quality proteins** that are efficiently used by the body and are used as reference standards against which other proteins can be compared. Meat protein is of high protein quality, whereas several proteins from plants used as major food sources are relatively deficient in certain essential amino acids, eg, tryptophan and lysine in maize (corn), lysine in wheat, and methionine in some beans. In a mixed diet, a deficiency of an amino acid in one protein is made up by its abundance in another; such proteins are described as complementary; eg, the protein of wheat and beans combined provides a satisfactory amino acid intake. Under such circumstances, a greater *total* amount of protein must be consumed to satisfy requirements. Amino acids that are not incorporated into new protein and are unnecessary for immediate re-

Table 53–4. Estimated protein and amino acid requirements and intakes in humans.*

	Requirement (mg/kg body weight/d)			Intake (g/d)	
	Infant (4–6 months)	Child (10–12 years)	Adult	Adult (70 kg) Allowance*	Estimated US Adult Intake†
Protein	2000	1400	800	56	101
Animal	. . .	. . .	. . .	. . .	71
Vegetable	. . .	. . .	. . .	. . .	30
Essential amino acids					
Histidine	33	?	10	0.70	?
Isoleucine	83	28	12	0.84	5.3
Leucine	135	42	16	1.12	8.2
Lysine	99	44	12	0.84	6.7
Methionine (and cysteine)	49	22	10	0.70	2.1
Phenylalanine (and tyrosine)	141	22	16	1.12	4.7
Threonine	68	28	8	0.56	4.1
Tryptophan	21	4	3	0.21	1.2
Valine	92	25	14	0.98	5.7

*Data from *Recommended Dietary Allowances,* 9th ed. Food and Nutrition Board, National Research Council—National Academy of Sciences, 1980.
†Data from Munro HN, Crim M: The proteins and amino acids. In: Goodhart RS, Shils ME: *Modern Nutrition in Health and Disease,* 6th ed. Lea & Febiger, 1980.

quirements cannot be stored and are rapidly degraded, and the nitrogen is excreted as urea and other products.

B. Energy Intake: The energy derived from carbohydrate and fat affects protein requirements because it spares the use of protein as an energy source. To use expensive (high-quality) dietary protein efficiently and to reduce requirements for it to a minimum, it is necessary to ensure adequate provision of energy from nonprotein sources, some of which should be carbohydrate in order to spare protein from gluconeogenesis.

C. Physical Activity: Physical activity increases nitrogen retention from dietary protein.

Protein-Energy Malnutrition

Protein-energy malnutrition includes a range of disorders of starvation and malnutrition that involve other nutrients such as vitamins and minerals in addition to protein deficiency. In severe form, it occurs in growing children, usually under 5 years of age, in poor areas of Asia, Africa, and South America. Two extreme forms are recognized, **marasmus** and **kwashiorkor,** with other conditions intermediate between the 2 extremes. In marasmus, there is generalized wasting due to deficiency of both energy and protein, whereas in kwashiorkor, which is characterized by edema, there is a deficiency in both the quantity and quality of protein although energy intake may be adequate.

CARBOHYDRATE REQUIREMENTS

Glucose is specifically required by many tissues but does not have to be provided as such in the diet, since other dietary carbohydrates are readily converted to glucose, either during digestion (eg, starch) or subsequently in the liver (eg, fructose, galactose; see pp 182 and 183). Glucose is also formed from the glycerol moiety of fats and from glucogenic amino acids by gluconeogenesis (see p 172). However, a minimum daily intake of carbohydrate (50–100 g) is recommended in humans to prevent **ketosis** (see p 254) and loss of muscle protein. The major foodstuffs containing carbohydrates are described in Chapter 14.

FIBER REQUIREMENTS

Dietary fiber denotes all plant cell wall components that cannot be digested by an animal's own enzymes, eg, cellulose, hemicellulose, lignin, gums, pectins, and pentosans. In herbivores such as ruminants, fiber (mainly as cellulose) is a major source of energy after it is digested by microorganisms.

In humans, a high-fiber diet exerts beneficial effects by aiding water retention during passage of food along the gut and thereby producing larger, softer feces. A high-fiber diet is associated with reduced incidence of **diverticulosis, cancer of the colon, cardiovascular disease,** and **diabetes mellitus.** The more insoluble fibers such as cellulose and lignin found in wheat bran are beneficial with regard to colonic func-

tion, whereas the more soluble fibers found in legumes and fruit, eg, gums and pectins, lower blood cholesterol, possibly by binding bile acids and dietary cholesterol. The soluble fibers also slow stomach emptying, and they delay and attenuate the postprandial rise in blood glucose, with consequent reduction in insulin secretion. This effect is beneficial to diabetics and to dieters because it reduces the rebound fall in blood glucose that stimulates appetite.

LIPID REQUIREMENTS

Although lipid frequently provides a significant proportion of the dietary requirement for energy, this is not an essential function.

Apart from increasing the palatability of food and producing a feeling of satiety, dietary lipid has 2 essential functions in mammalian nutrition. It acts as the dietary vehicle for the **fat-soluble vitamins,** and it supplies **essential polyunsaturated fatty acids** that the body is unable to synthesize. Three polyunsaturated fatty acids have been recognized as essential in the diet of at least some animals: **linoleic acid** ($\omega6$, 18:2), α-**linolenic acid** ($\omega3$, 18:3), and **arachidonic acid** ($\omega6$, 20:4). These acids are found in the lipids of plant and animal food (see Table 15–2). See also Chapter 24 for a discussion of their metabolism.

In humans, arachidonic acid is generally formed from linoleic acid and is not essential if sufficient linoleic acid is present in the diet. Debate continues about whether α-linolenic acid is truly essential in humans (but see p 212). **Deficiency of linoleic acid** is rare but may occur in infants on skim milk diets and in patients fed lipid-free diets intravenously.

A principal function of essential fatty acids is to serve as precursors of **leukotrienes, prostaglandins,** and **thromboxanes** (see Fig 24–6), which function as "local hormones." A dietary intake in which 1–2% of the total energy requirement is supplied as essential fatty acid prevents clinical deficiency.

Lipids and disease. Numerous studies have shown a correlation between **coronary heart disease,** blood cholesterol, and the consumption of fat, particularly of saturated fat (see Chapter 27). High fat consumption is also associated with cancer of the breast and colon. The main source of saturated fat in the human diet is the meat of ruminants, dairy products, and hard margarine. Cholesterol is found only in foods of animal origin and particularly in egg yolk.

VITAMIN REQUIREMENTS

Vitamins are organic nutrients which are required in small quantities for normal metabolism and which cannot be synthesized by the body in adequate amounts. Humans require either milligram or microgram quantities of each vitamin per day. Vitamins carry out specific biochemical functions. A summary of these functions, deficiency syndromes, and dietary sources is presented in Table 53–5 for the **water-soluble vitamins** and in Table 53–6 for the **fat-soluble vitamins.**

Water-soluble vitamins include the vitamin B complex and ascorbic acid (vitamin C). Water-soluble vitamins are absorbed into the hepatic portal vein, and any surplus is excreted in the urine. There is thus little storage of the free vitamin, which in most instances needs to be continually supplied in the diet. Some storage of folic acid occurs in the liver. Depletion may take several months for ascorbic acid and several years for B_{12} (also stored in the liver). Excess intake is generally well tolerated except for side effects occurring with large doses of nicotinic acid, ascorbic acid, or pyridoxine.

Absorption of vitamin B_{12} requires the presence of

Table 53–5. Essential water-soluble vitamins: Summary of major characteristics.

Vitamin	Coenzymes	Biochemical or Physiologic Function[1]	Deficiency Syndrome or Symptoms[2] (and Associated Diet)	Sources[3]
Niacin (nicotinic acid, nicotinamide)	Nicotinamide adenine dinucleotide (NAD); nicotinamide adenine dinucleotide phosphate (NADP).	Electron (hydrogen) transfer reactions carried out by dehydrogenase enzymes, eg, pyruvate dehydrogenase, glyceraldehyde-3-phosphate dehydrogenase.	Pellagra (milled corn).	Protein foods containing tryptophan, in addition to niacin sources in note[3].
Thiamin (vitamin B_1)	Thiamin pyrophosphate (TPP).	Oxidative decarboxylation of α-keto acids (pyruvate and α-ketoglutarate dehydrogenases) and 2-keto sugars (transketolases).	Beriberi (milled rice); Wernicke-Korsakoff syndrome (alcohol). Antagonized by thiaminase in raw fish.	
Riboflavin (vitamin B_2)	Flavin adenine dinucleotide (FAD); flavin mononucleotide (FMN).	Electron (hydrogen) transfer reactions (eg, acyl-CoA dehydrogenase).	Cheilosis.	
Pantothenic acid	CoA.	Acyl transfer reactions involving CoA or fatty acid synthase complex.		
Vitamin B_6, pyridoxine, pyridoxal, pyridoxamine	Pyridoxal phosphate.	Transamination and decarboxylation via Schiff base (many aminotransferase and decarboxylase enzymes).	Low serum levels are associated with pregnancy and oral contraceptive agents. Antagonized by isoniazid, penicillamine, and other drugs.	
Biotin	N-Carboxybiotinyl lysine.	CO_2 transfer reactions of carboxylase coenzymes (eg, pyruvate carboxylase, acetyl-CoA carboxylase).	Induced by avidin, a protein in raw egg whites, or by antibiotic therapy.	Synthesized by intestinal microorganisms.
Vitamin B_{12} (cobalamin)	Methylcobalamin; 5'-deoxyadenosyl cobalamin.	Methylation of homocysteine to methionine; conversion of methylmalonyl-CoA to succinyl-CoA.	Megaloblastic anemia, methylmalonic aciduria, peripheral neuropathy (strict vegetarian diet). Pernicious anemia induced by lack of intrinsic factor.	Animal foods (eg, meat).
Folic acid (folacin)	Derivatives of tetrahydrofolic acid.	One-carbon transfer reactions, eg, purine nucleotide and thymidylate synthesis.	Megaloblastic anemia.	
Ascorbic acid (vitamin C)	Unknown.	Antioxidant; collagen biosynthesis; tyrosine catabolism (?).	Scurvy (lack of fresh fruits and vegetables).	Fresh fruits (especially citrus) and vegetables.

[1]The metabolism of most water-soluble vitamins is similar. They are absorbed in the intestine, stored bound to enzymes and transport proteins, and excreted in urine when plasma levels exceed kidney thresholds. The one notable exception is vitamin B_{12}, which requires intrinsic factor (synthesized by gastric parietal cells) for absorption in the distal ileum, is stored in milligram amounts in the liver, and is excreted in bile (and reabsorbed via the enterohepatic circulation) as well as in urine.

[2]Excess intake of water-soluble vitamins is not usually toxic. Exceptions: Excess nicotinic acid—but not nicotinamide—causes vascular dilatation of skin ("flushing"); megadose intake of ascorbic acid has been reported to produce diarrhea, oxalate kidney stones, and a variety of other toxic symptoms; high-dose pyridoxine (5 g/d) has caused sensory ataxia, sensory nerve dysfunction, and axonal degeneration in a few cases. Deficiency of these vitamins affects actively metabolizing tissues; symptoms usually include disorders of the digestive and nervous systems, skin, and blood cells.

[3]Unless otherwise stated, a varied intake of adequate amounts of foods from the following groups will meet nutritional requirements for water-soluble vitamins: whole-grain cereals, legumes, leafy green vegetables, meat, and dairy products.

Table 53–6. Essential fat-soluble vitamins: Summary of major characteristics.

Vitamin/ Provitamin[1]	Metabolism[2]	Active Metabolite: Physiologic Function	Deficiency Syndrome or Symptoms	Toxicity Syndrome or Symptoms	Sources[3]
Vitamin A Provitamin: β-carotene Vitamin: retinol	Transported in lymph as retinyl esters in blood bound to retinol-binding protein and prealbumin.	11-*Cis* retinal is a constituent of rhodopsin and other light-receptor pigments. Unknown metabolites (retinoic acid?) are required for growth and differentiation of epithelial, nervous, and bone tissues.	Children: poor dark adaptation, xerosis, keratomalacia, growth failure, death. Adults: night blindness, xeroderma.	Hypervitaminosis A: headache, dizziness, nausea, skin sloughing, bone pain.	Highly pigmented vegetables (containing carotenes), fortified margarine.
Vitamin D Provitamins: ergosterol (plants, yeast) and 7-dehydrocholesterol (skin) Vitamins D_2 (ergocalciferol) and D_3 (cholecalciferol)	Provitamins converted to vitamins by ultraviolet irradiation. Vitamins hydroxylated in liver to 25-hydroxyvitamin D and in kidney to 1,25-dihydroxyvitamin D and other metabolites.	1,25-Dihydroxyvitamin D_3 is major hormonal regulator of bone mineral (calcium and phosphorus) metabolism.	Children: rickets. Adults: osteomalacia.	Hypervitaminosis D: hypercalcemia, hypercalciuria, nephrocalcinosis.	Fortified milk: sunlight on skin.
Vitamin E Tocopherols, tocotrienols	Generally unknown.	Active metabolite unknown. Functions as an antioxidant.	Children: anemia in premature infants. Adults: no known syndrome.	Undefined. Megadose intake reported to induce blurred vision, headaches.	Vegetable seed oils are major source.
Vitamin K K_1 (phylloquinone), K_2 (menaquinone), others	Generally undefined.	Active metabolite unknown but probably hydroquinone derivative. Activates blood clotting factors II, VII, IX, and X by γ-carboxylating glutamic acid residues; also carboxylates bone and kidney proteins.	Infants: hemorrhagic disease of newborn. Adults: defective blood clotting. Deficiency symptoms can be produced by coumarin anticoagulants and by antibiotic therapy.	Can be induced by water-dispersible analogs: hemolytic anemia, liver damage.	Synthesized by intestinal bacteria.

[1]The fat-soluble vitamins are insoluble in water but dissolve in fats and oils. They are relatively stable to normal cooking temperatures but are slowly inactivated by ultraviolet light and by oxidation.
[2]Absorption of fat-soluble vitamins requires dietary fat and bile; malabsorption or biliary obstruction leads to deficiency. Transport is via lipoproteins or specific transport proteins. Storage is mainly in liver, some in adipose tissue. These vitamins are excreted in bile and either reabsorbed via the enterohepatic circulation or excreted in feces. Some metabolites may be excreted in urine.
[3]Food sources of all fat-soluble vitamins include leafy green vegetables, vegetable seed oils, and fat-containing meat and dairy products, in addition to the specific sources listed here.

intrinsic factor, a glycoprotein secreted by the stomach. Gastrectomy patients may show deficiency symptoms after about 5 years unless they receive parenteral vitamin B_{12} supplementation. Vitamin B_{12} is not found in plant food but is synthesized by microorganisms, some of which are found in the gut flora of herbivores, particularly ruminants whose meat and milk products contain the vitamin. Animal foods are therefore the main sources of this vitamin. Vegans risk a dietary deficiency unless they receive appropriate supplements. Since many of the water-soluble vitamins occur in the same foods, deficiency caused by a single vitamin is rare, and most patients present with multiple vitamin B deficiencies.

The **fat-soluble vitamins** (vitamins A, D, E, and K) are found in food lipids of both plant and animal origin. They are digested with fat and absorbed by the intestine and incorporated into chylomicrons. They are consequently transported mainly in chylomicron remnants, initially to the liver, which is a major store of vitamins A, D, and K. Adipose tissue is the major storage source of vitamin E. Fat-soluble vitamins are not excreted in the urine and, if taken in excess, are toxic (particularly vitamins A and D).

Deficiency of the fat-soluble vitamins (D, causing **rickets**) occurs mainly in children. However, it can also occur in adults (**osteomalacia**), particularly in association with **defects in lipid absorption.**

Some diseases concerned with cofactor metabolism that respond to treatment with specific vitamins are listed in Table 53–7.

Table 53–7. Vitamin-responsive syndromes. Examples of specific defects in vitamin cofactor metabolism that can be corrected by vitamin therapy, usually requiring very large doses.*

Vitamin	Disease	Biochemical Defect
Biotin	Propionic acidemia	Propionyl-CoA carboxylase
Vitamin B_{12}	Methylmalonic aciduria	Formation of cobamide coenzyme
Folic acid	Folate malabsorption	Folic acid transport
Niacin	Hartnup disease	Tryptophan transport
Pyridoxine (vitamin B_6)	Infantile convulsions Cystathioninuria Homocystinuria	Glutamic acid decarboxylase (?) Cystathioninase Cystathionine synthase
Thiamin	Hyperalaninemia Thiamin-responsive lactic acidosis	Pyruvate decarboxylase Hepatic pyruvate carboxylase

*From: Herman RH, Stifel FB, Greene HL: Vitamin-deficient states and other related diseases. In: *Disorders of the Gastrointestinal Tract; Disorders of the Liver; Nutritional Disorders.* Dietschy JM (editor). Grune & Stratton, 1976.

MINERAL REQUIREMENTS

The minerals required for physiologic functions may be divided arbitrarily into 2 groups: (1) **macrominerals,** which are required in amounts greater than 100 mg/d, and (2) **microminerals (trace elements),** which are required in amounts less than 100 mg/d. Mineral requirements are summarized in Table 53–1. Table 53–8 provides a further summary of the properties of the macrominerals, and Table 53–9 does the same for the microminerals.

DIETARY RECOMMENDATIONS

All nutritional requirements must be met to prevent deficiency diseases and ill health. Ignorance or poor economic conditions are almost always the underlying cause of failure to satisfy nutritional requirements. On the other hand, certain common diseases are associated with *excess* intake of nutrients. **Obesity** generally reflects excess intake of energy and is often associated

Table 53–8. Essential macrominerals: Summary of major characteristics.

Elements	Functions	Metabolism[1]	Deficiency Disease or Symptoms	Toxicity Disease or Symptoms[2]	Sources[3]
Calcium	Constituent of bones, teeth; regulation of nerve, muscle function.	Absorption requires calcium-binding protein. Regulated by vitamin D, parathyroid hormone, calcitonin, etc.	Children: rickets. Adults: osteomalacia. May contribute to osteoporosis.	Occurs with excess absorption due to hypervitaminosis D or hypercalcemia due to hyperparathyroidism, or idiopathic hypercalcemia.	Dairy products, beans, leafy vegetables.
Phosphorus	Constituent of bones, teeth, ATP, phosphorylated metabolic intermediates. Nucleic acids.	Control of absorption unknown (vitamin D?). Serum levels regulated by kidney reabsorption.	Children: rickets. Adults: osteomalacia.	Low serum Ca^{2+}:P_i ratio stimulates secondary hyperthyroidism; may lead to bone loss.	Phosphate food additives.
Sodium	Principal cation in extracellular fluid. Regulates plasma volume, acid-base balance, nerve and muscle function, Na^+/K^+-ATPase.	Regulated by aldosterone.	Unknown on normal diet; secondary to injury or illness.	Hypertension (in susceptible individuals).	Table salt; salt added to prepared food.
Potassium	Principal cation in intracellular fluid; nerve and muscle function, Na^+/K^+-ATPase.	Also regulated by aldosterone.	Occurs secondary to illness, injury, or diuretic therapy; muscular weakness, paralysis, mental confusion.	Cardiac arrest, small bowel ulcers.	
Chloride	Fluid and electrolyte balance; gastric fluid.		Infants fed salt-free formula. Secondary to vomiting, diuretic therapy, renal disease.		Table salt.
Magnesium	Constituent of bones, teeth; enzyme cofactor (kinases, etc).		Secondary to malabsorption or diarrhea, alcoholism.	Depressed deep tendon reflexes and respiration.	Leafy green vegetables (containing chlorophyll).

[1] In general, minerals require carrier proteins for absorption. Absorption is rarely complete; it is affected by other nutrients and compounds in the diet (eg, oxalates and phytates that chelate divalent cations). Transport and storage also require special proteins. Excretion occurs in feces (unabsorbed minerals) and in urine, sweat, and bile.
[2] Excess mineral intake produces toxic symptoms. Unless otherwise specified, symptoms include nonspecific nausea, diarrhea, and irritability.
[3] Mineral requirements are met by a varied intake of adequate amounts of whole-grain cereals, legumes, leafy green vegetables, meat, and dairy products.

with the development of **non-insulin-dependent diabetes mellitus. Atherosclerosis** and **coronary heart disease** are associated with diets high in total fat and saturated fat. **Cancer** of the breast, colon, and prostate correlates with high fat intake. High incidence of **cerebrovascular disease** and **hypertension** is associated with a high salt intake.

Many committees throughout the world have investigated the composition of human diets and made recommendations for improvement. These may be summarized as follows: (1) If overweight, total energy intake should be reduced to achieve optimum weight. (2) There should be a general shift away from fat consumption to consumption of more carbohydrate. (3) A greater proportion of carbohydrate should be consumed as complex carbohydrates and less as sugars. (4) A greater proportion of dietary fat should be in the form of polyunsaturated fat and less as saturated fat. (5) Consumption of cholesterol and salt should be reduced. (6) Dietary fiber should be increased.

Table 53–9. Essential microminerals (trace elements): Summary of major characteristics.

Elements	Functions	Metabolism	Deficiency Disease or Symptoms	Toxicity Disease or Symptoms[1]	Sources[2]
Chromium	Trivalent chromium, a constituent of "glucose tolerance factor."		Impaired glucose tolerance; secondary to parenteral nutrition.		
Cobalt	Constituent of vitamin B_{12}.	As for vitamin B_{12}.	Vitamin B_{12} deficiency.		Foods of animal origin.
Copper	Oxidase enzymes: cytochrome c oxidase, ferroxidase, etc.	Transported by albumin; bound to ceruloplasmin.	Anemia (hypochromic, microcytic); secondary to malnutrition, Menke's syndrome.	Rare; secondary to Wilson's disease.	
Iodine	Thyroxine, triiodothyronine.	Stored in thyroid as thyroglobulin.	Children: cretinism. Adults: goiter and hypothyroidism, myxedema.	Thyrotoxicosis, goiter.	Iodized salt, seafood.
Iron	Heme enzymes (hemoglobin, cytochromes, etc.).	Transported as transferrin; stored as ferritin or hemosiderin; excreted in sloughed cells and by bleeding.	Anemia (hypochromic, microcytic).	Siderosis; hereditary hemochromatosis.	Iron cookware.
Manganese	Hydrolase, decarboxylase, and transferase enzymes. Glycoprotein and proteoglycan synthesis.		Unknown in humans.	Inhalation poisoning produces psychotic symptoms and parkinsonism.	
Molybdenum	Oxidase enzymes (xanthine oxidase).		Secondary to parenteral nutrition.		
Selenium	Glutathione peroxidase.	Synergistic antioxidant with vitamin E.	Marginal deficiency when soil content is low; secondary to parenteral nutrition, protein-energy malnutrition.	Megadose supplementation induces hair loss, dermatitis, and irritability.	
Zinc	Cofactor of many enzymes: lactic dehydrogenase, alkaline phosphatase, carbonic anhydrase, etc.		Hypogonadism, growth failure, impaired wound healing, decreased taste and smell acuity; secondary to acrodematitis enteropathica, parenteral nutrition.	Gastrointestinal irritation, vomiting.	
Fluoride[3]	Increases hardness of bones and teeth.		Dental caries; osteoporosis(?).	Dental fluorosis.	Drinking water.

[1]Excess mineral intake produces toxic symptoms. Unless otherwise specified, symptoms include nonspecific nausea, diarrhea, and irritability.
[2]Mineral requirements are met by a varied intake of adequate amounts of whole-grain cereals, legumes, leafy green vegetables, meat, and dairy products.
[3]Fluoride is essential for rat growth. While not proved to be strictly essential for human nutrition, fluorides have a well-defined role in prevention and treatment of dental caries.

DIGESTION

Most foodstuffs are ingested in forms that are unavailable to the organism, since they cannot be absorbed from the digestive tract until they have been broken down into smaller molecules. This disintegration of the naturally occurring foodstuffs into assimilable forms constitutes the process of digestion.

The chemical changes incident to digestion are accomplished with the aid of hydrolase enzymes of the digestive tract that catalyze the hydrolysis of native proteins to amino acids, of starches to monosaccharides, and of triacylglycerols to monoacylglycerols, glycerol, and fatty acids. In the course of these digestive reactions, the minerals and vitamins of the foodstuffs are also made more assimilable.

A systematic account of the nature and functions of the gastrointestinal hormones is given in Chapter 52.

DIGESTION IN THE ORAL CAVITY

Salivary Digestion

Saliva, secreted by the salivary glands, consists of about 99.5% water. It acts as a lubricant for mastication and for swallowing. Adding water to dry food provides a medium in which food molecules can dissolve and in which hydrolases can initiate digestion. Mastication subdivides the food, increasing its solubility and surface area for enzyme attack. The saliva is also a vehicle for the excretion of certain drugs (eg, ethanol and morphine), of inorganic ions such as K^+, Ca^{2+}, HCO_3^-, thiocyanate (SCN^-), and iodine, and of immunoglobulins (IgA).

The pH of saliva is usually about 6.8, although it may vary on either side of neutrality.

Salivary amylase is capable of bringing about the hydrolysis of starch and glycogen to maltose; this is of little significance in the body because of the short time it can act on the food. Salivary amylase is readily inactivated at pH 4.0 or less, so that digestive action on food in the mouth will soon cease in the acid environment of the stomach. In many animals, a salivary amylase is entirely absent. A **lingual lipase** secreted by the dorsal surface of the tongue (Ebner's glands) has been reported.

DIGESTION IN THE STOMACH

The gastric secretion is known as **gastric juice.** It is a clear, pale yellow fluid of 0.2–0.5% HCl, with a pH of about 1.0. The gastric juice is 97–99% water. The remainder consists of mucin and inorganic salts, the digestive enzymes (pepsin and rennin), and a lipase.

A. Hydrochloric Acid: The parietal cells are the source of gastric HCl, which originates according to the reactions shown in Fig 53–1. The process is similar to that of the "chloride shift" described for the red blood cell. There is also a resemblance to the renal tubular mechanisms for secretion of H^+, wherein the source of H^+ is also the **carbonic anhydrase**-catalyzed formation of H_2CO_3 from H_2O and CO_2. An alkaline urine often follows the ingestion of a meal ("alkaline tide"), as a result of the formation of bicarbonate in the process of hydrochloric acid secretion. Secretion of H^+ into the lumen is an active process driven by a membrane-located K^+/ATPase, which, unlike the Na^+/K^+-ATPase, is ouabain-insensitive. HCO_3^- passes into the plasma in exchange for Cl^-, which is coupled to the secretion of H^+ into the lumen.

As a result of contact with gastric HCl, proteins are denatured; ie, the tertiary protein structure is lost as a result of the destruction of hydrogen bonds. This allows the polypeptide chain to unfold, making it more

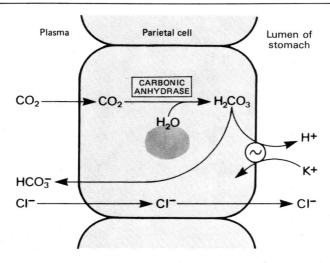

Figure 53–1. Production of gastric hydrochloric acid. $\odot$, K^+-ATPase.

accessible to the actions of proteolytic enzymes (proteases). The low pH also has the effect of destroying most microorganisms entering the gastrointestinal tract.

B. Pepsin: The chief digestive function of the stomach is the initiation of protein digestion. Pepsin is produced in the chief cells as the inactive zymogen, **pepsinogen.** This is activated to pepsin by H^+, which splits off a protective polypeptide to expose active pepsin; and by pepsin, which rapidly activates further molecules of pepsinogen (**autocatalysis**). Pepsin transforms denatured protein into proteoses and then peptones, which are large polypeptide derivatives. Pepsin is an **endopeptidase,** since it hydrolyzes peptide bonds within the main polypeptide structure rather than adjacent to N- or C-terminal residues, which is characteristic of **exopeptidases.** It is specific for peptide bonds formed by aromatic amino acids (eg, tyrosine) or dicarboxylic amino acids (eg, glutamate).

C. Rennin (Chymosin, Rennet): This enzyme causes the coagulation of milk. It is important in the digestive processes of infants, because it prevents the rapid passage of milk from the stomach. In the presence of calcium, rennin changes irreversibly the casein of milk to a **paracasein** which is then acted on by pepsin. Rennin is said to be absent from the stomach of adults. It is used in the making of cheese.

D. Lipase: The heat of the stomach is important in liquidizing the bulk of dietary lipids; emulsification takes place aided by peristaltic contractions. Although the stomach does contain a lipase capable of hydrolyzing triacylglycerols of short and medium chain length, the lipolytic action of gastric juice is not important. However, the **lingual lipase** can continue its activity at the low pH of the stomach, where because of the retention time of 2–4 hours, about 30% of dietary triacylglycerol may be digested. Lingual lipase is more active on triacylglycerol having shorter-chain fatty acids and is more specific for the ester linkage in the *sn*-3 position rather than position 1. Milk fat contains short- and medium-chain fatty acids, which tend to be esterified in the *sn*-3 position. Therefore, milk fat seems to be a particularly good substrate for this enzyme. The released hydrophilic short-chain fatty acids are absorbed via the stomach wall and enter the portal vein, whereas longer-chain fatty acids dissolve in the fat droplets.

PANCREATIC & INTESTINAL DIGESTION

The stomach contents, or **chyme,** are intermittently introduced during digestion into the duodenum through the pyloric valve. The pancreatic and bile ducts open into the duodenum at a point very close to the pylorus. The alkaline content of pancreatic and biliary secretions neutralizes the acid of the chyme and changes the pH of this material to the alkaline side; this shift of pH is necessary for the activity of the enzymes contained in pancreatic and intestinal juice, but **it inhibits further action of pepsin.**

Bile

In addition to many functions in intermediary metabolism, the liver, by producing bile, plays an important role in digestion. The gallbladder, a saccular organ attached to the hepatic duct, stores a certain amount of the bile produced by the liver between meals. During digestion, the gallbladder contracts and supplies bile rapidly to the small intestine by way of the common bile duct. The pancreatic secretions mix with the bile, since they empty into the common duct shortly before its entry into the duodenum.

A. Composition of Bile: The composition of hepatic bile differs from that of gallbladder bile. As shown in Table 53–10, the latter is more concentrated.

B. Functions of Bile:

1. Emulsification–The bile salts have considerable ability to lower surface tension. This enables them to emulsify fats in the intestine and to dissolve fatty acids and water-insoluble soaps. The presence of bile in the intestine is an important adjunct to accomplish the digestion and absorption of fats as well as the absorption of the fat-soluble vitamins A, D, E, and K. When fat digestion is impaired, other foodstuffs are also poorly digested, since the fat covers the food particles and prevents enzymes from attacking them. Under these conditions, the activity of the intestinal bacteria causes considerable putrefaction and production of gas.

2. Neutralization of acid–In addition to its functions in digestion, the bile, having a pH slightly above 7, neutralizes the acid chyme from the stomach and prepares it for intestinal digestion.

3. Excretion–Bile is an important vehicle for bile acid and cholesterol excretion, but it also removes many drugs, toxins, bile pigments, and various inorganic substances such as copper, zinc, and mercury.

Table 53–10. The composition of hepatic and of gallbladder bile.

	Hepatic Bile (as secreted)		Bladder Bile
	Percent of Total Bile	Percent of Total Solids	Percent of Total Bile
Water	97.00	. . .	85.92
Solids	2.52	. . .	14.08
Bile acids	1.93	36.9	9.14
Mucin and pigments	0.53	21.3	2.98
Cholesterol	0.06	2.4	0.26
Esterified and non-esterified fatty acids	0.14	5.6	0.32
Inorganic salts	0.84	33.3	0.65
Specific gravity	1.01	. . .	1.04
pH	7.1–7.3	. . .	6.9–7.7

4. Cholesterol solubility in bile; formation of gallstones—Free cholesterol is insoluble in water; consequently, it is incorporated into a phosphatidylcholine-bile salt micelle (see p 141). Indeed, phosphatidylcholine, the predominant phospholipid in bile, is itself insoluble in aqueous systems but can be dissolved by bile salts in micelles. The large quantities of cholesterol present in the bile of humans are solubilized in these water-soluble mixed micelles, allowing cholesterol to be transported in bile via the biliary tract to the intestine. However, the actual solubility of cholesterol in bile depends on the relative proportions of bile salt, phosphatidylcholine, and cholesterol. The solubility also depends on the water content of bile. This is especially important in dilute hepatic bile.

Using triangular coordinates (Fig 53–2), Redinger and Small were able to determine the maximum solubility of cholesterol in human gallbladder bile. Reference to the figure indicates that any triangular point falling above the line ABC would represent a bile whose composition is such that cholesterol is either supersaturated or precipitated.

It is believed that at some time during the life of a patient with gallstones there is formed an abnormal bile that has become supersaturated with cholesterol.

With time, various factors such as infection, for example, serve as seeding agents to cause the supersaturated bile to precipitate the excess cholesterol as crystals. Unless the newly formed crystals are promptly excreted into the intestine with the bile, the crystals will grow to form stones. When the activities of key enzymes in bile acid formation were measured in the livers of patients with gallstones, cholesterol synthesis was elevated but bile acid synthesis was reduced, causing liver cholesterol concentrations to increase. It seems that decreased 7α-hydroxylase activity leads to a diminished enterohepatic bile acid pool that signals the liver to produce more cholesterol. The bile then becomes overloaded with cholesterol, which is unable to dissolve completely in the mixed micelles.

The above information concerning cholesterol solubility has been used in attempts to dissolve gallstones or to prevent their further formation. Chenodeoxycholic acid appears to offer specific medical treatment for asymptomatic radiolucent gallstones in functioning gallbladders because of its specific inhibition of HMG-CoA reductase in the liver, with consequent reduction in cholesterol synthesis.

5. Bile pigment metabolism—The origin of the bile pigments from hemoglobin is discussed on p 328.

Digestion by Pancreatic Secretion

Pancreatic secretion is a nonviscid watery fluid that is similar to saliva in its content of water and contains some protein and other organic and inorganic compounds—mainly Na^+, K^+, HCO_3^-, and Cl^-, but Ca^{2+}, Zn^{2+}, HPO_4^{2-}, and SO_4^{2-} are also present in small amounts. The pH of pancreatic secretion is distinctly alkaline, 7.5–8.0 or higher.

Many enzymes are found in pancreatic secretion; some are secreted as zymogens.

A. Trypsin, Chymotrypsin, and Elastase: The proteolytic action of pancreatic secretion is due to the 3 **endopeptidases** trypsin, chymotrypsin, and elastase, which attack proteins and polypeptides released from the stomach to produce polypeptides, peptides, or both. Trypsin is specific for peptide bonds of basic amino acids, and chymotrypsin is specific for peptide bonds containing uncharged amino acid residues such as aromatic amino acids, whereas elastase, in spite of its name, has rather broad specificity in attacking bonds next to small amino acid residues such as glycine, alanine, and serine. All 3 enzymes are secreted as zymogens. Activation of **trypsinogen** is due to another proteolytic enzyme, **enterokinase,** secreted by the intestinal mucosa. This hydrolyzes a lysine peptide bond in the zymogen, releasing a small polypeptide that allows the molecule to unfold as active trypsin. Once trypsin is formed, it will attack not only additional molecules of trypsinogen but also the other zymogens in the pancreatic secretion, **chymotrypsinogen, proelastase,** and **procarboxypeptidase,** liberating chymotrypsin, elastase, and **carboxypeptidase,** respectively.

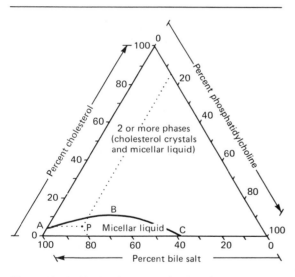

Figure 53–2. Method for presenting 3 major components of bile (bile salts, phosphatidylcholine, and cholesterol) on triangular coordinates. Each component is expressed as a percentage mole of total bile salt, phosphatidylcholine, and cholesterol. Line ABC represents maximum solubility of cholesterol in varying mixtures of bile salt and phosphatidylcholine. Point P represents normal bile composition, containing 5% cholesterol, 15% phosphatidylcholine, and 80% bile salt, and falls within the zone of a single phase of micellar liquid. Bile having a composition falling above the line would contain excess cholesterol in either supersaturated or precipitated form (crystals or liquid crystals). (Reproduced, with permission, from Redinger RN, Small DM: Bile composition, bile salt metabolism, and gallstones. *Arch Intern Med* 1972;**130**:620. Copyright © 1972. American Medical Association.)

B. Carboxypeptidase: The further attack on the polypeptides produced by the action of endopeptidases is carried on by the exopeptidase carboxypeptidase, which attacks the carboxy-terminal peptide bond, liberating single amino acids.

C. Amylase: The starch-splitting action of pancreatic secretion is due to a pancreatic **α-amylase.** It is similar in action to salivary amylase, hydrolyzing starch and glycogen to maltose, maltotriose [three α-glucose residues linked by $\alpha(1\rightarrow4)$ bonds], and a mixture of branched $(1\rightarrow6)$ oligosaccharides (α-limit dextrins), nonbranched oligosaccharides, and some glucose.

D. Lipase: The pancreatic lipase acts at the oil-water interface of the finely emulsified lipid droplets formed by mechanical agitation in the gut in the presence of the products of lingual lipase activity, bile salts, **colipase** (a protein present in pancreatic secretion), **phospholipids,** and **phospholipase A$_2$** (also present in the pancreatic secretion). Phospholipase A$_2$ and colipase are secreted in pro- forms and require activation by tryptic hydrolysis of specific peptide bonds. Ca^{2+} is necessary for phospholipase A$_2$ activity. A limited hydrolysis of the ester bond in the 2 position of the phospholipid by phospholipase A$_2$ (see Fig 25–5) results in the binding of lipase to the substrate interface and a rapid rate of hydrolysis of triacylglyc-

erol. Colipase binds to the bile salt-triacylglycerol/water interface, providing a high-affinity anchor for the lipase. The complete hydrolysis of triacylglycerols produces glycerol and fatty acids. However, the second and third fatty acids are hydrolyzed from the triacylglycerols with increasing difficulty. Pancreatic lipase is virtually specific for the hydrolysis of primary ester linkages, ie, at positions 1 and 3 of triacylglycerols. During fat digestion, the aqueous or "micellar phase" contains mixed **disklike micelles** and **liposomes** of bile salts saturated with lipolytic products (see Fig 15–34).

Because of the difficulty of hydrolysis of the secondary ester linkage in the triacylglycerol, it is probable that the digestion of triacylglycerol proceeds by removal of the terminal fatty acids to produce 2-monoacylglycerol. Since this last fatty acid is linked by a secondary ester bond, its removal requires isomerization to a primary ester linkage. This is a relatively slow process; as a result, **2-monoacylglycerols are major end products of triacylglycerol digestion,** and less than one-fourth of the ingested triacylglycerol is completely broken down to glycerol and fatty acids (Fig 53–3).

E. Cholesteryl Ester Hydrolase (Cholesterol Esterase): Under the conditions within the lumen of the intestine, the enzyme catalyzes the hydrolysis of

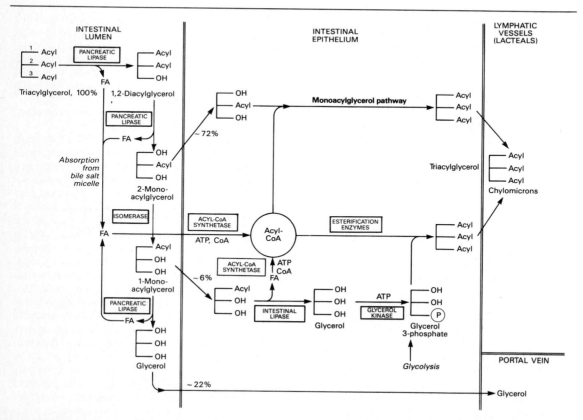

Figure 53–3. Digestion and absorption of triacylglycerols. FA, long-chain fatty acid. (Modified from Mattson FH, Volpenheim RA: The digestion and absorption of triglycerides. *J Biol Chem* 1964;239:2772.)

cholesteryl esters, which are thus absorbed from the intestine in a nonesterified, free form.

F. Ribonuclease (RNase) and Deoxyribonuclease (DNase) have been prepared from pancreatic tissue (see Chapters 38 and 39).

G. Phospholipase A₂: Phospholipase A_2 hydrolyzes the ester bond in the 2 position of glycerophospholipids of both biliary and dietary origins to form lysophospholipids.

Digestion by Intestinal Secretion

The intestinal juice secreted by the glands of Brunner and of Lieberkühn also contains digestive enzymes, including the following:

(1) **Aminopeptidase,** which is an exopeptidase attacking peptide bonds next to N-terminal amino acids of polypeptides and oligopeptides; and **dipeptidases** of various specificity, some of which may be within the intestinal epithelium. The latter complete digestion of dipeptides to free amino acids.

(2) Specific **disaccharidases** and **oligosaccharidases,** ie, α-glucosidase (maltase), which removes single glucose residues from $\alpha(1\rightarrow4)$ linked oligosaccharides and disaccharides, starting from the nonreducing ends; **isomaltase (α-dextrinase),** which hydrolyzes $1 \rightarrow 6$ bonds in α-limit dextrins; **β-galactosidase (lactase)** for removing galactose from lactose; **sucrase** for hydrolyzing sucrose; and **trehalase** for hydrolyzing trehalose.

(3) A **phosphatase,** which removes phosphate from certain organic phosphates such as hexosephosphates, glycerophosphate, and the nucleotides derived from the diet and the digestion of nucleic acids by nucleases.

(4) **Polynucleotidases,** which split nucleic acids into nucleotides.

(5) **Nucleosidases** (nucleoside phosphorylases) catalyze the phosphorolysis of nucleosides to give the free nitrogen base plus a pentose phosphate.

(6) The intestinal secretion is also said to contain a **phospholipase** that attacks phospholipids to produce glycerol, fatty acids, phosphoric acid, and bases such as choline.

The Major Products of Digestion

The final result of the action of the digestive enzymes described is to reduce the foodstuffs of the diet to forms that can be absorbed and assimilated. These end products of digestion are, for carbohydrates, the monosaccharides (principally glucose); for proteins, the amino acids; for triacylglycerol, the fatty acids, glycerol, and monoacylglycerols; and for nucleic acids, the nucleobases, nucleosides, and pentoses.

The plant cell wall polysaccharides and lignin of the diet that cannot be digested by mammalian enzymes constitute **dietary fiber** and make up the bulk of the residues from digestion. Fiber performs an important function in adding bulk to the diet and is discussed earlier in this chapter.

Table 53–11 summarizes the major digestive processes.

ABSORPTION FROM THE GASTROINTESTINAL TRACT

There is little absorption from the stomach, although considerable gastric absorption from ethanol is possible.

The small intestine is the main digestive and absorptive organ. About 90% of the ingested foodstuffs are absorbed in the course of passage through the small intestine, and water is absorbed at the same time. Considerably more water is absorbed after the foodstuffs pass into the large intestine, so that the contents, which were fluid in the small intestine, gradually become more solid in the colon.

There are 2 pathways for the transport of materials absorbed by the intestine: the **hepatic portal system,** which leads directly to the liver, and the **lymphatic vessels,** which lead to the blood by way of the thoracic duct.

ABSORPTION OF CARBOHYDRATES

The products of carbohydrate digestion are absorbed from the jejunum into the blood of the portal venous system in the form of monosaccharides, chiefly the hexoses (glucose, fructose, mannose, and galactose) and pentose sugars. The oligosaccharides (compounds derived from starches that yield 3–10 monosaccharide units upon hydrolysis) and the disaccharides are hydrolyzed by appropriate enzymes derived from the mucosal surfaces of the small intestine, which may include pancreatic amylase adsorbed onto the mucosa. There is little free disaccharidase activity in the intestinal lumen. Most of the activity is associated with small "knobs" on the brush border of the intestinal epithelial cell.

Two mechanisms are responsible for the absorption of monosaccharides: **active transport** against a concentration gradient and **simple diffusion.** However, the absorption of some sugars does not fit clearly into one or the other of these mechanisms. The molecular configurations that seem necessary for active transport, which are present in glucose and galactose, are the following: the OH on carbon 2 should have the same configuration as in glucose, a pyranose ring should be present, and a methyl or substituted methyl group should be present on carbon 5. Fructose is absorbed more slowly than glucose and galactose. Its absorption appears to proceed by diffusion with the concentration gradient.

ABSORPTION OF GLUCOSE

The brush border of the enterocyte contains several transporter systems, some very similar to those of the renal brush border membranes, which specialize in the uptake of the different amino acids and sugars. A carrier has been postulated which binds both glucose and Na^+ at separate sites and which transports them both through the plasma membrane of the intestinal cell. It is envisaged that both glucose and Na^+ are released into the cytosol, allowing the carrier to take up more "cargo." The Na^+ is transported down its concentration gradient and at the same time causes the carrier to transport glucose against its concentration gradient. The free energy required for this active transport is obtained from the hydrolysis of ATP linked to a sodium pump that expels Na^+ from the cell in exchange for K^+ (Fig 53–4). The active transport of glucose is inhibited by ouabain (a cardiac glycoside), an inhibitor of the sodium pump, and by phlorhizin, a known inhibitor of glucose reabsorption in the kidney tubule. The ratio of Na^+:glucose transported varies but may be consistent with 2 carriers—a 1:1 and a 3:1 carrier operating in parallel. There is also an Na^+-independent carrier of glucose.

Table 53–11. Summary of digestive processes.

Source of Secretion and Stimulus for Secretion	Enzyme	Method of Activation and Optimal Conditions for Activity	Substrate	End Products or Action
Salivary glands: Secrete saliva in reflex response to presence of food in oral cavity.	Salivary amylase	Chloride ion necessary. pH 6.6–6.8.	Starch Glycogen	Maltose plus 1:6 glucosides (oligosaccharides) plus maltotriose.
Lingual glands	Lingual lipase	pH range 2.0–7.5; optimal, 4.0–4.5.	Short-chain primary ester link at sn-3	Fatty acids plus 1,2-diacylglycerols.
Stomach glands: Chief cells and parietal cells secrete gastric juice in response to reflex stimulation and action of gastrin.	Pepsin A (fundus) Pepsin B (pylorus)	Pepsinogen converted to active pepsin by HCl. pH 1.0–2.0.	Protein	Peptides.
	Rennin	Calcium necessary for activity. pH 4.0.	Casein of milk	Coagulates milk.
Pancreas: Presence of acid chyme from the stomach activates duodenum to produce (1) secretin, which hormonally stimulates flow of pancreatic juice; (2) cholecystokinin, which stimulates the production of enzymes.	Trypsin	Trypsinogen converted to active trypsin by enterokinase of intestine at pH 5.2–6.0. Autocatalytic at pH 7.9.	Protein Peptides	Polypeptides. Dipeptides.
	Chymotrypsin	Secreted as chymotrypsinogen and converted to active form by trypsin. pH 8.0.	Protein Peptides	Same as trypsin. More coagulating power for milk.
	Elastase	Secreted as proelastase and converted to active form by trypsin.	Protein Peptides	Polypeptides. Dipeptides.
	Carboxypeptidase	Secreted as procarboxypeptidase, activated by trypsin.	Polypeptides at the free carboxyl end of the chain	Lower peptides. Free amino acids.
	Pancreatic amylase	pH 7.1.	Starch Glycogen	Maltose plus 1:6 glucosides (oligosaccharides) plus maltotriose.
	Lipase	Activated by bile salts, phospholipids, colipase. pH 8.0.	Primary ester linkages of triacylglycerol	Fatty acids, monoacylglycerols, diacylglycerols, glycerol.
	Ribonuclease		Ribonucleic acid	Nucleotides.
	Deoxyribonuclease		Deoxyribonucleic acids	Nucleotides.
	Cholesteryl ester hydrolase	Activated by bile salts.	Cholesteryl esters	Free cholesterol plus fatty acids.
	Phospholipase A_2	Secreted as proenzyme, activated by trypsin and Ca^{2+}.	Phospholipids	Fatty acids, lysophospholipids.
Liver and gallbladder: Cholecystokinin, a hormone from the intestinal mucosa—and possibly also gastrin and secretin—stimulate the gallbladder and secretion of bile by the liver.	(Bile salts and alkali)		Fats—also neutralize acid chyme	Fatty acid-bile salt conjugates and finely emulsified neutral fat-bile salt micelles and liposomes.

Table 53–11 (cont'd). Summary of digestive processes.

Source of Secretion and Stimulus for Secretion	Enzyme	Method of Activation and Optimal Conditions for Activity	Substrate	End Products or Action
Small intestine: Secretions of Brunner's glands of the duodenum and glands of Lieberkühn.	Aminopeptidase		Polypeptides at the free amino end of the chain	Lower peptides. Free amino acids.
	Dipeptidases		Dipeptides	Amino acids.
	Sucrase	pH 5.0–7.0.	Sucrose	Fructose, glucose.
	Maltase	pH 5.8–6.2.	Maltose	Glucose.
	Lactase	pH 5.4–6.0.	Lactose	Glucose, galactose.
	Trehalase		Trehalose	Glucose.
	Phosphatase	pH 8.6.	Organic phosphates	Free phosphate.
	Isomaltase or 1:6 glucosidase		1:6 glucosides	Glucose.
	Polynucleotidase		Nucleic acid	Nucleotides.
	Nucleosidases (nucleoside phosphorylases)		Purine or pyrimidine nucleosides	Purine or pyrimidine bases, pentose phosphate.

Hydrolysis of polysaccharides, oligosaccharides, and disaccharides is rapid; therefore, the absorptive mechanisms for glucose and fructose are quickly saturated. A conspicuous exception is the hydrolysis of lactose, which proceeds at only half the rate for sucrose, accounting for the fact that digestion of lactose does not lead to saturation of the transport mechanisms for glucose and galactose.

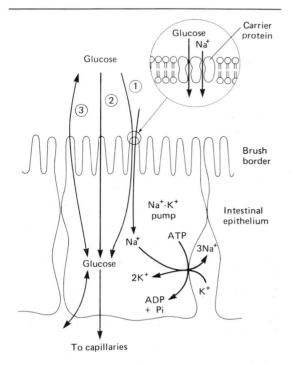

Figure 53–4. Transport of glucose across the intestinal epithelium. Active glucose transport is coupled to the Na⁺-K⁺ pump, 1, or to an Na⁺-independent system, 2. Diffusion is represented by 3.

Defects in Digestion & Absorption of Carbohydrates

A. Lactase Deficiency: Intolerance to lactose, the sugar of milk, may be attributable to a deficiency of lactase. The syndrome should not be confused with intolerance to milk resulting from a sensitivity to milk proteins, usually to the β-lactoglobulin. The signs and symptoms of lactose intolerance are the same regardless of the cause. These include abdominal cramps, diarrhea, and flatulence. They are attributed to accumulation of lactose, which is osmotically active, so that it holds water, and to the fermentative action on the sugar of the intestinal bacteria which produce gases and other products that serve as intestinal irritants.

There are 3 types of lactase deficiency:

1. Inherited lactase deficiency–In this syndrome, which is relatively rare, symptoms of intolerance develop very **soon after birth.** The feeding of a lactose-free diet results in disappearance of the symptoms. Occasionally, infants who appear to be able to digest and absorb lactose nonetheless develop severe symptoms after ingestion of milk or lactose. The occurrence of lactose in the urine is a prominent feature of this syndrome, which appears to be attributable to an effect of lactose on the intestine.

2. Secondary low-lactase activity–Because digestion of lactose is limited even in normal humans, intolerance to milk is not uncommon as a consequence of intestinal diseases. Examples are tropical and nontropical (celiac) sprue, kwashiorkor, colitis, and gastroenteritis. The disorder may be noted also after surgery for peptic ulcer.

3. Primary low-lactase activity–This is a relatively common syndrome, particularly among nonwhite populations in the USA as well as other parts of the world. Since intolerance to lactose was not a feature of the early life of adults with this disorder, it is presumed to represent a gradual decline in activity of lactase in susceptible individuals.

B. Sucrase Deficiency: There is an inherited deficiency of the disaccharidases sucrase and isomaltase. These 2 deficiencies coexist, because sucrase and isomaltase occur together as a complex enzyme. Symptoms occur in early childhood and are the same as those described in lactase deficiency.

C. Disacchariduria: An increase in the excretion of disaccharides may be observed in some patients with disaccharidase deficiencies. As much as 300 mg or more of disaccharide may be excreted in the urine of these people and in patients with intestinal damage (eg, sprue).

D. Monosaccharide Malabsorption: There is a congenital condition in which glucose and galactose are absorbed only slowly, owing to a defect in the carrier mechanism. Because fructose is not absorbed via the carrier, its absorption is normal.

ABSORPTION OF LIPIDS

The 2-monoacylglycerols, fatty acids, and small amounts of 1-monoacylglycerols leave the oil phase of the lipid emulsion and diffuse into the mixed micelles and liposomes consisting of bile salts, phosphatidylcholine, and cholesterol, furnished by the bile (Fig 53–2). Because the micelles are soluble, they allow the products of digestion to be transported through the aqueous environment of the intestinal lumen to the brush border of the mucosal cells, where they are absorbed into the intestinal epithelium. The bile salts pass on to the ileum, where most are absorbed into the **enterohepatic circulation** (see Chapter 27). Phospholipids of dietary and biliary origin (eg, phosphatidylcholine) are hydrolyzed by phospholipase A_2 of the pancreatic secretion to fatty acids and lysophospholipids, which are also absorbed from the micelles. Cholesteryl esters are hydrolyzed by cholesteryl ester hydrolase of the pancreatic juice, and the free cholesterol, together with most of the biliary cholesterol, is absorbed through the brush border after transportation in the micelles. Over 98% of dietary lipid is normally absorbed.

Within the intestinal wall, 1-monoacylglycerols are further hydrolyzed to produce free glycerol and fatty acids by a lipase, which is distinct from pancreatic lipase, whereas 2-monoacylglycerols may be reconverted to triacylglycerols via the **monoacylglycerol pathway** (Fig 53–3). The utilization of fatty acids for resynthesis of triacylglycerols requires first their "activation."

It is likely that the synthesis of triacylglycerols proceeds in the intestinal mucosa in a manner similar to that which takes place in other tissues, as described on p 218. The absorbed lysophospholipids, together with much of the absorbed cholesterol, are also reacylated with acyl-CoA to regenerate phospholipids and cholesteryl esters.

The free glycerol released in the intestinal lumen is not reutilized but passes directly to the portal vein. However, the glycerol released within the intestinal cells can be reutilized for triacylglycerol synthesis by activation by ATP to glycerol 3-phosphate. Thus, **all long-chain fatty acids absorbed in intestinal wall mucosal cells are ultimately utilized in the re-formation of triacylglycerols.**

Triacylglycerols, having been synthesized in the intestinal mucosa, are not transported to any extent in the portal venous blood. Instead, the great majority of absorbed lipids, including phospholipids, cholesteryl esters, cholesterol, and fat-soluble vitamins, generate chylomicrons that form a milky fluid, the **chyle,** that is collected by the lymphatic vessels of the abdominal region and passed to the systemic blood via the thoracic duct (see also Fig 26–3).

The majority of absorbed fatty acids of more than 10 carbon atoms in length, irrespective of the form in which they are absorbed, are found as esterified fatty acids in the lymph of the thoracic duct. Fatty acids with carbon chains **shorter than 10–12 carbons** are transported in the portal venous blood as unesterified (free) fatty acids.

Of the plant sterols (phytosterols), none are absorbed from the intestine except activated ergosterol (provitamin D).

Chyluria is an abnormality in which the patient excretes milky urine because of the presence of an abnormal connection between the urinary tract and the lymphatic drainage system of the intestine, a so-called chylous fistula. In a similar abnormality, **chylothorax,** there is an abnormal connection between the pleural space and the lymphatic drainage of the small intestine that results in the accumulation of milky pleural fluid. Feeding triacylglycerols in which the fatty acids are of medium chain length (< 12 carbons) in place of dietary fat results in a disappearance of chyluria. In chylothorax, the use of triacylglycerol with short-chain fatty acids results in the appearance of clear pleural fluid.

ABSORPTION OF AMINO ACIDS & PROTEIN

Under normal circumstances, the dietary proteins are almost completely digested to their constituent amino acids, and these end products of protein digestion are then rapidly absorbed from the intestine into the portal blood. It is possible that some hydrolysis, eg, of dipeptides, is completed in the intestinal wall. Animals may be successfully maintained when a complete amino acid mixture is fed to them.

The natural (L) isomer—but not the D isomer—of an amino acid is actively transported across the intestine from the mucosa to the serosa; vitamin B_6 (pyridoxal phosphate) may be involved in this transfer. This active transport of the L-amino acids is energy-dependent, as evidenced by the fact that 2,4-dinitrophenol, the uncoupler of oxidative phosphorylation (see p 113), inhibits transport. Amino acids are transported through the brush border by a multiplicity of carriers, many having Na^+-dependent mechanisms similar to the glucose carrier system (Fig 53–4). Of the Na^+-dependent carriers, there is a neutral amino acid carrier, a phenylalanine and methionine carrier, and a carrier specific for imino acids such as proline and hydroxyproline. Na^+-independent carriers specializing in the transport of neutral and lipophilic amino acids (eg, phenylalanine and leucine) or of cationic amino acids (eg, lysine) have been characterized.

Clinical aspects. Individuals in whom an immunologic response to ingested protein occurs must be able to absorb some unhydrolyzed protein, since digested protein is nonantigenic. This is not entirely undocumented, since the antibodies of the colostrum are known to be available to the infant.

There is increasing support for the hypothesis that the basic defect in **nontropical sprue** is located within the mucosal cells of the intestine and permits the polypeptides resulting from the peptic and tryptic digestion of gluten, the principal protein of wheat, not only to exert a local harmful effect within the intestine but also to be absorbed into the circulation and thus to elicit the production of antibodies. It has been established that circulating antibodies to wheat gluten or its fractions are frequently present in patients with nontropical sprue. The harmful entity is a polypeptide composed of 6 or 7 amino acids of which glutamine and proline must be present to ensure the harmful properties of the peptide.

These observations on a disease entity that is undoubtedly the adult analog of **celiac disease** in children advance the possibility that protein fragments of larger molecular size than amino acids are absorbed from the intestine under certain conditions.

Tables 53–12 and 53–13 summarize the sites of intestinal absorption of some common nutrients and some disorders resulting from malabsorption, respectively.

INTESTINAL PUTREFACTION & FERMENTATION

Most ingested food is absorbed from the small intestine. The residue passes into the large intestine. Here considerable absorption of water takes place, and the semiliquid intestinal contents gradually become more solid. During this period, considerable bacterial activity occurs. By fermentation and putrefaction, the

Table 53–12. Site of absorption of nutrients.

Site	Nutrient
Jejunum	Glucose and other monosaccharides; some disaccharides Monoacylglycerols, fatty acids, glycerol, cholesterol Amino acids, peptides Vitamins, folate Electrolytes, iron, calcium, water
Ileum	Bile acids Vitamin B_{12} Electrolytes Water

Table 53–13. Summary of disturbances due to malabsorption.

Sign or Symptom	Substance Malabsorbed
Anemia	Iron, vitamin B_{12}, folate
Edema	Products of protein digestion
Tetany	Calcium, magnesium, vitamin D
Osteoporosis	Calcium, products of protein digestion, vitamin D
Milk intolerance	Lactose
Bleeding, bruising	Vitamin K
Steatorrhea (fatty stools)	Lipids and fat-soluble vitamins
Hartnup disease (defect in intestinal neutral amino acid carrier)	Neutral amino acids

bacteria produce various gases, such as CO_2, methane, hydrogen, nitrogen, and hydrogen sulfide, as well as acetic, lactic, and butyric acids. The bacterial decomposition of phosphatidylcholine may produce choline and related toxic amines such as neurine.

Choline   Neurine

Fate of Amino Acids

Many amino acids undergo decarboxylation as a result of the action of intestinal bacteria to produce toxic amines (ptomaines).

Such decarboxylation reactions produce cadaverine from lysine; agmatine from arginine; tyramine from tyrosine; putrescine from ornithine; and histamine from histidine. Many of these amines are powerful vasopressor substances.

The amino acid tryptophan undergoes a series of reactions to form indole and methylindole (skatole), the substances particularly responsible for the odor of feces.

Indole   Skatole

The sulfur-containing amino acid cysteine undergoes a series of transformations to form mercaptans such as ethyl and methyl mercaptan as well as H_2S.

$$
\begin{array}{ll}
CH_3 & \\
| & CH_3SH \\
CH_2SH &
\end{array}
$$

Ethyl mercaptan   Methyl mercaptan

$$
CH_3SH \xrightarrow{[2H]} CH_4 + H_2S
$$

Methyl mercaptan   Methane and hydrogen sulfide

The large intestine is a source of considerable quantities of ammonia, presumably as a product of the putrefactive activity on nitrogenous substrates by the intestinal bacteria. This ammonia is absorbed into the portal circulation, but under normal conditions it is rapidly removed from the blood by the liver. In **liver disease,** this function of the liver may be impaired, in which case the concentration of ammonia in the peripheral blood will rise to toxic levels. It is believed that ammonia intoxication may play a role in the genesis of hepatic coma in some patients. The oral administration of neomycin has been shown to reduce the quantity of ammonia delivered from the intestine to the blood, owing to the antibacterial action of the drug. The feeding of high-protein diets to patients suffering from advanced liver disease, or the occurrence of gastrointestinal hemorrhage in such patients, may contribute to the development of ammonia intoxication. Neomycin is also beneficial under these circumstances.

Intestinal Bacteria

The intestinal flora may comprise as much as 25% of the dry weight of the feces. In herbivora, whose diet consists largely of cellulose, the intestinal or ruminal bacteria are essential to digestion, since they decompose the polysaccharide and make it available for absorption. In addition, these symbiotic bacteria accomplish the synthesis of essential amino acids and vitamins. In humans, although the intestinal flora is not as important as in the herbivora, nevertheless some nutritional benefit is derived from bacterial activity in the synthesis of certain vitamins, particularly vitamins K and B_{12}, and possibly other members of the B complex, which are made available to the body.

REFERENCES

American Cancer Society: Nutrition and cancer: Cause and prevention. (Special report.) *CA* 1984;**34:**121.

Börgstrom B: The micellar hypothesis of fat absorption: Must it be revisited? *Scand J Gastroenterol* 1985;**20:**389.

Crapo PA: Simple versus complex carbohydrate use in the diabetic diet. *Annu Rev Nutr* 1985;**5:**95.

Nestle M: *Nutrition in Clinical Practice.* Jones Medical Publications, 1985.

Passmore R, Eastwood MA: *Davidson and Passmore Human Nutrition and Dietetics,* 8th ed. Churchill Livingstone, 1986.

Steven BR, Kaunitz JD, Wright EM: Intestinal transport of amino acids and sugars. *Annu Rev Physiol* 1984;**46:**417.

Tso P: Gastrointestinal digestion and absorption of lipid. *Adv Lipid Res* 1985;**21:**143.

Woo R, Daniels-Kush R, Horton ES: Regulation of energy balance. *Annu Rev Nutr* 1985;**5:**411.

Glycoproteins & Proteoglycans

54

Robert K. Murray, MD, PhD

INTRODUCTION

Glycoproteins are proteins that have oligosaccharide (glycan) chains covalently attached to their polypeptide backbones. **Glycosaminoglycans** are polysaccharides built up of repeating disaccharide units, generally composed of an amino sugar (either glucosamine or galactosamine, which may or may not be sulfated) and a uronic acid (glucuronic acid or iduronic acid). Glycosaminoglycans were formerly called **mucopolysaccharides.** They usually occur covalently attached to protein; the complex of one or more glycosaminoglycans with protein is called a **proteoglycan. Glycoconjugates** and **complex carbohydrates** are equivalent terms used to denote molecules containing one or more carbohydrate chains covalently linked to protein or lipid; glycoproteins, proteoglycans, and glycolipids all fall within these categories.

BIOMEDICAL IMPORTANCE

Almost all the plasma proteins of humans, except albumin, are glycoproteins. Many proteins of cellular membranes (see Chapter 42) contain substantial amounts of carbohydrate. A number of the blood group substances are glycoproteins, whereas others are glycosphingolipids. Certain hormones (eg, chorionic gonadotropin) are glycoproteins. Cancer is increasingly recognized as a disorder resulting from abnormal genetic regulation (see Chapter 57). The major problem in cancer is metastasis, the phenomenon whereby cancer cells leave their tissue of origin (eg, the breast), migrate through the bloodstream to some distant site in the body (eg, the brain), and grow there in a completely unregulated manner, with catastrophic results for the affected individual. Many cancer researchers think that alterations in the structures of glycoconjugates on the surfaces of cancer cells are responsible, at least in part, for the phenomenon of metastasis. A number of diseases (the mucopolysaccharidoses) are due to deficiencies of the activities of specific lysosomal enzymes that degrade particular glycosaminoglycans; as a consequence, one or more of them accumulate in the tissues, resulting in various signs and symptoms. Hurler's syndrome is one such condition.

GLYCOPROTEINS

OCCURRENCE & FUNCTIONS

Glycoproteins occur in most organisms, from bacteria to humans. Many animal viruses also contain glycoproteins, some of which have been much investigated, in part because they are suitable for biosynthetic studies.

Numerous proteins of diverse functions are glycoproteins (Table 54–1); their carbohydrate content ranges from 1% to over 85% by weight. The precise roles oligosaccharide chains play in the functions of glycoproteins are still not clear despite extensive research; some suggested roles are listed in Table 54–2.

Table 54–1. Some functions served by glycoproteins.

Structural molecules
Cell walls
Collagen, elastin
Fibrins
Bone matrix
Lubricants and protective agents
Mucins
Mucous secretions
Transport molecules for
Vitamins
Lipids
Minerals and trace elements
Immunologic molecules
Immunoglobins
Histocompatibility antigens
Complement
Interferon
Hormones
Chorionic gonadotropin
Thyrotropin (TSH)
Enzymes
Proteases
Nucleases
Glycosidases
Hydrolases
Clotting factors
Cell attachment/recognition sites
Cell-cell
Virus-cell
Bacterium-cell
Hormone receptors
Antifreeze in antarctic fishes
Lectins

Table 54–2. Some functions of the oligosaccharide chains of glycoproteins.*

Modulate physicochemical properties, eg, solubility, viscosity, charge, and denaturation
Protect against proteolysis, from inside and outside the cell
Affect proteolytic processing of precursor proteins to smaller products
Are involved in biologic activity, eg, of hCG
Affect insertion into membranes, intracellular migration, sorting, and secretion
Affect embryonic development and differentiation
May affect sites of metastasis selected by cancer cells

*Adapted from Schachter H: Biosynthetic controls that determine the branching and heterogeneity of protein-bound oligosaccharides. *Biochem Cell Biol* 1986;**64**:163.

INFORMATIONAL CONTENT OF OLIGOSACCHARIDE CHAINS

An enormous number of glycosidic linkages can be generated between sugars. For example, 3 different hexoses may be linked to each other to form over 1000 different trisaccharides. The conformations of the sugars in oligosaccharide chains vary depending upon their linkages and proximity to other molecules with which the oligosaccharides may interact. A widely held belief is that oligosaccharide chains encode considerable biologic information and that this depends upon their constituent sugars, their sequences, and their conformations. There is also evidence that certain drugs and toxins interact with specific oligosaccharide chains of cell surface glycoconjugates (eg, cholera toxin interacts with that of the ganglioside G_{M1}; see Chapter 15). In addition, mannose 6-phosphate residues appear to target newly synthesized lysosomal enzymes to that organelle (see below).

DETECTION, PURIFICATION, & STRUCTURAL ANALYSIS OF GLYCOPROTEINS

Detection

Glycoproteins can be resolved from complex mixtures (eg, cell membranes) by **SDS-PAGE** (see Chapter 2) and detected by staining with the **periodic acid-Schiff (PAS) reagent,** which detects aldehyde groups in sugars produced by the action of periodic acid. They can also be detected by showing, through the use of **SDS-PAGE-radioautography,** that they incorporate radioactivity when cells or tissues are incubated with suitable radioactive sugars.

Purification & Structural Analysis

As for other molecules, it is first necessary to purify glycoproteins before attempting to determine their structures. This can generally be accomplished using conventional methods of protein purification. Affinity columns using lectins (see below) have also proved valuable in the purification of certain glycoproteins. Their carbohydrate composition is determined follow-

ing acid hydrolysis using analyses by **gas-liquid chromatography-mass spectrometry (GLC-MS).** Various methods have been used in the past to determine the detailed structure of their oligosaccharide chains. At present, the combined use of GLC-MS and **high-resolution NMR spectrometry** often makes possible complete determination of the structures (ie, sugar composition, details of linkages and of their anomeric natures) of the glycan chains of most glycoconjugates. Details of the linkages between the sugars of glycoproteins (which are beyond the scope of this chapter) are of fundamental importance in determining the structures and functions of these molecules.

SUGARS PRESENT IN GLYCOPROTEINS

Although about 200 monosaccharides are found in nature, fewer than a dozen are present in the oligosac-

Table 54–3. The principal sugars found in human glycoproteins. The structures of most of the sugars listed are illustrated in Chapter 14.

Sugar	Type	Abbreviation	Comments
Galactose	Hexose	Gal	Often found subterminal to NeuAc in N-linked glycoproteins. Also found in core trisaccharide of proteoglycans.
Glucose	Hexose	Glc	Present during the biosynthesis of N-linked glycoproteins but not usually present in mature glycoproteins.
Mannose	Hexose	Man	Common sugar in N-linked glycoproteins.
N-acetyl-neuraminic acid	Sialic acid (9 C atoms)	NeuAc	Often the terminal sugar in both N- and O-linked glycoproteins. Other types of sialic acid are also found, but NeuAc is the major species found in humans.
Fucose	Deoxy-hexose	Fuc	May be external in both N- and O-linked glycoproteins or linked to the GlcNAc residue attached to Asn in N-linked species.
N-Acetyl-galactos-amine	Amino-hexose	GalNAc	Present in both N- and O-linked glycoproteins.
N-Acetylglu-cosamine	Amino-hexose	GlcNAc	The sugar attached to the polypeptide chain via Asn in N-linked glycoproteins; also found at other sites in the oligosaccharides of these proteins.

charide chains of glycoproteins (Table 54–3). Most of these sugars were described in Chapter 14. N-Acetyl-neuraminic acid (NeuAc) is found at the termini of oligosaccharide chains, usually attached to subterminal galactose (Gal) or N-acetylgalactosamine (GalNAc) residues. The other sugars listed are generally found in more internal positions.

NUCLEOTIDE SUGARS

The first nucleotide sugar to be reported was uridine diphosphate glucose (UDPGlc). The common nucleotide sugars involved in the biosynthesis of glycoproteins are listed in Table 54–3; the reasons some contain UDP and others guanosine diphosphate (GDP) or cytidine monophosphate (CMP) are obscure. Many but not all of the glycosylation reactions involved in the biosynthesis of glycoproteins utilize these compounds (see below). The anhydro nature of the linkage between the phosphate group and the sugars is of the high-energy, high-group-transfer-potential type. The sugars of these compounds are thus "activated" and can be transferred to suitable acceptors provided appropriate transferases are available.

The nucleotide sugars are formed in the cytosol, generally from reactions involving the corresponding nucleoside triphosphate. Formation of uridine diphosphate galactose (UDPGal) requires the following 2 reactions:

$$\text{UTP + Glucose 1-phosphate} \xrightarrow[\text{PYROPHOSPHORYLASE}]{\text{UDPGlc}} \text{UDPGlc + Pyrophosphate}$$

$$\text{UDPGlc} \xleftrightarrow[\text{EPIMERASE}]{\text{UDPGlc}} \text{UDPGal}$$

Because many glycosylation reactions occur within the lumen of the Golgi apparatus, carrier systems (permeases, transporters) transport nucleotide sugars across the Golgi membrane. Systems transporting UDPGal, GDP-Man, and CMP-NeuAc into the cisternae of the Golgi apparatus have been described. They are antiport systems; ie, the influx of one molecule of nucleotide sugar is balanced by the efflux of one molecule of the corresponding nucleotide (eg, UMP, GMP, or CMP) formed from the nucleotide sugars. This mechanism ensures an adequate concentration of each nucleotide sugar inside the Golgi. UMP is formed from UDPGal in the above process as follows:

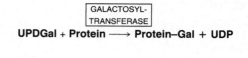

$$\text{UPDGal + Protein} \xrightarrow[]{\text{GALACTOSYL-TRANSFERASE}} \text{Protein–Gal + UDP}$$

$$\text{UDP} \xrightarrow[]{\substack{\text{NUCLEOSIDE}\\\text{DIPHOSPHATE}\\\text{PHOSPHATASE}}} \text{UMP + P}_i$$

EXOGLYCOSIDASES & ENDOGLYCOSIDASES

A number of **glycosidases** of defined specificity have proved useful in examining structural and functional aspects of glycoproteins (Table 54–4). These enzymes act at either external (exoglycosidases) or internal (endoglycosidases) positions of oligosaccharide chains. Examples of exoglycosidases are **neuraminidases** and **galactosidases;** their sequential use removes terminal NeuAc and subterminal Gal residues from most glycoproteins. **Endoglycosidases F** and **H** are examples of the latter class; these enzymes cleave the oligosaccharide chains at specific GlcNAc residues close to the polypeptide backbone (ie, at internal sites) and are thus useful in producing large oligosaccharide chains for structural analyses. A glycoprotein can be treated with one or more of the above glycosidases to analyze the effects on its biologic behavior of removal of specific sugars.

Experiments performed by Ashwell and his colleagues in the early 1970s strongly suggested that exposure of Gal residues by treatment with neuraminidase led to the rapid disappearance of ceruloplasmin from the plasma. Further work demonstrated that liver cells contain a receptor which recognizes the galactosyl moiety of many desialylated (ie, free of NeuAc) plasma proteins and leads to their endocytosis. This work clearly indicated that an individual sugar could play an important role in governing at least one of the biologic properties (ie, time of residence in the circulation) of certain glycoproteins.

Table 54–4. Some glycosidases used to study the structure and functions of glycoproteins. The enzymes are available from a variety of sources and are often specific for certain types of glycosidic linkages and also for their anomeric natures. The sites of action of endoglycosidases F and H are shown in Fig 54–4. F acts on both high-mannose and complex oligosaccharide chains, whereas H acts on the former.

Enzymes	Type
Neuraminidases	Exoglycosidase
Galactosidases	Exoglycosidase
Endoglycosidase F	Endoglycosidase
Endoglycosidase H	Endoglycosidase

LECTINS

Lectins are proteins of plant origin that bind one or more specific sugars. Their precise roles in plants are still under investigation. Many lectins have been purified, and more than 40 are commercially available (Table 54–5). Lectins have found a number of uses in biochemical research, including the following: (1) **In purification and analysis of glycoproteins.** Lectins such as concanavalin A (Con A) can be attached covalently to inert supporting media such as Sepharose. The resulting Sepharose-Con A may be used for the purification of glycoproteins that contain oligosaccharide chains which interact with Con A. (2) **As tools for probing the surfaces of cells.** Since lectins recognize specific sugars, they can be used to probe, at a general level, the sugar residues exposed on the surface membranes of cells. Numerous studies have been performed using lectins to compare the surfaces of normal and cancer cells (see Chapter 57). A general finding has been that smaller amounts of certain lectins are required to cause agglutination of tumor cells than of normal cells. This suggests that the organization or structure of a number of glycoproteins on the surfaces of tumor cells may be different from that found in normal cells. (3) **To generate mutant cells lacking certain enzymes of oligosaccharide synthesis.** When mammalian cells in tissue culture are exposed to appropriate concentrations of certain lectins (eg, Con A), most of them are killed, but a few resistant cells survive. Such cells are often found to lack certain enzymes involved in oligosaccharide synthesis. The cells are resistant, presumably because they do not produce one or more surface glycoproteins that would interact with the lectin used. The use of such mutant cells has been important in elucidating a number of aspects of glycoprotein biosynthesis.

CLASSIFICATION OF GLYCOPROTEINS

Glycoproteins can be divided into 4 classes, based on the nature of the linkage between their polypeptide chains and their oligosaccharide chains: (1) those containing a Ser (or Thr)-GalNAc linkage (Fig 54–1); (2) proteoglycans containing a Ser-Xyl linkage; (3) collagens containing a hydroxylysine (Hyl)-Gal linkage; and (4) glycoproteins containing an Asn-GlcNAc linkage (Fig 54–1).

Table 54–5. Three lectins and the sugars with which they interact. In most cases, lectins show specificity for the anomeric nature of the glycosidic linkage (α or β); this is not indicated in the table.

Lectin	Abbreviation	Sugars
Concanavalin A	Con A	Man and Glc
Soybean lectin		Gal and GalNAc
Wheat germ agglutinin	WGA	GlcNac and NeuAc

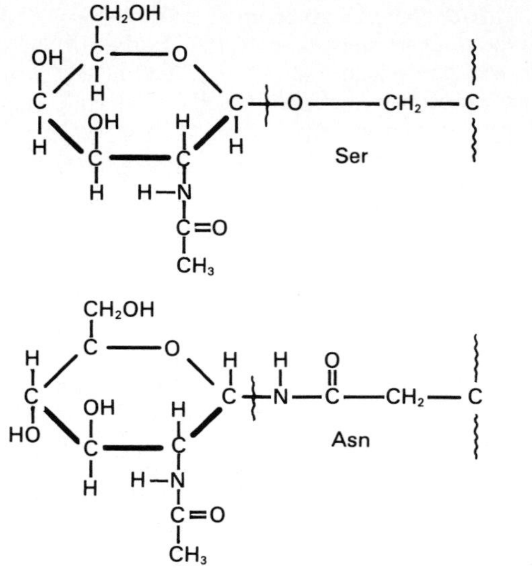

Figure 54–1. Linkage of N-acetylgalactosamine to serine and of N-acetylglucosamine to asparagine.

Classes 1, 2, and 3 are joined to the corresponding amino acids via **O-glycosidic linkages** (ie, a linkage involving an OH in the side chain of an amino acid and a sugar residue). The fourth class involves an **N-glycosidic linkage** (ie, a linkage involving the N of the amide group of asparagine and a sugar residue). Because classes 2 and 3 are relatively uncommon, the term **O-linked glycoproteins** is often used, as it is here, to refer only to members of class 1. The members of class 4 are called **N-linked glycoproteins.** Other minor classes exist. The number of oligosaccharide chains attached to one protein can vary from one to 30 or more, with the sugar chains ranging from 2 or 3 residues in length to much larger structures. Some glycoproteins contain *both* N- and O-glycosidic linkages.

O-Linked Glycoproteins

Most of the O-glycosidic links occur to the free OH groups of Ser and Thr residues of the polypeptide in a tripeptide sequence of Asn-Y-Ser(Thr), where Y is an amino acid other than aspartate. This specific tripeptide sequence is very common in proteins, but *not* every such sequence is glycosylated. The decision to glycosylate such Ser and Thr residues is based on the protein conformation surrounding that tripeptide as it emerges through the endoplasmic reticulum.

A. Occurrence and Structure of Oligosaccharide Chains of O-Linked Glycoproteins: Many glycoproteins of this class are found in **mucins.** However, O-glycosidic linkages are also found in some membrane and circulating glycoproteins. As indicated above, the most common sugar attached directly to the Ser or Thr residue is GalNac (Fig 54–1). Usually a Gal or a NeuAc residue is attached to the

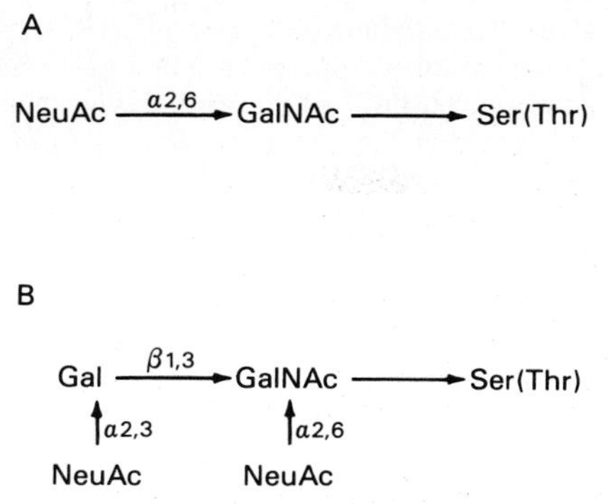

A

$$NeuAc \xrightarrow{\alpha 2,6} GalNAc \longrightarrow Ser(Thr)$$

B

$$Gal \xrightarrow{\beta 1,3} GalNAc \longrightarrow Ser(Thr)$$

$\uparrow \alpha 2,3$ $\uparrow \alpha 2,6$

NeuAc   NeuAc

Figure 54–2. Structures of 2 O-linked oligosaccharides found in *(A)* submaxillary mucins and *(B)* fetuin and in the sialoglycoprotein of the membrane of human red blood cells. (Modified and reproduced, with permission, from Lennarz WJ: *The Biochemistry of Glycoproteins and Proteoglycans.* Plenum Press, 1980.)

GalNAc. The structures of 2 typical oligosaccharide chains of glycoproteins of this class are shown in Fig 54–2. Many variations are found.

B. Biosynthesis of O-Linked Glycoproteins: The polypeptide chains of these and other glycoproteins are encoded by mRNA species; because most glycoproteins are membrane-bound or secreted, they are generally translated on membrane-bound polyribosomes (see Chapter 40). The sugars of the oligosaccharide chains of the O-glycosidic type of glyco-

protein are built up by the **stepwise donation of sugars from nucleotide sugars,** such as UDPGalNAc, UDPGal, and CMP-NeuAc. The enzymes catalyzing this type of reaction are membrane-bound **glycoprotein glycosyltransferases.** The synthesis of each such enzyme is controlled by one specific gene. Generally, synthesis of one specific type of linkage requires the activity of a correspondingly specific transferase (ie, the "one linkage, one glycosyl transferase" hypothesis). The enzymes catalyzing addition of the inner sugar residues are located in the endoplasmic reticulum, and addition of the first sugars occurs during translation (ie, cotranslational modification of the protein occurs). The enzymes adding the terminal sugars (such as NeuAc) are located in the Golgi apparatus.

N-Linked Glycoproteins

This class is distinguished by the presence of the Asn-GlcNAc linkage (Fig 54–1). It is the major class of glycoproteins and has been much studied, since the most readily accessible glycoproteins (eg, plasma proteins) mainly belong to this group. It includes both membrane-bound and circulating glycoproteins. The principal difference between this and the previous class, apart from the nature of the amino acid to which the oligosaccharide chain is attached (mainly Asn versus Ser), concerns their biosynthesis.

A. Classes of N-linked Glycoproteins and Their Structures: There are 3 major classes of N-linked glycoproteins: **complex, hybrid,** and **high-mannose** (Fig 54–3). Each type shares a common pentasaccharide ($[Man]_3 [GlcNAc]_2$, shown within the

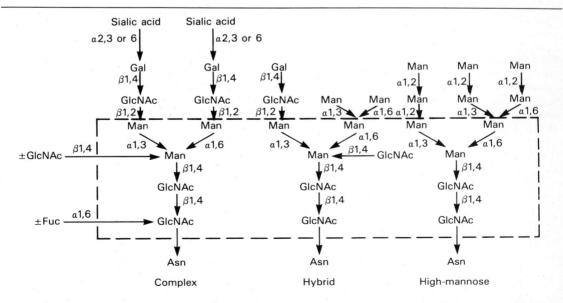

Figure 54–3. Structures of the major types of asparagine-linked oligosaccharides. The boxed area encloses the pentasaccharide core common to all N-linked glycoproteins. (Reproduced, with permission, from Kornfeld R, Kornfeld S: Assembly of asparagine-linked oligosaccharides. *Annu Rev Biochem* 1985;54:631.)

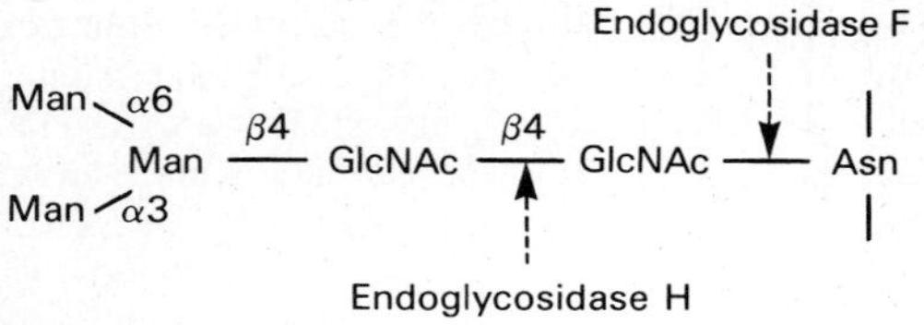

Endoglycosidase F

```
Man  α6
        β4          β4
      Man ── GlcNAc ── GlcNAc ── Asn
Man  α3
```

Endoglycosidase H

Figure 54–4. Schematic diagram of the pentasaccharide core common to all N-linked glycoproteins and to which various outer chains of oligosaccharides may be attached. The sites of action of endoglycosidases F and H are also indicated.

boxed area in Fig 54–3 and also depicted in Fig 54–4), but they differ in their outer branches. The presence of the common pentasaccharide is explained by the fact that all 3 classes share an initial common mechanism of biosynthesis. Glycoproteins of the complex type generally contain terminal NeuAc residues and underlying Gal and GlcNAc residues, the latter often constituting the disaccharide lactosamine. (The presence of repeating lactosamine units, [Galβ1-4GlcNAcβ1-3]$_n$, characterizes a fourth class of glycoprotein, the **polylactosamine** class; it is important because the I/i blood group substances belong to it.) The majority of complex-type oligosaccharides contain 2 (Fig 54–2), 3, or 4 outer branches, but structures containing 5 branches have also been described. The oligosaccharide branches are often referred to as antennae, so that bi-, tri-, tetra-, and penta-antennary structures may all be found. A bewildering number of chains of the complex type exist, and that indicated in Fig 54–3 is only one of many. Other complex chains may terminate in Gal or Fuc. High-mannose oligosaccharides typically have 2–6 additional Man residues linked to the pentasaccharide core. Hybrid molecules contain features of both of the 2 other classes.

B. Overview of The Biosynthesis of N-Linked Glycoproteins: Leloir and his colleagues described the occurrence of an **oligosaccharide-pyrophosphoryl-dolichol (oligosaccharide-P-P-Dol),** which subsequent research showed to play a key role in the biosynthesis of N-linked glycoproteins. The oligosaccharide chain of this compound generally has the

structure (Glc)$_3$(Man)$_9$(GlcNAc)$_2$–R (R = P-P-Dol). The sugars of this compound are first assembled on the pyrophosphoryl-dolichol backbone, and the oligosaccharide chain is then transferred en bloc to suitable Asn residues of acceptor apoglycoproteins during their synthesis on membrane-bound polyribosomes. To form an oligosaccharide chain of the **complex** type, the Glc residues and 6 of the Man residues are removed, forming the core pentasaccharide (Man)$_3$(GlcNac)$_2$– (Fig 54–4). The sugars characteristic of complex chains (GlcNAc, Gal, NeuAc) are then added by the action of individual glycosyltransferases, mostly located in the Golgi apparatus. To form **high-mannose** chains, only the Glc residues plus or minus certain of the peripheral Man residues are removed. The phenomenon whereby the glycan chains of N-linked glycoproteins are first partially degraded and then in some cases rebuilt is referred to as oligosaccharide processing. Rather surprisingly, the initial steps involved in the biosynthesis of the N-linked glycoproteins differ markedly from those involved in the biosynthesis of the O-linked glycoproteins. The former involves the oligosaccharide-P-P-Dol; the latter, as described earlier, does not.

The process can be broken down into 2 stages: (1) assembly and transfer of oligosaccharide-P-P-dolichol, and (2) processing of the oligosaccharide chain.

1. Assembly and transfer of oligosaccharide-P-P-dolichol–Polyisoprenol compounds exist in both bacteria and eukaryotic tissues. They participate in the synthesis of bacterial cell walls and in the biosynthesis of N-linked glycoproteins. The polyisoprenol used in eukaryotic tissues is **dolichol,** which is, next to rubber, the longest naturally occurring hydrocarbon made up of a single repeating unit. Dolichol is made up of 17–20 repeating isoprenoid units (Fig 54–5).

Before it participates in the biosynthesis of oligosaccharide-P-P-Dol, dolichol must first be phosphorylated to form dolichol phosphate (Dol-P) in a reaction catalyzed by **dolichol kinase** and using ATP as the phosphate donor.

GlcNAc-pyrophosphoryl-dolichol (GlcNAc-P-P-Dol) is the key lipid that acts as an **acceptor** for other sugars in the assembly of oligosaccharide-P-P-Dol. It is synthesized in the membranes of the endo-

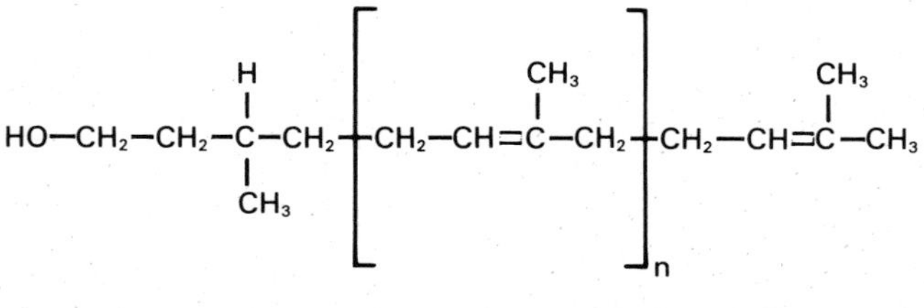

```
                    H                     CH₃              CH₃
                    |                     |                |
HO—CH₂—CH₂—C—CH₂—[ CH₂—CH=C—CH₂—]CH₂—CH=C—CH₃
                    |
                    CH₃                                    n
```

Figure 54–5. The structure of dolichol. The phosphate in dolichol phosphate is attached to the primary alcohol group at the left-hand end of the molecule. The group within the brackets is an isoprene unit (n = 17–20 isoprenoid units).

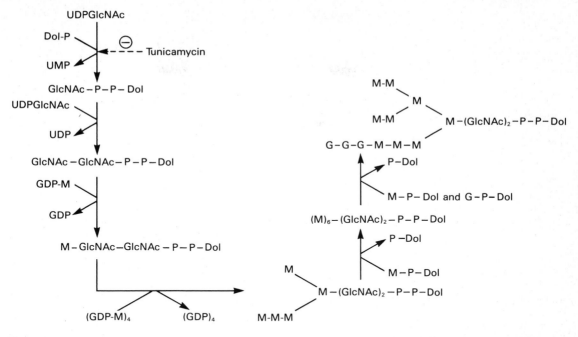

Figure 54–6. Pathway of biosynthesis of dolichol-P-P-oligosaccharide. The specific linkages formed are indicated in Fig 54–7. Note that the internal mannose residues are donated by GDP-mannose, whereas the more external mannose residues and the glucose residues are donated by dolichol-P-mannose and dolichol-P-glucose. UDP, uridine diphosphate; Dol, dolichol; P, phosphate; UMP, uridine monophosphate; GDP, guanosine diphosphate; M, mannose; G, glucose.

plasmic reticulum from Dol-P and UDPGlcNAc in the following reaction:

Dol-P + UDPGlcNAc ⟶ GlcNAc-P-P-Dol + UMP

The above reaction—which is the first step in the assembly of oligosaccharide-P-P-Dol—and the other later reactions are summarized in Fig 54–6.

The essential features of the subsequent steps in the assembly of oligosaccharide-P-P-dolichol are as follows:

a. A second GlcNAc residue is added to the first, again using UDPGlcNAc as the donor.

b. Five Man residues are added, using GDP-mannose as the donor.

c. Four additional Man residues are next added, using Dol-P-Man as the donor. Dol-P-Man is formed by the following reaction:

Dol-P + GDP-Man ⟶ Dol-P-Man + GDP

d. Finally, the **3 peripheral glucose residues** are donated by Dol-P-Glc, which is formed in a reaction analogous to that just presented, except that Dol-P and UDPGlc are the substrates.

It should be noted that the first 7 sugars (2 GlcNAc and 5 Man residues) are donated by nucleotide sugars,

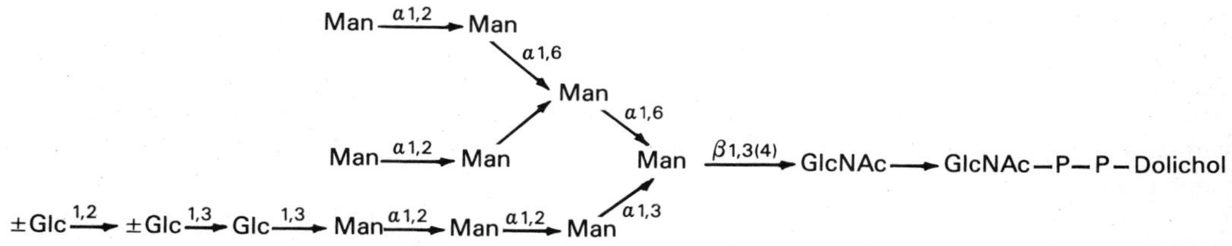

Man $\xrightarrow{\alpha 1,2}$ Man
$\searrow \alpha 1,6$
Man
$\searrow \alpha 1,6$
Man $\xrightarrow{\alpha 1,2}$ Man
Man $\xrightarrow{\beta 1,3(4)}$ GlcNAc ⟶ GlcNAc—P—P—Dolichol
$\nearrow$
$\pm$ Glc $\xrightarrow{1,2}$ $\pm$ Glc $\xrightarrow{1,3}$ Glc $\xrightarrow{1,3}$ Man $\xrightarrow{\alpha 1,2}$ Man $\xrightarrow{\alpha 1,2}$ Man $\xrightarrow{\alpha 1,3}$

Figure 54–7. Structure of dolichol-P-P-oligosaccharide. (Reproduced, with permission, from Lennarz WJ: *The Biochemistry of Glycoproteins and Proteoglycans.* Plenum Press, 1980.)

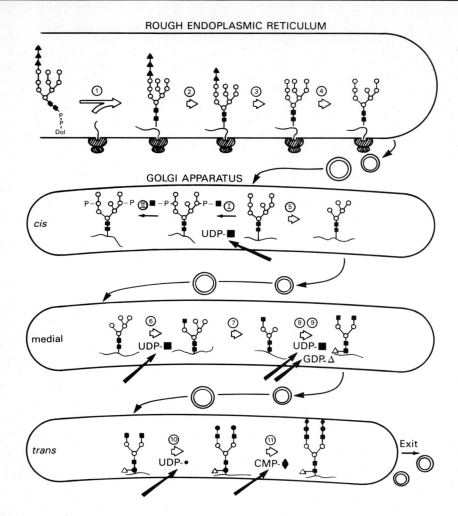

Figure 54–8. Schematic pathway of oligosaccharide processing. The reactions are catalyzed by the following enzymes: 1, oligosaccharyltransferase; 2, α-glucosidase I; 3, α-glucosidase II; 4, endoplasmic reticulum α1,2 mannosidase; I, N-acetylglucosaminylphosphotransferase; II, N-acetylglucosamine-1-phosphodiester α-N-acetylglucosaminidase; 5, Golgi apparatus α-mannosidase I; 6, N-acetylglucosaminyltransferase I; 7, Golgi apparatus α-mannosidase II; 8, N-acetylglucosaminyltransferase II; 9, fucosyltransferase; 10, galactosyltransferase; 11, sialyltransferase. ■, N-acetylglucosamine; ○, mannose; ▲, glucose; △, fucose; ●, galactose; ◆, sialic acid. (Reproduced, with permission, from Kornfeld R, Kornfeld S: Assembly of asparagine-linked oligosaccharides. Annu Rev Biochem 1985;54:631.)

whereas the last 7 sugars (4 Man and 3 Glc residues) added are donated by dolichol sugars. The net result of the above is assembly of the compound illustrated in Fig 54–7 and referred to in shorthand as $(Glc)_3 (Man)_9 (GlcNAc)_2$-P-P-Dol.

The oligosaccharide linked to dolichol-P-P is transferred en bloc to form an N-glycosidic bond with one or more specific Asn residues of an acceptor protein emerging from the luminal surface of the membrane of the endoplasmic reticulum. The reaction is catalyzed by an **"oligosaccharide transferase,"** a membrane-associated enzyme. The transferase will recognize and transfer any glycolipid with the general structure R–$(GlcNAc)_2$-P-P-Dol. Glycosylation occurs at the

Asn residue of an Asn-X-Ser/Thr tripeptide sequence, where X is any amino acid except possibly proline or aspartate. A tripeptide site contained within a β turn is favored. Only about one-third of the Asn residues that are potential acceptor sites are actually glycosylated. The acceptor proteins are of both the secretory and integral membrane class. Cytosolic proteins are rarely glycosylated. The transfer reaction and subsequent processes in the glycosylation of N-linked glycoproteins, along with their subcellular locations, are depicted in Fig 54–8.

The other product of the oligosaccharide transferase reaction is dolichol-P-P, which is subsequently converted to dolichol-P by a phosphatase. The dolichol-P can serve again as an acceptor for the synthesis of another molecule of oligosaccharide-P-P-dolichol.

2. Processing of the oligosaccharide chain—

a. Early phase—The various reactions involved are indicated in Fig 54–8. The oligosaccharide transferase catalyzes reaction 1 (see above). Reactions 2 and 3 involve the removal of the terminal Glc residue by glucosidase I and of the next 2 Glc residues by glucosidase II, respectively. In the case of **high-mannose** glycoproteins, the process may stop here, or up to 4 Man residues may also be removed. However, to form **complex** chains, additional steps are necessary, as follows. Four external Man residues are removed in reactions 4 and 5 by 2 different mannosidases. In reaction 6, a GlcNAc residue is added to one of the Man residues by GlcNAc transferase I. The action of this latter enzyme permits the occurrence of reaction 7, a reaction catalyzed by yet another mannosidase (Golgi α-mannosidase II) and which results in a reduction of the Man residues to the core number of 3 (Fig 54–4).

An important additional pathway is indicated in reactions I and II of Fig 54–8. This involves enzymes destined for **lysosomes.** Such enzymes are targeted to the lysosomes by a specific chemical marker. In reaction I, a residue of GlcNAc-1-P is added to carbon 6 of one or more specific Man residues of these enzymes. The reaction is catalyzed by a GlcNAc phosphotransferase, which uses UDPGlcNAc as the donor and generates UMP as the other product:

GlcNAc PHOSPHOTRANSFERASE

UDPGlcNAc + Man–Protein $\longrightarrow$

GlcNAc-1-P-6-Man–Protein + UMP

In reaction II the GlcNAc is removed by the action of a phosphodiesterase, leaving the Man residues phosphorylated in the 6 position:

PHOSPHODIESTERASE

GlcNAc-1-P-6-Man–Protein $\longrightarrow$

P-6-Man–Protein + GlcNAc

A Man-6-P receptor, located in the Golgi apparatus, binds the Man-6-P residue of these enzymes and directs them to the lysosomes. Fibroblasts from patients with I-cell disease (see below) are severely deficient in the activity of the GlcNAc phosphotransferase.

b. Late phase—To assemble a typical complex oligosaccharide chain, additional sugars must be added to the structure formed in reaction 7. Hence, in reaction 8, a second GlcNAc is added to the peripheral Man residue of the other arm of the bi-antennary structure shown in Fig 54–8; the enzyme catalyzing this step is GlcNAc transferase II. Reactions 9, 10, and 11 involve the addition of Fuc, Gal, and NeuAc residues at the sites indicated, in reactions catalyzed by fucosyl, galactosyl, and sialyl transferases, respectively.

Subcellular sites. As indicated in Fig 54–8, the endoplasmic reticulum and the Golgi apparatus are the major sites involved in glycosylation processes. Addition of the oligosaccharide occurs in the rough endoplasmic reticulum during or after translation. Removal of the Glc and some of the peripheral Man residues also occurs in the endoplasmic reticulum. The Golgi apparatus is composed of *cis,* medial, and *trans* cisternae; these can be separated by appropriate centrifugation procedures. Vesicles containing glycoproteins appear to bud off in the endoplasmic reticulum and are transported to the *cis* Golgi. Various studies have shown that the enzymes involved in glycoprotein processing show differential locations in the cisternae of the Golgi. As indicated in Fig 54–8, Golgi α-mannosidase I (catalyzing reaction 5) is located mainly in the *cis* Golgi, whereas GlcNAc transferase I (catalyzing reaction 6) appears to be located in the medial Golgi, and the fucosyl, galactosyl, and sialyl transferases (catalyzing reactions 9, 10, and 11) are located in the *trans* Golgi.

Regulation of the Glycosylation of Glycoproteins

It is evident that glycosylation of glycoproteins is a complex process, involving a large number of enzymes. One index of its complexity is that 7 distinct GlcNAc transferases involved in glycoprotein biosynthesis have been reported, and a number of others are theoretically possible. Multiple species of the other glycosyltransferases (eg, sialyltransferases) also exist. Controlling factors of the first stage (ie, oligosaccharide assembly and transfer) of N-linked glycoprotein biosynthesis include (1) the presence of suitable acceptor sites in proteins, (2) the tissue level of Dol-P, and (3) the activity of the oligosaccharide transferase.

Factors involved in the regulation of oligosaccharide processing include the following: (1) The activities in cells of the various hydrolases and transferases involved in processing are important determining factors of the types of oligosaccharide chains (eg, complex or high-mannose) formed. Obviously, if a particular transferase is not present in a tissue, the corresponding sugar linkage will not be synthesized. (2) Certain enzymes can act only if another enzyme has acted previously. For example, prior action of GlcNAc transferase I (Fig 54–8) is necessary for the action of Golgi α-mannosidase II. (3) The activities of various transferases can wax and wane systematically during development, explaining in part how different oligosaccharides may be formed during different stages of an organism's life cycle. (4) The fact that the various glycosyltransferases have different intracellular locations plays a role in regulation. For example, if a protein is destined for insertion into the membranes of the endoplasmic reticulum (eg, HMG-CoA reductase), it will never travel to the Golgi cisternae to encounter the various processing enzymes located there. Accordingly, it is not surprising to discover that HMG-CoA reductase is a high-mannose type of glycoprotein. (5) Another important factor is protein conformation. Closely related viral glycoproteins exhibit different types of oligosaccharide chains when grown in the same cells. The best explanation appears to be that these proteins must exhibit different conformations that affect the extent of processing. (6) There are

Table 54–6. Three inhibitors of enzymes involved in the glycosylation of glycoproteins and their sites of action.

Inhibitor	Site of Action
Tunicamycin	Inhibits the enzyme catalyzing addition of GlcNAc to dolichol-P, the first step in the biosynthesis of oligosaccharide-P-P-dolichol
Deoxynojirimycin	Inhibitor of glucosidases I and II
Swainsonine	Inhibitor of mannosidase II

considerable differences in the profiles of processing enzymes displayed by cells from different species. The oligosaccharides of one particular Sindbis virus vary (eg, from high-mannose to complex) depending upon the host cell in which the virus is grown. (7) An area of great current interest involves analysis of the activities of glycoprotein-processing enzymes in various types of cancer cells. There is some evidence that cancer cells synthesize different oligosaccharide chains (eg, they often exhibit greater branching) from those made in control cells; correlating the activity of a particular processing enzyme with the metastatic properties of some cancer cells (see Chapter 57) is of great interest.

Inhibitors of Processes Involved in Glycosylation

A number of compounds are known to inhibit various reactions involved in glycoprotein processing. Tunicamycin, deoxynojirimycin, and swainsonine are 3 such agents. The reactions that they inhibit are indicated in Table 54–6. These agents can be used experimentally to inhibit various stages of glycoprotein biosynthesis and to study the effects of specific alterations upon the process. For instance, if cells are grown in the presence of tunicamycin, no glycosylation of their normally N-linked glycoproteins will occur. In certain cases, this has been shown to increase the susceptibility of these proteins to proteolysis. Inhibition of glycosylation does not appear to have a consistent effect upon the secretion of glycoproteins that are normally secreted.

I-CELL DISEASE

As explained above, Man-6-P serves as a chemical marker to target lysosomal enzymes to that organelle. Analyses of cultured fibroblasts derived from patients with I-cell (inclusion cell) disease played a large part in revealing the above role of Man-6-P. I-cell disease is a rare condition in which cultured cells lack almost all of the normal lysosomal enzymes; the lysosomes thus accumulate many different types of undegraded molecules. However, samples of serum from patients with the disease contain very high activities of lysosomal enzymes, suggesting that lysosomal enzymes are being synthesized but failing to reach their proper in-

tracellular destination. Cultured cells from patients with the disease can take up exogenously added lysosomal enzymes obtained from normal subjects, indicating that the cells contain a normal receptor for uptake of lysosomal enzymes. In addition, it suggests that lysosomal enzymes from patients with I-cell disease might lack a recognition marker. Further studies revealed that lysosomal enzymes from normal individuals carried the Man-6-P recognition marker described above. Cultured cells from patients with I-cell disease were then found to be deficient in the activity of the GlcNAc phosphotransferase, explaining how the lysosomal enzymes failed to acquire the Man-6-P marker. Thus, biochemical investigations of this disease not only led to elucidation of its basis but also contributed significantly to knowledge of how newly synthesized proteins are targeted to specific organelles, in this case the lysosome.

BLOOD GROUP ANTIGENS

The blood group antigens are oligosaccharides of specific medical interest. Their structures and syntheses will be described in detail. In 1900, **Landsteiner** described the **ABO blood groups.** Today, there are more than 20 blood group systems expressing more than 160 distinct antigens. The most commonly studied blood groups are those of the ABH(O) and the Lewis (Le) systems. These erythrocyte antigens are linked to specific membrane proteins by O-glycosidic bonds in which GalNAc is the most proximal sugar residue. The specific oligosaccharides forming these antigens occur in 3 forms: (1) as glycosphingolipids and glycoproteins on the surfaces of erythrocytes and other cells, (2) as oligosaccharides in milk and urine, and (3) as oligosaccharides attached to mucins secreted in the gastrointestinal, genitourinary, and respiratory tracts.

There are 4 independent gene systems related to the expressions of these oligosaccharide antigens (Table 54–7); only the first 3 will be discussed here.

Table 54–7. The 4 independent gene systems responsible for the expressions of the ABH(O) and Lewis (Le) blood group antigens.

Genetic Locus	Alleles
H	H, h
Secretor	Se, se
ABO	A, B, O
Lewis	Le, le

THE H LOCUS

The **H locus** codes for the fucosyltransferase in hematopoietic tissues that attaches a fucose residue in $\alpha 1,2$- linkage to a Gal residue, itself attached in either $\beta 1,4$- or $\beta 1,3$- linkage to an oligosaccharide. The fucosyltransferase catalyzes

$$\text{GDP-}\beta\text{-Fuc + Gal-}\beta\text{–R} \longrightarrow \text{Fuc-}\alpha\text{1,2-Gal-}\beta\text{-R + GDP}$$

The product, Fuc-α1,2-Gal-β-R, is a **precursor** for the formation of both the A and B oligosaccharide antigens of erythrocytes. The h allele of the H locus codes for an inactive fucosyltransferase; therefore, individuals with the **hh genotype cannot generate this necessary precursor** of the A and B erythrocyte blood antigens. Accordingly, hh genotypic persons will be erythrocyte blood type O even though they may possess genes for the active A or active B glycosyltransferases described below.

THE SECRETOR LOCUS

The **secretor locus (Se)** codes for a specific Fuc transferase in secretory organs, such as the exocrine glands, but *not* in the erythrocytes. Accordingly—for example—individuals with the SeSe or Sese genotype will generate the A and B antigen **precursor** in the exocrine glands that form saliva and will be secretors of the A or B antigens (or both) when the A- or B-specific transferases are present (Table 54–8). Individuals who are sese genotype will **not secrete A or B** antigens, but if they possess an H allele and A or B alleles, their **erythrocytes will express** the A, B, or both antigens.

THE ABO LOCUS

The ABO locus codes for 2 specific transferases that act to transfer specific Gal moieties to the Fuc-α1,2-Gal-β-R precursor oligosaccharide formed by the action of the H or Se allele-coded fucosyltransferase. The A-specific transferase carries out the following reaction:

Table 54–8. A, B antigen expression.

Genotyopes			Phenotypes	
ABO Locus	H Locus	Secretor Locus	Erythrocytes	Secretions
OO	Any	Any	O	O
A and/or B	HH or Hh	SeSe or Sese	A and/or B	A and/or B
A and/or B	HH or Hh	sese	A and/or B	O
A and/or B	hh	Any	O	O

$$\text{UDP-}\alpha\text{-GalNAc + Fuc-}\alpha\text{1,2-Gal-}\beta\text{–R} \longrightarrow$$
$$\text{GalNAc-}\alpha\text{1,3-(Fuc-}\alpha\text{1,2)Gal-}\beta\text{–R + UDP}$$

The B allele-specific transferase catalyzes the reaction

$$\text{UDP-}\alpha\text{-Gal + Fuc-}\alpha\text{1,2-Gal-}\beta\text{–R} \longrightarrow$$
$$\text{Gal-}\alpha\text{1,3-(Fuc-}\alpha\text{1,2)Gal-}\beta\text{–R + UDP}$$

Accordingly, persons possessing an A allele will attach a **GalNAc moiety to the precursor generated by the 1,2-Fuc transferase,** and individuals possessing a B allele will transfer a **Gal moiety to the same precursor.** Individuals possessing both an A allele and a B allele will generate both oligosaccharides, ie, one with a GalNAc and another with a Gal moiety on the nonreducing terminus. Individuals lacking both A and B alleles (OO homozygotes) will not attach either GalNAc or Gal to the precursor. The anti-A antiserum originally described by Landsteiner recognizes the specific oligosaccharide with the GalNAc nonreducing terminus. The anti-B antiserum recognizes the closely related oligosaccharide with the Gal nonreducing terminus. When neither GalNAc nor Gal is at the reducing terminus of this oligosaccharide, it will not be recognized by either anti-A or anti-B antisera, and the blood group antigen is said to be **type O.** It can be seen that individuals with the hh genotype and thus incapable of attaching the Fuc moiety to the appropriate Gal-β–R oligosaccharide would be incapable of expressing the A or the B antigen determinant and thus also would be considered to be of the O type blood group.

PROTEOGLYCANS & GLYCOSAMINOGLYCANS

The proteoglycans and the glycoproteins are molecules consisting of proteins to which oligosaccharide or polysaccharide chains are covalently attached. The distinction between proteoglycans and glycoproteins is based on the chemical nature of the attached polysaccharides. In proteoglycans, each polysaccharide consists of **repeating disaccharide units in which D-glucosamine or D-galactosamine is always present.** Each disaccharide unit in the proteoglycan polysaccharides (with the exception of keratan sulfate) contains a uronic acid, **L-glucuronic acid (GlcUA)** or its 5-epimer, **L-iduronic acid (IdUA).** With the exception of hyaluronic acid, all polysaccharides of proteoglycans contain **sulfate groups** either as O-esters or N-sulfate (in heparin and heparan sulfate).

The linkage of the proteoglycan polysaccharides to their polypeptide chain is one of 3 types:

(1) An O-glycosidic bond between the Xyl and Ser, a bond that is **unique to proteoglycans.**

(2) An O-glycosidic bond between GalNAc and Ser(Thr), present in keratan sulfate II.

(3) An N-glycosylamine bond between GlcNAc and the amide nitrogen of Asn.

The formation of the polysaccharide chains occurs in pathways quite similar to those responsible for the attachment and growth of the oligosaccharide chains of glycoproteins. A UDP-Xyl transferase attaches the Xyl of the nucleotide sugar to Ser to form the Xyl-Ser O-glycosidic bond. The formation of the O-glycosidic bond between GalNAc and Ser (or Thr) probably occurs by a similar UDPGalNAc transferase. The N-glycosidic bond between GlcNAc and the amide nitrogen of Asn almost certainly involves the lipid-linked polysaccharide, dolichol-P-P polysaccharide, which, as discussed above, is responsible for the transfer of a preformed oligo- or polysaccharide in the formation of glycoproteins. However, the details of this reaction in the synthesis of proteoglycans have not been established.

The chain elongation process involves the nucleotide sugars acting as donors. The reactions are governed primarily by the substrate specificities of the specific glycosyltransferases. Again, the "one enzyme, one linkage" relationship seems to hold. The specificity of these reactions is dependent upon the nucleotide sugar donor, the acceptor oligosaccharide, and the anomeric configuration and position of the linkage. The enzyme systems involved in this chain elongation are capable of high-fidelity reproduction of complex polysaccharides.

The termination of polysaccharide chain growth seems to result from (1) capping effects of **sialylation** by the specific sialyltransferases; (2) **sulfation,** particularly at the 4- positions of the sugars, and (3) the progression of the particular polysaccharide **away from the site** in the membrane where the catalysis occurs.

After formation of the polysaccharide chain, there occur numerous chemical **modifications,** such as the introduction of sulfate groups onto GalNAc moieties of chondroitin sulfate and dermatan sulfate and the epimerization of GlcUA to IdUA residues in heparin and heparan sulfate.

An important aspect of the metabolism of proteoglycans is their **degradation.** Inherited defects in the degradation of the polysaccharide chains of proteoglycans lead to the group of diseases known as **mucopolysaccharidoses** and **mucolipidoses,** discussed below and in Chapter 14. These catabolic defects have allowed the study of specific degradation enzymes and their substrates. There exists a battery of exoglycosidases that act in a stepwise manner to remove the sulfate moieties and glycosyl groups. In addition, there are normally present endoglycosidases with different specificities. For example, **hyaluronidase** is a widely distributed enzyme that cleaves N-acetylhexosamine linkages in hyaluronic acid and chondroitin sulfates.

There are **7 types of polysaccharides** (glycosaminoglycans) found attached to the proteins of proteoglycans. Six of them are structurally related and

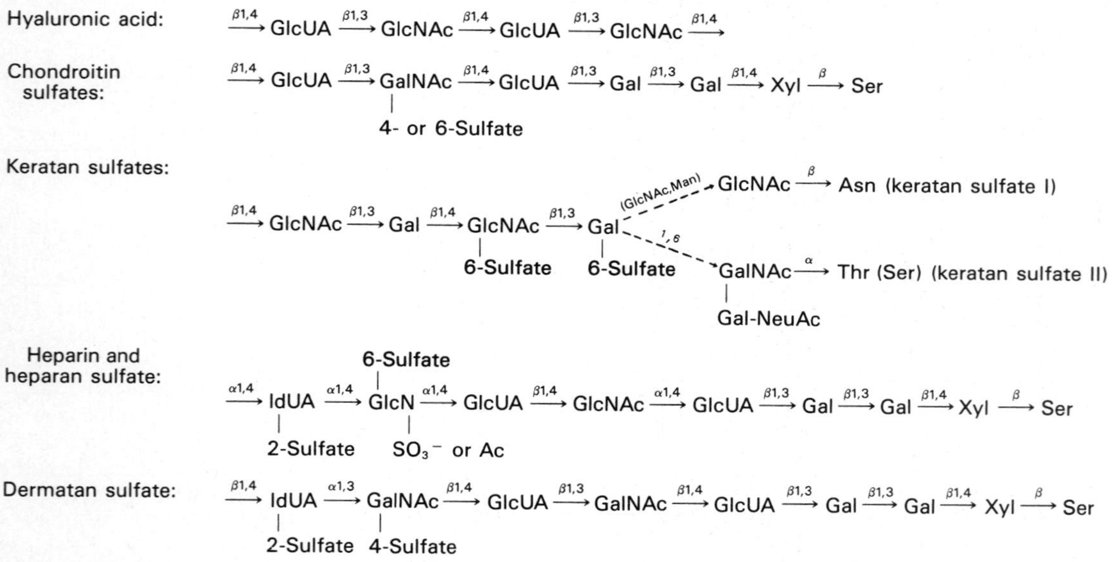

Figure 54–9. Summary of structures of proteoglycans and glycosaminoglycans. (GlcUA, D-glucuronic acid; IdUA, L-iduronic acid; GlcN, D-glucosamine; GalN, D-galactosamine; Ac, acetyl; Gal, D-galactose; Xyl, D-xylose; Ser, L-serine; Thr, L-threonine; Asn, L-asparagine; Man, D-mannose; NeuAc, N-acetylneuraminic acid.) The summary structures are qualitative representations only and do not reflect, for example, the uronic acid composition of hybrid polysaccharides such as heparin and dermatan sulfate, which contain both L-iduronic and D-glucuronic acid. Neither should it be assumed that the indicated substituents are always present, eg, whereas most iduronic acid residues in heparin carry a 2-sulfate group, a much smaller proportion of these residues are sulfated in dermatan sulfate. (Slightly modified and reproduced, with permission, from Lennarz WJ: The Biochemistry of Glycoproteins and Proteoglycans. Plenum Press, 1980.)

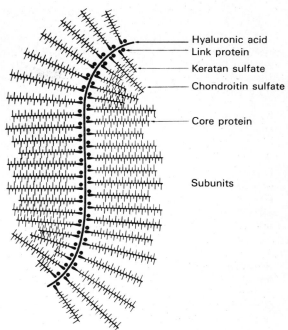

Figure 54–11. Schematic representation of proteoglycan aggregate. (Reproduced, with permission, from Lennarz WJ: *The Biochemistry of Glycoproteins and Proteoglycans.* Plenum Press, 1980.)

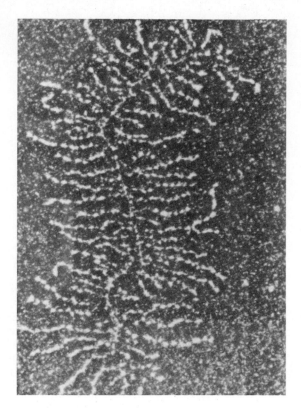

Figure 54–10. Darkfield electron micrograph of a proteoglycan aggregate of intermediate size in which the proteoglycan subunits and filamentous backbone are particularly well extended. (Reproduced, with permission, from Rosenberg L, Hellman W, Kleinschmidt AK: Electron microscopic studies of proteoglycan aggregates from bovine articular cartilage. *J Biol Chem* 1975;**250**:1877.)

contain **alternating uronic acid and hexosamine residues** in repeating disaccharide units; the exception is keratan sulfate, which lacks a uronic acid. All except hyaluronic acid contain **sulfated sugars** and are attached covalently to proteins. The 7 types of polysaccharides can be distinguished by their **monomer composition,** their **glycosidic linkage,** and the amount and location of their **sulfate substituents.**

All of the glycosaminoglycans are **polyanions,** since they have acidic sulfate or carboxyl groups of uronic acids present throughout their structures. Many of their functions result from this particular characteristic.

The structures of the 7 glycosaminoglycans of the proteoglycan molecules are summarized in Fig 54–9.

Hyaluronic Acid

Hyaluronic acid consists of an unbranched chain of repeating disaccharide units containing **GlcUA** and **GlcNAc.** There is no firm evidence that hyaluronic acid is linked to a protein molecule, as are other connective tissue polysaccharides, but it is probably synthesized as a proteoglycan, as are the other glycosaminoglycans. Hyaluronic acid is present in bacteria and widely distributed among various animal organisms and tissues, including synovial fluid, the vitreous body of the eye, and loose connective tissue.

Chondroitin Sulfates

Chondroitin sulfates are proteoglycans that are a very prominent component of cartilage. The polysaccharide is linked to protein by the Xyl-Ser O-glycosidic bond. The structure of the chondroitin sulfates is summarized in Fig 54–9. The repeating disaccharide unit is very similar to that of hyaluronic acid, except that the hexosamine is **GalNAc** rather than GlcNAc. However, in both the chondroitin sulfates and hyaluronic acid, the uronic acid is **GlcUA,** and the bond positions and anomeric configurations are the same. In the chondroitin sulfates, the GalNAc carries a sulfate substituent in the 4- or 6- position. As a rule, both 4- and 6-sulfate substituents are present in the same molecule but not on the same monosaccharide residue. There is on the average about one sulfate substituent per disaccharide unit. Each chain of polysaccharide consists of about 40 repeating disaccharide units and thus has a molecular weight of about 20,000. Many such chains are attached to a single protein molecule, generating high-molecular-weight proteoglycans. For instance, the molecular weight of nasal cartilage chondroitin sulfate is approximately 2.5×10^6.

The chondroitin sulfates associate tightly with hyaluronic acid with the aid of 2 **"link proteins"** to generate very large aggregates in connective tissue. These aggregates can be observed in the electron microscope (Fig 54–10) and are diagrammatically presented in Fig 54–11.

The link proteins are strongly hydrophobic and interact both with hyaluronic acid and the proteoglycan.

The chondroitin sulfates contain 6 types of intersaccharide linkages and thus are synthesized by 6 different glycosyltransferase enzymes, one for each type of linkage. In addition, there are 2 types of sulfate esters, one on the 4- and another on the 6- position. Two sulfotransferases carry out these esterifications. The sulfate-containing substrate for these sulfotransferases is 3'-phosphoadenosine-5'-phosphosulfate (PAPS).

Keratan Sulfate I & Keratan Sulfate II

As shown in Fig 54–9, the keratan sulfates consist of repeating **Gal-GlcNAc** disaccharide units and contain sulfates on the 6- position of GlcNAc residues and occasionally on the Gal 6- position. The polysaccharide of keratan sulfate I is attached to its polypeptide chain by a **GlcNAc-Asn** bond. It is abundant in the cornea.

Keratan sulfate II is a skeletal proteoglycan present along with chondroitin sulfate, attached to hyaluronic acid in loose connective tissue. Its polysaccharide chains are attached to its polypeptide chain by a **GalNAc-Thr(Ser)** linkage.

Heparin

Heparin is a classic proteoglycan in which several polysaccharide chains are linked to a common protein core. However, heparin is found stored in granules of mast cells and thus occurs **intracellularly.** Heparin has several other unique structural and functional features, including some that are of medical importance. Fig 54–12 illustrates some characteristic features of the heparin structure. The repeating disaccharide unit contains **glucosamine** (GlcN) and a **uronic acid.** Most of the amino groups of the GlcN residues are **N-sulfated,** but a few are **acetylated.** The GlcN also carries a C_6 sulfate ester.

Approximately 90% of the uronic acid residues are **IdUA;** only 10% are GlcUA. Initially, all of the uronic acids are GlcUA, but, as described below, a 5-epimerase converts approximately 90% of the GlcUA residues to IdUA residues after the polysaccharide is formed. The IdUA residues are frequently sulfated at the 2- position.

The protein molecule of the heparin proteoglycan is unique, consisting exclusively of **serine and glycine** residues. Approximately two-thirds of the serine residues contain polysaccharide chains, usually with molecular weights of 5000–15,000 but occasionally as high as 100,000.

The polysaccharide chains of the heparin proteoglycans undergo a specific sequence of modification after the polymerization. This sequence of modifications occurs in the following way:

(1) The primary product is not sulfated but fully N-acetylated and thus is a polymer of $(GlcUA-GlcNAc)_n$.

(2) Approximately 50% of the GlcNAc residues are N-deacetylated.

(3) The free amino groups of the GlcN are sulfated; subsequently, there is further deacetylation of about half of the remaining GlcNAc residues.

(4) The N-sulfated polymer then becomes a substrate for the 5-epimerase that converts approximately 90% of the GlcUA residues to IdUA.

(5) The recently formed IdUA residues are then O-sulfated on their C_2 positions.

(6) The modification is then completed by O-sulfation of the C_6 positions of GlcN units.

Heparan Sulfate

Heparan sulfate is present throughout cell surfaces as a proteoglycan and is **extracellular.** The polypeptide backbone of the heparan sulfate proteoglycan has a typical amino acid complement, unlike that for heparin. In the process of modification of its polysaccharide chains, there is less deacetylation of the GlcNAc residues, and thus it contains fewer N-sulfates. Because the 5-epimerase (as described above for the modification of heparin) requires the N-sulfate substituents on its substrate, heparan sulfate contains a lower content of IdUA but more GlcUA than does heparin. Accordingly, GlcUA is the predominant uronic

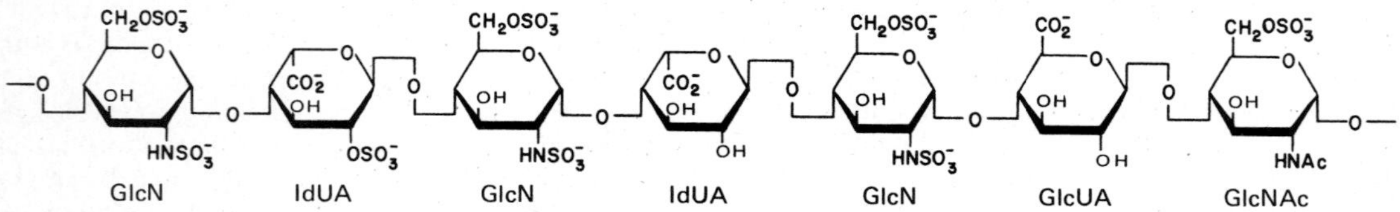

Figure 54–12. Structure of heparin. The polymer section illustrates structural features typical of heparin; however, the sequence of variously substituted repeating disaccharide units has been arbitrarily selected. In addition, non-O-sulfated or 3-O-sulfated glucosamine residues may also occur. (Modified, redrawn, and reproduced, with permission, from Lindahl U et al: Structure and biosynthesis of heparinlike polysaccharides. *Fed Proc* 1977;**36**:19.)

acid in heparan sulfate, while IdUA is the predominant uronic acid in heparin.

Dermatan Sulfate

Dermatan sulfate is a proteoglycan widely distributed in animal tissues. Structurally, it resembles both chondroitin sulfates and heparan sulfate. Its structure is similar to that of chondroitin sulfate except that in place of a GlcUA in beta-1,3-linkage to GalNAc, dermatan sulfate contains an **IdUA** in an alpha-1,3-linkage to **GalNAc.** Formation of the IdUA occurs, as in heparin and heparan sulfate, by the 5-epimerization of **GlcUA.** As in the formation of heparin, the epimerization reaction is coupled tightly to the sulfation of hexosamine. Thus, the dermatan sulfate contains 2 types of repeating disaccharide units: IdUA-GalNAc and GlcUA-GalNAc.

DEGRADATION OF THE POLYSACCHARIDE MOIETIES OF GLYCOPROTEINS & PROTEOGLYCANS

Our understanding of the degradative pathways for glycoproteins, proteoglycans, and glycosaminoglycans has been greatly aided by discoveries of the specific enzyme deficiencies of inborn errors of human metabolism. Two groups of diseases whose study has contributed greatly are the mucopolysaccharidoses and the mucolipidoses. (For many years, what we now call proteoglycans were called mucopolysaccharides.) Table 54–9 lists the biochemical defects in the mucopolysaccharidoses, mucolipidoses, and related disorders.

Degradation of the polysaccharide chains is carried out by **endoglycosidases, exoglycosidases,** and **sulfatases.** In each case, the enzymes exhibit substrate specificities that allow one to deduce which of the polysaccharide chains will be subject to degradation by the particular glycosidase or sulfatase.

Hyaluronidase is a widely distributed endoglycosidase that cleaves hexosaminidic linkages. From hyaluronic acid, the hyaluronidase will generate a tetrasaccharide with the structure $(GlcUA-\beta 1,3-GlcNAc-\beta 1,4)_2$. Hyaluronidase acts on both **hyaluronic acid** and **chondroitin sulfate.** The tetrasaccharide described above can be further degraded by a β-glucuronidase and β-N-acetylhexosaminidase.

β-Glucuronidase is an exoglycosidase that removes both GlcUA and IdUA from nonreducing termini of tetrasaccharides or larger polysaccharides. In general, the disaccharides are poor substrates for β-glucuronidase. β-Glucuronidase, itself a glycoprotein, is localized in both **lysosomes** and **microsomes** of many mammalian cells. Its substrates include **dermatan sulfate, heparan sulfate, chondroitin sulfate,** and **hyaluronic acid.** In inherited β-glucuronidase deficiency in humans, dermatan sulfate, heparan sulfate, and chondroitin sulfate compounds are excreted in the urine, but hyaluronic acid is not. Apparently, there are other degradative pathways that can degrade the tetrasaccharide produced from hyaluronic acid by hyaluronidase.

β-D-Acetylhexosaminidase is an exoglycosidase present in many mammalian tissues. It cleaves from the nonreducing termini of polysaccharides GlcNAc and GalNAc when in beta- linkage. The substrates for the β-D-acetylhexosaminidase include **gangliosides** and **chondroitin sulfates, hyaluronic acid, dermatan sulfates,** and **keratan sulfates** I and II. There are 2 isozymes of β-D-acetylhexosaminidase. The **A isozyme** consists of 2 different types of subunits, the alpha subunit and the beta subunit $(\alpha\beta)_n$, while the **B isozyme** consists of only beta subunits $(\beta\beta)_n$. In **Tay-Sachs disease,** the alpha subunit is defective, and thus only the A isozyme is inactive. In **Sandoff disease,** the beta subunit is defective, resulting in a deficiency of both A and B isozymes.

β-Galactosidases exist in several forms in animal tissues. Both chondroitin sulfate and keratan sulfate contain β-galactosides and thus are substrates for the **acid galactosidases.** In the deficiency of acid β-galactosidase, both keratan sulfate and glycoprotein fragments accumulate, along with the G_{M1} gangliosides. (See Chapter 25.)

α-L-Iduronidase is a lysosomal hydrolase that removes IdUA from the nonreducing terminus of polysaccharide chains. This enzyme is deficient in **Hurler's syndrome.**

Mammalian tissues contain heparin and heparan sulfate-specific endoglycosidases, particularly an endoglucuronidase that exists in liver, intestinal mucosa, platelets, and lysosomes.

A large series of specific **sulfatases** exist for the removal of the sulfate substituents. There are 3 arylsulfatases: A, B, and C. **Arylsulfatase A** degrades the Gal-3-sulfate from sulfatides. **Arylsulfatase B** removes the 4-sulfate from chondroitin sulfate and dermatan sulfate. However, patients with inherited deficiency of 4-sulfatase (**Maroteaux-Lamy syndrome**) spill only dermatan sulfate in urine. Distinct from arylsulfatases A and B is an enzyme that cleaves the 6-sulfate from GalNAc-6-sulfate. This sulfatase is deficient in patients with **Morquio's syndrome.** It normally will cleave the sulfate groups from both Gal-6-sulfate and GalNAc-6-sulfate. Thus, patients with Morquio's syndrome excrete both keratan 6-sulfate and chondroitin 6-sulfate.

A deficiency of **N-acetylglucosamine-6-sulfatase** has been observed in mucopolysaccharidosis A. This enzyme can utilize as a substrate GlcNAc-6-sulfate and Glc-6-sulfate.

Iduronate sulfatase is a specific exoenzyme that will cleave the C_2 sulfate from an IdUA residue at the nonreducing end of heparin, heparan sulfate, and dermatan sulfate. This enzyme is present normally in serum, lymphocytes, fibroblasts, and amniotic fluid. The inherited deficiency of iduronate sulfatase causes **Hunter's syndrome.**

Table 54–9. Biochemical defects and diagnostic tests in mucopolysaccharidoses, mucolipidoses, and related disorders.*

Name	Alternate Designation	Enzymatic Defect	Material for Enzyme Assay	Abnormal ^{35}S-Mucopolysaccharide Level in Fibroblasts	Urinary Metabolites
Mucopolysaccharidoses					
Hurler, Scheie, Hurler/Scheie	MPS I	α-L-Iduronidase	Fibroblasts, leukocytes, tissues, amniotic fluid cells	+	DS, HS
Hunter	MPS II	Iduronate sulfatase	Serum, fibroblasts, leukocytes, tissues, amniotic fluid cells, amniotic fluid	+	DS, HS
Sanfilippo A	MPS III A	HS N-sulfatase (sulfamidase)	Fibroblasts, leukocytes, tissues, amniotic fluid cells	±	HS (±)
Sanfilippo B	MPS III B	α-N-acetylglucosaminidase	Serum, fibroblasts, leukocytes, tissues, amniotic fluid cells	+	HS
Sanfilippo C	MPS III C	Acetyltransferase	Fibroblasts	+	HS
Morquio	MPS IV	N-acetylgalactosamine 6-sulfatase	Fibroblasts	−	KS
Morquio-like	None	β-Galactosidase	Fibroblasts	−	KS
Maroteaux-Lamy	MPS VI	N-acetylgalactosamine 4-sulfatase (arylsulfatase B)	Fibroblasts, leukocytes, tissues, amniotic fluid cells	+	DS
β-Glucuronidase deficiency	MPS VII	β-Glucuronidase	Serum, fibroblasts, leukocytes, amniotic fluid cells	+	DS, HS (±)
Unnamed disorder	MPS VIII	N-acetylglucosamine 6-sulfatase	Fibroblasts	+	HS, KS
Mucolipidoses and related disorders					
Sialidosis	ML I	Sialidase (neuraminidase)	Fibroblasts, leukocytes	−	GF
I-cell disease	ML II	UDP-N-acetylglucosamine: glycoprotein N-acetylglucosaminylphosphotransferase (acid hydrolases thus lack phosphomannosyl residue)	Serum, fibroblasts, amniotic fluid cells	+	GF
Pseudo-Hurler polydystrophy	ML III	As for ML II but deficiency is incomplete	Serum, fibroblasts, amniotic fluid cells	±	GF
Multiple sulfatase deficiency	None	Arylsulfatase A and other sulfatases	Serum, fibroblasts, leukocytes, tissues, amniotic fluid cells	+	DS, HS
Mannosidosis	None	α-Mannosidase	Serum, fibroblasts, leukocytes, amniotic fluid cells	−	GF
Fucosidosis	None	α-L-Fucosidase	Serum, fibroblasts, leukocytes, amniotic fluid cells	−	GF

MPS = mucopolysaccharidosis; ML = mucolipidosis; DS = dermatan sulfate; KS = keratan sulfate; HS = heparan sulfate; GF = glycoprotein fragments.

*Reproduced, with permission, from DiNatale P, Neufeld EF: The biochemical diagnosis of mucopolysaccharidoses, mucolipidosis and related disorders. In: *Perspectives in Inherited Metabolic Diseases*. Vol 2. Barra B et al (editors). Editones Ermes (Milan), 1979.

A specific **α-N-acetylglucosaminidase** can remove the specific alpha-linked GlcNAc residues present in heparin and heparan sulfate. The enzyme is normally present in fibroblasts but is missing in **Sanfilippo B syndrome.**

Heparin sulfamidase (heparan-N-sulfatase) is present in spleen, lung, and ileum. This enzyme is capable of removing a sulfate from GlcN-sulfates at a nonreducing terminus of heparin and heparan sulfate. It is deficient in **Sanfilippo A syndrome.** When the

sulfates are removed, an α-glucosamine (GlcN), a free amino group, remains. This GlcN is not a substrate for the α-N-acetylglucosaminidase described above. The enzyme α-glucosamine:N-acetyltransferase **reacetylates** the free amino group of GlcN at a nonreducing terminus, using acetyl-CoA as the acetyl donor, and thereby renders its product susceptible to the action of the above-described α-N-acetylglucosaminidase. The acetyltransferase activity is absent in **Sanfilippo C syndrome.**

FUNCTIONAL ASPECTS OF GLYCOSAMINOGLYCANS & PROTEOGLYCANS

The binding between glycosaminoglycans and other extracellular macromolecules contributes significantly to the structural organization of connective tissue matrix. Glycosaminoglycans can interact with extracellular macromolecules, plasma proteins, cell surface components, and intracellular macromolecules.

The binding of glycosaminoglycans is generally **electrostatic** in character because of their remarkable polyanionic nature. However, some binding interactions are more specific. Generally, the glycosaminoglycans containing IdUA, such as dermatan sulfate and heparan sulfate, bind proteins with greater affinities than do those containing GlcUA as their only uronic acid constituent.

Interactions With Extracellular Macromolecules

All glycosaminoglycans except those that lack sulfate groups (hyaluronate) or carboxyl groups (keratan sulfates) bind electrostatically to collagen at neutral pH. The presence of IdUA promotes tighter binding, and the proteoglycans interact more strongly than the corresponding glycosaminoglycans. Between 2 and 5 polysaccharide chains bind to each collagen monomer. The soluble collagens (types I, II, and III) all bind chondroitin sulfate proteoglycan.

Chondroitin sulfate and heparan sulfate bind specifically to elastin.

As mentioned above, chondroitin sulfate and keratan sulfate chains in their respective proteoglycans aggregate with the aid of link proteins with hyaluronic acid. As many as 100 proteoglycan molecules may bind to one hyaluronate molecule.

Interactions With Plasma Proteins

The intima of the arterial wall contains hyaluronate and chondroitin sulfate, dermatan sulfate, and heparan sulfate proteoglycans. Of these proteoglycans, **dermatan sulfate binds plasma lipoproteins.** In addition, dermatan sulfate appears to be the major glycosaminoglycan **synthesized by arterial smooth muscle cells.** As these smooth muscle cells are those that proliferate at the atherosclerotic lesion in arterial vessels, dermatan sulfate may play a significant role in development of the atherosclerotic plaque.

Heparin, although synthesized and stored in mast cells, is always in close proximity to blood vessels. Heparin, with its high negative charge density (due to the IdUA and sulfate residues), interacts strongly with several plasma components. Heparin specifically binds clotting factors IX and XI. More important in the **anticoagulant activity** of heparin is its interaction with a plasma alpha$_2$ glycoprotein called **antithrombin III.** The 1:1 stoichiometric binding of heparin to antithrombin III greatly accelerates the ability of the latter to inactivate serine proteases, particularly

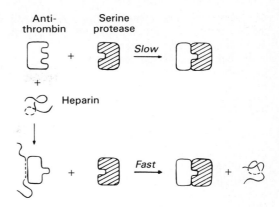

Figure 54–13. Schematic representation of inactivation by antithrombin of serine proteases (eg, thrombin) participating in the coagulation mechanism. Heparin is believed to accelerate the inactivation by binding to antithrombin, thereby inducing a conformational change in the antithrombin molecule that facilitates its interaction with thrombin. (Binding of heparin to thrombin as well cannot be excluded.) The interaction requires a specific binding site (– – –) in the polysaccharide chain. (Reproduced, with permission, from Lennarz WJ: *The Biochemistry of Glycoproteins and Proteoglycans.* Plenum Press, 1980.)

thrombin. The binding of heparin to Lys residues in antithrombin III appears to induce a conformational change that favors the binding of antithrombin III to the serine proteases. Such a scheme is diagrammatically depicted in Fig 54–13.

Commercially available heparin contains 2 components—a high-affinity heparin and a low-affinity heparin—both of which seem to bind to the same site of antithrombin III molecules. However, high-affinity heparin has an anticoagulant activity about 10 times higher than that of low-affinity material, and its binding constant is similarly greater. The N-desulfation or modification of the IdUA residues of heparin reduces its anticoagulant activity.

Heparan sulfate, the structure of which resembles that of heparin, is also capable of accelerating the action of antithrombin III, but it is much less potent than heparin.

Heparin can bind specifically to **lipoprotein lipase** present in capillary walls and cause a release of that triglyceride-degrading enzyme into the circulation. Similarly, hepatic lipase binds heparin and is released into the circulation, but it binds heparin with a lower affinity than does the lipoprotein lipase. Lipoprotein lipase will bind partially N-desulfated heparin better than antithrombin III will.

Glycosaminoglycans & Cell Surface Molecules

Heparin is capable of associating with many cell types, including blood platelets, arterial endothelial cells, and liver cells. Chondroitin sulfate, dermatan sulfate, and heparan sulfate bind to independent sites on surfaces of cells such as fibroblasts. At those sites, the glycosaminoglycans and proteoglycans are taken up by fibroblasts and degraded.

Hyaluronate is deposited by cells as they grow on the plastic substrata of culture dishes. In addition, hyaluronate appears to be involved in the cell-cell adhesion processes so important during the growth and development of metazoan organisms.

Some proteoglycans appear to serve as **receptors** and **carriers** for macromolecules, including the lipoproteins, lipases, and, as described above, antithrombin. Proteoglycans seem to be involved in the regulation of cell growth, the mediation of cell-cell communication, and the shielding of cell surface receptors.

Glycosaminoglycans & Intracellular Macromolecules

In addition to interacting with the enzymes involved in their biosynthesis and degradation, proteoglycans and their glycosaminoglycan components have effects on protein synthesis and intranuclear functions. Heparin particularly seems to have an effect on chromatin structure and can activate DNA polymerase activities in vitro. It is not clear how physiologic these actions are. Glycosaminoglycans are found in significant quantities in nuclei from different cell types, and in fact there is some suggestion that heparan sulfate has some role in the embryonic development of sea urchins.

Various lysosomal acid hydrolase activities can be affected in negative or positive ways by chondroitin sulfates, dermatan sulfates, and heparin. The acid hydrolases in lysosomes may be naturally complexed with glycosaminoglycans to provide a protected and inactive form.

Numerous storage or secretory granules such as the chromaffin granules in adrenal medulla, the prolactin secretory granules in the pituitary gland, and the basophilic granules in mast cells contain sulfated glycosaminoglycans. The glycosaminoglycan-peptide complexes that occur in these granules may play a role in the release of biogenic amines.

REFERENCES

Buckwalter JA, Rosenberg LC: Electron microscopic studies of cartilage proteoglycans. *J Biol Chem* 1982;**257**:9830.

DiNatale P, Neufeld EF: The biochemical diagnosis of mucopolysaccharidoses, mucolipidosis and related disorders. In: *Perspectives in Inherited Metabolic Diseases*. Vol 2. Barra B et al (editors). Editiones Ermes (Milan), 1979.

Höök M et al: Cell-surface glycosaminoglycans. *Annu Rev Biochem* 1984;**53**:847.

Hubbard SC, Ivatt RJ: Synthesis and processing of asparagine-linked oligosaccharides. *Annu Rev Biochem* 1981;**50**:555.

Jaques LB: Heparin: An old drug with a new paradigm. *Science* 1979;**206**:528.

Kornfeld R, Kornfeld S: Assembly of asparagine-linked oligosaccharides. *Annu Rev Biochem* 1985;**54**:631.

Kornfeld S, Sly WS: Lysosomal storage defects. *Hosp Pract* (Aug) 1985;**20**:71.

Lennarz WJ: *The Biochemistry of Glycoproteins and Proteoglycans*. Plenum Press, 1980.

Poole AR et al: Proteoglycans from bovine nasal cartilage: Immunochemical studies of link protein. *J Biol Chem* 1980; **255**:9295.

Reitman ML et al: Fibroblasts from patients with I-cell disease and pseudo-Hurler polydystrophy are deficient in uridine 5'-diphosphate-N-acetylglucosamine:glycoprotein N-acetylglucosaminylphosphotransferase activity. *J Clin Invest* 1981; **67**:1574.

Schachter H: Biosynthetic controls that determine the branching and heterogeneity of protein-bound oligosaccharides. *Biochem Cell Biol* 1986;**64**:163.

Snider MD, Rogers OC: Transmembrane movement of oligosaccharide-lipids during glycoprotein synthesis. *Cell* 1984;**36**:753.

Wedgwood JF, Strominger JL: Enzymatic activities in cultured human lymphocytes that dephosphorylate dolichyl pyrophosphate and dolichyl phosphate. *J Biol Chem* 1980;**255**:1120.

Blood Plasma & Clotting

55

Robert K. Murray, MD, PhD

INTRODUCTION

Blood is a tissue that circulates in what is virtually a closed system of blood vessels. It consists of solid elements—the red and white blood cells and the platelets—suspended in a liquid medium, the **plasma.** As indicated below, blood—and plasma in particular—perform many functions that are absolutely critical for the maintenance of health.

Once the blood has clotted (coagulated), as discussed below, the remaining liquid phase is called **serum.** Serum lacks the clotting factors (including fibrinogen) that are normally present in plasma but have been consumed during the process of coagulation. Serum does contain some degradation products of clotting factors—products that have been generated during the coagulation process and thus are *not* normally present in plasma.

BIOMEDICAL IMPORTANCE

The fundamental role that the blood plays in the maintenance of homeostasis and the ease with which blood can be obtained have meant that the study of its constituents has been of central importance in the development of biochemistry and clinical biochemistry. Hemoglobin, albumin, the immunoglobulins, and the various clotting factors are among the most studied proteins. Changes in the amounts of various plasma proteins occur in many diseases and can be monitored by electrophoresis. Alterations of the activities of certain plasma enzymes are of diagnostic use in a number of pathologic conditions (see the Appendix). Hemorrhagic and thrombotic states can pose serious medical emergencies; their rational management requires a clear understanding of the bases of blood clotting and fibrinolysis.

FUNCTIONS OF BLOOD

The functions of blood—all except specific cellular ones such as oxygen transport and cell-mediated immunologic defense—are carried out by plasma and its constituents. They are as follows: (1) respiration—transport of oxygen from the lungs to the tissues and of CO_2 from the tissues to the lungs; (2) nutrition—transport of absorbed food materials; (3) excretion—transport of metabolic wastes to the kidneys, lungs, skin, and intestines for removal; (4) maintenance of normal acid-base balance in the body; (5) regulation of water balance through the effects of blood on the exchange of water between the circulating fluid and the tissue fluid; (6) regulation of body temperature by the distribution of body heat; (7) defense against infection by the white cells and the circulating antibodies; (8) transport of hormones and regulation of metabolism; and (9) transport of metabolites.

Plasma consists of water, electrolytes, metabolites, nutrients, proteins, and hormones. The water and electrolyte composition of plasma is practically the same as that of all extracellular fluids.

THE PLASMA PROTEINS

The total protein of the plasma is about 7–7.5 g/dL. Thus, the plasma proteins comprise the major part of the solids of the plasma. The proteins of the plasma are actually a very complex mixture that includes not only simple proteins but also mixed or conjugated proteins such as glycoproteins and various types of lipoproteins.

The separation of individual proteins from a complex mixture is frequently accomplished by the use of various solvents or electrolytes (or both) to remove different protein fractions in accordance with their solubility characteristics. This is the basis of the so-called salting-out methods commonly utilized in the determination of protein fractions in the clinical laboratory. Thus, it is customary to separate the proteins of the plasma into 3 major groups—fibrinogen, albumin, and globulin—by the use of varying concentrations of sodium or ammonium sulfate.

Blood plasma is by definition an intravascular fluid. On the arterial side of the circulation, the intravascular hydrostatic pressure generated by the heart and large vessels is 20–25 mm Hg greater than the hydrostatic pressure in the tissue spaces. In order to prevent too much intravascular fluid from being forced into the extravascular tissue spaces, the hydrostatic pressure is opposed by an **intravascular colloid osmotic pressure** generated by the plasma proteins.

Albumin

Of the 3 major plasma protein groups, albumin is present in the highest mass concentration. Albumin

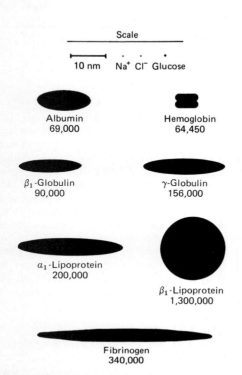

Figure 55–1. Relative dimensions and molecular weights of protein molecules in the blood (Oncley).

also has the lowest molecular weight of the major protein molecules in plasma (Fig 55–1); thus, **albumin is the largest contributor to the intravascular colloid osmotic pressure.** Albumin is synthesized in the liver and consists of a single chain of 610 amino acids. In addition to contributing to the colloid osmotic pressure, albumin also acts as a **carrier molecule** for bilirubin, fatty acids, trace elements, and many drugs. Some of its ligand binding sites are highly specific and saturable, while others are much less so. The major effect of low serum albumin concentration (hypoalbuminemia), which occurs frequently in liver and kidney disease, is soft tissue edema due to the diminished intravascular colloid osmotic pressure.

Globulins

As described in Chapter 5, globulins are protein molecules that are insoluble in plain water but soluble in salt water. The serum globulins are a heterogeneous, complex mixture of protein molecules that are frequently designated as α-, β-, or γ-globulins, sometimes with number designations as well, all based on their electrophoretic mobility (Fig 55–2). A more rational classification is based on their structure or function.

The **glycoproteins** contain covalently bound oligosaccharide moieties (see Chapter 54) and are found principally in the α_1- and α_2-globulin fractions. Among the glycoproteins are many specific molecules with specific functions, some understood and others not.

Lipoproteins contain lipids, usually noncovalently bound to the protein molecule (see Chapter 26). The lipoproteins migrate with the α-globulins or the

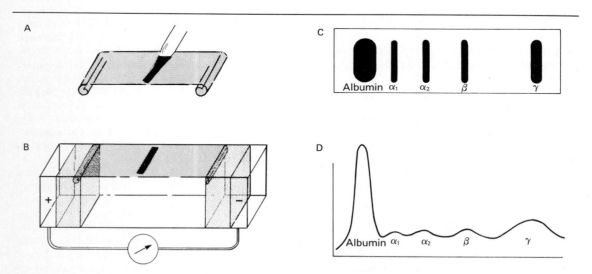

Figure 55–2. Technique of cellulose acetate zone electrophoresis. *A:* Small amount of serum or other fluid is applied to cellulose acetate strip. *B:* Electrophoresis of sample in electrolyte buffer is performed. *C:* Separated protein bands are visualized in characteristic position after being stained. *D:* Densitometer scanning from cellulose acetate strip converts bands to characteristic peaks of albumin, α_1-globulin, α_2-globulin, β-globulin, and γ-globulin. (Reproduced, with permission, from Stites DP, Stobo JD, Wells JV [editors]: *Basic & Clinical Immunology,* 6th ed. Appleton & Lange, 1987.)

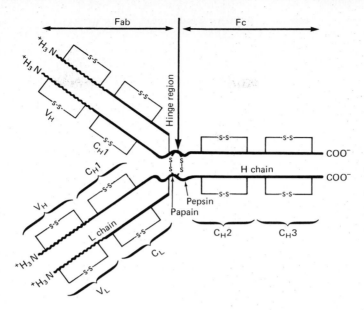

Figure 55–3. A simplified model for an IgG human antibody molecule showing the 4-chain basic structure and domains. V indicates variable region; C, the constant region; and the vertical arrow, the hinge region. Thick lines represent H and L chains; thin lines represent disulfide bonds. (Modified and reproduced, with permission, from Stites DP, Stobo JD, Wells JV [editors]: *Basic & Clinical Immunology,* 6th ed. Appleton & Lange, 1987.)

β-globulins. The higher the fat content and the lower the protein content of a lipoprotein, the lower its specific gravity. The lipoproteins act as **carrier molecules** for many different types of lipids and lipid-soluble molecules that are not soluble in the plasma water.

Some **metal-binding proteins** such as transferrin have the properties of globulins and act as carriers for trace elements.

Plasma normally contains a number of specific **enzyme molecules** such as phosphatases, lipases, lactate dehydrogenase, amylase, and ferroxidase (ceruloplasmin). In addition, as tissues break down or their membranes leak, intracellular enzymes can be released into the intravascular space, and their catalytic activities may serve as qualitative or quantitative indexes of tissue damage. Determinations of serum transaminases, creatine kinases, and acid phosphatases are particularly useful in clinical medicine.

Polypeptide **hormones** circulate in plasma. Hormones such as the hydrophobic steroids and 1,25-dihydroxyvitamin D_3 circulate in plasma bound to specific carrier molecules.

Fibrinogen, the precursor of fibrin that forms blood clots, and **immunoglobulins,** which constitute the effector arm of the humoral immunity system, are important plasma proteins discussed in detail below.

The plasma lipoproteins are described in Chapter 26.

Immunoglobulins

Immunoglobulins, or **antibodies,** are synthesized in B lymphocytes or their derivatives, plasma cells, and with remarkable specificity **bind to antigenic sites** on other molecules.

All immunoglobulin molecules consist of 2 identical light (L) chains (MW 23,000) and 2 identical heavy (H) chains (MW 53,000–75,000) held together as a tetramer (L_2H_2) by disulfide bonds (Fig 55–3). Each chain can be divided conceptually into specific **domains,** or regions, that have structural and functional significance. The half of the **light (L) chain** toward the carboxyl terminus is referred to as the **constant region** (C_L), while the amino-terminal half is the **variable region** of the light chain (V_L). Approximately one-quarter of the **heavy (H) chain** at the amino terminus is referred to as its variable region (V_H), and the other three-quarters of the heavy chain are referred to as the constant regions (C_H1, C_H2, C_H3) of that H chain. The portion of the immunoglobulin molecule that **binds the specific antigen** is formed by the amino-terminal portions (variable regions) of both the H and L chains—ie, the V_H and V_L **domains.** The domains of the protein chains do not simply exist as linear sequences of amino acids but form globular regions with secondary and tertiary structure in order to effect binding of specific antigens.

As depicted in Fig 55–3, digestion of an immunoglobulin by the enzyme papain produces 2 antigen-binding fragments **(Fab)** and one crystallizable fragment **(Fc).** The area in which papain cleaves the immunoglobulin molecule—ie, the region between the C_H1 and C_H2 domains—is referred to as the **hinge region.**

There are 2 general types of light chains, kappa (κ)

Table 55–1. Properties of human immunoglobulin chains.*

Designation	H Chains					L Chains		Secretory Component	J Chain
	γ IgG	α IgA	μ IgM	δ IgD	ϵ IgE	κ All classes	λ All classes	SC IgA	J IgA, IgM
Classes in which chains occur									
Subclasses or subtypes	1,2,3,4	1,2	1,2	. . .	. . .	. . .	1,2,3,4	. . .	. . .
Allotypic variants	Gm(1)–(25)	A2m(1), (2)	. . .	. . .	. . .	Km(1)–(3)†	. . .	. . .	. . .
Molecular weight (approximate)	50,000‡	55,000	70,000	62,000	70,000	23,000	23,000	70,000	15,000
V region subgroups	$V_H I – V_H IV$					$V_\kappa I – V_\kappa IV$	$V_\lambda I – V_\lambda VI$		
Carbohydrate (average percentage)	4	10	15	18	18	0	0	16	8
Number of oligosaccharides	1	2 or 3	5	?	5	0	0	?	1

*Reproduced, with permission, from Stites DP, Stobo JD, Wells JV (editors): *Basic & Clinical Immunology,* 6th ed. Appleton & Lange, 1987.
†Formerly Inv(1)–(3).
‡60,000 for γ3.

and lambda (λ), which can be distinguished on the basis of structural differences in their C_L regions (Table 55–1). A given immunoglobulin molecule always contains two κ or two λ light chains, **never a mixture of κ and λ**. In humans, the κ chains are more frequent than λ chains in immunoglobulin molecules.

Five classes of H chains have been found in humans, and these classes can be distinguished by differences in their C_H regions (Table 55–1). The 5 classes of H chains are designated γ, α, μ, δ, and ϵ and vary in molecular weight from 50,000 to 70,000 (Table 55–1). The μ and ϵ chains each have four C_H domains rather than the usual 3. The type of H chain determines the class of immunoglobulin and thus its effector function. There are 5 immunoglobulin classes: **IgG, IgA, IgM, IgD,** and **IgE**. As shown in Table 55–1, many of the H chain classes can be further divided into subclasses on the basis of subtle structural differences in the C_H regions.

The variable regions of immunoglobulin molecules consist of the V_L and V_H domains and are quite heterogeneous. In fact, no 2 variable regions from different humans have been found to have identical amino acid sequences. However, there are discernible patterns between the regions from different individuals, and these shared patterns have been divided into 3 main groups based on the degree of amino acid sequence homology. There is a V_K group for kappa L chains, a V_λ group for lambda L chains, and a V_H group for the H chains. At higher resolution, there are even subgroups within each of these 3 groups.

Thus, within the variable regions there are some positions that are relatively invariable to account for the groups and subgroups. Upon comparing variable regions from different light chains of the same group or subgroup or different heavy chains from the same group or subgroup, it is apparent that there are **hypervariable regions** interspersed between the relatively invariable (subgroup-determining) positions (Fig 55–4). L chains have 3 hypervariable regions (in V_L), and H chains have 4 (in V_H).

The constant regions of the immunoglobulin molecules, particularly the $C_H 2$ and $C_H 3$ (and $C_H 4$ of IgM and IgE), which constitute the Fc fragment, are responsible for the class-specific effector functions of the different immunoglobulin molecules (Table 55–2). Some immunoglobulins such as immune IgG exist only in the basic tetrameric structure, while others such as IgA and IgM can exist as higher-order polymers of 2, 3 (IgA) or 5 (IgM) tetrameric units (Fig 55–5).

The L chains and H chains are synthesized as separate molecules and are subsequently assembled within the B cell or plasma cell into mature immunoglobulin molecules, all of which are **glycoproteins** (Table 55–1).

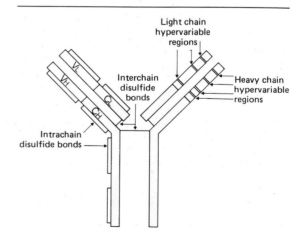

Figure 55–4. Schematic model of an IgG molecule showing approximate positions of the hypervariable regions in heavy and light chains. (Modified and reproduced, with permission, from Stites DP, Stobo JD, Wells JV [editors]: *Basic & Clinical Immunology,* 6th ed. Appleton & Lange, 1987.)

Table 55–2. Properties of human immunoglobulins.*

	IgG	IgA	IgM	IgD	IgE
H chain class	γ	α	μ	δ	ϵ
H chain subclass	$\gamma 1, \gamma 2, \gamma 3, \gamma 4$	$\alpha 1, \alpha 2$	$\mu 1, \mu 2$		
L chain type	κ and λ	κ and λ	κ and λ	κ and λ	κ and λ
Molecular formula	$\gamma_2 L_2$	$\alpha_2 L_2\dagger$ or $(\alpha_2 L_2)_2 SC\S J\ddagger$	$(\alpha_2 L_2)_5 J\ddagger$	$\delta_2 L_2$	$\epsilon_2 L_2$
Sedimentation coefficient (S)	6–7	7	19	7–8	8
Molecular weight (approximate)	150,000	160,000† 400,000**	900,000	180,000	190,000
Electrophoretic mobility (average)	γ	Fast γ to β	Fast γ to β	Fast γ	Fast γ
Complement fixation (classic)	+	0	++++	0	0
Serum concentration (approximate; mg/dL)	1000	200	120	3	0.05
Placental transfer	+	0	0	0	0
Reaginic activity	?	0	0	0	++++
Antibacterial lysis	+	+	+++	?	?
Antiviral activity	+	+++	+	?	?

*Reproduced, with permission, from Stites DP, Stobo JD, Wells JV (editors): *Basic & Clinical Immunology,* 6th ed. Appleton & Lange, 1987.
†For monomeric serum IgA.
‡J chain.
§Secretory component.
**For secretory IgA.

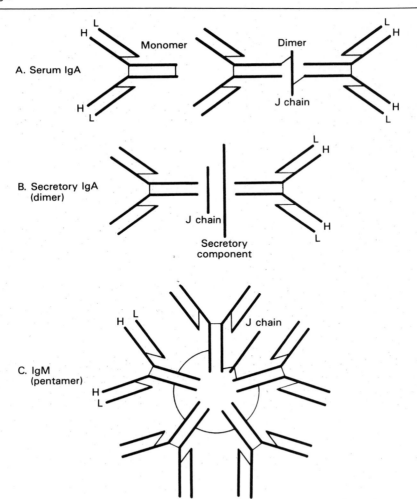

Figure 55–5. Highly schematic illustration of polymeric human immunoglobulins. Polypeptide chains are represented by thick lines; disulfide bonds linking different polypeptide chains are represented by thin lines. (Reproduced, with permission, from Stites DP, Stobo JD, Wells JV [editors]: *Basic & Clinical Immunology,* 6th ed. Appleton & Lange, 1987.)

Each immunoglobulin light chain is the product of at least 3 separate structural genes: a variable region (V_L) gene, a joining region (J) gene (bearing no relationship to the J chain of IgA or IgM), and a constant region (C_L) gene. Each heavy chain is the product of at least 4 different genes; a variable region (V_H) gene, a diversity region (D) gene, a joining region (J) gene, and a constant region (C_H) gene. Thus, the "one gene, one protein" concept is invalid. The molecular mechanisms responsible for the generation of the single immunoglobulin chains from multiple structural genes are discussed in Chapters 38 and 41.

Each person is capable of generating antibodies directed against perhaps 1 million different antigens. The generation of such immense **antibody diversity** appears to depend upon the **combinations of the various structural genes** contributing to the formation of each immunoglobulin chain and upon a high frequency of **somatic mutational events** in the rearranged V_H and V_L genes.

In most humoral immune responses, antibodies with identical specificity but of different classes are generated in a specific chronologic order in response to the immunogen (immunizing antigen). A single type of immunoglobulin light chain can combine with an antigen-specific μ chain to generate a specific IgM molecule. Subsequently, the same antigen-specific light chain combines with a γ chain with an identical V_H region to generate an IgG molecule with antigen specificity identical to that of the original IgM molecule. Subsequently, the same light chain can combine with an α heavy chain, again containing the identical V_H region, to form an IgA molecule with identical antigen specificity. These 3 classes (IgM, IgG, and IgA) of immunoglobulin molecules against the same antigen have identical variable domains of both their light (V_L) chains and heavy (V_H) chains and are said to share an **idiotype.** The different class **isotypes** are determined by the C_H regions combined with the same antigen-specific V_H region. One aspect of the genetic regulatory mechanisms responsible for the switching of the C_H region gene is discussed in Chapter 41.

Disorders of immunoglobulins include increased production of specific classes of immunoglobulins or even specific immunoglobulin molecules, the latter by clonal tumors of plasma cells called **myelomas. Hypogammaglobulinemia** may be restricted to a single class of immunoglobulin molecules (eg, IgA or IgG) or may involve underproduction of all classes of immunoglobulins (IgA, IgD, IgE, IgG, and IgM). The disorders of immunoglobulin levels are almost without exception due to disordered rates of immunoglobulin production or secretion, for which there can be many causes.

BLOOD CLOTTING

Hemostasis is the cessation of bleeding that follows traumatic interruption of vascular integrity. There are 4 phases to hemostasis. The **first phase** is constric-

tion of the injured vessel to diminish blood flow distal to the injury. The **second phase** consists of formation of a loose **platelet plug,** or white thrombus, at the site of injury. **Collagen** exposed at the site of injury acts as a binding site for platelets, which, in response to binding collagen, undergo disruption of their internal structure and release thromboxane and **ADP.** These induce other platelets to adhere to those bound to collagen, forming the loose and temporary platelet plug. This phase of hemostasis is measured by determining the **bleeding time.** The **third phase** is the formation of the red thrombus (blood clot). The **fourth phase** is the partial or complete **dissolution** of the clot.

There are 3 types of thrombi, or clots. The white thrombus is composed of platelets and fibrin and is relatively poor in erythrocytes. It forms at the site of an injury or abnormal vessel wall, particularly in areas of rapid blood flow (arteries). A second type of thrombus is a disseminated fibrin deposit in very small vessels (capillaries).

The **red thrombus** is the third type of clot and consists of red cells and fibrin. The red thrombus morphologically resembles the clot formed in a test tube. It can form in vivo in areas of **retarded blood flow** without any abnormal vascular wall, or it may form at the site of the **injury** or **abnormal vessel wall** in conjunction with the initiating platelet plug. Initiation of the clot formation in response to tissue injury is carried out by the **extrinsic pathway** of clotting. The initiation of the pure red thrombus in an area of restricted blood flow or in response to an abnormal vessel wall without tissue injury is carried out by the **intrinsic pathway.** The intrinsic and extrinsic pathways converge in a **final common pathway**—the activation of prothrombin to thrombin and the thrombin-catalyzed conversion of fibrinogen to the fibrin clot.

The Conversion of Fibrinogen to Fibrin by Thrombin

Fibrinogen* (factor I; see Fig 55–1 and Table 55–3) is a soluble plasma glycoprotein, 46 nm in length, with a molecular weight of 340,000, which consists of 6 polypeptide chains synthesized in liver. The 6 chains are two Aα chains, two Bβ chains, and two γ chains, making the structure Aα_2 Bβ_2 γ_2. The Bβ and γ chains contain Asn-linked complex oligosaccharides. All 3 genes (Aα, Bβ, and γ) are genetically linked and coordinately regulated in humans. The ends of the fiber-shaped fibrinogen molecule are **highly negatively charged,** the negative charges being contributed by a large number of aspartate and glutamate residues in the A portion of the Aα chains and the B portion of the Bβ chains (Fig 55–6). In addition, the B portion of the Bβ chains contains the unusual negatively charged tyrosine O-sulfate residue. These negatively charged termini of the fibrinogen molecules

---

*Except for fibrinogen and prothrombin (and their activated products) and Ca^{2+}, all clotting factors will be referred to by their designated roman numerals (Table 55–3).

Table 55-3. Numerical system for nomenclature of blood clotting factors. The numbers have no relationship to the order in which the factors act.

Factor	Name
I	Fibrinogen
II	Prothrombin
IV	Calcium
V	Labile factor, proaccelerin, accelerator (Ac-) globulin
VII	Proconvertin, serum prothrombin conversion accelerator (SPCA), cothromboplastin, autoprothrombin I
VIII	Antihemophilic factor, antihemophilic globulin (AHG)
IX	Plasma thromboplastin component (PTC) (Christmas factor)
X	Stuart-Prower factor
XI	Plasma thromboplastin antecedent (PTA)
XII	Hageman factor
XIII	Laki-Lorand factor (LLF)

not only contribute to its water solubility but also repulse the termini of other fibrinogen molecules, thereby preventing aggregation.

Thrombin is a 34,000-MW serine **protease** that consists of 2 polypeptide chains and hydrolyzes four Arg-Gly peptide bonds in fibrinogen (Fig 55–6). These 4 peptide bonds are the 2 between the A and α portions of the 2 Aα chains and the 2 between the B and β portions of the Bβ chains. Removal of the A and B portions of the fibrinogen molecule releases these negatively charged **fibrinopeptides** and generates the **fibrin monomer,** which has the subunit structure ($\alpha\beta\gamma$)$_2$. These long insoluble fibrin monomers spontaneously associate in a regularly staggered array to form the insoluble **fibrin polymer clot.** The A and B fibrinopeptides consist of only 18 amino acid residues; thus, the fibrin monomer retains 97% of the amino acid residues of fibrinogen. It is the formation of this fibrin polymer that traps red cells, platelets, and other components to form the red thrombus or the white thrombus (platelet plug). The initial fibrin clot is a rather weak one, held together only by the noncovalent staggered array of insoluble fibrin monomers.

Thrombin, in addition to converting fibrinogen to fibrin, also converts factor XIII to active factor XIII (XIII$_a$). Factor XIII$_a$ is a **transglutaminase.** The transglutaminase covalently **cross-links** fibrin monomers by forming a specific isopeptide bond between the γ-carboxyl group of glutamine and the ϵ-amino group of Lys (Fig 55–7). This strengthening of the initial fibrin clot contributes to retraction of the clot that can be observed in the test tube. Individuals with an inherited deficiency of factor XIII have a bleeding tendency, because they cannot form a stable fibrin clot.

The activity of thrombin must be carefully controlled in order to avoid the formation of uncalled for, potentially catastrophic blood clots. This control is exerted by 2 mechanisms. One is the existence of a thrombin antagonist called antithrombin III (see below). The second mechanism involves the synthesis and circulation of a catalytically **inactive thrombin zymogen, prothrombin.** Prothrombin, or factor II, is synthesized in the liver and contains the vitamin K-dependent γ-carboxyglutamate (Gla) residues. Prothrombin is a 72,000-MW single-chain glycoprotein; its primary and secondary structures are represented in Fig 55–8. The amino-terminal region of prothrombin, indicated by 1 in Fig 55–8, contains up to fourteen Gla residues. The dotted line represents a disulfide bridge between region A and region B of prothrombin. The serine-dependent active protease site is indicated by the arrowhead.

The **activation of prothrombin occurs on the platelet** and requires platelet anionic phospholipid, Ca^{2+}, factor V$_a$, and factor X$_a$. The phospholipids on the internal side of the platelet plasma membrane must be exposed as a result of the collagen-induced platelet disruption and degranulation. These phospholipids bind Ca^{2+} and prothrombin, the latter at the Gla-containing N-terminal region. The platelets also contain factor V, which, when activated as factor V$_a$, binds to specific receptors in the platelet membrane (Fig 55–9). Factor V$_a$ acts as a receptor for factor X$_a$, which in turn binds prothrombin in the F-1·2 region (Fig 55–8). Factor X$_a$ is a serine protease also and cleaves the catalytically inactive prothrombin at the

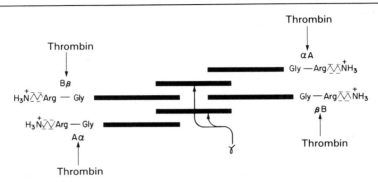

Figure 55–6. Diagrammatic representation of fibrinogen, its (AαB$\beta\gamma$)$_2$ structure, charged termini, and the sites of thrombin cleavage (arrows) of four Arg-Gly peptide bonds.

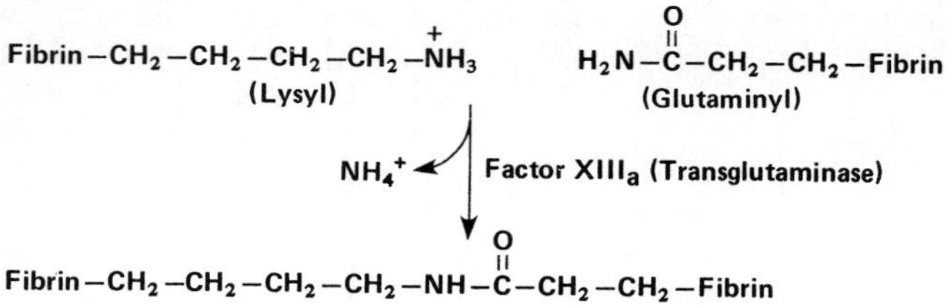

$$\underset{\text{(Lysyl)}}{\text{Fibrin}-CH_2-CH_2-CH_2-CH_2-\overset{+}{N}H_3} \qquad \underset{\text{(Glutaminyl)}}{H_2N-\overset{\overset{\displaystyle O}{\parallel}}{C}-CH_2-CH_2-\text{Fibrin}}$$

$$NH_4^+ \longleftarrow \quad \text{Factor XIII}_a \text{ (Transglutaminase)}$$

$$\text{Fibrin}-CH_2-CH_2-CH_2-CH_2-NH-\overset{\overset{\displaystyle O}{\parallel}}{C}-CH_2-CH_2-\text{Fibrin}$$

Figure 55–7. Cross-linking of fibrin monomers by activated factor XIII.

sites indicated in Fig 55–8, and the amino portion of prothrombin is released. The disulfide bridge holds together the thrombin A and B polypeptides that have been generated by the X_a cleavages.

The bridging of the phospholipid via Ca^{2+} to the Gla residues of prothrombin accelerates the activation of prothrombin 50- to 100-fold, apparently as a result of providing a high local concentration of the prothrombin and factor X_a (Fig 55–9). Factor V_a adds about a 350-fold acceleration, also as the result of locally concentrating factor X_a.

Factor V_a, which is generated by thrombin, is also subsequently **inactivated by thrombin,** thereby providing a means of limiting the activation of prothrombin to thrombin.

Prothrombin can also be activated by staphylocoagulase as a result of a simple conformational alteration not involving cleavage of the molecule.

Activation of Factor X

Activation of factor X occurs at the conceptual site where the intrinsic and extrinsic pathways join to form **the final common pathway** (Fig 55–10). Factor X is a zymogen (MW 55,000) of a serine protease and contains Gla residues. As in prothrombin, the Gla residues of factor X are responsible for the calcium-mediated binding of factor X to the acidic phospholipids of platelet membranes. In order to convert factor X to factor X_a, an Arg-Ile bond must be cleaved by still another serine protease. There are 2 serine proteases capable of cleaving the specific Arg-Ile bond of factor X.

The Extrinsic Pathway for Generating Factor X_a

Factor VII$_a$ operates exclusively in the extrinsic pathway in conjunction with tissue factor to cleave this Arg-Ile bond and generate X_a. This extrinsic pathway is very **rapid in response to tissue injury.** The precursor of factor VII$_a$ is factor VII, another Gla-containing glycoprotein synthesized in the liver. Factor VII can be cleaved by thrombin or factor X_a. Factor VII is a zymogen but has rather high endogenous activity. The **tissue factor** necessary to accelerate the attack of factor VII or VII$_a$ on factor X is abundant in **placenta, lung,** and **brain.**

While there is approximately 3 mg of fibrinogen in 1 mL of plasma, there is only 0.01 mg of factor X per milliliter of plasma. This requires that the clotting system provide the amplification. Conversion of factor X to X_a is an autocatalytic process and therefore an **amplification system.** In this group of reactions, it is difficult to know which came first, the chicken or the egg—II$_a$ (thrombin) or X_a (Fig 55–10).

The Intrinsic Pathway for Generating Factor X_a

The intrinsic pathway for the generating of X_a commences with the exposure of **prekallikrein,** high-molecular-weight **kininogen, factor XII,** and **factor XI** to an activating surface, perhaps collagen in vivo (Fig 55–11). Glass or kaolin will provide an activat-

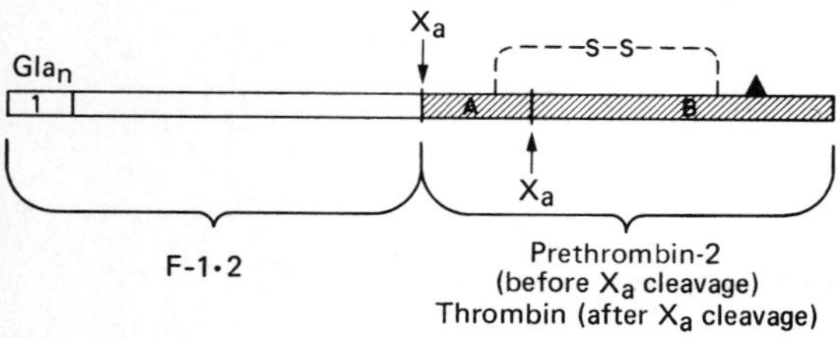

Figure 55–8. Diagrammatic representation of prothrombin. The amino terminus is to the left; region 1 contains all the Gla residues. The sites of cleavage by factor X_a are shown and the products named. The site of the catalytically active serine residue is indicated by ▲. The A and B chains of active thrombin (shaded) are held together by the disulfide bridge.

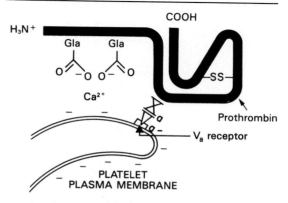

Figure 55–9. Diagrammatic representation of the binding of factors V_a, X_a, Ca^{2+}, and prothrombin to the platelet plasma membrane.

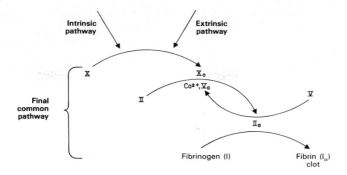

Figure 55–10. Relationship between the intrinsic, extrinsic, and final common pathways of blood clotting.

ing surface for in vitro tests of the intrinsic pathway. Exposure to the activating surface makes factor XII more labile to proteolysis by kallikrein. Factor XII_a is generated by kallikrein and attacks prekallikrein to generate more kallikrein, setting up a reciprocal activation. Factor XII_a releases bradykinin from high-molecular-weight kininogen and activates factor XI to XI_a. Factor IX, a Gla-containing zymogen, is activated in a 2-step reaction by factor XI_a. Factor IX_a, in the presence of calcium and acid phospholipids, slowly **activates factor X by cleaving the same Arg-Ile bond** that factor VII_a of the extrinsic system hydrolyzes. The factor IX_a-catalyzed activation of factor X is accelerated about 500-fold by the presence of **factor VIII** or VIII_a. Factor VIII probably requires activation by minute quantities of thrombin to form factor VIII_a. **Factor VIII is not a protease** but probably serves as a receptor for factor IX_a as it cleaves the Arg-Ile bond of factor X. The **intrinsic pathway is slow,** because it involves many factors operating in a **cascade mechanism** to generate factor X_a (Fig 55–11).

Table 55–4 lists many of the inherited deficiencies of the clotting system in humans. The most common deficiency is that of factor VIII, which produces a disease known as **hemophilia A.** The X chromosome-linked deficiency of factor VIII has played a major role in the history of the royal families of Europe.

Individuals with **von Willebrand's disease** have an autosomal dominant defect in **platelet adherence** and a deficiency of factor VIII clotting activity (and

antigenic material). Individuals with hemophilia A lack only factor VIII clotting activity and have normal platelet adherence. The platelet adherence factor (von Willebrand factor) is a large glycoprotein (MW > 200,000) synthesized in vascular endothelial cells and megakaryocytes (platelet precursor cells). It exists in plasma and platelets tightly associated with the factor VIII molecule. Platelet surfaces seem to possess a glycoprotein receptor for the factor VIII/von Willebrand factor complex. Functionally, the von Willebrand factor also probably stabilizes factor VIII procoagulant activity. Von Willebrand disease may be an inherited defect in a specific oligosaccharide moiety on the von Willebrand glycoprotein factor. The abnormal oligosaccharide may prevent normal platelet adherence and destabilize factor VIII. Hemophilia A is a factor VIII protein defect that interrupts its clotting function but does not adversely affect the platelet adherence function of the von Willebrand factor. Factor VIII is a 2300-amino-acid glycoprotein that has homology with ceruloplasmin and factor V. It is made in the liver, spleen, and kidney.

Clotting Tests

The reader is referred to a textbook of physiology or hematology for discussion of the various tests used to assess blood coagulation.

Anticoagulants

Three naturally occurring antithrombin activities

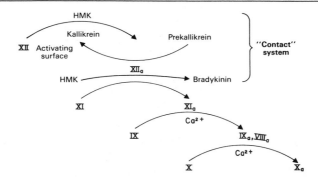

Figure 55–11. The intrinsic pathway for activating factor X to X_a. HMK is high-molecular-weight kininogen.

Table 55–4. Hemorrhagic disorders and their abnormalities.

Factor	Disorder	Bleeding Time	Clotting Time	Activated Partial Thromboplastin Time	Prothrombin Time
I	Afibrinogenemia	Variable	Infinite	Infinite	Infinite
II	Hypoprothrombinemia	Normal	Normal to long	Variable	Long
V	Parahemophilia	Normal	Long	Long	Long
VII	Factor VII deficiency	Normal	Normal	Normal	Long
VIII	Hemophilia A	Normal	Normal to long	Long	Normal
VIII	Von Willebrand's disease	Long	Variable	Variable	Normal
IX	Christmas disease, hemophilia B	Normal	Normal to long	Long	Normal
X	Stuart factor deficiency	Normal	Normal to long	Long	Long
XI	PTA deficiency	Variable	Normal to long	Long	Normal
XII	Hageman trait	Normal	Long	Long	Normal
XIII	Fibrin-stabilizing factor deficiency	Normal	Normal	Normal	Normal
Prekallikrein	Fletcher trait	Normal	Long	Long	Normal
High-molecular-weight kininogen	Fitzgerald trait	Normal	Long	Long	Normal

exist in normal plasma. Alpha$_1$-antitrypsin contributes only a minor antithrombin activity, but a specific α_2-globulin is responsible for about 25% of the antithrombin activity present in plasma. The α_2-globulin forms an irreversible complex with thrombin and other proteases and thereby prevents their binding to their natural (proteinaceous) substrates. The α_2-globulin is referred to as alpha$_2$ plasmin inhibitor, because it also inactivates plasmin, a serine protease with fibrinolytic activity discussed below.

The major antithrombin activity is contributed by antithrombin III. **Antithrombin III** has some endogenous activity but is greatly **activated** by the presence of **heparin,** a strongly anionic proteoglycan (see Chapter 54). Heparin probably binds to a specific cationic site of antithrombin III, inducing a conformational change that promotes the binding of antithrombin III to **all serine proteases,** including trypsin, chymotrypsin, and plasmin. In the clotting system, antithrombin III will **inhibit the activity of thrombin, IX$_a$, X$_a$, XI$_a$, and XII$_a$.** Individuals with inherited deficiencies of antithrombin III are prone to develop frequent and severe widespread clots, providing evidence that antithrombin III has physiologic functions and that the **clotting system in humans is normally very dynamic.**

Heparin is frequently used in clinical medicine to inhibit clotting. Its major anticoagulant action depends upon its activation of antithrombin III, which in turn inhibits the serine proteases described above. In addition, heparin in low doses appears to coat the endothelial lining of vessels and perhaps thereby reduces the activation of the intrinsic pathway. The anticoagulant effects of heparin can be antagonized by the use of strongly cationic polypeptides such as protamine to compete with the antithrombin III cationic region for the binding of the polyanionic heparin.

The **coumarin** drugs inhibit the vitamin K-dependent carboxylation of Glu to Gla residues at the amino-terminal regions of factors II, VII, IX, and X. These factors, all of which are synthesized in the liver, are dependent upon the Gla residues for maturation and thus normal function in the intrinsic, extrinsic, and final common pathways. The coumarin drugs seem to inhibit reduction of the quinone derivatives of vitamin K to the active hydroquinone forms. Thus, the administration of vitamin K will bypass the coumarin-induced block and allow maturation of the Gla-dependent clotting factors in the liver to occur. Reversal of coumarin effects by vitamin K takes place over 12–24 hours, whereas reversal of the anticoagulant effects of heparin by protamine is practically instantaneous, because of the nature of the antagonistic mechanisms.

Fibrinolysis

As described above, there is ample evidence that the blood clotting system is normally in a dynamic steady state in which fibrin clots are constantly being laid down and subsequently dissolved. **Plasmin** is a serine protease capable of digesting both fibrinogen and fibrin as well as factors V and VIII, complement, and various polypeptide hormones. Plasmin exists normally in plasma in a proenzyme or inactive form, **plasminogen.** Plasminogen activators of various types are found in most body tissues. Tissue plasminogen activator is a serine protease that is catalytically inactive until exposed to fibrin. Upon exposure to fibrin, plasminogen activator cleaves plasminogen to generate plasmin. When plasmin digests fibrin, the plasminogen activator is no longer active and proteolysis ceases, providing a well-regulated fibrinolytic process. There is considerable therapeutic interest in the use of tissue plasminogen activator (TPA), produced by recombinant DNA technology, to restore coronary arterial patency and thus reduce myocardial damage following acute coronary thrombosis. The urine contains the proteolytic enzyme **urokinase,** which is also a serine protease and can cleave plasminogen at 2 sites, activating the protease activity of plasmin.

Plasminogen normally coprecipitates with fibrin and thus is **incorporated into fibrin deposits.** When activated, the plasmin in clots digests the fibrin to soluble fragments, dissolving the clot. Cross-linked fibrin clots are less sensitive to dissolution by plasmin.

There are a number of disorders, including cancers and shock, in which the concentrations of plasminogen activators increase. In addition, the antiplasmin activities contributed by alpha$_1$-antitrypsin and alpha$_2$ plasmin inhibitor may be impaired in diseases such as cirrhosis of the liver. Some bacterial products, such as **streptokinase,** are capable of activating plasminogen without cleavage and may be responsible for the diffuse hemorrhage sometimes observed in patients with disseminated bacterial infections.

REFERENCES

Deykin D: Thrombogenesis. *N Engl J Med* 1967;**276:**622.

Genton E et al: Platelet-inhibiting drugs in the prevention of clinical thrombotic disease. (2 parts.) *N Engl J Med* 1975; **293:**1236, 1296.

George JN, Nurden AT, Phillips DR: Molecular defects in interactions of platelets with the vessel wall. *N Engl J Med* 1984;**311:**1084.

Gitschier J et al: Characterization of the human factor VIII gene. *Nature* 1984;**312:**326.

Heimark RL et al: Surface activation of blood coagulation, fibrinolysis and kinin formation. *Nature* 1980;**286:**456.

Jackson CM, Nemerson Y: Blood coagulation. *Annu Rev Biochem* 1980;**49:**767.

Kane WH et al: Factor V$_a$-dependent binding of factor X$_a$ to human platelets. *J Biol Chem* 1980;**255:**1170.

McKee PA: Hemostasis and disorders of blood coagulation. In: *The Metabolic Basis of Inherited Disease,* 5th ed. Stanbury JB et al (editors). McGraw-Hill, 1983.

Stenflo J, Suttie JW: Vitamin K-dependent formation of gamma-carboxyglutamic acid. *Annu Rev Biochem* 1977;**46:**157.

Stites DP, Stobo JD, Wells JV: *Basic & Clinical Immunology,* 6th ed. Appleton & Lange, 1987.

Weiss HJ: Platelet physiology and abnormalities of platelet function. (2 parts.) *N Engl J Med* 1975;**293:**531, 580.

56

Contractile & Structural Proteins

Victor W. Rodwell, PhD

INTRODUCTION

Protein molecules in biologic systems may serve primary functions other than catalysis. The regulatory, signal transmission, and recognition functions of protein molecules have been described in earlier chapters. Protein molecules also provide important transducing and structural functions to biologic systems. Some of these latter roles, which are dependent upon the fibrous nature of specific protein molecules, are reviewed in this chapter.

BIOMEDICAL IMPORTANCE

Our understanding of the molecular basis of major genetic diseases of structural and contractile proteins received significant impetus in 1986 with the successful cloning of the gene for Duchenne type muscular dystrophy, an achievement that holds high promise for therapy for this disease. While many molecular diseases of proteins arise from mutations in the structural genes that code for that protein (eg, hemoglobin), proteins subject to posttranslational modification present additional sites for genetic deficiency diseases. For example, several human diseases result from genetic defects in processing of immature collagen precursors. Of these, many reflect defects in cross-linking of these fibers, with resulting failure to form a stable collagen triple helix. Not all collagen diseases are, however, genetic. Hydroxyprolyl and hydroxylysyl residues arise by posttranslational hydroxylation of peptide-bound prolyl and lysyl residues in a reaction that requires ascorbate (vitamin C) as an essential cofactor. These hydroxylated residues function both as glycosylation sites and in cross-linking of fibers. The clinical presentation of softened connective tissue typical of classical scurvy thus is readily comprehended in terms of precise biochemical events.

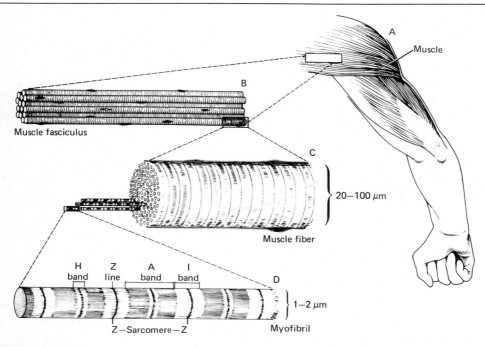

Figure 56–1. The structure of voluntary muscle. (Drawing by Sylvia Colard Keene. Reproduced, with permission, from Bloom W, Fawcett DW: *A Textbook of Histology,* 10th ed. Saunders, 1975.)

MUSCLE

Muscle is the major biochemical transducer (machine) that converts potential (chemical) energy into kinetic (mechanical) energy. Muscle is the largest single tissue in the human body, comprising somewhat less than 25% of body mass at birth, more than 40% of body mass in the young adult, and somewhat less than 30% in the aged adult.

An effective **chemical-mechanical transducer** must meet several requirements: (1) There must exist a constant supply of chemical energy. In vertebrate muscle, ATP and creatine phosphate are the forms of chemical energy. (2) There must be a means of regulating the mechanical activity—ie, the speed, duration, and force of contraction in the case of muscle. (3) The machine must be connected to an operator, a requirement met in biologic systems by the nervous system. (4) If it is to be used more than once, there must be a way of returning the machine to its original state.

Muscle is only a pulling machine, not a pushing machine. Therefore, a given muscle must be antagonized by another group of muscles or another force such as gravity or elastic recoil.

In vertebrate organisms, the above requirements and the specific needs of the organisms are met by the existence of 3 types of muscles: skeletal muscle, cardiac muscle, and smooth muscle. Both **skeletal** and **cardiac muscle** appear **striated** upon microscopic observation; **smooth muscle** is **nonstriated.** Although skeletal muscle is under **voluntary** nervous control, the control of both cardiac and smooth muscle is **involuntary.**

Muscle Structure

Striated muscle is composed of multinucleated muscle fiber cells surrounded by an electrically excitable membrane, the **sarcolemma.** When an individual muscle fiber cell, which may extend the entire length of the muscle, is examined microscopically, it will be found to contain a bundle of many **myofibrils** arranged in parallel; these are embedded in a type of intracellular fluid termed the **sarcoplasm.** Within this fluid is contained glycogen, the high-energy compounds ATP and phosphocreatine, and the enzymes of glycolysis.

The **sarcomere** is the functional unit of muscle. It is repeated along the axis of a fibril at distances of

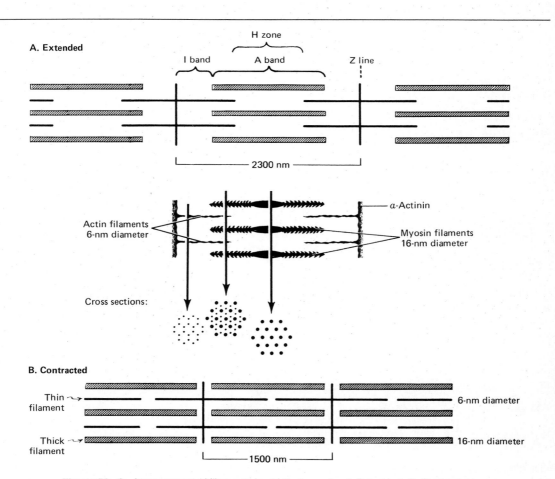

Figure 56–2. Arrangement of filaments in striated muscle. *A:* Extended. *B:* Contracted.

1500–2300 nm (Fig 56–1). When the myofibril is examined by electron microscopy, alternating dark and light bands (A bands and I bands) can be observed. The central region of the A band (the H zone) appears less dense than the rest of the band. The I band is bisected by a very dense and narrow Z line. These structural details are illustrated in Fig 56–2.

The striated appearance of voluntary and cardiac muscles in light microscopic studies results from their high degree of organization in which most muscle fiber cells are aligned so that their sarcomeres are in parallel register (Fig 56–1).

When **cross sections** of a myofibril are examined in an electron micrograph, it appears that each myofibril is constructed of 2 types of longitudinal filaments. One type (the thick filament), confined to the A band, contains chiefly the protein **myosin.** These filaments are about 16 nm in diameter and arranged in cross section as a hexagonal array (Fig 56–2). The other filament (thin filament) lies in the I band and extends also into the A band but not into the H zone of the A band (Fig 56–2). The thin filaments are about 6 nm in diameter.

They contain the proteins **actin, tropomyosin,** and **troponin.** In the A band, the thin filaments are arranged around the thick (myosin) filament as a secondary hexagonal array. Thus, as shown in Fig 56–2, each thin filament lies symmetrically between 3 thick filaments, and each thick filament is surrounded symmetrically by 6 thin filaments.

The thick and thin filaments interact via cross-bridges that emerge at intervals of 14 nm along the thick filaments. As depicted in Fig 56–2, the cross-bridges or "arrowheads" on the thick filaments have opposite polarities at the 2 ends of the filaments. The 2 poles of the filaments are separated by a 150-nm segment (the M band) that is free of projections.

When muscle contracts, there is no change in the lengths of the thick filaments or of the thin filaments, but the **H zone and the I bands shorten.** Thus, the **arrays of interdigitating filaments must slide past one another during muscle contraction.** The **cross-bridges generate and sustain the tension.** The tension developed during muscle contraction is proportionate to the filament overlap and thereby the number

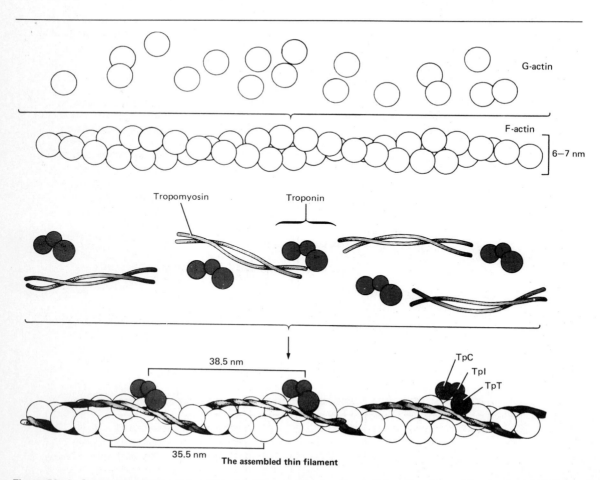

Figure 56–3. Schematic representation of the thin filament, showing the spatial configuration of the 3 major protein components—actin, tropomyosin, and troponin.

of cross-bridges. Each cross-bridge head is connected to the thick filament via a flexible fibrous segment that can bend outward from the thick filament to accommodate the interfilament spacing.

The Proteins of Muscle

The mass of a fresh muscle fibril is made up of 75% water and more than 20% protein. The 2 major muscle proteins are actin and myosin.

Monomeric (globular) actin (G-actin) is a 43,000-MW globular protein that comprises 25% of muscle protein by weight. At physiologic ionic strength and in the presence of magnesium, G-actin **polymerizes** noncovalently to form an insoluble double helical filament called F-actin (Fig 56–3). The **F-actin** fiber is 6–7 nm thick and has a pitch or repeating structure every 35.5 nm. Neither G- nor F-actin exhibits any catalytic activity.

In striated muscle, there are 4 other proteins that are minor in terms of their mass contribution but important in terms of their function. **Tropomyosin** is a fibrous molecule that consists of 2 chains, alpha and beta, that attach to the F-actin in the groove between the 2 polymers (Fig 56–3). **Tropomyosin is present in all muscle** and musclelike structures. The **troponin** system is unique to **striated muscle** and consists of 3 separate proteins. **Troponin T (TpT)** binds to tropomyosin as well as the other 2 troponin components (Fig 56–3). **Troponin I (TpI)** inhibits the F-actin–myosin interaction and also binds to the other components of troponin. **Troponin C (TpC)** is a calcium-binding protein that has a primary and secondary structure as well as a function quite analogous to **calmodulin,** a protein widely spread in nature. Four molecules of calcium ion are bound per molecule of troponin C or calmodulin, and both protein molecules have a molecular weight of 17,000. The thin filament of striated muscle consists of F-actin, tropomyosin, and the 3 components of troponin: TpC, TpI, and TpT (Fig 56–3). The repeat distance of the tropomyosin and troponin system is 38.5 nm.

Myosin contributes 55% of muscle protein by weight and forms the thick filaments. Myosin is an asymmetric hexamer with a molecular weight of 460,000. The myosin has a **fibrous portion** consisting of 2 intertwined helices, each with a **globular head** portion attached at one end (Fig 56–4). The **hexamer** consists of one pair of heavy chains (MW 200,000) and 2 pairs of light chains (MW 15,000–27,000). Skeletal muscle myosin exhibits **ATP-hydrolyzing (ATPase) activity** and binds to F-actin, an insoluble molecule.

Much has been learned from studies of the partial digestion products of myosin. When myosin is digested with trypsin, 2 myosin fragments (meromyosins) are generated. Light meromyosin (LMM) consists of aggregated, insoluble α-helical fibers (Fig 56–5). LMM exhibits no ATPase activity and will not bind to F-actin.

Heavy meromyosin (HMM) is a 340,000-MW soluble protein that has both a fibrous portion and a

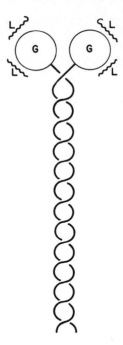

Figure 56–4. Diagram of a myosin molecule showing the 2 intertwined α helices (fibrous portion), the globular region (G), and the light chains (L).

globular portion (Fig 56–5). HMM exhibits **ATPase activity** and **binds to F-actin.** The digestion of HMM with papain generates 2 subfragments, S-1 and S-2. The S-2 is fibrous in character, exhibits no ATPase activity, and does not bind to F-actin.

S-1 has a molecular weight of 115,000, exhibits **ATPase activity,** and in the absence of ATP will **bind to and decorate actin with "arrowheads"** (Fig 56–6). Although both S-1 and HMM exhibit ATPase activity, that **catalytic activity is accelerated 100- to 200-fold by the addition of F-actin.** As discussed below, F-actin greatly enhances the rate at which myosin ATPase releases its products, ADP and P_i. Thus, although F-actin does not affect the hydrolysis step per se, its ability to **promote the release of the ATPase products** greatly accelerates the overall rate of catalysis.

α-Actinin is a protein molecule found in the Z line to which the ends of the F-actin molecules of the thin filaments attach (Fig 56–2).

The Molecular Function of Muscle

The question of how the structure and function of muscle are related can be rephrased in biochemical terms: **How can ATP hydrolysis produce macroscopic movement?** It should be apparent from the above discussion that muscle contraction consists of the **cyclic attachment and detachment of the globular head portion of myosin to the F-actin filament.** The attachment is followed by a change in the actin-

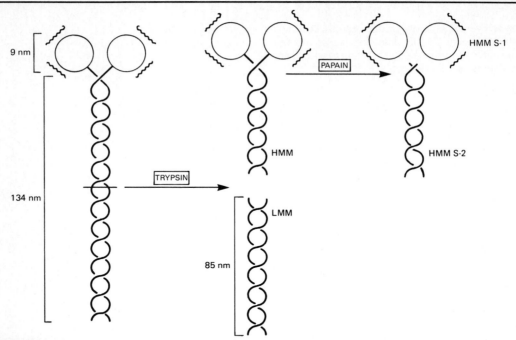

Figure 56–5. Enzymatic cleavage of myosin. HMM, heavy meromyosin; LMM, light meromyosin; S-1, subfragment 1; S-2, subfragment 2.

myosin interaction, so that the actin filaments and the myosin filaments slide past one another. The energy is supplied indirectly by ATP, which is hydrolyzed. ATP hydrolysis by the myosin ATPase is greatly accelerated by the binding of the myosin head to F-actin. The biochemical cycle of muscle contraction consists of 5 steps (Fig 56–7): (1) The myosin head alone can hydrolyze ATP to ADP + P_i, but it cannot release the products of this hydrolysis. Thus, the hydrolysis of ATP by the myosin head alone is stoichiometric rather than catalytic. (2) The myosin head containing ADP and P_i can rotate freely through large angles in order to locate and bind to F-actin, making an angle of about 90 degrees with the fiber axis. This interaction (3) promotes the release of ADP and P_i from the actin-myosin

complex. Because the conformation of lowest energy for the actomyosin bond is 45 degrees, the myosin changes its angle from 90 degrees to about 45 degrees by **pulling the actin** (10–15 nm) toward the center of the sarcomere. (4) A new ATP molecule binds to the myosin-F actin complex. Myosin-ATP has a poor affinity for actin, and thus the myosin (ATP) **head is released** (5) from the F-actin. This last step is **relaxation,** a process clearly **dependent upon the binding of ATP** to the actin-myosin complex. The ATP is again hydrolyzed by the myosin head, but without releasing ADP + P_i, to continue the cycle.

It should be clear that **ATP dissociates the myosin head from the thin filament and powers the contraction.** The efficiency of this contraction is about

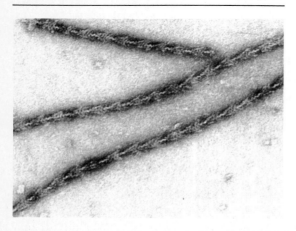

Figure 56–6. The decoration of actin filaments with the S-1 fragments of myosin to form "arrowheads." (Courtesy of Professor James Spudich, Stanford University.)

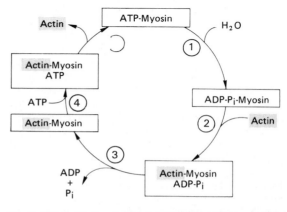

Figure 56–7. The hydrolysis of ATP drives the cyclic association and dissociation of actin and myosin in 5 reactions described in the text. (Modified from Stryer L: *Biochemistry,* 2nd ed. Freeman, 1981.)

50%; that of the internal combustion engine is less than 20%.

The Regulation of Muscle Contraction & Relaxation

The contraction of muscles from all sources occurs by the general mechanism described immediately above. Muscles from different organisms and from different cells and tissues within the same organism may have different molecular mechanisms responsible for the regulation of their contraction and relaxation. In all systems, **Ca²⁺ plays a key regulatory role.** There are 2 general mechanisms of regulation of muscle contraction: actin-based and myosin-based.

Actin-Based Regulation

Actin-based regulation of muscle occurs in vertebrate skeletal and cardiac muscles, both **striated.** In the general mechanism described above, the only potentially limiting factor in the cycle of muscle contraction might be ATP, not a seemingly ideal regulatory molecule, since it is required as the immediate energy source for contraction. The skeletal muscle system is inhibited at rest and is deinhibited to activate contraction. The **inhibitor of striated muscle is the troponin system,** which is bound to tropomyosin and F-actin in the thin filament (Fig 56–3). In striated muscle, there is no control of contraction (or ATPase as a biochemical indicator of contraction) unless the tropomyosin-troponin systems are present along with the actin and myosin filaments. As described above, tropomyosin lies along the groove of F-actin, and the 3 components of troponin—TpT, TpI, and TpC—are bound to the F-actin-tropomyosin complex. TpI prevents binding of the myosin head to its F-actin attachment site either by altering the conformation of F-actin via the tropomyosin molecules or by simply rolling tropomyosin into a position that directly blocks the sites on F-actin to which the myosin heads attach. Either way prevents the acceleration of the myosin ATPase that is mediated by binding of the myosin head to F-actin. Hence, the TpI system blocks the contraction cycle at step 2 of Fig 56–7. This accounts for the inhibited state of relaxed striated muscle.

The excitation of muscle contraction is mediated by Ca^{2+}. In resting muscle sarcoplasm, the concentration of Ca^{2+} is $10^{-7} - 10^{-8}$ mol/L. Calcium is sequestered in the sarcoplasmic reticulum, a network of fine membranous sacs, by an active transport system utilizing a Ca^{2+}-binding protein called calsequestrin. The sarcomere is surrounded by an **excitable membrane** that has transverse (T) channels closely associated with the sarcoplasmic reticulum. When the sarcomere membrane is excited, such as by the occupation of an acetylcholine receptor by acetylcholine, **Ca^{2+} is rapidly released** into the sarcoplasm from the sarcoplasmic reticulum. The Ca^{2+} concentration in sarcoplasm rapidly rises to 10^{-5} mol/L. The Ca^{2+} binding sites on TpC in the thin filament are quickly occupied by Ca^{2+}. The $TpC \cdot 4Ca^{2+}$ interacts with TpI and TpT to alter this interaction with tropomyosin. Accord-

ingly, tropomyosin simply moves out of the way or alters the F-actin conformation so that the myosin head $ADP-P_i$ can interact with F-actin to start the contraction cycle.

Relaxation occurs when (1) sarcoplasm Ca^{2+} falls below 10^{-7} mol/L owing to its resequestration in the sarcoplasmic reticulum by an energy-dependent Ca^{2+} pump; (2) $TpC \cdot 4Ca^{2+}$ loses its Ca^{2+}; (3) troponin, via its interaction with tropomyosin, inhibits further myosin head-F-actin interaction; and (4) in the presence of ATP, the myosin head detaches from the F-actin to induce relaxation. Thus, **Ca^{2+} controls muscle contraction by an allosteric mechanism** mediated in muscle by TpC, TpI, TpT, tropomyosin, and F-actin.

In cardiac muscle, the extracellular fluid is a major source of Ca^{2+} for excitation. In the absence of Ca^{2+} in the bathing extracellular fluid, cardiac muscle will cease contracting (beating) within 1 minute; skeletal muscle can contract for hours without extracellular Ca^{2+}.

The loss of ATP in the sarcoplasm has 2 major effects: (1) The Ca^{2+} pump in the sarcoplasmic reticulum ceases to maintain the low sarcoplasm Ca^{2+} concentration. Thus, the interaction of the myosin heads with F-actin is promoted. (2) The ATP-dependent detachment of myosin heads from F-actin cannot occur, and "rigor mortis" sets in.

Muscle contraction is not an all-or-none phenomenon, as anyone who can turn these pages will recognize. Muscle contraction is a delicate dynamic balance of the attachment and detachment of myosin heads to F-actin. The system is subject to fine regulation via the nervous system.

Myosin-Based Regulation of Contraction

As described above, all muscles contain actin, myosin, and tropomyosin, but **only vertebrate striated muscles contain the troponin system.** Thus, the mechanisms of regulating contraction must differ in various contractile systems.

Smooth muscles have molecular structures very similar to those in striated muscle, but the sarcomeres are not aligned in such a way as to generate the striated appearance. Smooth muscles contain α-actinin and tropomyosin molecules, as do skeletal muscles. They do not have the troponin system, and the light chains of smooth muscle myosin molecules differ from those of striated muscle myosin. However, like striated muscle, **smooth muscle contraction is regulated by Ca^{2+}.**

When smooth muscle myosin is bound to F-actin in the absence of other muscle proteins such as tropomyosin, there is no detectable ATPase activity. This absence of ATPase is quite unlike the situation described for striated muscle myosin and F-actin, which has abundant ATPase activity. Smooth muscle myosin contains a light chain (p-light chain) that prevents the binding of the myosin head to F-actin. The p-light chain must be phosphorylated before it allows

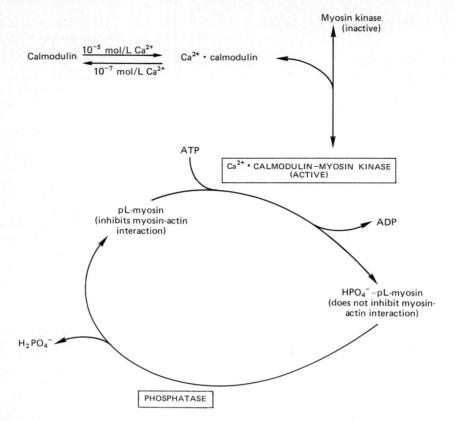

Figure 56–8. Regulation of smooth muscle contraction by Ca^{2+}. (Adapted from Adelstein RS, Eisenberg R: Regulation and kinetics of actin-myosin ATP interaction. *Annu Rev Biochem* 1980;**49**:921.)

F-actin to activate myosin ATPase. The **phosphorylation of p-light chain commences the attachment-detachment contraction cycle of smooth muscle.**

In smooth muscle sarcoplasm, there exists a **myosin light chain kinase. The myosin light chain kinase activity is calcium-dependent.** The Ca^{2+} activation of myosin light chain kinase requires binding of **calmodulin·4Ca^{2+}** to its 105,000-MW kinase subunit (Fig 56–8). The calmodulin·4Ca^{2+}-activated light chain kinase phosphorylates the p-light chain, which then ceases to inhibit the myosin-F-actin interaction. The contraction cycle then begins (Fig 56–8).

Relaxation of smooth muscle occurs when (1) sarcoplasm Ca^{2+} falls below 10^{-7} mol/L. The Ca^{2+} dissociates from calmodulin, which in turn dissociates from the myosin light chain kinase, (2) inactivating the kinase. (3) No new phosphates are attached to the p-light chain, and light chain protein phosphatase, which is continually active and calcium-independent, removes the existing phosphates from the p-light chain. (4) Dephosphorylated myosin p-light chain then inhibits the binding of myosin heads to F-actin and the ATPase activity. (5) The myosin head detaches from the F-actin in the presence of ATP, but it cannot reattach because of the presence of dephosphorylated p-light chain; hence, relaxation occurs.

Table 56–1 summarizes and compares the regulation of actin-myosin interactions (activation of myosin ATPase) in striated and smooth muscles.

The myosin light chain kinase is not directly affected or activated by cAMP. However, the usual cAMP-activated protein kinase (see Chapter 44) can phosphorylate the myosin light chain kinase (*not* the p-light chain itself). The phosphorylated myosin light chain kinase exhibits a significantly lower affinity for calmodulin·Ca^{2+} and thus is less sensitive to activation. Accordingly, an increase in cAMP dampens the contraction response of smooth muscle to a given elevation of sarcoplasm Ca^{2+}. This molecular mechanism can explain the relaxing effect of β-adrenergic stimulation on smooth muscle. The **phenothiazines,** widely used antipsychotic drugs, bind to calmodulin and prevent its attachment to calcium-dependent enzymes. Phenothiazines also relax smooth muscle.

Striated muscle from mollusks such as the scallop exhibits a myosin-based regulation of contraction. Like myosin and F-actin from smooth muscle, that from scallops also exhibits no ATPase, an effect of the inhibitory properties of the "regulatory" light chain of scallop myosin. The inhibition of scallop actin-myosin interaction is relieved when Ca^{2+} binds directly to a specific site on the myosin molecule. This regulation

Table 56–1. Actin-myosin interactions in striated and smooth muscle.

	Striated Muscle	Smooth Muscle (and Nonmuscle Cells)
Proteins of muscle filaments	Actin Myosin (hexamer) Tropomyosin Troponin (TpI, TpT, TpC)	Actin Myosin (hexamer)* Tropomyosin
Spontaneous interaction of F-actin and myosin *alone* (spontaneous activation of myosin ATPase by F-actin)	Yes	No
Inhibitor of F-actin-myosin interaction (inhibitor of F-actin-dependent activation of ATPase)	Troponin system (TpI)	Unphosphorylated myosin p-light chain
Contraction activated by	Ca^{2+}	Ca^{2+}
Direct effect of Ca^{2+}	$4Ca^{2+}$ bind to TpC	$4Ca^{2+}$ bind to calmodulin
Effect of protein-bound Ca^{2+}	$TpC \cdot 4Ca^{2+}$ antagonizes TpI inhibition of F-actin-myosin interaction (allows F-actin activation of ATPase).	$Calmodulin \cdot 4Ca^{2+}$ activates myosin light chain kinase that phosphorylates myosin p-light chain. The phosphorylated p-light chain no longer inhibits F-actin-myosin interaction (allows F-actin activation of ATPase).

*Light chains of myosin are different in striated and smooth muscles.

does not require covalent modification of myosin or the addition of a separate protein such as calmodulin or TpC to be Ca^{2+}-dependent.

Phosphorylation of Muscle Proteins

As described above, the phosphorylation of the light chain of smooth muscle myosin alleviates its inhibitory effect on the actin-myosin interaction and thereby commences the contraction cycle. Thus, phosphorylation is required for the actin-myosin interaction of smooth muscle.

One of the pairs of light chains of skeletal muscle myosin can also be phosphorylated, but this has no effect on the actin-activated ATPase of myosin, as it does on smooth muscle myosin. It has been proposed that the phosphate on the myosin light chains may form a chelate with the Ca^{2+} bound to the tropomyosin-TpC-actin complex, leading to an increased rate of formation of cross-bridges between the myosin heads and actin.

Some recent evidence suggests that phosphorylation of myosin heavy chains is a prerequisite for their assembly into the thick filaments in skeletal muscle, smooth muscle, and nonmuscle cells (see below).

The TpI and a peptide component of the sarcoplasmic reticulum Ca^{2+} pump in cardiac muscle can be phosphorylated by cAMP-dependent protein kinase. There is a rough correlation between the phosphorylation of TpI and the increased contraction of cardiac muscle induced by catecholamines. This mechanism may account for the inotropic effects (increased contractility) of the β-adrenergic compounds on the heart.

Muscle Metabolism

The ATP required as the constant energy source for the contraction-relaxation cycle of muscle can be generated by glycolysis, oxidative phosphorylation, cre-

atine phosphate, or two ADP molecules. The ATP stores in skeletal muscle are short-lived during contraction, providing energy probably for less than 1 second of contraction. In **slow skeletal muscle,** which has abundant O_2 stores in myoglobin, **oxidative phosphorylation** is the major source of ATP regeneration. **Fast** skeletal muscles regenerate ATP from glycolysis, mainly.

Phosphagens such as creatine phosphate prevent the rapid depletion of ATP by providing a readily available high-energy phosphate, which is all that is necessary to re-form ATP from ADP. Creatine phosphate is formed from ATP and creatine at times when the muscle is relaxed and ATP demands are not so great. The enzyme catalyzing the phosphorylation of creatine is creatine phosphokinase (CPK), a muscle-specific enzyme with clinical utility in the detection of acute or chronic disorders of muscle.

Skeletal muscle sarcoplasm contains large **glycogen** stores, located in granules close to the I bands. The release of glucose from glycogen is dependent upon a specific muscle glycogen phosphorylase enzyme (see Chapter 19). In order to generate glucose 6-phosphate for glycolysis in skeletal muscle, the glycogen phosphorylase b must be activated to phosphorylase a. The activation requires phosphorylation of phosphorylase b by the enzyme phosphorylase b kinase (see Chapter 19). Ca^{2+} promotes the activation of phosphorylase b kinase, also by phosphorylation. Thus, Ca^{2+} not only activates muscle contraction but also activates a pathway to provide the necessary source of energy, ATP. Muscle glycogen phosphorylase b is missing in a specific disorder of muscle (McArdle's disease), a form of glycogen storage disease.

ATP is also available from oxidative phosphorylation in muscle tissue, a process dependent upon a constant oxygen supply. Muscles that have high oxygen

Table 56–2. Characteristics of fast and slow skeletal muscle.

	Fast Skeletal Muscle	Slow Skeletal Muscle
Myosin ATPase	High	Low
Energy utilization	High	Low
Color	White	Red
Myoglobin	No	Yes
Contraction rate	Fast	Slow
Duration	Short	Prolonged

demands as a result of sustained contraction (such as to maintain posture) have the ability to store oxygen in **myoglobin** (see Chapter 6). Because of the heme moiety to which oxygen is bound in myoglobin, muscles containing myoglobin are red, as compared to white skeletal muscle. Table 56–2 compares some of the properties of fast (or white) skeletal muscle with slow (or red) skeletal muscle.

Myoadenylate kinase, an enzyme present in muscle, catalyzes the formation of one ATP molecule and one AMP from two ADP molecules. This reaction is shown in Fig 56–9 coupled with the hydrolysis of ATP by myosin ATPase during muscle contraction. The relationships between these various sources of ATP and its consumption during muscle contraction are also depicted.

In humans, skeletal muscle protein is the major nonfat source of stored energy. This explains the very large losses of muscle mass, particularly in adults, resulting from prolonged caloric undernutrition.

The study of tissue protein breakdown in vivo is difficult, because amino acids released during intracellular breakdown of proteins can be extensively reutilized for protein synthesis within the cell, or the amino acids may be transported to other organs where they enter anabolic pathways. However, actin and myosin are methylated following synthesis of their peptide bonds, producing **3-methylhistidine**. During intracellular breakdown of actin and myosin, 3-methylhistidine is released and excreted into the urine. When labeled material was administered to rats and humans, it was found that the urinary output of the methylated amino acid provides a reliable index of the rate of myofibrillar protein breakdown in the musculature of rats or human subjects. The fractional rate of muscle protein breakdown is not significantly different in the elderly as compared with young adults, but since muscle mass is less in the elderly, this tissue contributes less to the whole body protein breakdown that occurs with aging in humans.

As noted above, skeletal muscle is the major reserve of protein in the body. Also, this tissue is highly active in the degradation of certain amino acids as well as in the synthesis of others. In mammals, muscle appears to be the primary site of catabolism of the branched-chain amino acids. It oxidizes leucine to CO_2 and converts the carbon skeletons of aspartate, asparagine, glutamate, isoleucine, and valine into intermediates of the tricarboxylic acid cycle. The capacity of muscles to degrade branched-chain amino acids increases 3- to 5-fold during fasting and in diabetes.

Muscle also synthesizes and releases large amounts of alanine and glutamine. These compounds are synthesized utilizing amino groups that are generated in the breakdown of branched-chain amino acids, and the amino nitrogen is then transferred to α-ketoglutarate and to pyruvate by transamination. Glycolysis from exogenous glucose is the source of almost all of the pyruvate for synthesis of alanine. These reactions con-

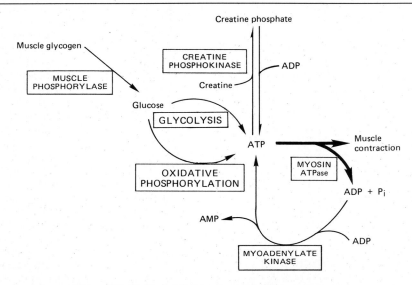

Figure 56–9. The multiple sources of ATP in muscle.

stitute the so-called glucose-alanine cycle, wherein alanine from muscle is utilized in hepatic gluconeogenesis while at the same time bringing amino groups to the liver for removal as urea.

The carbon skeletons of the amino acids that are degraded in muscle and enter the tricarboxylic acid cycle in muscle are converted mostly to glutamine and to pyruvate, which itself is further oxidized or converted to lactate. It thus appears that in fasting or the postabsorptive state, muscle releases most amino acids coming from net protein breakdown except for isoleucine, valine, glutamate, aspartate, and asparagine, which are used to contribute to the formation of glutamine, which itself is released for use by other tissues.

For many years, it has been observed that working muscle releases ammonia. It is now known that the immediate souce of ammonia in skeletal muscle is AMP, which is deaminated to IMP, catalyzed by adenylate deaminase. IMP may be converted back to AMP by reactions utilizing aspartate and catalyzed by adenylosuccinate synthetase and adenylosuccinase (see Chapter 35).

CELL MOTILITY & THE CYTOSKELETON

It is apparent that nonmuscle cells perform mechanical work, including self-propulsion, morphogenesis, cleavage, endocytosis, exocytosis, intracellular transport, and changing cell shape. These cellular functions are carried out by an extensive intracellular network of filamentous structures constituting the **cytoskeleton.** As will be shown, the cell cytoplasm is not a structural sac of fluid, as once thought. Essentially all eukaryotic cells contain 3 types of filamentous structures: **actin filaments** (7–9.5 nm in diameter), **microtubules** (25 nm), and **intermediate filaments** (10–12 nm). Each of these types of filaments can be distinguished biochemically and electron microscopically by special techniques.

Nonmuscle Actin

The G-actin protein isolated from nonmuscle cells has a molecular weight of about 43,000 and contains, as does muscle actin (α-actin), N-methylhistidyl residues. In the presence of magnesium and potassium chloride, this actin will spontaneously polymerize to form the double helical **F-actin filaments** like those seen in muscle. There are at least 2 types of actin in nonmuscle cells: β-actin and γ-actin. Both types can coexist in the same cell and probably even copolymerize in the same filament. In the cellular cytoplasm, actin forms **microfilaments** of 7–9.5 nm that frequently exist as bundles of tangled-appearing meshwork. The bundles of microfilaments are prominent just underlying the plasma membrane of resting cells and are there referred to as **stress fibers.** These stress fibers will decorate with the S-1 portion of myosin to reveal their double helical character (Fig 56–10). The stress fibers disappear as cell motility increases or

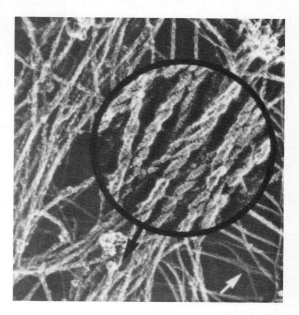

Figure 56–10. Replica of a freeze-dried cytoskeleton that was exposed to the myosin subfragment 1 (S-1) before quick-freezing. Nearly all the filaments in the lengthwise bundles, and many of the intervening filaments, have been thickened and converted into ropelike double helices (see *inset*). However, some of the filaments that travel by themselves, in between the bundles, remain totally undecorated (arrow); these are presumably intermediate filaments. × 70,000; *Inset*, × 200,000. (Reproduced, with permission, from Heuser JE, Kirschner MW: Filament organization revealed in platinum replicas of freeze-dried cytoskeletons. *J Cell Biol* 1980;**86:**212.)

upon the malignant transformation of the cell by chemicals or oncogenic viruses.

Microfilaments are also tightly packed in a meshwork pattern underlying the leading edge or "ruffle" of a motile cell (Fig 56–11). Actin microfilaments are found in all cellular microprojections such as filopodia and microvilli. For instance, the microvilli of intestinal mucosal cells contain 20–30 actin microfilaments arranged longitudinally within the microvilli as diagrammed in Fig 56–12. These microfilaments will decorate with myosin S-1, demonstrating a uniform polarity (Fig 56–12). At the base of the microvilli, myosin filaments exist and are capable of pulling together the actin filaments projecting into the microvilli. The contraction process does not involve any change of length of actin or myosin and thus must occur, as in muscle, by the sliding filament mechanism of Huxley. As in smooth muscle, the activation of the actin-myosin interaction and thereby contraction is mediated by phosphorylation of the myosin light chain.

Actin and myosin are also both found between the spindle poles and the chromosomes and along the cleavage furrow of mitotic telophase.

Actin microfilaments are associated with other

musclelike proteins in nonmuscle cells. **α-Actinin** is present at the plasma membrane sites to which microfilaments attach, such as the tips of microvilli. The geodesic domes—cytoskeletal scaffolding surrounding the nuclei of eukaryotic cells—consist of actin, α-actinin, and tropomyosin. α-Actinin is also found along actin microfilaments themselves.

As described above, **myosin** is found in association with actin microfilaments at the bases of microvilli. Myosin is also found along the actin fibers but as filaments thinner and shorter than in muscle. They seem to play a role in maintenance of the filamentous character of actin.

Tropomyosin, as mentioned above, participates in the formation of the geodomes surrounding nuclei. Tropomyosin along actin microfilaments seems to serve a structural rather than a motility function.

The regulation of nonmuscle actin function seems to depend upon several specialized proteins. **Profilin** prevents the polymerization of G-actin even in the presence of the proper concentrations of magnesium and potassium chloride. **Filamin** promotes the formation of an actin microfilament meshwork. **Tropomyosin** promotes the formation of bundles of actin stress fibers. **α-Actinin** promotes the attachment of actin microfilaments to membranes, substratum, and other cell organelles. **Cytochalasin** is a naturally occurring peptide that breaks microfilaments and prevents their polymerization. It is frequently used as a diagnostic test for the existence or function of microfilaments.

The actual motility of cells appears to be led by the **ruffle membrane,** or lamellipodium, that contains fingerlike projections called filopodia. The ruffle attaches at its tip to the substratum via the filopodia, and the cell then seems to pull in its rear margins. The ruffle releases and folds back over the top of the cell as new filopodia attach to the substratum (Fig 56–13).

Microtubules

Microtubules are an integral component of the cellular cytoskeleton. They consist of cytoplasmic tubes

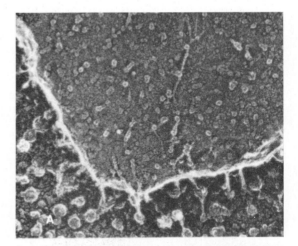
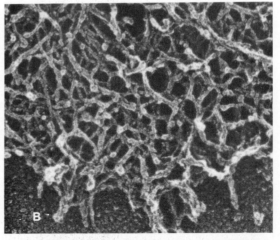
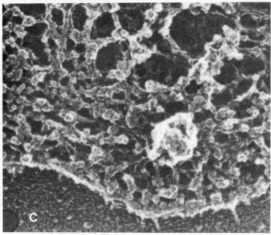

Figure 56–11 (at right). Three moderately high powered views of ruffles or lamellipodia from fibroblasts that were fixed while whole (in **A**), were extracted with Triton before fixation (in **B**), or extracted with Triton after fixation (in **C**). In **A**, the plasma membrane is intact, and no internal structure can be seen. In **B**, the plasma membrane has been removed and an underlying web of "kinky" filaments revealed. In other experiments, these filaments decorate with S-1, but they are much more concentrated and much more extensively interdigitated than actin in other regions of the cell. In **C**, the plasma membrane has again been removed, but only after the cell was fixed with aldehyde. The delicate meshwork of underlying filaments appears coarser after the chemical fixation. **A,** × 140,000; **B** and **C,** × 115,000. (Reproduced, with permission, from Heuser JE, Kirschner MW: Filament organization revealed in platinum replicas of freeze-dried cytoskeletons. *J Cell Biol* 1980;**86:**212.)

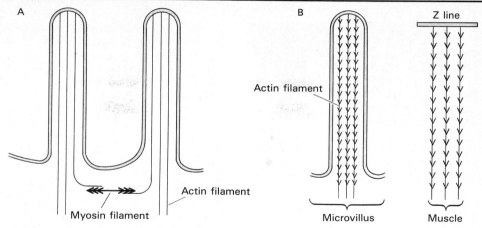

Figure 56–12. Microvilli are tiny cytoplasmic protrusions that extend out from the epithelial cells lining the small intestine, greatly increasing the surface for the absorption of nutrients. Microvilli contain both actin and myosin filaments and are known to contract much like muscle cells, and so they provide a convincing example of nonmuscle movement mediated by sliding filaments of actin and myosin. As is shown in **A**, bundles of actin filaments project upward inside each microvillus; the myosin filaments are localized at the base of the microvilli. In **B**, the orientation of the actin filaments was determined by treating the microvilli with isolated head fragments from muscle myosin, termed heavy meromyosin; these fragments retain the ability to bind to actin filaments. When the head fragments are applied to muscle cells, they form "arrowhead" complexes with the actin filaments that point in the direction of the filaments. When heavy meromyosin was added to microvilli, the head fragments formed arrowhead complexes with the actin filaments that pointed downward from the attachment sites in the tips of the microvilli. The actin filaments within the microvilli are therefore analogous to the actin filament arrays of muscle cells. (Reproduced, with permission, from Lazarides E, Revel JP: The molecular basis of cell movement. *Sci Am* [May] 1979;**240**:100.)

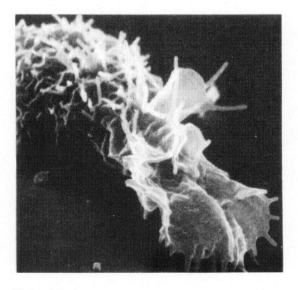

Figure 56–13. Individual cells in tissue culture are depicted. The delicate feathery structure at the bottom right is a "ruffle," or lamellipodium, which marks the leading edge of the cell. A cell is shown from an oblique angle as it moves across the substrate, extending its ruffle to form new adhesions. (Reproduced, with permission, from Lazarides E, Revel JP: The molecular basis of cell movement. *Sci Am* [May] 1979;**240**:100.)

25 nm in diameter and of indefinite length. Microtubules are necessary for the formation and function of the **mitotic spindle** and thus are present in all eukaryotic cells. Microtubules carry out a number of other cellular functions. They are responsible for the intracellular movement of endocytotic and exocytotic vesicles. They form the major structural component of **cilia and flagella.** Microtubules are a major protein component of **axons and dendrites,** where they maintain the structure and participate in the axoplasmic flow of material along these neuronal processes.

Microtubules are cylinders of 13 longitudinally arranged **protofilaments,** each consisting of dimers of α-**tubulin** and β-**tubulin** (Fig 56–14). α-Tubulin (MW 53,000) and β-tubulin (MW 55,000) are closely related protein molecules. The tubulin dimers assemble into protofilaments and subsequently into sheets and then cylinders, as depicted in Fig 56–15. The assembly of tubulin into microtubules requires two **GTP** molecules per tubulin dimer. Two proteins termed high-molecular-weight (HMW) protein and Tau promote the formation of microtubules but are not required for assembly. Calmodulin and phosphorylation may both play roles in microtubule assembly.

A number of particularly important alkaloids can prevent microtubule assembly. These include colchicine and its derivative demecolcine (used for treatment of acute gouty arthritis), vinblastine (a *Vinca* alkaloid used for treating cancer), and griseofulvin (an antifungal agent).

Microtubules "grow" with a polarity from specific sites (centrioles) within cells. On each chromatid of a chromosome (see Chapter 37) there exists a kinetochore that serves as a point of origin for microtubular

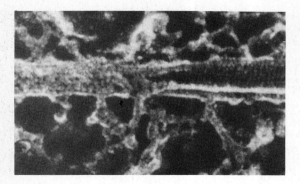

Figure 56–14. High magnification of a microtubule that was fractured and deep etched after quick-freezing. The left half of the field illustrates the outer surface of the microtubule, which displays longitudinal bands of bumps spaced 55 nm apart, which may represent the microtubule's protofilaments. To the right, the microtubule is fractured open to reveal its inner luminal walls, which display characteristic oblique striations separated by 40 nm. The reticulum surrounding the microtubule is thought to be unpolymerized tubulin and microtubule-associated proteins. (Reproduced, with permission, from Heuser JE, Kirschner MW: Filament organization revealed in platinum replicas of freeze-dried cytoskeletons. *J Cell Biol* 1980; 86:212.)

growth. Many abnormalities of chromosomal segregation result from abnormal structure or function of kinetochores. The centrosome, which is at the center of the mitotic poles, also nucleates microtubular formation. The movement of chromosomes during anaphase of mitosis is dependent upon microtubules, but the molecular mechanism has not been delineated.

At the base of all eukaryotic **flagella and cilia** is a structure called the **basal body;** it is identical to the centriole and acts as a nucleation center for the formation of the 9-doublet array of microtubules in the flagella and cilia. These microtubular structures are specialized for motility. Each member of a doublet shares a common wall of 3 protofilaments with its partner, and the doublets are connected by a flexible protein, **nexin.** Movement is effected by the sliding of the doublets past one another causing distortion of the cilium in waves. Connected to one of the doublets in a cilium is a large protein, **dynein,** which possesses an ATPase necessary for the microtubular doublet sliding movement.

Intermediate Filaments

Recent studies have confirmed the existence of an intracellular fibrous system of filaments with an axial periodicity of 21 nm and 8–10 nm in diameter that are

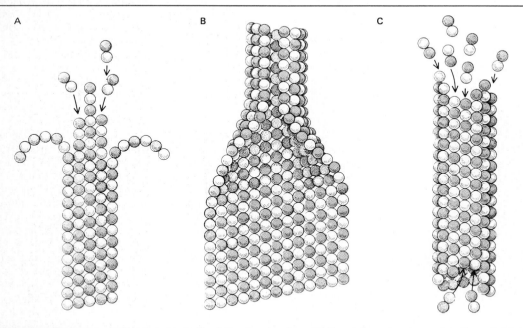

Figure 56–15. Assembly of microtubules in the laboratory begins with 2 protein molecules, α-tubulin and β-tubulin, which are globular molecules (probably more ovoid than these highly schematic spheres). The tubulins form dimers, or double molecules. If the dimers are present in a high enough concentration, they associate to form various intermediate structures, including double rings, spirals, and stacked rings; the equilibrium is biased in favor of either the isolated dimers or the intermediate structures, depending on the conditions. The next steps are not well established. It seems that the rings or spirals open up to form strands, called protofilaments, of linearly associated dimers, which assemble side by side in a sheet (*A*); sometimes the ends of protofilaments curve. When a sheet is wide enough, it forms a tube, perhaps by curling up (*B*). Once a short tube has formed (*C*), it is lengthened by the addition of dimers preferentially at one end. (Reproduced, with permission, from Dustin P: Microtubules. *Sci Am* [Aug] 1980;243:67.)

Table 56–3. Classes of intermediate filaments and their distributions.

Proteins	MW (Thousands)	Diameter (nm)	Distributions
Keratin type I and type II (tonofilaments)	40–65 (10–20 major proteins)	8	Epithelial cells (never cells of mesenchymal origin).
Desmin	50–55	10	Muscle (Z lines).
Vimentin	52	10	Mesenchymal and nonmesenchymal cells, eg, muscle, glial cells, epithelial cells.
Neurofilament	200 150 70	10	Neurons.
Glial filament	51	10	Glial cells.

distinct from microfilaments (6 nm in diameter) and microtubules (23 nm in diameter). There are 6 major classes of these filaments that share an antigenic determinant and exhibit diameters **intermediate** in size between actin microfilaments and microtubules. Each intermediate filament consists of biochemically and immunologically distinct subunits. Intermediate filaments seem to form relatively **stable components** of the cytoskeleton, not undergoing rapid assembly and disassembly and not disappearing during mitosis as do actin and many microtubular filaments. Table 56–3 summarizes some properties and distributions of intermediate filaments.

There are 2 types of **keratin,** I and II, comprising 10–20 different polypeptides that are as different from one another as they are from the other 4 classes of intermediate filament proteins—desmin, vimentin, neurofilament, and glial filament. These latter 4 classes have a high degree of homology among them. A keratin filament will contain at least 2 different keratin polypeptides, whereas the other 4 classes of intermediate filaments are homopolymers. Each of the intermediate filaments consists of 4 α-helical domains separated from one another by regions of β-pleated sheets and flanked on both ends by nonhelical terminal domains. The nonhelical terminal domains are involved in end-to-end extension of protofilaments and side-to-side interprotofibrillar interactions. The ends of the microfibrillar keratins can be cross-linked through disulfide bonding to form insoluble filaments, such as those characteristic of wool.

It is apparent that some of the intermediate filaments, particularly those of muscle and mesenchymal origin, coexist in numerous tissues.

COLLAGEN

Collagen, the major macromolecule of connective tissues, is the most common protein in the animal world. It provides an extracellular framework for all metazoan animals and exists in virtually every animal tissue. There are at least **5 distinct types** of collagen in mammalian tissues; thus, they exist as a family of molecules sharing many properties. The most definitive property of collagen molecules is their **triple helix,** a coiled structure of 3 polypeptide subunits. Each

polypeptide subunit, or alpha chain, is twisted into a **left-handed helix of 3 residues per turn.** Three of these left-handed helices are then wound to a right-handed superhelix to form a stiff rodlike molecule 1.4 nm in diameter and about 300 nm long. These triple helical molecules—unique to collagen—are then associated bilaterally and longitudinally into fibrils (Fig 56–16). The arrangement of collagen fibrils involves longitudinal staggering slightly less than one-quarter the length of the triple helix. Between the end of one triple helix and the beginning of the next is a gap that may provide a site for deposition of hydroxyapatite crystals in bone formation. Collagen fibrils range from 10 to 100 nm in diameter and are visible by microscopy as banded structures in the extracellular matrix of connective tissues.

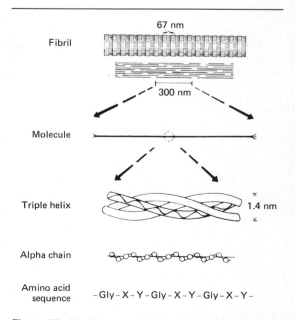

Figure 56–16. Molecular features of collagen structure from primary sequence up to the fibril. (Slightly modified and reproduced, with permission, from Eyre DR: Collagen: Molecular diversity in the body's protein scaffold. *Science* 1980;207:1315. Copyright © 1980 by the American Association for the Advancement of Science.)

Table 56-4. Genetically distinct vertebrate collagens. At least 5 different molecules containing 7 genetically distinct α chains are present in higher animals.*

Type	Molecular Formula	Native Polymer	Tissue Distribution	Distinctive Features
I	$[\alpha 1(I)]_2\alpha 2$	Fibril	Skin, tendon, bone, dentin, fascia; most abundant.	Low content of hydroxylysine; few sites of hydroxylysine glycosylation; broad fibrils.
II	$[\alpha 1(II)]_3$	Fibril	Cartilage, nucleus pulposus, notochord, vitreous body.	High content of hydroxylysine; heavily glycosylated; usually thinner fibrils than type I.
III	$[\alpha 1(III)]_3$	Fibril	Skin, uterus, blood vessels; "reticulin" fibers generally.	High content of hydroxyproline; low content of hydroxylysine; few sites of hydroxylysine glycosylation; interchain disulfides between cysteines at the carboxyl end of the helix; long carboxyl telopeptide.
IV	$[\alpha 1(IV)]_3$ (tentative, under dispute)	Basement lamina	Kidney glomeruli, lens capsule; Descemet's membrane; basement laminae of all epithelial and endothelial cells?	Very high content of hydroxylysine; almost fully glycosylated; relatively rich in 3-hydroxyproline; low alanine content; retains procollagen extension pieces.
V	$\alpha A(\alpha B)_2$ or $(\alpha A)_3$ and $(\alpha B)_3$	Unknown	Widespread in small amounts; basement laminae of blood vessels and smooth muscle cells; exoskeleton of fibroblasts and other mesenchymal cells?	High content of hydroxylysine; heavily glycosylated; low alanine content; fails to form native fibrils in vitro.

*Reproduced, with permission, from Eyre DR: Collagen: Molecular diversity in the body's protein scaffold. *Science* 1980;**207**:1315. Copyright © 1980 by the American Association for the Advancement of Science.

The other striking characteristic of the collagen molecule is that **glycine constitutes every third residue** in the triple helical portion of each alpha chain. Glycine is the only amino acid small enough to exist in the limited space available down the central core of the triple helical molecule; thus, the central core of the triple helical molecule consists of glycine residues provided by each of the 3 alpha subunits. This repeating structure can be represented by $(Gly-X-Y)_n$, where X and Y are amino acids other than glycine.

In mammalian collagen, about 100 of the X positions are **proline,** and 100 of the Y positions are **4-hydroxyproline.** These rigid imino acids limit rotation of the polypeptide backbone and thus increase the stability of the triple helix. The hydroxyproline residues contribute additional stability to the collagen triple helix by forming more intramolecular hydrogen bonds mediated through extra water molecules. Collagen also contains 3-hydroxyproline in some X positions and 5-hydroxylysine in Y positions.

The collagen triple helix is stabilized by multiple interchain cross-links between lysyl and hydroxylysyl residues. The chemical nature of these cross-links is described below. Mature collagen is a glycoprotein containing saccharides attached in O-glycosidic linkage to the hydroxylysine residues.

A summary of the vertebrate collagens, their tissue distributions, and distinctive features is presented in Table 56-4.

The Synthesis of Collagen

Collagen is an extracellular protein but is synthesized as an intracellular precursor molecule that undergoes posttranslational modification before becoming a mature collagen fibril. Like all secreted proteins, the precursor of collagen is processed as it passes through the endoplasmic reticulum and Golgi complex prior to appearing extracellularly. The earliest collagen precursor is a **preprocollagen** that contains a leader or signal sequence of approximately 100 amino acids at its amino terminus. Preprocollagen is generated by ribosomes attached to the endoplasmic reticulum. As the signal sequence penetrates into the vesicular space of the endoplasmic reticulum, the leader sequence is cleaved off and the amino-terminal end of **procollagen** continues to protrude into the endoplasmic reticular space. At this site, **prolyl 4-hydroxylase** and **lysyl hydroxylase** act on proline or lysine residues, respectively, in the Y position of the $(Gly-X-Y)_n$ peptide. A prolyl 3-hydroxylase acts on prolyl residues in the X position immediately preceding a 4-hydroxyproline in the Y position.

Table 56-5. Order and location of processing the collagen precursor (containing repeating structure $[Gly-X-Y]_n$).

Intracellular (endoplasmic reticulum)
(1) Cleavage of signal peptide.
(2) 4-Hydroxylation of Y-prolyl residues.
(3) 3-Hydroxylation of X-prolyl, where Y = 4-hydroxyprolyl residue.
(4) 5-Hydroxylation of Y-lysyl residues.
(5) Glycosylation of hydroxylysyl residues.
(6) Formation of intrachain and interchain S−S bonds.
(7) Formation of triple helix procollagen.

Extracellular
(1) Cleavage of NH_2-terminal propeptide (MW 20,000).
(2) Cleavage of COOH-terminal propeptide (MW 30,000−35,000).
(3) Formation of immature collagen fibrils.
(4) Oxidation of lysyl, hydroxylysyl, glycosylated hydroxylysyl residue to aldehydes.
(5) Cross-linking of chains and helical molecules of fibrils via Schiff bases and aldol condensations.

The procollagen molecule contains at its amino terminus a 20,000-MW peptide and at its carboxyl terminus a 30,000- to 35,000-MW peptide, neither of which is present in mature collagen. Both of these propeptides contain **cysteine** residues. While the amino-terminal propeptide collagen forms only intrachain disulfide bonds, the carboxy-terminal peptides form both intrachain and interchain disulfide bonds. Following the formation of these disulfide bonds, the procollagen molecules assemble as the triple helix.

After formation of the triple helix, further hydroxylation of prolyl and lysyl residues *cannot* occur. The glycosyltransferase activities that attach glucose or galactose to hydroxylysine residues also require that the procollagen alpha chains be nonhelical.

Following this intracellular processing, the glycosylated procollagen molecule reaches the outside of the cell by way of the Golgi apparatus. Extracellular **procollagen aminoprotease** and **procollagen car-** **boxyprotease** remove the amino-terminal and carboxy-terminal propeptides, respectively. The newly formed collagen molecules have approximately 1000 amino acids per chain and spontaneously assemble into **collagen fibrils** that are indistinguishable from the mature fibrils found in tissues.

These fibrils, however, do not have the tensile strength of mature collagen fibrils until they are **crosslinked by a series of covalent bonds.** The extracellular copper-containing enzyme lysyl oxidase oxidatively deaminates the ϵ-amino groups of certain lysyl and hydroxylysyl residues of collagen, yielding reactive aldehydes. The aldehydes can form Schiff bases with ϵ-amino groups of other lysines or hydroxylysines or even glycosylated hydroxylysines. These Schiff bases are chemically rearranged and provide stable covalent cross-links such as new peptide bonds or secondary amine bridges. The aldehyde component derived from a hydroxylysine forms a more stable

Table 56–6. Consequences of molecular defects in 4 heritable diseases of collagen.*

Disease	Defect†	Consequences
Osteogenesis imperfecta Type I	Proα1(I)°	Half normal amount of type I collagen
	Proα2(I)^S	Probably abnormal fibrils
	Other (unidentified)	
Type II	Proα1(I)^S	Unstable triple helix; increased synthesis of proα1(III)
	Proα2(I)^S and proα2(I)°	Uncertain
	Other (unidentified)	
Type III	Proα2(I)^{CX}	Synthesis of proα1(I) trimers
	Proα1(I)^{CX} or proα2(I)^{CX}	Increased mannose in C-propeptide and decreased solubility of type I procollagen
	Other (unidentified)	
Variant with characteristics of Ehlers-Danlos syndrome	Proα2(I)^S	Resistance to procollagen N-proteinase and persistence of pNcollagen‡
Marfan's syndrome	Proα2(I)^L	Probably abnormal cross-linking
	Other (unidentified)	
Ehlers-Danlos syndrome Type IV	Proα1(III)$^+$	Marked decrease in type III collagen
	Proα1(III)^{SM}	Unstable triple helix
	Proα1(III)^X	Decreased secretion rate of type III collagen
	Other (unidentified)	
Type VI	Lysine hydroxylase deficiency	Hydroxylysine-deficient collagen and defective cross-linking
	Other (unidentified)	
Type VII	Procollagen N-proteinase deficiency	Persistence of pNcollagen‡
	Proα2(I)^X	Resistance to procollagen N-proteinase and persistence of pNcollagen‡
Type IX	Defective copper metabolism	Lysine oxidase deficiency and defective cross-linking
Menkes' syndrome	Defective copper metabolism	Lysine oxidase deficiency and defective cross-linking

*Reproduced, with permission, from Prockop DJ, Kivirikko KI: Heritable diseases of collagen. *N Engl J Med* 1984;**311:**376.
†Proα1(I)°, proα2(I)°, and proα1(III)$^+$, nonfunctional or inefficiently functioning alleles for proα chains; proα1(I)^S and proα2(I)^S, shortened proα chains; proα2(I)^L, lengthened proα chains; proα1(I)^{CX} and proα2(I)^{CX}, mutations altering the structure of the C-propeptides of proα chains; proα1(III)^{SM}, an altered proα1(III) chain that migrates slowly in electrophoretic gels; and proα1(III)^X and proα2(I)^X, poorly defined mutations altering the structure of proα chains.
‡Intermediate in the conversion of procollagen that contains the N-propeptides but not the C-propeptides.

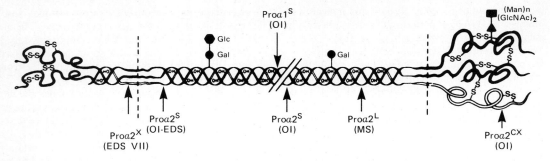

Figure 56–17. Approximate locations of mutations in the structure of type I procollagen. EDS, Ehlers-Danlos syndrome; MS, Marfan's syndrome; OI, osteogenesis imperfecta. For other abbreviations, see second footnote to Table 56–6. (Reproduced, with permission, from Prockop DJ, Kivirikko KI: Heritable diseases of collagen. *N Engl J Med* 1984;311:376.)

cross-link than does the aldehyde derived from a lysyl residue. Aldol bridges also provide intramolecular cross-links.

The intracellular and extracellular processing of the collagen precursor molecule is summarized in Table 56–5.

The same cells that secrete collagen also secrete **fibronectin,** a large glycoprotein present on cell surfaces, in the extracellular matrix, and in blood. Fibronectin binds to aggregating procollagen fibers and alters the kinetics of fibril formation in the pericellular matrix. Associated with fibronectin and procollagen in this matrix are the proteoglycans heparan sulfate and chondroitin sulfate (see Chapter 54).

Cartilage is an extracellular matrix in which collagen contributes **tensile strength** and proteoglycans are responsible for its remarkable **resilience.**

Inherited Defects of Collagen & Its Assembly

The inherited diseases that result in abnormal collagen constitute an increasing number of variants of at least 4 different types of syndromes: **osteogenesis imperfecta, Marfan's syndrome, Ehlers-Danlos syndrome,** and **Menkes' (kinky-hair) syndrome.** Originally, these diseases were defined on the basis of having similar phenotypes, but as the understanding of collagen structure and function increases, it is becoming apparent that similar molecular defects may present dissimilar clinical syndromes and vice versa.

Many of the defects include defective processing of collagen precursors owing to abnormalities within the precursor or, in some cases, abnormalities of the processing enzyme per se. In many cases, understanding the nature of the molecular defect allows one to predict whether the disease is recessively or dominantly inherited. So far, only defects in type I and type III procollagen molecules have been recognized among the inherited diseases of collagen.

Table 56–6 summarizes defects and their consequences in some inherited human diseases of collagen, and Fig 56–17 depicts sites at which defects in procollagen type I are recognized.

REFERENCES

Adelstein RS, Eisenberg R: Regulation and kinetics of actin-myosin ATP interaction. *Annu Rev Biochem* 1980;**49:**921.

Adelstein RS et al: Phosphorylation of muscle contractile proteins. *Fed Proc* 1980;**39:**1544.

Barany M, Barany K: Phosphorylation of the myofibrillar proteins. *Annu Rev Physiol* 1980;**42:**275.

Bornstein P, Sage H: Structurally distinct collagen types. *Annu Rev Biochem* 1980;**49:**957.

Caplan A: Cartilage. *Sci Am* (Oct) 1984;**250:**84.

Clark M, Spudich JA: Nonmuscle contractile proteins: The role of actin and myosin in cell motility and shape determination. *Annu Rev Biochem* 1977;**46:**797.

DeCrombrugghe B, Pastan I: Structure and regulation of a collagen gene. *Trends Biochem Sci* 1982;**7:**11.

Dustin P: Microtubules. *Sci Am* (Aug) 1980;**243:**67.

Eyre DR et al: Cross-linking in collagen and elastin. *Annu Rev Biochem* 1984;**53:**717.

Fuchs E, Hanukoglu I: Unraveling the structure of intermediate filaments. *Cell* 1983;**34:**332.

Heuser JE, Kirschner MW: Filament organization revealed in platinum replicas of freeze-dried cytoskeletons. *J Cell Biol* 1980;**86:**212.

Kleinman HK, Klebe RJ, Martin GR: Role of collagenous matrices in the adhesion and growth of cells. *J Cell Biol* 1981;**88:**473.

Lazarides E: Intermediate filaments: A chemically heterogeneous, developmentally regulated class of proteins. *Annu Rev Biochem* 1982;**51:**219.

Lazarides E, Revel JP: The molecular basis of cell movement. *Sci Am* (May) 1979;**240:**100.

Murray JM, Weber A: The cooperative action of muscle proteins. *Sci Am* (Feb) 1974;**230:**59.

Prockop DJ, Kivirikko KI: Heritable diseases of collagen. *N Engl J Med* 1984;**311:**376.

Sandberg LB, Soskel NT, Leslie JG: Elastin structure, biosynthesis, and relation to disease states. *N Engl J Med* 1981; **304:**566.

Cancer, Oncogenes, & Growth Factors

57

Robert K. Murray, MD, PhD

INTRODUCTION

Cancer cells are characterized by 3 properties: (1) diminished or unrestrained control of growth; (2) invasion of local tissues, and (3) spread, or metastasis, to other parts of the body. Cells of benign tumors also show diminished control of growth but do not invade local tissue or spread to other parts of the body.

In this chapter, we discuss some biochemical aspects of cancer. The key issues are to explain in biochemical terms the uncontrolled growth of cancer cells and their ability to invade and metastasize. The genes controlling growth and interactions with other normal cells are apparently abnormal in structure or regulation in cancer cells. Information on how cell growth—both normal and pathologic—is controlled is limited, and knowledge of specific genes involved in growth regulation is even more meager. Little is known as yet about the biochemical basis of metastasis, so that coverage of this topic will be brief. At least some types of cancer (eg, certain leukemias) can be regarded as examples of abnormal differentiation. Again, astonishingly little is known about the molecular basis of differentiation. However, many workers in this area think that further research on oncogenes and growth factors will provide insight into the nature of the disturbed control of growth, of differentiation (where applicable), and of cell-cell interaction exhibited by cancer cells. Thus, both of these topics will be discussed in some detail.

BIOMEDICAL IMPORTANCE

Cancer is the second most common cause of death in the USA after cardiovascular disease. Humans of all ages develop cancer, and a wide variety of organs are affected. The incidence of many cancers increases with age, so that as people live longer, more will develop the disease. Apart from individual suffering, the economic burden to society is immense.

BIOCHEMICAL LABORATORY TESTS & CANCER

Biochemical laboratory tests are helpful in management of patients with cancer. Many cancers are as-

Table 57–1. Clinically useful tumor markers.*

Marker	Associated Cancer
Carcinoembryonic antigen (CEA)	Colon, lung, breast, pancreas
Alpha-fetoprotein (AFP)	Liver, germ cell
Human chorionic gonadotropin (hCG)	Trophoblast, germ cell
Calcitonin (CT)	Thyroid (medullary carcinoma)
Prostatic acid phosphatase (PAP)	Prostate

*Adapted, with permission, from McIntire KR: Tumor markers: How useful are they? *Hosp Pract* (Dec) 1984;**19:**55.

sociated with the abnormal production of enzymes, proteins, and hormones (see Progression of Tumors, below), which can be measured in plasma or serum. These molecules are known as **tumor markers.** Measurement of some tumor markers is now an integral part of management of some types of cancer (Table 57–1). Applications of tumor markers in diagnosis and management of cancer are listed in Table 57–2. Three major conclusions have emerged from the study of tumor markers (McIntire, 1984): (1) No single marker is useful for all types of cancer or for all patients with a given type of cancer. For this reason, the use of a battery of tumor markers is sometimes advantageous. (2) Markers are most often detected in advanced stages of cancer rather than early stages, when they would be more helpful. (3) Of the uses of markers

Table 57–2. Applications of tumor markers.*

Detection: Screening in asymptomatic persons.
Diagnosis: Differentiating malignant from benign conditions.
Monitoring: Predicting effect of therapy and detecting recurrent cancer.
Classification: Choosing therapy and predicting tumor behavior (prognosis).
Staging: Defining extent of disease.
Localization: Nuclear scanning of injected radioactive antibodies.
Therapy: Cytotoxic agents directed to marker-containing cells.

*Adapted, with permission, from McIntire KR: Tumor markers: How useful are they? *Hosp Pract* (Dec) 1984;**19:**55.

listed in Table 57–2, the most successful have been the monitoring of responses to therapy and the detection of early recurrence.

CAUSES OF CANCER

Agents causing cancer fall into 3 broad groups: radiant energy, chemical compounds, and viruses.

RADIANT ENERGY

Ultraviolet rays, x-rays, and γ-rays are mutagenic and carcinogenic. These rays damage DNA in several ways. Ultraviolet radiation may cause pyrimidine dimers to form. Apurinic or apyrimidinic sites may form by elimination of corresponding bases. Single- and double-strand breaks or cross-linking of strands may occur. Damage to DNA is presumed to be the basic mechanism of carcinogenicity with radiant energy, but the details are unclear. Repair of DNA is discussed in Chapter 38. Apart from direct effects on DNA, x-rays and γ-rays cause **free radicals** to form in tissues. The resultant ·OH, superoxide, and other radicals can interact with DNA and other macromolecules, leading to molecular damage and thereby probably contributing to carcinogenic effects of radiant energy.

CHEMICAL CARCINOGENS

A wide variety of chemical compounds are carcinogenic (Table 57–3); the structures of 3 of the most widely studied are shown in Fig 57–1. Most of the compounds listed in Table 57–3 have been tested by administration to rodents or other animals. However, many substances are associated with the development of cancer in humans. It is estimated that up to 80% of human cancers are caused by environmental factors, principally chemicals. Exposure to such compounds can occur because of a person's **occupation** (eg, benzene, asbestos); **diet** (eg, aflatoxin B_1, which is produced by the mold *Aspergillus flavus* and sometimes found as a contaminant of peanuts and other foodstuffs); **life-style** (eg, cigarette smoking); or in other ways (eg, certain therapeutic **drugs** can be carcino-

Table 57–3. Some chemical carcinogens.

Class	Compound
Polycyclic aromatic hydrocarbons	Benzo[a]pyrene, dimethylbenzanthracene
Aromatic amines	2-Acetylaminofluorene, N-methyl-4-aminoazobenzene (MAB)
Nitrosamines	Dimethylnitrosamine, diethylnitrosamine
Various drugs	Alkylating agents (eg, cyclophosphamide), diethylstilbestrol
Naturally occurring compounds	Dactinomycin, aflatoxin B_1
Inorganic compounds	Arsenic, asbestos, beryllium, cadmium, chromium

genic). We shall present only a few important generalizations that have emerged from the study of chemical carcinogenesis.

Structure

Both organic and inorganic molecules may be carcinogenic (Table 57–3). The diversity of these compounds indicates that they do not possess one common structural feature that confers carcinogenicity.

Action

The organic carcinogens have been the most thoroughly studied. Some, such as nitrogen mustard and β-propiolactone, have been found to interact directly with target molecules **(direct carcinogens)**, but others require prior metabolism to become carcinogenic **(procarcinogens)**. The process whereby one or more enzyme-catalyzed reactions convert procarcinogens to active carcinogens is called **metabolic activation.** Any intermediate compounds formed are **proximate carcinogens,** and the final compound that reacts with cellular components (eg, DNA) is the **ultimate carcinogen.** The sequence is thus:

Procarcinogen → Proximate carcinogen A →

Proximate carcinogen B →

Ultimate carcinogen

The procarcinogen itself is not a chemically reactive species, whereas the ultimate carcinogen is often highly reactive. At least 2 reactions are required to

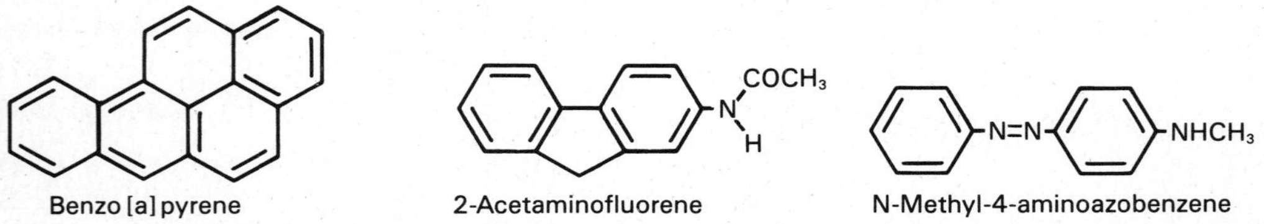

Benzo [a] pyrene   2-Acetaminofluorene   N-Methyl-4-aminoazobenzene

Figure 57–1. Structures of 3 important experimentally used chemical carcinogens.

convert the procarcinogen 2-acetylaminofluorene (2-AAF) to the ultimate carcinogen, the sulfate ester of N-hydroxy-AAF. An important generalization is that ultimate carcinogens are usually **electrophiles** (ie, molecules deficient in electrons), which readily attack nucleophilic (electron-rich) groups in DNA, RNA, and proteins.

Monooxygenases & Transferases

The metabolism of procarcinogens and other xenobiotics involves monooxygenases and transferases. The enzymes responsible for metabolic activation of procarcinogens are principally heme-containing monooxygenases of the cytochrome P-450 type located in the endoplasmic reticulum. These are the same enzymes that are involved in the metabolism of other xenobiotics, such as drugs and environmental pollutants (eg, polychlorinated biphenyls [PCBs]). The monooxygenases catalyze the hydroxylation of various procarcinogens and other xenobiotics using molecular oxygen as the source of oxygen and NADPH as a reducing source.

$$R\text{–}H + O_2 + NADPH + H^+ \rightarrow R\text{–}OH + H_2O + NADP$$

Because one atom of oxygen is introduced into the product and the other into water, these enzymes were formerly called mixed-function oxidases. In the case of drugs, such hydroxylation reactions may be responsible for converting the drug to an active or an inactive form. Either way, these reactions play a central role in the metabolism of many drugs. At least 6—and perhaps many more—such monooxygenases are present in the endoplasmic reticulum of human liver, each with broad and overlapping specificities for their substrates. The particular monooxygenase involved in the metabolism of polycyclic aromatic hydrocarbons is named cytochrome P-448, or aromatic hydrocarbon hydroxylase. The reactions catalyzed by these monooxygenases are called **phase 1 reactions of xenobiotic metabolism;** these reactions generally introduce a hydroxyl group into a compound, making it more polar and preparing it for subsequent excretion. In **phase 2 reactions of xenobiotic metabolism,** the hydroxylated xenobiotics are conjugated with various moieties (eg, glucuronate, sulfate, acetate, glutathione). These reactions generally lessen (detoxify) the reactivity of the compounds involved, further increase their polarity, and ready them for excretion, mainly in urine. In some instances, conjugation may actually increase the biologic activity or chemical reactivity of a molecule. The enzymes catalyzing the above conjugation reactions (eg, glucuronyl, sulfate, acetyl, and glutathione transferases) are usually cytosolic in location, although some are also present in the endoplasmic reticulum. Glucuronyl, sulfate, and acetyl transferases use UDP-glucuronate, 3′-phosphoadenosine-5′-phosphosulfate (PAPS), and acetyl-CoA as the donors of the glucuronyl, sulfate, and acetyl groups, respectively. The various glutathione transferases use glutathione itself as the donor.

Factors Affecting Enzymes Metabolizing Xenobiotics

Various factors affect the activities of the enzymes metabolizing xenobiotics. Such factors include the following: (1) The activities of these enzymes may differ substantially among **species.** (2) There are also significant differences in enzyme activities among individuals, many of which appear to be due to **genetic factors.** (3) The activities of some of these enzymes vary according to **age** and **sex.** (4) Intake of various xenobiotics, such as phenobarbital, PCBs, or certain hydrocarbons, can also increase the activities of a number of these enzymes—a process known as **enzyme induction.** For instance, inhalation of hydrocarbons from cigarette smoking during pregnancy appears to induce the activity of cytochrome P-448 in the placenta, thus altering the amounts of certain metabolites of hydrocarbons to which the fetus is exposed. (5) Metabolites of certain **drugs** can inhibit the activities of xenobiotic-metabolizing enzymes. The above considerations help explain the often appreciable differences in the carcinogenicity of chemicals among different species and individuals of the same species.

Covalent Binding

When chemical carcinogens are administered to animals or placed in cultured cells, it can be shown (eg, by using radioactive carcinogens) that they or their derivatives generally bind covalently to cellular macromolecules, including DNA, RNA, and proteins. The chemical nature of the adducts formed by interaction of certain ultimate carcinogens with their target molecules have been determined. Most interest has focused on products formed with DNA. Carcinogens have been found to interact with the purine, pyrimidine, or phosphodiester groups of DNA. The most common site of attack is guanine, and the addition of various carcinogens to the N_2, N_3, N_7, O_6, and O_8 atoms of this base has been observed.

Damage to DNA

The covalent interaction of direct carcinogens or ultimate carcinogens with DNA can result in several types of damage; this damage can be repaired, as described in Chapter 38.

Despite the existence of repair systems, certain modifications of DNA by chemical carcinogens persist for relatively long periods of time. It is possible that these persistent unrepaired lesions are of special importance in generating mutations critical to carcinogenesis.

Mutagens

Most chemical carcinogens are mutagens. This has been demonstrated using the Ames assay (see below) and other tests. At a molecular level, transitions, transversions, and other types of mutation (see Chapters 38 and 40) have been shown to occur following exposure of certain bacteria to ultimate carcinogens. It has been assumed that some types of cancer are due to mutations in somatic cells that affect key regulatory processes. Direct evidence of this has now been obtained (see Oncogenes, below).

Since testing the carcinogenicity of chemicals in animals is slow and expensive, assays for screening the potential carcinogenicity of chemical compounds have been developed. Many are based on detection of the mutagenicity of chemical carcinogens. Such assays are more rapid and less expensive than detecting tumors in animals. None is ideal, since the ultimate test of a carcinogen is to show that it causes tumors in animals. However, one assay based on detecting mutagenicity, the **Ames assay,** has proved useful in screening for potential carcinogens. This assay uses a specially constructed strain of *Salmonella typhimurium* that has a mutation (His^-) in a gene that codes for one of the enzymes involved in the synthesis of histidine. Thus, these particular salmonellae cannot synthesize histidine, which must be present in the medium for growth to occur. When a mutation caused by a carcinogen occurs at the site of the His^- mutation, the latter mutation can restore its reading sequence, converting it to His^+. The progeny from bacteria containing such a reverse mutation can now synthesize histidine and thus grow in a medium lacking it. Such salmonellae can be detected as readily observable and quantifiable colonies growing on agar plates.

One problem with the use of bacteria in mutagenicity tests is that they do not contain the spectrum of monooxygenases found in higher animals. Thus, if a compound requires activation to become a mutagenic or carcinogenic species, this may not occur when bacteria are used. Ames circumvented this problem by incubating the agents to be tested in a postmitochondrial supernatant of rat liver (the S-9 fraction, which is the supernatant fraction after centrifuging a rat liver homogenate at 9000 g for a suitable period of time). The S-9 fraction contains most of the various monooxygenases and other enzymes required to activate potential mutagens and carcinogens.

The Ames assay identifies approximately 90% of known carcinogens. It is becoming routine to test newly synthesized chemicals by this assay, particularly if they are to be introduced commercially or widely used in industry. Compounds giving a positive reaction should undergo further testing, including assessment of carcinogenicity in animals.

Initiation & Promotion

In certain organs such as skin and liver, it has been shown that carcinogenesis can be divided into at least 2 stages. The classic example is skin. Typically, identical areas of the skin of a group of mice are painted once with benzo[a]pyrene. If no other subsequent treatment is used, no skin tumors develop (Fig 57–2). However, if the application of benzo[a]pyrene is followed by several applications of croton oil, many tumors subsequently develop. Applications of croton oil alone (ie, no pretreatment with benzo[a]pyrene) do not result in skin tumors. Many other variants of this basic protocol have been carried out, permitting the following conclusions: (1) The stage of carcinogenesis

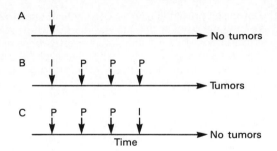

Figure 57–2. Diagrammatic representation of the stages of initiation and promotion of chemical carcinogenesis in skin. *A:* One application of the initiator (eg, benzo[a]pyrene) is made to the skin of a number of mice. *B:* The application of the initiator is followed by a number of applications of a promoter (eg, croton oil) at, for instance, weekly intervals. *C:* The promoter is applied first, and then the initiator is applied. Benign tumors of the skin (papillomas) may appear in about 100 days; malignant tumors (carcinomas) take about 1 year to appear. Many other informative variants of this protocol have been performed, all substantiating the basic concepts of initiation and promotion. I, initiator; P, promoter.

caused by application of benzo[a]pyrene is called initiation; this stage appears to be rapid and irreversible. It is presumed to involve an irreversible modification of DNA, perhaps resulting in one or more mutations. Benzo[a]pyrene is thus called an **initiating agent.** (2) The second, much slower (ie, months or years) stage of carcinogenesis, resulting from application of croton oil, is called promotion. Croton oil is thus a promoting agent, or **promoter.** Promoters are incapable of causing initiation. (3) Most carcinogens are capable of acting as both initiating and promoting agents.

A large number of compounds, including phenobarbital and saccharin, can act as promoters in various organs. The active agent of croton oil is a mixture of phorbol esters. The most active phorbol ester is 12-0-tetradecanoylphorbol-13-acetate (TPA), which has numerous effects. The most interesting finding has been that protein kinase C can act as a receptor for TPA. Stimulation of the activity of this enzyme by interaction with TPA may result in the phosphorylation of a number of membrane proteins, leading to effects on transport and other functions. This important result ties in the action of certain tumor promoters to the field of transmembrane signaling (see Growth Factors, below). Many tumor promoters appear to act by causing alterations of gene expression, but the precise mechanisms by which promoters influence the initiated cell to become a tumor cell remain to be determined.

Role of DNA

DNA is the critical target molecule in carcinogenesis. The following facts support this conclusion. (1) Cancer cells beget cancer cells, ie, the essential changes responsible for cancer are transmitted from mother to daughter cells. This is consistent with the behavior of DNA. (2) Both irradiation and chemical carcinogens damage DNA and are capable of causing mutations in DNA. (3) Many tumor cells exhibit abnormal chromosomes. (4) Transfection experiments (see below) indicate that purified DNA (oncogenes) from cancer cells can transform normal cells into (potential) cancer cells. However, epigenetic factors may also play a role in carcinogenesis.

ONCOGENIC VIRUSES

Oncogenic viruses contain either DNA or RNA as their genome (Table 57–4). Only a few important features of the major members of these 2 classes will be described here.

Polyomavirus and **SV40 viruses** have played an important role in the development of current ideas about viral oncogenesis. They are both small (containing a genome of about 5 kb), and their circular genomes code for only about 5–6 proteins. Under certain circumstances, infection of appropriate cells with these viruses can result in malignant transformation. Specific viral proteins are known to be involved. In the case of SV40, these proteins (often called **antigens,** because they were detected by immunologic methods) are known as T ("large T") and t ("small t"), and in the case of polyomavirus, they are known as T, mid-T, and t. (T refers to the fact that the first of these proteins was detected in a tumor.) How these proteins cause malignant transformation is still under investigation; the T antigens are known to bind tightly to DNA and cause alterations in gene expression. These proteins show cooperative effects, suggesting that alteration of more than one reaction or process is required for transformation.

Some types of adenovirus are known to cause transformation of certain animal cells. There is considerable interest in the Epstein-Barr virus, since it is associated with Burkitt's lymphoma and nasopharyngeal carcinoma in humans. Herpes simplex virus (type 2) has been associated with cancer of the cervix, and hepatitis B virus may be associated with some cases of liver cancer in humans.

Since much of the knowledge of oncogenes obtained in recent years has emerged from the study of RNA-containing tumor viruses, the subsequent discussion of oncogenes contains frequent references to these viruses.

TRANSFORMATION

When cultured cells are infected with certain oncogenic viruses, they may undergo malignant transformation. The most important morphologic and biochemical changes occurring upon transformation are listed in Table 57–5. These changes affect cell shape, motility, adhesiveness to the culture dish, growth, and a number of biochemical processes. They are interpreted as reflecting the primary processes that cause—and the secondary changes that result from—conversion from the normal to the malignant state. The ability to approximately equate transformation with acquisition of malignant properties has been of tremendous importance in cancer research. However, acquisition by cells of the changes collectively known as transformation does not necessarily mean that such cells will display the same biologic properties as tumor cells in vivo; cells must yield tumors when injected into a suitable host animal.

ONCOGENES

Oncogenes are genes capable of causing cancer. Their discovery has had a major impact on research on the fundamental mechanisms involved in carcinogenesis. Oncogenes were first recognized as unique genes of tumor-causing viruses that are responsible for the process of transformation (**viral oncogenes).**

Table 57–4. Some important tumor viruses.

Class	Members
DNA viruses	
Papovavirus	Polyomavirus, SV40 virus, papillomavirus
Adenovirus	Adenoviruses 12, 18, and 31
Herpesvirus	Epstein-Barr virus, herpes simplex type 2 virus
Hepadnavirus	Hepatitis B virus
RNA viruses	
Retrovirus type C	Murine sarcoma and leukemia viruses, avian sarcoma and leukemia viruses, human T cell leukemia viruses I and II
Retrovirus type B	Mouse mammary tumor virus

Table 57–5. Some changes shown by cultured cells which suggest that malignant transformation has occurred (eg, after infection by an oncogenic virus). The crucial test of malignancy is the ability of cells to grow into tumors in vivo.

Alterations of morphology: Transformed cells often have a much rounder shape than control cells.
Increased cell density (loss of contact inhibition of growth): Transformed cells often form multilayers, while control cells usually form a monolayer.
Loss of anchorage dependence: Transformed cells can grow without attachment to the surface of the culture dish and will often grow in agar.
Loss of contact inhibition of movement: Transformed cells grow over one another, while normal cells stop moving when they come into contact with each other.
A variety of biochemical changes, including an increased rate of glycolysis, alterations of the cell surface (eg, changes in the composition of glycoproteins or glycosphingolipids), and secretion of certain proteases.
Alterations of cytoskeletal structures, such as actin filaments.
Diminished requirement for growth factors and, often, increased secretion of certain growth factors into the surrounding medium.

Oncogenes of Rous Sarcoma Virus

Analyses of the oncogene of the Rous sarcoma virus and its product have been particularly revealing. The genome of this retrovirus contains 4 genes named *gag, pol, env,* and *src.* This can be shown schematically as follows:

The *gag* gene codes for group-specific antigens of the virus, *pol* for the reverse transcriptase that characterizes retroviruses, and *env* for certain glycoproteins of the viral envelope. A **protein-tyrosine kinase** was shown to be the product of *src* (ie, the sarcoma-causing gene) that is responsible for transformation. This finding was of fundamental importance. It revealed a specific biochemical mechanism (ie, abnormal phosphorylation of a number of proteins) that could explain, at least in part, how a tumor virus could cause the **pleiotropic effects** of transformation. The **critical cell proteins,** whose abnormal phosphorylation presumably leads to transformation, are still to be defined. **Vinculin,** a protein found in focal adhesion plaques (structures involved in intercellular adhesion), is one candidate. The abnormal phosphorylation of vinculin in focal adhesion plaques could help explain the rounding-up of cells and their diminished adhesion to the substratum and to one another observed during transformation (Table 57–5). Certain **glycolytic enzymes** appear to be target proteins for the *src* protein-tyrosine kinase; this is in keeping with the observation that transformed cells often show increased rates of glycolysis. The product of *src* may also catalyze phosphorylation of **phosphatidylinositol** to phosphatidylinositol mono- and bisphosphate. When phosphatidylinositol 4,5-bisphosphate is hydrolyzed by the action of phospholipase C, 2 second messengers are released: inositol triphosphate and diacylglycerol (see Chapter 44). The first compound mediates release of Ca^{2+} from intracellular sites of storage (eg, the endoplasmic reticulum). Diacylglycerol stimulates the activity of the plasma membrane-bound protein kinase C, which in turn phosphorylates a number of proteins, some of which may be components of ion pumps. Specifically, it has been proposed that mild alkalinization of the cell, brought about by activation of an Na^+/H^+ antiport system (see Chapter 42), could play a role in stimulating mitosis. Thus, the product of *src* may affect a large number of cellular processes by its ability to phosphorylate various target proteins and enzymes and by stimulating the pathway of synthesis of the polyphosphoinositides.

Protein-Tyrosine Kinases in Normal & Transformed Cells

The observation that Rous sarcoma virus contained a protein-tyrosine kinase stimulated much research on the phosphorylation of tyrosine. It is now known that many normal cells contain protein-tyrosine kinase activity. The amount of phosphotyrosine in most normal cells is low but is usually elevated in cells transformed by an oncogenic virus containing a protein-tyrosine kinase, although the amount is still relatively small ($\sim$ 1% of the total phosphoamino acids [mainly phosphoserine, phosphothreonine, and phosphotyrosine] in such cells). Certain receptors (eg, for epidermal growth factor, insulin, and platelet-derived growth factor) found in both normal and transformed cells have protein-tyrosine activities that are stimulated upon interaction with their ligands (see Growth Factors, below). Protein-tyrosine kinase activities thus play important roles in both normal and transformed cells.

Oncogenes of Other Retroviruses

In addition to the oncogenes of Rous sarcoma virus, approximately 20 oncogenes of other retroviruses have been recognized. About half of the products of these viral oncogenes are protein kinases, mostly of the tyrosine type. Some viral oncogenes are listed in Table 57–6, along with their products. While some of those listed encode protein kinases, the remainder encode various other proteins with interesting

Table 57–6. Some oncogenes of retroviruses.*

Oncogene	Retrovirus	Origin	Oncogene Product	Subcellular Location
abl	Abelson murine leukemia virus	Mouse	Protein-tyrosine kinase	Plasma membrane
erb-B	Avian erythroblastosis virus	Chicken	Truncated EGF receptor	Plasma membrane
fes	Feline sarcoma virus	Cat	Protein-tyrosine kinase	Plasma membrane
fos	Murine sarcoma virus	Mouse	?	Nucleus
myc	Myelocytoma virus 29	Chicken	DNA-binding protein	Nucleus
sis	Simian sarcoma virus	Monkey	Truncated PDGF (B chain)	Membranes ? Secreted
src	Rous sarcoma virus	Chicken	Protein-tyrosine kinase	Plasma membrane

Tyr, tyrosine; EGF, epidermal growth factor; PDGF, platelet-derived growth factor.
*Modified and reproduced, with permission, from Franks LM, Teich NM (editors): *Introduction to the Cellular and Molecular Biology of Cancer.* Oxford Univ Press, 1986.

biologic activities. The product of the *erb*-B gene of avian erythroblastosis virus is a truncated form of the receptor for epidermal growth factor, and that of the *sis* oncogene of simian sarcoma virus is a truncated B chain of platelet-derived growth factor. The product of the oncogene *(fms)* of one type of viral isolate of feline sarcoma virus is a macrophage colony-stimulating factor. On the other hand, the product of the *myc* oncogene, originally found in chicken myelocytoma viruses, is a DNA-binding protein, which may affect the control of mitosis. The product of the *ras* oncogene of murine sarcoma viruses binds GTP, has GTPase activity, and appears to be related to the proteins that regulate the activity of the important plasma membrane enzyme, adenylate cyclase (see Chapter 44).

Proto-oncogenes

A key issue raised by the discovery of viral oncogenes relates to their origin. Use of nucleic acid hybridization (see Chapter 36) revealed that normal cells contained DNA sequences similar—if not identical—to those of the viral oncogenes. Thus, the viruses apparently incorporated cellular genes into their genomes during their passages through cells. The retention of such genes in their genomes indicated that they must confer a selective advantage on the affected viruses, presumably related to the altered growth properties of transformed cells.

The cellular sequences were found to be conserved in a wide range of eukaryotic cells, suggesting that they were important components of normal cells. In addition, mRNA species and proteins derived from these normal sequences could be detected at various stages of their development or life cycles. The genes present in normal cells were thus designated as proto-oncogenes, and their products are believed to play important roles in normal differentiation and other cellular processes.

Oncogenes From Tumor Cells

Experiments using DNA extracted from tumors have also provided evidence for the existence of oncogenes. The method used for detecting such cellular oncogenes is called **gene transfer** or **DNA transfection.** It depends on the fact that certain genes present in tumors can cause transformation of "normal" cultured cells. DNA is isolated from tumor cells and added to recipient cells, often a line of mouse fibroblasts known as NIH/3T3 cells. DNA isolated from tumor cells is precipitated with calcium phosphate (to facilitate endocytosis) and added to NIH/3T3 cells in tissue culture. The cells are observed microscopically over a period of 1–2 weeks for formation of foci of transformed cells. If transformation occurs, the NIH/3T3 cells change their morphology from flat to rounded cells that grow in characteristic foci. The procedure is repeated several times using DNA extracted from the transformed cells, thus reducing the amount of DNA not involved in transformation that was transfected and facilitating identification (eg, by the Southern blotting technique, using a suitable probe [see

Chapter 36]) of the specific gene involved. Some 20 or so different cellular oncogenes have been recognized in this manner, with a number of them being related to the *ras* oncogene of murine sarcoma viruses. These cellular oncogenes either are identical to normal genes or show very small structural differences from their normal counterparts (see below). In the former case, the regulation of their expression may be abnormal in cancer cells.

Abbreviations for Cellular & Viral Oncogenes

The abbreviation c-*onc* (cellular oncogene, eg, c-*ras*) is used to designate an oncogene present in tumor cells. The species present in normal cells—ie, its proto-oncogene—can be conveniently referred to as the corresponding c-*onc* proto-oncogene (eg, the c-*ras* proto-oncogene). Similarly, a viral oncogene is designated v-*onc* (viral oncogene, eg, v-*ras*), with its proto-oncogene being referred to as a v-*onc* proto-oncogene (eg, the v-*ras* proto-oncogene).

Mechanisms by Which Proto-oncogenes Become Oncogenes

Five mechanisms will be discussed that alter the expression or structure of proto-oncogenes and thus participate in their becoming oncogenes. For the sake of convenience, the process whereby transcription of a gene is increased (from zero or a relatively low level) will be designated as **activation.** Familiarity with the mechanisms involved in activation is crucial for understanding contemporary thinking about carcinogenesis.

A. Promoter Insertion: Certain retroviruses lack oncogenes (eg, avian leukemia viruses) but may cause cancer over a longer period of time—months rather than days—than those which do contain oncogenes. As for other retroviruses, when these particular viruses infect cells, a DNA copy (cDNA) of their RNA genome is synthesized by reverse transcriptase, and the cDNA is integrated into the host genome. The integrated double-stranded cDNA is called a **provirus.** The cDNA copies of retroviruses are flanked at both ends by sequences named **long terminal repeats,** as are certain **transposons ("jumping genes")** found in bacteria and plants (see Chapter 38). The long terminal repeat sequences appear to be important in the mechanism of proviral integration, and they can act as promoters of transcription (see Chapter 39). Following infection of chicken B lymphocytes by certain avian leukemia viruses, their proviruses become integrated near the *myc* gene. The *myc* gene is activated by an upstream, adjacent viral long terminal repeat acting as a promoter, resulting in transcription of both the corresponding *myc* mRNA and translation of its product in such cells (Fig 57–3). A B cell tumor ensues, although the precise role of the products of the *myc* gene in the overall process is not clear. Similar events occur following infection of various cells with other retroviruses.

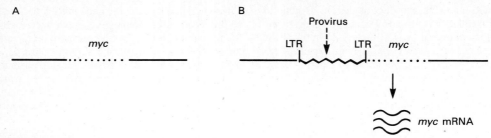

Figure 57–3. Schematic repesentation of how promoter insertion may activate a proto-oncogene. *A:* Normal chicken chromosome, showing an inactive *myc* gene. *B:* An avian leukemia virus has integrated in the chromosome in its proviral form, adjacent to the *myc* gene. Its right-hand long terminal repeat (LTR), containing a strong promoter, lies just upstream of the *myc* gene and activates that gene, resulting in transcription of *myc* mRNA. For simplicity, only one strand of DNA is depicted and other details have been omitted.

B. Enhancer Insertion: In some cases, the provirus is inserted downstream from the *myc* gene, or upstream from it but oriented in the reverse direction; nevertheless, the *myc* gene becomes activated (Fig 57–4). Such activation cannot be due to promoter insertion, since a promoter sequence must be upstream of the gene whose transcription it increases and the sequence must be in the correct 5′ to 3′ direction. Instead, enchancer sequences (see Chapters 39 and 41) present in the long terminal repeat sequences of the retroviruses under consideration appear to be involved.

The above 2 mechanisms—promoter and enhancer insertion—commonly operate in viral carcinogenesis. Proto-oncogenes other than *myc* are also probably involved.

C. Chromosomal Translocations: As mentioned earlier, many tumor cells exhibit chromosomal abnormalities. One type of chromosomal change seen in cancer cells is translocation. The basis of a translocation is that a piece of one chromosome is split off and then joined to another chromosome. If the second chromosome donates material to the first, the translocation is said to be **reciprocal.** Characteristic translocations are found in a number of tumor cells. One im-

portant translocation is the **Philadelphia chromosome,** involving chromosomes 9 and 22 and occurring in chronic granulocytic leukemia.

Burkitt's lymphoma is a fast-growing cancer of human B lymphocytes. In certain cases, an example of a reciprocal translocation (Fig 57–5) is found that has illuminated the mechanisms of activation of potential cellular oncogenes. Chromosomes 8 and 14 are involved. The segment of chromosome 8 that breaks off and moves to chromosome 14 contains the *myc* gene. As shown in Fig 57–6, the transposition places the previously inactive *myc* gene under the influence of enhancer sequences in the genes coding for the heavy chains of immunoglobulins. This juxtaposition results in activation of transcription of the *myc* gene. Apparently, synthesis of greatly increased amounts of the DNA-binding protein coded for by the *myc* gene acts to "drive" or "force" the cell toward becoming malignant, perhaps by an effect on the regulation of mitosis. This mechanism is similar to enhancer insertion, except that chromosomal translocation (rather than integration of provirus) is responsible for placing the proto-oncogene (ie, *myc*) under the influence of an enhancer.

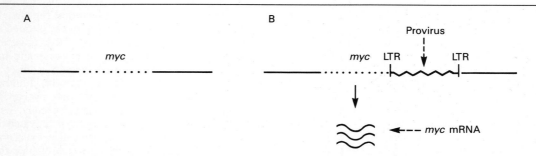

Figure 57–4. Schematic representation showing how enhancer insertion may activate a proto-oncogene. *A:* Normal chicken chromosome, showing an inactive *myc* gene. *B:* An avian leukemia virus has integrated in the chromosome in its proviral form, adjacent to the *myc* gene. However, in this instance, the site of integration is just downstream of the *myc* gene and it cannot act as a promoter (Fig 57–6). Instead, a certain proviral sequence acts as an enhancer element, leading to activation of the upstream *myc* gene and its transcription. For simplicity, only one strand of DNA is depicted and other details have been omitted.

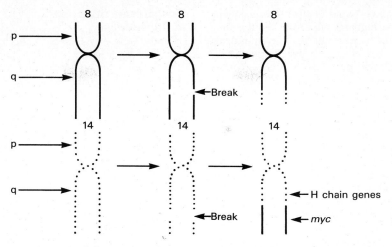

Figure 57–5. Schematic representation of the reciprocal translocation involved in Burkitt's lymphoma. The chromosomes involved are 8 and 14. A segment from the end of the q arm of chromosome 8 breaks off and moves to chromosome 14. The reverse process moves a small segment from the q arm of chromosome 14 to chromosome 8. The *myc* gene is contained in the small piece of chromosome 8 that was transferred to chromosome 14; it is thus placed next to genes transcribing the heavy chains of immunoglobulin molecules and itself becomes activated.

D. Gene Amplification: Amplification of certain genes (see Chapter 38) is found in a number of tumors. One method of bringing about gene amplification in tumors is by administration of the anticancer drug methotrexate, an inhibitor of the enzyme dihydrofolate reductase. Tumor cells can become resistant to the action of this drug. The basis of this phenomenon is that the gene for dihydrofolate reductase becomes amplified, resulting in an increase of the activity of the enzyme (up to 400-fold). The amplified genes, measuring up to 1000 kb or more in length, may be detected as **homogeneously staining regions** on a specific chromosome. Alternatively, they are detected as **double-minute chromosomes,** which are minichromosomes lacking centromeres. The precise relationship of homogeneously staining regions to double-minute chromosomes is under investigation. Certain cellular oncogenes can also be amplified in like manner and are thus activated. There is evidence suggesting that increased amounts of the products of certain oncogenes (such as c-*ras*) produced by gene amplification may play a role in the progression of tumor cells to a more malignant state (see Progression of Tumors, below).

E. Single-Point Mutation: The v-*ras* oncogene was originally detected in certain murine (ie, rat and mouse) retroviruses. Its product, a protein (p21) of MW 21,000, appears to be related to the G proteins that modulate the activity of adenylate cyclase (see above and Chapter 44) and thus play a key role in cellular responses to many hormones and drugs. Analyses by DNA sequencing of the c-*ras* proto-oncogene

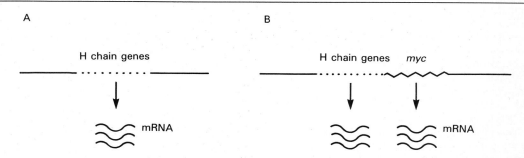

Figure 57–6. Schematic representation showing how the translocation involved in Burkitt's lymphoma may activate the *myc* proto-oncogene. **A:** A small segment of chromosome 14 prior to the translocation. The segment shown contains the genes encoding regions of heavy chains of immunoglobulins. **B:** Following the translocation, the previously inactive *myc* gene is placed under the influence of enhancer sequences in the genes encoding the heavy chains and is thus activated, resulting in transcription. For simplicity, only one strand of DNA is depicted and other details have been omitted.

from normal human cells and of the c-*ras* oncogene from a cancer of the human bladder showed that they differed solely in one base, resulting in an amino acid substitution at position 12 of p21. This intriguing result has been confirmed by analyses of c-*ras* genes from other human tumors. In each case the results were consistent; the gene isolated from the tumor exhibited only a single-point mutation, in comparison with the c-*ras* proto-oncogene from normal cells. The position of the mutation varied somewhat, so that other amino acid substitutions were observed. These mutations in p21 appear to affect its conformation and to diminish its activity as a GTPase. The lower activity of GTPase could result in chronic stimulation of the activity of adenylate cyclase, which normally is diminished when GDP is formed from GTP (see Chapter 44). The resulting stimulation of the activity of adenylate cyclase can result in a number of effects on cellular metabolism exerted by the increased amount of cAMP affecting the activities of various cAMP-dependent protein kinases. These events may assist in tipping the balance of cellular metabolism toward a state favoring transformation or its maintenance.

General Comments on Activation of Oncogenes

Of the 5 mechanisms described above, the first 4 (promoter insertion, enhancer insertion, chromosome translocation, and gene amplification) involve an increase in amount of the product of an oncogene due to increased transcription but no alteration of the structure of the product of the oncogene. Thus, it appears that increased amounts of the product of an oncogene may be sufficient to push a cell toward becoming malignant. The fifth mechanism, single-point mutation, involves a change in the structure of the product of the oncogene but not necessarily any change in its amount. This finding implies that the presence of a structurally abnormal key regulatory protein in a cell may also be sufficient to tip the scales toward cancer.

When considering the role of oncogenes in cancer, it is important to bear in mind that oncogenes have been isolated from only about 15% of human tumors. It is likely that their activation in at least some cases may be only a secondary occurrence associated with transformation, rather than a causal event. Their involvements in chemically induced experimental tumors are just beginning to be explored. However, recent work has shown that activation of c-*ras* in rat mammary cancers induced by nitrosomethylurea was apparently due to a specific $G \rightarrow A$ transition type of mutation, demonstrating that oncogenes are probably involved in chemical carcinogenesis. In addition, because a single dose of nitrosomethylurea was used (without any promoter), the above mutation may be an important event in the initiation stage of chemical carcinogenesis. Much further research is needed to examine the possible involvement of oncogenes in the phenomena of initiation, promotion, tumor progression, and metastasis.

Mechanisms of Action of Oncogenes

There are at least 3 mechanisms by which the products of oncogenes may stimulate growth (Fig 57–7). (1) **They may act on key intracellular pathways involved in growth control,** uncoupling them from the need for an exogenous stimulus. Relevant examples (described above) are the product of *src* acting as a protein-tyrosine kinase, the product of *ras* acting to stimulate the activity of adenylate cyclase, and the product of *myc* acting as a DNA-binding protein. Each of these could affect the control of mitosis, the first 2 by events involving phosphorylation of key regulatory proteins. A major deficiency in our knowledge of cell growth is that remarkably little is known about molecular aspects of the regulation of mitosis, even in normal cells. (2) **The products of oncogenes may also imitate the action of a polypeptide growth factor** or (3) **imitate an occupied receptor for a growth factor** (see below).

POLYPEPTIDE GROWTH FACTORS

The subject of growth factors is one of the most rapidly developing in contemporary biomedical science. A variety of such factors have been isolated and partly characterized (Table 57–7). Until recently, only very small quantities of most growth factors were available for study. However, the genes for a number of growth factors have now been cloned, confirming their separate identities and making usable amounts available through recombinant DNA technology. The growth factors known to date affect many different types of cells, eg, cells from the blood, nervous system, mesenchymal tissues, and epithelial tissues. They exert a mitogenic response on their target cells, although special conditions may be necessary to demonstrate this, such as depriving cells in culture of serum so that they have become quiescent prior to exposure to a growth factor. **Platelet-derived growth factor (PDGF),** released from the α granules of platelets, probably plays a role in normal wound healing. Various growth factors appear to play key roles in regulating differentiation of stem cells to form various types of mature hematopoietic cells. Growth inhibitory factors also exist (eg, **transforming growth factor [TGF-β]** may exert inhibitory effects on the growth of certain cells). Thus, chronic exposure to increased amounts of a growth factor or to decreased amounts of a growth inhibitory factor could alter the balance of cellular growth.

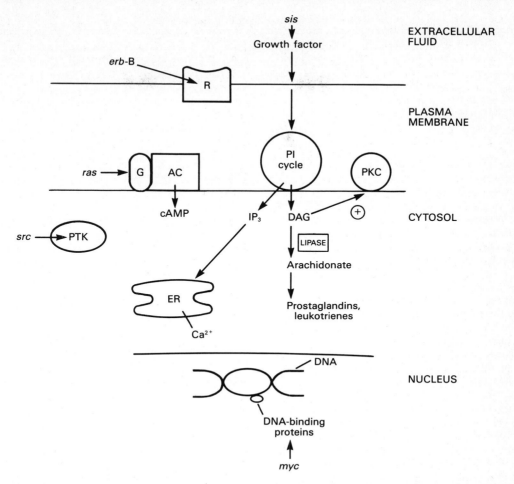

Figure 57–7. Schematic representation of mechanisms by which the products of certain oncogenes may alter cellular metabolism and thereby stimulate growth. Cyclic AMP can affect a number of cellular processes by activating cAMP-dependent protein kinases. Protein-tyrosine kinases and protein kinase C both can affect the activities of a number of target proteins. Ca^{2+} has numerous cellular effects, as do prostaglandins and leukotrienes produced from arachidonate. R, receptor; G, G protein; AC, adenylate cyclase; cAMP, cyclic AMP; PI, phosphatidylinositol; PKC, protein kinase C; PTK, protein-tyrosine kinase; IP_3, inositol triphosphate; DAG, diacylglycerol; ER, endoplasmic reticulum.

ENDOCRINE, PARACRINE, & AUTOCRINE ACTIONS OF GROWTH FACTORS

Growth factors may operate in 3 general ways (see also Chapter 43): (1) Their effects may be endocrine; that is to say, like hormones, they may be synthesized elsewhere in the body and pass in the circulation to their target cells. (2) They may be synthesized in certain cells and secreted from them to affect neighboring cells. However, the cells that synthesize the growth factor are not themselves affected, because they lack suitable receptors. This mode of action is called paracrine. (3) Certain growth factors can affect the cells that synthesize them. This third mode of action is called autocrine. For instance, a factor may be secreted and then attach to its cell of origin, provided

that cell possesses appropriate receptors. Alternatively, if a certain amount of the factor is not secreted, its presence inside the cell may directly stimulate various processes.

BIOCHEMICAL MECHANISMS OF ACTION OF GROWTH FACTORS

Relatively little is known about how growth factors operate at the molecular level. Like polypeptide hormones (see Chapter 44), they must transmit a message across the plasma membrane to the interior of the cell (**transmembrane signal transduction**). In the case of growth factors, the message will ultimately affect one or more processes involved in mitosis. Most growth factors have high-affinity protein receptors on the

Table 57–7. Some polypeptide growth factors.*

Growth Factor	Source	Function
Epidermal growth factor (EGF)	Mouse salivary gland	Stimulates growth of many epidermal and epithelial cells
Erythropoietin	Kidney, urine	Regulates development of early erythropoietic cells
Insulinlike growth factors I and II (IGF-I and IGF-II, also named somatomedins C and A)	Serum	Stimulate sulfate incorporation into cartilage, are mitogenic for chondrocytes, and exert insulinlike effects on many cells
Interleukin-1 (IL-1)	Conditioned media	Stimulates production of IL-2
Interleukin-2 (IL-2)	Conditioned media	Stimulates growth of T cells
Nerve growth factor (NGF)	Mouse salivary gland	Tropic effect on sympathetic and certain sensory neurons
Platelet-derived growth factor (PDGF)	Platelets	Stimulates growth of mesenchymal and glial cells
Transforming growth factor (TGF-α)	Conditioned media of transformed or tumor cells	Is similar to EGF
Transforming growth factor (TGF-β)	Kidney, platelets	Exerts both stimulatory and inhibitory effects on certain cells

*Modified and reproduced, with permission, from Franks LM, Teich NM (editors): *Introduction to the Cellular and Molecular Biology of Cancer.* Oxford Univ Press, 1986.

plasma membrane of target cells. The genes for the receptors for **epidermal growth factor (EGF)** and insulin have been cloned and models of the structures of the receptors constructed. They have short membrane-spanning segments and external and cytoplasmic domains of varying lengths. The ligands bind to the external domains. A number of receptors (eg, for EGF, insulin, and PDGF) have been found to exhibit protein-tyrosine kinase activities, reminiscent of the product of the v-*src* gene (see above). This kinase activity, located in the cytoplasmic domains, causes autophosphorylation of the receptor protein and also phosphorylates other target proteins. The receptor-ligand complexes are subjected to endocytosis in coated vesicles (see LDL receptor, Chapter 27); it is not yet clear whether the receptors recirculate to the cell surface. The precise events resulting in transmembrane signaling are still under investigation and differ among the various factors. The case of PDGF will be described as an example. Phospholipase C is stimulated following exposure of cells to PDGF, resulting in hydrolysis of phosphatidylinositol 4,5-bisphosphate to form inositol triphosphate and diacylglycerol (see v-*src*, above, and Fig 44–5). These 2 second messengers can effect intracellular release of Ca^{2+} and stimulation of the activity of protein kinase C, respectively, thus affecting a large number of cellular reactions. The subsequent hydrolysis of diacylglycerol by phospholipase A_2, liberating arachidonic acid, can also result in the production of prostaglandins and leukotrienes, which themselves may exert many biologic activities (see Chapter 24). Exposure of cells to PDGF can result in rapid (minutes to 1–2 hours) activation of certain cellular proto-oncogenes (eg, c-*myc* and c-*fos*). It seems likely that gene activation, whether of normal genes or proto-oncogenes, is involved in the action of most growth factors.

GROWTH FACTORS & ONCOGENES

The products of several oncogenes are either growth factors or parts of the receptors for growth factors (Table 57–6).

The B chain of PDGF contains 109 amino acids. It seems likely that the B chain is biologically active as a homodimer, without involvement of the A chain. The discovery that v-*sis* encodes 100 of the 109 amino acids of the B chain of PDGF revealed a direct relationship between oncogenes and growth factors. It also suggested that autocrine stimulation by PDGF—supplying a chronic mitogenic stimulus—could be an important factor in the mechanism of transformation of cells by v-*sis*. Indeed, many cultured tumor cells are known to secrete growth factors into their surrounding media and also to possess receptors for these molecules.

Sequence analysis of v-*erb*-B revealed that it encoded a truncated form of the receptor for EGF, with much of the external domain of the receptor being deleted but the protein-tyrosine kinase activity being retained. It has been suggested that the abnormal form of the receptor for EGF encoded by v-*erb*-B may be continuously active when present in cells, simulating an occupied receptor. As in the case of autocrine stimulation by PDGF, this could result in a chronic mitogenic signal, "driving" cells toward the transformed state.

Transforming growth factor (TGF-β) was originally thought to be a positive growth factor, since it caused fibroblasts to behave as if they were transformed. TFG-β is now known to inhibit the growth of most cell types, except fibroblasts. It will inhibit the growth of the monkey kidney cells that synthesize it. TGF-β may activate the *sis* gene in fibroblasts. How it produces its inhibitory effects on other cells remains to be established. There is other evidence that certain genes code for products that slow cell growth and also that tumor-suppressive genes exist. Thus, the control of cell growth is very complex, and both positive and negative regulatory factors are involved.

PROGRESSION OF TUMORS

Once a cell becomes a tumor cell, the composition and behavior of its progeny do not remain static. Instead, there is a tendency for malignancy to increase. This is manifested by increasingly abnormal karyotypes, increasing rates of growth, and increasing tendency to invade and metastasize. The important phenomenon of progression appears to reflect a fundamental instability of the genome of tumor cells. The activation of additional oncogenes may be involved. Cells with faster rates of growth have a selective advantage.

It is important to distinguish the biochemical profiles of cells that have just been transformed from those of fast-growing, highly malignant tumor cells. The former cells may show few differences from normal cells, apart from key changes leading to cancer (eg, activation of one or more oncogenes) and those changes generally associated with transformation (Table 57–5). Analyses of such cells can disclose the key biochemical alterations that result in transformation. The biochemical profile of highly malignant cells may be very different from that of normal cells. Many changes in enzyme profile and other biochemical parameters have occurred (Table 57–8), some of which are secondary to rapid growth rate and others probably due to chromosomal instability. These fast-growing cells tend to maximize the anabolic processes involved in growth (eg, DNA and RNA synthesis), cut down on catabolic functions (eg, catabolism of pyrimidines), and dispense with the differentiated functions shown by their normal ancestors. In other words, they are concentrating almost exclusively upon growth. They also show biochemical changes that reflect altered gene regulation, such as the synthesis of certain fetal proteins (some of which can be used as tumor markers; see Table 57–1) and the inappropriate manufacture of growth factors or hor-mones. Analyses of such cells are unlikely to reveal the initial key events responsible for transformation, partly because these events are obscured by a myriad of changes secondary to progression. However, knowledge of the biochemical profile of such cells is extremely important in choosing chemotherapy, as it is exactly this type of cell that the oncologist is usually called upon to deal with.

METASTASIS

Metastasis is the spread of cancer cells from a primary site of origin to other tissues where they grow as secondary tumors, and it is the major problem presented by the disease. Metastasis is a complex phenomenon to analyze in humans, and knowledge of its biochemical basis is quite restricted. Because it reflects a failure in cell-cell interaction, much attention has naturally focused on comparisons of the biochemistry of the surfaces of normal and malignant cells. Many changes have been documented at the surfaces of malignant cells (Table 57–9), although not all are directly relevant to the problem of metastasis.

At present, considerable research is devoted to developing suitable animal model systems for the study of metastasis. Many studies also are being done to uncover the possible roles of certain proteases (eg, type 4 collagenase) and of certain glycoproteins and glycosphingolipids of the cell surface in the phenomenon of metastasis. For instance, it is possible that changes in the oligosaccharide chains of cell glycoproteins (secondary to alterations of activities of specific glycoprotein glycosyltransferases) may be critical in permitting metastasis to occur. Elucidation of the biochemical mechanisms involved in metastasis could provide a basis for the rational development of more effective anticancer therapies.

Table 57–8. Biochemical changes often found in fast-growing tumor cells.

Increased activity of ribonucleotide reductase.
Increased synthesis of RNA and DNA.
Decreased catabolism of pyrimidines.
Increased rates of aerobic and anaerobic glycolysis.
Alterations of isozyme profiles, often to a fetal pattern.
Synthesis of fetal proteins (eg, carcinoembryonic antigen).
Loss of differentiated biochemical functions (eg, diminished synthesis of specialized proteins).
Inappropriate synthesis of certain growth factors and hor-mones.

Table 57–9. Some changes that have been detected at the surfaces of malignant cells.*

Alterations of permeability
Alterations in transport properties
Diminished adhesion
Increased agglutinability by many lectins
Alterations of the activities of a number of enzymes (eg, certain proteases)
Alterations of surface charge
Appearance of new antigens
Loss of certain antigens
Alterations of the oligosaccharide chains of glycoproteins
Changes of glycolipid constituents

*Adapted from Robbins JC, Nicolson GL: Surfaces of normal and transformed cells. In: *Cancer: A Comprehensive Treatise.* Vol 4. Becker FF (editor). Plenum Press, 1975.

REFERENCES

Barbacid M: Oncogenes and human cancer: Cause or consequence? *Carcinogenesis* 1986;**7**:1037.

Bishop JM: The molecular genetics of cancer. *Science* 1987;**235**:305.

Croce CM, Klein G: Chromosome translocations and human cancer. *Sci Am* (March) 1985;**252**:54.

Darnell J, Lodish H, Baltimore D: *Molecular Cell Biology.* Scientific American Books, 1986.

Farmer PB, Walker JM (editors): *The Molecular Basis of Cancer.* Croom Helm Ltd., 1985.

Franks LM, Teich N: *Introduction to the Cellular and Molecular Biology of Cancer.* Oxford Univ Press, 1986.

Hunter T: Oncogenes and growth control. *Trends Biochem Sci* 1985;**10**:275.

Hunter T, Cooper JA: Protein-tyrosine kinases. *Annu Rev Biochem* 1985;**54**:897.

Massague J: The transforming growth factors. *Trends Biochem Sci* 1985;**10**:237.

McIntire KR: Tumor markers: How useful are they? *Hosp Pract* (Dec) 1984;**19**:55.

Nicolson GL: Cell surface molecules and tumor metastasis: Regulation of metastatic phenotypic diversity. *Exp Cell Res* 1984;**150**:3.

Pitot HC: *Fundamentals of Oncology,* 3rd ed. Marcel Dekker, 1986.

Sporn MB, Roberts AB: Autocrine growth factors and cancer. *Nature* 1985;**313**:745.

Tannock IF, Hill R (editors): *The Basic Science of Oncology.* Pergamon Press, 1987.

Weinberg RA: The action of oncogenes in the cytoplasm and nucleus. *Science* 1985;**230**:770.

Appendix*

CHEMICAL CONSTITUENTS OF BLOOD & BODY FLUIDS

Validity of Numerical Values in Reporting Laboratory Results

The value reported from a clinical laboratory after determination of the concentration or amount of a substance in a specimen represents the best value obtainable with the method, reagents, instruments, and technical personnel involved in obtaining and processing the material.

Accuracy is the degree of agreement of the determination with the "true" value (eg, the known concentration in a control sample). **Precision** denotes the reproducibility of the analysis and is expressed in terms of variation among several determinations on the same sample. **Reliability** is a measure of the congruence of accuracy and precision.

Precision is not absolute but is subject to variation inherent in the complexity of the method, the stability of reagents, the accuracy of the primary standard, the sophistication of the equipment, and the skill of the technical personnel. Each laboratory should maintain data on precision (reproducibility) that can be expressed statistically in terms of the standard deviation from the mean value obtained by repeated analyses of the same sample. For example, the precision in determination of cholesterol in serum in a good laboratory may be the mean value ± 5 mg/dL. The 95% confidence limits are ± 2 SD, or ± 10 mg/dL. Thus, any value reported is "accurate" within a range of 20 mg/dL. Thus, the reported value 200 mg/dL means that the true value lies between 190 and 210 mg/dL. For the determination of serum potassium with a variance of 1 SD of ± 0.1 mmol/L, values ± 0.2 mmol could be obtained on the same specimen. A report of 5.5 could represent at best the range 5.3—5.7 mmol/L. That is, the 2 results—5.3 and 5.7 mmol/L—might be obtained on analysis of the same sample and still be within the limits of precision of the test.

Physicians should obtain from the laboratory the values for the variation of a given determination as a basis for deciding whether one reported value represents a change from another on the same patient.

Interpretation of Laboratory Tests

Normal values are those that fall within 2 standard deviations from the mean value for the normal population. This normal range encompasses 95% of the population. Many factors may affect values and influence the normal range; by the same token, various factors may produce values that are normal under the prevailing conditions but outside the 95% limits determined under other circumstances. These factors include **age; race; sex; environment; posture; diurnal** and **other cyclic variations; fasting** or **postprandial state, foods eaten; drugs;** and **level of exercise.**

Normal or reference values vary with the method employed, the laboratory, and conditions of collection and preservation of specimens. The normal values established by individual laboratories should be clearly expressed to ensure proper interpretation.

Interpretation of laboratory results must always be related to the condition of the patient. A low value may be the result of deficit or of dilution of the substance measured, eg, low serum sodium. Deviation from normal may be associated with a specific disease or with some drug consumed by the subject—eg, elevated serum uric acid levels may occur in patients with gout or may be due to treatment with chlorothiazides or with antineoplastic agents. The reader should consult an appropriate text for lists of drugs interfering with chemical tests.

Values may be influenced by the method of collection of the specimen. Inaccurate collection of a 24-hour urine specimen, variations in concentration of the randomly collected urine specimen, hemolysis in a blood sample, addition of an inappropriate anticoagulant, and contaminated glassware or other apparatus are examples of causes of erroneous results.

Note: Whenever an unusual or abnormal result is obtained, all possible sources of error must be considered before responding with therapy based on the laboratory report. Laboratory medicine is a specialty, and experts in the field should be consulted whenever results are unusual or in doubt.

*Reproduced, with permission, from Krupp MA, Schroeder SA, Tierney LM Jr: *Current Medical Diagnosis & Treatment 1987*. Appleton & Lange, 1987.

Effect of Meals & Posture on Concentration of Substances in Blood

A. Meals: The usual normal values for blood tests have been determined by assay of "fasting" specimens collected after 8–12 hours of abstinence from food. With few exceptions, water is usually permitted as desired.

Few routine tests are altered from usual fasting values if blood is drawn 3–4 hours after breakfast. When blood is drawn 3–4 hours after lunch, values are more likely to vary from those of the true fasting state (ie, as much as +31% for AST [SGOT], −5% for lactate dehydrogenase, and lesser variations for other substances). Valid measurement of triglyceride in serum or plasma requires abstinence from food for 10–14 hours.

B. Posture: Plasma volume measured in a person who has been supine for several hours is 12–15% greater than in a person who has been up and about or standing for an hour or so. It follows that measurements performed on blood obtained after the subject has been lying down for an hour or more will yield lower values than when blood has been obtained after the same subject has been upright. An intermediate change apparently occurs with sitting.

Values in the same subject change when position changes from supine to standing as follows: increase in total protein, albumin, calcium, potassium, phosphate, cholesterol, triglyceride, AST (SGOT), the phosphatases, total thyroxine, hematocrit, erythrocyte count, and hemoglobin. The greatest change occurs in concentration of total protein and enzymes (+11%) and calcium (+3 to +4%). In a series of studies, change from the upright to the supine position resulted in the following decreases: total protein, −0.5 g; albumin, −0.4 to −0.6 g; calcium, −0.4 mg; cholesterol, −10 to −25 mg; total thyroxine, −0.8 to −1.8 μg; and hematocrit, −4 to −9%, reflecting hemodilution as interstitial fluid reenters the circulation.

A tourniquet applied for 1 minute instead of 3 minutes produced the following changes in reported values: total protein, +5%; iron, +6.7%; cholesterol, +5%, AST (SGOT), +9.3%; and bilirubin, +8.4%. Decreases were observed for potassium, −6%; and creatinine, −2.3%.

Validity of Laboratory Tests*

The clinical value of a test is related to its specificity and sensitivity and the incidence of the disease in the population tested.

*This section is an abridged verson of an article by Krieg AF, Gambino R, Galen RS: Why are clinical laboratory tests performed? When are they valid? *JAMA* 1975;**233**:76. Reprinted from the Journal of the American Medical Association. Copyright 1975, American Medical Association. See also Galen RS, Gambino SR: *Beyond Normality: The Predictive Value and Efficiency of Medical Diagnosis*. Wiley, 1975.

Sensitivity means percentage of positive results in patients with the disease. The test for phenylketonuria is highly sensitive: a positive result is obtained in all who have the disease (100% sensitivity). The carcinoembryonic antigen (CEA) test has low specificity: only 72% of those with carcinoma of the colon provide a positive result when the disease is extensive, and only 20% are positive with early disease. Lower sensitivity occurs in the early stages of many diseases—in contrast to the higher sensitivity in well-established disease.

Specificity means percentage of negative results among people who do not have the disease. The test for phenylketonuria is highly specific: 99.9% of normal individuals give a negative result. In contrast, the CEA test for carcinoma of the colon has a variable specificity: about 3% of nonsmoking individuals give a false-positive result (97% specificity), whereas 20% of smokers give a false-positive result (80% specificity). The overlap of serum thyroxine levels between hyperthyroid patients and those taking oral contraceptives or those who are pregnant is an example of a change in specificity from that prevailing in a different set of individuals.

The **predictive value** of a positive test defines the percentage of positive results that are true positives. This is related fundamentally to the incidence of the disease. In a group of patients on a urology service, the incidence of renal disease is higher than in the general population, and the serum creatinine level will have a higher predictive value in that group than for the general population.

Formulas for definitions:

$$\text{Sensitivity} = \frac{\text{True positive}}{\text{True positive} + \text{false negative}} \times 100$$

$$\text{Specificity} = \frac{\text{True negative}}{\text{True negative} + \text{false positive}} \times 100$$

Predictive value

$$= \frac{\text{True positive}}{\text{True positive} + \text{false positive}} \times 100$$

Before ordering a test, attempt to determine whether test sensitivity, specificity, and predictive value are adequate to provide useful information. To be useful, the result should influence diagnosis, prognosis, or therapy; lead to a better understanding of the disease process; and benefit the patient.

SI Units (*Système International d'Unités*)

A "coherent" system of measurement has been developed by an international organization designated the General Conference of Weights and Measures. An adaptation has been tentatively recommended by the Commission on Quantities and Units of the Section on Clinical Chemistry, International Union of Pure and Applied Chemistry. SI units are in use in some European countries, and the conversion to SI will continue if the system proves to be helpful in understanding physiologic mechanisms.

Eight fundamental measurable properties of matter (with authorized abbreviations shown in parentheses) were selected for clinical use.

length: metre (m)
mass: kilogram (kg)
amount of substance: mole (mol)
time: second (s)
thermodynamic temperature: kelvin (K)
electric current: ampere (A)
luminous intensity: candela (cd)
catalytic activity: katal (kat)

Derived from these are the following measurable properties:

mass concentration: kilogram/litre (kg/L)
mass fraction: kilogram/kilogram (kg/kg)
volume fraction: litre/litre (L/L)
volume: cubic metre (m^3); for clinical use, the unit will be the litre (L)
substance concentration: mole/litre (mol/L)
molality: mole/kilogram (mol/kg)
mole fraction: mole/mole (mol/mol)
pressure: pascal (Pa) = newton/m^2

Decimal factors are as follows:

Number	Name	Symbol
10^{12}	tera	T
10^9	giga	G
10^6	mega	M
10^3	kilo	k
10^2	hecto	h
10^1	deca	da
10^{-1}	deci	d
10^{-2}	centi	c
10^{-3}	milli	m
10^{-6}	micro	μ
10^{-9}	nano	n
10^{-12}	pico	p
10^{-15}	femto	f
10^{-18}	atto	a

"Per"—eg, "per second"—is often written as the negative exponent. Per second thus becomes $\cdot s^{-1}$; per meter squared, $\cdot m^{-2}$; per kilogram, $\cdot kg^{-1}$. *Example:* cm/s = $cm \cdot s^{-1}$; $g/m^2 = g \cdot m^{-2}$; etc.

In anticipation that the SI system may be adopted in the USA in the next several years, values are reported here in the traditional units with equivalent SI units following in parentheses.

COMMON CLINICAL VALUES IN TRADITIONAL & SI MEASUREMENTS*

Albumin, Serum or Plasma:

See Proteins, Serum or Plasma.

Aminotransferases, Serum:

Normal (varies with method): AST (SGOT), 6–25 IU/L at 30 °C; SMA, 10–40 IU/L at 37 °C; SMAC, 0–41 IU/L at 37 °C. ALT (SGPT), 3–26 IU/L at 30 °C; SMAC, 0–45 IU/L at 37 °C.

A. Precautions: Avoid hemolysis. Remove serum from clot promptly.

B. Physiologic Basis: Aspartate aminotransferase (AST; SGOT), alanine aminotransferase (ALT; SGPT), and lactic dehydrogenase are intracellular enzymes involved in amino acid or carbohydrate metabolism. These enzymes are present in high concentrations in muscle, liver, and brain. Elevations of concentrations of these enzymes in the blood indicate nercrosis or disease, especially of these tissues.

C. Interpretation:

1. Elevated after myocardial infarction (especially AST); acute infectious hepatitis (ALT usually elevated more than AST); cirrhosis of the liver (AST usually elevated more than ALT); and metastatic or primary liver neoplasm. Elevated in transudates associated with neoplastic involvement of serous cavities. AST is elevated in muscular dystrophy, dermatomyositis, and paroxysmal myoglobinuria.

2. Decreased with pyridoxine (vitamin B_6) deficiency (often as a result of repeated hemodialysis), renal insufficiency, and pregnancy.

Ammonia, Blood:

Normal (Conway): 10–110 μg/dL whole blood. (SI: 12–65 μmol/L.)

A. Precautions: Do not use anticoagulants containing ammonia. Suitable anticoagulants include potassium oxalate, edetate calcium disodium, and heparin that is ammonia-free. The determination should be done immediately after drawing blood. If the blood is kept in an ice-water bath, it may be held for up to 1 hour.

B. Physiologic Basis: Ammonia present in the blood is derived from 2 principal sources: (1) In the large intestine, putrefactive action of bacteria on nitrogenous materials releases significant quantities of ammonia. (2) In the process of protein metabolism, ammonia is liberated. Ammonia entering the portal vein or the systemic circulation is rapidly converted to urea in the liver. Liver insufficiency may result in an increase in blood ammonia concentration, especially if protein consumption is high or if there is bleeding into the bowel.

C. Interpretation: Blood ammonia is elevated in hepatic insufficiency or with liver bypass in the form of a portacaval shunt, particularly if protein intake is high or if there is bleeding into the bowel.

Amylase, Serum:

Normal (varies with method): 80–180 Somogyi

---

*The values listed in this section and in the following section have been gleaned from many sources. Values will vary with method and individual laboratory.

units/dL serum. (One Somogyi unit equals amount of enzyme that will produce 1 mg of reducing sugar from starch at pH 7.2.) 0.8–3.2 IU/L.

A. Precautions: If storage for more than 1 hour is necessary, blood or serum must be refrigerated.

B. Physiologic Basis: Normally, small amounts of amylase (diastase), molecular weight about 50,000, originating in the pancreas and salivary glands, are present in the blood. Inflammatory disease of these glands or obstruction of their ducts results in regurgitation of large amounts of enzyme into the blood and increased excretion via the kidney.

C. Interpretation:

1. Elevated in acute pancreatitis, pseudocyst of the pancreas, obstruction of pancreatic ducts (carcinoma, stone, stricture, duct sphincter spasm after morphine), and mumps. Occasionally elevated in renal insufficiency, in diabetic acidosis, and in inflammation of the pancreas from a perforating peptic ulcer. Rarely, combination of amylase with an immunoglobulin produces elevated serum amylase activity (macroamylasemia) because the large molecular complex (molecular weight at least 160,000) is not filtered by the glomerulus.

2. Decreased in acute and chronic hepatitis, in pancreatic insufficiency, and, occasionally, in toxemia of pregnancy.

Amylase, Urine:

Normal (varies with method): 40–250 Somogyi units/h.

A. Precautions: If the determination is delayed more than 1 hour after collecting the specimen, urine must be refrigerated.

B. Physiologic Basis: See Amylase, Serum. If renal function is adequate, amylase is rapidly excreted in the urine. A timed urine specimen (ie, 2, 6, or 24 hours) should be collected and the rate of excretion determined.

C. Interpretation: Elevation of the concentration of amylase in the urine occurs in the same situations in which serum amylase concentration is elevated. Urinary amylase concentration remains elevated for up to 7 days after serum amylase levels have returned to normal following an attack of pancreatitis. Thus, the determination of urinary amylase may be useful if the patient is seen late in the course of an attack of pancreatitis. An elevated serum amylase with normal or low urine amylase excretory rate may be seen in the presence of renal insufficiency or with macroamylasemia.

Bicarbonate, Serum or Plasma:

Normal: 24–28 meq/L. (SI: 24–28 mmol/L.)

A. Precautions: Plasma or serum should be separated from blood cells and kept refrigerated in stoppered tubes.

B. Physiologic Basis: Bicarbonate-carbonic acid buffer is one of the most important buffer systems in maintaining normal pH of body fluids. Bicarbonate

and pH determinations on arterial whole blood serve as a basis for assessing "acid-base balance."

C. Interpretation:

1. Elevated in–

a. Metabolic alkalosis (arterial blood pH increased) due to ingestion of large quantities of sodium bicarbonate, protracted vomiting of acid gastric juice, or accompanying potassium deficit.

b. Respiratory acidosis (arterial blood pH decreased) due to inadequate elimination of CO_2 (leading to elevated P_{CO_2}) because of pulmonary emphysema, poor diffusion in alveolar membrane disease, heart failure with pulmonary congestion or edema, or ventilatory failure due to any cause, including oversedation, narcotics, or inadequate artificial respiration.

2. Decreased in–

a. Metabolic acidosis (arterial blood pH decreased) due to diabetic ketoacidosis, lactic acidosis, starvation, persistent diarrhea, renal insufficiency, ingestion of excess acidifying salts or methanol, or salicylate intoxication.

b. Respiratory alkalosis (arterial blood pH increased) due to hyperventilation (decreased P_{CO_2}).

Bilirubin, Serum:

Normal: Total, 1.1–1.2 mg/dL (SI: 3.5–19 μmol/L). Direct (glucuronide), 0.1–0.4 mg/dL. Indirect (unconjugated), 0.2–0.7 mg/dL. (SI: direct, up to 7 μmol/L; indirect, up to 12 μmol/L.)

A. Precautions: The fasting state is preferred to avoid turbidity of serum. For optimal stability of stored serum, samples should be frozen and stored in the dark.

B. Physiologic Basis: Destruction of hemoglobin yields bilirubin, which is conjugated in the liver to the diglucuronide and excreted in the bile. Bilirubin accumulates in the plasma when liver insufficiency exists, biliary obstruction is present, or the rate of hemolysis increases. Rarely, abnormalities of enzyme systems involved in bilirubin metabolism in the liver (eg, absence of glucuronyl transferase) result in abnormal bilirubin concentrations.

C. Interpretation:

1. Direct and indirect forms of serum bilirubin are elevated in acute or chronic hepatitis; biliary tract obstruction (cholangiolar, hepatic, or common ducts); toxic reactions to many drugs, chemicals, and toxins; and Dubin-Johnson and Rotor's syndromes.

2. Indirect serum bilirubin is elevated in hemolytic diseases or reactions and absence or deficiency of glucuronyl transferase, as in Gilbert's disease and Crigler-Najjar syndrome.

3. Direct and total bilirubin can be significantly elevated in normal and jaundiced subjects by fasting 24–48 hours (in some instances even 12 hours) or by prolonged caloric restriction.

Calcium, Serum:

Normal: Total, 8.5–10.3 mg/dL or 4.2–5.2 meq/L. Ionized, 4.2–5.2 mg/dL or 2.1–2.6 meq/L. (SI: total, 2.1–2.6 mmol/L; ionized, 1.05–1.3 mmol/L.)

A. Precautions: Glassware must be free of calcium. The patient should be fasting. Serum should be promptly separated from the clot.

B. Physiologic Basis: Endocrine, renal, gastrointestinal, and nutritional factors normally provide for precise regulation of calcium concentration in plasma and other body fluids. Since some calcium is bound to plasma protein, especially albumin, determination of the plasma albumin concentration is necessary before the clinical significance of abnormal serum calcium levels can be interpreted accurately.

C. Interpretation:

1. Elevated in hyperparathyroidism, secretion of parathyroidlike hormone by malignant tumors, vitamin D excess, milk-alkali syndrome, osteolytic disease such as multiple myeloma, invasion of bone by metastatic cancer, Paget's disease of bone, Boeck's sarcoid, immobilization, and familial hypocalciuria. Occasionally elevated with hyperthyroidism and with ingestion of thiazide drugs.

2. Decreased in hypoparathyroidism, vitamin D deficiency (rickets, osteomalacia), renal insufficiency, hypoproteinemia, malabsorption syndrome (sprue, ileitis, celiac disease, pancreatic insufficiency), severe pancreatitis with pancreatic necrosis, and pseudohypoparathyroidism.

Calcium, Urine (Daily Excretion):

Ordinarily there is a moderate continuous urinary calcium excretion of 50–150 mg/24 h, depending upon the intake. (SI: 1.2–3.7 mmol/24 h.)

A. Procedure: The patient should remain on a diet free of milk and cheese for 3 days prior to testing; for quantitative testing, a neutral ash diet containing about 150 mg calcium per day is given for 3 days. Quantitative calcium excretion studies may be made on a carefully timed 24–hour urine specimen.

B. Interpretation: On the quantitative diet, a normal person excretes 125 ± 50 mg (1.8–4.4 mmol) of calcium per 24 hours. In hyperparathyroidism, the urinary calcium excretion usually exceeds 200 mg/24 h (5 mmol/d). Urinary calcium excretion is almost always elevated when serum calcium is high.

Ceruloplasmin & Copper, Serum:

Normal: Ceruloplasmin, 25–43 mg/dL (SI: 1.7–2.9 μmol/L); copper, 100–200 μg/dL (SI: 16–31 μmol/L).

A. Precautions: None.

B. Physiologic Basis: About 5% of serum copper is loosely bound to albumin and 95% to ceruloplasmin, an oxidase enzyme that is an α_2 globulin with a blue color. In Wilson's disease, serum copper and ceruloplasmin are low and urinary copper is high.

C. Interpretation:

1. Elevated in pregnancy, hyperthyroidism, infection, aplastic anemia, acute leukemia, Hodgkin's disease, cirrhosis of the liver, and with use of oral contraceptives.

2. Decreased in Wilson's disease (accompanied by increased urinary excretion of copper), malabsorp-

tion, nephrosis, and copper deficiency that may accompany total parenteral nutrition.

Chloride, Serum or Plasma:

Normal: 96–106 meq/L. (SI: 96–106 mmol/L.)

A. Precautions: Determination with whole blood yields lower results than those obtained using serum or plasma as the specimen. Always use serum or plasma.

B. Physiologic Basis: Chloride is the principal inorganic anion of the extracellular fluid. It is important in maintenance of acid-base balance even though it exerts no buffer action. When chloride as HCl or NH_4Cl is lost, alkalosis follows; when chloride is retained or ingested, acidosis follows. Chloride (with sodium) plays an important role in control of osmolarity of body fluids.

C. Interpretation:

1. Elevated in renal insufficiency (when Cl intake exceeds excretion), nephrosis (occasionally), renal tubular acidosis, hyperparathyroidism (occasionally), ureterosigmoid anastomosis (reabsorption from urine in gut), dehydration (water deficit), and overtreatment with saline solution.

2. Decreased in gastrointestinal disease with loss of gastric and intestinal fluids (vomiting, diarrhea, gastrointestinal suction), renal insufficiency (with salt deprivation), overtreatment with diuretics, chronic respiratory acidosis (emphysema), diabetic acidosis, excessive sweating, adrenal insufficiency (NaCl loss), hyperadrenocorticism (chronic K^+ loss), and metabolic alkalosis ($NaHCO_3$ ingestion; K^+ deficit).

Chloride, Urine:

Urine chloride content varies with dietary intake, acid-base balance, endocrine "balance," body stores of other electrolytes, and water balance. Relationships and responses are so variable and complex that there is little clinical value in urine chloride determinations other than in balance studies.

Cholesterol, Serum or Plasma:

Normal: 150–280 mg/dL. (SI: 3.9–7.2 mmol/L.) See Table 1.

A. Precautions: The fasting state is preferred.

B. Physiologic Basis: Cholesterol concentra-

Table 1. Lipidemia: Ranges of population (USA) for serum concentrations of cholesterol (C), triglyceride (TG), low-density lipoprotein cholesterol (HDL-C), and high-density lipoprotein cholesterol (HDL-C).*

Age	C (mg/dL)	TG (mg/dL)	LDL-C (mg/dL) Upper Limit	HDL-C (mg/dL)	
				Male	Female
< 29	120–240	10–140	170	45 ± 12	55 ± 12
30–39	140–270	10–150	190		
40–49	150–310	10–160	190		
> 49	160–330	10–190	210		

*Reproduced, with permission, from Krupp MA et al: *Physician's Handbook,* 21st ed. Lange, 1985.

tions are determined by metabolic functions, which are influenced by heredity, nutrition, endocrine function, and integrity of vital organs such as the liver and kidney. Cholesterol metabolism is intimately associated with lipid metabolism.

C. Interpretation:

1. Elevated in familial hypercholesterolemia (xanthomatosis), hypothyroidism, poorly controlled diabetes mellitus, nephrotic syndrome, chronic hepatitis, biliary cirrhosis, obstructive jaundice, hypoproteinemia (idiopathic, with nephrosis or chronic hepatitis), and lipidemia (idiopathic, familial).

2. Decreased in acute hepatitis and Gaucher's disease. Occasionally decreased in hyperthyroidism, acute infections, anemia, malnutrition, and apolipoprotein deficiency.

Creatine Phosphokinase (CPK), Serum:

Normal (varies with method): 10–50 IU/L at 30 °C.

A. Precautions: The enzyme is unstable, and the red cell content inhibits enzyme activity. Serum must be removed from the clot promptly. If assay cannot be done soon after drawing blood, serum must be frozen.

B. Physiologic Basis: CPK splits creatine phosphate in the presence of ADP to yield creatine and ATP. Skeletal and heart muscle and brain are rich in the enzyme.

C. Interpretation:

1. Elevated in the presence of muscle damage such as with myocardial infarction, trauma to muscle, muscular dystrophies, polymyositis, severe muscular exertion (jogging), hypothyroidism, and cerebral infarction (necrosis). Following myocardial infarction, serum CPK concentration increases rapidly (within 3–5 hours), and it remains elevated for a shorter time after the episode (2 or 3 days) than does AST or LDH.

2. Not elevated in pulmonary infarction or parenchymal liver disease.

Creatine Phosphokinase Isoenzymes, Serum:

See Table 2.

A. Precautions: As for CPK Normal (see above).

B. Physiologic Basis: CPK consists of 3 proteins separable by electrophoresis. Skeletal muscle is characterized by isoenzyme MM, myocardium by isoenzyme MB, and brain by isoenzyme BB.

C. Interpretation: CPK isoenzymes are increased in serum. CPK-MM is elevated in injury to skeletal muscle, myocardial muscle, and brain; in muscle disease (eg, dystrophies, hypothyroidism, dermatomyositis, polymyositis); in rhabdomyolysis; and after severe exercise. CPK-MB is elevated soon (within 2–4 hours) after myocardial infarction and for up to 72 hours afterward (high levels are prolonged with extension of infarct or new infarction); also elevated in extensive rhabdomyolysis or muscle injury, severe muscle disease, Reye's syndrome, or Rocky Mountain spotted fever. CPK-BB is occasionally elevated in severe shock, in some carcinomas (especially oat cell carcinoma or carcinoma of the ovary, breast, or prostate), or in biliary atresia.

Creatine, Urine (24 Hours):

Normal: See Table 3.

A. Precautions: Collection of the 24-hour specimen must be accurate. The specimen may be refrigerated or preserved with 10 mL of toluene or 10 mL of 5% thymol in chloroform.

B. Physiologic Basis: Creatine is an important constituent of muscle, brain, and blood; in the form of creatine phosphate, it serves as a source of high-energy phosphate. Normally, small amounts of creatine are excreted in the urine, but in states of elevated catabolism and in the presence of muscular dystrophies, the rate of excretion is increased.

C. Interpretation:

1. Elevated in muscular dystrophies such as progressive muscular dystrophy, myotonia atrophica, and myasthenia gravis; muscle wasting, as in acute poliomyelitis, amyotrophic lateral sclerosis, and myositis manifested by muscle wasting; starvation and cachectic states; hyperthyroidism; and febrile diseases.

2. Decreased in hypothyroidism, amyotonia congenita, and renal insufficiency.

Creatinine, Serum or Plasma:

Normal: 0.7–1.5 mg/dL. (SI: 60–132 μmol/L.)

A. Precautions: None.

B. Physiologic Basis: Endogenous creatinine is excreted by filtration through the glomerulus and by tubular secretion at a rate about 20% greater than clearance of inulin. The Jaffe reaction measures chro-

Table 2. Creatine kinase isoenzymes.

Isoenzyme	Normal Levels % of Total
(Fastest) Fraction 1, BB	0
Fraction 2, MB	0–3
(Slowest) Fraction 3, MM	97–100

Table 3. Urine creatine and creatinine, normal values (24 hours).*

	Creatine	Creatinine
Newborn	4.5 mg/kg	10 mg/kg
1–7 months	8.1 mg/kg	12.8 mg/kg
2–3 years	7.9 mg/kg	12.1 mg/kg
4–4½ years	4.5 mg/kg	14.6 mg/kg
9–9½ years	2.5 mg/kg	18.1 mg/kg
11–14 years	2.7 mg/kg	20.1 mg/kg
Adult male	0–50 mg	25 mg/kg
Adult female	0–100 mg	21 mg/kg

mogens other than creatinine in the plasma. Because the chromogens are not passed into the urine, the measurement of creatinine in the urine is about 20% less than chromogen plus creatinine in plasma, providing, fortuitously, a compensation for the amount secreted. Thus, inulin and creatinine clearances for clinical purposes are comparable and creatinine clearance is an acceptable measure of glomerular filtration rate—except that with advancing renal failure, creatinine clearance exceeds inulin clearance owing to the secretion of creatinine by remaining renal tubules.

C. Interpretation: Creatinine is elevated in acute or chronic renal insufficiency, urinary tract obstruction, and impairment of renal function induced by some drugs. Materials other than creatinine may react to give falsely high results with the alkaline picrate method (Jaffe reaction): acetoacetate, acetone, β-hydroxybutyrate, α-ketoglutarate, pyruvate, glucose, bilirubin, hemoglobin, urea, and uric acid. Values below 0.7 mg/dL are of no known significance.

Creatinine, Urine:
Normal: See Table 3.

Glucose, Serum or Plasma:
Normal: Fasting "true" glucose, 65–110 mg/dL. (SI: 3.6–6.1 mmol/L.)

A. Precautions: If determination is delayed beyond 1 hour, sodium fluoride, about 3 mg/mL blood, should be added to the specimen. The filtrates may be refrigerated for up to 24 hours. Errors in interpretation may occur if the patient has eaten sugar or received glucose solution parenterally just prior to the collection of what is thought to be a "fasting" specimen.

B. Physiologic Basis: The glucose concentration in extracellular fluid is normally closely regulated, with the result that a source of energy is available to tissues, and no glucose is excreted in the urine. Hyperglycemia and hypoglycemia are nonspecific signs of abnormal glucose metabolism.

C. Interpretation:

1. Elevated in diabetes, hyperthyroidism, adrenocortical hyperactivity (cortical excess), hyperpituitarism, and hepatic disease (occasionally).

2. Decreased in hyperinsulinism, adrenal insufficiency, hypopituitarism, hepatic insufficiency (occasionally), functional hypoglycemia, and by hypoglycemic agents.

γ-Glutamyl Transpeptidase or Transferase, Serum:
Normal: Males, < 30 mU/mL at 30 °C. Females, < 25 mU/mL at 30 °C. Adolescents, < 50 mU/mL at 30 °C.

A. Precautions: Avoid hemolysis.

B. Physiologic Basis: γ-Glutamyl transferase (GGT) is an extremely sensitive indicator of liver disease. Levels are often elevated when transaminases and alkaline phosphatase are normal, and it is considered more specific than both for identifying liver impairment due to alcoholism.

The enzyme is present in liver, kidney, and pancreas and transfers C-terminal glutamic acid from a peptide to other peptides or L-amino acids. It is induced by alcohol.

C. Interpretation: Elevated in acute infectious or toxic hepatitis, chronic and subacute hepatitis, cirrhosis of the liver, intrahepatic or extrahepatic obstruction, primary or metastatic liver neoplasms, and liver damage due to alcoholism. It is elevated occasionally in congestive heart failure and rarely in postmyocardial infarction, pancreatitis, and pancreatic carcinoma.

Iron, Serum:
Normal: 50–175 μg/dL. (SI: 9–31.3 μmol/L.)

A. Precautions: Syringes and needles must be iron-free. Hemolysis of blood must be avoided. The serum must be free of hemoglobin.

B. Physiologic Basis: Because of diurnal variation with highest values in the morning, fasting morning blood specimens are desirable. Iron concentration in the plasma is determined by several factors, including absorption from the intestine; storage in intestine, liver, spleen, and marrow; breakdown or loss of hemoglobin; and synthesis of new hemoglobin.

C. Interpretation:

1. Elevated in hemochromatosis, hemosiderosis (multiple transfusions, excess iron administration), hemolytic disease, pernicious anemia, and hypoplastic anemias. Often elevated in viral hepatitis. Spuriously elevated if patient has received parenteral iron during the 2–3 months prior to determination.

2. Decreased in iron deficiency; with infections, nephrosis, and chronic renal insufficiency; and during periods of active hematopoiesis.

Iron-Binding Capacity, Serum:
Normal: Total, 250–410 μg/dL. (SI: 45–73 μmol/L.) Percent saturation, 20–55%.

A. Precautions: None.

B. Physiologic Basis: Iron is transported as a complex of the metal-binding globulin transferrin (siderophilin). Normally, this transport protein carries an amount of iron that represents about 30–40% of its capacity to combine with iron.

C. Interpretation of Total Iron-Binding Capacity:

1. Elevated in iron deficiency anemia, with use of oral contraceptives, in late pregnancy, and in infants. Occasionally elevated in hepatitis.

2. Decreased in association with decreased plasma proteins (nephrosis, starvation, cancer), chronic inflammation, and hemosiderosis (transfusions, thalassemia).

D. Interpretation of Saturation of Transferrin:

1. Elevated in iron excess (iron poisoning, hemolytic disease, thalassemia, hemochromatosis, pyridoxine deficiency, nephrosis, and, occasionally, hepatitis).

2. Decreased in iron deficiency, chronic infection, cancer, and late pregnancy.

Lactate Dehydrogenase (LDH), Serum, Serous Fluids, Spinal Fluid, or Urine:

Normal (varies with method): Serum, 55–140 IU/L at 30 °C; SMA, 100–225 IU/L at 37 °C; SMAC, 60–200 IU/L at 37 °C. Serous fluids, lower than serum. Spinal fluid, 15–75 units (Wroblewski); 6.3–30 IU/L. Urine, less than 8300 units/8 h (Wroblewski).

A. Precautions: Any degree of hemolysis must be avoided because the concentration of LDH within red blood cells is 100 times that in normal serum. Heparin and oxalate may inhibit enzyme activity. Remove serum from clot promptly.

B. Physiologic Basis: LDH catalyzes the interconversion of lactate and pyruvate in the presence of NADH or $NADH_2$. It is distributed generally in body cells and fluids.

C. Interpretation: Elevated in all conditions accompanied by tissue necrosis, particularly those involving acute injury of the heart, red cells, kidney, skeletal muscle, liver, lung, and skin. Marked elevations accompany hemolytic anemias, the anemias of vitamin B_{12} and folate deficiency, and polycythemia rubra vera. The course of rise in concentration over 3–4 days followed by a slow decline during the following 5–7 days may be helpful in confirming the presence of a myocardial infarction; however, pulmonary infarction, neoplastic disease, and megaloblastic anemia must be excluded. Although elevated during the acute phase of infectious hepatitis, enzyme activity is seldom increased in chronic liver disease.

Lactate Dehydrogenase (LDH) Isoenzymes, Serum:

Normal: See Table 4.

A. Precautions: As for LDH (see above).

B. Physiologic Basis: LDH consists of 5 separable proteins, each made of tetramers of 2 types, or subunits, H and M. The 5 isoenzymes can be distinguished by kinetics, electrophoresis, chromatography, and immunologic characteristics. By electrophoretic separation, the mobility of the isoenzymes corresponds to serum proteins α_1, α_2, β, γ_1, and γ_2. These are usually numbered 1 (fastest moving), 2, 3, 4, and 5 (slowest moving). Isoenzyme 1 is present in high concentrations in heart muscle (tetramer H H H H) and in erythrocytes and kidney cortex; isoenzyme 5, in skeletal muscle (tetramer M M M M) and liver.

C. Interpretation: In myocardial infarction, the

Table 4. Lactate dehydrogenase isoenzymes.

	Isoenzyme	Percentage of Total (and Range)
(Fastest)	1 (α_1)	28 (15–30)
	2 (α_2)	36 (22–50)
	3 (β)	23 (15–30)
	4 (γ_1)	6 (0–15)
(Slowest)	5 (γ_2)	6 (0–15)

α isoenzymes are elevated—particularly LDH 1—to yield a ratio of LDH 1:LDH 2 of greater than 1. Similar α isoenzyme elevations occur in renal cortex infarction and with hemolytic anemias.

LDH 5 and 4 are relatively increased in the presence of acute hepatitis, acute muscle injury, dermatomyositis, and muscular dystrophies.

Lipase, Serum:

Normal: 0.2–1.5 units.

A. Precautions: None. The specimen may be stored under refrigeration up to 24 hours prior to the determination.

B. Physiologic Basis: A low concentration of fat-splitting enzyme is present in circulating blood. In the presence of pancreatitis, pancreatic lipase is released into the circulation in higher concentrations, which persist, as a rule, for a longer period than does the elevated concentration of amylase.

C. Interpretation: Serum lipase is elevated in acute or exacerbated pancreatitis and in obstruction of pancreatic ducts by stone or neoplasm.

Magnesium, Serum:

Normal: 1.8–3 mg/dL or 1.5–2.5 meq/L. (SI: 0.75–1.25 mmol/L.)

A. Precautions: None.

B. Physiologic Basis: Magnesium is primarily an intracellular electrolyte. In extracellular fluid, it affects neuromuscular irritability and response. Magnesium deficit may exist with little or no change in extracellular fluid concentrations. Low magnesium levels in plasma have been associated with tetany, weakness, disorientation, and somnolence.

C. Interpretation:

1. Elevated in renal insufficiency and in overtreatment with magnesium salts.

2. Decreased in chronic diarrhea, acute loss of enteric fluids, starvation, chronic alcoholism, chronic hepatitis, hepatic insufficiency, excessive renal loss (diuretics), and inadequate replacement with parenteral nutrition. May be decreased in and contribute to persistent hypocalcemia in patients with hypoparathyroidism) and when large doses of vitamin D and calcium are being administered.

Phosphatase, Acid, Serum:

Normal values vary with method: 0.1–0.63 Sigma units. (SI: 36–175 nmol/s/L.)

A. Precautions: Do not draw blood for assay for 24 hours after prostatic massage or instrumentation. For methods that measure enzyme activity, complete the determination promptly, since activity declines quickly; avoid hemolysis. For immunoassay methods, the enzyme is stable for 3–4 days when the serum is refrigerated or frozen.

B. Physiologic Basis: Phosphatases active at pH 4.9 are present in high concentration in the prostate gland, erythrocytes, platelets, reticuloendothelial cells, liver, spleen, and kidney. A variety of isoenzymes have been found in these tissues and serum and account for different activities operating against different substrates.

C. Interpretation: In the presence of carcinoma of the prostate, the prostatic fraction of acid phosphatase may be increased in the serum, particularly if the cancer has spread beyond the capsule of the gland or has metastasized. Palpation of the prostate will produce a transient increase. Total acid phosphatase may be increased in Gaucher's disease, malignant tumors involving bone, renal disease, hepatobiliary disease, diseases of the reticuloendothelial system, and thromboembolism. Fever may cause spurious elevations.

Phosphatase, Alkaline, Serum:

Normal (varies with method): Bessey-Lowry, children, 2.8–6.7 units; Bessey-Lowry, adults, 0.8–2.3 units. Adults, King-Armstrong, 5–13 units: 24–71 IU/L at 30 °C; SMA, 30–85 IU/L at 37 °C; SMAC, 30–115 IU/L at 37 °C.

A. Precautions: Serum may be kept in the refrigerator for 24–48 hours, but values may increase slightly (10%). The specimen will deteriorate if it is not refrigerated. Do not use fluoride or oxalate.

B. Physiologic Basis: Alkaline phosphatase is present in high concentration in growing bone, in bile, and in the placenta. In serum, it consists of a mixture of isoenzymes not yet clearly defined. The isoenzymes may be separated by electrophoresis; liver alkaline phosphatase migrates faster than bone and placental alkaline phosphatase, which migrate together.

C. Interpretation:

1. Elevated in–

a. Children (normal growth of bone).

b. Osteoblastic bone disease–Hyperparathyroidism, rickets and osteomalacia, neoplastic bone disease (osteosarcoma, metastatic neoplasms), ossification as in myositis ossificans, Paget's disease (osteitis deformans), and Boeck's sarcoid.

c. Hepatic duct or cholangiolar obstruction due to stone, stricture, or neoplasm.

d. Hepatic disease resulting from drugs such as chlorpromazine and methyltestosterone.

e. Pregnancy.

2. Decreased in hypothyroidism and in growth retardation in children.

Phosphorus, Inorganic, Serum:

Normal: Children, 4–7 mg/dL. (SI: 1.3–2.3 mmol/L.) Adults, 3–4.5 mg/dL. (SI: 1–1.5 mmol/L.)

A. Precautions: Glassware cleaned with phosphate cleansers must be thoroughly rinsed. The fasting state is necessary to avoid postprandial depression of phosphate associated with glucose transport and metabolism.

B. Physiologic Basis: The concentration of inorganic phosphate in circulating plasma is influenced by parathyroid gland function, action of vitamin D, intestinal absorption, renal function, bone metabolism, and nutrition.

C. Interpretation:

1. Elevated in renal insufficiency, hypoparathyroidism, and hypervitaminosis D.

2. Decreased in hyperparathyroidism, hypovitaminosis D (rickets, osteomalacia), malabsorption syndrome (steatorrhea), ingestion of antacids that bind phosphate in the gut, starvation or cachexia, chronic alcoholism (especially with liver disease), hyperalimentation with phosphate-poor solutions, carbohydrate administration (especially intravenously), renal tubular defects, use of thiazide diuretics, acid-base disturbances, diabetic ketoacidosis (especially during recovery), and genetic hypophosphatemia. Occasionally decreased during pregnancy and with hypothyroidism.

Potassium, Serum or Plasma:

Normal: 3.5–5 meq/L. (SI: 3.5–5 mmol/L.)

A. Precautions: Avoid hemolysis, which releases erythrocyte potassium: Serum must be separated promptly from the clot, or plasma from the red cell mass, to prevent diffusion of potassium from erythrocytes. Platelets and leukocytes are rich in potassium; if these elements are present in large numbers (eg, in thrombocytosis or leukemia), the potassium released during coagulation will increase potassium concentration in serum. If these are sources of artifact, plasma from heparinized blood should be employed.

B. Physiologic Basis: Potassium concentration in plasma determines neuromuscular and muscular irritability. Elevated or decreased concentrations impair the capability of muscle tissue to contract.

C. Interpretation: (See Precautions, above.)

1. Elevated in renal insufficiency (especially in the presence of increased rate of protein or tissue breakdown); adrenal insufficiency (especially hypoaldosteronism); hyporeninemic hypoaldosteronism; use of spironolactone; too rapid administration of potassium salts, especially intravenously; and use of triamterene or phenformin.

2. Decreased in–

a. Inadequate intake (starvation).

b. Inadequate absorption or unusual enteric losses–Vomiting, diarrhea, malabsorption syndrome, or use of sodium polystyrene sulfonate resin.

c. Unusual renal loss–Secondary to hyperadrenocorticism (especially hyperaldosteronism) and to adrenocorticosteroid therapy, metabolic alkalosis, use of diuretics such as chlorothiazide and its derivatives and the mercurials; renal tubular defects such as the de Toni-Fanconi syndrome and renal tubular acidosis; treatment with antibiotics that are excreted as anions (carbenicillin, ticarcillin); use of phenothiazines, amphotericin B, and drugs with high sodium content; and use of degraded tetracycline.

d. Abnormal redistribution between extracellular and intracellular fluids–Familial periodic paralysis or testosterone administration.

Proteins, Serum or Plasma (Includes Fibrinogen):

Normal: See Interpretation, below.

A. Precautions: Serum or plasma must be free of hemolysis. Since fibrinogen is removed in the process of coagulation of the blood, fibrinogen determinations cannot be done on serum.

B. Physiologic Basis: Concentration of protein determines colloidal osmotic pressure of plasma. The concentration of protein in plasma is influenced by the nutritional state, hepatic function, renal function, occurrence of disease such as multiple myeloma, and metabolic errors. Variations in the fractions of plasma proteins may signify specific disease.

C. Interpretation:

1. Total protein, serum–Normal: 6–8 g/dL. (SI: 60–80 g/L.) See albumin and globulin fractions, below, and Table 5.

2. Albumin, serum or plasma–Normal: 3:5–5.5 g/dL. (SI: 33–55 g/L.)

a. Elevated in dehydration, shock, hemoconcentration, and administration of large quantities of concentrated albumin "solution" intravenously.

b. Decreased in malnutrition, malabsorption syndrome, acute or chronic glomerulonephritis, nephrosis, acute or chronic hepatic insufficiency, neoplastic diseases, and leukemia.

3. Globulin, serum or plasma–Normal: 2–3.6 g/dL. (SI: 20–36 g/L.) (See Tables 6 and 7.)

a. Elevated in hepatic disease, infectious hepatitis, cirrhosis of the liver, biliary cirrhosis, and hemochromatosis; disseminated lupus erythematosus; plasma cell myeloma; lymphoproliferative disease; sarcoidosis; and acute or chronic infectious diseases, particularly lymphogranuloma venereum, typhus, leishmaniasis, schistosomiasis, and malaria.

b. Decreased in malnutrition, congenital agammaglobulinemia, acquired hypogammaglobulinemia, and lymphatic leukemia.

4. Fibrinogen, plasma–Normal: 0.2–0.6 g/dL. (SI: 2–6 g/L.)

a. Elevated in glomerulonephritis, nephrosis (occasionally), and infectious diseases.

b. Decreased in disseminated intravascular coagulation (accidents of pregnancy such as placental ablation, amniotic fluid embolism, and violent labor; meningococcal meningitis; metastatic carcinoma of the prostate and occasionally of other organs; and leukemia), acute and chronic hepatic insufficiency, and congenital fibrinogenopenia.

Table 5. Protein fractions as determined by electrophoresis.

	Percentage of Total Protein
Albumin	52–68
α_1 globuliin	2.4–4.4
α_2 globulin	6.1–10.1
β globulin	8.5–14.5
γ globulin	10–21

Table 6. Gamma globulins by immunoelectrophoresis.

IgA	90–450 mg/dL
IgG	700–1500 mg/dL
IgM	40–250 mg/dL
IgD	0.3–40 mg/dL
IgE	0.006–0.16 mg/dL

Sodium, Serum or Plasma:

Normal: 136–145 meq/L. (SI: 136–145 mmol/L.)

A. Precautions: Clean glassware completely.

B. Physiologic Basis: Sodium constitutes about 140 of the 155 meq of cation in plasma. With its associated anions it provides the bulk of osmotically active solute in the plasma, thus affecting the distribution of body water significantly. A shift of sodium into cells or a loss of sodium from the body results in a decrease of extracellular fluid volume, affecting circulation, renal function, and nervous system function.

C. Interpretation:

1. Elevated in dehydration (water deficit), central nervous system trauma or disease, and hyperadrenocorticism with hyperaldosteronism or corticosterone or corticosteroid excess.

2. Decreased in adrenal insufficiency; in renal insufficiency, especially with inadequate sodium intake; in renal tubular acidosis; as a physiologic response to trauma or burns (sodium shift into cells); in unusual losses via the gastrointestinal tract, as with acute or chronic diarrhea or with intestinal obstruction or fistula; and in unusual sweating with inadequate sodium replacement. In some patients with edema associated with cardiac or renal disease, serum sodium concentration is low even though total body sodium content is greater than normal; water retention (excess ADH) and abnormal distribution of sodium between intracellular and extracellular fluid contribute to this paradoxic situation. Hyperglycemia occasionally results in shift of intracellular water to the extracellular space, producing a dilutional hyponatremia. (Artifact: When measured by the flame photometer, serum or plasma sodium will be decreased in the presence of hy-

Table 7. Some constituents of globulins.

Globulin	Representative Constituents
α_1	Thyroxine-binding globulin Transcortin Glycoprotein Lipoprotein Antitrypsin
α_2	Haptoglobin Glycoprotein Macroglobulin Ceruloplasmin
β	Transferrin Lipoprotein Glycoprotein
γ	γG γD γM γE γA

perlipidemia or hyperglobulinemia; in these disorders, the volume ordinarily occupied by water is taken up by other substances, and the serum or plasma will thus be "deficient" in water and electrolytes. In the presence of hyperglycemia, serum sodium concentration will be reduced by 1.6 meq/L per 100 mg/dL glucose above 200 mg/dL because of shifts of water into extracellular fluid.)

Thyroxine (T₄), Total (TT₄), Serum:

Normal: Radioimmunoassay (RIA), 5–12 μg/dL (SI: 65–156 nmol/L); competitive binding protein (CPB) (Murphy-Pattee), 4–11 μg/dL (SI: 51–142 nmol/L).

A. Precautions: None.

B. Physiologic Basis: The total thyroxine level does not necessarily reflect the physiologic hormonal effect of thyroxine. Levels of thyroxine vary with the concentration of the carrier proteins (thyroxine-binding globulin and prealbumin), which are readily altered by physiologic conditions such as pregnancy and by a variety of diseases and drugs. Any interpretation of the significance of total T_4 depends upon knowing the concentration of carrier protein either from direct measurement or from the result of the erythrocyte or resin uptake of triiodothyronine (T_3) (see below). It is the concentration of free T_4 and of T_3 that determines hormonal activity.

C. Interpretation:

1. Elevated in hyperthyroidism, with elevation of thyroxine-binding proteins, and at times with active thyroiditis or acromegaly.

2. Decreased in hypothyroidism (primary or secondary) and with decreased concentrations of thyroxine-binding proteins.

Thyroxine, Free, Serum:

Normal (equilibrium dialysis): 0.8–2.4 ng/dL. (SI: 0.01–0.03 nmol/L.) May be estimated from measurement of total thyroxine and resin T_3 uptake.

A. Precautions: None.

B. Physiologic Basis: The metabolic activity of T_4 is related to the concentration of free T_4. T_4 is apparently largely converted to T_3 in peripheral tissue. (T_3 is also secreted by the thyroid gland.) Both T_4 and T_3 seem to be active hormones.

C. Interpretation:

1. Elevated in hyperthyroidism and at times with active thyroiditis.

2. Decreased in hypothyroidism.

Thyroxine-Binding Globulin (TBG), Serum:

Normal (radioimmunoassay): 2–4.8 mg/dL.

A. Precautions: None.

B. Physiologic Basis: TBG is the principal carrier protein for T_4 and T_3 in the plasma. Variations in concentration of TBG are accompanied by corresponding variations in concentration of T_4 with intrinsic adjustments that maintain the physiologically active free hormones at proper concentration for euthyroid function. The inherited abnormalities of TBG concentration appear to be X-linked.

C. Interpretation:

1. Elevated in pregnancy, in infectious hepatitis, and in hereditary increase in TBG concentration.

2. Decreased in major depleting illness with hypoproteinemia (globulin), nephrotic syndrome, cirrhosis of the liver, active acromegaly, estrogen deficiency, and hereditary TBG deficiency.

Triiodothyronine (T₃) Uptake, Serum; Resin (RT₃U) or Thyroxine-Binding Globulin Assessment (TBG Assessment):

Normal: RT₃U, as percentage of uptake of ¹²⁵I-T₃ by resin, 25–36%; RT₃U ratio (TBG assessment) expressed as ratio of binding of ¹²⁵I-T₃ by resin in test serum/pooled normal serum, 0.85–1.15.

A. Precautions: None.

B. Physiologic Basis: When serum thyroxine-binding proteins are normal, more TBG binding sites will be occupied by T_4 in T_4 hyperthyroidism, and fewer binding sites will be occupied in hypothyroidism. ¹²⁵I-labeled T_3 added to serum along with a secondary binder (resin, charcoal, talc, etc) is partitioned between TBG and the binder. The binder is separated from the serum, and the radioactivity of the binder is measured for the RT₃U test. Since the resin takes up the non-TBG-bound radioactive T_3, its activity varies inversely with the numbers of available TBG sites, ie, RT₃U is increased if TBG is more nearly saturated by T_4 and decreased if TBG is less well saturated by T_4.

C. Interpretation:

1. RT₃U and RT₃U ratio are increased when available sites are decreased, as in hyperthyroidism, acromegaly, nephrotic syndrome, severe hepatic cirrhosis, and hereditary TBG deficiency.

2. RT₃U and RT₃U ratio are decreased when available TBG sites are increased, as in hypothyroidism, pregnancy, the newborn, infectious hepatitis, and hereditary increase in TBG.

Transaminases:

See Aminotransferases, above.

Triglycerides, Serum:

Normal: < 165 mg/dL. (SI: < 1.65 g/L.) See also Table 1.

A. Precautions: Subject must be in a fasting state (preferably for at least 16 hours). The determination may be delayed if the serum is promptly separated from the clot and refrigerated.

B. Physiologic Basis: Dietary fat is hydrolyzed in the small intestine, absorbed and resynthesized by the mucosal cells, and secreted into lacteals in the form of chylomicrons. Triglycerides in the chylomicrons are cleared from the blood by tissue lipoprotein lipase (mainly adipose tissue), and the split products are absorbed and stored. Free fatty acids derived mainly from adipose tissue are precursors of the endogenous triglycerides produced by the liver. Transport of endogenous triglycerides is in association with β-lipoproteins, the very low density lipoproteins. In order to ensure measurement of endogenous triglycerides, blood must be drawn in the postabsorptive state.

C. Interpretation: Concentration of triglycerides, cholesterol, and lipoprotein fractions (very low density, low-density, and high-density) is interpreted collectively. Disturbances in normal relationships of these lipid moieties may be primary or secondary in origin.

1. Elevated (hyperlipoproteinemia)–

a. Primary–Type I hyperlipoproteinemia (exogenous hyperlipidemia), type II hyperbetalipoproteinemia, type III broad beta hyperlipoproteinemia, type IV hyperlipoproteinemia (endogenous hyperlipidemia), and type V hyperlipoproteinemia (mixed hyperlipidemia).

b. Secondary–Hypothyroidism, diabetes mellitus, nephrotic syndrome, chronic alcoholism with fatty liver, ingestion of contraceptive steroids, biliary obstruction, and stress.

2. Decreased (hypolipoproteinemia)–

a. Primary–Tangier disease (α-lipoprotein deficiency), abetalipoproteinemia, and a few rare, poorly defined syndromes.

b. Secondary–Malnutrition, malabsorption, and, occasionally, with parenchymal liver disease.

Urea Nitrogen & Urea, Blood, Plasma, or Serum:

Normal: Blood urea nitrogen, 8–25 mg/dL (SI: 2.9–8.9 mmol/L). Urea, 21–53 mg/dL (SI: 3.5–9 mmol/L).

A. Precautions: *Do not use* ammonium oxalate or "double oxalate" as anticoagulant, for the ammonia will be measured as urea.

B. Physiologic Basis: Urea, an end product of protein metabolism, is excreted by the kidney. The urea concentration in the glomerular filtrate is the same as in the plasma. Tubular reabsorption of urea varies inversely with rate of urine flow. Thus, urea is a less useful measure of glomerular filtration than is creatinine, which is not reabsorbed. Blood urea nitrogen varies directly with protein intake and inversely with the rate of excretion of urea.

C. Interpretation:

1. Elevated in–

a. Renal insufficiency–Nephritis, acute and chronic; acute renal failure (tubular necrosis); and urinary tract obstruction.

b. Increased nitrogen metabolism associated with diminished renal blood flow or impaired renal function–Dehydration (from any cause) and upper gastrointestinal bleeding (combination of increased protein absorption from digestion of blood plus decreased renal blood flow).

c. Decreased renal blood flow–Shock, adrenal insufficiency, and, occasionally, congestive heart failure.

2. Decreased in hepatic failure, nephrosis not complicated by renal insufficiency, and cachexia.

Uric Acid, Serum or Plasma:

Normal: Males, 3–9 mg/dL (SI: 0.18–0.53 mmol/L); females, 2.5–7.5 mg/dL (SI: 0.15–0.45 mmol/L).

A. Precautions: If plasma is used, lithium oxalate should be used as the anticoagulant; potassium oxalate may interfere with the determination.

B. Physiologic Basis: Uric acid, an end product of nucleoprotein metabolism, is excreted by the kidney. Gout, a genetically transmitted metabolic error, is characterized by an increased plasma or serum uric acid concentration, an increase in total body uric acid, and deposition of uric acid in tissues. An increase in uric acid concentration in plasma and serum may accompany increased nucleoprotein catabolism (blood dyscrasias, therapy with antileukemic drugs), use of thiazide diuretics, or decreased renal excretion.

C. Interpretation:

1. Elevated in gout, preeclampsia-eclampsia, leukemia, polycythemia, therapy with antileukemic drugs and a variety of other agents, renal insufficiency, glycogen storage disease (type I), Lesch-Nyhan syndrome (X-linked hypoxanthine-guanine phosphoribosyltransferase deficit), and Down's syndrome. The incidence of hyperuricemia is greater in Filipinos than in whites.

2. Decreased in acute hepatitis (occasionally), treatment with allopurinol, and treatment with probenecid.

Uric Acid, Urine:

Normal: 350–600 mg/24 h on a standard purine-free diet. (SI: 2.1–3.6 mmol/24 h.) Normal urinary uric acid/creatinine ratio for adults is 0.21–0.59; maximum of 0.75 for 24-hour urine while on purine-free diet.

A. Precautions: Diet should be free of high-purine foods prior to and during 24-hour urine collection. Strenuous activity may be associated with elevated purine excretion.

B. Physiologic Basis: Elevated serum uric acid may result from overproduction or diminished excretion.

C. Interpretation:

1. Elevated renal excretion occurs in about 25–30% of cases of gout due to increased purine synthesis. Excess uric acid synthesis and excretion are associated with myeloproliferative disorders. Lesch-Nyhan syndrome (hypoxanthine-guanine phosphoribosyltransferase deficit) and some cases of glycogen storage disease are associated with uricosuria.

2. Decreased in renal insufficiency, in some case of glycogen storage disease (type I), and in any metabolic defect producing either lactic acidemia or β-hydroxybutyric acidemia. Salicylates in doses of less than 2–3 g/d may produce renal retention of uric acid.

NORMAL LABORATORY VALUES

(Blood [B], Plasma [P], Serum [S], Urine [U])

BLOOD, PLASMA, OR SERUM CHEMICAL CONSTITUENTS (Values vary with method used.)

Acetone and acetoacetate: [S] 0.3–2 mg/dL (3–20 mg/L).

α-Amino acid nitrogen: [S, fasting] 3–5.5 mg/dL (2.2–3.9 mmol/L).

Aminotransferases:
Aspartate aminotransferase (AST; SGOT) (varies with method used): [S] 6–25 IU/L at 30 °C; SMA, 10–40 IU/L at 37 °C; SMAC, 0–41 IU/L at 37 °C.
Alanine aminotransferase (ALT: SGPT) (varies with method used): [S] 3–26 IU/L at 30 °C; SMAC, 0–45 IU/L at 37 °C.

Ammonia: [B] < 110 μg/dL (< 65 μmol/L) (diffusion method). Do not use anticoagulant containing ammonium oxalate.

Amylase: [S] 80–180 units/dL (Somogyi). Values vary with method used.

$α_1$-Antitrypsin: [S] > 180 mg/dL.

Ascorbic acid: [P] 0.4–1.5 mg/dL (23–85 μmol/L).

Base, total serum: [S] 145–160 meq/L (145–160 mmol/L).

Bicarbonate: [S] 24–28 meq/L (24–28 mmol/L).

Bilirubin: [S] Total, 0.2–1.2 mg/dL (3.5–20.5 μmol/L). Direct (conjugated), 0.1–0.4 mg/dL (< 7 μmol/L). Indirect, 0.2–0.7 mg/dL (< 12 μmol/L).

Calcium: [S] 8.5–10.3 mg/dL (2.1–2.6 mmol/L). Values vary with albumin concentration.

Calcium, ionized: [S] 4.25–5.25 mg/dL; 2.1–2.6 meq/L (1.05–1.3 mmol/L).

β-Carotene: [S, fasting] 50–300 μg/dL (0.9–5.58 μmol/L).

Ceruloplasmin: [S] 25–43 mg/dL (1.7–2.9 μmol/L).

Chloride: [S or P] 96–106 meq/L (96–106 mmol/L).

Cholesterol: [S or P] 150–265 mg/dL (3.9–6.85 mmol/L). (See Lipid fractions.) Values vary with age.

Cholesteryl esters: [S] 65–75% of total cholesterol.

CO_2 content: [S or P] 24–29 meq/L (24–29 mmol/L).

Complement: [S] C3 ($β_{1C}$), 90–250 mg/dL. C4 ($β_{1E}$), 10–60 mg/dL. Total (CH_{50}), 75–160 mg/dL.

Copper: [S or P] 100–200 μg/dL (16–31 μmol/L).

Cortisol: [P] 8:00 AM, 5–25 μg/dL (138–690 nmol/L); 8:00 PM, < 10 μg/dL (275 nmol/L).

Creatine phosphokinase (CPK): [S] 10–50 IU/L at 30 °C. Values vary with method used.

Creatine phosphokinase isoenzymes: See Table 2.

Creatinine: [S or P] 0.7–1.5 mg/dL (62–132 μmol/L).

Cyanocobalamin: [S] 200 pg/mL (148 pmol/L).

Epinephrine: [P] Supine, < 100 pg/mL (< 550 pmol/L).

Ferritin: [S] Adult women, 20–120 ng/mL; men, 30–300 ng/mL. Child to 15 years, 7–140 ng/mL.

Folic acid: [S] 2–20 ng/mL (4.5–45 nmol/L). [RBC] > 140 ng/mL (> 318 nmol/L).

Glucose: [S or P] 65–110 mg/dL (3.6–6.1 mmol/L).

Haptoglobin: [S] 40–170 mg of hemoglobin-binding capacity.

Iron: [S] 50–175 μg/dL (9–31.3 μmol/L).

Iron-binding capacity: [S] Total, 250–410 μg/dL (44.7–73.4 μmol/L). Percent saturation, 20–55%.

Lactate: [B, special handling] Venous, 4–16 mg/dL (0.44–1.8 mmol/L).

Lactate dehydrogenase (LDH): (Varies with method.) [S] 55–140 IU/L at 30 °C; SMA, 100–225 IU/L at 37 °C; SMAC, 60–200 IU/L at 37 °C.

Lipase: [S] < 150 U/L.

Lipid fractions: [S or P] Desirable levels: HDL cholesterol, > 40 mg/dL; LDL cholesterol, < 180 mg/dL; VLDL cholesterol, < 40 mg/dL. (To convert to mmol/L, multiply by 0.026.)

Lipids, total: [S] 450–1000 mg/dL (4.5–10 g/L).

Magnesium: [S or P] 1.8–3 mg/dL (0.75–1.25 mmol/L).

Norepinephrine: [P] Supine, < 500 pg/mL (< 3 nmol/L).

Osmolality: [S] 280–296 mosm/kg water (280–296 mmol/kg water).

Oxygen:
Capacity: [B] 16–24 vol%. Values vary with hemoglobin concentration.
Arterial content: [B] 15–23 vol%. Values vary with hemoglobin concentration.
Arterial % saturation: 94–100% of capacity.
Arterial P_{O2} (P_{aO2}): 80–100 mm Hg (10.67–13.33 kPa) (sea level). Values vary with age.

P_{aCO2}: [B, arterial] 35–45 mm Hg (4.7–6 kPa).

pH (reaction): [B, arterial] 7.35–7.45 (H^+ 44.7–45.5 nmol/L).

Phosphatase, acid: [S] 1–5 units (King-Armstrong), 0.1–0.63 units (Bessey-Lowry).

Phosphatase, alkaline: [S] Adults, 5–13 units (King-Armstrong), 0.8–2.3 (Bessey-Lowry); SMA, 30–85 IU/L at 37 °C; SMAC, 30–115 IU/L at 37 °C.

Phospholipid phosphorus: [S] 5–12 mg/dL (1.45–2 g/L).

Phosphorus, inorganic: [S, fasting] 3–4.5 mg/dL (1–1.5 mmol/L).

Potassium: [S or P] 3.5–5 meq/L (3.5–5 mmol/L).

Protein:
Total: [S] 6–8 g/dL (60–80 g/L).
Albumin: [S] 3.5–5.5 g/dL (35–55 g/L).
Globulin: [S] 2–3.6 g/dL (20–36 g/L).
Fibrinogen: [P] 0.2–0.6 g/dL (2–6 g/L).
Separation by electrophoresis: See Table 5.

Prothrombin clotting time: [P] By control.

Pyruvate: [B] 0.6–1 mg/dL (70–114 μmol/L).

Serotonin: [B] 5.0–20 μg/dL (0.2–1.14 μmol/L).

Sodium: [S or P] 136–145 meq/L (136–145 mmol/L).

Specific gravity: [B] 1.056 (varies with hemoglobin and protein concentration). [S] 1.0254–1.0288 (varies with protein concentration).

Sulfate: [S or P] As sulfur, 0.5–1.5 mg/dL (156–468 μmol/L).

Transferrin: [S] 200–400 mg/dL (23–45 μmol/L).

Triglycerides: [S] < 165 mg/dL (1.9 mmol/L). (See Lipid fractions.)

Urea nitrogen: [S or P] 8–25 mg/dL (2.9–8.9 mmol/L). Do not use anticoagulant containing ammonium oxalate.

Uric acid: [S or P] Men, 3–9 mg/dL (0.18–0.54 mmol/L); women, 2.5–7.5 mg/dL (0.15–0.45 mmol/L).

Vitamin A: [S] 15–60 μg/dL (0.53–2.1 μmol/L).

Vitamin B_{12}: [S] > 200 pg/mL (> 148 pmol/L).

Vitamin D: [S] Cholecalciferol (D_3): 25-Hydroxycholecalciferol, 8–55 ng/mL (19.4–137 nmol/L); 1,25-dihydroxycholecalciferol, 26–65 pg/mL (62–155 pmol/L); 24,25-dihydroxycholecalciferol, 1–5 ng/mL (2.4–12 nmol/L).

Volume, blood (Evans blue dye method): Adults, 2990–6980 mL. Women, 46.3–85.5 mL/kg; men, 66.2–97.7 mL/kg.

Zinc: [S] 50–150 μg/dL (7.65–22.95 μmol/L).

HORMONES, SERUM OR PLASMA

Pituitary:
Growth hormone (GH): [S] Adults, 1–10 ng/mL (46–465 pmol/L) (by RIA).
Thyroid-stimulating hormone (TSH): [S] < 10 μU/mL.
Follicle-stimulating hormone (FSH): [S] Prepubertal, 2–12 mIU/mL; adult men, 1–15 mIU/mL; adult women, 1–30 mIU/mL; castrate or postmenopausal, 30–200 mIU/mL (by RIA).
Luteinizing hormone (LH): [S] Prepubertal, 2–12 mIU/mL; adult men, 1–15 mIU/mL; adult women, < 30 mIU/mL; castrate or postmenopausal, > 30 mIU/mL.
Corticotropin (ACTH): [P] 8:00–10:00 AM, up to 100 pg/mL (22 pmol/L).
Prolactin: [S] 1–25 ng/mL (0.4–10 nmol/L).
Somatomedin C: [P] 0.4–2 U/mL.
Antidiuretic hormone (ADH; vasopressin): [P] Serum osmolality 285 mosm/kg, 0–2 pg/mL; > 290 mosm/kg, 2–12 + pg/mL.

Adrenal:
Aldosterone: [P] Supine, normal salt intake, 2–9

ng/dL (56–250 pmol/L); increased when upright.

Cortisol: [S] 8:00 AM, 5–20 μg/dL (0.14–0.55 μmol/L), 8:00 PM, < 10 μg/dL (0.28 μmol/L).

Deoxycortisol: [S] After metyrapone, > 7 μg/dL (> 0.2 μmol/L).

Dopamine: [P] < 135 pg/mL.

Epinephrine: [P] < 0.1 ng/mL (< 0.55 nmol/L).

Norepinephrine: [P] < 0.5 μg/L (< 3 nmol/L).

See also Miscellaneous Normal Values.

Thyroid:

Thyroxine, free (FT$_4$): [S] 0.8–2.4 ng/dL (10–30 pmol/L).

Thyroxine, total (TT$_4$): [S] 5–12 μg/dL (65–156 nmol/L) (by RIA).

Thyroxine-binding globulin capacity: [S] 12–28 μg T$_4$/dL (150–360 nmol T$_4$/dL).

Triiodothyronine (T$_3$): [S] 80–220 ng/dL (1.2–3.3 nmol/L).

Reverse triiodothyronine (rT$_3$): [S] 30–80 ng/dL (0.45–1.2 nmol/L).

Triiodothyronine uptake (RT$_3$U); [S] 25–36%; as TBG assessment (RT$_3$U ratio), 0.85–1.15.

Calcitonin: [S] < 100 pg/mL (< 29.2 pmol/L).

Parathyroid: Parathyroid hormone levels vary with method and antibody. Correlate with serum calcium.

Islets:

Insulin: [S] 4–25 μU/mL (29–181 pmol/L).

C peptide: [S] 0.9–4.2 ng/mL.

Glucagon: [S, fasting] 20–100 pg/mL.

Stomach:

Gastrin: [S, special handling] Up to 100 pg/mL (47 pmol/L). Elevated, > 200 pg/mL.

Pepsinogen I: [S] 25–100 ng/mL.

Kidney:

Renin activity: [P, special handling] Normal sodium intake: Supine, 1–3 ng/mL/h; standing, 3–6 ng/mL/h. Sodium depleted: Supine, 2–6 ng/mL/h; standing, 3–20 ng/mL/h.

Gonad:

Testosterone, free: [S] Men, 10–30 ng/dL; women, 0.3–2 ng/dL. (1 ng/dL = 0.035 nmol/L.)

Testosterone, total: [S] Prepubertal, < 100 ng/dL; adult men, 300–1000 ng/dL; adult women, 20–80 ng/dL; luteal phase, up to 120 ng/dL.

Estradiol (E$_2$): [S, special handling] Men, 12–34 pg/mL; women, menstrual cycle 1–10 days, 24–68 pg/mL; 11–20 days, 50–300 pg/mL;

21–30 days, 73–149 pg/mL (by RIA). (1 pg/mL = 3.6 pmol/L.)

Progesterone: [S] Follicular phase, 0.2–1.5 mg/mL; luteal phase, 6–32 ng/mL; pregnancy, > 24 ng/mL; men, < 1 ng/mL. (1 ng/mL = 3.2 nmol/L.)

Placenta:

Estriol (E$_3$): [S] Men and nonpregnant women, < 0.2 μg/dL (< 7 nmol/L) (by RIA).

Chorionic gonadotropin: [S] Beta subunit: Men, < 9 mIU/mL; pregnant women after implantation, > 10 mIU/mL.

MISCELLANEOUS NORMAL VALUES

Adrenal hormones and metabolites:

Aldosterone: [U] 2–26 μg/24 h (5.5–72 nmol). Values vary with sodium and potassium intake.

Catecholamines: [U] Total, < 100 μg/24 h. Epinephrine, < 10 μg/24 h (< 55 nmol); norepinephrine, < 100 μg/24 h (< 591 nmol). Values vary with method used.

Cortisol, free: [U] 20–100 μg/24 h (0.55–2.76 μmol).

11,17-Hydroxycorticoids: [U] Men, 4–12 mg/24 h; women, 4–8 mg/24 h. Values vary with method used.

17-Ketosteroids: [U] Under 8 years, 0–2 mg/24 h; adolescents, 2–20 mg/24 h. Men, 10–20 mg/24 h; women, 5–15 mg/24 h. Values vary with method used. (1 mg = 3.5 μmol.)

Metanephrine: [U] < 1.3 mg/24 h (< 6.6 μmol) or < 2.2 μg/mg creatinine. Values vary with method used.

Vanillylmandelic acid (VMA): [U] Up to 7 mg/24 h (< 35 μmol).

Fecal fat: Less than 30% dry weight.

Lead: [U] < 80 μg/24 h (< 0.4 μmol/d).

Porphyrins:

Delta-aminolevulinic acid: [U] 1.5–7.5 mg/24 h (11.4–57.2 μmol).

Coproporphyrin: [U] < 230 μg/24 h (< 345 nmol).

Uroporphyrin: [U] < 50 μg/24 h (< 60 nmol).

Porphobilinogen: [U] < 2 mg/24 h (< 8.8 μmol).

Urobilinogen: [U] 0–2.5 mg/24 h (< 4.23 μmol).

Urobilinogen, fecal: 40–280 mg/24 h (68–474 μmol).

REFERENCES

Friedman RB et al: Effects of diseases on clinical laboratory tests. *Clin Chem* 1980;**26(Suppl 4):**1D.

Hansten PD, Lybecker LA: Drug effects on laboratory tests. In: *Basic & Clinical Pharmacology,* 2nd ed. Katzung BG (editor). Lange, 1984.

Lippert H, Lehmann HP: *SI Units in Medicine: An Introduction to the International System of Units With Conversion Tables and Normal Ranges.* Urban & Schwarzenberg, 1978.

Lundberg GD, Iverson C, Radulescu G (editors): Now read this: The SI units are here. (Editorial.) *JAMA* 1986;**255:**2329.

Powsner EK: SI quantities and units for American medicine. *JAMA* 1984;**252:**1737.

Scully RE et al: Normal reference laboratory values: Case records of the Massachusetts General Hospital. *N Engl J Med* 1986;**314:**39.

Sonnenwirth AC, Jarett L: *Gradwohl's Laboratory Methods and Diagnosis,* 8th ed. Vols 1 and 2. Mosby, 1980.

ABBREVIATIONS ENCOUNTERED IN BIOCHEMISTRY

A (Å)	Angstrom units(s) (10^{-10} m, 0.1 nm)
AA	Amino acid
α-AA	α-Amino acid
ACTH	Adrenocorticotropic hormone, adreno-corticotropin, corticotropin
Acyl-CoA	An acyl derivative of coenzyme A (eg, butyryl-CoA)
ADH	Alcohol dehydrogenase
ADH	Antidiuretic hormone (vasopressin)
AHG	Antihemophilic globulin
Ala	Alanine
ALA	Aminolevulinic acid
AMP	Adenosine monophosphate
Arg	Arginine
Asn	Asparagine
Asp	Aspartic acid
ATP	Adenosine triphosphate
BAL	Dimercaprol (British anti-lewisite)
cAMP	3′,5′-Cyclic adenosine monophosphate, cyclic AMP
CBG	Corticosteroid-binding globulin
CBZ	Carbobenzoxy
CCCP	m-Chlorocarbonyl cyanide phenyl-hydrazone
CCK(PZ)	Cholecystokinin (pancreozymin)
CDP	Cytidine diphosphocholine
Cer	Ceramide
cGMP	3′,5′-Guanosine monophosphate, cyclic GMP
CI	Chain-initiating
CMP	Cytidine monophosphate; 5′-phosphoribosyl cytosine
CoA·SH	Free (uncombined) coenzyme A. A pantothenic acid-containing nucleotide that functions in the metabolism of fatty acids, ketone bodies, acetate, and amino acids

$$\text{CoA·S·}\overset{\text{O}}{\underset{}{\text{C}}}\text{·CH}_3$$

CoA·S·C·CH₃	Acetyl-CoA, "activated acetate." The form in which acetate is "activated" by combination with coenzyme A for participation in various reactions
CPK	Creatine phosphokinase
CRH (CRF)	Corticotropin-releasing hormone
CRP	C-reactive protein
CTP	Cytidine triphosphate
Cys	Cysteine
D-	Dextrorotatory
D₂ (vitamin)	Ergocalciferol
D₃ (vitamin)	Cholecalciferol
1,25(OH)₂-D₃	1,25-Dihydroxycholecalciferol
dA	Deoxyadenosine
dC	Deoxycytosine
dG	Deoxyguanosine
DNA	Deoxyribonucleic acid
DNP	Dinitrophenol
Dopa	3,4-Dihydroxyphenylalanine
DPG	Diphosphoglycerate (bisphosphoglycerate)
DPN	Diphosphopyridine nucleotide (now replaced by NAD)
dT	Deoxythymidine

dTMP	Deoxythymidine 5′-monophosphate
dUMP	Deoxyribose uridine 5′-phosphate
E	Enzyme (also Enz)
E.C.	Enzyme code number (IUB) system
EDTA	Ethylenediaminetetraacetic acid. A reagent used to chelate divalent metals
Enz	Enzyme (also E)
Eq	Equivalent
eu	Enzyme unit
FAD	Flavin adenine dinucleotide (oxidized)
FADH₂	Flavin adenine dinucleotide (reduced)
FDA	Food & Drug Administration
FFA	Free fatty acids
Figlu	Formiminoglutamic acid
FMN	Flavin mononucleotide
FP	Flavoprotein
FSF	Fibrin stabilizing factor
FSH	Follicle-stimulating hormone
FSHRH (FSHRH)	Follicle-stimulating hormone-releasing hormone
g	Gram(s)
g	Gravity
Gal	Galactose
GalNAc	N-Acetylgalactosamine
GDP	Guanosine diphosphate
GFR	Glomerular filtration rate
GH	Growth hormone
GHRH (GHRF)	Growth hormone-releasing hormone
GHRIH (GHRIF)	Growth hormone release-inhibiting hormone (somatostatin)
GLC	Gas-liquid chromatography
Glc	Glucose
GlcNAc	N-Acetylglucosamine
GlcUA	Glucuronic acid
Gln	Glutamine
Glu	Glutamic acid
Gly	Glycine
GMP	Guanosine monophosphate
GnRH	Gonadotropin-releasing hormone
GTP	Guanosine triphosphate
Hb	Hemoglobin
hCG	Human chorionic gonadotropin
hCS	Human chorionic somatomammotropin
HDL	High-density lipoproteins
H₂folate	Dihydrofolate
H₄folate	Tetrahydrofolate
His	Histidine
HMG-CoA	β-Hydroxy-β-methylglutaryl-CoA
Hyl	Hydroxylysine
Hyp	4-Hydroxyproline
ICD	Isocitric dehydrogenase
IDL	Intermediate-density lipoproteins
IDP	Inosine diphosphate
IF	Initiation factor (for protein synthesis)
Ile	Isoleucine
IMP	Inosine monophosphate; hypoxanthine ribonucleotide
INH	Isonicotinic acid hydrazide (isoniazid)
ITP	Inosine triphosphate
ITyr	Monoiodotyrosine
I₂Tyr	Diiodotyrosine

IU	International unit(s)
IUB	International Union of Biochemistry
α-KA	α-Keto acid
kcal	Kilocalorie (calorie)
α-KG	α-Ketoglutarate
kJ	Kilojoule
K_m	Substrate concentration producing half-maximal velocity (Michaelis constant)
L-	Levorotatory
LCAT	Lecithin:cholesterol acyltransferase
LD	Lactate dehydrogenase (see also LDH)
LDH	Lactic dehydrogenase
LDL	Low-density lipoproteins
Leu	Leucine
LH	Luteinizing hormone
LHRH (LHRF)	Luteinizing hormone-releasing hormone
LLF	Laki-Lorand factor
LTH	Luteotropic hormone
Lys	Lysine
M	Molar
MAO	Monoamine oxidase
MCH	Mean corpuscular hemoglobin
MCHC	Mean corpuscular hemoglobin concentration
MCV	Mean corpuscular volume
Met	Methionine
mol	Mole(s)
MRF	Melanocyte releasing factor
MRH	Melanocyte releasing hormone
MRIH	Melanocyte release-inhibiting hormone
mRNA	Messenger RNA
MSH	Melanocyte-stimulating hormone
MW	Molecular weight
NAD	Nicotinamide adenine dinucleotide (oxidized)
NADH	Nicotinamide adenine dinucleotide (reduced)
NADP	Nicotinamide adenine dinucleotide phosphate (oxidized)
NADPH	Nicotinamide adenine dinucleotide phosphate (reduced)
NDP	Any nucleoside diphosphate
NeuAc	N-acetylneuraminic acid
NTP	Any nucleoside triphosphate
OA	Oxaloacetic acid
OD	Optical density
P_i	Inorganic phosphate (orthophosphate)
PCV	Packed cell volume
Phe	Phenylalanine
PIH (PIF)	Prolactin release-inhibiting hormone
PL	Pyridoxal
PLP	Pyridoxal phosphate
PP_i	Inorganic pyrophosphate
PRH (PRF)	Prolactin releasing hormone
PRIH (PRIF)	Prolactin release-inhibiting hormone
PRL	Prolactin
Pro	Proline

PRPP	5-Phosphoribosyl-1-pyrophosphate
PTA	Plasma thromboplastin antecedent
PTC	Plasma thromboplastin component
RBC	Red blood cell
RDA	Recommended daily allowance
RE	Retinol equivalents
RNA	Ribonucleic acid
RQ	Respiratory quotient
rRNA	Ribosomal RNA
S (Sf) units	Svedberg units of flotation
SDA	Specific dynamic action
SDS	Sodium dodecyl sulfate
Ser	Serine
SGOT	Serum glutamic oxaloacetic transaminase
SGPT	Serum glutamic pyruvic transaminase
SH	Sulfhydryl
SLR	*Streptococcus lactis* R
SPCA	Serum prothrombin conversion accelerator
SRIH (SRIF)	Somatostatin (growth hormone release-inhibiting hormone)
sRNA	Soluble RNA (same as tRNA, which term is preferred)
STP	Standard temperature and pressure (273° absolute, 760 mm Hg)
T_3	Triiodothyronine
T_4	Tetraiodothyronine; thyroxine
TEBG	Testosterone-estrogen-binding globulin
TG	Triacylglycerols (formerly called triglycerides)
Thr	Threonine
TLC	Thin layer chromatography
Tm_{Ca}	Tubular maximum for calcium
Tm_G	Tubular maximum for glucose
TPN	Triphosphopyridine nucleotide (now replaced by NADP)
TRH (TRF)	Thyrotropin-releasing hormone
Tris	Tris(hydroxymethyl)aminomethane, a buffer
tRNA	Transfer RNA (see also sRNA)
Trp	Tryptophan
TSH	Thyroid-stimulating hormone; thyrotropin
Tyr	Tyrosine
UDP	Uridine diphosphate
UDPGal	Uridine diphosphate galactose
UDPGlc	Uridine diphosphate glucose
UDPGlcUA	Uridine diphosphoglucuronic acid
UDPGluc	Uridine diphosphoglucuronic acid
UMP	Uridine monophosphate; uridine-5'-phosphate; uridylic acid
UTP	Uridine triphosphate
V_{max}	Maximal velocity
Val	Valine
VHDL	Very high density lipoproteins
VLDL	Very low density lipoproteins
VMA	Vanillylmandelic acid
vol%	Volumes percent

Index

A antigen, 598
  expression of, 598
A band, 620
A chain, 548
Abbreviations used in biochemistry, **665–666**
Abetalipoproteinemia, 230, 251
ABH(O) blood group system, 598
ABO blood groups, 598
ABO locus, and blood group antigens, 599
ABP. *See* Androgen-binding protein.
Absorption and digestion from gastrointestinal tract, **571–588**
Acanthosis nigricans, 470, 471
Acetaldehyde, 289
2-Acetaminofluorene, 636
Acetate, 207, 208, 291
Acetic acid, 131
Acetoacetate, 255, 257, 258, 262, 291, 292, 303
  formation of, 257
  serum, normal values, 661
Acetoacetyl enzyme, 205, 206
Acetoacetyl-CoA, 255, 256, 257, 282
Acetoacetyl-CoA synthetase, 257
Acetone, 254, 255
  serum, normal values, 661
Acetone bodies, 254
Acetyl (acyl)-malonyl enzyme, 206
Acetyl hexosamines, in glycoproteins, 128
Acetyl-carnitine, 199
Acetylcholine, 471, 623
Acetylcholine receptor, 623
Acetyl-CoA, 155, 188, 205, 207, 291, 302, 303, 304
  amino acids forming, 290
  in carbohydrate metabolism, 189
  catabolism of, **149–157**
  in cholesterol metabolism, 241
  and fatty acid oxidation, 199, 202, 203
  in food interconversion, 262
  in ketosis, 255
  in Krebs cycle, 152
  from oxidation of pyruvate, 163
  from threonine and glycine, 289
Acetyl-CoA carboxylase, 188, 207, 559
  and fatty acid biosynthesis, 202
  in IUB classification, 51
  quaternary structure of, 39
N-Acetylgalactosamine (GalNAc), 591, 592
  in glycoproteins, 128
N-Acetylgalactosamine sulfate, 128

N-Acetylglucosamine (GlcNAc), 128, 185, 592
  in glycoproteins, 128
N-Acetylglucosamine 1-phosphate, 185
N-Acetylglucosamine 6-phosphate, 185
N-Acetylglucosamine 6-sulfatase, 604
α-N-Acetylglucosaminidase, 603, 604
N-Acetylglutamate, 278
N-Acetylglutamate-γ-semialdehyde, 270
β-D-Acetylhexosaminidase, 603
N-Acetylmannosamine 6-phosphate, 185
N-Acetyl-5-methoxyserotonin, 313
N-Acetylneuraminic acid (NeuAc), 128, 129, 223, 224, 591
N-Acetylneuraminic acid 9-phosphate, 185
N-Acetylserotonin, 313
N-Acetyltyrosine, 292
Acid(s). *See also specific acid.*
  catalysis by, 78
  monoethenoid, 131
  monounsaturated, 131
  pantothenic, 53
  polyethenoid, 131
  polyunsaturated, 131
Acid maltase, 171
Acidemia
  aminobutyric, 316
  propionic, 305, 577
Aciduria
  dicarboxylic, 202
  methylmalonic, 174, 305, 577
Aconitase, 151
  in Krebs cycle, 152
Aconitate hydratase, 151
ACP. *See* Acyl carrier protein.
Acrodermatitis enteropathica, 213
2-Acroleyl-3-aminofumarate, 297
Acromegaly, 487
ACTH (adrenocorticotropic hormone), 238, 464, 475, 482, 483, 491
  human, structure of, 592
Actin, 620
  globular, 621
  monomeric, 621
  nonmuscle, 627, 628
α-Actin, 627
β-Actin, 627
γ-Actin, 627
Actin filaments, 627
  arrowhead formation of, 621
α-Actinin, 621, 628

Actin-myosin interactions, in striated and smooth muscle, 625
Active dihydroxyacetone, 177
Active glycolaldehyde, 177
Acyl carrier protein (ACP), 204, 206
Acyl enzyme, 205
Acyl(acetyl)-malonyl enzyme, 205
Acylcarnitine, 199
Acyl-CoA, 237, 255, 261
  in acylglycerol metabolism, 218
  in fatty acid oxidation, 198, 199
Acyl-CoA dehydrogenase, 103, 199, 201
Acyl-CoA synthetase, 98, 174, 198, 201, 218, 236, 237
  in mitochondria, 116
Acyl-CoA thioesters, 300, 301
1-Acyldihydroxyacetone phosphate, 219, 221
1-Acyldihydroxyacetone phosphate reductase, 219
Acylglycerol(s), 130
  biosynthesis of, 218
1-Acylglycerol 3-phosphate, 219
1-Acylglycerol 3-phosphate acyltransferase, 219
Acyltransferase, 221
  classification of, 51
Addison's disease, 493, 522
Adenine, 311, 336, 337, 362, 395
  in DNA, 377, 395, 401
  phosphoribosylation of, 349
  in RNA, 382
Adenine nucleotide(s), interconversion of, 98
Adenine nucleotide transporter, 118
Adenine phosphoribosyltransferase, 349
Adenosine, 300, 337, 339, 344, 345
  derivatives of, 337, 340
  phosphorylation of, to AMP, 350
  syn and anti configurations of, 338
Adenosine deaminase, deficiency of, 358
Adenosine diphosphate (ADP), 95, 96, 110, 188, 189, 191, 339, 625, 626
  structure of, 339
Adenosine kinase, 349, 350, 351
Adenosine monophosphate (AMP), 96, 188, 337, 339, 348, 627
  derivatives of, 342
  structure of, 339
Adenosine 3'-monophosphate, 339
Adenosine 5'-monophosphate, 339

Adenosine 3'-phosphate-5'-phospho-
  sulfate, 340
Adenosine triphosphate (ATP), 95,
  96, 110, 159, 164, 168, 172, 188,
  207, 218, 220, 235, 244, 273,
  299, 339, 340, 619, 626
  in actin-myosin reactions, 622
  in coupling, 95
  hydrolysis of, free energy of, 95
  magnesium complexes of, 95
  structure of, 339
S-Adenosylhomocysteine, 299
S-Adenosylmethionine, 96, 235, 299,
  310, 311, 312, 316, 340
  formation of, 299, 340
S-Adenosylmethionine decarboxylase,
  310, 312
Adenylate cyclase, 168, 169, 170,
  238, 340
Adenylate cyclase system, 475
Adenylate deaminase, 627
Adenylate kinase, 98, 191
  in mitochondria, 116
Adenylic acid, 338
Adenylosuccinase, 347, 348
Adenylosuccinate synthase, 348
ADH. See Antidiuretic hormone.
Adipic acids, 200
Adipose tissue, 239
  brown, in thermogenesis, 239
  in lipolysis, 238
  metabolic interrelationships of,
    with liver and extrahepatic
    tissue, 261
  metabolism of, 236, 237
ADP. See Adenosine diphosphate.
(ADP-ribose)$_n$, 402
Adrenal cortex, 239, 511
  hormones of, 511
    biosynthesis of, 512
    classification and mechanism of
      action of, 520
    metabolic effects of, 518
    pathophysiology of, 522
    regulation of synthesis of, 516
    secretion, transport, and
      metabolism of, 515
  zones of, 511
Adrenal hormones, 511–529
  normal values, 662
Adrenal hyperplasia, congenital, 523
Adrenal insufficiency
  primary, 522
  secondary, 523
Adrenal medulla, 524
  pathophysiology of, 529
β-Adrenergic compounds, inotropic
  effects on heart, 625
α$_2$-Adrenergic hormones, 475
β-Adrenergic hormones, 475
Adrenergic receptors, actions of, 528
α-Adrenergic receptors, 528
β-Adrenergic receptors, 528
β-Adrenergic stimulation of smooth
  muscle, 624
Adrenocorticotropic hormone. See
  ACTH.
Adrenogenital syndrome, 523
Aerobic organisms, respiration in,
  major pathway of, 150

Aglycone, 123
Agmatine, 587
Agonist(s), 520, 521
  partial, 520, 521
Ala. See Alanine.
Alanine (Ala), 16, 20, 156, 188, 262,
  265, 267, 272, 275, 276, 286,
  307, 626
  biosynthesis of, 267
  conversion of, 307
  isoelectric zwitterionic structure of,
    15
  in kidney, 276
  in liver and gut, 275
  in muscle, 275
α-Alanine, 307
β-Alanine, 19, 307
Alanine transaminase, 272
Alanine-glucose cycle, 276
Alanine-pyruvate transaminase, 272
β-Alanyl dipeptides, 307
Alanyl-valine, 20
Albinism, 314
  ocular, 314
  oculocutaneous, 314
    tyrosinase-negative, 314
    tyrosinase-positive, 314
Albinoidism, oculocutaneous, 314
Albumin, 32, 607
  serum, 218, 229
Albumin-FFA, 227
Alcohol(s). See also specific alcohol.
  with lipid molecules, 133
  polyhydric, 120
Alcohol dehydrogenase, 103, 235
  in IUB classification, 51
Alcoholism, 213, 235, 251
Aldehyde(s)
  fatty, 133
Aldehyde dehydrogenase, 103, 236,
  289
Aldehyde-lyases, classification of,
  51
Aldolase, 159, 160, 178
  in IUB classification, 51
Aldolase A, 159, 182
Aldolase B, 159, 182
Aldose(s), 120
  D series, structural relations of, 124
Aldose reductase, 182, 183
Aldose-ketose isomerism, 122
Aldosterone, 464, 511, 513, 515,
  516, 522
Aldosteronism
  primary, 523
  secondary, 523
Alkali reserve, 255
Alkaline tide, 579
Alkaptonuria, 291, 292
1-Alkenyl, 2-acyl glycerol 3-phospho-
  ethanolamine plasmalogen, 221
1-Alkyl, 2-acyl glycerol, 221
1-Alkyl, 2-acyl glycerol 3-phosphate,
  221
1-Alkyl, 2-acyl glycerol 3-phospho-
  ethanolamine, 221
1-Alkyldihydroxyacetone phosphate,
  221
1-Alkylglycerol 3-phosphate, 221
Allantoin, 102, 352, 353

Allopurinol, 342, 359
  structure of, 343
Allosteric effectors, 87
  of allosteric enzymes, 89
Allosteric enzyme(s), 89
  kinetics of, 89
  models of, 90
Allosteric inhibitor, 89
Allosteric proteins, 187
Allosteric regulation, 87
  of aspartokinase, 88
  of pyrimidine biosynthesis, 355
Allosteric site, 87
  allosteric enzymes, 88
Alu family, in DNA, 393
Amenorrhea, 487
Ames assay, 638
Amination, 274
Amino acid(s), 13–20
  absorption of, 586
  in acetyl-CoA formation, 290
  aspartate family of, 84
    biosynthesis of, 270
  biosynthesis of, 265–270
  branched-chain, 84, 302, 303
  catabolism of, 271, 281–305
  chemical reactions of, 19
  in collagen, 632
  color reactions of, 19
  conversion of, to specialized prod-
    ucts, 306–318
  and enzymes, 83
  gluconeogenic, 276
  glycogenic, 281
  in gut, 275
  interorgan exchange of
    in fed state, 276
    in postabsorptive state, 275
  in intestines, 275
  ionic forms of, 14
  ketogenic, 281
  in α-ketoglutarate formation, 283
  in kidney, 275
  in liver, 275
  metabolism of, 144, 145
  in muscle, 275
  with non-α-amino groups, 19
  nutritionally essential, 270
    biosynthesis of
      from aspartate, 270
    requirements in humans, 572,
      573
  nutritionally nonessential, 267
  in oxaloacetate formation, 282
  and peptides, 24–31
    determination of, 27
  pI of, 15
  properties of, 19
  in proteins, identification of, 21
  in pyruvate formation, 285
  reaction with 1-fluoro-2,4-dini-
    trobenzene, 28
  requirements in humans, 265
    in A and B chains of insulin, 38
    garbled, 419
  solubility of, 19
  in splanchnic tissues, 275
  structures of, 17
  in succinyl-CoA formation, 299
  sulfur-containing, 287

Amino acid(s) (cont'd)
 sulfur-containing (cont'd)
  inborn errors of metabolism of, 289
 synthesis of, from amphibolic intermediates, 265
 weak acid groups of, 14
α-Amino acid(s), 267
 not present in proteins, 18
L-α-Amino acid(s)
 conversion of carbon skeletons to
   amphibolic intermediates, 281
 nonpolar/polar classification of, 15
 in proteins, 15, 16, 17
L-Amino acid dehydrogenase, 102
Amino acid nitrogen, 272
 catabolism of, **271–280**
 serum, normal values, 661
L-Amino acid oxidase, 102
Amino acid residues
 abbreviations for, 3-letter and one-
   letter, 25
 and alpha-helix formation, 34, 35
 and helix formation, 35
Amino acid sequences
 at catalytic sites, 66
2-Amino *cis, cis* muconate semialdehyde, 297
Amino sugars, 124
 metabolism of, 184, 185
Aminoacyl-tRNA, 416, 423
Aminoacyl-tRNA synthetase, 416
γ-Aminobutyrate (GABA), 19, 308,
   316, 317
 metabolism of, 317
2-Amino-3-hydroxybenzoyl pyruvate,
   298
Aminoimidazole carboxylate ribosyl-
   5-phosphate, 347
Aminoimidazole ribosyl 5-phosphate,
   347
Aminoimidazole ribosylphosphate
   synthetase, 346
Aminoimidazole succinyl carboxam-
   ide ribosyl-5-phosphate, 347
Aminoimidazole succinyl carboxam-
   ide ribosylphosphate (SAICAR),
   346
β-Aminoisobutyrate, 19, 304, 355, 356
β-Aminoisobutyric acid. *See* β-
   Aminoisobutyrate.
β-Aminoisobutyric aciduria, 359
α-Amino-β-ketoadipate, 306, 307, 320
δ-Aminolevulinate, 306, 320, 321
 dehydrase, 321
 synthase, 320, 321
Aminolipids, 130
2-Amino-6-oxypurine, 336
Aminopeptidases, 54, 583, 584
6-Aminopurine, 336
Aminopyrine, 106, 182
Aminotransferase(s), 272
 normal values, 661
 serum, normal values, 651
Ammonia, 274, 588
 blood, normal values, 651, 661
 excretion of, 272
 formation of, 274
 intoxication due to, 274
 release of, by muscle, 627
 transport of, 274

Ammonium ion, 274
Ammonotelic, defined, 272
Ammonotelic organisms, 272
Amobarbital, in oxidative phosphory-
   lation, 112
AMP. *See* Adenosine monophosphate.
5'-AMP, 168, 193
Amphibolic intermediates, 281, 282
Amphibolic pathways, 142
Amphipathic molecule, defined, 447
Ampicillin resistance gene, 366
Amylase, 579, 582
 serum, normal values, 661
 serum or urine, normal values, 651,
   652
Amylo-(1→6)-glucosidase, 167
Amylopectin, 125, 127
Amylopectinosis, 177
Amylose, 125, 127
Amylo-(1→4)→(1→6)-transglucosi-
   dase, 166
Anabolic pathways, 142
 flow of metabolites through, 83
Anabolism, defined, 94
Anaerobic dehydrogenases, in oxida-
   tion, 101
Andersen's disease, 171
Androgen(s), 471, 516, 536
 actions of, mechanism of, 534
 biosynthesis of, 515, 531
Androgen resistance, steps involved
   in, 536
Androgen-binding protein (ABP), 533
Androgenic steroids, 511
Androstenedione, 511, 513, 515
Androsterone, 532
Anemia, 571
 due to malabsorption, 587
Angiotensin(s), 193
 formation and metabolism of, 517
Angiotensin I, 517
Angiotensin II, 475, 482, 517
Angiotensin III, 518
Angiotensinases, 518
Angiotensin-converting enzyme, 518
Angiotensinogen, 517
Anhydride(s), 96
-anoic, in fatty acid nomenclature,
   130
Anomer(s), α and β, 122
Anomeric carbon atom, 122
Anserine, 308, 309
Antagonist(s), 520, 521
Anterior pituitary gland, 196
 hormones of, 483
Antibodies, 612
 class switching of, 439, 449
 IgG, human, model of, 609
 plasma, 609
Anticoagulants, 615
Anticodon loop of RNA, 385
Anticodon region, 415
Antidiuretic hormone (ADH), 471,
   475, 482, 493, 494
Antigen(s), 609
 and antibodies, 438, 439
 blood group, 598
 H-Y, 544
Antigen-binding fragments, 609
Anti-Lepore hemoglobins, 393

Antimycin A, and oxidative phospho-
   rylation, 112
Antioxidants, 138
Antiparallel β-pleated sheet, 35
Antipyrine, 182
Antithrombin, inactivation of serine
   proteases by, 605
Antithrombin III, 616
Antithyroid drugs, thiourea class of,
   499
 structure of, 498
Antiport transport system, 456
α_1-Antitrypsin, serum, normal values,
   661
Apo-B receptor, 231
Apo-D, 232
Apo-E, 251
Apo-E receptor, 231
Apoenzymes, 50, 103
Apolipoprotein(s), 227, 231
Apoprotein(s), 227
 arginine-rich, 228
 of human plasma lipoproteins,
   228
 nomenclature of, 227
Arabinose, 128
 in glycoproteins, 128
D-Arabinose, 123
Arabinosyl cytosine, 342
 structure of, 343
AraC, 342
Arachidic acid, 131
Arachidonate. *See* Arachidonic acid.
Arachidonic acid, 132, 210, 211, 213,
   214, 215, 216, 574
 conversion of linoleic acid to, 211,
   213
Arg. *See* Arginine.
Argentaffinoma, 312
Arginase, 277, 278, 279
 in amino acid metabolism, 86
 regulation of, 85
Arginine (Arg), 17, 262, 265, 270,
   283, 310, 587
 biosynthesis of, 270
 cleavage of, to ornithine and urea,
   277
 conversion of, 310
 metabolism of, 309
L-Arginine, 278, 317
 catabolism of, 284
Arginine-rich apoprotein, 228
Argininosuccinase, 277, 278, 279
Argininosuccinate, synthesis of, 277
Argininosuccinate synthase, 279
Argininosuccinic acid, 18
Argininosuccinic acid synthase, 278
Argininosuccinicaciduria, 279
Aromatic L-amino acid decarboxylase,
   309
Arsenate, 161
Arsenic, 235
Arsenite, 164
 and citric acid cycle, 152
 and Krebs cycle, 152
1-Arseno-3-phosphoglycerate, 161
Arthritis, gouty, 357
Arylsulfatase A, 224, 225, 603, 604
Arylsulfatase B, 603, 604
L-Ascorbate, 181

Ascorbic acid, 180, 575
  major characteristics of, 575
  plasma, normal values, 661
-ase, in enzyme nomenclature, 50
Asn. *See* Asparagine.
Asp. *See* Aspartate.
Asparaginase, 283
L-Asparaginase, 274
Asparagine (Asn), 16, 265, 267, 282, 576
  biosynthesis of, 267
L-Asparagine, 283
  catabolism of, to amphibolic intermediates, 283
Asparagine synthetase, 266
Aspartate (Asp), 16, 155, 156, 262, 265, 267, 282
  biosynthesis of, 267
  protionic equilibria of, 15
L-Aspartate, 283
Aspartate family of amino acids, 84
  biosynthesis of, 270
Aspartate transaminase, quaternary structure of, 39
Aspartate transcarbamoylase, 86, 88, 89, 353, 355
  allosteric regulation of, 88
  in purine or pyrimidine metabolism, 86
Aspartate transcarbamoylase reaction, 89
Aspartic acid. *See* Aspartate.
Aspartokinase, allosteric regulation of, 88
A-specific transferase, 599
Aspirin, 215
  and G6PD deficiency, 179
Asthma, 471
Ataxia-telangiectasia, DNA in, 402
ATCase. *See* Aspartate transcarbamoylase.
ATCase reaction, 89
Atheromas, 251
Atherosclerosis, 250
ATLC. *See* chromatography, adsorption thin-layer.
Atom(s)
  carbon
    anomeric, 122
    asymmetric, 120
    numbering of, 512
ATP. *See* Adenosine triphosphate.
ATP synthase
  membrane-located, in oxidative phosphorylation, 113
  proton-translocating, 116
ATP/ADP cycle, 96
ATPase, 621, 622
  quaternary structure of, 39
ATP-citrate lyase, 155, 188, 207, 239
Atractyloside, 112, 118
Autocrine function, 463
Automated Edman technique for determination of polypeptide structures, 28
Automated ion exchange chromatography, 22
Auto-oxidation, 138
Auto-oxidizable flavoproteins, 273
Autoradiography, defined, 375

Axial ratios, 32
Axons, 629
Axoplasmic flow, 629
8-Azaguanine, 342
Azaserine, 348
Azathioprine, 342, 343
6-Azauridine, 342, 360
6-Azauridylate, 360

B antigen, 599
  expression of, 599
B chain, 548
B vitamins, 53
Bacteria
  decarboxylation due to, 587
  intestinal, 588
  symbiotic, 588
Bacteriophage, defined, 375
Bacteriophage lambda, 432
Bam HI, 60
Banding, of chromosome complement, 391
Barbital, 180
Barbiturates, 182
  in oxidative phosphorylation, 112
Baroreceptors, 494, 517
Basal metabolic rate, 572
Base(s)
  adenine, in DNA, 378, 395
  catalysis by, 78, 79
  conjugate, 14
  cytosine, in DNA, 378, 395
  guanine, in DNA, 378, 395
  pairing of, in DNA, 362, 380
  thymine, in DNA, 378, 395
  total serum, normal values, 661
  uracil, in DNA, 401
Behenic acid, 131
Benign prostatic hypertrophy, 535
Benzopyrene, 636
Benzoate, 307
Benzoquinoneacetate, 292
Benzoyl-CoA, 307
Benzphetamine, 106
Benzpyrene, 106
Bernard, Claude, 82
Beta anomers, 122
Betaine, 268
Betaine aldehyde, 268
Bicarbonate, 46
  serum, normal values, 661
  serum or plasma, normal values, 652
Bilayer, 448
Bile, 248, 580
  bilirubin secretion into, 330
  cholesterol solubility in, 581
  composition of, 580, 581
Bile acids, 248, 580
  biosynthesis and degradation of, 248, 249
Bile pigments
  formation of, 328
  and porphyrins, **319–334**
  structure of, 331
Bile salts, 248, 581, 584
Biliary tree, obstruction of, 332
Bilirubin
  conjugation of, 328
    with glucuronic acid, 330

Bilirubin (cont'd)
  metabolism of, in intestine, 330
  secretion of, into bile, 330
  serum, normal values, 652, 661
  uptake of, by liver, 328
Bilirubin diglucuronide, 330
  structure of, 329
Bilirubin-urobilinogen cycle, 333
Biliverdin IX-α, 328
Biliverdin reductase, 328
Biochemical thermodynamics, 93
Bioenergetics, **93–99**
  of coupled reactions, 97
  defined, 93
Biosynthetic pathway(s)
  branched or multiple-branched, 84, 85
  in branched sites of repression, 84, 85
Biotin, 172, 174, 202, 577
  major characteristics of, 575
Biotin-dependent enzymes, 174
1,3-Bisphosphoglycerate, 96, 97, 159, 160, 173
2,3-Bisphosphoglycerate (DPG), 46, 47, 161, 162
  in erythrocytes, 87, 162
  binding of, 46, 47
Bisphosphoglycerate mutase, 162
2,3-Bisphosphoglycerate phosphatase, 162
Bleeding, due to malabsorption, 587
Bleeding time, 616
BLI. *See* Bombesinlike immunoreactivity.
Blood
  buffering capacity of, 46
  clotting of, **612–617**
  coagulation of, **612–617**
    final common pathway of, 614
    tissue factor in, 614
  functions of, 607
  protein molecules in, dimensions and molecular weights of, 608
Blood cells, red, sickle, 48
Blood glucose, 194
Blood group antigens, 598
Blood plasma, **607–612**
  lipids of, 226
Blood volume, normal values, 662
Blot transfer(s), 368, 369
BMR. *See* Basal metabolic rate.
Bohr effect, 46
  at submolecular level, 46
Bomb calorimeter, 572
Bombesin, 566, 568
Bombesinlike immunoreactivity (BLI), 565
Bond(s)
  covalent
    formation of, 50, 63
    rupture of, 63
  disulfide, 33
  electrostatic, 34
  high-energy, 154
  hydrogen, 34
  peptide, 33
  salt, 34
  stability of, 34
Bone mineralization, 502

Boston, hemoglobin M, 419
Bovine proteases, 66
Bovine ribonuclease, structure of, 38
Bradykinin, 25, 518
Brain, amino acids in, 275
Branched-chain ketonuria, 303
Branching, 165
Branching enzyme, 166, 167
  absence of, 171
Broad beta disease, 251
Brown adipose tissue, in thermogenesis, 239
Bruising, due to malabsorption, 587
Brunner's glands, 583, 584
B-specific transferase, 599
"Bubbles, replication," 398
Buffering capacity of blood, 46
Burkitt's lymphoma, 642
Butyric acid, 131
Butyryl-CoA, 206

C peptide, properties of, 551
C_H gene, 612
C_H region, 609
C_L gene, 612
C_L region, 609
C_{17-20} lyase, 531
Ca^{2+}-calmodulin-sensitive phosphorylase b kinase, 169
CAA. *See* Carbamoyl aspartic acid.
Cadaverine, 587
  structure of, 310
Caffeine, 238, 336, 337
Calcification, ectopic, 502
Calcitonin (CT), 475, 503, 510
Calcitriol, 471, 503, 507
Calcium, 479, 584
  homeostasis of, 503
    hormones involved in, 503
  and hormone action, 479, 480
  metabolism of, and hormonal regulation, **502–510**
  serum, normal values, 661
  serum or urine, normal values, 652, 653
Calcium ion, in muscle contraction and relaxation, 623
Calcium-binding protein (CBP), 509
Calmodulin, 169, 193, 479, 621, 624, 625, 629
Calmodulin-dependent protein kinase, 170
Caloric homeostasis, 186
Calorie balance, 571
Calorimeter, bomb, 572
Calorimetry, of food, 572
Calsequestrin, 623
cAMP (cyclic adenosine monophosphate; cyclic AMP), 168, 169, 170, 216, 238, 261, 335, 340
  destruction of, by phosphodiesterase, 340
  formation of, from ATP, 340
  as the second messenger, 475
Cancer, 598, **635–648**
  and biochemistry, 635
  causes of, 636
CAP. *See* Carbamoyl phosphate.

CAP. *See* Catabolite gene activator protein.
Capric acid, 131
Caproic acid, 131
Caprylic acid, 131
Carbamoyl aspartic acid (CAA), 354
Carbamoyl phosphate (CAP), 96, 278, 353, 354
  synthesis of, 277
Carbamoyl phosphate synthase, 277, 278, 279, 353, 354, 356
N-Carbamoyl-β-alanine, 356
N-Carbamoyl-β-aminoisobutyrate, 356
Carbohydrate(s), **120–129**
  absorption of, 583
    defects in, 585
  blood glucose as source, 194
  of cell membranes, 129
  complex, 589
  conversion into fat, 260
  dietary, requirement for, 574
  energy available in, 572
  interconversion of, 262
  metabolism of, 143
    economics of, 260
    regulation of, **186–197**
      at cellular and enzymatic level, 189
  oxidation of, energetics of, 164
  storage of, 165
Carbohydrate excess syndrome, 239
Carbomycin, 125
Carbon atoms
  anomeric, 122
  asymmetric, 120, 121
  numbering of, 512
Carbon dioxide (CO_2), 172, 285
  blood transport of, 46
  serum or plasma, normal values, 661
Carbon dioxide-biotin enzyme complex, formation of, 174
Carbon monoxide, 101
  in oxidative phosphorylation, 112
Carbon skeletons
  of amino acids, catabolism of, **281–305**
  of common L-α-amino acids, conversion of, 281
Carbon tetrachloride, 235
Carbonic acid, formation and dissociation of, 46
Carbonic anhydrase, in carbonic acid formation, 46
Carbon-oxygen lyases, classification of, 51
Carboxin, 112
Carboxylate ion, 14
Carboxypeptidase(s), 54, 581, 582, 584
Carcinogen(s)
  chemical, 636, 638
  direct, 636
  proximal, 636
  ultimate, 636
Carcinoid, 312
Cardiac glycosides, 124
Cardiac muscle, 619
Cardiolipin, 134, 219
  biosynthesis of, 219, 220
  in mitochondria, 116

Carnitine, 199
  deficiency of, 200
  in fatty acid metabolism, 199
Carnitine acetyltransferase, 199
Carnitine acyltransferase, 199
Carnitine palmitoyltransferase, 199, 258, 259
  deficiency of, 202
Carnosinase deficiency, 308
Carnosine, 308, 309
β-Carotene
  major characteristics of, 576
  serum, normal values, 661
Carrier proteins, 467
Cartilage, 634
Casein, of milk, 584
Catabolic pathways, 142
  flow of metabolites through, 82, 83
Catabolism, defined, 94
Catabolite gene activator protein (CAP), 430
Catabolite gene activator protein-cAMP complex, 431
Catabolite regulatory protein (CRP), 476
Catabolite repression, 84, 430
Catalase(s), 101, 105, 273, 321
  in IUB classification, 51
  in oxidative reactions, 105
  quaternary structure of, 39
Catalysis, 50
  by acid, 78, 79
  by base, 78, 79
  metal ions in, 80
Catalyst(s)
  defined, 50
  in formation of productive transition states, 62
  nonprotein, 50
  protein, enzymes as, 50
  reaction-specific, enzymes as, 50
Catalytic efficiency of enzymes, regulation of, 85
  covalent modification in, 90
Catalytic site(s), 64
  amino acid sequences at, 66
  of specific enzymes, 65
Cataract, diabetic, 183
Catechol estrogens, 537
Catechol oxidase, 102
Catecholamine(s), 193, 464
  biosynthesis of, 524, 525, 526
    regulation of, 527
  classification and mechanism of action of, 528
  metabolism of, 526
  storage and release of, 526
Catecholamine hormones, 524
Catechol-O-methyltransferase (COMT), 527
Cathepsin B, 505
Cathepsin D, 505
CBG. *See* Corticosteroid-binding globulin.
CBP. *See* Calcium-binding protein.
CCCP. *See* m-Chlorocarbonyl cyanide phenylhydrazone.
CCK. *See* Cholecystokinin.
cDNA, defined, 375
CDP. *See* Cytidine diphosphate.

CDP-choline, 219, 220, 223
CDP-diacylglycerol, 219, 220
CDP-diacylglycerol inositol trans-
    ferase, 219, 220
CDP-ethanolamine, 220
Celiac sprue, 585
Cell(s)
    composition of, and extracellular
        fluid, 446
    contact and communication be-
        tween, 462
    eukaryotic, compartmentation of, 86
    follicular, 489
    granulosa, 536
    idealized, in steady state, 83
    intestinal, 230
        mucosal, microvilli of, 627
    juxtaglomerular, 517
    Leydig, 530
    membrane(s) of, carbohydrates of,
        129
    motility of, 627
    normal environment of, mainte-
        nance of, 445
    parenchymal, hepatic, 229
    plasma, 609
    red, sickle, 48
    Sertoli, 489, 530
    in steady-state system, 82
Cell cycle, DNA synthesis during,
    400
Cellobiose, 126
Cellulose, 125
Centromere, defined, 391
Cephalin, 134, 220
Cer. See Ceramide.
Ceramidase, deficiency of, 224
Ceramide, 135, 222, 223, 224
Ceramide lactosidase, deficiency of,
    224
Ceramide lactoside lipidosis, 224
Cerebrohepatorenal syndrome, 202
Cerebronic acid, 136, 223
Cerebroside(s), 183, 223, 447
    biosynthesis of, 223
Ceruloplasmin, serum, normal values,
    653, 661
Cetyl alcohol, 133
cGMP (cyclic guanosine monophos-
    phate; cyclic GMP), 335, 341,
    479
    structure of, 341
Chain breaks, in DNA, 401
Chain elongation, microsomal system
    for, 208
Chemical hypothesis, of oxidative
    phosphorylation, 113
Chemical reactions
    coupled, bioenergetics of, 97
    endergonic, defined, 94
    energy barriers for, 62
    enzyme-catalyzed, 186, 187
        control of, 187
        test for, 37
    equilibrium and nonequilibrium,
        286
    exergonic, defined, 94
    kinetic (collision) theory for, 62
    reversible, 82
Chemiosmotic theory, 113, 240

Chenodeoxycholic acid, 581
Chenodeoxycholyl-CoA, 249
Chitin, 128
Chloride major characteristics
    of, 577
Chloride, normal values, 653, 661
Chlorobutanol, 180
m-Chlorocarbonyl cyanide phenylhy-
    drazone (CCCP), in oxidative
    phosphorylation, 112
Chloroform, 235
Chlorophyll, 319
Cholecalciferol, major characteristics
    of, 576
Cholecystokinin (CCK), 564, 565,
    566, 568, 584
Δ7,24-Cholestadienol, 243
Cholestasis, 248
Cholesterol, 130, 137, 222, 229, 231,
    232, 234, 241, 244, 250, 251,
    447, 513, 574
    in bile, 580, 581
Cholesterol, biosynthesis of, 241, 243
    excretion of, 248
    free, 232
        in blood plasma, 226
    and polyunsaturated fatty acids, 250
    serum or plasma, normal values,
        653, 661
    side-chain cleavage of, 513
    solubility of, in bile, 581
    synthesis of, 241
    total, in blood plasma, 226
    transport of, 245, 246
Cholesterol desmolase deficiency, 523
Cholesterol esterase, 582
Cholesterol transport, reverse, 233
Cholesterolemia, 251
Cholesteryl ester(s), 130, 222, 231,
    232, 234, 250
    serum, normal values, 661
Cholesteryl ester hydrolase, 582, 584
Cholesteryl ester storage disease, 251
Cholesteryl ester transfer protein, 232,
    248
Cholestyramine, 250
Cholic acid, 248
Choline, 134, 135, 221, 222, 235,
    268, 587
    active, 220
Choline acyltransferase, in IUB
    classification, 51
Choline kinase, 219
Cholyl-CoA, 249
Chondroitin, 185
Chondroitin sulfate(s), 180, 601, 603
    structure of, 600
Chondroitin 4-sulfate, 128
Choriocarcinoma, 544
Chorionic gonadotropin, 488
Chorionic somatomammotropin, 484,
    487
Chromaffin granules, 526
Chromatid exchanges, sister, 395
    in haploid genome, 391
    nucleoproteins within, 391
Chromatin, 387
    "active," 389, 484
    30-nm fiber of, 389
    10-nm fibril of, 389

Chromatogram, 2-dimensional, of
    natural amino acids, 21
Chromatography, 20
    adsorption thin-layer (ATLC), 22
    affinity technique of, 56
    for enzyme isolation, 55
        dye ligand, 56
        hydrophobic ligand, 56
    gas-liquid (GLC), 139
    gel filtration, 56
    ion exchange, automated, 22
    normal partition, 21
    paper, 2-dimensional, 21
    partition thin-layer (PTLC), 21
    phase relationships for, 20
    reversed phase high-pressure liquid,
        27
    reversed phase partition, 21
    thin-layer, 21, 140
        separation of major lipid classes
            by, 139
Chromium, major characteristics of,
    578
Chromosomal sex, 544
Chromosome(s), 391
    double-minute, 643
    integration of, 393
    interphase, chromatin fibers in, 389
    metaphase, 391
    Philadelphia, 642
    polytene, 391
    recombination of, 393
    sex, 544
Chromosome walking, 374, 375
Chronotropic effect, 528
Chyle, 586
Chylomicron(s), 229, 261, 576, 582,
    586
    catabolism of, 230
    formation of, 229
        and secretion of, by intestinal
            cells, 230
    metabolic fate of, 231
    nascent, 231
Chylomicron remnant, 228, 231, 232,
    251
Chylothorax, 586
Chyluria, 586
Chyme, 580
Chymosin, 580
Chymotrypsin, 54, 76, 581, 584
    in IUB classification, 51
α-Chymotrypsin, 77
π-Chymotrypsin, 77
Chymotrypsin A, 66
Chymotrypsin B, 66
Chymotrypsinogen, 581, 584
Cilia, 629
Circulation, 248
Cirrhosis, 213, 235
Cis-, defined, 512
Cis-aconitate, 151, 152
Δ3-Cis-Δ6-cis-dienoyl-CoA, 203
Cis-prenyl transferase, 242
Cis-trans isomerases, classification
    of, 51
Cis-trans isomerism, in unsaturated
    fatty acids, 132
Δ3-Cis-Δ2-trans-enoyl-CoA iso-
    merase, 202, 203

Cistron, defined, 429
Citrate, 149, 155, 188, 262, 282
    cycle of, 282
    in citric acid cycle, 152
    and oxaloacetate, 151
Citrate cleavage enzyme, in lipid
    metabolism, 86
Citrate synthase, 151, 152
Citrate transporter, 208
Citric acid cycle, 97, **149–157**, 164,
        173, 201, 207, 237, 256, 260,
        262
    amphibolic role of, 154
    deamination in, 154
    energetics of, 153
    in fatty acid synthesis from glucose,
        156
    gluconeogenesis in, 154
    high-energy phosphate bonds in, 153
    reactions of, 151
    regulation of, 194
    role of vitamins in, 153
    transamination in, 154
Citrulline, 18, 279
    synthesis of, 277
L-Citrulline, 278
Citrullinemia, 279
CK. *See* Creatine phosphokinase.
Class switching, of antibodies, 439,
        440
Clathrin, 460
CLIP. *See* Corticotropinlike interme-
        diate lobe peptide.
Clofibrate, 200, 250
Clomiphene citrate, 542
Clone, defined, 375
Clot, fibrin polymer, 613
Clotting, blood, **612–617**
Clotting factors
    IX and XI, 605
    in plasma and serum, 613
Clotting tests, 615
Clotting time, 616
CNBr. *See* Cyanogen bromide.
CO_2. *See* Carbon dioxide.
CoA. *See* Coenzyme A.
CoA transferase, 257
Coagulation, blood, **612–617**
    final common pathway of, 615
    tissue factor in, 614
Coagulation factors, in IUB
    classification, 51
Cobalamin, major characteristics of,
        575
Cobalt, major characteristics of,
        578
Cocoa, 336
Codon(s), 415
    nonsense, 415, 418
    in protein synthesis, 414
    tryptophan, 436
Coenzyme(s), 50
    as B-vitamin derivatives, 53
    classification of, 52
    concentration of, effective, 86
    as derivatives of AMP, 53
    as group transfer reagents, 51
    as second substrates, 51
Coenzyme A (CoA), 154, 188, 205
    synthesis of, 155

Coenzyme Q, 109
Coffee, and free fatty acid levels, 238,
        250
Colestipol, 250
Colipase, 582
Colitis, 585
Collagen(s), 631
    and glycosaminoglycans, 605
    inherited defects of, 634
    molecular features of, 631
    synthesis of, 632
    vertebrate, 632
Collagen fibrils, 631, 633
Collagen precursor, order and location
        of, processing of, 632
    hydroxyproline in, 268
Collagenase, type 4, 647
Collision theory for chemical reac-
        tions, 62
Combustion, heats of, for food
        molecules, 572
Compactin, 241, 250
Competitive inhibition of enzymes,
        classic, Lineweaver-Burk plot of,
        73
    graphic evaluation of, constants of,
        72
Complement, serum, normal values,
        661
Compound B, 513
Compound E, 513
Compound F, 513
Compound S, 513
COMT. *See* Catechol-O-methyltrans-
        ferase.
c-*onc* (cellular oncogene), 641
Concanavalin A (Con A), 129, 592
Concerted feedback inhibition, 88
Condensing enzyme, 206
Conformational change, defined, 64
Congenital nephrogenic diabetes in-
        sipidus, 471
Connexons, 462
Conn's syndrome, 523
Constant region, 609
Constant region gene, 612
Constitutive expression of gene,
        429
Constitutive heterochromatin, 390
Contractile and structural proteins,
        618–634
Contraction, muscle, ATP in, 622
Converter proteins, 91
α-COOH group in proteins,
        modification of, 37
Convulsions, infantile, 577
Cooperativity phenomenon, in al-
        losteric enzymes, 90
Coordinate induction of enzymes, 84
Coordinate repression of enzymes, 84
Copper, major characteristics of, 578
    in oxidases, 101, 102
    serum, normal values, 653
    serum or plasma, normal values,
        661
Coproporphyria, hereditary, 327
Coproporphyrin(s), 320
Coproporphyrinogen oxidase, 321,
        327
Coprostanol, 137, 248

Coprosterol, 137, 248
Corepressor, 84
Cori cycle, 195
Cori's disease, 171
Coronary heart disease, 213, 250, 574
Corticosteroid-binding globulin
        (CBG), 516, 537, 538
Corticosterone, 511, 513
Corticotropin. *See* ACTH.
Corticotropinlike intermediate lobe
        peptide (CLIP), 490
Corticotropin-releasing hormone
        (CRH), 475, 482, 483, 491
Cortisol, 482, 511
    biosynthesis of, 516
    plasma, normal values, 661
Cortisone (compound E), 513
Corynebacterium diphtheriae, 426
Cos site, in DNA cloning, 366
Cosmid,
    defined, 375
    in DNA cloning, 366
Cosubstrate, 51
    NAD^+ as, 51
Cotransport transport system, 456
Coumarin drugs, 616
Coupled reactions, bioenergetics of,
        97
Coupling, in bioenergetics, 94
Covalent bonds
    formation of, 50, 63
    rupture of, 63
Covalent modification, 91
CPK. *See* Creatine phosphokinase.
CRABP. *See* Cellular retinoic acid-
        binding protein.
CRBP. *See* Cellular retinol-binding
        protein.
Creatine, 96, 97, 316
    biosynthesis of, 307, 317
    urine, normal values, 654
Creatine phosphate, 95, 96, 97, 317,
        619, 625
Creatine phosphokinase (CPK),
        625
    serum, normal values, 654, 661
Creatinine, 316, 317
    biosynthesis of, 317
    normal values, 654, 655, 661
Cretinism, 501
CRH. *See* Corticotropin-releasing hor-
        mone.
Crigler-Najjar syndrome, 332
Cro, defined, 432
Cro protein, 433
Crohn's disease, 213
Cross-linkage, in DNA, 401
CRP. *See* Catabolite regulatory
        protein.
Crystallography, x-ray, in determina-
        tion of protein structure, 38
CT. *See* Calcitonin.
CTP. *See* Cytidine triphosphate.
CTP synthase, 354
CTP-phosphatidate cytidyl trans-
        ferase, 219
Culture, stepdown, 85
Cushing's syndrome, 492, 523
Cyanide, 101
    in oxidative phosphorylation, 112

Cyanocobalamin, serum, normal values, 661
Cyanogen bromide (CNBr), 29
Cyclase, 243
Cyclic adenosine monophosphate. See cAMP.
3',5'-Cyclic adenylic acid, 169
Cyclic AMP. See cAMP.
Cyclic GMP. See cGMP.
Cyclic guanosine monophosphate. See cGMP.
Cyclic 3',5'-nucleotide phosphodiesterase, 238
Cyclooxygenase, 215
Cyclooxygenase pathway, 215
Cyclopentanoperhydrophenanthrene nucleus, 512
Cyproterone acetate, 534
Cys. See Cysteine.
Cystathionine, 269, 300
Cystathioninuria, 289, 577
Cysteine (Cys), 16, 262, 265, 269, 287, 290, **309**, 588
 biosynthesis of, 269
 catabolism of
   direct oxidative pathway of, 287, 288
   transaminase (3-mercaptopyruvate) pathway of, 287, 288
 conversion of, 309
 human requirements of, 573
 in muscle, 275
 synthesis of, 269
L-Cysteine, 269, 287, 288
Cysteine residue, 205
L-Cysteine sulfinate, 288
Cysteine sulfinic acid, 18
Cysteine thiol, 204
5-S-Cysteinyl-dopa, 315
5-S-Cysteinyl-dopaquinone, 315
Cystic fibrosis, 213
Cystine, 286
L-Cystine, 287
Cystine reductase reaction, 287
Cystine storage disease, 288
Cystine-lysinuria, 287
Cystinosis, 288, 289
Cystinuria, 287
Cytarabine, 342
 structure of, 343
Cytidine, 337, 341, 344, 355
Cytidine diphosphate (CDP), 341
Cytidine diphosphocholine, 220
Cytidine diphosphoethanolamine, 220
Cytidine monophosphate, 591
Cytidine triphosphate (CTP), 88, 98, 220, 341
Cytochalasin, 628
Cytochrome(s), 103, 319
 in respiratory chain, 103
Cytochrome aa$_3$, 101
Cytochrome b, 103, 109
Cytochrome b$_5$, 103, 211
Cytochrome c, 103, 319
Cytochrome oxidase, 101, 110, 242
Cytochrome P-450, 103, 200
Cytochrome P-450$_{scc}$, 512, 513
Cytochrome system, 100
Cytokinins, 242

Cytosine, 335, 336, 337, 355, 356, 362, 395
 derivatives of, 341
 in DNA, 379, 395, 402
 in RNA, 382
 structure of, 337
Cytosine-ribose, 341
Cytoskeleton, 627

D gene, 612
dAMP. See Deoxyadenosine monophosphate.
DBH. See Dopamine β-hydroxylase.
D-binding protein, 508
DD. See Dopa decarboxylase.
Deacylase, 153, 256
Deamination, 154
 oxidative, 273
Debranching, 166
Debranching enzyme, 166, 167
 absence of, 171
Decanoic acid, 131
Decarboxylase, α-keto acid, 303
Decarboxylation of pyruvate, oxidative, 163
Degradation, 166
L-Dehydroascorbate, 181
7-Dehydrocholesterol, major characteristics of, 576
24-Dehydrocholesterol, 243
Dehydroepiandrosterone (DHEA), 511, 513, 515
Dehydroepiandrosterone pathway, 531
Dehydrogenase(s)
 aerobic, 101, 102
 anaerobic, 103
 dependent on nicotinamide coenzymes, 103
 NADH, 110
 NADH- or NADPH-dependent, 55
 quantitative analysis of, 54
18-Dehydrogenase, deficiency of, 523
Dehydrogenation reactions, 94
Deiodinase, 497
Demyelinating diseases, 223
Denaturation
 of DNA molecule, 379
 of proteins, 37
Dendrites, 629
Deoxy sugars, 124
2'-Deoxyadenosine monophosphate (dAMP), 337, 338
Deoxyadenosine nucleotides of DNA, 377
5'-Deoxyadenosylcobalamin, 305
Deoxyadenylate, 377
2'-Deoxyadenylic acid, 338
Deoxycholic acid, 248, 249
11-Deoxycorticosterone (DOC), 513
11-Deoxycortisol (compound S), 513, 515
Deoxycytidine, 355
Deoxycytidine kinase, 355
Deoxycytidine nucleotides of DNA, 377
Deoxycytidylate, 377
6-Deoxy-L-galactose, 128
Deoxyguanosine nucleotides of DNA, 377

Deoxyguanylate, 377
2-Deoxy-D-ribofuranose, 126
Deoxyribonuclease, 583, 584
 defined, 413
Deoxyribonucleic acid. See DNA.
Deoxyribonucleoside triphosphate, in DNA synthesis, 396
Deoxyribonucleotides, discontinuous polymerization of, 397
Dephospho-coenzyme A, 155
Depurination of DNA, 401
Derepression, 84
Dermatan sulfate, 603, 605
 structure of, 600
Desaturase, 221
"Desensitization," 506
14-Desmethyl lanosterol, 244, 243
Desmin, 631
Desmosterol, 243
Desulfinase, 288
Dexamethasone, 513
Dextrins, 120, 125
DHOA. See Dihydroorotic acid.
DHT. See Dihydrotestosterone.
DHU. See Dihydrouracil.
Diabetes, 251
 fructose in, 182
 insipidus, 494
   acquired nephrogenic, 494
   congenital nephrogenic, 471
   hereditary nephrogenic, 494
 mellitus, 197, 200, 229, 234, 239, 250, 251, 255, 553, 560
   insulin-dependent (IDDM) (type I), 560
   non-insulin-dependent (NIDDM) (type II), 251, 471, 560
 sorbital in, 183
Diacylglycerol, 133, 238, 480
1,2-Diacylglycerol, 220, 234, 582
Diacylglycerol acyltransferase, 219, 220
Diacylglycerol lipase, 238
1,2-Diacylglycerol phosphate, 219
1,4-Diaminobutane, structure of, 310
α,ε-Diamino-δ-hydroxycaproate, 270
1,5-Diaminopentane, structure of, 310
1,3-Diaminopropane, structure of, 310
O-Diazoacetyl-L-serine, 348
Diazonorleucine, 348
(6-Diazo-5-oxo)-L-norleucine, 348
Dicarboxylate transporter, 118
Dienoic acids, 132
Dietary fiber, 583
Dietary recommendations, 577
Dietary triacylglycerol, 231
Diethylstilbestrol, 542
Diffusion
 facilitated, 456, 457, 458
 kinetics of, 457
 passive, 455
DIFP. See Diisopropylphosphofluoridate.
Digestion, 344, **579–583**
 defects in, 585
 and gallbladder, 584
 in intestine, 580, 584
 and liver, 584
 major products of, 583

Digestion (cont'd)
  in mouth, 579, 584
  and pancreas, 580, 584
  process of, summary of, 584
  salivary, 579, 584
  in stomach, 579, 584
Digitalis, 124
Diglycerides, 133
Dihomo-γ-linolenyl-CoA, 213
Dihydrobiopterin, 269
Dihydrolipoyl dehydrogenase, 103, 163
Dihydrolipoyl transacetylase, 163
Dihydroorotase, 86, 353
  in purine or pyrimidine metabolism, 86
Dihydroorotate dehydrogenase, 353
Dihydroorotic acid (DHOA), 354
Dihydropicolinate, 270
Dihydrosphingosine reductase, 222
Dihydrotestosterone (DHT), 531, 532, 534, 536, 545
Dihydrothymine, 356
Dihydrouracil (DHU), 356
Dihydrouracil loop of RNA, 385
Dihydroxyacetone, 120, 124
  active, 177
Dihydroxyacetone phosphate, 159, 160, 173, 174, 182, 218, 219, 220
  pathway of, 219
Dihydroxyacetone phosphate acyl-transferase, 219
1,25-Dihydroxycholecalciferol, 464
5,6-Dihydroxyindole, 315
3,4-Dihydroxyphenylalanine (dopa), 18, 315, 316, 525
Diiodotyrosine, 496, 497
Diisopropylphosphofluoridate (DIFP), 76
Dimer(s)
  defined, 36
  of histones, 388
Dimercaprol, 112
Dimethyladenine, 336
3,3-Dimethylallyl pyrophosphate, 242, 244
Dimethylglycine, 268
1,3-Dimethylxanthine, 336, 337
3,7-Dimethylxanthine, 336, 337
Dinitrocresol, in oxidative phosphorylation, 112
Dinitrophenol, in oxidative phosphorylation, 112
Dioxygenases, 105
2,4-Dioxy-5-methylpyrimidine, 336
2,6-Dioxypurine, 336
2,4-Dioxypyrimidine, 336
Dipalmityl lecithin, 134
Dipeptidases, 583, 584
Dipeptides, 584
o-Diphenols, 102
Diphosphate galactose, 183
Diphosphatidylglycerol, 134, 220
Diphosphoglycerate. See 2,3-Bisphosphoglycerate.
Diphtheria toxin, 426
Direct repeats, in DNA, 392
Disaccharides, 120, **125**
  structures of, 126

Disacchariduria, 586
Disialogangliosides, 224
Dissociation constant, for enzyme-substrate complex, 71
1,2-Distearopalmitin, 133
1,3-Distearopalmitin, 133
Disulfide(s), "mixed," of cysteine and homocysteine, structure of, 288
Disulfide bonds, of proteins, 33
DIT. See Diiodotyrosine.
Diversity region gene, 612
DL mixture, defined, 121
DNA (deoxyribonucleic acid)
  A form of, 378
  B form of, 378
    model of, 379
  base pairing in, 380
  basic features of, 362
  blunt-ended, 364
  defined, 375
  chemical nature of, 377
  chimeric molecules of, 365
  cloning of, 366
  code of, genetic mutation altering, 83
  coding strand of, 378, 404
  damage to, 401
    by chemical carcinogens, 637
  degradation and repair of, 401
  deletions, insertions, and rearrangements of, 373
  denaturation of, 379
  depurination of, 401
  features of some structures of, 380
  5' flanking-sequence, 364
  fragments of, method of screening recombinants for, 368
  function of, **377**, 381
  and ionizing radiation, 401
  "jumping," 394
  melting of, 379
  melting temperature of, 379
  molecule of, grooves of, 380
  noncoding, 392
  noncoding strand of, 378, 404
  nonrepetitive, 392
  organization and replication of, **387–403**
  phosphates in, 379
  polymorphisms of, 372
  rearrangement of, in immunoglobulin, 438
  recombinant, defined, 376
  recombinant technology with. See Recombinant DNA technology.
  relaxed, 380
  repair of, 401
  repetitive sequences in, 392
  replication of, 395
  restriction map of, 364
  role of, in carcinogenesis, 638
  sequencing of, 369
    by Maxam and Gilbert method, 370, 371
    by Sanger method, 370
  single-base alteration in, 401
  sticky-ended, 364

DNA (cont'd)
  defined, 376
  structures of, 378
  supercoiled, 380, 400
  synthesis of, 395
    initiation of, 395, 396
    regulation of, 400
    semidiscontinuous, 397, 398
    unscheduled, 402
  synthesizing RNA from, 404
  template of, 378
  template strand of, 395
  transposition of, defined, 394
  underwound, 380
  unique-sequence, 392
  Watson and Crick model of, 379
  Z form of, 378
DNA ligases, 398
DNA polymerase
  mitochondrial, 398
  RNA-dependent, 383
DNA polymerase I, 395
DNA transfection, 641
DNA-dependent RNA polymerase, 404
  nomenclature and localization of, 406
DNA-dependent RNA polymerase II, 408
DNA-dependent RNA polymerase III, 409
DNase, 583
DNase I, 390
DOC. See 11-Deoxycorticosterone.
Dolichol, 133, 138, 244, 594, 595
  biosynthesis of, 242
Dolichol kinase, 594
Dolichol-P-P polysaccharide, in proteoglycan synthesis, 600
Domains
  of chromatin fibers, 389
    looped, 391
  of immunoglobulins, defined, 609
Dopa, 18, 315, 316, 525
Dopa decarboxylase (DD), 314, 525
Dopachrome, 315
Dopamine, 316, 483, 524, 525, 527
Dopamine β-hydroxylase (DBH), 526
Dopaquinonone, 315
Double helix, left-handed, of Z-DNA, 378
Down-regulation, 468, 506
DPG. See 2,3-Bisphosphoglycerate.
Dubin-Johnson syndrome, 332
Dwarfism, 487
  GH-deficient, 487
  Laron type, 487
Dynamic equilibrium (Schoenheimer), 85
Dynein, 630
Dysbetalipoproteinemia, familial, 251
Dyslipoproteinemias, 251

E⁰′ **(redox potential), 100**
E₂. See 17β-Estradiol.
Eadie-Hofstee plot, 70
E.C. (enzyme code) number, in enzyme classification, 50
Eck fistula, 274

Ectopic hormone syndromes, 464
Edema, due to malabsorption, 587
Edman reaction, 29
Edman reagent, 29
Edman technique, automated, for determination of polypeptide structures, 28
EFA. *See* Essential fatty acids.
Effectors, allosteric, 87, 89
EGF. *See* Epidermal growth factor.
Ehlers-Danlos syndrome, 634
Ehrlich's diazo reagent, 331
Ehrlich's reaction, 331
Eicosanoids, 131, 210, 213
Eicosapentaenoic acid, 210
5,8,11,14,17-Eicosapentaenoate, 214
5,8,11,14-Eicosatetraenoate, 214
Eicosatrienoate (dihomo γ-linolenate), 211
8,11,14-Eicosatrienoate, 214
Elaidic acids, 132
Elastase, 581, 584
Elastin, 605
Electrochemical potential difference, in oxidative phosphorylation, 113
Electron photomicrography, in determination of protein structure, 39
Electron-transferring flavoprotein, 103, 109, 199
Electrophiles, 637
Electrophoresis, 22
  polyacrylamide gel, 39
    in enzyme isolation, 56
  separation of plasma lipoproteins by, 227
  zone, cellulose acetate, 608
Electrostatic bonds, 34
Elongase, 208
Elongation factors, in protein synthesis, 423
Embden-Meyerhof pathway. *See* Glycolysis, pathway of.
Emulsification, as function of bile system, 580
Emulsions, 141
  formation of, 141
Endergonic reactions, 94
  defined, 94
Endocrine glands, derivation and location of, 463
Endocrine system
  characteristics of, **463–471**
  diversity of, 463
Endocytosis, 459, 460, 461
Endocytotic vesicles, 629
Endoglycosidase(s), 591, 594, 603
Endonuclease(s), 364
  defined, 375, 413
  restriction, 413
    results of digestion of, 364
    and sequence specificities, 365
Endopeptidase, 580
Endoplasmic reticulum
  rough, 230, 414
  smooth, 230
Endorphins, 492
Energetics, 91
  of interconvertible enzymes, 91

Energy
  barriers for, for chemical reactions, 62
  capture of, 95, 97
  conservation of, 97
  dietary, requirements for, 571
  essential nutritional requirements of, 572
  expenditure of, 571
  free, 93
    changes in, and formation and decay of transition states, 61
    of hydrolysis
      of ATP, 95
      of organophosphates, 95
    intake of
      and protein requirements, 574
      recommended amounts, 572
    in major food sources, 572
  transduction of, 95
Energy equilibrium, 571
Enhancer(s), 364, 642
Enhancer elements, 440
Enkephalins, 565
-enoic, in fatty acid nomenclature, 130
Enolase, 160, 161
Enolphosphates, 96
p-Enolpyruvate, 262
Enoyl reductase, 204
Enoyl-CoA, 203
Δ2-Enoyl-CoA hydratase, 199, 201, 203
Enteroglucagon, 565
Enterohepatic circulation, 248
Enterokinase, 581
Enthalpy, 93
  defined, 93
Entropy, 93
  defined, 93
env gene, 640
Enzymatic analysis, by coupled assay, 54
Enzyme(s). *See also specific enzyme*.
  activity of
    altered by glucagon, 562
    altered by insulin, 559, 562
    inhibition of, 72
    modulators of, 73
    quantitative measurement of, 54
    regulation of, **82–92**
  adaptation of, to environmental stimulus, 86
  adaptive, 188
  allosteric, 89
    inhibition of, 89
    models of, 90
  angiotensin-converting, 518
  basal level of, 84
  catalytic efficiency of
    covalent modification in regulation of, 90
    regulation of, 85
  catalytic sites of, 64
  of citric acid cycle, in mitochondria, 116
  classification of, 50
  in clinical diagnosis, 58
  code number for, 50
  and coenzymes, in oxidation and reduction, 100

Enzyme(s) (cont'd)
  compartmentation of, 86
  concentration of, 68
  condensing, 206
  constitutive, 84
  coordinate induction of, 84
  coordinate repression of, 84
  debranching, 166, 167
    absence of, 171
  degradation of, to amino acids, 83
  denaturation of, 66
  as general acid or as general base catalysts, 78
  general properties of, **50–60**
  glycolytic, 640
  inducible, 84
  interconvertible, 90
    analogies to feedback inhibition, 91
  intracellular distribution of, 56
  isolation of, 55
  kinetic properties of, **61–81**
  as markers of membranes, 449
  metal ions in, 87
  multiple, 88
  nomenclature of, 50
  of β-oxidation of fatty acids, in mitochondria, 116
  and pH, 66
  planar representation of, 53
  plasma
    functional, 58
    nonfunctional, 58
  primary structure of, 83
  as protein catalysts, 50
  purification of, 55, 56
  quantity of, regulation of, 83
  as reaction-specific catalysts, 50
  in recombinant DNA research, 366
  regulated
    allosteric sites on, evidence for, 89
    covalent modification of, 91
  regulation of
    available options for, 83
    catalytic efficiency of, 85
  regulatory, 87, 188
  restriction, 364
    defined, 376
  in rupture and formation of covalent bonds, 63
  specificity of, 53
  synthesis of
    from amino acids, 83
    induction of, 83
    repression and derepression as control of, 84
  and temperature, 66
  "three-point attachment" of, 53
  turnover of, 85
  unnecessary, 84
Enzyme induction, 637
Enzyme units, defined, 54
Enzyme-bridge complexes, 80
Enzyme-catalyzed reaction, 187
  control of, 187
  test for, 37
Enzyme-metal-substrate complexes, 79
Enzyme-substrate complex, 68
  dissociation constant for, 71

Epidermal growth factor (EGF), 557, 640, 646
Epimer(s), of glucose, 122
Epimerase, 177, 178, 183
Epinephrine, 168, 169, 170, 188, 193, 196, 238, 314, 316, 524, 525, 527
  plasma, normal values, 661
Epoxidase, 243
Equilibrium, dynamic (Schoen-heimer), 85
Equilibrium constant, 67
Equilibrium reactions, 186
erb-B gene, 640, 641
Ergocalciferol, major characteristics of, 576
Ergosterol, 137, 586
  major characteristics of, 576
Ergothioneine, 309
  structure of, 308
Erythrocruorins, 319
Erythrocytes, glycolysis in, 162
Erythro-γ-hydroxyglutamate, 290
Erythro-γ-hydroxy-L-glutamate, 290
Erythromycin, 125
Erythrose, 120
D-Erythrose, 124
Erythrose 4-phosphate, 177, 178
Erythrulose, 120
Escherichia coli, 83, 430
Essential fructosuria, 182
Essential pentosuria, 180
Ester(s), phorbol, tumor-promoting, 557
Esterification, 236
Estradiol, 464, 513
17β-Estradiol (E₂), 536
Estriol, 513, 537
Estrogen(s), 536, 537
  biosynthesis of, 537
  catechol, 537
Estrogen receptor, 543
Estrone, 513
Ethanol, 234
  energy available in, 572
Ethanolamine, 219
  active, 220
Ether, and plasmalogens, biosynthesis of, 221
Ether lipids, 219
17α-Ethinyl estradiol, 542
Ethionine, 234
Ethyl mercaptan, 588
N-Ethylmaleimide, 118
Etiocholanolone, 513, 532
Euchromatin, 390
Euglobulins, 32
Eukaryotes, regulation of RNA pro-cessing in, 437
Eukaryotic cells, compartmentation of, 86
Eumelanins, 314, 315
Excision-repair, of DNA, 401
Excitable membrane, of sarcomere, 623
Excretion, as function of bile system, 580
Exergonic reaction(s), 94, 95
  defined, 94
Exocytosis, 461

Exocytotic vesicles, 629
Exoglycosidase(s), 591, 603
Exon(s), 363, 414
  defined, 375, 392, 409
Exonuclease(s), 364, 413
  defined, 376
Exopeptidase, 580
Extracellular fluid, 445
Extrahepatic tissues, 255, 257
  metabolic interrelationships of, with liver and adipose tissue, 261
Extramitochondrial system for de novo synthesis of fatty acids, 202

F₁ subunit, 113, 115, 116
F₀, 113, 115
Fab, 609
Fabry's disease, 224
F-actin, 621
  defined, 621
F-actin filaments, 629
Factors, blood clotting, 613
  disorders of, 616
Facultative heterochromatin, 390
FAD. *See* Flavin adenine dinucleotide (oxidized).
FADH₂. *See* Flavin adenine dinucle-otide (reduced).
Familial alpha-lipoprotein deficiency, 251
Familial dysbetalipoproteinemia, 251
Familial hypercholesterolemia, 251
Familial hyperlipoproteinemia, 251
Familial hypertriacylglycerolemia, 251
Familial hypobetalipoproteinemia, 251
Familial lipoprotein lipase deficiency, 251
Fanconi's anemia, DNA in, 402
Farber's disease, 224
Farnesyl pyrophosphate, 242, 244
Fasciculus, muscle, 618
Fast skeletal muscle, ATP regenera-tion in, 625
Fasting, 165, 208
Fat(s), 130, 281
  carbohydrate conversion into, 260
  energy available in, 572
  interconversion of, 262
  mobilization of, 236
Fatty acid(s), **130**, 188, 199, 200, 231, 260, 262
  active, 201
  in bile, 580
  C₂₄, 223
  essential (EFA), 212, 216
    deficiencies of, 212, 235
    functions of, 212
    metabolism of, 210
  essential nutritional requirements of, 572
  free (FFA), 198, 199, 226, 229, 236, 237, 254, 256, 258, 262
    in plasma, 226
    level of, 236
  long-chain
    biosynthesis of, 205
    oxidation of, regulation of, 259

Fatty acid(s) (cont'd)
  monounsaturated, synthesis of, 211
  nomenclature of, 130
  nonessential, 210
  nonesterified, 229
    in plasma, 226
  odd number of carbon atoms in, 200
  oxidation of, 198
    carnitine in, 198
    energetics of, 200
    peroxisomal, 200
    regulation of, **254**
    α-oxidation of, 200
    β-oxidation of, 199
    ω-oxidation of, 200
  oxidation and biosynthesis of, **198–209**
  polyunsaturated, 250
    biosynthesis of, 212
    essential, 574
  saturated, 131
    biosynthesis of, 202
  synthesis of, 155, 202, 234
    regulation of, 253
  unesterified, 229
  unsaturated, 131, 132
    isomerism in, 132
    metabolism of, 210
    oxidation of, 202, 203
Fatty acid oxidase, 199
Fatty acid pool, free, 236
Fatty acid synthase, 188
  in lipid metabolism, 86
  multienzyme complex of, 204, 205
  quaternary structure of, 39
Fatty acid-binding protein, 198, 229
Fatty aldehydes, 133
Fava beans, and G6PD deficiency, 179
Favism, 185
Fc (crystallizable fragment), 609
Feedback control, 465
  negative, 465
  positive, 465
Feedback inhibition, 87, 90
  cooperative, 88
  cumulative, 88
  kinetics of, 89
Feedback loops, multiple, 88
Feedback regulation, 90
Feeding, continuous, 250
Fermentation, 587
  in yeast, 158
Ferritin, serum, normal values, 661
Ferrochelatase, 323, 328
FeS. *See* Iron-sulfur protein.
Fetal hemoglobin, 47
FFA. *See* Fatty acids, free.
FGF. *See* Fibroblast growth factor.
Fiber(s)
  dietary, 583
    requirement for, 574
  essential nutritional requirements of, 572
  stress, 627
Fibrin monomer, 613
Fibrin polymer clot, 613
Fibrinogen, 609
Fibrinolysis, 616
Fibroblast(s), 231, 232

Fibroblast growth factor (FGF), 557
Fibronectin, 634
"Fight or flight" response, 520, 524
Figlu. *See* N-Formiminoglutamate.
Filamin, 628
Filopodia, 627, 628
Fischer, Emil, 64
Fischer formula, shorthand, for porphyrins, 321
Fischer template hypothesis for enzymes, 64
Fistula, Eck, 274
Flagella, 629
Flavin adenine dinucleotide (FAD) (oxidized), 102, 199
    structure of, 102
Flavin adenine dinucleotide (FADH₂) (reduced), 153
    oxidation of, 152
Flavin mononucleotide (FMN), 102, 103
Flavoprotein(s), 100, 101, 110, 164, 201, 222
    electron-transferring, 103, 199
Flavoprotein dehydrogenases, 103
Flora, intestinal, 588
Fluid mosaic model, of membrane structure, 451
Fluorescamine, 19
Fluoride, major characteristics of, 578
Fluoroacetate, 151
    and citric acid cycle, 152
    and Krebs cycle, 152
Fluoroacetyl-CoA, 151
Fluorocitrate, 151
9α-Fluorocortisol, 513
5-Fluorouracil, 342
Flux-generating reaction, 186
FMN. *See* Flavin mononucleotide.
fms oncogene, 641
Folacin, major characteristics of, 575
Folate, malabsorption of, 577
Folic acid, 577
    major characteristics of, 575
    serum, normal values, 661
Follicle-stimulating hormone (FSH), 464, 475, 488, 489, 533
Follicular cells, 489
Follicular phase, 540
Food(s)
    calorimetry of, 572
    energy in, 572
    interconversion of, 260
Forbes' disease, 171
Formaldehyde, 268
N-Formiminoglutamate (Figlu), 285
Formylglycinamide ribosyl 5-phosphate, 347
Formylglycinamide ribosylphosphate synthetase, 346
N-Formylkynurenine, 297, 298
N-Formylmethionyl-tRNA, 422
Formyltransferase, 346, 347
Fragments, antigen-binding, 609
Frame shift mutations, 419
Free energy, 93
    changes in, and formation and decay of transition states, 61
    of hydrolysis
        of ATP, 95

Free energy change, standard, 93
Free fatty acids. *See* Fatty acids, free.
D-Fructofuranose, 121
Fructokinase, 182
D-Fructopyranose, 122
Fructosan, 125
Fructose, 120, 180, 182, 188, 235, 583, 584
    metabolism of, 180, **182,** 194
D-Fructose, 123, 182
Fructose 1,6-bisphosphate, 159, 160, 173, 182, 183, 188
Fructose diphosphatase, quaternary structure of, 39
Fructose intolerance, hereditary, 182
Fructose 1,6-phosphatase, in carbohydrate metabolism, 86
Fructose 1-phosphate, 182, 183
Fructose 6-phosphate, 96, 159, 160, 173, 177, 178, 179, 182, 185, 207
Fructose-1,6-bisphosphatase, 172, 173, 178, 179, 188
    deficiency of, 183
Fructose-2,6-bisphosphatase, 192
Fructosuria, essential, 182
FSH. *See* Follicle-stimulating hormone.
1,2-Fuc transferase, 599
Fucose, 128, 228
L-Fucose, 124
    in glycoproteins, 128
α-Fucosidase, deficiency of, 224
Fucosidosis, 224
Fucosyltransferase, 599
Fumarase, 152, 153
Fumarate, 72, 152, 173, 262, 277, 278, 282, 291, 292, 348
Fumarate hydratase, 153
Fumarylacetoacetate, 291
Fumarylacetoacetate hydrolase, 291
Furan ring, 121
Furanose ring structure, 121
"Fusion genes," 475
Futile cycle, 193

G₁, 400
G₂, 400
GABA. *See* γ-Aminobutyrate.
G-actin, 621, 627
    defined, 621
gag gene, 640
Galactokinase, 183
Galactorrhea, 487
Galactosamine, 184
D-Galactosamine, 125
    in proteoglycans, 599
Galactose, 128, 183, 228, 583, 584, 591
    conversion to glucose, 184
    in glycoproteins, 128
    metabolism of, 180, 183, 194
α-D-Galactose, 123
D-Galactose, 123, 124
Galactose 1-phosphate, 184
Galactose tolerance test, 183
Galactosemia, 183
Galactose-1-phosphate uridyl transferase, 183
    deficiency of, 183

Galactosidase(s), 591
β-Galactosidase(s), 83, 603
    deficiency of, 224
    G_{M1}, 224
    as hydrolyzing agent, 430
    in IUB classification, 51
Galactoside, defined, 123
Galactosylceramide, 135, 223
Galactosyltransferase, 591
Gallbladder, and digestion, 584
Gallstones, formation of, 581
GalNAc (N-acetylgalactosamine), 598
Ganglionic blockade, 239
Ganglioside(s), 135, 223, 447, 603
    biosynthesis of, 224
    higher, 224
    simple, 224
Gangliosidosis, generalized, 224
GAP. *See* GnRH-associated peptide.
Gap junctions, 462
Gap 1, 400
Gap 2, 400
GAR. *See* Glycinamide ribotide.
Gastric inhibitory polypeptide (GIP), 552, 564, 565, 566, 567
Gastric juice, 579
Gastrin, 564, 565, 566, 568, 584
Gastrin family, 566
Gastrin-cholecystokinin family, 568
Gastrinomas, 568
Gastrin-releasing peptide (GRP), 568
Gastroenteritis, 585
Gastrointestinal hormones, **564–568**
Gastrointestinal peptides, amino acid sequences of, 566
Gastrointestinal tract
    absorption from, 583
    hormones of, **564–568**
Gaucher's disease, 224
GDP. *See* Guanosine diphosphate.
GDP-mannose, 595
Gemfibrozil, 250
Gene(s), 377
    A, 430
    amplification of, 642
        during development, 438
    coding regions on, 363
    constant region of, 612
    conversion of, 395
    cro, 432
    deletions and insertions in, 420
    eukaryotic, pathway of expression of, 363
        transcription unit of, 363
    expression of
        coordinate, 430
        effect of insulin on, 559
        rates of, 428
        regulation of, **428–443**
    fusion of, 475
    α-globin, 373
    β-globin, 372, 373
        and genetic disorders, 372
        mutations in, causing β-thalassemia, 373
        structural alterations of, 372
    human, localization of, 370
    i, 430
    in immunoglobulin light chains, 612

Gene(s) (cont'd)
  inducible, defined, 429
  joining region of, 612
  libraries of, 367
  mapping of, 370
  mutations of, 417
    effect of, 417
  normal variations of, 372
  organization of, 363
  processed, 394
  in protein synthesis, 414
  repressor, 432
  therapy using, 375
  transcription of, 364
    steroid regulation of, 474
  variable region of, 612
  variations in, causing disease, 372
  y, 430
  z, 430
Gene conversion, 395
Gene transfer, 641
Genetic code, 414
  and protein synthesis, **414–427**
Genetic engineering, 362
Genetic material, alteration and rearrangement of, 393
Genetic mutation, and DNA code, 83
Genome
  genetic organization of, 392
  haploid, human, 391
Geometric isomerism, 132
Geranyl pyrophosphate, 242, 244
GH. *See* Growth hormone.
GHRH. *See* Growth hormone-releasing hormone.
GHRIH (growth hormone release-inhibiting hormone). *See* Somatostatin.
Gigantism, 487
Gilbert and Maxam method of DNA sequencing, 371
Gilbert's disease, 332
GIP. *See* Gastric inhibitory polypeptide.
Gland(s). *See specific gland.*
GLC. *See* Gas-liquid chromatography.
GlcNAc phosphotransferase, 597
GlcNAc-pyrophosphoryl-dolichol, 594
GlcUA (glucuronic acid), 180, 599
Glial filament, 631
Glicentin, 568
Gln. *See* Glutamine.
Globin gene, mutations in, 418
Globulin(s), 32, 608
Glu. *See* Glutamic acid.
Glucagon, 169, 188, 196, 237, 475, 547, 550, 561, 566, 568
  amino acid sequence of, 561
  biosynthesis and metabolism of, 561
  chemistry of, 561
  effect on phosphorylase, 169
  physiologic effects of, 562
  and rate of enzyme synthesis, 86
  regulation of secretion of, 562
Glucan, 125
Glucan transferase, 166
α-(1→4)→α-(1→4) Glucan transferase, 167

Glucocorticoid(s), 188, 238, 441, 511
  classification and mechanism of action of, 520
  metabolic effects of, 518
  and rate of enzyme synthesis, 86
Glucocorticoid receptor, 521
Glucocorticoid steroids, 511
Glucocorticoid synthesis, 515
  regulation of, 516
α-D-Glucofuranose, 121
Glucogenic amino acids, 174
Glucogenic compounds, as source of blood glucose, 194
Glucokinase, 159, 160, 162, 173, 182, 188, 195
Gluconeogenesis, 152, 159, **172–175,** 188, 190, 555
  enzymes of, 188
  in liver, 192
    major pathways of, 173
    metabolic pathways involved in, 172
  regulation of, 189
Gluconolactone hydrolase, 176, 177
α-D-Glucopyranose, 121
β-D-Glucopyranose, 122
Glucosamine, 126, 184, 185, 228
  sulfated, 128
D-Glucosamine, 124
  in proteoglycans, 599
Glucosamine 1-phosphate, 185
Glucosamine 6-phosphate, 185
Glucosan, 125
Glucose, 120, 158, 164, 165, 172, 173, 174, 180, 185, 228, 237, 256, 260, 261, 583, 584, 585
  absorption of, 583
  basal requirement for, 172
  blood, 194
    concentration of, 194
    regulation of, 194
  catabolism of, 165
  dietary requirement for, 574
  entry into muscle cells, 554
  epimerization of, 123
  and fatty acid synthesis, 156
  D- and L-isomerism of, 121
  and lipogenesis, 253
  metabolism of, 194, 236
    pathways of, 180
  moiety of, epimerization of, to galactose, 223
  mutarotation of, 122
  pyranose and furanose forms of, 121
  renal threshold for, 196
  serum or plasma, normal values, 655, 661
  structure of, 120
  tolerance for, 197, 252
  transport of, 459, 585
D-Glucose, 121, 123, 124
α-D-Glucose, 121, 123, 160
L-Glucose, 121
Glucose 1,6-bisphosphate, 96, 165
Glucose effect, 84
Glucose oxidase, 103
Glucose 1-phosphate, 96, 165, 167, 168, 181

Glucose 6-phosphate, 96, 160, 166, 173, 176, 177, 178, 181, 182, 185, 188, 195, 207, 237, 261
Glucose tolerance test, 196
Glucose transporter(s), 554
  human, proposed model for, 449
Glucose-alanine cycle, 194, 195, 276, 627
Glucose-fatty acid cycle, 263
Glucose-6-phosphatase, 57, 166, 167, 173, 174, 182, 184, 188
  deficiency of, 357
Glucose-6-phosphate dehydrogenase (G6PD), 176, 177, 178, 188, 239
  in carbohydrate metabolism, 86
  deficiency of, 179
  in muscle tissue, 179
1,6-Glucosidase, 584
Glucosides, defined, 123
1,6-Glucosides, 584
Glucosylceramide, 135, 223, 224
Glucosyltransferase, 165
Glucuronate, 180
D-Glucuronate, 123, 181
α-D-Glucuronate, 125
Glucuronic acid, 180, 599
β-Glucuronic acid, 128
β-Glucuronidases, 330, 603
Glutamate, 155, 262, 265, 267, 270, 272, 282, 283
  in muscle, 275
L-Glutamate, 268, 272, 273, 283, 284, 285, 296, 317
L-Glutamate decarboxylase, 317
Glutamate dehydrogenase, 274, 278
L-Glutamate dehydrogenase, 273
Glutamate dehydrogenase reaction, 266
Glutamate formiminotransferase, 285
Glutamate semialdehyde dehydrogenase, 284
Glutamate transaminase, 273
Glutamate-α-ketoglutarate transaminase, 272
L-Glutamate-γ-semialdehyde, 268, 284
Glutamic acid (Glu), 17, 265, 266
Glutamic dehydrogenase, in IUB classification, 51
Glutaminase, 274, 283
Glutaminase reaction, 274
Glutamine (Gln), 17, 265, 267, 274, 275, 283, 353, 354, 626, 627
  biosynthesis of, 267
  in kidney, 275
  in liver and gut, 275
  in muscle, 275
L-Glutamine, 283
  catabolism of, to amphibolic intermediates, 283
Glutamine synthetase, 266, 274
  in IUB classification, 51
  quaternary structure of, 39
Glutamine synthetase reaction, 266, 275
γ-Glutamyl-cysteinyl-glycine, 26
γ-Glutamyl transpeptidase, serum, normal values, 655
Glutaryl-CoA, 296
Glutathione, 26, 291, 315

Glutathione peroxidase, 105, 139, 179
Glutathione reductase, 179
Glutathione-insulin transhydrogenase, 553
Gly. *See* Glycine.
D-Glyceraldehyde, 121, 124, 182
L-Glyceraldehyde, 121
Glyceraldehyde 3-phosphate, 159, 160, 173, 176, 177, 178, 182, 207
  oxidation of, 161
Glyceraldehyde-3-phosphate dehydrogenase, 103, 160, 161, 207
Glycerol, 130, 172, 173, 174, 218, 219, 231, 261, 263, 582
Glycerol ether, biosynthesis of, 220
Glycerol kinase, 173, 174, 218, 219, 236
Glycerol 3-phosphate, 96, 173, 174, 220, 237, 261, 262, 582, 586
sn-Glycerol 3-phosphate, 218, 219, 220, 221
Glycerol phosphate pathway, 219
Glycerol-3-phosphate acyltransferase, 219, 220
Glycerol-3-phosphate dehydrogenase, 174, 218
Glycerophosphate acyltransferase, in mitochondria, 116
α-Glycerophosphate dehydrogenase, in carbohydrate metabolism, 86
Glycerophosphate shuttle, 116
Glycerophospholipid(s), 130
Glycerose, 120
D-Glycerose, 121, 124
L-Glycerose, 121
Glycerylphosphocholine, 221
Glycerylphosphocholine hydrolase, 221
Glycinamide kinosynthetase, 346
Glycinamide ribosyl 5-phosphate, 347
Glycinamide ribosylphosphate, 346
Glycinamide ribosylphosphate formyl-transferase, 346
Glycinamide ribotide (GAR), 346
Glycine (Gly), 16, 262, 265, 275, 285, 289, 306, **307,** 317
  biosynthesis of, 267
  cleavage of, 289
  conjugation with, 307
  conversion of, 307
  formation of, from choline, 268
    conjugates of, 307
  reversible cleavage of, 286
  synthesis of, 267
Glycinuria, 286
Glycochenodeoxycholic acid, 248, 249
Glycocholic acid, 248, 249
Glycoconjugate(s), 589
Glycocyamine, 316, 317
Glycogen, 125, 158, 160, 175, 168, 172, 173, 182, 185, 188, 260, 262, 263, 281, 584, 625
  in liver, as source of blood glucose, 194
  metabolism of, **165–171**
    in liver, 190
    regulation of, 192
  molecule of, 127

Glycogen primer, 165, 166
Glycogen storage diseases, 170
Glycogen synthase, 165, 166, 167, 174, 559
  activation and inactivation of, 170
  in muscle, 170
Glycogen synthase a, 170, 188
Glycogen synthase b, 170
Glycogen synthase kinases, 170
Glycogen synthase system, in rat, 188
Glycogenesis, **165,** 188, 193
  control mechanisms of, 167
  and glycogenolysis, 166
Glycogenic amino acids, 281
Glycogenolysis, 165, **166,** 193
  cAMP-independent, 193
  control mechanisms of, 167
Glycogenosis, 170, 171
  myophosphorylase deficiency, 171
Glycolaldehyde, 177, 181
  active, 177
Glycolate, 181
  in oxalate formation, 180
Glycolipid(s), 130, **135,** 183, 223
Glycolysis, 97, **158–164,** 189, 190, 625
  aerobic, 158
  anaerobic, 158
  in liver, 192
  pathway of, 158, 159, 160, 176, 178
  regulation of, 189
  sequence of reactions in, 159
  summary of, 158
Glycophorin, 129
Glycoprotein(s), 128, 184, 185, **589–598,** 603
  classification of, 592
  complex, 593
  defined, 128, 589
  degradation of
    disorders of, 603
    polysaccharide moieties of, 603
  detection, purification, and structural analysis of, 590
  functions of, 589
  glycosylation of, regulation of, 597
  high-mannose, 593
  hybrid, 593
  and metastasis, 647
  N-linked, 593, 594
  O-linked, 592
  oligosaccharide chains of, 590
  polylactosamine, 594
  sugars in, 590
Glycoprotein hormones, 488
α-D-Glycopyranose, 122
Glycosaminoglycans, 184, 589, 599, 605, 606
  and cell surface molecules, 605
  and intracellular macromolecules, 606
  structures of, 600
Glycosidase(s), 591
Glycosides, 123
  cardiac, 124
Glycosidic bond, as acetal link, 123
N-Glycosidic bonds, in RNA, 382
N-Glycosidic linkages, of glyco-proteins, 592

O-Glycosidic linkage(s)
  of glycoproteins, 592
  in proteoglycans, 599
Glycosphingolipid(s), 130, 135, 184, 223, 447
  functions of, 223
  and metastasis, 647
Glycosuria, 196, 553, 554
  renal, 197
N-Glycosylamine links, in proteogly-cans, 600
Glycosylation, 597
  inhibitors of, 598
Glycosyltransferase(s), 597
  in proteoglycan synthesis, 600
G_{M1}, 136
G_{M3}, 136
GMP. *See* Guanosine monophos-phate.
GnRH. *See* Gonadotropin-releasing hormone.
GnRH-associated peptide (GAP), 483, 484
Goiter, 501
Golgi complex, 230
Gonad(s)
  hormones of, **530–546**
    normal values, 663
    ovarian, **536–544**
    and sexual differentiation, 544
    testicular, **530–535**
  primitive, 544
Gonadal dysgenesis, 543
Gonadal sex, 544
Gonadotropin(s), 489
Gonadotropin-releasing hormone (GnRH), 482, 483, 489, 533
GOT. *See* Glutamic oxaloacetic transaminase.
Gout, 352, 356, 358
Gouty arthritis, 357
G6PD. *See* Glucose 6-phosphate de-hydrogenase.
GPT. *See* Glutamic pyruvic transami-nase.
Gramicidin S, 26
Gratuitous inducers, 84, 431
Graves' disease, 470, 471, 501
Griseofulvin, 328
Group I hormones, mechanism of ac-tion of, 472
Group II hormones, mechanism of ac-tion of, 475
Group specificity of enzymes, 54
Group transfer potential, 96
Group transfer reactions, 50
Growth factor(s)
  and oncogenes, 646
  polypeptide, **644,** 646
Growth hormone (GH), 238, 484, 485
  feedback system regulating, 486
  gene for, structure of, 484
  structure of, 485
Growth hormone release-inhibiting hormone (GHRIH). *See* Somato-statin.
Growth hormone-prolactin-chorionic somatomammotropin group, 484
Growth hormone-releasing hormone (GHRH), 483

GRP. *See* Gastrin-releasing peptide.
GTP. *See* Guanosine triphosphate.
Guanase, 345, 352
Guanidoacetate, 317
Guanine, 336, 345, 349, 350, 362, 395
  in DNA, 379, 395
  phosphoribosylation of, 350
  in RNA, 382
Guanosine, 337, 345, 351
  derivatives of, 341
Guanosine diphosphate (GDP), 591
Guanosine monophosphate (GMP), 348, 349, 350
Guanosine triphosphate (GTP), 98, 154, 172, 629
Guanylate cyclase, 341, 479
L-Gulonate, 123, 180, 181
Gynecomastia, 487
Gyrase, bacterial, in DNA, 381

H chains, 609, 610
H locus, and blood group antigens, 599
H protomer, 58
H zone, 620
Hairpin, defined, 376
Haploid genome, human, 391
Haptoglobin, serum, normal values, 661
Hartnup disease, 299, 308, 571, 577
  due to malabsorption, 587
hCG. *See* Human chorionic gonadotropin.
hCS. *See* Human chorionic somatomammotropin.
HDL. *See* High-density lipoproteins.
Heart disease, coronary, 250
Heavy chains, 439, 609, 610
Heavy meromyosin (HMM), 621
Helix, triple, of collagen, 631
α Helix, 34, 35, 621
  structure of protein, 34, 35
Hematoporphyrin, absorption spectrum of, 324
Heme, 40, 42, 45, 319
  biosynthesis of
    pathway of, 326
    regulation of, 323
  catabolism of, 328
  formation of, 323
  synthesis of, 306
    regulation of, 324
Heme a, 242
Heme oxygenase system, 328, 329
Heme synthase, 323
Hemiacetal, defined, 122
Hemiketal, defined, 122
Hemin, 328
Hemoglobin(s), 40, **43**, 163, 319
  A, 43, 418
  Bristol, 418
  Chesapeake, 47
  fetal, 47
  Hikari, 418, 419
  human, mutant, 47
  Lepore, 393, 394
  M, 47
    Boston, 419

Hemoglobin(s) (cont'd)
  Milwaukee, 418
  S, 47, 48, 418, 419
  sickle, 48
  Sydney, 418
Hemoproteins, iron-containing, 103
Hemorrhagic disorders, 616
Hemostasis, 612
Heparan sulfate, 602, 604, 605
  structure of, 600
Heparan-N-sulfatase, 604
Heparin, 128, 230, 600
  anticoagulant activity of, 605
  structure of, 600, 602
Heparin sulfamidase, 604
Heparin-releasable hepatic lipase, 230, 233
Hepatic coma, 274
Hepatic parenchymal cells, 229
Hepatic portal system, 229
Hepatic sinusoids, 229
Hepatorenal syndrome, 213
Heptapeptide, 25
Heptoses, 120
Hereditary fructose intolerance, 182
HETE. *See* Hydroxyeicosatetraenoate.
Heterochromatin, 390
  constitutive, 390
  facultative, 390
Heteropolysaccharides, 120
Hexaenoic acids, 132
Hexamethonium, 239
Hexokinase, 98, 159, 160, 162, 173, 182, 184, 188, 195
  coupled assay for, 55
  in IUB classification, 51
Hexosamines, **124**, 184
Hexosaminidase A and B, deficiency of, 224
Hexosans, 120
Hexose(s), 120
  in glycoproteins, 128
  metabolism of, 180
Hexose monophosphate shunt. *See* Pentose phosphate pathway.
HGPRTase. *See* Hypoxanthine-guanine phosphoribosyl transferase.
HHT. *See* Hydroxyheptadecatrienoate.
Hibernation, 239
High-density lipoproteins (HDL), 227, 228, 229, 231, 232, 250, 251
  metabolism of, 232, 233
  nascent, 232
High-energy bonds, 154, 166
  sulfur, 161
  thio ester, 164
High-energy phosphate, **95**, 150, 151, 161, 162
High-fat diets, 234
High-fat feeding, 255
High-voltage electrophoresis (HVE), 22
Hill equation, 71, 72
Hind III, 60
Hinge region, 609
Hippurate, biosynthesis of, 307
His. *See* Histidine.
Histamine, 309

Histidase, 285
  in amino acid metabolism, 86
Histidine (His), 17, 65, 66, 262, 265, 270, 283, **309**, 587
  biosynthesis of, 270
    in microorganisms, 270
  catabolism of, metabolic disorders of, 284
  compounds related to, 308
  human requirements of, 573
L-Histidine, catabolism of, to α-ketoglutarate, 285
Histidine decarboxylase, 309
Histoenzymology, 56
Histone(s), 32, 363
  in chromatin, 387
HMG. *See* β-Hydroxy-β-methylglutaryl.
HMG-CoA. *See* β-Hydroxy-β-methylglutaryl-CoA.
HMG-CoA lyase, 257
HMG-CoA reductase, 241, 244, 559
  feedback regulation of, 90
  in lipid metabolism, 86
HMG-CoA synthase, 241, 255, 257
HMM. *See* Heavy meromyosin.
hnRNA. *See* RNA, heterogeneous nuclear.
Holoenzyme, 50
Homeostasis, 82
  of calcium, 503
    hormones involved in, 503
  caloric, 186
  of phosphate, 504
Homocarnosine, 308
Homocarsinosis, 308
Homocysteine, 18, 288
L-Homocysteine, 269, 299
Homocystinuria(s), 289, 577
Homogentisate, 291
  conversion to fumarate and acetoacetate, 292
Homogentisate dioxygenase, 106
Homogentisate oxidase, 291, 292
Homopolymer tailing, 365
Homopolysaccharides, 120
Homoserine, 18
L-Homoserine, 269, 300
  conversion to α-ketobutyrate, 300
Hormone(s). *See also specific hormone.*
  action of, **472–481**
    and calcium, 479, 480
    and phosphatidylinositides, 479, 480
  adrenal, **511–529**
  α_2-adrenergic, 475
  β-adrenergic, 475
  adrenocorticotropic. *See* ACTH.
  anterior pituitary, 483
  antidiuretic, 471, 475, 482, 493, 494
  autocrine function of, 463
  biosynthesis and modification of, 464
  catecholamine, 524
  classification of, 472, 473
  corticotropin-releasing, 491
  follicle-stimulating, 464, 488, 489
  gastrointestinal, **564–568**

Hormone(s) (cont'd)
  glycoprotein, 488
  gonadal, **530–546**
  gonadotropin-releasing, 489
  group I, mechanism of action of, 472
  group II, mechanism of action of, 475
  growth, 238, 484, 485
    feedback system regulating, 486
    gene for, structure of, 484
    structure of, 485
  growth hormone-prolactin-chorionic somatomammotropin group, 484
  hypothalamic, 482
  hypothalamic releasing, structures of, 483
  hypothalamic-hypophyseal, 482, 483
  lactogenic, 487
  luteinizing, 464, 475, 488, 489, 533, 540
  melanocyte-stimulating, 26, 238, 475, 493
  ovarian, **536–544**
  pancreatic, **547–563**
  paracrine function of, 463
  parathyroid, 471, 475, 502, 503, 504
  peripheral conversion of, 464
  pituitary, **482–495**
  placental, and pregnancy, 540
  posterior pituitary, 493
  preproparathyroid, 503
  progestational, 251
  proparathyroid, 503
  receptors for, 465, 466, 467
    and disease, 470, 471
    recognition and coupling domains of, 466
    regulation of, 468
    structure of, 468
  regulation of
    by Ca^{2+}, 478
    and calcium metabolism, **502–510**
  serum or plasma, normal values, 662
  somatotropin release-inhibiting, 483
  steroid. *See* Steroids.
  testicular, **530–535**
  thyroid, 196, **496–501**
  thyroid-stimulating, 238, 464, 471, 475, 488, 490, 494
Hormone responsive element, 474
Hormone systems, characteristics of, **463–471**
Hormone-sensitive lipase, 238
Hormone-sensitive triacylglycerol lipase a and b, 238
Hpa I, 60
HPETE. *See* Hydroperoxyeicosatetraenoate.
H_2S, in oxidative phosphorylation, 112
Human chorionic gonadotropin (hCG), 464, 475, 489, 540
Human chorionic somatomammotropin (hCS), 484, 487, 540

Humoral immune responses, 612
Hunter's syndrome, 603
Hurler's syndrome, 589, 603
Huxley, sliding filament mechanism of, 627
HVE. *See* High-voltage electrophoresis.
H-Y antigen, 544
Hyaluronic acid, 125, 128, 185, 601, 603
  structure of, 600
Hyaluronidase, 600, 603
Hybridization, 362, 368, 369
  colony, 369
  defined, 376
  in situ, 370
  plaque, 369
Hydatidiform mole, 544
Hydratase, 204, 205, 208
Hydrochloric acid, gastric, 579
  production of, 579
Hydrocortisone, 464
Hydrogen bonds, 34
Hydrogen sulfide, 101, 588
Hydrogenation reactions, 94
Hydrolases, classification of, 51
Hydrolysis, 13
  by proteolytic enzymes, enzyme degradation involving, 85
Hydrolytic activity of phospholipase, sites of, 222
Hydrolytic reactions, 51
Hydroperoxidases, 101, 103
Hydrophobic interactions, 34
Hydrostatic pressure, tissue, 607
D(-)-3-Hydroxyacyl enzyme, 205
D(-)-β-Hydroxyacyl-CoA, 203
L(+)-3-Hydroxyacyl-CoA, 201, 203
3-Hydroxyacyl-CoA dehydrogenase, 199
L(+)-3-Hydroxyacyl-CoA dehydrogenase, 201
D(-)-β-Hydroxyacyl-CoA epimerase, 203
3-Hydroxyanthranilate, 297, 298
3-Hydroxyanthranilate dioxygenase, 106
3-Hydroxyanthranilate oxidase, 297
Hydroxyapatite, 502, 631
D(-)-3-Hydroxybutyrate, 254, 256, 257
γ-Hydroxybutyrate, 317
3-Hydroxybutyrate dehydrogenase, 116
D(-)-3-Hydroxybutyrate dehydrogenase, 255, 256
25-Hydroxycholecalciferol, 464
7α-Hydroxycholesterol, 249
Hydroxycinnamate, 118
γ-Hydroxy-L-glutamate-γ-semialdehyde, 290
5-Hydroxyindole-3-acetic acid, 313
β-Hydroxyisobutyrate, 301, 304
β-Hydroxyisobutyryl-CoA, 301, 304
Hydroxykynurenine, 298
3-Hydroxykynurenine, 297, 298
Hydroxylamine, 29
Hydroxylase, 106, 243
18-Hydroxylase, 515
  deficiency of, 523

21-Hydroxylase, 515
  deficiency of, 523
1α-Hydroxylase, 508
7α-Hydroxylase, 249
17α-Hydroxylase, 531
11β-Hydroxylase, 515
  deficiency of, 523
Hydroxylysine (Hyl), 16
  biosynthesis of, 270
5-Hydroxymethylcytosine, 336
β-Hydroxy-β-methylglutaryl (HMG), 242
β-Hydroxy-β-methylglutaryl-CoA (HMG-CoA), 242, 244, 255, 256, 257, 301, 303
p-Hydroxyphenylacetate, 293
p-Hydroxyphenyllactate, 293
p-Hydroxyphenylpyruvate, 291, 292, 293
p-Hydroxyphenylpyruvate hydroxylase, 291, 292
Hydroxyproline (Hyp), 16, 262, 265, 273, 290
  biosynthesis of, 268
  catabolism of, metabolic disorders of, 290
4-Hydroxy-L-proline, 290
15-Hydroxyprostaglandin dehydrogenase, 216
4-Hydroxypyrazolopyrimidine, 342, 343
3β-Hydroxysteroid dehydrogenase (3β-OHSD), 514, 531
17β-Hydroxysteroid dehydrogenase (17β-OHSD), 531
5-Hydroxytryptamine, 312, 313
5-Hydroxytryptophan, 312
Hyl. *See* Hydroxylysine.
Hyp. *See* Hydroxyproline.
Hyperalaninemia, 577
Hyper-β-alaninemia, 307
Hyperalphalipoproteinemia, 252
Hyperammonemia, 279
Hyperargininemia, 279, 283
Hyperbetalipoproteinemia, 251
Hyperbilirubinemia, 331
  conjugated, 332
  toxic, 332
  unconjugated, 331
Hyperbilirubinemic toxic encephalopathy, 332
Hypercholesterolemia, familial, 251
Hyperchromicity of denaturation, in DNA molecule, 379
Hyperglycemia, 197, 553, 554
Hyperhydroxyprolinemia, 290
Hyperketonemia, 255
Hyperlipoproteinemia
  type I, 251
  type II, 251
  type III, 229, 251
  type IV, 251
  type V, 251
Hyperlysinemia, 295, 296
Hypermethioninemia, 289
Hyperoxaluria, 286
Hyperphenylalaninemia, 293
Hyperplasia, congenital adrenal, 523
Hyperprolinemia, 283
Hypertension, renin-dependent, 518

Hyperthyroidism, 471, 501
Hypertriacylglycerolemia, 251
Hypertrophy, benign prostatic, 535
Hyperuricemia, 356
    classification of, 358
Hypervalinemia, 303
Hypervariable regions, in heavy and
    light chains, 610
Hyperventilation syndrome, 502
Hypobetalipoproteinemia, 251
Hypoglycemia, 171, 196, 260
    convulsions in, 197
Hypoglycin, 202
Hypogonadism
    primary, 535, 543
    secondary, 535, 543
Hypolipidemic drugs, 250
Hypolipoproteinemia, 251
Hypomelanosis, 314
Hypoparathyroidism, 506
Hypothalamic hormones, 482
Hypothalamic releasing hormones,
    structures of, 483
Hypothalamic-hypophyseal hor-
    mones, 483
Hypothyroidism, 250, 251, 501
Hypouricemia, 357, 358
Hypoxanthine, 336, 344, 345, 350
    derivatives of, 341
    in DNA, 401
    and guanine, phosphoribosylation
        of, to form IMP and GMP, 350
Hypoxanthine-guanine phosphoribo-
    syltransferase (HGPRTase), 350,
    351, 358

I band, 619, 620
I⁻. See Iodide.
¹³¹I, 499, 501
i gene, 430
IAP. See Intermittent acute porphyria.
ICD. See Isocitrate dehydrogenase.
I-cell disease, 598
Icterus, 331
IDDM. See Insulin-dependent dia-
    betes mellitus.
Idiotype, 612
IDL. See Intermediate-density lipo-
    proteins.
β-L-Iduronate, 125
Iduronate sulfatase, 604
Iduronic acid, sulfated, 128
L-Iduronic acid, 599
α-L-Iduronidase, 603
IgA, 610, 611
    secretory, 611
    serum, 611
IgD, 610
IgE, 610
IGF. See Insulinlike growth factor.
IGF-I. See Insulinlike growth factor I.
IGF-II. See Insulinlike growth factor
    II.
IgG, 610
IgM, 441, 610, 611
Ile. See Isoleucine.
Imidazole, 205
Imidazole ring, closure of, 346
4-Imidazolone-5-propionate, 284, 285

Imidazolonepropionate hydrolase,
    285
Imino acids, 17
Immune deficiency, 358
Immune responses, humoral, 612
Immunoelectrophoresis, 227
Immunoglobulin(s), 438, 609
    chains, properties of, 610
    classes, 610
    domains, 609
    in plasma, 609
    polymeric, 611
    properties of, 611
    regions of, 609
Immunoglobulin A (IgA), 610, 611
    secretory, 611
    serum, 611
Immunoglobulin D (IgD), 610
Immunoglobulin E (IgE), 610
Immunoglobulin G (IgG), 610
Immunoglobulin gene, rearrangement
    of, 438
Immunoglobulin light chain, 438
Immunoglobulin M (IgM), 441, 610,
    611
IMP. See Inosine monophosphate.
IMP cyclohydrolase, 346
IMP dehydrogenase, 348
Indefinite response, 428
Indole, 587
    derivatives in urine, 314
Indoleacetate, 299
α-N(Indole-3-acetyl)glutamine, 299
Indole-5-6-quinone, 315
"Induced fit" model of catalytic site,
    64
Inducers, gratuitous, 84, 431
Inducible enzyme, basal level of, 84
Inducible gene, defined, 429
Induction
    as control of enzyme synthesis, 83
    coordinate, of enzymes, 84
Inhibin, 489, 534
Inhibition
    of allosteric enzymes, kinetics of,
        89
    feedback, 87, 90
        cooperative, 88
        cumulative, 88
        multivalent, 88
    noncompetitive
        irreversible, 73
        reversible, 73
Inhibitor-1, 168, 170
Initial velocity, 68
    measured, 69
Initiating agent, 638
Initiation, 638
Initiation factors in protein synthesis,
    422
Injury, tissue, response to, 614
Inorganic phosphate (P$_i$), 95, 161, 188
Inorganic pyrophosphatase, 98, 165
Inosine, 344, 345, 351
Inosine monophosphate (IMP), 346,
    350, 627
    conversion to AMP and GMP, 348
    interconversion to adenosine nucle-
        otides and guanosine nucle-
        otides, 352

Inosine triphosphate (ITP), 172
Inosinic acid, 341, 346
Inositol, 135
Inositol triphosphate, 461
Inotropic effect, 528
Insert, defined, 376
Insertions, 419
Insulin, 169, 170, 188, 195, 237,
        238, 464, 471, 547, 548, 550
    A and B chains of, 38, 548
    biosynthesis of, 549
    chemistry of, 548
    deficiency of
        major symptoms of, 553
        metabolic consequences of, 556
        pathophysiology of, 554
    effect on gene expression, 559
    human, structure of, 548
    and lipogenesis, 254
    mechanism of action of, 557
    metabolism of, 553
    nonhuman, variations in structure
        of, 548
    pathophysiology of, 560
    physiologic effects of, 553
    primary structure of, 38
    secretion of, regulation of, 552, 553
Insulin receptor, 557, 561
Insulin-dependent diabetes mellitus
        (IDDM), 560
Insulinlike growth factor (IGF), 557,
        561
Insulinlike growth factor I (IGF-I),
        482, 486, 561
Insulinlike growth factor II (IGF-II),
        486, 561
Interconversion of major foodstuffs,
        262
Interconvertible enzymes, 90
    analogies to feedback inhibition, 91
Intermediate filaments, 627
    classes and distributions of, 631
Intermediate-density lipoproteins
        (IDL), 228, 231
Intermittent acute porphyria, 325
Intermittent branched-chain ketonuria,
        303
Internal milieu (Claude Bernard), 82
International Union of Biochemistry
        (IUB) system, for classification
        of enzymes, 50
Interphase chromosomes, chromatin
        fibers in, 389
Intestinal cells, 230
Intestinal digestion, 580, 584
Intestinal mucosal cells, microvilli of,
        627
Intestinal putrefaction, 587
Intestinal secretions, digestion by, 583
Intestines, amino acids in, 275
Intracellular distribution of enzymes, 56
Intracellular fluid, 445
Intrinsic factor, 576
Intron(s), 28, 363, 414
    defined, 376, 392, 409
Inulin, 125
Invert sugar, 125
Invertase, 84
Iodide, metabolism of, 497, 498
    model of, 498

Iodine
  and biosynthesis of thyroid hormones, 496
  deficiency of, 497
  major characteristics of, 578
  radioactive, 496
Iodoacetate, 160, 161
5′-Iodo-2′-deoxyuridine, 342
o-Iodosobenzene, 29
Iodothyronyl, 497
Iodotyrosyls, coupling of, 499
Ion(s), metal, in enzymes, catalytic and structural roles of, 87
Ionic forms of amino acids, 14
Ionizing radiation, and DNA, 401
Ionophores
  action of, 119
  in membrane transport, 456
Iproniazid, 312
Iron, 103
  major characteristics of, 578
  nonheme, 109
  serum, normal values, 655, 661
Iron-binding capacity, serum, normal values, 655, 661
Iron-containing hemoproteins, 103
Iron-sulfur protein (FeS), 109
Iron-sulfur-protein complex, 110
Islets of Langerhans, 547
  cell types in, 547
  hormones of, normal values, 663
Isoalloxazine, reduction of, 102
Isobutyryl-CoA, 302
Isocitrate, 151, 152, 207
Isocitrate dehydrogenase (ICD), 151, 152, 205, 207
Isocitrate dehydrogenase reaction, 207
Isoelectric pH (pI), 15
Isolation of enzymes, 55
Isoleucine (Ile), 16, 262, 265, 298, 299, 300, 302
  biosynthesis of, 270
  in brain, 275
  catabolism of, 298, 299, 301, 302
  human requirements of, 573
Isomaltase, 584
Isomer, optical, 121
Isomerase(s), 582
  classification of, 51
$\Delta^{5,4}$ Isomerase, 514, 531
Isomerism, 120
  aldose-ketose, 122
  in unsaturated fatty acids, 132
  geometric, 132
D-Isomerism, 121
L-Isomerism, 121
Isomerization reactions, 50
Isopentenyladenine, 242
Isopentenylpyrophosphate, 242, 244
Isopentenylpyrophosphate isomerase, 242
Isoprenoid unit(s), 244
  active, 244
Isotypes, 612
Isovaleric acidemia, 301, 304
Isovaleryl-CoA, 302
Isovaleryl-CoA dehydrogenase, 305

Isozymes, 57
  diagnostic significance of, 57
  physical basis for, 58
ITP. *See* Inosine triphosphate.
IUB. *See* International Union of Biochemistry.

J chain, 610
J gene, 612
Jamaican vomiting sickness, 202
Jaundice, 331
  cholestatic, 333
  chronic idiopathic, 332
  congenital nonhemolytic, 332
  hemolytic, 333
  obstructive, 333
Joining region gene, 612
"Jumping genes," 641
Juxtaglomerular cells, 517

K_d, relationship to K_m, 71
K_{deg}, 83
K_{eq}, 68
K_m, relationship to K_d, 71
K_m value, 69
K_s in enzymes, 83
Kallidin, 25
Karyotype, human, 391
K^+-ATPase, 579
Keratan sulfates, 602, 603
  structure of, 600
Keratin, 631
Kernicterus, 332
α-Keto acid, 267
α-Keto acid decarboxylase, 303
Ketoacidosis, 255
3-Ketoacyl enzyme, 205, 206
Ketoacyl reductase, 204
3-Ketoacyl reductase, 205
Ketoacyl synthase, 204
3-Ketoacyl synthase, 205, 206
3-Ketoacyl-CoA, 199, 201
3-Ketoacyl-CoA reductase, 208
3-Ketoacyl-CoA synthase, 208
α-Ketoadipate, 295, 297
α-Ketoarginine, 309
α-Ketobutyrate, 298, 299, 300
Ketogenesis, **254**, 255
  in liver, 255
    pathways of, 256
  regulation of, 258
Ketogenic amino acids, 281
α-Ketoglutaraldehyde, 306
α-Ketoglutarate, 152, 155, 156, 207, 262, 266, 275, 282, 283, 296, 317
  amino acids forming, 283
  oxidative decarboxylation of, 151
α-Ketoglutarate dehydrogenase, 151, 152
α-Ketoglutarate transporter, 118, 207
3-Keto-L-gulonate, 180, 181
2-Keto-L-gulonolactone, 181
α-Keto-γ-hydroxyglutarate, 290
α-Ketoisocaproate, 302
Ketoisomerase, 178
α-Ketoisovalerate, 302

α-Keto-β-methylvalerate, 302
Ketone bodies, 130, 188, 254, 255, 256, 258, 260, 261, 263
  formation, utilization, and excretion of, 256
  interrelationships of, 255
Ketonemia, 255, 258
Ketonuria, 255
  intermittent branched-chain, 303
3-Keto-6-phosphogluconate, 176
Ketose(s), 120
  examples of, 124
Ketosis, 171, 200, 255, 258
  assessing severity of, 258
  in cattle, 235, 255
  in vivo, 260
3-Ketosphinganine, 222
3-Ketosphingosine reductase, 222
β-Ketothiolase, 199, 201
Khorana, 415
Kidney
  amino acids in, 275
  hormones of, normal values, 663
Kinase(s), 242
  calcium-calmodulin-dependent, 477
  calcium-phospholipid-dependent, 477
  cAMP-dependent, 477
  casein, 477
  cGMP-dependent, 477
  dsRNA-dependent eIF-2α, 477
  epidermal growth factor-dependent tyrosine, 477
  hemin-dependent eIF-2α, 477
  insect cyclic nucleotide-dependent, 477
  insulin-dependent tyrosine, 477
  myosin light chain, 477
  phosphorylase, 477
  protease-activated, 477
  protein, 476
    cAMP-dependent, 168, 169, 170, 187, 188, 189, 192, 238, 477
    cAMP-independent, 477
    cGMP-dependent, 479
    purified, 477
  pyruvate dehydrogenase, 477
  rhodopsin, 477
  viral tyrosine, 477
Kinetic properties of enzymes, **61–81**
Kinetic (collision) theory for chemical reactions, 62
Kinetics
  of feedback inhibition, 89
  sigmoid saturation, 71
  "stop-flow," 75
Kinetochores, 629
Kininogen, high-molecular-weight, 614
  disorders of, 615, 616
Kornberg, Arthur, 395
Koshland, "induced fit" model of, 64
Krabbe's disease, 224
Krebs, 259
Krebs cycle, 149
Kupffer cells, 505
Kwashiorkor, 267, 271, 571, 574, 585

Kynurenic acid, 298
Kynureninase, 297, 298
Kynurenine, 297, 298
Kynurenine formylase, 297, 298
Kynurenine hydroxylase, 297
Kynurenine-anthranilate pathway, 298

L chains, 609, 610
Lac operon, 430, 431
Lac repressor, 430
Lactase, 584
 deficiency of, 585
Lactate, 154, 158, 162, 172, 173,
  174, 262
 blood, normal values, 661
L(+)-Lactate, 160
Lactate dehydrogenase (LDH), 160,
  162, 172, 288, 317
 activity of, detection of, 57
 isozymes of, 57, 58
 normal values, 656
 quaternary structure of, 39
 serum, normal values, 661
Lactation, 541
Lacteals, 229, 582
Lactic acid cycle, 194, 195
Lactic acidosis, 164
 responsive to thiamin, 577
Lactogen, placental, 484, 487
Lactogenic hormone, 487
Lactose, 120, 126, 183, 584
 metabolism of, 430
 synthesis of, 183, 184
Lactose operon, 430, 431
Lactose synthase, 183, 184
Lactotropes, 487
Lambda
 bacteriophage, 432
 life cycle of, 434
Lambda repressor, 432
Lambda repressor protein, 433
Lamellipodia, 628, 629
Landsteiner, 598
Lanosterol, 243, 244
Lauric acid, 131
LCAT. *See* Lecithin:cholesterol acyl-
  transferase.
LDH. *See* Lactate dehydrogenase.
LDL. *See* Low-density lipoproteins.
Le blood group system, 598
Lead, 235
 urine, normal values, 663
Leader sequence, of proteins, 452
Lecithin, 134, 220, 581
 metabolism of, 221
Lecithin:cholesterol acyltransferase
  (LCAT), 222, 232, 247, 248
 deficiency of, 232, 252
Lectin(s), 129, 592
Left-handed double helix, of Z-DNA,
  378
Leprechaunism, 560
Lesch-Nyhan syndrome, 357
Leu. *See* Leucine.
Leucine (Leu), 16, 262, 265, 270,
  290, 299, 302
 in brain, 275
 catabolism of, 301, 302
 human requirements of, 573

Leucodopachrome, 315
Leu-enkephalin, 568
Leukodystrophies, 224
Leukotriene(s) (LT), 131, 210, 213,
  214, 216
Leukotriene A4, 132
Lewis blood group system, 598
Leydig cells, 530
LH. *See* Luteinizing hormone.
Library
 cDNA, 367
 complete genomic, composition of,
  368
 defined, 376
 genomic, 367
Lieberkühn, glands of, 583,
  584
Ligases
 classification of, 51
 DNA, 398
Ligation, defined, 376
Light chain(s), 609, 610
 coding of, 438
 protein phosphatase, 624
p-Light chain, 623
Light meromyosin (LMM), 621
Lignoceric acid, 131, 223
Limit dextrinosis, 171
Linear polypeptides, resolution into, 28
Lineweaver-Burk plot, 70, 73
Lingual glands, 584
Lingual lipase, 579, 584
"Link proteins," 602
Linoleic acid, 210, 211, 212, 574
 conversion to arachidonate, 213
 oxidation of, 203
α-Linoleic acid, 574
α-Linolenic acid, 211
γ-Linolenic acid, 211
γ-Linolenyl-CoA, 213
Linoleyl-CoA, 203, 213
Lipase, 579, 580, 582, 584
 hormone-sensitive, 236
 lingual, 579, 584
 lipoprotein, 229, 230, 231, 233,
  234, 251, 605
  activity of, cofactors for, 230
  in catabolism of chylomicrons, 230
 serum, normal values, 656, 661
Lipid(s), **130–141**
 absorption of, 586
 alcohols in, 133
 amphipathic, 140, 141, 447
 bilayer of, 447, 448
 of blood plasma, 226
 classification of, 130
 complex, 130
 cycle of, 263
 derived, 130
 dietary
  and disease, 574
  requirement for, 574
 ether, 219
 membrane, 446
  diagram of, 447
  formation of, 141
 metabolism of, 144
  economics of, 260
  liver in, 233
  regulation of, 253

Lipid(s) (cont'd)
 neutral, 130
 peroxidation of, 138
 physical properties of, 133
 rancidity of, 138
 total
  in blood plasma, 226
  serum, normal values, 662
Lipid bilayer membranes, permeability coefficients in, 448
Lipid fractions, serum or plasma, normal values, 662
Lipid membrane
 diagram of, 447
 organization of, 447
Lipid nephrosis, 250
Lipid peroxides, 235
Lipid-linked oligosaccharide precursor, structure of, 595
Lipidoses, 223
 summary of, 224
Lipoate, 103
Lipogenesis, 188, 236, 237, 239, 253
 hepatic, direct inhibition of, by free
  fatty acids, 253
 provision of acetyl-CoA and
  NADPH for, 207
Lipoid, defined, 130
Lipolysis, 236, 238
Lipoprotein(s), 130, 261, 608
 high-density, 228, 229, 230, 231,
  232, 233, 250
 intermediate-density, 228
 low-density, 227, 228, 229, 232,
  246, 248, 251
 nascent, 230
 plasma, 226
  composition of, 227
  metabolism of, 229
 very low density, 227, 228, 229,
  232, 246, 248, 250, 251
  catabolism of, 230
  formation of, 229
  and secretion of, by hepatic
   cells, 230
  metabolic fate of, 231
α-Lipoprotein, 229
 deficiency, familial, 251
β-Lipoprotein, 229, 246
Lipoprotein lipase, 229, 230, 231,
  233, 237, 251, 605
 activity of, cofactors for, 230
 in catabolism of chylomicrons, 230
Lipoprotein-X, 248
Liposomes, 141, 450
 formation of, 141
Lipotropic factors and fatty livers, 235
Lipotropin (LPH), 475
β-Lipotropin (β-LPH), 492
 primary structure of, 26
Lipoxygenase, 139
Lipoxygenase pathway, 216
Lithium, 494
Lithocholic acid, 248, 249
Liver, 193, 231, 256
 amino acids in, 275
 in catabolism of chylomicrons, 230
 and digestion, 584
 diseases of, parenchymal, 248
 fatty, and lipotropic factors, 235

Liver (cont'd)
   and glycogen, 165
      as source of blood glucose, 194
   in lipid metabolism, 231, 233
   long-chain fatty acid oxidation in, 259
   metabolic interrelationships of, with adipose and extrahepatic tissues, 261
Liver peroxisomes, polyamine oxidase in, 312
Liver phosphorylase, 183
LMM. *See* Light meromyosin.
"Lock and key" model of catalytic site, 64
Long interspersed repeats, in DNA, 392
Long terminal repeats (LTR), in DNA, 392, 641
Long-chain acyl-CoA, 188
Long-chain fatty acid, oxidation of, regulation of, 259
Loop(s), of chromatin fibers, 389
Looped domains, of chromatin fibers, 391
Low-density lipoproteins (LDL), 227, 228, 229, 230, 231, 232, 251, 252
   in membranes, 460
   metabolism of, 232
   receptor for, 246
Low-energy phosphates, 96
LPH. *See* Lipotropin.
β-LPH. *See* β-Lipotropin.
LT. *See* Leukotriene.
LTH. *See* Luteotropic hormone.
LTR. *See* Long terminal repeats.
Lung(s), 256
Lung surfactant, 220
Luteal phase, 540
Luteinizing hormone (LH), 464, 475, 488, 489, 533, 540
Luteotropic hormone (LTH), 487
Lyase(s)
   $C_{17\text{-}20}$, 531
   classification of, 51
Lymphatic system, 229
Lymphocytes, 231, 232
Lys. *See* Lysine.
Lysine (Lys), 17, 265, 270, 273, 290, 295
   biosynthesis of, by yeast, 270
   catabolism of, metabolic disorders of, 295
   human requirements of, 573
L-Lysine, 296
   conversion of to α-aminoadipate and α-ketoadipate, 295
Lysogenic pathway, of lambda DNA, 432
Lysolecithin, 135, 221, 222, 232
Lysophosphatidate, 219
Lysophosphatidate acyltransferase, 220
Lysophospholipase, 221
Lysophospholipids, 135, 221
Lysozyme, catalytic site of, 65
Lytic pathway, of lambda DNA, 432
Lytic reactions, 50
D-Lyxose, 123, 124

M protomer, 58
Macromineral(s), 577
   essential nutritional requirements of, 572
Macromolecular complexes, regulatory implications of, 86
Macromolecules, cross-membrane movement of, 459
Magnesium, 160
   dietary, major characteristics of, 577
   serum, normal values, 656
   serum or plasma, normal values, 662
Magnesium fluoride, 160
Magnesium ions, 177
Malabsorption, disturbances due to, summary of, 587
Malabsorption syndrome, 571
Malate, 153, 172, 173, 207
L-Malate, 152
Malate dehydrogenase, 57, 152, 153, 207
Malate shuttle, 117
Maleylacetoacetate, 291, 292
Maleylacetoacetate *cis, trans* isomerase, 291
Malic enzyme, 188, 205, 206, 207
Malnutrition, protein-energy, 574
Malonate, 72
   and citric acid cycle, 152
Malonyl-CoA, 202, 204, 206, 207, 208, 258, 259
   biosynthesis of, 204
Maltase, 584
Maltose, 120, 126, 584
Malonyl-CoA, 202
Mammary gland, development of, 541
Mammotropin, 487
Manganese
   dietary, major characteristics of, 578
   in glycolysis, 162
Mannosamine, 184
D-Mannosamine, 125
Mannose, 122, 128, 228
   in glycoproteins, 128
D-Mannose, 123, 124
α-D-Mannose, 123
MAO. *See* Monoamine oxidase.
Maple syrup urine disease, 303
Marasmus, 267, 574
Marfan's syndrome, 634
Maroteaux-Lamy syndrome, 603
Matthaei, 415
Maxam and Gilbert method of DNA sequencing, 370, 371
McArdle's syndrome, 171
Medroxyprogesterone acetate, 543
Medullary thyroid carcinoma, 510
Melanin, 314
Melanocyte-stimulating hormone (MSH), 26, 238, 475, 493
α-Melanocyte-stimulating hormone (α-MSH), 238, 492, 493
β-Melanocyte-stimulating hormone (β-MSH), 26, 238, 492, 493
γ-Melanocyte-stimulating hormone (γ-MSH), 493

Melanogenesis, 493
Melanoprotein, 314
Melanosomes, 314
Melatonin, 313
   biosynthesis and metabolism of, 313
Membrane(s), **445–462**
   amphipathic nature of, 447
   artificial, 450
   assembly of, **452–455**
      models for, 454
   asymmetry of, 449
   carbohydrates of, 129
   defined, 445
   enzymatic markers of, 449
   excitable, of sarcomere, 623
   fluidity of, 451
   lipid, permeability coefficients in, 448
   lipids in, 446
      composition of, 446
      organization of, 447
   lipid bilayer of, 447
   movement across, 455, 456
   permeability of, 448
   protein, integral and peripheral, 449
   proteins of, 448
      insertion of, 453
   protein:lipid ratio in, 446
   self-assembly of, 450
   signal transmission across, 461
   specialized functions of, **455–462**
   structure of, **446–452**
      fluid mosaic model for, 451
   transport systems of, 456, 457
Membrane trigger hypothesis, 454
Membrane-located ATP synthetase, in oxidative phosphorylation, 113
Menaquinone(s), major characteristics of, 576
Menkes' (kinky-hair) syndrome, 634
Menopause, 542
Menstrual cycle, 539
   hormonal and physiologic changes during, 539
Mercaptans, 588
6-Mercaptopurine, 342, 348
3-Mercaptopyruvate-cysteine disulfiduria, 289
Meromyosin
   heavy, 621, 622
   light, 621, 622
Mesobilirubinogen, structure of, 331
Messenger, second, 472
Messenger RNA. *See* mRNA.
Mestranol, 542
Met. *See* Methionine.
Metabolic pathways
   basic, 142
   general principles of regulation of, 186
Metabolic regulation, 82
Metabolism
   defined, 94
   intermediary
      overview of, **142–148**
      at subcellular level, 147
      at tissue and organ level, 145

Metabolites, unidirectional flow of, 82
Metachromatic leukodystrophy, 224
Metal ions
    in catalysis, 80
    catalytic and structural roles of, in
        enzymes, 87
    in enzyme action, 79
Metal-activated enzymes, 79
Metal-bridge complexes, 80
Metalloenzymes, 79
Metalloflavoproteins, 102, 109
Metallothionein, 440, 441
Metanephrine, 527
Metaphase chromosomes, 391
Metastasis, 647
Met-enkephalin, 568
Methacrylyl-CoA, 302, 304
    catabolism of, 304
Methane, 588
Methemoglobinemia, 47
N^5,N^{10}-Methenyltetrahydrofolate, 346
Methimazole, 498
Methionine (Met), 16, 235, 262, 265,
        298, 299, **309**, 311, 340
    active, 299
    biosynthesis of, 270
    catabolism of, 298
    metabolic disorders of, 300
    conversion of, to propionyl-CoA,
        300
    human requirements of, 573
L-Methionine, 299, 300
Methionine malabsorption syndrome,
    289
Methionyl-tRNA, 422
3-Methoxy-4-hydroxymandelic acid,
    527
5-Methoxyindole-3-acetic acid, 313
5-Methoxytryptamine, 313
Methyl mercaptan, 588
Methyl pentose, in glycoproteins, 128
Methyl xanthines, 238
    in foodstuffs, structures of, 337
α-Methylacetoacetyl-CoA, 302, 304
N-Methyl-4-aminoazobenzene, 636
2-Methyl-3-aminopropanoate, 19
Methylases, DNA, site-specific, 364
α-Methylbutyryl-CoA, 302
β-Methylcrotonyl-CoA, 301, 302
    carboxylation of, 301
    catabolism of, 303
5-Methylcytosine, 336
Methylene blue, 101
Methylene bridges, 321
Methylene-interrupted bonds, 211
N^5,N^{10}-Methylenetetrahydrofolate,
    286, 346
β-Methylglutaconyl-CoA, 301, 303
7-Methylguanine, 336
3-Methylhistidine, 626
α-Methyl-β-hydroxybutyryl-CoA,
    302, 304
Methylindole, 587
Methylmalonate, 304
Methylmalonate semialdehyde, 301,
    304
Methylmalonyl-CoA, 298, 304
D-Methylmalonyl-CoA, 174
L-Methylmalonyl-CoA, 174
Methylmalonyl-CoA isomerase, 174

Methylmalonyl-CoA racemase, 174
N-Methylnicotinamide, 298
Methylthioadenosine, 311
Mevalonate, 241, 242, 244
Mevalonate kinase, 242
Mevalonate 5-phosphate, 242
Mevalonate 3-phospho-5-pyrophos-
    phate, 242
Mevalonate 5-pyrophosphate, 242
Mevinolin, 241, 250
Mg-ATP complex, 159
Micelle(s), 140, 581
    cross section of, 448
    diagram of, 448
    formation of, 141
Michaelis constant, 69
Michaelis-Menten equation, 69
Michaelis-Menten model, limitations
    of, 71
Microfilaments, 627
Micromineral(s), 577
    essential nutritional requirements
        of, 572
Microsomal chain elongation system,
    208, 212
Microsomal desaturase system, 211,
    212
Microsomes
    cytochrome P-450 hydroxylase
        cycle in, 106
    electron transport chain in, 106
Microtubules, 627, 628, 630
    assembly of, 629, 630
    drugs inhibiting, 629
    sites of, 629
Microvilli, 627, 629
MIF. *See* Mullerian inhibiting
    factor.
Milk intolerance, 577, 587
Mineral(s)
    dietary
        major characteristics of, 577
        requirements for, 577
    essential nutritional requirements
        of, 572
Mineralization of bone, 502
Mineralocorticoid(s), 511
    classification and mechanisms of
        action of, 522
Missense mutations of genes, 418
    types of, 419
MIT. *See* Monoiodotyrosine.
Mitochondria
    energy-linked ion transport in, 117
    respiratory chain in, 108
    organization of, 108
    transport systems of, **116–119**
Mitochondrial ATPase, quaternary
    structure of, 39
Mitochondrial DNA polymerase, 398
Mitochondrial glycerol-3-phosphate
    dehydrogenase, 103
Mitochondrial glycine synthase com-
    plex, 286
Mitochondrial membrane(s), 212
    structure of, 115
    transporter systems in, 118
Mitochondrial system, 202
Mitochondrial transporter systems,
    118

Mitotic spindle, 629
Mixture, racemic, 121
Mobilization of fat, 236
Moderately repetitive sequences, in
    DNA, 392
Modulators
    of catalytic activity, 74
    molecule, 74
Molecular sieves, filtration through,
    39
Molecular weight of proteins, deter-
    mination of, 38
Molecule, chimeric, defined, 375
Molybdenum, 103
    major characteristics of, 578
Monoacyl glycerophosphate acyltrans-
    ferase, in mitochondria, 116
Monoacylglycerol, 133, 219
1-Monoacylglycerol, 582
2-Monoacylglycerol, 219, 582,
    586
Monoacylglycerol acyltransferase,
    219, 220
Monoacylglycerol hydrolase, 230
Monoacylglycerol pathway, 219, 220,
    582
Mono(ADP-ribose), 402
Monoamine oxidase (MAO), 101,
    102, 312, 527
    in mitochondria, 116
Monoenoic acids, 131, 132
Monoethenoid acids, 131
Monoglycerides, 133
Monoiodotyrosine (MIT), 496,
    497
3-Monoiodotyrosine, 18
Monomers, 36
Monooxygenase(s), 106, 637
    microsomal cytochrome P-450 sys-
        tems, 106
    mitochondrial cytochrome P-450
        systems, 106
Monophenols, 102
Monosaccharides, 120, **122**
    absorption of, 583
    malabsorption of, 586
Monosialoganglioside, 224
Monounsaturated acids, 131
Monounsaturated fatty acids, synthe-
    sis of, 211
Morphine, 106
Morquio's syndrome, 603
Motilin, 565, 566, 568
Mouth, digestion in, 579, 584
mRNA (messenger RNA), 363, 383,
    414
    amplification of, during develop-
        ment, 438
    polycistronic, 430
    regulated by insulin, 560
    in RNA processing, 411
    splicing patterns of, 442
    stability of, 442
    translation of, differential, 442
    trinucleotide code of, 83
    triplet code of, 83
mRNA transcript, 384
MSH. *See* Melanocyte-stimulating
    hormone.
Mst II, 59

Mucin, 579
  in bile, 580
Mucolipidoses, 600, 603
  biochemical defects in, 604
Mucopolysaccharides, 128
Mucopolysaccharidoses, 589, 600, 603
  biochemical defects in, 604
Mucoproteins, 128
Müllerian duct system, 545
Müllerian inhibiting factor
  (MIF), 545
Multienzyme complex, fatty acid syn-
  thase, 204
Multiple endocrine neoplasia (MEN)
  syndromes, 464
Multiple feedback loops, in metabolic
  regulation, 88
Multiple sclerosis, 223
Multiple sulfatase deficiency, 225
Multiplication-stimulating activity
  (MSA). See Insulinlike growth
  factor II.
Multiplicity, enzyme, 88
Multivalent feedback inhibition, 88
Multivalent repression, 85
Muscle, 619
  amino acids in, 275
  ATP in, sources of, 626
  cardiac, 619
  contraction of, 158, 620
    ATP in, 623
    myosin-based regulation of, 623
    and relaxation of, regulation of,
      623
  fiber, 618
  glycogen in, 165
  involuntary control of, 619
  molecular function of, 621
  nonstriated, 619
  phosphorylase in, 169
  proteins of, 621
  relaxation of, 623, 624
  skeletal, 619
    fast and slow, characteristics of,
      626
    glycolysis in, 158
  smooth, 619
    β-adrenergic stimulation of, 624
    contraction of, 623
    and striated muscle, actin-myosin
      interactions in, 625
  striated, 619
    arrangements of filaments in, 619
    contraction of
      actin-based regulation of, 623
      allosteric mechanism of, con-
        trol of, 623
      in mollusks, 624
      and smooth muscle, actin-myosin
        interactions in, 625
  structure of, 619
  voluntary control of, 619
Muscle fasciculus, 618
Muscle glycogen phosphorylase, 625
Muscle metabolism, 625
Muscle proteins, phosphorylation of,
  625
Mutagens, 637
Mutarotation, 122
  of glucose, 122

Mutations, of genes, 417
  altering DNA code, 83
  base-substitution, in protein synthe-
    sis, 417
  constitutive, defined, 429
  defined, 393
  effect of, 417
  frame shift, 419
  missense, 418
  point, 372
  in protein synthesis, 417
  single-point, 643
  transition, 417
  transversion, 417
Myasthenia gravis, 470, 471
myc oncogene, 641, 642
Myoadenylate kinase, 626
Myofibrils, 618
Myoglobin(s), 40, **41**, 319, 626
  oxygen dissociation curve for, 44
Myo-inositol, 135
Myo-inositol 1,4,5-P_3, 480
Myokinase, 98, 348
Myophosphorylase deficiency glycogenesis, 171
Myosin, 620, 621, 628
pL-Myosin, 624
Myosin filament, 629
Myosin light chain kinase, 624
Myristic acid, 131

NAD. *See* Nicotinamide adenine dinucleotide (oxidized).
NAD^+, 159
  absorption spectrum of, 54
  anaerobic regeneration of, represen-
    tative mechanisms for, 51
  as cosubstrate, 51
  synthesis of, 104
NADH. *See* Nicotinamide adenine
  dinucleotide (reduced).
NADH dehydrogenase, 103, 110
$NADH/NAD^+$ ratio, 236
NAD-linked dehydrogenases, in oxidative pathways, 103
NADP. *See* Nicotinamide adenine
  dinucleotide phosphate (oxidized).
$NAD(P)^+$, 53
NADPH. *See* Nicotinamide adenine
  dinucleotide phosphate (reduced).
NADPH generators, 205
NADP-linked dehydrogenases, in re-
  ductive syntheses, 103
Na^+/K^+-ATPase, 124
Na^+/K^+-ATPase pump, 458, 450
Neomycin, 250, 588
Nerve impulses, transmission of, 461
Nervonic acid, 223
Nervous system
  parasympathetic, 524
  sympathetic, 239, 524
NeuAc. *See* N-Acetylneuraminic acid.
Neural crest, and origin of endocrine
  cell types, 463
Neuraminic acid, 129
Neuraminidases, 591
Neurine, 587
Neurofilament, 631

Neurophysins, 493
Neurotensin, 565, 568
Neurotransmitters, 463
Nexin, 630
NGF. *See* Nerve growth factor.
α-NH_2 group in proteins, modification
  of, 37
Niacin, 577
  major characteristics of, 575
Nick translation, defined, 376
Nick-sealing reactions on DNA, 399
Nicotinamide, 53, 103
  coenzymes of, mechanism of oxida-
    tion of, 105
  major characteristics of, 575
Nicotinamide adenine dinucleotide
  (NAD) (oxidized), 103, 158,
    188, 200
  synthesis of, 104
Nicotinamide adenine dinucleotide
  (NADH) (reduced), 106, 153,
    158, 188
  absorption spectrum of, 54
  extramitochondrial, oxidation of,
    117
  oxidation of, 152
Nicotinamide adenine dinucleotide
  phosphate (NADP) (oxidized),
    103
Nicotinamide adenine dinucleotide
  phosphate (NADPH) (reduced),
    106, 179, 202, 206, 207, 208,
    211, 220
Nicotinamide coenzymes, dehydroge-
  nases dependent on, 103
Nicotine, from cigarette smoking, 250
Nicotinic acid, 238, 251
  major characteristics of, 575
NIDDM. *See* Non-insulin-dependent
  diabetes mellitus.
Niemann-Pick disease, 224
Nigericin, 119
NIH/3T3 cells, 641
Ninhydrin, 19
Nirenberg, Marshall, 415
Nitrogen, amino acid
  catabolism of, **271–280**
  removal from amino acids, 271
  as source of, 306
Nitrogen balance, 271, 573
Nitrogen equilibrium, 271
p-Nitrophenylacetate, 75, 76
NMN. *See* Nicotinate mononucle-
  otide.
Non-α functional groups in proteins,
  modifications of, 38
Noncoding DNA, 392
Noncompetitive inhibition
  irreversible, 73
  reversible, 73
Nonequilibrium reactions, 186
Nonheme iron, 109
Nonhistone proteins, in chromatin,
  387
Non-insulin-dependent diabetes melli-
  tus (NIDDM), 251, 471, 560
Nonmuscle actin, 627
Nonprotein catalysts, 50
Nonsense codon, 415, 418
Nonstriated muscle, 619

Norepinephrine, 169, 237, 238, 239, 314, 316, 524, 525, 527
  liberation of, through sympathetic activity, 239
  plasma, normal values, 662
Norethindrone, 543
Northern blot, 376
Northern blot transfer, 369
Nuclease(s), 413
  restriction, in chimeric DNA molecules, 367
Nucleic acid, structure and function of, **377–386**
Nucleobases, synthetic analogs of, 341
Nucleoplasmin, 388
Nucleoproteins, within chromatids, 391
Nucleosidase(s), 583, 584
Nucleoside(s), 344
  and nucleotides, 337
  nomenclature of, 339
Nucleoside diphosphate kinase, 98, 348
Nucleoside diphosphate phosphatase, 591
Nucleoside monophosphate, conversion of, to nucleoside diphosphates and nucleoside triphosphates, 348
Nucleoside monophosphate kinase, 98, 348
Nucleosomes, in chromatin, 388
Nucleotidase, 344
Nucleotide(s), **335–343,** 344
  naturally occurring, 340
  and nucleosides, 337
    nomenclature of, 339
  synthetic analogs of, 341
Nutrient(s), absorption of, site of, 587
Nutrition, **571–578**
  essential requirements in, 572

O antigen, 598
O blood group, 599
OA. See Oxaloacetic acid.
Obesity, 250, 252, 471, 571
Octamer, of histone, 388
Octanoic acid, 131
Octoses, 120
3β-OHSD. See 3β-Hydroxysteroid dehydrogenase.
17β-OHSD. See 17β-Hydroxysteroid dehydrogenase.
-oic, in fatty acid nomenclature, 130
Oil(s), 250
Okazaki pieces, 398
  formation of, 397
Oleic acid, 132, 210, 211, 212
  geometric isomerism of, 132
Oleyl-CoA, 211
Oligomers, defined, 36
Oligomycin, 112
Oligosaccharidases, 583
Oligosaccharide(s), 120
  asparagine-linked, 593
  in blood group antigens, 598
  O-linked, 593

Oligosaccharide(s) (cont'd)
  processing of, 594, 596
Oligosaccharide transferase, 596
Oligosaccharide-pyrophosphoryl-dolichol, 594, 598
OMP. See Orotidine monophosphate.
Oncogene(s), **639,** 640, 645
  abbreviations for, 641
  and growth factors, 646
  from tumor cells, 641
One cistron, one subunit concept, 429
One gene, one enzyme concept, 429
Operator locus, 430
Operator region, 432
Operon, 84
  lac, 430, 431, 432
    constitutive expression of, 432
    induction of, 430
    repression and derepression of, 431
    structural and regulatory genes of, positional relationships of, 430
Opioids, 475
Optical activity, 121
Optical isomer, 121
Optical specificity of enzymes, 53
OR (right operator) region, 432
Ordered binding of substrates, 78
Organophosphates, hydrolysis of, free energy of, 96
Ornithine, 18, 26, 262, **310,** 587
  metabolism of, 310
L-Ornithine, 284, 311
Ornithine decarboxylase, 310
Ornithine transaminase, quaternary structure of, 39
Ornithine transcarbamoylase, 278, 279, 359
L-Ornithine transcarbamoylase, 277
Orotate phosphoribosyltransferase, 354, 359, 360
Orotic acid, 234, 354
Orotic aciduria, 359, 360
  hereditary, 359
Orotidine monophosphate (OMP), 353, 354
Orotidylate, 353
Orotidylate decarboxylase, 359
Orotidylic acid decarboxylase, 354
Osmolality, serum, normal values, 662
Osmoreceptors, 494
Osmotic pressure, 607
Osteodystrophy, renal, 510
Osteogenesis imperfecta, 634
Osteomalacia, 507, 509, 576
Osteoporosis, 520, 542, 571
  due to malabsorption, 587
Ouabain, 124, 125, 584
Ova, 536
Ovaries
  hormones of, **536–544**
    biosynthesis and metabolism of, 536
    mechanism of action of, 543
    pathophysiology of, 543
    regulation and physiologic actions of, 539
  polycystic, 489

Oxalate, 181
Oxaloacetate, 149, 150, 153, 154, 156, 172, 173, 207, 262, 266, 282
  amino acids forming, 282
  catalytic role of, 149
Oxaloacetic acid (OA), 234, 354
Oxalocrotonate, 297
Oxalosis, from xylitol, 180
Oxalosuccinate, 151, 152
Oxidase(s), 101
  mixed-function, 106, 637
  in oxidation, 101
Oxidation, 100
  biologic, **100–119**
  defined, 100
  of fatty acids, 198
    energetics of, 200
  peroxisomal fatty acid, 200
  and phosphorylation, uncoupling of, 161
  of protein, 572
  and reduction, enzymes and coenzymes involved in, 100
β-Oxidation, 199, 203, 237
Oxidation-reduction equilibria, 100
Oxidation-reduction loops, 113
Oxidative deamination, 273
Oxidative decarboxylation
  of pyruvate, 163
Oxidative phosphorylation. See Phosphorylation, oxidative.
Oxidoreductases, 54, 100
  classification of, 51
Oxidoreductions, 50
2-Oxy-4-aminopyrimidine, 336
Oxygen
  blood, normal values, 662
  consumption of, 572
Oxygen transferases, 105
Oxygenases, 101, 105
  true, 105
Oxygenated sterol, 244
Oxyhemoglobin, 163
6-Oxypurine, 336
Oxytocin, 193, 493

P (peptidyl) site, 422
P_{aCO_2}, blood, normal values, 662
PAGE. See Polyacrylamide gel electrophoresis.
Palindrome, defined, 376
Palmitate, 202, 205, 206, 207
  fate of, 206
Palmitic acid, 131
Palmitol-CoA, 222
Palmitoleic acid, 210, 211
Palmityl, 206
Pancreas
  hormones of, **547–563**
  structure and function of, 547
    endocrine function, 547
    exocrine function, 547
Pancreatic amylase, 584
Pancreatic digestion, 580
Pancreatic lipase, 582
Pancreatic polypeptide (PP), 547, 550, 563, 565

Pancreatic secretion, 581
  digestion by, 581
Pantothenic acid, 53, **154**
  major characteristics of, 575
  in synthesis of coenzyme A, 155
Papain, 621
Paper chromatography, 20
  2-dimensional, 21
PAPS. *See* Phosphoadenosinephosphosulfate.
Paracrine function, 463
Parallel β-pleated sheet, 35
Parasympathetic nervous system, 524
Parathyroid gland, hormones of, normal values, 663
Parathyroid hormone (PTH), 464, 471, 475, 502, **503**, 504
Parenchymal cells, hepatic, 229
Parenchymal liver disease, 248
Parietal cells, 579
Partial agonists, 520, 521
Pasteur effect, 191
pBR322, 366
PDGF. *See* Platelet-derived growth factor.
PDH. *See* Pyruvate dehydrogenase.
Pentachlorophenol, in oxidative phosphorylation, 112
Pentaenoic acids, 132
Pentagastrin, 568
Pentosans, 120
Pentose(s), 120, 122, 123, 180
  in glycoproteins, 128
Pentose phosphate pathway, 143, **175–179**, 180, 181, 188, 189, 205, 207, 237, 254
  flow chart of, 178
  metabolic significance of, 179
  nonoxidative phase of, 177
  oxidative phase of, 177
Pentose sugars, 583
Pentosuria, essential, 180
PEPCK. *See* Phosphoenolpyruvate carboxykinase.
Pepsin, in IUB classification, 51
Pepsin A and B, 580, 584
Pepsinogen, 580, 584
Peptide(s), **24–31**
  amino acids in, determination of, 27
  conformation of, in solution, 25
  defined, 24
  endorphins, 492
  gastrin-releasing, 568
  gastrointestinal, amino acid sequences of, 566
  in gut and central nervous system, 565
  ionic forms of, 25
  β-lipotropin, 492
  physiologically active, 25
  primary structure of, 25
    changes in, physiologic consequences of, 25
    defined, 25
  pro-opiomelanocortin family of, 490
    functions of, 491
  separatory techniques for, 26
  structural representation of, 24
  synthesis of, by automated techniques, 30

Peptide bond(s), 19, 24, 33
Peptidyl site, 422
Peptidyltransferase, 423
Peptidyl-tRNA, 422
Peripheral conversion, of hormones, 464
Permeability coefficients, in lipid bilayer membranes, 448
Peroxidase, 101, 103, 215
Peroxidation, of lipids, 138
Peroxisomal fatty acid oxidation, 200
Peroxisomes, 105
PG. *See* Prostaglandin.
PG₁ series of prostaglandins, defined, 131
PG₂ series of prostaglandins, 213, 215
  defined, 131
PG₃ series of prostaglandins, 215
  defined, 131
PGE₁. *See* Prostaglandin E₁.
PGE₂. *See* Prostaglandin E₂.
PGE₃. *See* Prostaglandin E₃.
PGF₂α. *See* Prostaglandin F₂α.
PGH. *See* Prostaglandin H.
PGI. *See* Prostacyclin.
PGI₂. *See* Prostacyclin.
pH, 66
  effect on enzyme activity, 67
  isoelectric (pI), 15
  (reaction), blood, normal values, 662
Phage, in DNA cloning, 366
Phagocytosis, 459
Phagolysosomes, 497
Phasing, defined, 389
Phe. *See* Phenylalanine.
Phenolase, 102
Phenotypic sex, 545
Phenylacetate, 294, 295
Phenylacetic acid, 294
Phenylacetylglutamine, 294, 295
Phenylalanine (Phe), 17, 262, 265, 290, 292, 294, 295
  carbon atoms of, catabolic fate of, 293
  catabolism of, metabolic disorders of, 292
  human requirements of, 573
  ultraviolet absorption spectrum of, 20
D-Phenylalanine, 26
L-Phenylalanine, 269, 293, 294
Phenylalanine hydroxylase complex, 270
Phenylalanine hydroxylase reaction, 269
Phenylethanolamine-N-methyltransferase (PNMT), 526
Phenylisothiocyanate, 30
Phenylketonuria (PKU), 293
Phenyllactate, 294, 295
Phenyllactic acid, 294
Phenylpyruvate, 294, 295
Phenylpyruvic acid, 294
Pheochromocytomas, 529
Pheomelanins, 314, 315
Phlorhizin, 197
Phosphagens, 97, 625

Phosphatase(s), 583, 584, 624
  acid or alkaline, serum, normal values, 656, 657, 662
  phosphoprotein, 478
Phosphate(s)
  in DNA, 379
  high-energy, **95,** 96, 150, 151, 161
    as "energy currency" of cell, 97
    transfer of, 96, 97
  inorganic (Pᵢ), 95, 188
  low-energy, 96
Phosphate cycles, and interchange of adenine nucleotides, 98
Phosphate homeostasis, 504
Phosphate transporter, 118
Phosphatidate phosphohydrolase, 219, 220, 258
Phosphatidic acid, 134, 446
Phosphatidylcholine, 134, 219, 220
  metabolism of, 221
3-Phosphatidylcholine, 134
Phosphatidylethanolamine, 134, 135, 219, 220
3-Phosphatidylethanolamine, 134
Phosphatidylglycerols, 134
Phosphatidylinositides, and hormone action, 480
Phosphatidylinositol, 134, 219, 220, 640
3-Phosphatidylinositol, 135
Phosphatidylinositol 4,5-bisphosphate, structure of, 461
Phosphatidylserine, 134, 219, 220
3-Phosphatidylserine, 135
Phosphoadenosine-phosphosulfate (PAPS), 223
3′-Phosphoadenosine-5′-phosphosulfate, 223
Phosphocholine, 219, 220
Phosphocholine diacylglycerol transferase, 220, 221
Phosphocholine-cytidyl transferase, 219
Phosphodiester bonds, in RNA, 382
3′,5′-Phosphodiester bridges, of DNA and RNA, 377, 382
Phosphodiesterase, 169, 170, 238, 340, 478, 559, 597
Phosphoenolpyruvate, 96, 97, 154, 160, 161, 172, 173, 185
Phosphoenolpyruvate carboxykinase (PEPCK), 154, 172, 173, 188, 519, 560, 562
  distribution of, 172
Phosphoesterases, 344
Phosphoethanolamine, 219, 220
Phosphofructokinase, 159, 160, 162, 173, 182, 188, 191, 192
Phosphofructokinase-1, 159
6-Phosphofructo-1-kinase, 192
6-Phosphofructo-2-kinase, 192
6-Phosphofructo-2-kinase/fructose-2,6-bisphosphatase, 559
Phosphoglucomutase, 165, 166, 167, 181, 184, 185
6-Phosphogluconate, 176, 178
6-Phosphogluconate dehydrogenase, 176, 177, 178, 188
  in muscle tissue, 179
6-Phosphogluconolactone, 176, 177

2-Phosphoglycerate, 160, 161, 173, 182
3-Phosphoglycerate, 96, 97, 160, 173
Phosphoglycerate kinase, 160, 161
Phosphoglycerate mutase, 160, 161, 162
Phosphoglycerides, membrane, 446, 447
Phosphoguanidines, 96
Phosphohexose isomerase, 159, 160, 178
Phosphohydrolase, 221
Phosphoinositides, and hormone action, 479
Phospholipase(s), 583
sites of hydrolytic activity of, 222
Phospholipase A$_1$, 221, 222
Phospholipase A$_2$, 221, 222, 582, 583, 584, 586
in mitochondria, 116
Phospholipase B, 221
Phospholipase C, 222
Phospholipase D, 222
Phospholipid(s), 130, **134,** 215, 220, 230, 232, 234, 256, 446, 582
degradation of, 221
glycerol ether, 220
phosphorus, serum, normal values, 662
and sphingolipids, in disease, 223
surface, 232
total, in blood plasma, 226
and triacylglycerols, biosynthesis of, 219
turnover of, 221
Phosphomevalonate kinase, 242
4′-Phosphopantetheine, 204, 205
Phosphoprotein phosphatases, 478
5-Phosphoribosyl-1-pyrophosphate (PRPP), 346, 347, 349, 350, 351, 353, 355, 358
5-Phosphoribosylamine, 347
synthesis of, 346
Phosphoribosylpyrophosphate amidotransferase, 346
Phosphorolysis, 161
Phosphorus, 235
inorganic, serum, normal values, 657, 662
major characteristics of, 577
Phosphorylase, 166, 169, 184, 188, 559
activation and inactivation of, 167, 169
in muscle, 168, 169, 625
Phosphorylase a, 167, 168, 169, 187
Phosphorylase b, 168, 169
Phosphorylase a kinase, 168, 169
Phosphorylase b kinase, 168, 169
Phosphorylase kinase, 169, 170, 559
Ca^{2+}/calmodulin-sensitive, 169
Phosphorylation, 91, 629
of ADP to ATP, 199
multisite, 193
of muscle proteins, 625
oxidative, 96, 97, **108–116,** 149, 150, 152, 164, 625
ATP from, 625
electrochemical potential difference in, 113

Phosphorylation (cont'd)
oxidative (cont'd)
mechanisms of, 113
chemical hypothesis of, 113
chemiosmotic theory of, 113
sites of, 91
at substrate level, 161
Phosphorylation-dephosphorylation, 91
Phosphorylation-dephosphorylation protein, 559
O-Phosphoserine, 309
Phosphosphingolipids, 222
Phosphotriose isomerase, 159, 160, 178
Photolysis, 508
Photomicrography, electron, in determination of protein structure, 39
Photosensitivity, cutaneous, 328
Phylloquinone, major characteristics of, 576
Phytanic acid, 200
Phytosterols, 586
P$_i$ (inorganic phosphate), 95, 188
pI. See Isoelectric pH.
Piericidin A, in oxidative phosphorylation, 112
Pigments in bile, 580
Pinocytosis, 460
absorptive, 460
adsorptive, 460
fluid-phase, 460
receptor-mediated, 460
reverse, 229
Pituitary gland
and fat mobilization, 239
hormones of, **482–495**
anterior, 483
normal values, 662
posterior, 493
pK$_a$, defined, 14
PKU. See Phenylketonuria.
Placenta, 183
hormones of, normal values, 663
Placental lactogen, 484, 487
Plasma, **607–612**
antibodies in, 609
cholesterol in, 245, 246
constituents of, 607
immunoglobulins in, 609
lipids in, 226
lipoproteins in, 226
block in production of, 235
disorders of, 251
metabolism of, 229
proteins in, 607
seminal, 183
Plasma cells, 609
Plasma lipoproteins, 605
Plasmalogen(s)
and ether lipids, biosynthesis of, 221
structure of, 135
Plasmalogenic diacylglycerol, 220
Plasmid
defined, 376
in DNA cloning, 366
Plasmin, in IUB classification, 51
Plasminogen, 617
Plasminogen activator, tissue, 617

Platelet-activating factor, 221
Platelet-derived growth factor (PDGF), 557, 640, 644, 646
β-Pleated sheet, 35
antiparallel, 35
parallel, 35
p-Light chain, defined, 623
pL-myosin, 624
PNMT. See Phenylethanolamine-N-methyltransferase.
pol gene, 640
Poly(A) addition sites, 442
Poly(A) "tail" of mRNAs, 385
Polyacrylamide gel electrophoresis (PAGE), 27, 39
for enzyme isolation, 56
Poly(ADP-ribose) polymerase, 402
Polyamines, 309, 310
biosynthesis of, 310
catabolism of, 312
structures of, 310
Polycystic ovaries, 489
Polycystic ovary syndrome, 543
Polycythemia, 48
Polydipsia, 553
Polyenoic acids, 131
Polyethenoid acids, 131
Polyhydric alcohols, 120
Polymerase, mitochondrial DNA, 398
Polymerase alpha, 398
Polymerase beta, 398
Polymerase gamma, 398
Polynucleotidases, 344, 583, 584
Polyol dehydrogenase, 183
Polyol pathway, 183
Polyomavirus, 639
Polypeptide(s), 584
cleavage of, for automated sequencing, 29
defined, 24
gastric inhibitory, 552, 564, 565, 567
linear, resolution into, 28
pancreatic, 547, 550, 563, 564
primary structure of, determination of, 25, **28**
structures of, determination of, automated Edman technique for, 28
vasoactive intestinal, 565, 568
Polypeptide chain(s)
fully extended, dimensions of, 33
order of, 36
ordered conformations of, 34
Polyphenol oxidase, 102
Polyprenoid compounds, 138
Polyribosome, 414, 425
Polysaccharide(s), 120, **125**
degradation of, 603
structure of, 128
Polysaccharide chains, in proteoglycans, 599
Polysome, 425
Polytene chromosomes, 391
Polyunsaturated acids, 131
Polyunsaturated fatty acids, 250
Polyuria, 553
POMC. See Pro-opiomelanocortin.
Pompe's disease, 171
Porphin molecule, 319

Porphobilinogen, 321
 biosynthesis of, 320
 biosynthesis of porphyrin derivatives from, 323
 conversion to uroporphyrinogens,
  322
Porphyria(s), 325
 acquired, 328
 classification of, 325
 congenital erythropoietic, 327
 cutanea tarda, 327
 excretion of porphyrins in, 326
 intermittent acute, 325
 variegate, 327
Porphyrin(s)
 and bile pigments, **319–334**
 biosynthesis of, 321, 323
 chemistry of, 324
 fluorescence of, 324
 and porphyrin precursors, upper
  limits of excretory values of,
  325
 structure of, 321
 tests for, 325
 urine, normal values, 663
Porphyrinogens, 324
Portacaval shunts, 274
Postabsorptive state, exercise in, 255
Posterior pituitary hormones, 493
Potassium, major characteristics of,
 577
 serum or plasma, normal values,
  657, 662
PP. See Pancreatic polypeptide.
PP$_i$ (inorganic pyrophosphate), 98
Precursor proteins, 464
Prednisolone, 513
Prednisone, 513
Pregnancy
 hormone levels during, 540
 and placental hormones, 540
 toxemia of, in sheep, 235, 255
Pregnanediol, 513
Pregnanetriol, 513
Pregnenolone, 513
Prekallikrein, 614
 disorders of, 616
Pre-β-lipoproteins, 229
Pre-mRNA splicing, proposed path-
 way for, 411
Preprocollagen, 632
Preproparathyroid hormone (pre-
 proPTH), 503
PreproPTH. See Preproparathyroid
 hormone.
PRH. See Prolactin-releasing hor-
 mone.
Pribnow box, 406
PRIH. See Prolactin release-inhibiting
 hormone.
Primaquine, and G6PD deficiency,
 179
Primary structure of proteins, 36
 determination of, 37
Primer, glycogen, 165, 166
PRL. See Prolactin.
Pro. See Proline.
Probe(s), 367
 defined, 376
Probucol, 251

Procarboxypeptidase, 581, 587
Procarcinogen(s), 636
Processed genes, 394
Prochymotrypsin, conversion of, to
 chymotrypsin, 77
Procollagen, 632, 633
 defects of, 634
Product feedback repression, 84
Proelastase, 581
Proenzymes, 77, 85
Profilin, 628
Progestational hormones, 251
Progesterone, 513, 536, 537, 538,
 540
 biosynthesis of, 538
Progesterone pathway, 531
Progestins, 536, 537, 538
Prohormones, 426
Proinsulin, 38, 464
 cleavage of, 550
 human, structure of, 550
 properties of, 551
Prokaryotes, regulation of RNA pro-
 cessing in, 430
Prolactin (PRL), 484, 487
 ovine, structure of, 488
Prolactin release-inhibiting hormone
 (PRIH), 483
Prolactin-releasing hormone, 487
Proline (Pro), 17, 262, 265, 267, 273,
 283
 biosynthesis of, 267
  from glutamate, 268
 catabolism of, metabolic disorders
  of, 283
 in kidney, 275
 metabolism of, 309
L-Proline, 268
 catabolism of, 284
Proline dehydrogenase, 284
Proline hydroxylase reaction, 269
Promoter(s), 431, 638, 641, 642
 in RNA synthesis, 404
 in transcription control, 440
Promoter element, 473
Promoter recognition specificity of
 RNA polymerase, 405
Promoter site, 430
Promotion, 638
Pro-opiomelanocortin (POMC), 464
Pro-opiomelanocortin gene, 490
Pro-opiomelanocortin peptide(s),
 functions of, 491
Pro-opiomelanocortin peptide family,
 490
Proparathyroid hormone (proPTH),
 503
Propionate, 131, 173, 174, 206, 262,
 355
 metabolism of, 174
  defects of, 305
 in odd number of fatty acids, 174
Propionic acid. See Propionate.
Propionyl-CoA, 174, 200, 206, 262,
 299, 300, 302, 304
Propionyl-CoA carboxylase, 174
 quaternary structure of, 39
Proproteins, 77, 426
ProPTH. See Proparathyroid hormone.

Propylthiouracil, 498
Prostacyclin (PGI; PGI$_2$), 131, 215
Prostaglandin(s) (PG), 131, 210, 213,
 214, 215, 216
 defined, 131
 PG$_1$ series, 131
 PG$_2$ series, 131, 215
 PG$_3$ series, 131, 215
Prostaglandin E$_1$ (PGE$_1$), 214, 239
Prostaglandin E$_2$ (PGE$_2$), 214
Prostaglandin E$_3$ (PGE$_3$), 214
Prostaglandin endoperoxide synthase,
 213
Prostaglandin F$_{2\alpha}$ (PGF$_{2\alpha}$), 557
Prostaglandin H (PGH), 215
Prostanoids, 131, 215
Prostatic hypertrophy, benign, 535
Protamines, 32
Protease(s), 29
 insulin-specific, 553
 and metastasis, 647
Protease V8, *Staphylococcus aureus*,
 29
Protein(s), **32–39**, 584
 absorption of, 586
 acyl carrier, 204, 206
 α-amino acids not present in, 18
 androgen-binding, 533
 carrier, 467
  sterol, 244
 catabolite gene activator, 430
 catabolite regulatory, 476
 in chromatin, 387
 classification of, 32
 contractile and structural, **618–634**
 converter, 91
 cotranslational insertion of, 453
 cro, 433
  defined, 432
 degradation to amino acids, 83
 denaturation of, 37
 determining molecular weight of,
  38
 dietary
  energy available in, 572
  quality of, 573
  requirements for, 573
 disulfide bonds of, 33
 fibrous, 32
 function of, 32
 globular, 32
 gluconeogenesis from, 263
 α helix structure of, 34
 heme, **40–49**
 human requirements of, 573
 integral membrane, 450
 iron-sulfur, 109
 lambda repressor, 432, 433
 leader sequence of, 452
 "link," 402
 membrane, 448
  insertion of, 453
  integral and peripheral, 449
 in membrane assembly, 454
 molecular weight of, determination
  of, 38
 of muscle, 621
  phosphorylation of, 625
 non-α functional groups in, 38
 nonhistone, in chromatin, 387

Protein(s) (cont'd)
oxidation of, 572
phosphorylation-dephosphorylation, 559
plasma, **607–612**
precursor, 464
primary structure of, 36
determination of, 37
processing of, 425
production of, in recombinant DNA technology, 370
random coil conformation of, 35
repressor, 433
serum or plasma, normal values, 658, 662
simple, 32
solubility of, 32
sterol carrier, 244
structural, **618–634**
structure of, 36
bonds responsible for, 33
α helix, 34
orders of, 36
primary, 36
determination of, 37
quaternary, 36
determination of, 38
secondary, 36
determination of, 38
tertiary, 36
determination of, 38
3-dimensional, 33
subunits of, 36
synthesis of, 422
from amino acids, 83
and genetic code, **414–427**
inhibitors of, 426
initiation of, 421
peptide elongation in, 423
process of, 422
termination process of, 424, 425
transcortin, 515
transfer, 248
transport (carrier), 467
Protein catalysts, enzymes as, 50
Protein kinase(s), 168, 170, 476, 559
cAMP-activated, 238
cAMP-dependent, 168, 169, 170, 188, 192, 238, 477
cAMP-independent, 477
cGMP-dependent, 479
purified, 477
Protein phosphatase-1, 168, 169, 170, 192, 193
Protein-energy malnutrition, 574
Protein-iron-sulfur complex, 110
Protein-tyrosine kinase(s), 640
Proteoglycan(s), 128, 183, **599**, 600, 603
and cell surface molecules, 605
defined, 128, 589
degradation of, 600
disorders of, 603, 604
polysaccharide moieties of, 603
and extracellular macromolecules, 605
functional aspects of, 605
and intracellular macromolecules, 606
and plasma proteins, 605

Proteoglycan(s) (cont'd)
sialylation in, 600
sialyltransferases in, 600
structure of, 600
Proteoglycan aggregate, 601
Proteolytic enzymes, enzyme degradation involving hydrolysis by, 85
Prothrombin, 613
Protocoproporphyria hereditaria, 327
Protofilaments, 629
Protoheme, 105
Protomers
H, 58
M, 58
protein, 37
Proton translocation, 114, 116
Proton-translocating ATP synthetase, 116
Proton-translocating oxidation/reduction (o/r) loop, 114
Proto-oncogene(s), 557, 641, 642
Protoporphyria, 328
Protoporphyrin, addition of iron to form heme, 320
Protoporphyrin III, 321
Protoporphyrinogen III, 321
Protoporphyrinogen oxidase, 323, 327
Provirus, 641
PRPP. See 5-Phosphoribosyl-1-pyrophosphate
PRPP amidotransferase, 346
PRPP glutamyl amidotransferase, 347
PRPP synthetase, 347, 358
Pseudocholinesterase, in IUB classification, 51
Pseudogene(s), 394
defined, 376
Pseudoglobulins, 32
Pseudogout, 357
Pseudohypoparathyroidism, 471, 506
Pseudouridine, 339
PTH. See Parathyroid hormone.
PTLC. See Chromatography, partition thin-layer.
Ptomaines, 587
"Puffs," defined, 391
Purification, enzyme, 55, 56, 57
Purine(s), 335, 344, 346, 362
biosynthesis of, from ribose 5-phosphate and ATP, regulation of, 347
catabolism of, 352
metabolism of
clinical disorders of, 356
inherited disorders of, and associated enzyme abnormalities, 358
synthesis of, 307
de novo, 351
Purine bases, 102
structures of, 336
Purine nucleoside phosphorylase, 345, 350, 351
deficiency of, 358
Purine 5'-nucleotidase, 349, 351
Purine nucleotide(s)
biosynthesis of, 344
metabolism of, **344–361**
Purine nucleotide cycle, 341

Purine ribonucleotides, reduction of, regulation of, 352
Purine ring, nitrogen and carbon atoms of, sources of, 346
Purine salvage
cycles of, 351
pathways of, 349
Puromycin, 234
and tyrosyl-tRNA, comparative structures of, 426
Putrefaction, intestinal, 587
Putrescine, 309, 311, 587
structure of, 310
Pygmies, 487
Pyran ring, 121
Pyranose ring structure, 121
Pyridoxal, major characteristics of, 575
Pyridoxal phosphate, 169, 222, 272, 297, 298
Pyridoxamine, major characteristics of, 575
Pyridoxine, 577
major characteristics of, 575
Pyrimidine(s), 335, 344, **353**
biosynthesis of, 353
regulation of, 355
catabolism of, 355
metabolism of, clinical disorders of, 359
Pyrimidine nucleoside kinase reactions, 355
Pyrimidine nucleotide(s)
biosynthetic pathway for, 354
metabolism of, **344–361**
synthesis of, control of, 357
Pyrimidine salvage pathways, 355
Pyrimidine-pyrimidine dimers, ultraviolet light in formation of, 401
Pyroglutamic acid, 26
Pyroglutamylhistidylprolinamide, 26
Pyrophosphatase, inorganic, 98, 165
Pyrophosphate, dimethylallyl, 244
farnesyl, 242, 244
geranyl, 242, 244
Pyrophosphomevalonate decarboxylase, 242
Pyrrole, 40
Δ^2-Pyrrolidine-5-carboxylate, 268
L-Δ^1-Pyrroline-3-hydroxy-5-carboxylate, 290
Pyruvate, 97, 154, 155, 156, 158, 160, 162, 163, 164, 172, 173, 182, 185, 188, 207, 259, 282, 287, 288, 289, 290
alanine formed from, 266
amino acids forming, 285
blood, normal values, 662
conversion of alanine and serine to, 287
oxidation of, **158–164**
oxidative decarboxylation of, 163
Pyruvate carboxylase, 154, 172, 173, 188, 189
Pyruvate dehydrogenase (PDH), 155, 156, 163, 164, 188, 190, 207, 289, 559
regulation of, 191
Pyruvate kinase, 160, 162, 173, 188, 190, 559

Q, defined, **109**
Q_{10}, defined, 66
Quaternary structure
 of proteins, 36
  determination of, 38
 of selected enzymes, 39
Questran, 250
o-Quinones, 102

R_f value, defined, **21**
Racemic mixture, 121
 defined, 121
Radiant energy, as cause of cancer, 636
Radiation, ionizing, and DNA, 401
Rancidity of lipids, 138
Random binding of substrates, 78
Random coil conformation of
  proteins, 35
ras oncogene, 641, 643, 644
Reactant(s)
 concentration of, effect of, 67
Reaction(s), chemical. *See also*
  specific reaction.
 coupled, bioenergetics of, 97
 endergonic, defined, 94
 energy barriers for, 62
 enzyme-catalyzed, 187
  control of, 187
  test for, 37
 equilibrium and nonequilibrium,
  186
 exergonic, defined, 94
 kinetic (collision) theory for, 62
 reversible, 82
Reaction profiles, 61
Reaction-specific catalysts, enzymes
  as, 50
recA, defined, 435
Receptor(s)
 adrenergic, actions of, 528
 α-adrenergic, 475, 528
 β-adrenergic, 475, 528
 adrenergic hormone, classification
  of, 528
 baroreceptor, 517
 estrogen, 543
 glucocorticoid, 521
 hormone, 466, 467
  and disease, 471
  recognition and coupling domains
   of, 466
  regulation of, 468
  structure of, 468
 IGF-I, 561
 IGF-II, 561
 insulin, 557, 558, 559, 560
 spare, 468
 testosterone/DHT, 535
Receptor-effector coupling, 467
Receptosomes, 460
Recombinant DNA, defined, 376
Recombinant DNA technology,
  362–376
 concepts used in, 364
 glossary of terms related to, 375
 in molecular analysis of disease,
  372
 practical applications of, 370
 and prenatal diagnosis, 373

Recombination, defined, 393
Red cells, sickle, 48
Redox potential, 100
Redox state of tissue, 162, 254
Reductase, 221
Reduction, defined, 100
Refsum's disease, 200
Region, operator, 432
Regulation
 allosteric, 87
 of aspartokinase, allosteric patterns
  of, 88
 of catalytic efficiency of enzymes,
  85
 of enzyme activity, **82–92**
 of enzyme quantity, 83
 of gene expression, **428–443**
 metabolic, scope of, 83
 negative, by genes, 428
 positive, by genes, 428
Regulator
 negative, 430
 positive, 431
Regulatory enzymes, 87, 188
Releasing factors, 425
Remnant removal disease, 251
Renal glycosuria, 197
Renal lithiasis, 358
Renal osteodystrophy, 510
Renal threshold for glucose, 196
Renin
 inhibitors of, 517
 stimulators of, 517
Renin-angiotensin system, 517
Rennet, 580
Rennin, 579, 580, 584
Repeats, in DNA, 392
Repetitive sequences, in DNA, 392
"Replication bubbles," 398
Repression, 84, 355
 coordinate, 84
 and derepression, as control of en-
  zyme synthesis, 84
 multivalent, 88
 product feedback, 84
Repressor gene, 432, 433
Repressor protein, lambda, 432, 433
Reserpine, 239
Respiration, rate of, conditions limit-
  ing, 111
Respiratory chain, 101, **108,** 110, 199
 configuration of oxidation/reduction
  (o/r) loops in, 114
 dysfunction of, 119
 in energy capture, 110
 inhibitors of, 111
Respiratory control, 94, 111
 ADP in, 112
 in mitochondria, 112
 states of, 111
Respiratory distress syndrome, 134,
  220
Respiratory quotient (RQ), 229
Restriction endonucleases, 413
 diagnostic applications of, 59
Restriction fragment length polymorphism (RFLP), 59, 374
Reticulum, endoplasmic
 rough, 230, 414
 smooth, 230

Retinene isomerase, in IUB
  classification, 51
Retinol, major characteristics of, 576
Retrovirus(es), oncogenes of, 640
Reverse cholesterol transport, 233
Reverse T_3. *See* 3,3′,5′-Triiodothy-
  ronine.
Reverse transcriptase, 400
Reversible reactions, 82
Reye's syndrome, 213, 360
RFLP. *See* Restriction fragment
  length polymorphism.
Rho factor, defined, 406
Riboflavin, 53, 102, 103
 major characteristics of, 575
Riboflavin phosphate, 102
Ribonuclease(s), 38, 65, 583, 584
 bovine
  pancreatic, diagram of main
   chain folding of, 36
  structure of, 38
 catalytic site of, 65
 defined, 413
 pancreatic, 36
 primary structure of, 38
 structure of, 65
Ribonucleic acid. *See* RNA.
Ribonucleoside(s), structures of, 338
Ribonucleoside diphosphates, reduc-
  tion of, to 2′ deoxyribonucleoside
  diphosphates, 349
Ribonucleotide reductase, 349, 354
Ribose, 120
 generation of, 179
 in RNA, 382
D-Ribose, 124
Ribose 1,5-bisphosphate, 176, 179
Ribose 1-phosphate, 176
Ribose 5-phosphate, 176, 177, 347
Ribose-5-phosphate ketoisomerase,
  177
Ribosomal RNA. *See* rRNA.
Ribosome(s), 414
 defined, 385
 mammalian, components of, 386
Ribulose, 120
D-Ribulose, 123, 124
Ribulose 5-phosphate, 177, 178
Ribulose 5-phosphate 3-epimerase,
  176, 177
Richner-Hanhart syndrome, 292
Ricinoleic acid, 132
Rickets, 507, 509, 576
 vitamin D-resistant (type II), 471
Rigor mortis, 623
Ring structures
 furanose, 121
 pyranose, 121
RME. *See* Resting metabolic expenditure.
RNA (ribonucleic acid), **382,** 414
 biologic function of, 383
 chemical nature of, 382
 in chromatin, 387
 heterogeneous nuclear (hnRNA),
  385, 392, 414
 messenger. *See* mRNA.
 processing of, 409
  differential, 441
  regulation of, 437

RNA (ribonucleic acid) (cont'd)
  ribosomal. *See* rRNA.
  secondary structure of, 383
  segment of, 382
  small nuclear, 383
  small stable, 386
  splicing, 363
  structure of, 383
  synthesis and processing of,
    404–413
  transcript of, 383
  transfer. *See* tRNA.
RNA polymerase, promoter recognition specificity of, 405
RNA polymerase II, 392, 406
RNA primer, 396
  of deoxynucleoside triphosphate, 395
RNA transcript, 382
RNA-dependent DNA polymerase, 383
RNase, 583
RNase H, 400
Rotenone, and oxidative phosphorylation, 112
Rough endoplasmic reticulum, 230, 414
Rous sarcoma virus, 640
RQ. *See* Respiratory quotient.
rRNA (ribosomal RNA), 383, 385
  amplification of, during development, 438
  in RNA processing, 413
rT₃. *See* 3,3′,5′-Triiodothyronine.
Ruffle(s), 627, 628, 629
Ruffle membrane, 628

**S (synthetic) phases of life span of
  cell, 400**
S-1. *See* Subfragment 1, of heavy meromyosin.
S-2. *See* Subfragment 2, of heavy meromyosin.
Saccharopine, 296
SAICAR. *See* Aminoimidazole succinyl carboxamide ribosylphosphate.
Saliva, constituents of, 579
Salivary amylase, 579, 584
Salivary digestion, 579
Salivary glands, 579, 584
Salmonella typhimurium, 84
Salt, inorganic, in bile, 580
Salt bonds, 34
Salting-out methods in determination of protein fractions, 607
Salt-links, in deoxyhemoglobin, 44
Sandoff disease, 603
Sanfilippo syndrome, 604
Sanger, Frederick, 28
  reagent of, 28
Sanger method of DNA sequencing, 370
Sarcolemma, 619
Sarcomere, 619, 623
Sarcoplasm, 619
Sarcoplasmic reticulum, 623
Sarcosine, 268
Saturated fatty acids, 131

Sau I, 59
Scleroproteins, 32
Sclerosis, multiple, 223
scRNA. *See* Small cytoplasmic RNA.
SDS-PAGE, 590
SDS-PAGE-radioautography, 590
Second messenger(s), 472
Second substrates, coenzymes as, 51
Secondary structure of proteins, 36
  determination of, 38
Secretin, 564, 565, 566, 567, 584
Secretin family, 566, 567
Secretor locus, and blood group antigens, 599
Secretory component, 610
Secretory IgA, 611
D-Sedoheptulose, 124
Sedoheptulose 7-phosphate, 176, 177
Sedormid, 328
Selenium, 105
  major characteristics of, 578
  metabolism of, vitamin E in, 139
Seminal plasma, 183
Semiquinone, 103
Sephadex, 55
Ser. *See* Serine.
Serine (Ser), 16, 65, 219, 222, 262, 265, 267, 268, 286, 287, **309**
  biosynthesis of, 267
  conversion of, 287, 309
  in kidney, 275
  in liver and gut, 275
  in muscle, 275
L-Serine, 269, 285, 286, 287, 289
Serine dehydratase, 286, 287
  in amino acid metabolism, 86
Serine dehydratase reaction, 287
Serine hydroxymethyltransferase reaction, 268, 286
Serine proteases, inactivation by antithrombin, 605
Serotonin, 312, 313, 568
  blood, normal values, 662
Sertoli cells, 489, 530
Serum IgA, 611
Sex
  chromosomal, 544
  gonadal, 544
  phenotypic, 545
Sex chromosomes, 544
Sex hormone-binding globulin (SHBG), 532, 537, 538
Sexual differentiation, hormonal involvement in, 544
Sf (Svedberg) units, defined, 226
SGOT. *See* Serum glutamic oxaloacetic transaminase.
SGPT. *See* Serum glutamic pyruvic transaminase.
SH groups. *See* Sulfhydryl groups.
SHBG. *See* Sex hormone-binding globulin.
Short interspersed repeats, in DNA, 392
Shuttle mechanisms, in enzyme compartmentation, 86
SI units, 650
Sialic acids, 128, 184, 185, 228
  in glycoproteins, 128
  structure of, 129

Sialylation, in proteoglycans, 600
Sialyltransferase(s), in proteoglycans, 600
Sickle cell disease
  pedigree analysis of, 373, 374
  point mutations in, 372
Sigma factor, 405
Sigmoid saturation kinetics, 71
Signal, defined, 376
Signal hypothesis, 453, 454
Signal recognition particle (SRP), 453
Silencer(s), 364
Simian virus 40 (SV4O) enhancer elements, 440
Single-base alteration, in DNA, 401
Sinusoids, hepatic, 229
sis oncogene, 640, 641
Sister chromatid exchanges, 395
Sitosterols, 138, 246, 250
Sjögren-Larsson syndrome, 213
Skatole, 587
Skeletal muscle, 619
  fast and slow, characteristics of, 626
  glycolysis in, 158
Sliding filament mechanism of Huxley, 627
Slow skeletal muscle, ATP regeneration in, 625
Slow-reacting substance of anaphylaxis (SRS-A), 210, 217
Small nuclear ribonucleoprotein particles, 386
Small nuclear RNA, 383
Small stable RNA, 386
Smooth muscle, 619
  β-adrenergic stimulation of, 624
  contraction of, 623
  and striated muscle, actin-myosin interactions in, 625
-*sn*- (stereochemical numbering) system, defined, 133
snRNA. *See* Small nuclear RNA.
Snurps, 386
Sodium
  major characteristics of, 577
  serum or plasma, normal values, 658, 662
Sodium pump, 584
Solids in bile, 580
Somatomedin C, 482, 486, 561
Somatostatin, 475, 483, 486, 500, 547, 550, 562, 565, 566, 568
  amino acid sequence of, 562
Somatotropes, 485
Somatotropin release-inhibiting hormone (SRIH). *See* Somatostatin.
Sorbitol, 183
  in diabetes, 183
  metabolism of, 183
Sorbitol dehydrogenase, 182, 183
Sorbitol intolerance, 183
Sorbitol pathway, 183
Soret band, 324
Southern blot, 376
Southern blot transfer, 369
Spare receptors, 468
Specific gravity, blood or serum, normal values, 662

Specificity, of enzymes, 53
Spectrometry
  mass, 590
  NMR, 590
Spermatogenesis, 533
Spermatogonia, 530
Spermatozoa, 530
Spermidine, 309
  biosynthesis of, 311
  structure of, 310
Spermine, 309, 310
  biosynthesis of, 311
  structure of, 310
Sphinganine, 222
Sphingenine, 222
Sphingolipid(s), 222
  and phospholipids, in disease, 223
Sphingolipidoses, 223
Sphingomyelin(s), 135, 222, 446
  biosynthesis of, 223
  in blood plasma, 226
  structure of, 135
Sphingomyelinase, deficiency of, 224
Sphingophospholipids, 130
Sphingosine, 135, 222, 223, 224
  biosynthesis of, 222
Splanchnic tissues, amino acids in,
  275
Splice junctions, sequences at, 410
Splicing, defined, 376
Splicosome, 417
Squalene, 243, 244
  biosynthesis of, 242
Squalene synthetase, 242
src gene, 640, 646
SRIH. *See* Somatostatin.
SRS-A. *See* Slow-reacting substance
  of anaphylaxis.
Standard free energy change, defined,
  93
Staphylococcus aureus, 307
Staphylococcus aureus protease V8,
  29
Starch(es), 125, 584
  structure of, 127
Starvation, 200, 235, 255, 263, 571,
  574
Steady-state system, cell in, 82
Stearic acid, 131
Steatorrhea, due to malabsorption,
  587
Stein-Leventhal syndrome, 489, 543
Stepdown culture, 85
Stercobilin, structure of, 331
Stercobilinogen, structure of, 331
Stereochemical aspects of steroids,
  137
Stereochemical numbering (*-sn-*) sys-
  tem, defined, 133
Stereoisomer(s), 120
  conformations of, 137
Stereoisomerism, 512
Steroid(s), 130, **136.** *See also specific
  steroid hormone.*
  classification of, by glucocorticoid
    action, 521
  glucocorticoid, 511
  inactive, 520, 521
  mineralocorticoid, 511
  names of, trivial and chemical, 513

Steroid(s) (cont'd)
  nomenclature and chemistry of,
    512
  stereochemical aspects of, 137
Steroid sulfates, 223
Steroidogenesis, 533
Sterols, 447
  oxygenated, 244
Sterol carrier protein, 244
Stomach
  digestion in, 579, 584
  hormones of, normal values, 663
Streptococcus faecalis, 307
Streptokinase, 617
Streptomycin(s), 124, 125
Stress, 250
Stress fibers, 627
"Stress response," 520
Striated muscle, 619
  arrangements of filaments in, 619
  contraction of
    actin-based regulation of, 623
    in mollusks, 624
  and smooth muscle, actin-myosin
    interactions in, 625
Strophanthus, 124
Structural proteins, **618–634**
Suberic acids, 200
Subfragment 1 (S-1), of heavy
  meromyosin, 621, 622
Subfragment 2 (S-2), of heavy
  meromyosin, 621
Substance P, 565, 566, 568
Substrate(s)
  binding of, 78
  and coenzymes, effective concen-
    trations of, 86
  concentration of, 68
  second, coenzymes as, 51
Substrate analog inhibition, 72
Substrate cycle, 193
Substrate-bridge complexes, 80
Subunits, protein, 36
Succinate, 72, 152, 306, 317
Succinate dehydrogenase, 103, 152,
  153, 194
  in mitochondria, 116
Succinate semialdehyde, 306, 317
Succinate thiokinase, 151, 152
Succinate-glycine cycle, 306, 307
Succinyl-CoA, 96, 152, 173, 174,
  257, 262, 282, 298, 299
  amino acids forming, 299
Succinyl-CoA synthetase, 151
Succinyl-CoA-acetoacetate-CoA
  transferase, 151, 257
Sucrase, 584
  deficiency of, 586
Sucrose, 120, 126, 235, 584
Sucrose density gradient centrifugation, in determining molecular
  weight of protein, 38
Sugar(s). *See also specific sugar.*
  amino, 124
  deoxy, 124
  deoxyribose, in DNA, 379
  invert, 125
  nucleotide, 591
  pentose, 583
  sulfated, 601

"Sugar" diabetes, 197
Sulfatase(s), 603
Sulfatase deficiency, multiple, 225
Sulfate
  active, 223
  serum or plasma, normal values,
    662
Sulfate groups, 599
Sulfated sugars, 601
Sulfatide(s), 135, 223
  biosynthesis of, 223
Sulfation, in proteoglycan synthesis,
  600
Sulfation factor, 486
Sulfhydryl (SH) groups, 161
β-Sulfinylpyruvate, 288
Sulfituria, 289
Sulfocysteinuria, 289
Sulfo(galacto)glycerolipids, 223
Sulfogalactosylceramide, 223
Sulfolipids, 130
Sulfonamides, and G6PD deficiency,
  179
Sulfonylurea compounds, 553
Sulfur-containing amino acids, 287
Sulfur-protein-iron complex, 110
Sunlight, sensitivity to, DNA in, 402
Supercoiling of DNA, 400
Superoxide dismutase, 107
Superoxide metabolism, 107
Surfactant, lung, 220
SV4O enhancer elements. *See* Simian
  virus 40 enhancer elements.
SV40 virus, 639
Svedberg (Sf) units, defined, 226
Sympathetic nervous system, 239,
  524
Sympathoadrenal system, 524
Symport transport system, 456

T₃. *See* Triiodothyronine.
T₄. *See* Thyroxine.
Tandem, defined, 376
Tangier disease, 251
Target gland, 465
Tarui's disease, 171
Taurine, 19
Taurochenodeoxycholic acid, 248,
  249
Taurocholic acid, 248, 249
Tautomerism, 336
Tay-Sachs disease, 224, 603
TBG. *See* Thyroid hormone-binding
  globulin *and* Thyroxine-binding
  globulin.
TBPA. *See* Thyroxine-binding pre-
  albumin.
T·ψ·C loop. *See* Thymidine-pseudo-
  uridine-cytidine loop.
Tea, 336
TEBG. *See* Testosterone-estrogen-
  binding globulin.
Telomeres, 390
Temperature
  and chemical reactions, 62
  and energy expenditure, 573
  in enzyme-catalyzed reactions, 66
  and enzymes, 66
  transition, defined, 450

"Template" model of catalytic site, 64
Termination, 425
Termination signal(s), transcriptional, 406, 415
Tertiary structure of proteins, 36
  determination of, 38
Testes
  hormones of, **530–535**
    biosynthesis and metabolism of, 530
    mechanism of action of, 535
    pathophysiology of, 535
    physiologic action of, 534
    and regulation of testicular function, 533
Testicular feminization syndrome, 471, 535
Testosterone, 513, 515, 530, 531, 532, 533, 536, 545
  biosynthesis of, 530
Testosterone/DHT receptor, 535
Testosterone-estrogen-binding globulin (TEBG), 532
Tetany, 571
  due to malabsorption, 587
Tetracycline resistance gene, 366
Tetraenoic acids, 132
Tetrahydrobiopterin, 270
Tetraiodothyronine. *See* Thyroxine.
3,5,3′,5′-Tetraiodothyronine. *See* Thyroxine.
Tetramer
  defined, 36
  of histones, 388
Tetrapyrroles, 40
Tetroses, 120, 122
TH. *See* Tyrosine hydroxylase.
Thalassemia(s), 49
α-Thalassemia, 373
β-Thalassemia, 373
Theobromine, 336, 377
Theophylline, 238, 336, 337
Thermodynamics
  biochemical, 93
  laws of, 93
Thermogenesis, in brown adipose tissue, 239
Thermogenic effect, 572
Thermogenin, 240
Thiamin, 53, **153**, 154, 177, 577
  deficiency of, 164
  major characteristics of, 574
  structure of, 154
Thiamin diphosphate, 163, 177
  as coenzyme in transketolase, 177
Thick filaments, 620
Thin filaments, 620
Thioesterase, 204, 205, 206
6-Thioguanine, 342
Thiokinase. *See* Acyl-CoA synthetase.
Thiolase, 199, 241, 255, 256, 257
Thiolpyruvate, 288
Thiolytic cleavage, 199
Thiophorase, 151
Thioredoxin, 349
Thioredoxin reductase, 349
Thiouracil, 498
Thiourea, 498

Thiourea class of antithyroid drugs, 498
Thoracic duct, 229
Thr. *See* Threonine.
Three-point attachment, in enzyme-substrate binding, 53
Threonine (Thr), 16, 262, 265, 273, 290, **309**
  biosynthesis of, 270
  human requirements of, 573
L-Threonine, 289
Threonine aldolase, 289, 290
Threonine dehydrase, 84
D-Threose, 124
Thrombin, 66, 612, 613
Thromboxane(s) (TX), 131, 210, 213, 214, 215
Thromboxane A₂, 131, 215
Thrombus
  red, 612
  white, 612
Thymidine, 355
³H-Thymidine, 344
Thymidine kinase, 355
Thymidine monophosphate (TMP), 354
Thymidine nucleotides, of DNA, 377
Thymidine-pseudouridine-cytidine (T·ψ·C) loop, of RNA, 385
Thymidylate, 354, 377
Thymidylate synthase, 354
Thymidylic acid, 339
Thymine, 335, 336, 337, 356, 362, 395
  in DNA, 379, 395
Thymine-thymine dimer, ultraviolet light-induced, 401
Thyroglobulin, metabolism of, 497
Thyroid gland, medullary carcinoma of, 510
Thyroid hormone(s), 196, **496–501**
  biosynthesis of, 496
  mechanism of action of, 500
  normal values, 663
  pathophysiology of, 501
  structure of, 496
  synthesis and release of, regulation of, 499
  transport and metabolism of, 499
Thyroid-stimulating hormone (TSH), 238, 464, 465, 471, 475, 482, 483, 488, 490
Thyroid-stimulating IgG (TSI), 501
Thyronine, 496
Thyroperoxidase, 499
Thyrotropin-releasing hormone (TRH), 26, 482, 483
Thyroxine (T₄), 18, 464, 482, 496
  serum, normal values, 659
Thyroxine-binding globulin (TBG), 499
Thyroxine-binding prealbumin (TBPA), 499
Tiglyl-CoA, 302
  catabolism of, 304
TLC. *See* Chromatography, thin-layer.
T_m, of DNA, 379
TMP. *See* Thymidine monophosphate.

Tocopherol(s), 139
  major characteristics of, 576
α-Tocopherol, 107, 139
Tolbutamide, 553
Tonofilaments, 631
Tophi, 356
Topoisomerases, DNA, 381, 399, 400
TpC. *See* Troponin C.
TpI. *See* Troponin I.
TpT. *See* Troponin T.
Trace elements
  deficiency of, 578
  essential nutritional requirements of, 572
  major characteristics of, 578
Transacylase, 204, 206
Transaldolase, 177
Transaminases, 272, 283, 285, 288, 359
  alanine and glutamate, 272, 273
Transamination, 154, 272, 299
Δ²-*Trans*-Δ⁶-*cis*-dienoyl-CoA, 203
Transcortin, 515
Transcriptase, reverse, 383, 400
Transcription
  control of, 440
  defined, 376
  reverse, defined, 376
Transcription unit, 404
Transcriptional signals, in RNA synthesis, 406
Transcriptionally active chromatin, 390
Transcriptionally inactive chromatin, 390
Transducer, chemical-mechanical, requirements of, 619
Trans-fatty acids, 212
Transfection, DNA, 641
Transfer protein, 248
Transfer RNA. *See* tRNA.
Transferase(s), 167, 637
  classification of, 51
  terminal, defined, 376
Transferrin, serum, normal values, 662
Transformation
  malignant, 639
  pleiotropic effects of, 640
Transforming growth factor, 644
Transgenic, defined, 376
Transhydrogenase, energy-linked, 112
Transient response, 428
Transition mutations, 417
  and transversion mutations, diagrammatic representation of, 417
Transition state
  decay of, 61
  formation of, 61
  catalysts in, 62
Transition temperature, defined, 450
Transketolase, 176, 177, 178, 179
Translation, genetic, 414
  defined, 376
Translocation, 423
  chromosomal, 642, 643
  of protons, 114, 116
  reciprocal, 642, 643

Transmembrane channels, 455
Transmembrane signal transduction, 645
Transmethylation, 235
Trans-methylglutaconate shunt, 244
Trans-3-methylglutaconate-CoA, 244
Transport, active, in membranes, 456, 458
Transport (carrier) proteins, 467
Transporter systems, mitochondrial, 118
Transposition of DNA, defined, 394
Transposons, 641
Trans-prenyl transferase, 242
Transversion mutations, 417
Trehalase, 583, 584
Trehalose, 126, 583, 584
TRH. *See* Thyrotropin-releasing hormone.
Triacylglycerol(s), **133**, 219, 229, 231, 234, 235, 236, 238, 251, 256, 261, 262, 582, 583
  in blood plasma, 226
  catabolism of, 215
  dietary, 231
  digestion and absorption of, 582
  nomenclature of, 133
  and phospholipids, biosynthesis of, 219
  stores of, lipolysis and reesterification of, 236
  structure of, 133
  synthesis of, 234
Triacyl-*sn*-glycerol, 133
Triacylglycerol lipase, 559
Triacylglycerolemia, 251
Triamcinolone, 513
Tricarboxylate transporter, 207
Tricarboxylic acid cycle, 149
Trichochromes, 314, 315
Trichorrhexis nodosa, 279
Trienoic acids, 132
2′,3′,5′-Triester of RNA, 382
Triglyceride(s), 133
  serum, normal values, 659, 662
Triiodothyronine (T₃), 18, 482, 496, 497
  serum, normal values, 659
3,3′,5′-Triiodothyronine (reverse T₃; rT₃), 496
3,5,3′-Triiodothyronine, 496
1,3,7-Trimethylxanthine, 336, 337
Trinucleotide code of mRNA, 83
Triokinase, 182, 183
Triose, 120, 122
Triose phosphate, 159, 161
Triosephosphate isomerase, in IUB classification, 51
Tripeptide(s), 24
Triple helix, of collagen, 631
Triplet code of mRNA, 83
Trisialogangliosides, 225
tRNA (transfer RNA), 385
  aminoacyl, 385
  arms of, 385
  function of, 416
  molecules of
    abnormally functioning, 420
    suppressor, 420
      in protein synthesis, 420

tRNA (transfer RNA) (cont'd)
  in protein synthesis, 416
  in RNA processing, 412
Tropomyosin, 620, 621, 628
Tropomyosin-troponin systems, 623
Troponin, 620, 621
Troponin C (TpC), 621
Troponin I (TpI), 621
Troponin system, 623
Troponin T (TpT), 621
Trp. *See* Tryptophan.
Trypsin, 29, 66, 581, 584, 622
  in IUB classification, 51
Trypsinogen, 581
Tryptophan (Trp), 17, 262, 265, 290, 298, 299, **312**, 587
  catabolism of, metabolic disorders of, 299
  conversion of, 312
  human requirements of, 573
  ultraviolet absorption spectrum of, 20
L-Tryptophan, 297
  catabolism of, 297
Tryptophan codons, 436
L-Tryptophan dioxygenase, 106
Tryptophan operon, attenuation in, 435
Tryptophan oxygenase, 85, 298
Tryptophan pyrrolase, 84, 85, 106, 297, 298, 321
TSH. *See* Thyroid-stimulating hormone.
TSI. *See* Thyroid-stimulating IgG.
α-Tubulin, defined, 629, 630
β-Tubulin, defined, 629, 630
Tumor(s), progression of, 647
Tumor markers, 635
Tumor-promoting phorbol esters, 557
Tunicamycin, 595, 597
Turner's syndrome, 543
Two-base alteration, in DNA, 401
TX. *See* Thromboxane.
TX₃, 215
TXA₂, 215
Tyr. *See* Tyrosine.
Tyramine, 102, 293, 587
Tyrocidin, 26
Tyrosinase, 102, 292, 315
Tyrosine (Tyr), 16, 17, 262, 265, 290, 292, 315, 316, 525, 587
  alternative catabolites of, 293
  biosynthesis of, 269
  catabolism of,
    intermediates in, 291
    metabolic disorders of, 292
  conversion of, 316
  human requirements of, 573
  iodination of, 499
  transamination of, 292
  ultraviolet absorption spectrum of, 20
L-Tyrosine, 269, 291
Tyrosine hydroxylase (TH), 314, 525
Tyrosine kinase, 559
Tyrosine transaminase, 291
Tyrosine-α-ketoglutarate transaminase, 292
Tyrosine-α-ketoglutaric transaminase, 84

Tyrosinemia, 291, 292
  neonatal, 291, 292
Tyrosinosis, 292
Tyrosinyl-tRNA, 426

Ubiquinone, 109, 115, 244
  biosynthesis of, 242
Ubiquitin, 387
UDP. *See* Uridine diphosphate.
UDP glucosamine, 185
UDP glucuronate, 181
UDP glucuronyltransferase, 330
UDP-N-acetylgalactosamine, 185
UDP-N-acetylglucosamine, 185
UDPGal. *See* Uridine diphosphate galactose.
UDPGalNAc transferase, in proteoglycan synthesis, 600
UDPGlc. *See* Uridine diphosphate glucose.
UDPGlc dehydrogenase, 180, 181
UDPGlc epimerase, 591
UDPGlc pyrophosphorylase, 165, 166, 181, 184, 591
UDPGlcUA (uridine diphosphoglucuronic acid), 341
UDP-Xyl (uridine diphosphate xylose) transferase, in proteoglycan synthesis, 600
Ultracentrifugation
  in determining molecular weight of protein, 38
  to separate lipoproteins, 226
Ultraviolet absorption spectrum, of tryptophan, tyrosine, and phenylalanine, 20
Ultraviolet light
  and DNA, 401
  in formation of pyrimidine-pyrimidine dimers, 401
UMP (uridine monophosphate; uridylic acid). *See* Uridine monophosphate.
Uncouplers, 112
Unequal crossover, 393
Unidirectional flow of metabolites, 82
Uniport transport system, 456
Δ²-Unsaturated acyl-CoA, 201, 203
2,3-Unsaturated acyl-CoA reductase, 208
α,β-Unsaturated acyl-CoA thioesters, 301
Unsaturated fatty acids, 131, 132
  isomerism in, 132
  metabolism of, 210
  oxidation of, 202, 203
"Up-regulation," 468
Uracil, 335, 355, 356
  derivatives of, 341
  in DNA, 402
  in RNA, 382
Uracil DNA glycosylase, 402
Urate pool, miscible, 352
Urea, 272, 278
  biosynthesis of, 277, 278
    regulation of, 278
  normal values, 660

Urea cycle
  disorders involving, 278
  reactions of, 277
Urea nitrogen, normal values, 660, 662
β-Ureidoisobutyrate, 356
β-Ureidopropionic acid, 356
Ureotelic, defined, 272
Ureotelic organisms, 272, 344
Uric acid, 102, 336, 344
  conversion of, to allantoin, 353
  excretion of, 272
  generation of, from purine nu-
    cleosides, 345
  normal values, 660, 662
Uricase, 101, 102
Uricotelic, defined, 272
Uricotelic organisms, 272, 344
Uridine, 337, 355
Uridine diphosphate (UDP), 165, 166,
    184
Uridine diphosphate galactose (UDP-
    Gal), 165, 183, 184, 223, 224, 591
Uridine diphosphate glucose
    (UDPGlc), 96, 165, 166, 180,
      181, 183, 184, 223, 224,
      341, 591
Uridine diphospho- sugars, 341
Uridine diphosphogalactose
    epimerase, 223
Uridine diphosphogalactose 4-
    epimerase, 184
Uridine diphosphoglucuronic acid,
    341
Uridine monophosphate (UMP;
    uridylic acid), 339, 353, 354
Uridine nucleotide derivatives, 341
Uridine triphosphate (UTP), 98, 165,
    166, 180, 341
Uridine-cytidine kinase, 355
Uridyl transferase, 183
Uridylate, 353
Uridylic acid. See Uridine monophos-
    phate.
Urine, 256
  indole derivatives in, 314
Urobilinogen(s), 330
  normal values, 663
  in urine, 333
Urobilinogen cycle, intrahepatic, 330
Urobilinogen-bilirubin cycle, 333
Urocanase, 285
Urocanate, 285
Uronic acid pathway, of glucose
    metabolism, 180
Uroporphyrin, and coproporphyrins,
    320
Uroporphyrin III, 319
Uroporphyrinogen(s)
  decarboxylation to coproporphyri-
    nogens in cytosol, 322
Uroporphyrinogen III cosynthase, 321
Uroporphyrinogen decarboxylase,
    321, 327
Uroporphyrinogen I synthase, 321,
    326
UTP. See Uridine triphosphate.

V_H gene, 612
V_H region, 609

V_L gene, 612
V_L region, 609
Val. See Valine.
Valeric acid, 131
Valine (Val), 16, 20, 24, 262, 265,
    298, 299, 302, 303
  biosynthesis of, 270
  in brain, 275
  catabolism of, 298, 301, 302
  human requirements of, 573
Valinomycin, 119
Van den Bergh reaction, 331
Vanillylmandelic acid (VMA),
    527
Variable region, 609
Variable region gene, 612
Vasoactive intestinal polypeptide
    (VIP), 482, 566, 568
Vasopressin, 193, 237, 238, 493,
    494, 555
Vector(s)
  cloning, common, 366
  defined, 376
  expression, 367
Vertebrate collagens, 632
Very low density lipoproteins
    (VLDL), 227, 228, 229, 230,
      231, 232, 234, 235, 237,
      246, 248, 250, 251
  formation of, 234
  nascent, 231
  remnant, 231
  secretion of, 250
Vimentin, 631
Vinculin, 640
Vinyl ether, 220
VIP. See Vasoactive intestinal
    polypeptide.
VIPomas, 568
Virus(es), oncogenic, **639**
Vitamin(s)
  deficiencies of, 235
    diseases due to, 577
  dietary, requirements for, 574
  disease responsive to, 577
  fat-soluble, 576
    essential nutritional requirements
      of, 572
  water-soluble, major characteristics
      of, 575
    essential nutritional requirements
      of, 572
Vitamin A, major characteristics of,
    576
  serum, normal values, 662
Vitamin B, 53
  derivatives of, coenzymes as, 53
Vitamin B complex, 575
Vitamin B_1, major characteristics of,
    575
Vitamin B_2, major characteristics of,
    575
Vitamin B_6, 298, 577
  major characteristics of, 575
Vitamin B_{12}, 174, 577
  absorption of, 575
  deficiency of, 174
  major characteristics of, 575
  metabolism of, defects of, 305
  serum, normal values, 662

Vitamin C, major characteristics of,
    575
Vitamin D, 503, 508, 576
  deficiency of, 509, 576
  major characteristics of, 576
  serum, normal values, 662
Vitamin D_2, major characteristics of,
    576
Vitamin D_3
  formation and hydroxylation of,
    507
  major characteristics of, 576
Vitamin D-resistant rickets, type II,
    471
Vitamin E, 107, 576
  deficiency of, 235
  major characteristics of, 576
Vitamin K, major characteristics of,
    576
Vitamin K_1, major characteristics of,
    576
Vitamin K_2, major characteristics of,
    576
Vitamin nucleotides, 341
VLDL. See Very low density lipo-
    proteins.
β-VLDL, 229
VMA. See Vanillylmandelic acid.
Von Gierke's disease, 170, 357
von Willebrand factor, 615
von Willebrand's disease, 615, 616
v-*onc* (viral oncogene), 641

Water
  in bile, 580
  body, 445
  essential nutritional requirements
    of, 572
  extracellular, 445
  intracellular, 445
Water-soluble vitamins, major charac-
    teristics of, 575
Watson and Crick model of DNA, 379
Waxes, 130
Western blot, 376
Western blot transfer, 369
Wilson's disease, 308, 309
Wobble, 415, 416
Wolffian duct system, 545
Wolman's disease, 251

Xanthine, 336, 344, 345, 352
Xanthine dehydrogenase, 102
Xanthine lithiasis, 358
Xanthine oxidase, 102, 345, 352
  deficiency of, 358
  in purine or pyrimidine metabolism,
    86
Xanthinuria, 358
Xanthomas, 251
Xanthosine monophosphate (XMP),
    348
Xanthurenate, 298
  formation of, in vitamin B_6
    deficiency, 298
Xenobiotic(s), 637
Xeroderma pigmentosum, DNA in, 402
Xerophthalmia, 571

XMP. *See* Xanthosine monophosphate.
X-ray crystallography, 38
Xylitol, 180, 181
Xylose, 128, 180
  in glycoproteins, 128
D-Xylose, 123, 124
D-Xylulose, 124, 180, 181
L-Xylulose, 180, 181
Xylulose 1-phosphate, in oxalate formation, 180

D-Xylulose 1-phosphate, 181
Xylulose 5-phosphate, 176, 177, 178
D-Xylulose 5-phosphate, 180, 181

Yeast, fermentation in, 158

Z gene, defined, 430
Z line, 619, 620

Z-DNA (zigzag DNA), 379
  left-handed double helix of, 378
Zellweger's syndrome, 202
Zinc, 103
  major characteristics of, 578
  serum, normal values, 662
Z-protein, 198, 229
Zwitterion, 15
Zymogens, 77
Zymosterol, 243

BUSINESS REPLY MAIL
FIRST CLASS PERMIT NO. 150 E. NORWALK, CT

POSTAGE WILL BE PAID BY ADDRESSEE

APPLETON & LANGE

DEPARTMENT B
25 VAN ZANT STREET
EAST NORWALK, CT 06855